Trapezoid: A four-sided figure with one pair of parallel sides

Area: $A = \frac{1}{2}h(b_1 + b_2)$

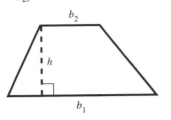

Parallelogram: A four-sided figure with opposite sides parallel

Area: $A = bh$

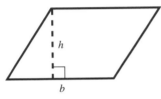

Rectangle: A four-sided figure with four right angles

Area: $A = LW$

Perimeter: $P = 2L + 2W$

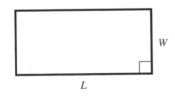

Rhombus: A four-sided figure with four equal sides

Perimeter: $P = 4a$

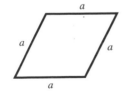

Square: A four-sided figure with four equal sides and four right angles

Area: $A = s^2$

Perimeter: $P = 4s$

Circle

Area: $A = \pi r^2$

Circumference: $C = 2\pi r$

Diameter: $d = 2r$

Value of pi: $\pi \approx 3.14$

Sphere

Volume: $V = \frac{4}{3}\pi r^3$

Surface Area: $S = 4\pi r^2$

Right Circular Cone

Volume: $V = \frac{1}{3}\pi r^2 h$

Lateral Surface Area: $S = \pi r \sqrt{r^2 + h^2}$

Right Circular Cylinder

Volume: $V = \pi r^2 h$

Lateral Surface Area: $S = 2\pi rh$

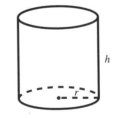

Rectangular Solid

Volume: $V = LWH$

Surface Area:
$A = 2LW + 2WH + 2LH$

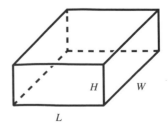

Elementary and Intermediate

Algebra

Fourth edition

Mark Dugopolski

Southeastern Louisiana University

Mc Graw Hill

Connect
Learn
Succeed™

ELEMENTARY AND INTERMEDIATE ALGEBRA, FOURTH EDITION

Published by McGraw-Hill, a business unit of The McGraw-Hill Companies, Inc., 1221 Avenue of the Americas, New York, NY 10020. Copyright © 2012 by The McGraw-Hill Companies, Inc. All rights reserved. Previous editions © 2009, 2006, and 2002. No part of this publication may be reproduced or distributed in any form or by any means, or stored in a database or retrieval system, without the prior written consent of The McGraw-Hill Companies, Inc., including, but not limited to, in any network or other electronic storage or transmission, or broadcast for distance learning.

Some ancillaries, including electronic and print components, may not be available to customers outside the United States.

This book is printed on acid-free paper.

1 2 3 4 5 6 7 8 9 0 RJE/RJE 1 0 9 8 7 6 5 4 3 2 1

ISBN 978–0–07–338435–1
MHID 0–07–338435–6

ISBN 978–0–07–735329–2 (Annotated Instructor's Edition)
MHID 0–07–735329–3

Vice President, Editor-in-Chief: *Marty Lange*
Vice President, EDP: *Kimberly Meriwether David*
Senior Director of Development: *Kristine Tibbetts*
Editorial Director: *Stewart K. Mattson*
Sponsoring Editor: *Mary Ellen Rahn*
Developmental Editor: *Adam Fischer*
Marketing Manager: *Peter A. Vanaria*
Lead Project Manager: *Peggy J. Selle*
Senior Buyer: *Sandy Ludovissy*
Senior Media Project Manager: *Jodi K. Banowetz*
Designer: *Tara McDermott*
Cover Designer: *Greg Nettles/Squarecrow Design*
Cover Image: © *Bryan Mullennix/Alamy*
Lead Photo Research Coordinator: *Carrie K. Burger*
Compositor: *Glyph International*
Typeface: *10.5/12 Times Roman*
Printer: *R. R. Donnelley*

All credits appearing on page or at the end of the book are considered to be an extension of the copyright page.

Photo Credits: Page 75: © Vol. 141/Corbis; p. 82: © Reuters/Corbis; p. 150 (top): © George Disario/Corbis; p. 168: © Vol. 166/Corbis; p. 193 (bottom): © Ann M. Job/AP/Wide World Photos; p. 246 (bottom left): © Fancy Photography/Veer RF; p. 253: © Michael Keller/Corbis; p. 255: © DV169/Digital Vision; p. 476: © Vol. 128/Corbis; p. 557: © Daniel Novisedlak/Flickr/Getty RF; p. 810: © Vol. 168/Corbis; p. 839: © Stockdisc/Digital Vision RF. All other photos © PhotoDisc/Getty RF.

Library of Congress Cataloging-in-Publication Data

Dugopolski, Mark.
   Elementary and intermediate algebra / Mark Dugopolski.—4th ed.
     p. cm.
   Includes index.
ISBN 978–0–07–338435–1—ISBN 0–07–338435–6 and ISBN 978–0–07–735329–2—ISBN
0–07–735329–3 (annotated instructor's edition) (hard copy: alk. paper) 1. Algebra—Textbooks. I. Title.
   QA152.3.D84 2012
   512.9—dc22   2010024307

www.mhhe.com

About the Author

Mark Dugopolski was born and raised in Menominee, Michigan. He received a degree in mathematics education from Michigan State University and then taught high school mathematics in the Chicago area. While teaching high school, he received a master's degree in mathematics from Northern Illinois University. He then entered a doctoral program in mathematics at the University of Illinois in Champaign, where he earned his doctorate in topology in 1977. He was then appointed to the faculty at Southeastern Louisiana University, where he taught for 25 years. He is now professor emeritus of mathematics at SLU. He is a member of MAA and AMATYC. He has written many articles and numerous mathematics textbooks. He has a wife and two daughters. When he is not working, he enjoys gardening, hiking, bicycling, jogging, tennis, fishing, and motorcycling.

In loving memory of my parents,
Walter and Anne Dugopolski

McGraw-Hill Connect Mathematics McGraw-Hill conducted in-depth research to create a new and improved learning experience that meets the needs of today's students and instructors. The result is a reinvented learning experience rich in information, visually engaging, and easily accessible to both instructors and students. McGraw-Hill's Connect is a Web-based assignment and assessment platform that helps students connect to their coursework and prepares them to succeed in and beyond the course.

Connect Mathematics enables math instructors to create and share courses and assignments with colleagues and adjuncts with only a few clicks of the mouse. All exercises, learning objectives, videos, and activities are directly tied to text-specific material.

1 *You and your students want a fully integrated online homework and learning management system all in one place.*

McGraw-Hill and Blackboard Inc. Partnership

▶ McGraw-Hill has partnered with Blackboard Inc. to offer the deepest integration of digital content and tools with Blackboard's teaching and learning platform.

▶ **Life simplified.** Now, all McGraw-Hill content (text, tools, & homework) can be accessed directly from within your Blackboard course. All with one sign-on.

▶ **Deep integration.** McGraw-Hill's content and content engines are seamlessly woven within your Blackboard course.

▶ **No more manual synching!** Connect assignments within Blackboard automatically (and instantly) feed grades directly to your Blackboard grade center. No more keeping track of two gradebooks!

Do More

2 *Your students want an assignment page that is easy to use and includes lots of extra resources for help.*

Efficient Assignment Navigation

▶ Students have access to immediate feedback and help while working through assignments.

▶ Students can view detailed step-by-step solutions for each exercise.

Connect. Learn. Succeed.

 3 Your students want an interactive eBook rich with integrated functionality.

Integrated Media-Rich eBook

▶ A Web-optimized eBook is seamlessly integrated within ConnectPlus Mathematics for ease of use.

▶ Students can access videos, images, and other media in context within each chapter or subject area to enhance their learning experience.

▶ Students can highlight, take notes, or even access shared instructor highlights/notes to learn the course material.

▶ The integrated eBook provides students with a cost-saving alternative to traditional textbooks.

4 You want a more intuitive and efficient assignment creation process to accommodate your busy schedule.

Assignment Creation Process

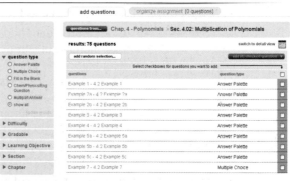

▶ Instructors can select textbook-specific questions organized by chapter, section, and objective.

▶ Drag-and-drop functionality makes creating an assignment quick and easy.

▶ Instructors can preview their assignments for efficient editing.

5 You want a gradebook that is easy to use and provides you with flexible reports to see how your students are performing.

Flexible Instructor Gradebook

▶ Based on instructor feedback, Connect Mathematics' straightforward design creates an intuitive, visually pleasing grade management environment.

▶ View scored work immediately and track individual or group performance with various assignment and grade reports.

www.mcgrawhillconnect.com

FROM THE AUTHOR

I would like to thank the many students and faculty that have used my books over the years. You have provided me with excellent feedback that has assisted me in writing a better, more student-focused book in each edition. Your comments are always taken seriously, and I have adjusted my focus on each revision to satisfy your needs.

Understandable Explanations

I originally undertook the task of writing my own book for the elementary and intermediate algebra course so I could explain mathematical concepts to students in language they would understand. Most books claim to do this, but my experience with a variety of texts had proven otherwise. What students and faculty will find in my book are **short, precise explanations** of terms and concepts that are written in **understandable language.** For example, when I introduce the Commutative Property of Addition, I make the concrete analogy that "the price of a hamburger plus a Coke is the same as the price of a Coke plus a hamburger," a mathematical fact in their daily lives that students can readily grasp. Math doesn't need to remain a mystery to students, and students reading my book will find other analogies like this one that connect abstractions to everyday experiences.

Detailed Examples Keyed to Exercises

My experience as a teacher has taught me two things about examples: they need to be detailed, and they need to help students do their homework. As a result, users of my book will find abundant examples with every step carefully laid out and explained where necessary so that students can follow along in class if the instructor is demonstrating an example on the board. Students will also be able to read them on their own later when they're ready to do the exercise sets. I have also included a **double cross-referencing** system between my examples and exercise sets so that, no matter which one students start with, they'll see the connection to the other. All examples in this edition refer to specific exercises by ending with a phrase such as "Now do Exercises 11–18" so that students will have the opportunity for immediate practice of that concept. If students work an exercise and find they are stumped on how to finish it, they'll see that for that group of exercises they're directed to a specific example to follow as a model. Either way, students will find my book's examples give them the guidance they need to succeed in the course.

Varied Exercises and Applications

A third goal of mine in writing this book was to give students **more variety** in the kinds of exercises they perform than I found in other books. Students won't find an intimidating page of endless drills in my book, but instead will see exercises in manageable groups with specific goals. They will also be able to augment their math proficiency using different formats (true/false, written response, multiple-choice) and different methods (discussion, collaboration, calculators). Not only is there an abundance of skill-building exercises, I have also researched a wide variety of **realistic applications** using **real data** so that those "dreaded word problems" will be seen as a useful and practical extension of what students have learned. Finally, every chapter ends with **critical thinking exercises** that go beyond numerical computation and call on students to employ their intuitive problem-solving skills to find the answers to mathematical puzzles in **fun and innovative** ways. With all of these resources to choose from, I am sure that instructors will be comfortable adapting my book to fit their course, and that students will appreciate having a text written for their level and to stimulate their interest.

Listening to Student and Instructor Concerns

McGraw-Hill has given me a wonderful resource for making my textbook more responsive to the immediate concerns of students and faculty. In addition to sending my manuscript out for review by instructors at many different colleges, several times a year McGraw-Hill holds symposia and focus groups with math instructors where the emphasis is *not* on selling products but instead on the **publisher listening** to the needs of faculty and their students. These encounters have provided me with a wealth of ideas on how to improve my chapter organization, make the page layout of my books more readable, and fine-tune exercises in every chapter. Consequently, students and faculty will feel comfortable using my book because it incorporates their specific suggestions and anticipates their needs. These events have particularly helped me in the shaping of the fourth edition.

Improvements in the Fourth Edition
OVERALL
- All Warm-Up exercise sets have been rewritten and now include a combination of fill-in-the-blank and true/false exercises. This was done to put a greater emphasis on vocabulary.
- Using a graphing calculator with this text is still optional. However, more Calculator Close-Ups and more graphing calculator required exercises have been included throughout the text for those instructors who prefer to emphasize graphing calculator use.
- Every chapter now includes a Mid-Chapter Quiz. This quiz can be used to assess student progress in the chapter.
- Numerous applications have been updated and rewritten.
- All Enriching Your Mathematical Word Power exercise sets have been expanded and rewritten as fill-in-the-blank exercises.
- All Making Connections exercise sets have been expanded so that they present a more comprehensive cumulative review.
- Teaching Tips are now included throughout the text, along with many new Helpful Hints.

CHAPTER 1

- New material on equivalent fractions and reducing fractions
- Exercise sets: 9 updated and rewritten applications

CHAPTER 2

- Functions are now introduced in the context of formulas.
- New material on the language of functions
- The language of functions and function notation are now used more extensively throughout the text.
- New material on the simple interest formula, perimeter, and original price applications
- New definition of a function and new caution box for the formula and function section
- Three updated and rewritten examples to reflect functions in the context of formulas
- Exercise sets: 10 updated and rewritten applications
- End of chapter: revised and updated summary, review exercises, and chapter test

CHAPTER 3

- New material on graphing ordered pairs and ordered pairs as solutions to equations
- Simplified introduction to graphing a linear equation in two variables
- New material on graphing a line using intercepts
- Improved definitions of intercepts and slope intercept form
- New material on function notation and applications
- Two updated examples and a new caution box
- Revised and updated exercise sets for Sections 3.1, 3.3, and 3.4
- Revised and updated Math at Work feature
- Exercise sets: 5 updated and rewritten applications
- End of chapter: revised and updated review exercises and chapter test

CHAPTER 4

- Section 4.2, Negative Exponents and Scientific Notation, has been split into two sections— Section 4.2, Negative Exponents, and Section 4.3, Scientific Notation.
- Three revised examples and new study tips
- New material on using rules for negative exponents
- New material on scientific notation, including "Combining Numbers and Words" and "Applications"
- New material on polynomial functions

- Exercise sets: revised Sections 4.2 and 4.3 to reflect new organization
- End of chapter: revised and updated review exercises and chapter test

CHAPTER 5

- Section 5.2 has been rewritten with more emphasis on factoring by grouping.
- Section 5.2 has been reorganized so that factoring by grouping comes before special products.
- Section 5.5 has been simplified by eliminating division in factoring.
- New material on factoring applications, factoring by grouping, and the Pythagorean Theorem
- New strategy box for factoring a four-term polynomial by grouping
- New strategy box for factoring $x^2 + bx + c$ by grouping
- Three new examples and four revised examples
- Rewritten explanation on factoring $ax^2 + bx + c$ with $a = 1$
- New explanation of the sum of two squares prime polynomial
- New strategy and explanation for factoring sum and difference of cubes
- Revised strategy for factoring polynomials completely
- Exercise sets: revised Sections 5.1, 5.5, and 5.6 to reflect new organization
- End of chapter: revised and updated review exercises

CHAPTER 6

- Updated Section 6.1 by including rational functions
- New explanation on rational functions and domain of a rational function

CHAPTER 7

- Section 7.1, Solving Systems by Graphing and Substitution, has been split into two sections— Section 7.1, The Graphing Method, and Section 7.2, The Substitution Method.
- Five updated examples
- New summary of the methods for solving systems of equations
- Exercise sets: revised Sections 7.3 and 7.4
- End of chapter: revised and updated review exercises and chapter test

CHAPTER 9

- New material on roots and variables
- New presentation of perfect squares, cubes, and fourth powers
- New material on radical functions and domain of radical functions
- Four revised applications and one revised example
- Exercise sets: revised Section 9.2
- End of chapter: revised and updated review exercises and chapter test

CHAPTER 10

- Simplified Section 10.5 to focus exclusively on quadratic inequalities
- New definition of quadratic inequalities
- New strategies for solving a quadratic inequality graphically and with the Test-Point Method
- Four new examples on solving quadratic inequalities graphically and with the Test-Point Method
- The sign-graph method of solving quadratic and rational inequalities has been removed and replaced with the more intuitive graphical method. The Test-Point Method is also presented.
- New material on quadratic functions
- Improved figures to help clarify graphing examples
- New material using function notation with quadratics
- Exercise sets: revised Section 10.5
- End of chapter: revised and updated review exercises and chapter test

CHAPTER 11

- Section 11.3, Transformations of Graphs, has been rearranged in a more natural order.

- New material on transformations of graphs, horizontal translation, and multiple transformations
- Solving polynomial inequalities by the graphical method and Test-Point Method has been added to Section 11.4 after graphs of polynomial functions.
- New material and two new examples on solving polynomial inequalities
- Solving rational inequalities by the graphical method and Test Points has been added to Section 11.5 after the graphs of rational functions are discussed.
- Rational inequalities have been moved to Section 11.5 where the graphs of rational functions are discussed.
- New material on rational inequalities, along with two new examples for solving graphically and with test points
- Exercise sets: revised Sections 11.3, 11.4 and 11.5.
- End of chapter: revised and updated summary, review exercises, and chapter test

CHAPTER 12

- New definition of domain
- New material on exponential and logarithmic functions

CHAPTER 13

- Simplified material on parabolas in Section 13.2

CHAPTER 14

- Two updated applications.

Manuscript Review Panels

Teachers and academics from across the country reviewed the various drafts of the manuscript to give feedback on content, design, pedagogy, and organization. This feedback was summarized by the book team and used to guide the direction of the text. I would like to thank the following professors for their participation in making this fourth edition.

Seth Daugherty, *Saint Louis CC–Forest Park*

Shing So, *University of Central Missouri*

Elsie Newman, *Owens Community College*

Patrick Ward, *Illinois Central College*

Sean Stewart, *Owens Community College*

Sharon Robertson, *University of Tennessee–Martin*

Randall Castleton, *University of Tennessee–Martin*

David Ray, *University of Tennessee–Martin*

Roland Trevino, *San Antonio College*

Irma Bakenhus, *San Antonio College*

Larry Green, *Lake Tahoe Community College*

Pinder Naidu, *Kennesaw State University*

Fereja Tahir, *Illinois Central College*

Brooke Lee, *San Antonio College*

Timothy McKenna, *University of Michigan–Dearborn*

Jean Peterson, *University of Wisconsin–Oshkosh*

Amy Young, *Navarro College*

Jenell Sargent, *Tennessee State University*

Mark Brenneman, *Mesa Community College*

Litsa St Amand, *Mesa Community College*

Jeff Igo, *University of Michigan–Dearborn*

Gerald Busald, *San Antonio College*

Bobbie Jo Hill, *Coastal Bend College*

Mary Kay Best, *Coastal Bend College*

Mary Frey, *Cincinnati State Tech and Community College*

Kimberly Bonacci, *Indiana University–Southeast*

Kenneth Thompson, *East Central Community College*

Joseph Sedlacek, *Kirkwood Community College*

Mildred Vernia, *Indiana University–Southeast*

Manuel Sanders, *University of South Carolina*

Randell Simpson, *Temple College*

Debra Pharo, *Northwestern Michigan College*

Carmen Buhler, *Minneapolis Community and Tech College*

Mary Peddycoart, *Kingwood College*

Tim McBride, *Spartanburg Technical College*

Glenn Robert Jablonski, *Triton College*

Derek Martinez, *Central New Mexico Community College*

Dennis Reissig, *Suffolk County Community College*

Charles Patterson, *Louisiana Tech University*

Rhoderick Fleming, *Wake Technical Community College*

Toni McCall, *Angelina College*

Suzanne Doviak, *Old Dominican University*

Judith Atkinson, *University of Alaska–Fairbanks*

Stephen Drake, *Northwestern Michigan College*

Michael Price, *University of Oregon*

Donald Munsey, *Louisiana Delta Community College*

Peggy Blanton, *Isothermal Community College*

Paul Diehl, *Indiana University Southeast*

Jinhua Tao, *University of Central Missouri*

Rajalakshmi Baradwaj, *University of Maryland–Baltimore County*

Jinfeng Wei, *Maryville University*

Dale Vanderwilt, *Dordt College*

Lori Wall, *University of England–Biddeford*

Hossein Behforooz, *Utica College*

Jan Butler, *CCC Online*

Paul Jones, *University of Cincinnati*

Joan Brown, *Eastern New Mexico University*

Kimberly Caldwell, *Volunteer State Community College*

Wendy Conway, *Oakland Community College–Highland Lakes*

Robert Diaz, *Fullerton College*

David French, *Tidewater Community College*

Teresa Houston, *East Mississippi Community College–Scooba*

Carla Monticelli, *Camden County College*

Madhu Motha, *Butler County Community College*

Chris Reisch, *Jamestown Community College*

Jill Rafael, *Sierra College*

Dan Rothe, *Alpena Community College*

Richard Rupp, *Del Mar College*

Kristina Sampson, *Lone Star College*

John Squires, *Chattanooga State Tech*

Jane Thompson, *Waubonsee Community College*

Richard Watkins, *Tidewater Community College*

Jackie Wing, *Angelina College*

I also want to express my sincere appreciation to my wife, Cheryl, for her invaluable patience and support.

Mark Dugopolski
Ponchatoula, Louisiana

> "I was 'gripped' by the examples and introductions to the topics. These were interesting, current, and nicely written. Students will find these motivating to learn the material."
>
> *Timothy McKenna,*
> *University of Michigan–Dearborn*

Chapter Opener ❯
Each chapter opener features a real-world situation that can be modeled using mathematics. The application then refers students to a specific exercise in the chapter's exercise sets.

⌄

Chapter **5**

Factoring

The sport of skydiving was born in the 1930s soon after the military began using parachutes as a means of deploying troops. Today, skydiving is a popular sport around the world.

With as little as 8 hours of ground instruction, first-time jumpers can be ready to make a solo jump. Without the assistance of oxygen, skydivers can jump from as high as 14,000 feet and reach speeds of more than 100 miles per hour as they fall toward the earth. Jumpers usually open their parachutes between 2000 and 3000 feet and then gradually glide down to their landing area. If the jump and the parachute are handled correctly, the landing can be as gentle as jumping off two steps.

Making a jump and floating to earth are only part of the sport of skydiving. For example, in an activity called "relative work skydiving," a team of as many as 920 free-falling skydivers join together to make geometrically shaped formations. In a related exercise called "canopy relative work," the team members form geometric patterns after their parachutes or canopies have opened. This kind of skydiving takes skill and practice, and teams are not always successful in their attempts.

The amount of time a skydiver has for a free fall depends on the height of the jump and how much the skydiver uses the air to slow the fall.

5.1 Factoring Out Common Factors

5.2 Special Products and Grouping

5.3 Factoring the Trinomial $ax^2 + bx + c$ with $a = 1$

5.4 Factoring the Trinomial $ax^2 + bx + c$ with $a \neq 1$

5.5 Difference and Sum of Cubes and a Strategy

5.6 Solving Quadratic Equations by Factoring

In Exercises 85 and 86 of Section 5.6 we find the amount of time that it takes a skydiver to fall from a given height.

85. Skydiving. If there were no air resistance, then the height (in feet) above the earth for a skydiver t seconds after jumping from an airplane at 10,000 feet would be given by

$$h(t) = -16t^2 + 10,000.$$

a) Find the time that it would take to fall to earth with no air resistance; that is, find t for which $h(t) = 0$. A skydiver actually gets about twice as much free fall time due to air resistance.

b) Use the accompanying graph to determine whether the skydiver (with no air resistance) falls farther in the first 5 seconds or the last 5 seconds of the fall.

c) Is the skydiver's velocity increasing or decreasing as she falls?

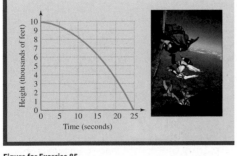

Figure for Exercise 85

In This Section

The In This Section listing gives a preview of the topics to be covered in the section. These subsections have now been numbered for easier reference. In addition, these subsections are listed in the relevant places in the end-of-section exercises.

5.1 Factoring Out Common Factors

In Chapter 4, you learned how to multiply a monomial and a polynomial. In this section, you will learn how to reverse that multiplication by finding the greatest common factor for the terms of a polynomial and then factoring the polynomial.

In This Section

⟨1⟩ Prime Factorization of Integers
⟨2⟩ Greatest Common Factor
⟨3⟩ Greatest Common Factor for Monomials
⟨4⟩ Factoring Out the Greatest Common Factor
⟨5⟩ Factoring Out the Opposite of the GCF
⟨6⟩ Applications

⟨1⟩ Prime Factorization of Integers

To **factor** an expression means to write the expression as a product. For example, if we start with 12 and write $12 = 4 \cdot 3$, we have factored 12. Both 4 and 3 are **factors** or **divisors** of 12. There are other factorizations of 12:

$$12 = 2 \cdot 6 \qquad 12 = 1 \cdot 12 \qquad 12 = 2 \cdot 2 \cdot 3 = 2^2 \cdot 3$$

The one that is most useful to us is $12 = 2^2 \cdot 3$, because it expresses 12 as a product of *prime numbers*.

Examples

Examples refer directly to exercises, and those exercises in turn refer back to that example. This **double cross-referencing** helps students connect examples to exercises no matter which one they start with.

EXAMPLE 5

Factoring completely

Factor each polynomial completely.

a) $4x^3 + 14x^2 + 6x$

b) $12x^2y + 6xy + 6y$

Solution

a) $4x^3 + 14x^2 + 6x = 2x(2x^2 + 7x + 3)$ Factor out the GCF, $2x$.

$= 2x(2x + 1)(x + 3)$ Factor $2x^2 + 7x + 3$.

Check by multiplying.

b) $12x^2y + 6xy + 6y = 6y(2x^2 + x + 1)$ Factor out the GCF, $6y$.

To factor $2x^2 + x + 1$ by the *ac* method, we need two numbers with a product of 2 and a sum of 1. Because there are no such numbers, $2x^2 + x + 1$ is prime and the factorization is complete.

Now do Exercises 75–84

⟨3⟩ Factoring Completely

Factor each polynomial completely. See Examples 5 and 6.

75. $81w^3 - w$ **76.** $81w^3 - w^2$

77. $4w^2 + 2w - 30$ **78.** $2x^2 - 28x + 98$

79. $27 + 12x^2 + 36x$ **80.** $24y + 12y^2 + 12$

81. $6w^2 - 11w - 35$ **82.** $8y^2 - 14y - 15$

83. $3x^2z - 3zx - 18z$ **84.** $a^2b + 2ab - 15b$

85. $9x^3 - 21x^2 + 18x$ **86.** $-8x^3 + 4x^2 - 2x$

87. $a^2 + 2ab - 15b^2$ **88.** $a^2b^2 - 2a^2b - 15a^2$

89. $2x^2y^2 + xy^2 + 3y^2$ **90.** $18x^2 - 6x + 6$

"I really appreciate how the examples correlate with the homework sections. These specific examples are helpful to students that go onto college algebra and pre-calc math classes."

Sean Stewart, Owens Community College

"The worked out examples are clearly explained, no step is left out, and they progress in a fashion that eases the student from very basic to the somewhat complex."

Larry Green, Lake Tahoe Community College

Math at Work

The Math at Work feature appears in each chapter to reinforce the book's theme of real applications in the everyday world of work.

> "Dugopolski uses language and context appropriate for the level of student for whom the text is written without sacrificing mathematical rigor or precision."
>
> *Irma Bakenhus,*
> *San Antonio College*

Math *at Work*   Kayak Design

Kayaks have been built by the Aleut and Inuit people for the past 4000 years. Today's builders have access to materials and techniques unavailable to the original kayak builders. Modern kayakers incorporate hydrodynamics and materials technology to create designs that are efficient and stable. Builders measure how well their designs work by calculating indicators such as prismatic coefficient, block coefficient, and the midship area coefficient, to name a few.

Even the fitting of a kayak to the paddler is done scientifically. For example, the formula

$$PL = 2 \cdot BL + BS\left(0.38 \cdot EE + 1.2\sqrt{\left(\frac{BW}{2} - \frac{SW}{2}\right)^2 + (SL)^2}\right)$$

can be used to calculate the appropriate paddle length. *BL* is the length of the paddle's blade. *BS* is a boating style factor, which is 1.2 for touring, 1.0 for river running, and 0.95 for play boating. *EE* is the elbow to elbow distance with the paddler's arms straight out to the sides. *BW* is the boat width and *SW* is the shoulder width. *SL* is the spine length, which is the distance measured in a sitting position from the chair seat to the top of the paddler's shoulder. All lengths are in centimeters.

The degree of control a kayaker exerts over the kayak depends largely on the body contact with it. A kayaker wears the kayak. So the choice of a kayak should hinge first on the right body fit and comfort and second on the skill level or intended paddling style. So designing, building, and even fitting a kayak is a blend of art and science.

Strategy Boxes

The strategy boxes provide a handy reference for students to use when they review key concepts and techniques to prepare for tests and homework. They are now directly referenced in the end-of-section exercises where appropriate.

Strategy for Factoring Polynomials Completely

1. Factor out the GCF (with a negative coefficient if necessary).
2. When factoring a binomial, check to see whether it is a difference of two squares, a difference of two cubes, or a sum of two cubes. *A sum of two squares does not factor.*
3. When factoring a trinomial, check to see whether it is a perfect square trinomial.
4. If the polynomial has four terms, try factoring by grouping.
5. When factoring a trinomial that is not a perfect square, use the *ac* method or the trial-and-error method.
6. Check to see whether any of the factors can be factored again.

Margin Notes

Margin notes include **Helpful Hints,** which give advice on the topic they're adjacent to; **Calculator Close-Ups,** which provide advice on using calculators to verify students' work; and **Teaching Tips,** which are especially helpful in programs with new instructors who are looking for alternate ways to explain and reinforce material.

‹ Helpful Hint ›

Some students grow up believing that the only way to solve an equation is to "do the same thing to each side." Then along come quadratic equations and the zero factor property. For a quadratic equation, we write an equivalent compound equation that is not obtained by "doing the same thing to each side."

‹ Teaching Tip ›

Show students how to make up a problem like this example: If $x = 5$, then $(5 - 2)(5 + 7) = 36$. So one of the solutions to $(x - 2)(x + 7) = 36$ is 5. Now solve it to find both solutions.

‹ Calculator Close-Up ›

Your calculator can add signed numbers. Most calculators have a key for subtraction and a different key for the negative sign.

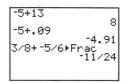

You should do the exercises in this section by hand and then check with a calculator.

Guided Tour Features and Supplements

Exercises
Section exercises are preceded by true/false **Warm-Ups,** which can be used as quizzes or for class discussion.

Warm-Ups ▼

Fill in the blank.

1. To _____ means to write as a product.
2. A _____ number is an integer greater than 1 that has no factors besides itself and 1.
3. The _____ of two numbers is the largest number that is a factor of both.
4. All factoring can be checked by _____ the factors.

True or false?

5. There are only nine prime numbers.
6. The prime factorization of 32 is $2^3 \cdot 3$.
7. The integer 51 is a prime number.
8. The GCF for 12 and 16 is 4.
9. The GCF for $x^5y^3 - x^4y^7$ is x^4y^3.
10. We can factor out $2xy$ or $-2xy$ from $2x^2y - 6xy^2$.

Getting More Involved
concludes the exercise set with **Discussion, Writing, Exploration,** and **Cooperative Learning** activities for well-rounded practice in the skills for that section.

Applications

Solve each problem.

99. *Approach speed.* The formula $1211.1L = CA^2S$ is used to determine the approach speed for landing an aircraft, where L is the gross weight of the aircraft in pounds, C is the coefficient of lift, S is the surface area of the wings in square feet (ft^2), and A is approach speed in feet per second. Find A for the Piper Cheyenne, which has a gross weight of 8700 lb, a coefficient of lift of 2.81, and a wing surface area of 200 ft^2.

100. *Time to swing.* The period T (time in seconds for one complete cycle) of a simple pendulum is related to the length L (in feet) of the pendulum by the formula $8T^2 = \pi^2L$. If a

Getting More Involved

103. *Discussion*

Which of the following equations is not a quadratic equation? Explain your answer.
a) $\pi x^2 - \sqrt{5}x - 1 = 0$ b) $3x^2 - 1 = 0$
c) $4x + 5 = 0$ d) $0.009x^2 = 0$

104. *Exploration*

Solve $x^2 - 4x + k = 0$ for $k = 0, 4, 5,$ and 10.
a) When does the equation have only one solution?
b) For what values of k are the solutions real?
c) For what values of k are the solutions imaginary?

Calculator Exercises
Optional calculator exercises provide students with the opportunity to use scientific or graphing calculators to solve various problems.

Video Exercises
A video icon indicates an exercise that has a video walking through how to solve it.

33. $(y - 1)(y + 1)$ 34. $(p + 2)(p - 2)$
35. $(3x - 8)(3x + 8)$ 36. $(6x + 1)(6x - 1)$
37. $(r + s)(r - s)$ 38. $(b - y)(b + y)$
39. $(8y - 3a)(8y + 3a)$ 40. $(4u - 9v)(4u + 9v)$
41. $(5x^2 - 2)(5x^2 + 2)$ 42. $(3y^2 + 1)(3y^2 - 1)$

74. $(0.1y + 0.5)^2$
75. $(a + b)^3$
76. $(2a - 3b)^3$
77. $(1.5x + 3.8)^2$
78. $(3.45a - 2.3)^2$
79. $(3.5t - 2.5)(3.5t + 2.5)$
80. $(4.5h + 5.7)(4.5h - 5.7)$

Mid-Chapter Quiz
Mid-Chapter Quizzes give students an earlier chance check their progress through the chapter allowing them to identify what past skills they need to practice as they move forward in their class.

Mid-Chapter **Quiz** | Sections 6.1 through 6.4 | Chapter 6

Reduce to lowest terms.

1. $\dfrac{36}{84}$ 2. $\dfrac{8x - 2}{8}$

3. $\dfrac{w^2 - 1}{2w + 2}$ 4. $\dfrac{2a^2 - 10a + 12}{6 - 3a}$

Perform the indicated operation.

5. $\dfrac{6}{7} \cdot \dfrac{21}{10}$ 6. $\dfrac{3xy^2}{5z} \cdot \dfrac{8x^2z^3}{8y^4}$

13. $\dfrac{5}{6} - \dfrac{5}{21}$ 14. $\dfrac{4}{ab^3} + \dfrac{5}{a^2b}$

15. $\dfrac{3x}{x + 1} + \dfrac{x}{x^2 + 2x + 1}$ 16. $\dfrac{y}{y + 5} - \dfrac{y}{y + 2}$

17. $\dfrac{1}{a} + \dfrac{1}{b} + \dfrac{1}{c}$

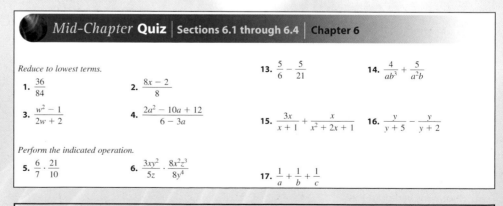

"This text is very well written with good, detailed examples. It offers plenty of practice exercises in each section including several real world applications."

Randall Casleton, University of Tennessee–Martin

Wrap-Up
The extensive and varied review in the chapter Wrap-Up will help students prepare for tests. First comes the **Summary** with key terms and concepts illustrated by examples; then **Enriching Your Mathematical Word Power** enables students to test their recall of new terminology in a fill-in-the-blank format.

Chapter 5 Wrap-Up

Summary

Factoring		Examples
Prime number	A positive integer larger than 1 that has no integral factors other than 1 and itself	2, 3, 5, 7, 11
Prime polynomial	A polynomial that cannot be factored is prime.	$x^2 + 3$ and $x^2 - x + 5$ are prime.

Enriching Your Mathematical Word Power

Fill in the blank.

1. A _____ number is an integer greater than 1 that has no integral factors other than itself and 1.

2. An integer larger than 1 that is not prime is _____.

3. A polynomial that has no factors is a _____ polynomial.

4. Writing a polynomial as a product is _____.

5. Writing a polynomial as a product of primes is factoring _____.

8. The trinomial $a^2 + 2ab + b^2$ is a perfect _____ trinomial.

9. The polynomial $a^3 + b^3$ is a ____ of two cubes.

10. The polynomial $a^3 - b^3$ is a _____ of two cubes.

11. A _____ equation is an equation of the form $ax^2 + bx + c = 0$.

12. According to the ____ factor property, if $ab = 0$ then $a = 0$ or $b = 0$.

Review Exercises

5.1 Factoring Out Common Factors
Find the prime factorization for each integer.

1. 144
2. 121
3. 58
4. 76
5. 150
6. 200

43. $r^2 - 4r - 60$
44. $x^2 + 13x + 40$
45. $y^2 - 6y - 55$
46. $a^2 + 6a - 40$
47. $u^2 + 26u + 120$
48. $v^2 - 22v - 75$

Next come **Review Exercises,** which are first linked back to the section of the chapter that they review, and then the exercises are mixed without section references in the **Miscellaneous** section.

Miscellaneous

Find all real and imaginary solutions to each equation.

63. $w^2 + 4 = 0$
64. $w^2 + 9 = 0$
65. $a^4 + 6a^2 + 8 = 0$
66. $b^4 + 13b^2 + 36 = 0$
67. $m^4 - 16 = 0$
68. $t^4 - 4 = 0$
69. $16b^4 - 1 = 0$
70. $b^4 - 81 = 0$
71. $x^3 + 1 = 0$
72. $x^3 - 1 = 0$
73. $x^3 + 8 = 0$

Photo for Exercise 81

Chapter Test
The test gives students additional practice to make sure they're ready for the real thing, with **all** answers provided at the back of the book and **all** solutions available in the Student's Solutions Manual.

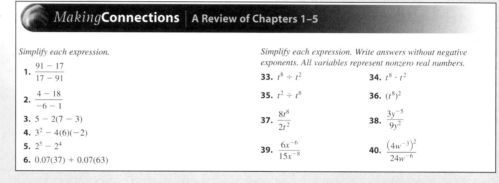

Chapter 5 Test

Give the prime factorization for each integer.

1. 66 **2.** 336

Find the greatest common factor (GCF) for each group.

3. 48, 80 **4.** 42, 66, 78

5. $6y^2$, $15y^3$ **6.** $12a^2b$, $18ab^2$, $24a^3b^3$

Factor each polynomial completely.

7. $5x^2 - 10x$ **8.** $6x^2y^2 + 12xy^2 + 12y^2$

Solve each equation.

22. $x^2 + 6x + 9 = 0$ **23.** $2x^2 + 5x - 12 = 0$

24. $3x^3 = 12x$ **25.** $(2x - 1)(3x + 5) = 5$

26. $\frac{1}{8}x^2 - \frac{3}{4}x + 1 = 0$ **27.** $0.3x^2 - 1.7x + 1 = 0$

The **Making Connections** feature following the Chapter Test is a cumulative review of all chapters up to and including the one just finished, helping to tie the course concepts together for students on a regular basis.

*Making***Connections** | A Review of Chapters 1–5

Simplify each expression.

1. $\frac{91 - 17}{17 - 91}$

2. $\frac{4 - 18}{-6 - 1}$

3. $5 - 2(7 - 3)$

4. $3^2 - 4(6)(-2)$

5. $2^5 - 2^4$

6. $0.07(37) + 0.07(63)$

Simplify each expression. Write answers without negative exponents. All variables represent nonzero real numbers.

33. $t^8 \div t^2$ **34.** $t^8 \cdot t^2$

35. $t^2 \div t^8$ **36.** $(t^8)^2$

37. $\frac{8t^8}{2t^2}$ **38.** $\frac{3y^{-5}}{9y^2}$

39. $\frac{6x^{-6}}{15x^{-8}}$ **40.** $\frac{(4w^{-3})^2}{24w^{-6}}$

Critical Thinking
The Critical Thinking section that concludes every chapter encourages students to think creatively to solve unique and intriguing problems and puzzles.

*Critical***Thinking** | For Individual or Group Work | **Chapter 6**

These exercises can be solved by a variety of techniques, which may or may not require algebra. So be creative and think critically. Explain all answers. Answers are in the Instructor's Edition of this text.

1. *Equilateral triangles.* Consider the sequence of three equilateral triangles shown in the accompanying figure.

  a) How many equilateral triangles are there in (a) of the accompanying figure?
  b) How many equilateral triangles congruent to the one in (a) can be found in (b) of the accompanying figure? How many are found in (c)?
  c) Suppose the sequence of equilateral triangles shown in (a), (b), and (c) is continued. How many equilateral triangles [congruent to the one in (a)] could be found in the *n*th such figure?

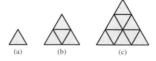

(a)   (b)   (c)

Figure for Exercise 1

4. *Eyes and feet.* A rancher has some sheep and ostriches. His young daughter observed that the animals have a total of 60 eyes and 86 feet. How many animals of each type does the rancher have?

Photo for Exercise 4

> "The critical thinking exercises at the end of the chapter are a good way to help students learn to work in groups and to write mathematically. Having to explain how and why you worked out a solution reinforces the thinking and writing skills necessary to be successful in today's world."
> *Mark Brenneman,*
> *Mesa Community College*

SUPPLEMENTS

Multimedia Supplements

MCGRAW-HILL HIGHER EDUCATION AND BLACKBOARD HAVE TEAMED UP.

Blackboard, the Web-based course-management system, has partnered with McGraw-Hill to better allow students and faculty to use online materials and activities to complement face-to-face teaching. Blackboard features exciting social learning and teaching tools that foster more logical, visually impactful and active learning opportunities for students. You'll transform your closed-door classrooms into communities where students remain connected to their educational experience 24 hours a day.

This partnership allows you and your students access to McGraw-Hill's Connect™ and Create™ right from within your Blackboard course—all with one single sign-on.

Not only do you get single sign-on with Connect™ and Create™, you also get deep integration of McGraw-Hill content and content engines right in Blackboard. Whether you're choosing a book for your course or building Connect™ assignments, all the tools you need are right where you want them—inside of Blackboard.

Gradebooks are now seamless. When a student completes an integrated Connect™ assignment, the grade for that assignment automatically (and instantly) feeds your Blackboard grade center.

McGraw-Hill and Blackboard can now offer you easy access to industry leading technology and content, whether your campus hosts it, or we do. Be sure to ask your local McGraw-Hill representative for details.

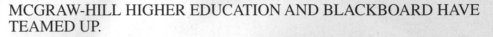

www.mcgrawhillconnect.com

McGraw-Hill conducted in-depth research to create a new and improved learning experience that meets the needs of today's students and instructors. The result is a reinvented learning experience rich in information, visually engaging, and easily accessible to both instructors and students.

McGraw-Hill's Connect is a Web-based assignment and assessment platform that helps students connect to their coursework and prepares them to succeed in and beyond the course. Connect Mathematics enables math instructors to create and share courses and assignments with colleagues and adjuncts with only a few clicks of the mouse. All exercises, learning objectives, videos, and activities are directly tied to text-specific material.

- Students have access to immediate feedback and help while working through assignments.

- A Web-optimized eBook is seamlessly integrated within ConnectPlus Mathematics.
- Instructors can select textbook-specific questions organized by chapter, section, and objective.
- Connect Mathematics' straightforward design creates and intuitive, visually pleasing grade management environment.

Instructors: To access Connect, request registration information from your McGraw-Hill sales representative.

Computerized Test Bank (CTB) Online (Instructors Only)

Available through Connect, this **computerized test bank,** utilizing Wimba Diploma® algorithm-based testing software, enables users to create customized exams quickly. This user-friendly program enables instructors to search for questions by topic, format, or difficulty level; to edit existing questions or to add new ones; and to scramble questions and answer keys for multiple versions of the same test. Hundreds of text-specific open-ended and multiple-choice questions are included in the question bank. Sample chapter tests in Microsoft Word® and PDF formats are also provided.

Online Instructor's Solutions Manual (Instructors Only)

Available on Connect, the Instructor's Solutions Manual provides comprehensive, **worked-out solutions** to all exercises in the text. The methods used to solve the problems in the manual are the same as those used to solve the examples in the textbook.

Video Lectures Available Online

In the videos, qualified teachers work through selected exercises from the textbook, following the solution methodology employed in the text. The video series is available online as an assignable element of Connect. The videos are closed-captioned for the hearing impaired, are subtitled in Spanish, and meet the Americans with Disabilities Act Standards for Accessible Design. Instructors may use them as resources in a learning center, for online courses, and/or to provide extra help for students who require extra practice.

ALEKS®   www.ALEKS.com

ALEKS (**A**ssessment and **LE**arning in **K**nowledge **S**paces) is a dynamic online learning system for mathematics education, available over the Web 24/7. ALEKS assesses students, accurately determines their knowledge, and then guides them to the material that they are most ready to learn. With a variety of reports, Textbook Integration Plus, quizzes, and homework assignment capabilities, ALEKS offers flexibility and ease of use for instructors.

- ALEKS uses artificial intelligence to determine exactly what each student knows and is ready to learn. ALEKS remediates student gaps and provides highly efficient learning and improved learning outcomes.
- ALEKS is a comprehensive curriculum that aligns with syllabi or specified textbooks. Used in conjunction with a McGraw-Hill text, students also receive links to text-specific videos, multimedia tutorials, and textbook pages.
- Textbook Integration Plus enables ALEKS to be automatically aligned with syllabi or specified McGraw-Hill textbooks with instructor-chosen dates, chapter goals, homework, and quizzes.
- ALEKS with AI-2 gives instructors increased control over the scope and sequence of student learning. Students using ALEKS demonstrate a steadily increasing mastery of the content of the course.
- ALEKS offers a dynamic classroom management system that enables instructors to monitor and direct student progress toward mastery of course objectives. See: www.aleks.com

Printed Supplements

Annotated Instructor's Edition (Instructors Only)

This ancillary contains answers to all exercises in the text. These answers are printed in a special color for ease of use by the instructor and are located on the appropriate pages throughout the text.

Student's Solutions Manual

The Student's Solutions Manual provides comprehensive, **worked-out solutions** to all of the odd-numbered section exercises and all exercises in the Mid-Chapter Quizzes, Chapter Tests, and Making Connections. The steps shown in the solutions match the style of solved examples in the textbook.

Contents

1 Chapter

Real Numbers and Their Properties   1

2 Chapter

Linear Equations and Inequalities in One Variable   85

Linear Equations in Two Variables and Their Graphs   169

Exponents and Polynomials   255

11 Chapter

Functions   689

12 Chapter

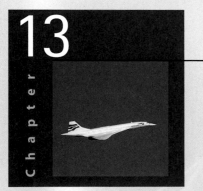

Exponential and Logarithmic Functions   787

13 Chapter

Nonlinear Systems and the Conic Sections   839

14 Sequences and Series (Available online at www.mhhe.com/dugopolski)   899

Appendix   A-1

Answers to Selected Exercises   A-59

Index   I-1

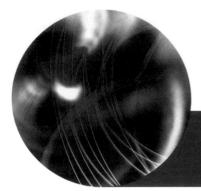

Applications Index

Real Numbers and Their Properties

It has been said that baseball is the "great American pastime." All of us who have played the game or who have only been spectators believe we understand the game. But do we realize that a pitcher must aim for an invisible three-dimensional target that is about 20 inches wide by 23 inches high by 17 inches deep and that a pitcher must throw so that the batter has difficulty hitting the ball? A curve ball may deflect 14 inches to skim over the outside corner of the plate, or a knuckle ball can break 11 inches off center when it is 20 feet from the plate and then curve back over the center of the plate.

The batter is trying to hit a rotating ball that can travel up to 120 miles per hour and must make split-second decisions about shifting his weight, changing his stride, and swinging the bat. The size of the bat each batter uses depends on his strengths, and pitchers in turn try to capitalize on a batter's weaknesses.

Millions of baseball fans enjoy watching this game of strategy and numbers. Many watch their favorite teams at the local ballparks, while others cheer for the home team on television. Of course, baseball fans are always interested in which team is leading the division and the number of games that their favorite team is behind the leader. Finding the number of games behind for each team in the division involves both arithmetic and algebra. Algebra provides the formula for finding games behind, and arithmetic is used to do the computations.

In Exercise 95 of Section 1.6 we will find the number of games behind for each team in the American League East.

| 1.1 | **The Real Numbers** |

The numbers that we use in algebra are called the real numbers. We start the discussion of the real numbers with some simpler sets of numbers.

⟨1⟩ The Integers

The most fundamental collection or **set** of numbers is the set of **counting numbers** or **natural numbers.** Of course, these are the numbers that we use for counting. The set of natural numbers is written in symbols as follows.

The Natural Numbers

$$\{1, 2, 3, \ldots\}$$

Braces, { }, are used to indicate a set of numbers. The three dots after 1, 2, and 3, which are read "and so on," mean that the pattern continues without end. There are infinitely many natural numbers.

The natural numbers, together with the number 0, are called the **whole numbers.** The set of whole numbers is written as follows.

The Whole Numbers

$$\{0, 1, 2, 3, \ldots\}$$

Figure 1.1

Although the whole numbers have many uses, they are not adequate for indicating losses or debts. A debt of \$20 can be expressed by the negative number -20 (negative twenty). See Fig. 1.1. When a thermometer reads 10 degrees below zero on a Fahrenheit scale, we say that the temperature is $-10°$F. See Fig. 1.2. The whole numbers together with the negatives of the counting numbers form the set of **integers.**

The Integers

$$\{\ldots, -3, -2, -1, 0, 1, 2, 3, \ldots\}$$

Degrees
Fahrenheit

Figure 1.2

⟨2⟩ The Rational Numbers

In arithmetic, we discuss and perform operations with specific numbers. In algebra, we like to make more general statements about numbers. In making general statements, we often use letters to represent numbers. A letter that is used to represent a number is called a **variable** because its value may vary. For example, we might say that a and b are integers. This means that a and b could be any of the infinitely many possible integers. They could be different integers or they could even be the same integer. We will use variables to describe the next set of numbers.

The set of **rational numbers** consists of all possible ratios of the form $\frac{a}{b}$, where a and b are integers, except that b is not allowed to be 0. For example,

$$\frac{1}{2}, \quad \frac{-9}{8}, \quad \frac{6}{1}, \quad \frac{150}{-70}, \quad \frac{2}{4}, \quad \frac{-9}{-1}, \quad \text{and} \quad \frac{0}{2}$$

are rational numbers. These numbers are not all in their simplest forms. We usually write 6 instead of $\frac{6}{1}$, $\frac{1}{2}$ instead of $\frac{2}{4}$, and 0 instead of $\frac{0}{2}$. A ratio such as $\frac{5}{0}$ does not represent any number. So we say that it is **undefined.** Any integer is a rational number because it could be written with a denominator of 1 as we did with 6 or $\frac{6}{1}$. Don't be concerned about how to simplify all of these ratios now. You will learn how to simplify all of them when we study fractions and signed numbers later in this chapter.

We cannot make a nice list of rational numbers like we did for the natural numbers, the whole numbers, and the integers. So we write the set of rational numbers in symbols using **set-builder notation** as follows.

The Rational Numbers

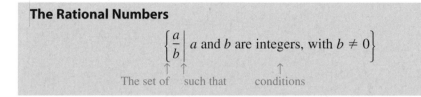

$$\left\{ \frac{a}{b} \,\middle|\, a \text{ and } b \text{ are integers, with } b \neq 0 \right\}$$

The set of   such that   conditions

We read this notation as "the set of all numbers of the form $\frac{a}{b}$, where a and b are integers, with b not equal to 0."

If you divide the denominator into the numerator, then you can convert a rational number to decimal form. As a decimal, every rational number either repeats indefinitely $\left(\frac{1}{3} = 0.\overline{3} = 0.333 \ldots \right)$ or terminates $\left(\frac{1}{8} = 0.125 \right)$. The line over the 3 indicates that it repeats forever. The part that repeats can have more digits than the display of your calculator. In this case you will have to divide by hand to do the conversion. For example, try converting $\frac{11}{17}$ to a repeating decimal.

‹3›   The Number Line

The number line is a diagram that helps us visualize numbers and their relationships to each other. A number line is like the scale on the thermometer in Fig. 1.2. To construct a number line, we draw a straight line and label any convenient point with the number 0. Now we choose any convenient length and use it to locate other points. Points to the right of 0 correspond to the positive numbers, and points to the left of 0 correspond to the negative numbers. Zero is neither positive nor negative. The number line is shown in Fig. 1.3.

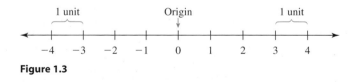

Figure 1.3

The numbers corresponding to the points on the line are called the **coordinates** of the points. The distance between two consecutive integers is called a **unit** and is the same for any two consecutive integers. The point with coordinate 0 is called the **origin.** The numbers on the number line increase in size from left to right. *When we compare the size of any two numbers, the larger number lies to the right of the smaller on the number line.* Zero is larger than any negative number and smaller than any positive number.

EXAMPLE   **1**   ## Comparing numbers on a number line

Determine which number is the larger in each given pair of numbers.

a) $-3, 2$ b) $0, -4$ c) $-2, -1$

Solution

a) The larger number is 2, because 2 lies to the right of -3 on the number line. In fact, any positive number is larger than any negative number.

b) The larger number is 0, because 0 lies to the right of -4 on the number line.

c) The larger number is -1, because -1 lies to the right of -2 on the number line.

Now do Exercises 1–12

The set of integers is illustrated or *graphed* in Fig. 1.4 by drawing a point for each integer. The three dots to the right and left below the number line and the blue arrows indicate that the numbers go on indefinitely in both directions.

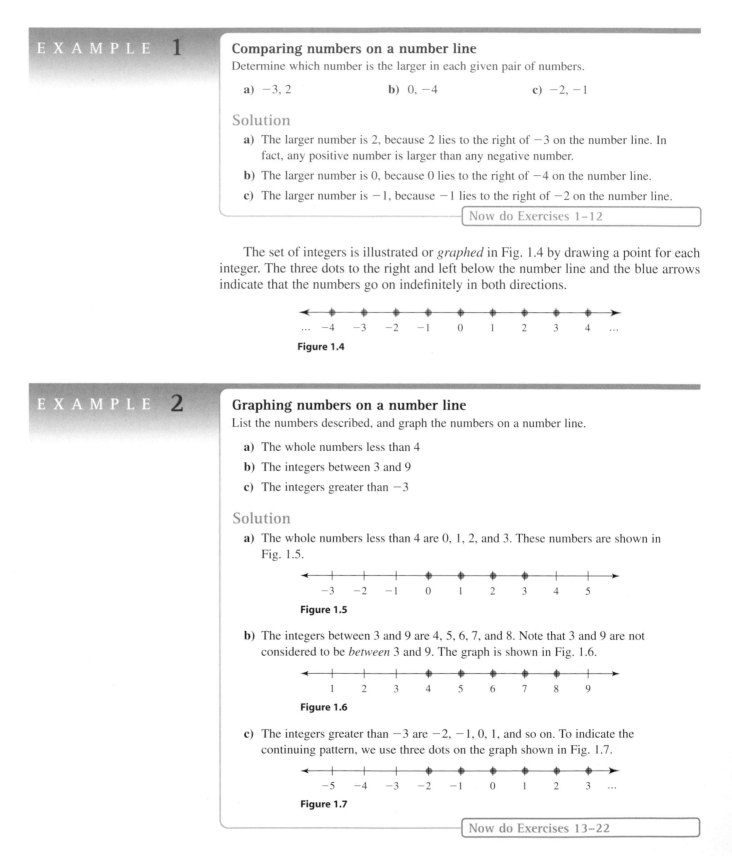

Figure 1.4

EXAMPLE   **2**   ## Graphing numbers on a number line

List the numbers described, and graph the numbers on a number line.

a) The whole numbers less than 4

b) The integers between 3 and 9

c) The integers greater than -3

Solution

a) The whole numbers less than 4 are 0, 1, 2, and 3. These numbers are shown in Fig. 1.5.

Figure 1.5

b) The integers between 3 and 9 are 4, 5, 6, 7, and 8. Note that 3 and 9 are not considered to be *between* 3 and 9. The graph is shown in Fig. 1.6.

Figure 1.6

c) The integers greater than -3 are $-2, -1, 0, 1$, and so on. To indicate the continuing pattern, we use three dots on the graph shown in Fig. 1.7.

Figure 1.7

Now do Exercises 13–22

⟨4⟩ The Real Numbers

For every rational number there is a point on the number line. For example, the number $\frac{1}{2}$ corresponds to a point halfway between 0 and 1 on the number line, and $-\frac{5}{4}$ corresponds to a point one and one-quarter units to the left of 0, as shown in Fig. 1.8. Since there is a correspondence between numbers and points on the number line, the points are often referred to as numbers.

Figure 1.8

⟨ **Calculator Close-Up** ⟩

A calculator can give rational approximations for irrational numbers such as $\sqrt{2}$ and π.

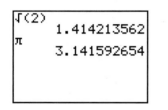

The calculator screens in this text may differ from the screen of the calculator model you use. If so, you may have to consult your manual to get the desired results.

The set of numbers that corresponds to *all* points on a number line is called the set of **real numbers** or *R*. A graph of the real numbers is shown on a number line by shading all points as in Fig. 1.9. All rational numbers are real numbers, but there are points on the number line that do not correspond to rational numbers. Those real numbers that are not rational are called **irrational.** An irrational number cannot be written as a ratio of integers. It can be shown that numbers such as $\sqrt{2}$ (the square root of 2) and π (Greek letter pi) are irrational. The number $\sqrt{2}$ is a number that can be multiplied by itself to obtain $2\,(\sqrt{2} \cdot \sqrt{2} = 2)$. The number π is the ratio of the circumference and diameter of any circle. Irrational numbers are not as easy to represent as rational numbers. That is why we use symbols such as $\sqrt{2}$, $\sqrt{3}$, and π for irrational numbers. When we perform computations with irrational numbers, we sometimes use rational approximations for them. For example, $\sqrt{2} \approx 1.414$ and $\pi \approx 3.14$. The symbol $\approx$ means "is approximately equal to." Note that not all square roots are irrational. For example, $\sqrt{9} = 3$, because $3 \cdot 3 = 9$. We will deal with irrational numbers in greater depth when we discuss roots in Chapter 9.

Figure 1.9

Figure 1.10 summarizes the sets of numbers that make up the real numbers, and shows the relationships between them.

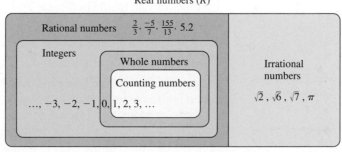

Figure 1.10

EXAMPLE **3**

Types of numbers

Determine whether each statement is true or false.

a) Every rational number is an integer.

b) Every counting number is an integer.

c) Every irrational number is a real number.

Solution

a) False. For example, $\frac{1}{2}$ is a rational number that is not an integer.

b) True, because the integers consist of the counting numbers, the negatives of the counting numbers, and zero.

c) True, because the rational numbers together with the irrational numbers form the real numbers.

> Now do Exercises 23–34

⟨5⟩ Intervals of Real Numbers

Retailers often have a sale for a certain *interval* of time. Between 6 A.M. and 8 A.M. you get a 20% discount. A **bounded** or finite interval of real numbers is the set of real numbers that are between two real numbers, which are called the **endpoints** of the interval. The endpoints may or may not belong to an interval. **Interval notation** is used to represent intervals of real numbers. In interval notation, parentheses are used to indicate that the endpoints do not belong to the interval and brackets indicate that the endpoints do belong to the interval. The following box shows the four types of finite intervals for two real numbers a and b, where a is less than b.

Finite Intervals		
Verbal Description	**Interval Notation**	**Graph**
The set of real numbers between a and b	(a, b)	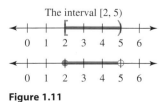 a b
The set of real numbers between a and b inclusive	$[a, b]$	a b
The set of real numbers greater than a and less than or equal to b	$(a, b]$	a b
The set of real numbers greater than or equal to a and less than b	$[a, b)$	a b

The interval [2, 5)

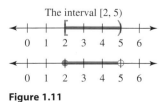

Figure 1.11

Note how the parentheses and brackets are used on the graph and in the interval notation. It is also common to draw the graph of an interval of real numbers using an open circle for an endpoint that does not belong to the interval and a closed circle for an endpoint that belongs to the interval. For example, see the graphs of the interval [2, 5) in Fig. 1.11. In this text, graphs of intervals will be drawn with parentheses and brackets so that they agree with interval notation.

E X A M P L E   **4**

Interval notation for finite intervals

Write the interval notation for each interval of real numbers and graph the interval.

  a) The set of real numbers greater than 3 and less than or equal to 5

  b) The set of real numbers between 0 and 4 inclusive

  c) The set of real numbers greater than or equal to -1 and less than 4

  d) The set of real numbers between -2 and -1

Solution

  a) The set of real numbers greater than 3 and less than or equal to 5 is written in interval notation as (3, 5] and graphed in Fig. 1.12.

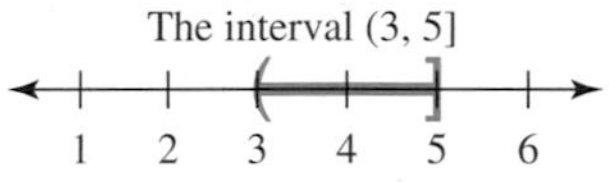

The interval (3, 5]

Figure 1.12

  b) The set of real numbers between 0 and 4 inclusive is written in interval notation as [0, 4] and graphed in Fig. 1.13.

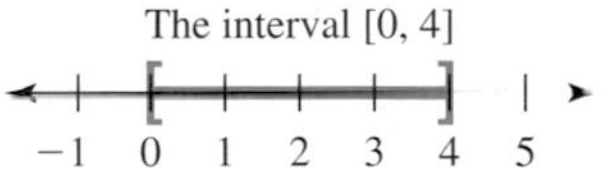

The interval [0, 4]

Figure 1.13

  c) The set of real numbers greater than or equal to -1 and less than 4 is written in interval notation as $[-1, 4)$ and graphed in Fig. 1.14.

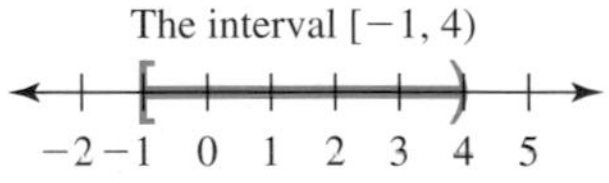

The interval $[-1, 4)$

Figure 1.14

  d) The set of real numbers between -2 and -1 is written in interval notation as $(-2, -1)$ and graphed in Fig. 1.15.

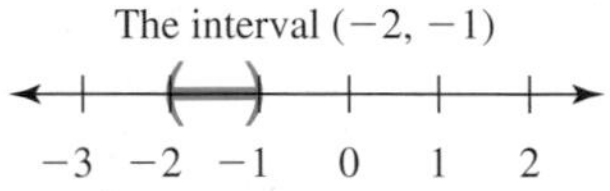

The interval $(-2, -1)$

Figure 1.15

Now do Exercises 35–40

Some sales never end. After 8 A.M. all merchandise is 10% off. An **unbounded** or **infinite interval** of real numbers is missing at least one endpoint. It may extend infinitely far to the right or left on the number line. In this case the infinity symbol ∞ is used as an endpoint in the interval notation. Note that parentheses are always used next to ∞ or $-\infty$ in interval notation, because ∞ is not a number. It is just used to indicate that there is no end to the interval. The following box shows the five types of infinite intervals for a real number a.

Infinite Intervals

Verbal Description	Interval Notation	Graph
The set of real numbers greater than a	(a, ∞)	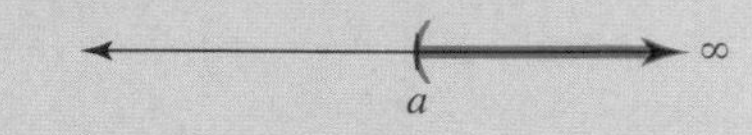
The set of real numbers greater than or equal to a	$[a, \infty)$	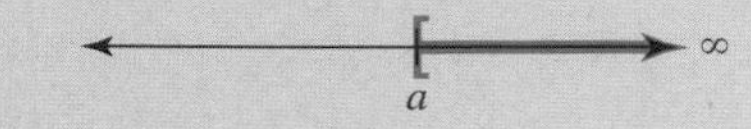
The set of real numbers less than a	$(-\infty, a)$	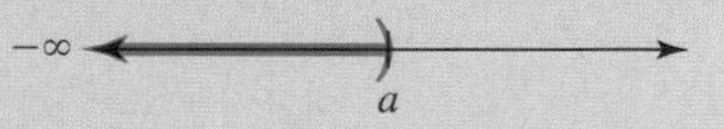
The set of real numbers less than or equal to a	$(-\infty, a]$	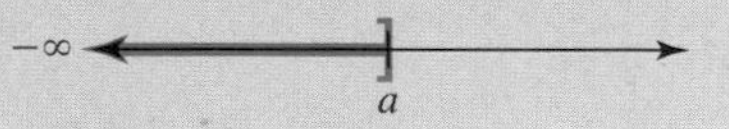
The set of all real numbers	$(-\infty, \infty)$	

E X A M P L E   5

Interval notation for infinite intervals

Write each interval of real numbers in interval notation and graph it.

a) The set of real numbers greater than or equal to 3

b) The set of real numbers less than -2

c) The set of real numbers greater than 2.5

Solution

a) The set of real numbers greater than or equal to 3 is written in interval notation as $[3, \infty)$ and graphed in Fig. 1.16.

The interval $[3, \infty)$

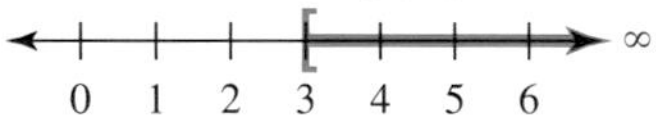

Figure 1.16

b) The set of real numbers less than -2 is written in interval notation as $(-\infty, -2)$ and graphed in Fig. 1.17.

The interval $(-\infty, -2)$

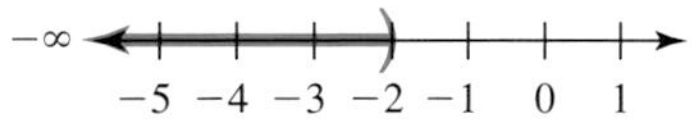

Figure 1.17

c) The set of real numbers greater than 2.5 is written in interval notation as $(2.5, \infty)$ and graphed in Fig. 1.18.

The interval $(2.5, \infty)$

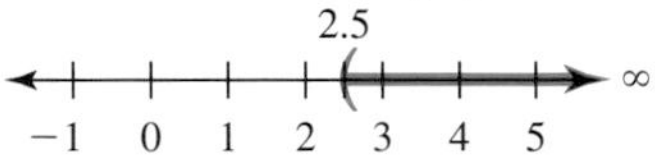

Figure 1.18

Now do Exercises 41–46

⟨6⟩ Absolute Value

The concept of absolute value will be used to define the basic operations with real numbers in Section 1.3. The **absolute value** of a number is the number's distance from 0 on the number line. For example, the numbers 5 and -5 are both five units away from 0 on the number line. So the absolute value of each of these numbers is 5. See Fig. 1.19. We write $|a|$ for "the absolute value of a." So,

$$|5| = 5 \qquad \text{and} \qquad |-5| = 5.$$

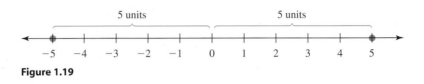

Figure 1.19

The notation $|a|$ represents distance, and distance is never negative. So $|a|$ is greater than or equal to zero for any real number a.

EXAMPLE 6

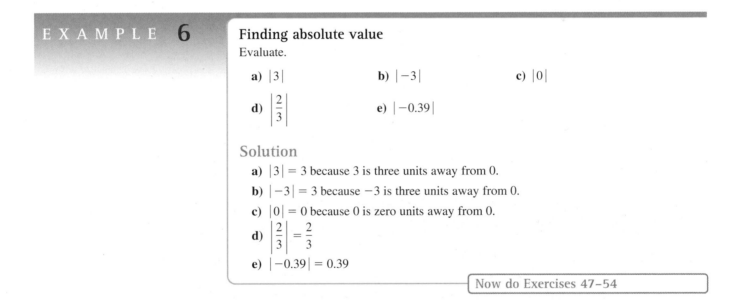

Finding absolute value

Evaluate.

a) $|3|$ b) $|-3|$ c) $|0|$

d) $\left|\dfrac{2}{3}\right|$ e) $|-0.39|$

Solution

a) $|3| = 3$ because 3 is three units away from 0.

b) $|-3| = 3$ because -3 is three units away from 0.

c) $|0| = 0$ because 0 is zero units away from 0.

d) $\left|\dfrac{2}{3}\right| = \dfrac{2}{3}$

e) $|-0.39| = 0.39$

Now do Exercises 47–54

Two numbers that are located on opposite sides of zero and have the same absolute value are called **opposites** of each other. The numbers 5 and -5 are opposites of each other. We say that the opposite of 5 is -5 and the opposite of -5 is 5. The symbol "$-$" is used to indicate "opposite" as well as "negative." When the negative sign is used before a number, it should be read as "negative." When it is used in front of parentheses or a variable, it should be read as "opposite." For example, $-(5) = -5$ means "the opposite of 5 is negative 5," and $-(-5) = 5$ means "the opposite of negative 5 is 5." Zero does not have an opposite in the same sense as nonzero numbers. Zero is its own opposite. We read $-(0) = 0$ as the "the opposite of zero is zero."

In general, $-a$ means "the opposite of a." If a is positive, $-a$ is negative. If a is negative, $-a$ is positive. Opposites have the following property.

> **Opposite of an Opposite**
>
> For any real number a,
>
> $$-(-a) = a.$$

Remember that we have defined $|a|$ to be the distance between 0 and a on the number line. Using opposites, we can give a symbolic definition of absolute value.

> **Absolute Value**
>
> $$|a| = \begin{cases} a & \text{if } a \text{ is positive or zero} \\ -a & \text{if } a \text{ is negative} \end{cases}$$

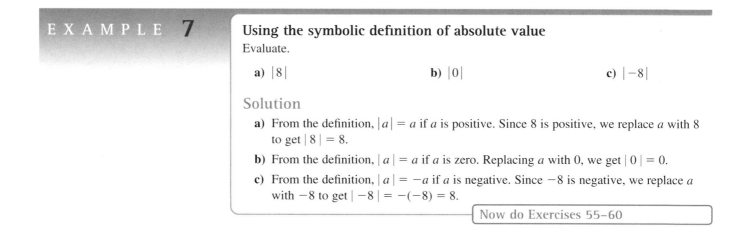

E X A M P L E   7

Using the symbolic definition of absolute value
Evaluate.

a) $|8|$ b) $|0|$ c) $|-8|$

Solution

a) From the definition, $|a| = a$ if a is positive. Since 8 is positive, we replace a with 8 to get $|8| = 8$.

b) From the definition, $|a| = a$ if a is zero. Replacing a with 0, we get $|0| = 0$.

c) From the definition, $|a| = -a$ if a is negative. Since -8 is negative, we replace a with -8 to get $|-8| = -(-8) = 8$.

> Now do Exercises 55–60

Warm-Ups ▼

Fill in the blank.

1. The set of _____ is $\{\ldots, -3, -2, -1, 0, 1, 2, 3, \ldots\}$.

2. The set of _____ numbers is $\{1, 2, 3, \ldots\}$.

3. Every _____ number can be expressed as a ratio of integers.

4. _____ and _____ decimal numbers are rational numbers.

5. A decimal number that does not repeat and does not terminate is _____.

6. The rationals together with the irrationals form the set of _____ numbers.

7. The ratio of the _____ and diameter of any circle is π.

8. The _____ of a number is its distance form 0 on a number line.

True or false?

9. The natural numbers and the counting numbers are the same.

10. Zero is a counting number.

11. Zero is an irrational number.

12. The opposite of negative 3 is positive 3.

13. The absolute value of 4 is -4.

14. The real number π is in the interval $(3, 4)$.

15. The interval $(4, 9)$ contains 8.

16. The interval $(2, 6)$ contains 6.

17. The interval $[3, 5]$ contains 3.

18. The interval $(9, \infty)$ contains 88 trillion.

‹ 3 › **The Number Line**

Determine which number is the larger in each given pair of numbers. See Example 1.

1. 0, 6   **2.** 7, 4

3. −3, 6   **4.** 7, −10

5. 0, −6   **6.** −8, 0

7. −3, −2   **8.** −5, −8

9. −12, −15   **10.** −13, −7

11. −2.9, −2.1   **12.** 2.1, 2.9

List the numbers described and graph them on a number line. See Example 2.

13. The counting numbers smaller than 6

14. The natural numbers larger than 4

15. The whole numbers smaller than 5

16. The integers between −3 and 3

17. The whole numbers between −5 and 5

18. The integers smaller than −1

19. The counting numbers larger than −4

20. The natural numbers between −5 and 7

21. The integers larger than $\frac{1}{2}$

22. The whole numbers smaller than $\frac{7}{4}$

‹ 4 › **The Real Numbers**

Determine whether each statement is true or false. Explain your answer. See Example 3.

23. Every integer is a rational number.

24. Every counting number is a whole number.

25. Zero is a counting number.

26. Every whole number is a counting number.

27. The ratio of the circumference and diameter of a circle is an irrational number.

28. Every rational number can be expressed as a ratio of integers.

29. Every whole number can be expressed as a ratio of integers.

30. Some of the rational numbers are integers.

31. Some of the integers are natural numbers.

32. There are infinitely many rational numbers.

33. Zero is an irrational number.

34. Every irrational number is a real number.

‹ 5 › **Intervals of Real Numbers**

Write each interval of real numbers in interval notation and graph it. See Example 4.

35. The set of real numbers between 0 and 1

36. The set of real numbers between 2 and 6

37. The set of real numbers between −2 and 2 inclusive

38. The set of real numbers between −3 and 4 inclusive

39. The set of real numbers greater than 0 and less than or equal to 5

40. The set of real numbers greater than or equal to −1 and less than 6

Write each interval of real numbers in interval notation and graph it. See Example 5.

41. The set of real numbers greater than 4

42. The set of real numbers greater than 2

43. The set of real numbers less than or equal to −1

44. The set of real numbers less than or equal to −4

45. The set of real numbers greater than or equal to 0

46. The set of real numbers greater than or equal to 6

⟨6⟩ Absolute Value

Determine the values of the following. See Examples 6 and 7.

47. $|-6|$ **48.** $|4|$
49. $|0|$ **50.** $|2|$
51. $|7|$ **52.** $|-7|$
53. $|-9|$ **54.** $|-2|$
55. $|-45|$ **56.** $|-30|$
57. $\left|\dfrac{3}{4}\right|$ **58.** $\left|-\dfrac{1}{2}\right|$
59. $|-5.09|$ **60.** $|0.00987|$

Select the smaller number in each given pair of numbers.

61. $-16, 9$ **62.** $-12, -7$
63. $-\dfrac{5}{2}, -\dfrac{9}{4}$ **64.** $\dfrac{5}{8}, \dfrac{6}{7}$

65. $|-3|, 2$ **66.** $|-6|, 0$
67. $|-4|, 3$ **68.** $|5|, -4$

Which number in each given pair has the larger absolute value?

69. $-5, -9$ **70.** $-12, -8$
71. $16, -9$ **72.** $-12, 7$

Determine which number in each pair is closer to 0 on the number line.

73. $-4, -5$ **74.** $-8.1, 7.9$
75. $-2.01, -1.99$ **76.** $2.01, 1.99$

77. $-75, 74$ **78.** $-75, -74$

What is the distance on the number line between 0 and each of the following numbers?

79. 5.25 **80.** 4.2 **81.** -40
82. -33 **83.** $-\dfrac{1}{2}$ **84.** $-\dfrac{1}{3}$

Consider the following nine integers:

$$-4, -3, -2, -1, 0, 1, 2, 3, 4$$

85. Which of these integers has an absolute value equal to 3?

86. Which of these integers has an absolute value equal to 0?

87. Which of these integers has an absolute value greater than 2?

88. Which of these integers has an absolute value greater than 1?

89. Which of these integers has an absolute value less than 2?

90. Which of these integers has an absolute value less than 4?

Miscellaneous

Write the interval notation for the interval of real numbers shown in each graph.

91.
92.
93.
94.

95.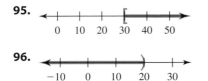

A number line marked at 0, 10, 20, 30, 40, 50 with a bracket starting at 30 and an arrow extending right.

96.

A number line marked at −10, 0, 10, 20, 30 with a parenthesis at 20 and an arrow extending left.

True or false? Explain your answer.

97. If we add the absolute values of −3 and −5, we get 8.

98. If we multiply the absolute values of −2 and 5, we get 10.

99. The absolute value of any negative number is greater than 0.

100. The absolute value of any positive number is less than 0.

101. The absolute value of −9 is larger than the absolute value of 6.

102. The absolute value of 12 is larger than the absolute value of −11.

Getting More Involved

103. *Exploration*

a) Find a rational number between $\frac{1}{3}$ and $\frac{1}{4}$.

b) Find a rational number between -3.205 and -3.114.

c) Find a rational number between $\frac{2}{3}$ and 0.6667.

d) Explain how to find a rational number between any two given rational numbers.

104. *Discussion*

Suppose that a is a negative real number. Determine whether each of the following is positive or negative, and explain your answer.

a) $-a$ b) $|-a|$ c) $-|a|$ d) $-(-a)$ e) $-|-a|$

105. *Discussion*

Determine whether each number listed in the following table is a member of each set listed on the side of the table. For example, $\frac{1}{2}$ is a real number and a rational number. So check marks are placed in those two cells of the table.

	$\frac{1}{2}$	-2	π	$\sqrt{3}$	$\sqrt{9}$	6	0	$-\frac{7}{3}$
Real	✓							
Irrational								
Rational	✓							
Integer								
Whole								
Counting								

1.2   Fractions

In This Section

⟨1⟩ Equivalent Fractions

⟨2⟩ Multiplying Fractions

⟨3⟩ Unit Conversion

⟨4⟩ Dividing Fractions

⟨5⟩ Adding and Subtracting Fractions

⟨6⟩ Fractions, Decimals, and Percents

⟨7⟩ Applications

In this section and Sections 1.3 and 1.4 we will discuss operations performed with real numbers. We begin by reviewing operations with fractions. Note that this section on fractions is not an entire arithmetic course. We are simply reviewing selected fraction topics that will be used in this text.

⟨1⟩ Equivalent Fractions

If a pizza is cut into 3 equal pieces and you eat 2 of them, then you have eaten $\frac{2}{3}$ of the pizza. We read $\frac{2}{3}$ as "two-thirds." The rational number $\frac{2}{3}$ is a *fraction*. Any rational number that is not an integer is a **fraction**. The top number is the **numerator** and the bottom number is the **denominator**.

If a pizza is cut into 6 equal pieces and you eat 4 of them, then you have eaten $\frac{4}{6}$ (four-sixths) of the pizza. Figure 1.20 shows that $\frac{4}{6}$ of a pizza is the same amount

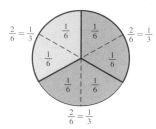

Figure 1.20

as $\frac{2}{3}$ of a pizza. So $\frac{4}{6}$ is **equal** or **equivalent** to $\frac{2}{3}$. Every fraction can be written in infinitely many equivalent forms. Consider the following equivalent form of $\frac{2}{3}$:

$$\frac{2}{3} = \frac{4}{6} = \frac{6}{9} = \frac{8}{12} = \frac{10}{15} = \cdots$$

The three dots
mean "and so on."

Notice that each equivalent form of $\frac{2}{3}$ can be obtained by multiplying the numerator and denominator by the same nonzero number. For example,

$$\frac{2}{3} = \frac{2 \cdot 5}{3 \cdot 5} = \frac{10}{15}.$$ The raised dot indicates multiplication.

Converting a fraction into an equivalent fraction with a larger denominator is called **building up** the fraction. As we have just seen, $\frac{2}{3}$ is built up to $\frac{10}{15}$ by multiplying its numerator and denominator by 5.

Building Up Fractions

If $b \neq 0$ and $c \neq 0$, then

$$\frac{a}{b} = \frac{a \cdot c}{b \cdot c}.$$

Multiplying the numerator and denominator of a fraction by a nonzero number changes the fraction's appearance but not its value.

E X A M P L E   1

Building up fractions

Build up each fraction so that it is equivalent to the fraction with the indicated denominator.

a) $\frac{3}{4} = \frac{?}{28}$ **b)** $\frac{5}{3} = \frac{?}{30}$

Solution

a) Because $4 \cdot 7 = 28$, we multiply both the numerator and denominator by 7:

$$\frac{3}{4} = \frac{3 \cdot 7}{4 \cdot 7} = \frac{21}{28}$$

b) Because $3 \cdot 10 = 30$, we multiply both the numerator and denominator by 10:

$$\frac{5}{3} = \frac{5 \cdot 10}{3 \cdot 10} = \frac{50}{30}$$

Now do Exercises 1–12

The method for building up fractions shown in Example 1 will be used again on rational expressions in Chapter 6. So it is good to use this method and show the details. The same goes for the method of reducing fractions that is coming next.

If we convert a fraction to an equivalent fraction with a smaller denominator, we are **reducing** the fraction. For example, to reduce $\frac{10}{15}$, we *factor* 10 as $2 \cdot 5$ and 15 as $3 \cdot 5$, and then **divide out** or **cancel** the common factor 5.

$$\frac{10}{15} = \frac{2 \cdot \cancel{5}}{3 \cdot \cancel{5}} = \frac{2}{3}$$

The fraction $\frac{2}{3}$ cannot be reduced further because the numerator 2 and the denominator 3 have no factors (other than 1) in common. So we say that $\frac{2}{3}$ is in **lowest terms.**

> **Reducing Fractions**
> If $b \neq 0$ and $c \neq 0$, then
> $$\frac{a \cdot c}{b \cdot c} = \frac{a}{b}.$$

CAUTION Reducing a fraction changes its appearance, but not its value. The fraction $\frac{2}{3}$ is *not* smaller than $\frac{10}{15}$.

E X A M P L E **2**

Reducing fractions

Reduce each fraction to lowest terms.

a) $\frac{15}{24}$ **b)** $\frac{42}{30}$ **c)** $\frac{13}{26}$ **d)** $\frac{35}{7}$

Solution

For each fraction, factor the numerator and denominator and then divide by the common factor:

a) $\frac{15}{24} = \frac{3 \cdot 5}{3 \cdot 8} = \frac{5}{8}$ **b)** $\frac{42}{30} = \frac{7 \cdot \cancel{6}}{5 \cdot \cancel{6}} = \frac{7}{5}$

c) $\frac{13}{26} = \frac{1 \cdot \cancel{13}}{2 \cdot \cancel{13}} = \frac{1}{2}$ The number 1 in the numerator is essential.

d) $\frac{35}{7} = \frac{5 \cdot \cancel{7}}{1 \cdot \cancel{7}} = \frac{5}{1} = 5$

> Now do Exercises 13–28

‹ **Calculator Close-Up** ›

To reduce a fraction to lowest terms using a graphing calculator, display the fraction and use the fraction feature.

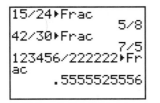

If the fraction is too complicated, the calculator will return a decimal equivalent instead of reducing it.

Strategy for Obtaining Equivalent Fractions

Equivalent fractions can be obtained by multiplying or dividing the numerator and denominator by the same nonzero number.

‹2› Multiplying Fractions

Suppose a pizza is cut into three equal pieces. If you eat $\frac{1}{2}$ of one piece, you have eaten $\frac{1}{6}$ of the pizza. See Fig. 1.21. You can obtain $\frac{1}{6}$ by multiplying $\frac{1}{2}$ and $\frac{1}{3}$:

$$\frac{1}{2} \cdot \frac{1}{3} = \frac{1 \cdot 1}{2 \cdot 3} = \frac{1}{6}$$

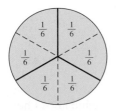

Figure 1.21

This example illustrates the definition of multiplication of fractions. To multiply two fractions, we multiply their numerators and multiply their denominators.

> **Multiplication of Fractions**
>
> If $b \neq 0$ and $d \neq 0$, then
>
> $$\frac{a}{b} \cdot \frac{c}{d} = \frac{a \cdot c}{b \cdot d}.$$

We can multiply the numerators and the denominators and then reduce, as in Example 3(a) or we can reduce before multiplying as in Example 3(b) and (c). It is usually simpler to reduce before multiplying.

EXAMPLE 3

Multiplying fractions

Find each product.

a) $\dfrac{2}{3} \cdot \dfrac{5}{8}$ b) $\dfrac{1}{3} \cdot \dfrac{3}{4}$ c) $\dfrac{4}{5} \cdot \dfrac{15}{22}$

Solution

a) First multiply the numerators and the denominators, and then reduce:

$$\frac{2}{3} \cdot \frac{5}{8} = \frac{10}{24}$$

$$= \frac{2 \cdot 5}{2 \cdot 12} \qquad \text{Factor the numerator and denominator.}$$

$$= \frac{5}{12} \qquad \text{Divide out the common factor 2.}$$

b) Reduce before multiplying:

$$\frac{1}{3} \cdot \frac{3}{4} = \frac{1}{\cancel{3}} \cdot \frac{\cancel{3}}{4} = \frac{1}{4}$$

c) Factor the numerators and denominators, and then divide out the common factors before multiplying:

$$\frac{4}{5} \cdot \frac{15}{22} = \frac{2 \cdot 2}{\cancel{5}} \cdot \frac{3 \cdot \cancel{5}}{\cancel{2} \cdot 11} = \frac{6}{11}$$

> Now do Exercises 29–40

‹ Calculator Close-Up ›

A graphing calculator can multiply fractions and get fractional answers using the fraction feature.

```
2/3*5/8►Frac
           5/12
1/3*3/4►Frac
            1/4
4/5*15/22►Frac
           6/11
```

Multiplication of fractions can help us better understand the idea of building up fractions. For example, we have already seen that multiplying $\frac{2}{3}$ by 5 in its numerator and denominator builds it up to $\frac{10}{15}$:

$$\frac{2}{3} = \frac{2 \cdot 5}{3 \cdot 5} = \frac{10}{15}$$

We can get this same result by multiplying $\frac{2}{3}$ by 1, using $\frac{5}{5}$ for 1:

$$\frac{2}{3} = \frac{2}{3} \cdot 1 = \frac{2}{3} \cdot \frac{5}{5} = \frac{10}{15}$$

So building up a fraction is equivalent to multiplying it by 1, which does not change its value.

⟨3⟩ Unit Conversion

Most measurements can be expressed in a variety of units. For example, distance could be in miles or kilometers. Converting from one unit of measurement to another can always be done by multiplying by a conversion factor expressed as a fraction. (Some common conversion factors can be found on the inside back cover of this text.) This method is called **cancellation of units,** because the units cancel just like the common factors cancel in multiplication of fractions.

EXAMPLE 4

Unit conversion

a) Convert 6 yards to feet.

b) Convert 12 miles to kilometers.

c) Convert 60 miles per hour to feet per second.

Solution

a) Because 3 feet = 1 yard, multiplying by $\frac{3\ \text{feet}}{1\ \text{yard}}$ is equivalent to multiplying by 1. Notice how yards cancels and the result is feet.

$$6\ \text{yd} = 6\ \cancel{\text{yd}} \cdot \frac{3\ \text{ft}}{1\ \cancel{\text{yd}}} = 18\ \text{ft}$$

b) There are two ways to convert 12 miles to kilometers using the conversion factors given on the inside back cover:

$$12\ \text{mi} = 12\ \cancel{\text{mi}} \cdot \frac{1.609\ \text{km}}{1\ \cancel{\text{mi}}} \approx 19.31\ \text{km}$$

$$12\ \text{mi} = 12\ \cancel{\text{mi}} \cdot \frac{1\ \text{km}}{0.6214\ \cancel{\text{mi}}} \approx 19.31\ \text{km}$$

Notice that in the second method we are also multiplying by a fraction that is equivalent to 1, but we actually divide 12 by 0.6214.

c) Convert 60 miles per hour to feet per second as follows:

$$60\ \text{mi/hr} = \frac{60\ \cancel{\text{mi}}}{1\ \cancel{\text{hr}}} \cdot \frac{5280\ \text{ft}}{1\ \cancel{\text{mi}}} \cdot \frac{1\ \cancel{\text{hr}}}{60\ \cancel{\text{min}}} \cdot \frac{1\ \cancel{\text{min}}}{60\ \text{sec}} = 88\ \text{ft/sec}$$

> Now do Exercises 41–52

⟨4⟩ Dividing Fractions

Suppose that a pizza is cut into three pieces. If one piece is divided between two people $\left(\frac{1}{3} \div 2\right)$, then each of these two people gets $\frac{1}{6}$ of the pizza. Of course $\frac{1}{3}$ times $\frac{1}{2}$ is also $\frac{1}{6}$. So dividing by 2 is equivalent to multiplying by $\frac{1}{2}$. In symbols:

$$\frac{1}{3} \div 2 = \frac{1}{3} \div \frac{2}{1} = \frac{1}{3} \cdot \frac{1}{2} = \frac{1}{6}$$

The pizza example illustrates the general rule for dividing fractions.

Division of Fractions

If $b \neq 0$, $c \neq 0$, and $d \neq 0$, then

$$\frac{a}{b} \div \frac{c}{d} = \frac{a}{b} \cdot \frac{d}{c}.$$

In general if $m \div n = p$, then n is called the **divisor** and p (the result of the division) is called the **quotient** of m and n. We also refer to $m \div n$ and $\frac{m}{n}$ as the quotient of m and n. So in words, *to find the quotient of two fractions we invert the divisor and multiply.*

EXAMPLE 5	**Dividing fractions** Find the indicated quotients.

a) $\dfrac{1}{3} \div \dfrac{7}{6}$ **b)** $\dfrac{2}{3} \div 5$ **c)** $\dfrac{3}{8} \div \dfrac{3}{2}$

‹ **Calculator Close-Up** ›

When the divisor is a fraction on a graphing calculator, it must be in parentheses. A different result is obtained without using parentheses. Note that when the divisor is a whole number, parentheses are not necessary.

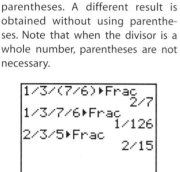

Try these computations on your calculator.

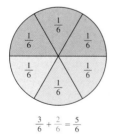

$$\frac{3}{6} + \frac{2}{6} = \frac{5}{6}$$

Figure 1.22

Solution

In each case we invert the divisor (the number on the right) and multiply.

a) $\dfrac{1}{3} \div \dfrac{7}{6} = \dfrac{1}{3} \cdot \dfrac{6}{7}$ Invert the divisor.

$\qquad = \dfrac{1}{3} \cdot \dfrac{2 \cdot 3}{7}$ Reduce.

$\qquad = \dfrac{2}{7}$ Multiply.

b) $\dfrac{2}{3} \div 5 = \dfrac{2}{3} \div \dfrac{5}{1} = \dfrac{2}{3} \cdot \dfrac{1}{5} = \dfrac{2}{15}$

c) $\dfrac{3}{8} \div \dfrac{3}{2} = \dfrac{3}{8} \cdot \dfrac{2}{3} = \dfrac{3 \cdot 1}{4 \cdot 2} \cdot \dfrac{2}{3} = \dfrac{1}{4}$

> Now do Exercises 53–62

‹5› Adding and Subtracting Fractions

To understand addition and subtraction of fractions, again consider the pizza that is cut into six equal pieces as shown in Fig. 1.22. If you eat $\frac{3}{6}$ and your friend eats $\frac{2}{6}$, together you have eaten $\frac{5}{6}$ of the pizza. Similarly, if you remove $\frac{1}{6}$ from $\frac{6}{6}$, you have $\frac{5}{6}$ left. To add or subtract fractions with identical denominators, we add or subtract their numerators and write the result over the common denominator.

Addition and Subtraction of Fractions

If $b \neq 0$, then

$$\frac{a}{b} + \frac{c}{b} = \frac{a+c}{b} \qquad \text{and} \qquad \frac{a}{b} - \frac{c}{b} = \frac{a-c}{b}.$$

An **improper fraction** is a fraction in which the numerator is larger than the denominator. For example, $\frac{7}{6}$ is an improper fraction. A **mixed number** is a natural number plus a fraction, with the plus sign removed. For example, $1\frac{1}{6}$ (or $1 + \frac{1}{6}$) is a mixed number. Since $1 + \frac{1}{6} = \frac{6}{6} + \frac{1}{6} = \frac{7}{6}$, we have $1\frac{1}{6} = \frac{7}{6}$.

EXAMPLE **6**

A good way to remember that you need common denominators for addition is to think of a simple example. If you own $\frac{1}{3}$ share of a car wash and your spouse owns $\frac{1}{3}$, then together you own $\frac{2}{3}$ of the business.

Adding and subtracting fractions
Perform the indicated operations.

a) $\frac{1}{7} + \frac{2}{7}$ **b)** $\frac{7}{10} - \frac{3}{10}$

Solution

a) $\frac{1}{7} + \frac{2}{7} = \frac{3}{7}$ **b)** $\frac{7}{10} - \frac{3}{10} = \frac{4}{10} = \frac{2 \cdot 2}{2 \cdot 5} = \frac{2}{5}$

> Now do Exercises 63–66

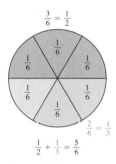

$$\frac{3}{6} = \frac{1}{2}$$

$$\frac{2}{6} = \frac{1}{3}$$

$$\frac{1}{2} + \frac{1}{3} = \frac{5}{6}$$

Figure 1.23

CAUTION Do not add the denominators when adding fractions: $\frac{1}{7} + \frac{2}{7} \neq \frac{3}{14}$.

To add or subtract fractions with different denominators, we must convert them to equivalent fractions with the same denominator and then add or subtract. For example, to add $\frac{1}{2}$ and $\frac{1}{3}$, we build up each fraction to a denominator of 6. See Fig. 1.23. Since $\frac{1}{2} = \frac{3}{6}$ and $\frac{1}{3} = \frac{2}{6}$, we have

$$\frac{1}{2} + \frac{1}{3} = \frac{3}{6} + \frac{2}{6} = \frac{5}{6}.$$

The smallest number that is a multiple of the denominators of two or more fractions is called the **least common denominator (LCD)**. So 6 is the LCD for $\frac{1}{2}$ and $\frac{1}{3}$. Note that we obtained the LCD 6 by examining Fig. 1.23. We must have a more systematic way.

The procedure for finding the LCD is based on factors. For example, to find the LCD for the denominators 6 and 9, factor 6 and 9 as $6 = 2 \cdot 3$ and $9 = 3 \cdot 3$. To obtain a multiple of both 6 and 9 the number must have two 3's as factors and one 2. So the LCD for 6 and 9 is $2 \cdot 3 \cdot 3$ or 18. If any number is omitted from $2 \cdot 3 \cdot 3$, we will not have a multiple of both 6 and 9. So each factor found in either 6 or 9 appears in the LCD the maximum number of times that it appears in either 6 or 9. The general strategy follows.

The *least* common denominator is *greater than* or equal to all of the denominators, because they must all divide into the LCD.

Strategy for Finding the LCD

1. Factor each denominator completely.
2. Determine the maximum number of times each distinct factor occurs in any denominator.
3. The LCD is the product of all of the distinct factors, where each factor is used the maximum number of times from step 2.

Note that a **prime number** is a number 2 or larger that has no factors other than itself and 1. If a denominator is prime (such as 2, 3, 5, 7, 11), then we do not factor it. A number is **factored completely** when it is written as a product of prime numbers.

E X A M P L E   **7**

Adding and subtracting fractions

Perform the indicated operations.

a) $\dfrac{3}{4} + \dfrac{1}{6}$ b) $\dfrac{1}{3} - \dfrac{1}{12}$

c) $\dfrac{7}{12} + \dfrac{5}{18}$ d) $2\dfrac{1}{3} + \dfrac{5}{9}$

Solution

a) First factor the denominators as $4 = 2 \cdot 2$ and $6 = 2 \cdot 3$. Since 2 occurs twice in 4 and once in 6, it appears twice in the LCD. Since 3 appears once in 6 and not at all in 4, it appears once in the LCD. So the LCD is $2 \cdot 2 \cdot 3$ or 12. Now build up each denominator to 12:

$$\frac{3}{4} + \frac{1}{6} = \frac{3 \cdot 3}{4 \cdot 3} + \frac{1 \cdot 2}{6 \cdot 2} \qquad \text{Build up each denominator to 12.}$$

$$= \frac{9}{12} + \frac{2}{12} \qquad \text{Simplify.}$$

$$= \frac{11}{12} \qquad \text{Add.}$$

b) The denominators are 12 and 3. Factor 12 as $12 = 2 \cdot 6 = 2 \cdot 2 \cdot 3$. Since 3 is a prime number, we do not factor it. Since 2 occurs twice in 12 and not at all in 3, it appears twice in the LCD. Since 3 occurs once in 3 and once in 12, 3 appears once in the LCD. The LCD is $2 \cdot 2 \cdot 3$ or 12. So we must build up $\frac{1}{3}$ to have a denominator of 12:

$$\frac{1}{3} - \frac{1}{12} = \frac{1 \cdot 4}{3 \cdot 4} - \frac{1}{12} \qquad \text{Build up the first fraction to the LCD.}$$

$$= \frac{4}{12} - \frac{1}{12} \qquad \text{Simplify.}$$

$$= \frac{3}{12} \qquad \text{Subtract.}$$

$$= \frac{1}{4} \qquad \text{Reduce to lowest terms.}$$

c) Since $12 = 2 \cdot 6 = 2 \cdot 2 \cdot 3$ and $18 = 2 \cdot 9 = 2 \cdot 3 \cdot 3$, the factor 2 appears twice in the LCD and the factor 3 appears twice in the LCD. So the LCD is $2 \cdot 2 \cdot 3 \cdot 3$ or 36:

$$\frac{7}{12} + \frac{5}{18} = \frac{7 \cdot 3}{12 \cdot 3} + \frac{5 \cdot 2}{18 \cdot 2} \qquad \text{Build up each denominator to 36.}$$

$$= \frac{21}{36} + \frac{10}{36} \qquad \text{Simplify.}$$

$$= \frac{31}{36} \qquad \text{Add.}$$

d) To perform addition with the mixed number $2\frac{1}{3}$, first convert it into an improper fraction: $2\frac{1}{3} = 2 + \frac{1}{3} = \frac{6}{3} + \frac{1}{3} = \frac{7}{3}$.

Now do Exercises 67–78

<Calculator Close-Up>

You can check these results with a graphing calculator. Note how a graphing calculator handles mixed numbers.

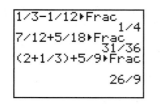

$$2\frac{1}{3} + \frac{5}{9} = \frac{7}{3} + \frac{5}{9} \qquad \text{Write } 2\frac{1}{3} \text{ as an improper fraction.}$$

$$= \frac{7 \cdot 3}{3 \cdot 3} + \frac{5}{9} \qquad \text{The LCD is 9.}$$

$$= \frac{21}{9} + \frac{5}{9} \qquad \text{Simplify.}$$

$$= \frac{26}{9} \qquad \text{Add.}$$

Note that $\frac{1}{3} + \frac{5}{9} = \frac{3}{9} + \frac{5}{9} = \frac{8}{9}$. Then add on the 2 to get $2\frac{8}{9}$, which is the same as $\frac{26}{9}$.

⟨6⟩ Fractions, Decimals, and Percents

In the decimal number system, fractions with a denominator of 10, 100, 1000, and so on are written as decimal numbers. For example,

$$\frac{3}{10} = 0.3, \qquad \frac{25}{100} = 0.25, \qquad \text{and} \qquad \frac{5}{1000} = 0.005.$$

Fractions with a denominator of 100 are often written as percents. Think of the percent symbol (%) as representing the denominator of 100. For example,

$$\frac{25}{100} = 25\%, \qquad \frac{5}{100} = 5\%, \qquad \text{and} \qquad \frac{300}{100} = 300\%.$$

Example 8 illustrates further how to convert from any one of the forms (fraction, decimal, percent) to the others.

⟨ Helpful Hint ⟩

Recall the *place value* for decimal numbers:

tenths
⌐hundredths
⌐thousandths
⌐ten thousandths
0.2635

So $0.2635 = \frac{2635}{10,000}$.

E X A M P L E   **8**

Changing forms

Convert each given fraction, decimal, or percent into its other two forms.

a) $\frac{1}{5}$ **b)** 6% **c)** 0.1

Solution

a) $\frac{1}{5} = \frac{1 \cdot 20}{5 \cdot 20} = \frac{20}{100} = 20\%$ and $\frac{1}{5} = \frac{1 \cdot 2}{5 \cdot 2} = \frac{2}{10} = 0.2$

So $\frac{1}{5} = 0.2 = 20\%$. Note that a fraction can also be converted to a decimal by dividing the denominator into the numerator with long division.

b) $6\% = \frac{6}{100} = 0.06$ and $\frac{6}{100} = \frac{2 \cdot 3}{2 \cdot 50} = \frac{3}{50}$

So $6\% = 0.06 = \frac{3}{50}$.

c) $0.1 = \frac{1}{10} = \frac{1 \cdot 10}{10 \cdot 10} = \frac{10}{100} = 10\%$

So $0.1 = \frac{1}{10} = 10\%$.

⟨ Calculator Close-Up ⟩

A calculator can convert fractions to decimals and decimals to fractions. The calculator shown here converts the terminating decimal 0.333333333333 into 1/3 even though 1/3 is a repeating decimal with infinitely many threes after the decimal point.

Now do Exercises 79–90

‹7› Applications

The dimensions for lumber used in construction are usually given in fractions. For example, a two-by-four (2 × 4) stud used in framing walls is actually $1\frac{1}{2}$ in. by $3\frac{1}{2}$ in. by $92\frac{5}{8}$ in. A two-by-twelve (2 × 12) floor joist with a width of $1\frac{1}{2}$ in. and height of $11\frac{1}{2}$ in. comes in various lengths, usually 8, 10, 12, 14, and 16 feet. In Example 9 we find the height of a wall.

EXAMPLE 9

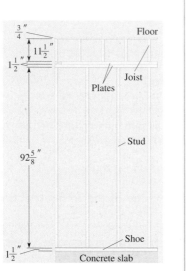

Figure 1.24

Framing a two-story house

In framing a two-story house, a carpenter uses a 2 × 4 shoe, a wall stud, two 2 × 4 plates, then 2 × 12 floor joists, and a $\frac{3}{4}$-in. plywood floor, before starting the second level. Use the dimensions in Fig. 1.24 to find the total height of the framing shown.

Solution

We can find the total height using multiplication and addition:

$$3 \cdot 1\frac{1}{2} + 92\frac{5}{8} + 11\frac{1}{2} + \frac{3}{4} = 4\frac{1}{2} + 92\frac{5}{8} + 11\frac{1}{2} + \frac{3}{4}$$

$$= 4\frac{4}{8} + 92\frac{5}{8} + 11\frac{4}{8} + \frac{6}{8}$$

$$= 107\frac{19}{8}$$

$$= 107 + \frac{16}{8} + \frac{3}{8} = 107 + 2 + \frac{3}{8} = 109\frac{3}{8}$$

The total height of the framing shown is $109\frac{3}{8}$ in.

> Now do Exercises 115–118

Warm-Ups ▼

Fill in the blank.

1. _____ fractions are identical when they are reduced to lowest terms.

2. A fraction in lowest terms has no common _____ (greater than 1) in the numerator and denominator.

3. _____ denominators are required for addition and subtraction of fractions.

4. We can convert a fraction to a decimal by dividing the _____ into the _____.

5. We can convert a percent into a fraction by _____ by 100 and deleting the percent symbol.

True or false?

6. $\dfrac{8}{12} = \dfrac{4}{6}$

7. $\dfrac{1}{2} \cdot \dfrac{2}{3} = \dfrac{1}{3}$

8. $\dfrac{1}{2} \cdot \dfrac{3}{5} = \dfrac{3}{10}$

9. $\dfrac{1}{2} \div 3 = \dfrac{1}{6}$

10. $5 \div \dfrac{1}{2} = 10$

11. $\dfrac{1}{2} + \dfrac{1}{4} = \dfrac{2}{6}$

12. $2 - \dfrac{1}{2} = \dfrac{3}{2}$

‹1› Equivalent Fractions

Build up each fraction or whole number so that it is equivalent to the fraction with the indicated denominator. See Example 1.

1. $\dfrac{3}{4} = \dfrac{?}{8}$

2. $\dfrac{5}{7} = \dfrac{?}{21}$

3. $\dfrac{8}{3} = \dfrac{?}{12}$

4. $\dfrac{7}{2} = \dfrac{?}{8}$

5. $5 = \dfrac{?}{2}$

6. $9 = \dfrac{?}{3}$

7. $\dfrac{3}{4} = \dfrac{?}{100}$

8. $\dfrac{1}{2} = \dfrac{?}{100}$

9. $\dfrac{3}{10} = \dfrac{?}{100}$

10. $\dfrac{2}{5} = \dfrac{?}{100}$

11. $\dfrac{5}{3} = \dfrac{?}{42}$

12. $\dfrac{5}{7} = \dfrac{?}{98}$

Reduce each fraction to lowest terms. See Example 2.

13. $\dfrac{3}{6}$

14. $\dfrac{2}{10}$

15. $\dfrac{12}{18}$

16. $\dfrac{30}{40}$

17. $\dfrac{15}{5}$

18. $\dfrac{39}{13}$

19. $\dfrac{50}{100}$

20. $\dfrac{5}{1000}$

21. $\dfrac{200}{100}$

22. $\dfrac{125}{100}$

23. $\dfrac{18}{48}$

24. $\dfrac{34}{102}$

25. $\dfrac{26}{42}$

26. $\dfrac{70}{112}$

27. $\dfrac{84}{91}$

28. $\dfrac{121}{132}$

‹2› Multiplying Fractions

Find each product. See Example 3.

29. $\dfrac{2}{3} \cdot \dfrac{5}{9}$

30. $\dfrac{1}{8} \cdot \dfrac{1}{8}$

31. $\dfrac{1}{3} \cdot 15$

32. $\dfrac{1}{4} \cdot 16$

33. $\dfrac{3}{4} \cdot \dfrac{14}{15}$

34. $\dfrac{5}{8} \cdot \dfrac{12}{35}$

35. $\dfrac{2}{5} \cdot \dfrac{35}{26}$

36. $\dfrac{3}{10} \cdot \dfrac{20}{21}$

37. $\dfrac{1}{2} \cdot \dfrac{6}{5}$

38. $\dfrac{1}{2} \cdot \dfrac{3}{5}$

39. $\dfrac{1}{2} \cdot \dfrac{1}{3}$

40. $\dfrac{3}{16} \cdot \dfrac{1}{7}$

‹3› Unit Conversion

Perform the indicated unit conversions. See Example 4. Round approximate answers to the nearest hundredth. Answers can vary slightly depending on the conversion factors used.

41. Convert 96 feet to inches.

42. Convert 33 yards to feet.

43. Convert 14.22 miles to kilometers.

44. Convert 33.6 kilometers to miles.

45. Convert 13.5 centimeters to inches.

46. Convert 42.1 inches to centimeters.

47. Convert 14.2 ounces to grams.

48. Convert 233 grams to ounces.

49. Convert 40 miles per hour to feet per second.

50. Convert 200 feet per second to miles per hour.

51. Convert 500 feet per second to kilometers per hour.

52. Convert 230 yards per second to miles per minute.

⟨4⟩ Dividing Fractions

Find each quotient. See Example 5.

53. $\dfrac{3}{4} \div \dfrac{1}{4}$

54. $\dfrac{2}{3} \div \dfrac{1}{2}$

55. $\dfrac{1}{3} \div 5$

56. $\dfrac{3}{5} \div 3$

57. $5 \div \dfrac{5}{4}$

58. $8 \div \dfrac{2}{3}$

59. $\dfrac{6}{10} \div \dfrac{3}{4}$

VIDEO **60.** $\dfrac{2}{3} \div \dfrac{10}{21}$

61. $\dfrac{3}{16} \div \dfrac{5}{2}$

62. $\dfrac{1}{8} \div \dfrac{5}{16}$

⟨5⟩ Adding and Subtracting Fractions

Find each sum or difference. See Examples 6 and 7. See Strategy for Finding the LCD box on page 19.

63. $\dfrac{1}{4} + \dfrac{1}{4}$

64. $\dfrac{1}{10} + \dfrac{1}{10}$

65. $\dfrac{5}{12} - \dfrac{1}{12}$

66. $\dfrac{17}{14} - \dfrac{5}{14}$

67. $\dfrac{1}{2} - \dfrac{1}{4}$

68. $\dfrac{1}{3} + \dfrac{1}{6}$

69. $\dfrac{1}{3} + \dfrac{1}{4}$

70. $\dfrac{1}{2} + \dfrac{3}{5}$

71. $\dfrac{3}{4} - \dfrac{2}{3}$

72. $\dfrac{4}{5} - \dfrac{3}{4}$

VIDEO **73.** $\dfrac{1}{6} + \dfrac{5}{8}$

74. $\dfrac{3}{4} + \dfrac{1}{6}$

75. $\dfrac{5}{24} - \dfrac{1}{18}$

76. $\dfrac{3}{16} - \dfrac{1}{20}$

77. $3\dfrac{5}{6} + \dfrac{5}{16}$

78. $5\dfrac{3}{8} - \dfrac{15}{16}$

⟨6⟩ Fractions, Decimals, and Percents

Convert each given fraction, decimal, or percent into its other two forms. See Example 8.

79. $\dfrac{3}{5}$

80. $\dfrac{19}{20}$

81. 9%

82. 60%

83. 0.08

84. 0.4

85. $\dfrac{3}{4}$

86. $\dfrac{5}{8}$

87. 2%

88. 120%

89. 0.01

90. 0.005

Perform the indicated operations.

91. $\dfrac{3}{8} \div \dfrac{1}{8}$

92. $\dfrac{7}{8} \div \dfrac{3}{14}$

93. $\dfrac{3}{4} \cdot \dfrac{28}{21}$

94. $\dfrac{5}{16} \cdot \dfrac{3}{10}$

95. $\dfrac{7}{12} + \dfrac{5}{32}$

96. $\dfrac{2}{15} + \dfrac{8}{21}$

97. $\dfrac{5}{24} - \dfrac{1}{15}$

98. $\dfrac{9}{16} - \dfrac{1}{12}$

99. $3\dfrac{1}{8} + \dfrac{15}{16}$

100. $5\dfrac{1}{4} - \dfrac{9}{16}$

101. $7\dfrac{2}{3} \cdot 2\dfrac{1}{4}$

102. $6\dfrac{1}{2} \div \dfrac{7}{2}$

103. $\dfrac{1}{2} + \dfrac{1}{3} + \dfrac{1}{4}$

104. $\dfrac{1}{2} + \dfrac{1}{3} - \dfrac{1}{6}$

105. $\dfrac{1}{2} \cdot \dfrac{1}{2} \cdot \dfrac{1}{2}$

106. $\dfrac{2}{3} \cdot \dfrac{2}{3} \cdot \dfrac{2}{3}$

Fill in the blank so that each equation is correct.

107. $\dfrac{1}{4} + \underline{} = \dfrac{5}{8}$

108. $\dfrac{1}{3} + \underline{} = \dfrac{4}{9}$

109. $\dfrac{5}{16} - \underline{} = \dfrac{1}{8}$

110. $\dfrac{3}{5} - \underline{} = \dfrac{1}{10}$

111. $\dfrac{4}{9} \cdot \underline{} = \dfrac{8}{27}$

112. $\dfrac{3}{8} \cdot \underline{} = \dfrac{3}{4}$

113. $\dfrac{2}{3} \div \underline{} = \dfrac{4}{3}$

114. $\dfrac{1}{15} \div \underline{} = \dfrac{1}{5}$

⟨7⟩ **Applications**

Solve each problem. See Example 9.

115. *Planned giving.* Marie's will specifies that one-sixth of her estate will go to Tulane University and one-thirty-second will go to the Humane Society. What is the total portion of her estate that will go to these two organizations?

116. *Diversification.* Helen has $\frac{1}{5}$ of her portfolio in U.S. stocks, $\frac{1}{8}$ of her portfolio in European stocks, and $\frac{1}{10}$ of her portfolio in Japanese stocks. The remainder is invested in municipal bonds. What fraction of her portfolio is invested in municipal bonds? What percent is invested in municipal bonds?

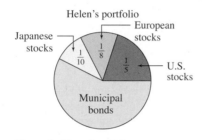

Figure for Exercise 116

117. *Concrete patio.* A contractor plans to pour a concrete rectangular patio.

a) Use the table to find the approximate volume of concrete in cubic yards for a 9 ft by 12 ft patio that is 4 inches thick.

b) Find the exact volume of concrete in cubic feet and cubic yards for a patio that is $12\frac{1}{2}$ feet long, $8\frac{3}{4}$ feet wide, and 4 inches thick.

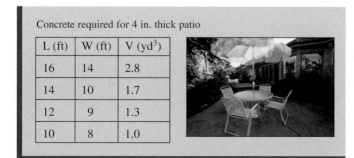

Concrete required for 4 in. thick patio

L (ft)	W (ft)	V (yd³)
16	14	2.8
14	10	1.7
12	9	1.3
10	8	1.0

Figure for Exercise 117

118. *Bundle of studs.* A lumber yard receives 2×4 studs in a bundle that contains 25 rows (or layers) of studs with 20 studs in each row. A 2×4 stud is actually $1\frac{1}{2}$ in. by $3\frac{1}{2}$ in. by $92\frac{5}{8}$ in. Find the cross-sectional area of a bundle in square inches. Find the volume of a bundle in cubic feet. (The formula $V = LWH$ gives the volume of a rectangular solid.) Round approximate answers to the nearest tenth.

Getting More Involved

119. *Writing*

Find an example of a real-life situation in which it is necessary to add two fractions.

120. *Cooperative learning*

Write a step-by-step procedure for adding two fractions with different denominators. Give your procedure to a classmate to try out on some addition problems. Refine your procedure as necessary.

121. *Fraction puzzle.* A wheat farmer in Manitoba left his L-shaped farm (shown in the diagram) to his four daughters. Divide the property into four pieces so that each piece is exactly the same size and shape.

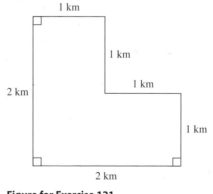

Figure for Exercise 121

Math *at Work*   **Stock Price Analysis**

Stock market analysts use mathematics daily to evaluate the potential success of a stock based on its financial statements and its current performance. Each analyst has a philosophy of investing. If an analyst is working for a mutual fund that specializes in retirement investing for clients with a lengthy time horizon, the analyst may recommend higher-risk stocks. If the client base is older and has a shorter time horizon, the analyst may recommend more secure investments.

There are hundreds of ratios and formulas that a stock market analyst uses to estimate the value of a stock. Two popular ones are the capital asset pricing model (CAPM) and the price/earnings ratio (P/E). The CAPM is used to assess the price of a stock in relation to general movements in the stock market, whereas the P/E ratio is used to compare the price of one stock to others in the same industry.

Using CAPM a stock's price P is determined by $P = A + BM$, where A is the stock's variance, B is the stock's fluctuation in relation to the market, and M is the market level. For example, a stock trading at \$10.50 on the New York Stock Exchange has a variance of 3.24 and fluctuation of 0.001058 using the Dow Jones Industrial Average. If the Dow is at 13,125, then $P = 3.24 + 0.001058(13,125) \approx 17.13$. So the stock is worth \$17.13 and is a good buy at \$10.50. If the company has earned \$1.53 per share, then P/E = $10.50/1.53 \approx 6.9$. If other stocks in the same industry have higher P/E ratios, then this stock is a good buy.

Since there are hundreds of ways to analyze a stock and all analysts have access to the same data, the analysts must decide which data are most important. The analyst must also look beyond data and formulas to determine whether to buy a stock.

1.3   Addition and Subtraction of Real Numbers

In This Section

⟨1⟩ **Addition of Two Negative Numbers**

⟨2⟩ **Addition of Numbers with Unlike Signs**

⟨3⟩ **Subtraction of Signed Numbers**

⟨4⟩ **Applications**

In arithmetic we add and subtract only positive numbers and zero. In Section 1.1 we introduced the concept of absolute value of a number. Now we will use absolute value to extend the operations of addition and subtraction to the real numbers. We will work only with rational numbers in this chapter. You will learn to perform operations with irrational numbers in Chapter 9.

⟨1⟩ Addition of Two Negative Numbers

A good way to understand positive and negative numbers is to *think of the positive numbers as assets and the negative numbers as debts.* For this illustration we can think of assets simply as cash. For example, if you have \$3 and \$5 in cash, then your total cash is \$8. You get the total by adding two positive numbers.

Think of debts as unpaid bills such as the electric bill or the phone bill. If you have debts of \$70 and \$80, then your total debt is \$150. You can get the total debt by adding negative numbers:

$$
\underset{\substack{\uparrow \\ \$70 \text{ debt}}}{(-70)} \quad \underset{\substack{\uparrow \\ \text{plus}}}{+} \quad \underset{\substack{\uparrow \\ \$80 \text{ debt}}}{(-80)} \quad = \quad \underset{\substack{\uparrow \\ \$150 \text{ debt}}}{-150}
$$

We think of this addition as adding the absolute values of -70 and -80 ($70 + 80 = 150$), and then putting a negative sign on that result to get -150. These examples illustrate the following rule.

> **Sum of Two Numbers with Like Signs**
>
> To find the sum of two numbers with the same sign, add their absolute values. The sum has the same sign as the given numbers.

E X A M P L E 1 **Adding numbers with like signs**
Perform the indicated operations.

a) $23 + 56$

b) $(-12) + (-9)$

c) $(-3.5) + (-6.28)$

d) $\left(-\dfrac{1}{2}\right) + \left(-\dfrac{1}{4}\right)$

Solution

a) The sum of two positive numbers is a positive number: $23 + 56 = 79$.

b) The absolute values of -12 and -9 are 12 and 9, and $12 + 9 = 21$. So,

$$(-12) + (-9) = -21.$$

c) Add the absolute values of -3.5 and -6.28, and put a negative sign on the sum. Remember to line up the decimal points when adding decimal numbers:

$$\begin{array}{r} 3.50 \\ \underline{6.28} \\ 9.78 \end{array}$$

So $(-3.5) + (-6.28) = -9.78$.

d) $\left(-\dfrac{1}{2}\right) + \left(-\dfrac{1}{4}\right) = \left(-\dfrac{2}{4}\right) + \left(-\dfrac{1}{4}\right) = -\dfrac{3}{4}$

> Now do Exercises 1–10

⟨2⟩ Addition of Numbers with Unlike Signs

If you have a debt of \$5 and have only \$5 in cash, then your debts equal your assets (in absolute value), and your net worth is \$0. **Net worth** is the total of debts and assets. Symbolically,

$$-5 \quad + \quad 5 \quad = \quad 0.$$

$$\underset{\substack{\uparrow \\ \text{\$5 debt}}}{} \qquad \underset{\substack{\uparrow \\ \text{\$5 cash}}}{} \qquad \underset{\substack{\uparrow \\ \text{Net worth}}}{}$$

For any number a, a and its opposite, $-a$, have a sum of zero. For this reason, a and $-a$ are called **additive inverses** of each other. Note that the words "negative," "opposite," and "additive inverse" are often used interchangeably.

Additive Inverse Property

For any number a,

$$a + (-a) = 0 \qquad \text{and} \qquad (-a) + a = 0.$$

EXAMPLE 2

Finding the sum of additive inverses

Evaluate.

a) $34 + (-34)$

b) $-\dfrac{1}{4} + \dfrac{1}{4}$

c) $2.97 + (-2.97)$

Solution

a) $34 + (-34) = 0$

b) $-\dfrac{1}{4} + \dfrac{1}{4} = 0$

c) $2.97 + (-2.97) = 0$

> Now do Exercises 11–14

< **Helpful Hint** >

We use the illustrations with debts and assets to make the rules for adding signed numbers understandable. However, in the end the carefully written rules tell us exactly how to perform operations with signed numbers, and we must obey the rules.

To understand the sum of a positive and a negative number that are not additive inverses of each other, consider the following situation. If you have a debt of $6 and $10 in cash, you may have $10 in hand, but your net worth is only $4. Your assets exceed your debts (in absolute value), and you have a positive net worth. In symbols,

$$-6 + 10 = 4.$$

Note that to get 4, we actually subtract 6 from 10.

If you have a debt of $7 but have only $5 in cash, then your debts exceed your assets (in absolute value). You have a negative net worth of $-$2. In symbols,

$$-7 + 5 = -2.$$

Note that to get the 2 in the answer, we subtract 5 from 7.

As you can see from these examples, the sum of a positive number and a negative number (with different absolute values) may be either positive or negative. These examples help us to understand the rule for adding numbers with unlike signs and different absolute values.

Sum of Two Numbers with Unlike Signs (and Different Absolute Values)

To find the sum of two numbers with unlike signs (and different absolute values), subtract their absolute values.

- The answer is positive if the number with the larger absolute value is positive.
- The answer is negative if the number with the larger absolute value is negative.

EXAMPLE   **3**

Adding numbers with unlike signs

Evaluate.

a) $-5 + 13$ **b)** $6 + (-7)$ **c)** $-6.4 + 2.1$

d) $-5 + 0.09$ **e)** $\left(-\dfrac{1}{3}\right) + \left(\dfrac{1}{2}\right)$ **f)** $\dfrac{3}{8} + \left(-\dfrac{5}{6}\right)$

‹ Calculator Close-Up ›

Your calculator can add signed numbers. Most calculators have a key for subtraction and a different key for the negative sign.

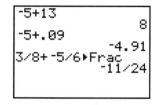

You should do the exercises in this section by hand and then check with a calculator.

Solution

a) The absolute values of -5 and 13 are 5 and 13. Subtract them to get 8. Since the number with the larger absolute value is 13 and it is positive, the result is positive:

$$-5 + 13 = 8$$

b) The absolute values of 6 and -7 are 6 and 7. Subtract them to get 1. Since -7 has the larger absolute value, the result is negative:

$$6 + (-7) = -1$$

c) Line up the decimal points and subtract 2.1 from 6.4.

$$\begin{array}{r} 6.4 \\ -2.1 \\ \hline 4.3 \end{array}$$

Since 6.4 is larger than 2.1, and 6.4 has a negative sign, the sign of the answer is negative. So $-6.4 + 2.1 = -4.3$.

d) Line up the decimal points and subtract 0.09 from 5.00.

$$\begin{array}{r} 5.00 \\ -0.09 \\ \hline 4.91 \end{array}$$

Since 5.00 is larger than 0.09, and 5.00 has the negative sign, the sign of the answer is negative. So $-5 + 0.09 = -4.91$.

e) $\left(-\dfrac{1}{3}\right) + \left(\dfrac{1}{2}\right) = \left(-\dfrac{2}{6}\right) + \left(\dfrac{3}{6}\right) = \dfrac{1}{6}$

f) $\dfrac{3}{8} + \left(-\dfrac{5}{6}\right) = \dfrac{9}{24} + \left(-\dfrac{20}{24}\right) = -\dfrac{11}{24}$

> Now do Exercises 15–24

‹3› Subtraction of Signed Numbers

Each subtraction problem with signed numbers is solved by doing an equivalent addition problem. So before attempting subtraction of signed numbers be sure that you understand addition of signed numbers.

We can think of subtraction as removing debts or assets, and addition as receiving debts or assets. Removing a debt means the debt is forgiven. If you owe your mother $20 and she tells you to forget it, then that debt is removed and your net worth has gone up by $20. Paying off a debt is not the same. Paying off a debt does not affect your net worth. If you lose your wallet, which contains $50, then that asset is removed. When your electric bill arrives, you have received a debt. When you get your paycheck, you have received an asset.

How does removing debts or assets affect your net worth? Suppose that your net worth is $100. Losing $30 or receiving a phone bill for $30 has the same effect. Your net worth goes down to $70.

$$100 \quad - \quad 30 \quad = \quad 100 \quad + \quad (-30)$$

$$\uparrow \qquad \uparrow \qquad\qquad \uparrow \qquad \uparrow$$

Remove   Cash   Receive   Debt

Removing an asset (cash) is equivalent to receiving a debt.

Suppose you have $15 but owe a friend $5. Your net worth is only $10. If the debt of $5 is canceled or forgiven, your net worth will go up to $15, the same as if you received $5 in cash. In symbols,

$$10 \quad - \quad (-5) \quad = \quad 10 \quad + \quad 5.$$

$$\uparrow \qquad \uparrow \qquad\qquad \uparrow \qquad \uparrow$$

Remove   Debt   Receive   Cash

Removing a debt is equivalent to receiving cash.

Notice that each subtraction problem is equivalent to an addition problem in which we add the opposite of what we want to subtract. In other words, *subtracting a number is the same as adding its opposite.*

Subtraction of Real Numbers

For any real numbers a and b,

$$a - b = a + (-b).$$

EXAMPLE 4

Subtracting signed numbers
Perform each subtraction.

a) $-5 - 3$ **b)** $5 - (-3)$

c) $-5 - (-3)$ **d)** $\dfrac{1}{2} - \left(-\dfrac{1}{4}\right)$

e) $-3.6 - (-5)$ **f)** $0.02 - 8$

Solution

To do *any* subtraction, we can change it to addition of the opposite.

a) $-5 - 3 = -5 + (-3) = -8$

b) $5 - (-3) = 5 + (3) = 8$

c) $-5 - (-3) = -5 + 3 = -2$

d) $\dfrac{1}{2} - \left(-\dfrac{1}{4}\right) = \dfrac{2}{4} + \dfrac{1}{4} = \dfrac{3}{4}$

e) $-3.6 - (-5) = -3.6 + 5 = 1.4$

f) $0.02 - 8 = 0.02 + (-8) = -7.98$

Now do Exercises 25–52

⟨4⟩ Applications

EXAMPLE 5

Net worth

A couple has $18,000 in credit card debt, $2000 in their checking account, and $6000 in a 401(k). The mortgage balance on their $180,000 house is $170,000. Their two cars are worth a total of $19,000, but the loan balances on them total $23,000. Find their net worth.

Solution

Net worth is the total of all debts and assets. To find it, subtract the debts from the assets:

$$2000 + 6000 + 180{,}000 + 19{,}000 - 18{,}000 - 170{,}000 - 23{,}000 = -4000$$

The net worth is $-$4000.

Now do Exercises 99–102

Warm-Ups ▼

Fill in the blank.

1. If the sum of two numbers is zero, then the numbers are _____ or _____.

2. The sum of two numbers with opposite signs and the same absolute value is _____.

3. When adding two numbers with opposite signs, we _____ their absolute values and use the sign of the number with the larger absolute value.

4. Subtraction is defined in terms of additions as $a - b =$ _____.

True or false?

5. $-9 + 8 = -1$
6. $-2 + (-4) = -6$
7. $0 - 7 = -7$
8. $5 - (-2) = 3$
9. $-5 - (-2) = -7$
10. The additive inverse of -3 is 0.
11. If b is negative, then $-b$ is positive.
12. The sum of a positive number and a negative number is a negative number.

Exercises

‹1› Addition of Two Negative Numbers

Perform the indicated operation. See Example 1.

1. $3 + 10$

2. $81 + 19$

3. $(-3) + (-10)$

4. $(-81) + (-19)$

5. $-3 + (-5)$

6. $-7 + (-2)$

7. $-0.25 + (-0.9)$

8. $-0.8 + (-2.35)$

9. $\left(-\dfrac{1}{3}\right) + \left(-\dfrac{1}{6}\right)$

10. $\dfrac{2}{3} + \dfrac{1}{12}$

‹2› Addition of Numbers with Unlike Signs

Evaluate. See Examples 2 and 3.

11. $-8 + 8$

12. $20 + (-20)$

13. $-\dfrac{17}{50} + \dfrac{17}{50}$

14. $\dfrac{12}{13} + \left(-\dfrac{12}{13}\right)$

15. $-7 + 9$

16. $10 + (-30)$ 📀 VIDEO

17. $7 + (-13)$

18. $-8 + 20$ 📀 VIDEO

19. $8.6 + (-3)$

20. $-9.5 + 12$

21. $3.9 + (-6.8)$

22. $-5.24 + 8.19$

23. $\dfrac{1}{4} + \left(-\dfrac{1}{2}\right)$

24. $-\dfrac{2}{3} + 2$

‹3› Subtraction of Signed Numbers

Fill in the parentheses to make each statement correct. See Example 4.

25. $8 - 2 = 8 + (?)$

26. $3.5 - 1.2 = 3.5 + (?)$

27. $4 - 12 = 4 + (?)$

28. $\dfrac{1}{2} - \dfrac{5}{6} = \dfrac{1}{2} + (?)$

29. $-3 - (-8) = -3 + (?)$

30. $-9 - (-2.3) = -9 + (?)$

31. $8.3 - (-1.5) = 8.3 + (?)$

32. $10 - (-6) = 10 + (?)$

Perform the indicated operation. See Example 4.

33. $6 - 10$

34. $3 - 19$

35. $-3 - 7$

36. $-3 - 12$

37. $5 - (-6)$

38. $5 - (-9)$

39. $-6 - 5$

40. $-3 - 6$

41. $\dfrac{1}{4} - \dfrac{1}{2}$

42. $\dfrac{2}{5} - \dfrac{2}{3}$

43. $\dfrac{1}{2} - \left(-\dfrac{1}{4}\right)$

44. $\dfrac{2}{3} - \left(-\dfrac{1}{6}\right)$

45. $10 - 3$

46. $13 - 3$

47. $1 - 0.07$

48. $0.03 - 1$

49. $7.3 - (-2)$

50. $-5.1 - 0.15$

51. $-0.03 - 5$

52. $0.7 - (-0.3)$

Miscellaneous

Perform the indicated operations. Do not use a calculator.

53. $-5 + 8$

54. $-6 + 10$

55. $-6 + (-3)$

56. $(-13) + (-12)$ 📀 VIDEO

57. $-80 - 40$

58. $44 - (-15)$

59. $61 - (-17)$ 📀 VIDEO

60. $-19 - 13$ 📀 VIDEO

61. $(-12) + (-15)$

62. $-12 + 12$

63. $13 + (-20)$

64. $15 + (-39)$

65. $-102 - 99$

66. $-94 - (-77)$

67. $-161 - 161$

68. $-19 - 88$

69. $-16 + 0.03$

70. $0.59 + (-3.4)$

71. $0.08 - 3$

72. $1.8 - 9$

73. $-3.7 + (-0.03)$

74. $0.9 + (-1)$

75. $-2.3 - (-6)$

76. $-7.08 - (-9)$

77. $\dfrac{3}{4} + \left(-\dfrac{3}{5}\right)$

78. $-\dfrac{1}{3} + \dfrac{3}{5}$

79. $-\dfrac{1}{12} - \left(-\dfrac{3}{8}\right)$

80. $-\dfrac{1}{17} - \left(-\dfrac{1}{17}\right)$

Fill in the parentheses so that each equation is correct.

81. $-5 + (\quad) = 8$

82. $-9 + (\quad) = 22$

83. $12 + (\quad) = 2$

84. $13 + (\quad) = -4$

85. $10 - (\quad) = -4$

86. $14 - (\quad) = -8$

87. $6 - (\quad) = 10$

88. $3 - (\quad) = 15$

89. $-4 - (\quad) = -1$

90. $-11 - (\quad) = 2$

Use a calculator to perform the indicated operations.

91. $45.87 + (-49.36)$

92. $-0.357 + (-3.465)$

93. $0.6578 + (-1)$

94. $-2.347 + (-3.5)$

95. $-3.45 - 45.39$

96. $9.8 - 9.974$

97. $-5.79 - 3.06$

98. $0 - (-4.537)$

⟨4⟩ Applications

Solve each problem. See Example 5.

99. *Overdrawn.* Willard opened his checking account with a deposit of $97.86. He then wrote checks and had other charges as shown in his account register. Find his current balance.

Deposit		97.86
Wal-Mart	27.89	
Kmart	42.32	
ATM cash	25.00	
Service charge	3.50	
Check printing	8.00	

Figure for Exercise 99

100. *Net worth.* Melanie's house is worth $125,000, but she still owes $78,422 on her mortgage. She has $21,236 in a savings account and has $9477 in credit card debt. She owes $6131 to the credit union and figures that her cars and other household items are worth a total of $15,000. What is Melanie's net worth?

101. *Falling temperatures.* At noon the temperature in Montreal was 5°C. By midnight the mercury had fallen 12°. What was the temperature at midnight?

102. *Bitter cold.* The overnight low temperature in Milwaukee was -13°F for Monday night. The temperature went up 20° during the day on Tuesday and then fell 15° to reach Tuesday night's overnight low temperature.

a) What was the overnight low Tuesday night?

b) Judging from the accompanying graph, was the average low for the week above or below 0°F?

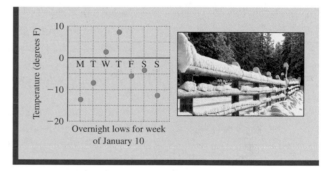

Figure for Exercise 102

Getting More Involved

103. *Writing*

What does absolute value have to do with adding signed numbers? Can you add signed numbers without using absolute value?

104. *Discussion*

Why do we learn addition of signed numbers before subtraction?

105. *Discussion*

Aimee and Joni are traveling south in separate cars on Interstate 5 near Stockton. While they are speaking to each other on cellular telephones, Aimee gives her location as mile marker x and Joni gives her location as mile marker y. Which of the following expressions gives the distance between them? Explain your answer.

a) $y - x$ **b)** $x - y$
c) $|x - y|$ **d)** $|y - x|$
e) $|x| + |y|$

1.4   Multiplication and Division of Real Numbers

In This Section

⟨1⟩ **Multiplication of Real Numbers**

⟨2⟩ **Division of Real Numbers**

⟨3⟩ **Division by Zero**

In this section, we will complete the study of the four basic operations with real numbers.

⟨1⟩ Multiplication of Real Numbers

The result of multiplying two numbers is referred to as the **product** of the numbers. The numbers multiplied are called **factors.** In algebra we use a raised dot between the factors to indicate multiplication, or we place symbols next to one another to indicate multiplication. Thus, $a \cdot b$ or ab are both referred to as the product of a and b. When multiplying numbers, we may enclose them in parentheses to make the meaning clear. To write 5 times 3, we may write it as $5 \cdot 3$, $5(3)$, $(5)3$, or $(5)(3)$. In multiplying a number and a variable, no sign is used between them. Thus, $5x$ is used to represent the product of 5 and x.

Multiplication is just a short way to do repeated additions. Adding together five 3's gives

$$3 + 3 + 3 + 3 + 3 = 15.$$

So we have the multiplication fact $5 \cdot 3 = 15$. Adding together five -3's gives

$$(-3) + (-3) + (-3) + (-3) + (-3) = -15.$$

So we should have $5(-3) = -15$. Receiving five debts of $3 each is the same as a $15 debt. If you have five debts of $3 each and they are forgiven, then you have gained $15. So we should have $(-5)(-3) = 15$.

These examples illustrate the rule for multiplying signed numbers.

⟨ **Helpful Hint** ⟩

The product of two numbers with like signs is positive, but the product of three numbers with like signs can be positive or negative. For example,

$$2 \cdot 2 \cdot 2 = 8$$

and

$$(-2)(-2)(-2) = -8.$$

Product of Signed Numbers

To find the product of two nonzero real numbers, multiply their absolute values.

- The product is *positive* if the numbers have *like* signs.
- The product is *negative* if the numbers have *unlike* signs.

E X A M P L E **1**

Multiplying signed numbers

Evaluate each product.

a) $(-2)(-3)$ **b)** $3(-6)$ **c)** $-5 \cdot 10$

d) $\left(-\dfrac{1}{3}\right)\left(-\dfrac{1}{2}\right)$ **e)** $(-0.02)(0.08)$ **f)** $(-300)(-0.06)$

‹ **Calculator Close-Up** ›

Try finding the products in Example 1 with your calculator.

```
(-2)(-3)
                    6
3(-6)
                   -18
-5*10
                   -50
```

Solution

a) First find the product of the absolute values:

$$|-2| \cdot |-3| = 2 \cdot 3 = 6$$

Because -2 and -3 have the same sign, we get $(-2)(-3) = 6$.

b) First find the product of the absolute values:

$$|3| \cdot |-6| = 3 \cdot 6 = 18$$

Because 3 and -6 have unlike signs, we get $3(-6) = -18$.

c) $-5 \cdot 10 = -50$ Unlike signs, negative result

d) $\left(-\dfrac{1}{3}\right)\left(-\dfrac{1}{2}\right) = \dfrac{1}{6}$ Like signs, positive result

e) When multiplying decimals, we total the number of decimal places in the factors to get the number of decimal places in the product. Thus,

$$(-0.02)(0.08) = -0.0016.$$

f) $(-300)(-0.06) = 18$ Like signs, positive result

Now do Exercises 1–12

‹2› Division of Real Numbers

We say that $10 \div 2 = 5$ because $5 \cdot 2 = 10$. This example illustrates how division is defined in terms of multiplication.

Division of Real Numbers

If a, b, and c are any real numbers with $b \neq 0$, then

$$a \div b = c \qquad \text{provided that} \qquad c \cdot b = a.$$

Using the definition of division, we can make the following table:

Positive quotient $\Big\{$	$10 \div 2 = 5$ because $5 \cdot 2 = 10$
	$-10 \div (-2) = 5$ because $5(-2) = -10$
Negative quotient $\Big\{$	$10 \div (-2) = -5$ because $-5(-2) = 10$
	$-10 \div 2 = -5$ because $-5 \cdot 2 = -10$

Notice that in this table, the quotient for two numbers with the same sign is positive and the quotient for two numbers with opposite signs is negative. These examples illustrate the rule for dividing signed numbers. The rule for dividing signed numbers is similar to that for multiplying signed numbers because of the definition of division.

Division of Signed Numbers

To find the quotient of two nonzero real numbers, divide their absolute values.

- The quotient is *positive* if the two numbers have *like* signs.
- The quotient is *negative* if the two numbers have *unlike* signs.

Zero divided by any nonzero real number is zero.

E X A M P L E 2

Dividing signed numbers
Evaluate.

a) $(-8) \div (-4)$ b) $(-8) \div 8$ c) $8 \div (-4)$

d) $-4 \div \dfrac{1}{3}$ e) $-2.5 \div 0.05$ f) $0 \div (-6)$

⟨ **Helpful Hint** ⟩

Do not use negative numbers in long division. To find $-378 \div 7$, divide 378 by 7:

$$\begin{array}{r} 54 \\ 7)\overline{378} \\ 35 \\ \hline 28 \\ 28 \\ \hline 0 \end{array}$$

Since a negative divided by a positive is negative,
$$-378 \div 7 = -54.$$

Solution

a) $(-8) \div (-4) = \dfrac{-8}{-4} = 2$ Same sign, positive result

b) $(-8) \div 8 = \dfrac{-8}{8} = -1$ Unlike signs, negative result

c) $8 \div (-4) = \dfrac{8}{-4} = -2$ Unlike signs, negative result

d) $-4 \div \dfrac{1}{3} = -4 \cdot \dfrac{3}{1}$ Invert and multiply.

      $= -4 \cdot 3$

      $= -12$

e) $-2.5 \div 0.05 = \dfrac{-2.5}{0.05}$ Write in fraction form.

      $= \dfrac{-2.5 \cdot 100}{0.05 \cdot 100}$ Multiply by 100 to eliminate the decimals.

      $= \dfrac{-250}{5}$ Simplify.

      $= -50$ Divide.

f) $0 \div (-6) = \dfrac{0}{-6} = 0$ Zero divided by a nonzero number is zero.

Now do Exercises 13–26

Division can also be indicated by a fraction bar. For example,

$$24 \div 6 = \frac{24}{6} = 4.$$

If signed numbers occur in a fraction, we use the rules for dividing signed numbers. For example,

$$\frac{-9}{3} = -3, \qquad \frac{9}{-3} = -3, \qquad \frac{-1}{2} = \frac{1}{-2} = -\frac{1}{2}, \qquad \text{and} \qquad \frac{-4}{-2} = 2.$$

Note that if one negative sign appears in a fraction, the fraction has the same value whether the negative sign is in the numerator, in the denominator, or in front of the fraction. If the numerator and denominator of a fraction are both negative, then the fraction has a positive value.

⟨3⟩ Division by Zero

Why do we exclude division by zero from the definition of division? If we write $10 \div 0 = c$, we need to find a number c such that $c \cdot 0 = 10$. This is impossible. If we write $0 \div 0 = c$, we need to find a number c such that $c \cdot 0 = 0$. In fact, $c \cdot 0 = 0$ is true for any value of c. Having $0 \div 0$ equal to any number would be confusing in doing computations. Thus, $a \div b$ is defined only for $b \neq 0$. Quotients such as

$$8 \div 0, \qquad 0 \div 0, \qquad \frac{8}{0}, \qquad \text{and} \qquad \frac{0}{0}$$

are said to be **undefined.**

E X A M P L E **3**	**Division involving zero** Evaluate. If the operation is undefined, say so.

a) $0 \div 1$ **b)** $\dfrac{3}{4} \div 0$

c) $\dfrac{-12}{0}$ **d)** $\dfrac{0}{-9}$

Solution

a) The result of 0 divided by a nonzero number is zero. So $0 \div 1 = 0$.

b) Since division by zero is not allowed, $\dfrac{3}{4} \div 0$ is an undefined operation.

c) Since division by zero is not allowed, $\dfrac{-12}{0}$ is undefined.

d) The result of 0 divided by a nonzero number is zero. So $\dfrac{0}{-9} = 0$.

Now do Exercises 27–34

Warm-Ups ▼

Fill in the blank.

1. The result of multiplication is a _____ .
2. To find the product of two signed numbers multiply their _____ and use a negative sign if the original numbers have opposite signs.
3. To find the _____ of two signed numbers divide their absolute values and use a negative sign if the original numbers have opposite signs.
4. Division is defined in terms of _____ as $a \div b = c$ provided $c \cdot b = a$ and $b \neq 0$.

True or false?

5. The product of 7 and y is $7y$.
6. The product of -2 and 5 is -10.
7. The quotient of x and 3 is $x \div 3$ or $\frac{x}{3}$.
8. $0 \div 6$ is undefined.
9. $-9 \div (-3) = 3$
10. $6 \div (-2) = -3$
11. $(-0.2)(0.2) = -0.4$
12. $0 \div 0 = 0$

1.4

Exercises

⟨ **Study Tips** ⟩

- If you don't know how to get started on the exercises, go back to the examples. Read the solution in the text, and then cover it with a piece of paper and see if you can solve the example.
- If you need help, don't hesitate to get it. If you don't straighten out problems in a timely manner, you can get hopelessly lost.

⟨1⟩ **Multiplication of Real Numbers**

Evaluate. See Example 1.

1. $-3 \cdot 9$
2. $6(-4)$
3. $(-12)(-11)$
4. $(-9)(-15)$
5. $-\dfrac{3}{4} \cdot \dfrac{4}{9}$
6. $\left(-\dfrac{2}{3}\right)\left(-\dfrac{6}{7}\right)$
7. $0.5(-0.6)$
8. $(-0.3)(0.3)$
9. $(-12)(-12)$
10. $(-11)(-11)$
11. $-3 \cdot 0$
12. $0(-7)$

⟨2⟩ **Division of Real Numbers**

Evaluate. See Example 2.

13. $8 \div (-8)$
14. $-6 \div 2$
15. $(-90) \div (-30)$
16. $(-20) \div (-40)$

17. $\dfrac{44}{-66}$
18. $\dfrac{-33}{-36}$
19. $\left(-\dfrac{2}{3}\right) \div \left(-\dfrac{4}{5}\right)$
20. $-\dfrac{1}{3} \div \dfrac{4}{9}$
21. $0 \div \left(-\dfrac{1}{3}\right)$
22. $0 \div 43.568$
23. $40 \div (-0.5)$
24. $3 \div (-0.1)$
25. $-0.5 \div (-2)$
26. $-0.75 \div (-0.5)$

⟨3⟩ **Division by Zero**

Evaluate. If the operation is undefined, say so. See Example 3.

27. $0 \div 125$
28. $0 \div (-99)$
29. $\dfrac{-125}{0}$
30. $\dfrac{3.5}{0}$
31. $\dfrac{1}{2} \div 0$
32. $0.236 \div 0$
33. $\dfrac{0}{2}$
34. $\dfrac{0}{-5}$

Miscellaneous

Perform the indicated operations.

35. $(25)(-4)$ **36.** $(5)(-4)$

37. $(-3)(-9)$ **38.** $(-51) \div (-3)$

39. $-9 \div 3$ **40.** $86 \div (-2)$

41. $20 \div (-5)$ **42.** $(-8)(-6)$

43. $(-6)(5)$ **44.** $(-18) \div 3$

45. $(-57) \div (-3)$ **46.** $(-30)(4)$

47. $(0.6)(-0.3)$ **48.** $(-0.2)(-0.5)$

49. $(-0.03)(-10)$ **50.** $(0.05)(-1.5)$

51. $(-0.6) \div (0.1)$ **52.** $8 \div (-0.5)$

53. $(-0.6) \div (-0.4)$ **54.** $(-63) \div (-0.9)$

55. $-\dfrac{12}{5}\left(-\dfrac{55}{6}\right)$ **56.** $-\dfrac{9}{10} \cdot \dfrac{4}{3}$

57. $-2\dfrac{3}{4} \div 8\dfrac{1}{4}$ **58.** $-9\dfrac{1}{2} \div \left(-3\dfrac{1}{6}\right)$

Use a calculator to perform the indicated operations. Round approximate answers to two decimal places.

59. $(0.45)(-365)$ **60.** $8.5 \div (-0.15)$

61. $(-52) \div (-0.034)$ **62.** $(-4.8)(5.6)$

Fill in the parentheses so that each equation is correct.

63. $-5 \cdot (\quad) = 60$ **64.** $-9 \cdot (\quad) = 54$

65. $12 \cdot (\quad) = -96$ **66.** $11 \cdot (\quad) = -44$

67. $24 \div (\quad) = -4$ **68.** $51 \div (\quad) = -17$

69. $-36 \div (\quad) = 36$ **70.** $-48 \div (\quad) = 6$

71. $-40 \div (\quad) = -8$ **72.** $-13 \div (\quad) = -1$

Perform the indicated operations. Use a calculator to check.

73. $(-4)(-4)$ **74.** $-4 - 4$

75. $-4 + (-4)$ **76.** $-4 \div (-4)$

77. $-4 + 4$ **78.** $-4 \cdot 4$

79. $-4 - (-4)$ **80.** $0 \div (-4)$

81. $0.1 - 4$ **82.** $(0.1)(-4)$

83. $(-4) \div (0.1)$ **84.** $-0.1 - 4$

85. $(-0.1)(-4)$ **86.** $-0.1 + 4$

87. $|-0.4|$ **88.** $|0.4|$

89. $\dfrac{-0.06}{0.3}$ **90.** $\dfrac{2}{-0.04}$

91. $\dfrac{3}{-0.4}$ **92.** $\dfrac{-1.2}{-0.03}$

93. $-\dfrac{1}{5} + \dfrac{1}{6}$ **94.** $-\dfrac{3}{5} - \dfrac{1}{4}$

95. $\left(-\dfrac{3}{4}\right)\left(\dfrac{2}{15}\right)$ **96.** $-1 \div \left(-\dfrac{1}{4}\right)$

Use a calculator to perform the indicated operations. Round approximate answers to three decimal places.

97. $\dfrac{45.37}{6}$ **98.** $(-345) \div (28)$

99. $(-4.3)(-4.5)$ **100.** $\dfrac{-12.34}{-3}$

101. $\dfrac{0}{6.345}$ **102.** $0 \div (34.51)$

103. $199.4 \div 0$ **104.** $\dfrac{23.44}{0}$

Applications

105. *Big loss.* Ford Motor Company's profit for 2008 was $-\$14.6$ billion. Find the rate in dollars per minute (to the nearest dollar) at which Ford was "making" money in 2008.

106. *Negative divided by a positive.* In 2009, the national debt was $11.27 trillion dollars and the U.S. population was 306.2 million people. Find the amount of the debt per person to the nearest dollar.

Getting More Involved

107. *Discussion*

If you divide $0 among five people, how much does each person get? If you divide $5 among zero people, how much does each person get? What do these questions illustrate?

108. *Discussion*

What is the difference between the nonnegative numbers and the positive numbers?

109. *Writing*

Why do we learn multiplication of signed numbers before division?

110. *Writing*

Try to rewrite the rules for multiplying and dividing signed numbers without using the idea of absolute value. Are your rewritten rules clearer than the original rules?

Mid-Chapter **Quiz** | Sections 1.1 through 1.4 | **Chapter 1**

Graph each set of numbers on a number line.

1. The set of integers between 4 and 8

2. The real numbers between 4 and 8

3. The whole numbers less than or equal to 3

4. The real numbers less than or equal to 3

Perform the indicated operations.

5. $\dfrac{1}{2} + \dfrac{1}{8}$

6. $\dfrac{5}{6} - \dfrac{3}{4}$

7. $\dfrac{3}{5} \cdot \dfrac{1}{6}$

8. $\dfrac{2}{3} \div \dfrac{8}{9}$

9. $-19 + 33$

10. $-6 - (-5)$

11. $-12(3)$

12. $-15 \div (-3)$

13. $-12 + (-5)$

14. $-60 - 40$

15. $-11(-6)$

16. $56 \div (-8)$

17. $-\dfrac{3}{4} + \dfrac{3}{5}$

18. $\dfrac{3}{7} - \dfrac{5}{6}$

19. $\left(-\dfrac{3}{8}\right) \cdot \left(-\dfrac{2}{15}\right)$

20. $0 \div \left(-\dfrac{3}{4}\right)$

Miscellaneous.

21. What is the distance between 0 and -5 on the number line?

22. Evaluate $|-8|$, $|8|$, and $|0|$.

23. Write $\dfrac{1}{4}$ as a decimal and a percent.

24. Convert 50 yards per minute to feet per second.

25. Convert $\dfrac{3}{8}$ to an equivalent fraction with a denominator of 32.

26. Reduce $\dfrac{24}{36}$ to lowest terms.

27. What is $44 \div 0$?

1.5   **Exponential Expressions and the Order of Operations**

In This Section

⟨1⟩ **Arithmetic Expressions**

⟨2⟩ **Exponential Expressions**

⟨3⟩ **The Order of Operations**

⟨4⟩ **Applications**

In Sections 1.3 and 1.4, you learned how to perform operations with a pair of real numbers to obtain a third real number. In this section, you will learn to evaluate expressions involving several numbers and operations.

⟨1⟩ Arithmetic Expressions

The result of writing numbers in a meaningful combination with the ordinary operations of arithmetic is called an **arithmetic expression** or simply an **expression.** Consider the expressions

$$(3 + 2) \cdot 5 \qquad \text{and} \qquad 3 + (2 \cdot 5).$$

The parentheses are used as **grouping symbols** and indicate which operation to perform first. Because of the parentheses, these expressions have different values:

$$(3 + 2) \cdot 5 = 5 \cdot 5 = 25$$
$$3 + (2 \cdot 5) = 3 + 10 = 13$$

Absolute value symbols and fraction bars are also used as grouping symbols. The numerator and denominator of a fraction are treated as if each is in parentheses.

E X A M P L E   **1**

‹ **Calculator Close-Up** ›

One advantage of a graphing calculator is that you can enter an entire
expression on its display and then
evaluate it. If your calculator does not
allow built-up form for fractions, then
you must use parentheses around
the numerator and denominator as
shown here.

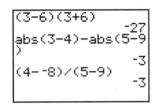

Using grouping symbols

Evaluate each expression.

    a) $(3 - 6)(3 + 6)$ **b)** $|3 - 4| - |5 - 9|$ **c)** $\dfrac{4 - (-8)}{5 - 9}$

Solution

  a) $(3 - 6)(3 + 6) = (-3)(9)$ Evaluate within parentheses first.
                   $= -27$ Multiply.

  b) $|3 - 4| - |5 - 9| = |-1| - |-4|$ Evaluate within absolute value symbols.
                   $= 1 - 4$ Find the absolute values.
                   $= -3$ Subtract.

  c) $\dfrac{4 - (-8)}{5 - 9} = \dfrac{12}{-4}$ Evaluate the numerator and denominator.

            $= -3$ Divide.

> Now do Exercises 1–12

‹2› Exponential Expressions

An arithmetic expression with repeated multiplication can be written by using exponents. For example,

$$2 \cdot 2 \cdot 2 = 2^3 \qquad \text{and} \qquad 5 \cdot 5 = 5^2.$$

The 3 in 2^3 is the number of times that 2 occurs in the product $2 \cdot 2 \cdot 2$, while the 2 in 5^2 is the number of times that 5 occurs in $5 \cdot 5$. We read 2^3 as "2 cubed" or "2 to the third power." We read 5^2 as "5 squared" or "5 to the second power." In general, an expression of the form a^n is called an **exponential expression** and is defined as follows.

Exponential Expression

For any counting number n,

$$a^n = \underbrace{a \cdot a \cdot a \cdot \cdots \cdot a}_{n \text{ factors}}.$$

We call a the **base** and n the **exponent.**

The expression a^n is read "a to the nth power." If the exponent is 1, it is usually omitted. For example, $9^1 = 9$.

E X A M P L E   **2**

Using exponential notation

Write each product as an exponential expression.

    a) $6 \cdot 6 \cdot 6 \cdot 6 \cdot 6$ **b)** $(-3)(-3)(-3)(-3)$ **c)** $\dfrac{3}{2} \cdot \dfrac{3}{2} \cdot \dfrac{3}{2}$

Solution

  a) $6 \cdot 6 \cdot 6 \cdot 6 \cdot 6 = 6^5$

  b) $(-3)(-3)(-3)(-3) = (-3)^4$

  c) $\dfrac{3}{2} \cdot \dfrac{3}{2} \cdot \dfrac{3}{2} = \left(\dfrac{3}{2}\right)^3$

> Now do Exercises 13–20

EXAMPLE   3

Writing an exponential expression as a product

Write each exponential expression as a product without exponents.

a) y^6 **b)** $(-2)^4$ **c)** $\left(\dfrac{5}{4}\right)^3$ **d)** $(-0.1)^2$

Solution

a) $y^6 = y \cdot y \cdot y \cdot y \cdot y \cdot y$

b) $(-2)^4 = (-2)(-2)(-2)(-2)$

c) $\left(\dfrac{5}{4}\right)^3 = \dfrac{5}{4} \cdot \dfrac{5}{4} \cdot \dfrac{5}{4}$

d) $(-0.1)^2 = (-0.1)(-0.1)$

Now do Exercises 21–26

To evaluate an exponential expression, write the base as many times as indicated by the exponent, and then multiply the factors from left to right.

EXAMPLE   4

Evaluating exponential expressions

Evaluate.

a) 3^3 **b)** $(-2)^3$ **c)** $\left(\dfrac{2}{3}\right)^4$ **d)** $(0.4)^2$

‹ **Calculator Close-Up** ›

You can use the power key for any power. Most calculators also have an x^2 key that gives the second power. Note that parentheses must be used when raising a fraction to a power.

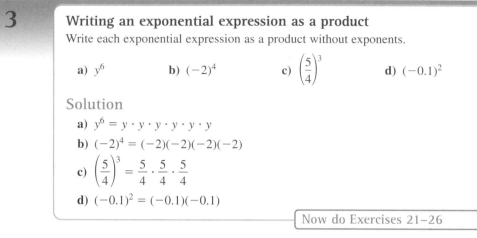

```
(-2)^3
               -8
(2/3)^4►Frac
            16/81
.4²
             .16
```

Solution

a) $3^3 = 3 \cdot 3 \cdot 3 = 9 \cdot 3 = 27$

b) $(-2)^3 = (-2)(-2)(-2)$

$\qquad = 4(-2)$

$\qquad = -8$

c) $\left(\dfrac{2}{3}\right)^4 = \dfrac{2}{3} \cdot \dfrac{2}{3} \cdot \dfrac{2}{3} \cdot \dfrac{2}{3}$

$\qquad = \dfrac{4}{9} \cdot \dfrac{2}{3} \cdot \dfrac{2}{3}$

$\qquad = \dfrac{8}{27} \cdot \dfrac{2}{3}$

$\qquad = \dfrac{16}{81}$

d) $(0.4)^2 = (0.4)(0.4) = 0.16$

Now do Exercises 27–42

CAUTION Note that $3^3 \neq 9$. We do not multiply the exponent and the base when evaluating an exponential expression.

Be especially careful with exponential expressions involving negative numbers. An exponential expression with a negative base is written with parentheses around the base as in $(-2)^4$:

$$(-2)^4 = (-2)(-2)(-2)(-2) = 16$$

To evaluate $-(2^4)$, use the base 2 as a factor four times, and then find the opposite:

$$-(2^4) = -(2 \cdot 2 \cdot 2 \cdot 2) = -(16) = -16$$

We often omit the parentheses in $-(2^4)$ and simply write -2^4. So,

$$-2^4 = -(2^4) = -16.$$

To evaluate $-(-2)^4$, use the base -2 as a factor four times, and then find the opposite:

$$-(-2)^4 = -(16) = -16$$

EXAMPLE **5**

Evaluating exponential expressions involving negative numbers
Evaluate.

a) $(-10)^4$ b) -10^4

c) $-(-0.5)^2$ d) $-(5 - 8)^2$

Solution

a) $(-10)^4 = (-10)(-10)(-10)(-10)$ Use -10 as a factor four times.

      $= 10,000$

b) $-10^4 = -(10^4)$ Rewrite using parentheses.

      $= -(10,000)$ Find 10^4.

      $= -10,000$ Then find the opposite of 10,000.

c) $-(-0.5)^2 = -(-0.5)(-0.5)$ Use -0.5 as a factor two times.

      $= -(0.25)$

      $= -0.25$

d) $-(5 - 8)^2 = -(-3)^2$ Evaluate within parentheses first.

      $= -(9)$ Square -3 to get 9.

      $= -9$ Take the opposite of 9 to get -9.

> Now do Exercises 43–50

CAUTION   Be careful with -10^4 and $(-10)^4$. It is tempting to evaluate these two the same. However, we have agreed that $-10^4 = -(10^4)$, where the exponent is applied only to positive 10. The negative sign is handled last. So $-10^4 = -10,000$, a negative number. Likewise, $-1^2 = -1$, $-2^2 = -4$, and $-3^4 = -81$.

‹ Helpful Hint ›

"Please Excuse My Dear Aunt Sally" (PEMDAS) is often used as a memory aid for the order of operations. Do Parentheses, Exponents, Multiplication and Division, then Addition and Subtraction. Multiplication and division have equal priority. The same goes for addition and subtraction.

⟨3⟩ **The Order of Operations**

When we evaluate expressions, operations within grouping symbols are always performed first. For example,

$$(3 + 2) \cdot 5 = (5) \cdot 5 = 25 \qquad \text{and} \qquad (2 \cdot 3)^2 = 6^2 = 36.$$

To make expressions look simpler, we often omit some or all parentheses. In this case, we must agree on the order in which to perform the operations. We agree to do multiplication before addition and exponential expressions before multiplication. So,

$$3 + 2 \cdot 5 = 3 + 10 = 13 \qquad \text{and} \qquad 2 \cdot 3^2 = 2 \cdot 9 = 18.$$

We state the complete **order of operations** in the following box.

Order of Operations

1. Evaluate expressions within grouping symbols first. Parentheses and brackets are grouping symbols. Absolute value bars and fraction bars indicate grouping and an operation.
2. Evaluate each exponential expression (in order from left to right).
3. Perform multiplication and division (in order from left to right).
4. Perform addition and subtraction (in order from left to right).

Multiplication and division have equal priority in the order of operations. If both appear in an expression, they are performed in order from left to right. The same holds for addition and subtraction. For example,

$$8 \div 4 \cdot 3 = 2 \cdot 3 = 6 \quad \text{and} \quad 9 - 3 + 5 = 6 + 5 = 11.$$

EXAMPLE 6

Using the order of operations
Evaluate each expression.

a) $2^3 \cdot 3^2$ b) $2 \cdot 5 - 3 \cdot 4 + 4^2$ c) $2 \cdot 3 \cdot 4 - 3^3 + \dfrac{8}{2}$

Solution

a) $2^3 \cdot 3^2 = 8 \cdot 9$ Evaluate exponential expressions before multiplying.
$= 72$

b) $2 \cdot 5 - 3 \cdot 4 + 4^2 = 2 \cdot 5 - 3 \cdot 4 + 16$ Exponential expressions first
$= 10 - 12 + 16$ Multiplication second
$= 14$ Addition and subtraction from left to right

c) $2 \cdot 3 \cdot 4 - 3^3 + \dfrac{8}{2} = 2 \cdot 3 \cdot 4 - 27 + \dfrac{8}{2}$ Exponential expressions first
$= 24 - 27 + 4$ Multiplication and division second
$= 1$ Addition and subtraction from left to right

> Now do Exercises 51–66

⟨ **Calculator Close-Up** ⟩

Most calculators follow the same order of operations shown here. Evaluate these expressions with your calculator.

```
2^3*3²
                72
2*5-3*4+4²
                14
2*3*4-3^3+8/2
                 1
```

When grouping symbols are used, we perform operations within grouping symbols first. The order of operations is followed within the grouping symbols.

EXAMPLE 7

Grouping symbols and the order of operations
Evaluate.

a) $3 - 2(7 - 2^3)$ b) $3 - |7 - 3 \cdot 4|$ c) $\dfrac{9 - 5 + 8}{-5^2 - 3(-7)}$

Solution

a) $3 - 2(7 - 2^3) = 3 - 2(7 - 8)$ Evaluate within parentheses first.
$= 3 - 2(-1)$
$= 3 - (-2)$ Multiply.
$= 5$ Subtract.

b) $3 - |7 - 3 \cdot 4| = 3 - |7 - 12|$ Evaluate within the absolute value symbols first.

$\qquad\qquad\qquad\quad = 3 - |-5|$

$\qquad\qquad\qquad\quad = 3 - 5$ Evaluate the absolute value.

$\qquad\qquad\qquad\quad = -2$ Subtract.

c) $\dfrac{9 - 5 + 8}{-5^2 - 3(-7)} = \dfrac{12}{-25 + 21} = \dfrac{12}{-4} = -3$ Numerator and denominator are treated as if in parentheses.

> Now do Exercises 67–80

When grouping symbols occur within grouping symbols, we evaluate within the innermost grouping symbols first and then work outward. In this case, brackets [ ] can be used as grouping symbols along with parentheses to make the grouping clear.

E X A M P L E 8

Grouping within grouping
Evaluate each expression.

a) $6 - 4[5 - (7 - 9)]$

b) $-2|3 - (9 - 5)| - |-3|$

Solution

a) $6 - 4[5 - (7 - 9)] = 6 - 4[5 - (-2)]$ Innermost parentheses first

$\qquad\qquad\qquad\qquad\quad = 6 - 4[7]$ Next evaluate within the brackets.

$\qquad\qquad\qquad\qquad\quad = 6 - 28$ Multiply.

$\qquad\qquad\qquad\qquad\quad = -22$ Subtract.

b) $-2|3 - (9 - 5)| - |-3| = -2|3 - 4| - |-3|$ Innermost grouping first

$\qquad\qquad\qquad\qquad\qquad\quad = -2|-1| - |-3|$ Evaluate within the first absolute value.

$\qquad\qquad\qquad\qquad\qquad\quad = -2 \cdot 1 - 3$ Evaluate absolute values.

$\qquad\qquad\qquad\qquad\qquad\quad = -2 - 3$ Multiply.

$\qquad\qquad\qquad\qquad\qquad\quad = -5$ Subtract.

> Now do Exercises 81–88

‹ **Calculator Close-Up** ›

Graphing calculators can handle grouping symbols within grouping symbols. Since parentheses must occur in pairs, you should have the same number of left parentheses as right parentheses. You might notice other grouping symbols on your calculator, but they may or may not be used for grouping. See your manual.

```
6-4(5-(7-9))
              -22
-2abs(3-(9-5))-a
bs(-3)
               -5
```

‹4› Applications

E X A M P L E 9

Doubling your bet
A strategy among gamblers is to double your bet and bet again after a loss. The only problem with this strategy is that you might run out of money before you get a win. A gambler loses $100 and employs this strategy. He keeps losing, six times in a row. His seventh bet will be $100 \cdot 2^6$ dollars.

a) Find the amount of the seventh bet.

b) Find the total amount lost on the first six bets.

Solution

a) By the order of operations, $100 \cdot 2^6 = 100 \cdot 64 = 6400$. So the seventh bet is \$6400.

b) Now find the total of the first six losses:
$$100 + 100 \cdot 2 + 100 \cdot 2^2 + 100 \cdot 2^3 + 100 \cdot 2^4 + 100 \cdot 2^5$$
$$= 100 + 200 + 400 + 800 + 1600 + 3200$$
$$= 6300$$

So the gambler has lost a total of \$6300 on the first six bets.

Now do Exercises 121–124

Warm-Ups ▼

Fill in the blank.

1. An _____ is the result of writing numbers in a meaningful combination with the ordinary operations of arithmetic.

2. _____ symbols indicate the order in which operations are performed.

3. An _____ expression is an expression of the form a^n.

4. The _____ tells us the order in which to perform operations when grouping symbols are omitted.

True or false?

5. $(-3)^2 = 6$

6. $(5 - 3) \cdot 2 = 4$

7. $5 - 3 \cdot 2 = 4$

8. $|5 - 6| = |5| - |6|$

9. $5 + 6 \cdot 2 = (5 + 6)2$

10. $(2 + 3)^2 = 2^2 + 3^2$

11. $5 - 3^3 = 8$

12. $(5 - 3)^3 = 8$

13. $\dfrac{6 - 6}{2} = 0$

1.5   Exercises

⟨ **Study Tips** ⟩

• Take notes in class. Write down everything that you can. As soon as possible after class, rewrite your notes. Fill in details and make corrections.
• If your instructor takes the time to work an example, it is a good bet that your instructor expects you to understand the concepts involved.

⟨1⟩ **Arithmetic Expressions**

Evaluate each expression. See Example 1.

1. $(4 - 3)(5 - 9)$

2. $(5 - 7)(-2 - 3)$

3. $|3 + 4| - |-2 - 4|$

4. $|-4 + 9| + |-3 - 5|$

5. $\dfrac{7 - (-9)}{3 - 5}$

6. $\dfrac{-8 + 2}{-1 - 1}$

7. $(-6 + 5)(7)$

8. $-6 + (5 \cdot 7)$

9. $(-3 - 7) - 6$

10. $-3 - (7 - 6)$

11. $-16 \div (8 \div 2)$

12. $(-16 \div 8) \div 2$

⟨2⟩ Exponential Expressions

Write each product as an exponential expression. See Example 2.

13. $4 \cdot 4 \cdot 4 \cdot 4$

14. $1 \cdot 1 \cdot 1 \cdot 1 \cdot 1$

15. $(-5)(-5)(-5)(-5)$

16. $(-7)(-7)(-7)$

17. $(-y)(-y)(-y)$

18. $x \cdot x \cdot x \cdot x \cdot x$

19. $\frac{3}{7} \cdot \frac{3}{7} \cdot \frac{3}{7} \cdot \frac{3}{7} \cdot \frac{3}{7}$

20. $\frac{y}{2} \cdot \frac{y}{2} \cdot \frac{y}{2} \cdot \frac{y}{2}$

Write each exponential expression as a product without exponents. See Example 3.

21. 5^3

22. $(-8)^4$

23. b^2

24. $(-a)^5$

25. $\left(-\frac{1}{2}\right)^5$

26. $\left(-\frac{13}{12}\right)^3$

Evaluate each exponential expression. See Examples 4 and 5.

27. 3^4

28. 5^3

29. 0^9

30. 0^{12}

31. $(-5)^4$

32. $(-2)^5$

33. $(-6)^3$

34. $(-12)^2$

35. $(10)^5$

36. $(-10)^6$

37. $(-0.1)^3$

38. $(-0.2)^2$

39. $\left(\frac{1}{2}\right)^3$

40. $\left(\frac{2}{3}\right)^3$

41. $\left(-\frac{1}{2}\right)^2$

42. $\left(-\frac{2}{3}\right)^2$

43. -8^2

44. -7^2

45. -8^4

46. -7^4

47. $-(7-10)^3$

48. $-(6-9)^4$

49. $(-2^2) - (3^2)$

50. $(-3^4) - (-5^2)$

⟨3⟩ The Order of Operations

Evaluate each expression. See Example 6.

51. $20 \div 2 \cdot 5$

52. $30 \div 6 \cdot 5$

53. $11 - 6 + 5$

54. $8 - 2 + 4$

55. $3^2 \cdot 2^2$

56. $5 \cdot 10^2$

57. $-3 \cdot 2 + 4 \cdot 6$

58. $-5 \cdot 4 - 8 \cdot 3$

59. $(-3)^3 + 2^3$

60. $3^2 - 5(-1)^3$

61. $-21 + 36 \div 3^2$

62. $-18 - 9^2 \div 3^3$

63. $-3 \cdot 2^3 - 5 \cdot 2^2$

64. $2 \cdot 5 - 3^2 + 4 \cdot 0$

65. $\frac{-8}{2} + 2 \cdot 3 \cdot 5 - 2^3$

66. $-4 \cdot 2 \cdot 6 - \frac{12}{3} + 3^3$

Evaluate each expression. See Example 7.

67. $(-3 + 4^2)(-6)$

68. $-3 \cdot (2^3 + 4) \cdot 5$

69. $(-3 \cdot 2 + 6)^3$

70. $5 - 2(-3 + 2)^3$

71. $2 - 5(3 - 4 \cdot 2)$

72. $(3 - 7)(4 - 6 \cdot 2)$

73. $3 - 2 \cdot |5 - 6|$

74. $3 - |6 - 7 \cdot 3|$

75. $(3^2 - 5) \cdot |3 \cdot 2 - 8|$

76. $|4 - 6 \cdot 3| + |6 - 9|$

77. $\frac{3 - 4 \cdot 6}{7 - 10}$

78. $\frac{6 - (-8)^2}{-3 - (-1)}$

79. $\frac{7 - 9 - 3^2}{9 - 7 - 3}$

80. $\frac{3^2 - 2 \cdot 4}{-30 + 2 \cdot 4^2}$

Evaluate each expression. See Example 8.

81. $3 + 4[9 - 6(2 - 5)]$

82. $9 + 3[5 - (3 - 6)^2]$

83. $6^2 - [(2 + 3)^2 - 10]$

84. $3[(2 - 3)^2 + (6 - 4)^2]$

85. $4 - 5 \cdot |3 - (3^2 - 7)|$

86. $2 + 3 \cdot |4 - (7^2 - 6^2)|$

87. $-2|3 - (7 - 3)| - |-9|$

88. $[3 - (2 - 4)][3 + |2 - 4|]$

Evaluate each expression. Use a calculator to check.

89. $1 + 2^3$

90. $(1 + 2)^3$

91. $(-2)^2 - 4(-1)(3)$

92. $(-2)^2 - 4(-2)(-3)$

93. $4^2 - 4(1)(-3)$

94. $3^2 - 4(-2)(3)$

95. $(-11)^2 - 4(5)(0)$

96. $(-12)^2 - 4(3)(0)$

97. $-5^2 - 3 \cdot 4^2$

98. $-6^2 - 5(-3)^2$

99. $[3 + 2(-4)]^2$

100. $[6 - 2(-3)]^2$

101. $|-1| - |-1|$

102. $4 - |1 - 7|$

103. $\dfrac{4 - (-4)}{-2 - 2}$

104. $\dfrac{3 - (-7)}{3 - 5}$

105. $3(-1)^2 - 5(-1) + 4$

106. $-2(1)^2 - 5(1) - 6$

107. $5 - 2^2 + 3^4$

108. $5 + (-2)^2 - 3^2$

109. $-2 \cdot |9 - 6^2|$

110. $8 - 3|5 - 4^2 + 1|$

111. $-3^2 - 5[4 - 2(4 - 9)]$

112. $-2[(3 - 4)^3 - 5] + 7$

113. $1 - 5|5 - (9 + 1)|$

114. $|6 - 3 \cdot 7| + |7 - (5 - 2)|$

Use a calculator to evaluate each expression. Round approximate answers to four decimal places.

115. $3.2^2 - 4(3.6)(-2.2)$

116. $(-4.5)^2 - 4(-2.8)(-4.6)$

117. $(5.6)^3 - [4.7 - (-3.3)^2]$

118. $9.8^3 - [1.2 - (4.4 - 9.6)^2]$

119. $\dfrac{3.44 - (-8.32)}{6.89 - 5.43}$

120. $\dfrac{-4.56 - 3.22}{3.44 - (-6.26)}$

⟨4⟩ **Applications**

Solve each problem. See Example 9.

121. *Gambler's ruin.* A gambler bets $5 and loses. He doubles his bet and loses again. He continues this pattern, losing eight times in a row. His ninth bet will be $5 \cdot 2^8$ dollars.

    a) Calculate the amount of the ninth bet.

    b) What is the total amount lost on the first eight bets?

122. *Big profits.* Big Bulldog Motorcycles showed a profit of $50,000 in its first year of operation. The company plans to double the profit each year for the next 9 years.

    a) What will be the amount of the profit in the tenth year?

    b) What will be the total amount of profit for the first 10 years of business?

123. *Population of the United States.* In 2009 the population of the United States was 306.2 million (U.S. Census Bureau, www.census.gov). If the population continues to grow at an annual rate of 1.05%, then the population in the year 2020 will be $306.2(1.0105)^{11}$ million.

    a) Evaluate the expression to find the predicted population in 2020 to the nearest tenth of a million people.

    b) Use the accompanying graph to estimate the year in which the population will reach 400 million people.

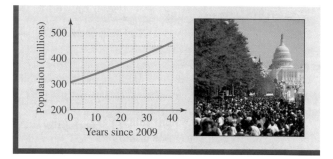

Figure for Exercise 123

124. *Population of Mexico.* In 2009 the population of Mexico was 110.3 million. If Mexico's population continues to grow at an annual rate of 1.43%, then the population in 2020 will be $110.3(1.0143)^{11}$ million.

    a) Find the predicted population in 2020 to the nearest tenth of a million people.

    b) Use the result of Exercise 123 to determine whether the United States or Mexico will have the greater increase in population between 2009 and 2020.

Getting More Involved

125. *Discussion*

How do the expressions $(-5)^3$, $-(5^3)$, -5^3, $-(-5)^3$, and $-1 \cdot 5^3$ differ?

126. *Discussion*

How do the expressions $(-4)^4$, $-(4^4)$, -4^4, $-(-4)^4$, and $-1 \cdot 4^4$ differ?

1.6 Algebraic Expressions

In This Section

In Section 1.5, you studied arithmetic expressions. In this section, you will study expressions that are more general—expressions that involve variables.

⟨1⟩ Identifying Algebraic Expressions

Variables (or letters) are used to represent numbers. With variables we can express ideas better than we can with numbers alone. For example, we know that $3 + 3 = 2(3)$, $4 + 4 = 2(4)$, $5 + 5 = 2(5)$, and so on. But using the variable x we can say that $x + x = 2x$ is true for any real number x.

The result of combining numbers and variables with the ordinary operations of arithmetic (in some meaningful way) is called an **algebraic expression** or simply an **expression.** Some examples of algebraic expressions are

$$x + x, \qquad 2x, \qquad \pi r^2, \qquad b^2 - 4ac, \qquad \text{and} \qquad \frac{a - b}{c - d}.$$

Expressions are often named by the last operation to be performed in the expression. For example, the expression $x + 2$ is a **sum** because the only operation in the expression is addition. The expression $a - bc$ is referred to as a **difference** because subtraction is the last operation to be performed. The expression $3(x - 4)$ is a **product,** while $\frac{3}{x - 4}$ is a **quotient.** The expression $(a + b)^2$ is a **square** because the addition is performed before the square is found.

E X A M P L E   **1**

Naming expressions

Identify each expression as either a sum, difference, product, quotient, or square.

a) $3(x + 2)$ **b)** $b^2 - 4ac$

c) $\dfrac{a - b}{c - d}$ **d)** $(a - b)^2$

Solution

a) In $3(x + 2)$ we add before we multiply. So this expression is a product.

b) By the order of operations the last operation to perform in $b^2 - 4ac$ is subtraction. So this expression is a difference.

c) The last operation to perform in this expression is division. So this expression is a quotient.

d) In $(a - b)^2$ we subtract before we square. This expression is a square.

> Now do Exercises 1–12

⟨ **Helpful Hint** ⟩

Sum, difference, product, and quotient are nouns. They are used as names for expressions. Add, subtract, multiply, and divide are verbs. They indicate an action to perform.

⟨2⟩ Translating Algebraic Expressions

Algebra is useful because it can be used to solve problems. Since problems are often communicated verbally, we must be able to translate verbal expressions into algebraic expressions and translate algebraic expressions into verbal expressions. Consider the following examples of verbal expressions and their corresponding algebraic expressions.

Verbal Expressions and Corresponding Algebraic Expressions	
Verbal Expression	**Algebraic Expression**
The sum of $5x$ and 3	$5x + 3$
The product of 5 and $x + 3$	$5(x + 3)$
The sum of 8 and $\dfrac{x}{3}$	$8 + \dfrac{x}{3}$
The quotient of $8 + x$ and 3	$\dfrac{8 + x}{3}$, $(8 + x)/3$, or $(8 + x) \div 3$
The difference of 3 and x^2	$3 - x^2$
The square of $3 - x$	$(3 - x)^2$

Note that the word "difference" must be used carefully. To be consistent, we say that the difference between a and b is $a - b$. So the difference between 10 and 12 is $10 - 12$ or -2. However, outside of a textbook most people would say that the difference in age between a 10-year-old and a 12-year-old is 2, not -2. Users of the English language do not follow precise rules like we follow in mathematics. Of course, in mathematics we must make our mathematics and our English sentences perfectly clear. So we try to avoid using "difference" in an ambiguous or vague manner. (We will study verbal and algebraic expressions further in Section 2.5.)

Example 2 shows how the terms sum, difference, product, quotient, and square are used to describe expressions.

E X A M P L E   2

Algebraic expressions to verbal expressions

Translate each algebraic expression into a verbal expression. Use the word sum, difference, product, quotient, or square.

a) $\dfrac{3}{x}$ b) $2y + 1$ c) $3x - 2$ d) $(a - b)(a + b)$ e) $(a + b)^2$

Solution

a) The quotient of 3 and x b) The sum of $2y$ and 1
c) The difference of $3x$ and 2   d) The product of $a - b$ and $a + b$
e) The square of the sum $a + b$

Now do Exercises 13–22

E X A M P L E   3

Verbal expressions to algebraic expressions

Translate each verbal expression into an algebraic expression.

a) The quotient of $a + b$ and 5   b) The difference of x^2 and y^2
c) The product of π and r^2 d) The square of the difference $x - y$

> **Solution**
>
> **a)** $\dfrac{a+b}{5}$, $(a+b) \div 5$, or $(a+b)/5$ **b)** $x^2 - y^2$
>
> **c)** πr^2 **d)** $(x-y)^2$

Now do Exercises 23–38

⟨3⟩ Evaluating Algebraic Expressions

The value of an algebraic expression depends on the values given to the variables. For example, the value of $x - 2y$ when $x = -2$ and $y = -3$ is found by replacing x and y by -2 and -3, respectively:

$$x - 2y = -2 - 2(-3) = -2 - (-6) = 4$$

If $x = 1$ and $y = 2$, the value of $x - 2y$ is found by replacing x by 1 and y by 2, respectively:

$$x - 2y = 1 - 2(2) = 1 - 4 = -3$$

Note that we use the order of operations when evaluating an algebraic expression.

E X A M P L E   4

Evaluating algebraic expressions

Evaluate each expression using $a = 3$, $b = -2$, and $c = -4$.

    a) $2a + b - c$ **b)** $(a - b)(a + b)$ **c)** $b^2 - 4ac$ **d)** $\dfrac{-a^2 - b^2}{c - b}$

Solution

a) $\begin{aligned} 2a + b - c &= 2(3) + (-2) - (-4) &&\text{Replace } a \text{ by 3, } b \text{ by } -2, \text{ and } c \text{ by } -4. \\ &= 6 - 2 + 4 &&\text{Multiply and remove parentheses.} \\ &= 8 &&\text{Addition and subtraction last} \end{aligned}$

b) $\begin{aligned} (a - b)(a + b) &= [3 - (-2)][3 + (-2)] &&\text{Replace.} \\ &= [5][1] &&\text{Simplify within the brackets.} \\ &= 5 &&\text{Multiply.} \end{aligned}$

c) $\begin{aligned} b^2 - 4ac &= (-2)^2 - 4(3)(-4) &&\text{Replace.} \\ &= 4 - (-48) &&\text{Square } -2, \text{ and then multiply before subtracting.} \\ &= 52 &&\text{Subtract.} \end{aligned}$

d) $\dfrac{-a^2 - b^2}{c - b} = \dfrac{-3^2 - (-2)^2}{-4 - (-2)} = \dfrac{-9 - 4}{-2} = \dfrac{13}{2}$

Now do Exercises 39–62

Mathematical notation is readily available in scientific word processors. However, on Internet pages or in e-mail, multiplication is often written with a star (*), fractions are written with a slash (/), and exponents with a caret (^). For example, $\dfrac{x+y}{2x^3}$ is written as $(x + y)/(2*x^3)$. If the numerator or denominator contains more than one symbol, it is best to enclose them in parentheses to avoid confusion. An expression such as $1/2x$ is confusing. If your class evaluates it for $x = 4$, some students will probably assume that it is $1/(2x)$ and get $1/8$, and some will assume that it is $(1/2)x$ and get 2.

⟨4⟩ Equations

An **equation** is a statement of equality of two expressions. For example,

$$11 - 5 = 6, \qquad x + 3 = 9, \qquad 2x + 5 = 13, \qquad \text{and} \qquad \frac{x}{2} - 4 = 1$$

are equations. In an equation involving a variable, any number that gives a true statement when we replace the variable by the number is said to **satisfy** the equation and is called a **solution** or **root** to the equation. For example, 6 is a solution to $x + 3 = 9$ because $6 + 3 = 9$ is true. Because $5 + 3 = 9$ is false, 5 is not a solution to the equation $x + 3 = 9$. We have **solved** an equation when we have found all solutions to the equation. You will learn how to solve certain equations in Chapter 2.

E X A M P L E   5

Satisfying an equation

Determine whether the given number is a solution to the equation following it.

a) $6, 3x - 7 = 9$ **b)** $-3, \dfrac{2x - 4}{5} = -2$ **c)** $-5, -x - 2 = 3(x + 6)$

Solution

a) Replace x by 6 in the equation $3x - 7 = 9$:

$$3(6) - 7 = 9$$
$$18 - 7 = 9$$
$$11 = 9 \quad \text{False}$$

The number 6 is not a solution to the equation $3x - 7 = 9$.

b) Replace x by -3 in the equation $\dfrac{2x - 4}{5} = -2$:

$$\frac{2(-3) - 4}{5} = -2$$
$$\frac{-10}{5} = -2$$
$$-2 = -2 \quad \text{True}$$

The number -3 is a solution to the equation.

c) Replace x by -5 in $-x - 2 = 3(x + 6)$:

$$-(-5) - 2 = 3(-5 + 6)$$
$$5 - 2 = 3(1)$$
$$3 = 3 \quad \text{True}$$

The number -5 is a solution to the equation $-x - 2 = 3(x + 6)$.

> Now do Exercises 63–76

Just as we translated verbal expressions into algebraic expressions, we can translate verbal sentences into algebraic equations. In an algebraic equation we use the equality symbol ($=$). Equality is indicated in words by phrases such as "is equal to," "is the same as," or simply "is."

E X A M P L E   6

Writing equations

Translate each sentence into an equation.

a) The sum of x and 7 is 12.

b) The product of 4 and x is the same as the sum of y and 5.

c) The quotient of $x + 3$ and 5 is equal to -1.

Solution

a) $x + 7 = 12$ **b)** $4x = y + 5$ **c)** $\dfrac{x + 3}{5} = -1$

Now do Exercises 77–84

⟨5⟩ Applications

Algebraic expressions are used to describe or **model** real-life situations. We can evaluate an algebraic expression for many values of a variable to get a collection of data. A graph (picture) of this data can give us useful information. For example, a forensic scientist can use a graph to estimate the length of a person's femur from the person's height.

EXAMPLE 7

Reading a graph

A forensic scientist uses the expression $69.1 + 2.2F$ as an estimate of the height in centimeters of a male with a femur of length F centimeters (National Space Biomedical Research Institute, www.nsbri.org).

a) If the femur of a male skeleton measures 50.6 cm, then what was the person's height?

b) Use the graph shown in Fig. 1.25 to estimate the length of a femur for a person who is 150 cm tall.

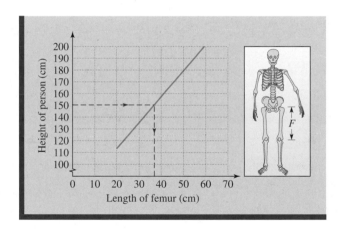

Figure 1.25

Solution

a) To find the height of the person, we use $F = 50.6$ in the expression $69.1 + 2.2F$:

$$69.1 + 2.2(50.6) \approx 180.4$$

So the person was approximately 180.4 cm tall.

b) To find the length of a femur for a person who is 150 cm tall, first locate 150 cm on the height scale of the graph in Fig. 1.25. Now draw a horizontal line to the graph and then a vertical line down to the length scale. So the length of a femur for a person who is 150 cm tall is approximately 37 cm.

Now do Exercises 93–98

Warm-Ups ▼

Fill in the blank.

1. An _____ is the result of writing numbers and variables in a meaningful combination with the ordinary operations of arithmetic.

2. An algebraic expression is a _____ if the last operation to be performed is addition.

3. An algebraic expression is a _____ if the last operation to be performed is multiplication.

4. An algebraic expression is a _____ if the last operation to be performed is division.

5. An algebraic expression is a _____ if the last operation to be performed is subtraction.

6. An _____ is a sentence that expresses equality between two algebraic expressions.

True or false?

7. The expression $2x + 3y$ is a sum.

8. The expression $5(y - 9)$ is a difference.

9. The expression $2(x + 3y)$ is a product.

10. The expression $\dfrac{6}{x} + 7$ is a quotient.

11. If $x = -2$, then the value of $2x + 4$ is 0.

12. If $a = -3$, then $a^3 - 5 = 22$.

13. The number 5 is a solution to $2x - 3 = 13$.

14. The number -2 is a solution to $3x - 5 = x - 9$.

1.6 Exercises

‹ **Study Tips** ›

• The review exercises at the end of this chapter are keyed to the sections in this chapter. If you have trouble with the review exercises, go back and study the corresponding section.

• Work the sample test at the end of this chapter to see if you are ready for your instructor's chapter test. Your instructor might not ask the same questions, but you will get a good idea of your test readiness.

‹1› Identifying Algebraic Expressions

Identify each expression as a sum, difference, product, quotient, square, or cube. See Example 1.

1. $a^3 - 1$

2. $b(b - 1)$

3. $(w - 1)^3$

4. $m^2 + n^2$

5. $3x + 5y$

6. $\dfrac{a - b}{b - a}$

7. $\dfrac{u}{v} - \dfrac{v}{u}$

8. $(s - t)^2$

9. $3(x + 5y)$

10. $a - \dfrac{a}{2}$

11. $\left(\dfrac{2}{z}\right)^2$

12. $(2q - p)^3$

‹2› Translating Algebraic Expressions

Use the term sum, difference, product, quotient, square, or cube to translate each algebraic expression into a verbal expression. See Example 2.

13. $x^2 - a^2$

14. $a^3 + b^3$

15. $(x - a)^2$

16. $(a + b)^3$

17. $\dfrac{x - 4}{2}$

18. $2(x - 3)$

19. $\dfrac{x}{2} - 4$

20. $2x - 3$

21. $(ab)^3$

22. $a^3 b^3$

Translate each verbal expression into an algebraic expression. Do not simplify. See Example 3.

23. The sum of 8 and y

24. The sum of $8x$ and $3y$

25. The product of $5x$ and z

26. The product of $x + 9$ and $x + 12$

27. The difference of 8 and $7x$

28. The difference of a^3 and b^3

29. The quotient of 6 and $x + 4$

30. The quotient of $x - 7$ and $7 - x$

31. The square of $a + b$

32. The cube of $x - y$

33. The sum of the cube of x and the square of y

34. The quotient of the square of a and the cube of b

35. The product of 5 and the square of m

36. The difference of the square of m and the square of n

37. The square of the sum of s and t

38. The cube of the difference of a and b

⟨3⟩ **Evaluating Algebraic Expressions**

Evaluate each expression using $a = -1$, $b = 2$, and $c = -3$. See Example 4.

39. $-(a - b)$

40. $b - a$

41. $-b^2 + 7$

42. $-c^2 - b^2$

43. $c^2 - 2c + 1$

44. $b^2 - 2b + 4$

45. $a^3 - b^3$

46. $b^3 - c^3$

47. $(a - b)(a + b)$

48. $(a - c)(a + c)$

49. $b^2 - 4ac$

50. $a^2 - 4bc$

51. $\dfrac{a - c}{a - b}$

52. $\dfrac{b - c}{b + a}$

53. $\dfrac{2}{a} + \dfrac{6}{b} - \dfrac{9}{c}$

54. $\dfrac{c}{a} + \dfrac{6}{b} - \dfrac{b}{a}$

55. $a \div |-a|$

56. $|a| \div a$

57. $|b| - |a|$

58. $|c| + |b|$

59. $-|-a - c|$

60. $-|-a - b|$

61. $(3 - |a - b|)^2$

62. $(|b + c| - 2)^3$

⟨4⟩ **Equations**

Determine whether the given number is a solution to the equation following it. See Example 5.

63. $2, \ 3x + 7 = 13$

64. $-1, \ -3x + 7 = 10$

65. $-2, \ \dfrac{3x - 4}{2} = 5$

66. $-3, \ \dfrac{-2x + 9}{3} = 5$

67. $-2, \ -x + 4 = 6$

68. $-9, \ -x + 3 = 12$

69. $4, \ 3x - 7 = x + 1$

70. $5, \ 3x - 7 = 2x + 1$

71. $3, \ -2(x - 1) = 2 - 2x$

72. $-8, \ x - 9 = -(9 - x)$

73. $8, \ \dfrac{x}{x - 8} = 0$

74. $3, \ \dfrac{x - 3}{x + 3} = 0$

75. $-6, \ \dfrac{x + 6}{x + 6} = 1$

76. $9, \ \dfrac{9}{x - 9} = 0$

Translate each sentence into an equation. See Example 6.

77. The sum of $5x$ and $3x$ is $8x$.

78. The sum of $\dfrac{y}{2}$ and 3 is 7.

79. The product of 3 and $x + 2$ is equal to 12.

80. The product of -6 and $7y$ is equal to 13.

81. The quotient of x and 3 is the same as the product of x and 5.

82. The quotient of $x + 3$ and $5y$ is the same as the product of x and y.

83. The square of the sum of a and b is equal to 9.

84. The sum of the squares of a and b is equal to the square of c.

Miscellaneous

Fill in the tables with the appropriate values for the given expressions.

85.

x	$2x - 3$
-2	
-1	
0	
1	
2	

86.

x	$-\dfrac{1}{2}x + 4$
-4	
-2	
0	
2	
4	

87.

a	a^2	a^3	a^4
2			
$\dfrac{1}{2}$			
10			
0.1			

88.

b	$\dfrac{1}{b}$	$\dfrac{1}{b^2}$	$\dfrac{1}{b^3}$
3			
$\dfrac{1}{3}$			
10			
0.1			

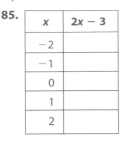*Use a calculator to find the value of $b^2 - 4ac$ for each of the following choices of a, b, and c.*

89. $a = 4.2$, $b = 6.7$, $c = 1.8$

90. $a = -3.5$, $b = 9.1$, $c = 3.6$

91. $a = -1.2$, $b = 3.2$, $c = 5.6$

92. $a = 2.4$, $b = -8.5$, $c = -5.8$

⟨5⟩ Applications

Solve each problem. See Example 7.

93. *Forensics.* A forensic scientist uses the expression $81.7 + 2.4T$ to estimate the height in centimeters of a male with a tibia of length T centimeters. If a male skeleton has a tibia of length 36.5 cm, then what was the height of the person?

Use the accompanying graph to estimate the length of a tibia for a male with a height of 180 cm.

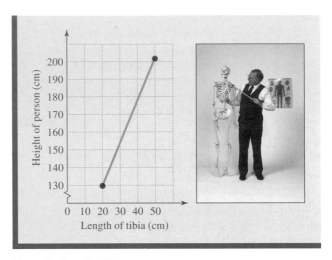

Figure for Exercise 93

94. *Forensics.* A forensic scientist uses the expression $72.6 + 2.5T$ to estimate the height in centimeters of a female with a tibia of length T centimeters. If a female skeleton has a tibia of length 32.4 cm, then what was the height of the person? Find the length of your tibia in centimeters, and use the expression from this exercise or the previous exercise to estimate your height.

95. *Games behind.* In baseball a team's standing is measured by its percentage of wins and by the number of games it is behind the leading team in its division. The expression

$$\frac{(X - x) + (y - Y)}{2}$$

gives the number of games behind for a team with x wins and y losses, where the division leader has X wins and Y losses. The table shown gives the won-lost records for the American League East on July 3, 2006

	W	L	Pct	GB
Boston	50	29	0.633	—
NY Yankees	46	33	0.582	?
Toronto	46	35	0.568	?
Baltimore	38	45	0.458	?
Tampa Bay	35	47	0.427	?

Table for Exercise 95

(www.espn.com). Fill in the column for the games behind (GB).

96. *Fly ball.* The approximate distance in feet that a baseball travels when hit at an angle of 45° is given by the expression

$$\frac{(v_0)^2}{32}$$

where v_0 is the initial velocity in feet per second. If Barry Bonds of the Giants hits a ball at a 45° angle with an initial velocity of 120 feet per second, then how far will the ball travel? Use the accompanying graph to estimate the initial velocity for a ball that has traveled 370 feet.

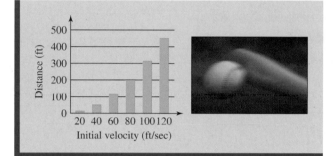

Figure for Exercise 96

97. *Football field.* The expression $2L + 2W$ gives the perimeter of a rectangle with length L and width W. What is the perimeter of a football field with length 100 yards and width 160 feet?

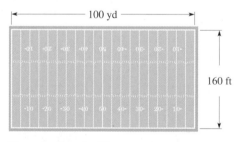

Figure for Exercise 97

98. *Crop circles.* The expression πr^2 gives the area of a circle with radius r. How many square meters of wheat were destroyed when an alien ship made a crop circle of diameter 25 meters in the wheat field at the Southwind Ranch? Round to the nearest tenth. Find π on your calculator.

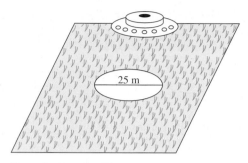

Figure for Exercise 98

Getting More Involved

99. *Writing*

Explain why the square of the sum of two numbers is different from the sum of the squares of two numbers.

100. *Cooperative learning*

The sum of the integers from 1 through n is $\frac{n(n + 1)}{2}$. The sum of the squares of the integers from 1 through n is $\frac{n(n + 1)(2n + 1)}{6}$. The sum of the cubes of the integers from 1 through n is $\frac{n^2(n + 1)^2}{4}$. Use the appropriate expressions to find the following values.

a) The sum of the integers from 1 through 30.

b) The sum of the squares of the integers from 1 through 30.

c) The sum of the cubes of the integers from 1 through 30.

d) The square of the sum of the integers from 1 through 30.

e) The cube of the sum of the integers from 1 through 30.

1.7   Properties of the Real Numbers

Everyone knows that the price of a hamburger plus the price of a Coke is the same as the price of a Coke plus the price of a hamburger. But do you know that this example illustrates the commutative property of addition? The properties of the real numbers are commonly used by anyone who performs the operations of arithmetic. In algebra we must have a thorough understanding of these properties.

⟨1⟩ The Commutative Properties

We get the same result whether we evaluate $3 + 5$ or $5 + 3$. This example illustrates the commutative property of addition. The fact that $4 \cdot 6$ and $6 \cdot 4$ are equal illustrates the commutative property of multiplication.

Commutative Property of Addition

For any real numbers a and b,

$$a + b = b + a.$$

Commutative Property of Multiplication

For any real numbers a and b,

$$ab = ba.$$

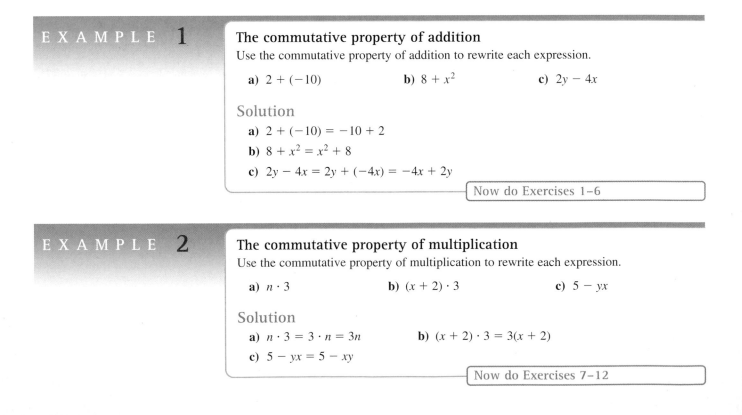

EXAMPLE   **1**

The commutative property of addition
Use the commutative property of addition to rewrite each expression.

  a) $2 + (-10)$ **b)** $8 + x^2$ **c)** $2y - 4x$

Solution

  a) $2 + (-10) = -10 + 2$
  b) $8 + x^2 = x^2 + 8$
  c) $2y - 4x = 2y + (-4x) = -4x + 2y$

Now do Exercises 1–6

EXAMPLE   **2**

The commutative property of multiplication
Use the commutative property of multiplication to rewrite each expression.

  a) $n \cdot 3$ **b)** $(x + 2) \cdot 3$ **c)** $5 - yx$

Solution

  a) $n \cdot 3 = 3 \cdot n = 3n$ **b)** $(x + 2) \cdot 3 = 3(x + 2)$
  c) $5 - yx = 5 - xy$

Now do Exercises 7–12

Addition and multiplication are commutative operations, but what about subtraction and division? Since $5 - 3 = 2$ and $3 - 5 = -2$, subtraction is not commutative. To see that division is not commutative, try dividing \$8 among 4 people and \$4 among 8 people.

⟨2⟩ The Associative Properties

⟨ **Helpful Hint** ⟩

In arithmetic we would probably write $(2 + 3) + 7 = 12$ without thinking about the associative property. In algebra, we need the associative property to understand that

$$(x + 3) + 7 = x + (3 + 7)$$
$$= x + 10.$$

Consider the computation of $2 + 3 + 6$. Using the order of operations, we add 2 and 3 to get 5 and then add 5 and 6 to get 11. If we add 3 and 6 first to get 9 and then add 2 and 9, we also get 11. So,

$$(2 + 3) + 6 = 2 + (3 + 6).$$

We get the same result for either order of addition. This property is called the **associative property of addition.** The commutative and associative properties of addition are the reason that a hamburger, a Coke, and French fries cost the same as French fries, a hamburger, and a Coke.

We also have an **associative property of multiplication.** Consider the following two ways to find the product of 2, 3, and 4:

$$(2 \cdot 3)4 = 6 \cdot 4 = 24$$
$$2(3 \cdot 4) = 2 \cdot 12 = 24$$

We get the same result for either arrangement.

Associative Property of Addition

For any real numbers a, b, and c,

$$(a + b) + c = a + (b + c).$$

Associative Property of Multiplication

For any real numbers a, b, and c,

$$(ab)c = a(bc).$$

EXAMPLE **3**

Using the properties of multiplication

Use the commutative and associative properties of multiplication and exponential notation to rewrite each product.

a) $(3x)(x)$ **b)** $(xy)(5yx)$

Solution

a) $(3x)(x) = 3(x \cdot x) = 3x^2$

b) The commutative and associative properties of multiplication allow us to rearrange the multiplication in any order. We generally write numbers before variables, and we usually write variables in alphabetical order:

$$(xy)(5yx) = 5xxyy = 5x^2y^2$$

Now do Exercises 13–18

Consider the expression

$$3 - 9 + 7 - 5 - 8 + 4 - 13.$$

According to the accepted order of operations, we could evaluate this by computing from left to right. However, using the definition of subtraction, we can rewrite this expression as addition:

$$3 + (-9) + 7 + (-5) + (-8) + 4 + (-13)$$

The commutative and associative properties of addition allow us to add these numbers in any order we choose. It is usually faster to add the positive numbers, add the negative numbers, and then combine those two totals:

$$3 + 7 + 4 + (-9) + (-5) + (-8) + (-13) = 14 + (-35) = -21$$

Note that by performing the operations in this manner, we must subtract only once. There is no need to rewrite this expression as we have done here. We can sum the positive numbers and the negative numbers from the original expression and then combine their totals.

EXAMPLE 4

Using the properties of addition
Evaluate.

    a) $3 - 7 + 9 - 5$ **b)** $4 - 5 - 9 + 6 - 2 + 4 - 8$

Solution

    a) First add the positive numbers and the negative numbers:

$$3 - 7 + 9 - 5 = 12 + (-12)$$
$$= 0$$

  b) $4 - 5 - 9 + 6 - 2 + 4 - 8 = 14 + (-24)$
$$= -10$$

Now do Exercises 19–26

It is certainly not essential that we evaluate the expressions of Example 4 as shown. We get the same answer by adding and subtracting from left to right. However, in algebra, just getting the answer is not always the most important point. Learning new methods often increases understanding.

Even though addition is associative, subtraction is not an associative operation. For example, $(8 - 4) - 3 = 1$ and $8 - (4 - 3) = 7$. So,

$$(8 - 4) - 3 \neq 8 - (4 - 3).$$

We can also use a numerical example to show that division is not associative. For instance, $(16 \div 4) \div 2 = 2$ and $16 \div (4 \div 2) = 8$. So,

$$(16 \div 4) \div 2 \neq 16 \div (4 \div 2).$$

⟨3⟩ The Distributive Property

If four men and five women pay $3 each for a movie, there are two ways to find the total amount spent:

$$3(4 + 5) = 3 \cdot 9 = 27$$
$$3 \cdot 4 + 3 \cdot 5 = 12 + 15 = 27$$

<Helpful Hint>

To visualize the distributive property, we can determine the number of circles shown here in two ways:

o o o o   o o o o o
o o o o   o o o o o
o o o o   o o o o o

There are $3 \cdot 9$ or 27 circles, or there are $3 \cdot 4$ circles in the first group and $3 \cdot 5$ circles in the second group for a total of 27 circles.

Since we get $27 either way, we can write

$$3(4 + 5) = 3 \cdot 4 + 3 \cdot 5.$$

We say that the multiplication by 3 is *distributed* over the addition. This example illustrates the **distributive property.**

Consider the following expressions involving multiplication and subtraction:

$$5(6 - 4) = 5 \cdot 2 = 10$$
$$5 \cdot 6 - 5 \cdot 4 = 30 - 20 = 10$$

Since both expressions have the same value, we can write

$$5(6 - 4) = 5 \cdot 6 - 5 \cdot 4.$$

Multiplication by 5 is distributed over each number in the parentheses. This example illustrates that multiplication distributes over subtraction.

Distributive Property

For any real numbers a, b, and c,

$$a(b + c) = ab + ac \qquad \text{and} \qquad a(b - c) = ab - ac.$$

We can use the distributive property to remove parentheses. If we start with $4(x + 3)$ and write

$$4(x + 3) = 4x + 4 \cdot 3 = 4x + 12,$$

we are using it to multiply 4 and $x + 3$ or to remove the parentheses. We wrote the product $4(x + 3)$ as the sum $4x + 12$.

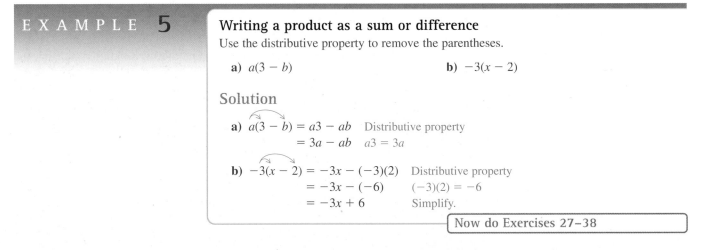

EXAMPLE 5

Writing a product as a sum or difference
Use the distributive property to remove the parentheses.

a) $a(3 - b)$ **b)** $-3(x - 2)$

Solution

a) $a(3 - b) = a3 - ab$ Distributive property
         $= 3a - ab$ $a3 = 3a$

b) $-3(x - 2) = -3x - (-3)(2)$ Distributive property
         $= -3x - (-6)$ $(-3)(2) = -6$
         $= -3x + 6$ Simplify.

Now do Exercises 27–38

When we write a number or an expression as a product, we are **factoring.** If we start with $3x + 15$ and write

$$3x + 15 = 3x + 3 \cdot 5 = 3(x + 5),$$

we are using the distributive property to factor $3x + 15$. We factored out the common factor 3.

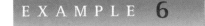

Writing a sum or difference as a product
Use the distributive property to factor each expression.

    a) $7x - 21$ **b)** $5a + 5$

Solution

    a) $7x - 21 = 7x - 7 \cdot 3$ Write 21 as $7 \cdot 3$.
                    $= 7(x - 3)$ Distributive property
    b) $5a + 5 = 5a + 5 \cdot 1$ Write 5 as $5 \cdot 1$.
                 $= 5(a + 1)$ Factor out the common factor 5.

> Now do Exercises 39–50

⟨4⟩ The Identity Properties

The numbers 0 and 1 have special properties. Multiplication of a number by 1 does not change the number, and addition of 0 to a number does not change the number. That is why 1 is called the **multiplicative identity** and 0 is called the **additive identity.**

Additive Identity Property

For any real number a,
$$a + 0 = 0 + a = a.$$

Multiplicative Identity Property

For any real number a,
$$a \cdot 1 = 1 \cdot a = a.$$

⟨5⟩ The Inverse Properties

The idea of additive inverses was introduced in Section 1.3. Every real number a has an **additive inverse** or **opposite,** $-a$, such that $a + (-a) = 0$. Every nonzero real number a also has a **multiplicative inverse** or **reciprocal,** written $\frac{1}{a}$, such that $a \cdot \frac{1}{a} = 1$. Note that the sum of additive inverses is the additive identity and that the product of multiplicative inverses is the multiplicative identity.

Additive Inverse Property

For any real number a, there is a unique number $-a$ such that
$$a + (-a) = 0.$$

Multiplicative Inverse Property

For any nonzero real number a, there is a unique number $\frac{1}{a}$ such that
$$a \cdot \frac{1}{a} = 1.$$

We are already familiar with multiplicative inverses for rational numbers. For example, the multiplicative inverse of $\frac{2}{3}$ is $\frac{3}{2}$ because
$$\frac{2}{3} \cdot \frac{3}{2} = \frac{6}{6} = 1.$$

EXAMPLE 7

Multiplicative inverses

Find the multiplicative inverse of each number.

a) 5   **b)** 0.3   **c)** $-\dfrac{3}{4}$ **d)** 1.7

Solution

a) The multiplicative inverse of 5 is $\frac{1}{5}$ because

$$5 \cdot \frac{1}{5} = 1.$$

b) To find the reciprocal of 0.3, we first write 0.3 as a ratio of integers:

$$0.3 = \frac{3}{10}$$

The multiplicative inverse of 0.3 is $\frac{10}{3}$ because

$$\frac{3}{10} \cdot \frac{10}{3} = 1.$$

c) The reciprocal of $-\frac{3}{4}$ is $-\frac{4}{3}$ because

$$\left(-\frac{3}{4}\right)\left(-\frac{4}{3}\right) = 1.$$

d) First convert 1.7 to a ratio of integers:

$$1.7 = 1\frac{7}{10} = \frac{17}{10}$$

The multiplicative inverse is $\frac{10}{17}$.

> Now do Exercises 51–62

‹ **Calculator Close-Up** ›

You can find multiplicative inverses with a calculator as shown here.

```
1/.3▶Frac
             10/3
1/(-3/4)▶Frac
             -4/3
1/1.7▶Frac
            10/17
```

When the divisor is a fraction, it must be in parentheses.

‹6› Identifying the Properties

Zero has a property that no other number has. Multiplication involving zero always results in zero.

Multiplication Property of Zero

For any real number a,

$$0 \cdot a = 0 \qquad \text{and} \qquad a \cdot 0 = 0.$$

EXAMPLE 8

Identifying the properties

Name the property that justifies each equation.

a) $5 \cdot 7 = 7 \cdot 5$ **b)** $4 \cdot \frac{1}{4} = 1$

c) $1 \cdot 864 = 864$ **d)** $6 + (5 + x) = (6 + 5) + x$

e) $3x + 5x = (3 + 5)x$ **f)** $6 + (x + 5) = 6 + (5 + x)$

g) $\pi x^2 + \pi y^2 = \pi(x^2 + y^2)$ **h)** $325 + 0 = 325$

i) $-3 + 3 = 0$ **j)** $455 \cdot 0 = 0$

Solution

a) Commutative property of multiplication   b) Multiplicative inverse property
c) Multiplicative identity property   d) Associative property of addition
e) Distributive property   f) Commutative property of addition
g) Distributive property   h) Additive identity property
i) Additive inverse property   j) Multiplication property of 0

> Now do Exercises 63–82

Warm-Ups ▼

Fill in the blank.

1. According to the _____ property of addition, $a + b = b + a$ for any real numbers a and b.

2. According to the _____ property, $a(b + c) = ab + ac$ for any real numbers a, b, and c.

3. According to the _____ property of addition, $a + (b + c) = (a + b) + c$ for any real numbers a, b, and c.

4. _____ is the process of writing a number or expression as a product.

5. The number 0 is the _____ identity.

6. The number 1 is the _____ identity.

True or false?

7. $99 + (36 + 78) = (99 + 36) + 78$

8. $24 \div (4 \div 2) = (24 \div 4) \div 2$

9. $9 - (4 - 3) = (9 - 4) - 3$

10. $156 + 387 = 387 + 156$

11. $156 \div 387 = 387 \div 156$

12. $5x + 5 = 5(x + 1)$ for any real number x.

13. The multiplicative inverse of 0.02 is 50.

14. The additive inverse of 0 is 0.

15. The number -2 is a solution to $3x - 5 = x - 9$.

1.7 Exercises

‹ **Study Tips** ›

- Don't stay up all night cramming for a test. Prepare for a test well in advance and get a good night's sleep before a test.
- Do your homework on a regular basis so that there is no need to cram.

‹1› **The Commutative Properties**

Use the commutative property of addition to rewrite each expression. See Example 1.

1. $9 + r$ 2. $t + 6$ 3. $3(2 + x)$

4. $P(1 + rt)$ 5. $4 - 5x$ 6. $b - 2a$

Use the commutative property of multiplication to rewrite each expression. See Example 2.

7. $x \cdot 6$ 8. $y \cdot (-9)$ 9. $(x - 4)(-2)$

10. $a(b + c)$ 11. $4 - y \cdot 8$ 12. $z \cdot 9 - 2$

⟨2⟩ The Associative Properties

Use the commutative and associative properties of multiplication and exponential notation to rewrite each product. See Example 3.

13. $(4w)(w)$ **14.** $(y)(2y)$ **15.** $3a(ba)$

16. $(x \cdot x)(7x)$ **17.** $(x)(9x)(xz)$ **18.** $y(y \cdot 5)(wy)$

Evaluate by finding first the sum of the positive numbers and then the sum of the negative numbers. See Example 4.

19. $8 - 4 + 3 - 10$
20. $-3 + 5 - 12 + 10$
21. $8 - 10 + 7 - 8 - 7$
22. $6 - 11 + 7 - 9 + 13 - 2$
23. $-4 - 11 + 7 - 8 + 15 - 20$
24. $-8 + 13 - 9 - 15 + 7 - 22 + 5$
25. $-3.2 + 2.4 - 2.8 + 5.8 - 1.6$
26. $5.4 - 5.1 + 6.6 - 2.3 + 9.1$

⟨3⟩ The Distributive Property

Use the distributive property to remove the parentheses. See Example 5.

27. $3(x - 5)$ **28.** $4(b - 1)$
29. $a(2 + t)$ **30.** $b(a + w)$
31. $-3(w - 6)$  **32.** $-3(m - 5)$
33. $-4(5 - y)$ **34.** $-3(6 - p)$
35. $-1(a - 7)$ **36.** $-1(c - 8)$
37. $-1(t + 4)$ **38.** $-1(x + 7)$

Use the distributive property to factor each expression. See Example 6.

39. $2m + 12$ **40.** $3y + 6$
41. $4x - 4$ **42.** $6y + 6$
43. $4y - 16$ **44.** $5x + 15$
45. $4a + 8$ **46.** $7a - 35$
47. $x + xy$ **48.** $a - ab$
49. $6a - 2b$ **50.** $8a + 2c$

⟨5⟩ The Inverse Properties

Find the multiplicative inverse (reciprocal) of each number. See Example 7.

51. $\dfrac{1}{2}$ **52.** $\dfrac{1}{3}$ **53.** -5

54. -6 **55.** 7 **56.** 8

57. 1 **58.** -1 **59.** -0.25

60. 0.75 **61.** 2.5 **62.** 3.5

⟨6⟩ Identifying the Properties

Name the property that justifies each equation. See Example 8.

63. $3 \cdot x = x \cdot 3$
64. $x + 5 = 5 + x$
65. $2(x - 3) = 2x - 6$
66. $a(bc) = (ab)c$
67. $-3(xy) = (-3x)y$
68. $3(x + 1) = 3x + 3$
69. $4 + (-4) = 0$
70. $1.3 + 9 = 9 + 1.3$
71. $x^2 \cdot 5 = 5x^2$
72. $0 \cdot \pi = 0$
73. $1 \cdot 3y = 3y$
74. $(0.1)(10) = 1$
75. $2a + 5a = (2 + 5)a$
76. $3 + 0 = 3$
77. $-7 + 7 = 0$
78. $1 \cdot b = b$
79. $(2346)0 = 0$
80. $4x + 4 = 4(x + 1)$
81. $ay + y = y(a + 1)$
82. $ab + bc = b(a + c)$

Complete each equation, using the property named.

83. $a + y =$ ____, commutative property of addition
84. $6x + 6 =$ ____, distributive property
85. $5(aw) =$ ____, associative property of multiplication

86. $x + 3 =$ ____, commutative property of addition
87. $\dfrac{1}{2}x + \dfrac{1}{2} =$ ____, distributive property

88. $-3(x - 7) =$ ____, distributive property
89. $6x + 15 =$ ____, distributive property
90. $(x + 6) + 1 =$ ____, associative property of addition

91. $4(0.25) =$ ____, multiplicative inverse property
92. $-1(5 - y) =$ ____, distributive property
93. $0 = 96($____$)$, multiplication property of zero
94. $3 \cdot ($____$) = 3$, multiplicative identity property

95. $0.33($____$) = 1$, multiplicative inverse property

96. $-8(1) =$ ____, multiplicative identity property

Getting More Involved

97. *Writing*

The perimeter of a rectangle is the sum of twice the length and twice the width. Write in words another way to find the perimeter that illustrates the distributive property.

98. *Discussion*

Eldrid bought a loaf of bread for $2.50 and a gallon of milk for $4.31. Using a tax rate of 5%, he correctly figured that the tax on the bread would be 13 cents and the tax on the milk would be 22 cents, for a total of $7.16. However, at the cash register he was correctly charged $7.15. How

could this happen? Which property of the real numbers is in question in this case?

99. *Exploration*

Determine whether each of the following pairs of tasks are "commutative." That is, does the order in which they are performed produce the same result?

a) Put on your coat; put on your hat.
b) Put on your shirt; put on your coat.

Find another pair of "commutative" tasks and another pair of "noncommutative" tasks.

1.8   Using the Properties to Simplify Expressions

In This Section

⟨1⟩ **Using the Properties in Computation**
⟨2⟩ **Combining Like Terms**
⟨3⟩ **Products and Quotients**
⟨4⟩ **Removing Parentheses**
⟨5⟩ **Applications**

The properties of the real numbers can be helpful when we are doing computations. In this section we will see how the properties can be applied in arithmetic and algebra.

⟨1⟩ Using the Properties in Computation

The properties of the real numbers can often be used to simplify computations. For example, to find the product of 26 and 200, we can write

$$
\begin{aligned}
(26)(200) &= (26)(2 \cdot 100) \\
&= (26 \cdot 2)(100) \\
&= 52 \cdot 100 \\
&= 5200.
\end{aligned}
$$

It is the associative property that allows us to multiply 26 by 2 to get 52, and then multiply 52 by 100 to get 5200.

E X A M P L E   1

Using the properties
Use the appropriate property to aid you in evaluating each expression.

a) $347 + 35 + 65$ **b)** $3 \cdot 435 \cdot \dfrac{1}{3}$ **c)** $6 \cdot 28 + 4 \cdot 28$

Solution

a) Notice that the sum of 35 and 65 is 100. So apply the associative property as follows:

$$
\begin{aligned}
347 + (35 + 65) &= 347 + 100 \\
&= 447
\end{aligned}
$$

b) Use the commutative and associative properties to rearrange this product. We can then do the multiplication quickly:

$$3 \cdot 435 \cdot \frac{1}{3} = 435\left(3 \cdot \frac{1}{3}\right) \quad \text{Commutative and associative properties}$$

$$= 435 \cdot 1 \qquad \text{Multiplicative inverse property}$$

$$= 435 \qquad \text{Multiplicative identity property}$$

c) Use the distributive property to rewrite this expression.

$$6 \cdot 28 + 4 \cdot 28 = (6 + 4)28$$

$$= 10 \cdot 28$$

$$= 280$$

Now do Exercises 1–16

⟨2⟩ Combining Like Terms

An expression containing a number or the product of a number and one or more variables raised to powers is called a **term.** For example,

$$-3, \qquad 5x, \qquad -3x^2y, \qquad a, \qquad \text{and} \qquad -abc$$

are terms. The number preceding the variables in a term is called the **coefficient.** In the term $5x$, the coefficient of x is 5. In the term $-3x^2y$ the coefficient of x^2y is -3. In the term a, the coefficient of a is 1 because $a = 1 \cdot a$. In the term $-abc$ the coefficient of abc is -1 because $-abc = -1 \cdot abc$. If two terms contain the same variables with the same exponents, they are called **like terms.** For example, $3x^2$ and $-5x^2$ are like terms, but $3x^2$ and $-5x^3$ are not like terms.

Using the distributive property on an expression involving the sum of like terms allows us to combine the like terms as shown in Example 2.

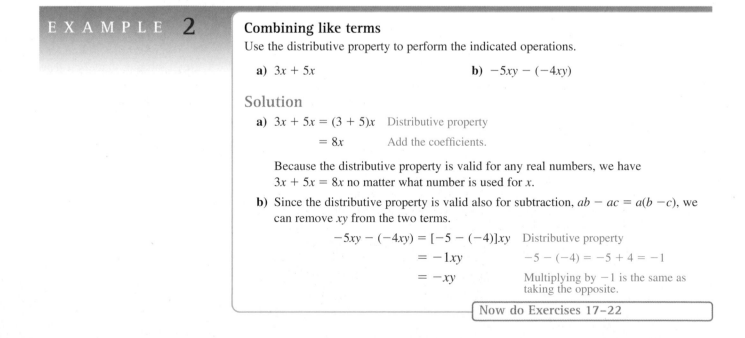

EXAMPLE   **2**

Combining like terms

Use the distributive property to perform the indicated operations.

a) $3x + 5x$ **b)** $-5xy - (-4xy)$

Solution

a) $3x + 5x = (3 + 5)x$ Distributive property

$\qquad = 8x$ Add the coefficients.

Because the distributive property is valid for any real numbers, we have $3x + 5x = 8x$ no matter what number is used for x.

b) Since the distributive property is valid also for subtraction, $ab - ac = a(b - c)$, we can remove xy from the two terms.

$$-5xy - (-4xy) = [-5 - (-4)]xy \quad \text{Distributive property}$$

$$= -1xy \qquad\qquad -5 - (-4) = -5 + 4 = -1$$

$$= -xy \qquad\qquad \text{Multiplying by } -1 \text{ is the same as taking the opposite.}$$

Now do Exercises 17–22

Of course, we do not want to write out all of the steps shown in Example 2 every time we combine like terms. We can combine like terms as easily as we can add or subtract their coefficients.

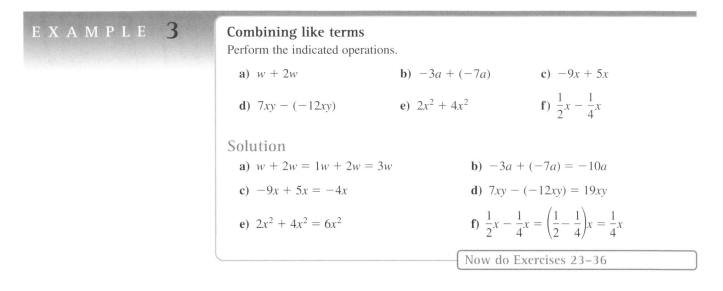

EXAMPLE 3

Combining like terms

Perform the indicated operations.

a) $w + 2w$ b) $-3a + (-7a)$ c) $-9x + 5x$

d) $7xy - (-12xy)$ e) $2x^2 + 4x^2$ f) $\dfrac{1}{2}x - \dfrac{1}{4}x$

Solution

a) $w + 2w = 1w + 2w = 3w$ b) $-3a + (-7a) = -10a$

c) $-9x + 5x = -4x$ d) $7xy - (-12xy) = 19xy$

e) $2x^2 + 4x^2 = 6x^2$ f) $\dfrac{1}{2}x - \dfrac{1}{4}x = \left(\dfrac{1}{2} - \dfrac{1}{4}\right)x = \dfrac{1}{4}x$

Now do Exercises 23–36

CAUTION There are no like terms in expressions such as

$$2 + 5x, \qquad 3xy + 5y, \qquad 3w + 5a, \qquad \text{and} \qquad 3z^2 + 5z.$$

The terms in these expressions cannot be combined.

⟨3⟩ Products and Quotients

To **simplify** an expression means to perform operations, combine like terms, and get an equivalent expression that looks simpler. However, *simplify* is *not* a precisely defined term. An expression that uses fewer symbols is usually considered simpler, but we should not be too picky with this idea. Simplifying $2x + 3x$ we get $5x$, but we would not say that $\frac{x}{2}$ is simpler than $\frac{1}{2}x$. Some would say that $2ax + 2ay$ is simpler than $2a(x + y)$ because the parentheses have been removed. However, there are seven symbols in each expression, and five operations indicated in $2ax + 2ay$ with only three in $2a(x + y)$. If you are asked to write $2a(x + y)$ as a sum or to remove the parentheses rather than to simplify it, then the answer is clearly $2ax + 2ay$.

In Example 4 we use the associative property of multiplication to simplify some products.

EXAMPLE 4

Finding products

Simplify.

a) $3(5x)$ b) $2\left(\dfrac{x}{2}\right)$

c) $(4x)(6x)$ d) $(-2a)(4b)$

Solution

a) $3(5x) = (3 \cdot 5)x$ Associative property of multiplication

$\qquad = (15)x$ Multiply.

$\qquad = 15x$ Remove unnecessary parentheses.

b) $2\left(\dfrac{x}{2}\right) = 2\left(\dfrac{1}{2} \cdot x\right)$ Multiplying by $\dfrac{1}{2}$ is the same as dividing by 2.

$\qquad = \left(2 \cdot \dfrac{1}{2}\right)x$ Associative property of multiplication

$\qquad = 1 \cdot x$ Multiplicative inverse property

$\qquad = x$ Multiplicative identity property

c) $(4x)(6x) = 4 \cdot 6 \cdot x \cdot x$ Commutative and associative properties

$\qquad = 24x^2$ Definition of exponent

d) $(-2a)(4b) = -2 \cdot 4 \cdot a \cdot b = -8ab$

> Now do Exercises 37–46

Note that $\dfrac{x}{2}$ is equivalent to $\dfrac{1}{2} \cdot x$ in Example 4(b) because division is defined as multiplication by the reciprocal of the divisor. In general, $\dfrac{x}{b}$ is equivalent to $\dfrac{1}{b} \cdot x$.

CAUTION Be careful with expressions such as $3(5x)$ and $3(5 + x)$. In $3(5x)$, we multiply 5 by 3 to get $3(5x) = 15x$. In $3(5 + x)$, both 5 and x are multiplied by the 3 to get $3(5 + x) = 15 + 3x$.

In Example 4 we showed how the properties are used to simplify products. However, in practice we usually do not write out any steps for these problems—we can write just the answer.

E X A M P L E   **5**

Finding products quickly

Find each product.

a) $(-3)(4x)$ **b)** $(-4a)(-7a)$ **c)** $(-3a)\left(\dfrac{b}{3}\right)$ **d)** $6 \cdot \dfrac{x}{2}$

Solution

a) $-12x$ **b)** $28a^2$ **c)** $-ab$ **d)** $3x$

> Now do Exercises 47–52

In Section 1.2 we found the quotient of two numbers by inverting the divisor and then multiplying. Since $a \div b = a \cdot \dfrac{1}{b}$, any quotient can be written as a product.

E X A M P L E   **6**

Simplifying quotients

Simplify.

a) $\dfrac{10x}{5}$ **b)** $\dfrac{4x + 8}{2}$

Solution

a) Since dividing by 5 is equivalent to multiplying by $\frac{1}{5}$, we have

$$\frac{10x}{5} = \frac{1}{5}(10x) = \left(\frac{1}{5} \cdot 10\right)x = (2)x = 2x.$$

Note that you can simply divide 10 by 5 to get 2.

b) Since dividing by 2 is equivalent to multiplying by $\frac{1}{2}$, we have

$$\frac{4x + 8}{2} = \frac{1}{2}(4x + 8)$$

$$= \frac{1}{2} \cdot 4x + \frac{1}{2} \cdot 8 \quad \text{Distributive property}$$

$$= 2x + 4.$$

Note that both 4 and 8 are divided by 2. So we could have written

$$\frac{4x + 8}{2} = \frac{4x}{2} + \frac{8}{2} = 2x + 4 \quad \text{or} \quad \frac{4x + 8}{2} = \frac{\cancel{2}(2x + 4)}{\cancel{2}} = 2x + 4.$$

> Now do Exercises 53–64

CAUTION It is not correct to divide only one term in the numerator by the denominator. For example,

$$\frac{4 + 7}{2} \neq 2 + 7$$

because $\frac{4 + 7}{2} = \frac{11}{2}$ and $2 + 7 = 9$.

⟨ **Calculator Close-Up** ⟩

A negative sign in front of parentheses changes the sign of every term inside the parentheses.

```
-(5-3)
                -2
-1(5-3)
                -2
-5+3
                -2
```

⟨**4**⟩ **Removing Parentheses**

In Section 1.7 we used the distributive property to multiply a sum or difference by -1 and remove the parentheses. For example,

$$-1(a + 5) = -a - 5 \quad \text{and} \quad -1(-x - 2) = x + 2.$$

If -1 is replaced with a negative sign, the parentheses are removed in the same manner because multiplying a number by -1 is equivalent to finding its opposite. So,

$$-(a + 5) = -1(a + 5) = -a - 5 \quad \text{and} \quad -(-x - 2) = -1(-x - 2) = x + 2.$$

If a subtraction sign precedes the parentheses, it is removed in the same manner also, because subtraction is defined as addition of the opposite. So,

$$3a - (a + 5) = 3a - a - 5 = 2a - 5 \quad \text{and} \quad 5x - (-x - 2) = 5x + x + 2 = 6x + 2.$$

If parentheses are preceded by a negative sign or a subtraction symbol, the signs of all terms within the parentheses are changed when the parentheses are removed.

EXAMPLE **7**

Removing parentheses with opposites and subtraction
Remove the parentheses and combine the like terms.

a) $-(x - 4) + 5x - 1$ b) $-(-5 - y) + 2y - 6$

c) $10 - (x + 3)$ d) $3x - 6 - (2x - 4)$

Solution

The procedure is the same for each part: change the signs of each term in parentheses and then combine like terms.

a) $-(x - 4) + 5x - 1 = -x + 4 + 5x - 1$
$$= 4x + 3$$

b) $-(-5 - y) + 2y - 6 = 5 + y + 2y - 6$
$$= 3y - 1$$

c) $10 - (x + 3) = 10 - x - 3$
$$= -x + 7$$

d) $3x - 6 - (2x - 4) = 3x - 6 - 2x + 4$
$$= x - 2$$

Now do Exercises 65–80

Some parentheses are used for emphasis or clarity and are unnecessary. They can be removed without changing anything. For example,

$$(2x + 3) + (x - 4) = 2x + 3 + x - 4 - 3x - 1.$$

In Example 8, we simplify more algebraic expressions, some of which contain unnecessary parentheses.

EXAMPLE **8**

Simplifying algebraic expressions
Simplify each expression.

a) $(-2x + 3) + (5x - 7)$ b) $(-3x + 6x) + 5(4 - 2x)$

c) $-2x(3x - 7) - 3(x - 6)$ d) $x - 0.02(x + 500)$

Solution

a) $(-2x + 3) + (5x - 7) = -2x + 3 + 5x - 7$ Remove unnecessary parentheses.
$$= 3x - 4$$ Combine like terms.

b) $(-3x + 6x) + 5(4 - 2x) = -3x + 6x + 20 - 10x$ Distributive property
$$= -7x + 20$$ Combine like terms.

c) $-2x(3x - 7) - 3(x - 6) = -6x^2 + 14x - 3x + 18$ Distributive property

$= -6x^2 + 11x + 18$ Combine like terms.

d) $x - 0.02(x + 500) = 1x - 0.02x - 10$ Distributive property

$= 0.98x - 10$ Combine like terms.

> Now do Exercises 81–98

⟨5⟩ Applications

E X A M P L E   9

Perimeter of a rectangle

Find an algebraic expression for the perimeter of the rectangle shown here and then find the perimeter if $x = 15$ inches.

Solution

The perimeter of any figure is the sum of the lengths of its sides:

$$2(x) + 2(2x + 1) = 2x + 4x + 2$$
$$= 6x + 2$$

So $6x + 2$ is an algebraic expression for the perimeter. If $x = 15$ inches, then the perimeter is $6(15) + 2$ or 92 inches.

> Now do Exercises 115–118

Warm-Ups ▼

Fill in the blank.

1. An expression containing a number or the product of a number and one or more variables raised to powers is a _____.

2. _____ terms are terms with the same variables and the same exponents.

3. The number preceding the variables(s) in a term is the _____.

4. To _____ an expression we combine like terms and perform operations to get an equivalent expression that looks simpler.

5. Multiplying a number by -1 changes the _____ of the number.

True or false?

6. The expressions $3x^2y$ and $5xy^2$ are like terms.

7. The coefficient in $-7ab^3$ is -7.

8. The expression $6 + 2(x + 6)$ simplified is $2x + 18$.

9. $-1(x - 4) = -x + 4$ for any real number x.

10. $(3a)(4a) = 12a$ for any real number a.

11. $b + b = b^2$ for any real number b.

12. $3(5 \cdot 2) = 15 \cdot 6$

⟨ **Study Tips** ⟩

- When you get a test back, don't simply file it in your notebook. Rework all of the problems that you missed.
- Being a full-time student is a full-time job. A successful student spends 2 to 4 hours studying outside of class for every hour spent in the classroom.

⟨1⟩ Using the Properties in Computation

Use the appropriate properties to evaluate the expressions. See Example 1.

1. $35(200)$

2. $15(300)$

3. $\dfrac{4}{3}(0.75)$

4. $5(0.2)$

5. $256 + 78 + 22$

6. $12 + 88 + 376$

7. $35 \cdot 3 + 35 \cdot 7$

8. $98 \cdot 478 + 2 \cdot 478$

9. $18 \cdot 4 \cdot 2 \cdot \dfrac{1}{4}$

10. $19 \cdot 3 \cdot 2 \cdot \dfrac{1}{3}$

11. $(120)(300)$

12. $150 \cdot 200$

13. $12 \cdot 375(-6 + 6)$

14. $354^2(-2 \cdot 4 + 8)$

15. $78 + 6 + 8 + 4 + 2$

16. $-47 + 12 - 6 - 12 + 6$

⟨2⟩ Combining Like Terms

Combine like terms where possible. See Examples 2 and 3.

17. $5w + 6w$

18. $4a + 10a$

19. $4x - x$

20. $a - 6a$

21. $2x - (-3x)$

22. $2b - (-5b)$

23. $-3a - (-2a)$

24. $-10m - (-6m)$

25. $-a - a$

26. $a - a$

27. $10 - 6t$

28. $9 - 4w$

29. $3x^2 + 5x^2$

30. $3r^2 + 4r^2$

31. $-4x + 2x^2$

32. $6w^2 - w$

33. $5mw^2 - 12mw^2$

34. $4ab^2 - 19ab^2$

35. $\dfrac{1}{3}a + \dfrac{1}{2}a$

36. $\dfrac{3}{5}b - b$

⟨3⟩ Products and Quotients

Simplify the following products or quotients. See Examples 4–6.

37. $3(4h)$

38. $2(5h)$

39. $6b(-3)$

40. $-3m(-1)$

41. $(-3m)(3m)$

42. $(2x)(-2x)$

43. $(-3d)(-4d)$

44. $(-5t)(-2t)$

45. $(-y)(-y)$

46. $y(-y)$

47. $-3a(5b)$

48. $-7w(3r)$

49. $-3a(2 + b)$

50. $-2x(3 + y)$

51. $-k(1 - k)$

52. $-t(t - 1)$

53. $\dfrac{3y}{3}$

54. $\dfrac{-9t}{9}$

55. $\dfrac{-15y}{5}$

56. $\dfrac{-12b}{2}$

57. $2\left(\dfrac{y}{2}\right)$

58. $6\left(\dfrac{m}{3}\right)$

59. $8y\left(\dfrac{y}{4}\right)$

60. $10\left(\dfrac{2a}{5}\right)$

61. $\dfrac{6a - 3}{3}$

62. $\dfrac{-8x + 6}{2}$

63. $\dfrac{-9x + 6}{-3}$

64. $\dfrac{10 - 5x}{-5}$

⟨4⟩ Removing Parentheses

Simplify each expression by removing the parentheses and combining like terms. See Example 7.

65. $-(5x + 1) + 7x$

66. $-(7a + 3) + 8a$

67. $-(-c + 4) + 5c - 9$

68. $-(-y + 4) + 9 + 4y$

69. $-(7b - 2) - 1$

70. $-(a - 1) - 9$

71. $-(-w - 4) - 8 + w$

72. $-(-y - 3) - 9y - 1$

73. $x - (3x - 1)$ **74.** $4x - (2x - 5)$

75. $5 - (y - 3)$ **76.** $8 - (m - 6)$

77. $2m + 3 - (m + 9)$

78. $7 - 8t - (2t + 6)$

79. $-3 - (-w + 2)$

80. $-5x - (-2x + 9)$

Simplify the following expressions by combining like terms. See Example 8.

81. $3x + 5x + 6 + 9$

82. $2x + 6x + 7 + 15$

83. $(-2x + 3) + (7x - 4)$

84. $(-3x + 12) + (5x - 9)$

85. $3a - 7 - (5a - 6)$

86. $4m - 5 - (m - 2)$

87. $2(a - 4) - 3(-2 - a)$

88. $2(w + 6) - 3(-w - 5)$

89. $3x(2x - 3) + 5(2x - 3)$

90. $2a(a - 5) + 4(a - 5)$

91. $-b(2b - 1) - 4(2b - 1)$

92. $-2c(c - 8) - 3(c - 8)$

93. $-5m + 6(m - 3) + 2m$

94. $-3a + 2(a - 5) + 7a$

95. $5 - 3(x + 2) - 6$

96. $7 + 2(k - 3) - k + 6$

97. $x - 0.05(x + 10)$

98. $x - 0.02(x + 300)$

Simplify each expression.

99. $3x - (4 - x)$ **100.** $2 + 8x - 11x$

101. $y - 5 - (-y - 9)$ **102.** $a - (b - c - a)$

103. $7 - (8 - 2y - m)$ **104.** $x - 8 - (-3 - x)$

105. $\frac{1}{2}(10 - 2x) + \frac{1}{3}(3x - 6)$

106. $\frac{1}{2}(x - 20) - \frac{1}{5}(x + 15)$

107. $\frac{1}{2}(3a + 1) - \frac{1}{3}(a - 5)$

108. $\frac{1}{4}(6b + 2) - \frac{2}{3}(3b - 2)$

109. $0.2(x + 3) - 0.05(x + 20)$

110. $0.08x + 0.12(x + 100)$

111. $2k + 1 - 3(5k - 6) - k + 4$

112. $2w - 3 + 3(w - 4) - 5(w - 6)$

113. $-3m - 3[2m - 3(m + 5)]$

114. $6h + 4[2h - 3(h - 9) - (h - 1)]$

⟨5⟩ **Applications**

Solve each problem. See Example 9.

115. *Perimeter of a corral.* The perimeter of a rectangular corral that has width x feet and length $x + 40$ feet is $2(x) + 2(x + 40)$. Simplify the expression for the perimeter. Find the perimeter if $x = 30$ feet.

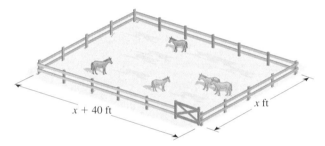

Figure for Exercise 115

116. *Perimeter of a mirror.* The perimeter of a rectangular mirror that has a width of x inches and a length of $x + 16$ inches is $2(x) + 2(x + 16)$ inches. Simplify the expression for the perimeter. Find the perimeter if $x = 14$ inches.

117. *Married filing jointly.* The value of the expression

$$9350 + 0.25(x - 67,900)$$

is the 2009 federal income tax for a married couple filing jointly with a taxable income of x dollars, where x is over \$67,900 but not over \$137,050 (Internal Revenue Service, www.irs.gov).

a) Simplify the expression.

b) Use the expression to find the amount of tax for a couple with a taxable income of \$80,000.

c) Use the accompanying graph to estimate the 2009 federal income tax for a couple with a taxable income of \$200,000

d) Use the accompanying graph to estimate the taxable income for a couple who paid \$80,000 in federal income tax.

Figure for Exercise 117

118. *Marriage penalty eliminated.* The value of the expression

$$4675 + 0.25(x - 33{,}950)$$

is the 2009 federal income tax for a single taxpayer with taxable income of x dollars, where x is over \$33,950 but not over \$82,250.

a) Simplify the expression.
b) Find the amount of tax for a single taxpayer with taxable income of \$40,000.
c) Who pays more, a married couple with a joint taxable income of \$80,000 or two single taxpayers with

taxable incomes of \$40,000 each? See Exercise 117.

Getting More Involved

119. *Discussion*

What is wrong with the way in which each of the following expressions is simplified?

a) $4(2 + x) = 8 + x$
b) $4(2x) = 8 \cdot 4x = 32x$
c) $\dfrac{4 + x}{2} = 2 + x$
d) $5 - (x - 3) = 5 - x - 3 = 2 - x$

120. *Discussion*

An instructor asked his class to evaluate the expression $1/2x$ for $x = 5$. Some students got 0.1; others got 2.5. Which answer is correct and why?

Summary

The Real Numbers		Examples
Counting or natural numbers	$\{1, 2, 3, \ldots\}$	
Whole numbers	$\{0, 1, 2, 3, \ldots\}$	
Integers	$\{\ldots, -3, -2, -1, 0, 1, 2, 3, \ldots\}$	
Rational numbers	$\left\{\dfrac{a}{b} \mid a \text{ and } b \text{ are integers with } b \neq 0\right\}$	$\dfrac{3}{2}, 5, -6, 0$
Irrational numbers	$\{x \mid x \text{ is a real number that is not rational}\}$	$\sqrt{2}, \sqrt{3}, \pi$
Real numbers	The set of real numbers consists of all rational numbers together with all irrational numbers.	
Intervals of real numbers	If a is less than b, then the set of real numbers between a and b is written as (a, b). The set of real numbers between a and b inclusive is written as $[a, b]$.	The notation $(1, 9)$ represents the real numbers between 1 and 9. The notation $[1, 9]$ represents the real numbers between 1 and 9 inclusive.

Fractions		Examples
Reducing fractions	$\dfrac{a \cdot c}{b \cdot c} = \dfrac{a}{b}$	$\dfrac{4}{6} = \dfrac{2 \cdot 2}{2 \cdot 3} = \dfrac{2}{3}$
Building up fractions	$\dfrac{a}{b} = \dfrac{a \cdot c}{b \cdot c}$	$\dfrac{3}{8} = \dfrac{3 \cdot 5}{8 \cdot 5} = \dfrac{15}{40}$
Multiplying fractions	$\dfrac{a}{b} \cdot \dfrac{c}{d} = \dfrac{ac}{bd}$	$\dfrac{2}{3} \cdot \dfrac{4}{5} = \dfrac{8}{15}$
Dividing fractions	$\dfrac{a}{b} \div \dfrac{c}{d} = \dfrac{a}{b} \cdot \dfrac{d}{c}$	$\dfrac{2}{3} \div \dfrac{4}{5} = \dfrac{2}{3} \cdot \dfrac{5}{4} = \dfrac{10}{12} = \dfrac{5}{6}$
Adding or subtracting fractions	$\dfrac{a}{b} + \dfrac{c}{b} = \dfrac{a + c}{b}$ $\dfrac{a}{b} - \dfrac{c}{b} = \dfrac{a - c}{b}$	$\dfrac{1}{5} + \dfrac{2}{5} = \dfrac{3}{5}$ $\dfrac{3}{5} - \dfrac{2}{5} = \dfrac{1}{5}$

Least common denominator	The smallest number that is a multiple of all denominators.	$\dfrac{1}{4} + \dfrac{1}{6} = \dfrac{3}{12} + \dfrac{2}{12} = \dfrac{5}{12}$

Operations with Real Numbers

		Examples
Absolute value	$\lvert a \rvert = \begin{cases} a & \text{if } a \text{ is positive or zero} \\ -a & \text{if } a \text{ is negative} \end{cases}$	$\lvert 3 \rvert = 3, \lvert 0 \rvert = 0$ $\lvert -3 \rvert = 3$
Sum of two numbers with like signs	Add their absolute values. The sum has the same sign as the given numbers.	$-3 + (-4) = -7$
Sum of two numbers with unlike signs (and different absolute values)	Subtract the absolute values of the numbers. The answer is positive if the number with the larger absolute value is positive. The answer is negative if the number with the larger absolute value is negative.	$-4 + 7 = 3$ $-7 + 4 = -3$
Sum of opposites	The sum of any number and its opposite is 0.	$-6 + 6 = 0$
Subtraction of signed numbers	$a - b = a + (-b)$ Subtract any number by adding its opposite.	$3 - 5 = 3 + (-5) = -2$ $4 - (-3) = 4 + 3 = 7$
Product or quotient	Like signs $\leftrightarrow$ Positive result Unlike signs $\leftrightarrow$ Negative result	$(-3)(-2) = 6$ $(-8) \div 2 = -4$
Definition of exponents	For any counting number n, $a^n = \underbrace{a \cdot a \cdot a \cdots \cdots a.}_{n \text{ factors}}$	$2^3 = 2 \cdot 2 \cdot 2 = 8$ $(-5)^2 = 25$ $-5^2 = -(5^2) = -25$
Order of operations	No parentheses or absolute value present: 1. Exponential expressions 2. Multiplication and division 3. Addition and subtraction With parentheses or absolute value: First evaluate within each set of parentheses or absolute value, using the order of operations.	$5 + 2^3 = 13$ $2 + 3 \cdot 5 = 17$ $4 + 5 \cdot 3^2 = 49$ $(2 + 3)(5 - 7) = -10$ $2 + 3\lvert 2 - 5 \rvert = 11$

Properties of the Real Numbers

		Examples
	For any real numbers a, b, and c	

Commutative property of
     Addition
     Multiplication
	$a + b = b + a$
	 $a \cdot b = b \cdot a$
		$5 + 7 = 7 + 5$
		 $6 \cdot 3 = 3 \cdot 6$

Associative property of		
Addition	$a + (b + c) = (a + b) + c$	$1 + (2 + 3) = (1 + 2) + 3$
Multiplication	$a \cdot (b \cdot c) = (a \cdot b) \cdot c$	$2(3 \cdot 4) = (2 \cdot 3)4$
Distributive properties	$a(b + c) = ab + ac$	$2(3 + x) = 6 + 2x$
	$a(b - c) = ab - ac$	$-2(x - 5) = -2x + 10$
Additive identity property	$a + 0 = a$ and $0 + a = a$	$5 + 0 = 0 + 5 = 5$
	Zero is the additive identity.	
Multiplicative identity property	$1 \cdot a = a$ and $a \cdot 1 = a$	$7 \cdot 1 = 1 \cdot 7 = 7$
	One is the multiplicative identity.	
Additive inverse property	For any real number a, there is a number $-a$ (additive inverse or opposite) such that $a + (-a) = 0$ and $-a + a = 0.$	$3 + (-3) = 0$ $-3 + 3 = 0$
Multiplicative inverse property	For any nonzero real number a there is a number $\frac{1}{a}$ (multiplicative inverse or reciprocal) such that $a \cdot \frac{1}{a} = 1$ and $\frac{1}{a} \cdot a = 1.$	$3 \cdot \frac{1}{3} = 1$ $\frac{1}{3} \cdot 3 = 1$
Multiplication property of 0	$a \cdot 0 = 0$ and $0 \cdot a = 0$	$5 \cdot 0 = 0$ $0(-7) = 0$

Enriching Your Mathematical Word Power

Fill in the blank.

1. The numbers $\{\ldots, -3, -2, -1, 0, 1, 2, 3, \ldots\}$ are the _____ .

2. The numbers $\{1, 2, 3, 4, \ldots\}$ are the _____ or counting numbers.

3. The numbers $\{0, 1, 2, 3, 4, \ldots\}$ are the _____ numbers.

4. The real numbers that can be expressed as a ratio of two integers are the _____ numbers.

5. The real numbers that cannot be expressed as a ratio of two integers are the _____ numbers.

6. A _____ is an expression containing a number or the product of a number and one or more variables raised to powers.

7. Terms that have the same variables with the same exponents are _____ terms.

8. A letter that is used to represent some numbers is a _____ .

9. A _____ is a rational number that is not an integer.

10. A fraction is _____ by dividing out common factors of the numerator and denominator.

11. A fraction is in _____ terms if the numerator and denominator have no common factors.

12. If a is a real number, then $-a$ is the _____ inverse of a.

13. The _____ of operations is the order in which operations are to be performed in the absence of grouping symbols.

14. The least common multiple of the denominators is the _____ common denominator.

15. The _____ value of a number is its distance from 0 on the number line.

16. The number 0 is the _____ identity.

17. The number 1 is the _____ identity.

18. In the division $a \div b = c$, b is the _____ and c is the _____ .

19. A(n) _____ number is a natural number 2 or larger that has no factors other than itself and 1.

20. A(n) _____ fraction has a larger numerator than denominator.

Review Exercises

1.1 The Real Numbers
Which of the numbers $-\sqrt{5}$, -2, 0, 1, 2, 3.14, π, *and* 10 *are*

1. whole numbers?

2. natural numbers?

3. integers?

4. rational numbers?

5. irrational numbers?

6. real numbers?

True or false? Explain your answer.

7. Every whole number is a rational number.

8. Zero is not a rational number.

9. The counting numbers between -4 and 4 are -3, -2, -1, 0, 1, 2, and 3.

10. There are infinitely many integers.

11. The set of counting numbers smaller than the national debt is infinite.

12. The decimal number 0.25 is a rational number.

13. Every integer greater than -1 is a whole number.

14. Zero is the only number that is neither rational nor irrational.

Graph each set of numbers.

15. The set of integers between -3 and 3

16. The set of natural numbers between -3 and 3

17. The set of real numbers between -1 and 4

18. The set of real numbers between -2 and 3 inclusive

Write the interval notation for each interval of real numbers.

19. The set of real numbers between 4 and 6 inclusive

20. The set of real numbers greater than 2 and less than 5

21. The set of real numbers greater than or equal to -30

22. The set of real numbers less than 50

1.2 Fractions
Perform the indicated operations.

23. $\dfrac{1}{3} + \dfrac{3}{8}$ **24.** $\dfrac{2}{3} - \dfrac{1}{4}$ **25.** $\dfrac{3}{5} \cdot 10$

26. $\dfrac{3}{5} \div 10$ **27.** $\dfrac{2}{5} \cdot \dfrac{15}{14}$ **28.** $7 \div \dfrac{1}{2}$

29. $4 + \dfrac{2}{3}$ **30.** $\dfrac{7}{12} - \dfrac{1}{4}$

31. $\dfrac{1}{2} + \dfrac{1}{3} + \dfrac{1}{4}$ **32.** $\dfrac{3}{4} \div 9$

1.3 Addition and Subtraction of Real Numbers
Evaluate.

33. $-5 + 7$ **34.** $-9 + (-4)$

35. $35 - 48$ **36.** $-3 - 9$

37. $-12 + 5$ **38.** $-12 - 5$

39. $-12 - (-5)$ **40.** $-9 - (-9)$

41. $-0.05 + 12$ **42.** $-0.03 + (-2)$

43. $-0.1 - (-0.05)$ **44.** $-0.3 + 0.3$

45. $\dfrac{1}{3} - \dfrac{1}{2}$ **46.** $-\dfrac{2}{3} + \dfrac{1}{4}$

47. $-\dfrac{1}{3} + \left(-\dfrac{2}{5}\right)$ **48.** $\dfrac{1}{3} - \left(-\dfrac{1}{4}\right)$

1.4 Multiplication and Division of Real Numbers
Evaluate.

49. $(-3)(5)$ **50.** $(-9)(-4)$

51. $(-8) \div (-2)$ **52.** $50 \div (-5)$

53. $\dfrac{-20}{-4}$

54. $\dfrac{30}{-5}$

55. $\left(-\dfrac{1}{2}\right)\left(-\dfrac{1}{3}\right)$

56. $8 \div \left(-\dfrac{1}{3}\right)$

57. $-0.09 \div 0.3$

58. $4.2 \div (-0.3)$

59. $(0.3)(-0.8)$

60. $0 \div (-0.0538)$

61. $(-5)(-0.2)$

62. $\dfrac{1}{2}(-12)$

1.5 Exponential Expressions and the Order of Operations
Evaluate.

63. $3 + 7(9)$

64. $(3 + 7)9$

65. $(3 + 4)^2$

66. $3 + 4^2$

67. $3 + 2 \cdot |5 - 6 \cdot 4|$

68. $3 - (8 - 9)$

69. $(3 - 7) - (4 - 9)$

70. $3 - 7 - 4 - 9$

71. $-2 - 4(2 - 3 \cdot 5)$

72. $3^2 - 7 + 5^2$

73. $3^2 - (7 + 5)^2$

74. $|4 - 6 \cdot 3| - |7 - 9|$

75. $\dfrac{-3 - 5}{2 - (-2)}$

76. $\dfrac{1 - 9}{4 - 6}$

77. $\dfrac{6 + 3}{3} - 5 \cdot 4 + 1$

78. $\dfrac{2 \cdot 4 + 4}{3} - 3(1 - 2)$

1.6 Algebraic Expressions
Let $a = -1$, $b = -2$, and $c = 3$. Find the value of each algebraic expression.

79. $b^2 - 4ac$

80. $a^2 - 4b$

81. $(c - b)(c + b)$

82. $(a + b)(a - b)$

83. $a^2 + 2ab + b^2$

84. $a^2 - 2ab + b^2$

85. $a^3 - b^3$

86. $a^3 + b^3$

87. $\dfrac{b + c}{a + b}$

88. $\dfrac{b - c}{2b - a}$

89. $|a - b|$

90. $|b - a|$

91. $(a + b)c$

92. $ac + bc$

Determine whether the given number is a solution to the equation following it.

93. $4, 3x - 2 = 10$

94. $1, 5(x + 3) = 20$

95. $-6, \dfrac{3x}{2} = 9$

96. $-30, \dfrac{x}{3} - 4 = 6$

97. $15, \dfrac{x + 3}{2} = 9$

98. $1, \dfrac{12}{2x + 1} = 4$

99. $4, -x - 3 = 1$

100. $7, -x + 1 = 6$

1.7 Properties of the Real Numbers
Name the property that justifies each statement.

101. $a(x + y) = ax + ay$

102. $3(4y) = (3 \cdot 4)y$

103. $(0.001)(1000) = 1$

104. $xy = yx$

105. $0 + y = y$

106. $325 \cdot 1 = 325$

107. $3 + (2 + x) = (3 + 2) + x$

108. $2x - 6 = 2(x - 3)$

109. $5 \cdot 200 = 200 \cdot 5$

110. $3 + (x + 2) = (x + 2) + 3$

111. $-50 + 50 = 0$

112. $43 \cdot 59 \cdot 82 \cdot 0 = 0$

113. $12 \cdot 1 = 12$

114. $3x + 1 = 1 + 3x$

1.8 Using the Properties to Simplify Expressions
Simplify by combining like terms.

115. $3a + 7 - (4a - 5)$

116. $2m + 6 - (m - 2)$

117. $2a(3a - 5) + 4a$

118. $3a(a - 5) + 5a(a + 2)$

119. $3(t - 2) - 5(3t - 9)$

120. $2(m + 3) - 3(3 - m)$

121. $0.1(a + 0.3) - (a + 0.6)$

122. $0.1(x + 0.3) - (x - 0.9)$

123. $0.05(x - 20) - 0.1(x + 30)$

124. $0.02(x - 100) + 0.2(x - 50)$

125. $5 - 3x(-5x - 2) + 12x^2$

126. $7 - 2x(3x - 7) - x^2$

127. $-(a - 2) - 2 - a$

128. $-(w - y) - 3(y - w)$

129. $x(x + 1) + 3(x - 1)$

130. $y(y - 2) + 3(y + 1)$

Miscellaneous

Evaluate each expression. Use a calculator to check.

131. $752(-13) + 752(13)$ **132.** $75 - (-13)$

133. $|15 - 23|$ **134.** $4^2 - 6^2$

135. $-6^2 + 3(5)$ **136.** $(0.03)(-200)$

137. $\dfrac{2}{5} + \dfrac{1}{10}$ **138.** $\dfrac{2 + 1}{5 + 10}$

139. $(0.05) \div (-0.1)$ **140.** $(4 - 9)^2 + (2 \cdot 3 - 1)^2$

141. $2\left(-\dfrac{1}{2}\right)^2 + \left(-\dfrac{1}{2}\right) - 1$ **142.** $\left(-\dfrac{6}{7}\right)\left(\dfrac{21}{26}\right)$

Simplify each expression if possible.

143. $\dfrac{2x + 4}{2}$ **144.** $4(2x)$

145. $4 + 2x$ **146.** $4(2 + x)$

147. $4 \cdot \dfrac{x}{2}$ **148.** $4 - (x - 2)$

149. $-4(x - 2)$ **150.** $(4x)(2x)$

151. $4x + 2x$ **152.** $2 + (x + 4)$

153. $4 \cdot \dfrac{x}{4}$ **154.** $4 \cdot \dfrac{3x}{2}$

155. $2 \cdot x \cdot 4$ **156.** $4 - 2(2 - x)$

157. $2(x - 4) - x(x - 4)$

158. $-x(2 - x) - 2(2 - x)$

159. $\dfrac{1}{2}(x - 4) - \dfrac{1}{4}(x - 2)$

160. $\dfrac{1}{4}(x + 2) - \dfrac{1}{2}(x - 4)$

Fill in the tables with the appropriate values for the given expressions.

161.

x	$-\dfrac{1}{3}x + 1$
-6	
-3	
0	
3	
6	

162.

x	$\dfrac{1}{2}x + 3$
-4	
-2	
0	
2	
4	

163.

a	a^2	a^3	a^4
5			
-4			

164.

b	$\dfrac{1}{b}$	$\dfrac{1}{b^2}$	$\dfrac{1}{b^3}$
-3			
$-\dfrac{1}{2}$			

Applications

Solve each problem.

165. *High-income bracket.* The expression

$$108{,}216 + 0.35(x - 372{,}950)$$

represents the amount for the 2009 federal income tax in dollars for a single taxpayer with x dollars of taxable income, where x is over \$372,950.

a) Simplify the expression.

b) Use the graph on the next page to estimate the amount of tax for a single taxpayer with a taxable income of \$500,000.

c) Find the amount of tax for MLB player Alex Rodriguez for 2009. At \$28 million he was the highest paid baseball player that year.

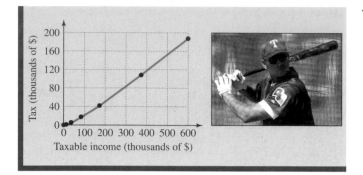

Figure for Exercise 165

166. *Married filing jointly.* The expression

$$26{,}638 + 0.28(x - 137{,}050)$$

represents the amount for the 2009 federal income tax in dollars for a married couple with x dollars of taxable income, where x is over \$137,050 but not over \$208,850.
a) Simplify the expression.
b) Find the amount of tax for Mr. and Mrs. Smith who teach at a college and have a taxable income of \$145,341.

Chapter 1 Test

Which of the numbers -3, $-\sqrt{3}$, $-\dfrac{1}{4}$, 0, $\sqrt{5}$, π, and 8 are

1. whole numbers?

2. integers?

3. rational numbers?

4. irrational numbers?

Evaluate each expression.

5. $6 + 3(-9)$

6. $(-2)^2 - 4(-2)(-1)$

7. $\dfrac{-3^2 - 9}{3 - 5}$

8. $-5 + 6 - 12 + 4$

9. $0.05 - 1$

10. $(5 - 9)(5 + 9)$

11. $(878 + 89) + 11$

12. $6 + |3 - 5(2)|$

13. $8 - 3|7 - 10|$

14. $(839 + 974)[3(-4) + 12]$

15. $974(7) + 974(3)$

16. $-\dfrac{2}{3} + \dfrac{3}{8}$

17. $(-0.05)(400)$

18. $\left(-\dfrac{3}{4}\right)\left(\dfrac{2}{9}\right)$

19. $13 \div \left(-\dfrac{1}{3}\right)$

Graph each set of numbers.

20. The set of whole numbers less than 5

21. The set of real numbers less than or equal to 4

Write the interval notation for each interval of real numbers.

22. The real numbers greater than 2

23. The real numbers greater than or equal to 3 and less than 9

Identify the property that justifies each equation.

24. $2(x + 7) = 2x + 14$

25. $48 \cdot 1000 = 1000 \cdot 48$

26. $2 + (6 + x) = (2 + 6) + x$

27. $-348 + 348 = 0$

28. $1 \cdot (-6) = -6$

29. $0 \cdot 388 = 0$

Use the distributive property to write each sum or difference as a product.

30. $3x + 30$

31. $7w - 7$

Simplify each expression.

32. $6 + 4x + 2x$

33. $6 + 4(x - 2)$

34. $5x - (3 - 2x)$

35. $x + 10 - 0.1(x + 25)$

36. $2a(4a - 5) - 3a(-2a - 5)$

37. $\dfrac{6x + 12}{6}$

38. $8 \cdot \dfrac{t}{2}$

39. $(-9xy)(-6xy)$

40. $\dfrac{1}{2}(3x + 2) - \dfrac{1}{4}(3x - 2)$

Evaluate each expression if a = −2, b = 3, and c = 4.

41. $b^2 - 4ac$

42. $\dfrac{a - b}{b - c}$

43. $(a - c)(a + c)$

Determine whether the given number is a solution to the equation following it.

44. $-2, 3x - 4 = 2$

45. $13, \dfrac{x + 3}{8} = 2$

46. $-3, -x + 5 = 8$

Solve the problem.

47. A forensic scientist uses the expression

$$80.405 + 3.660R - 0.06(A - 30)$$

to estimate the height in centimeters for a male with a radius (bone in the forearm) of length R centimeters and age A in years, where A is over 30. Simplify the expression. Use the expression to estimate the height of an 80-year-old male with a radius of length 25 cm.

Critical Thinking | For Individual or Group Work | Chapter 1

These exercises can be solved by a variety of techniques, which may or may not require algebra. So be creative and think critically. Explain all answers. Answers are in the Instructor's Edition of this text.

1. *Dividing evenly.* Suppose that you have a three-ounce glass, a five-ounce glass, and an eight-ounce glass, as shown in the accompanying figure. The two smaller glasses are empty, but the largest glass contains eight ounces of milk. How can you divide the milk into two equal parts by using only these three glasses as measuring devices?

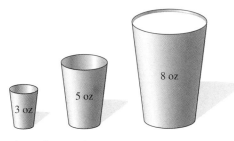

Figure for Exercise 1

2. *Totaling one hundred.* Start with the sequence of digits 123456789. Place any number of plus or minus signs between the digits in the sequence so that the value of the resulting expression is 100. For example, we could write

$$123 - 45 + 6 + 78 - 9,$$

but the value is not 100.

3. *More hundreds.* We can easily find an expression whose value is 6 using only 2's. For example, $2^2 + 2 = 6$. Find an expression whose value is 100 using only 3's. Only 4's, and so on.

4. *Forming triangles.* It is possible to draw three straight lines through a capital M to form nine nonoverlapping triangles. Try it.

5. *The right time.* Starting at 12 noon determine the number of times in the next 24 hours for which the hour and minute hands on a clock form a right angle.

6. *Perfect power.* One is the smallest positive integer that is a perfect square, a perfect cube, and a perfect fifth power. What is the next larger positive integer that is a perfect square, a perfect cube, and a perfect fifth power?

Photo for Exercise 5

7. *Summing the digits.* The sum of all of the digits that are used in writing the integers from 29 through 32 is

$$2 + 9 + 3 + 0 + 3 + 1 + 3 + 2$$

or 23. Find the sum of all of the digits that are used in writing the integers from 1 through 1000 without using a calculator.

8. *Integral rectangles.* Find all rectangles whose sides are integers and the numerical value for the area is equal to the numerical value for the perimeter.

9. *Big square.* Find the exact area of a square that is 111,111,111 feet on each side.

10. *Spelling bee.* If you spell out the counting numbers starting at 1, then what is the first counting number for which you will use the letter "a"?

Linear Equations and Inequalities in One Variable

Some ancient peoples chewed on leaves to cure their headaches. Thousands of years ago, the Egyptians used honey, salt, cedar oil, and sycamore bark to cure illnesses. Currently, some of the indigenous people of North America use black birch as a pain reliever.

Today, we are grateful for modern medicine and the seemingly simple cures for illnesses. From our own experiences we know that just the right amount of a drug can work wonders but too much of a drug can do great harm. Even though physicians often prescribe the same drug for children and adults, the amount given must be tailored to the individual. The portion of a drug given to children is usually reduced on the basis of factors such as the weight and height of the child. Likewise, older adults frequently need a lower dosage of medication than what would be prescribed for a younger, more active person.

Various algebraic formulas have been developed for determining the proper dosage for a child and an older adult.

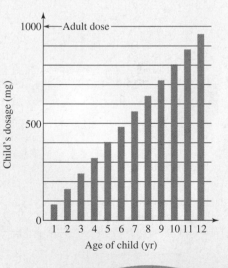

In Exercises 91 and 92 of Section 2.4 you will see two formulas that are used to determine a child's dosage by using the adult dosage and the child's age.

2.1 The Addition and Multiplication Properties of Equality

In This Section

⟨1⟩ The Addition Property of Equality

⟨2⟩ The Multiplication Property of Equality

⟨3⟩ Variables on Both Sides

⟨4⟩ Applications

In Section 1.6, an **equation** was defined as a statement that two expressions are equal. A **solution** to an equation is a number that can be used in place of the variable to make the equation a true statement. The **solution set** is the set of all solutions to an equation. Equations with the same solution set are **equivalent equations**. To **solve** an equation means to find all solutions to the equation. In this section you will learn systematic procedures for solving equations.

⟨1⟩ The Addition Property of Equality

If two workers have equal salaries and each gets a $1000 raise, then they will have equal salaries after the raise. If two people are the same age now, then in 5 years they will still be the same age. If you add the same number to two equal quantities, the results will be equal. This idea is called the *addition property of equality*:

> **The Addition Property of Equality**
>
> Adding the same number to both sides of an equation does not change the solution to the equation. In symbols, $a = b$ and
> $$a + c = b + c$$
> are equivalent equations.

Consider the equation $x = 5$. The only possible number that could be used in place of x to get a true statement is 5, because $5 = 5$ is true. So the solution set is $\{5\}$. We say that x in $x = 5$ is **isolated** because it occurs only once in the equation and it is by itself. The variable in $x - 3 = -7$ is not isolated. In Example 1, we solve $x - 3 = -7$ by using the addition property of equality to isolate the variable.

E X A M P L E **1**

⟨ Helpful Hint ⟩

Think of an equation like a balance scale. To keep the scale in balance, what you add to one side you must also add to the other side.

Adding the same number to both sides

Solve $x - 3 = -7$.

Solution

Because 3 is subtracted from x in $x - 3 = -7$, adding 3 to each side of the equation will isolate x:

$$x - 3 = -7$$
$$x - 3 + 3 = -7 + 3 \quad \text{Add 3 to each side.}$$
$$x + 0 = -4 \quad \text{Simplify each side.}$$
$$x = -4 \quad \text{Zero is the additive identity.}$$

Since -4 satisfies the last equation, it should also satisfy the original equation because all of the previous equations are equivalent. Check that -4 satisfies the original equation by replacing x by -4:

$$x - 3 = -7 \quad \text{Original equation}$$
$$-4 - 3 = -7 \quad \text{Replace } x \text{ by } -4.$$
$$-7 = -7 \quad \text{Simplify.}$$

Since $-4 - 3 = -7$ is correct, $\{-4\}$ is the solution set to the equation.

Now do Exercises 1–8

Note that enclosing the solutions to an equation in braces is not absolutely necessary. It is simply a formal way of stating the answer. At times we may simply state that the solution to the equation is -4.

The equations that we work with in this section and Sections 2.2 and 2.3 are called linear equations. The name comes from the fact that similar equations in two variables that we will study in Chapter 3 have graphs that are straight lines.

> **Linear Equation**
>
> A **linear equation in one variable** x is an equation that can be written in the form
>
> $$ax = b$$
>
> where a and b are real numbers and $a \neq 0$.

An equation such as $2x = 3$ is a linear equation. We also refer to equations such as

$$x + 8 = 0, \quad 2x + 5 = 9 - 5x, \quad \text{and} \quad 3 + 5(x - 1) = -7 + x$$

as linear equations, because these equations could be written in the form $ax = b$ using the properties of equality.

In Example 1, we used addition to isolate the variable on the left-hand side of the equation. Once the variable is isolated, we can determine the solution to the equation. Because subtraction is defined in terms of addition, we can also use subtraction to isolate the variable.

EXAMPLE 2

Subtracting the same number from both sides
Solve $9 + x = -2$.

Solution

We can remove the 9 from the left side by adding -9 to each side or by subtracting 9 from each side of the equation:

$$9 + x = -2$$
$$9 + x - 9 = -2 - 9 \qquad \text{Subtract 9 from each side.}$$
$$x = -11 \qquad \text{Simplify each side.}$$

Check that -11 satisfies the original equation by replacing x by -11:

$$9 + x = -2 \qquad \text{Original equation}$$
$$9 + (-11) = -2 \qquad \text{Replace } x \text{ by } -11.$$

Since $9 + (-11) = -2$ is correct, $\{-11\}$ is the solution set to the equation.

> Now do Exercises 9–18

Our goal in solving equations is to isolate the variable. In Examples 1 and 2, the variable was isolated on the left side of the equation. In Example 3, we isolate the variable on the right side of the equation.

EXAMPLE **3**

Isolating the variable on the right side

Solve $\frac{1}{2} = -\frac{1}{4} + y$.

Solution

We can remove $-\frac{1}{4}$ from the right side by adding $\frac{1}{4}$ to both sides of the equation:

$$\frac{1}{2} = -\frac{1}{4} + y$$

$$\frac{1}{2} + \frac{1}{4} = -\frac{1}{4} + y + \frac{1}{4} \quad \text{Add } \frac{1}{4} \text{ to each side.}$$

$$\frac{3}{4} = y \qquad\qquad \frac{1}{2} + \frac{1}{4} = \frac{2}{4} + \frac{1}{4} = \frac{3}{4}$$

Check that $\frac{3}{4}$ satisfies the original equation by replacing y by $\frac{3}{4}$:

$$\frac{1}{2} = -\frac{1}{4} + y \quad \text{Original equation}$$

$$\frac{1}{2} = -\frac{1}{4} + \frac{3}{4} \quad \text{Replace } y \text{ by } \frac{3}{4}.$$

$$\frac{1}{2} = \frac{2}{4} \qquad\quad \text{Simplify.}$$

Since $\frac{1}{2} = \frac{2}{4}$ is correct, $\left\{\frac{3}{4}\right\}$ is the solution set to the equation.

> **Now do Exercises 19–26**

⟨2⟩ The Multiplication Property of Equality

To isolate a variable that is involved in a product or a quotient, we need the multiplication property of equality.

The Multiplication Property of Equality

Multiplying both sides of an equation by the same nonzero number does not change the solution to the equation. In symbols, for $c \neq 0$, $a = b$ and

$$ac = bc$$

are equivalent equations.

We specified that $c \neq 0$ in the multiplication property of equality because multiplying by 0 can change the solution to an equation. For example, $x = 4$ is satisfied only by 4, but $0 \cdot x = 0 \cdot 4$ is true for any real number x.

In Example 4, we use the multiplication property of equality to solve an equation.

E X A M P L E   4

Multiplying both sides by the same number

Solve $\frac{z}{2} = 6$.

Solution

We isolate the variable z by multiplying each side of the equation by 2.

$$\frac{z}{2} = 6 \qquad \text{Original equation}$$

$$2 \cdot \frac{z}{2} = 2 \cdot 6 \qquad \text{Multiply each side by 2.}$$

$$1 \cdot z = 12 \qquad \text{Because } 2 \cdot \frac{z}{2} = 2 \cdot \frac{1}{2}z = 1z$$

$$z = 12 \qquad \text{Multiplicative identity}$$

Because $\frac{12}{2} = 6$, $\{12\}$ is the solution set to the equation.

> **Now do Exercises 27–34**

Because dividing by a number is the same as multiplying by its reciprocal, the multiplication property of equality allows us to divide each side of the equation by any nonzero number.

E X A M P L E   5

Dividing both sides by the same number

Solve $-5w = 30$.

Solution

Since w is multiplied by -5, we can isolate w by dividing by -5:

$$-5w = 30 \qquad \text{Original equation}$$

$$\frac{-5w}{-5} = \frac{30}{-5} \qquad \text{Divide each side by } -5.$$

$$1 \cdot w = -6 \qquad \text{Because } \frac{-5}{-5} = 1$$

$$w = -6 \qquad \text{Multiplicative identity}$$

We could also solve this equation by multiplying each side by $-\frac{1}{5}$:

$$-\frac{1}{5} \cdot -5w = -\frac{1}{5} \cdot 30$$

$$1 \cdot w = -6$$

$$w = -6$$

Because $-5(-6) = 30$, $\{-6\}$ is the solution set to the equation.

> **Now do Exercises 35–44**

In Example 6, the coefficient of the variable is a fraction. We could divide each side by the coefficient as we did in Example 5, but it is easier to multiply each side by the reciprocal of the coefficient.

EXAMPLE   **6**   **Multiplying by the reciprocal**

Solve $\frac{4}{5}p = 40$.

Solution

Multiply each side by $\frac{5}{4}$, the reciprocal of $\frac{4}{5}$, to isolate p on the left side.

$$\frac{4}{5}p = 40$$

$$\frac{5}{4} \cdot \frac{4}{5}p = \frac{5}{4} \cdot 40 \quad \text{Multiply each side by } \frac{5}{4}.$$

$$1 \cdot p = 50 \qquad \text{Multiplicative inverses}$$

$$p = 50 \qquad \text{Multiplicative identity}$$

Because $\frac{4}{5} \cdot 50 = 40$, we can be sure that the solution set is $\{50\}$.

> Now do Exercises 45–52

< **Helpful Hint** >

You could solve this equation by multiplying each side by 5 to get $4p = 200$, and then dividing each side by 4 to get $p = 50$.

If the coefficient of the variable is an integer, we usually divide each side by that integer, as we did in solving $-5w = 30$ in Example 5. Of course, we could also solve that equation by multiplying each side by $-\frac{1}{5}$. If the coefficient of the variable is a fraction, we usually multiply each side by the reciprocal of the fraction as we did in solving $\frac{4}{5}p = 40$ in Example 6. Of course, we could also solve that equation by dividing each side by $\frac{4}{5}$. If $-x$ appears in an equation, we can multiply by -1 to get x or divide by -1 to get x, because $-1(-x) = x$ and $\frac{-x}{-1} = x$.

EXAMPLE   **7**   **Multiplying or dividing by −1**

Solve $-h = 12$.

Solution

This equation can be solved by multiplying each side by -1 or dividing each side by -1. We show both methods here. First replace $-h$ with $-1 \cdot h$:

Multiplying by −1   **Dividing by −1**

$$-h = 12 \qquad\qquad\qquad -h = 12$$

$$-1(-1 \cdot h) = -1 \cdot 12 \qquad\qquad \frac{-1 \cdot h}{-1} = \frac{12}{-1}$$

$$h = -12 \qquad\qquad\qquad h = -12$$

Since $-(-12) = 12$, the solution set is $\{-12\}$.

> Now do Exercises 53–60

⟨3⟩ Variables on Both Sides

In Example 8, the variable occurs on both sides of the equation. Because the variable represents a real number, we can still isolate the variable by using the addition property of equality. Note that it does not matter whether the variable ends up on the right side or the left side.

E X A M P L E   **8**

‹ Helpful Hint ›

It does not matter whether the variable ends up on the left or right side of the equation. Whether we get $y = -9$ or $-9 = y$ we can still conclude that the solution is -9.

Subtracting an algebraic expression from both sides
Solve $-9 + 6y = 7y$.

Solution

The expression $6y$ can be removed from the left side of the equation by subtracting $6y$ from both sides.

$$-9 + 6y = 7y$$
$$-9 + 6y - 6y = 7y - 6y \quad \text{Subtract } 6y \text{ from each side.}$$
$$-9 = y \quad\quad\quad\quad \text{Simplify each side.}$$

Check by replacing y by -9 in the original equation:

$$-9 + 6(-9) = 7(-9)$$
$$-63 = -63$$

The solution set to the equation is $\{-9\}$.

Now do Exercises 61–68

‹4› Applications

In Example 9, we use the multiplication property of equality in an applied situation.

E X A M P L E   **9**

Comparing populations
In the 2000 census, Georgia had $\frac{2}{3}$ as many people as Illinois (U.S. Bureau of Census, www.census.gov). If the population of Georgia was 8 million, then what was the population of Illinois?

Solution

If p represents the population of Illinois, then $\frac{2}{3}p$ represents the population of Georgia. Since the population of Georgia was 8 million, we can write the equation $\frac{2}{3}p = 8$. To find p, solve the equation:

$$\frac{2}{3}p = 8$$

$$\frac{3}{2} \cdot \frac{2}{3}p = \frac{3}{2} \cdot 8 \quad \text{Multiply each side by } \frac{3}{2}.$$

$$p = 12 \quad \text{Simplify.}$$

So the population of Illinois was 12 million in 2000.

Now do Exercises 89–94

Warm-Ups ▼

Fill in the blank.

1. An _____ is a sentence that expresses the equality of two algebraic expressions.

2. The _____ is the set of all solutions to an equation.

3. A number _____ an equation if the equation is true when the variable is replaced by the number.

4. Equations that have the same solution set are _____.

5. A _____ equation in one variable has the form $ax = b$, with $a \neq 0$.

6. According to the _____, adding the same number to both sides of an equation does not change the solution set.

True or false?

7. The solution to $x - 5 = 5$ is 10.

8. The equation $\frac{x}{2} = 4$ is equivalent to $x = 8$.

9. To solve $\frac{3}{4}y = 12$, we should multiply each side by $\frac{3}{4}$.

10. The equation $\frac{x}{7} = 4$ is equivalent to $\frac{1}{7}x = 4$.

11. The equations $5x = 0$ and $4x = 0$ are equivalent.

12. To isolate t in $2t = 7 + t$, we subtract t from each side.

13. The solution set to $2x - 3 = x - 1$ is $\{4\}$.

2.1 Exercises

⟨ **Study Tips** ⟩

- Get to know your classmates whether you are an online student or in a classroom.
- Talk about what you are learning. Verbalizing ideas helps you get them straight in your mind.

⟨ 1 ⟩ **The Addition Property of Equality**

Solve each equation. Show your work and check your answer. See Example 1.

1. $x - 6 = -5$
2. $x - 7 = -2$
3. $-13 + x = -4$
4. $-8 + x = -12$
5. $y - \frac{1}{2} = \frac{1}{2}$
6. $y - \frac{1}{4} = \frac{1}{2}$
7. $w - \frac{1}{3} = \frac{1}{3}$
8. $w - \frac{1}{3} = \frac{1}{2}$

Solve each equation. Show your work and check your answer. See Example 2.

9. $x + 3 = -6$
10. $x + 4 = -3$ VIDEO

11. $12 + x = -7$
12. $19 + x = -11$
13. $t + \frac{1}{2} = \frac{3}{4}$
14. $t + \frac{1}{3} = 1$
15. $\frac{1}{19} + m = \frac{1}{19}$
16. $\frac{1}{3} + n = \frac{1}{2}$
17. $a + 0.05 = 6$
18. $b + 4 = -0.7$

Solve each equation. Show your work and check your answer. See Example 3.

19. $2 = x + 7$
20. $3 = x + 5$
21. $-13 = y - 9$
22. $-14 = z - 12$
23. $0.5 = -2.5 + x$
24. $0.6 = -1.2 + x$
25. $\frac{1}{8} = -\frac{1}{8} + r$
26. $\frac{1}{6} = -\frac{1}{6} + h$

⟨2⟩ The Multiplication Property of Equality

Solve each equation. Show your work and check your answer. See Example 4.

27. $\dfrac{x}{2} = -4$

28. $\dfrac{x}{3} = -6$

29. $0.03 = \dfrac{y}{60}$

30. $0.05 = \dfrac{y}{80}$

31. $\dfrac{a}{2} = \dfrac{1}{3}$

32. $\dfrac{b}{2} = \dfrac{1}{5}$

33. $\dfrac{1}{6} = \dfrac{c}{3}$

34. $\dfrac{1}{12} = \dfrac{d}{3}$

Solve each equation. Show your work and check your answer. See Example 5.

35. $-3x = 15$

36. $-5x = -20$

37. $20 = 4y$

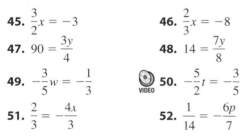 **38.** $18 = -3a$

39. $2w = 2.5$

40. $-2x = -5.6$

41. $5 = 20x$

42. $-3 = 27d$

43. $5x = \dfrac{3}{4}$

44. $3x = -\dfrac{2}{3}$

Solve each equation. Show your work and check your answer. See Example 6.

45. $\dfrac{3}{2}x = -3$

46. $\dfrac{2}{3}x = -8$

47. $90 = \dfrac{3y}{4}$

48. $14 = \dfrac{7y}{8}$

49. $-\dfrac{3}{5}w = -\dfrac{1}{3}$

50. $-\dfrac{5}{2}t = -\dfrac{3}{5}$

51. $\dfrac{2}{3} = -\dfrac{4x}{3}$

52. $\dfrac{1}{14} = -\dfrac{6p}{7}$

Solve each equation. Show your work and check your answer. See Example 7.

53. $-x = 8$

54. $-x = 4$

55. $-y = -\dfrac{1}{3}$

56. $-y = -\dfrac{7}{8}$

57. $3.4 = -z$

58. $4.9 = -t$

59. $-k = -99$

60. $-m = -17$

⟨3⟩ Variables on Both Sides

Solve each equation. Show your work and check your answer. See Example 8.

61. $4x = 3x - 7$

62. $3x = 2x + 9$

63. $9 - 6y = -5y$

64. $12 - 18w = -17w$

65. $-6x = 8 - 7x$

66. $-3x = -6 - 4x$

67. $\dfrac{1}{2}c = 5 - \dfrac{1}{2}c$

68. $-\dfrac{1}{2}h = 13 - \dfrac{3}{2}h$

Miscellaneous

Use the appropriate property of equality to solve each equation.

69. $12 = x + 17$

70. $-3 = x + 6$

71. $\dfrac{3}{4}y = -6$

72. $\dfrac{5}{9}z = -10$

73. $-3.2 + x = -1.2$

74. $t - 3.8 = -2.9$

75. $2a = \dfrac{1}{3}$

76. $-3w = \dfrac{1}{2}$

77. $-9m = 3$

78. $-4h = -2$

79. $-b = -44$

80. $-r = 55$

81. $\dfrac{2}{3}x = \dfrac{1}{2}$

82. $\dfrac{3}{4}x = \dfrac{1}{3}$

83. $-5x = 7 - 6x$

84. $-\dfrac{1}{2} + 3y = 4y$

85. $\dfrac{5a}{7} = -10$

86. $\dfrac{7r}{12} = -14$

87. $\dfrac{1}{2}v = -\dfrac{1}{2}v + \dfrac{3}{8}$

88. $\dfrac{1}{3}s + \dfrac{7}{9} = \dfrac{4}{3}s$

⟨4⟩ Applications

Solve each problem by writing and solving an equation. See Example 9.

89. *Births to teenagers.* In 2006 there were 41.8 births per 1000 females 15 to 19 years of age (National Center for Health Statistics, www.cdc.gov/nchs). This birth rate is $\dfrac{2}{3}$ of the birth rate for teenagers in 1991.

   a) Write an equation and solve it to find the birth rate for teenagers in 1991.

   b) Use the accompanying graph to estimate the birth rate to teenagers in 2000.

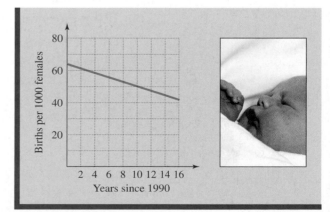

Figure for Exercise 89

90. *World grain demand.* Freeport McMoRan projects that in 2015 world grain supply will be 2.1 trillion metric tons and the supply will be only $\frac{3}{4}$ of world grain demand. What will world grain demand be in 2015?

Photo for Exercise 90

91. *Advancers and decliners.* On Thursday, $\frac{2}{3}$ of the stocks traded on the New York Stock Exchange advanced in price. If 1918 stocks advanced, then how many stocks were traded on that day?

92. *Births in the United States.* In 2009, two-fifths of all births in the United States were to unmarried women (National Center for Health Statistics, www.cdc.gov/nchs). If there were 1,707,600 births to unmarried women, then how many births were there in 2009?

93. *College students.* At Springfield College 40% of the students are male. If there are 1200 males, then how many students are there at the college?

94. *Credit card revenue.* Seventy percent of the annual revenue for a credit card company comes from interest and penalties. If the amount for interest and penalties was $210 million, then what was the annual revenue?

2.2   Solving General Linear Equations

In This Section

⟨1⟩ Equations of the Form
    $ax + b = 0$

⟨2⟩ Equations of the Form
    $ax + b = cx + d$

⟨3⟩ Equations with Parentheses

⟨4⟩ Applications

All of the equations that we solved in Section 2.1 required only a single application of a property of equality. In this section you will solve equations that require more than one application of a property of equality.

⟨1⟩ Equations of the Form $ax + b = 0$

To solve an equation of the form $ax + b = 0$ we might need to apply both the addition property of equality and the multiplication property of equality.

EXAMPLE 1

⟨ Helpful Hint ⟩

If we divide each side by 3 first, we must divide each term on the left side by 3 to get $r - \frac{5}{3} = 0$. Then add $\frac{5}{3}$ to each side to get $r = \frac{5}{3}$. Although we get the correct answer, we usually save division to the last step so that fractions do not appear until necessary.

Using the addition and multiplication properties of equality
Solve $3r - 5 = 0$.

Solution

To isolate r, first add 5 to each side, and then divide each side by 3.

$$3r - 5 = 0 \qquad \text{Original equation}$$
$$3r - 5 + 5 = 0 + 5 \qquad \text{Add 5 to each side.}$$
$$3r = 5 \qquad \text{Combine like terms.}$$
$$\frac{3r}{3} = \frac{5}{3} \qquad \text{Divide each side by 3.}$$
$$r = \frac{5}{3} \qquad \text{Simplify.}$$

Checking $\frac{5}{3}$ in the original equation gives

$$3 \cdot \frac{5}{3} - 5 = 5 - 5 = 0.$$

So $\left\{\frac{5}{3}\right\}$ is the solution set to the equation.

Now do Exercises 1–6

CAUTION   In solving $ax + b = 0$, we usually use the addition property of equality first and the multiplication property last. Note that this is the reverse of the order of operations (multiplication before addition), because we are undoing the operations that are done in the expression $ax + b$.

EXAMPLE 2

Using the addition and multiplication properties of equality

Solve $-\frac{2}{3}x + 8 = 0$.

Solution

To isolate x, first subtract 8 from each side, and then multiply each side by $-\frac{3}{2}$.

$$-\frac{2}{3}x + 8 = 0 \qquad \text{Original equation}$$

$$-\frac{2}{3}x + 8 - 8 = 0 - 8 \qquad \text{Subtract 8 from each side.}$$

$$-\frac{2}{3}x = -8 \qquad \text{Combine like terms.}$$

$$-\frac{3}{2}\left(-\frac{2}{3}x\right) = -\frac{3}{2}(-8) \qquad \text{Multiply each side by } -\frac{3}{2}.$$

$$x = 12 \qquad \text{Simplify.}$$

Checking 12 in the original equation gives

$$-\frac{2}{3}(12) + 8 = -8 + 8 = 0.$$

So $\{12\}$ is the solution set to the equation.

Now do Exercises 7–14

⟨2⟩ Equations of the Form $ax + b = cx + d$

In solving equations, our goal is to isolate the variable. We use the addition property of equality to eliminate unwanted terms. Note that it does not matter whether the variable ends up on the right or left side. For some equations, we will perform fewer steps if we isolate the variable on the right side.

E X A M P L E   3

Isolating the variable on the right side

Solve $3w - 8 = 7w$.

Solution

To eliminate the $3w$ from the left side, we can subtract $3w$ from both sides.

$$3w - 8 = 7w \qquad \text{Original equation}$$

$$3w - 8 - 3w = 7w - 3w \qquad \text{Subtract } 3w \text{ from each side.}$$

$$-8 = 4w \qquad \text{Simplify each side.}$$

$$-\frac{8}{4} = \frac{4w}{4} \qquad \text{Divide each side by 4.}$$

$$-2 = w \qquad \text{Simplify.}$$

To check, replace w with -2 in the original equation:

$$3w - 8 = 7w \qquad \text{Original equation}$$

$$3(-2) - 8 = 7(-2)$$

$$-14 = -14$$

Since -2 satisfies the original equation, the solution set is $\{-2\}$.

> **Now do Exercises 15–22**

You should solve the equation in Example 3 by isolating the variable on the left side to see that it takes more steps. In Example 4, it is simplest to isolate the variable on the left side.

E X A M P L E   4

Isolating the variable on the left side

Solve $\frac{1}{2}b - 8 = 12$.

Solution

To eliminate the 8 from the left side, we add 8 to each side.

$$\frac{1}{2}b - 8 = 12 \qquad \text{Original equation}$$

$$\frac{1}{2}b - 8 + 8 = 12 + 8 \qquad \text{Add 8 to each side.}$$

$$\frac{1}{2}b = 20 \qquad \text{Simplify each side.}$$

$$2 \cdot \frac{1}{2}b = 2 \cdot 20 \qquad \text{Multiply each side by 2.}$$

$$b = 40 \qquad \text{Simplify.}$$

To check, replace b with 40 in the original equation:

$$\frac{1}{2}b - 8 = 12 \quad \text{Original equation}$$

$$\frac{1}{2}(40) - 8 = 12$$

$$12 = 12$$

Since 40 satisfies the original equation, the solution set is {40}.

> Now do Exercises 23–30

In Example 5, both sides of the equation contain two terms.

E X A M P L E   5

Solving $ax + b = cx + d$

Solve $2m - 4 = 4m - 10$.

Solution

First, we decide to isolate the variable on the left side. So we must eliminate the 4 from the left side and eliminate $4m$ from the right side:

$$2m - 4 = 4m - 10$$

$$2m - 4 + 4 = 4m - 10 + 4 \quad \text{Add 4 to each side.}$$

$$2m = 4m - 6 \quad \text{Simplify each side.}$$

$$2m - 4m = 4m - 6 - 4m \quad \text{Subtract } 4m \text{ from each side.}$$

$$-2m = -6 \quad \text{Simplify each side.}$$

$$\frac{-2m}{-2} = \frac{-6}{-2} \quad \text{Divide each side by } -2.$$

$$m = 3 \quad \text{Simplify.}$$

To check, replace m by 3 in the original equation:

$$2m - 4 = 4m - 10 \quad \text{Original equation}$$

$$2 \cdot 3 - 4 = 4 \cdot 3 - 10$$

$$2 = 2$$

Since 3 satisfies the original equation, the solution set is {3}.

> Now do Exercises 31–38

⟨3⟩ Equations with Parentheses

Equations that contain parentheses or like terms on the same side should be simplified as much as possible before applying any properties of equality.

E X A M P L E   **6**   **Simplifying before using properties of equality**
Solve $2(q - 3) + 5q = 8(q - 1)$.

Solution

First remove parentheses and combine like terms on each side of the equation.

$$2(q - 3) + 5q = 8(q - 1) \qquad \text{Original equation}$$
$$2q - 6 + 5q = 8q - 8 \qquad \text{Distributive property}$$
$$7q - 6 = 8q - 8 \qquad \text{Combine like terms.}$$
$$7q - 6 + 6 = 8q - 8 + 6 \qquad \text{Add 6 to each side.}$$
$$7q = 8q - 2 \qquad \text{Combine like terms.}$$
$$7q - 8q = 8q - 2 - 8q \qquad \text{Subtract } 8q \text{ from each side.}$$
$$-q = -2$$
$$-1(-q) = -1(-2) \qquad \text{Multiply each side by } -1.$$
$$q = 2 \qquad \text{Simplify.}$$

To check, we replace q by 2 in the original equation and simplify:

$$2(q - 3) + 5q = 8(q - 1) \qquad \text{Original equation}$$
$$2(2 - 3) + 5(2) = 8(2 - 1) \qquad \text{Replace } q \text{ by 2.}$$
$$2(-1) + 10 = 8(1)$$
$$8 = 8$$

Because both sides have the same value, the solution set is {2}.

Now do Exercises 39–46

‹ Calculator Close-Up ›

You can check an equation by entering the equation on the home screen as shown here. The equal sign is in the TEST menu.

When you press ENTER, the calculator returns the number 1 if the equation is true or 0 if the equation is false. Since the calculator shows a 1, we can be sure that 2 is the solution.

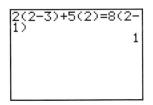

Linear equations can vary greatly in appearance, but there is a strategy that you can use for solving any of them. The following strategy summarizes the techniques that we have been using in the examples. Keep it in mind when you are solving linear equations.

Strategy for Solving Equations

1. Remove parentheses by using the distributive property and then combine like terms to simplify each side as much as possible.
2. Use the addition property of equality to get like terms from opposite sides onto the same side so that they can be combined.
3. The multiplication property of equality is generally used last.
4. Check that the solution satisfies the original equation.

⟨4⟩ Applications

Linear equations occur in business situations where there is a fixed cost and a per item cost. A mail-order company might charge $3 plus $2 per CD for shipping and handling. A lawyer might charge $300 plus $65 per hour for handling your lawsuit. AT&T might charge 5 cents per minute plus $2.95 for long distance calls. Example 7 illustrates the kind of problem that can be solved in this situation.

E X A M P L E   7

Long-distance charges

With AT&T's One Rate plan you are charged 5 cents per minute plus $2.95 for long-distance service for one month. If a long-distance bill is $4.80, then what is the number of minutes used?

Solution

Let x represent the number of minutes of calls in the month. At $0.05 per minute, the cost for x minutes is the product $0.05x$ dollars. Since there is a fixed cost of $2.95, an expression for the total cost is $0.05x + 2.95$ dollars. Since the total cost is $4.80, we have $0.05x + 2.95 = 4.80$. Solve this equation to find x.

$$0.05x + 2.95 = 4.80$$

$$0.05x + 2.95 - 2.95 = 4.80 - 2.95 \qquad \text{Subtract 2.95 from each side.}$$

$$0.05x = 1.85 \qquad \text{Simplify.}$$

$$\frac{0.05x}{0.05} = \frac{1.85}{0.05} \qquad \text{Divide each side by 0.05.}$$

$$x = 37 \qquad \text{Simplify.}$$

So the bill is for 37 minutes.

Now do Exercises 87–94

Warm-Ups ▼

Fill in the blank.

1. To solve $-x = 8$ we use the _____ property of equality.

2. To solve $x + 5 = 9$ we use the _____ property of equality.

3. To solve $3x - 7 = 11$ we apply the _____ property of equality and then the _____ property of equality.

True or false?

4. The solution set to $4x - 3 = 3x$ is $\{3\}$.

5. The equation $2x + 7 = 8$ is equivalent to $2x = 1$.

6. To solve $3x - 5 = 8x + 7$, you could add 5 to each side and then subtract $8x$ from each side.

7. To solve $5 - 4x = 9 + 7x$, you could subtract 9 from each side and then subtract $7x$ from each side.

8. The equation $-n = 9$ is equivalent to $n = -9$.

9. The equation $-y = -7$ is equivalent to $y = 7$.

10. The solution to $7x = 5x$ is 0.

11. To isolate y in $3y - 7 = 6$, you could divide each side by 3 and then add 7 to each side.

Exercises

‹ **Study Tips** ›

- Don't simply work exercises to get answers. Keep reminding yourself of what you are actually doing.
- Look for the big picture. Where have we come from? Where are we going next? When will the picture be complete?

‹1› Equations of the Form $ax + b = 0$

Solve each equation. Show your work and check your answer. See Examples 1 and 2.

1. $5a - 10 = 0$ **2.** $8y + 24 = 0$

3. $-3y - 6 = 0$ **4.** $-9w - 54 = 0$

5. $3x - 2 = 0$ **6.** $5y + 1 = 0$

7. $\frac{1}{2}w - 3 = 0$ **8.** $\frac{3}{8}t + 6 = 0$

9. $-\frac{2}{3}x + 8 = 0$ **10.** $-\frac{1}{7}z - 5 = 0$

11. $-m + \frac{1}{2} = 0$ **12.** $-y - \frac{3}{4} = 0$

13. $3p + \frac{1}{2} = 0$ **14.** $9z - \frac{1}{4} = 0$

‹2› Equations of the Form $ax + b = cx + d$

Solve each equation. See Examples 3 and 4.

15. $6x - 8 = 4x$ **16.** $9y + 14 = 2y$

17. $4z = 5 - 2z$ **18.** $3t = t - 3$

19. $4a - 9 = 7$ **20.** $7r + 5 = 47$

21. $9 = -6 - 3b$ **22.** $13 = 3 - 10s$

23. $\frac{1}{2}w - 4 = 13$ **24.** $\frac{1}{3}q + 13 = -5$

25. $6 - \frac{1}{3}d = \frac{1}{3}d$ **26.** $9 - \frac{1}{2}a = \frac{1}{4}a$

27. $2w - 0.4 = 2$ **28.** $10h - 1.3 = 6$

29. $x = 3.3 - 0.1x$ **30.** $y = 2.4 - 0.2y$

Solve each equation. See Example 5.

31. $3x - 3 = x + 5$ **32.** $9y - 1 = 6y + 5$

33. $4 - 7d = 13 - 4d$ **34.** $y - 9 = 12 - 6y$

35. $c + \frac{1}{2} = 3c - \frac{1}{2}$ **36.** $x - \frac{1}{4} = \frac{1}{2} - x$

37. $\frac{2}{3}a - 5 = \frac{1}{3}a + 5$ **38.** $\frac{1}{2}t - 3 = \frac{1}{4}t - 9$

‹3› Equations with Parentheses

Solve each equation. See Example 6.

39. $5(a - 1) + 3 = 28$

40. $2(w + 4) - 1 = 1$

41. $2 - 3(q - 1) = 10 - (q + 1)$

42. $-2(y - 6) = 3(7 - y) - 5$

43. $2(x - 1) + 3x = 6x - 20$

44. $3 - (r - 1) = 2(r + 1) - r$

45. $2\left(y - \frac{1}{2}\right) = 4\left(y - \frac{1}{4}\right) + y$

46. $\frac{1}{2}(4m - 6) = \frac{2}{3}(6m - 9) + 3$

Miscellaneous

Solve each equation. Show your work and check your answer. See the Strategy for Solving Equations box on page 98.

47. $2x = \frac{1}{3}$ **48.** $3x = \frac{6}{11}$

49. $5t = -2 + 4t$ **50.** $8y = 6 + 7y$

51. $3x - 7 = 0$ **52.** $5x + 4 = 0$

53. $-x + 6 = 5$ **54.** $-x - 2 = 9$

55. $-9 - a = -3$ **56.** $4 - r = 6$

57. $2q + 5 = q - 7$ **58.** $3z - 6 = 2z - 7$

59. $-3x + 1 = 5 - 2x$ **60.** $5 - 2x = 6 - x$

61. $-12 - 5x = -4x + 1$ **62.** $-3x - 4 = -2x + 8$

63. $3x + 0.3 = 2 + 2x$ **64.** $2y - 0.05 = y + 1$

65. $k - 0.6 = 0.2k + 1$ **66.** $2.3h + 6 = 1.8h - 1$

67. $0.2x - 4 = 0.6 - 0.8x$ **68.** $0.3x = 1 - 0.7x$

69. $-3(k - 6) = 2 - k$ **70.** $-2(h - 5) = 3 - h$

71. $2(p + 1) - p = 36$ **72.** $3(q + 1) - q = 23$

73. $7 - 3(5 - u) = 5(u - 4)$

74. $v - 4(4 - v) = -2(2v - 1)$

75. $4(x + 3) = 12$ **76.** $5(x - 3) = -15$

77. $\dfrac{w}{5} - 4 = -6$

78. $\dfrac{q}{2} + 13 = -22$

79. $\dfrac{2}{3}y - 5 = 7$

80. $\dfrac{3}{4}u - 9 = -6$

81. $4 - \dfrac{2n}{5} = 12$

82. $9 - \dfrac{2m}{7} = 19$

83. $-\dfrac{1}{3}p - \dfrac{1}{2} = \dfrac{1}{2}$

84. $-\dfrac{3}{4}z - \dfrac{2}{3} = \dfrac{1}{3}$

85. $3.5x - 23.7 = -38.75$

86. $3(x - 0.87) - 2x = 4.98$

‹4› Applications

Solve each problem. See Example 7.

87. *The practice.* A lawyer charges $300 plus $65 per hour for a divorce. If the total charge for Bill's divorce was $1405, then for what number of hours did the lawyer work on the case?

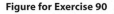 **88.** *The plumber.* Tamika paid $165 to her plumber for a service call. If her plumber charges $45 plus $40 per hour for a service call, then for how many hours did the plumber work?

89. *Celsius temperature.* If the air temperature in Quebec is 68° Fahrenheit, then the solution to the equation $\frac{9}{5}C + 32 = 68$ gives the Celsius temperature of the air. Find the Celsius temperature.

90. *Fahrenheit temperature.* Water boils at 212°F.

a) Use the accompanying graph to determine the Celsius temperature at which water boils.

b) Find the Fahrenheit temperature of hot tap water at 70°C by solving the equation

$$70 = \frac{5}{9}(F - 32).$$

Figure for Exercise 90

91. *Rectangular patio.* If the rectangular patio in the accompanying figure has a length that is 3 feet longer than its width and a perimeter of 42 feet, then the width can be found by solving the equation $2x + 2(x + 3) = 42$. What is the width?

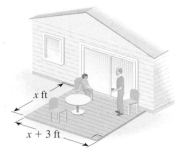

Figure for Exercise 91

92. *Perimeter of a triangle.* The perimeter of the triangle shown in the accompanying figure is 12 meters. Determine the values of x, $x + 1$, and $x + 2$ by solving the equation

$$x + (x + 1) + (x + 2) = 12.$$

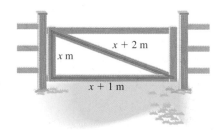

Figure for Exercise 92

93. *Cost of a car.* Jane paid 9% sales tax and a $150 title and license fee when she bought her new Saturn for a total of $16,009.50. If x represents the price of the car, then x satisfies $x + 0.09x + 150 = 16,009.50$. Find the price of the car by solving the equation.

94. *Cost of labor.* An electrician charged Eunice $29.96 for a service call plus $39.96 per hour for a total of $169.82 for installing her electric dryer. If n represents the number of hours for labor, then n satisfies

$$39.96n + 29.96 = 169.82.$$

Find n by solving this equation.

2.3 More Equations

In this section we will solve more equations of the type that we solved in Sections 2.1 and 2.2. However, some equations in this section will contain fractions or decimal numbers. Some equations will have infinitely many solutions, and some will have no solution.

⟨1⟩ Equations Involving Fractions

We solved some equations involving fractions in Sections 2.1 and 2.2. Here, we will solve equations with fractions by eliminating all fractions in the first step. All of the fractions will be eliminated if we multiply each side by the least common denominator.

EXAMPLE 1

⟨ **Helpful Hint** ⟩

Note that the fractions in Example 1 will be eliminated if you multiply each side of the equation by any number divisible by both 2 and 3. For example, multiplying by 24 yields

$$12y - 24 = 8y + 24$$
$$4y = 48$$
$$y = 12.$$

Multiplying by the least common denominator

Solve $\dfrac{y}{2} - 1 = \dfrac{y}{3} + 1$.

Solution

The least common denominator (LCD) for the denominators 2 and 3 is 6. Since both 2 and 3 divide into 6 evenly, multiplying each side by 6 will eliminate the fractions:

$$6\left(\frac{y}{2} - 1\right) = 6\left(\frac{y}{3} + 1\right) \qquad \text{Multiply each side by 6.}$$

$$6 \cdot \frac{y}{2} - 6 \cdot 1 = 6 \cdot \frac{y}{3} + 6 \cdot 1 \quad \text{Distributive property}$$

$$3y - 6 = 2y + 6 \qquad \text{Simplify: } 6 \cdot \tfrac{y}{2} = 3y$$

$$3y = 2y + 12 \qquad \text{Add 6 to each side.}$$

$$y = 12 \qquad \text{Subtract } 2y \text{ from each side.}$$

Check 12 in the original equation:

$$\frac{12}{2} - 1 = \frac{12}{3} + 1$$

$$5 = 5$$

Since 12 satisfies the original equation, the solution set is {12}.

> Now do Exercises 1–18

CAUTION You can multiply each side of the equation in Example 1 by 6 to clear the fractions and get an equivalent equation, but multiplying an expression by a number to clear the fraction is not allowed. For example, multiplying the expression $\frac{1}{6}x + \frac{2}{3}$ by 6 to simplify it will change its value when x is replaced with a number.

⟨2⟩ **Equations Involving Decimals**

When an equation involves decimal numbers, we can work with the decimal numbers or we can eliminate all of the decimal numbers by multiplying both sides by 10, or 100, or 1000, and so on. Multiplying a decimal number by 10 moves the decimal point one place to the right. Multiplying by 100 moves the decimal point two places to the right, and so on.

E X A M P L E **2**

An equation involving decimals
Solve $0.3p + 8.04 = 12.6$.

Solution

The largest number of decimal places appearing in the decimal numbers of the equation is two (in the number 8.04). Therefore, we multiply each side of the equation by 100 because multiplying by 100 moves decimal points two places to the right:

$$0.3p + 8.04 = 12.6 \qquad \text{Original equation}$$

$$100(0.3p + 8.04) = 100(12.6) \qquad \text{Multiplication property of equality}$$

$$100(0.3p) + 100(8.04) = 100(12.6) \qquad \text{Distributive property}$$

$$30p + 804 = 1260$$

$$30p + 804 - 804 = 1260 - 804 \qquad \text{Subtract 804 from each side.}$$

$$30p = 456$$

$$\frac{30p}{30} = \frac{456}{30} \qquad \text{Divide each side by 30.}$$

$$p = 15.2$$

You can use a calculator to check that

$$0.3(15.2) + 8.04 = 12.6.$$

The solution set is {15.2}.

> Now do Exercises 19–28

⟨ **Helpful Hint** ⟩

After you have used one of the properties of equality on each side of an equation, be sure to simplify all expressions as much as possible before using another property of equality.

E X A M P L E **3**

Another equation with decimals
Solve $0.5x + 0.4(x + 20) = 13.4$.

Solution

First use the distributive property to remove the parentheses:

$$0.5x + 0.4(x + 20) = 13.4 \qquad \text{Original equation}$$

$$0.5x + 0.4x + 8 = 13.4 \qquad \text{Distributive property}$$

$$10(0.5x + 0.4x + 8) = 10(13.4) \qquad \text{Multiply each side by 10.}$$

$$5x + 4x + 80 = 134 \qquad \text{Simplify.}$$

$$9x + 80 = 134 \qquad \text{Combine like terms.}$$

$$9x + 80 - 80 = 134 - 80 \qquad \text{Subtract 80 from each side.}$$

$$9x = 54 \qquad \text{Simplify.}$$

$$x = 6 \qquad \text{Divide each side by 9.}$$

Check 6 in the original equation:

$$0.5(6) + 0.4(6 + 20) = 13.4 \quad \text{Replace } x \text{ by 6.}$$
$$3 + 0.4(26) = 13.4$$
$$3 + 10.4 = 13.4$$

Since both sides of the equation have the same value, the solution set is {6}.

Now do Exercises 29–32

CAUTION If you multiply each side by 10 in Example 3 before using the distributive property, be careful how you handle the terms in parentheses:

$$10 \cdot 0.5x + 10 \cdot 0.4(x + 20) = 10 \cdot 13.4$$
$$5x + 4(x + 20) = 134$$

It is not correct to multiply 0.4 by 10 *and also* to multiply $x + 20$ by 10.

⟨3⟩ Simplifying the Process

It is very important to develop the skill of solving equations in a systematic way, writing down every step as we have been doing. As you become more skilled at solving equations, you will probably want to simplify the process a bit. One way to simplify the process is by writing only the result of performing an operation on each side. Another way is to isolate the variable on the side where the variable has the larger coefficient, when the variable occurs on both sides. We use these ideas in Example 4 and in future examples in this text.

EXAMPLE 4

Simplifying the process
Solve each equation.

a) $2a - 3 = 0$

b) $2k + 5 = 3k + 1$

Solution

a) Add 3 to each side, and then divide each side by 2:

$$2a - 3 = 0$$
$$2a = 3 \quad \text{Add 3 to each side.}$$
$$a = \frac{3}{2} \quad \text{Divide each side by 2.}$$

Check that $\frac{3}{2}$ satisfies the original equation. The solution set is $\left\{\frac{3}{2}\right\}$.

b) For this equation we can get a single k on the right by subtracting $2k$ from each side. (If we subtract $3k$ from each side, we get $-k$, and then we need another step.)

$$2k + 5 = 3k + 1$$
$$5 = k + 1 \quad \text{Subtract } 2k \text{ from each side.}$$
$$4 = k \quad \text{Subtract 1 from each side.}$$

Check that 4 satisfies the original equation. The solution set is {4}.

Now do Exercises 33–48

⟨4⟩ Identities, Conditional Equations, and Inconsistent Equations

It is easy to find equations that are satisfied by any real number that we choose as a replacement for the variable. For example, the equations

$$x \div 2 = \frac{1}{2}x, \qquad x + x = 2x, \qquad \text{and} \qquad x + 1 = x + 1$$

are satisfied by all real numbers. The equation

$$\frac{5}{x} = \frac{5}{x}$$

is satisfied by any real number except 0 because division by 0 is undefined.

All of these equations are called *identities*. Remember that the solution set for an identity is not always the entire set of real numbers. There might be some exclusions because of undefined expressions.

> **Identity**
>
> An equation that is satisfied by every real number for which both sides are defined is called an **identity.**

We cannot recognize that the equation in Example 5 is an identity until we have simplified each side.

E X A M P L E   5

Solving an identity

Solve $7 - 5(x - 6) + 4 = 3 - 2(x - 5) - 3x + 28$.

Solution

We first use the distributive property to remove the parentheses:

$$7 - 5(x - 6) + 4 = 3 - 2(x - 5) - 3x + 28$$
$$7 - 5x + 30 + 4 = 3 - 2x + 10 - 3x + 28$$
$$41 - 5x = 41 - 5x \quad \text{Combine like terms.}$$

This last equation is true for any value of x because the two sides are identical. So the solution set to the original equation is the set of all real numbers or R.

> Now do Exercises 49–50

CAUTION   If you get an equation in which both sides are identical, as in Example 5, there is no need to continue to simplify the equation. If you do continue, you will eventually get $0 = 0$, from which you can still conclude that the equation is an identity.

The statement $2x + 4 = 10$ is true only on condition that we choose $x = 3$. The equation $x^2 = 4$ is satisfied only if we choose $x = 2$ or $x = -2$. These equations are called conditional equations.

Conditional Equation

A **conditional equation** is an equation that is satisfied by at least one real number but is not an identity.

Every equation that we solved in Sections 2.1 and 2.2 is a conditional equation.

It is easy to find equations that are false no matter what number we use to replace the variable. Consider the equation

$$x = x + 1.$$

If we replace x by 3, we get $3 = 3 + 1$, which is false. If we replace x by 4, we get $4 = 4 + 1$, which is also false. Clearly, there is no number that will satisfy $x = x + 1$. Other examples of equations with no solutions include

$$x = x - 2, \qquad x - x = 5, \qquad \text{and} \qquad 0 \cdot x + 6 = 7.$$

Inconsistent Equation

An equation that has no solution is called an **inconsistent equation.**

The solution set to an inconsistent equation has no members. The set with no members is called the **empty set,** and it is denoted by the symbol $\varnothing$.

EXAMPLE 6

Solving an inconsistent equation
Solve $2 - 3(x - 4) = 4(x - 7) - 7x$.

Solution
Use the distributive property to remove the parentheses:

$$
\begin{array}{ll}
2 - 3(x - 4) = 4(x - 7) - 7x & \text{The original equation} \\
2 - 3x + 12 = 4x - 28 - 7x & \text{Distributive property} \\
14 - 3x = -28 - 3x & \text{Combine like terms on each side.} \\
14 - 3x + 3x = -28 - 3x + 3x & \text{Add } 3x \text{ to each side.} \\
14 = -28 & \text{Simplify.}
\end{array}
$$

The last equation is not true for any x. So the solution set to the original equation is the empty set, $\varnothing$. The equation is inconsistent.

 Now do Exercises 51–68

Keep the following points in mind when solving equations.

Recognizing Identities and Inconsistent Equations

If you are solving an equation and you get
1. an equation in which both sides are identical, the original equation is an identity.
2. an equation that is false, the original equation is an inconsistent equation.

The solution set to an identity is the set of all real numbers for which both sides of the equation are defined. The solution set to an inconsistent equation is the empty set, $\varnothing$.

⟨5⟩ Applications

EXAMPLE 7

Discount

Olivia got a 6% discount when she bought a new Xbox. If she paid $399.50 and x is the original price, then x satisfies the equation $x - 0.06x = 399.50$. Solve the equation to find the original price.

Solution

We could multiply each side by 100, but in this case, it might be easier to just work with the decimals:

$$x - 0.06x = 399.50$$
$$0.94x = 399.50 \qquad 1.00 - 0.06 = 0.94$$
$$x = \frac{399.50}{0.94} = 425 \quad \text{Divide each side by 0.94.}$$

Check that $425 - 0.06(425) = 399.50$. The original price was $425.

Now do Exercises 87–90

Warm-Ups ▼

Fill in the blank.

1. If an equation involves fractions, we multiply each side by the _____ of all of the fractions.

2. If an equation involves decimals, we _____ each side by a power of 10 to eliminate all decimals.

3. An _____ is satisfied by all numbers for which both sides are defined.

4. A _____ equation has at least one solution but is not an identity.

5. An _____ equation has no solution.

True or false?

6. To solve $\frac{1}{2}x - \frac{1}{3} = x + \frac{1}{6}$, multiply each side by 6.

7. The equation $0.2x + 0.03x = 8$ is equivalent to $20x + 3x = 8$.

8. The equation $5a + 3 = 0$ is inconsistent.

9. The equation $2t = t$ is a conditional equation.

10. The equation $w - 0.1w = 0.9w$ is an identity.

11. The equation $\frac{x}{x} = 1$ is an identity.

‹1› Equations Involving Fractions

Solve each equation by first eliminating the fractions.
See Example 1.

1. $\dfrac{x}{4} - \dfrac{3}{10} = 0$

2. $\dfrac{x}{15} + \dfrac{1}{6} = 0$

3. $3x - \dfrac{1}{6} = \dfrac{1}{2}$

4. $5x + \dfrac{1}{2} = \dfrac{3}{4}$

5. $\dfrac{x}{2} + 3 = x - \dfrac{1}{2}$

6. $13 - \dfrac{x}{2} = x - \dfrac{1}{2}$

7. $\dfrac{x}{2} + \dfrac{x}{3} = 20$

8. $\dfrac{x}{2} - \dfrac{x}{3} = 5$

9. $\dfrac{w}{2} + \dfrac{w}{4} = 12$

10. $\dfrac{a}{4} - \dfrac{a}{2} = -5$

11. $\dfrac{3z}{2} - \dfrac{2z}{3} = -10$

12. $\dfrac{3m}{4} + \dfrac{m}{2} = -5$

13. $\dfrac{1}{3}p - 5 = \dfrac{1}{4}p$

14. $\dfrac{1}{2}q - 6 = \dfrac{1}{5}q$

15. $\dfrac{1}{6}v + 1 = \dfrac{1}{4}v - 1$

16. $\dfrac{1}{15}k + 5 = \dfrac{1}{6}k - 10$

17. $\dfrac{1}{2}x + \dfrac{1}{3} = \dfrac{1}{4}x$

18. $\dfrac{1}{3}x - \dfrac{2}{5}x = \dfrac{5}{6}$

‹2› Equations Involving Decimals

Solve each equation by first eliminating the decimal numbers.
See Examples 2 and 3.

19. $x - 0.2x = 72$

20. $x - 0.1x = 63$

21. $0.3x + 1.2 = 0.5x$

22. $0.4x - 1.6 = 0.6x$

23. $0.02x - 1.56 = 0.8x$

24. $0.6x + 10.4 = 0.08x$

25. $0.1a - 0.3 = 0.2a - 8.3$

26. $0.5b + 3.4 = 0.2b + 12.4$

27. $0.05r + 0.4r = 27$

28. $0.08t + 28.3 = 0.5t - 9.5$

29. $0.05y + 0.03(y + 50) = 17.5$

30. $0.07y + 0.08(y - 100) = 44.5$

31. $0.1x + 0.05(x - 300) = 105$

32. $0.2x - 0.05(x - 100) = 35$

‹3› Simplifying the Process

Solve each equation. If you feel proficient enough, try simplifying the process, as described in Example 4.

33. $2x - 9 = 0$

34. $3x + 7 = 0$

35. $-2x + 6 = 0$

36. $-3x - 12 = 0$

37. $\dfrac{z}{5} + 1 = 6$

38. $\dfrac{s}{2} + 2 = 5$

39. $\dfrac{c}{2} - 3 = -4$

40. $\dfrac{b}{3} - 4 = -7$

41. $3 = t + 6$

42. $-5 = y - 9$

43. $5 + 2q = 3q$

44. $-4 - 5p = -4p$

45. $8x - 1 = 9 + 9x$

46. $4x - 2 = -8 + 5x$

47. $-3x + 1 = -1 - 2x$

48. $-6x + 3 = -7 - 5x$

‹4› Identities, Conditional Equations, and Inconsistent Equations

Solve each equation. Identify each as a conditional equation, an inconsistent equation, or an identity. See Examples 5 and 6. See Recognizing Identities and Inconsistent Equations on page 107.

49. $x + x = 2x$

50. $2x - x = x$

51. $a - 1 = a + 1$

52. $r + 7 = r$

53. $3y + 4y = 12y$

54. $9t - 8t = 7$

55. $-4 + 3(w - 1) = w + 2(w - 2) - 1$

56. $4 - 5(w + 2) = 2(w - 1) - 7w - 4$

57. $3(m + 1) = 3(m + 3)$

58. $5(m - 1) - 6(m + 3) = 4 - m$

59. $x + x = 2$

60. $3x - 5 = 0$

61. $2 - 3(5 - x) = 3x$

62. $3 - 3(5 - x) = 0$

63. $(3 - 3)(5 - z) = 0$

64. $(2 \cdot 4 - 8)p = 0$

65. $\dfrac{0}{x} = 0$

66. $\dfrac{2x}{2} = x$

67. $x \cdot x = x^2$

68. $\dfrac{2x}{2x} = 1$

Miscellaneous

Solve each equation.

69. $3x - 5 = 2x - 9$

70. $5x - 9 = x - 4$

71. $x + 2(x + 4) = 3(x + 3) - 1$

72. $u + 3(u - 4) = 4(u - 5)$

73. $23 - 5(3 - n) = -4(n - 2) + 9n$

74. $-3 - 4(t - 5) = -2(t + 3) + 11$

75. $0.05x + 30 = 0.4x - 5$

76. $x - 0.08x = 460$

77. $-\dfrac{2}{3}a + 1 = 2$

78. $-\dfrac{3}{4}t = \dfrac{1}{2}$

79. $\dfrac{y}{2} + \dfrac{y}{6} = 20$

80. $\dfrac{3w}{5} - 1 = \dfrac{w}{2} + 1$

81. $0.09x - 0.2(x + 4) = -1.46$

82. $0.08x + 0.5(x + 100) = 73.2$

83. $436x - 789 = -571$

84. $0.08x + 4533 = 10x + 69$

85. $\dfrac{x}{344} + 235 = 292$

86. $34(x - 98) = \dfrac{x}{2} + 453.5$

⟨5⟩ **Applications**

Solve each problem. See Example 7.

87. *Sales commission.* Danielle sold her house through an agent who charged 8% of the selling price. After the commission was paid, Danielle received $117,760. If x is the selling price, then x satisfies

$$x - 0.08x = 117,760.$$

Solve this equation to find the selling price.

88. *Raising rabbits.* Before Roland sold two female rabbits, half of his rabbits were female. After the sale, only one-third of his rabbits were female. If x represents his original number of rabbits, then

$$\frac{1}{2}x - 2 = \frac{1}{3}(x - 2).$$

Solve this equation to find the number of rabbits that he had before the sale.

89. *Eavesdropping.* Reginald overheard his boss complaining that his federal income tax for 2009 was $60,531.

  a) Use the accompanying graph to estimate his boss's taxable income for 2009.

  b) Find his boss's exact taxable income for 2009 by solving the equation

$$46,742 + 0.33(x - 208,850) = 60,531.$$

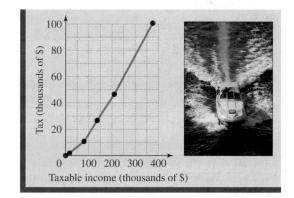

Figure for Exercise 89

90. *Federal taxes.* According to Bruce Harrell, CPA, the federal income tax for a class C corporation is found by solving a linear equation. The reason for the equation is that the amount x of federal tax is deducted before the state tax is figured, and the amount of state tax is deducted before the federal tax is figured. To find the amount of federal tax for a corporation with a taxable income of $200,000, for which the federal tax rate is 25% and the state tax rate is 10%, Bruce must solve

$$x = 0.25[200,000 - 0.10(200,000 - x)].$$

Solve the equation for Bruce.

2.4   Formulas and Functions

In this section, you will learn to rewrite formulas using the same properties of equality that we used to solve equations. You will also learn how to find the value of one of the variables in a formula when we know the value of all of the others.

⟨1⟩ Solving for a Variable

Most drivers know the relationship between distance, rate, and time. For example, if you drive 70 mph for 3 hours, then you will travel 210 miles. At 60 mph a 300-mile trip will take 5 hours. If a 400-mile trip took 8 hours, then you averaged 50 mph. The relationship between distance D, rate R, and time T is expressed by the formula

$$D = R \cdot T.$$

A **formula** or **literal equation** is an equation involving two or more variables.

To find the time for a 300-mile trip at 60 mph, you are using the formula in the form $T = \frac{D}{R}$. The process of rewriting a formula for one variable in terms of the others is called **solving for a certain variable.** To solve for a certain variable, we use the same techniques that we use in solving equations.

E X A M P L E   1

Solving for a certain variable

Solve the formula $D = RT$ for T.

Solution

Since T is multiplied by R, dividing each side of the equation by R will isolate T:

$$D = RT \qquad \text{Original formula}$$

$$\frac{D}{R} = \frac{R \cdot T}{R} \qquad \text{Divide each side by } R.$$

$$\frac{D}{R} = T \qquad \text{Divide out (or cancel) the common factor } R.$$

$$T = \frac{D}{R} \qquad \text{It is customary to write the single variable on the left.}$$

> Now do Exercises 1–12

The formula $C = \frac{5}{9}(F - 32)$ is used to find the Celsius temperature for a given Fahrenheit temperature. If we solve this formula for F, then we have a formula for finding Fahrenheit temperature for a given Celsius temperature.

E X A M P L E   2

Solving for a certain variable

Solve the formula $C = \frac{5}{9}(F - 32)$ for F.

Solution

We could apply the distributive property to the right side of the equation, but it is simpler to proceed as follows:

$$C = \frac{5}{9}(F - 32)$$

$$\frac{9}{5}C = \frac{9}{5} \cdot \frac{5}{9}(F - 32) \qquad \text{Multiply each side by } \frac{9}{5}, \text{ the reciprocal of } \frac{5}{9}.$$

$$\frac{9}{5}C = F - 32 \qquad \text{Simplify.}$$

$$\frac{9}{5}C + 32 = F - 32 + 32 \qquad \text{Add 32 to each side.}$$

$$\frac{9}{5}C + 32 = F \qquad \text{Simplify.}$$

The formula is usually written as $F = \frac{9}{5}C + 32$.

> **Now do Exercises 13–18**

⟨2⟩ The Language of Functions

The formula $D = RT$ is a *rule* for determining the distance D from the rate R and the time T. (In words, the rule is to multiply the rate and time to obtain the distance.) We say that $D = RT$ expresses D as *a function of R and T* and that the formula is a *function*. Distance is a function of rate and time. The formula $T = \frac{D}{R}$ can be used to determine the time from the distance and rate. So this formula expresses time as a function of distance and rate. The formula $C = \frac{5}{9}(F - 32)$ expresses the Celsius temperature C as a function of the Fahrenheit temperature F. The formula $F = \frac{9}{5}C + 32$ expresses the Fahrenheit temperature as a function of the Celsius temperature.

> **Function**
>
> **A function** is a rule for determining uniquely the value of one variable a from the value(s) of one or more other variable(s). We say that a **is a function of** the other variable(s).

If y is a function of x, then there is only one y-value for any given x-value. The plus or minus symbol, $\pm$, is sometimes used in a formula as in $y = \pm x$. In this case, there are two possible y-values for a given x-value. Since y is not uniquely determined by x, y is *not* a function of x.

E X A M P L E   **3**

⟨ Helpful Hint ⟩

If we simply wanted to solve $x + 2y = 6$ for y, we could have written

$$y = \frac{6 - x}{2} \quad \text{or} \quad y = \frac{-x + 6}{2}.$$

However, in Example 3 we requested the form $y = mx + b$. This form is a popular form that we will study in detail in Chapter 3.

Expressing y as a function of x

Find a formula that expresses y as a function of x if $x + 2y = 6$. Write the answer in the form $y = mx + b$ where m and b are real numbers.

Solution

$$x + 2y = 6 \qquad \text{Original equation}$$

$$2y = 6 - x \qquad \text{Subtract } x \text{ from each side.}$$

$$\frac{1}{2} \cdot 2y = \frac{1}{2}(6 - x) \qquad \text{Multiply each side by } \tfrac{1}{2}.$$

$$y = 3 - \frac{1}{2}x \qquad \text{Distributive property}$$

$$y = -\frac{1}{2}x + 3 \qquad \text{Rearrange to get } y = mx + b \text{ form.}$$

The formula $y = -\frac{1}{2}x + 3$ expresses y as a function of x.

> **Now do Exercises 19–28**

Notice that in Example 3 we multiplied each side of the equation by $\frac{1}{2}$, and so we multiplied each term on the right-hand side by $\frac{1}{2}$. Instead of multiplying by $\frac{1}{2}$, we could have divided each side of the equation by 2. We would then divide each term on the right side by 2. This idea is illustrated in Example 4.

E X A M P L E   4

Expressing y as a function of x

Find a formula that expresses y as a function of x if $2x - 3y = 9$. Write the answer in the form $y = mx + b$ where m and b are real numbers.

Solution

$$2x - 3y = 9 \qquad \text{Original equation}$$

$$-3y = -2x + 9 \qquad \text{Subtract } 2x \text{ from each side.}$$

$$\frac{-3y}{-3} = \frac{-2x + 9}{-3} \qquad \text{Divide each side by } -3.$$

$$y = \frac{-2x}{-3} + \frac{9}{-3} \qquad \text{By the distributive property, each term is divided by } -3.$$

$$y = \frac{2}{3}x - 3 \qquad \text{Simplify.}$$

The formula $y = \frac{2}{3}x - 3$ expresses y as a function of x.

> Now do Exercises 29–40

Note that in Example 4 we wrote the answer as $y = \frac{2}{3}x - 3$ rather than $y = \frac{2}{3}x + (-3)$. If the form $y = mx + b$ is requested, we may use a subtraction symbol in place of the addition symbol when b is negative.

When solving for a variable that appears more than once in the equation, we must combine the terms to obtain a single occurrence of the variable. *When a formula has been solved for a certain variable, that variable will not occur on both sides of the equation.*

E X A M P L E   5

Solving for a variable that appears on both sides

Find a formula that expresses x as a function of b and d if $5x - b = 3x + d$.

Solution

First get all terms involving x onto one side and all other terms onto the other side:

$$5x - b = 3x + d \qquad \text{Original formula}$$

$$5x - 3x - b = d \qquad \text{Subtract } 3x \text{ from each side.}$$

$$5x - 3x = b + d \qquad \text{Add } b \text{ to each side.}$$

$$2x = b + d \qquad \text{Combine like terms.}$$

$$x = \frac{b + d}{2} \qquad \text{Divide each side by 2.}$$

The formula solved for x is $x = \frac{b + d}{2}$. The formula $x = \frac{b + d}{2}$ expresses x as a function of b and d.

> Now do Exercises 41–48

CAUTION If we simply add b to both sides and then divide by 5 in Example 5, we get $x = \frac{3x + b + d}{5}$. Since x appears on both sides, this formula is *not* solved for x and does *not* express x as a function of b and d.

⟨3⟩ Finding the Value of a Variable

In many situations, we know the values of all variables in a formula except one. We use the formula to determine the unknown value.

 EXAMPLE 6

Finding the value of a variable in a formula
If $2x - 3y = 9$, find y when $x = 6$.

Solution

Method 1: First solve the equation for y. Because we have already solved this equation for y in Example 4, we will not repeat that process in this example. We have

$$y = \frac{2}{3}x - 3.$$

Now replace x by 6 in this equation:

$$y = \frac{2}{3}(6) - 3$$

$$= 4 - 3 = 1$$

So when $x = 6$, we have $y = 1$.

Method 2: First replace x by 6 in the original equation, and then solve for y:

$$2x - 3y = 9 \qquad \text{Original equation}$$
$$2 \cdot 6 - 3y = 9 \qquad \text{Replace } x \text{ by 6.}$$
$$12 - 3y = 9 \qquad \text{Simplify.}$$
$$-3y = -3 \qquad \text{Subtract 12 from each side.}$$
$$y = 1 \qquad \text{Divide each side by } -3.$$

So when $x = 6$, we have $y = 1$.

> **Now do Exercises 49–58**

It usually does not matter which method from Example 6 is used. However, if you want many y-values, it is best to have the equation solved for y. For example, completing the y-column in the following table is straightforward if you have a formula that expresses y as a function of x:

$$y = \frac{2}{3}x - 3$$

x	y
0	
3	
6	

$$y = \frac{2}{3}(0) - 3 = -3$$
$$y = \frac{2}{3}(3) - 3 = -1$$
$$y = \frac{2}{3}(6) - 3 = 1$$

x	y
0	-3
3	-1
6	1

⟨4⟩ Applications

Example 7 involves the simple interest formula $I = Prt$, where I is the amount of interest, P is the principal or the amount invested, r is the annual interest rate, and t is the time in years. The amount of interest is a function of the principal, rate, and time. The interest rate is usually expressed as a percent, which must be converted to a decimal for computations.

E X A M P L E   7

Finding the simple interest rate

The principal is $400 and the time is 2 years. Find the simple interest rate for each of the following amounts of interest: $120, $60, $30.

Solution

First solve the formula $I = Prt$ for r:

$$Prt = I \qquad \text{Simple interest formula}$$

$$\frac{Prt}{Pt} = \frac{I}{Pt} \qquad \text{Divide each side by } Pt.$$

$$r = \frac{I}{Pt} \qquad \text{Simplify.}$$

Now insert the values for P, t, and the three amounts of interest:

$$r = \frac{120}{400 \cdot 2} = 0.15 = 15\% \qquad \text{Move the decimal point two places to the left.}$$

$$r = \frac{60}{400 \cdot 2} = 0.075 = 7.5\%$$

$$r = \frac{30}{400 \cdot 2} = 0.0375 = 3.75\%$$

If the amount of interest is $120, $60, or $30, then the simple interest rate is 15%, 7.5%, or 3.75%, respectively.

> Now do Exercises 67–70

⟨ **Helpful Hint** ⟩

All interest computation is based on simple interest. However, depositors do not like to wait 2 years to get interest as in Example 7. More often the time is $\frac{1}{12}$ year or $\frac{1}{365}$ year. Simple interest computed every month is said to be compounded monthly. Simple interest computed every day is said to be compounded daily.

In Example 8, we use the formula for the perimeter of a rectangle, $P = 2L + 2W$, which can be found inside the front cover of this book. The perimeter P is a function of the length L and the width W. For geometric problems it is usually best to draw a diagram as we do in Example 8.

E X A M P L E   8

Using a geometric formula

The perimeter of a rectangle is 36 feet. If the width is 6 feet, then what is the length?

Solution

First, put the given information on a diagram as shown in Fig. 2.1. Substitute the given values into the formula for the perimeter of a rectangle and then solve for L. (We could solve for L first and then insert the given values.)

$$P = 2L + 2W \qquad \text{Perimeter of a rectangle}$$

$$36 = 2L + 2 \cdot 6 \qquad \text{Substitute 36 for } P \text{ and 6 for } W.$$

$$36 = 2L + 12 \qquad \text{Simplify.}$$

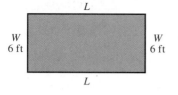

Figure 2.1

$$24 = 2L \qquad \text{Subtract 12 from each side.}$$
$$12 = L \qquad \text{Divide each side by 2.}$$

Check: If $L = 12$ and $W = 6$, then $P = 2(12) + 2(6) = 36$ feet. So we can be certain that the length is 12 feet.

> **Now do Exercises 71–74**

If L is the list price or original price of an item and r is the rate of discount, then the amount of discount is the product of the list price and the rate of discount, rL. The sale price S is the list price minus the amount of discount. So $S = L - rL$. The sale price S is a function of the list price L and the rate of discount r. The rate of discount is usually expressed as a percent, which must be converted to a decimal for computations.

E X A M P L E   9

Finding the original price

What was the original price of a stereo that sold for $560 after a 20% discount?

Solution

Express 20% as the decimal 0.20 or 0.2, and use the formula $S = L - rL$:

$$\text{Selling price} = \text{list price} - \text{amount of discount}$$
$$560 = L - 0.2L$$
$$10(560) = 10(L - 0.2L) \qquad \text{Multiply each side by 10.}$$
$$5600 = 10L - 2L \qquad \text{Remove the parentheses.}$$
$$5600 = 8L \qquad \text{Combine like terms.}$$
$$\frac{5600}{8} = \frac{8L}{8} \qquad \text{Divide each side by 8.}$$
$$700 - L$$

Since 20% of $700 is $140 and $700 − $140 = $560, we can be sure that the original price was $700. Note that if the discount is 20%, then the selling price is 80% of the list price. So we could have started with the equation $560 = 0.80L$.

> **Now do Exercises 75–80**

Warm-Ups ▼

Fill in the blank.

1. An equation with two or more variables is a _____ or _____ equation.
2. To _____ for a variable means to find an equivalent equation in which the variable is isolated.
3. If $D = RT$, then D is a _____ of R and T.
4. The formula $P = 2L + 2W$ is the formula for the _____ of a rectangle.
5. The formula $A = LW$ is the formula for the _____ of a rectangle.
6. The formula $C = \pi d$ is the formula for the _____ of a circle.

True or false?

7. The formula $D = R \cdot T$ solved for T is $T \cdot R = D$.
8. The formula $a - b = 3a - m$ solved for a is $a = 3a - m + b$.
9. The formula $A = LW$ solved for L is $L = \frac{A}{W}$.
10. The perimeter of a rectangle is the product of its length and width.
11. If $x = -1$ and $y = -3x + 6$, then $y = 9$.

Exercises

‹ **Study Tips** ›

- When studying for an exam, start by working the exercises in the Chapter Review. They are grouped by section so that you can go back and review any topics that you have trouble with.
- Never leave an exam early. Most papers turned in early contain careless errors that could be found and corrected. Every point counts.

‹1› Solving for a Variable

Solve each formula for the specified variable. See Examples 1 and 2.

1. $D = RT$ for R

2. $A = LW$ for W

3. $C = \pi D$ for D

4. $F = ma$ for a

5. $I = Prt$ for P

6. $I = Prt$ for t

7. $F = \dfrac{9}{5}C + 32$ for C

8. $y = \dfrac{3}{4}x - 7$ for x

9. $A = \dfrac{1}{2}bh$ for h

10. $A = \dfrac{1}{2}bh$ for b

11. $P = 2L + 2W$ for L

12. $P = 2L + 2W$ for W

13. $A = \dfrac{1}{2}(a + b)$ for a

14. $A = \dfrac{1}{2}(a + b)$ for b

15. $S = P + Prt$ for r

16. $S = P + Prt$ for t

17. $A = \dfrac{1}{2}h(a + b)$ for a

18. $A = \dfrac{1}{2}h(a + b)$ for b

‹2› The Language of Functions

In each case find a formula that expresses y as a function of x. See Examples 3 and 4.

19. $x + y = -9$

20. $3x + y = -5$

21. $x + y - 6 = 0$

22. $4x + y - 2 = 0$

23. $2x - y = 2$

24. $x - y = -3$

25. $3x - y + 4 = 0$

26. $-2x - y + 5 = 0$

27. $x + 2y = 4$

28. $3x + 2y = 6$

29. $2x - 2y = 1$

30. $3x - 2y = -6$

31. $y + 2 = 3(x - 4)$

32. $y - 3 = -3(x - 1)$

33. $y - 1 = \dfrac{1}{2}(x - 2)$

34. $y - 4 = -\dfrac{2}{3}(x - 9)$

35. $\dfrac{1}{2}x - \dfrac{1}{3}y = -2$

36. $\dfrac{x}{2} + \dfrac{y}{4} = \dfrac{1}{2}$

37. $y - 2 = \dfrac{3}{2}(x + 3)$

38. $y + 4 = \dfrac{2}{3}(x - 2)$

39. $y - \dfrac{1}{2} = -\dfrac{1}{4}\left(x - \dfrac{1}{2}\right)$

40. $y + \dfrac{1}{2} = -\dfrac{1}{3}\left(x + \dfrac{1}{2}\right)$

Solve each equation for x. See Example 5.

41. $5x + a = 3x + b$

42. $2c - x = 4x + c - 5b$

43. $4(a + x) - 3(x - a) = 0$

44. $-2(x - b) - (5a - x) = a + b$

45. $3x - 2(a - 3) = 4x - 6 - a$

46. $2(x - 3w) = -3(x + w)$

47. $3x + 2ab = 4x - 5ab$

48. $x - a = -x + a + 4b$

⟨3⟩ Finding the Value of a Variable

*For each equation that follows, find y given that x = 2.
See Example 6.*

49. $y = 3x - 4$ **50.** $y = -2x + 5$

51. $3x - 2y = -8$ **52.** $4x + 6y = 8$

53. $\dfrac{3x}{2} - \dfrac{5y}{3} = 6$ **54.** $\dfrac{2y}{5} - \dfrac{3x}{4} = \dfrac{1}{2}$

55. $y - 3 = \dfrac{1}{2}(x - 6)$ **56.** $y - 6 = -\dfrac{3}{4}(x - 2)$

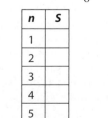 **57.** $y - 4.3 = 0.45(x - 8.6)$

58. $y + 33.7 = 0.78(x - 45.6)$

Fill in the tables using the given formulas.

59. $y = -3x + 30$

x	y
−10	
0	
10	
20	
30	

60. $y = 4x - 20$

x	y
−10	
−5	
0	
5	
10	

61. $F = \dfrac{9}{5}C + 32$

C	F
−10	
−5	
0	
40	
100	

62. $C = \dfrac{5}{9}(F - 32)$

F	C
−40	
14	
32	
59	
86	

63. $T = \dfrac{400}{R}$

R (mph)	T (hr)
10	
20	
40	
80	
100	

64. $R = \dfrac{100}{T}$

T (hr)	R (mph)
1	
5	
20	
50	
100	

65. $S = \dfrac{n(n + 1)}{2}$

n	S
1	
2	
3	
4	
5	

66. $S = \dfrac{n(n + 1)(2n + 1)}{6}$

n	S
1	
2	
3	
4	
5	

⟨4⟩ Applications

Solve each of the following problems. Some geometric formulas that may be helpful can be found inside the front cover of this text. See Examples 7–9.

67. *Finding the rate.* A loan of $5000 is made for 3 years. Find the interest rate for simple interest amounts of $600, $700, and $800.

68. *Finding the rate.* A loan of $1000 is made for 7 years. Find the interest rate for simple interest amounts of $420, $455, and $472.50.

69. *Finding the time.* Kathy paid $500 in simple interest on a loan of $2500. If the annual interest rate was 5%, then what was the time?

70. *Finding the time.* Robert paid $240 in simple interest on a loan of $1000. If the annual interest rate was 8%, then what was the time?

71. *Finding the length.* The area of a rectangle is 28 square yards. Find the length if the width is 2 yards, 3 yards, or 4 yards.

72. *Finding the width.* The area of a rectangle is 60 square feet. Find the width if the length is 10 feet, 16 feet, or 18 feet.

73. *Finding the length.* If it takes 600 feet of wire fencing to fence a rectangular feed lot that has a width of 75 feet, then what is the length of the lot?

74. *Finding the depth.* If it takes 500 feet of fencing to enclose a rectangular lot that is 104 feet wide, then how deep is the lot?

75. *Finding MSRP.* What was the manufacturer's suggested retail price (MSRP) for a Lexus SC 430 that sold for $54,450 after a 10% discount?

76. *Finding MSRP.* What was the MSRP for a Hummer H1 that sold for $107,272 after an 8% discount?

77. *Finding the original price.* Find the original price if there is a 15% discount and the sale price is $255.

78. *Finding the list price.* Find the list price if there is a 12% discount and the sale price is $4400.

79. *Rate of discount.* Find the rate of discount if the discount is $40 and the original price is $200.

80. *Rate of discount.* Find the rate of discount if the discount is $20 and the original price is $250.

81. *Width of a football field.* The perimeter of a football field in the NFL, excluding the end zones, is 920 feet. How wide is the field? See the figure on the next page.

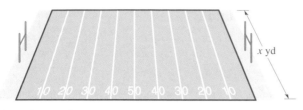

Figure for Exercise 81

82. *Perimeter of a frame.* If a picture frame is 16 inches by 20 inches, then what is its perimeter?

83. *Volume of a box.* A rectangular box measures 2 feet wide, 3 feet long, and 4 feet deep. What is its volume?

84. *Volume of a refrigerator.* The volume of a rectangular refrigerator is 20 cubic feet. If the top measures 2 feet by 2.5 feet, then what is the height?

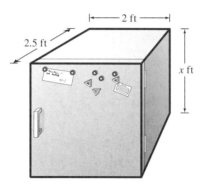

Figure for Exercise 84

85. *Radius of a pizza.* If the circumference of a pizza is 8π inches, then what is the radius?

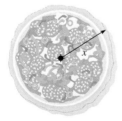

Figure for Exercise 85

86. *Diameter of a circle.* If the circumference of a circle is 4π meters, then what is the diameter?

87. *Height of a banner.* If a banner in the shape of a triangle has an area of 16 square feet with a base of 4 feet, then what is the height of the banner?

88. *Length of a leg.* If a right triangle has an area of 14 square meters and one leg is 4 meters in length, then what is the length of the other leg?

Figure for Exercise 87

89. *Length of the base.* A trapezoid with height 20 inches and lower base 8 inches has an area of 200 square inches. What is the length of its upper base?

90. *Height of a trapezoid.* The end of a flower box forms the shape of a trapezoid. The area of the trapezoid is 300 square centimeters. The bases are 16 centimeters and 24 centimeters in length. Find the height.

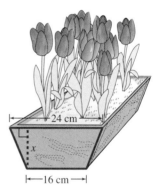

Figure for Exercise 90

91. *Fried's rule.* Doctors often prescribe the same drugs for children as they do for adults. The formula

$$d = 0.08aD$$

(Fried's rule) expresses the child's dosage d as a function of the adult dosage D and the child's age a.

a) If a doctor prescribes 1000 milligrams of acetaminophen for an adult, then how many milligrams would he prescribe for an 8-year-old child?

b) If a doctor uses Fried's rule to prescribe 200 milligrams of a drug to a child when he would prescribe 600 milligrams to an adult, then how old is the child?

c) Use the accompanying bar graph to determine the age at which a child would get the same dosage as an adult.

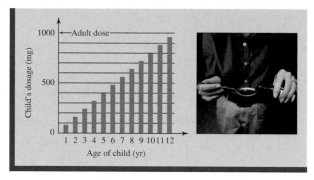

Figure for Exercise 91

92. *Cowling's rule.* Cowling's rule is another function for determining the child's dosage of a drug. For this rule, the formula

$$d = \frac{D(a + 1)}{24}$$

expresses the child's dosage d as a function of the adult dosage D and the child's age a.

a) If a doctor prescribes 1000 milligrams of acetaminophen for an adult, then how many milligrams would she prescribe for an eight-year-old child using Cowling's rule?

b) If a doctor uses Cowling's rule to prescribe 200 milligrams of a drug to a child when she would prescribe 600 milligrams to an adult, then how old is the child?

93. *Administering vancomycin.* A patient is to receive 750 milligrams (desired dose) of the antibiotic vancomycin. However, vancomycin comes in a solution containing 1000 milligrams (available dose) of vancomycin per 5 milliliters (quantity) of solution. The amount of solution to be given to the patient is a function of the desired dose, the available dose, and the quantity, given by the formula

$$\text{Amount} = \frac{\text{desired dose}}{\text{available dose}} \times \text{quantity}.$$

Find the amount of the solution that should be administered to the patient.

94. *International communications.* The global investment in telecom infrastructure since 1990 can be modeled by the function

$$I = 7.5t + 115,$$

where I is in billions of dollars and t is the number of years since 1990 (*Fortune*, www.fortune.com).

a) Use the formula to find the global investment in 2000.

b) Use the accompanying graph to estimate the year in which the global investment will reach $300 billion.

c) Use the formula to find the year in which the global investment will reach $300 billion.

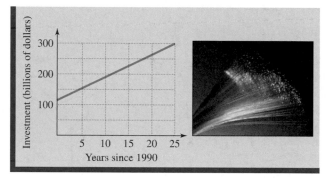

Figure for Exercise 94

95. *The 2.4-meter rule.* A 2.4-meter sailboat is a one-person boat that is about 13 feet in length, has a displacement of about 550 pounds, and a sail area of about 81 square feet. To compete in the 2.4-meter class, a boat must satisfy the formula

$$2.4 = \frac{L + 2D - F\sqrt{S}}{2.37},$$

where L = length, F = freeboard, D = girth, and S = sail area. Solve the formula for L.

Photo for Exercise 95

Mid-Chapter **Quiz** | Sections 2.1 through 2.4 | Chapter 2

Solve each equation.

1. $x + 9 = -12$

2. $\frac{3}{4}m = \frac{1}{2}$

3. $-9x = 5 - 10x$

4. $4a - 3 = 0$

5. $8w - 5 = 6w + 4$

6. $4(a + 3) + 8 = 48$

7. $6 - 3(x + 2) = 4(x - 7)$

8. $\frac{3}{2}x + \frac{1}{6} = \frac{2}{3}$

9. $0.8x + 120 = x - 70$

10. $0.09x + 3.4 = 0.4x + 65.4$

Identify each equation as a conditional equation, an inconsistent equation, or an identity.

11. $7x - 12x = -5x$

12. $7x - 12x = -5$

13. $7x - 12x = 6x$

14. $7x - 12x = -5x + 4$

Solve each equation for x.

15. $ax + b = c$

16. $5(x - a) = 2(x - b)$

Miscellaneous.

17. What was the original price of a car that sold for $13,904 after a 12% discount?

18. If the perimeter of a rectangle is 48 yards and the length is 15 yards, then what is the width?

19. If $x = 8$ and $3x - 4y = 12$, then what is y?

20. If the principal is $4000, the simple interest is $640, and the time is 2 years, then what is the simple interest rate?

2.5 Translating Verbal Expressions into Algebraic Expressions

In This Section

⟨1⟩ **Writing Algebraic Expressions**

⟨2⟩ **Pairs of Numbers**

⟨3⟩ **Consecutive Integers**

⟨4⟩ **Using Formulas**

⟨5⟩ **Writing Equations**

You translated some verbal expressions into algebraic expressions in Section 1.6; in this section you will study translating in more detail.

⟨1⟩ Writing Algebraic Expressions

The following box contains a list of some frequently occurring verbal expressions and their equivalent algebraic expressions.

Translating Words into Algebra

	Verbal Phrase	Algebraic Expression
Addition:	The sum of a number and 8	$x + 8$
	Five is added to a number	$x + 5$
	Two more than a number	$x + 2$
	A number increased by 3	$x + 3$
Subtraction:	Four is subtracted from a number	$x - 4$
	Three less than a number	$x - 3$
	The difference between 7 and a number	$7 - x$
	A number decreased by 2	$x - 2$

Multiplication:	The product of 5 and a number	$5x$
	Twice a number	$2x$
	One-half of a number	$\frac{1}{2}x$
	Five percent of a number	$0.05x$
Division:	The ratio of a number to 6	$\frac{x}{6}$
	The quotient of 5 and a number	$\frac{5}{x}$
	Three divided by some number	$\frac{3}{x}$

EXAMPLE 1

Writing algebraic expressions

Translate each verbal expression into an algebraic expression.

a) The sum of a number and 9

b) Eighty percent of a number

c) A number divided by 4

d) The result of a number subtracted from 5

e) Three less than a number

Solution

a) If x is the number, then the sum of x and 9 is $x + 9$.

b) If w is the number, then eighty percent of the number is $0.80w$.

c) If y is the number, then the number divided by 4 is $\frac{y}{4}$.

d) If z is the number, then the result of subtracting z from 5 is $5 - z$.

e) If a is the number, then 3 less than a is $a - 3$.

Now do Exercises 1–12

‹ **Helpful Hint** ›

We know that x and $10 - x$ have a sum of 10 for any value of x. We can easily check that fact by adding:

$$x + 10 - x = 10$$

In general, it is not true that x and $x - 10$ have a sum of 10, because

$$x + x - 10 = 2x - 10.$$

For what value of x is the sum of x and $x - 10$ equal to 10?

‹2› **Pairs of Numbers**

There is often more than one unknown quantity in a problem, but a relationship between the unknown quantities is given. For example, if one unknown number is 5 more than another unknown number, we can use x to represent the smaller one and $x + 5$ to represent the larger one. If we use x to represent the larger unknown number, then $x - 5$ represents the smaller. Either way is correct.

If two numbers differ by 5, then one of them is 5 more than the other. So x and $x + 5$ can also be used to represent two numbers that differ by 5. Likewise, x and $x - 5$ could represent two numbers that differ by 5.

How would you represent two numbers that have a sum of 10? If one of the numbers is 2, the other is certainly $10 - 2$, or 8. Thus, if x is one of the numbers, then $10 - x$ is the other. The expressions

$$x \quad \text{and} \quad 10 - x$$

have a sum of 10 for any value of x.

EXAMPLE 2

Algebraic expressions for pairs of numbers

Write algebraic expressions for each pair of numbers.

  a) Two numbers that differ by 12

  b) Two numbers with a sum of -8

Solution

  a) The expressions x and $x - 12$ represent two numbers that differ by 12. We can check by subtracting:

$$x - (x - 12) = x - x + 12 = 12$$

  Of course, x and $x + 12$ also differ by 12 because $x + 12 - x = 12$.

  b) The expressions x and $-8 - x$ have a sum of -8. We can check by addition:

$$x + (-8 - x) = x - 8 - x = -8$$

Now do Exercises 13–22

Pairs of numbers occur in geometry in discussing measures of angles. You will need the following facts about degree measures of angles.

Degree Measures of Angles

Two angles are called **complementary** if the sum of their degree measures is 90°.

Two angles are called **supplementary** if the sum of their degree measures is 180°.

The sum of the degree measures of the three angles of any triangle is 180°.

For complementary angles, we use x and $90 - x$ for their degree measures. For supplementary angles, we use x and $180 - x$. Complementary angles that share a common side form a right angle. Supplementary angles that share a common side form a straight angle or straight line.

EXAMPLE 3

Degree measures

Write algebraic expressions for each pair of angles shown.

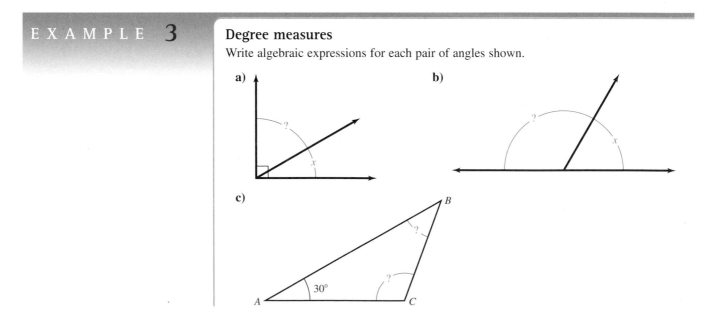

Solution

a) Since the angles shown are complementary, we can use x to represent the degree measure of the smaller angle and $90 - x$ to represent the degree measure of the larger angle.

b) Since the angles shown are supplementary, we can use x to represent the degree measure of the smaller angle and $180 - x$ to represent the degree measure of the larger angle.

c) If we let x represent the degree measure of angle B, then $180 - x - 30$, or $150 - x$, represents the degree measure of angle C.

> Now do Exercises 23–26

⟨3⟩ Consecutive Integers

Note that each integer is one larger than the previous integer. For example, if $x = 5$, then $x + 1 = 6$ and $x + 2 = 7$. So if x is an integer, then x, $x + 1$, and $x + 2$ represent three consecutive integers. Each even (or odd) integer is two larger than the previous even (or odd) integer. For example, if $x = 6$, then $x + 2 = 8$, and $x + 4 = 10$. If $x = 7$, then $x + 2 = 9$, and $x + 4 = 11$. So x, $x + 2$, and $x + 4$ represent three consecutive even integers if x is even and three consecutive odd integers if x is odd.

CAUTION The expressions x, $x + 1$, and $x + 3$ do not represent three consecutive odd integers no matter what x represents.

E X A M P L E   4

Expressions for integers

Write algebraic expressions for the following unknown integers.

a) Two consecutive integers, the smallest of which is w.

b) Three consecutive even integers, the smallest of which is z.

c) Four consecutive odd integers, the smallest of which is y.

Solution

a) Each integer is 1 larger than the preceding integer. So if w represents the smallest of two consecutive integers, then w and $w + 1$ represent the integers.

b) Each even integer is 2 larger than the preceding even integer. So if z represents the smallest of three consecutive even integers, then z, $z + 2$, and $z + 4$ represent the three consecutive even integers.

c) Each odd integer is 2 larger than the preceding odd integer. So if y represents the smallest of four consecutive odd integers, then y, $y + 2$, $y + 4$, and $y + 6$ represent the four consecutive odd integers.

> Now do Exercises 27–34

The following box contains a summary of some common verbal phrases and algebraic expressions for pairs of numbers.

Summary of Algebraic Expressions for Pairs of Numbers	
Verbal Phrase	**Algebraic Expressions**
Two numbers that differ by 5	x and $x + 5$
Two numbers with a sum of 6	x and $6 - x$
Two consecutive integers	x and $x + 1$
Two consecutive even integers	x and $x + 2$
Two consecutive odd integers	x and $x + 2$
Complementary angles	x and $90 - x$
Supplementary angles	x and $180 - x$

⟨4⟩ Using Formulas

In writing expressions for unknown quantities, we often use standard formulas such as those given inside the front cover of this book.

EXAMPLE 5

Writing algebraic expressions using standard formulas

Find an algebraic expression for

a) the distance if the rate is 30 miles per hour and the time is T hours.

b) the discount if the rate is 40% and the original price is p dollars.

Solution

a) Using the formula $D = RT$, we have $D = 30T$. So $30T$ is an expression that represents the distance in miles.

b) Since the discount is the rate times the original price, an algebraic expression for the discount is $0.40p$ dollars.

> Now do Exercises 35–58

⟨5⟩ Writing Equations

To solve a problem using algebra, we describe or **model** the problem with an equation. In this section we write the equations only, and in Section 2.6 we write and solve them. Sometimes we must write an equation from the information given in the problem, and sometimes we use a standard model to get the equation. Some standard models are shown in the following box.

Uniform Motion Model	
Distance = Rate · Time	$D = R \cdot T$
Percentage Models	
What number is 5% of 40?	$x = 0.05 \cdot 40$
Ten is what percent of 80?	$10 = x \cdot 80$
Twenty is 4% of what number?	$20 = 0.04 \cdot x$
Selling Price and Discount Model	
Discount = Rate of discount · Original price	$d = r \cdot L$
Selling Price = Original price − Discount	$S = L - r \cdot L$

Real Estate Commission Model

Commission = Rate of commission · Selling price

Amount for owner = Selling price − Commission

Geometric Models for Perimeter

Perimeter of any figure = the sum of the lengths of the sides

Rectangle: $P = 2L + 2W$ Square: $P = 4s$

Geometric Models for Area

Rectangle: $A = LW$ Square: $A = s^2$

Parallelogram: $A = bh$ Triangle: $A = \frac{1}{2}bh$

More geometric formulas can be found inside the front cover of this text.

E X A M P L E   6

Writing equations

Identify the variable and write an equation that describes each situation.

a) Find two numbers that have a sum of 14 and a product of 45.

b) A coat is on sale for 25% off the list price. If the sale price is $87, then what is the list price?

c) What percent of 8 is 2?

d) The value of x dimes and $x - 3$ quarters is $2.05.

‹ **Helpful Hint** ›

At this point we are simply learning to write equations that model certain situations. Don't worry about solving these equations now. In Section 2.6 we will solve problems by writing an equation and solving it.

Solution

a) Let $x =$ one of the numbers and $14 - x =$ the other number. Since their product is 45, we have

$$x(14 - x) = 45.$$

b) Let $x =$ the list price and $0.25x =$ the amount of discount. We can write an equation expressing the fact that the selling price is the list price minus the discount:

$$\text{List price} - \text{discount} = \text{selling price}$$
$$x - 0.25x = 87$$

c) If we let x represent the percentage, then the equation is $x \cdot 8 = 2$, or $8x = 2$.

d) The value of x dimes at 10 cents each is $10x$ cents. The value of $x - 3$ quarters at 25 cents each is $25(x - 3)$ cents. We can write an equation expressing the fact that the total value of the coins is 205 cents:

$$\text{Value of dimes} + \text{value of quarters} = \text{total value}$$
$$10x + 25(x - 3) = 205$$

Now do Exercises 59–84

CAUTION   The value of the coins in Example 6(d) is either 205 cents or 2.05 dollars. If the total value is expressed in dollars, then all of the values must be expressed in dollars. So we could also write the equation as

$$0.10x + 0.25(x - 3) = 2.05.$$

Warm-Ups ▼

Fill in the blank.

1. Words such as "sum," "plus," "increased by," and "more than" indicate _____.

2. Words such as "product," "twice," and "percent of" indicate _____.

3. _____ angles have degree measures with a sum of 90°.

4. _____ angles have degree measures with a sum of 180°.

5. Distance is the _____ of rate and time.

6. We can use x and $x + 2$ to represent consecutive _____ or consecutive _____ integers.

True or false?

7. For any value of x, x and $x + 6$ differ by 6.

8. For any value of a, a and $10 - a$ have a sum of 10.

9. If Jack ran x miles per hour for 3 hours, then he ran $3x$ miles.

10. If Jill ran x miles per hour for 10 miles, then she ran $10x$ hours.

11. Three consecutive odd integers can be represented by x, $x + 1$, and $x + 3$.

12. The value in cents of n nickels and d dimes is $0.05n + 0.10d$

2.5 Exercises

⟨ **Study Tips** ⟩

• Almost everything that we do in algebra can be redone by another method or checked. So don't close your mind to a new method or checking. The answers will not always be in the back of the book.

• When you take a test, work the problems that are easiest for you first. This will build your confidence. Make sure that you do not forget to answer a question.

⟨ 1 ⟩ **Writing Algebraic Expressions**

Translate each verbal expression into an algebraic expression. See Example 1. See Translating Words into Algebra box on pages 120–121.

1. The sum of a number and 3

2. Two more than a number

3. Three less than a number

4. Four subtracted from a number

5. The product of a number and 5

6. Five divided by some number

7. Ten percent of a number

8. Eight percent of a number

9. The ratio of a number and 3

10. The quotient of 12 and a number

11. One-third of a number

12. Three-fourths of a number

⟨ 2 ⟩ **Pairs of Numbers**

Write algebraic expressions for each pair of numbers. See Example 2.

13. Two numbers with a difference of 15

14. Two numbers that differ by 9

15. Two numbers with a sum of 6

16. Two numbers with a sum of 5

17. Two numbers such that one is 3 larger than the other

18. Two numbers such that one is 8 smaller than the other

19. Two numbers such that one is 5% of the other

20. Two numbers such that one is 40% of the other

21. Two numbers such that one is 30% more than the other

22. Two numbers such that one is 20% smaller than the other

Each of the following figures shows a pair of angles. Write algebraic expressions for the degree measures of each pair of angles. See Example 3.

23.

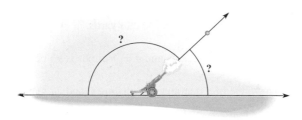

Figure for Exercise 23

24.

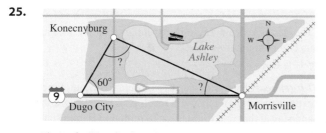

Figure for Exercise 24

25.

Figure for Exercise 25

26.

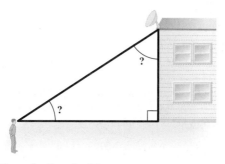

Figure for Exercise 26

⟨3⟩ **Consecutive Integers**

Write algebraic expressions for the following unknown integers. See Example 4.

27. Two consecutive even integers, the smallest of which is *n*

28. Two consecutive odd integers, the smallest of which is *x*

29. Two consecutive integers

30. Three consecutive even integers

31. Three consecutive odd integers

32. Three consecutive integers

33. Four consecutive even integers

34. Four consecutive odd integers

⟨4⟩ **Using Formulas**

Find an algebraic expression for the quantity in italics using the given information. See Example 5.

35. The *distance,* given that the rate is *x* miles per hour and the time is 3 hours

36. The *distance,* given that the rate is *x* + 10 miles per hour and the time is 5 hours

37. The *discount,* given that the rate is 25% and the original price is *q* dollars

38. The *discount,* given that the rate is 10% and the original price is *t* yen

39. The *time,* given that the distance is *x* miles and the rate is 20 miles per hour

40. The *time,* given that the distance is 300 kilometers and the rate is *x* + 30 kilometers per hour

41. The *rate,* given that the distance is *x* − 100 meters and the time is 12 seconds

42. The *rate,* given that the distance is 200 feet and the time is *x* + 3 seconds

43. The *area* of a rectangle with length *x* meters and width 5 meters

44. The *area* of a rectangle with sides *b* yards and *b* − 6 yards

45. The *perimeter* of a rectangle with length *w* + 3 inches and width *w* inches

46. The *perimeter* of a rectangle with length *r* centimeters and width *r* − 1 centimeters

47. The *width* of a rectangle with perimeter 300 feet and length *x* feet

48. The *length* of a rectangle with area 200 square feet and width *w* feet

49. The *length* of a rectangle, given that its width is *x* feet and its length is 1 foot longer than twice the width

50. The *length* of a rectangle, given that its width is *w* feet and its length is 3 feet shorter than twice the width

51. The *area* of a rectangle, given that the width is x meters and the length is 5 meters longer than the width

52. The *perimeter* of a rectangle, given that the length is x yards and the width is 10 yards shorter

53. The *simple interest*, given that the principal is $x + 1000$, the rate is 18%, and the time is 1 year

54. The *simple interest*, given that the principal is $3x$, the rate is 6%, and the time is 1 year

55. The *price per pound* of peaches, given that x pounds sold for $16.50

56. The *rate per hour* of a mechanic who gets $480 for working x hours

57. The *degree measure* of an angle, given that its complementary angle has measure x degrees

58. The *degree measure* of an angle, given that its supplementary angle has measure x degrees

⟨5⟩ Writing Equations

Identify the variable and write an equation that describes each situation. Do not solve the equation. See Example 6.

59. Two numbers differ by 5 and have a product of 8.

60. Two numbers differ by 6 and have a product of -9.

61. Herman's house sold for x dollars. The real estate agent received 7% of the selling price and Herman received $84,532.

62. Gwen sold her car on consignment for x dollars. The saleswoman's commission was 10% of the selling price and Gwen received $6570.

63. What percent of 500 is 100?

64. What percent of 40 is 120?

65. The value of x nickels and $x + 2$ dimes is $3.80.

66. The value of d dimes and $d - 3$ quarters is $6.75.

67. The sum of a number and 5 is 13.

68. Twelve subtracted from a number is -6.

69. The sum of three consecutive integers is 42.

70. The sum of three consecutive odd integers is 27.

71. The product of two consecutive integers is 182.

72. The product of two consecutive even integers is 168.

73. Twelve percent of Harriet's income is $3000.

74. If 9% of the members buy tickets, then we will sell 252 tickets to this group.

75. Thirteen is 5% of what number?

76. Three hundred is 8% of what number?

77. The length of a rectangle is 5 feet longer than the width, and the area is 126 square feet.

78. The length of a rectangle is 1 yard shorter than twice the width, and the perimeter is 298 yards.

79. The value of n nickels and $n - 1$ dimes is 95 cents.

80. The value of q quarters, $q + 1$ dimes, and $2q$ nickels is 90 cents.

81. The measure of an angle is 38° smaller than the measure of its supplementary angle.

82. The measure of an angle is 16° larger than the measure of its complementary angle.

83. *Target heart rate.* For a cardiovascular workout, fitness experts recommend that you reach your target heart rate and stay at that rate for at least 20 minutes (HealthStatus, www.healthstatus.com). To find your target heart rate, find the sum of your age and your resting heart rate, and then subtract that sum from 220. Find 60% of that result and add it to your resting heart rate.

    a) Write an equation with variable r expressing the fact that the target heart rate for 30-year-old Bob is 144.

    b) Judging from the accompanying graph, does the target heart rate for a 30-year-old increase or decrease as the resting heart rate increases?

84. *Adjusting the saddle.* The saddle height on a bicycle should be 109% of the rider's inside leg measurement L (www.harriscyclery.com). See the figure. Write an equation expressing the fact that the saddle height for Brenda is 36 in.

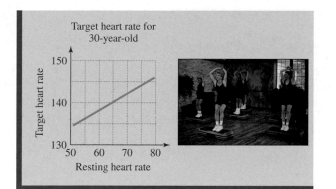

Target heart rate for 30-year-old

Figure for Exercise 83

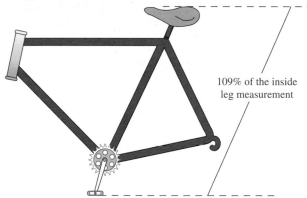

109% of the inside leg measurement

Figure for Exercise 84

Miscellaneous

Translate each verbal expression into an algebraic expression. Do not simplify.

85. The sum of 6 and x

86. w less than 12

87. m increased by 9

88. q decreased by 5

89. t multiplied by 11

90. 10 less than the square of y

91. 5 times the difference between x and 2

92. The sum of two-thirds of k and 1

93. m decreased by the product of 3 and m

94. 7 increased by the quotient of x and 2

95. The ratio of 8 more than h and h

96. The product of 5 and the total of r and 3

97. 5 divided by the difference between y and 9

98. The product of n and the sum of n and 6

99. The quotient of 8 less than w and twice w

100. 3 more than one-third of the square of b

101. 9 less than the product of v and -3

102. The total of 4 times the cube of t and the square of b

103. x decreased by the quotient of x and 7

104. Five-eighths of the sum of y and 3

105. The difference between the square of m and the total of m and 7

106. The product of 13 and the total of t and 6

107. x increased by the difference between 9 times x and 8

108. The quotient of twice y and 8

109. 9 less than the product of 13 and n

110. The product of s and 5 more than s

111. 6 increased by one-third of the sum of x and 2

112. x decreased by the difference between $5x$ and 9

113. The sum of x divided by 2 and x

114. Twice the sum of 6 times n and 5

Given that the area of each figure is 24 square feet, use the dimensions shown to write an equation expressing this fact. Do not solve the equation.

115.

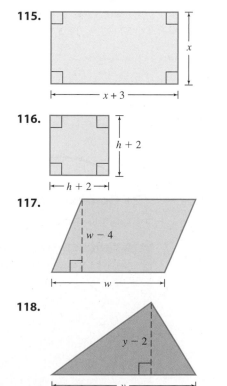

x

$x + 3$

116.

$h + 2$

$h + 2$

117.

$w - 4$

w

118.

$y - 2$

y

2.6 | **Number, Geometric, and Uniform Motion Applications**

In This Section

⟨1⟩ **Number Problems**

⟨2⟩ **General Strategy for Solving Verbal Problems**

⟨3⟩ **Geometric Problems**

⟨4⟩ **Uniform Motion Problems**

In this section, we apply the ideas of Section 2.5 to solving problems. Many of the problems can be solved by using arithmetic only and not algebra. However, remember that we are not just trying to find the answer; we are trying to learn how to apply algebra. So even if the answer is obvious to you, set the problem up and solve it by using algebra as shown in the examples.

⟨1⟩ Number Problems

Algebra is often applied to problems involving time, rate, distance, interest, or discount. **Number problems** do not involve any physical situation; we simply find some numbers that satisfy some given conditions. These problems can provide good practice for solving more complex problems.

EXAMPLE   1

⟨ **Helpful Hint** ⟩

Making a guess can be a good way to get familiar with the problem. For example, let's guess that the answers to Example 1 are 20, 21, and 22. Since $20 + 21 + 22 = 63$, these are not the correct numbers. But now we realize that we should use x, $x + 1$, and $x + 2$ and that the equation should be

$$x + x + 1 + x + 2 = 48.$$

A consecutive integer problem

The sum of three consecutive integers is 48. Find the integers.

Solution

If x represents the smallest of the three consecutive integers, then x, $x + 1$, and $x + 2$ represent the three consecutive integers. Since the sum of x, $x + 1$, and $x + 2$ is 48, we write that fact as an equation and solve it:

$$x + (x + 1) + (x + 2) = 48$$
$$3x + 3 = 48 \quad \text{Combine like terms.}$$
$$3x = 45 \quad \text{Subtract 3 from each side.}$$
$$x = 15 \quad \text{Divide each side by 3.}$$
$$x + 1 = 16 \quad \text{If } x \text{ is 15, then } x + 1 \text{ is 16 and } x + 2 \text{ is 17.}$$
$$x + 2 = 17$$

Because $15 + 16 + 17 = 48$, the three consecutive integers that have a sum of 48 are 15, 16, and 17.

Now do Exercises 1–8

⟨2⟩ General Strategy for Solving Verbal Problems

You should use the following steps as a guide for solving problems.

Strategy for Solving Problems

1. Read the problem as many times as necessary. Guessing the answer and checking it will help you understand the problem.
2. If possible, draw a diagram to illustrate the problem.
3. Choose a variable and *write* what it represents.
4. Write algebraic expressions for any other unknowns in terms of that variable.
5. Write an equation that describes the situation.

6. Solve the equation.
7. Answer the original question.
8. Check your answer in the original problem (not the equation).

⟨3⟩ Geometric Problems

For geometric problems, always draw the figure and label it. Common geometric formulas are given in Section 2.5 and inside the front cover of this text. The **perimeter** of any figure is the sum of the lengths of all of the sides of the figure. The perimeter for a square is given by $P = 4s$, for a rectangle $P = 2L + 2W$, and for a triangle $P = a + b + c$. You can use these formulas or simply remember that the sum of the lengths of all sides is the perimeter.

EXAMPLE 2

⟨ **Helpful Hint** ⟩
──────────────
To get familiar with the problem, guess that the width is 50 ft. Then the length is $2 \cdot 50 - 1$ or 99. The perimeter would be

$$2(50) + 2(99) = 298,$$

which is too small. But now we realize that we should let x be the width, $2x - 1$ be the length, and we should solve

$$2x + 2(2x - 1) = 748.$$

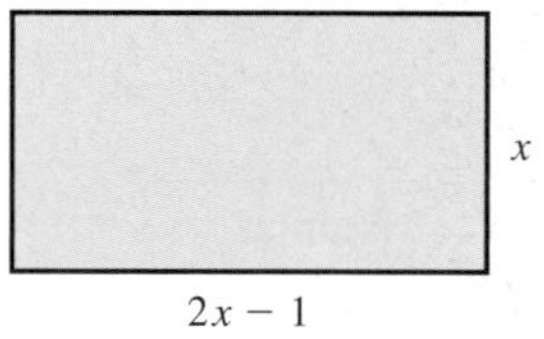

x

$2x - 1$

Figure 2.2

A perimeter problem

The length of a rectangular piece of property is 1 foot less than twice the width. If the perimeter is 748 feet, find the length and width.

Solution

Let $x =$ the width. Since the length is 1 foot less than twice the width, $2x - 1 =$ the length. Draw a diagram as in Fig. 2.2. We know that $2L + 2W = P$ is the formula for the perimeter of a rectangle. Substituting $2x - 1$ for L and x for W in this formula yields an equation in x:

$$2L + 2W = P$$
$$2(2x - 1) + 2(x) = 748 \quad \text{Replace } L \text{ by } 2x - 1 \text{ and } W \text{ by } x.$$
$$4x - 2 + 2x = 748 \quad \text{Remove the parentheses.}$$
$$6x - 2 = 748 \quad \text{Combine like terms.}$$
$$6x = 750 \quad \text{Add 2 to each side.}$$
$$x = 125 \quad \text{Divide each side by 6.}$$

If $x = 125$, then $2x - 1 = 2(125) - 1 = 249$. Check by computing the perimeter:

$$P = 2L + 2W = 2(249) + 2(125) = 748$$

So the width is 125 feet and the length is 249 feet.

Now do Exercises 9–14

Example 3 involves the degree measures of angles. For this problem, the figure is given.

EXAMPLE 3

Complementary angles

In Fig. 2.3, the angle formed by the guy wire and the ground is 3.5 times as large as the angle formed by the guy wire and the antenna. Find the degree measure of each of these angles.

Solution

Let $x =$ the degree measure of the smaller angle, and let $3.5x =$ the degree measure of the larger angle. Since the antenna meets the ground at a 90° angle, the sum of the degree

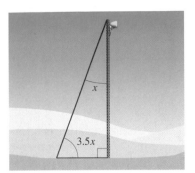

Figure 2.3

measures of the other two angles of the right triangle is 90°. (They are complementary angles.) So we have the following equation:

$$x + 3.5x = 90$$
$$4.5x = 90 \quad \text{Combine like terms.}$$
$$x = 20 \quad \text{Divide each side by 4.5.}$$
$$3.5x = 70 \quad \text{Find the other angle.}$$

Check: 70° is 3.5 · 20° and 20° + 70° = 90°. So the smaller angle is 20°, and the larger angle is 70°.

Now do Exercises 15–16

⟨4⟩ Uniform Motion Problems

Problems involving motion at a constant rate are called **uniform motion problems.** In uniform motion problems, we often use an average rate when the actual rate is not constant. For example, you can drive all day and average 50 miles per hour, but you are not driving at a constant 50 miles per hour.

EXAMPLE **4**

Finding the rate

Bridgette drove her car for 2 hours on an icy road. When the road cleared up, she increased her speed by 35 miles per hour and drove 3 more hours, completing her 255-mile trip. How fast did she travel on the icy road?

Solution

It is helpful to draw a diagram and then make a table to classify the given information. Remember that $D = RT$.

⟨ **Helpful Hint** ⟩

To get familiar with the problem, guess that she traveled 20 mph on the icy road and 55 mph (20 + 35) on the clear road. Her total distance would be

$$20 \cdot 2 + 55 \cdot 3 = 205 \text{ mi.}$$

Of course this is not correct, but now you are familiar with the problem.

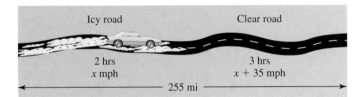

	Rate	Time	Distance
Icy road	$x \dfrac{\text{mi}}{\text{hr}}$	2 hr	$2x$ mi
Clear road	$x + 35 \dfrac{\text{mi}}{\text{hr}}$	3 hr	$3(x + 35)$ mi

The equation expresses the fact that her total distance traveled was 255 miles:

Icy road distance + clear road distance = total distance

$$2x + 3(x + 35) = 255$$
$$2x + 3x + 105 = 255$$
$$5x + 105 = 255$$
$$5x = 150$$
$$x = 30$$
$$x + 35 = 65$$

If she drove at 30 miles per hour for 2 hours on the icy road, she went 60 miles. If she drove at 65 miles per hour for 3 hours on the clear road, she went 195 miles. Since $60 + 195 = 255$, we can be sure that her speed on the icy road was 30 mph.

Now do Exercises 17–20

In the next uniform motion problem we find the time.

E X A M P L E   5

Finding the time

Pierce drove from Allentown to Baker, averaging 55 miles per hour. His journey back to Allentown using the same route took 3 hours longer because he averaged only 40 miles per hour. How long did it take him to drive from Allentown to Baker? What is the distance between Allentown and Baker?

Solution

Draw a diagram and then make a table to classify the given information. Remember that $D = RT$.

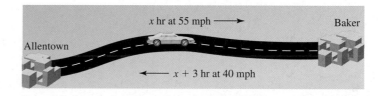

	Rate	Time	Distance
Going	$55 \frac{mi}{hr}$	x hr	$55x$ mi
Returning	$40 \frac{mi}{hr}$	$x + 3$ hr	$40(x + 3)$ mi

We can write an equation expressing the fact that the distance either way is the same:

$$\text{Distance going} = \text{distance returning}$$
$$55x = 40(x + 3)$$
$$55x = 40x + 120$$
$$15x = 120$$
$$x = 8$$

The trip from Allentown to Baker took 8 hours. The distance between Allentown and Baker is $55 \cdot 8$, or 440 miles.

Now do Exercises 21–22

Warm-Ups ▼

Fill in the blank.

1. _____ motion is motion at a constant rate.
2. When solving a _____ problem you should draw a figure and label it.
3. If x and $x + 10$ are _____ angles, then $x + x + 10 = 90$.
4. If x and $x - 45$ are _____ angles, then $x + x - 45 = 180$.
5. If x is an even integer, then $x + 2$ is an _____ integer.
6. If x is an odd integer, then $x + 2$ is an ____ integer.

True or false?

7. The first step in solving a word problem is to write the equation.
8. You should always write down what the variable represents.
9. Diagrams and tables are used as aids in solving word problems.
10. If x is an odd integer, then $x + 1$ is also an odd integer.
11. The degree measures of two complementary angles can be represented by x and $90 - x$.
12. The degree measures of two supplementary angles can be represented by x and $x + 180$.

2.6 Exercises

⟨ Study Tips ⟩

- Make sure you know how your grade in this course is determined. How much weight is given to tests, homework, quizzes, and projects? Does your instructor give any extra credit?
- You should keep a record of all of your scores and compute your own final grade.

⟨1⟩ Number Problems

Show a complete solution to each problem. See Example 1.

1. *Consecutive integers.* Find two consecutive integers whose sum is 79.
2. *Consecutive odd integers.* Find two consecutive odd integers whose sum is 56.
3. *Consecutive integers.* Find three consecutive integers whose sum is 141.
4. *Consecutive even integers.* Find three consecutive even integers whose sum is 114.
5. *Consecutive odd integers.* Two consecutive odd integers have a sum of 152. What are the integers?

6. *Consecutive odd integers.* Four consecutive odd integers have a sum of 120. What are the integers?
7. *Consecutive integers.* Find four consecutive integers whose sum is 194.
8. *Consecutive even integers.* Find four consecutive even integers whose sum is 340.

⟨3⟩ Geometric Problems

Show a complete solution to each problem. See Examples 2 and 3. See the Strategy for Solving Problems box on pages 130–131.

9. *Olympic swimming.* If an Olympic swimming pool is twice as long as it is wide and the perimeter is 150 meters, then what are the length and width?

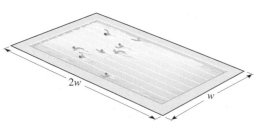

Figure for Exercise 9

10. *Wimbledon tennis.* If the perimeter of a tennis court is 228 feet and the length is 6 feet longer than twice the width, then what are the length and width?

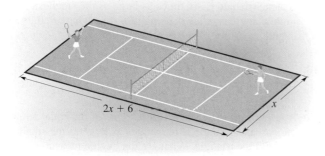

Figure for Exercise 10

11. *Framed.* Julia framed an oil painting that her uncle gave her. The painting was 4 inches longer than it was wide, and it took 176 inches of frame molding. What were the dimensions of the picture?

12. *Industrial triangle.* Geraldo drove his truck from Indianapolis to Chicago, then to St. Louis, and then back to Indianapolis. He observed that the second side of his triangular route was 81 miles short of being twice as long as the first side and that the third side was 61 miles longer than the first side. If he traveled a total of 720 miles, then how long is each side of this triangular route?

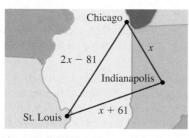

Figure for Exercise 12

13. *Triangular banner.* A banner in the shape of an isosceles triangle has a base that is 5 inches shorter than either of the equal sides. If the perimeter of the banner is 34 inches, then what is the length of the equal sides?

14. *Border paper.* Dr. Good's waiting room is 8 feet longer than it is wide. When Vincent wallpapered Dr. Good's waiting room, he used 88 feet of border paper. What are the dimensions of Dr. Good's waiting room?

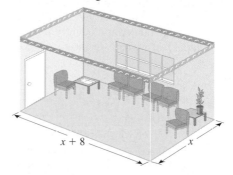

Figure for Exercise 14

15. *Roof truss design.* An engineer is designing a roof truss as shown in the accompanying figure. Find the degree measure of the angle marked w.

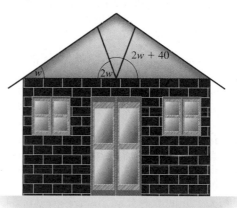

Figure for Exercise 15

16. *Another truss.* Another truss is shown in the accompanying figure. Find the degree measure of the angle marked z.

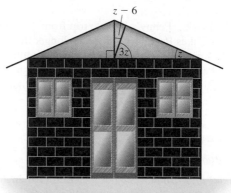

Figure for Exercise 16

⟨4⟩   Uniform Motion Problems

Show a complete solution to each problem. See Examples 4 and 5.

17. *Highway miles.*   Bret drove for 4 hours on the freeway, and then decreased his speed by 20 miles per hour and drove for 5 more hours on a country road. If his total trip was 485 miles, then what was his speed on the freeway?

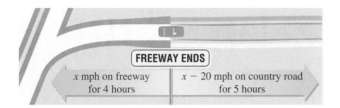

FREEWAY ENDS

x mph on freeway for 4 hours   *x* − 20 mph on country road for 5 hours

Figure for Exercise 17

18. *Walking and running.*   On Saturday morning, Lynn walked for 2 hours and then ran for 30 minutes. If she ran twice as fast as she walked and she covered 12 miles altogether, then how fast did she walk?

19. *Driving all night.*   Kathryn drove her rig 5 hours before dawn and 6 hours after dawn. If her average speed was 5 miles per hour more in the dark and she covered 630 miles altogether, then what was her speed after dawn?

20. *Commuting to work.*   On Monday, Roger drove to work in 45 minutes. On Tuesday he averaged 12 miles per hour more, and it took him 9 minutes less to get to work. How far does he travel to work?

21. *Head winds.*   A jet flew at an average speed of 640 mph from Los Angeles to Chicago. Because of head winds the jet averaged only 512 mph on the return trip, and the return trip took 48 minutes longer. How many hours was the flight from Chicago to Los Angeles? How far is it from Chicago to Los Angeles?

22. *Ride the Peaks.*   Penny's bicycle trip from Colorado Springs to Pikes Peak took 1.5 hours longer than the return trip to Colorado Springs. If she averaged 6 mph on the way to Pikes Peak and 15 mph for the return trip, then how long was the ride from Colorado Springs to Pikes Peak?

Miscellaneous

Solve each problem.

23. *Perimeter of a frame.*   The perimeter of a rectangular frame is 64 in. If the width of the frame is 8 in. less than the length, then what are the length and width of the frame?

24. *Perimeter of a box.*   The width of a rectangular box is 20% of the length. If the perimeter is 192 cm, then what are the length and width of the box?

25. *Isosceles triangle.*   An isosceles triangle has two equal sides. If the shortest side of an isosceles triangle is 2 ft less than one of the equal sides and the perimeter is 13 ft, then what are the lengths of the sides?

26. *Scalene triangle.*   A scalene triangle has three unequal sides. The perimeter of a scalene triangle is 144 m. If the first side is twice as long as the second side and the third side is 24 m longer than the second side, then what are the measures of the sides?

27. *Angles of a scalene triangle.*   The largest angle in a scalene triangle is six times as large as the smallest. If the middle angle is twice the smallest, then what are the degree measures of the three angles?

28. *Angles of a right triangle.*   If one of the acute angles in a right triangle is 38°, then what are the degree measures of all three angles?

29. *Angles of an isosceles triangle.*   One of the equal angles in an isosceles triangle is four times as large as the smallest angle in the triangle. What are the degree measures of the three angles?

30. *Angles of an isosceles triangle.*   The measure of one of the equal angles in an isosceles triangle is 10° larger than twice the smallest angle in the triangle. What are the degree measures of the three angles?

31. *Super Bowl score.*   The 1977 Super Bowl was played in the Rose Bowl in Pasadena. In that football game the Oakland Raiders scored 18 more points than the Minnesota Vikings. If the total number of points scored was 46, then what was the final score for the game?

32. *Top payrolls.*   Payrolls for the three highest paid baseball teams (the Yankees, Mets, and Cubs) for 2009 totaled $485 million (www.usatoday.com). If the team payroll for the Yankees was $52 million greater than the payroll for the Mets and the payroll for the Mets was $14 million greater than the payroll for the Cubs, then what was the 2009 payroll for each team?

33. *Idabel to Lawton.*   Before lunch, Sally drove from Idabel to Ardmore, averaging 50 mph. After lunch she continued on to Lawton, averaging 53 mph. If her driving time after lunch was 1 hour less than her driving time before lunch and the total trip was 256 miles, then how many hours did she drive before lunch? How far is it from Ardmore to Lawton?

34. *Norfolk to Chadron.*   On Monday, Chuck drove from Norfolk to Valentine, averaging 47 mph. On Tuesday, he continued on to Chadron, averaging 69 mph. His driving time on Monday was 2 hours longer than his driving time on Tuesday. If the total distance from Norfolk to Chadron is 326 miles, then how many hours did he drive on Monday? How far is it from Valentine to Chadron?

35. *Golden oldies.* Joan Crawford, John Wayne, and James Stewart were born in consecutive years (*Doubleday Almanac*). Joan Crawford was the oldest of the three, and James Stewart was the youngest. In 1950, after all three had their birthdays, the sum of their ages was 129. In what years were they born?

36. *Leading men.* Bob Hope was born 2 years after Clark Gable and 2 years before Henry Fonda (*Doubleday Almanac*). In 1951, after all three of them had their birthdays, the sum of their ages was 144. In what years were they born?

37. *Trimming a garage door.* A carpenter used 30 ft of molding in three pieces to trim a garage door. If the long piece was 2 ft longer than twice the length of each shorter piece, then how long was each piece?

Figure for Exercise 37

38. *Fencing dog pens.* Clint is constructing two adjacent rectangular dog pens. Each pen will be three times as long as it is wide, and the pens will share a common long side. If Clint has 65 ft of fencing, what are the dimensions of each pen?

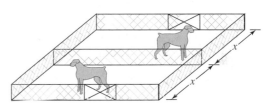

Figure for Exercise 38

2.7 Discount, Investment, and Mixture Applications

In This Section

⟨1⟩ **Discount Problems**
⟨2⟩ **Commission Problems**
⟨3⟩ **Investment Problems**
⟨4⟩ **Mixture Problems**

In this section, we continue our study of applications of algebra. The problems in this section involve percents.

⟨1⟩ **Discount Problems**

When an item is sold at a discount, the amount of the discount is usually described as being a percentage of the original price. The percentage is called the **rate of discount.** Multiplying the rate of discount and the original price gives the amount of the discount.

E X A M P L E   **1**   **Finding the original price**

Ralph got a 12% discount when he bought his new 2010 Corvette Coupe. If the amount of his discount was $6606, then what was the original price of the Corvette?

Solution

Let x represent the original price. The discount is found by multiplying the 12% rate of discount and the original price:

$$\text{Rate of discount} \cdot \text{original price} = \text{amount of discount}$$

$$0.12x = 6606$$

$$x = \frac{6606}{0.12} \quad \text{Divide each side by 0.12.}$$

$$x = 55{,}050$$

To check, find 12% of $55,050. Since $0.12 \cdot 55{,}050 = 6606$, the original price of the Corvette was $55,050.

| Now do Exercises 1–2 |

E X A M P L E   **2**

Finding the original price

When Susan bought her new car, she also got a discount of 12%. She paid $17,600 for her car. What was the original price of Susan's car?

⟨ **Helpful Hint** ⟩

To get familiar with the problem, guess that the original price was $30,000. Then her discount is 0.12(30,000) or $3600. The price she paid would be $30,000 − 3600$ or $26,400, which is incorrect.

Solution

Let x represent the original price for Susan's car. The amount of discount is 12% of x, or $0.12x$. We can write an equation expressing the fact that the original price minus the discount is the price Susan paid.

$$\text{Original price} - \text{discount} = \text{sale price}$$

$$x - 0.12x = 17{,}600$$

$$0.88x = 17{,}600 \qquad 1.00x - 0.12x = 0.88x$$

$$x = \frac{17{,}600}{0.88} \qquad \text{Divide each side by 0.88.}$$

$$x = 20{,}000$$

Check: 12% of $20,000 is $2400, and $20{,}000 - \$2400 = \$17{,}600$. The original price of Susan's car was $20,000.

| Now do Exercises 3–4 |

⟨2⟩ Commission Problems

A salesperson's commission for making a sale is often a percentage of the selling price. **Commission problems** are very similar to other problems involving percents. The commission is found by multiplying the rate of commission and the selling price.

E X A M P L E   **3**

Real estate commission

Sarah is selling her house through a real estate agent whose commission rate is 7%. What should the selling price be so that Sarah can get the $83,700 she needs to pay off the mortgage?

Solution

Let x be the selling price. The commission is 7% of x (not 7% of $83,700). Sarah receives the selling price less the sales commission:

$$\text{Selling price} - \text{commission} = \text{Sarah's share}$$
$$x - 0.07x = 83,700$$
$$0.93x = 83,700 \quad {\scriptstyle 1.00x - 0.07x = 0.93x}$$
$$x = \frac{83,700}{0.93}$$
$$x = 90,000$$

Check: 7% of $90,000 is $6300, and $90,000 − $6300 = $83,700. So the house should sell for $90,000.

Now do Exercises 5–8

⟨3⟩ Investment Problems

The interest on an investment is a percentage of the investment, just as the sales commission is a percentage of the sale amount. However, in **investment problems** we must often account for more than one investment at different rates. So it is a good idea to make a table, as in Example 4.

E X A M P L E   4

Diversified investing

Ruth Ann invested some money in a certificate of deposit with an annual yield of 9%. She invested twice as much in a mutual fund with an annual yield of 10%. Her interest from the two investments at the end of the year was $232. How much was invested at each rate?

Solution

When there are many unknown quantities, it is often helpful to identify them in a table. Since the time is 1 year, the amount of interest is the product of the interest rate and the amount invested.

	Interest Rate	Amount Invested	Interest for 1 Year
CD	9%	x	$0.09x$
Mutual fund	10%	$2x$	$0.10(2x)$

Since the total interest from the investments was $232, we can write the following equation:

$$\text{CD interest} + \text{mutual fund interest} = \text{total interest}$$
$$0.09x + 0.10(2x) = 232$$
$$0.09x + 0.20x = 232$$
$$0.29x = 232$$
$$x = \frac{232}{0.29}$$
$$x = 800$$
$$2x = 1600$$

⟨ Helpful Hint ⟩

To get familiar with the problem, guess that she invested $1000 at 9% and $2000 at 10%. Then her earnings in 1 year would be

$$0.09(1000) + 0.10(2000)$$

or $290, which is close but incorrect.

To check, we find the total interest:

$$0.09(800) + 0.10(1600) = 72 + 160$$
$$= 232$$

So Ruth Ann invested $800 at 9% and $1600 at 10%.

Now do Exercises 9–12

⟨4⟩ Mixture Problems

Mixture problems are concerned with the result of mixing two quantities, each of which contains another substance. Notice how similar the following mixture problem is to the last investment problem.

EXAMPLE 5

⟨ **Helpful Hint** ⟩

To get familiar with the problem, guess that we need 100 gal of 4% milk. Mixing that with 80 gal of 1% milk would produce 180 gal of 2% milk. Now the two milks separately have

$$0.04(100) + 0.01(80)$$

or 4.8 gal of fat. Together the amount of fat is 0.02(180) or 3.6 gal. Since these amounts are not equal, our guess is incorrect.

Mixing milk

How many gallons of milk containing 4% butterfat must be mixed with 80 gallons of 1% milk to obtain 2% milk?

Solution

It is helpful to draw a diagram and then make a table to classify the given information.

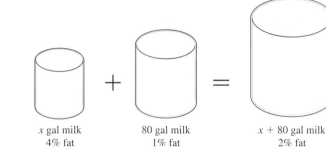

x gal milk   80 gal milk $x + 80$ gal milk
4% fat   1% fat   2% fat

	Percentage of Fat	Amount of Milk	Amount of Fat
4% milk	4%	x	$0.04x$
1% milk	1%	80	$0.01(80)$
2% milk	2%	$x + 80$	$0.02(x + 80)$

The equation expresses the fact that the total fat from the first two types of milk is the same as the fat in the mixture:

Fat in 4% milk + fat in 1% milk = fat in 2% milk

$$0.04x + 0.01(80) = 0.02(x + 80)$$

$0.04x + 0.8 = 0.02x + 1.6$	Simplify.
$100(0.04x + 0.8) = 100(0.02x + 1.6)$	Multiply each side by 100.
$4x + 80 = 2x + 160$	Distributive property
$2x + 80 = 160$	Subtract $2x$ from each side.
$2x = 80$	Subtract 80 from each side.
$x = 40$	Divide each side by 2.

To check, calculate the total fat:

$$2\% \text{ of } 120 \text{ gallons} = 0.02(120) = 2.4 \text{ gallons of fat}$$
$$0.04(40) + 0.01(80) = 1.6 + 0.8 = 2.4 \text{ gallons of fat}$$

So we mix 40 gallons of 4% milk with 80 gallons of 1% milk to get 120 gallons of 2% milk.

Now do Exercises 13–16

In mixture problems, the solutions might contain fat, alcohol, salt, or some other substance. We always assume that the substance neither appears nor disappears in the process. For example, if there are 3 grams of salt in one glass of water and 2 grams in another, then there are exactly 5 grams in a mixture of the two.

Warm-Ups ▼

Fill in the blank.

1. The _____ of discount is a percentage.
2. The _____ is the amount by which a price is reduced.
3. The _____ of the original price and the rate of discount is the discount.
4. A _____ helps us to organize information given in a word problem.
5. An interest _____ is a percentage.

True or false?

6. If Jim gets a 12% commission for selling a $1000 Wonder Vac, then his commission is $120.
7. If Bob earns a 5% commission on an $80,000 motorhome sale, then Bob earns $400.
8. If Sue gets a 20% discount on a TV with a list price of x dollars, then Sue pays $0.8x$ dollars.
9. If you get a 6% discount on a Chevy Volt for which the MSRP is x dollars, then your discount is $0.6x$ dollars.

Exercises 2.7

⟨ **Study Tips** ⟩

• Find out what kinds of help are available for commuting students, online students, and on-campus students.
• Sometimes a minor issue can be resolved very quickly and you can get back on the path to success.

⟨ 1 ⟩   **Discount Problems**

Show a complete solution to each problem. See Examples 1 and 2.

1. **Close-out sale.** At a 25% off sale, Jose saved $80 on a 19-inch Panasonic TV. What was the original price of the television?

2. **Nice tent.** A 12% discount on a Walrus tent saved Melanie $75. What was the original price of the tent?

3. **Circuit city.** After getting a 20% discount, Robert paid $320 for a Pioneer CD player for his car. What was the original price of the CD player?

4. **Chrysler Sebring.** After getting a 15% discount on the price of a new Chrysler Sebring convertible, Helen paid $27,000. What was the original price of the convertible to the nearest dollar?

⟨2⟩ **Commission Problems**

Show a complete solution to each problem. See Example 3.

5. ***Selling price of a home.*** Kirk wants to get $115,000 for his house. The real estate agent gets a commission equal to 8% of the selling price for selling the house. What should the selling price be?

Photo for Exercise 5

6. ***Horse trading.*** Gene is selling his palomino at an auction. The auctioneer's commission is 10% of the selling price. If Gene still owes $810 on the horse, then what must the horse sell for so that Gene can pay off his loan?

7. ***Sales tax collection.*** Merilee sells tomatoes at a roadside stand. Her total receipts including the 7% sales tax were $462.24. What amount of sales tax did she collect?

8. ***Toyota Corolla.*** Gwen bought a new Toyota Corolla. The selling price plus the 8% state sales tax was $15,714. What was the selling price?

⟨3⟩ **Investment Problems**

Show a complete solution to each problem. See Example 4.

9. ***Wise investments.*** Wiley invested some money in the Berger 100 Fund and $3000 more than that amount in the Berger 101 Fund. For the year he was in the fund, the 100 Fund paid 18% simple interest and the 101 Fund paid 15% simple interest. If the income from the two investments totaled $3750 for 1 year, then how much did he invest in each fund?

10. ***Loan shark.*** Becky lent her brother some money at 8% simple interest, and she lent her sister twice as much at twice the interest rate. If she received a total of 20 cents interest, then how much did she lend to each of them?

11. ***Investing in bonds.*** David split his $25,000 inheritance between Fidelity Short-Term Bond Fund with an annual yield of 5% and T. Rowe Price Tax-Free Short-Intermediate Fund with an annual yield of 4%. If his total income for 1 year on the two investments was $1140, then how much did he invest in each fund?

12. ***High-risk funds.*** Of the $50,000 that Natasha pocketed on her last real estate deal, $20,000 went to charity. She invested part of the remainder in Dreyfus New Leaders Fund with an annual yield of 16% and the rest in Templeton Growth Fund with an annual yield of 25%. If she made $6060 on these investments in 1 year, then how much did she invest in each fund?

⟨4⟩ **Mixture Problems**

Show a complete solution to each problem. See Example 5.

13. ***Mixing milk.*** How many gallons of milk containing 1% butterfat must be mixed with 30 gallons of milk containing 3% butterfat to obtain a mixture containing 2% butterfat?

Figure for Exercise 13

14. ***Acid solutions.*** How many gallons of a 5% acid solution should be mixed with 30 gallons of a 10% acid solution to obtain a mixture that is 8% acid?

15. ***Alcohol solutions.*** Gus has on hand a 5% alcohol solution and a 20% alcohol solution. He needs 30 liters of a 10% alcohol solution. How many liters of each solution should he mix together to obtain the 30 liters?

16. ***Adjusting antifreeze.*** Angela needs 20 quarts of 50% antifreeze solution in her radiator. She plans to obtain this by mixing some pure antifreeze with an appropriate amount of a 40% antifreeze solution. How many quarts of each should she use?

Figure for Exercise 16

Miscellaneous

Solve each problem.

17. ***Registered voters.*** If 60% of the registered voters of Lancaster County voted in the November election and 33,420 votes were cast, then how many registered voters are there in Lancaster County?

Photo for Exercise 17

18. Tough on crime. In a random sample of voters, 594 respondents said that they favored passage of a $33 billion crime bill. If the number in favor of the crime bill was 45% of the number of voters in the sample, then how many voters were in the sample?

19. Ford Taurus. At an 8% sales tax rate, the sales tax on Peter's new Ford Taurus was $1200. What was the price of the car?

20. Taxpayer blues. Last year, Faye paid 24% of her income to taxes. If she paid $9600 in taxes, then what was her income?

21. Making a profit. A retail store buys shirts for $8 and sells them for $14. What percent increase is this?

22. Monitoring AIDS. If 28 new AIDS cases were reported in Landon County this year and 35 new cases were reported last year, then what percent decrease in new cases is this?

23. High school integration. Wilson High School has 400 students, of whom 20% are African American. The school board plans to merge Wilson High with Jefferson High. This one school will then have a student population that is 44% African American. If Jefferson currently has a student population that is 60% African American, then how many students are at Jefferson?

24. Junior high integration. The school board plans to merge two junior high schools into one school of 800 students in which 40% of the students will be Caucasian. One of the schools currently has 58% Caucasian students; the other has only 10% Caucasian students. How many students are in each of the two schools?

25. Hospital capacity. When Memorial Hospital is filled to capacity, it has 18 more people in semiprivate rooms (two patients to a room) than in private rooms. The room rates are $200 per day for a private room and $150 per day for a semiprivate room. If the total receipts for rooms is $17,400 per day when all are full, then how many rooms of each type does the hospital have?

26. Public relations. Memorial Hospital is planning an advertising campaign. It costs the hospital $3000 each time a television ad is aired and $2000 each time a radio ad is aired. The administrator wants to air 60 more television ads than radio ads. If the total cost of airing the ads is $580,000, then how many ads of each type will be aired?

27. Mixed nuts. Cashews sell for $4.80 per pound, and pistachios sell for $6.40 per pound. How many pounds of pistachios should be mixed with 20 pounds of cashews to get a mixture that sells for $5.40 per pound?

28. Premium blend. Premium coffee sells for $6.00 per pound, and regular coffee sells for $4.00 per pound. How many pounds of each type of coffee should be blended to obtain 100 pounds of a blend that sells for $4.64 per pound?

29. Nickels and dimes. Candice paid her library fine with 10 coins consisting of nickels and dimes. If the fine was $0.80, then how many of each type of coin did she use?

30. Dimes and quarters. Jeremy paid for his breakfast with 36 coins consisting of dimes and quarters. If the bill was $4.50, then how many of each type of coin did he use?

31. Cooking oil. Crisco Canola Oil is 7% saturated fat. Crisco blends corn oil that is 14% saturated fat with Crisco Canola Oil to get Crisco Canola and Corn Oil, which is 11% saturated fat. How many gallons of corn oil must Crisco mix with 600 gallons of Crisco Canola Oil to get Crisco Canola and Corn Oil?

32. Chocolate ripple. The Delicious Chocolate Shop makes a dark chocolate that is 35% fat and a white chocolate that is 48% fat. How many kilograms of dark chocolate should be mixed with 50 kilograms of white chocolate to make a ripple blend that is 40% fat?

33. Hawaiian Punch. Hawaiian Punch is 10% fruit juice. How much water would you have to add to one gallon of Hawaiian Punch to get a drink that is 6% fruit juice?

34. Diluting wine. A restaurant manager has 2 liters of white wine that is 12% alcohol. How many liters of white grape juice should he add to get a drink that is 10% alcohol?

35. Bargain hunting. A smart shopper bought 5 pairs of shorts and 8 tops for a total of $108. If the price of a pair of shorts was twice the price of a top, then what was the price of each type of clothing?

36. VCRs and CDs. The manager of a stereo shop placed an order for $10,710 worth of VCRs at $120 each and CD players at $150 each. If the number of VCRs she ordered was three times the number of CD players, then how many of each did she order?

2.8 Inequalities

⟨ **Helpful Hint** ⟩

A good way to learn inequality symbols is to notice that the inequality symbol always points at the smaller number. This observation will help you read an inequality such as $-2 < x$. Reading right to left, we say that x is greater than -2. It is usually easier to understand an inequality if you read the variable first.

In Chapter 1, we defined inequality in terms of the number line. One number is greater than another number if it lies to the right of the other number on the number line. In this section, you will study inequality in greater depth.

⟨1⟩ Inequalities

The symbols used to express inequality and their meanings are given in the following box.

Inequality Symbols

Symbol	Meaning
$<$	Is less than
$\leq$	Is less than or equal to
$>$	Is greater than
$\geq$	Is greater than or equal to

The statement $a < b$ means that a is to the left of b on the number line as shown in Fig. 2.4. The statement $c > d$ means that c is to the right of d on the number line, as shown in Fig. 2.5. Of course, $a < b$ has the same meaning as $b > a$. The statement $a \leq b$ means that either a is to the left of b or a corresponds to the same point as b on the number line. The statement $a \leq b$ has the same meaning as the statement $b \geq a$.

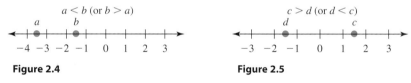

Figure 2.4   Figure 2.5

E X A M P L E **1**

Verifying inequalities

Determine whether each of the following statements is correct.

a) $3 < 4$ **b)** $-1 < -2$ **c)** $-2 \leq 0$

d) $0 \geq 0$ **e)** $2(-3) + 8 > 9$ **f)** $(-2)(-5) \leq 10$

Solution

a) Locate 3 and 4 on the number line shown in Fig. 2.6. Because 3 is to the left of 4 on the number line, $3 < 4$ is correct.

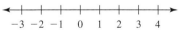

Figure 2.6

b) Locate -1 and -2 on the number line shown in Fig. 2.6. Because -1 is to the right of -2, on the number line, $-1 < -2$ is not correct.

c) Because -2 is to the left of 0 on the number line, $-2 \leq 0$ is correct.

d) Because 0 is equal to 0, $0 \geq 0$ is correct.

e) Simplify the left side of the inequality to get $2 > 9$, which is not correct.

f) Simplify the left side of the inequality to get $10 \leq 10$, which is correct.

Now do Exercises 1–16

⟨ **Calculator Close-Up** ⟩

A graphing calculator can determine whether an inequality is correct. Use the inequality symbols from the TEST menu to enter the inequality.

```
3<4
            1
0≥0
            1
2( -3)+8>9
            0
```

When ENTER is pressed, the calculator returns a 1 if the inequality is correct or a 0 if the inequality is incorrect.

⟨2⟩ Graphing Inequalities

If a is a fixed real number, then any real number x located to the right of a on the number line satisfies $x > a$. The set of real numbers located to the right of a on the number line is the solution set to $x > a$. This solution set is written in set-builder notation as $\{x \mid x > a\}$, or more simply in interval notation as (a, ∞). We **graph the inequality** by graphing the solution set (a, ∞). Recall from Chapter 1 that a bracket means that an endpoint is included in an interval and a parenthesis means that an endpoint is not included in an interval. Remember also that ∞ is not a number. It simply indicates that there is no end to the interval.

EXAMPLE 2

⟨ **Helpful Hint** ⟩

A person in debt has a negative net worth. If Bob's net worth is −$8000 and Mary's net worth is −$3000, then Bob certainly has the greater debt, but we write

$$-8000 < -3000$$

because −8000 lies to the left of −3000 on the number line.

Figure 2.7

Graphing inequalities

State the solution set to each inequality in interval notation and sketch its graph.

a) $x < 5$ **b)** $-2 < x$ **c)** $x \geq 10$

Solution

a) All real numbers less than 5 satisfy $x < 5$. The solution set is the interval $(-\infty, 5)$ and the graph of the solution set is shown in Fig. 2.7.

b) The inequality $-2 < x$ indicates that x is greater than -2. The solution set is the interval $(-2, \infty)$ and the graph of the inequality is shown in Fig. 2.8.

c) All real numbers greater than or equal to 10 satisfy $x \geq 10$. The solution set is the interval $[10, \infty)$ and the graph is shown in Fig. 2.9.

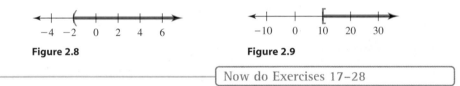

Figure 2.8   **Figure 2.9**

> Now do Exercises 17–28

⟨3⟩ Graphing Compound Inequalities

A statement involving more than one inequality is a **compound inequality.** We will study one type of compound inequality here and see other types in Section 8.1.

If a and b are real numbers and $a < b$, then the compound inequality

$$a < x < b$$

means that $a < x$ *and* $x < b$. Reading x first makes $a < x < b$ clearer:

"x is greater than a *and* x is less than b."

If x is greater than a and less than b, then x is between a and b. So the solution set to $a < x < b$ is the interval (a, b).

EXAMPLE 3

Graphing compound inequalities

State the solution set to each inequality in interval notation and sketch its graph.

a) $2 < x < 3$ **b)** $-2 \leq x < 1$

Solution

a) All real numbers between 2 and 3 satisfy $2 < x < 3$. The solution set is the interval $(2, 3)$, and the graph of the solution set is shown in Fig. 2.10 on the next page.

b) The real numbers that satisfy $-2 \leq x < 1$ are between -2 and 1, including -2 but not including 1. So the solution set is the interval $[-2, 1)$, and the graph of this compound inequality is shown in Fig. 2.11.

Figure 2.10   **Figure 2.11**

Now do Exercises 29–36

CAUTION   We write $a < x < b$ only if $a < b$, and we write $a > x > b$ only if $a > b$. Similar rules hold for $\leq$ and $\geq$. So $4 < x < 9$ and $-6 \geq x \geq -8$ are correct uses of this notation, but $5 < x < 2$ is not correct. Also, the inequalities should *not* point in opposite directions as in $5 < x > 7$.

⟨4⟩ Checking Inequalities

In Examples 2 and 3 we determined the solution sets to some inequalities. In Section 2.9, more complicated inequalities will be solved by using steps similar to those used for solving equations. In Example 4, we determine whether a given number satisfies an inequality of the type that we will be solving in Section 2.9.

EXAMPLE 4

Checking inequalities
Determine whether the given number satisfies the inequality following it.

a) $0, 2x - 3 \leq -5$ **b)** $-4, x - 5 > 2x + 1$ **c)** $\frac{13}{3}, 6 < 3x - 5 < 14$

⟨ **Calculator Close-Up** ⟩

To check 13/3 in
$$6 < 3x - 5 < 14$$
we check each part of the compound inequality separately.

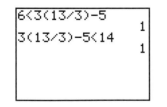

Because both parts of the compound inequality are correct, 13/3 satisfies the compound inequality.

Solution

a) Replace x by 0 in the inequality and simplify:
$$2x - 3 \leq -5$$
$$2 \cdot 0 - 3 \leq -5$$
$$-3 \leq -5 \quad \text{Incorrect}$$

Since this last inequality is incorrect, 0 is not a solution to the inequality.

b) Replace x by -4 and simplify:
$$x - 5 > 2x + 1$$
$$-4 - 5 > 2(-4) + 1$$
$$-9 > -7 \quad \text{Incorrect}$$

Since this last inequality is incorrect, -4 is not a solution to the inequality.

c) Replace x by $\frac{13}{3}$ and simplify:
$$6 < 3x - 5 < 14$$
$$6 < 3 \cdot \frac{13}{3} - 5 < 14$$
$$6 < 13 - 5 < 14$$
$$6 < 8 < 14 \quad \text{Correct}$$

Since 8 is greater than 6 and less than 14, this inequality is correct. So $\frac{13}{3}$ satisfies the original inequality.

Now do Exercises 47–64

⟨5⟩ **Writing Inequalities**

Inequalities occur in applications, just as equations do. Certain verbal phrases indicate inequalities. For example, if you must be at least 18 years old to vote, then you can vote if you are 18 or older. The phrase "at least" means "greater than or equal to." If an elevator has a capacity of at most 20 people, then it can hold 20 people or fewer. The phrase "at most" means "less than or equal to."

E X A M P L E   **5**

Writing inequalities

Write an inequality that describes each situation.

a) Lois plans to spend at most $500 on a washing machine including the 9% sales tax.

b) The length of a certain rectangle must be 4 meters longer than the width, and the perimeter must be at least 120 meters.

c) Fred made a 76 on the midterm exam. To get a B, the average of his midterm and his final exam must be between 80 and 90.

Solution

a) If x is the price of the washing machine, then $0.09x$ is the amount of sales tax. Since the total must be less than or equal to $500, the inequality is

$$x + 0.09x \leq 500.$$

b) If W represents the width of the rectangle, then $W + 4$ represents the length. Since the perimeter $(2W + 2L)$ must be greater than or equal to 120, the inequality is

$$2(W) + 2(W + 4) \geq 120.$$

c) If we let x represent Fred's final exam score, then his average is $\frac{x + 76}{2}$.

To indicate that the average is between 80 and 90, we use the compound inequality

$$80 < \frac{x + 76}{2} < 90.$$

> Now do Exercises 73–85

CAUTION   In Example 5(b) you are given that L is 4 meters longer than W. So $L = W + 4$, and you can use $W + 4$ in place of L. If you knew only that L was longer than W, then you would know only that $L > W$.

Warm-Ups ▼

Fill in the blank.

1. The symbols $<$, $\leq$, $>$, and $\geq$ are _____ symbols.

2. To graph $x \geq a$ on a number line we use a _____ at a.

3. To graph $x < a$ on a number line we use a _____ at a.

4. A _____ inequality involves more than one inequality.

5. If $a < x < b$, then x is _____ a and b.

True or false?

6. $-2 \leq -2$

7. $-5 < -6 < 7$

8. $-3 < -2 < -1$

9. The inequalities $x < 7$ and $7 > x$ have the same graph.

10. The graph of $x < -3$ includes the point at -3.

11. The number -3 is a solution to $-2 < x$.

Exercises

> **‹ Study Tips ›**
>
> - Be careful not to spend too much time on a single problem when taking a test. If a problem seems to be taking too much time, you might be on the wrong track. Be sure to finish the test.
> - Before you take a test on this chapter, work the test given in this book at the end of this chapter. This will give you a good idea of your test readiness.

‹1› Inequalities

Determine whether each of the following statements is true. See Example 1.

1. $-5 < -8$

2. $-6 > -3$

3. $-3 < 5$

4. $-6 < 0$

5. $4 \leq 4$

6. $-3 \geq -3$

7. $-6 > -5$

8. $-2 < -9$

9. $-4 \leq -3$

10. $-5 \geq -10$

11. $(-3)(4) - 1 < 0 - 3$

12. $2(4) - 6 \leq -3(5) + 1$

13. $-4(5) - 6 \geq 5(-6)$

14. $4(8) - 30 > 7(5) - 2(17)$

15. $7(4) - 12 \leq 3(9) - 2$

16. $-3(4) + 12 \leq 2(3) - 6$

‹2› Graphing Inequalities

State the solution set to each inequality in interval notation and sketch its graph. See Example 2.

17. $x \leq 3$

18. $x \leq -7$

19. $x > -2$

20. $x > 4$

21. $-1 > x$

22. $0 > x$

23. $-2 \leq x$

24. $-5 \geq x$

25. $x \geq \dfrac{1}{2}$

26. $x \geq -\dfrac{2}{3}$

27. $x \leq 5.3$

28. $x \leq -3.4$

‹3› Graphing Compound Inequalities

State the solution set to each inequality in interval notation and sketch its graph. See Example 3.

29. $-3 < x < 1$

30. $0 < x < 5$

31. $3 \leq x \leq 7$

32. $-3 \leq x \leq -1$

33. $-5 \leq x < 0$

34. $-2 < x \le 2$

35. $40 < x \le 100$

36. $0 \le x < 600$

For each graph, write the corresponding inequality and the solution set to the inequality using interval notation.

37.
$-3\,-2\,-1\;\;0\;\;1\;\;2\;\;3\;\;4\;\;5\;\;6\;\;7$

38.
$-4\,-3\,-2\,-1\;\;0\;\;1\;\;2\;\;3\;\;4\;\;5\;\;6$

39.
$-6\,-5\,-4\,-3\,-2\,-1\;\;0\;\;1\;\;2\;\;3\;\;4$

40.
$-5\,-4\,-3\,-2\,-1\;\;0\;\;1\;\;2\;\;3\;\;4\;\;5$

41.
$-5\,-4\,-3\,-2\,-1\;\;0\;\;1\;\;2\;\;3\;\;4\;\;5$

42.
$-5\,-4\,-3\,-2\,-1\;\;0\;\;1\;\;2\;\;3\;\;4\;\;5$

43.
$-6\;\;-4\;\;-2\;\;\;0\;\;\;2\;\;\;4\;\;\;6\;\;\;8$

44.
$-5\,-4\,-3\,-2\,-1\;\;0\;\;1\;\;2\;\;3\;\;4\;\;5$

45.
$-5\,-4\,-3\,-2\,-1\;\;0\;\;1\;\;2\;\;3\;\;4\;\;5$

46.
$-5\,-4\,-3\,-2\,-1\;\;0\;\;1\;\;2\;\;3\;\;4\;\;5$

⟨4⟩ Checking Inequalities

Determine whether the given number satisfies the inequality following it. See Example 4.

47. $-9,\ -x > 3$

48. $5,\ -3 < -x$

49. $-2,\ 5 \le x$

50. $4,\ 4 \ge x$

51. $-6,\ 2x - 3 > -11$

52. $4,\ 3x - 5 < 7$

53. $3,\ -3x + 4 > -7$

54. $-4,\ -5x + 1 > -5$

55. $0,\ 3x - 7 \le 5x - 7$

56. $0,\ 2x + 6 \ge 4x - 9$

57. $2.5,\ -10x + 9 \le 3(x + 3)$

58. $1.5,\ 2x - 3 \le 4(x - 1)$

59. $-7,\ -5 < x < 9$

60. $-9,\ -6 \le x \le 40$

61. $-2,\ -3 \le 2x + 5 \le 9$

62. $-5,\ -3 < -3x - 7 \le 8$

63. $-3.4,\ -4.25x - 13.29 < 0.89$

64. $4.8,\ 3.25x - 14.78 \le 1.3$

For each inequality, determine which of the numbers -5.1, 0, and 5.1 satisfies the inequality.

65. $x > -5$

66. $x \le 0$

67. $5 < x$

68. $-5 > x$

69. $5 < x < 7$

70. $5 < -x < 7$

71. $-6 < -x < 6$

72. $-5 \le x - 0.1 \le 5$

⟨5⟩ Writing Inequalities

Write an inequality to describe each situation. Do not solve. See Example 5.

73. *Sales tax.* At an 8% sales tax rate, Susan paid more than $1500 sales tax when she purchased her new Camaro. Let p represent the price of the Camaro.

74. *Internet shopping.* Carlos paid less than $1000 including $40 for shipping and 9% sales tax when he bought his new computer. Let p represent the price of the computer.

75. *Fine dining.* At Burger Brothers the price of a hamburger is twice the price of an order of French fries, and the price of a Coke is $0.25 more than the price of the fries. Burger Brothers advertises that you can get a complete meal (burger, fries, and Coke) for under $2.00. Let p represent the price of an order of fries.

76. *Cats and dogs.* Willow Creek Kennel boards only cats and dogs. One Friday night there were twice as many dogs as cats in the kennel and at least 30 animals spent the night there. Let d represent the number of dogs.

77. *Barely passing.* Travis made 44 and 72 on the first two tests in algebra and has one test remaining. The average on the three tests must be at least 60 for Travis to pass the course. Let s represent his score on the last test.

78. *Ace the course.* Florence made 87 on her midterm exam in psychology. The average of her midterm and her final must be at least 90 to get an A in the course. Let *s* represent her score on the final.

79. *Coast to coast.* On Howard's recent trip from Bangor to San Diego, he drove for 8 hours each day and traveled between 396 and 453 miles each day. Let *R* represent his average speed for each day.

80. *Mother's Day present.* Bart and Betty are looking at color televisions that range in price from $399.99 to $579.99. Bart can afford more than Betty and has agreed to spend $100 more than Betty when they purchase this gift for their mother. Let *b* represent Betty's portion of the gift.

81. *Positioning a ladder.* Write an inequality in the variable *x* for the degree measure of the angle at the base of the ladder shown in the figure, given that the angle at the base must be between 60° and 70°.

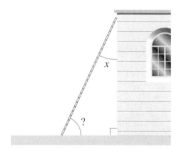

Figure for Exercise 81

82. *Building a ski ramp.* Write an inequality in the variable *x* for the degree measure of the smallest angle of the triangle shown in the figure, given that the degree measure of the smallest angle is at most 30°.

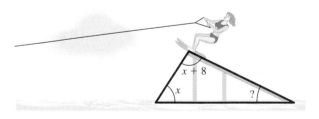

Figure for Exercise 82

83. *Maximum girth.* United Parcel Service defines the girth of a box as the sum of the length, twice the width, and twice the height. The maximum girth that UPS will ship is 130 in.

a) If a box has a length of 45 in. and a width of 30 in., then what inequality must be satisfied by the height?

b) The accompanying graph shows the girth of a box with a length of 45 in., a width of 30 in., and height of *h* in. Use the graph to estimate the maximum height that is allowed for this box.

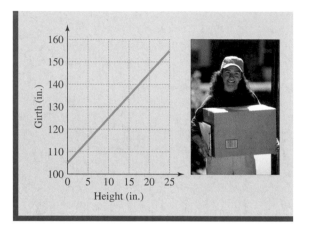

Figure for Exercise 83

84. *Batting average.* Near the end of the season a professional baseball player has 93 hits in 317 times at bat for an average of 93/317 or 0.293. He gets a $1 million bonus if his season average is over 0.300. He estimates that he will bat 20 more times before the season ends. Let *x* represent the number of hits in the last 20 at bats of the season.

a) Write an inequality that must be satisfied for him to get the bonus.

b) Use the accompanying graph to estimate the number of hits in 337 at bats that will put his average over 0.300.

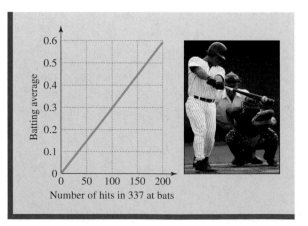

Figure for Exercise 84

Solve.

85. ***Bicycle gear ratios.*** The gear ratio r for a bicycle is defined by the formula

$$r = \frac{Nw}{n},$$

where N is the number of teeth on the chainring (by the pedal), n is the number of teeth on the cog (by the wheel), and w is the wheel diameter in inches (*Cycling,* Burkett and Darst). The accompanying chart gives uses for the various gear ratios. A bicycle with a 27-inch-diameter wheel has 50 teeth on the chainring and 17 teeth on the cog. Find the gear ratio and indicate what this gear ratio is good for.

Ratio	Use
$r > 90$	hard pedaling on level ground
$70 < r \leq 90$	moderate effort on level ground
$50 < r \leq 70$	mild hill climbing
$35 < r \leq 50$	long hill climbing with load

Figure for Exercise 85

Math *at Work*   Body Mass Index

Medical professionals say that two-thirds of all Americans are overweight and excess weight has about the same effect on life expectancy as smoking. How can you tell if you are overweight or normal? Body mass index (BMI) can help you decide. To determine BMI divide your weight in kilograms by the square of your height in meters. Don't know your weight and height in the metric system? Then use the formula $BMI = 703W/H^2$, where W is your weight in pounds and H is your height in inches.

If $23 < BMI < 25$, then you are probably not overweight. If $BMI \geq 26$, then you are probably overweight and are statistically likely to have a lower life expectancy. According to the National Heart, Lung, and Blood Institute, you are overweight if $25 < BMI < 29.9$ and obese if $BMI \geq 30$. If your BMI is between 17 and 22, your life span might be longer than average. Men are usually happy with a BMI between 23 and 25 and women like to see their BMI between 20 and 22. However, BMI does not distinguish between muscle and fat and can wrongly suggest that a person with a short muscular build is overweight. Also, the BMI does not work well for children, because normal varies with age.

If you want to learn more about body mass index or don't want to do the calculations yourself, then check out any of the numerous websites that discuss BMI and even have online BMI calculators. Just do a search for body mass index.

2.9   Solving Inequalities and Applications

In This Section

⟨1⟩ **Rules for Inequalities**

⟨2⟩ **Solving Inequalities**

⟨3⟩ **Applications**

To solve equations, we write a sequence of equivalent equations that ends in a very simple equation whose solution is obvious. In this section, you will learn that the procedure for solving inequalities is the same. However, the rules for performing operations on each side of an inequality are slightly different from the rules for equations.

⟨1⟩ Rules for Inequalities

Equivalent inequalities are inequalities that have exactly the same solutions. Inequalities such as $x > 3$ and $x + 2 > 5$ are equivalent because any number that is larger than 3 certainly satisfies $x + 2 > 5$ and any number that satisfies $x + 2 > 5$ must certainly be larger than 3.

We can get equivalent inequalities by performing operations on each side of an inequality just as we do for solving equations. If we start with the inequality $6 < 10$ and add 2 to each side, we get the true statement $8 < 12$. Examine the results of performing the same operation on each side of $6 < 10$.

Perform these operations on each side:

	Add 2	Subtract 2	Multiply by 2	Divide by 2
Start with $6 < 10$	$8 < 12$	$4 < 8$	$12 < 20$	$3 < 5$

All of the resulting inequalities are correct. Now if we repeat these operations using -2, we get the following results.

Perform these operations on each side:

	Add -2	Subtract -2	Multiply by -2	Divide by -2
Start with $6 < 10$	$4 < 8$	$8 < 12$	$-12 > -20$	$-3 > -5$

Notice that the direction of the inequality symbol is the same for all of the results except the last two. When we multiplied each side by -2 and when we divided each side by -2, we had to reverse the inequality symbol to get a correct result. These tables illustrate the rules for solving inequalities.

> **Addition Property of Inequality**
>
> If we add the same number to each side of an inequality, we get an equivalent inequality. If $a < b$, then $a + c < b + c$.

The addition property of inequality also enables us to subtract the same number from each side of an inequality because subtraction is defined in terms of addition.

> **Multiplication Property of Inequality**
>
> If we multiply each side of an inequality by the same *positive* number, we get an equivalent inequality. If $a < b$ and $c > 0$, then $ac < bc$. If we multiply each side of an inequality by the same *negative* number and *reverse the inequality symbol*, we get an equivalent inequality. If $a < b$ and $c < 0$, then $ac > bc$.

The multiplication property of inequality also enables us to divide each side of an inequality by a nonzero number because division is defined in terms of multiplication. So if we multiply or divide each side by a negative number, the inequality symbol is reversed.

⟨ **Helpful Hint** ⟩

You can think of an inequality like a seesaw that is out of balance.

$50 > 20$

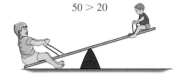

If the same weight is added to or subtracted from each side, it will remain in the same state of imbalance.

⟨ **Helpful Hint** ⟩

Changing the signs of numbers changes their relative position on the number line. For example, 3 lies to the left of 5 on the number line, but -3 lies to the right of -5. So $3 < 5$, but $-3 > -5$. Since multiplying and dividing by a negative cause sign changes, these operations reverse the inequality.

EXAMPLE 1

Writing equivalent inequalities

Write the appropriate inequality symbol in the blank so that the two inequalities are equivalent.

   a) $x + 3 > 9$, x _____ 6 **b)** $-2x \leq 6$, x _____ -3

Solution

   a) If we subtract 3 from each side of $x + 3 > 9$, we get the equivalent inequality $x > 6$.

   b) If we divide each side of $-2x \leq 6$ by -2, we get the equivalent inequality $x \geq -3$.

> Now do Exercises 1–10

CAUTION   We use the properties of inequality just as we use the properties of equality. However, when we multiply or divide each side by a negative number, we must reverse the inequality symbol.

⟨2⟩ Solving Inequalities

To solve inequalities, we use the properties of inequality to isolate x on one side.

EXAMPLE 2

Isolating the variable on the left side

Solve the inequality $4x - 5 > 19$. State the solution set using interval notation and sketch its graph.

Solution

$$4x - 5 > 19 \qquad \text{Original inequality}$$
$$4x - 5 + 5 > 19 + 5 \qquad \text{Add 5 to each side.}$$
$$4x > 24 \qquad \text{Simplify.}$$
$$x > 6 \qquad \text{Divide each side by 4.}$$

Since the last inequality is equivalent to the first, it has the same solution set as the first. So the solution set to $4x - 5 > 19$ is $(6, \infty)$. The graph is shown in Fig. 2.12.

> Now do Exercises 11–12

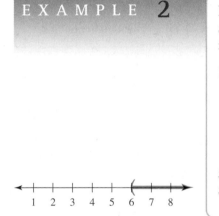

Figure 2.12

⟨ Calculator Close-Up ⟩

You can use the TABLE feature of a graphing calculator to numerically support the solution to the inequality $4x - 5 > 19$ in Example 2. Use the Y = key to enter the equation $y_1 = 4x - 5$.

Next, use TBLSET to set the table so that the values of x start at 4.5 and the change in x is 0.5.

```
TABLE SETUP
 TblStart=4.5
 ΔTbl=.5
Indpnt: Auto Ask
Depend: Auto Ask
```

Finally, press TABLE to see lists of x-values and the corresponding y-values.

Notice that when x is larger than 6, y_1 (or $4x - 5$) is larger than 19. The table verifies or supports the algebraic solution, but it should not replace the algebraic method.

```
Plot1 Plot2 Plot3
\Y1■4X-5
\Y2=
\Y3=
\Y4=
\Y5=
\Y6=
\Y7=
```

Remember that $5 < x$ is equivalent to $x > 5$. So the variable can be isolated on the right side of an inequality as shown in Example 3.

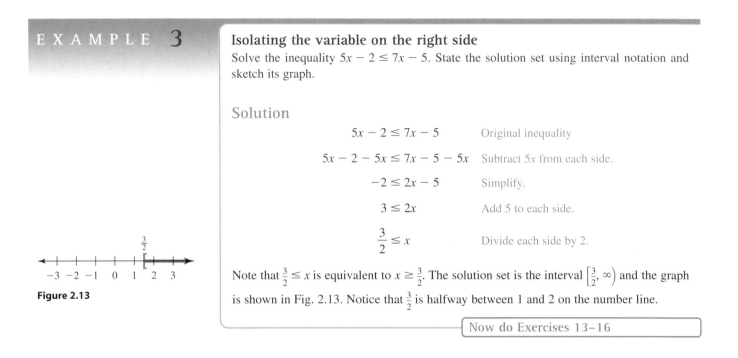

E X A M P L E   **3**

Isolating the variable on the right side

Solve the inequality $5x - 2 \le 7x - 5$. State the solution set using interval notation and sketch its graph.

Solution

$$5x - 2 \le 7x - 5 \qquad \text{Original inequality}$$

$$5x - 2 - 5x \le 7x - 5 - 5x \qquad \text{Subtract } 5x \text{ from each side.}$$

$$-2 \le 2x - 5 \qquad \text{Simplify.}$$

$$3 \le 2x \qquad \text{Add 5 to each side.}$$

$$\frac{3}{2} \le x \qquad \text{Divide each side by 2.}$$

Note that $\frac{3}{2} \le x$ is equivalent to $x \ge \frac{3}{2}$. The solution set is the interval $\left[\frac{3}{2}, \infty\right)$ and the graph is shown in Fig. 2.13. Notice that $\frac{3}{2}$ is halfway between 1 and 2 on the number line.

Figure 2.13

Now do Exercises 13–16

Rewriting $\frac{3}{2} \le x$ as $x \ge \frac{3}{2}$ in Example 3 is not "reversing the inequality." Multiplying or dividing each side of $\frac{3}{2} \le x$ by a negative number would reverse the inequality. For example, multiplying by -1 yields $-\frac{3}{2} \ge -x$. In Example 4, we divide each side of an inequality by a negative number and reverse the inequality symbol.

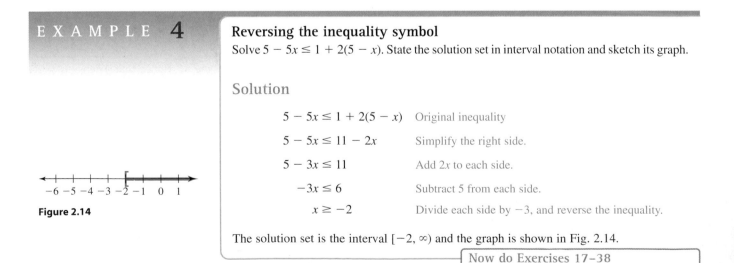

E X A M P L E   **4**

Reversing the inequality symbol

Solve $5 - 5x \le 1 + 2(5 - x)$. State the solution set in interval notation and sketch its graph.

Solution

$$5 - 5x \le 1 + 2(5 - x) \qquad \text{Original inequality}$$

$$5 - 5x \le 11 - 2x \qquad \text{Simplify the right side.}$$

$$5 - 3x \le 11 \qquad \text{Add } 2x \text{ to each side.}$$

$$-3x \le 6 \qquad \text{Subtract 5 from each side.}$$

$$x \ge -2 \qquad \text{Divide each side by } -3, \text{ and reverse the inequality.}$$

The solution set is the interval $[-2, \infty)$ and the graph is shown in Fig. 2.14.

Figure 2.14

Now do Exercises 17–38

We can use the rules for solving inequalities on the compound inequalities that we studied in Section 2.8.

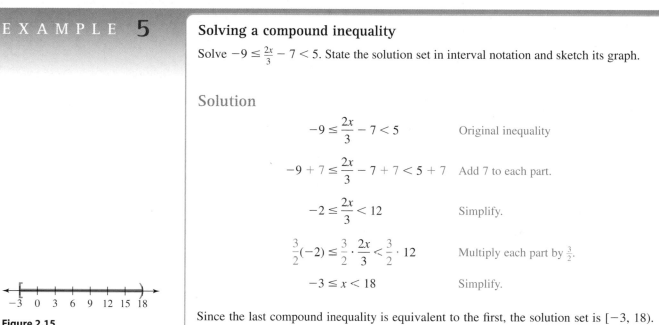

EXAMPLE 5

Solving a compound inequality

Solve $-9 \le \frac{2x}{3} - 7 < 5$. State the solution set in interval notation and sketch its graph.

Solution

$$-9 \le \frac{2x}{3} - 7 < 5 \qquad \text{Original inequality}$$

$$-9 + 7 \le \frac{2x}{3} - 7 + 7 < 5 + 7 \qquad \text{Add 7 to each part.}$$

$$-2 \le \frac{2x}{3} < 12 \qquad \text{Simplify.}$$

$$\frac{3}{2}(-2) \le \frac{3}{2} \cdot \frac{2x}{3} < \frac{3}{2} \cdot 12 \qquad \text{Multiply each part by } \tfrac{3}{2}.$$

$$-3 \le x < 18 \qquad \text{Simplify.}$$

Since the last compound inequality is equivalent to the first, the solution set is $[-3, 18)$. The graph is shown in Fig. 2.15.

Figure 2.15

Now do Exercises 39–42

CAUTION There are many negative numbers in Example 5, but the inequality was not reversed, since we did not multiply or divide by a negative number. An inequality is reversed only if you multiply or divide by a negative number.

EXAMPLE 6

Reversing inequality symbols in a compound inequality

Solve $-3 \le 5 - x \le 5$. State the solution set in interval notation and sketch its graph.

Solution

$$-3 \le 5 - x \le 5 \qquad \text{Original inequality}$$

$$-3 - 5 \le 5 - x - 5 \le 5 - 5 \qquad \text{Subtract 5 from each part.}$$

$$-8 \le -x \le 0 \qquad \text{Simplify.}$$

$$(-1)(-8) \ge (-1)(-x) \ge (-1)(0) \qquad \text{Multiply each part by } -1,$$
$$\text{reversing the inequality symbols.}$$

$$8 \ge x \ge 0$$

Figure 2.16

It is customary to write $8 \geq x \geq 0$ with the smallest number on the left:

$$0 \leq x \leq 8$$

Since the last compound inequality is equivalent to the first, the solution set is [0, 8]. The graph is shown in Fig. 2.16.

Now do Exercises 43–52

⟨3⟩ Applications

Example 7 shows how inequalities can be used in applications.

EXAMPLE 7

Averaging test scores

Mei Lin made a 76 on the midterm exam in history. To get a B, the average of her midterm and her final exam must be between 80 and 90. For what range of scores on the final exam will she get a B?

Solution

Let x represent the final exam score. Her average is then $\frac{x+76}{2}$. The inequality expresses the fact that the average must be between 80 and 90:

$$80 < \frac{x+76}{2} < 90$$

$$2(80) < 2\left(\frac{x+76}{2}\right) < 2(90) \qquad \text{Multiply each part by 2.}$$

$$160 < x + 76 < 180 \qquad \text{Simplify.}$$

$$160 - 76 < x + 76 - 76 < 180 - 76 \qquad \text{Subtract 76 from each part.}$$

$$84 < x < 104 \qquad \text{Simplify.}$$

The last inequality indicates that Mei Lin's final exam score must be between 84 and 104.

Now do Exercises 59–74

> **⟨ Helpful Hint ⟩**
>
> Remember that all inequality symbols in a compound inequality must point in the same direction. We usually have them all point to the left so that the numbers are increasing in size as you go from left to right in the inequality.

Warm-Ups ▼

Fill in the blank.

1. _____ inequalities have the same solution set.
2. According to the _____ property of inequality, adding the same number to both sides of an inequality produces an equivalent inequality.
3. According to the _____ property of inequality, the inequality symbol is reversed when multiplying by a negative number and not reversed when multiplying by a positive number.

True or false?

4. The inequality $2x > 18$ is equivalent to $x > 9$.
5. The inequality $x - 5 > 0$ is equivalent to $x < 5$.
6. The inequality $-2x \leq 6$ is equivalent to $x \geq -3$.
7. The statement "x is at most 7" is written as $x < 7$.
8. The statement "x is not more than 85" is written as $x \leq 85$.
9. The inequality $-3 > x > -9$ is equivalent to $-9 < x < -3$.

‹1› Rules for Inequalities

Write the appropriate inequality symbol in the blank so that the two inequalities are equivalent. See Example 1.

1. $x + 7 > 0$
   $x \_\_ -7$

2. $x - 6 < 0$
   $x \_\_ 6$

3. $9 \le 3w$
   $w \_\_ 3$

4. $10 \ge 5z$
   $z \_\_ 2$

5. $-x < 8$
   $x \_\_ -8$

6. $-x \ge -3$
   $x \_\_ 3$

7. $-4k < -4$
   $k \_\_ 1$

8. $-9t > 27$
   $t \_\_ -3$

9. $-\frac{1}{2}y \ge 4$
   $y \_\_ -8$

10. $-\frac{1}{3}x \le 4$
   $x \_\_ -12$

‹2› Solving Inequalities

Solve each inequality. State the solution set in interval notation and sketch its graph. See Examples 2–4.

11. $x + 3 > 0$

12. $x + 9 \le -8$

13. $-3 < w - 1$

14. $9 > w - 12$

15. $8 > 2b$

16. $35 < 7b$

17. $-8z \le 4$

18. $-4y \ge -10$

19. $3y - 2 < 7$

20. $2y - 5 > -9$

21. $3 - 9z \le 6$

22. $5 - 6z \ge 13$

23. $6 > -r + 3$

24. $6 \le 12 - r$

25. $5 - 4p > -8 - 3p$

26. $7 - 9p > 11 - 8p$

27. $-\frac{5}{6}q \ge -20$

28. $-\frac{2}{3}q \ge -4$

29. $1 - \frac{1}{4}t \ge \frac{1}{8}$

30. $\frac{1}{6} - \frac{1}{3}t > 0$

31. $0.1x + 0.35 > 0.2$

32. $1 - 0.02x \le 0.6$

33. $2x + 5 < x - 6$

34. $3x - 4 < 2x + 9$

35. $x - 4 < 2(x + 3)$

36. $2x + 3 < 3(x - 5)$

37. $0.52x - 35 < 0.45x + 8$

38. $8455(x - 3.4) > 4320$

Solve each compound inequality. State the solution set in interval notation and sketch its graph. See Examples 5 and 6.

39. $5 < x - 3 < 7$

40. $2 < x - 5 < 6$

41. $3 < 2v + 1 < 10$

42. $-3 < 3v + 4 < 7$

43. $-4 \leq 5 - k \leq 7$

44. $2 \leq 3 - k \leq 8$

45. $-2 < 7 - 3y \leq 22$

46. $-1 \leq 1 - 2y < 3$

47. $5 < \dfrac{2u}{3} - 3 < 17$

48. $-4 < \dfrac{3u}{4} - 1 < 11$

49. $-2 < \dfrac{4m - 4}{3} \leq \dfrac{2}{3}$

50. $0 \leq \dfrac{3 - 2m}{2} < 9$

51. $0.02 < 0.54 - 0.0048x < 0.05$

52. $0.44 < \dfrac{34.55 - 22.3x}{124.5} < 0.76$

Solve each inequality. State the solution set in interval notation and sketch its graph.

53. $\dfrac{1}{2}x - 1 \leq 4 - \dfrac{1}{3}x$

54. $\dfrac{y}{4} - \dfrac{5}{12} \geq \dfrac{y}{3} + \dfrac{1}{4}$

55. $\dfrac{1}{2}\left(x - \dfrac{1}{4}\right) > \dfrac{1}{4}\left(6x - \dfrac{1}{2}\right)$

56. $-\dfrac{1}{2}\left(z - \dfrac{2}{5}\right) < \dfrac{2}{3}\left(\dfrac{3}{4}z - \dfrac{6}{5}\right)$

57. $\dfrac{1}{3} < \dfrac{1}{4}x - \dfrac{1}{6} < \dfrac{7}{12}$

58. $-\dfrac{3}{5} < \dfrac{1}{5} - \dfrac{2}{15}w < -\dfrac{1}{3}$

⟨3⟩   **Applications**

Solve each of the following problems by using an inequality. See Example 7.

59. *Boat storage.* The length of a rectangular boat storage shed must be 4 meters more than the width, and the perimeter must be at least 120 meters. What is the range of values for the width?

60. *Fencing a garden.* Elka is planning a rectangular garden that is to be twice as long as it is wide. If she can afford to buy at most 180 feet of fencing, then what are the possible values for the width?

Photo for Exercise 60

61. *Car shopping.* Harold Ivan is shopping for a new car. In addition to the price of the car, there is a 5% sales tax and a $144 title and license fee. If Harold Ivan decides that he will spend less than $9970 total, then what is the price range for the car?

62. Car selling. Ronald wants to sell his car through a broker who charges a commission of 10% of the selling price. Ronald still owes $11,025 on the car. Ronald must get enough to at least pay off the loan. What is the range of the selling price?

63. Microwave oven. Sherie is going to buy a microwave in a city with an 8% sales tax. She has at most $594 to spend. In what price range should she look?

64. Dining out. At Burger Brothers the price of a hamburger is twice the price of an order of French fries, and the price of a Coke is $0.40 more than the price of the fries. Burger Brothers advertises that you can get a complete meal (burger, fries, and Coke) for under $4.00. What is the price range of an order of fries?

65. Averaging test scores. Tilak made 44 and 72 on the first two tests in algebra and has one test remaining. For Tilak to pass the course, the average on the three tests must be at least 60. For what range of scores on his last test will Tilak pass the course?

66. Averaging income. Helen earned $400 in January, $450 in February, and $380 in March. To pay all of her bills, she must average at least $430 per month. For what income in April would her average for the 4 months be at least $430?

67. Going for a C. Professor Williams gives only a midterm exam and a final exam. The semester average is computed by taking $\frac{1}{3}$ of the midterm exam score plus $\frac{2}{3}$ of the final exam score. To get a C, Stacy must have a semester average between 70 and 79 inclusive. If Stacy scored only 48 on the midterm, then for what range of scores on the final exam will Stacy get a C?

68. Different weights. Professor Williamson counts his midterm as $\frac{2}{3}$ of the grade and his final as $\frac{1}{3}$ of the grade. Wendy scored only 48 on the midterm. What range of scores on the final exam would put Wendy's average between 70 and 79 inclusive? Compare to the previous exercise.

69. Average driving speed. On Halley's recent trip from Bangor to San Diego, she drove for 8 hours each day and traveled between 396 and 453 miles each day. In what range was her average speed for each day of the trip?

70. Driving time. On Halley's trip back to Bangor, she drove at an average speed of 55 mph every day and traveled between 330 and 495 miles per day. In what range was her daily driving time?

71. Sailboat navigation. As the sloop sailed north along the coast, the captain sighted the lighthouse at points A and B as shown in the figure. If the degree measure of the angle at the lighthouse is less than 30°, then what are the possible values for x?

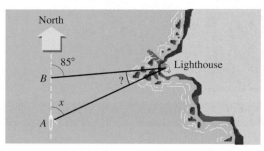

Figure for Exercise 71

72. Flight plan. A pilot started at point A and flew in the direction shown in the diagram for some time. At point B she made a 110° turn to end up at point C, due east of where she started. If the measure of angle C is less than 85°, then what are the possible values for x?

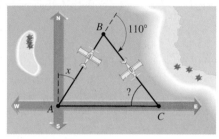

Figure for Exercise 72

73. Bicycle gear ratios. The gear ratio r for a bicycle is defined by the formula

$$r = \frac{Nw}{n},$$

where N is the number of teeth on the chainring (by the pedal), n is the number of teeth on the cog (by the wheel), and w is the wheel diameter in inches (www.sheldonbrown.com/gears).

a) If the wheel has a diameter of 27 in. and there are 12 teeth on the cog, then for what number of teeth on the chainring is the gear ratio between 60 and 80?

b) If a bicycle has 48 teeth on the chainring and 17 teeth on the cog, then for what diameter wheel is the gear ratio between 65 and 70?

c) If a bicycle has a 26-in.-diameter wheel and 40 teeth on the chainring, then for what number of teeth on the cog is the gear ratio less than 75?

74. Virtual demand. The weekly demand (the number bought by consumers) for the Acme Virtual Pet is given by the formula

$$d = 9000 - 60p$$

where p is the price for each in dollars.

a) What is the demand when the price is $30 each?

b) In what price range will the demand be above 6000?

Chapter 2   Wrap-Up

Summary

Equations		**Examples**
Linear equation	An equation of the form $ax = b$ with $a \neq 0$	$3x = 7$
Identity	An equation that is satisfied by every number for which both sides are defined	$x + x = 2x$
Conditional equation	An equation that has at least one solution but is not an identity	$5x - 10 = 0$
Inconsistent equation	An equation that has no solution	$x = x + 1$
Equivalent equations	Equations that have exactly the same solutions	$2x + 1 = 5$ $2x = 4$
Properties of equality	If the same number is added to or subtracted from each side of an equation, the resulting equation is equivalent to the original equation.	$x - 5 = -9$ $x = -4$
	If each side of an equation is multiplied or divided by the same nonzero number, the resulting equation is equivalent to the original equation.	$9x = 27$ $x = 3$
Solving equations	1. Remove parentheses by using the distributive property and then combine like terms to simplify each side as much as possible. 2. Use the addition property of equality to get like terms from opposite sides onto the same side so that they may be combined. 3. The multiplication property of equality is generally used last. 4. Check that the solution satisfies the original equation.	$2(x - 3) = -7 + 3(x - 1)$ $2x - 6 = -10 + 3x$ $-x - 6 = -10$ $-x = -4$ $x = 4$ *Check:* $2(4 - 3) = -7 + 3(4 - 1)$ $2 = 2$

Formulas and Functions		**Examples**
Formula	An equation involving two or more variables	$D = RT$
Solving for a specified variable	Rewrite the formula so that the specified variable is isolated on the left side and does not occur on the right side.	Solve for R. $R = \dfrac{D}{T}$

| Functions | A function is a rule for determining uniquely the value of one variable a from the value(s) of one or more other variable(s). | D is a function of R and T. $D = RT$ is a function. |

Applications

Steps in solving applied problems	1. Read the problem.
	2. If possible, draw a diagram to illustrate the problem.
	3. Choose a variable and write down what it represents.
	4. Represent any other unknowns in terms of that variable.
	5. Write an equation that describes the situation.
	6. Solve the equation.
	7. Answer the original question.
	8. Check your answer by using it to solve the original problem (not the equation).

Inequalities

Examples

| Properties of inequality | Addition, subtraction, multiplication, and division may be performed on each side of an inequality, just as we do in solving equations, with one exception. When multiplying or dividing by a negative number, the inequality symbol is reversed. | $-3x + 1 > 7$ $-3x > 6$ $x < -2$ |

Enriching Your Mathematical Word Power

Fill in the blank.

1. A(n) _____ is a sentence that expresses the equality of two algebraic expressions.

2. A(n) _____ equation has the form $ax = b$ with $a \neq 0$.

3. An _____ is satisfied by all real numbers for which both sides are defined.

4. A _____ equation has at least one solution but is not an identity.

5. A(n) _____ equation has no solutions.

6. Equations that have the same solution are _____ equations.

7. An equation involving two or more variables is a(n) _____ equation or _____ .

8. If the value of y can be determined from the value of x, then y is a _____ of x.

9. Angles whose degree measures total $90°$ are _____ angles.

10. Angles whose degree measures total $180°$ are _____ angles.

11. Motion at a constant rate is _____ motion.

12. A statement that uses $<$, $\leq$, $>$, or $\geq$ is an _____ .

13. Inequalities that have the same solution set are _____ .

Review Exercises

2.1 The Addition and Multiplication Properties of Equality

Solve each equation and check your answer.

1. $x - 23 = 12$

2. $14 = 18 + y$

3. $\frac{2}{3}u = -4$

4. $-\frac{3}{8}r = 15$

5. $-5y = 35$

6. $-12 = 6h$

7. $6m = 13 + 5m$

8. $19 - 3n = -2n$

2.2 Solving General Linear Equations

Solve each equation and check your answer.

9. $2x - 5 = 9$

10. $5x - 8 = 38$

11. $3p - 14 = -4p$

12. $36 - 9y = 3y$

13. $2z + 12 = 5z - 9$

14. $15 - 4w = 7 - 2w$

15. $2(h - 7) = -14$

16. $2(t - 7) = 0$

17. $3(w - 5) = 6(w + 2) - 3$

18. $2(a - 4) + 4 = 5(9 - a)$

2.3 More Equations

Solve each equation. Identify each equation as a conditional equation, an inconsistent equation, or an identity.

19. $2(x - 7) - 5 = 5 - (3 - 2x)$

20. $2(x - 7) + 5 = -(9 - 2x)$

21. $2(w - w) = 0$

22. $2y - y = 0$

23. $\dfrac{3r}{3r} = 1$

24. $\dfrac{3t}{3} = 1$

25. $\dfrac{1}{2}a - 5 = \dfrac{1}{3}a - 1$

26. $\dfrac{1}{2}b - \dfrac{1}{2} = \dfrac{1}{4}b$

27. $0.06q + 14 = 0.3q - 5.2$

28. $0.05(z + 20) = 0.1z - 0.5$

29. $0.05(x + 100) + 0.06x = 115$

30. $0.06x + 0.08(x + 1) = 0.41$

Solve each equation.

31. $2x + \dfrac{1}{2} = 3x + \dfrac{1}{4}$

32. $5x - \dfrac{1}{3} = 6x - \dfrac{1}{2}$

33. $\dfrac{x}{2} - \dfrac{3}{4} = \dfrac{x}{6} + \dfrac{1}{8}$

34. $\dfrac{1}{3} - \dfrac{x}{5} = \dfrac{1}{2} - \dfrac{x}{10}$

35. $\dfrac{5}{6}x = -\dfrac{2}{3}$

36. $-\dfrac{2}{3}x = \dfrac{3}{4}$

37. $-\dfrac{1}{2}(x - 10) = \dfrac{3}{4}x$

38. $-\dfrac{1}{3}(6x - 9) = 23$

39. $3 - 4(x - 1) + 6 = -3(x + 2) - 5$

40. $6 - 5(1 - 2x) + 3 = -3(1 - 2x) - 1$

41. $5 - 0.1(x - 30) = 18 + 0.05(x + 100)$

42. $0.6(x - 50) = 18 - 0.3(40 - 10x)$

2.4 Formulas and Functions

Solve each equation for x.

43. $ax + b = 0$

44. $mx + e = t$

45. $ax - 2 = b$

46. $b = 5 - x$

47. $LWx = V$

48. $3xy = 6$

49. $2x - b = 5x$

50. $t - 5x = 4x$

In each case find a formula that expresses y as a function of x. Write the answer in the form y = mx + b where m and b are real numbers.

51. $5x + 2y = 6$

52. $5x - 3y + 9 = 0$

53. $y - 1 = -\dfrac{1}{2}(x - 6)$

54. $y + 6 = \dfrac{1}{2}(x + 8)$

55. $\dfrac{1}{2}x + \dfrac{1}{4}y = 4$

56. $-\dfrac{x}{3} + \dfrac{y}{2} = 1$

Find the value of y in each formula if x = −3.

57. $y = 3x - 4$

58. $2x - 3y = -7$

59. $5xy = 6$

60. $3xy - 2x = -12$

61. $y - 3 = -2(x - 4)$

62. $y + 1 = 2(x - 5)$

Fill in the tables using the given formulas.

63. $y = -5x + 10$

x	y
−1	
0	
1	
2	
3	

64. $y = 2x - 4$

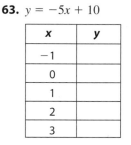

x	y
0	
1	
2	
3	
4	

65. $y = \dfrac{2}{3}x - 1$

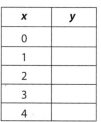

x	y
−3	
0	
3	
6	

66. $y = 10x + 100$

x	y
−20	
−10	
0	
10	

2.5 Translating Verbal Expressions into Algebraic Expressions

Translate each verbal expression into an algebraic expression.

67. The sum of a number and 9

68. The product of a number and 7

69. Two numbers that differ by 8

70. Two numbers with a sum of 12

71. Sixty-five percent of a number

72. One-half of a number

Identify the variable, and write an equation that describes each situation. Do not solve the equation.

73. One side of a rectangle is 5 feet longer than the other, and the area is 98 square feet.

74. One side of a rectangle is one foot longer than twice the other side, and the perimeter is 56 feet.

75. By driving 10 miles per hour slower than Jim, Barbara travels the same distance in 3 hours as Jim does in 2 hours.

76. Gladys and Ned drove 840 miles altogether, with Gladys averaging 5 miles per hour more in her 6 hours at the wheel than Ned did in his 5 hours at the wheel.

77. The sum of three consecutive even integers is 90.

78. The sum of two consecutive odd integers is 40.

79. The three angles of a triangle have degree measures of t, $2t$, and $t - 10$.

80. Two complementary angles have degree measures p and $3p - 6$.

2.6–7 Applications

Solve each problem.

81. *Odd integers.* If the sum of three consecutive odd integers is 237, then what are the integers?

82. *Even integers.* Find two consecutive even integers that have a sum of 450.

83. *Driving to the shore.* Lawanda and Betty both drive the same distance to the shore. By driving 15 miles per hour faster than Betty, Lawanda can get there in 3 hours while Betty takes 4 hours. How fast does each of them drive?

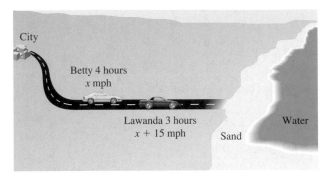

City

Betty 4 hours
x mph

Lawanda 3 hours
x + 15 mph

Water

Sand

Figure for Exercise 83

84. *Rectangular lot.* The length of a rectangular lot is 50 feet more than the width. If the perimeter is 500 feet, then what are the length and width?

85. *Combined savings.* Wanda makes $6000 more per year than her husband does. Wanda saves 10% of her income for retirement, and her husband saves 6%. If together they save $5400 per year, then how much does each of them make per year?

86. *Layoffs looming.* American Products plans to lay off 10% of its employees in its aerospace division and 15% of its employees in its agricultural division. If altogether 12% of the 3000 employees in these two divisions will be laid off, then how many employees are in each division?

2.8 Inequalities

Determine whether the given number is a solution to the inequality following it.

87. $3, -2x + 5 \leq x - 6$

88. $-2, 5 - x > 4x + 3$

89. $-1, -2 \leq 6 + 4x < 0$

90. $0, 4x + 9 \geq 5(x - 3)$

For each graph write the corresponding inequality and the solution set to the inequality using interval notation.

91.
$$\text{---}+++(+++++++ \rightarrow$$
$$-2\,{-}1\;\;0\;\;1\;\;2\;\;3\;\;4\;\;5\;\;6\;\;7\;\;8$$

92.
$$\leftarrow +++++++) +++\rightarrow$$
$$-5{-}4{-}3{-}2{-}1\;\;0\;\;1\;\;2\;\;3\;\;4\;\;5$$

93.
$$\leftarrow ++++[+++++++ \rightarrow$$
$$-2{-}1\;\;0\;\;1\;\;2\;\;3\;\;4\;\;5\;\;6\;\;7\;\;8$$

94.
$$\leftarrow +++++(+)++++\rightarrow$$
$$-2{-}1\;\;0\;\;1\;\;2\;\;3\;\;4\;\;5\;\;6\;\;7\;\;8$$

95.
$$\leftarrow ++[++++)++\rightarrow$$
$$-5{-}4{-}3{-}2{-}1\;\;0\;\;1\;\;2\;\;3\;\;4\;\;5$$

96.
$$\leftarrow ++++++++] +++\rightarrow$$
$$-6{-}5{-}4{-}3{-}2{-}1\;\;0\;\;1\;\;2\;\;3\;\;4$$

97.
$$\leftarrow +++++++)+++\rightarrow$$
$$-8{-}7{-}6{-}5{-}4{-}3{-}2{-}1\;\;0\;\;1\;\;2$$

98.
$$\leftarrow ++++[++++)++\rightarrow$$
$$-5{-}4{-}3{-}2{-}1\;\;0\;\;1\;\;2\;\;3\;\;4\;\;5$$

2.9 Solving Inequalities and Applications

Solve each inequality. State the solution set in interval notation and sketch its graph.

99. $x + 2 > 1$

100. $x - 3 > 7$

101. $3x - 5 < x + 1$

102. $5x - 5 > 9 - 2x$

103. $-\dfrac{3}{4}x \geq 3$

104. $-\dfrac{2}{3}x \leq 10$

105. $3 - 2x < 11$

106. $5 - 3x > 35$

107. $-3 < 2x - 1 < 9$

108. $2 \leq 3x + 2 < 8$

109. $0 \leq 1 - 2x < 5$

110. $-5 < 3 - 4x \leq 7$

111. $-1 \leq \dfrac{2x - 3}{3} \leq 1$

112. $-3 < \dfrac{4 - x}{2} < 2$

113. $\dfrac{1}{3} < \dfrac{1}{3} + \dfrac{x}{2} < \dfrac{5}{6}$

114. $-\dfrac{3}{8} \leq -\dfrac{1}{4}x + \dfrac{1}{8} < \dfrac{5}{8}$

Miscellaneous

Use an equation, inequality, or formula to solve each problem.

115. *Plasma TV discount.* Nexus got a 14% discount when he bought a new plasma television. If the amount of the discount was $392, then what was the original price of the television?

116. *Laptop discount.* Zeland got a 12% discount on a new laptop computer. If he paid $1166 for the laptop, then what was the original price?

117. *Rug commission.* Caroline sold an antique rug through a broker who got 8% of the selling price as a commission. If Caroline got $7820 for the rug after the broker's commission, then what was the selling price of the rug?

118. *Buyer's premium.* Brittany paid $95,920 for a 1966 Mustang at a classic car auction where there is a 9% buyer's premium. This means that the buyer pays the bid price plus 9% of the bid price. What was the bid price?

119. *Long-term yields.* The annual yield on a 30-year treasury bond is 5.375%. Use the simple interest formula to find the amount of interest earned during the first year on a $10,000 bond.

120. *High interest rate.* Eddie wrote a $280 check to a check holding company, which gave him $260 in cash. After two weeks, the company will cash his $280 check. Find the annual simple interest rate for this loan. Note that the time is a fraction of a year.

121. *Combined videos.* The owners of ABC Video discovered that they had no movies in common with XYZ Video and bought XYZ's entire stock. Although XYZ had 200 titles, they had no children's movies, while 60% of ABC's titles were children's movies. If 40% of the movies in the combined stock are children's movies, then how many movies did ABC have before the merger?

122. *Living comfortably.* Gary has figured that he needs to take home $30,400 a year to live comfortably. If the government gets 24% of Gary's income, then what must his income be for him to live comfortably?

123. *Bracing a gate.* The diagonal brace on a rectangular gate forms an angle with the horizontal side with degree measure x and an angle with the vertical side with degree measure $2x - 3$. Find x.

124. *Digging up the street.* A contractor wants to install a pipeline connecting point A with point C on opposite sides of a road as shown in the figure below. To save money, the contractor has decided to lay the pipe to point B and then under the road to point C. Find the measure of the angle marked x in the figure.

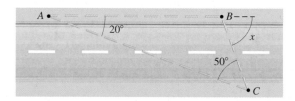

Figure for Exercise 124

125. *Perimeter of a triangle.* One side of a triangle is 1 foot longer than the shortest side, and the third side is twice as long as the shortest side. If the perimeter is less than 25 feet, then what is the range of the length of the shortest side?

126. *Restricted hours.* Alana makes $5.80 per hour working in the library. To keep her job, she must make at least $116 per week; but to keep her scholarship, she must not earn more than $145 per week. What is the range of the number of hours per week that she may work?

Chapter 2 Test

Solve each equation.

1. $-10x - 6 + 4x = -4x + 8$

2. $5(2x - 3) = x + 3$

3. $-\dfrac{2}{3}x + 1 = 7$

4. $x + 0.06x = 742$

5. $x - 0.03x = 0.97$

6. $6x - 7 = 0$

7. $\dfrac{1}{2}x - \dfrac{1}{3} = \dfrac{1}{4}x + \dfrac{1}{6}$

8. $2(x + 6) = 2x - 5$

9. $x + 7x = 8x$

Solve for the indicated variable.

10. $2x - 3y = 9$ for y

11. $m = aP - w$ for a

For each graph write the corresponding inequality and the solution set to the inequality using interval notation.

12.
$$-5\ -4\ -3\ -2\ -1\ \ 0\ \ 1\ \ 2\ \ 3\ \ 4\ \ 5$$

13.
$$-2\ -1\ \ 0\ \ 1\ \ 2\ \ 3\ \ 4\ \ 5\ \ 6\ \ 7\ \ 8$$

Solve each inequality. State the solution set in interval notation and sketch its graph.

14. $4 - 3(w - 5) < -2w$

15. $1 < \dfrac{1 - 2x}{3} < 5$

16. $1 < 3x - 2 < 7$

17. $-\dfrac{2}{3}y < 4$

Write a complete solution to each problem.

18. The perimeter of a rectangle is 72 meters. If the width is 8 meters less than the length, then what is the width of the rectangle?

19. a) What formula expresses the area of a triangle A as a function of its base b and height h?

    b) Find a formula that expresses the height of a triangle as a function of its area and base.

    c) If the area of a triangle is 54 square inches and the base is 12 inches, then what is the height?

20. How many liters of a 20% alcohol solution should Maria mix with 50 liters of a 60% alcohol solution to obtain a 30% solution?

21. Brandon gets a 40% discount on loose diamonds where he works. The cost of the setting is $250. If he plans to spend at most $1450, then what is the price range (list price) of the diamonds that he can afford?

22. If the degree measure of the smallest angle of a triangle is one-half of the degree measure of the second largest angle and one-third of the degree measure of the largest angle, then what is the degree measure of each angle?

Making **Connections** │ A Review of Chapters 1–2

Simplify each expression.

1. $3x + 5x$

2. $3x \cdot 5x$

3. $\dfrac{4x + 2}{2}$

4. $5 - 4(3 - x)$

5. $3x + 8 - 5(x - 1)$

6. $(-6)^2 - 4(-3)2$

7. $3^2 \cdot 2^3$

8. $4(-7) - (-6)(3)$

9. $-2x \cdot x \cdot x$

10. $(-1)(-1)(-1)(-1)(-1)$

Evaluate each expression if $x = -2$ and $y = 3$.

11. $5x + 4x$

12. $9x$

13. $(y - x)(y + x)$

14. $y^2 - x^2$

15. $(x - y)^2$

16. $x^2 - 2xy + y^2$

17. $(2x + y)^2$

18. $4x^2 + 4xy + y^2$

Write the interval notation for each set.

19. The real numbers less than 2

20. The real numbers greater than -6

21. The real numbers greater than or equal to 5

22. The real numbers less than or equal to -1

23. The real numbers between 2 and 6 inclusive

24. The real numbers greater than 4 and less than 8

Perform the following operations.

25. $\dfrac{1}{2} + \dfrac{1}{6}$

26. $\dfrac{1}{2} - \dfrac{1}{3}$

27. $\dfrac{5}{3} \cdot \dfrac{1}{15}$

28. $\dfrac{2}{3} \cdot \dfrac{5}{6}$

29. $6 \cdot \left(\dfrac{5}{3} + \dfrac{1}{2}\right)$

30. $15\left(\dfrac{2}{3} - \dfrac{2}{15}\right)$

31. $4 \cdot \left(\dfrac{x}{2} + \dfrac{1}{4}\right)$

32. $12\left(\dfrac{5}{6}x - \dfrac{3}{4}\right)$

Find the solution set to each equation or inequality.

33. $x - \dfrac{1}{2} = \dfrac{1}{6}$

34. $x + \dfrac{1}{3} = \dfrac{1}{2}$

35. $x - \dfrac{1}{2} > \dfrac{1}{6}$

36. $x + \dfrac{1}{3} \le \dfrac{1}{2}$

37. $\dfrac{3}{5}x = \dfrac{1}{15}$

38. $\dfrac{3}{2}x = \dfrac{5}{6}$

39. $-\dfrac{3}{5}x \le \dfrac{1}{15}$

40. $-\dfrac{3}{2}x > \dfrac{5}{6}$

41. $\dfrac{5}{3}x + \dfrac{1}{2} = 1$

42. $\dfrac{2}{3}x - \dfrac{2}{15} = 2$

43. $\dfrac{x}{2} + \dfrac{1}{4} = \dfrac{1}{2}$

44. $\dfrac{5}{6}x - \dfrac{3}{4} = \dfrac{5}{12}$

45. $3x + 5x = 8$

46. $3x + 5x = 8x$

47. $3x + 5x = 7x$

48. $3x + 5 = 8$

49. $3x + 5x > 7x$

50. $3x + 5x > 8x$

51. $3x + 1 = 7$

52. $5 - 4(3 - x) = 1$

53. $3x + 8 = 5(x - 1)$

54. $x - 0.05x = 190$

55. $5 - 3x < 11$

56. $19 \le 3 + 8x$

57. $0 \le \dfrac{x + 3}{5} \le 3$

58. $1 < \dfrac{7 - x}{12} < 4$

Solve the problem.

59. *Linear Depreciation.* In computing income taxes, a company is allowed to depreciate a $20,000 computer system over five years. Using *linear depreciation,* the value V of the computer system at any year t from 0 through 5 is given by

$$V = C - \dfrac{C - S}{5}t,$$

where C is the initial cost of the system and S is the scrap value of the system.

a) What is the value of the computer system after two years if its scrap value is $4000?

b) If the value of the system after three years is claimed to be $14,000, then what is the scrap value of the company's system?

c) If the accompanying graph models the depreciation of the system, then what is the scrap value of the system?

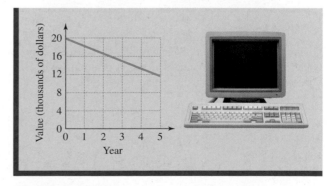

Figure for Exercise 59

Critical **Thinking** | For Individual or Group Work | Chapter 2

These exercises can be solved by a variety of techniques, which may or may not require algebra. So be creative and think critically. Explain all answers. Answers are in the Instructor's Edition of this text.

1. **Visible squares.** How many squares are visible in each of the following diagrams?

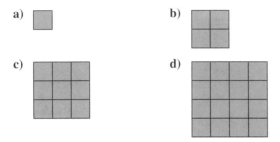

2. **Baker's dilemma.** A baker needs 8 cups of flour. He sends his apprentice to the flour bin with a scoop that holds 6 cups and a scoop that holds 11 cups. How can the apprentice measure 8 cups of flour with these scoops?

Photo for Exercise 2

3. **Totaling one hundred.** Start with the sequence of digits 987654321. Place any number of plus or minus signs between the digits in the sequence so that the value of the resulting expression is 100. For example,

$$98 + 7 - 6 - 5 + 4 + 3 - 2 + 1 = 100.$$

4. **Four threes.** Check out these equations:

$$\frac{3 \cdot 3}{3 \cdot 3} = 1, \frac{3}{3} + \frac{3}{3} = 2, (3 - 3)3 + 3 = 3.$$

Using exactly four 3's write arithmetic expressions whose values are 4, 5, 6, and so on. How far can you go?

5. **Palindrome time.** A palindrome is a sequence of words or numbers that reads the same forward or backward. For example, "A TOYOTA" is a palindrome and 14341 is a palindromic number. How many times per day does a digital clock display a palindromic number? Of course the answer depends on the format in which the digital clock displays the time. First, state precisely the type of digital clock display you are using, and then count the palindromic numbers for that type of display.

6. **Reversible products.** Find the product of 32 and 46. Now reverse the digits and find the product of 23 and 64. The products are the same. Does this happen with any pair of two-digit numbers? Find two other pairs of two-digit numbers (with different digits) that have this property.

7. **Running late.** Alice, Bea, Carl, and Don all have an 8 o'clock class. Alice's watch is 8 minutes fast, but she thinks it is 4 minutes slow. Bea's watch is 8 minutes slow, but she thinks it is 8 minutes fast. Carl's watch is 4 minutes slow, but he thinks it is 8 minutes fast. Don's watch is 4 minutes fast, but he thinks it is 8 minutes slow. Each student leaves so they will get to class at exactly 8 o'clock. Each student assumes the correct time is what they think it is by their watch. Who is late to class and by how much?

8. **Automorphic numbers.** Automorphic numbers are integers whose squares end in the given integer. Since $1^2 = 1$ and $6^2 = 36$, both 1 and 6 are automorphic. Find the next four automorphic numbers.

3

Linear Equations in Two Variables and Their Graphs

If you pick up any package of food and read the label, you will find a long list that usually ends with some mysterious looking names. Many of these strange elements are food additives. A food additive is a substance or a mixture of substances other than basic foodstuffs that is present in food as a result of production, processing, storage, or packaging. They can be natural or synthetic and are categorized in many ways: preservatives, coloring agents, processing aids, and nutritional supplements, to name a few.

Food additives have been around since prehistoric humans discovered that salt would help to preserve meat. Today, food additives can include simple ingredients such as red color from Concord grape skins, calcium, or an enzyme.

Throughout the centuries there have been lively discussions on what is healthy to eat. At the present time the food industry is working to develop foods that have less cholesterol, fats, and other unhealthy ingredients.

Although they frequently have different viewpoints, the food industry and the Food and Drug Administration (FDA) are working to provide consumers with information on a healthier diet. Recent developments such as the synthetically engineered tomato stirred great controversy, even though the FDA declared the tomato safe to eat.

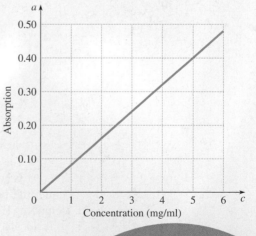

In Exercise 87 of Section 3.4 you will see how a food chemist uses a linear equation in testing the concentration of an enzyme in a fruit juice.

3.1   Graphing Lines in the Coordinate Plane

In Chapter 1 you learned to graph numbers on a number line. We also used number lines to illustrate the solution to inequalities in Chapter 2. In this section, you will learn to graph pairs of numbers in a coordinate system made up of a pair of number lines. We will use this coordinate system to illustrate the solution to equations and inequalities in two variables.

⟨1⟩ Graphing Ordered Pairs

A GPS unit uses longitude and latitude to locate points on the earth. In mathematics we also use pairs of real numbers to describe the locations of points in a plane. We position two number lines at a right angle as shown in Fig. 3.1. The horizontal number line is the **x-axis** and the vertical number line is the **y-axis.** The point at which the axes intersect is the **origin.** The axes divide the **coordinate plane** or **xy-plane** into four quadrants The quadrants are numbered as shown in Fig. 3.1. The quadrants do not include any points on the axes. The system is called the **rectangular coordinate system** or the **Cartesian coordinate system.** It is named after the French mathematician René Descartes (1596–1650).

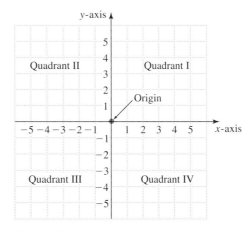

Figure 3.1

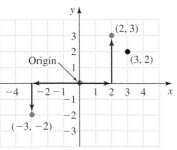

Figure 3.2

Every point in the plane in Fig. 3.1 corresponds to a pair of numbers. For example, the point corresponding to the pair (2, 3) is found by starting at the origin and moving 2 units to the right (in the x direction) and then 3 units up (in the y direction). To locate (3, 2) start at the origin and go 3 units to the right and 2 units up. To locate $(-3, -2)$ start at the origin and go 3 units to the left and then 2 units down. All three points are shown in Fig. 3.2.

Note that (3, 2) and (2, 3) correspond to different points in Fig. 3.2. Since the order of the numbers in the pair makes a difference, a pair of numbers in parentheses is called an **ordered pair.** The first number in an ordered pair is the **x-coordinate,** and the second number is the **y-coordinate.** Locating a point in the xy-plane that corresponds to an ordered pair is called **plotting** or **graphing** the point.

E X A M P L E   1

Plotting points

Plot the points $(2, 5)$, $(-1, 4)$, $(-3, -4)$, and $(3, -2)$.

Solution

To locate $(2, 5)$, start at the origin, move two units to the right, and then move up five units. To locate $(-1, 4)$, start at the origin, move one unit to the left, and then move up four units. All four points are shown in Fig. 3.3.

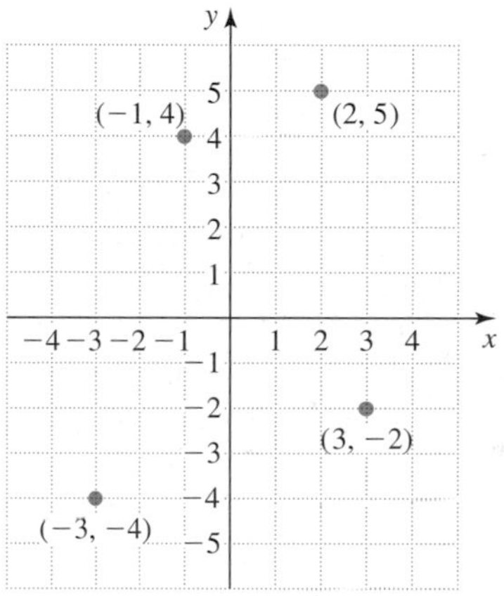

‹ **Helpful Hint** ›

In this chapter, you will be doing a lot of graphing. Using graph paper will help you understand the concepts and help you recognize errors. For your convenience, a page of graph paper can be found on page 250 of this text. Make as many copies of it as you wish.

Figure 3.3

Now do Exercises 1–28

CAUTION In Chapter 1 the notation $(2, 5)$ was used to represent an interval of real numbers. Now it represents an ordered pair of real numbers. The context should always make it clear what we are referring to.

‹2› Ordered Pairs as Solutions to Equations

An equation in two variables such as $y = 2x - 1$ is satisfied if we choose a value for x and a value for y that make it true. If $x = 2$ and $y = 3$, then $y = 2x - 1$ becomes

$$
\begin{array}{cc}
y & x \\
\downarrow & \downarrow \\
\end{array}
$$
$$
3 = 2(2) - 1
$$
$$
3 = 3.
$$

Because the last statement is true, the ordered pair $(2, 3)$ **satisfies the equation** or is a **solution to the equation.** The x-value is always written first and the y-value second.

CAUTION The ordered pair $(3, 2)$ does not satisfy $y = 2x - 1$, because for $x = 3$ and $y = 2$, we have

$$
2 \neq 2(3) - 1.
$$

In Section 2.4 we said that an equation such as $y = 2x - 1$ expresses y as a function of x because it uniquely determines y from any chosen x-value. For this reason we call x the **independent variable** and y the **dependent variable.** We usually use a function to determine the value of the dependent variable from the value of the independent variable. However, for a function of the form $y = mx + b$ we can find either coordinate when given the other, as shown in Example 2.

E X A M P L E   **2**   **Finding solutions to an equation**

Each of the following ordered pairs is missing one coordinate. Complete each ordered pair so that it satisfies the equation $y = -3x + 4$.

    a) (2,)   **b)** (, −5)   **c)** (0,)

Solution

    a) The x-coordinate of (2,) is 2. Let $x = 2$ in the equation $y = -3x + 4$:

$$y = -3 \cdot 2 + 4$$
$$= -6 + 4$$
$$= -2$$

The ordered pair $(2, -2)$ satisfies the equation.

    b) The y-coordinate of (, −5) is −5. Let $y = -5$ in the equation $y = -3x + 4$:

$$-5 = -3x + 4$$
$$-9 = -3x$$
$$3 = x$$

The ordered pair $(3, -5)$ satisfies the equation.

    c) Replace x by 0 in the equation $y = -3x + 4$:

$$y = -3 \cdot 0 + 4 = 4$$

So $(0, 4)$ satisfies the equation.

> Now do Exercises 29–44

⟨3⟩ Graphing a Linear Equation in Two Variables

In Chapter 2 we defined a linear equation in one variable as an equation of the form $ax = b$, where $a \neq 0$. A linear equation in two variables is defined similarly:

Linear Equation in Two Variables

A **linear equation in two variables** is an equation of the form

$$Ax + By = C,$$

where A and B are not both zero.

    Consider the linear equation $-2x + y = -1$. If we solve it for y, we get $y = 2x - 1$. If we choose any real number for x, we can use $y = 2x - 1$ to compute a corresponding y-value. So there are infinitely many ordered pairs that satisfy the equation. To get a better understanding of the solution set to a linear equation in two variables, we often graph all of the ordered pairs in the solution set. *The graph of the solution set to a linear equation in two variables is a straight line*, as shown in Example 3.

EXAMPLE 3

You can make a table of values for x and y with a graphing calculator. Enter the equation $y = 2x - 1$ using $Y =$ and then press TABLE.

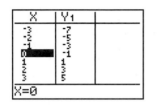

‹ **Helpful Hint** ›

The graph of a linear equation in one variable consists of a single point on a number line. The graph of a linear equation in two variables consists of a line in a coordinate plane.

Graphing an equation

Graph the equation $y = 2x - 1$ in the coordinate plane.

Solution

To find ordered pairs that satisfy $y = 2x - 1$, we arbitrarily select some x-coordinates and calculate the corresponding y-coordinates:

$$\text{If } x = -3, \quad \text{then } y = 2(-3) - 1 = -7.$$
$$\text{If } x = -2, \quad \text{then } y = 2(-2) - 1 = -5.$$
$$\text{If } x = -1, \quad \text{then } y = 2(-1) - 1 = -3.$$
$$\text{If } x = 0, \quad \text{then } y = 2(0) - 1 = -1.$$
$$\text{If } x = 1, \quad \text{then } y = 2(1) - 1 = 1.$$
$$\text{If } x = 2, \quad \text{then } y = 2(2) - 1 = 3.$$
$$\text{If } x = 3, \quad \text{then } y = 2(3) - 1 = 5.$$

We can make a table for these results as follows:

x	-3	-2	-1	0	1	2	3
$y = 2x - 1$	-7	-5	-3	-1	1	3	5

The ordered pairs $(-3, -7)$, $(-2, -5)$, $(-1, -3)$, $(0, -1)$, $(1, 1)$, $(2, 3)$, and $(3, 5)$ are graphed in Fig. 3.4. Draw a straight line through these points, as shown in Fig. 3.5. The line in Fig. 3.5 is the graph of the solution set to $y = 2x - 1$. The arrows on the ends of the line indicate that it goes indefinitely in both directions.

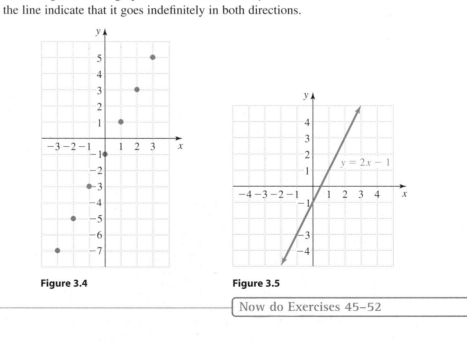

Figure 3.4

Figure 3.5

Now do Exercises 45–52

A linear equation in two variables is an equation of the form $Ax + By = C$, where A and B are not both zero. Note that we can have $A = 0$ if $B \neq 0$, and we can have $B = 0$ with $A \neq 0$. So equations such as $x = 8$ and $y = 2$ are linear equations.

Equations such as $x - y - 5 = 0$ and $y = 2x + 3$ are also called linear equations because they could be rewritten in the form $Ax + By = C$. Equations such as $y = 2x^2$ or $y = \frac{5}{x}$ are not linear equations.

EXAMPLE 4

Graphing an equation

Graph the equation $3x + y = 2$. Plot at least five points.

Solution

It is easier to make a table of ordered pairs if we express y as a function of x. So subtract $3x$ from each side to get $y = -3x + 2$. Now select some values for x and then calculate the corresponding y-coordinates:

$$\text{If } x = -2, \quad \text{then } y = -3(-2) + 2 = 8.$$
$$\text{If } x = -1, \quad \text{then } y = -3(-1) + 2 = 5.$$
$$\text{If } x = 0, \quad \text{then } y = -3(0) + 2 = 2.$$
$$\text{If } x = 1, \quad \text{then } y = -3(1) + 2 = -1.$$
$$\text{If } x = 2, \quad \text{then } y = -3(2) + 2 = -4.$$

The following table shows these five ordered pairs:

x	-2	-1	0	1	2
$y = -3x + 2$	8	5	2	-1	-4

Plot $(-2, 8)$, $(-1, 5)$, $(0, 2)$, $(1, -1)$, and $(2, -4)$. Draw a line through them, as shown in Fig. 3.6.

Now do Exercises 53–56

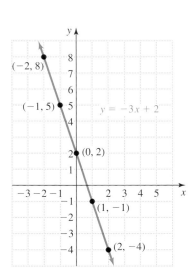

Figure 3.6

‹ Calculator Close-Up ›

To graph $y = -3x + 2$, enter the equation using the Y = key:

Next, set the viewing window (WINDOW) to get the desired view of the graph. Xmin and Xmax indicate the minimum and maximum x-values used for the graph, and likewise for Ymin and Ymax. Xscl and Yscl (scale) give the distance between tick marks on the respective axes.

Press GRAPH to get the graph:

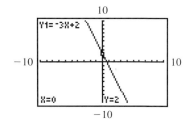

Even though the graph is not really "straight," it is consistent with the graph of $y = -3x + 2$ in Fig. 3.6.

EXAMPLE 5

Horizontal and vertical lines

Graph each linear equation.

a) $y = 4$ **b)** $x = 3$

Solution

a) The equation $y = 4$ is a simplification of $0 \cdot x + y = 4$. So if y is replaced with 4, then we can use any real number for x. For example, $(-1, 4)$ satisfies $0 \cdot x + y = 4$ because $0(-1) + 4 = 4$ is correct. The following table shows five ordered pairs that satisfy $y = 4$.

x	-2	-1	0	1	2
$y = 4$	4	4	4	4	4

Figure 3.7 shows a horizontal line through these points.

b) The equation $x = 3$ is a simplification of $x + 0 \cdot y = 3$. So if x is replaced with 3, then we can use any real number for y. For example, $(3, -2)$ satisfies $x + 0 \cdot y = 3$ because $3 + 0(-2) = 3$ is correct. The following table shows five ordered pairs that satisfy $x = 3$.

$x = 3$	3	3	3	3	3
y	-2	-1	0	1	2

Figure 3.8 shows a vertical line through these points.

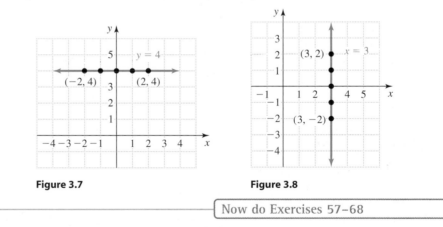

Figure 3.7   **Figure 3.8**

Now do Exercises 57–68

‹ **Calculator Close-Up** ›

You cannot graph the vertical line $x = 3$ on most graphing calculators. The only equations that can be graphed are ones in which y is written in terms of x.

CAUTION If $x = 3$ occurs in the context of equations in a single variable, then $x = 3$ has only one solution, 3. In the context of equations in two variables, $x = 3$ is assumed to be a simplified form of $x + 0 \cdot y = 3$, and it has infinitely many solutions (all of the ordered pairs on the line in Fig. 3.8).

All of the equations we have considered so far have involved single-digit numbers. If an equation involves large numbers, then we must change the scale on the x-axis, the y-axis, or both to accommodate the numbers involved. The change of scale is arbitrary, and the graph will look different for different scales.

EXAMPLE   **6**

Adjusting the scale

Graph the equation $y = 20x + 500$. Plot at least five points.

Solution

The following table shows five ordered pairs that satisfy the equation.

x	-20	-10	0	10	20
$y = 20x + 500$	100	300	500	700	900

To fit these points onto a graph, we change the scale on the x-axis to let each division represent 10 units and change the scale on the y-axis to let each division represent 200 units. The graph is shown in Fig. 3.9.

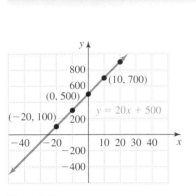

Figure 3.9

Now do Exercises 69–74

⟨4⟩ Graphing a Line Using Intercepts

For many lines, the easiest points to locate are the points where the line crosses the axes.

> **Intercepts**
>
> The **x-intercept** is the point at which a line crosses the x-axis.
> The **y-intercept** is the point at which a line crosses the y-axis.

The second coordinate of the x-intercept is 0, and the first coordinate of the y-intercept is 0. If a line has two distinct intercepts, they can be used as two points that determine the location of the line.

EXAMPLE   **7**

Graphing a line using intercepts

Graph the equation $2x - 3y = 6$ by using the x- and y-intercepts.

Solution

To find the x-intercept, let $y = 0$ in the equation $2x - 3y = 6$:

$$2x - 3 \cdot 0 = 6$$
$$2x = 6$$
$$x = 3$$

The x-intercept is $(3, 0)$. To find the y-intercept, let $x = 0$ in $2x - 3y = 6$:

$$2 \cdot 0 - 3y = 6$$
$$-3y = 6$$
$$y = -2$$

⟨ **Helpful Hint** ⟩

You can find the intercepts for $2x - 3y = 6$ using the *cover-up method*. Cover up $-3y$ with your pencil, and then solve $2x = 6$ mentally to get $x = 3$ and an x-intercept of $(3, 0)$. Now cover up $2x$ and solve $-3y = 6$ to get $y = -2$ and a y-intercept of $(0, -2)$.

‹ Calculator Close-Up ›

To graph $2x - 3y = 6$ on a calculator you must solve for y. In this case, $y = (2/3)x - 2$.

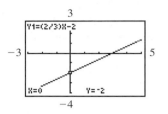

Since the calculator graph appears to be the same as the graph in Fig. 3.10, it supports the conclusion that Fig. 3.10 is correct.

The y-intercept is $(0, -2)$. Locate the intercepts and draw a line through them, as shown in Fig. 3.10. To check, find one additional point that satisfies the equation, say $(6, 2)$, and see whether the line goes through that point.

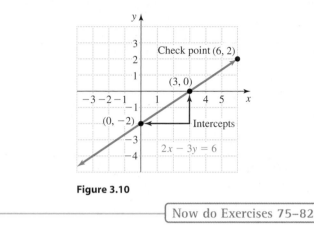

Figure 3.10

Now do Exercises 75–82

‹5› Function Notation and Applications

An equation of the form $y = mx + b$ expresses y as a function of x, and it is called a **linear function.** Linear functions occur in many real-life situations. For example, if the monthly cost C of a cell phone is \$50 plus 10 cents per minute, then $C = 50 + 0.10n$ where n is the number of minutes used. We may also write $C(n)$ in place of C. This notation is called **function notation.** We read $C(n)$ as "the cost of n minutes" or simply "C of n." Using function notation is very convenient for identifying more than one cost. For example, to express the fact that the cost for 100 minutes is \$60, 200 minutes is \$70, and 300 minutes is \$80 we can simply write

$$C(100) = \$60, \quad C(200) = \$70, \quad \text{and} \quad C(300) = \$80.$$

EXAMPLE **8**

House plans

An architect uses the function $C(x) = 30x + 900$ to determine the cost C for drawing house plans, where x is the number of copies of the plan that the client receives.

a) Find $C(5)$, $C(6)$, and $C(7)$.

b) Find the intercepts and interpret them.

c) Graph the function.

d) Does the cost increase or decrease as x increases?

Solution

a) Replace x by 5, 6, and 7 in the equation $C(x) = 30x + 900$:

$$C(5) = 30(5) + 900 = 1050$$
$$C(6) = 30(6) + 900 = 1080$$
$$C(7) = 30(7) + 900 = 1110$$

So the cost of 5 plans is \$1050, the cost of 6 plans is \$1080, and the cost of 7 plans is \$1110.

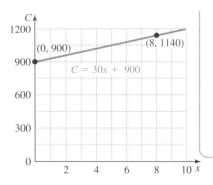

Figure 3.11

b) If $x = 0$, then $C(0) = 30(0) + 900 = 900$. The C-intercept is $(0, 900)$. The cost is $900 for the labor involved in drawing the plans, even if you get no copies of the plan. If $C(x) = 0$, then $30x + 900 = 0$ or $x = -30$. So the x-intercept is $(-30, 0)$, but in this situation the x-intercept is meaningless. The number of plans can't be negative.

c) The graph goes through $(0, 900)$ and $(8, 1140)$ as shown in Fig. 3.11. Since negative values of x are meaningless, the graph is drawn in the first quadrant only.

d) As x increases, the cost increases.

Now do Exercises 83–88

EXAMPLE **9**

Ticket demand

The demand for tickets to see the Ice Gators play hockey can be modeled by the equation $d = 8000 - 100p$, where d is the number of tickets sold and p is the price per ticket in dollars.

a) How many tickets will be sold at $20 per ticket?

b) Find the intercepts and interpret them.

c) Graph the linear equation.

d) What happens to the demand as the price increases?

Solution

a) If tickets are $20 each, then $d = 8000 - 100 \cdot 20 = 6000$. So at $20 per ticket, the demand will be 6000 tickets.

b) Replace d with 0 in the equation $d = 8000 - 100p$ and solve for p:

$$0 = 8000 - 100p$$

$$100p = 8000 \quad \text{Add } 100p \text{ to each side.}$$

$$p = 80 \quad \text{Divide each side by 100.}$$

If $p = 0$, then $d = 8000 - 100 \cdot 0 = 8000$. So the intercepts are $(0, 8000)$ and $(80, 0)$. If the tickets are free, the demand will be 8000 tickets. At $80 per ticket, no tickets will be sold.

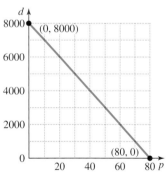

Figure 3.12

c) Graph the line using the intercepts $(0, 8000)$ and $(80, 0)$ as shown in Fig. 3.12. The line is graphed in the first quadrant only, because negative values for demand or price are meaningless.

d) When the tickets are free, the demand is high. As the price increases, the demand goes down. At $80 per ticket, there will be no demand.

Now do Exercises 89–92

Note that $d = 8000 - 100p$ is a *model* for the demand in Example 9. A model car has only some of the features of a real car, and the same is true here. For instance, the line in Fig. 3.12 contains infinitely many points. But there is really only a finite number of possibilities for price and demand, because we cannot sell a fraction of a ticket.

Warm-Ups ▼

Fill in the blank.

1. The point at the intersection of the x- and y-axis is the
   _____.

2. Every point in the coordinate plane corresponds to an
   _____ of real numbers.

3. The point at which a line crosses the x-axis is the
   _____.

4. The point at which a line crosses the y-axis is the
   _____.

5. The graph of $y = 5$ is a _____ line.

6. The graph of $x = 3$ is a _____ line.

7. A _____ equation in two variables has the form
   $Ax + By = C$ where A and B are not both zero.

True or false?

8. The point $(2, 4)$ satisfies $2y - 3x = -8$.

9. If $(1, 5)$ satisfies an equation, then $(5, 1)$ does also.

10. The origin is in quadrant I.

11. The point $(4, 0)$ is on the y-axis.

12. The graph of $x + 0 \cdot y = 9$ is the same as the graph
    of $x = 9$.

13. The y-intercept for $x + 2y = 5$ is $(5, 0)$.

14. The point $(5, -3)$ is in quadrant III.

Exercises 3.1

⟨ **Study Tips** ⟩

• It is a good idea to work with others, but don't be misled. Working a problem with help is not the same as working a problem on your own.
• Math is personal. Make sure that you can do it.

⟨1⟩ **Graphing Ordered Pairs**

*Plot the points on a rectangular coordinate system.
See Example 1.*

1. $(1, 5)$

2. $(4, 3)$

3. $(-2, 1)$

 4. $(-3, 5)$

5. $\left(3, -\dfrac{1}{2}\right)$

6. $\left(2, -\dfrac{1}{3}\right)$

7. $(-2, -4)$

8. $(-3, -5)$

9. $(0, 3)$

10. $(0, -2)$

11. $(-3, 0)$

12. $(5, 0)$

13. $(\pi, 1)$

14. $(-2, \pi)$

15. $(1.4, 4)$

16. $(-3, 0.4)$

*For each point, name the quadrant in which it lies or the axis
on which it lies.*

17. $(-3, 45)$

18. $(-33, 47)$

19. $(-3, 0)$

20. $(0, -9)$

21. $(-2.36, -5)$

22. $(89.6, 0)$

23. $(3.4, 8.8)$ **24.** $(4.1, 44)$ **25.** $\left(-\dfrac{1}{2}, 50\right)$

26. $\left(-6, -\dfrac{1}{2}\right)$ **27.** $(0, -99)$ **28.** $(\pi, 0)$

⟨2⟩ Ordered Pairs as Solutions to Equations

Complete each ordered pair so that it satisfies the given equation. See Example 2.

29. $y = 3x + 9$: $(0, \ \ )$, $(\ \ , 24)$, $(2, \ \ )$

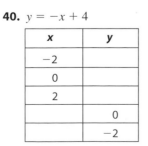**30.** $y = 2x + 5$: $(8, \ \ )$, $(-1, \ \ )$, $(\ \ , -1)$

31. $y = -3x - 7$: $(0, \ \ )$, $\left(\dfrac{1}{3}, \ \ \right)$, $(\ \ , -5)$

32. $y = -5x - 3$: $(-1, \ \ )$, $\left(-\dfrac{1}{2}, \ \ \right)$, $(\ \ , -2)$

33. $y = 1.2x + 54.3$: $(0, \ \ )$, $(10, \ \ )$, $(\ \ , 54.9)$

34. $y = 1.8x + 22.6$: $(1, \ \ )$, $(-10, \ \ )$, $(\ \ , 22.6)$

35. $2x - 3y = 6$: $(3, \ \ )$, $(\ \ , -2)$, $(12, \ \ )$

36. $3x + 5y = 0$: $(-5, \ \ )$, $(\ \ , -3)$, $(10, \ \ )$

37. $0 \cdot y + x = 5$: $(\ \ , -3)$, $(\ \ , 5)$, $(\ \ , 0)$

38. $0 \cdot x + y = -6$: $(3, \ \ )$, $(-1, \ \ )$, $(4, \ \ )$

Use the given equations to find the missing coordinates in the following tables.

39. $y = -2x + 5$

x	y
−2	
0	
2	
	−3
	−7

40. $y = -x + 4$

x	y
−2	
0	
2	
	0
	−2

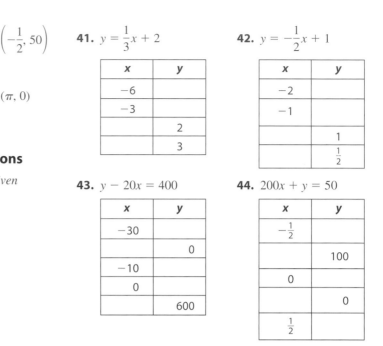

41. $y = \dfrac{1}{3}x + 2$

x	y
−6	
−3	
	2
	3

42. $y = -\dfrac{1}{2}x + 1$

x	y
−2	
−1	
	1
	$\dfrac{1}{2}$

43. $y - 20x = 400$

x	y
−30	
	0
−10	
0	
	600

44. $200x + y = 50$

x	y
$-\dfrac{1}{2}$	
	100
0	
	0
$\dfrac{1}{2}$	

⟨3⟩ Graphing a Linear Equation in Two Variables

Graph each equation. Plot at least five points for each equation. Use graph paper. See Examples 3–5. If you have a graphing calculator, use it to check your graphs when possible.

45. $y = x + 1$ **46.** $y = x - 1$

47. $y = 2x + 1$ **48.** $y = 3x - 1$

49. $y = 3x - 2$ **50.** $y = 2x + 3$

51. $y = x$

52. $y = -x$

63. $x + 2y = 4$

64. $x - 2y = 6$

65. $x - 3y = 6$

66. $x + 4y = 5$

53. $y = 1 - x$

54. $y = 2 - x$

67. $y = 0.36x + 0.4$

68. $y = 0.27x - 0.42$

55. $y = -2x + 3$

56. $y = -3x + 2$

Graph each equation. Plot at least five points for each equation. Use graph paper. See Example 6. If you have a graphing calculator, use it to check your graphs.

57. $y = -3$

58. $y = 2$

69. $y = x + 1200$

70. $y = 2x - 3000$

59. $x = 2$

60. $x = -4$

71. $y = 50x - 2000$

72. $y = -300x + 4500$

61. $2x + y = 5$

62. $3x + y = 5$

73. $y = -400x + 2000$

74. $y = 500x + 3$

⟨4⟩ Graphing a Line Using Intercepts

For each equation, state the x-intercept and y-intercept. Then graph the equation using the intercepts and a third point. See Example 7.

75. $3x + 2y = 6$

76. $2x + y = 6$

77. $x - 4y = 4$

78. $-2x + y = 4$

79. $y = \dfrac{3}{4}x - 9$

80. $y = -\dfrac{1}{2}x + 5$

81. $\dfrac{1}{2}x + \dfrac{1}{4}y = 1$

82. $\dfrac{1}{3}x - \dfrac{1}{2}y = 3$

⟨5⟩ Function Notation and Applications

Solve each problem. See Examples 8 and 9.

83. *Plumbing charges.* The cost of a plumber's service call is a function of the number of hours spent on the job. The linear function

$$C(n) = 50n + 90$$

is used to determine the cost C from the number of hours n.

a) Find $C(0)$ and $C(2)$.

b) If the cost was \$440, then how many hours were spent on the job?

84. *Moving day.* The one-day cost of renting a truck for a local move is a function of the number of miles put on the truck. The cost C in dollars is determined from the mileage m by the linear function

$$C(m) = 0.42m + 39.$$

a) Find the cost for 66 miles.

b) If the cost of the truck was \$54.96, then how many miles were driven?

85. *Social Security.* The percentage of full benefit that you receive from Social Security is a function of the age at which you retire. The linear function

$$p(a) = 8a - 436$$

determines the percentage of full benefit p from the retirement age a for ages 67 through 70.

a) Find $p(67)$ and $p(68)$.

b) If a person receives 124% of full benefit, then at what age did the person retire?

c) If full benefit is \$14,000 per year for Bob Jones, then how much does he get per year if he retires at age 69?

86. *Retiring early.* If you retire before the full retirement age of 67, you get less than full benefit. For ages 64 through 67 the linear function

$$p(a) = \frac{20}{3}a - \frac{1040}{3}$$

determines the percentage of full benefit p for retirement age a.

a) Find $p(64)$ and $p(66)$.

b) If a person receives $86\frac{2}{3}$% of full benefit, then at what age did the person retire?

c) If full benefit is \$12,300 per year for Sue Smith, then how much does she get per year if she retires at age 65?

87. *Medicaid spending.* The cost C in billions of dollars for federal Medicaid (health care for the poor) can be modeled by the linear function

$$C(n) = 11.5n + 319,$$

where n is the number of years since 2007 (Health Care Financing Administration, www.hcfa.gov).

a) Find $C(0)$, $C(1)$, and $C(2)$.

b) What was the cost in 2010?

c) In what year will the cost reach $400 billion?

d) Graph the equation for *n* ranging from 0 through 20.

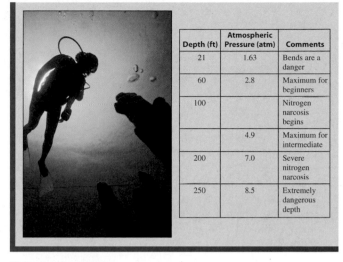

Depth (ft)	Atmospheric Pressure (atm)	Comments
21	1.63	Bends are a danger
60	2.8	Maximum for beginners
100		Nitrogen narcosis begins
	4.9	Maximum for intermediate
200	7.0	Severe nitrogen narcosis
250	8.5	Extremely dangerous depth

Figure for Exercise 89

88. **Dental services.** The national cost *C* in billions of dollars for dental services can be modeled by the linear function

$$C(n) = 3.1n + 62.5,$$

where *n* is the number of years since 2000 (Health Care Financing Administration, www.hcfa.gov).

a) Find $C(2)$, $C(4)$, and $C(8)$.

b) Find and interpret the *C*-intercept for the line.

c) Find and interpret the *n*-intercept for the line.

d) In what year will the cost of dental services reach $120 billion?

e) Graph the line for *n* ranging from 0 through 20.

90. **Demand equation.** Helen's Health Foods usually sells 400 cans of ProPac Muscle Punch per week when the price is $5 per can. After experimenting with prices for some time, Helen has determined that the weekly demand can be found by using the equation

$$d = 600 - 40p,$$

where *d* is the number of cans and *p* is the price per can.

a) Will Helen sell more or less Muscle Punch if she raises her price from $5?

b) What happens to her sales every time she raises her price by $1?

c) Graph the equation.

89. **Hazards of depth.** The accompanying table shows the depth below sea level and atmospheric pressure. The equation

$$A = 0.03d + 1$$

expresses the atmospheric pressure in terms of the depth *d*.

a) Find the atmospheric pressure at the depth where nitrogen narcosis begins.

b) Find the maximum depth for intermediate divers.

c) Graph the equation for *d* ranging from 0 to 250 feet.

d) What is the maximum price that she can charge and still sell at least one can?

91. **Advertising blitz.** Furniture City in Toronto had $24,000 to spend on advertising a year-end clearance sale. A 30-second radio ad costs $300, and a 30-second local television ad costs $400. To model this situation, the advertising manager wrote the equation $300x + 400y = 24,000$. What do *x* and *y* represent? Graph the equation. How many solutions are there to the equation, given that the number of ads of each type must be a whole number?

Graphing Calculator Exercises

Graph each straight line on your graphing calculator using a viewing window that shows both intercepts. Answers may vary.

93. $2x + 3y = 1200$ **94.** $3x - 700y = 2100$

92. *Material allocation.* A tent maker had 4500 square yards of nylon tent material available. It takes 45 square yards of nylon to make an 8 × 10 tent and 50 square yards to make a 9 × 12 tent. To model this situation, the manager wrote the equation $45x + 50y = 4500$. What do *x* and *y* represent? Graph the equation. How many solutions are there to the equation, given that the number of tents of each type must be a whole number?

95. $200x - 300y = 6$ **96.** $300x + 5y = 20$

97. $y = 300x - 1$ **98.** $y = 300x - 6000$

Math *at Work*   Predicting the Future

No one knows what the future may bring, but everyone plans for and tries to predict the future. Stock market analysts predict the profits of companies, pollsters predict the outcomes of elections, and urban planners predict sizes of cities. These predictions of the future are often based on the trends of the past.

Consider the accompanying table, which shows the population of the United States in millions for each census year from 1950 through 2010. It certainly appears that the population is going up, and it would be a safe bet to predict that the population in 2020 will be somewhat larger than 309 million. We get a different perspective if we look at the accompanying graph of the population data. Not only does the graph show an increasing population, it shows the population increasing in a linear manner. Now we can make a prediction based on the line that appears to fit the data. The equation of this line, the *regression line,* is $y = 2.55x - 4820$, where *x* is the year and *y* is the population. The equation of the regression line can be found with a computer or graphing calculator. Now if $x = 2020$, then $y = 2.35(2020) - 4820 \approx 331$. So we can predict 331 million people in 2020.

Year	Population (millions)
1950	152
1960	180
1970	204
1980	227
1990	249
2000	279
2010	309

3.2 Slope

In Section 3.1 you learned that the graph of a linear equation is a straight line. In this section, we will continue our study of lines in the coordinate plane.

⟨**1**⟩ Slope

If a highway rises 6 feet in a horizontal run of 100 feet, then the grade is $\frac{6}{100}$ or 6%. See Fig. 3.13. The grade is a measurement of the steepness of the road. A road with an 8% grade rises 8 feet in a horizontal run of 100 feet, and it is steeper than a road with a 6% grade. We use exactly the same idea to measure the steepness of a line in a coordinate system, but the measurement is called **slope** rather than grade. For the line in Fig. 3.14 the y-coordinate increases by 2 units and the x-coordinate increases by 3 units as you move from (1, 1) to (4, 3). So its slope is $\frac{2}{3}$.

In general, the change in y-coordinate is the **rise** and the change in x-coordinate is the **run.** The letter m is often used for slope.

Figure 3.13

> **Slope**
>
> $$m = \text{slope} = \frac{\text{rise}}{\text{run}} = \frac{\text{change in } y\text{-coordinate}}{\text{change in } x\text{-coordinate}}$$

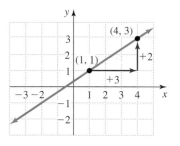

Figure 3.14

Signed numbers are not used to describe the grade of a road, but they are used for lines in a coordinate system. If the y-coordinate increases (moving upward) as you move from one point on the line to another, the rise is positive. If it decreases (moving downward), the rise is negative. The same goes for the run. If the x-coordinate increases (moving to the right), then the run is positive, and if it decreases (moving to the left), the run is negative. Using signed numbers for the rise and run causes the slope to be positive or negative, as shown in Example 1.

EXAMPLE 1

Finding the slope of a line

Find the slope of each blue line by going from point A to point B.

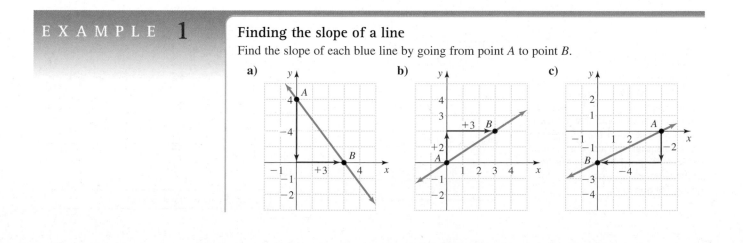

Solution

a) The coordinates of point A are $(0, 4)$, and the coordinates of point B are $(3, 0)$. Going from A to B, the change in y is -4, and the change in x is $+3$. So,

$$m = \frac{-4}{3} = -\frac{4}{3}.$$

Note that it does not matter whether you move from A to B or from B to A. Moving from B to A, the y-coordinate increases by 4 units (rise $+4$) and the x-coordinate decreases by 3 units (run -3). So rise over run is $\frac{+4}{-3}$ or $-\frac{4}{3}$.

b) Going from A to B, the rise is 2, and the run is 3. So,

$$m = \frac{2}{3}.$$

c) Going from A to B, the rise is -2, and the run is -4. So,

$$m = \frac{-2}{-4} = \frac{1}{2}.$$

> Now do Exercises 1–4

CAUTION   The change in y is always in the numerator, and the change in x is always in the denominator.

The ratio of rise to run is the ratio of the lengths of the two legs of any right triangle whose hypotenuse is on the line. As long as one leg is vertical and the other is horizontal, all such triangles for a certain line have the same shape. These triangles are similar triangles. The ratio of the length of the vertical side to the length of the horizontal side for any two such triangles is the same number. So we get the same value for the slope no matter which two points of the line are used to calculate it or in which order the points are used.

EXAMPLE 2

Finding slope

Find the slope of the line shown here using points A and B, points A and C, and points B and C.

Solution

Using A and B, we get

$$m = \frac{\text{rise}}{\text{run}} = \frac{1}{4}.$$

Using A and C, we get

$$m = \frac{\text{rise}}{\text{run}} = \frac{2}{8} = \frac{1}{4}.$$

Using B and C, we get

$$m = \frac{\text{rise}}{\text{run}} = \frac{1}{4}.$$

‹ **Helpful Hint** ›

It is good to think of what the slope represents when x and y are measured quantities rather than just numbers. For example, if the change in y is 50 miles and the change in x is 2 hours, then the slope is 25 mph (or 25 miles per 1 hour). So the slope is the amount of change in y for a change of one in x.

> Now do Exercises 5–12

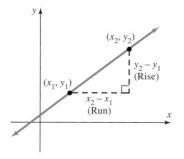

Figure 3.15

⟨2⟩ Slope Using Coordinates

One way to obtain the rise and run is from a graph. The rise and run can also be found by using the coordinates of two points on the line as shown in Fig. 3.15.

Coordinate Formula for Slope

The slope of the line containing the points (x_1, y_1) and (x_2, y_2) is given by

$$m = \frac{y_2 - y_1}{x_2 - x_1},$$

provided that $x_2 - x_1 \neq 0$.

The small lowered numbers following x and y in the slope formula are **subscripts.** We read x_1 as "x sub one" or simply "x one." We think of (x_1, y_1) as the x- and y-coordinates of the first point and (x_2, y_2) as the x- and y-coordinates of the second point.

E X A M P L E   **3**

Using coordinates to find slope
Find the slope of each of the following lines.

a) The line through $(0, 5)$ and $(6, 3)$

b) The line through $(-3, -4)$ and $(-5, -2)$

c) The line through $(-4, 2)$ and the origin

Solution

a) If $(x_1, y_1) = (0, 5)$ and $(x_2, y_2) = (6, 3)$, then

$$m = \frac{y_2 - y_1}{x_2 - x_1}$$

$$= \frac{3 - 5}{6 - 0} = \frac{-2}{6} = -\frac{1}{3}.$$

If $(x_1, y_1) = (6, 3)$ and $(x_2, y_2) = (0, 5)$, then

$$m = \frac{y_2 - y_1}{x_2 - x_1}$$

$$= \frac{5 - 3}{0 - 6} = \frac{2}{-6} = -\frac{1}{3}.$$

Note that it does not matter which point is called (x_1, y_1) and which is called (x_2, y_2). In either case, the slope is $-\frac{1}{3}$.

b) Let $(x_1, y_1) = (-3, -4)$ and $(x_2, y_2) = (-5, -2)$:

$$m = \frac{y_2 - y_1}{x_2 - x_1}$$

$$= \frac{-2 - (-4)}{-5 - (-3)}$$

$$= \frac{2}{-2} = -1$$

c) Let $(x_1, y_1) = (0, 0)$ and $(x_2, y_2) = (-4, 2)$:

$$m = \frac{2 - 0}{-4 - 0} = \frac{2}{-4} = -\frac{1}{2}$$

Now do Exercises 13–26

CAUTION   Order matters. If you divide $y_2 - y_1$ by $x_1 - x_2$, your slope will have the wrong sign. However, you will get the correct slope regardless of which point is called (x_1, y_1) and which is called (x_2, y_2).

Because division by zero is undefined, slope is undefined if $x_2 - x_1 = 0$ or $x_2 = x_1$. The x-coordinates of two distinct points on a line are equal only if the points are on a vertical line. *So slope is undefined for vertical lines.* The concept of slope does not exist for a vertical line.

Any two points on a horizontal line have equal y-coordinates. So for points on a horizontal line we have $y_2 - y_1 = 0$. Since $y_2 - y_1$ is in the numerator of the slope formula, *the slope for any horizontal line is zero.* We never refer to a line as having "no slope," because in English "no" can mean zero or does not exist.

E X A M P L E   4

Slope for vertical and horizontal lines

Find the slope of the line through each pair of points.

a) $(2, 1)$ and $(2, -3)$

b) $(-2, 2)$ and $(4, 2)$

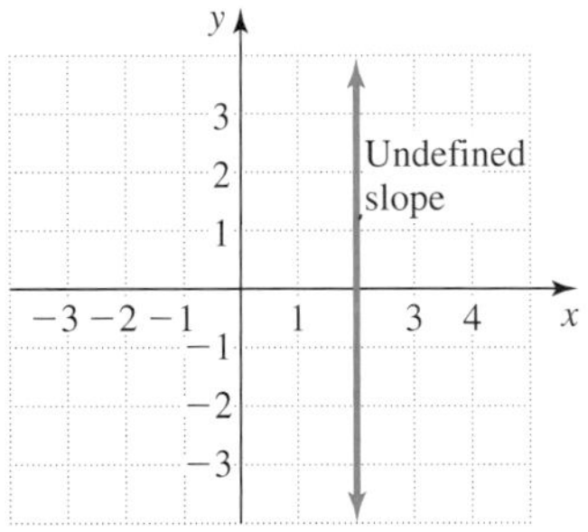

Solution

a) The points $(2, 1)$ and $(2, -3)$ are on the vertical line shown in Fig. 3.16. Since slope is undefined for vertical lines, this line does not have a slope. Using the slope formula we get

$$m = \frac{-3 - 1}{2 - 2} = \frac{-4}{0}.$$

Since division by zero is undefined, we can again conclude that slope is undefined for the vertical line through the given points.

b) The points $(-2, 2)$ and $(4, 2)$ are on the horizontal line shown in Fig. 3.17. Using the slope formula we get

$$m = \frac{2 - 2}{-2 - 4} = \frac{0}{-6} = 0.$$

So the slope of the horizontal line through these points is 0.

Now do Exercises 27–32

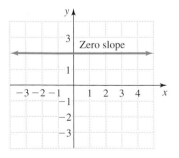

Figure 3.16

Vertical line

Horizontal line

Figure 3.17

Note that for a line with *positive slope,* the y-values increase as the x-values increase. For a line with *negative slope,* the y-values decrease as the x-values increase. See Fig. 3.18. As the absolute value of the slope increases, the line gets steeper.

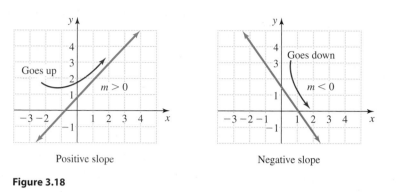

Figure 3.18

⟨3⟩ Graphing a Line Given a Point and Slope

To graph a line from its equation we usually make a table of ordered pairs and then draw a line through the points or we use the intercepts. In Example 5 we will graph a line using one point and the slope. From the slope we find additional points by using the rise and the run.

EXAMPLE   5

Graphing a line given a point and its slope
Graph each line.

a) The line through $(2, 1)$ with slope $\frac{3}{4}$

b) The line through $(-2, 4)$ with slope -3

Solution

a) First locate the point $(2, 1)$. Because the slope is $\frac{3}{4}$, we can find another point on the line by going up three units and to the right four units to get the point $(6, 4)$, as shown in Fig. 3.19. Now draw a line through $(2, 1)$ and $(6, 4)$. Since $\frac{3}{4} = \frac{-3}{-4}$, we could have obtained the second point by starting at $(1, 2)$ and going down 3 units and to the left 4 units.

⟨ **Calculator Close-Up** ⟩

When we graph a line, we usually draw a graph that shows both intercepts, because they are important features of the graph. If the intercepts are not between -10 and 10, you will have to adjust the window to get a good graph. The viewing window that has x- and y-values ranging from a minimum of -10 to a maximum of 10 is called the *standard viewing window*.

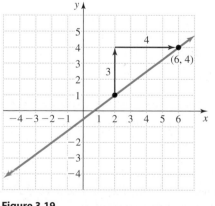

Figure 3.19

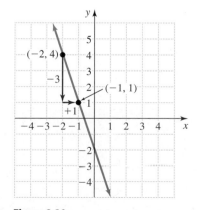

Figure 3.20

b) First locate the point $(-2, 4)$. Because the slope is -3, or $\frac{-3}{1}$, we can locate another point on the line by starting at $(-2, 4)$ and moving down three units and then one unit to the right to get the point $(-1, 1)$. Now draw a line through $(-2, 4)$ and $(-1, 1)$ as shown in Fig. 3.20. Since $\frac{-3}{1} = \frac{3}{-1}$, we could have obtained the second point by starting at $(-2, 4)$ and going up 3 units and to the left 1 unit.

> Now do Exercises 33–38

⟨4⟩ Parallel Lines

Two lines in a coordinate plane that do not intersect are **parallel.** Consider the two lines with slope $\frac{1}{3}$ shown in Fig. 3.21. At the y-axis these lines are 4 units apart, measured vertically. A slope of $\frac{1}{3}$ means that you can forever rise 1 and run 3 to get to another point on the line. So the lines will always be 4 units apart vertically, and they will never intersect. This example illustrates the following fact.

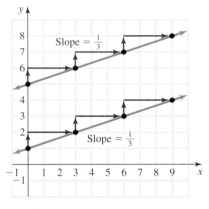

Figure 3.21

Parallel Lines
Two lines with slopes m_1 and m_2 are parallel if and only if $m_1 = m_2$.

For lines that have slope, the slopes can be used to determine whether the lines are parallel. The only lines that do not have slope are vertical lines. Of course, any two vertical lines are parallel.

E X A M P L E   6

Graphing parallel lines
Draw a line through the point $(-2, 1)$ with slope $\frac{1}{2}$ and a line through $(3, 0)$ with slope $\frac{1}{2}$.

Solution

Because slope is the ratio of rise to run, a slope of $\frac{1}{2}$ means that we can locate a second point of the line by starting at $(-2, 1)$ and going up one unit and to the right two units.

For the line through $(3, 0)$ we start at $(3, 0)$ and go up one unit and to the right two units. See Fig. 3.22.

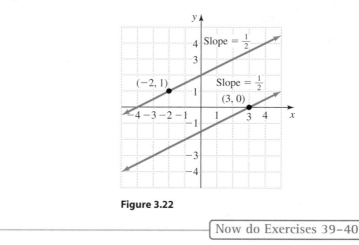

Figure 3.22

Now do Exercises 39–40

⟨5⟩ Perpendicular Lines

Figure 3.23 shows line l_1 with positive slope m_1. The rise m_1 and the run 1 are the sides of a right triangle. If l_1 and the triangle are rotated 90° clockwise, then l_1 will coincide with line l_2, and the slope of l_2 can be determined from the triangle in its new position. Starting at the point of intersection, the run for l_2 is m_1 and the rise is -1 (moving downward). So if m_2 is the slope of l_2, then $m_2 = -\frac{1}{m_1}$. *The slope of l_2 is the opposite of the reciprocal of the slope of l_1.* This result can be stated also as $m_1 m_2 = -1$ or as follows.

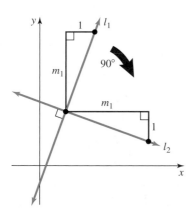

Figure 3.23

> **Perpendicular Lines**
>
> Two lines with slopes m_1 and m_2 are perpendicular if and only if
> $$m_1 = -\frac{1}{m_2}.$$

Notice that we cannot compare slopes of horizontal and vertical lines to see if they are perpendicular because slope is undefined for vertical lines. However, you should just remember that any horizontal line is perpendicular to any vertical line and vice versa.

EXAMPLE **7**

Graphing perpendicular lines

Draw two lines through the point $(-1, 2)$, one with slope $-\frac{1}{3}$ and the other with slope 3.

Solution

Because slope is the ratio of rise to run, a slope of $-\frac{1}{3}$ means that we can locate a second point on the line by starting at $(-1, 2)$ and going down one unit and to the right three units.

<Helpful Hint>

The relationship between the slopes of perpendicular lines can also be remembered as

$$m_1 \cdot m_2 = -1.$$

For example, lines with slopes -3 and $\frac{1}{3}$ are perpendicular because

$$-3 \cdot \frac{1}{3} = -1.$$

For the line with slope 3, we start at $(-1, 2)$ and go up three units and to the right one unit. See Fig. 3.24.

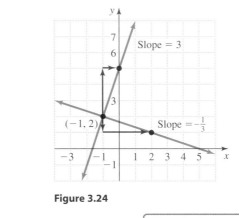

Figure 3.24

Now do Exercises 41–48

E X A M P L E   **8**

Parallel, perpendicular, or neither

Determine whether the lines l_1 and l_2 are parallel, perpendicular, or neither.

  a) l_1 goes through $(1, 2)$ and $(4, 8)$, l_2 goes through $(0, 3)$ and $(1, 5)$.

  b) l_1 goes through $(-2, 5)$ and $(3, 7)$, l_2 goes through $(8, 4)$ and $(6, 9)$.

  c) l_1 goes through $(0, 4)$ and $(-1, 6)$, l_2 goes through $(7, 7)$ and $(4, 4)$.

Solution

  a) The slope of l_1 is $\frac{8-2}{4-1}$ or 2. The slope of l_2 is $\frac{5-3}{1-0}$ or 2. Since the slopes are equal, the lines are parallel.

  b) The slope of l_1 is $\frac{7-5}{3-(-2)}$ or $\frac{2}{5}$. The slope of l_2 is $\frac{4-9}{8-6}$ or $-\frac{5}{2}$. Since one slope is the opposite of the reciprocal of the slope of the other, the lines are perpendicular.

  c) The slope of l_1 is $\frac{6-4}{-1-0}$ or -2. The slope of l_2 is $\frac{7-4}{7-4}$ or 1. Since $-2 \neq 1$ and $-2 \neq -\frac{1}{1}$, the lines are neither parallel nor perpendicular.

Now do Exercises 49–56

⟨6⟩ Applications

The slope of a line is the ratio of the rise and the run. If the rise is measured in dollars and the run in days, then the slope is measured in dollars per day (dollars/day). The slope is the amount of increase or decrease in dollars for *one* day. The slope of a line is the rate at which the dependent variable is increasing or decreasing. It is the amount of change in the dependent variable per a change of one unit in the independent variable. In some cases, the slope is a fraction, but whole numbers sound better for interpretation. For example, the birth rate at a hospital of $\frac{1}{3}$ birth/day might sound better stated as one birth per three days.

EXAMPLE **9**

Interpreting slope as a rate of change

A car goes from 60 mph to 0 mph in 120 feet after applying the brakes.

a) Find and interpret the slope of the line shown here.

b) What is the velocity at a distance of 80 feet?

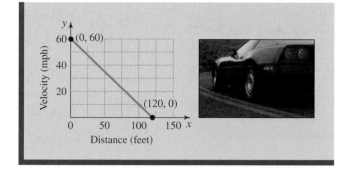

Solution

a) Find the slope of the line through (0, 60) and (120, 0):

$$m = \frac{60 - 0}{0 - 120} = -0.5$$

Because the vertical axis is miles per hour and the horizontal axis is feet, the slope is −0.5 mph/ft, which means the car is losing 0.5 mph of velocity for every foot it travels after the brakes are applied.

b) If the velocity is decreasing 0.5 mph for every foot the car travels, then in 80 feet the velocity goes down 0.5(80) or 40 mph. So the velocity at 80 feet is 60 − 40 or 20 mph.

Now do Exercises 57–60

EXAMPLE **10**

Finding points when given the slope

Assume that the base price of a new Jeep Wrangler is increasing $300 per year. Find the data that are missing from the table.

Year (x)	Price (y)
2001	$15,600
2002	
2003	
	$18,300
	$20,100

Solution

The price in 2002 is $15,900 and in 2003 it is $16,200 because the slope is $300 per year. The rise in price from $16,200 to $18,300 is $2100, which takes 7 years at $300 per year. So in 2010 the price is $18,300. The rise from $18,300 to $20,100 is $1800, which takes 6 years at $300 per year. So in 2016 the price is $20,100.

Now do Exercises 61–62

Warm-Ups ▼

Fill in the blank.

1. The _____ of a line is the ratio of its rise and run.
2. ____ is the change in y-coordinates and ____ is the change in x-coordinates.
3. Slope is undefined for _____ lines.
4. _____ lines have zero slope.
5. Lines with _____ slope are rising as you go from left to right.
6. Lines with _____ slope are falling as you go from left to right.
7. If m_1 and m_2 are the slopes of _____ lines, then $m_1 m_2 = -1$.
8. If m_1 and m_2 are the slopes of _____ lines, then $m_1 = m_2$.

True or false?

9. Slope is a measurement of the steepness of a line.
10. Every line in the coordinate plane has a slope.
11. The line through (1, 1) and the origin has slope 1.
12. A line with slope 2 is perpendicular to a line with slope -0.5.
13. The slope of the line through (0, 3) and (4, 0) is $\dfrac{3}{4}$.
14. Two different lines can't have the same slope.
15. The line through (1, 3) and (−5, 3) has zero slope.

3.2 Exercises

⟨ **Study Tips** ⟩

- Don't expect to understand a topic the first time you see it. Learning mathematics takes time, patience, and repetition.
- Keep reading the text, asking questions, and working problems. Someone once said, "All math is easy once you understand it."

⟨1⟩ **Slope**

In Exercises 1–12, find the slope of each line. See Examples 1 and 2.

1.   **2.**

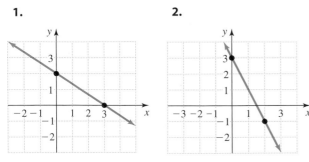

3.   **4.**

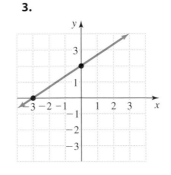
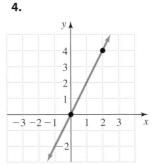

5.

6.

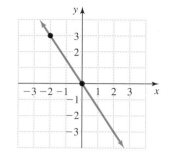

⟨2⟩ **Slope Using Coordinates**

Find the slope of the line that goes through each pair of points. See Examples 3 and 4.

13. (1, 2), (3, 6)   **14.** (2, 7), (3, 10)

15. (2, 5), (6, 10)   **16.** (5, 1), (8, 9)

17. (2, 4), (5, −1)   **18.** (3, 1), (6, −2)

19. (−2, 4), (5, 9)   **20.** (−1, 3), (3, 5)

21. (−2, −3), (−5, 1)   **22.** (−6, −3), (−1, 1)

23. (−3, 4), (3, −2)   **24.** (−1, 3), (5, −2)

25. $\left(\frac{1}{2}, 2\right), \left(-1, \frac{1}{2}\right)$ **26.** $\left(\frac{1}{3}, 2\right), \left(-\frac{1}{3}, 1\right)$

27. (2, 3), (2, −9)   **28.** (−3, 6), (8, 6)

29. (−2, −5), (9, −5)   **30.** (4, −9), (4, 6)

31. (0.3, 0.9), (−0.1, −0.3)   **32.** (−0.1, 0.2), (0.5, 0.8)

7.

8.

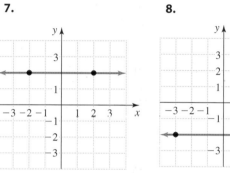

⟨3⟩ **Graphing a Line Given a Point and Slope**

Graph the line with the given point and slope. See Example 5.

33. The line through (1, 1) with slope $\frac{2}{3}$

9.

10.

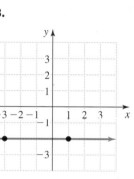

34. The line through (2, 3) with slope $\frac{1}{2}$

11.

12.

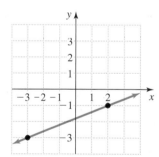
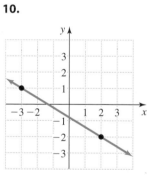
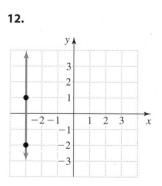
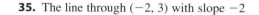

35. The line through (−2, 3) with slope −2

36. The line through $(-2, 5)$ with slope -1

37. The line through $(0, 0)$ with slope $-\frac{2}{5}$

38. The line through $(-1, 4)$ with slope $-\frac{2}{3}$

41. Draw l_1 through $(1, 2)$ with slope $\frac{1}{2}$, and draw l_2 through $(1, 2)$ with slope -2.

42. Draw l_1 through $(-2, 1)$ with slope $\frac{2}{3}$, and draw l_2 through $(-2, 1)$ with slope $-\frac{3}{2}$.

43. Draw any line l_1 with slope $\frac{3}{4}$. What is the slope of any line perpendicular to l_1? Draw any line l_2 perpendicular to l_1.

⟨4–5⟩ Parallel and Perpendicular Lines

Solve each problem. See Examples 6 and 7.

39. Draw line l_1 through $(1, -2)$ with slope $\frac{1}{2}$ and line l_2 through $(-1, 1)$ with slope $\frac{1}{2}$.

44. Draw any line l_1 with slope -1. What is the slope of any line perpendicular to l_1? Draw any line l_2 perpendicular to l_1.

40. Draw line l_1 through $(0, 3)$ with slope 1 and line l_2 through $(0, 0)$ with slope 1.

45. Draw l_1 through $(-2, -3)$ and $(4, 0)$. What is the slope of any line parallel to l_1? Draw l_2 through $(1, 2)$ so that it is parallel to l_1.

46. Draw l_1 through $(-4, 0)$ and $(0, 6)$. What is the slope of any line parallel to l_1? Draw l_2 through the origin and parallel to l_1.

47. Draw l_1 through $(-2, 4)$ and $(3, -1)$. What is the slope of any line perpendicular to l_1? Draw l_2 through $(1, 3)$ so that it is perpendicular to l_1.

48. Draw l_1 through $(0, -3)$ and $(3, 0)$. What is the slope of any line perpendicular to l_1? Draw l_2 through the origin so that it is perpendicular to l_1.

In each case, determine whether the lines l_1 and l_2 are parallel, perpendicular, or neither. See Example 8.

49. Line l_1 goes through $(3, 5)$ and $(4, 7)$. Line l_2 goes through $(11, 7)$ and $(12, 9)$.

50. Line l_1 goes through $(-2, -2)$ and $(2, 0)$. Line l_2 goes through $(-2, 5)$ and $(-1, 3)$.

51. Line l_1 goes through $(-1, 4)$ and $(2, 6)$. Line l_2 goes through $(2, -2)$ and $(4, 1)$.

52. Line l_1 goes through $(-2, 5)$ and $(4, 7)$. Line l_2 goes through $(2, 4)$ and $(3, 1)$.

53. Line l_1 goes through $(-1, 4)$ and $(4, 6)$. Line l_2 goes through $(-7, 0)$ and $(3, 4)$.

54. Line l_1 goes through $(1, 2)$ and $(1, -1)$. Line l_2 goes through $(4, 4)$ and $(3, 3)$.

55. Line l_1 goes through $(3, 5)$ and $(3, 6)$. Line l_2 goes through $(-2, 4)$ and $(-3, 4)$.

56. Line l_1 goes through $(-3, 7)$ and $(4, 7)$. Line l_2 goes through $(-5, 1)$ and $(-3, 1)$.

⟨6⟩ Applications

Solve each problem. See Examples 9 and 10.

57. *Super cost.* The average cost of a 30-second ad during the 1998 Super Bowl was $1.3 million, and in 2009 it was $3 million (www.adage.com).

   a) Find the slope of the line through $(1998, 1.3)$ and $(2009, 3)$ and interpret your result.

   b) Use the slope to estimate the average cost of an ad in 2002. Is your estimate consistent with the accompanying graph?

   c) Use the slope to predict the average cost in 2014.

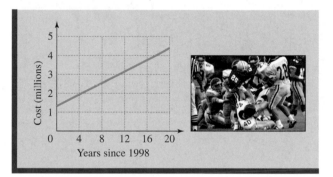

Figure for Exercise 57

58. *Retirement pay.* The annual Social Security benefit of a retiree depends on the age at the time of retirement. The accompanying graph gives the annual benefit for persons retiring at ages 62 through 70 in the year 2005 or later (Social Security Administration, www.ssa.gov). What is the annual benefit for a person who retires at age 64? At what retirement age does a person receive an annual benefit of $11,600? Find the slope of each line segment on the graph, and interpret your results. Why do people who postpone retirement until 70 years of age get the highest benefit?

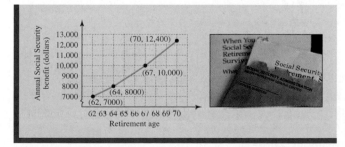

Figure for Exercise 58

59. *Increasing training.* The accompanying graph shows the percentage of U.S. workers receiving training by their employers. The percentage went from 5% in 1982 to 29% in 2006 (Department of Labor, www.dol.gov). Find the slope of this line. Interpret your result.

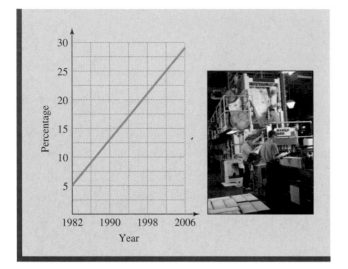

Figure for Exercise 59

60. *Saving for retirement.* Financial advisors at Fidelity Investments, Boston, use the accompanying table as a measure of whether a client is on the road to a comfortable retirement.

a) Graph these points and draw a line through them.

b) What is the slope of the line?
c) By what percentage of your salary should you be increasing your savings every year?

Age (*a*)	Years of Salary Saved (*y*)
35	0.5
40	1.0
45	1.5
50	2.0

Figure for Exercise 60

61. *Increasing salary.* An elementary school teacher gets a raise of $400 per year. Find the data that are missing from the accompanying table.

Year	Salary (dollars)
2000	
2002	28,900
	29,300
	32,900
2015	

62. *Declining population.* The population of Springfield is decreasing at a rate of 250 people per year. Find the data that are missing from the table.

Year	Population
	8400
2002	8150
2008	
	5900
	4900

Determine whether the points in each table lie on a straight line.

63.

x	*y*
4	10
7	19
11	31
17	49

64.

x	*y*
2	−4
4	−14
8	−34
13	−59

65.

x	*y*
−2	7
0	3
3	−3
9	−16

66.

x	*y*
−3	−12
0	2
2	10
6	26

3.3   Equations of Lines in Slope-Intercept Form

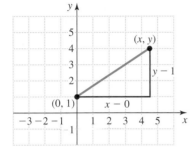

Figure 3.25

In Section 3.1 you learned that the graph of all solutions to a linear equation in two variables is a straight line. In this section, we start with a line or a description of a line and write an equation for the line. The equation of a line in any form is called a **linear equation in two variables.**

⟨1⟩ Slope-Intercept Form

Consider the line through $(0, 1)$ with slope $\frac{2}{3}$ shown in Fig. 3.25. If we use the points (x, y) and $(0, 1)$ in the slope formula, we get an equation that is satisfied by every point on the line:

$$\frac{y_2 - y_1}{x_2 - x_1} = m \quad \text{Slope formula}$$

$$\frac{y - 1}{x - 0} = \frac{2}{3} \quad \text{Let } (x_1, y_1) = (0, 1) \text{ and } (x_2, y_2) = (x, y).$$

$$\frac{y - 1}{x} = \frac{2}{3}$$

Now solve the equation for y:

$$x \cdot \frac{y - 1}{x} = \frac{2}{3} \cdot x \quad \text{Multiply each side by } x.$$

$$y - 1 = \frac{2}{3}x$$

$$y = \frac{2}{3}x + 1 \quad \text{Add 1 to each side.}$$

Because $(0, 1)$ is on the y-axis, it is called the **y-intercept** of the line. Note how the slope $\frac{2}{3}$ and the y-coordinate of the y-intercept $(0, 1)$ appear in $y = \frac{2}{3}x + 1$. For this reason, it is called the **slope-intercept form** of the equation of the line.

Slope-Intercept Form

The equation of the line with y-intercept $(0, b)$ and slope m is

$$y = mx + b.$$

Note that $y = mx + b$ is also called a linear function.

E X A M P L E   **1**

Using slope-intercept form

Write the equation of each line in slope-intercept form.

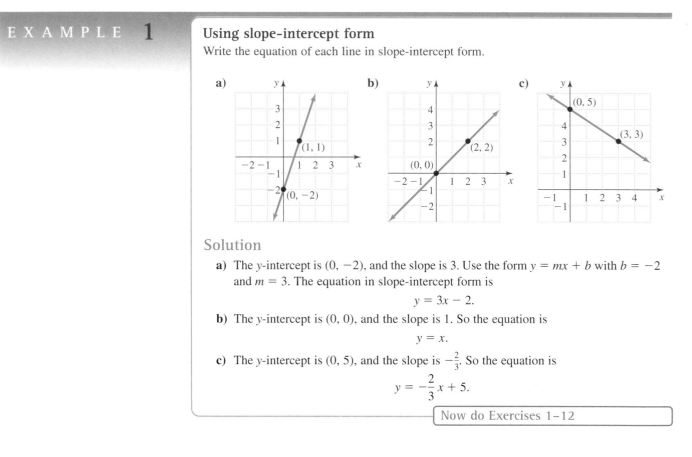

a)

b)

c)

Solution

  a) The y-intercept is $(0, -2)$, and the slope is 3. Use the form $y = mx + b$ with $b = -2$
  and $m = 3$. The equation in slope-intercept form is

$$y = 3x - 2.$$

  b) The y-intercept is $(0, 0)$, and the slope is 1. So the equation is

$$y = x.$$

  c) The y-intercept is $(0, 5)$, and the slope is $-\frac{2}{3}$. So the equation is

$$y = -\frac{2}{3}x + 5.$$

Now do Exercises 1–12

    The equation of a line may take many different forms. The easiest way to find the
slope and y-intercept for a line is to rewrite the equation in slope-intercept form.

E X A M P L E   **2**

Finding slope and y-intercept

Determine the slope and y-intercept of the line $3x - 2y = 6$.

Solution

Solve for y to get slope-intercept form:

$$3x - 2y = 6$$
$$-2y = -3x + 6$$
$$y = \frac{3}{2}x - 3$$

The slope is $\frac{3}{2}$, and the y-intercept is $(0, -3)$.

Now do Exercises 13–32

⟨2⟩ Standard Form

In Section 3.1 we defined a linear equation in two variables as an equation of the form
$Ax + By = C$, where A and B are not both zero. The form $Ax + By = C$ is called the
standard form of the equation of a line. It includes vertical lines such as $x = 6$ and

horizontal lines such as $y = 5$. *Every line has an equation in standard form.* Since slope is undefined for vertical lines, there is no equation in slope-intercept form for a vertical line. *Every nonvertical line has an equation in slope-intercept form.*

There is only one slope-intercept equation for a given line, but standard form is not unique. For example,

$$2x - 3y = 5, \quad 4x - 6y = 10, \quad x - \frac{3}{2}y = \frac{5}{2}, \quad \text{and} \quad -2x + 3y = -5$$

are all equations in standard form for the same line. When possible, we will write the standard form in which A is positive, and A, B, and C are integers with a greatest common factor of 1. So $2x - 3y = 5$ is the *preferred* standard form for this line.

In Example 2 we converted an equation in standard form to slope-intercept form. In Example 3, we convert an equation in slope-intercept form to standard form.

EXAMPLE 3

Converting to standard form

Write the equation of the line $y = \frac{2}{5}x + 3$ in standard form using only integers.

Solution

To get standard form, first subtract $\frac{2}{5}x$ from each side:

$$y = \frac{2}{5}x + 3$$

$$-\frac{2}{5}x + y = 3$$

$$-5\left(-\frac{2}{5}x + y\right) = -5 \cdot 3 \qquad \text{Multiply each side by } -5 \text{ to eliminate the fraction and get positive } 2x.$$

$$2x - 5y = -15$$

Now do Exercises 33–48

⟨3⟩ Using Slope-Intercept Form for Graphing

One way to graph a linear equation is to find several points that satisfy the equation and then draw a straight line through them. We can also graph a linear equation by using the y-intercept and the slope.

Strategy for Graphing a Line Using y-Intercept and Slope

1. Write the equation in slope-intercept form if necessary.
2. Plot the y-intercept.
3. Starting from the y-intercept, use the rise and run to locate a second point.
4. Draw a line through the two points.

EXAMPLE 4

Graphing a line using y-intercept and slope

Graph the line $2x - 3y = 3$.

‹ **Calculator Close-Up** ›

To check Example 4, graph $y = (2/3)x - 1$ on a graphing calculator as follows:

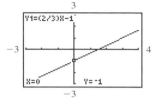

The calculator graph is consistent with the graph in Fig. 3.26.

Solution

First write the equation in slope-intercept form:

$$2x - 3y = 3$$
$$-3y = -2x + 3 \quad \text{Subtract } 2x \text{ from each side.}$$
$$y = \frac{2}{3}x - 1 \quad \text{Divide each side by } -3.$$

The slope is $\frac{2}{3}$, and the y-intercept is $(0, -1)$. A slope of $\frac{2}{3}$ means a rise of 2 and a run of 3. Start at $(0, -1)$ and go up two units and to the right three units to locate a second point on the line. Now draw a line through the two points. See Fig. 3.26 for the graph of $2x - 3y = 3$.

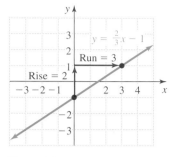

Figure 3.26

Now do Exercises 49–50

CAUTION   When using the slope to find a second point on the line, be sure to start at the y-intercept, not at the origin.

E X A M P L E   **5**

Graphing lines with y-intercept and slope

Graph each line.

a) $y = -3x + 4$ **b)** $2y - 5x = 0$

Solution

a) For $y = -3x + 4$ the slope is -3 and the y-intercept is $(0, 4)$. Because $-3 = \frac{-3}{1}$, the rise is -3 and the run is 1. First plot the y-intercept $(0, 4)$. To locate a second point on the line start at $(0, 4)$ and go down three units and to the right one unit. Draw a line through $(0, 4)$ and $(1, 1)$. See Fig. 3.27.

b) First solve the equation for y:

$$2y - 5x = 0$$
$$2y = 5x$$
$$y = \frac{5}{2}x$$

The slope is $\frac{5}{2}$ and the y-intercept is $(0, 0)$. Using a rise of five and a run of two from the origin yields the point $(2, 5)$. Draw a line through $(0, 0)$ and $(2, 5)$ as shown in Fig. 3.28.

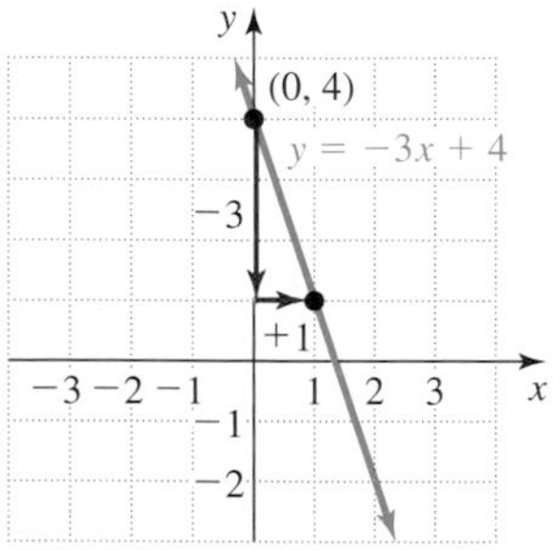

Figure 3.27

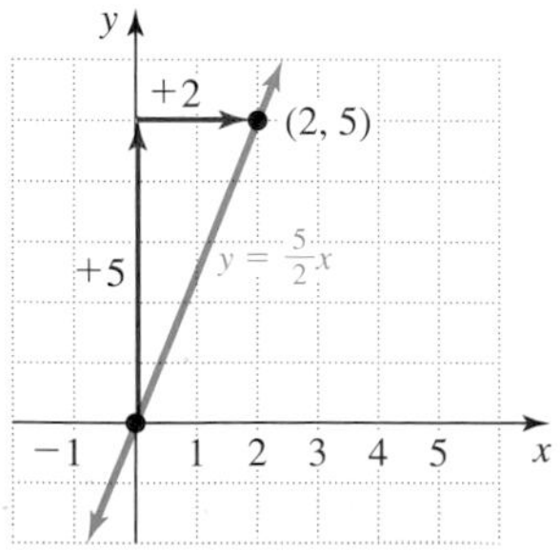

Figure 3.28

Now do Exercises 51–62

If your equation is in slope-intercept form, it is usually easiest to use the y-intercept and the slope to graph the line, as shown in Example 5. If your equation is in standard form, it is usually easiest to graph the line using the intercepts, as discussed in Section 3.1. These guidelines are summarized as follows.

The Method for Graphing Depends on the Form	
Slope-intercept form $y = mx + b$	Start at the y-intercept $(0, b)$ and count off the rise and run. This works best if b is an integer and m is rational.
Standard form $Ax + By = C$	Find the x-intercept by setting $y = 0$. Find the y-intercept by setting $x = 0$. Find one additional point as a check.

⟨4⟩ Writing the Equation for a Line

In Example 1 we wrote the equation of a line by finding its slope and y-intercept from a graph. In Example 6 we write the equation of a line from a description of the line.

EXAMPLE 6

Writing an equation

Write the equation in slope-intercept form for each line:

  a) The line through $(0, 3)$ that is parallel to the line $y = 2x - 1$

  b) The line through $(0, 4)$ that is perpendicular to the line $2x - 4y = 1$

Solution

  a) The line $y = 2x - 1$ has slope 2, and any line parallel to it has slope 2. So the equation of the line with y-intercept $(0, 3)$ and slope 2 is $y = 2x + 3$.

  b) First find the slope of $2x - 4y = 1$:

$$2x - 4y = 1$$

$$-4y = -2x + 1$$

$$y = \frac{1}{2}x - \frac{1}{4}$$

So $2x - 4y = 1$ has slope $\frac{1}{2}$ and the slope of any line perpendicular to $2x - 4y = 1$ is the opposite of the reciprocal of $\frac{1}{2}$ or -2. The equation of the line through the y-intercept $(0, 4)$ with slope -2 is $y = -2x + 4$.

Now do Exercises 71–84

⟨ **Calculator Close-Up** ⟩

If you use the same minimum and maximum window values for x and y, then the length of one unit on the x-axis is larger than on the y-axis because the screen is longer in the x-direction. In this case, perpendicular lines will not look perpendicular. The viewing window chosen here for the lines in Example 6 makes them look perpendicular.

Any viewing window proportional to this one will also produce approximately the same unit length on each axis. Some calculators

have a square feature that automatically makes the unit length the same on both axes.

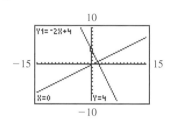

We have now seen four ways to find the slope of a line. These methods are summarized as follows:

Finding the Slope of a Line

1. Starting with a graph of a line, count the rise and run between two points and use $m = \dfrac{\text{rise}}{\text{run}}$.

2. Starting with the coordinates of two points on a line (x_1, y_1) and (x_2, y_2) use the formula $m = \dfrac{y_2 - y_1}{x_2 - x_1}$.

3. Starting with the equation of a line, rewrite it in the form $y = mx + b$ if necessary. The slope is m, the coefficient of x.

4. If a line with unknown slope m_1 is parallel or perpendicular to a line with known slope m_2, then use $m_1 = m_2$ for parallel lines or $m_1 = -\dfrac{1}{m_2}$ for perpendicular lines.

⟨5⟩ Applications

In Example 7 we see that the slope-intercept and standard forms are both important in applications.

EXAMPLE 7

Changing forms

A landscaper has a total of $800 to spend on bushes at $20 each and trees at $50 each. So if x is the number of bushes and y is the number of trees he can buy, then $20x + 50y = 800$. Write this equation in slope-intercept form. Find and interpret the y-intercept and the slope.

Solution

Write in slope-intercept form:

$$20x + 50y = 800$$
$$50y = -20x + 800$$
$$y = -\frac{2}{5}x + 16$$

The slope is $-\frac{2}{5}$ and the intercept is (0, 16). So he can get 16 trees if he buys no bushes and he loses $\frac{2}{5}$ of a tree for each additional bush that he purchases.

Now do Exercises 85–92

Warm-Ups ▼

Fill in the blank.

1. The _____ form is $y = mx + b$.

2. In $y = mx + b$, m is the _____ and (0, b) is the _____.

3. The _____ form of the equation of a line is $Ax + By = C$.

True or false?

4. There is only one line with y-intercept (0, 3) and slope $-\frac{4}{3}$.

5. The equation of the line through (1, 2) with slope 3 is $y = 3x + 2$.

6. The vertical line $x = -2$ has no y-intercept.

7. The line $y = x - 3$ is perpendicular to the line $y = 5 - x$.

8. The lines $y = 2x - 3$ and $y = 4x - 3$ are parallel.

9. The line $2y = 3x - 8$ has slope 3.

10. The line $x = 2$ is perpendicular to the line $y = 5$.

11. The line $y = x$ has no y-intercept.

12. The lines $x = -4$ and $x = 1$ are parallel.

Exercises 3.3

‹ **Study Tips** ›

• Finding out what happened in class and attending class are not the same. Attend every class and be attentive.
• Don't just take notes and let your mind wander. Use class time as a learning time.

‹ 1 › **Slope-Intercept Form**

Write an equation for each line. Use slope-intercept form if possible. See Example 1.

1.

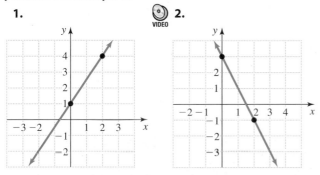

2.

3.

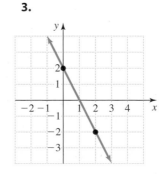

4.

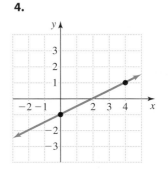

5.

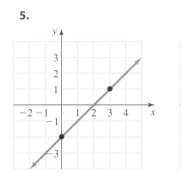

6.

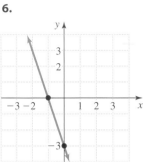

7.

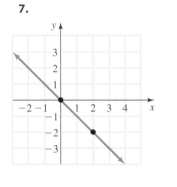

8.

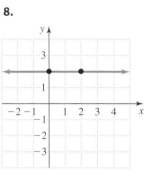

9.

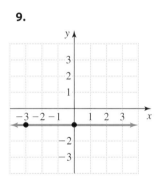

10.

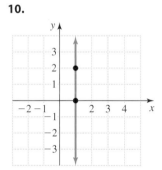

11.

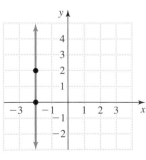

12.

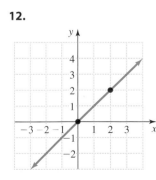

Find the slope and y-intercept for each line that has a slope and y-intercept. See Example 2.

13. $y = 3x - 9$

14. $y = -5x + 4$

15. $y = -\dfrac{1}{2}x + 3$

16. $y = \dfrac{1}{4}x + 2$

17. $y = 4$

18. $y = -5$

19. $y = x$

20. $y = -x$

21. $y = -3x$

22. $y = 2x$

23. $x + y = 5$

24. $x - y = 4$

25. $x - 2y = 4$

26. $x + 2y = 3$

 27. $2x - 5y = 10$

28. $2x + 3y = 9$

29. $2x - y + 3 = 0$

30. $3x - 4y - 8 = 0$

31. $x = -3$

32. $\dfrac{2}{3}x = 4$

⟨2⟩ **Standard Form**

Write each equation in standard form using only integers. See Example 3.

33. $y = -x + 2$

34. $y = 3x - 5$

35. $y = \dfrac{1}{2}x + 3$

36. $y = \dfrac{2}{3}x - 4$

37. $y = \dfrac{3}{2}x - \dfrac{1}{3}$

38. $y = \dfrac{4}{5}x + \dfrac{2}{3}$

39. $y = -\dfrac{3}{5}x + \dfrac{7}{10}$

40. $y = -\dfrac{2}{3}x - \dfrac{5}{6}$

41. $\dfrac{3}{5}x + 6 = 0$

42. $\dfrac{1}{2}x - 9 = 0$

43. $\dfrac{3}{4}y = \dfrac{5}{2}$

44. $\dfrac{2}{3}y = \dfrac{1}{9}$

45. $\dfrac{x}{2} = \dfrac{3y}{5}$

46. $\dfrac{x}{8} = -\dfrac{4y}{5}$

47. $y = 0.02x + 0.5$

48. $0.2x = 0.03y - 0.1$

⟨3⟩ Using Slope-Intercept Form for Graphing

Graph each line using its y-intercept and slope. See Examples 4 and 5. See the Strategy for Graphing a Line Using y-Intercept and Slope on page 201.

49. $y = 2x - 1$

50. $y = 3x - 2$

51. $y = -3x + 5$

52. $y = -4x + 1$

53. $y = \dfrac{3}{4}x - 2$

54. $y = \dfrac{3}{2}x - 4$

55. $2y + x = 0$

56. $2x + y = 0$

57. $3x - 2y = 10$

58. $4x + 3y = 9$

59. $4y + x = 8$

60. $y + 4x = 8$

61. $y - 2 = 0$

62. $y + 5 = 0$

In each case determine whether the lines are parallel, perpendicular, or neither.

63. $y = 3x - 4$
$y = 3x - 9$

64. $y = -5x + 7$
$y = \dfrac{1}{5}x - 6$

65. $y = 2x - 1$
$y = -2x + 1$

66. $y = x + 7$
$y = -x + 2$

67. $y = 3$
$y = -\dfrac{1}{3}$

68. $y = 3x + 2$
$y = \dfrac{1}{3}x - 4$

69. $y = -4x + 1$
$y = \dfrac{1}{4}x - 5$

70. $y = \dfrac{1}{3}x + \dfrac{1}{2}$
$y = \dfrac{1}{3}x - 2$

⟨4⟩ Writing the Equation for a Line

Write an equation in slope-intercept form, if possible, for each line. See Example 6.

71. The line through $(0, -4)$ with slope $\dfrac{1}{2}$

72. The line through $(0, 4)$ with slope $-\dfrac{1}{2}$

73. The line through $(0, 3)$ that is parallel to the line $y = 2x - 1$

74. The line through $(0, -2)$ that is parallel to the line

$$y = -\frac{1}{3}x + 6$$

75. The line through $(0, 6)$ that is perpendicular to the line

$$y = 3x - 5$$

76. The line through $(0, -1)$ that is perpendicular to the line
$y = x$

77. The line with y-intercept $(0, 3)$ that is parallel to the line
$2x + y = 5$

78. The line through the origin that is parallel to the line
$y - 3x = -3$

79. The line through $(2, 3)$ that runs parallel to the x-axis

80. The line through $(-3, 5)$ that runs parallel to the y-axis

81. The line through $(0, 4)$ that is perpendicular to

$$2x - 3y = 6$$

82. The line through $(0, -1)$ that is perpendicular to

$$2x - 5y = 10$$

83. The line through $(0, 4)$ and $(5, 0)$

84. The line through $(0, -3)$ and $(4, 0)$

⟨5⟩ Applications

Solve each problem. See Example 7.

85. *Labor cost.* An appliance repair service uses the formula
$C = 50n + 80$ to determine the labor cost for a service call,
where C is the cost in dollars and n is the number of hours.

   a) Find the cost of labor for $n = 0$, 1, and 2 hours.

   b) Find the slope and C-intercept for the line $C = 50n + 80$.

   c) Interpret the slope and C-intercept.

86. *Decreasing price.* World Auto uses the formula
$P = -3000n + 17{,}000$ to determine the wholesale price
for a used Ford Focus, where P is the price in dollars and n
is the age of the car in years.

   a) Find the price for a Focus that is 1, 2, or 3 years old.

   b) Find the slope and P-intercept for the line
$P = -3000n + 17{,}000$.

   c) Interpret the slope and P-intercept.

87. *Marginal cost.* A manufacturer plans to spend $150,000
on research and development for a new lawnmower and
then $200 to manufacture each mower. The function

$$C(n) = 200n + 150{,}000$$

gives the total cost in dollars of n mowers.

   a) Find $C(5000)$ and $C(5001)$.

   b) By how much did the one extra lawnmower increase
the cost in part (a)?

   c) The increase in cost in part (b) is called the *marginal
cost* of the 5001st mower. What is the marginal cost of
the 6001st mower?

   d) Find the average cost per mower when 100 mowers are
made. Average cost is total cost divided by the number
of mowers.

88. *Marginal revenue.* A defense attorney charges her client
$4000 plus $120 per hour. The function

$$R(n) = 120n + 4000$$

gives her revenue R in dollars for n hours of work.

   a) Find $R(100)$ and $R(101)$.

   b) By how much did the extra hour of work increase her
revenue in part (a)?

   c) The increase in revenue in part (a) is called the
marginal revenue for the 101st hour. What is the
marginal revenue for the 61st hour?

   d) Find the average revenue per hour when she works
10 hours.

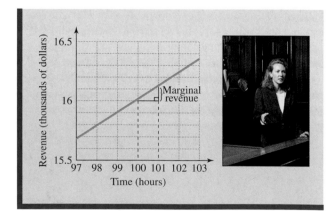

Figure for Exercise 88

89. *In-house training.* The accompanying graph shows the
percentage of U.S. workers receiving training by their

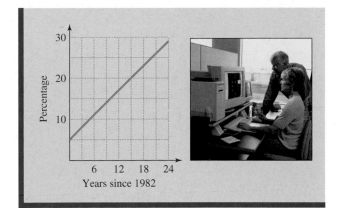

Figure for Exercise 89

employers (Department of Labor, www.dol.gov). The percentage went from 5% in 1982 to 29% in 2006.

a) Find and interpret the slope of the line.
b) Write the equation of the line in slope-intercept form.
c) What is the meaning of the *y*-intercept?
d) Use your equation to predict the percentage that will be receiving training in 2010.

90. *Single women.* The percentage of women in the 20–24 age group who have never married went from 55% in 1970 to 73% in 2000 (Census Bureau, www.census.gov). Let 1970 be year 0 and 2000 be year 30.

a) Find and interpret the slope of the line through the points (0, 55) and (30, 73).
b) Find the equation of the line in part (a).
c) What is the meaning of the *y*-intercept?
d) Use the equation to predict the percentage in 2010.
e) If this trend continues, then in what year will the percentage of women in the 20–24 age group who have never married reach 100%?

91. *Pansies and snapdragons.* A nursery manager plans to spend $100 on 6-packs of pansies at 50 cents per pack and snapdragons at 25 cents per pack. The equation $0.50x + 0.25y = 100$ can be used to model this situation.

a) What do *x* and *y* represent?

b) Graph the equation.

c) Write the equation in slope-intercept form.

d) What is the slope of the line?
e) What does the slope tell you?

92. *Pens and pencils.* A bookstore manager plans to spend $60 on pens at 30 cents each and pencils at 10 cents each. The equation $0.10x + 0.30y = 60$ can be used to model this situation.

a) What do *x* and *y* represent?

b) Graph the equation.

c) Write the equation in slope-intercept form.

d) What is the slope of the line?
e) What does the slope tell you?

Getting More Involved

Exploration

If $a \neq 0$ and $b \neq 0$, then $\frac{x}{a} + \frac{y}{b} = 1$ is called the double-intercept form for the equation of a line.

93. Find the *x*- and *y*-intercepts for $\frac{x}{2} + \frac{y}{3} = 1$.

94. Find the *x*- and *y*-intercepts for $\frac{x}{a} + \frac{y}{b} = 1$.

95. Write the equation of the line through (0, 5) and (9, 0) in double-intercept form.

96. Write the equation of the line through $\left(\frac{1}{2}, 0\right)$ and $\left(0, \frac{1}{3}\right)$ in standard form.

Graphing Calculator Exercises

Graph each pair of straight lines on your graphing calculator using a viewing window that makes the lines look perpendicular. Answers may vary.

97. $y = 12x - 100$, $y = -\dfrac{1}{12}x + 50$

98. $2x - 3y = 300$, $3x + 2y = -60$

Mid-Chapter **Quiz** | Sections 3.1 through 3.3 | **Chapter 3**

Use the given equation to find the missing coordinates in the given table.

1. $2x + 3y = 12$

x	y
−3	
	4
3	
	0

2. $y = \dfrac{3}{4}x + 3$

x	y
−4	
	12
8	
	−6

Graph each equation in the rectangular coordinate system.

3. $y = x$

4. $y = 5 - 3x$

5. $x = 4$

6. $y = 2$

7. $2x - 3y = 6$

8. $y = \dfrac{5}{3}x - 4$

Find the slope and y-intercept for each line.

9. $y = -5x + 2$

10. $y = 6$

11. $3x - 8y = 16$

Write each equation in standard form using only integers.

12. $y = -0.02x + 5$

13. $\dfrac{1}{2}y - \dfrac{1}{3}x - 9 = 0$

Find the equation in slope-intercept form for each line.

14. The line through $(0, 3)$ with slope $\dfrac{1}{3}$

15. The line through $(0, 6)$ that is parallel to $y = -5x + 12$

16. The line through $(0, -4)$ that is perpendicular to $3x - 5y = 9$

17. The line through $(4, 5)$ that is parallel to the *x*-axis

Miscellaneous.

18. Find the slope of the line through $(-3, 4)$ and $(1, -1)$.

19. Draw the graph of a line through the origin with slope $\frac{3}{4}$.

20. Is the line through $(-2, 1)$ and $(3, 5)$ parallel or perpendicular to the line through $(-4, 0)$ and $(0, -5)$?

3.4   The Point-Slope Form

In This Section

⟨**1**⟩ **Point-Slope Form**
⟨**2**⟩ **Parallel Lines**
⟨**3**⟩ **Perpendicular Lines**
⟨**4**⟩ **Applications**

In Section 3.3 we wrote the equation of a line given its slope and y-intercept. In this section, you will learn to write the equation of a line given the slope and *any* point on the line.

⟨1⟩ Point-Slope Form

Consider a line through the point $(4, 1)$ with slope $\frac{2}{3}$ as shown in Fig. 3.29. Because the slope can be found by using any two points on the line, we use $(4, 1)$ and an arbitrary point (x, y) in the formula for slope:

$$\frac{y_2 - y_1}{x_2 - x_1} = m \qquad \text{Slope formula}$$

$$\frac{y - 1}{x - 4} = \frac{2}{3} \qquad \text{Let } m = \frac{2}{3}, (x_1, y_1) = (4, 1), \text{ and } (x_2, y_2) = (x, y).$$

$$y - 1 = \frac{2}{3}(x - 4) \qquad \text{Multiply each side by } x - 4.$$

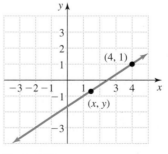

Figure 3.29

⟨ **Helpful Hint** ⟩

If a point (x, y) is on a line with slope m through (x_1, y_1), then

$$\frac{y - y_1}{x - x_1} = m.$$

Multiplying each side of this equation by $x - x_1$ gives us the point-slope form.

Note how the coordinates of the point $(4, 1)$ and the slope $\frac{2}{3}$ appear in the preceding equation. We can use the same procedure to get the equation of any line given one point on the line and the slope. The resulting equation is called the **point-slope form** of the equation of the line.

Point-Slope Form

The equation of the line through the point (x_1, y_1) with slope m is

$$y - y_1 = m(x - x_1).$$

EXAMPLE   1

Writing an equation given a point and a slope

Find the equation of the line through $(-2, 3)$ with slope $\frac{1}{2}$, and write it in slope-intercept form.

Solution

Because we know a point and the slope, we can use the point-slope form:

$$y - y_1 = m(x - x_1) \qquad \text{Point-slope form}$$

$$y - 3 = \frac{1}{2}[x - (-2)] \qquad \text{Substitute } m = \frac{1}{2} \text{ and } (x_1, y_1) = (-2, 3).$$

$$y - 3 = \frac{1}{2}(x + 2) \qquad \text{Simplify.}$$

$$y - 3 = \frac{1}{2}x + 1 \qquad \text{Distributive property}$$

$$y = \frac{1}{2}x + 4 \qquad \text{Slope-intercept form}$$

Alternate Solution

Replace m by $\frac{1}{2}$, x by -2, and y by 3 in the slope-intercept form:

$$y = mx + b \qquad \text{Slope-intercept form}$$

$$3 = \frac{1}{2}(-2) + b \qquad \text{Substitute } m = \frac{1}{2} \text{ and } (x, y) = (-2, 3).$$

$$3 = -1 + b \qquad \text{Simplify.}$$

$$4 = b$$

Since $b = 4$, we can write $y = \frac{1}{2}x + 4$.

Now do Exercises 1–18

The alternate solution to Example 1 is shown because many students have seen that method in the past. This does not mean that you should ignore the point-slope form. It is always good to know more than one method to accomplish a task. The good thing about using the point-slope form is that you immediately write down the equation and then you simplify it. In the alternate solution, the last thing you do is to write the equation.

The point-slope form can be used to find the equation of a line for *any* given point and slope. However, if the given point is the y-intercept, then it is simpler to use the slope-intercept form. Note that it is not necessary that the slope be given, because the slope can be found from any two points. So if we know two points on a line, then we can find the slope and use the slope with either one of the points in the point-slope form.

E X A M P L E   **2**

Writing an equation given two points
Find the equation of the line that contains the points $(-3, -2)$ and $(4, -1)$, and write it in standard form.

‹ **Calculator Close-Up** ›

Graph $y = (x + 3)/7 - 2$ to see that the line goes through $(-3, -2)$ and $(4, -1)$.

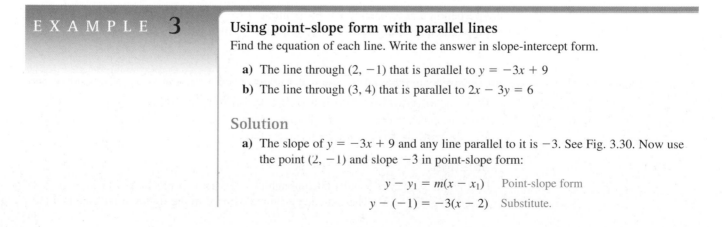

Note that the form of the equation does not matter on the calculator as long as it is solved for y.

Solution
First find the slope using the two given points:

$$m = \frac{-2 - (-1)}{-3 - 4} = \frac{-1}{-7} = \frac{1}{7}$$

Now use one of the points, say $(-3, -2)$, and slope $\frac{1}{7}$ in the point-slope form:

$$y - y_1 = m(x - x_1) \qquad \text{Point-slope form}$$

$$y - (-2) = \frac{1}{7}[x - (-3)] \qquad \text{Substitute.}$$

$$y + 2 = \frac{1}{7}(x + 3) \qquad \text{Simplify.}$$

$$7(y + 2) = 7 \cdot \frac{1}{7}(x + 3) \qquad \text{Multiply each side by 7.}$$

$$7y + 14 = x + 3$$

$$7y = x - 11 \qquad \text{Subtract 14 from each side.}$$

$$-x + 7y = -11 \qquad \text{Subtract } x \text{ from each side.}$$

$$x - 7y = 11 \qquad \text{Multiply each side by } -1.$$

The equation in standard form is $x - 7y = 11$. Using the other given point, $(4, -1)$, would give the same final equation in standard form. Try it.

Now do Exercises 19–38

‹**2**› **Parallel Lines**

In Section 3.2 you learned that parallel lines have the same slope. We will use this fact in Example 3.

E X A M P L E   **3**

Using point–slope form with parallel lines
Find the equation of each line. Write the answer in slope-intercept form.

a) The line through $(2, -1)$ that is parallel to $y = -3x + 9$

b) The line through $(3, 4)$ that is parallel to $2x - 3y = 6$

Solution

a) The slope of $y = -3x + 9$ and any line parallel to it is -3. See Fig. 3.30. Now use the point $(2, -1)$ and slope -3 in point-slope form:

$$y - y_1 = m(x - x_1) \qquad \text{Point-slope form}$$

$$y - (-1) = -3(x - 2) \qquad \text{Substitute.}$$

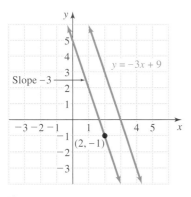

Figure 3.30

$$y + 1 = -3x + 6 \quad \text{Simplify.}$$

$$y = -3x + 5 \quad \text{Slope-intercept form}$$

Since $-1 = -3(2) + 5$ is correct, the line $y = -3x + 5$ goes through $(2, -1)$. It is certainly parallel to $y = -3x + 9$. So $y = -3x + 5$ is the desired equation.

b) Solve $2x - 3y = 6$ for y to determine its slope:

$$2x - 3y = 6$$

$$-3y = -2x + 6$$

$$y = \frac{2}{3}x - 2$$

So the slope of $2x - 3y = 6$ and any line parallel to it is $\frac{2}{3}$. Now use the point $(3, 4)$ and slope $\frac{2}{3}$ in the point-slope form:

$$y - y_1 = m(x - x_1) \quad \text{Point-slope form}$$

$$y - 4 = \frac{2}{3}(x - 3) \quad \text{Substitute.}$$

$$y - 4 = \frac{2}{3}x - 2 \quad \text{Simplify.}$$

$$y = \frac{2}{3}x + 2 \quad \text{Slope-intercept form}$$

Since $4 = \frac{2}{3}(3) + 2$ is correct, the line $y = \frac{2}{3}x + 2$ contains the point $(3, 4)$. Since $y = \frac{2}{3}x + 2$ and $y = \frac{2}{3}x - 2$ have the same slope, they are parallel. So the equation is $y = \frac{2}{3}x + 2$.

Now do Exercises 43–44

‹3› Perpendicular Lines

In Section 3.2 you learned that lines with slopes m and $-\frac{1}{m}$ (for $m \neq 0$) are perpendicular to each other. For example, the lines

$$y = -2x + 7 \qquad \text{and} \qquad y = \frac{1}{2}x - 8$$

are perpendicular to each other. In Example 4 we will write the equation of a line that is perpendicular to a given line and contains a given point.

EXAMPLE 4

Writing an equation given a point and a perpendicular line
Write the equation of the line that is perpendicular to $3x + 2y = 8$ and contains the point $(1, -3)$. Write the answer in slope-intercept form.

Solution

First graph $3x + 2y = 8$ and a line through $(1, -3)$ that is perpendicular to $3x + 2y = 8$ as shown in Fig. 3.31. The right angle symbol is used in the figure to indicate that the

< **Calculator Close-Up** >

Graph $y_1 = (2/3)x - 11/3$ and $y_2 = (-3/2)x + 4$ as shown:

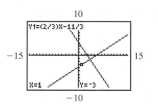

Because the lines look perpendicular and y_1 goes through $(1, -3)$, the graph supports the answer to Example 4.

lines are perpendicular. Now write $3x + 2y = 8$ in slope-intercept form to determine its slope:

$$3x + 2y = 8$$

$$2y = -3x + 8$$

$$y = -\frac{3}{2}x + 4 \quad \text{Slope-intercept form}$$

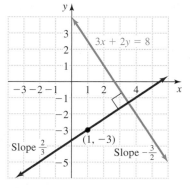

Figure 3.31

The slope of the given line is $-\frac{3}{2}$. The slope of any line perpendicular to it is $\frac{2}{3}$. Now we use the point-slope form with the point $(1, -3)$ and the slope $\frac{2}{3}$:

$$y - y_1 = m(x - x_1) \quad \text{Point-slope form}$$

$$y - (-3) = \frac{2}{3}(x - 1)$$

$$y + 3 = \frac{2}{3}x - \frac{2}{3}$$

$$y = \frac{2}{3}x - \frac{2}{3} - 3 \quad \text{Subtract 3 from each side.}$$

$$y = \frac{2}{3}x - \frac{11}{3} \quad \text{Slope-intercept form}$$

So $y = \frac{2}{3}x - \frac{11}{3}$ is the equation of the line that contains $(1, -3)$ and is perpendicular to $3x + 2y = 8$. Check that $(1, -3)$ satisfies $y = \frac{2}{3}x - \frac{11}{3}$.

Now do Exercises 45–54

< 4 > Applications

We use the point-slope form to find the equation of a line given two points on the line. In Example 5, we use that same procedure to find a linear equation that relates two variables in an applied situation.

EXAMPLE   **5**

Writing a formula given two points

A contractor charges $30 for installing 100 feet of pipe and $120 for installing 500 feet of pipe. To determine the charge, he uses a linear equation that gives the charge C in terms of the length L. Find the equation and find the charge for installing 240 feet of pipe.

Solution

Because C is determined from L, we let C take the place of the dependent variable y and let L take the place of the independent variable x. So the ordered pairs are in the form (L, C). We can use the slope formula to find the slope of the line through the two points $(100, 30)$ and $(500, 120)$ shown in Fig. 3.32.

$$m = \frac{120 - 30}{500 - 100} = \frac{90}{400} = \frac{9}{40}$$

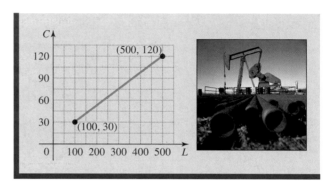

Figure 3.32

Now we use the point-slope form with the point $(100, 30)$ and slope $\frac{9}{40}$:

$$y - y_1 = m(x - x_1)$$

$$C - 30 = \frac{9}{40}(L - 100)$$

$$C - 30 = \frac{9}{40}L - \frac{45}{2}$$

$$C = \frac{9}{40}L - \frac{45}{2} + 30$$

$$C = \frac{9}{40}L + \frac{15}{2}$$

Note that $C = \frac{9}{40}L + \frac{15}{2}$ means that the charge is $\frac{9}{40}$ dollars/foot plus a fixed charge of $\frac{15}{2}$ dollars (or $7.50). We can now find C when $L = 240$:

$$C = \frac{9}{40} \cdot 240 + \frac{15}{2}$$

$$C = 54 + 7.5$$

$$C = 61.5$$

The charge for installing 240 feet of pipe is $61.50.

Now do Exercises 73–88

Warm-Ups ▼

Fill in the blank.

1. The _____ form is $y - y_1 = m(x - x_1)$.

2. Nonvertical _____ lines have equal slopes.

3. Lines with slopes m_1 and m_2 are _____ if $m_1 m_2 = -1$.

True or false?

4. It is impossible to find the equation of the line through $(2, 5)$ and $(-3, 1)$.

5. The point-slope form will not work for the line through $(3, 4)$ and $(3, 6)$.

6. The line through the origin with slope 1 is $y = x$.

7. The slope of $5x + y = 4$ is 5.

8. The slope of any line perpendicular to $y = 4x - 3$ is $-\dfrac{1}{4}$.

9. The slope of any line parallel to $x + y = 1$ is -1.

10. The line $2x - y = -1$ goes through $(-2, -3)$.

11. The lines $2x + y = 4$ and $y = -2x + 1$ are parallel.

12. The lines $y = x$ and $y = -x$ are perpendicular.

Exercises 3.4

‹ **Study Tips** ›

- When taking a test, put a check mark beside every problem that you have answered and checked. Spend any extra time working on unchecked problems.
- Make sure that you don't forget to answer any of the questions on a test.

‹ 1 › **Point-Slope Form**

Write each equation in slope-intercept form. See Example 1.

1. $x + y = 1$

2. $x - y = 1$

3. $y - 1 = 5(x + 2)$

4. $y + 3 = -3(x - 6)$

5. $3x - 4y = 80$

6. $2x + 3y = 90$

7. $y - \dfrac{1}{2} = \dfrac{2}{3}\left(x - \dfrac{1}{4}\right)$

8. $y + \dfrac{2}{3} = -\dfrac{1}{2}\left(x - \dfrac{2}{5}\right)$

Find the equation of the line that goes through the given point and has the given slope. Write the answer in slope-intercept form. See Example 1.

9. $(1, 2)$, 3

10. $(2, 5)$, 4

 11. $(2, 4)$, $\dfrac{1}{2}$

12. $(4, 6)$, $\dfrac{1}{2}$

13. $(2, 3)$, $\dfrac{1}{3}$

14. $(1, 4)$, $\dfrac{1}{4}$

15. $(-2, 5)$, $-\dfrac{1}{2}$

16. $(-3, 1)$, $-\dfrac{1}{3}$

17. $(-1, -7)$, -6 **18.** $(-1, -5)$, -8

Write each equation in standard form using only integers. See Example 2.

19. $y - 3 = 2(x - 5)$ **20.** $y + 2 = -3(x - 1)$

21. $y = \dfrac{1}{2}x - 3$ **22.** $y = \dfrac{1}{3}x + 5$

23. $y - 2 = \dfrac{2}{3}(x - 4)$ **24.** $y + 1 = \dfrac{3}{2}(x + 4)$

Find the equation of the line through each given pair of points. Write the answer in standard form using only integers. See Example 2.

25. $(1, 3)$, $(2, 5)$ **26.** $(2, 5)$, $(3, 9)$

27. $(1, 1)$, $(2, 2)$ **28.** $(-1, 1)$, $(1, -1)$

29. $(1, 2)$, $(5, 8)$ **30.** $(3, 5)$, $(8, 15)$

31. $(-2, -1)$, $(3, -4)$ **32.** $(-1, -3)$, $(2, -1)$

33. $(-2, 0)$, $(0, 2)$ **34.** $(0, 3)$, $(5, 0)$

35. $(2, 4)$, $(2, 6)$ **36.** $(-3, 5)$, $(-3, -1)$

37. $(-3, 9)$, $(3, 9)$ **38.** $(2, 5)$, $(4, 5)$

40.

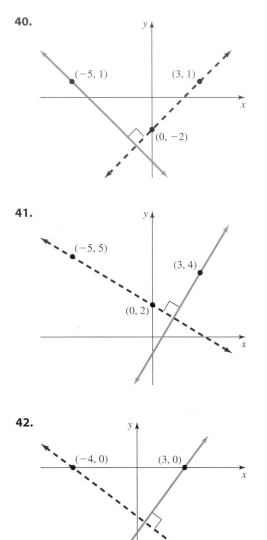

41.

42.

⟨2–3⟩ Parallel and Perpendicular Lines

The lines in each figure are perpendicular. Find the equation (in slope-intercept form) for the solid line.

39.

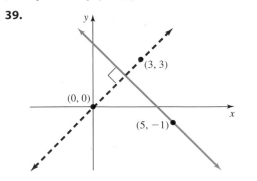

Find the equation of each line. Write each answer in slope-intercept form. See Examples 3 and 4.

43. The line is parallel to $y = x - 9$ and goes through the point $(7, 10)$.

44. The line is parallel to $y = -x + 5$ and goes through the point $(-3, 6)$.

45. The line contains the point $(3, 4)$ and is perpendicular to $y = 3x - 1$.

46. The line contains the point $(-2, 3)$ and is perpendicular to $y = 2x + 7$.

47. The line is perpendicular to $3x - 2y = 10$ and passes through the point $(1, 1)$.

48. The line is perpendicular to $x - 5y = 4$ and passes through the point $(-1, 1)$.

49. The line is parallel to $2x + y = 8$ and contains the point $(-1, -3)$.

50. The line is parallel to $-3x + 2y = 9$ and contains the point $(-2, 1)$.

51. The line goes through $(-1, 2)$ and is perpendicular to $3x + y = 5$.

52. The line goes through $(1, 2)$ and is perpendicular to $y = \frac{1}{2}x - 3$.

53. The line goes through $(2, 3)$ and is parallel to $-2x + y = 6$.

54. The line goes through $(1, 4)$ and is parallel to $x - 2y = 6$.

Miscellaneous

Find the equation of each line in the form $y = mx + b$ if possible.

55. The line through $(3, 2)$ with slope 0

56. The line through $(3, 2)$ with undefined slope

57. The line through $(3, 2)$ and the origin

58. The line through the origin that is perpendicular to $y = \frac{2}{3}x$

59. The line through the origin that is parallel to the line through $(5, 0)$ and $(0, 5)$

60. The line through the origin that is perpendicular to the line through $(-3, 0)$ and $(0, -3)$

61. The line through $(-30, 50)$ that is perpendicular to the line $x = 400$

62. The line through $(20, -40)$ that is parallel to the line $y = 6000$

63. The line through $(-5, -1)$ that is perpendicular to the line through $(0, 0)$ and $(3, 5)$

64. The line through $(3, 1)$ that is parallel to the line through $(-3, -2)$ and $(0, 0)$

For each line described here choose the correct equation from (a) through (h).

65. The line through $(1, 3)$ and $(2, 5)$

66. The line through $(1, 3)$ and $(5, 2)$

67. The line through $(1, 3)$ with no x-intercept

68. The line through $(1, 3)$ with no y-intercept

69. The line through $(1, 3)$ with x-intercept $(5, 0)$

70. The line through $(1, 3)$ with y-intercept $(0, -5)$

71. The line through $(1, 3)$ with slope -2

72. The line through $(1, 3)$ with slope $\frac{1}{2}$

    a) $x + 4y = 13$ **b)** $x = 1$

    c) $x - 2y = -5$ **d)** $y = 8x - 5$

    e) $y = 2x + 1$ **f)** $y = 3$

    g) $2x + y = 5$ **h)** $3x + 4y = 15$

⟨4⟩ Applications

Solve each problem. See Example 5.

73. *Automated tellers.* ATM volume reached 14.2 billion transactions in 2000 and 27.2 billion transactions in 2008 as shown in the accompanying graph. If 2000 is year 0 and 2008 is year 8, then the line goes through the points $(0, 14.2)$ and $(8, 27.2)$.

    a) Find and interpret the slope of the line.

    b) Write the equation of the line in slope-intercept form.

    c) Use your equation from part (b) to predict the number of transactions at automated teller machines in 2014.

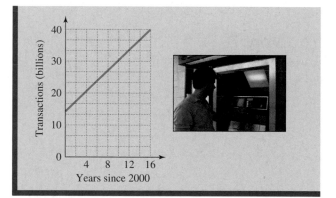

Figure for Exercise 73

74. *Direct deposit.* The percentage of workers receiving direct deposit of their paychecks went from 32% in 1994 to 71% in 2009 (www.directdeposit.com). Let 1994 be year 0 and 2009 be year 15.

    a) Write the equation of the line through $(0, 32)$ and $(15, 71)$ to model the growth of direct deposit.

    b) Use the graph on the next page to predict the year in which 100% of all workers will receive direct deposit of their paychecks.

    c) Use the equation from part (a) to predict the year in which 100% of all workers will receive direct deposit.

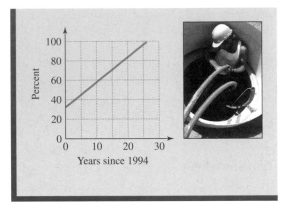

Figure for Exercise 74

75. *Gross domestic product.* The U.S. gross domestic product (GDP) per employed person increased from $62.7 thousand in 1996 to $93.8 thousand in 2007 (Bureau of Labor Statistics, www.bls.gov). Let 1996 be year 6 and 2007 be year 17.

a) Find the equation of the line through (6, 62.7) and (17, 93.8) to model the gross domestic product.

b) What do x and y represent in your equation?

c) Use the equation to predict the GDP per employed person in 2015.

d) Graph the equation.

76. *Age at first marriage.* The median age at first marriage for females increased from 24.5 years in 1995 to 26.0 years in 2007 (U.S. Census Bureau, www.census.gov). Let 1995 be year 5 and 2007 be year 17.

a) Find the equation of the line through (5, 24.5) and (17, 26.0).

b) What do x and y represent in your equation?

c) Interpret the slope of this line.

d) In what year will the median age be 30?

e) Graph the equation.

77. *Plumbing charges.* Pete worked 2 hours at Millie's house and charged her $70. He then worked 4 hours at Rosalee's house and charged her $110. Pete uses a linear function to determine the charge C from the number of hours n.

a) Find the linear function.

b) Find the charge for 7 hours at Fred's house.

c) If Pete charges Julio $270, then how many hours did he work for Julio?

78. *Interior angles.* The sum of the measures of the interior angles is 180° for a triangle and 360° for a square. The sum of the measures S for the interior angles of a regular polygon is a linear function of the number of sides n.

a) Find the linear function.

b) Find the sum of the measures of the interior angles of the stop sign in the accompanying figure.

c) If the sum of the measures of the interior angles of a regular polygon is 3240°, then how many sides does it have?

Figure for Exercise 78

79. *Shoe sizes.* If a child's foot is 7.75 inches long, then the child's shoe size is 13. If a child's foot is 5.75 inches long, then the child's shoe size is 7. The shoe size S is a linear function of the length of the foot L as shown in the accompanying figure.

a) Find the linear function.

b) Find the shoe size for a 6.25-inch foot.

c) Find the length of the foot for a child who wears a size 11.5 shoe.

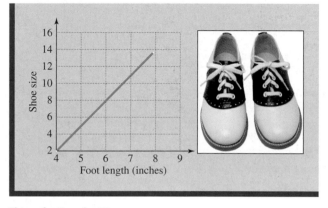

Figure for Exercise 79

1 sec
42 ft/sec

2 sec
74 ft/sec

Figure for Exercise 81

80. *Celsius to Fahrenheit.* Water freezes at 0°C or 32°F and boils at 100°C or 212°F. The Fahrenheit temperature F is a linear function of the Celsius temperature C.

a) Find the linear function.

b) Find the Fahrenheit temperature when the Celsius temperature is 45°.

c) Find the Celsius temperature when the Fahrenheit temperature is 68°.

81. *Velocity of a projectile.* A ball is thrown downward from the top of a tall building. Its velocity is 42 feet per second after 1 second and 74 feet per second after 2 seconds. The velocity v in feet per second is a linear function of time t in seconds.

a) Find the linear function. Use function notation.

b) Find $v(3.5)$.

c) Find t if $v(t) = 106$ feet per second

82. *Natural gas.* The cost of 1000 cubic feet of natural gas is $39, and the cost of 3000 cubic feet is $99. The cost C in dollars is a linear function of the amount used a in cubic feet.

a) Find the linear function. Use function notation.

b) Find $C(2400)$.

c) Find a if $C(a) = \$264$.

83. *Expansion joint.* An expansion joint on the Washington bridge has a width of 0.75 inch when the air temperature is 90°F and a width of 1.25 inches when the air temperature is 30°F. The width w in inches is a linear function of the temperature t in degrees Fahrenheit.

a) Find the linear function. Use function notation.

b) Find $w(80)$.

c) Find t when $w(t) = 1$ inch.

84. *Perimeter of a rectangle.* A rectangle has a fixed width and a variable length. The perimeter P in inches is a linear function of the length L in inches. When $L = 6.5$ inches, $P = 28$ inches. When $L = 10.5$ inches, $P = 36$ inches.

a) Find the linear function. Use function notation.

b) Find $P(40)$.

c) Find L if $P(L) = 215$ inches.

d) What is the width of the rectangle?

85. *Stretching a spring.* A weight of 3 pounds stretches a spring 1.8 inches beyond its natural length, and a weight of 5 pounds stretches the same spring 3 inches beyond its natural length. Let A represent the amount of stretch and w the weight. There is a linear equation that expresses A in terms of w. Find the equation, and find the amount that the spring will stretch with a weight of 6 pounds. See the figure on the next page.

86. *Velocity of a bullet.* A gun is fired straight upward. The bullet leaves the gun at 100 feet per second (time $t = 0$). After 2 seconds, the velocity of the bullet is 36 feet per second. There is a linear equation that gives the velocity v in terms of the time t. Find the equation and find the velocity after 3 seconds.

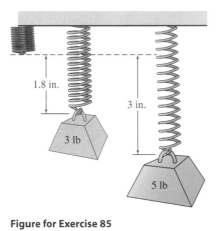

Figure for Exercise 85

87. *Enzyme concentration.* The amount of light absorbed by a certain liquid depends on the concentration of an enzyme in the liquid. A concentration of 2 milligrams per milliliter (mg/ml) produces an absorption of 0.16 and a concentration of 5 mg/ml produces an absorption of 0.40. There is a linear equation that expresses the absorption a in terms of the concentration c.

  a) Find the equation.
  b) What is the absorption when the concentration is 3 mg/ml?
  c) Use the graph to estimate the concentration when the absorption is 0.50.

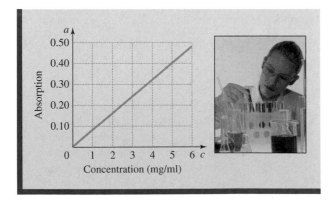

Figure for Exercise 87

88. *Basal energy requirement.* The basal energy requirement B is the number of calories that a person needs to maintain the life process. For a 28-year-old female with a height of 160 centimeters and a weight of 45 kilograms (kg), B is 1300 calories. If her weight increases to 50 kg, then B is 1365 calories. There is a linear equation that expresses B in

terms of her weight w. Find the equation, and find the basal energy requirement if her weight is 53.2 kg.

Getting More Involved

89. *Exploration*

Each linear equation in the following table is given in standard form $Ax + By = C$. In each case identify A, B, and the slope of the line.

Equation	A	B	Slope
$2x + 3y = 9$			
$4x - 5y = 6$			
$\frac{1}{2}x + 3y = 1$			
$2x - \frac{1}{3}y = 7$			

90. *Exploration*

Find a pattern in the table of Exercise 89, and write a formula for the slope of $Ax + By = C$, where $B \neq 0$.

Graphing Calculator Exercises

91. Graph each equation on a graphing calculator. Choose a viewing window that includes both the x- and y-intercepts. Use the calculator output to help you draw the graph on paper.

  a) $y = 20x - 300$
  b) $y = -30x + 500$
  c) $2x - 3y = 6000$

93. Graph $y = 0.5x + 0.8$ and $y = 0.5x + 0.7$ on a graphing calculator. Find a viewing window in which the two lines are separate.

94. Graph $y = 3x + 1$ and $y = -\frac{1}{3}x + 2$ on a graphing calculator. Do the lines look perpendicular? Explain.

92. Graph $y = 2x + 1$ and $y = 1.99x - 1$ on a graphing calculator. Are these lines parallel? Explain your answer.

3.5 Variation

In This Section

⟨1⟩ **Direct , Inverse, and Joint Variation**

⟨2⟩ **Finding the Variation Constant**

⟨3⟩ **Applications**

The linear equation $y = 5x$ can be used to determine the value of y from any given x-value. So y is a linear function of x, and $y = 5x$ is a linear function. As x varies so does y. This linear function and other functions are customarily expressed in terms of variation. In this section you will learn the language of variation and learn to write some functions from verbal descriptions.

⟨1⟩ Direct, Inverse, and Joint Variation

If you are averaging 60 miles per hour on the freeway, then the distance that you travel D is a function of the time T. Consider some possible values for T and D in the following table.

T (hours)	1	2	3	4	5	6
D (miles)	60	120	180	240	300	360

Since distance is the product of the rate and the time, the function $D = 60T$ can be used to determine the distance from the time. The graph of $D = 60T$ is shown in Fig. 3.33. Note that as T gets larger, so does D. So we say that D *varies directly* with T, or D is *directly proportional* to T. The constant rate of 60 miles per hour is the *variation constant* or *proportionality constant*. Notice that $D = 60T$ is simply a linear equation. We are just introducing some new terms to express an old idea.

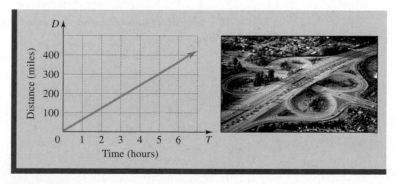

Figure 3.33

Direct Variation

The statement **"y varies directly as x"** or **"y is directly proportional to x"** means that

$$y = kx$$

for some constant k. The **variation constant** or **proportionality constant** k is a fixed nonzero real number.

CAUTION   Direct variation refers only to equations of the form $y = kx$ (lines through the origin). We do *not* refer to $y = 3x + 5$ as a direct variation.

If you plan to make a 400-mile trip by car, the time it will take is a function of your speed. Using the formula $D = RT$, we can write

$$T = \frac{400}{R}.$$

Consider the possible values for R and T given in the following table:

R (mph)	10	20	40	50	80	100
T (hours)	40	20	10	8	5	4

The graph of $T = \frac{400}{R}$ is shown in Fig. 3.34. As your rate increases, the time for the trip decreases. In this situation we say that the time is *inversely proportional* to the speed. Note that the graph of $T = \frac{400}{R}$ is not a straight line because $T = \frac{400}{R}$ is not a linear equation.

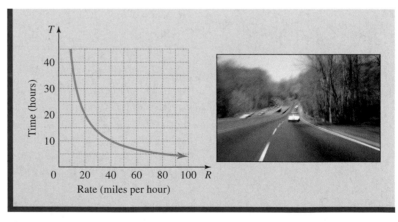

Figure 3.34

Inverse Variation

The statement **"y varies inversely as x"** or **"y is inversely proportional to x"** means that

$$y = \frac{k}{x}$$

for some nonzero constant of variation k.

CAUTION The constant of variation is usually positive because most physical examples involve positive quantities. However, the definitions of direct and inverse variation do not rule out a negative constant.

If the price of carpet is $30 per square yard, then the cost C of carpeting a rectangular room depends on the width W (in yards) and the length L (in yards). As the width or length of the room increases, so does the cost. We can write the cost as a function of the two variables L and W:

$$C = 30LW$$

We say that C *varies jointly* as L and W.

Joint Variation

The statement **"*y* varies jointly as *x* and *z*"** or **"*y* is jointly proportional to *x* and *z*"** means that

$$y = kxz$$

for some nonzero constant of variation k.

E X A M P L E   **1**   **Writing the formula**

Write a formula that expresses the relationship described in each statement. Use k as the variation constant.

a) a varies directly as t.

b) c is inversely proportional to m.

c) q varies jointly as x and y.

Solution

a) Since a varies directly as t, we have $a = kt$.

b) Since c is inversely proportional to m, we have $c = \dfrac{k}{m}$.

c) Since q varies jointly as x and y, we have $q = kxy$.

Now do Exercises 4–10

⟨2⟩ Finding the Variation Constant

If we know the values of all variables in a variation statement, we can find the value of the constant and write a formula using the value of the constant rather than an unknown constant k.

E X A M P L E   **2**   **Finding the variation constant**

Find the variation constant and write a formula that expresses the relationship described in each statement.

a) a varies directly as x, and $a = 10$ when $x = 2$.

b) w is inversely proportional to t, and $w = 10$ when $t = 5$.

c) m varies jointly as a and b, and $m = 24$ when $a = 2$ and $b = 3$.

Solution

a) Since a varies directly as x, we have $a = kx$. Since $a = 10$ when $x = 2$, we have $10 = k(2)$. Solve $2k = 10$ to get $k = 5$. So we can write the formula as $a = 5x$.

b) Since w is inversely proportional to t, we have $w = \frac{k}{t}$. Since $w = 10$ when $t = 5$, we have $10 = \frac{k}{5}$. Solve $\frac{k}{5} = 10$ to get $k = 50$. So we can write the formula $w = \frac{50}{t}$.

c) Since m varies jointly as a and b, we have $m = kab$. Since $m = 24$ when $a = 2$ and $b = 3$, we have $24 = k \cdot 2 \cdot 3$. Solve $6k = 24$ to get $k = 4$. So we can write the formula as $m = 4ab$.

> Now do Exercises 11-20

⟨3⟩ Applications

Examples 3, 4, and 5 illustrate applications of the language of variation.

E X A M P L E   3

A direct variation problem

Your electric bill at Middle States Electric Co-op varies directly with the amount of electricity that you use. If the bill for 2800 kilowatts of electricity is $196, then what is the bill for 4000 kilowatts of electricity?

Solution

Because the amount A of the electric bill varies directly as the amount E of electricity used, we have

$$A = kE$$

for some constant k. Because 2800 kilowatts cost $196, we have

$$196 = k \cdot 2800$$

or

$$0.07 = k.$$

So $A = 0.07E$. Now if $E = 4000$, we get

$$A = 0.07(4000) = 280.$$

The bill for 4000 kilowatts would be $280.

> Now do Exercises 21-22

⟨ **Helpful Hint** ⟩

In any variation problem you must first determine the general form of the relationship. Because this problem involves direct variation, the general form is $y = kx$.

E X A M P L E   4

An inverse variation problem

The volume of a gas in a cylinder is inversely proportional to the pressure on the gas. If the volume is 12 cubic centimeters when the pressure on the gas is 200 kilograms per square centimeter, then what is the volume when the pressure is 150 kilograms per square centimeter? See Fig. 3.35.

Solution

Because the volume V is inversely proportional to the pressure P, we have

$$V = \frac{k}{P}$$

$P = 200 \text{ kg/cm}^2$ $P = 150 \text{ kg/cm}^2$

$V = 12 \text{ cm}^3$ $V = ?$

Figure 3.35

for some constant k. Because $V = 12$ when $P = 200$, we can find k:

$$12 = \frac{k}{200}$$

$$200 \cdot 12 = 200 \cdot \frac{k}{200} \quad \text{Multiply each side by 200.}$$

$$2400 = k$$

Now to find V when $P = 150$, we can use the formula $V = \frac{2400}{P}$:

$$V = \frac{2400}{150} = 16$$

So the volume is 16 cubic centimeters when the pressure is 150 kilograms per square centimeter.

Now do Exercises 23–24

EXAMPLE 5

A joint variation problem

The cost of shipping a piece of machinery by truck varies jointly with the weight of the machinery and the distance that it is shipped. It costs $3000 to ship a 2500-lb milling machine a distance of 600 miles. Find the cost for shipping a 1500-lb lathe a distance of 800 miles.

Solution

Because the cost C varies jointly with the weight w and the distance d, we have

$$C = kwd$$

where k is the constant of variation. To find k, we use $C = 3000$, $w = 2500$, and $d = 600$:

$$3000 = k \cdot 2500 \cdot 600$$

$$\frac{3000}{2500 \cdot 600} = k \quad \text{Divide each side by } 2500 \cdot 600.$$

$$0.002 = k$$

Now use $w = 1500$ and $d = 800$ in the formula $C = 0.002wd$:

$$C = 0.002 \cdot 1500 \cdot 800$$

$$= 2400$$

So the cost of shipping the lathe is $2400.

Now do Exercises 25–26

⟨ **Helpful Hint** ⟩

Because the variation in this problem is joint, we know the general form is $y = kxz$, where k is the constant of variation.

CAUTION   The variation words (directly, inversely, or jointly) are never used to indicate addition or subtraction. We use multiplication in the formula unless we see the word "inversely." We use division for inverse variation.

Warm-Ups ▼

Fill in the blank.

1. If y varies _____ as x, then $y = kx$ for some constant k.

2. If y varies _____ as x, then $y = \dfrac{k}{x}$ for some constant k.

3. If y varies _____ as x and z, then $y = kxz$ for some constant k.

True or false

4. If $y = 5x$, then y is directly proportional to x.

5. If $y = \dfrac{6}{a}$, then y is inversely proportional to a.

6. If C varies jointly as h and t, then $C = ht$.

7. If y varies directly as x and $y = 8$ when $x = 2$, then the variation constant is 4.

8. If y varies inversely as x and $y = 8$ when $x = 2$, then the variation constant is $\dfrac{1}{4}$.

9. The amount of sales tax on a new car varies directly with the purchase price of the car.

10. If z varies inversely as w and $z = 10$ when $w = 2$, then $z = \dfrac{20}{w}$.

11. The time that it takes to travel a fixed distance varies inversely with the rate.

12. The distance that you can travel at a fixed rate varies directly with the time.

3.5 Exercises

‹ Study Tips ›

- Get in a habit of checking your work. Don't look in the back of the book for the answer until after you have checked your work.
- You will not always have an answer section for your problems.

‹1› Direct, Inverse, and Joint Variation

Write a formula that expresses the relationship described by each statement. Use k for the constant in each case. See Example 1.

1. T varies directly as h.
2. m varies directly as p.

3. y varies inversely as r.

4. u varies inversely as n.

5. R is jointly proportional to t and s.
6. W varies jointly as u and v.
7. i is directly proportional to b.
8. p is directly proportional to x.
9. A is jointly proportional to y and m.

10. t is inversely proportional to e.

‹2› Finding the Variation Constant

Find the variation constant, and write a formula that expresses the indicated variation. See Example 2.

11. y varies directly as x, and $y = 5$ when $x = 3$.

12. m varies directly as w, and $m = \dfrac{1}{2}$ when $w = \dfrac{1}{4}$.

13. A varies inversely as B, and $A = 3$ when $B = 2$.

14. c varies inversely as d, and $c = 5$ when $d = 2$.

15. m varies inversely as p, and $m = 22$ when $p = 9$.

16. s varies inversely as v, and $s = 3$ when $v = 4$.

17. A varies jointly as t and u, and $A = 24$ when $t = 6$ and $u = 2$.

18. N varies jointly as p and q, and $N = 720$ when $p = 3$ and $q = 2$.

19. T varies directly as u, and $T = 9$ when $u = 2$.

20. R varies directly as p, and $R = 30$ when $p = 6$.

⟨3⟩ Applications

Solve each variation problem. See Examples 3–5.

21. Y varies directly as x, and $Y = 100$ when $x = 20$. Find Y when $x = 5$.

22. n varies directly as q, and $n = 39$ when $q = 3$. Find n when $q = 8$.

23. a varies inversely as b, and $a = 3$ when $b = 4$. Find a when $b = 12$.

24. y varies inversely as w, and $y = 9$ when $w = 2$. Find y when $w = 6$.

25. P varies jointly as s and t, and $P = 56$ when $s = 2$ and $t = 4$. Find P when $s = 5$ and $t = 3$.

26. B varies jointly as u and v, and $B = 12$ when $u = 4$ and $v = 6$. Find B when $u = 5$ and $v = 8$.

27. *Aluminum flatboat.* The weight of an aluminum flatboat varies directly with the length of the boat. If a 12-foot boat weighs 86 pounds, then what is the weight of a 14-foot boat?

28. *Christmas tree.* The price of a Christmas tree varies directly with the height. If a 5-foot tree costs $20, then what is the price of a 6-foot tree?

29. *Sharing the work.* The time it takes to erect the big circus tent varies inversely as the number of elephants working on the job. If it takes four elephants 75 minutes, then how long would it take six elephants?

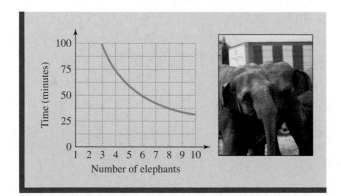

Figure for Exercise 29

30. *Gas laws.* The volume of a gas is inversely proportional to the pressure on the gas. If the volume is 6 cubic centimeters when the pressure on the gas is 8 kilograms per square centimeter, then what is the volume when the pressure is 12 kilograms per square centimeter?

31. *Steel tubing.* The cost of steel tubing is jointly proportional to its length and diameter. If a 10-foot tube with a 1-inch diameter costs $5.80, then what is the cost of a 15-foot tube with a 2-inch diameter?

32. *Sales tax.* The amount of sales tax varies jointly with the number of Cokes purchased and the price per Coke. If the sales tax on eight Cokes at 65 cents each is 26 cents, then what is the sales tax on six Cokes at 90 cents each?

33. *Approach speed.* The approach speed of an airplane is directly proportional to its landing speed. If the approach speed for a Piper Cheyenne is 90 mph with a landing speed of 75 mph, then what is the landing speed for an airplane with an approach speed of 96 mph?

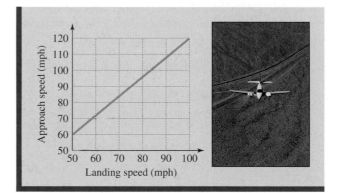

Figure for Exercise 33

34. *Ideal waist size.* According to Dr. Aaron R. Folsom of the University of Minnesota School of Public Health, your maximum ideal waist size is directly proportional to your hip size. For a woman with 40-inch hips, the maximum ideal waist size is 32 inches. What is the maximum ideal waist size for a woman with 35-inch hips?

35. *Sugar Pops.* The number of days that it takes to eat a large box of Sugar Pops varies inversely with the size of the family. If a family of three eats a box in 7 days, then how many days does it take a family of seven?

36. *Cost of CDs.* The cost for manufacturing a CD varies inversely with the number of CDs made. If the cost is $2.50 per CD when 10,000 are made, then what is the cost per CD when 100,000 are made?

37. *Carpeting.* The cost C of carpeting a rectangular living room with $20 per square yard carpet varies jointly with the length L and the width W. Fill in the missing entries in the following table.

Length (yd)	Width (yd)	Cost ($)
8	10	
10		2400
	14	3360

38. *Waterfront property.* At $50 per square foot, the price of a rectangular waterfront lot varies jointly with the length and width. Fill in the missing entries in the following table.

Length (ft)	Width (ft)	Cost ($)
60	100	
80		360,000
	150	750,000

Miscellaneous

Use the given formula to fill in the missing entries in each table, and determine whether b varies directly or inversely as a.

39. $b = \dfrac{300}{a}$

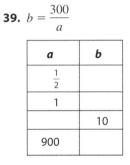

a	b
$\frac{1}{2}$	
1	
	10
900	

40. $b = \dfrac{500}{a}$

a	b
$\frac{1}{5}$	
1	
	10
1500	

41. $b = \dfrac{3}{4}a$

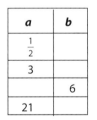

a	b
$\frac{1}{3}$	
8	
	9
20	

42. $b = \dfrac{2}{3}a$

a	b
$\frac{1}{2}$	
3	
	6
21	

For each table, determine whether y varies directly or inversely as x and find a formula for y in terms of x.

43.

x	y
2	7
3	10.5
4	14
5	17.5

44.

x	y
10	5
15	7.5
20	10
25	12.5

45.

x	y
2	10
4	5
10	2
20	1

46.

x	y
5	100
10	50
50	10
250	2

Solve each problem.

47. *Distance.* With the cruise control set at 65 mph, the distance traveled varies directly with the time spent traveling. Fill in the missing entries in the following table.

Time (hours)	1	2	3	4
Distance (miles)				

48. *Cost.* With gas selling for $1.60 per gallon, the cost of filling your tank varies directly with the amount of gas that you pump. Fill in the missing entries in the following table.

Amount (gallons)	5	10	15	20
Cost (dollars)				

49. *Time.* The time that it takes to complete a 400-mile trip varies inversely with your average speed. Fill in the missing entries in the following table.

Speed (mph)	20	40	50	
Time (hours)				2

50. *Amount.* The amount of gasoline that you can buy for $20 varies inversely with the price per gallon. Fill in the missing entries in the following table.

Price per gallon (dollars)	1	2	4	
Amount (gallons)				2

Getting More Involved

51. *Discussion*

If y varies directly as x, then the graph of the equation is a straight line. What is its slope? What is the y-intercept? If $y = 3x + 2$, then does y vary directly as x? Which straight lines correspond to direct variations?

52. *Writing*

Write a summary of the three types of variation. Include an example of each type that is not found in this text.

3.6   **Graphing Linear Inequalities in Two Variables**

In This Section

⟨1⟩ **Linear Inequalities**
⟨2⟩ **Graphing a Linear Inequality**
⟨3⟩ **The Test-Point Method**
⟨4⟩ **Applications**

You studied linear equations and inequalities in one variable in Chapter 2. In this section we extend the ideas of linear equations in two variables to study linear inequalities in two variables.

⟨1⟩ Linear Inequalities

If we replace the equals sign in any linear equation with any one of the inequality symbols $<$, $\leq$, $>$, or $\geq$, we have a *linear inequality*. For example, $x + y = 5$ is a linear equation and $x + y < 5$ is a linear inequality.

Linear Inequality in Two Variables

If A, B, and C are real numbers with A and B not both zero, then

$$Ax + By < C$$

is called a **linear inequality in two variables.** In place of $<$, we could have $\leq$, $>$, or $\geq$.

The inequalities

$$3x - 4y \leq 8, \qquad y > 2x - 3, \qquad \text{and} \qquad x - y + 9 < 0$$

are linear inequalities. Not all of these are in the form of the definition, but they could all be rewritten in that form.

A point (or ordered pair) is a solution to an inequality in two variables if the ordered pair satisfies the inequality.

E X A M P L E   **1**

Satisfying a linear inequality
Determine whether each point satisfies the inequality $2x - 3y \geq 6$.

   a) $(4, 1)$ **b)** $(3, 0)$ **c)** $(3, -2)$

Solution

   a) To determine whether $(4, 1)$ is a solution to the inequality, we replace x by 4 and y by 1 in the inequality $2x - 3y \geq 6$:

$$2(4) - 3(1) \geq 6$$
$$8 - 3 \geq 6$$
$$5 \geq 6 \quad \text{Incorrect}$$

   So $(4, 1)$ does not satisfy the inequality $2x - 3y \geq 6$.

   b) Replace x by 3 and y by 0:

$$2(3) - 3(0) \geq 6$$
$$6 \geq 6 \quad \text{Correct}$$

   So the point $(3, 0)$ satisfies the inequality $2x - 3y \geq 6$.

c) Replace x by 3 and y by -2:

$$2(3) - 3(-2) \geq 6$$
$$6 + 6 \geq 6$$
$$12 \geq 6 \quad \text{Correct}$$

So the point $(3, -2)$ satisfies the inequality $2x - 3y \geq 6$.

> Now do Exercises 1–8

⟨2⟩ Graphing a Linear Inequality

The solution set to an equation in one variable such as $x = 3$ is $\{3\}$. This single number divides the number line into two regions as shown in Fig. 3.36. Every number to the right of 3 satisfies $x > 3$, and every number to the left satisfies $x < 3$.

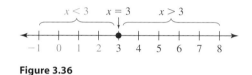

Figure 3.36

A similar situation occurs for linear equations in two variables. For example, the solution set to $y = x + 2$ consists of all ordered pairs on the line shown in Fig. 3.37. This line divides the coordinate plane into two regions. Every ordered pair above the line satisfies $y > x + 2$, and every ordered pair below the line satisfies $y < x + 2$. To see that this statement is correct, check a point such as $(3, 5)$, which is on the line. A point with a larger y-coordinate such as $(3, 6)$ is certainly above the line and satisfies $y > x + 2$. A point with a smaller y-coordinate such as $(3, 4)$ is certainly below the line and satisfies $y < x + 2$.

⟨ **Helpful Hint** ⟩

Why do we keep drawing graphs? When we solve $2x + 1 = 7$, we don't bother to draw a graph showing 3, because the solution set is so simple. However, the solution set to a linear inequality is an infinite set of ordered pairs. Graphing gives us a way to visualize the solution set.

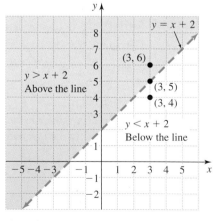

Figure 3.37

So the graph of a linear inequality consists of all ordered pairs that satisfy the inequality, and they all lie on one side of the boundary line. If the inequality symbol is $<$ or $>$, the line is not included and it is drawn dashed. If the inequality symbol is $\leq$ or $\geq$, the line is included and it is drawn solid.

Strategy for Graphing a Linear Inequality in Two Variables

1. Solve the inequality for y, and then graph $y = mx + b$.

> $y > mx + b$ is the region above the line.
> $y = mx + b$ is the line itself.
> $y < mx + b$ is the region below the line.

2. If the inequality involves only x, then graph the vertical line $x = k$.

> $x > k$ is the region to the right of the line.
> $x = k$ is the line itself.
> $x < k$ is the region to the left of the line.

3. If the inequality involves only y, then graph the horizontal line $y = k$.

> $y > k$ is the region above the line.
> $y = k$ is the line itself.
> $y < k$ is the region below the line.

Note that this case is included in part 1, but is restated for clarity.

CAUTION The symbol $>$ corresponds to "above" and the symbol $<$ corresponds to "below" only when the inequality is solved for y. You would certainly *not* shade below the line for $x - y < 0$, because $x - y < 0$ is equivalent to $y > x$. The graph of $y > x$ is the region above $y = x$.

EXAMPLE 2

Graphing a linear inequality

Graph each inequality.

a) $y < \dfrac{1}{3}x + 1$

b) $y \geq -2x + 3$

c) $2x - 3y < 6$

Solution

a) The set of points satisfying this inequality is the region below the line

$$y = \frac{1}{3}x + 1.$$

To show this region, we first graph the boundary line. The slope of the line is $\frac{1}{3}$, and the y-intercept is $(0, 1)$. We draw the line dashed because it is not part of the graph of $y < \frac{1}{3}x + 1$. In Fig. 3.38 on the next page, the graph is the shaded region.

b) Because the inequality symbol is $\geq$, every point on or above the line satisfies this inequality. We use the fact that the slope of this line is -2 and the y-intercept is $(0, 3)$ to draw the graph of the line. To show that the line $y = -2x + 3$ is included in the graph, we make it a solid line and shade the region above. See Fig. 3.39 on the next page.

⟨ **Helpful Hint** ⟩
_____

An inequality such as $y < \frac{1}{3}x + 1$ does not express y as a function of x, because there are infinitely many y-values for any given x-value. You can't determine y from x.

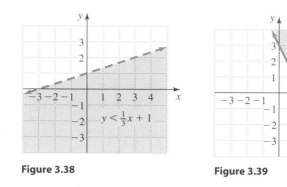

Figure 3.38   **Figure 3.39**

c) First solve for y:

$$2x - 3y < 6$$
$$-3y < -2x + 6$$
$$y > \frac{2}{3}x - 2 \quad \text{Divide by } -3 \text{ and reverse the inequality.}$$

To graph this inequality, we first graph the line with slope $\frac{2}{3}$ and y-intercept $(0, -2)$. We use a dashed line for the boundary because it is not included, and we shade the region above the line. Remember, "less than" means below the line and "greater than" means above the line only when the inequality is solved for y. See Fig. 3.40 for the graph.

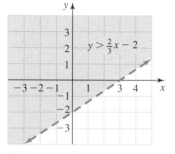

Figure 3.40

Now do Exercises 9–22

E X A M P L E   **3**

Horizontal and vertical boundary lines
Graph each inequality.

a) $y \leq 4$ **b)** $x > 3$

Solution

a) The line $y = 4$ is the horizontal line with y-intercept $(0, 4)$. We draw a solid horizontal line and shade below it as in Fig. 3.41.

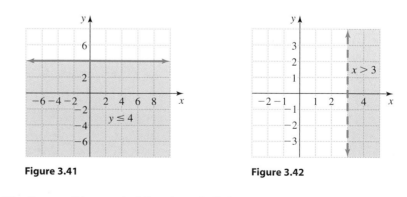

Figure 3.41   **Figure 3.42**

b) The line $x = 3$ is a vertical line through $(3, 0)$. Any point to the right of this line has an x-coordinate larger than 3. The graph is shown in Fig. 3.42.

Now do Exercises 23–26

⟨3⟩ The Test-Point Method

The graph of the linear equation $Ax + By = C$ separates the coordinate plane into two regions. All points in one region satisfy $Ax + By > C$, and all points in the other region satisfy $Ax + By < C$. To see which region satisfies which inequality we test a point in one of the regions. With this **test-point method** the form of the inequality does not matter and it does not matter how you graph the line. Here are the steps to follow.

Strategy for the Test-Point Method

To graph a linear inequality follow these steps.
1. Replace the inequality symbol with the equals symbol, and graph the resulting boundary line by using any appropriate method. Use a solid line for $\geq$ or $\leq$ and a dashed line for $>$ or $<$.
2. Select any point that is not on the line. Pick one with simple coordinates.
3. Check whether the selected point satisfies the inequality.
4. If the inequality is satisfied, shade the region containing the test point. If not, shade the other region.

E X A M P L E 4

The test-point method

Graph each inequality.

   a) $2x - 3y > 6$ **b)** $x - y \leq 0$

Solution

a) First graph the equation $2x - 3y = 6$ using the x-intercept $(3, 0)$ and the y-intercept $(0, -2)$ as shown in Fig. 3.43. Select a point on one side of the line, say $(0, 1)$, and check to see if it satisfies the inequality. Since $2(0) - 3(1) > 6$ is false, points on the other side of the line must satisfy the inequality. So shade the region on the other side of the line to get the graph of $2x - 3y > 6$, as shown in Fig. 3.44. The boundary line is dashed because the inequality symbol is $>$.

<⟨ **Helpful Hint** ⟩

Some people always like to choose $(0, 0)$ as the test point for lines that do not go through $(0, 0)$. The arithmetic for testing $(0, 0)$ is generally easier than for any other point.

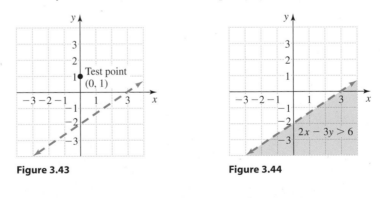

Figure 3.43   **Figure 3.44**

b) First graph $x - y = 0$. This line goes through $(1, 1)$, $(2, 2)$, $(3, 3)$, and so on. Select a point not on this line, say, $(1, 3)$, and test it in the inequality. Since $1 - 3 < 0$ is true, shade the region containing $(1, 3)$, as shown in Fig. 3.45. The boundary line is solid because the inequality symbol is $\leq$.

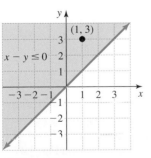

Figure 3.45

> Now do Exercises 33–44

The test-point method could be used on inequalities in one variable. For example, to solve $x > 2$ first replace the inequality symbol with equals, to get $x = 2$. The graph of $x = 2$ is a single point at 2 on the number line shown in Fig. 3.46. That point divides the number line into two regions. Every point in one region satisfies $x > 2$, and every point in the other satisfies $x < 2$. Selecting a test point such as 4 and checking that $4 > 2$ is correct tells us that the region to the right of 2 is the solution set to $x > 2$.

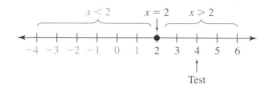

Figure 3.46

⟨4⟩ Applications

The values of variables used in applications are often restricted to nonnegative numbers. So solutions to inequalities in these applications are graphed in the first quadrant only.

E X A M P L E   5

Manufacturing tables

The Ozark Furniture Company can obtain at most 8000 board feet of oak lumber for making two types of tables. It takes 50 board feet to make a round table and 80 board feet to make a rectangular table. Write an inequality that limits the possible number of tables of each type that can be made. Draw a graph showing all possibilities for the number of tables that can be made.

Solution

If x is the number of round tables and y is the number of rectangular tables, then x and y satisfy the inequality

$$50x + 80y \leq 8000.$$

Now find the intercepts for the line $50x + 80y = 8000$:

$$50 \cdot 0 + 80y = 8000 \qquad\qquad 50x + 80 \cdot 0 = 8000$$
$$80y = 8000 \qquad\qquad\qquad 50x = 8000$$
$$y = 100 \qquad\qquad\qquad\quad x = 160$$

Draw the line through $(0, 100)$ and $(160, 0)$. Because $(0, 0)$ satisfies the inequality, the number of tables must be below the line. Since the number of tables cannot be negative, the number of tables made must be below the line and in the first quadrant as shown in Fig. 3.47. Assuming that Ozark will not make a fraction of a table, only points in Fig. 3.47 with whole-number coordinates are practical.

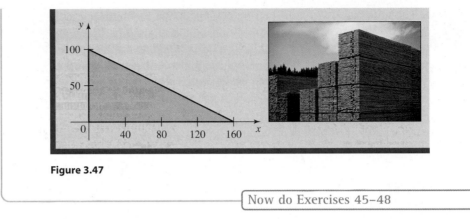

Figure 3.47

Now do Exercises 45–48

Graphical Summary of Equations and Inequalities

The graphs that follow summarize the different types of graphs that can occur for equations and inequalities in two variables. For these graphs m, b, and k are positive. Similar graphs could be made with negative numbers.

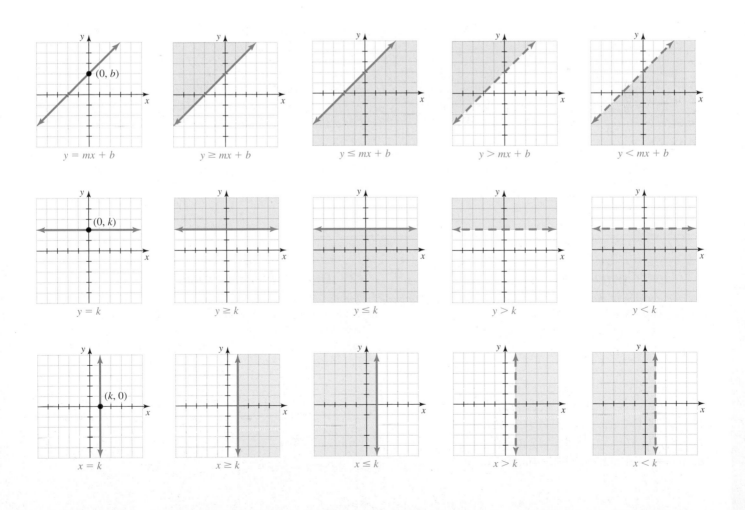

Warm-Ups ▼

Fill in the blank.

1. The inequality $Ax + By > C$ is a _____ inequality.

2. The boundary line to the graph of $Ax + By > C$ is drawn as a _____ line.

3. The boundary line to the graph of $Ax + By \geq C$ is drawn as a _____ line.

True or false?

4. The point $(1, 4)$ satisfies $y > 3x + 1$.

5. The point $(2, -3)$ satisfies $3x - 2y \geq 12$.

6. The graph of $y > x + 9$ is the region above $y = x + 9$.

7. The graph of $x < y + 2$ is the region below $x = y + 2$.

8. The graph of $x = 3$ is a single point on the x-axis.

9. The graph of $y \leq 5$ is the region below $y = 5$.

10. The graph of $x < 3$ is the region to the left of the line $x = 3$.

11. The point $(0, 0)$ is on the graph of $y \geq x$.

3.6

Exercises

‹ **Study Tips** ›

- Everyone knows that you must practice to be successful with musical instruments, foreign languages, and sports. Success in algebra also requires regular practice.
- As soon as possible after class find a quiet place to work on your homework. The longer you wait, the harder it is to remember what happened in class.

‹ 1 › **Linear Inequalities**

Determine which of the points following each inequality satisfy that inequality. See Example 1.

1. $x + y > 0$ $(0, 0), (3, -1), (-5, 4)$

2. $x + y \leq 0$ $(0, 0), (2, -1), (-6, 3)$

3. $x - y > 5$ $(2, 3), (-3, -9), (8, 3)$

4. $2x + y < 3$ $(-2, 6), (0, 3), (3, 0)$

5. $y \geq -2x + 5$ $(3, 0), (1, 3), (-2, 5)$

6. $y \leq -x + 6$ $(2, 0), (-3, 9), (-4, 12)$

7. $x > -3y + 4$ $(2, 3), (7, -1), (0, 5)$

8. $x < -y - 3$ $(1, 2), (-3, -4), (0, -3)$

‹ 2 › **Graphing a Linear Inequality**

Graph each inequality. See Examples 2 and 3. See the Strategy for Graphing a Linear Inequality in Two Variables box on page 233.

9. $y < x + 4$

10. $y < 2x + 2$

11. $y > -x + 3$　　　　VIDEO **12.** $y < -2x + 1$　　　　　**21.** $x - 2y + 4 \leq 0$　　　　**22.** $2x - y + 3 \geq 0$

13. $y > \dfrac{2}{3}x - 3$　　　　**14.** $y < \dfrac{1}{2}x + 1$　　　　**23.** $y \geq 2$　　　　VIDEO **24.** $y < 7$

15. $y \leq -\dfrac{2}{5}x + 2$　　　　**16.** $y \geq -\dfrac{1}{2}x + 3$　　　　**25.** $x > 9$　　　　VIDEO **26.** $x \leq 1$

17. $y - x \geq 0$　　　　**18.** $x - 2y \leq 0$　　　　**27.** $x + y \leq 60$　　　　**28.** $x - y \leq 90$

19. $x > y - 5$　　　　**20.** $2x < 3y + 6$　　　　**29.** $x \leq 100y$　　　　**30.** $y \geq 600x$

31. $3x - 4y \le 8$  **32.** $2x + 5y \ge 10$ **41.** $3x - 4y < -12$ **42.** $4x + 3y > 24$

43. $x < 5y - 100$ **44.** $-x > 70 - y$

⟨3⟩ The Test-Point Method

Graph each inequality using a test point. See Example 4. See the Strategy for the Test-Point Method box on page 235.

33. $2x - 3y < 6$ **34.** $x - 4y > 4$

⟨4⟩ Applications

Solve each problem. See Example 5.

45. *Storing the tables.* Ozark Furniture Company must store its oak tables before shipping. A round table is packaged in a carton with a volume of 25 cubic feet (ft^3), and a rectangular table is packaged in a carton with a volume of 35 ft^3. The warehouse has at most 3850 ft^3 of space available for these tables. Write an inequality that limits the possible number of tables of each type that can be stored, and graph the inequality in the first quadrant.

35. $x - 4y \le 8$ **36.** $3y - 5x \ge 15$

37. $y - \dfrac{7}{2}x \le 7$ **38.** $\dfrac{2}{3}x + 3y \le 12$

46. *Maple rockers.* Ozark Furniture Company can obtain at most 3000 board feet of maple lumber for making its classic and modern maple rocking chairs. A classic maple rocker requires 15 board feet of maple, and a modern rocker requires 12 board feet of maple. Write an inequality that limits the possible number of maple rockers of each type that can be made, and graph the inequality in the first quadrant.

39. $x - y < 5$ **40.** $y - x > -3$

47. *Pens and notebooks.* A student has at most $4 to spend on pens at $0.25 each and notebooks at $0.40 each. Write an inequality that limits the possibilities for the number of pens (x) and the number of notebooks (y) that can be purchased. Graph the inequality in the first quadrant.

48. *Enzyme concentration.* A food chemist tests enzymes for their ability to break down pectin in fruit juices (Dennis Callas, *Snapshots of Applications in Mathematics*). Excess pectin makes juice cloudy. In one test, the chemist measures the concentration of the enzyme, c, in milligrams per milliliter and the fraction of light absorbed by the liquid, a. If $a > 0.07c + 0.02$, then the enzyme is working as it should. Graph the inequality in the first quadrant.

Getting More Involved

49. *Discussion*

When asked to graph the inequality $x + 2y < 12$, a student found that $(0, 5)$ and $(8, 0)$ both satisfied $x + 2y < 12$. The student then drew a dashed line through these two points and shaded the region below the line. What is wrong with this method? Do all of the points graphed by this student satisfy the inequality?

50. *Writing*

Compare and contrast the two methods presented in this section for graphing linear inequalities. What are the advantages and disadvantages of each method? How do you choose which method to use?

3 Wrap-Up

Summary

Slope of a Line

Examples

Slope

The slope of the line through (x_1, y_1) and (x_2, y_2) is given by

$$m = \frac{y_2 - y_1}{x_2 - x_1}, \text{ provided that } x_2 - x_1 \neq 0.$$

$(0, 1), (3, 5)$

$$m = \frac{5 - 1}{3 - 0} = \frac{4}{3}$$

Slope is the ratio of the rise to the run for any two points on the line:

$$m = \frac{\text{change in } y}{\text{change in } x} = \frac{\text{rise}}{\text{run}}$$

Types of slope

Parallel lines

Nonvertical parallel lines have equal slopes.
Two vertical lines are parallel.

The lines $y = 3x - 9$ and $y = 3x + 7$ are parallel lines.

Perpendicular lines

Lines with slopes m and $-\frac{1}{m}$ are perpendicular.
Any vertical line is perpendicular to any horizontal line.

The lines $y = -5x + 7$ and $y = \frac{1}{5}x$ are perpendicular.

Equations of Lines

Examples

Slope-intercept form

The equation of the line with y-intercept $(0, b)$ and slope m is $y = mx + b$.

$y = 3x - 1$ has slope 3 and y-intercept $(0, -1)$.

Point-slope form

The equation of the line with slope m that contains the point (x_1, y_1) is $y - y_1 = m(x - x_1)$.

The line through $(2, -1)$ with slope -5 is $y + 1 = -5(x - 2)$.

Standard form

Every line has an equation of the form $Ax + By = C$, where A, B, and C are real numbers with A and B not both equal to zero.

$4x - 9y = 15$
$x = 5$ (vertical line)
$y = -7$ (horizontal line)

Linear function	An equation of the form $y = mx + b$	$C = 5n + 20$; C is a linear function of n.
Function notation	The independent variable is placed in parentheses after the dependent variable as in $A(x)$. (Read "A of x.")	$C(n) = 5n + 20$ $C(2) = 30$ $C(3) = 35$
Graphing a line using y-intercept and slope	1. Write the equation in slope-intercept form. 2. Plot the y-intercept. 3. Use the rise and run to locate a second point. 4. Draw a line through the two points.	

Variation		**Examples**
Direct	If $y = kx$, then y varies directly as x.	$D = 50T$
Inverse	If $y = \dfrac{k}{x}$, then y varies inversely as x.	$R = \dfrac{400}{T}$
Joint	If $y = kxz$, then y varies jointly as x and z.	$V = 6LW$

Linear Inequalities in Two Variables		**Examples**
Graphing the solution to an inequality in two variables	1. Solve the inequality for y, and then graph $y = mx + b$. $y > mx + b$ is the region above the line. $y = mx + b$ is the line itself. $y < mx + b$ is the region below the line.	$y > x + 3$ $y = x + 3$ $y < x + 3$
	Remember that "less than" means below the line and "greater than" means above the line only when the inequality is solved for y. 2. If the inequality involves only x, then graph the vertical line $x = k$. $x > k$ is the region to the right of the line. $x = k$ is the line itself. $x < k$ is the region to the left of the line.	$x > 5$ Region to right of vertical line $x = 5$
Test points	A linear inequality may also be graphed by graphing the equation and then testing a point to determine which region satisfies the inequality.	$x + y > 4$ $(0, 6)$ satisfies the inequality.

Enriching Your Mathematical Word Power

Fill in the blank.

1. The _____ of an equation is an illustration in the coordinate plane that shows all ordered pairs that satisfy the equation.

2. The _____ is the point at the intersection of the x and y-axes.

3. The first number in an ordered pair is the _____.

4. A point at which a graph intersects the y-axis is the _____.

5. The _____ variable corresponds to the first coordinate of an ordered pair.

6. The _____ variable corresponds to the second coordinate of an ordered pair.

7. The _____ of a line is the change in y-coordinates divided by the change in x-coordinates.

8. $Ax + By = C$ is the _____ form for the equation of a line.

9. The _____ -intercept form for the equation of a line is $y = mx + b$.

10. The _____ -slope form for the equation of a line is $y - y_1 = m(x - x_1)$.

11. If $y = mx + b$, then y is a _____ function of x.

12. The notation $C(x)$ is _____ notation.

13. An inequality of the form $Ax + By > 0$ is a _____ inequality in two variables.

14. If $y = kx$ for some constant k, then y varies _____ as x.

15. If $y = \dfrac{k}{x}$ for some constant k, then y varies _____ as x.

16. If $y = kxz$ for some constant k, then y varies _____ as x and z.

● Review Exercises

3.1 Graphing Lines in the Coordinate Plane

For each point, name the quadrant in which it lies or the axis on which it lies.

1. $(-2, 5)$

2. $(-3, -5)$

3. $(3, 0)$

4. $(9, 10)$

5. $(0, -6)$

6. $(0, \pi)$

7. $(1.414, -3)$

8. $(-4, 1.732)$

Complete the given ordered pairs so that each ordered pair satisfies the given equation.

9. $y = 3x - 5$: $(0, \ )$, $(-3, \ )$, $(4, \ )$

10. $y = -2x + 1$: $(9, \ )$, $(3, \ )$, $(-1, \ )$

11. $2x - 3y = 8$: $(0, \ )$, $(3, \ )$, $(-6, \ )$

12. $x + 2y = 1$: $(0, \ )$, $(-2, \ )$, $(2, \ )$

Sketch the graph of each equation by finding three ordered pairs that satisfy each equation.

13. $y = -3x + 4$

14. $y = 2x - 6$

15. $x + y = 7$

16. $x - y = 4$

3.2 Slope

Determine the slope of the line that goes through each pair of points.

17. $(0, 0)$ and $(1, 1)$

18. $(-1, 1)$ and $(2, -2)$

19. $(-2, -3)$ and $(0, 0)$

20. $(-1, -2)$ and $(4, -1)$

21. $(-4, -2)$ and $(3, 1)$

22. $(0, 4)$ and $(5, 0)$

3.3 Equations of Lines in Slope-Intercept Form

Find the slope and y-intercept for each line.

23. $y = 3x - 18$

24. $y = -x + 5$

25. $2x - y = 3$

26. $x - 2y = 1$

27. $4x - 2y - 8 = 0$

28. $3x + 5y + 10 = 0$

In each case express y as a function of x.

29. $x + y = 12$

30. $x - y = -6$

31. $3x - 4y = 20$

32. $2x + 5y = 10$

Sketch the graph of each equation.

33. $y = \frac{2}{3}x - 5$ **34.** $y = \frac{3}{2}x + 1$

35. $2x + y = -6$ **36.** $-3x - y = 2$

37. $y = -4$ **38.** $x = 9$

Determine the equation of each line. Write the answer in standard form using only integers as the coefficients.

39. The line through $(0, 4)$ with slope $\frac{1}{3}$

40. The line through $(-2, 0)$ with slope $-\frac{3}{4}$

41. The line through the origin that is perpendicular to the line $y = 2x - 1$

42. The line through $(0, 9)$ that is parallel to the line $3x + 5y = 15$

43. The line through $(3, 5)$ that is parallel to the x-axis

44. The line through $(-2, 4)$ that is perpendicular to the x-axis

3.4 The Point-Slope Form

Write each equation in slope-intercept form.

45. $y - 3 = \frac{2}{3}(x + 6)$ **46.** $y + 2 = -6(x - 1)$

47. $3x - 7y - 14 = 0$ **48.** $1 - x - y = 0$

49. $y - 5 = -\frac{3}{4}(x + 1)$ **50.** $y + 8 = -\frac{2}{5}(x - 2)$

Determine the equation of each line. Write the answer in slope-intercept form.

51. The line through $(-4, 7)$ with slope -2

52. The line through $(9, 0)$ with slope $\frac{1}{2}$

53. The line through the two points $(-2, 1)$ and $(3, 7)$

54. The line through the two points $(4, 0)$ and $(-3, -5)$

55. The line through $(3, -5)$ that is parallel to the line $y = 3x - 1$

56. The line through $(4, 0)$ that is perpendicular to the line $x + y = 3$

Solve each problem.

57. *Electric charge.* An electrician uses the linear function $C(n) = 44n + 25$ to determine the labor cost for a service call, where C is in dollars and n is the number of hours worked.

   a) Find $C(0)$, $C(2)$, and $C(8)$.
   b) Find the number of hours worked if the cost of the service call is $289.

58. *Spiral bound.* A print shop uses the linear function $C(x) = 0.12x + 5.36$ to determine the cost for making a spiral-bound report, where C is in dollars and x is the number of pages.

   a) Find $C(10)$, $C(20)$, and $C(30)$.
   b) Find the number of pages in a report for which the cost was $8.48.

59. *Rental charge.* The rental charge for an air hammer is $113 for 2 days and $209 for 5 days. The rental charge R is a linear function of the number of days n.

   a) Find the linear function. Use function notation.

   b) Find the rental charge for a four-day rental.
   c) If the rental charge was $465, then for how many days was the air hammer rented?

60. *Time on a treadmill.* After 2 minutes on a treadmill, Jenny has a heart rate of 82 beats per minute. After 3 minutes her heart rate is 86. Her heart rate h is a linear function of the time t in minutes.

   a) Find the linear function. Use function notation.

b) Find her heart rate after 10 minutes on the treadmill.

c) If her heart rate is 102 beats per minute, then how long has she been on the treadmill?

61. *Probability of rain.* If the probability of rain is 90%, then the probability that it does not rain is 10%. If the probability of rain is 80%, then the probability that it does not rain is 20%. The probability that it does not rain q is a linear function of the probability of rain p.

a) Find the linear function.

b) Use the accompanying graph to determine the probability of rain if the probability that it does not rain is 0.

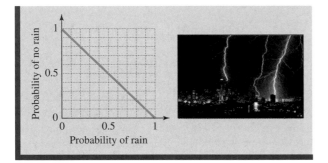

Figure for Exercise 61

62. *Social Security benefits.* If Lebron retires at age 62, 63, or 64, he will get an annual benefit of $7000, $7500, or $8000, respectively (Social Security Administration, www.ssa.gov). His benefit b is a linear function of age a for these years. Find the function.

63. *Predicting freshmen GPA.* A researcher who is studying the relationship between ACT score and grade point average for freshmen gathered the data shown in the accompanying table. Find the equation of the line in slope-intercept form that goes through these points.

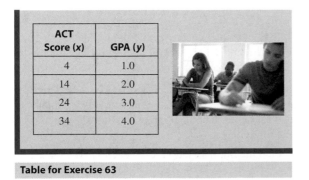

ACT Score (x)	GPA (y)
4	1.0
14	2.0
24	3.0
34	4.0

Table for Exercise 63

64. *Interest rates.* A credit manager rates each applicant for a car loan on a scale of 1 through 5 and then determines the interest rate from the accompanying table. Find the equation of the line in slope-intercept form that goes through these points.

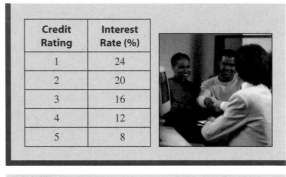

Credit Rating	Interest Rate (%)
1	24
2	20
3	16
4	12
5	8

Table for Exercise 64

3.5 Variation

Solve each variation problem.

65. Suppose y varies directly as w. If $y = 48$ when $w = 4$, then what is y when $w = 11$?

66. Suppose m varies directly as t. If $m = 13$ when $t = 2$, then what is m when $t = 6$?

67. If y varies inversely as v and $y = 8$ when $v = 6$, then what is y when $v = 24$?

68. If y varies inversely as r and $y = 9$ when $r = 3$, then what is y when $r = 9$?

69. Suppose y varies jointly as u and v, and $y = 72$ when $u = 3$ and $v = 4$. Find y when $u = 5$ and $v = 2$.

70. Suppose q varies jointly as s and t, and $q = 10$ when $s = 4$ and $t = 3$. Find q when $s = 25$ and $t = 6$.

71. *Taxi fare.* The cost of a taxi ride varies directly with the length of the ride in minutes. A 12-minute ride costs $9.00.

a) Write the cost C in terms of the length T of the ride.

b) What is the cost of a 20-minute ride?

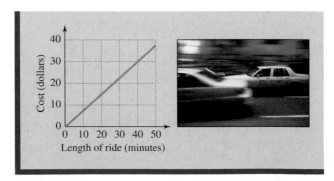

Figure for Exercise 71

c) Is the cost increasing or decreasing as the length of the ride increases?

76. $y \geq x - 6$

72. *Applying shingles.* The number of hours that it takes to apply 296 bundles of shingles varies inversely with the number of roofers working on the job. Three workers can complete the job in 40 hours.

a) Write the number of hours h in terms of the number n of roofers on the job.

77. $y \leq 8$

b) How long would it take five roofers to complete the job?

c) Is the time to complete the job increasing or decreasing as the number of workers increases?

78. $x \geq -6$

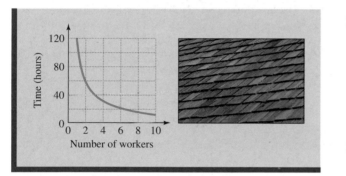

Figure for Exercise 72

79. $2x + 3y \leq -12$

3.6 Graphing Linear Inequalities in Two Variables

Graph each inequality.

80. $x - 3y < 9$

73. $y > \dfrac{1}{3}x - 5$

74. $y < \dfrac{1}{2}x + 2$

Miscellaneous

Write each equation in slope-intercept form.

81. $x - y = 1$

82. $x = 5 - y$

83. $2x + 4y = 16$

84. $3x + 5y = 10$

85. $y - 3 = 4(x - 2)$

86. $y + 6 = -3(x - 1)$

75. $y \leq -2x + 7$

87. $\dfrac{1}{2}x - \dfrac{1}{3}y = 12$

88. $\frac{2}{3}x + \frac{3}{4}y = 18$

Find the x- and y-intercepts for each line.

89. $x + y = 1$

90. $x - y = 6$

91. $3x + 4y = 12$

92. $5x + 6y = 30$

93. $y = 4x - 2$

94. $y = -3x - 1$

95. $\frac{3}{2}x + \frac{1}{3}y = 6$

96. $-\frac{2}{3}x + \frac{1}{4}y = 2$

Find the equation of each line in slope-intercept form.

97. The line through $(6, 0)$ with slope $\frac{1}{2}$

98. The line through $(-3, 0)$ with slope $\frac{2}{3}$

99. The line through $(2, 3)$ that is parallel to $y = 9$

100. The line through $(4, -5)$ that is perpendicular to $x = 1$

101. The line through $(-3, 0)$ and $(0, 9)$

102. The line through $(4, 0)$ and $(0, -6)$

103. The line through $(1, 1)$ and $(-2, 2)$

104. The line through $(-5, -3)$ and $(1, 1)$

105. The line through $(0, 2)$ that is perpendicular to $y = \frac{1}{4}x$

106. The line through $(0, 5)$ that is perpendicular to $y = -2x$

107. The line through $(1, 2)$ that is parallel to $3x - y = 0$

108. The line through $(2, -11)$ that is parallel to $y + 3x = 1$

Chapter 3 Test

For each point, name the quadrant in which it lies or the axis on which it lies.

1. $(-2, 7)$ **2.** $(-\pi, 0)$

3. $(3, -6)$ **4.** $(0, 1785)$

Find the slope of the line through each pair of points.

5. $(3, 3)$ and $(4, 4)$ **6.** $(-2, -3)$ and $(4, -8)$

Find the slope of each line.

7. The line $y = 3x - 5$ **8.** The line $y = 3$

9. The line $x = 5$ **10.** The line $2x - 3y = 4$

Write the equation of each line. Give the answer in slope-intercept form.

11. The line through $(0, 3)$ with slope $-\frac{1}{2}$

12. The line through $(-1, -2)$ with slope $\frac{3}{7}$

Write the equation of each line. Give the answer in standard form using only integers as the coefficients.

13. The line through $(2, -3)$ that is perpendicular to the line $y = -3x + 12$

14. The line through $(3, 4)$ that is parallel to the line $5x + 3y = 9$

Sketch the graph of each equation.

15. $y = \frac{1}{2}x - 3$ **16.** $2x - 3y = 6$

17. $y = 4$ **18.** $x = -2$

Graph each inequality.

19. $y > 3x - 5$ **20.** $x - y < 3$

21. $x - 2y \geq 4$

25. The demand for tickets to a play can be modeled by the equation $d = 1000 - 20p$, where d is the number of tickets sold and p is the price per ticket in dollars.

   a) How many tickets will be sold at $10 per ticket?
   b) Find the intercepts and interpret them.

   c) Find and interpret the slope, including units.

Solve each problem.

22. Julie's mail-order CD club charges a shipping and handling fee of $2.50 plus $0.75 per CD for each order shipped. Write the shipping and handling fee S in terms of the number n of CDs in the order.

23. The price in dollars p for a supreme pizza is determined by the function

$$p(n) = 0.50n + 12.75$$

where n is the number of toppings.
   a) Find $p(1)$, $p(3)$, and $p(10)$.

   b) Find the number of toppings on a pizza for which the price is $16.25.

24. A 10-ounce soft drink sells for 50 cents and a 16-ounce soft drink sells for 68 cents. The price P in cents is a linear function of the volume v in ounces.
   a) Find the linear function. Use function notation.

   b) Find the price of a 20-ounce soft drink.
   c) Find the number of ounces in a soft drink for which the price is $1.64.

26. The price P for a watermelon varies directly with its weight w.
   a) Write a formula for this variation.
   b) If the price of a 30-pound watermelon is $4.20, then what is the price of a 20-pound watermelon?

27. The number n of days that Jerry spends on the road is inversely proportional to the amount A of his sales for the previous month.
   a) Write a formula for this variation.

   b) Jerry spent 15 days on the road in February because his January sales amount was $75,000. If his August sales amount is $60,000, then how many days would he spend on the road in September?
   c) Does his road time increase or decrease as his sales increase?

28. The cost C for installing ceramic floor tile in a rectangular room varies jointly with the length L and width W of the room.
   a) Write a formula for this variation.
   b) The cost is $400 for a room that is 8 feet by 10 feet. What is the cost for a room that is 11 feet by 14 feet?

Graph Paper

Use these grids for graphing. Make as many copies of this page as you need. If you have access to a computer, you can download this page from www.mhhe.com/dugopolski and print it.

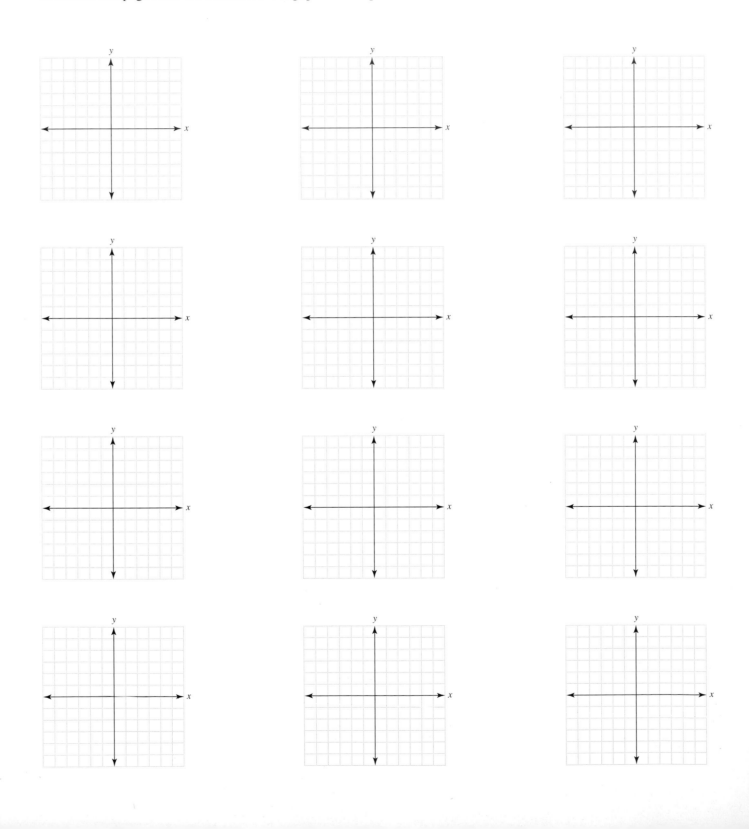

Making Connections | A Review of Chapters 1–3

Simplify each arithmetic expression.

1. $9 - 5 \cdot 2$

2. $-4 \cdot 5 - 7 \cdot 2$

3. $3^2 - 2^3$

4. $3^2 \cdot 2^3$

5. $(-4)^2 - 4(1)(5)$

6. $-4^2 - 4 \cdot 3$

7. $\dfrac{-5 - 9}{2 - (-2)}$

8. $\dfrac{6 - 3.6}{6}$

9. $\dfrac{1 - \frac{1}{2}}{4 - (-1)}$

10. $\dfrac{4 - (-6)}{1 - \frac{1}{3}}$

Simplify the given expression or solve the given equation, whichever is appropriate.

11. $4x - (-9x)$

12. $4(x - 9) - x$

13. $5(x - 3) + x = 0$

14. $5 - 2(x - 1) = x$

15. $\dfrac{1}{2} - \dfrac{1}{3}$

16. $\dfrac{1}{4} + \dfrac{1}{6}$

17. $\dfrac{1}{2}x - \dfrac{1}{3} = \dfrac{1}{4}x + \dfrac{1}{6}$

18. $\dfrac{2}{3}x + \dfrac{1}{5} = \dfrac{3}{5}x - \dfrac{1}{15}$

19. $\dfrac{4x - 8}{2}$

20. $\dfrac{-5x - 10}{-5}$

21. $\dfrac{6 - 2(x - 3)}{2} = 1$

22. $\dfrac{20 - 5(x - 5)}{5} = 6$

23. $-4(x - 9) - 4 = -4x$

24. $4(x - 6) = -4(6 - x)$

Solve each inequality. State the solution set using interval notation.

25. $2x - 3 > 6$

26. $5 - 3x < 7$

27. $51 - 2x \le 3x + 1$

28. $4x - 80 \ge 60 - 3x$

29. $-1 < 4 - 2x \le 5$

30. $1 - 2x \le x + 1 < 3 - 2x$

Solve each equation for y.

31. $3\pi y + 2 = t$

32. $x = \dfrac{y - b}{m}$

33. $3x - 3y - 12 = 0$

34. $2y - 3 = 9$

35. $\dfrac{y}{2} - \dfrac{y}{4} = \dfrac{1}{5}$

36. $0.6y - 0.06y = 108$

Solve each problem.

37. Which quadrant contains no points on the graph of $2y - 3x = 5$?

38. Which quadrants contain no points on the graph of $y = 22$?

39. Find the intercepts for the graph of $y = -3x + 6$.

40. Find the slope of the line that goes through $(2, -5)$ and $(-3, 10)$.

41. Find the slope of the line $5x + 12y = 36$.

42. Find the slope and y-intercept for the line $y = -4x + 7$.

43. Find the slope of any line that is perpendicular to $2x + 3y = 9$.

44. Find the slope of any line that is parallel to $5x - 10y = 1$.

Find the equation of each line in slope-intercept form.

45. The line through $(0, -12)$ with slope 5

46. The line through $(2, 3)$ with slope $\dfrac{1}{2}$

47. The line through $(4, 5)$ that is parallel to $y = 6$

48. The line through $(1, 6)$ that is parallel to $y = 3x - 8$

49. The line through $(-2, 7)$ that is perpendicular to $5x + 10y = 8$

50. The line through $(3, -5)$ and $(-3, 7)$

Solve.

51. *Financial planning.* Financial advisors at Fidelity Investments use the information in the accompanying graph as a guide for retirement investing.

   a) What is the slope of the line segment for ages 35 through 50?
   b) What is the slope of the line segment for ages 50 through 65?
   c) If a 38-year-old man is making $40,000 per year, then what percent of his income should he be saving?
   d) If a 58-year-old woman has an annual salary of $60,000, then how much should she have saved and how much should she be saving per year?

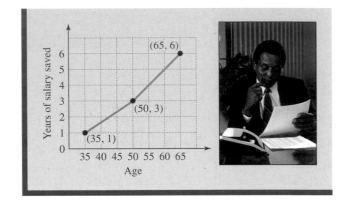

Figure for Exercise 51

Critical Thinking | For Individual or Group Work | Chapter 3

These exercises can be solved by a variety of techniques, which may or may not require algebra. So be creative and think critically. Explain all answers. Answers are in the Instructor's Edition of this text.

1. ***Share and share alike.*** A chocolate bar consists of two rows of small squares with four squares in each row as shown in part (a) of the accompanying figure. You want to share it with your friends.

   a) How many times must you break it to get it divided into 8 small squares?

   b) If the bar has 3 rows of 5 squares in each row as shown in part (b) of the accompanying figure, then how many breaks does it take to separate it into 15 small squares?

   c) If the bar is divided into *m* rows with *n* small squares in each row, then how many breaks does it take to separate it into *mn* small squares?

Photo for Exercise 4

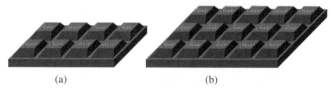

(a)   (b)

Figure for Exercise 1

2. ***Straight time.*** Starting at 8 A.M. determine the number of times in the next 24 hours for which the hour and minute hands on a clock form a 180° angle.

3. ***Dividing days by months.*** For how many days of the year do you get a whole number when you divide the day number by the month number? For example, for December 24, the result of 24 divided by 12 is 2.

4. ***Crossword fanatic.*** Ms. Smith loves to work the crossword puzzle in her daily newspaper. To keep track of her efforts, she gives herself 2 points for every crossword puzzle that she completes correctly and deducts 3 points for every crossword puzzle that she fails to complete or completes incorrectly. For the month of June her total score was zero. How many puzzles did she solve correctly in June?

5. ***Counting ones.*** If you write down the integers between 1 and 100 inclusive, then how many times will you write the number one?

6. ***Smallest sum.*** What is the smallest possible sum that can be obtained by adding five positive integers that have a product of 48?

7. ***Mind control.*** Each student in your class should think of an integer between 2 and 9 inclusive. Multiply your integer by 9. Think of the sum of the digits in your answer. Subtract 5 from your answer. Think of the letter in the alphabet that corresponds to the last answer. Think of a state that begins with that letter. Think of the second letter in the name of the state. Think of a large mammal that begins with that letter. Think of the color of that animal. What is the color that is on everyone's mind? Explain.

8. ***Four-digit numbers.*** How many four-digit whole numbers are there such that the thousands digit is odd, the hundreds digit is even, and all four digits are different? How many four-digit whole numbers are there such that the thousands digit is even, hundreds digit is odd, and all four digits are different?

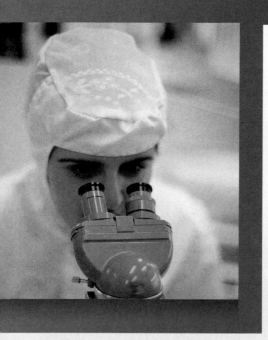

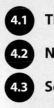

Exponents and Polynomials

The nineteenth-century physician and physicist Jean Louis Marie Poiseuille (1799–1869) is given credit for discovering a formula associated with the circulation of blood through arteries. Poiseuille's law, as it is known, can be used to determine the velocity of blood in an artery at a given distance from the center of the artery. The formula states that the flow of blood in an artery is faster toward the center of the blood vessel and is slower toward the outside. Blood flow can also be affected by a person's blood pressure, the length of the blood vessel, and the viscosity of the blood itself.

In later years, Poiseuille's continued interest in blood circulation led him to experiments to show that blood pressure rises and falls when a person exhales and inhales. In modern medicine, physicians can use Poiseuille's law to determine how much the radius of a blocked blood vessel must be widened to create a healthy flow of blood.

In this chapter, you will study polynomials, the fundamental expressions of algebra. Polynomials are to algebra what integers are to arithmetic. We use polynomials to represent quantities in general, such as perimeter, area, revenue, and the volume of blood flowing through an artery.

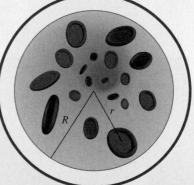

In Exercise 87 of Section 4.7, you will see Poiseuille's law represented by a polynomial.

4.1   The Rules of Exponents

In This Section

⟨1⟩ **The Product Rule for Exponents**

⟨2⟩ **Zero Exponent**

⟨3⟩ **The Quotient Rule for Exponents**

⟨4⟩ **The Power of a Power Rule**

⟨5⟩ **The Power of a Product Rule**

⟨6⟩ **The Power of a Quotient Rule**

⟨7⟩ **The Amount Formula**

We defined exponential expressions with positive integral exponents in Chapter 1. In this section, we will review that definition and then learn the rules for positive integral exponents.

⟨1⟩ The Product Rule for Exponents

Exponents were defined in Chapter 1 as a simple way of expressing repeated multiplication. For example,

$$x^1 = x, \quad y^2 = y \cdot y, \quad 5^3 = 5 \cdot 5 \cdot 5, \quad \text{and} \quad a^4 = a \cdot a \cdot a \cdot a.$$

To find the product of the exponential expressions x^3 and x^5 we could simply count the number of times x appears in the product:

$$x^3 \cdot x^5 = \overbrace{(x \cdot x \cdot x)}^{3 \text{ factors}}\overbrace{(x \cdot x \cdot x \cdot x \cdot x)}^{5 \text{ factors}} = x^8$$
$$\underbrace{}_{8 \text{ factors}}$$

Instead of counting to find that x occurs 8 times it is easier to add 3 and 5 to get 8. This example illustrates the **product rule for exponents.**

> **Product Rule for Exponents**
>
> If a is any real number, and m and n are positive integers, then
>
> $$a^m \cdot a^n = a^{m+n}.$$

CAUTION By the product rule $2^3 \cdot 2^2 = 2^5$. Note that $2^3 \cdot 2^2 \neq 4^5$ and $2^3 \cdot 2^2 \neq 2^6$. The bases are not multiplied in the product rule and neither are the exponents.

E X A M P L E   1

Using the product rule for exponents
Find the indicated products.

a) $2^3 \cdot 2^2$ **b)** $x^2 \cdot x^4 \cdot x$ **c)** $2y^3 \cdot 4y^8$ **d)** $-4a^2b^3(-3a^5b^9)$

Solution

a) $2^3 \cdot 2^2 = 2^5$ Product rule for exponents

$\quad\quad\quad = 32$ Simplify.

b) $x^2 \cdot x^4 \cdot x = x^2 \cdot x^4 \cdot x^1$ Product rule for exponents

$\quad\quad\quad\quad = x^7$

c) $2y^3 \cdot 4y^8 = (2)(4)y^3y^8$ Product rule for exponents

$\quad\quad\quad\quad = 8y^{11}$

d) $-4a^2b^3(-3a^5b^9) = (-4)(-3)a^2a^5b^3b^9$

$\quad\quad\quad\quad\quad\quad = 12a^7b^{12}$ Product rule for exponents

 Now do Exercises 1–12

⟨2⟩ Zero Exponent

A positive integer exponent indicates the number of times that the base is used as a factor. But that idea does not make sense for 0 as an exponent. To see what would make sense for the definition of 0 as an exponent, look at a table of the powers of 2:

2^6	2^5	2^4	2^3	2^2	2^1	2^0
64	32	16	8	4	2	?

The value of each expression in the table is one-half of the value of the preceding expression. So it would seem reasonable to define 2^0 to be half of 2 or 1. So the zero power of any nonzero real number is defined to be 1. We do not define the expression 0^0.

> **Zero Exponent**
>
> For any nonzero real number a,
> $$a^0 = 1.$$

Note that defining a^0 to be 1 is consistent with the product rule for exponents, because $a^0 \cdot a^n = 1 \cdot a^n = a^n$ and $a^0 \cdot a^n = a^{0+n} = a^n$. So the product rule is now valid for nonnegative integral exponents.

EXAMPLE 2

Using the definition of zero exponent

Simplify each expression. Assume that all variables represent nonzero real numbers.

a) 5^0 **b)** $(3xy)^0$ **c)** $b^0 \cdot b^9$ **d)** $2^0 + 3^0$

Solution

a) $5^0 = 1$ Definition of zero exponent

b) $(3xy)^0 = 1$ Definition of zero exponent

c) If we use the fact that $b^0 = 1$, then $b^0 \cdot b^9 = 1 \cdot b^9 = b^9$. If we use the product rule for exponents, then $b^0 \cdot b^9 = b^{0+9} = b^9$.

d) $2^0 + 3^0 = 1 + 1 = 2$ Definition of zero exponent

> Now do Exercises 13–22

⟨3⟩ The Quotient Rule for Exponents

To find the quotient of x^7 and x^3 we can write the quotient as a fraction and divide out or cancel the common factors. Then count the remaining factors:

$$x^7 \div x^3 = \frac{x^7}{x^3} = \frac{\cancel{x} \cdot \cancel{x} \cdot \cancel{x} \cdot x \cdot x \cdot x \cdot x}{\cancel{x} \cdot \cancel{x} \cdot \cancel{x}} = x^4$$

Instead of counting to find that there are four x's left, we can simply subtract 3 from 7 to get 4. This example illustrates the **quotient rule for exponents.**

Quotient Rule for Exponents

If a is a nonzero real number, and m and n are nonnegative integers (with $m \geq n$), then

$$\frac{a^m}{a^n} = a^{m-n}.$$

Note that $\frac{2^3}{2^3} = \frac{8}{8} = 1$, but we also have $\frac{2^3}{2^3} = 2^{3-3} = 2^0 = 1$. So the quotient rule is consistent with the definition of zero exponent. In this section, we will use the quotient rule only when $m \geq n$. The exponent in the numerator must be greater than or equal to the exponent in the denominator. In Section 4.2, we will define negative exponents and see that the quotient rule is valid also when $m < n$.

EXAMPLE 3

Using the quotient rule for exponents

Simplify each expression. Assume that all variables represent nonzero real numbers.

a) $x^7 \div x^4$ 　　b) $w^5 \div w^3$ 　　c) $\dfrac{2x^9}{-4x^3}$ 　　d) $\dfrac{6a^{12}b^6}{-3a^9b^6}$

Solution

a) $x^7 \div x^4 = x^{7-4}$ 　　　　Quotient rule for exponents
　　　$= x^3$ 　　　　　　　　Simplify.

b) $w^5 \div w^3 = w^{5-3}$ 　　　Quotient rule for exponents
　　　$= w^2$ 　　　　　　　Simplify.

c) $\dfrac{2x^9}{-4x^3} = -\dfrac{2}{4} \cdot \dfrac{x^9}{x^3} = -\dfrac{1}{2} \cdot x^{9-3}$ 　Quotient rule for exponents

　　　$= -\dfrac{1}{2} \cdot x^6$ 　　Simplify.

　　　$= -\dfrac{x^6}{2}$

d) $\dfrac{6a^{12}b^6}{-3a^9b^6} = \dfrac{6}{-3} \cdot \dfrac{a^{12}}{a^9} \cdot \dfrac{b^6}{b^6}$ 　Definition of fraction multiplication

　　　$= -2a^{12-9}b^{6-6}$ 　　Quotient rule for exponents
　　　$= -2a^3$ 　　　　　　$b^0 = 1$

Now do Exercises 23–34

< Helpful Hint >

Note that these rules of exponents are not absolutely necessary. We could simplify every expression here by using only the definition of exponents. However, these rules make it a lot simpler.

⟨4⟩ The Power of a Power Rule

The expression $(a^m)^n$ in which the mth power of a is raised to the nth power is called a **power of a power.** We can simplify a power of a power using the product rule:

$$(x^2)^4 = x^2 \cdot x^2 \cdot x^2 \cdot x^2 = x^8$$

Note that the exponent in the answer is the product of the two original exponents: $4 \cdot 2 = 8$. This example illustrates the **power of a power rule.**

Power of a Power Rule

If a is any real number, and m and n are positive integers, then

$$(a^m)^n = a^{mn}.$$

The power of a power rule is valid also if either of the exponents is zero. In that case, a must not be zero, because 0^0 is undefined.

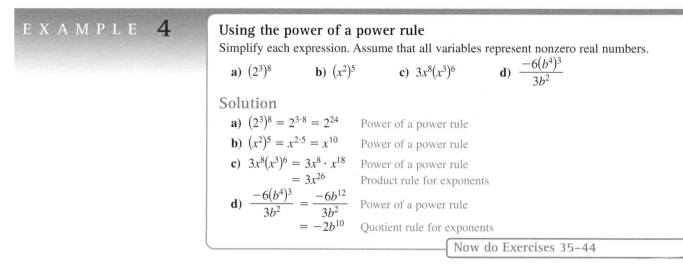

EXAMPLE 4

Using the power of a power rule
Simplify each expression. Assume that all variables represent nonzero real numbers.

   a) $(2^3)^8$ b) $(x^2)^5$ c) $3x^8(x^3)^6$ d) $\dfrac{-6(b^4)^3}{3b^2}$

Solution

   a) $(2^3)^8 = 2^{3\cdot8} = 2^{24}$ Power of a power rule

   b) $(x^2)^5 = x^{2\cdot5} = x^{10}$ Power of a power rule

   c) $3x^8(x^3)^6 = 3x^8 \cdot x^{18}$ Power of a power rule

             $= 3x^{26}$ Product rule for exponents

   d) $\dfrac{-6(b^4)^3}{3b^2} = \dfrac{-6b^{12}}{3b^2}$ Power of a power rule

                $= -2b^{10}$ Quotient rule for exponents

> Now do Exercises 35–44

⟨5⟩ The Power of a Product Rule

The expression $(ab)^n$ is a power of the product ab. We can simplify a power of a product using rules that we already know:

$$\overbrace{(5w^2)^3 = 5w^2 \cdot 5w^2 \cdot 5w^2}^{\text{3 factors of }5w^2} = 5^3 \cdot w^6 = 125w^6$$

Note that the exponent is applied to each factor of the product. So we have a new rule, the **power of a product rule,** which makes it easier to simplify this expression.

> **Power of a Product Rule**
>
> If a and b are real numbers, and n is any positive integer, then
>
> $$(ab)^n = a^n \cdot b^n.$$

The power of a power rule is valid also if $n = 0$. In that case, both a and b must be nonzero.

EXAMPLE 5

Using the power of a product rule
Simplify each expression. Assume that all variables represent nonzero real numbers.

   a) $(-2x)^3$ b) $(-3a^2)^4$ c) $(5x^3y^2)^3$

Solution

   a) $(-2x)^3 = (-2)^3x^3$ Power of a product rule

              $= -8x^3$ Simplify.

b) $(-3a^2)^4 = (-3)^4(a^2)^4$ Power of a product rule

$ = 81a^8$ Power of a power rule

c) $(5x^3y^2)^3 = 5^3(x^3)^3(y^2)^3$ Power of a product rule

$ = 125x^9y^6$ Power of a power rule

Now do Exercises 45–52

⟨6⟩ The Power of a Quotient Rule

The expression $\left(\dfrac{a}{b}\right)^n$ is a power of the quotient $\dfrac{a}{b}$. We can simplify a power of a quotient using the definition of exponents and the rule for multiplying fractions:

$$\left(\frac{x}{2}\right)^3 = \frac{x}{2} \cdot \frac{x}{2} \cdot \frac{x}{2} = \frac{x^3}{2^3}$$

Note that the exponent is applied to both the numerator and denominator. So we have a new rule, the **power of a quotient rule,** which makes it easier to simplify this expression.

Power of a Quotient Rule

If a and b are nonzero real numbers, and n is a nonnegative integer, then

$$\left(\frac{a}{b}\right)^n = \frac{a^n}{b^n}.$$

E X A M P L E   6

Using the power of a quotient rule

Simplify each expression. Assume that all variables represent nonzero real numbers.

a) $\left(\dfrac{y}{4}\right)^3$ **b)** $\left(-\dfrac{2x^2}{3y}\right)^4$ **c)** $\left(\dfrac{x^3}{y^5}\right)^4$

Solution

a) $\left(\dfrac{y}{4}\right)^3 = \dfrac{y^3}{4^3}$ Power of a quotient rule

$\phantom{\left(\dfrac{y}{4}\right)^3} = \dfrac{y^3}{64}$ Simplify.

b) $\left(-\dfrac{2x^2}{3y}\right)^4 = \dfrac{(-2x^2)^4}{(3y)^4}$ Power of a quotient rule

$\phantom{\left(-\dfrac{2x^2}{3y}\right)^4} = \dfrac{(-2)^4(x^2)^4}{3^4y^4}$ Power of a product rule

$\phantom{\left(-\dfrac{2x^2}{3y}\right)^4} = \dfrac{16x^8}{81y^4}$ Power of a power rule

c) $\left(\dfrac{x^3}{y^5}\right)^4 = \dfrac{(x^3)^4}{(y^5)^4}$ Power of a quotient rule

$\phantom{\left(\dfrac{x^3}{y^5}\right)^4} = \dfrac{x^{12}}{y^{20}}$ Power of a power rule

Now do Exercises 53–60

The five rules that we studied in this section are summarized as follows.

Rules for Nonnegative Integral Exponents

If a and b are nonzero real numbers, and m and n are nonnegative integers, then

1. $a^m a^n = a^{m+n}$ Product rule for exponents

2. $\dfrac{a^m}{a^n} = a^{m-n}$ Quotient rule for exponents ($m \geq n$)

3. $(a^m)^n = a^{mn}$ Power of a power rule

4. $(ab)^n = a^n b^n$ Power of a product rule

5. $\left(\dfrac{a}{b}\right)^n = \dfrac{a^n}{b^n}$ Power of a quotient rule

‹7› The Amount Formula

The amount of money invested is the **principal,** and the value of the principal after a certain time period is the **amount.** Interest rates are annual percentage rates.

Amount Formula

The amount A of an investment of P dollars with annual interest rate r compounded annually for n years is given by the formula

$$A = P(1 + r)^n.$$

E X A M P L E   7

Using the amount formula

A teacher invested \$10,000 in a bond fund that should have an average annual return of 6% per year for the next 20 years. What will be the amount of the investment in 20 years?

Solution

Use $n = 20$, $P = \$10{,}000$, and $r = 0.06$ in the amount formula:

$$A = P(1 + r)^n$$

$$A = 10{,}000(1 + 0.06)^{20}$$

$$= 10{,}000(1.06)^{20}$$

$$\approx 32{,}071.35$$

So the \$10,000 investment will amount to \$32,071.35 in 20 years.

Now do Exercises 83–88

Warm-Ups ▼

Fill in the blank.

1. According to the _____ rule, $a^m a^n = a^{m+n}$.

2. According to the _____ rule $\dfrac{a^m}{a^n} = a^{m-n}$.

3. According to the power of a _____ rule $(a^m)^n = a^{mn}$.

4. According to the power of a _____ rule $(ab)^m = a^m b^m$.

5. According to the power of a _____ rule $\left(\dfrac{a}{b}\right)^m = \dfrac{a^m}{b^m}$.

6. Any nonzero number to the power ____ is 1.

True or false?

7. $3^5 \cdot 3^6 = 3^{11}$

8. $2^3 \cdot 3^2 = 6^5$

9. $\dfrac{5^{13}}{5^{10}} = 125$

10. $(2^3)^2 = 64$

11. $(q^3)^5 = q^8$

12. $\dfrac{a^{12}}{a^4} = a^3$

13. $(2a^3)^4 = 8a^{12}$

14. $\left(\dfrac{m^3}{2}\right)^4 = \dfrac{m^{12}}{16}$

4.1 Exercises

‹ Study Tips ›

- Don't try to get everything done before you start studying. Since the average attention span for a task is only 20 minutes, it is better to study and take breaks from studying to do other duties.
- Your mood for studying should match the mood in which you are tested. Being too relaxed in studying will not match the increased anxiety that you feel during a test.

‹1› The Product Rule for Exponents

Find each product. See Example 1.

1. $3x^2 \cdot 9x^3$

2. $5x^7 \cdot 3x^5$

3. $2a^3 \cdot 7a^8$

4. $3y^{12} \cdot 5y^{15}$

5. $-6x^2 \cdot 5x^2$

6. $-2x^2 \cdot 8x^5$

7. $(-9x^{10})(-3x^7)$

8. $(-2x^2)(-8x^9)$

9. $-6st \cdot 9st$

10. $-12sq \cdot 3s$

11. $3wt \cdot 8w^7 t^6$

12. $h^8 k^3 \cdot 5h$

‹2› Zero Exponent

Simplify each expression. All variables represent nonzero real numbers. See Example 2.

13. 9^0

14. m^0

15. $(-2x^3)^0$

16. $(5a^3 b)^0$

17. $2 \cdot 5^0 - 5$

18. $-4^0 - 8^0$

19. $(2x - y)^0$

20. $(a^2 + b^2)^0$

21. $x^0 \cdot x^3$

22. $a^0 \cdot a^2$

‹3› The Quotient Rule for Exponents

Find each quotient. All variables represent nonzero real numbers. See Example 3.

23. $m^{18} \div m^6$

24. $a^{12} \div a^3$

25. $\dfrac{u^6}{u^3}$ 26. $\dfrac{w^{12}}{w^6}$

27. $b^3 \div b^3$ 28. $q^5 \div q^5$

29. $\dfrac{-6a^{10}}{2a^8}$ 30. $\dfrac{8m^{17}}{-2m^{13}}$

31. $\dfrac{8s^2 t^{13}}{-2st^5}$ 32. $\dfrac{-22v^3 w^9}{-11v^2 w^3}$

33. $\dfrac{-6x^8 y^4}{-3x^2 y^4}$ 34. $\dfrac{-51y^{16} z^3}{17y^9 z^3}$

⟨4⟩ The Power of a Power Rule

Simplify. All variables represent nonzero real numbers.
See Example 4.

35. $\left(x^2\right)^3$ 36. $\left(y^2\right)^4$

37. $2x^2 \cdot \left(x^2\right)^5$ 38. $\left(y^2\right)^6 \cdot 3y^5$

39. $\dfrac{\left(t^2\right)^5}{\left(t^3\right)^3}$ 40. $\dfrac{\left(r^4\right)^5}{\left(r^5\right)^3}$

41. $\dfrac{\left(x^3\right)^4}{\left(x^6\right)^2}$ 42. $\dfrac{\left(w^4\right)^6}{\left(w^2\right)^9}$

43. $\dfrac{-3x\left(x^5\right)^2}{6x^3\left(x^2\right)^4}$ 44. $\dfrac{-5y^4\left(y^5\right)^2}{15y^7\left(y^2\right)^3}$

⟨5⟩ The Power of a Product Rule

Simplify. All variables represent nonzero real numbers.
See Example 5.

45. $\left(xy^2\right)^3$ 46. $\left(wy^2\right)^6$

47. $\left(-2t^5\right)^3$ 48. $\left(-3r^3\right)^3$

49. $\left(-2x^2 y^5\right)^3$ 50. $\left(-3y^2 z^3\right)^3$

51. $\dfrac{\left(a^3 b^4 c^5\right)^4}{\left(a^2 b^3 c^4\right)^2}$ 52. $\dfrac{\left(2a^2 b^3\right)^6}{\left(4ab^3\right)^3}$

⟨6⟩ The Power of a Quotient Rule

Simplify. All variables represent nonzero real numbers.
See Example 6.

53. $\left(\dfrac{x}{2}\right)^3$ 54. $\left(\dfrac{y}{3}\right)^4$

55. $\left(\dfrac{a^4}{4}\right)^3$ 56. $\left(\dfrac{w^2}{2}\right)^3$

57. $\left(\dfrac{-2a^2}{b^3}\right)^4$ 58. $\left(\dfrac{-9r^3}{t^5}\right)^2$

59. $\left(\dfrac{2x^2 y^3}{-4y^2}\right)^3$ 60. $\left(\dfrac{3y^8}{2zy^2}\right)^4$

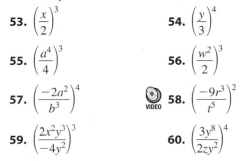

Miscellaneous

Simplify. All variables represent nonzero real numbers.

61. $5^2 \cdot 2^3$ 62. $10^3 \cdot 3^3$

63. $10^2 \cdot 10^4$ 64. $2^3 \cdot 2^4$

65. $\left(\dfrac{2^5}{2^3}\right)^3$ 66. $\left(\dfrac{3^3}{3}\right)^2$

67. $x^4 \cdot x^3$ 68. $x^5 \cdot x^8$

69. $x^0 \cdot x^5$ 70. $a^9 \cdot a^0$

71. $a^0 \cdot b^0$ 72. $a^0 + b^0$

73. $\left(a^8\right)^4$ 74. $\left(b^5\right)^8$

75. $\left(a^4 b^2\right)^3$ 76. $\left(x^2 t^4\right)^6$

77. $\dfrac{x^7}{x^4}$ 78. $\dfrac{m^{10}}{m^8}$

79. $\left(\dfrac{a^3}{b^4}\right)^3$ 80. $\left(\dfrac{t}{m^2}\right)^4$

81. $\left(2a^3 b\right)^2\left(3a^2 b^3\right)^2$ 82. $\left(-2x^2 y^3\right)^4\left(4xy^3\right)$

⟨7⟩ The Amount Formula

Solve each problem. See Example 7.

83. **CD investment.** Ernesto invested $25,000 in a CD that paid 5% compounded annually for 6 years. What was the value of his investment at the end of the sixth year?

84. **Venture capital.** Alberto invested $80,000 in his brother's restaurant. His brother did well and paid him back after 5 years with 10% interest compounded annually. What was the amount that Alberto received?

85. **Mutual fund.** Beryl invested $40,000 in a mutual fund that had an average annual return of 8%. What was the amount of his investment after 10 years?

86. **Savings account.** Helene put her $30,000 inheritance into a savings account at her bank and earned 2.2% compounded annually for 10 years. How much did she have after the tenth year?

87. **Long-term investing.** Sheila invested P dollars at annual rate r for 10 years. At the end of 10 years her investment was worth $P(1 + r)^{10}$ dollars. She then reinvested this money for another 5 years at annual rate r. At the end of the second time period her investment was worth $P(1 + r)^{10}(1 + r)^5$ dollars. Which rule of exponents can be used to simplify the expression? Simplify it.

88. **CD rollover.** Ronnie invested P dollars in a 2-year CD with an annual return of r. After the CD rolled over three times, its value was $P\left[(1 + r)^2\right]^3$ dollars. Which rule of exponents can be used to simplify the expression? Simplify it.

Getting More Involved

89. *Writing*

When we square a product, we square each factor in the product. For example, $(3b)^2 = 9b^2$. Explain why we cannot square a sum by simply squaring each term of the sum.

90. *Writing*

Explain why we defined 2^0 to be 1. Explain why $-2^0 \neq 1$.

4.2 Negative Exponents

In This Section

⟨1⟩ **Negative Integral Exponents**

⟨2⟩ **The Rules for Integral Exponents**

⟨3⟩ **The Present Value Formula**

We defined exponential expressions with positive integral exponents in Chapter 1 and learned five rules for exponents in Section 4.1. In this section we will define negative integral exponents and see that the rules from Section 4.1 can be applied to negative integral exponents also.

⟨1⟩ Negative Integral Exponents

A positive integral exponent indicates the number of times that the base is used as a factor. For example

$$x^2 = x \cdot x \qquad \text{and} \qquad a^3 = a \cdot a \cdot a.$$

If n is a positive integral exponent, a^n indicates that a is used as a factor n times. We define a^{-n} as the reciprocal of a^n. For example,

$$x^{-2} = \frac{1}{x^2} \qquad \text{and} \qquad a^{-3} = \frac{1}{a^3}.$$

> **Negative Integral Exponents**
>
> If a is a nonzero real number and n is a positive integer, then
> $$a^{-n} = \frac{1}{a^n}. \quad \text{(If } n \text{ is positive, } -n \text{ is negative.)}$$

E X A M P L E   1

Simplifying expressions with negative exponents

Simplify.

a) 2^{-5} **b)** $(-2)^{-5}$ **c)** -9^{-2} **d)** $\dfrac{2^{-3}}{3^{-2}}$

Solution

a) $2^{-5} = \dfrac{1}{2^5} = \dfrac{1}{32}$

b) $(-2)^{-5} = \dfrac{1}{(-2)^5}$ Definition of negative exponent

           $= \dfrac{1}{-32} = -\dfrac{1}{32}$

You can evaluate expressions with negative exponents on a calculator as shown here.

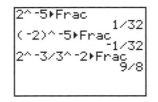

c) $-9^{-2} = -(9^{-2}) = -\dfrac{1}{9^2} = -\dfrac{1}{81}$

d) $\dfrac{2^{-3}}{3^{-2}} = 2^{-3} \div 3^{-2}$

$= \dfrac{1}{2^3} \div \dfrac{1}{3^2}$

$= \dfrac{1}{8} \div \dfrac{1}{9} = \dfrac{1}{8} \cdot \dfrac{9}{1} = \dfrac{9}{8}$

> Now do Exercises 1–10

CAUTION   A negative sign preceding an exponential expression is handled last for any exponents, resulting in a negative value for the expression:

$$-3^{-2} = -\dfrac{1}{3^2} = -\dfrac{1}{9}, \qquad -3^2 = -9, \qquad \text{and} \qquad -3^0 = -1.$$

If the base is negative, the value could be positive or negative:

$$(-2)^{-4} = \dfrac{1}{(-2)^4} = \dfrac{1}{16} \qquad \text{and} \qquad (-2)^{-3} = \dfrac{1}{(-2)^3} = \dfrac{1}{-8} = -\dfrac{1}{8}.$$

To evaluate a^{-n}, you can first find the nth power of a and then find the reciprocal. However, the result is the same if you first find the reciprocal of a and then find the nth power of the reciprocal. For example,

$$3^{-2} = \dfrac{1}{3^2} = \dfrac{1}{9} \qquad \text{or} \qquad 3^{-2} = \left(\dfrac{1}{3}\right)^2 = \dfrac{1}{3} \cdot \dfrac{1}{3} = \dfrac{1}{9}.$$

So the power and the reciprocal can be found in either order. If the exponent is -1, we simply find the reciprocal. For example,

$$5^{-1} = \dfrac{1}{5}, \qquad \left(\dfrac{1}{4}\right)^{-1} = 4, \qquad \text{and} \qquad \left(-\dfrac{3}{5}\right)^{-1} = -\dfrac{5}{3}.$$

Because $3^{-2} \cdot 3^2 = 1$, the reciprocal of 3^{-2} is 3^2, and we have

$$\dfrac{1}{3^{-2}} = 3^2.$$

Remember that if a negative sign in a negative exponent is deleted, then you must find a reciprocal. Four situations where this idea occurs are listed in the following box. Don't think of this as four more rules to be memorized. Remember the idea.

> **Rules for Negative Exponents**
>
> If a is a nonzero real number, and n is a positive integer, then
>
> **1.** $a^{-1} = \dfrac{1}{a}$ **2.** $\dfrac{1}{a^{-n}} = a^n$
>
> **3.** $a^{-n} = \left(\dfrac{1}{a}\right)^n$ **4.** $\left(\dfrac{a}{b}\right)^{-n} = \left(\dfrac{b}{a}\right)^n$

CAUTION   Note that a^{-1} is the multiplicative inverse of a and $-a$ is the additive inverse of a. For example, $2 \cdot 2^{-1} = 2 \cdot \dfrac{1}{2} = 1$ and $2 + (-2) = 0$.

With our definitions of the integral exponents, we get a nice pattern for the integral powers of 2 as shown in the following table. Whenever the exponent increases by 1, the value of the exponential expression is doubled.

2^{-5}	2^{-4}	2^{-3}	2^{-2}	2^{-1}	2^{0}	2^{1}	2^{2}	2^{3}	2^{4}	2^{5}
$\dfrac{1}{32}$	$\dfrac{1}{16}$	$\dfrac{1}{8}$	$\dfrac{1}{4}$	$\dfrac{1}{2}$	1	2	4	8	16	32

E X A M P L E   2

Using the rules for negative exponents

Simplify. Use only positive exponents in the answers.

a) $10^{-1} + 10^{-1}$ **b)** $\dfrac{2y^{-8}}{x^{-3}}$ **c)** 7^{-2} **d)** $\left(\dfrac{3}{4}\right)^{-3}$

Solution

a) $10^{-1} + 10^{-1} = \dfrac{1}{10} + \dfrac{1}{10} = \dfrac{2}{10} = \dfrac{1}{5}$ First rule for negative exponents

b) $\dfrac{2y^{-8}}{x^{-3}} = 2 \cdot y^{-8} \cdot \dfrac{1}{x^{-3}}$ $a \div b = a \cdot \dfrac{1}{b}$

$\qquad = 2 \cdot \dfrac{1}{y^{8}} \cdot x^{3}$ Second rule for negative exponents

$\qquad = \dfrac{2x^{3}}{y^{8}}$ Multiply.

Note that a negative exponent in the numerator or denominator can be changed to positive by simply relocating the expression.

c) $7^{-2} = \left(\dfrac{1}{7}\right)^{2} = \dfrac{1}{7} \cdot \dfrac{1}{7} = \dfrac{1}{49}$ Third rule for negative exponents

d) We can find the power and the reciprocal in either order:

$$\left(\frac{3}{4}\right)^{-3} = \left(\frac{4}{3}\right)^{3} = \frac{4}{3} \cdot \frac{4}{3} \cdot \frac{4}{3} = \frac{64}{27} \qquad \left(\frac{3}{4}\right)^{-3} = \left(\frac{27}{64}\right)^{-1} = \frac{64}{27}$$

Now do Exercises 11–20

In Example 2(b) the negative exponents were changed to positive by simply moving the expressions from numerator to denominator or denominator to numerator. We could do this so easily because there was no addition or subtraction involved in the expression. If an expression involves addition or subtraction, change all of the negative exponents to positive and then follow the order of operations. The numerator and denominator of an expression are evaluated before division is done.

E X A M P L E   3

Evaluating expressions with negative exponents

Evaluate each expression.

a) $\dfrac{2^{-1} + 2^{-1}}{2^{-1}}$ **b)** $\dfrac{2^{-1} - 2^{-2}}{3^{-1} - 4^{-1}}$

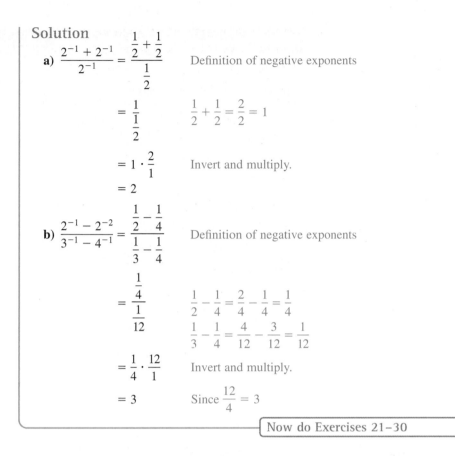

Solution

a) $\dfrac{2^{-1} + 2^{-1}}{2^{-1}} = \dfrac{\frac{1}{2} + \frac{1}{2}}{\frac{1}{2}}$ Definition of negative exponents

$= \dfrac{1}{\frac{1}{2}}$ $\dfrac{1}{2} + \dfrac{1}{2} = \dfrac{2}{2} = 1$

$= 1 \cdot \dfrac{2}{1}$ Invert and multiply.

$= 2$

b) $\dfrac{2^{-1} - 2^{-2}}{3^{-1} - 4^{-1}} = \dfrac{\frac{1}{2} - \frac{1}{4}}{\frac{1}{3} - \frac{1}{4}}$ Definition of negative exponents

$= \dfrac{\frac{1}{4}}{\frac{1}{12}}$ $\dfrac{1}{2} - \dfrac{1}{4} = \dfrac{2}{4} - \dfrac{1}{4} = \dfrac{1}{4}$

$\dfrac{1}{3} - \dfrac{1}{4} = \dfrac{4}{12} - \dfrac{3}{12} = \dfrac{1}{12}$

$= \dfrac{1}{4} \cdot \dfrac{12}{1}$ Invert and multiply.

$= 3$ Since $\dfrac{12}{4} = 3$

> Now do Exercises 21–30

CAUTION   Be careful changing negative exponents to positive when addition or subtraction is present: $\dfrac{1 + 2^{-3}}{5^{-2}} \neq \dfrac{1 + 5^{2}}{2^{3}}$.

⟨2⟩ The Rules for Integral Exponents

To find the product of y^{-2} and y^{-6} we could convert to positive exponents:

$$y^{-2} \cdot y^{-6} = \dfrac{1}{y^2} \cdot \dfrac{1}{y^6} = \dfrac{1}{y^8} = y^{-8}$$

To find the quotient of y^{-2} and y^{-6} we could again convert to positive exponents:

$$\dfrac{y^{-2}}{y^{-6}} = \dfrac{\frac{1}{y^2}}{\frac{1}{y^6}} = \dfrac{1}{y^2} \cdot \dfrac{y^6}{1} = \dfrac{y^6}{y^2} = y^{6-2} = y^4$$

⟨ **Calculator Close-Up** ⟩

You can use a calculator to demonstrate that the product rule for exponents holds when the exponents are negative numbers.

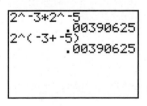
```
2^-3*2^-5
             .00390625
2^(-3+-5)
             .00390625
```

However, it is not necessary to convert to positive exponents. The exponent for the product is the sum of the exponents, and the exponent for the quotient is the difference: $-2 + (-6) = -8$ and $-2 - (-6) = 4$. These examples illustrate the fact that the product and quotient rules hold for negative exponents as well as positive exponents. In fact, *all five of the rules for exponents from Section 4.1 are valid for any integer exponents!*

The definitions and rules that we studied in this section and Section 4.1 are summarized as follows. Note the rules apply to any integers as exponents: positive, negative, or zero.

‹ **Helpful Hint** ›

The definitions of the different types of exponents are a really clever mathematical invention. The fact that we have rules for performing arithmetic with those exponents makes the notation of exponents even more amazing.

Rules for Integral Exponents

If a and b are nonzero real numbers, and m and n are integers, then

1. $a^{-n} = \dfrac{1}{a^n}$ Definition of negative exponent

2. $a^{-1} = \dfrac{1}{a}, \dfrac{1}{a^{-n}} = a^n, a^{-n} = \left(\dfrac{1}{a}\right)^n, \left(\dfrac{a}{b}\right)^{-n} = \left(\dfrac{b}{a}\right)^n$ Negative exponent rules

3. $a^0 = 1$ Definition of zero exponent

4. $a^m a^n = a^{m+n}$ Product rule for exponents

5. $\dfrac{a^m}{a^n} = a^{m-n}$ Quotient rule for exponents

6. $(a^m)^n = a^{mn}$ Power of a power rule

7. $(ab)^n = a^n b^n$ Power of a product rule

8. $\left(\dfrac{a}{b}\right)^n = \dfrac{a^n}{b^n}$ Power of a quotient rule

In Example 4, we use the product and quotient rules (rules 4 and 5) to simplify some expressions involving positive and negative exponents. Note that we specify that the answers are to be written without negative exponents. We do this to make the answers look simpler and so that there is only one correct answer. It is not wrong to use negative exponents in an answer.

EXAMPLE 4

Using the product and quotient rules with integral exponents

Simplify. Write answers without negative exponents. Assume that the variables represent nonzero real numbers.

a) $b^{-3}b^5$ **b)** $-3x^{-3} \cdot 5x^2$ **c)** $\dfrac{m^{-6}}{m^{-2}}$ **d)** $\dfrac{4x^{-6}y^5}{-12x^{-6}y^{-3}}$

Solution

a) $b^{-3}b^5 = b^{-3+5}$ Product rule for exponents

$= b^2$ Simplify.

b) $-3x^{-3} \cdot 5x^2 = -15x^{-3+2}$ Product rule for exponents

$= -15x^{-1}$ Simplify.

$= -15 \cdot \dfrac{1}{x}$ Definition of a negative exponent (to get answer without negative exponents)

$= -\dfrac{15}{x}$ Simplify.

c) $\dfrac{m^{-6}}{m^{-2}} = m^{-6-(-2)}$ Quotient rule for exponents

        $= m^{-4}$ Simplify.

        $= \dfrac{1}{m^4}$ Definition of negative exponent (to get answer without negative exponents)

d) $\dfrac{4x^{-6}y^5}{-12x^{-6}y^{-3}} = \dfrac{x^{-6-(-6)}y^{5-(-3)}}{-3} = \dfrac{x^0 y^8}{-3} = -\dfrac{y^8}{3}$

> Now do Exercises 31–46

In Example 5, we use the power rules (rules 6–8) to simplify some expressions involving positive and negative exponents.

EXAMPLE 5

Using the power rules with integral exponents

Simplify. Write answers without negative exponents. Assume that the variables represent nonzero real numbers.

a) $\left(a^{-3}\right)^2$ b) $\left(10x^{-3}\right)^{-2}$ c) $\left(\dfrac{4x^{-5}}{y^2}\right)^{-2}$

Solution

a) $\left(a^{-3}\right)^2 = a^{-3\cdot 2}$ Power of a power rule

        $= a^{-6}$ Simplify.

        $= \dfrac{1}{a^6}$ Definition of a negative exponent (to get answer without negative exponents)

b) $\left(10x^{-3}\right)^{-2} = 10^{-2}\left(x^{-3}\right)^{-2}$ Power of a product rule

        $= \dfrac{1}{10^2}x^6$ Power of a power rule

        $= \dfrac{x^6}{100}$ Simplify.

c) $\left(\dfrac{4x^{-5}}{y^2}\right)^{-2} = \dfrac{\left(4x^{-5}\right)^{-2}}{\left(y^2\right)^{-2}}$ Power of a quotient rule

        $= \dfrac{4^{-2}x^{10}}{y^{-4}}$ Power of a product and power of a power rule

        $= 4^{-2} \cdot x^{10} \cdot \dfrac{1}{y^{-4}}$ Because $\dfrac{a}{b} = a \cdot \dfrac{1}{b}$

        $= \dfrac{1}{4^2} \cdot x^{10} \cdot y^4$ Definition of a negative exponent

        $= \dfrac{x^{10}y^4}{16}$ Simplify.

> Now do Exercises 47–62

‹ Calculator Close-Up ›

You can use a calculator to demonstrate that the power of a power rule for exponents holds when the exponents are negative integers.

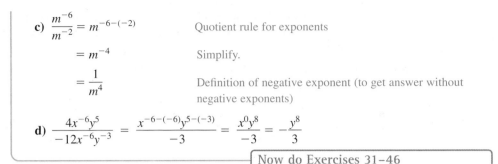

```
(3^-2)^-5
                59049
3^(-2*-5)
                59049
```

‹3› The Present Value Formula

In Section 4.1, we studied the amount formula $A = P(1 + r)^n$. If we are interested in the principal P that must be invested today to grow to a specified amount A in the future, then the principal is called the **present value** of the investment. We can find a formula for present value by solving the amount formula for P:

$$A = P(1 + r)^n \qquad \text{The amount formula}$$

$$P = \frac{A}{(1 + r)^n} \qquad \text{Divide each side by } (1 + r)^n.$$

$$P = A(1 + r)^{-n} \qquad \text{Definition of negative exponent}$$

Present Value Formula

The present value P that will amount to A dollars after n years with interest compounded annually at annual interest rate r is given by the formula

$$P = A(1 + r)^{-n}.$$

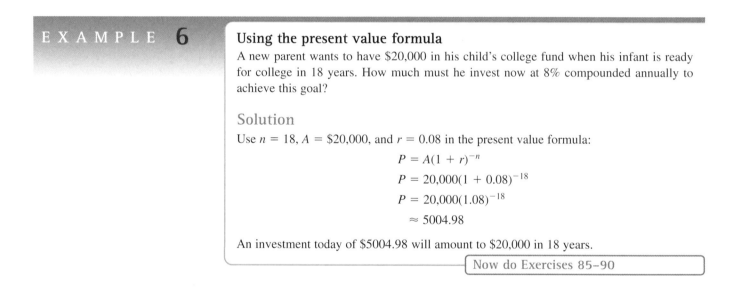

E X A M P L E   6

Using the present value formula

A new parent wants to have $20,000 in his child's college fund when his infant is ready for college in 18 years. How much must he invest now at 8% compounded annually to achieve this goal?

Solution

Use $n = 18$, $A = \$20,000$, and $r = 0.08$ in the present value formula:

$$P = A(1 + r)^{-n}$$
$$P = 20,000(1 + 0.08)^{-18}$$
$$P = 20,000(1.08)^{-18}$$
$$\approx 5004.98$$

An investment today of $5004.98 will amount to $20,000 in 18 years.

Now do Exercises 85–90

Warm-Ups ▼

Fill in the blank.

1. The expression a^{-n} is the _____ of a^n.

2. To evaluate 2^6 we use 2 as a _____ six times.

3. If n is positive, then a^{-n} has a _____ exponent.

True or false?

4. $10^{-2} = \dfrac{1}{10^2}$

5. $\left(-\dfrac{1}{5}\right)^{-1} = -5$

6. $3^{-2} \cdot 2^{-1} = 6^{-3}$

7. $\dfrac{3^{-2}}{3^{-1}} = \dfrac{1}{3}$

8. $(2^{-3})^{-2} = 64$

9. $-2^{-4} = \dfrac{1}{16}$

10. $5^{-3} = \dfrac{1}{5} \cdot \dfrac{1}{5} \cdot \dfrac{1}{5}$

Variables in all exercises represent nonzero real numbers. Write all answers without negative exponents.

‹1› Negative Integral Exponents

Evaluate each expression. See Example 1.

1. 3^{-1}

2. 3^{-3}

3. $(-2)^{-4}$

4. $(-3)^{-4}$

5. -4^{-2}

6. -2^{-4}

7. -3^{-3}

8. -5^{-3}

9. $\dfrac{5^{-2}}{10^{-2}}$

10. $\dfrac{3^{-4}}{6^{-2}}$

Simplify. See Example 2.

11. $6^{-1} + 6^{-1}$

12. $2^{-1} + 4^{-1}$

13. $\dfrac{10}{5^{-3}}$

14. $\dfrac{1}{25 \cdot 10^{-4}}$

15. $\dfrac{3a^{-3}}{b^{-9}}$

16. $\dfrac{6x^{-5}}{5y^{-1}}$

17. $\left(\dfrac{1}{b}\right)^{-7}$

18. $\left(\dfrac{1}{y}\right)^{-4}$

19. $\left(\dfrac{5}{2}\right)^{-3}$

20. $\left(\dfrac{4}{3}\right)^{-2}$

Evaluate. See Example 3.

21. $\dfrac{3^{-1} + 3^{-1}}{3^{-2}}$

22. $\dfrac{4^{-1}}{2^{-1} + 4^{-1}}$

23. $\dfrac{2^{-1} + 3^{-1}}{6^{-1} + 6^{-1}}$

24. $\dfrac{10^{-1} + 10^{-1}}{5^{-1} + 10^{-1}}$

25. $\dfrac{2^{-1} - 2^{-3}}{2^{-1} - 4^{-1}}$

26. $\dfrac{3^{-1} - 6^{-1}}{3^{-1} - 3^{-2}}$

27. $\dfrac{2 + 2^{-1}}{1 + 4^{-1}}$

28. $\dfrac{3 + 2^{-1}}{1 - 2^{-2}}$

29. $\dfrac{5 \cdot 3^{-2}}{6^{-1}}$

30. $\dfrac{3 \cdot 6^{-1}}{5 \cdot 10^{-1}}$

‹2› The Rules for Integral Exponents

Simplify. See Example 4.

31. $x^{-1} \cdot x^5$

32. $y^{-3} \cdot y^5$

33. $x^3 \cdot x \cdot x^{-7}$

34. $y \cdot y^{-8} \cdot y$

35. $y^{-3} \cdot y^{-5}$

36. $w^{-8} \cdot w^{-3}$

37. $-2x^2 \cdot 8x^{-6}$

38. $5y^5\left(-6y^{-7}\right)$

39. $b^3 \div b^9$

40. $q^5 \div q^7$

41. $\dfrac{-6a^6}{2a^8}$

42. $\dfrac{2m^{13}}{-8m^{17}}$

43. $\dfrac{u^{-5}}{u^3}$

44. $\dfrac{w^{-4}}{w^6}$

45. $\dfrac{8t^{-3}}{-2t^{-5}}$

46. $\dfrac{-22w^{-4}}{-11w^{-3}}$

Simplify. See Example 5.

47. $\left(y^{-3}\right)^4$

48. $\left(a^5\right)^{-3}$

49. $2x^{-3}\left(x^{-2}\right)^{-5}$

50. $3x^{16}\left(x^{-2}\right)^6$

51. $\dfrac{\left(b^3\right)^{-3}}{\left(b^{-2}\right)^5}$

52. $\dfrac{\left(a^{-3}\right)^{-3}}{\left(a^{-1}\right)^{-4}}$

53. $(2x)^{-4}$

54. $(3a)^{-3}$

55. $(xy^{-2})^{-3}$

56. $\left(a^{-3}b\right)^4$

57. $\left(\dfrac{x^{-4}}{9^{-1}}\right)^2$

58. $\left(\dfrac{w^{-3}}{2^{-2}}\right)^3$

59. $\left(\dfrac{-2m^{-3}}{n^{-2}}\right)^4$

60. $\left(\dfrac{-3b^{-1}}{a^4}\right)^{-2}$

61. $\left(\dfrac{6ab^{-2}}{3a^2b^{-4}}\right)^3$

62. $\left(\dfrac{2s^{-1}t^3}{6s^2t^{-4}}\right)^{-3}$

Miscellaneous

Simplify.

63. $2^{-1} + 2^{-1}$ **64.** $3^{-1} - 4^{-1}$

65. $-1^{-1} - 1^{-1}$ **66.** $-1^{-2} + 1^{-2}$

67. $(2^{-1})^{-1}$ **68.** $(2^{-2})^{-2}$

69. $4^{-2} \cdot 4^3$ **70.** $4^{-12} \cdot 4^{11}$

71. $5^{-5} \cdot 5^7$ **72.** $10^{-22} \cdot 10^{24}$

73. $(10^{-2})^{-3} \div 10^5$

74. $10^{-14} \div (10^{-3})^5$

75. $\left(\dfrac{1}{a^3}\right)^{-1}$ **76.** $\left(\dfrac{1}{x}\right)^{-1}$

77. x^{-5} **78.** $3b^{-6}$

79. $\dfrac{1}{5w^{-3}}$ **80.** $\dfrac{3}{6^{-1}d^{-6}}$

81. $\dfrac{a^{-2}c^3}{2b^{-5}}$ **82.** $\dfrac{2m^{-4}}{3^{-1}n^2}$

83. $\left(\dfrac{a^{-3}}{a^{-1}a^7}\right)^{-4}$ **84.** $\left(\dfrac{c^{-4}c^{-2}}{c^{-6}c^8}\right)^{-4}$

⟨3⟩ The Present Value Formula

Solve each problem. See Example 6.

85. *Saving for a car.* How much would Florence have to invest today at 6.2% compounded annually so that she would have $20,000 to buy a new car in 6 years?

86. *Saving for college.* Mr. Isaacs wants to have $60,000 in 18 years when little Debby will start college. How much would he have to invest today in high-yield bonds that pay 9% compounded annually to achieve his goal?

87. *Saving for retirement.* Nadine inherited a large sum of money and wants to make sure her son will have a comfortable retirement. How much should she invest today in Treasury bills paying 4.5% compounded annually so that her son will have $1,000,000 in 40 years when he retires?

88. *Saving for a boat.* Oscar has an account that is earmarked for a sailboat. He needs $200,000 for the boat when he retires in 10 years. If he averages 7% annually on this account, how much should he have in the account now so that his goal will be reached with no additional deposits?

89. *Present value.* Find the present value that will amount to $50,000 in 20 years at 8% compounded annually.

90. *Investing in stocks.* U.S. small company stocks have returned an average of 14.9% annually for the last 50 years (T. Rowe Price, www.troweprice.com). Find the amount invested today in small company stocks that would be worth $1 million in 50 years, assuming that small company stocks continue to return 14.9% annually for the next 50 years.

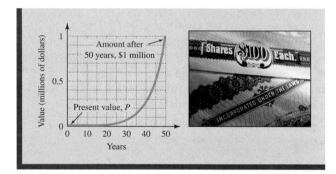

Figure for Exercise 90

Getting More Involved

91. *Exploration*

a) If $w^{-3} < 0$, then what can you say about w?

b) If $(-5)^m < 0$, then what can you say about m?

c) What restriction must be placed on w and m so that $w^m < 0$?

92. *Discussion*

Which of the following expressions is not equal to -1? Explain your answer.

a) -1^{-1} **b)** -1^{-2} **c)** $\left(-1^{-1}\right)^{-1}$

d) $(-1)^{-1}$ **e)** $(-1)^{-2}$

4.3 Scientific Notation

In This Section

⟨1⟩ **Converting Scientific to Standard Notation**

⟨2⟩ **Converting Standard to Scientific Notation**

⟨3⟩ **Combining Numbers and Words**

⟨4⟩ **Computations with Scientific Notation**

⟨5⟩ **Applications**

Many of the numbers occurring in science are either very large or very small. For example, the speed of light is 983,571,000 feet per second and 1 millimeter is equal to 0.000001 kilometer. Large numbers and small numbers can be written in a simpler way using *scientific notation*, which involves positive and negative integral exponents.

⟨1⟩ Converting Scientific to Standard Notation

In scientific notation the speed of light is 9.83571×10^8 feet per second and 1 millimeter is equal to 1×10^{-6} kilometer. In scientific notation there is always one digit to the left of the decimal point.

> **Scientific Notation**
>
> A number in **scientific notation** is written using the times symbol $\times$ in the form
> $$a \times 10^n$$
> where $1 \leq a < 10$ and n is a positive or negative integer.

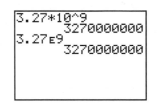
Scientific notation is based on multiplication by integral powers of 10. Multiplying a number by a positive power of 10 moves the decimal point to the right:

$$10(5.32) = 53.2$$
$$10^2(5.32) = 100(5.32) = 532$$
$$10^3(5.32) = 1000(5.32) = 5320$$

Multiplying by a negative power of 10 moves the decimal point to the left:

$$10^{-1}(5.32) = \frac{1}{10}(5.32) = 0.532$$
$$10^{-2}(5.32) = \frac{1}{100}(5.32) = 0.0532$$
$$10^{-3}(5.32) = \frac{1}{1000}(5.32) = 0.00532$$

So if n is a positive integer, multiplying by 10^n moves the decimal point n places to the right and multiplying by 10^{-n} moves it n places to the left.

To convert a number in scientific notation to standard notation, we simply multiply by the indicated power of 10, where multiplication is accomplished by moving the decimal point. We can use the following strategy.

> **Strategy for Converting to Standard Notation**
>
> **1.** Determine the number of places to move the decimal point by examining the exponent on the 10.
>
> **2.** Move to the right for a positive exponent and to the left for a negative exponent.

EXAMPLE   1

Converting to standard notation
Write in standard notation.

a) 7.02×10^6 b) 8.13×10^{-5} c) 9×10^{-6}

Solution

a) Because the exponent is positive, move the decimal point six places to the right:
$$7.02 \times 10^6 = 7020000. = 7,020,000$$

b) Because the exponent is negative, move the decimal point five places to the left:
$$8.13 \times 10^{-5} = 0.0000813$$

c) If the decimal point is not written, then it is assumed to be on the right of the number. So 9 and 9. are the same number. Because the exponent on 10 is -6 we move the decimal point 6 places to the left from this position:
$$9 \times 10^{-6} = 9. \times 10^{-6} = 0.000009$$

> Now do Exercises 1–14

⟨2⟩ Converting Standard to Scientific Notation

To convert a positive number to scientific notation, we just reverse the strategy for converting from scientific notation.

Strategy for Converting to Scientific Notation

1. Count the number of places (n) that the decimal must be moved so that it will follow the first nonzero digit of the number.

2. If the original number was larger than 10, use 10^n.

3. If the original number was smaller than 1, use 10^{-n}.

Remember that the scientific notation for a number larger than 10 will have a positive power of 10 and the scientific notation for a number between 0 and 1 will have a negative power of 10.

EXAMPLE   2

Converting to scientific notation
Write in scientific notation.

a) 7,346,200   b) 0.0000348   c) 135×10^{-12}

⟨ **Calculator Close-Up** ⟩

To convert to scientific notation, set the mode to scientific. In scientific mode all results are given in scientific notation.

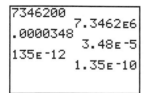

Solution

a) Because 7,346,200 is larger than 10, the exponent on the 10 will be positive:
$$7,346,200 = 7.3462 \times 10^6$$

b) Because 0.0000348 is smaller than 1, the exponent on the 10 will be negative:
$$0.0000348 = 3.48 \times 10^{-5}$$

c) There should be only one nonzero digit to the left of the decimal point:
$$135 \times 10^{-12} = 1.35 \times 10^2 \times 10^{-12} \quad \text{Convert 135 to scientific notation.}$$
$$= 1.35 \times 10^{-10} \quad \text{Product rule for exponents}$$

> Now do Exercises 15–24

⟨3⟩ Combining Numbers and Words

Large quantities are often expressed with a combination of a number and a word such as thousand, million, billion, or trillion. An expression such as "12 million" means 12 times one million. Such numbers can be converted to scientific notation or standard notation using

$$1 \text{ thousand} = 10^3, 1 \text{ million} = 10^6, 1 \text{ billion} = 10^9, \text{ and } 1 \text{ trillion} = 10^{12}.$$

EXAMPLE 3

Combining numbers and words
Write each number in scientific notation and standard notation.

   a) 327 thousand   **b)** 3788 million   **c)** 0.5 billion   **d** 16.5 trillion

Solution

a) $327 \text{ thousand} = 327 \times 10^3$ $1 \text{ thousand} = 10^3$

                       $= 3.27 \times 10^5$ Scientific notation

                       $= 327{,}000$ Standard notation

b) $3788 \text{ million} = 3788 \times 10^6$ $1 \text{ million} = 10^6$

                     $= 3.788 \times 10^9$ Scientific notation

                     $= 3{,}788{,}000{,}000$ Standard notation

c) $0.5 \text{ billion} = 0.5 \times 10^9$ $1 \text{ billion} = 10^9$

                 $= 5 \times 10^8$ Scientific notation

                 $= 500{,}000{,}000$ Standard notation

d) $16.5 \text{ trillion} = 16.5 \times 10^{12}$ $1 \text{ trillion} = 10^{12}$

                   $= 1.65 \times 10^{13}$ Scientific notation

                 $= 16{,}500{,}000{,}000{,}000$ Standard notation

Now do Exercises 25–30

⟨4⟩ Computations with Scientific Notation

An important feature of scientific notation is its use in computations. Numbers in scientific notation are nothing more than exponential expressions, and you have already studied operations with exponential expressions in Section 4.2. We use the same rules of exponents on numbers in scientific notation that we use on any other exponential expressions.

EXAMPLE 4

Using the rules of exponents with scientific notation
Perform the indicated computations. Write the answers in scientific notation.

   a) $(3 \times 10^6)(2 \times 10^8)$ **b)** $\dfrac{4 \times 10^5}{8 \times 10^{-2}}$ **c)** $(5 \times 10^{-7})^3$

Solution

   a) $(3 \times 10^6)(2 \times 10^8) = 3 \cdot 2 \cdot 10^6 \cdot 10^8 = 6 \times 10^{14}$

With a calculator's built-in scientific notation, some parentheses can be omitted as shown. Writing out the powers of 10 can lead to errors.

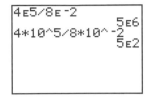

Try these computations with your calculator.

b) $\dfrac{4 \times 10^5}{8 \times 10^{-2}} = \dfrac{4}{8} \cdot \dfrac{10^5}{10^{-2}} = \dfrac{1}{2} \cdot 10^{5-(-2)}$ Quotient rule for exponents

$$= (0.5)10^7 \qquad \dfrac{1}{2} = 0.5$$

$$= 5 \times 10^{-1} \cdot 10^7 \qquad \text{Write 0.5 in scientific notation.}$$

$$= 5 \times 10^6 \qquad \text{Product rule for exponents}$$

c) $(5 \times 10^{-7})^3 = 5^3 (10^{-7})^3$ Power of a product rule

$$= 125 \cdot 10^{-21} \qquad \text{Power of a power rule}$$

$$= 1.25 \times 10^2 \times 10^{-21} \qquad 125 = 1.25 \times 10^2$$

$$= 1.25 \times 10^{-19} \qquad \text{Product rule for exponents}$$

Now do Exercises 31–42

E X A M P L E   **5**

Converting to scientific notation for computations

Perform these computations by first converting each number into scientific notation. Give your answer in scientific notation.

a) $(3,000,000)(0.0002)$ **b)** $(20,000,000)^3(0.0000003)$

Solution

a) $(3,000,000)(0.0002) = 3 \times 10^6 \cdot 2 \times 10^{-4}$ Scientific notation

$$= 6 \times 10^2 \qquad \text{Product rule for exponents}$$

b) $(20,000,000)^3(0.0000003) = (2 \times 10^7)^3(3 \times 10^{-7})$ Scientific notation

$$= 8 \times 10^{21} \cdot 3 \times 10^{-7} \qquad \text{Power of a product rule}$$

$$= 24 \times 10^{14}$$

$$= 2.4 \times 10^1 \times 10^{14} \qquad 24 = 2.4 \times 10^1$$

$$= 2.4 \times 10^{15} \qquad \text{Product rule for exponents}$$

Now do Exercises 43–50

‹5› **Applications**

E X A M P L E   **6**

Using scientific notation

a) The mean distance from Mars to the sun is 141.6×10^6 miles. Express this distance in feet. Use scientific notation rounded off with one digit to the right of the decimal point.

b) If the national debt is $\$1.8 \times 10^{13}$ and the population of the country is 3.5×10^8, then what is the debt per person? Express the answer in standard notation to the nearest thousand dollars.

Solution

a) There are 5280 feet in one mile. So we use a calculator to multiply 141.6×10^6 by 5280:

$$141.6 \times 10^6 \text{ miles} \cdot \frac{5280 \text{ feet}}{1 \text{ mile}} \approx 7.5 \times 10^{11} \text{ feet}$$

b) Use a calculator to divide the debt by the number of people:

$$\frac{\$1.8 \times 10^{13}}{3.5 \times 10^8 \text{ people}} \approx \$51,000 \text{ per person}$$

Now do Exercises 59–66

Warm-Ups ▼

Fill in the blank.

1. The number 1.2×10^{12} is written in _____ notation.

2. To convert _____ to standard notation, multiply by the appropriate power of 10.

3. To convert _____ to scientific notation, move the decimal point and use the appropriate power of 10.

True or false?

4. The number 12×10^4 is written in scientific notation.

5. The number 1×10^{55} is written in scientific notation.

6. $23.7 = 2.37 \times 10^{-1}$

7. $0.000036 = 3.6 \times 10^{-5}$

8. $(3 \times 10^{-9})^2 = 9 \times 10^{-18}$

9. $(2 \times 10^{-5})(4 \times 10^4) = 8 \times 10^{-20}$

10. $(1.8 \times 10^{12}) \div (3 \times 10^{-4}) = 6 \times 10^{15}$

Exercises 4.3

‹ **Study Tips** ›

- It is a good idea to review on a regular basis. Go back to a section that you have already studied and work some exercises.
- Every chapter of this text contains a Mid-Chapter Quiz. You can use these to review the first half of any chapter.

‹1› Converting Scientific to Standard Notation

Write each number in standard notation. See Example 1. See the Strategy for Converting to Standard Notation box on page 273.

1. 9.86×10^9
2. 4.007×10^4
3. 1.37×10^{-3}
4. 9.3×10^{-5}
5. 1×10^{-6}
6. 3×10^{-1}
7. 6×10^5
8. 8×10^6
9. 56×10^4
10. 286×10^5
11. 43.2×10^{-4}
12. 589.6×10^{-3}
13. 0.0067×10^3
14. 0.34×10^{-3}

‹2› Converting Standard to Scientific Notation

Write each number in scientific notation. See Example 2. See the Strategy for Converting to Scientific Notation box on page 274.

VIDEO
15. 9000
16. 5,298,000

17. 0.00078
18. 0.000214
19. 0.0000085
20. 0.015
21. 644,000,000
22. 5,670,000,000
23. 525×10^9
24. 0.0034×10^{-8}

‹3› Combining Numbers and Words

Write each number in scientific notation and standard notation.

25. 23 million
26. 344 million
27. 15 billion
28. 3478 billion
29. 13.6 trillion
30. 0.75 trillion

⟨4⟩ Computations with Scientific Notation

Perform the computations. Write answers in scientific notation. See Example 4.

31. $(3 \times 10^5)(2 \times 10^{-15})$

32. $(2 \times 10^{-9})(4 \times 10^{23})$

33. $\dfrac{4 \times 10^{-8}}{2 \times 10^{30}}$

34. $\dfrac{9 \times 10^{-4}}{3 \times 10^{-6}}$

35. $\dfrac{3 \times 10^{20}}{6 \times 10^{-8}}$

36. $\dfrac{1 \times 10^{-8}}{4 \times 10^7}$

37. $(3 \times 10^{12})^2$

38. $(2 \times 10^{-5})^3$

39. $(5 \times 10^4)^3$

40. $(5 \times 10^{14})^{-1}$

41. $(4 \times 10^{32})^{-1}$

42. $(6 \times 10^{11})^2$

Perform the following computations by first converting each number into scientific notation. Write answers in scientific notation. See Example 5.

43. $(4300)(2,000,000)$

44. $(40,000)(4,000,000,000)$

45. $(4,200,000)(0.00005)$

46. $(0.00075)(4,000,000)$

47. $(300)^3(0.000001)^5$

48. $(200)^4(0.0005)^3$

49. $\dfrac{(4000)(90,000)}{0.00000012}$

50. $\dfrac{(30,000)(80,000)}{(0.000006)(0.002)}$

Perform the following computations with the aid of a calculator. Write answers in scientific notation. Round to three decimal places.

51. $(6.3 \times 10^6)(1.45 \times 10^{-4})$

52. $(8.35 \times 10^9)(4.5 \times 10^3)$

53. $(5.36 \times 10^{-4}) + (3.55 \times 10^{-5})$

54. $(8.79 \times 10^8) + (6.48 \times 10^9)$

55. $\dfrac{(3.5 \times 10^5)(4.3 \times 10^{-6})}{3.4 \times 10^{-8}}$

56. $\dfrac{(3.5 \times 10^{-8})(4.4 \times 10^{-4})}{2.43 \times 10^{45}}$

57. $(3.56 \times 10^{85})(4.43 \times 10^{96})$

58. $(8 \times 10^{99}) + (3 \times 10^{99})$

⟨5⟩ Applications

Solve each problem.

59. *Distance to the sun.* The distance from the earth to the sun is 93 million miles. Express this distance in feet. (1 mile = 5280 feet.)

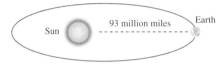

Figure for Exercise 59

60. *Speed of light.* The speed of light is 9.83569×10^8 feet per second. How long does it take light to travel from the sun to the earth? See Exercise 59.

61. *Warp drive, Scotty.* How long does it take a spacecraft traveling at 2×10^{35} miles per hour (warp factor 4) to travel 93 million miles?

62. *Area of a dot.* If the radius of a very small circle is 2.35×10^{-8} centimeters, then what is the circle's area?

63. *Circumference of a circle.* If the circumference of a circle is 5.68×10^9 feet, then what is its radius?

64. *Diameter of a circle.* If the diameter of a circle is 1.3×10^{-12} meters, then what is its radius?

65. *National debt.* In 2009 the national debt for the United States hit $\$1.2 \times 10^{13}$. If the population at that time was 308 million, then what was the amount of debt per person to the nearest thousand dollars?

66. *National debt.* In 1980 the national debt for the United States was $\$9.09 \times 10^{11}$. If the population at that time was 2.27×10^8, then what was the amount of the debt per person to the nearest thousand dollars?

Math *at* Work   Aerospace Engineering

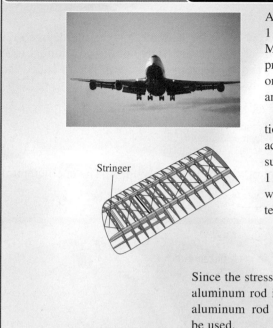

Stringer

Aircraft design is a delicate balance between weight and strength. Saving 1 pound of weight could save the plane's operators $5000 over 20 years. Mathematics is used to calculate the strength of each of a plane's parts and to predict when the material making up a part will fail. If calculations show that one kind of metal isn't strong enough, designers usually have to choose another material or change the design.

As an example, consider an aluminum stringer with a circular cross section. The stringer is used inside the wing of an airplane as shown in the accompanying figure. The aluminum rod has a diameter of 20 mm and will support a load of 5×10^4 Newtons (N). The maximum stress on aluminum is 1×10^8 Pascals (Pa), where 1 Pa $= 1$ N/m^2. To calculate the stress S on the rod we use $S = $ (load)/(cross-sectional area). Note that we must divide the diameter by 2 to get the radius and convert square millimeters to square meters:

$$S = \frac{L}{\pi r^2} = \frac{5 \times 10^4 \, \text{N}}{\pi (10 \, \text{mm})^2} \cdot \left(\frac{1000 \, \text{mm}}{1 \, \text{m}}\right)^2 \approx 1.6 \times 10^8 \, \text{Pa}$$

Since the stress is 1.6×10^8 Pa and the maximum stress on aluminum is 1×10^8 Pa, the aluminum rod is not strong enough. The design must be changed. The diameter of the aluminum rod could be increased, or stronger/lighter metal such as titanium could be used.

4.4   Addition and Subtraction of Polynomials

We first used polynomials in Chapter 1, but did not identify them as polynomials. Polynomials also occurred in the equations and inequalities of Chapter 2. In this section, we will define polynomials and begin a thorough study of polynomials.

In This Section

⟨1⟩   Polynomials

In Chapter 1 we defined a **term** as an expression containing a number or the product of a number and one or more variables raised to powers. If the number is 1 or the power is 1, we usually omit it, as in $1x^1 = x$. Some examples of terms are

$$4x^3, \quad -x^2y^3, \quad abc, \quad \text{and} \quad -2.$$

The number preceding the variable in a term is the **coefficient** of the variable or the coefficient of the term. The coefficients of the terms $4x^3$, $-x^2y^3$, and abc are 4, -1, and 1, respectively. The **degree** of a term in one variable is the power of the variable. So the degree of $4x^3$ is 3.

A **polynomial** is a single term or a finite sum of terms in which the powers of the variables are positive integers. If the coefficient of a term is negative, we use subtraction, as in $x^4 - 6y^4$ rather than $x^4 + (-6y^4)$. So,

$$4x^3 + 3x + 2, \quad a^2 + 2ab + b^2, \quad x^4 - 6y^4, \quad \text{and} \quad x$$

are polynomials. The **degree of a polynomial** in one variable is the highest degree of its terms. Consider the polynomial

$$4x^3 - 15x^2 + x - 2.$$

The degree of $4x^3$ is 3 and the degree of $-15x^2$ is 2. Since $x = x^1$, the degree of x is 1. Since $-2 = -2x^0$, the degree of -2 is 0. So the degree of the polynomial is 3. A single number is called a **constant,** and so the zero-degree term is also called the **constant term.** The degree of a polynomial consisting of a single number is 0.

$$4x^3 - 15x^2 + x - 2$$

Third-   Second-   First-   Zero-
degree   degree   degree   degree
term   term   term   term

In $4x^3 - 15x^2 + x - 2$, the coefficient of x^3 (or the term $4x^3$) is 4. The coefficient of x^2 is -15 and the coefficient of x is 1.

E X A M P L E   1

Identifying coefficients

Determine the coefficients of x^3 and x^2 in each polynomial:

  a) $x^3 + 5x^2 - 6$ **b)** $4x^6 - x^3 + x$

Solution

  a) Write the polynomial as $1 \cdot x^3 + 5x^2 - 6$ to see that the coefficient of x^3 is 1 and the coefficient of x^2 is 5.

  b) The x^2-term is missing in $4x^6 - x^3 + x$. Because $4x^6 - x^3 + x$ can be written as

$$4x^6 - 1 \cdot x^3 + 0 \cdot x^2 + x,$$

    the coefficient of x^3 is -1 and the coefficient of x^2 is 0.

> Now do Exercises 1–6

For simplicity we generally write polynomials in one variable with the exponents decreasing from left to right and the constant term last. So we write

$$x^3 - 4x^2 + 5x + 1 \qquad \text{rather than} \qquad -4x^2 + 1 + 5x + x^3.$$

When a polynomial is written with decreasing exponents, the coefficient of the first term is called the **leading coefficient.**

Certain polynomials are given special names. A **monomial** is a polynomial that has one term, a **binomial** is a polynomial that has two terms, and a **trinomial** is a polynomial that has three terms. For example, $3x^5$ is a monomial, $2x - 1$ is a binomial, and $4x^6 - 3x + 2$ is a trinomial.

E X A M P L E   2

Types of polynomials

Identify each polynomial as a monomial, binomial, or trinomial and state its degree.

  a) $5x^2 - 7x^3 + 2$ **b)** $x^{43} - x^2$ **c)** $5x$ **d)** -12

Solution

  a) The polynomial $5x^2 - 7x^3 + 2$ is a third-degree trinomial.

  b) The polynomial $x^{43} - x^2$ is a binomial with degree 43.

c) Because $5x = 5x^1$, this polynomial is a monomial with degree 1.

d) The polynomial -12 is a monomial with degree 0.

Now do Exercises 7–18

⟨2⟩ Evaluating Polynomials

A polynomial with a variable in it has no value until the variable is replaced with a number. Example 3 shows how to evaluate a polynomial.

EXAMPLE 3

Evaluating polynomials

a) Find the value of $-3x^4 - x^3 + 20x + 3$ when $x = 1$.

b) Find the value of $-3x^4 - x^3 + 20x + 3$ when $x = -2$.

Solution

a) Replace x by 1 in the polynomial:

$$-3x^4 - x^3 + 20x + 3 = -3(1)^4 - (1)^3 + 20(1) + 3$$
$$= -3 - 1 + 20 + 3$$
$$= 19$$

So the value of the polynomial is 19 when $x = 1$.

b) Replace x by -2 in the polynomial:

$$-3x^4 - x^3 + 20x + 3 = -3(-2)^4 - (-2)^3 + 20(-2) + 3$$
$$= -3(16) \quad (-8) - 40 + 3$$
$$= -48 + 8 - 40 + 3$$
$$= -77$$

So the value of the polynomial is -77 when $x = -2$.

Now do Exercises 19–26

If the value of a polynomial is used to determine the value of a second variable y, then we have a **polynomial function.** For example,

$$y = 3x + 5, \quad y = x^2 - 1, \quad y = x^3, \quad \text{and} \quad y = -3x^4 - x^3 + 20x + 3$$

are polynomial functions. First-degree polynomial functions like $y = 3x + 5$ are linear functions. We discussed them in Chapter 3. We use function notation here just as we used it with linear functions in Chapter 3. For example, let

$$P(x) = x^2 - 1 \quad \text{and} \quad Q(x) = -3x^4 - x^3 + 20x + 3.$$

Then $P(-2)$ (read "P of -2") is the value of the polynomial $x^2 - 1$ when $x = -2$ and

$$P(-2) = (-2)^2 - 1 = 3.$$

In Example 3(b) we found that if $x = -2$, then the value of $-3x^4 - x^3 + 20x + 3$ is -77. So $Q(-2) = -77$.

E X A M P L E   **4**

Evaluating polynomials using function notation

a) If $P(x) = -3x^4 - x^3 + 20x + 3$, find $P(1)$.

b) If $D(a) = a^3 - 5$, find $D(0)$, $D(1)$, and $D(2)$.

⟨ **Calculator Close-Up** ⟩

To evaluate the polynomial in Example 4(a) with a calculator, first use Y = to define the polynomial.

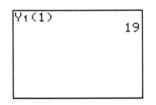

Then find $y_1(1)$.

Solution

a) To find $P(1)$, replace x by 1 in the formula for $P(x)$:

$$P(x) = -3x^4 - x^3 + 20x + 3$$
$$P(1) = -3(1)^4 - (1)^3 + 20(1) + 3$$
$$= 19$$

So $P(1) = 19$. The value of the polynomial when $x = 1$ is 19.

b) To find $D(0)$, $D(1)$, and $D(2)$ replace a with 0, 1, and 2:

$$D(0) = 0^3 - 5 = -5, \quad D(1) = 1^3 - 5 = -4, \quad D(2) = 2^3 - 5 = 3$$

So $D(0) = -5$, $D(1) = -4$, and $D(2) = 3$.

▸ **Now do Exercises 27–32**

⟨3⟩ Addition of Polynomials

You learned how to combine like terms in Chapter 1. Also, you combined like terms when solving equations in Chapter 2. Addition of polynomials is done simply by adding the like terms.

> **Addition of Polynomials**
>
> To add two polynomials, add the like terms.

Polynomials can be added horizontally or vertically, as shown in Example 5.

E X A M P L E   **5**

Adding polynomials

Perform the indicated operation.

a) $(x^2 - 6x + 5) + (-3x^2 + 5x - 9)$

b) $(-5a^3 + 3a - 7) + (4a^2 - 3a + 7)$

⟨ **Helpful Hint** ⟩

When we perform operations with polynomials and write the results as equations, those equations are identities. For example,

$$(x + 1) + (3x + 5) = 4x + 6$$

is an identity. This equation is satisfied by every real number.

Solution

a) The commutative and associative properties enable us to remove the parentheses and rearrange the terms with like terms next to each other:

$$(x^2 - 6x + 5) + (-3x^2 + 5x - 9) = x^2 - 3x^2 - 6x + 5x + 5 - 9$$
$$= -2x^2 - x - 4$$

Note that $x^2 - 3x^2 = (1 - 3)x^2 = -2x^2$ and $-6x + 5x = (-6 + 5)x = -x$ because of the distributive property. It is not necessary to write all of these details. You can simply pick out the like terms from each polynomial and combine them.

b) When adding vertically, we line up the like terms:

$$\begin{array}{r} -5a^3 \qquad\quad + 3a - 7 \\ 4a^2 - 3a + 7 \\ \hline -5a^3 + 4a^2 \qquad\qquad\quad \end{array}$$ Add.

▸ **Now do Exercises 33–46**

⟨4⟩ Subtraction of Polynomials

To add polynomials we add the like terms, and to subtract polynomials we subtract the like terms. However, since $a - b = a + (-b)$, it is usually simplest to change the signs of all terms in the second polynomial and then add.

> **Subtraction of Polynomials**
>
> To subtract two polynomials subtract the like terms, or change the signs of all terms in the second polynomial and then add.

Polynomials can be subtracted horizontally or vertically, as shown in Example 6. Vertical subtraction is used in the long division algorithm in Section 4.8.

EXAMPLE 6

Subtracting polynomials
Perform the indicated operation.

    a) $(x^2 - 5x - 3) - (4x^2 + 8x - 9)$ **b)** $(4y^3 - 3y + 2) - (5y^2 - 7y - 6)$

Solution

    a) $(x^2 - 5x - 3) - (4x^2 + 8x - 9) = x^2 - 5x - 3 - 4x^2 - 8x + 9$ Change signs.

                                 $= x^2 - 4x^2 - 5x - 8x - 3 + 9$ Rearrange.

                                   $= -3x^2 - 13x + 6$ Add.

    b) To subtract $5y^2 - 7y - 6$ from $4y^3 - 3y + 2$ vertically, we line up the like terms as we do for addition:

$$4y^3 \qquad\quad - 3y + 2$$
$$-\quad (5y^2 - 7y - 6)$$
$$\overline{}$$

    Now change the signs of $5y^2 - 7y - 6$ and add the like terms:

$$4y^3 \qquad\quad - 3y + 2$$
$$-5y^2 + 7y + 6$$
$$\overline{4y^3 - 5y^2 + 4y + 8}$$

> Now do Exercises 47–60

⟨ **Helpful Hint** ⟩

For subtraction, write the original problem and then rewrite it as addition with the signs changed. Many students have trouble when they write the original problem and then overwrite the signs. Vertical subtraction is essential for performing long division of polynomials in Section 4.8.

CAUTION When adding or subtracting polynomials vertically, be sure to line up the like terms.

In Example 7 we combine addition and subtraction of polynomials.

EXAMPLE 7

Adding and subtracting
Perform the indicated operations:

$$(2x^2 - 3x) + (x^3 + 6) - (x^4 - 6x^2 - 9)$$

Solution

Remove the parentheses and combine the like terms:

$$(2x^2 - 3x) + (x^3 + 6) - (x^4 - 6x^2 - 9) = 2x^2 - 3x + x^3 + 6 - x^4 + 6x^2 + 9$$
$$= -x^4 + x^3 + 8x^2 - 3x + 15$$

> Now do Exercises 77–84

⟨5⟩ **Applications**

Polynomials are often used to represent unknown quantities. In certain situations it is necessary to add or subtract such polynomials.

E X A M P L E   **8**

Profit from prints

Trey pays $60 per day for a permit to sell famous art prints in the Student Union Mall. Each print costs him $4, so the polynomial $C(x) = 4x + 60$ represents his daily cost in dollars for x prints sold. He sells the prints for $10 each. So the polynomial $R(x) = 10x$ represents his daily revenue for x prints sold. Find a polynomial $P(x)$ that represents his daily profit from selling x prints. Evaluate the profit polynomial for $x = 30$.

Solution

Because profit is revenue minus cost, we can subtract the corresponding polynomials to get a polynomial that represents the daily profit:

$$
\begin{aligned}
P(x) &= R(x) - C(x) \\
&= 10x - (4x + 60) \\
&= 10x - 4x - 60 \\
&= 6x - 60
\end{aligned}
$$

So the daily profit polynomial is $P(x) = 6x - 60$. Now evaluate this profit polynomial for $x = 30$:

$$
\begin{aligned}
P(30) &= 6(30) - 60 \\
&= 120
\end{aligned}
$$

So if Trey sells 30 prints, his profit is $120.

> Now do Exercises 85–94

Warm-Ups ▼

Fill in the blank.

1. A _____ of a polynomial is a single number or the product of a number and one or more variables raised to whole number powers.
2. The number preceding the variable in each term is the _____ of that term.
3. The _____ term is just a number.
4. A _____ is a single term or a finite sum of terms.
5. The _____ of a polynomial in one variable is the highest power of the variable in the polynomial.
6. A _____ is a polynomial with one term.
7. A _____ is a polynomial with two terms.
8. A _____ is a polynomial with three terms.

True or false?

9. The coefficient of x in $2x^2 - 4x + 7$ is 4.
10. The degree of the polynomial $x^2 + 5x - 9x^3$ is 2.
11. The coefficient of x in $x^2 - x$ is -1.
12. The degree of $x^2 - x$ is 2.
13. A binomial always has degree 2.
14. If $P(x) = 3x - 1$, then $P(5) = 14$.
15. For any value of x, $x^2 - 7x^2 = -6x^2$.
16. For any value of x, $(3x^2 - 8x) + (x^2 + 4x) = 4x^2 - 4x$.
17. For any value of x, $(3x^2 - 8x) - (x^2 + 4x) = 2x^2 - 12x$.

‹1› Polynomials

Determine the coefficients of x^3 and x^2 in each polynomial. See Example 1.

1. $-3x^3 + 7x^2$

2. $10x^3 - x^2$

3. $x^4 + 6x^2 - 9$

4. $x^5 - x^3 + 3$

5. $\dfrac{x^3}{3} + \dfrac{7x^2}{2} - 4$

6. $\dfrac{x^3}{2} - \dfrac{x^2}{4} + 2x + 1$

Identify each polynomial as a monomial, binomial, or trinomial and state its degree. See Example 2.

7. -1

8. 5

9. m^3

10. $3a^8$

11. $4x + 7$

12. $a + 6$

13. $x^{10} - 3x^2 + 2$

14. $y^6 - 6y^3 + 9$

15. $x^6 + 1$

16. $b^2 - 4$

17. $a^3 - a^2 + 5$

18. $-x^2 + 4x - 9$

‹2› Evaluating Polynomials

Evaluate each polynomial as indicated. See Examples 3 and 4.

19. Evaluate $x^2 + 1$ for $x = 3$.

20. Evaluate $x^2 - 1$ for $x = -3$.

21. Evaluate $2x^2 - 3x + 1$ for $x = -1$.

22. Evaluate $3x^2 - x + 2$ for $x = -2$.

23. Evaluate $\frac{1}{2}x^2 - x + 1$ for $x = \frac{1}{2}$.

24. Evaluate $3x^2 + \frac{1}{2}x - 1$ for $x = \frac{1}{3}$.

25. Evaluate $-3x^3 - x^2 + 3x - 4$ for $x = 3$.

26. Evaluate $-2x^4 - 3x^2 + 5x - 9$ for $x = 2$.

27. If $P(x) = x^2 - 4$, find $P(3)$.

28. If $P(x) = x^3 + 1$, find $P(2)$.

29. If $P(x) = 3x^4 - 2x^3 + 7$, find $P(-2)$.

30. If $P(x) = -2x^3 + 5x^2 - 12$, find $P(5)$.

31. If $P(x) = 1.2x^3 - 4.3x - 2.4$, find $P(1.45)$.

32. If $P(x) = -3.5x^4 - 4.6x^3 + 5.5$, find $P(-2.36)$.

‹3› Addition of Polynomials

Perform the indicated operation. See Example 5.

33. $(x - 3) + (3x - 5)$

34. $(x - 2) + (x + 3)$

35. $(q - 3) + (q + 3)$

36. $(q + 4) + (q + 6)$

37. $(3x + 2) + (x^2 - 4)$

38. $(5x^2 - 2) + (-3x^2 - 1)$

39. $(4x - 1) + (x^3 + 5x - 6)$

40. $(3x - 7) + (x^2 - 4x + 6)$

41. $(a^2 - 3a + 1) + (2a^2 - 4a - 5)$

42. $(w^2 - 2w + 1) + (2w - 5 + w^2)$

43. $(w^2 - 9w - 3) + (w - 4w^2 + 8)$

44. $(a^3 - a^2 - 5a) + (6 - a - 3a^2)$

45. $(5.76x^2 - 3.14x - 7.09) + (3.9x^2 + 1.21x + 5.6)$

46. $(8.5x^2 + 3.27x - 9.33) + (x^2 - 4.39x - 2.32)$

‹4› Subtraction of Polynomials

Perform the indicated operation. See Example 6.

47. $(x - 2) - (5x - 8)$

48. $(x - 7) - (3x - 1)$

49. $(m - 2) - (m + 3)$

50. $(m + 5) - (m + 9)$

51. $(2z^2 - 3z) - (3z^2 - 5z)$

52. $(z^2 - 4z) - (5z^2 - 3z)$

53. $(w^5 - w^3) - (-w^4 + w^2)$

54. $(w^6 - w^3) - (-w^2 + w)$

55. $(t^2 - 3t + 4) - (t^2 - 5t - 9)$

56. $(t^2 - 6t + 7) - (5t^2 - 3t - 2)$

57. $(9 - 3y - y^2) - (2 + 5y - y^2)$

58. $(4 - 5y + y^3) - (2 - 3y + y^2)$

59. $(3.55x - 879) - (26.4x - 455.8)$

60. $(345.56x - 347.4) - (56.6x + 433)$

Add or subtract the polynomials as indicated.
See Examples 5 and 6.

61. Add:

$3a - 4$
$\underline{a + 6}$

62. Add:

$2w - 8$
$\underline{w + 3}$

63. Subtract:

$3x + 11$
$\underline{-(5x + 7)}$

64. Subtract:

$4x + 3$
$\underline{-(2x + 9)}$

65. Add:

$a - b$
$\underline{a + b}$

66. Add:

$s - 6$
$\underline{s - 1}$

67. Subtract:

$-3m + 1$
$\underline{-(2m - 6)}$

68. Subtract:

$-5n + 2$
$\underline{-(3n - 4)}$

Add or subtract as indicated. Arrange the polynomials
vertically as in Exercises 61–68. See Examples 5 and 6.

69. Add $2x^2 - x - 3$ and $2x^2 + x + 4$.

70. Add $-x^2 + 4x - 6$ and $3x^2 - x - 5$.

71. Subtract $2a^3 + 4a^2 - 2a$ from $3a^3 - 5a^2 + 7$.

72. Subtract $b^3 - 4b - 2$ from $-2b^3 + 7b^2 - 9$.

73. $(x^2 - 3x + 6) - (x^2 - 3)$

74. $(x^4 - 3x^2 + 2) - (3x^4 - 2x)$

75. $(y^3 + 4y^2 - 6y - 5) + (y^3 + 3y - 9)$

76. $(q^2 - 4q + 9) + (-3q^3 - 7q + 5)$

Perform the indicated operations. See Example 7.

77. $(4m - 2) + (2m + 4) - (9m - 1)$

78. $(-5m - 6) + (8m - 3) - (-5m + 3)$

79. $(6y - 2) - (8y + 3) - (9y - 2)$

80. $(-5y - 1) - (8y - 4) - (y + 3)$

81. $(-x^2 - 5x + 4) + (6x^2 - 8x + 9) - (3x^2 - 7x + 1)$

82. $(-8x^2 + 5x - 12) + (-3x^2 - 9x + 18)$
$- (-3x^2 + 9x - 4)$

83. $(-6z^4 - 3z^3 + 7z^2) - (5z^3 + 3z^2 - 2) + (z^4 - z^2 + 5)$

84. $(-v^3 - v^2 - 1) - (v^4 - v^2 - v - 1) + (v^3 - 3v^2 + 6)$

⟨5⟩ Applications

Solve each problem. See Example 8.

85. *Water pumps.* Walter uses the polynomials $R(x) = 400x$ and $C(x) = 120x + 800$ to estimate his monthly revenue and cost in dollars for producing x water pumps per month.

   a) Write a polynomial $P(x)$ for his monthly profit.

   b) Find the monthly profit for $x = 50$.

86. *Manufacturing costs.* Ace manufacturing has determined that the cost of labor for producing x transmissions is $L(x) = 0.3x^2 + 400x + 550$ dollars, while the cost of materials is $M(x) = 0.1x^2 + 50x + 800$ dollars.

   a) Write a polynomial $T(x)$ that represents the total cost of materials and labor for producing x transmissions.

   b) Evaluate the total cost polynomial for $x = 500$.

   c) Find the cost of labor for 500 transmissions and the cost of materials for 500 transmissions.

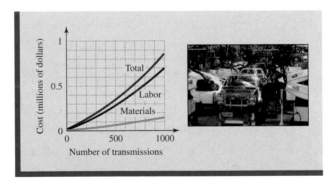

Figure for Exercise 86

87. *Perimeter of a triangle.* The shortest side of a triangle is x meters, and the other two sides are $3x - 1$ and $2x + 4$ meters. Write a polynomial $P(x)$ that represents the perimeter and then evaluate the perimeter polynomial if x is 4 meters.

88. *Perimeter of a rectangle.* The width of a rectangular playground is $2x - 5$ feet, and the length is $3x + 9$ feet. Write a polynomial $P(x)$ that represents the perimeter and then evaluate this perimeter polynomial if x is 4 feet.

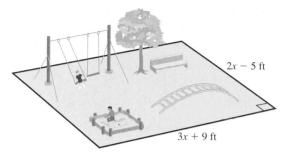

Figure for Exercise 88

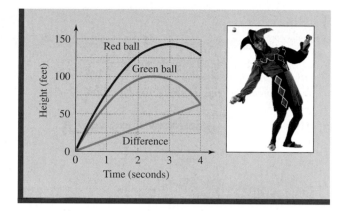

Figure for Exercise 92

🔊 **89.** *Total distance.* Hanson drove his rig at x mph for 3 hours and then increased his speed to $x + 15$ mph and drove 2 more hours. Write a polynomial $D(x)$ that represents the total distance that he traveled. Find $D(45)$.

90. *Before and after.* Jessica traveled $2x + 50$ miles in the morning and $3x - 10$ miles in the afternoon. Write a polynomial $T(x)$ that represents the total distance that she traveled. Find $T(20)$.

91. *Sky divers.* Bob and Betty simultaneously jump from two airplanes at different altitudes. Bob's altitude t seconds after leaving his plane is $-16t^2 + 6600$ feet. Betty's altitude t seconds after leaving her plane is $-16t^2 + 7400$ feet. Write a polynomial that represents the difference between their altitudes t seconds after leaving the planes. What is the difference between their altitudes 3 seconds after leaving the planes?

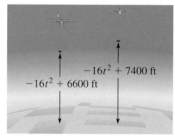

Figure for Exercise 91

92. *Height difference.* A red ball and a green ball are simultaneously tossed into the air. The red ball is given an initial velocity of 96 feet per second, and its height t seconds after it is tossed is $-16t^2 + 96t$ feet. The green ball is given an initial velocity of 80 feet per second, and its height t seconds after it is tossed is $-16t^2 + 80t$ feet.

a) Find a polynomial $D(t)$ that represents the difference in the heights of the two balls.
b) How much higher is the red ball 2 seconds after the balls are tossed?
c) In reality, when does the difference in the heights stop increasing?

93. *Total interest.* Donald received $0.08(x + 554)$ dollars interest on one investment and $0.09(x + 335)$ interest on another investment. Write a polynomial $T(x)$ that represents the total interest he received. What is the total interest if $x = 1000$?

94. *Total acid.* Deborah figured that the amount of acid in one bottle of solution is $0.12x$ milliliters and the amount of acid in another bottle of solution is $0.22(75 - x)$ milliliters. Find a polynomial $T(x)$ that represents the total amount of acid? What is the total amount of acid if $x = 50$?

Getting More Involved

95. *Discussion*

Is the sum of two natural numbers always a natural number? Is the sum of two integers always an integer? Is the sum of two polynomials always a polynomial? Explain.

96. *Discussion*

Is the difference of two natural numbers always a natural number? Is the difference of two rational numbers always a rational number? Is the difference of two polynomials always a polynomial? Explain.

97. *Writing*

Explain why the polynomial $2^4 - 7x^3 + 5x^2 - x$ has degree 3 and not degree 4.

98. *Discussion*

Which of the following polynomials does not have degree 2? Explain.

a) πr^2 b) $\pi^2 - 4$ c) $y^2 - 4$
d) $x^2 - x^4$ e) $a^2 - 3a + 9$

Mid-Chapter **Quiz** │ Sections 4.1 through 4.4 │ Chapter 4

Simplify. All variables represent nonzero real numbers. Use only positive exponents in your answers.

1. -24

2. $(-2)^4$

3. $2^3 \cdot 1^2 \cdot 5^0$

4. $x^3 \cdot x^2 \cdot x$

5. $\dfrac{a^9}{a^2}$

6. $\dfrac{b^3}{b^7}$

7. $(2a^5b^3)^5$

8. $\left(\dfrac{2}{w^4}\right)^3$

9. $\left(\dfrac{ab^6}{a^7b^4}\right)^3$

10. $(-3x^{-2}y^3)^{-3}$

11. $\left(\dfrac{a^{-3}}{3b^4}\right)^{-2}$

12. $\left(\dfrac{2w^{-1}y^4}{w^{-1}y^{-3}}\right)^{-4}$

13. $(2xy^{-3})^2 (3x^3y^{-4})^{-3}$

14. $(3.5 \times 10^3)(4 \times 10^{-4})$

15. $\dfrac{4.5 \times 10^{-4}}{9 \times 10^{-5}}$

Perform the indicated operations.

16. $(5x^2 - 3x) + (-8x^2 - 2x + 6)$

17. $(5x^2 - 3x) - (-8x^2 - 2x + 6)$

18. $-5x^3y + 2x^3y$

Miscellaneous.

19. Find the value of the polynomial $3x^3 - 5x^2 + 6x - 9$ when $x = -2$.

20. Find $P(-1)$ if $P(x) = -8x^4 + 9x^3 - 7x^2 + 5$.

4.5   Multiplication of Polynomials

In This Section

⟨1⟩ **Multiplying Monomials**

⟨2⟩ **Multiplying Polynomials**

⟨3⟩ **The Additive Inverse of a Polynomial**

⟨4⟩ **Applications**

You learned to multiply some polynomials in Chapter 1. In this section, you will learn how to multiply any two polynomials.

⟨1⟩ Multiplying Monomials

Monomials are the simplest polynomials. We learned to multiply monomials in Section 4.1 using the product rule for exponents.

E X A M P L E   1

Multiplying monomials

Find the indicated products.

  a) $2x^3 \cdot 3x^4$ **b)** $(-2ab^2)(-3ab^4)$ **c)** $(3a^2)^3$

Solution

  a) $2x^3 \cdot 3x^4 = 6x^7$ Product rule for exponents

  b) $(-2ab^2)(-3ab^4) = 6a^2b^6$ Product rule for exponents

  c) $(3a^2)^3 = 3^3(a^2)^3$ Power of a product rule

     $= 27a^6$ Power of a power rule

> Now do Exercises 1–16

CAUTION   Be sure to distinguish between adding and multiplying monomials. You can add like terms to get $3x^4 + 2x^4 = 5x^4$, but you cannot combine the terms in $3w^5 + 6w^2$. However, you can multiply any two monomials: $3x^4 \cdot 2x^4 = 6x^8$ and $3w^5 \cdot 6w^2 = 18w^7$. Note that the exponents are added, not multiplied.

⟨2⟩ Multiplying Polynomials

To multiply a monomial and a polynomial, we use the distributive property.

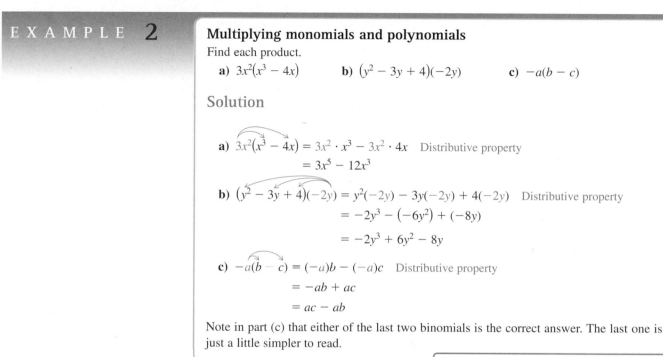

EXAMPLE **2**

Multiplying monomials and polynomials

Find each product.

a) $3x^2(x^3 - 4x)$ **b)** $(y^2 - 3y + 4)(-2y)$ **c)** $-a(b - c)$

Solution

a) $3x^2(x^3 - 4x) = 3x^2 \cdot x^3 - 3x^2 \cdot 4x$ Distributive property

$= 3x^5 - 12x^3$

b) $(y^2 - 3y + 4)(-2y) = y^2(-2y) - 3y(-2y) + 4(-2y)$ Distributive property

$= -2y^3 - (-6y^2) + (-8y)$

$= -2y^3 + 6y^2 - 8y$

c) $-a(b - c) = (-a)b - (-a)c$ Distributive property

$= -ab + ac$

$= ac - ab$

Note in part (c) that either of the last two binomials is the correct answer. The last one is just a little simpler to read.

Now do Exercises 17–30

Just as we use the distributive property to find the product of a monomial and a polynomial, we can use it to find the product of any two polynomials.

EXAMPLE **3**

Multiplying polynomials

Use the distributive property to find each product.

a) $(x + 2)(x + 5)$ **b)** $(x + 3)(x^2 + 2x - 7)$

Solution

a) First multiply each term of $x + 5$ by $x + 2$:

$(x + 2)(x + 5) = (x + 2)x + (x + 2)5$ Distributive property

$= x^2 + 2x + 5x + 10$ Distributive property

$= x^2 + 7x + 10$ Combine like terms.

b) First multiply each term of the trinomial by $x + 3$:

$(x + 3)(x^2 + 2x - 7) = (x + 3)x^2 + (x + 3)2x + (x + 3)(-7)$ Distributive property

$= x^3 + 3x^2 + 2x^2 + 6x - 7x - 21$ Distributive property

$= x^3 + 5x^2 - x - 21$ Combine like terms.

Now do Exercises 31–42

Examples 2 and 3 illustrate the following rule.

> **Multiplication of Polynomials**
> To multiply polynomials, multiply each term of one polynomial by every term of the other polynomial and then combine like terms.

⟨3⟩ The Additive Inverse of a Polynomial

The additive inverse of a is $-a$, because $a + (-a) = 0$. Since $-1 \cdot a = -a$, multiplying an expression by -1 produces that additive inverse of the expression. To find the additive inverse of $a - b$ multiply by -1:

$$-1(a - b) = -1 \cdot a - (-1)b = -a + b = b - a$$

By the distributive property, every term is multiplied by -1, causing every term to change sign. So the additive inverse (or opposite) of $a - b$ is $-a + b$ or $b - a$. In symbols,

$$-(a - b) = b - a$$

CAUTION   The additive inverse of $a + b$ is $-a - b$ *not* $a - b$.

The additive inverse of any polynomial can be found by multiplying each term by -1 or simply changing the sign of each term, as shown in Example 4.

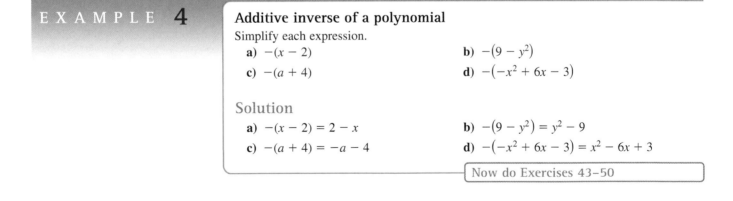

EXAMPLE   4

Additive inverse of a polynomial
Simplify each expression.
  a) $-(x - 2)$
  b) $-(9 - y^2)$
  c) $-(a + 4)$
  d) $-(-x^2 + 6x - 3)$

Solution
  a) $-(x - 2) = 2 - x$
  b) $-(9 - y^2) = y^2 - 9$
  c) $-(a + 4) = -a - 4$
  d) $-(-x^2 + 6x - 3) = x^2 - 6x + 3$

Now do Exercises 43–50

⟨4⟩ Applications

EXAMPLE   5

Multiplying polynomials
A parking lot is 20 yards wide and 30 yards long. If the college increases the length and width by the same amount to handle an increasing number of cars, then what polynomial represents the area of the new lot? What is the new area if the increase is 15 yards?

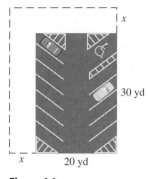

Figure 4.1

Solution

If x is the amount of increase in yards, then the new lot will be $x + 20$ yards wide and $x + 30$ yards long as shown in Fig. 4.1. Multiply the length and width to get the area:

$$(x + 20)(x + 30) = (x + 20)x + (x + 20)30$$
$$= x^2 + 20x + 30x + 600$$
$$= x^2 + 50x + 600$$

The polynomial $x^2 + 50x + 600$ represents the area of the new lot. If $x = 15$, then

$$x^2 + 50x + 600 = (15)^2 + 50(15) + 600 = 1575.$$

If the increase is 15 yards, then the area of the lot will be 1575 square yards.

Now do Exercises 71–80

Warm-Ups ▼

Fill in the blank.

1. To multiply a monomial and a binomial we use the _____ property.
2. The sum of two monomials is a _____ if the terms are not like terms.
3. To find the _____ of a polynomial we change the sign of every term in the polynomial.
4. When multiplying two monomials, we may need the _____ rule for exponents.

True or false?

5. For any value of x, $3x^3 \cdot 5x^4 = 15x^{12}$.
6. For any number x, $3x^2 \cdot 2x^7 = 5x^9$.
7. For any value of x, $-3x(5x - 7x^2) = -15x^2 + 21x^3$.
8. For any number x, $-2(3 - x) = 2x - 6$.
9. For any number x, $-(x - 7) = 7 - x$.
10. $37 - 83 = -(83 - 37)$

Exercises 4.5

‹ **Study Tips** ›

- Effective time management will allow adequate time for school, work, social life, and free time. However at times you will have to sacrifice to do well.
- Everyone has different attention spans. Start by studying 10 to 15 minutes at a time and then build up to longer periods. Be realistic. When you can no longer concentrate, take a break.

‹1› **Multiplying Monomials**

Find each product. See Example 1.

1. $3x^2 \cdot 9x^3$
2. $5x^7 \cdot 3x^5$
3. $2a^3 \cdot 7a^8$
4. $3y^{12} \cdot 5y^{15}$
5. $-6x^2 \cdot 5x^2$
6. $-2x^2 \cdot 8x^5$
7. $(-9x^{10})(-3x^7)$
8. $(-2x^2)(-8x^9)$
9. $-6st \cdot 9st$

10. $-12sq \cdot 3s$ **11.** $3wt \cdot 8w^7t^6$ **12.** $h^8k^3 \cdot 5h$

13. $(5y)^2$ **14.** $(6x)^2$

15. $(2x^3)^2$ **16.** $(3y^5)^2$

⟨2⟩ Multiplying Polynomials

Find each product. See Example 2.

17. $x(x + y^2)$
18. $x^2(x - y)$
19. $4y^2(y^5 - 2y)$
20. $6t^3(t^5 + 3t^2)$
21. $-3y(6y - 4)$
22. $-9y(y^2 - 1)$
23. $(y^2 - 5y + 6)(-3y)$
24. $(x^3 - 5x^2 - 1)7x^2$
25. $-x(y^2 - x^2)$
26. $-ab(a^2 - b^2)$
27. $(3ab^3 - a^2b^2 - 2a^3b)5a^3$
28. $(3c^2d - d^3 + 1)8cd^2$
29. $-\frac{1}{2}t^2v(4t^3v^2 - 6tv - 4v)$
30. $-\frac{1}{3}m^2n^3(-6mn^2 + 3mn - 12)$

Use the distributive property to find each product. See Example 3.

31. $(x + 1)(x + 2)$ **32.** $(x + 6)(x + 3)$

33. $(x - 3)(x + 5)$ **34.** $(y - 2)(y + 4)$

35. $(t - 4)(t - 9)$ **36.** $(w - 3)(w - 5)$

37. $(x + 1)(x^2 + 2x + 2)$ **38.** $(x - 1)(x^2 + x + 1)$

39. $(3y + 2)(2y^2 - y + 3)$ **40.** $(4y + 3)(y^2 + 3y + 1)$

41. $(y^2z - 2y^4)(y^2z + 3z^2 - y^4)$

42. $(m^3 - 4mn^2)(6m^4n^2 - 3m^6 + m^2n^4)$

⟨3⟩ The Additive Inverse of a Polynomial

Simplify each expression. See Example 4.

43. $-(3t - u)$ **44.** $-(-4 - u)$
45. $-(3x + y)$ **46.** $-(x - 5b)$
47. $-(-3a^2 - a + 6)$

48. $-(5b^2 - b - 7)$
49. $-(3w^2 + w - 6)$
50. $-(-4t^2 + t - 6)$

Miscellaneous

Perform the indicated operation.

51. $-3x(2x - 9)$ **52.** $-1(2 - 3x)$

53. $2 - 3x(2x - 9)$ **54.** $6 - 3(4x - 8)$

55. $(2 - 3x) + (2x - 9)$ **56.** $(2 - 3x) - (2x - 9)$

57. $(6x^6)^2$ **58.** $(-3a^3b)^2$
59. $3ab^3(-2a^2b^7)$ **60.** $-4xst \cdot 8xs$

61. $(5x + 6)(5x + 6)$ **62.** $(5x - 6)(5x - 6)$

63. $(5x - 6)(5x + 6)$ **64.** $(2x - 9)(2x + 9)$

65. $2x^2(3x^5 - 4x^2)$ **66.** $4a^3(3ab^3 - 2ab^3)$

67. $(m - 1)(m^2 + m + 1)$ **68.** $(a + b)(a^2 - ab + b^2)$

69. $(3x - 2)(x^2 - x - 9)$
70. $(5 - 6y)(3y^2 - y - 7)$

⟨4⟩ Applications

Solve each problem. See Example 5.

71. Office space. The length of a professor's office is x feet, and the width is $x + 4$ feet. Write a polynomial $A(x)$ that represents the area of the office. Find $A(10)$.

72. Swimming space. The length of a rectangular swimming pool is $2x - 1$ meters, and the width is $x + 2$ meters. Write a polynomial $A(x)$ that represents the area. Find $A(5)$.

73. Area. A roof truss is in the shape of a triangle with height of x feet and a base of $2x + 1$ feet. Write a polynomial $A(x)$ that represents the area of the triangle. Find $A(5)$. See the accompanying figure.

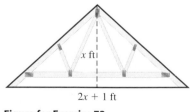

x ft

$2x + 1$ ft

Figure for Exercise 73

74. *Volume.* The length, width, and height of a box are x, $2x$, and $3x - 5$ inches, respectively. Write a polynomial $V(x)$ that represents its volume. Find $V(3)$.

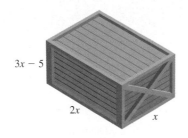

3x − 5

2x

x

Figure for Exercise 74

75. *Number pairs.* If two numbers differ by 5, then what polynomial represents their product?

76. *Number pairs.* If two numbers have a sum of 9, then what polynomial represents their product?

 77. *Area of a rectangle.* The length of a rectangle is $2.3x + 1.2$ meters, and its width is $3.5x + 5.1$ meters. What polynomial represents its area?

 78. *Patchwork.* A quilt patch cut in the shape of a triangle has a base of $5x$ inches and a height of $1.732x$ inches. What polynomial represents its area?

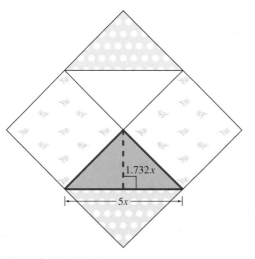

1.732x

5x

Figure for Exercise 78

79. *Total revenue.* At p dollars per ticket, a promoter expects to sell $40{,}000 - 1000p$ tickets to a concert.
   a) How many tickets will she sell at \$10 each?

   b) At \$10 per ticket, what is the total revenue?

   c) Find a polynomial $R(p)$ that represents the total revenue when tickets are p dollars each.

   d) Find $R(20)$, $R(30)$, and $R(35)$.

80. *Selling shirts.* If a vendor charges p dollars each for rugby shirts, then he expects to sell $2000 - 100p$ shirts at a tournament.
   a) Find a polynomial $R(p)$ that represents the total revenue when the shirts are p dollars each.

   b) Find $R(5)$, $R(10)$, and $R(20)$.

   c) Use the bar graph to determine the price that will give the maximum total revenue.

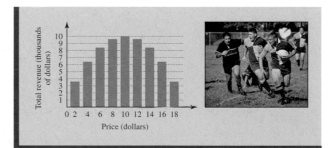

Figure for Exercise 80

Getting More Involved

 81. *Discussion*

Name all properties of the real numbers that are used in finding the following products:

   a) $-2ab^3c^2 \cdot 5a^2bc$ **b)** $(x^2 + 3)(x^2 - 8x - 6)$

82. *Discussion*

Find the product of 27 and 436 without using a calculator. Then use the distributive property to find the product $(20 + 7)(400 + 30 + 6)$ as you would find the product of a binomial and a trinomial. Explain how the two methods are related.

4.6   Multiplication of Binomials

In This Section

⟨1⟩ **The FOIL Method**

⟨2⟩ **Multiplying Binomials Quickly**

⟨3⟩ **Applications**

In Section 4.5, you learned to multiply polynomials. In this section, you will learn a rule that makes multiplication of binomials simpler.

⟨1⟩ The FOIL Method

We can use the distributive property to find the product of two binomials. For example,

$$(x + 2)(x + 3) = (x + 2)x + (x + 2)3 \quad \text{Distributive property}$$
$$= x^2 + 2x + 3x + 6 \quad \text{Distributive property}$$
$$= x^2 + 5x + 6 \quad \text{Combine like terms.}$$

There are four terms in $x^2 + 2x + 3x + 6$. The term x^2 is the product of the *first* terms of each binomial, x and x. The term $3x$ is the product of the two *outer* terms, 3 and x. The term $2x$ is the product of the two *inner* terms, 2 and x. The term 6 is the product of the last terms of each binomial, 2 and 3. We can connect the terms multiplied by lines as follows:

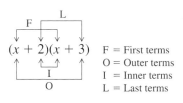

$$(x + 2)(x + 3)$$

F = First terms
O = Outer terms
I = Inner terms
L = Last terms

If you remember the word FOIL, you can get the product of the two binomials much faster than writing out all of the steps. This method is called the **FOIL method.** The name should make it easier to remember.

E X A M P L E   **1**

⟨ **Helpful Hint** ⟩

You may have to practice FOIL a while to get good at it. However, the better you are at FOIL, the easier you will find factoring in Chapter 5.

Using the FOIL method

Find each product.

a) $(x + 2)(x - 4)$ **b)** $(2x + 5)(3x - 4)$

c) $(a - b)(2a - b)$ **d)** $(x + 3)(y + 5)$

Solution

a)
$$\overset{\text{F O I L}}{(x + 2)(x - 4)} = x^2 - 4x + 2x - 8$$
$$= x^2 - 2x - 8 \quad \text{Combine like terms.}$$

b) $(2x + 5)(3x - 4) = 6x^2 - 8x + 15x - 20$
$$= 6x^2 + 7x - 20 \quad \text{Combine like terms.}$$

c) $(a - b)(2a - b) = 2a^2 - ab - 2ab + b^2$
$$= 2a^2 - 3ab + b^2$$

d) $(x + 3)(y + 5) = xy + 5x + 3y + 15 \quad \text{There are no like terms to combine.}$

> Now do Exercises 1–24

FOIL can be used to multiply any two binomials. The binomials in Example 2 have higher powers than those of Example 1.

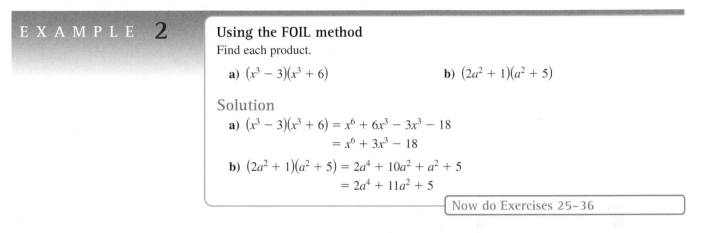

EXAMPLE **2**

Using the FOIL method
Find each product.

a) $(x^3 - 3)(x^3 + 6)$
b) $(2a^2 + 1)(a^2 + 5)$

Solution

a) $(x^3 - 3)(x^3 + 6) = x^6 + 6x^3 - 3x^3 - 18$
$$= x^6 + 3x^3 - 18$$

b) $(2a^2 + 1)(a^2 + 5) = 2a^4 + 10a^2 + a^2 + 5$
$$= 2a^4 + 11a^2 + 5$$

Now do Exercises 25–36

⟨2⟩ Multiplying Binomials Quickly

The outer and inner products in the FOIL method are often like terms, and we can combine them without writing them down. Once you become proficient at using FOIL, you can find the product of two binomials without writing anything except the answer.

EXAMPLE **3**

Using FOIL to find a product quickly
Find each product. Write down only the answer.

a) $(x + 3)(x + 4)$
b) $(2x - 1)(x + 5)$
c) $(a - 6)(a + 6)$

Solution

a) $(x + 3)(x + 4) = x^2 + 7x + 12$ Combine like terms: $3x + 4x = 7x$.

b) $(2x - 1)(x + 5) = 2x^2 + 9x - 5$ Combine like terms: $10x - x = 9x$.

c) $(a - 6)(a + 6) = a^2 - 36$ Combine like terms: $6a - 6a = 0$.

Now do Exercises 37–62

EXAMPLE **4**

Products of three binomials
Find each product.

a) $(b - 1)(b + 2)(b - 3)$
b) $\left(\frac{1}{2}x + 3\right)\left(\frac{1}{2}x - 3\right)(2x + 5)$

Solution

a) Use FOIL to find $(b - 1)(b + 2) = b^2 + b - 2$. Then use the distributive property to multiply $b^2 + b - 2$ and $b - 3$:

$(b - 1)(b + 2)(b - 3) = (b^2 + b - 2)(b - 3)$ FOIL

$$= (b^2 + b - 2)b + (b^2 + b - 2)(-3)$$ Distributive property

$$= b^3 + b^2 - 2b - 3b^2 - 3b + 6$$ Distributive property

$$= b^3 - 2b^2 - 5b + 6$$ Combine like terms.

b) $\left(\frac{1}{2}x + 3\right)\left(\frac{1}{2}x - 3\right)(2x + 5) = \left(\frac{1}{4}x^2 - 9\right)(2x + 5)$ FOIL

$= \frac{1}{2}x^3 + \frac{5}{4}x^2 - 18x - 45$ FOIL

Now do Exercises 63–70

⟨3⟩ Applications

EXAMPLE 5

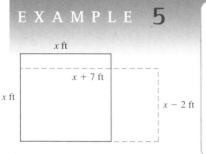

Figure 4.2

Area of a garden

Sheila has a square garden with sides of length x feet. If she increases the length by 7 feet and decreases the width by 2 feet, then what trinomial represents the area of the new rectangular garden?

Solution

The length of the new garden is $x + 7$ feet and the width is $x - 2$ feet as shown in Fig. 4.2. The area is $(x + 7)(x - 2)$ or $x^2 + 5x - 14$ square feet.

Now do Exercises 93–96

Warm-Ups ▼

Fill in the blank.

1. We can use the _____ property to multiply two binomials.
2. _____ stands for First, Outer, Inner, Last.
3. The _____ method gives the product of two binomials quickly.
4. The maximum number of terms that can result from the product of two binomials is _____.

True or false?

5. $(x + 3)(x + 2) = x^2 + 6$
6. $(x + 5)(x + 1) = x^2 + 5x + x + 5$
7. $(a + 3)(a - 2) = a^2 + a - 6$
8. $(y - 9)(y - 2) = y^2 - 11y - 18$
9. $(b^2 + 2)(b^2 - 5) = b^4 - 3b^2 - 10$
10. $(a - b)(c - d) = ac - bc + bd$

4.6 Exercises

⟨ Study Tips ⟩

• Set short-term goals and reward yourself for accomplishing them. When you have solved 10 problems, take a short break and listen to your favorite music.
• Study in a clean, comfortable, well-lit place, but don't get too comfortable. Study at a desk, not in bed.

⟨1⟩ The FOIL Method

Use FOIL to find each product. See Example 1.

1. $(x + 2)(x + 4)$
2. $(x + 3)(x + 5)$
3. $(a + 1)(a + 4)$
4. $(w + 3)(w + 6)$
5. $(x + 9)(x + 10)$
6. $(x + 5)(x + 7)$

7. $(2x + 1)(x + 3)$

8. $(3x + 2)(2x + 1)$

9. $(a - 3)(a + 2)$

10. $(b - 1)(b + 2)$

11. $(2x - 1)(x - 2)$

12. $(2y - 5)(y - 2)$

13. $(2a - 3)(a + 1)$

14. $(3x - 5)(x + 4)$

15. $(w - 50)(w - 10)$

16. $(w - 30)(w - 20)$

17. $(y - a)(y + 5)$

18. $(a + t)(3 - y)$

19. $(5 - w)(w + m)$

20. $(a - h)(b + t)$

21. $(2m - 3t)(5m + 3t)$

22. $(2x - 5y)(x + y)$

23. $(5a + 2b)(9a + 7b)$

24. $(11x + 3y)(x + 4y)$

Use FOIL to find each product. See Example 2.

25. $(x^2 - 5)(x^2 + 2)$

26. $(y^2 + 1)(y^2 - 2)$

27. $(h^3 + 5)(h^3 + 5)$

28. $(y^6 + 1)(y^6 - 4)$

29. $(3b^3 + 2)(b^3 + 4)$

30. $(5n^4 - 1)(n^4 + 3)$

31. $(y^2 - 3)(y - 2)$

32. $(x - 1)(x^2 - 1)$

33. $(3m^3 - n^2)(2m^3 + 3n^2)$

34. $(6y^4 - 2z^2)(6y^4 - 3z^2)$

35. $(3u^2v - 2)(4u^2v + 6)$

36. $(5y^3w^2 + z)(2y^3w^2 + 3z)$

⟨2⟩ Multiplying Binomials Quickly

Find each product. Try to write only the answer. See Example 3.

37. $(w + 2)(w + 1)$

38. $(q + 2)(q + 3)$

39. $(b + 4)(b + 5)$

40. $(y + 8)(y + 4)$

41. $(x - 3)(x + 9)$

42. $(m + 7)(m - 8)$

43. $(a + 5)(a + 5)$

44. $(t - 4)(t - 4)$

45. $(2x - 1)(2x - 1)$

46. $(3y + 4)(3y + 4)$

47. $(z - 10)(z + 10)$

48. $(3h - 5)(3h + 5)$

49. $(a + b)(a + b)$

50. $(x - y)(x - y)$

51. $(a - 1)(a - 2)$

52. $(b - 8)(b - 1)$

53. $(2x - 1)(x + 3)$

54. $(3y + 5)(y - 3)$

55. $(5t - 2)(t - 1)$

56. $(2t - 3)(2t - 1)$

57. $(h - 7)(h - 9)$

58. $(h - 7w)(h - 7w)$

59. $(h + 7w)(h + 7w)$

60. $(h - 7q)(h + 7q)$

61. $(2h^2 - 1)(2h^2 - 1)$

62. $(3h^2 + 1)(3h^2 + 1)$

Find each product. See Example 4.

63. $(a + 1)(a - 2)(a + 5)$

64. $(y - 1)(y + 3)(y - 4)$

65. $(h + 2)(h + 3)(h + 4)$

66. $(m - 1)(m - 3)(m - 5)$

67. $\left(\frac{1}{2}x + 4\right)\left(\frac{1}{2}x - 4\right)(4x - 8)$

68. $\left(\frac{1}{3}w - 3\right)\left(\frac{1}{3}w + 3\right)(w - 6)$

69. $\left(x + \frac{1}{2}\right)\left(x - \frac{1}{2}\right)(x + 8)$

70. $\left(x + \frac{1}{3}\right)\left(x - \frac{1}{3}\right)(x + 9)$

Miscellaneous

Perform the indicated operations.

71. $(x + 10)(x + 5)$

72. $(x + 4)(x + 8)$

73. $\left(x + \frac{1}{2}\right)\left(x + \frac{1}{2}\right)$

74. $\left(x + \frac{1}{3}\right)\left(x + \frac{1}{6}\right)$

75. $\left(4x + \frac{1}{2}\right)\left(2x + \frac{1}{4}\right)$

76. $\left(3x + \dfrac{1}{6}\right)\left(6x + \dfrac{1}{3}\right)$

77. $\left(2a + \dfrac{1}{2}\right)\left(4a - \dfrac{1}{2}\right)$

78. $\left(3b + \dfrac{2}{3}\right)\left(6b - \dfrac{1}{3}\right)$

79. $\left(\dfrac{1}{2}x - \dfrac{1}{3}\right)\left(\dfrac{1}{4}x + \dfrac{1}{2}\right)$

80. $\left(\dfrac{2}{3}t - \dfrac{1}{4}\right)\left(\dfrac{1}{2}t - \dfrac{1}{2}\right)$

81. $a(a + 3)(a + 4)$

82. $w(w + 5)(w + 9)$

83. $x^3(x + 6)(x + 7)$

84. $x^2(x^2 + 1)(x^2 + 8)$

85. $-2x^4(3x - 1)(2x + 5)$

86. $4xy^3(2x - y)(3x + y)$

87. $(x - 1)(x + 1)(x + 3)$

88. $(a - 3)(a + 4)(a - 5)$

89. $(3x - 2)(3x + 2)(x + 5)$

90. $(x - 6)(9x + 4)(9x - 4)$

91. $(x - 1)(x + 2) - (x + 3)(x - 4)$

92. $(k - 4)(k + 9) - (k - 3)(k + 7)$

⟨3⟩ **Applications**

Solve each problem. See Example 5.

93. *Area of a rug.* Find a trinomial $A(x)$ that represents the area of a rectangular rug whose sides are $x + 3$ feet and $2x - 1$ feet. Find $A(4)$.

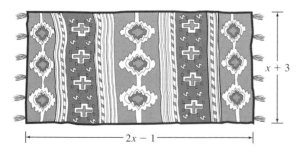

Figure for Exercise 93

94. *Area of a parallelogram.* Find a trinomial $A(x)$ that represents the area of a parallelogram whose base is $3x + 2$ meters and whose height is $2x + 3$ meters. Find $A(3)$.

95. *Area of a sail.* A sail is triangular in shape with a base of $2x - 1$ meters and a height of $4x - 4$ meters. Find a polynomial $A(x)$ that represents the area of the sail. Find $A(5)$.

96. *Area of a square.* A square has sides of length $3x + 1$ meters. Find a polynomial $A(x)$ that represents the area of the square. Find $A(1)$.

Getting More Involved

97. *Exploration*

Find the area of each of the four regions shown in the figure. What is the total area of the four regions? What does this exercise illustrate?

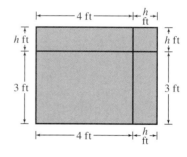

Figure for Exercise 97

98. *Exploration*

Find the area of each of the four regions shown in the figure. What is the total area of the four regions? What does this exercise illustrate?

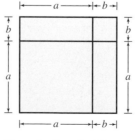

Figure for Exercise 98

4.7 Special Products

In This Section

⟨1⟩ **The Square of a Sum or Difference**

⟨2⟩ **Product of a Sum and a Difference**

⟨3⟩ **Higher Powers of Binomials**

⟨4⟩ **Applications**

In Section 4.6, you learned the FOIL method to make multiplying binomials simpler. In this section, you will learn rules for squaring binomials and for finding the product of a sum and a difference. These products are called **special products.**

⟨ Helpful Hint ⟩

To visualize the square of a sum, draw a square with sides of length $a + b$ as shown.

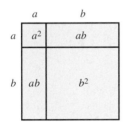

The area of the large square is $(a + b)^2$. You get the same area if you add the areas of the four smaller regions: $(a + b)^2 = a^2 + 2(ab) + b^2$.

⟨1⟩ The Square of a Sum or Difference

To compute $(a + b)^2$, the square of a sum, we can write it as $(a + b)(a + b)$ and use FOIL:

$$(a + b)^2 = (a + b)(a + b)$$
$$= a^2 + ab + ab + b^2$$
$$= a^2 + 2ab + b^2$$

So to square $a + b$, *we square the first term* (a^2), *add twice the product of the two terms* $(2ab)$, *and then add the square of the last term* (b^2). The square of a sum occurs so frequently that it is helpful to learn this new rule to find it. The rule for squaring a sum is given symbolically as follows.

The Square of a Sum

$$(a + b)^2 = a^2 + 2ab + b^2$$

EXAMPLE 1

Using the rule for squaring a sum

Find the square of each sum.

a) $(x + 3)^2$ **b)** $(2a + 5)^2$

Solution

a) $(x + 3)^2 = x^2 + 2(x)(3) + 3^2 = x^2 + 6x + 9$

         Square   Twice   Square
         of   the   of
         first   product   last

b) $(2a + 5)^2 = (2a)^2 + 2(2a)(5) + 5^2$
$$= 4a^2 + 20a + 25$$

> Now do Exercises 1–16

CAUTION Don't forget the middle term when squaring a sum. The square of $x + 3$ is $x^2 + 6x + 9$; it is not $x^2 + 9$. The equation $(x + 3)^2 = x^2 + 6x + 9$ is an identity. It is true for every real number x. The equation $(x + 3)^2 = x^2 + 9$ is true only if $x = 0$.

When we use FOIL to find $(a - b)^2$, we see that

$$(a - b)^2 = (a - b)(a - b)$$
$$= a^2 - ab - ab + b^2$$
$$= a^2 - 2ab + b^2.$$

So to square $a - b$, *we square the first term* (a^2), *subtract twice the product of the two terms* ($-2ab$), *and add the square of the last term* (b^2). The rule for squaring a difference is given symbolically as follows.

> **The Square of a Difference**
> $$(a - b)^2 = a^2 - 2ab + b^2$$

E X A M P L E   2

Using the rule for squaring a difference
Find the square of each difference.

a) $(x - 4)^2$ **b)** $(4b - 5y)^2$

Solution

a) $(x - 4)^2 = x^2 - 2(x)(4) + 4^2$
$$= x^2 - 8x + 16$$

b) $(4b - 5y)^2 = (4b)^2 - 2(4b)(5y) + (5y)^2$
$$= 16b^2 - 40by + 25y^2$$

| Now do Exercises 17–30 |

‹ **Helpful Hint** ›

Many students keep using FOIL to find the square of a sum or difference. However, learning the new rules for these special cases will pay off in the future.

‹2› Product of a Sum and a Difference

If we multiply the sum $a + b$ and the difference $a - b$ by using FOIL, we get

$$(a + b)(a - b) = a^2 - ab + ab - b^2$$
$$= a^2 - b^2.$$

The inner and outer products have a sum of 0. So *the product of the sum $a + b$ and the difference $a - b$ is equal to the difference of two squares $a^2 - b^2$.*

> **The Product of a Sum and a Difference**
> $$(a + b)(a - b) = a^2 - b^2$$

E X A M P L E   3

Product of a sum and a difference
Find each product.

a) $(x + 2)(x - 2)$

b) $(b + 7)(b - 7)$

c) $(3x - 5)(3x + 5)$

Solution

a) $(x + 2)(x - 2) = x^2 - 4$

b) $(b + 7)(b - 7) = b^2 - 49$

c) $(3x - 5)(3x + 5) = 9x^2 - 25$

| Now do Exercises 31–42 |

‹ **Helpful Hint** ›

You can use
$$(a + b)(a - b) = a^2 - b^2$$
to perform mental arithmetic tricks like
$$19 \cdot 21 = (20 - 1)(20 + 1)$$
$$= 400 - 1$$
$$= 399.$$
What is $29 \cdot 31$? $28 \cdot 32$?

⟨3⟩ Higher Powers of Binomials

To find a power of a binomial that is higher than 2, we can use the rule for squaring a binomial along with the method of multiplying binomials using the distributive property. Finding the second or higher power of a binomial is called **expanding the binomial** because the result has more terms than the original.

E X A M P L E   4

Higher powers of a binomial

Expand each binomial.

a) $(x + 4)^3$

b) $(y - 2)^4$

Solution

a) $(x + 4)^3 = (x + 4)^2(x + 4)$

$\qquad = (x^2 + 8x + 16)(x + 4)$ Square of a sum

$\qquad = (x^2 + 8x + 16)x + (x^2 + 8x + 16)4$ Distributive property

$\qquad = x^3 + 8x^2 + 16x + 4x^2 + 32x + 64$

$\qquad = x^3 + 12x^2 + 48x + 64$

b) $(y - 2)^4 = (y - 2)^2(y - 2)^2$

$\qquad = (y^2 - 4y + 4)(y^2 - 4y + 4)$

$\qquad = (y^2 - 4y + 4)(y^2) + (y^2 - 4y + 4)(-4y) + (y^2 - 4y + 4)(4)$

$\qquad = y^4 - 4y^3 + 4y^2 - 4y^3 + 16y^2 - 16y + 4y^2 - 16y + 16$

$\qquad = y^4 - 8y^3 + 24y^2 - 32y + 16$

> Now do Exercises 43–50

⟨4⟩ Applications

E X A M P L E   5

Area

a) A square patio has sides of length x feet. If the length and width are increased by 2 feet, then what trinomial represents the area of the larger patio?

b) A pizza parlor makes all of its pizzas 1 inch smaller in radius than advertised. If x is the advertised radius, then what trinomial represents the actual area?

Solution

a) The area of a square is given by $A = s^2$. Since the larger patio has sides of length $x + 2$ feet, its area is $(x + 2)^2$ or $x^2 + 4x + 4$ square feet.

b) The area of a circle is given by $A = \pi r^2$. If the advertised radius is x inches, then the actual radius is $x - 1$ inches. The actual area is $\pi(x - 1)^2$:

$$\pi(x - 1)^2 = \pi(x^2 - 2x + 1) = \pi x^2 - 2\pi x + \pi$$

So the actual area is $\pi x^2 - 2\pi x + \pi$ square inches. Since π is a number, this trinomial is a trinomial in one variable, x.

> Now do Exercises 81–92

Warm-Ups ▼

Fill in the blank.

1. The square of a sum, the square of a difference, and the product of a sum and a difference are the _____ products.

2. The product of a sum and a difference is equal to the _____ of two squares.

3. The _____ of a binomial is the square of the first term, plus twice the product of the terms, plus the square of the last term.

True or false?

4. $(2 + 3)^2 = 2^2 + 3^2$

5. For any value of x, $(x + 3)^2 = x^2 + 6x + 9$.

6. $(3 + 5)^2 = 9 + 30 + 25$

7. For any value of x, $(x - 6)(x + 6) = x^2 - 36$.

8. $(40 + 1)(40 - 1) = 1599$

9. $(49)(51) = 2499$

4.7 Exercises

‹ Study Tips ›

- We are all creatures of habit. When you find a place in which you study successfully, stick with it.
- Studying in a quiet place is better than studying in a noisy place. There are very few people who can listen to music or conversation and study effectively.

⟨1⟩ The Square of a Sum or Difference

Square each binomial. See Example 1.

1. $(x + 1)^2$
2. $(y + 2)^2$

3. $(y + 4)^2$
4. $(z + 3)^2$

5. $(m + 6)^2$
6. $(w + 7)^2$

7. $(a + 9)^2$

8. $(b + 10)^2$

9. $(3x + 8)^2$
10. $(2m + 7)^2$

11. $(s + t)^2$
12. $(x + z)^2$

13. $(2x + y)^2$
14. $(3t + v)^2$

15. $(2t + 3h)^2$
16. $(3z + 5k)^2$

Square each binomial. See Example 2.

17. $(p - 2)^2$
18. $(b - 5)^2$

19. $(a - 3)^2$
20. $(w - 4)^2$

21. $(t - 1)^2$
22. $(t - 6)^2$

23. $(3t - 2)^2$
24. $(5a - 6)^2$

25. $(s - t)^2$
26. $(r - w)^2$

27. $(3a - b)^2$
28. $(4w - 7)^2$

29. $(3z - 5y)^2$
30. $(2z - 3w)^2$

⟨2⟩ Product of a Sum and a Difference

Find each product. See Example 3.

31. $(a - 5)(a + 5)$
32. $(x - 6)(x + 6)$

33. $(y - 1)(y + 1)$ **34.** $(p + 2)(p - 2)$

35. $(3x - 8)(3x + 8)$ **36.** $(6x + 1)(6x - 1)$

37. $(r + s)(r - s)$ **38.** $(b - y)(b + y)$

39. $(8y - 3a)(8y + 3a)$ **40.** $(4u - 9v)(4u + 9v)$

41. $(5x^2 - 2)(5x^2 + 2)$ **42.** $(3y^2 + 1)(3y^2 - 1)$

⟨3⟩ Higher Powers of Binomials

Expand each binomial. See Example 4.

43. $(x + 1)^3$

44. $(y - 1)^3$

45. $(2a - 3)^3$

46. $(3w - 1)^3$

47. $(a - 3)^4$

48. $(2b + 1)^4$

49. $(a + b)^4$

50. $(2a - 3b)^4$

Miscellaneous

Find each product.

51. $(a - 20)(a + 20)$ **52.** $(1 - x)(1 + x)$

53. $(x + 8)(x + 7)$ **54.** $(x - 9)(x + 5)$

55. $(4x - 1)(4x + 1)$ **56.** $(9y - 1)(9y + 1)$

57. $(9y - 1)^2$ **58.** $(4x - 1)^2$

59. $(2t - 5)(3t + 4)$ **60.** $(2t + 5)(3t - 4)$

61. $(2t - 5)^2$ **62.** $(2t + 5)^2$

63. $(2t + 5)(2t - 5)$ **64.** $(3t - 4)(3t + 4)$

65. $(x^2 - 1)(x^2 + 1)$ **66.** $(y^3 - 1)(y^3 + 1)$

67. $(2y^3 - 9)^2$ **68.** $(3z^4 - 8)^2$

69. $(2x^3 + 3y^2)^2$ **70.** $(4y^5 + 2w^3)^2$

71. $\left(\dfrac{1}{2}x + \dfrac{1}{3}\right)^2$ **72.** $\left(\dfrac{2}{3}y - \dfrac{1}{2}\right)^2$

73. $(0.2x - 0.1)^2$

74. $(0.1y + 0.5)^2$

75. $(a + b)^3$

76. $(2a - 3b)^3$

77. $(1.5x + 3.8)^2$

78. $(3.45a - 2.3)^2$

79. $(3.5t - 2.5)(3.5t + 2.5)$

80. $(4.5h + 5.7)(4.5h - 5.7)$

⟨4⟩ Applications

Solve each problem. See Example 5.

81. *Area of a square.* Find a polynomial $A(x)$ that represents the area of the shaded region in the accompanying figure.

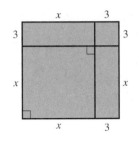

Figure for Exercise 81

82. *Area of a square.* Find a polynomial $A(x)$ that represents the area of the shaded region in the accompanying figure.

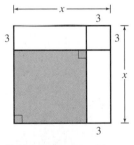

Figure for Exercise 82

83. *Shrinking garden.* Rose's garden is a square with sides of length x feet. Next spring she plans to make it rectangular by lengthening one side 5 feet and shortening the other side by 5 feet.

   a) Find a polynomial $A(x)$ that represents the new area.

   b) By how much will the area of the new garden differ from that of the old garden?

84. *Square lot.* Sam has a lot that he thought was a square, 200 feet by 200 feet. When he had it surveyed, he discovered that one side was x feet longer than he thought and the other side was x feet shorter than he thought.

a) Find a polynomial $A(x)$ that represents the new area.

b) Find $A(2)$.

c) If $x = 2$ feet, then how much less area does he have than he thought he had?

85. *Area of a circle.* Find a polynomial $A(b)$ that represents the area of a circle whose radius is $b + 1$. Use 3.14 for π.

86. *Comparing dart boards.* A small circular dart board has radius t inches and a larger one has a radius that is 3 inches larger.

a) Find a polynomial $D(t)$ that represents the difference in area between the two dart boards. Use 3.14 for π.

b) Find $D(4)$.

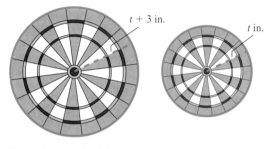

Figure for Exercise 86

87. *Poiseuille's law.* According to the nineteenth-century physician Jean Poiseuille, the velocity (in centimeters per second) of blood r centimeters from the center of an artery of radius R centimeters is given by

$$v = k(R - r)(R + r),$$

where k is a constant. Rewrite the formula using a special product rule.

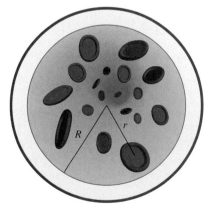

Figure for Exercise 87

88. *Going in circles.* A promoter is planning a circular race track with an inside radius of r feet and a width of w feet. The cost in dollars for paving the track is given by the formula

$$C = 1.2\pi[(r + w)^2 - r^2].$$

Use a special product rule to simplify this formula. What is the cost of paving the track if the inside radius is 1000 feet and the width of the track is 40 feet?

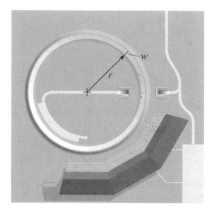

Figure for Exercise 88

89. *Compounded annually.* P dollars is invested at annual interest rate r for 2 years. If the interest is compounded annually, then the polynomial $P(1 + r)^2$ represents the value of the investment after 2 years. Rewrite this expression without parentheses. Evaluate the polynomial if $P = \$200$ and $r = 10\%$.

90. *Compounded semiannually.* P dollars is invested at annual interest rate r for 1 year. If the interest is compounded semiannually, then the polynomial $P\left(1 + \dfrac{r}{2}\right)^2$ represents the value of the investment after 1 year. Rewrite this expression without parentheses. Evaluate the polynomial if $P = \$200$ and $r = 10\%$.

91. *Investing in treasury bills.* An investment advisor uses the polynomial $P(1 + r)^{10}$ to predict the value in 10 years of a client's investment of P dollars with an average annual return r. The accompanying graph shows historic average annual returns for the last 20 years for various asset classes (T. Rowe Price, www.troweprice.com). Use the historical average return to predict the value in 10 years of an investment of $10,000 in U.S. treasury bills.

92. *Comparing investments.* How much more would the investment in Exercise 91 be worth in 10 years if the client

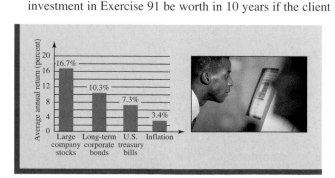

Figure for Exercises 91 and 92

invests in large company stocks rather than U.S. treasury bills?

Getting More Involved

93. *Writing*

What is the difference between the equations $(x + 5)^2 = x^2 + 10x + 25$ and $(x + 5)^2 = x^2 + 25$?

94. *Writing*

Is it possible to square a sum or a difference without using the rules presented in this section? Why should you learn the rules given in this section?

4.8 Division of Polynomials

In This Section

⟨1⟩ Dividing Monomials
⟨2⟩ Dividing a Polynomial by a Monomial
⟨3⟩ Dividing a Polynomial by a Binomial

You multiplied polynomials in Section 4.5. In this section, you will learn to divide polynomials.

⟨1⟩ Dividing Monomials

We actually divided some monomials in Section 4.1 using the quotient rule for exponents. We use the quotient rule here also. In Section 4.2, we divided expressions with positive and negative exponents. Since monomials and polynomials have nonnegative exponents only, we will not be using negative exponents here.

EXAMPLE 1

Dividing monomials

Find each quotient. All variables represent nonzero real numbers.

a) $\left(12x^5\right) \div \left(3x^2\right)$ **b)** $\dfrac{-4x^3}{2x^3}$ **c)** $\dfrac{-10a^2b^4}{-2a^2b^2}$

Solution

a) $\left(12x^5\right) \div \left(3x^2\right) = \dfrac{12x^5}{3x^2} = 4x^{5-2} = 4x^3$

The quotient is $4x^3$. Use the definition of division to check that $4x^3 \cdot 3x^2 = 12x^5$.

b) $\dfrac{-4x^3}{2x^3} = -2x^{3-3} = -2x^0 = -2 \cdot 1 = -2$

The quotient is -2. Use the definition of division to check that $-2 \cdot 2x^3 = -4x^3$.

c) $\dfrac{-10a^3b^4}{-2a^2b^2} = 5a^{3-2}b^{4-2} = 5ab^2$

The quotient is $5ab^2$. Check that $5ab^2(-2a^2b^2) = -10a^3b^4$.

Now do Exercises 1–18

If $a \div b = c$, then a is called the **dividend,** b is called the **divisor,** and c is called the **quotient.** We use these terms with division of real numbers or division of polynomials.

⟨2⟩ Dividing a Polynomial by a Monomial

We divided some simple polynomials by monomials in Chapter 1 using the distributive property. Now that we have the rules of exponents, we can use them to divide polynomials of higher degrees by monomials. Because of the distributive property, each term of the polynomial in the numerator is divided by the monomial from the denominator.

E X A M P L E   **2**

Dividing a polynomial by a monomial

Find the quotient.

  a) $(5x - 10) \div 5$ **b)** $(-8x^6 + 12x^4 - 4x^2) \div (4x^2)$

Solution

  a) By the distributive property, each term of $5x - 10$ is divided by 5:

$$\frac{5x - 10}{5} = \frac{5x}{5} - \frac{10}{5} = x - 2$$

  The quotient is $x - 2$. Check by multiplying: $5(x - 2) = 5x - 10$.

  b) By the distributive property, each term of $-8x^6 + 12x^4 - 4x^2$ is divided by $4x^2$:

$$\frac{-8x^6 + 12x^4 - 4x^2}{4x^2} = \frac{-8x^6}{4x^2} + \frac{12x^4}{4x^2} - \frac{4x^2}{4x^2}$$
$$= -2x^4 + 3x^2 - 1$$

  The quotient is $-2x^4 + 3x^2 - 1$. We can check by multiplying.

$$4x^2(-2x^4 + 3x^2 - 1) = -8x^6 + 12x^4 - 4x^2$$

> Now do Exercises 19–26

Because division by zero is undefined, we will always assume that the divisor is nonzero in any quotient involving variables. For example, the division in Example 3 is valid only if $4x^2 \neq 0$, or $x \neq 0$.

⟨3⟩ Dividing a Polynomial by a Binomial

Division of whole numbers is often done with a procedure called **long division.** For example, 253 is divided by 7 as follows:

$$
\begin{array}{r}
36 \quad \leftarrow \text{Quotient} \\
\text{Divisor} \rightarrow 7\overline{)253} \quad \leftarrow \text{Dividend} \\
\underline{21} \\
43 \\
\underline{42} \\
1 \quad \leftarrow \text{Remainder}
\end{array}
$$

Note that the remainder must be smaller than the divisor and

$$\text{dividend} = (\text{quotient})(\text{divisor}) + (\text{remainder}).$$

This fact is used to check. Since $253 = 36 \cdot 7 + 1$, the division was done correctly. Dividing each side of this last equation by "divisor" yields the equation

$$\frac{\text{dividend}}{\text{divisor}} = \text{quotient} + \frac{\text{remainder}}{\text{divisor}}.$$

There are two ways to express the result of dividing 253 by 7. One is to state that the quotient is 36 and the remainder is 1. The other is to write the equation

$$\frac{253}{7} = 36 + \frac{1}{7} = 36\frac{1}{7}.$$

If the division is done in a context where fractions are allowed, then $36\frac{1}{7}$ could be called the quotient. For example, dividing \$9 among 2 people results in \$$4\frac{1}{2}$ each.

However, dividing 9 people into groups of 2 to play tennis results in 4 groups with a remainder of 1 person.

To divide a polynomial by a binomial, we perform the division like long division of whole numbers. For example, to divide $x^2 - 3x - 10$ by $x + 2$, we get the first term of the quotient by dividing the first term of $x + 2$ into the first term of $x^2 - 3x - 10$. So divide x^2 by x to get x, and then multiply and subtract as follows:

1 Divide:
2 Multiply: $x + 2 \overline{)x^2 - 3x - 10}$
$\underline{x^2 + 2x}$
3 Subtract: $-5x$

$x^2 \div x = x$
$x \cdot (x + 2) = x^2 + 2x$
$-3x - 2x = -5x$

Now bring down -10 and continue the process. We get the second term of the quotient (see the following) by dividing the first term of $x + 2$ into the first term of $-5x - 10$. So divide $-5x$ by x to get -5:

1 Divide:
2 Multiply: $x + 2 \overline{)x^2 - 3x - 10}$
$\underline{x^2 + 2x}$
$-5x - 10$
$\underline{-5x - 10}$
3 Subtract: 0

$-5x \div x = -5$
Bring down -10.
$-5(x + 2) = -5x - 10$
$-10 - (-10) = 0$

So the quotient is $x - 5$, and the remainder is 0.

In Example 3 there is a term missing in the dividend. To account for the missing term we insert a term with a zero coefficient.

EXAMPLE 3

Dividing a polynomial by a binomial
Determine the quotient and remainder when $x^3 - 5x - 1$ is divided by $x - 4$.

Solution
Because the x^2-term in the dividend $x^3 - 5x - 1$ is missing, we write $0 \cdot x^2$ for it:

Place x^2 in the quotient because $x^3 \div x = x^2$.
Place $4x$ in the quotient because $4x^2 \div x = 4x$.
Place 11 in the quotient because $11x \div x = 11$.

$$x - 4 \overline{)x^3 + 0 \cdot x^2 - 5x - 1}$$

$x^2 + 4x + 11$

$\underline{x^3 - 4x^2}$ $x^2(x - 4) = x^3 - 4x^2$
$4x^2 - 5x$ $0 \cdot x^2 - (-4x^2) = 4x^2$
$\underline{4x^2 - 16x}$ $4x(x - 4) = 4x^2 - 16x$
$11x - 1$ $-5x - (-16x) = 11x$
$\underline{11x - 44}$ $11(x - 4) = 11x - 44$
43 $-1 - (-44) = 43$

So the quotient is $x^2 + 4x + 11$ and the remainder is 43. To check, multiply the quotient by divisor $x - 4$ and add the remainder to see if you get the dividend $x^3 - 5x - 1$:

$$(x - 4)(x^2 + 4x + 11) + 43 = x(x^2 + 4x + 11) - 4(x^2 + 4x + 11) + 43$$
$$= x^3 + 4x^2 + 11x - 4x^2 - 16x - 44 + 43$$
$$= x^3 - 5x - 1 \quad \text{The dividend}$$

> **Now do Exercises 27–30**

In Example 4, the terms of the dividend are not in order of decreasing exponents and there is a missing term.

E X A M P L E   4

Dividing a polynomial by a binomial
Divide $2x^3 - 4 - 7x^2$ by $2x - 3$, and identify the quotient and the remainder.

Solution

‹ **Helpful Hint** ›

Students usually have the most difficulty with the subtraction part of long division. So pay particular attention to that step and double check your work.

Rearrange the dividend as $2x^3 - 7x^2 - 4$. Because the x-term in the dividend is missing, we write $0 \cdot x$ for it:

$$
\begin{array}{r}
x^2 - 2x - 3 \\
2x - 3 \overline{\smash{)}\, 2x^3 - 7x^2 + 0 \cdot x - 4} \\
\underline{2x^3 - 3x^2} \\
-4x^2 + 0 \cdot x \\
\underline{-4x^2 + 6x} \\
-6x - 4 \\
\underline{-6x + 9} \\
-13
\end{array}
$$

$2x^3 \div (2x) = x^2$

$x^2(2x - 3) = 2x^3 - 3x^2$

$-7x^2 - (-3x^2) = -4x^2$

$-2x(2x - 3) = -4x^2 + 6x$

$0 \cdot x - 6x = -6x$

$-3(2x - 3) = -6x + 9$

$-4 - (9) = -13$

The quotient is $x^2 - 2x - 3$ and the remainder is -13. To check, multiply the quotient by the divisor $2x - 3$ and add the remainder -13 to see if you get the dividend $2x^3 - 7x^2 - 4$:

$$(2x - 3)(x^2 - 2x - 3) - 13 = 2x(x^2 - 2x - 3) - 3(x^2 - 2x - 3) - 13$$
$$= 2x^3 - 4x^2 - 6x - 3x^2 + 6x + 9 - 13$$
$$= 2x^3 - 7x^2 - 4 \qquad \text{The dividend}$$

> **Now do Exercises 31–44**

CAUTION   To avoid errors, always write the terms of the divisor and the dividend in descending order of the exponents and insert a zero for any term that is missing.

E X A M P L E   5

Rewriting algebraic fractions
Express $\dfrac{-3x}{x - 2}$ in the form

$$\text{quotient} + \frac{\text{remainder}}{\text{divisor}}.$$

Solution

Use long division to get the quotient and remainder:

$$
\begin{array}{r}
-3 \\
x - 2 \overline{)\,-3x + 0} \\
\underline{-3x + 6} \\
-6
\end{array}
$$

To check, multiply the divisor and quotient and add the remainder to see if you get the dividend $-3x$:

$$-3(x - 2) - 6 = -3x + 6 - 6 = -3x$$

Because the quotient is -3 and the remainder is -6, we can write

$$\frac{-3x}{x - 2} = -3 + \frac{-6}{x - 2}.$$

To check we must verify that $-3(x - 2) - 6 = -3x$.

> Now do Exercises 45–60

CAUTION When dividing polynomials by long division, we do not stop until the remainder is 0 or the degree of the remainder is smaller than the degree of the divisor. For example, we stop dividing in Example 5 because the degree of the remainder -6 is 0 and the degree of the divisor $x - 2$ is 1.

Warm-Ups ▼

Fill in the blank.

1. The _____ rule for exponents can be used when dividing monomials.

2. If $a \div b = c$, then a is the _____, b is the _____ and c is the _____.

3. The terms of a polynomial are written in _____ order of the exponents for long division.

4. The long division process stops when the degree of the remainder is _____ than the degree of the divisor.

True or false?

5. For any nonzero value of y, $y^{10} \div y^2 = y^5$.

6. For any value of x, $\dfrac{7x + 2}{7} = x + 2$.

7. For any value of x, $\dfrac{7x^2}{7} = x^2$.

8. If $3x^2 + 6$ is divided by 3, then the quotient is $x^2 + 2$.

9. The quotient times the remainder plus the dividend equals the divisor.

10. If the remainder is zero, then the quotient times the divisor is equal to the dividend.

Exercises

‹1› Dividing Monomials

Find each quotient. Try to write only the answer. See Example 1.

1. $\dfrac{x^8}{x^2}$ **2.** $\dfrac{y^9}{y^3}$

3. $\dfrac{w^{12}}{w^3}$ **4.** $\dfrac{m^{20}}{m^{10}}$

5. $\dfrac{a^{14}}{a^5}$ **6.** $\dfrac{b^{19}}{b^{12}}$

7. $\dfrac{6a^{12}}{2a^7}$ **8.** $\dfrac{30b^6}{3b^2}$

9. $a^9 \div a^3$ **10.** $b^{12} \div b^4$

11. $-12x^9 \div \left(3x^5\right)$ **12.** $-6y^{10} \div \left(-3y^5\right)$

13. $-6y^2 \div (6y)$ **14.** $-3a^2b \div (3ab)$

 **15.** $\dfrac{-6x^3y^2}{2x^2y^2}$ **16.** $\dfrac{-4h^2k^4}{-2hk^3}$

17. $\dfrac{-9x^5y^2}{3x^2y^2}$ **18.** $\dfrac{-12z^{10}y^2}{-2z^4y^2}$

‹2› Dividing a Polynomial by a Monomial

Find the quotients. See Example 2.

19. $\dfrac{3x - 6}{3}$

20. $\dfrac{5y - 10}{-5}$

21. $\dfrac{x^5 + 3x^4 - x^3}{x^2}$

 **22.** $\dfrac{6y^6 - 9y^4 + 12y^2}{3y^2}$

23. $\dfrac{-8x^2y^2 + 4x^2y - 2xy^2}{-2xy}$

24. $\dfrac{-9ab^2 - 6a^3b^3}{-3ab^2}$

25. $\left(x^2y^3 - 3x^3y^2\right) \div \left(x^2y\right)$

26. $\left(4h^5k - 6h^2k^2\right) \div \left(-2h^2k\right)$

‹3› Dividing a Polynomial by a Binomial

Complete each division and identify the quotient and remainder. See Example 3.

27. $x - 1 \overline{\smash{\big)}\, 2x - 3} \atop \underline{2x-2}$ with quotient 2 above

28. $x + 2 \overline{\smash{\big)}\, -3x + 4} \atop \underline{-3x-6}$ with quotient -3 above

29. $x - 3 \overline{\smash{\big)}\, x^2 + 2x + 1} \atop \underline{x^2-3x}$ with quotient x above

30. $x + 4 \overline{\smash{\big)}\, x^2 - 3x + 2} \atop \underline{x^2+4x}$ with quotient x above

Find the quotient and remainder for each division. Check by using the fact that dividend = (quotient)(divisor) + remainder. See Example 4.

31. $\left(x^2 + 5x + 13\right) \div (x + 3)$

32. $\left(x^2 + 3x + 6\right) \div (x + 3)$

33. $(2x) \div (x + 5)$

34. $(5x) \div (x - 1)$

35. $\left(a^3 + 4a - 3\right) \div (a - 2)$

36. $\left(w^3 + 2w^2 - 3\right) \div (w - 2)$

37. $\left(x^2 - 3x\right) \div (x + 1)$

38. $\left(3x^2\right) \div (x + 1)$

39. $\left(h^3 - 27\right) \div (h - 3)$

40. $\left(w^3 + 1\right) \div (w + 1)$

41. $\left(6x^2 - 13x + 7\right) \div (3x - 2)$

42. $\left(4b^2 + 25b - 3\right) \div (4b + 1)$

43. $\left(x^3 - x^2 + x - 2\right) \div (x - 1)$

44. $\left(a^3 - 3a^2 + 4a - 4\right) \div (a - 2)$

Write each expression in the form

$$quotient + \frac{remainder}{divisor}.$$

See Example 5.

45. $\dfrac{3x}{x - 5}$ **46.** $\dfrac{2x}{x - 1}$

47. $\dfrac{-x}{x + 3}$ **48.** $\dfrac{-3x}{x + 1}$

49. $\dfrac{x-1}{x}$ **50.** $\dfrac{a-5}{a}$

51. $\dfrac{3x+1}{x}$ **52.** $\dfrac{2y+1}{y}$

53. $\dfrac{x^2}{x+1}$ **54.** $\dfrac{x^2}{x-1}$

55. $\dfrac{x^2+4}{x+2}$

56. $\dfrac{x^2+1}{x-1}$

57. $\dfrac{x^3}{x-2}$

58. $\dfrac{x^3-1}{x+1}$

59. $\dfrac{x^3+3}{x}$ **60.** $\dfrac{2x^2+4}{2x}$

Miscellaneous

Find each quotient.

61. $-6a^3b \div (2a^2b)$

62. $-14x^7 \div (-7x^2)$

63. $-8w^9t^7 \div (-2w^4t^3)$

64. $-9y^7z^{11} \div (3y^3z^4)$

65. $(3a-12) \div (-3)$

66. $(-6z+3z^2) \div (-3z)$

67. $(3x^2-9x) \div (3x)$

68. $(5x^3+15x^2-25x) \div (5x)$

69. $(12x^4-4x^3+6x^2) \div (-2x^2)$

70. $(-9x^3+3x^2-15x) \div (-3x)$

71. $(t^2-5t-36) \div (t-9)$

72. $(b^2+2b-35) \div (b-5)$

73. $(6w^2-7w-5) \div (3w-5)$

74. $(4z^2+23z-6) \div (4z-1)$

75. $(8x^3+27) \div (2x+3)$

76. $(8y^3-1) \div (2y-1)$

77. $(t^3-3t^2+5t-6) \div (t-2)$

78. $(2u^3-13u^2-8u+7) \div (u-7)$

79. $(-6v^2-4+9v+v^3) \div (v-4)$

80. $(14y+8y^2+y^3+12) \div (6+y)$

Solve each problem.

81. *Area of a rectangle.* The area of a rectangular billboard is x^2+x-30 square meters. If the length is $x+6$ meters, find a binomial that represents the width.

x + 6 meters

Figure for Exercise 81

82. *Perimeter of a rectangle.* The perimeter of a rectangular backyard is $6x+6$ yards. If the width is x yards, find a binomial that represents the length.

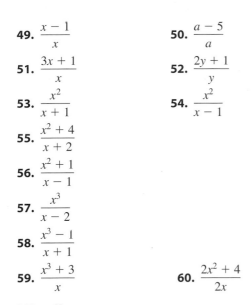

x yards

?

Figure for Exercise 82

Getting More Involved

83. *Exploration*

Divide x^3-1 by $x-1$, x^4-1 by $x-1$, and x^5-1 by $x-1$. What is the quotient when x^9-1 is divided by $x-1$?

84. *Exploration*

Divide a^3-b^3 by $a-b$ and a^4-b^4 by $a-b$. What is the quotient when a^8-b^8 is divided by $a-b$?

85. *Discussion*

Are the expressions $\dfrac{10x}{5x}$, $10x \div 5x$, and $(10x) \div (5x)$ equivalent? Before you answer, review the order of operations in Section 1.5 and evaluate each expression for $x=3$.

Wrap-Up

Summary

The Rules of Exponents   **Examples**

The following rules hold for any integers m and n, and nonzero real numbers a and b.

Zero exponent $a^0 = 1$ $(-3)^0 = 1, \quad -3^0 = -1$

Product rule for exponents $a^m \cdot a^n = a^{m+n}$ $a^2 \cdot a^3 = a^5, \quad b^{-5} \cdot b^3 = b^{-2}$

Quotient rule for exponents $\dfrac{a^m}{a^n} = a^{m-n}$ $x^8 \div x^2 = x^6, \quad \dfrac{y^{-3}}{y^{-7}} = y^4$

Power of a power rule $(a^m)^n = a^{mn}$ $(2^2)^3 = 2^6, \quad (w^{-3})^{-4} = w^{12}$

Power of a product rule $(ab)^n = a^n b^n$ $(2t)^3 = 8t^3, \quad (3t^{-2})^4 = 81t^{-8}$

Power of a quotient rule $\left(\dfrac{a}{b}\right)^n = \dfrac{a^n}{b^n}$ $\left(\dfrac{x}{3}\right)^3 = \dfrac{x^3}{27}, \quad \left(\dfrac{a^{-3}}{b^{-4}}\right)^{-2} = \dfrac{a^6}{b^8}$

Negative Exponents   **Examples**

Negative integral exponents

If n is a positive integer and a is a nonzero real number, then $a^{-n} = \dfrac{1}{a^n}$.

$3^{-2} = \dfrac{1}{3^2}, \quad x^{-5} = \dfrac{1}{x^5}$

Rules for negative exponents

If a is a nonzero real number and n is a positive integer, then $a^{-1} = \dfrac{1}{a}, \dfrac{1}{a^{-n}} = a^n, a^{-n} = \left(\dfrac{1}{a}\right)^n$,

and $\left(\dfrac{a}{b}\right)^{-n} = \left(\dfrac{b}{a}\right)^n$.

$5^{-1} = \dfrac{1}{5}, \quad \dfrac{1}{x^{-3}} = x^3$

$2^{-3} = \left(\dfrac{1}{2}\right)^3$

$\left(\dfrac{2}{3}\right)^{-3} = \left(\dfrac{3}{2}\right)^3$

Scientific Notation

		Examples
Converting from scientific notation	1. Find the number of places to move the decimal point by examining the exponent on the 10. 2. Move to the right for a positive exponent and to the left for a negative exponent.	$5.6 \times 10^3 = 5600$ $9 \times 10^{-4} = 0.0009$
Converting into scientific notation (positive numbers)	1. Count the number of places (n) that the decimal point must be moved so that it will follow the first nonzero digit of the number. 2. If the original number was larger than 10, use 10^n. 3. If the original number was smaller than 1, use 10^{-n}.	$304.6 = 3.046 \times 10^2$ $0.0035 = 3.5 \times 10^{-3}$

Polynomials

		Examples
Term	A number or the product of a number and one or more variables raised to powers	$5x^3$, $-4x$, 7
Polynomial	A single term or a finite sum of terms	$2x^5 - 9x^2 + 11$
Degree of a polynomial	The highest degree of any of the terms	Degree of $2x - 9$ is 1. Degree of $5x^3 - x^2$ is 3.
Naming a polynomial	A polynomial can be named with a letter such as P or $P(x)$ (function notation).	$P = x^2 - 1$ $P(x) = x^2 - 1$
Evaluating a polynomial	The value of a polynomial is the real number that is obtained when the variable (x) is replaced with a real number.	If $x = 3$, then $P = 8$, or $P(3) = 8$.

Adding, Subtracting, and Multiplying Polynomials

		Examples
Add or subtract polynomials	Add or subtract the like terms.	$(x + 1) + (x - 4) = 2x - 3$ $(x^2 - 3x) - (4x^2 - x)$ $= -3x^2 - 2x$
Multiply monomials	Use the product rule for exponents.	$-2x^5 \cdot 6x^8 = -12x^{13}$
Multiply polynomials	Multiply each term of one polynomial by every term of the other polynomial, and then combine like terms.	$(x - 1)(x^2 + 2x + 5)$ $= x(x^2 + 2x + 5) - 1(x^2 + 2x + 5)$ $= x^3 + 2x^2 + 5x - x^2 - 2x - 5$ $= x^3 + x^2 + 3x - 5$

Binomials   **Examples**

FOIL   A method for multiplying two binomials quickly $(x - 2)(x + 3) = x^2 + x - 6$

Square of a sum $(a + b)^2 = a^2 + 2ab + b^2$ $(x + 3)^2 = x^2 + 6x + 9$

Square of a difference $(a - b)^2 = a^2 - 2ab + b^2$ $(m - 5)^2 = m^2 - 10m + 25$

Product of a sum $(a - b)(a + b) = a^2 - b^2$ $(x + 2)(x - 2) = x^2 - 4$
and a difference

Dividing Polynomials   **Examples**

Dividing monomials   Use the quotient rule for exponents $8x^5 \div (2x^2) = 4x^3$

Divide a polynomial   Divide each term of the polynomial by the $\dfrac{3x^5 + 9x}{3x} = x^4 + 3$
by a monomial   monomial.

Divide a polynomial   If the divisor is a binomial, use $$\text{Divisor} \rightarrow x + 2\overset{\displaystyle x - 7 \leftarrow \text{Quotient}}{\overline{)x^2 - 5x - 4}} \leftarrow \text{Dividend}$$
by a binomial   long division. $$\underline{x^2 + 2x}$$
                          (quotient)(divisor) + (remainder) = dividend $$-7x - 4$$
                                                                           $$\underline{-7x - 14}$$
                                                                           $$10 \leftarrow \text{Remainder}$$

Enriching Your Mathematical Word Power

Fill in the blank.

1. A ____ is an expression containing one or more variables raised to whole number powers.

2. A _____ is a single term or a finite sum of terms.

3. The _____ of a polynomial is the highest degree of any of its terms.

4. The coefficient of the first term of a polynomial when it is written in order of decreasing exponents is the _____ coefficient.

5. A polynomial with one term is a _____.

6. A polynomial with two terms is a _____.

7. A polynomial with three terms is a _____.

8. The _____ method is a procedure for multiplying two binomials quickly.

9. The amount of money invested is the _____.

10. The value of the principal after a certain period of time is the _____.

11. The _____ value is the principal that must be invested today to grow to a specified amount in the future.

12. The expression $(a + b)^2$ is the _____ of a sum.

13. The expression $a^2 - b^2$ is the _____ of two squares.

14. If $a \div b = c$, then a is the _____, b is the _____, and c is the _____.

15. A notation for expressing large or small numbers using powers of 10 is _____ notation.

Review Exercises

4.1 The Rules of Exponents
Simplify each expression. Assume all variables represent nonzero real numbers.

1. $-5^0 + 3^0$

2. $-4^0 - 3^0$

3. $-3a^3 \cdot 2a^4$

4. $2y^{10}(-3y^{20})$

5. $\dfrac{-10b^5c^9}{2b^5c^3}$

6. $\dfrac{-30k^3y^9}{15k^3y^2}$

7. $\left(b^5\right)^6$

8. $\left(y^5\right)^8$

9. $\left(-2x^3y^2\right)^3$

10. $\left(-3a^4b^6\right)^4$

11. $\left(\dfrac{2a}{b^2}\right)^3$

12. $\left(\dfrac{3y^2}{2}\right)^3$

13. $\left(\dfrac{-6x^2y^5}{-3z^6}\right)^3$

14. $\left(\dfrac{-3a^4b^8}{6a^3b^{12}}\right)^4$

4.2 Negative Exponents

Simplify each expression. Assume all variables represent nonzero real numbers. Use only positive exponents in answers.

15. 2^{-3}

16. -2^{-4}

17. $\left(\dfrac{1}{7}\right)^{-1}$

18. $\left(\dfrac{1}{2}\right)^{-2}$

19. $x^5 \cdot x^{-8}$

20. $a^{-3}a^{-9}$

21. $\dfrac{a^{-8}}{a^{-12}}$

22. $\dfrac{a^{10}}{a^{-4}}$

23. $\left(x^{-3}\right)^4$

24. $\left(x^5\right)^{-10}$

25. $\left(2x^{-3}\right)^{-3}$

26. $\left(3y^{-5}\right)^2$

27. $\left(\dfrac{a}{3b^{-3}}\right)^{-2}$

28. $\left(\dfrac{a^{-2}}{5b}\right)^{-3}$

4.3 Scientific Notation

Write each number in standard notation.

29. 8.36×10^6

30. 3.4×10^7

31. 5.7×10^{-4}

32. 4×10^{-3}

33. 4.5 million

34. 34 trillion

35. 3561 thousand

36. 0.6 billion

Write each number in scientific notation.

37. 8,070,000

38. 90,000

39. 0.000709

40. 0.0000005

41. 1.2 trillion

42. 0.8 million

43. 500 thousand

44. 455.6 billion

Perform each computation without a calculator. Write the answer in scientific notation.

45. $\left(5\left(2 \times 10^4\right)\right)^3$

46. $\left(6\left(2 \times 10^{-3}\right)\right)^2$

47. $\dfrac{\left(2 \times 10^{-9}\right)\left(3 \times 10^7\right)}{5\left(6 \times 10^{-4}\right)}$

48. $\dfrac{\left(3 \times 10^{12}\right)\left(5 \times 10^4\right)}{30 \times 10^{-9}}$

49. $\dfrac{(4,000,000,000)(0.0000006)}{(0.000012)(2,000,000)}$

50. $\dfrac{(1200)(0.00002)}{0.0000004}$

4.4 Addition and Subtraction of Polynomials

Perform the indicated operations.

51. $(2w - 6) + (3w + 4)$

52. $(1 - 3y) + (4y - 6)$

53. $\left(x^2 - 2x - 5\right) - \left(x^2 + 4x - 9\right)$

54. $\left(3 - 5x - x^2\right) - \left(x^2 - 7x + 8\right)$

55. $\left(5 - 3w + w^2\right) + \left(w^2 - 4w - 9\right)$

56. $\left(-2t^2 + 3t - 4\right) + \left(t^2 - 7t + 2\right)$

57. $\left(4 - 3m - m^2\right) - \left(m^2 - 6m + 5\right)$

58. $\left(n^3 - n^2 + 9\right) - \left(n^4 - n^3 + 5\right)$

Find the following values.

59. Find the value of the polynomial $x^3 - 9x$ if $x = 3$.

60. Find the value of the polynomial $x^2 - 7x + 1$ if $x = 4$.

61. Suppose that $P(x) = x^3 - x^2 + x - 1$. Find $P(2)$.

62. Suppose that $Q(x) = x^2 - 6x - 8$. Find $Q(-3)$.

4.5 Multiplication of Polynomials

Perform the indicated operations.

63. $5x^2 \cdot \left(-10x^9\right)$

64. $3h^3t^2 \cdot 2h^2t^5$

65. $\left(-11a^7\right)^2$

66. $\left(12b^3\right)^2$

67. $x - 5(x - 3)$

68. $x - 4(x - 9)$

69. $5x + 3\left(x^2 - 5x + 4\right)$

70. $5 + 4x^2(x - 5)$

71. $3m^2\left(5m^3 - m + 2\right)$

72. $-4a^4\left(a^2 + 2a + 4\right)$

73. $(x - 5)\left(x^2 - 2x + 10\right)$

74. $(x + 2)\left(x^2 - 2x + 4\right)$

75. $\left(x^2 - 2x + 4\right)(3x - 2)$

76. $(5x + 3)\left(x^2 - 5x + 4\right)$

4.6 Multiplication of Binomials

Perform the indicated operations.

77. $(q - 6)(q + 8)$

78. $(w + 5)(w + 12)$

79. $(2t - 3)(t - 9)$

80. $(5r + 1)(5r + 2)$

81. $(4y - 3)(5y + 2)$

82. $(11y + 1)(y + 2)$

83. $(3x^2 + 5)(2x^2 + 1)$

84. $(x^3 - 7)(2x^3 + 7)$

4.7 Special Products

Perform the indicated operations. Try to write only the answers.

85. $(z - 7)(z + 7)$

86. $(a - 4)(a + 4)$

87. $(y + 7)^2$

88. $(a + 5)^2$

89. $(w - 3)^2$

90. $(a - 6)^2$

91. $(x^2 - 3)(x^2 + 3)$

92. $(2b^2 - 1)(2b^2 + 1)$

93. $(3a + 1)^2$

94. $(1 - 3c)^2$

95. $(4 - y)^2$

96. $(9 - t)^2$

4.8 Division of Polynomials

Find each quotient.

97. $-10x^5 \div (2x^3)$

98. $-6x^4y^2 \div (-2x^2y^2)$

99. $\dfrac{6a^5b^9c^6}{-3a^3b^7c^6}$

100. $\dfrac{-9h^7t^9r^2}{3h^5t^6r^2}$

101. $\dfrac{3x - 9}{-3}$

102. $\dfrac{7 - y}{-1}$

103. $\dfrac{9x^3 - 6x^2 + 3x}{-3x}$

104. $\dfrac{-8x^3y^5 + 4x^2y^4 - 2xy^3}{2xy^2}$

105. $(a - 1) \div (1 - a)$

106. $(t - 3) \div (3 - t)$

107. $(m^4 - 16) \div (m - 2)$

108. $(x^4 - 1) \div (x - 1)$

Find the quotient and remainder.

109. $(3m^3 - 9m^2 + 18m) \div (3m)$

110. $(8x^3 - 4x^2 - 18x) \div (2x)$

111. $(b^2 - 3b + 5) \div (b + 2)$

112. $(r^2 - 5r + 9) \div (r - 3)$

113. $(4x^2 - 9) \div (2x + 1)$

114. $(9y^3 + 2y) \div (3y + 2)$

115. $(x^3 + x^2 - 11x + 10) \div (x - 1)$

116. $(y^3 - 9y^2 + 3y - 6) \div (y + 1)$

Write each expression in the form

$$quotient + \frac{remainder}{divisor}.$$

117. $\dfrac{2x}{x - 3}$

118. $\dfrac{3x}{x - 4}$

119. $\dfrac{2x}{1 - x}$

120. $\dfrac{3x}{5 - x}$

121. $\dfrac{x^2 - 3}{x + 1}$

122. $\dfrac{x^2 + 3x + 1}{x - 3}$

123. $\dfrac{x^2}{x + 1}$

124. $\dfrac{-2x^2}{x - 3}$

Miscellaneous

Perform the indicated operations.

125. $(x + 3)(x + 7)$

126. $(k + 5)(k + 4)$

127. $(t - 3y)(t - 4y)$

128. $(t + 7z)(t + 6z)$

129. $(2x^3)^0 + (2y)^0$ **130.** $(4y^2 - 9)^0$

131. $(-3ht^6)^3$

132. $(-9y^3c^4)^2$

133. $(2w + 3)(w - 6)$

134. $(3x + 5)(2x - 6)$

135. $(3u - 5v)(3u + 5v)$

136. $(9x^2 - 2)(9x^2 + 2)$

137. $(3h + 5)^2$

138. $(4v - 3)^2$

139. $(x + 3)^3$

140. $(k - 10)^3$

141. $(-7s^2t)(-2s^3t^5)$

142. $-5w^3r^2 \cdot 2w^4r^8$

143. $\left(\dfrac{k^4m^2}{2k^2m^2}\right)^4$ **144.** $\left(\dfrac{-6h^3y^5}{2h^7y^2}\right)^4$

145. $(5x^2 - 8x - 8) - (4x^2 + x - 3)$

146. $(4x^2 - 6x - 8) - (9x^2 - 5x + 7)$

147. $(2x^2 - 2x - 3) + (3x^2 + x - 9)$

148. $(x^2 - 3x - 1) + (x^2 - 2x + 1)$

149. $(x + 4)(x^2 - 5x + 1)$

150. $(2x^2 - 7x + 4)(x + 3)$

151. $(x^2 + 4x - 12) \div (x - 2)$

152. $(a^2 - 3a - 10) \div (a - 5)$

Applications

Solve each problem.

153. *Roundball court.* The length of a basketball court is 44 feet more than its width w. Find polynomials $P(w)$ and $A(w)$ that represent its perimeter and area. Find $P(50)$ and $A(50)$.

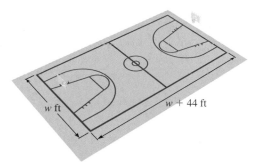

Figure for Exercise 153

154. *Badminton court.* The width of a badminton court is 24 feet less than its length x. Find polynomials $P(x)$ and $A(x)$ that represent its perimeter and area. Find $P(44)$ and $A(44)$.

155. *Smoke alert.* A retailer of smoke alarms knows that at a price of p dollars each, she can sell $600 - 15p$ smoke alarms per week. Find a polynomial $R(p)$ that represents the weekly revenue for the smoke alarms. Find the revenue for a week in which the price is $12 per smoke

alarm. Use the bar graph to find the price per smoke alarm that gives the maximum weekly revenue.

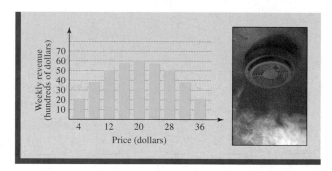

Figure for Exercise 155

156. *Boom box sales.* A retailer of boom boxes knows that at a price of q dollars each, he can sell $900 - 3q$ boom boxes per month. Find a polynomial $R(q)$ that represents the monthly revenue for the boom boxes. How many boom boxes will he sell if the price is $300 each?

157. *CD savings.* Valerie invested $12,000 in a CD that paid 6% compounded annually for 8 years. What was the value of her investment at the end of the eighth year?

158. *Risky business.* Tony invested $45,000 in Kirk's new business. If Kirk does well, he will pay Tony back in 5 years with interest at 5% compounded annually. If the business succeeds, then how much will Tony receive in 5 years?

159. *Saving for a house.* Newlyweds Michael and Leslie want to have $30,000 for a down payment on a house in 4 years. If they can earn 9% interest compounded annually, then how much would they have to have now to reach this goal?

160. *Opening a business.* Sandy wants to start a florist shop in 6 years and figures that she will need $20,000 to do it. If she can earn 7% interest compounded annually, then how much does she need now to reach this goal?

Chapter 4 Test

Use the rules of exponents to simplify each expression. Write answers without negative exponents.

1. $-5x^3 \cdot 7x^5$

2. $3x^3y \cdot (2xy^4)^2$

3. $-4a^6b^5 \div (2a^5b)$

4. $3x^{-2} \cdot 5x^7$

5. $\left(\dfrac{-2a}{b^2}\right)^5$

6. $\dfrac{-6a^7b^6c^2}{-2a^3b^8c^2}$

7. $\dfrac{6t^{-7}}{2t^9}$

8. $\dfrac{w^{-6}}{w^{-4}}$

9. $(-3s^{-3}t^2)^{-2}$

10. $(-2x^{-6}y)^3$

Convert to scientific notation.

11. 5,433,000

12. 0.0000065

Convert to standard notation.

13. 3.2×10^3

14. 8×10^{-5}

15. 3.5 billion

16. 12 trillion

Perform each computation by converting to scientific notation. Give answers in scientific notation.

17. (80,000)(0.000006)

18. $(0.0000003)^4$

Perform the indicated operations.

19. $(7x^3 - x^2 - 6) + (5x^2 + 2x - 5)$

20. $(x^2 - 3x - 5) - (2x^2 + 6x - 7)$

21. $\dfrac{6y^3 - 9y^2}{-3y}$

22. $(x - 2) \div (2 - x)$

23. $(x^3 - 2x^2 - 4x + 3) \div (x - 3)$

24. $3x^2(5x^3 - 7x^2 + 4x - 1)$

Find the products.

25. $(x + 5)(x - 2)$

26. $(3a - 7)(2a + 5)$

27. $(a - 7)^2$

28. $(4x + 3y)^2$

29. $(b - 3)(b + 3)$

30. $(3t^2 - 7)(3t^2 + 7)$

31. $(4x^2 - 3)(x^2 + 2)$

32. $(x - 2)(x + 3)(x - 4)$

Write each expression in the form

$$quotient + \frac{remainder}{divisor}.$$

33. $\dfrac{2x}{x - 3}$

34. $\dfrac{x^2 - 3x + 5}{x + 2}$

Solve each problem.

35. Find the value of the polynomial $x^3 - 5x + 1$ when $x = 3$.

36. Suppose that $P(x) = x^2 - 5x + 2$. Find $P(0)$ and $P(3)$.

37. Find the quotient and remainder when $x^2 - 5x + 9$ is divided by $x - 3$.

38. Subtract $3x^2 - 4x - 9$ from $x^2 - 3x + 6$.

39. The width of a pool table is x feet, and the length is 4 feet longer than the width. Find polynomials $A(x)$ and $P(x)$ that represent the area and perimeter of the pool table. Find $A(4)$ and $P(4)$.

40. If a manufacturer charges q dollars each for footballs, then he can sell $3000 - 150q$ footballs per week. Find a polynomial $R(q)$ that represents the revenue for one week. Find the weekly revenue if the price is $8 for each football.

41. Gordon got a $15,000 bonus and has decided to invest it in the stock market until he retires in 35 years. If he averages 9% return on the investment compounded annually, then how much will he have in 35 years?

Making **Connections** | **A Review of Chapters 1–4**

Evaluate each arithmetic expression.

1. $-16 \div (-2)$

2. $-16 \div \left(-\dfrac{1}{2}\right)$

3. $(-5)^2 - 3(-5) + 1$

4. $-5^2 - 4(-5) + 3$

5. $2^{15} \div 2^{10}$

6. $2^6 - 2^5$

7. $-3^2 \cdot 4^2$

8. $(-3 \cdot 4)^2$

9. $\left(\dfrac{1}{2}\right)^3 + \dfrac{1}{2}$

10. $\left(\dfrac{2}{3}\right)^2 - \dfrac{1}{3}$

11. $(5 + 3)^2$

12. $5^2 + 3^2$

13. $3^{-1} + 2^{-1}$

14. $2^{-2} - 3^{-2}$

15. $(30 - 1)(30 + 1)$

16. $(30 - 1) \div (1 - 30)$

Perform the indicated operations.

17. $(x + 3)(x + 5)$

18. $x + 3(x + 5)$

19. $-5t^3v \cdot 3t^2v^6$

20. $\left(-10t^3v^2\right) \div \left(-2t^2v\right)$

21. $\left(x^2 + 8x + 15\right) + (x + 5)$

22. $\left(x^2 + 8x + 15\right) - (x + 5)$

23. $\left(x^2 + 8x + 15\right) \div (x + 5)$

24. $\left(x^2 + 8x + 15\right)(x + 5)$

25. $\left(-6y^3 + 8y^2\right) \div \left(-2y^2\right)$

26. $\left(18y^4 - 12y^3 + 3y^2\right) \div \left(3y^2\right)$

Solve each equation.

27. $2x + 1 = 0$

28. $x - 7 = 0$

29. $\dfrac{3}{4}x - 3 = \dfrac{1}{2}$

30. $\dfrac{x}{2} - \dfrac{3}{4} = \dfrac{1}{8}$

31. $2(x - 3) = 3(x - 2)$

32. $2(3x - 3) = 3(2x - 2)$

33. $\dfrac{3}{11}x = \dfrac{1}{5}$

34. $x + \dfrac{1}{8} = \dfrac{9}{20}$

35. $0.35x = 0.4x + 2$

36. $\dfrac{0.05x - 9}{8} = 0.2(x - 25)$

37. $5 - 3(4x + 12) = 1 - 3(4x + 1)$

38. $5 - 3(4x + 12) = -12x - 31$

Solve.

39. Find the *x*-intercept for the line $y = 2x + 1$.

40. Find the *y*-intercept for the line $y = x - 7$.

41. Find the slope of the line $y = 2x + 1$.

42. Find the slope of the line that goes through $(0, 0)$ and $\left(\dfrac{1}{2}, \dfrac{1}{3}\right)$.

43. If $y = \dfrac{3}{4}x - 3$ and y is $\dfrac{1}{2}$, then what is x?

44. Find y if $y = \dfrac{x}{2} - \dfrac{3}{4}$ and x is $\dfrac{1}{2}$.

Solve each problem.

45. The perimeter of a rectangular field is 740 meters. If the width is 30 meters less than the length, then what is the length?

46. The area of a rectangular table top is 1200 square inches. If the length is 40 inches, then what is the width?

47. A diamond ring is on sale at 30% off the regular price. If the sale price is $3500, then what is the regular price?

48. A farmer has planted 4000 strawberry plants of which 12% are genetically modified. How many more genetically modified plants should be planted so that 20% of her strawberry plants are genetically modified plants?

Solve the problem.

49. ***Average cost.*** Pineapple Recording plans to spend $100,000 to record a new CD by the Woozies and $2.25 per CD to manufacture the disks. The polynomial $2.25n + 100,000$ represents the total cost in dollars for recording and manufacturing n disks. Find an expression that represents the average cost per disk by dividing the total cost by n. Find the average cost per disk for $n = 1000$, $100,000$, and $1,000,000$. What happens to the large initial investment of $100,000 if the company sells one million CDs?

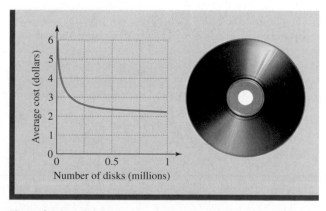

Figure for Exercise 49

*Critical*Thinking | For Individual or Group Work | Chapter 4

These exercises can be solved by a variety of techniques, which may or may not require algebra. So be creative and think critically. Explain all answers. Answers are in the Instructor's Edition of this text.

1. *Throwing darts.* A dart board contains a region worth 9 points and a region worth 4 points as shown in the accompanying figure. If you are allowed to throw as many darts as you wish, then what is the largest possible total score that you *cannot* get?

Figure for Exercise 1

2. *Counting squares.* A square checkerboard is made up of 36 alternately colored 1 inch by 1 inch squares.

a) What is the total number of squares that are visible on this checkerboard? (*Hint:* Count the 6 by 6 squares, then the 5 by 5 squares, and so on.)

b) How many squares are visible on a checkerboard that has 64 alternately colored 1 inch by 1 inch squares?

3. *Four fours.* Check out these equations:

$$\frac{4+4}{4+4} = 1, \quad \frac{4}{4} + \frac{4}{4} = 2, \quad 4 - 4^{4-4} = 3.$$

a) Using exactly four 4's write arithmetic expressions whose values are 4, 5, 6, and so on. How far can you go?

b) Repeat this exercise using four 5's, three 4's, and three 5's.

4. *Four coins.* Place four coins on a table with heads facing downward. On each move you must turn over exactly three coins. Count the number of moves it takes to get all four coins with heads facing upward. What is the minimum number of moves necessary to get all four heads facing upward?

5. *Snakes and iguanas.* A woman has a collection of snakes and iguanas. Her young son observed that the reptiles have a total of 50 eyes and 56 feet. How many reptiles of each type does the woman have?

Photo for Exercise 5

6. *Hungry bugs.* If it takes a colony of termites one day to devour a block of wood that is 2 inches wide, 2 inches long, and 2 inches high, then how long will it take them to devour a block of wood that is 4 inches wide, 4 inches long, and 4 inches high? Assume that they keep eating at the same rate.

7. *Ancient history.* This problem is from the second century. Four numbers have a sum of 9900. The second exceeds the first by one-seventh of the first. The third exceeds the sum of the first two by 300. The fourth exceeds the sum of the first three by 300. Find the four numbers.

8. *Related digits.* What is the largest four-digit number such that the second digit is one-fourth of the third digit, the third digit is twice the first digit, and the last digit is the same as the first digit?

Factoring

The sport of skydiving was born in the 1930s soon after the military began using parachutes as a means of deploying troops. Today, skydiving is a popular sport around the world.

With as little as 8 hours of ground instruction, first-time jumpers can be ready to make a solo jump. Without the assistance of oxygen, skydivers can jump from as high as 14,000 feet and reach speeds of more than 100 miles per hour as they fall toward the earth. Jumpers usually open their parachutes between 2000 and 3000 feet and then gradually glide down to their landing area. If the jump and the parachute are handled correctly, the landing can be as gentle as jumping off two steps.

Making a jump and floating to earth are only part of the sport of skydiving. For example, in an activity called "relative work skydiving," a team of as many as 920 free-falling skydivers join together to make geometrically shaped formations. In a related exercise called "canopy relative work," the team members form geometric patterns after their parachutes or canopies have opened. This kind of skydiving takes skill and practice, and teams are not always successful in their attempts.

The amount of time a skydiver has for a free fall depends on the height of the jump and how much the skydiver uses the air to slow the fall.

In Exercises 85 and 86 of Section 5.6 we find the amount of time that it takes a skydiver to fall from a given height.

5.1   Factoring Out Common Factors

In Chapter 4, you learned how to multiply a monomial and a polynomial. In this section, you will learn how to reverse that multiplication by finding the greatest common factor for the terms of a polynomial and then factoring the polynomial.

⟨1⟩   Prime Factorization of Integers

To **factor** an expression means to write the expression as a product. For example, if we start with 12 and write $12 = 4 \cdot 3$, we have factored 12. Both 4 and 3 are **factors** or **divisors** of 12. There are other factorizations of 12:

$$12 = 2 \cdot 6 \qquad 12 = 1 \cdot 12 \qquad 12 = 2 \cdot 2 \cdot 3 = 2^2 \cdot 3$$

The one that is most useful to us is $12 = 2^2 \cdot 3$, because it expresses 12 as a product of *prime numbers.*

> **Prime Number**
>
> A positive integer larger than 1 that has no positive integral factors other than itself and 1 is called a **prime number.**

The numbers 2, 3, 5, 7, 11, 13, 17, 19, and 23 are the first nine prime numbers. A positive integer larger than 1 that is not a prime is a **composite number.** The numbers 4, 6, 8, 9, 10, and 12 are the first six composite numbers. Every composite number is a product of prime numbers. The **prime factorization** for 12 is $2^2 \cdot 3$.

E X A M P L E   1

Prime factorization

Find the prime factorization for 36.

Solution

We start by writing 36 as a product of two integers:

$$36 = 2 \cdot 18 \qquad \text{Write 36 as } 2 \cdot 18.$$
$$= 2 \cdot 2 \cdot 9 \qquad \text{Replace 18 by } 2 \cdot 9.$$
$$= 2 \cdot 2 \cdot 3 \cdot 3 \qquad \text{Replace 9 by } 3 \cdot 3.$$
$$= 2^2 \cdot 3^2 \qquad \text{Use exponential notation.}$$

The prime factorization for 36 is $2^2 \cdot 3^2$.

Now do Exercises 1–6

⟨ **Helpful Hint** ⟩

The prime factorization of 36 can be found also with a *factoring tree:*

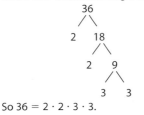

So $36 = 2 \cdot 2 \cdot 3 \cdot 3$.

For larger integers, it is better to use the method shown in Example 2 and to recall some divisibility rules. Even numbers are divisible by 2. If the sum of the digits of a number is divisible by 3, then the number is divisible by 3. Numbers that end in 0 or 5 are divisible by 5. Two-digit numbers with repeated digits (11, 22, 33, . . .) are divisible by 11.

EXAMPLE 2

‹ Helpful Hint ›

The fact that every composite number has a unique prime factorization is known as the fundamental theorem of arithmetic.

‹ Helpful Hint ›

Note that the division in Example 2 can be done also as follows:

$$
\begin{array}{r}
7 \\
5\overline{)35} \\
3\overline{)105} \\
2\overline{)210} \\
2\overline{)420}
\end{array}
$$

Factoring a large number

Find the prime factorization for 420.

Solution

Start by dividing 420 by the smallest prime number that will divide into it evenly (without remainder). The smallest prime divisor of 420 is 2.

$$
\begin{array}{r}
210 \\
2\overline{)420}
\end{array}
$$

Now find the smallest prime that will divide evenly into the quotient, 210. The smallest prime divisor of 210 is 2. Continue this procedure, as follows, until the quotient is a prime number:

$$
\begin{array}{r}
2\overline{)420} \\
2\overline{)210} \quad 420 \div 2 = 210 \\
3\overline{)105} \quad 210 \div 2 = 105 \\
5\overline{)35} \quad 105 \div 3 = 35 \\
7
\end{array}
$$

The product of all of the prime numbers in this procedure is 420:

$$420 = 2 \cdot 2 \cdot 3 \cdot 5 \cdot 7$$

So the prime factorization of 420 is $2^2 \cdot 3 \cdot 5 \cdot 7$. Note that it is not necessary to divide by the smallest prime divisor at each step. We get the same factorization if we divide by any prime divisor.

> Now do Exercises 7–12

‹2› Greatest Common Factor

The largest integer that is a factor of two or more integers is called the **greatest common factor (GCF)** of the integers. For example, 1, 2, 3, and 6 are common factors of 18 and 24. Because 6 is the largest, 6 is the GCF of 18 and 24. We can use prime factorizations to find the GCF. For example, to find the GCF of 8 and 12, we first factor 8 and 12:

$$8 = 2 \cdot 2 \cdot 2 = 2^3 \qquad 12 = 2 \cdot 2 \cdot 3 = 2^2 \cdot 3$$

We see that the factor 2 appears twice in both 8 and 12. So 2^2, or 4, is the GCF of 8 and 12. Notice that 2 is a factor in both 2^3 and $2^2 \cdot 3$ and that 2^2 is the smallest power of 2 in these factorizations. In general, we can use the following strategy to find the GCF.

Strategy for Finding the GCF for Positive Integers

1. Find the prime factorization for each integer.
2. The GCF is the product of the common prime factors using the smallest exponent that appears on each of them.

If two integers have no common prime factors, then their greatest common factor is 1, because 1 is a factor of every integer. For example, 6 and 35 have no common prime factors because $6 = 2 \cdot 3$ and $35 = 5 \cdot 7$. However, because $6 = 1 \cdot 6$ and $35 = 1 \cdot 35$, the GCF for 6 and 35 is 1.

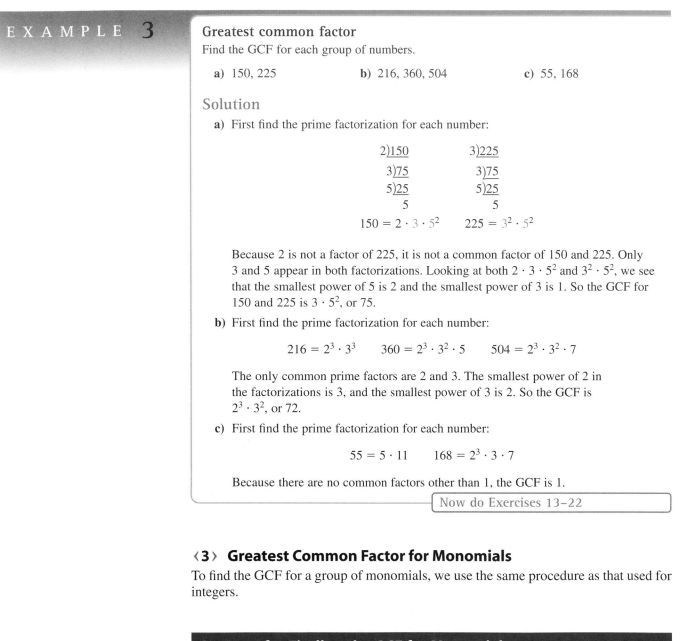

E X A M P L E   3

Greatest common factor

Find the GCF for each group of numbers.

a) 150, 225   **b)** 216, 360, 504   **c)** 55, 168

Solution

a) First find the prime factorization for each number:

$$
\begin{array}{cc}
2\overline{)150} & 3\overline{)225} \\
3\overline{)75} & 3\overline{)75} \\
5\overline{)25} & 5\overline{)25} \\
5 & 5
\end{array}
$$

$$150 = 2 \cdot 3 \cdot 5^2 \qquad 225 = 3^2 \cdot 5^2$$

Because 2 is not a factor of 225, it is not a common factor of 150 and 225. Only 3 and 5 appear in both factorizations. Looking at both $2 \cdot 3 \cdot 5^2$ and $3^2 \cdot 5^2$, we see that the smallest power of 5 is 2 and the smallest power of 3 is 1. So the GCF for 150 and 225 is $3 \cdot 5^2$, or 75.

b) First find the prime factorization for each number:

$$216 = 2^3 \cdot 3^3 \qquad 360 = 2^3 \cdot 3^2 \cdot 5 \qquad 504 = 2^3 \cdot 3^2 \cdot 7$$

The only common prime factors are 2 and 3. The smallest power of 2 in the factorizations is 3, and the smallest power of 3 is 2. So the GCF is $2^3 \cdot 3^2$, or 72.

c) First find the prime factorization for each number:

$$55 = 5 \cdot 11 \qquad 168 = 2^3 \cdot 3 \cdot 7$$

Because there are no common factors other than 1, the GCF is 1.

Now do Exercises 13–22

⟨3⟩ **Greatest Common Factor for Monomials**

To find the GCF for a group of monomials, we use the same procedure as that used for integers.

Strategy for Finding the GCF for Monomials

1. Find the GCF for the coefficients of the monomials.
2. Form the product of the GCF for the coefficients and each variable that is common to all of the monomials, where the exponent on each variable is the smallest power of that variable in any of the monomials.

EXAMPLE   **4**

Greatest common factor for monomials

Find the greatest common factor for each group of monomials.

    a) $15x^2, 9x^3$ **b)** $12x^2y^2, 30x^2yz, 42x^3y$

Solution

a) Since $15 = 3 \cdot 5$ and $9 = 3^2$, the GCF for 15 and 9 is 3. Since the smallest power of x in $15x^2$ and $9x^3$ is 2, the GCF is $3x^2$. If we write these monomials as

$$15x^2 = 5 \cdot 3 \cdot x \cdot x \qquad \text{and} \qquad 9x^3 = 3 \cdot 3 \cdot x \cdot x \cdot x,$$

we can see that $3x^2$ is the GCF.

b) Since $12 = 2^2 \cdot 3$, $30 = 2 \cdot 3 \cdot 5$, and $42 = 2 \cdot 3 \cdot 7$, the GCF for 12, 30, and 42 is $2 \cdot 3$ or 6. For the common variables x and y, 2 is the smallest power of x and 1 is the smallest power of y. So the GCF for the three monomials is $6x^2y$. Note that z is not in the GCF because it is not in all three monomials.

                               Now do Exercises 23–34

⟨4⟩   Factoring Out the Greatest Common Factor

In Chapter 4, we used the distributive property to multiply monomials and polynomials. For example,

$$6(5x - 3) = 30x - 18.$$

If we start with $30x - 18$ and write

$$30x - 18 = 6(5x - 3),$$

we have factored $30x - 18$. Because multiplication is the last operation to be performed in $6(5x - 3)$, the expression $6(5x - 3)$ is a product. Because 6 is the GCF for 30 and 18, we have **factored out** the GCF.

EXAMPLE   **5**

Factoring out the greatest common factor

Factor the following polynomials by factoring out the GCF.

    a) $25a^2 + 40a$ **b)** $6x^4 - 12x^3 + 3x^2$ **c)** $x^2y^5 + x^6y^3$

Solution

a) The GCF for the coefficients 25 and 40 is 5. Because the smallest power of the common factor a is 1, we can factor $5a$ out of each term:

$$25a^2 + 40a = 5a \cdot 5a + 5a \cdot 8$$
$$= 5a(5a + 8)$$

b) The GCF for 6, 12, and 3 is 3. We can factor x^2 out of each term, since the smallest power of x in the three terms is 2. So factor $3x^2$ out of each term as follows:

$$6x^4 - 12x^3 + 3x^2 = 3x^2 \cdot 2x^2 - 3x^2 \cdot 4x + 3x^2 \cdot 1$$
$$= 3x^2(2x^2 - 4x + 1)$$

Check by multiplying: $3x^2(2x^2 - 4x + 1) = 6x^4 - 12x^3 + 3x^2$.

c) The GCF for the numerical coefficients is 1. Both x and y are common to each term. Using the lowest powers of x and y, we get

$$x^2y^5 + x^6y^3 = x^2y^3 \cdot y^2 + x^2y^3 \cdot x^4$$
$$= x^2y^3(y^2 + x^4).$$

Check by multiplying.

> Now do Exercises 35–62

Because of the commutative property of multiplication, the common factor can be placed on either side of the other factor. So in Example 5, the answers could be written as $(5a + 8)5a$, $(2x^2 - 4x + 1)3x^2$, and $(y^2 + x^4)x^2y^3$.

CAUTION   If the GCF is one of the terms of the polynomial, then you must remember to leave a 1 in place of that term when the GCF is factored out. For example,

$$ab + b = a \cdot b + 1 \cdot b = b(a + 1).$$

You should always check your answer by multiplying the factors.

In Example 6, the greatest common factor is a binomial. This type of factoring will be used in factoring trinomials by grouping in Section 5.2.

E X A M P L E   6

A binomial factor

Factor out the greatest common factor.

a) $(a + b)w + (a + b)6$ **b)** $x(x + 2) + 3(x + 2)$

c) $y(y - 3) - (y - 3)$

Solution

a) The greatest common factor is $a + b$:

$$(a + b)w + (a + b)6 = (a + b)(w + 6)$$

b) The greatest common factor is $x + 2$:

$$x(x + 2) + 3(x + 2) = (x + 3)(x + 2)$$

c) The greatest common factor is $y - 3$:

$$y(y - 3) - (y - 3) = y(y - 3) - 1(y - 3)$$
$$= (y - 1)(y - 3)$$

> Now do Exercises 63–70

⟨5⟩ Factoring Out the Opposite of the GCF

The greatest common factor for $-4x + 2xy$ is $2x$. Note that you can factor out the GCF $(2x)$ or the opposite of the GCF $(-2x)$:

$$-4x + 2xy = 2x(-2 + y) \qquad -4x + 2xy = -2x(2 - y)$$

It is useful to know both of these factorizations. Factoring out the opposite of the GCF will be used in factoring by grouping in Section 5.2 and in factoring trinomials with negative leading coefficients in Section 5.4. Remember to check all factoring by multiplying the factors to see if you get the original polynomial.

EXAMPLE **7**

Factoring out the opposite of the GCF

Factor each polynomial twice. First factor out the greatest common factor, and then factor out the opposite of the GCF.

a) $3x - 3y$ b) $a - b$

c) $-x^3 + 2x^2 - 8x$

Solution

a) $3x - 3y = 3(x - y)$ Factor out 3.

$ = -3(-x + y)$ Factor out -3.

Note that the signs of the terms in parentheses change when -3 is factored out. Check the answers by multiplying.

b) $a - b = 1(a - b)$ Factor out 1, the GCF of a and b.

$ = -1(-a + b)$ Factor out -1, the opposite of the GCF.

We can also write $a - b = -1(b - a)$.

c) $-x^3 + 2x^2 - 8x = x(-x^2 + 2x - 8)$ Factor out x.

$ = -x(x^2 - 2x + 8)$ Factor out $-x$.

> Now do Exercises 71–86

CAUTION   Be sure to change the sign of each term in parentheses when you factor out the opposite of the greatest common factor.

⟨6⟩ Applications

EXAMPLE **8**

Area of a rectangular garden

The area of a rectangular garden is $x^2 + 8x + 15$ square feet. If the length is $x + 5$ feet, then what binomial represents the width?

Solution

Note that the area of a rectangle is the product of the length and width. Since $x^2 + 8x + 15 = (x + 5)(x + 3)$ and the length is $x + 5$ feet, the width must be $x + 3$ feet.

> Now do Exercises 87–90

Warm-Ups ▼

Fill in the blank.

1. To _____ means to write as a product.
2. A _____ number is an integer greater than 1 that has no factors besides itself and 1.
3. The _____ of two numbers is the largest number that is a factor of both.
4. All factoring can be checked by _____ the factors.

True or false?

5. There are only nine prime numbers.
6. The prime factorization of 32 is $2^3 \cdot 3$.
7. The integer 51 is a prime number.
8. The GCF for 12 and 16 is 4.
9. The GCF for $x^5 y^3 - x^4 y^7$ is $x^4 y^3$.
10. We can factor out $2xy$ or $-2xy$ from $2x^2 y - 6xy^2$.

‹1› Prime Factorization of Integers

Find the prime factorization of each integer. See Examples 1 and 2.

1. 18 **2.** 20

3. 52 **4.** 76

5. 98 **6.** 100

7. 216 **8.** 248

9. 460 **10.** 345

11. 924 **12.** 585

‹2› Greatest Common Factor

Find the greatest common factor for each group of integers. See Example 3. See the Strategy for Finding the GCF for Positive Integers box on page 323.

13. 8, 20 **14.** 18, 42

15. 36, 60 **16.** 42, 70

17. 40, 48, 88 **18.** 15, 35, 45

19. 76, 84, 100 **20.** 66, 72, 120

21. 39, 68, 77 **22.** 81, 200, 539

‹3› Greatest Common Factor for Monomials

Find the greatest common factor for each group of monomials. See Example 4. See the Strategy for Finding the GCF for Monomials box on page 324.

23. $6x, 8x^3$ **24.** $12x^2, 4x^3$

25. $12x^3, 4x^2, 6x$ **26.** $3y^5, 9y^4, 15y^3$

27. $3x^2y, 2xy^2$ **28.** $7a^2x^3, 5a^3x$

29. $24a^2bc, 60ab^2$ **30.** $30x^2yz^3, 75x^3yz^6$

31. $12u^3v^2, 25s^2t^4$ **32.** $45m^2n^5, 56a^4b^8$

33. $18a^3b, 30a^2b^2, 54ab^3$ **34.** $16x^2z, 40xz^2, 72z^3$

‹4› Factoring Out the Greatest Common Factor

Complete the factoring of each monomial.

35. $27x = 9(\quad)$ **36.** $51y = 3y(\quad)$

37. $24t^2 = 8t(\quad)$ **38.** $18u^2 = 3u(\quad)$

39. $36y^5 = 4y^2(\quad)$

40. $42z^4 = 3z^2(\quad)$

41. $u^4v^3 = uv(\quad)$

42. $x^5y^3 = x^2y(\quad)$

43. $-14m^4n^3 = 2m^4(\quad)$

44. $-8y^3z^4 = 4z^3(\quad)$

45. $-33x^4y^3z^2 = -3x^3yz(\quad)$

46. $-96a^3b^4c^5 = -12ab^3c^3(\quad)$

Factor out the GCF in each expression. See Example 5.

47. $2w + 4t$

48. $6y + 3$

49. $12x - 18y$

50. $24a - 36b$

51. $x^3 - 6x$

52. $10y^4 - 30y^2$

53. $5ax + 5ay$

54. $6wz + 15wa$

55. $h^5 + h^3$

56. $y^6 + y^5$

57. $-2k^7m^4 + 4k^3m^6$

58. $-6h^5t^2 + 3h^3t^6$

59. $2x^3 - 6x^2 + 8x$

60. $6x^3 + 18x^2 + 24x$

61. $12x^4t + 30x^3t - 24x^2t^2$

62. $15x^2y^2 - 9xy^2 + 6x^2y$

Factor out the GCF in each expression. See Example 6.

63. $(x - 3)a + (x - 3)b$

64. $(y + 4)3 + (y + 4)z$

65. $x(x - 1) - 5(x - 1)$

66. $a(a + 1) - 3(a + 1)$

67. $m(m + 9) + (m + 9)$

68. $(x - 2)x - (x - 2)$

69. $a(y + 1)^2 + b(y + 1)^2$

70. $w(w + 2)^2 + 8(w + 2)^2$

⟨5⟩ Factoring Out the Opposite of the GCF

First factor out the GCF, and then factor out the opposite of the GCF. See Example 7.

71. $8x - 8y$

72. $2a - 6b$

73. $-4x + 8x^2$

74. $-5x^2 + 10x$

75. $x - 5$

76. $a - 6$

77. $4 - 7a$

78. $7 - 5b$

79. $-24a^3 + 16a^2$

80. $-30b^4 + 75b^3$

81. $-12x^2 - 18x$

82. $-20b^2 - 8b$

83. $-2x^3 - 6x^2 + 14x$

 **84.** $-8x^4 + 6x^3 - 2x^2$

85. $4a^3b - 6a^2b^2 - 4ab^3$

86. $12u^5v^6 + 18u^2v^3 - 15u^4v^5$

⟨6⟩ Applications

Solve each problem by factoring. See Example 8.

87. *Uniform motion.* Helen traveled a distance of $20x + 40$ miles at 20 miles per hour on the Yellowhead Highway. Find a binomial that represents the time that she traveled.

88. *Area of a painting.* A rectangular painting with a width of x centimeters has an area of $x^2 + 50x$ square centimeters. Find a binomial that represents the length. See the accompanying figure.

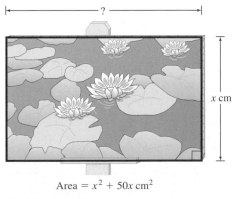

Area = $x^2 + 50x$ cm²

Figure for Exercise 88

89. *Tomato soup.* The amount of metal S (in square inches) that it takes to make a can for tomato soup depends on the radius r and height h:

$$S = 2\pi r^2 + 2\pi rh$$

a) Rewrite this formula by factoring out the greatest common factor on the right-hand side.

b) Let $h = 5$ in. and write a formula that expresses S in terms of r.

c) The accompanying graph shows S for r between 1 in. and 3 in. (with $h = 5$ in.). Which of these r-values gives the maximum surface area?

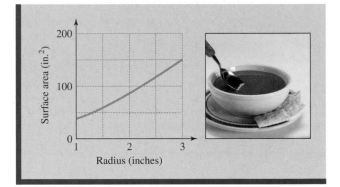

Figure for Exercise 89

90. *Amount of an investment.* The amount of an investment of P dollars for t years at simple interest rate r is given by $A = P + Prt$.

a) Rewrite this formula by factoring out the greatest common factor on the right-hand side.

b) Find A if \$8300 is invested for 3 years at a simple interest rate of 15%.

Getting More Involved

91. *Discussion*

Is the greatest common factor of $-6x^2 + 3x$ positive or negative? Explain.

92. *Writing*

Explain in your own words why you use the smallest power of each common prime factor when finding the GCF of two or more integers.

Math *at Work*   **Kayak Design**

Kayaks have been built by the Aleut and Inuit peoples for the past 4000 years. Today's builders have access to materials and techniques unavailable to the original kayak builders. Modern kayakers incorporate hydrodynamics and materials technology to create designs that are efficient and stable. Builders measure how well their designs work by calculating indicators such as prismatic coefficient, block coefficient, and the midship area coefficient, to name a few.

Even the fitting of a kayak to the paddler is done scientifically. For example, the formula

$$PL = 2 \cdot BL + BS \left(0.38 \cdot EE + 1.2 \sqrt{\left(\frac{BW}{2} - \frac{SW}{2} \right)^2 + (SL)^2} \right)$$

can be used to calculate the appropriate paddle length. BL is the length of the paddle's blade. BS is a boating style factor, which is 1.2 for touring, 1.0 for river running, and 0.95 for play boating. EE is the elbow to elbow distance with the paddler's arms straight out to the sides. BW is the boat width and SW is the shoulder width. SL is the spine length, which is the distance measured in a sitting position from the chair seat to the top of the paddler's shoulder. All lengths are in centimeters.

The degree of control a kayaker exerts over the kayak depends largely on the body contact with it. A kayaker wears the kayak. So the choice of a kayak should hinge first on the right body fit and comfort and second on the skill level or intended paddling style. So designing, building, and even fitting a kayak is a blend of art and science.

5.2   Special Products and Grouping

In This Section

‹ 1 › Factoring by Grouping

‹ 2 › Factoring a Difference of Two Squares

‹ 3 › Factoring a Perfect Square Trinomial

‹ 4 › Factoring Completely

In Section 5.1 you learned how to factor out the greatest common factor from all of the terms of a polynomial. In this section you will learn to factor a four-term polynomial by factoring out a common factor from the first two terms and then a common factor from the last two terms.

‹ 1 › Factoring by Grouping

The product of two binomials may have four terms. For example,

$$(x + a)(x + 3) = (x + a)x + (x + a)3$$
$$= x^2 + ax + 3x + 3a.$$

To factor $x^2 + ax + 3x + 3a$, we simply reverse the steps we used to find the product. Factor out the common factor x from the first two terms and the common factor 3 from the last two terms:

$$x^2 + ax + 3x + 3a = x(x + a) + 3(x + a) \qquad \text{Factor out } x \text{ and 3.}$$
$$= (x + 3)(x + a) \qquad\qquad \text{Factor out } x + a.$$

It does not matter whether you take out the common factor to the right or left. So $(x + a)(x + 3)$ is also correct and we could have factored as follows:

$$x^2 + ax + 3x + 3a = (x + a)x + (x + a)3$$
$$= (x + a)(x + 3)$$

This method of factoring is called **factoring by grouping.**

Strategy for Factoring a Four-Term Polynomial by Grouping

1. Factor out the GCF from the first group of two terms.
2. Factor out the GCF from the last group of two terms.
3. Factor out the common binomial.

EXAMPLE 1

Factoring by Grouping

Use grouping to factor each polynomial.

a) $xy + 2y + 5x + 10$ **b)** $x^2 + wx + x + w$

Solution

a) The first two terms have a common factor of y, and the last two terms have a common factor of 5:

$$xy + 2y + 5x + 10 = y(x + 2) + 5(x + 2) \quad \text{Factor out } y \text{ and } 5.$$
$$= (y + 5)(x + 2) \quad\quad\quad \text{Factor out } x + 2.$$

Check by using FOIL.

b) The first two terms have a common factor of x, and the last two have a common factor of 1:

$$x^2 + wx + x + w = x(x + w) + 1(x + w) \quad \text{Factor out } x \text{ and } 1.$$
$$= (x + 1)(x + w) \quad\quad\quad \text{Factor out } x + w.$$

Check by using FOIL.

> Now do Exercises 1–10

For some four-term polynomials it is necessary to rearrange the terms before factoring out the common factors.

EXAMPLE 2

Factoring by Grouping with Rearranging

Use grouping to factor each polynomial.

a) $mn + 4m + m^2 + 4n$ **b)** $ax + b + bx + a$

Solution

a) We can factor out m from the first two terms to get $m(n + 4)$, but we can't get another factor of $n + 4$ from the last two terms. By rearranging the terms we can factor by grouping:

$$mn + 4m + m^2 + 4n = m^2 + mn + 4m + 4n \quad \text{Rearrange terms.}$$
$$= m(m + n) + 4(m + n) \quad \text{Factor out } m \text{ and } 4.$$
$$= (m + 4)(m + n) \quad\quad\quad \text{Factor out } m + n.$$

b) $ax + b + bx + a = ax + bx + a + b$ Rearrange terms.

$= x(a + b) + 1(a + b)$ Factor out x and 1.

$= (x + 1)(a + b)$ Factor out $a + b$.

> Now do Exercises 11–18

Note that there are several rearrangements that will allow us to factor the polynomials in Example 2. For example, $m^2 + 4m + mn + 4n$ would also work for Example 2(a).

We saw in Section 5.1 that you could factor out a common factor with a positive sign or a negative sign. For example, we can factor $-2x + 10$ as $2(-x + 5)$ or $-2(x - 5)$. We use this technique in Example 3.

EXAMPLE 3

Factoring by Grouping with Negative Signs
Use grouping to factor each polynomial.

    a) $2x^2 - 3x - 2x + 3$ **b)** $ax + 3y - 3x - ay$

Solution

  a) We can factor out x from the first two terms and 1 from the last two terms:

$$2x^2 - 3x - 2x + 3 = x(2x - 3) + 1(-2x + 3)$$

However, we didn't get a common binomial. We can get a common binomial if we factor out -1 from the last two terms:

$$2x^2 - 3x - 2x + 3 = x(2x - 3) - 1(2x - 3) \quad \text{Factor out } x \text{ and } -1.$$
$$= (x - 1)(2x - 3) \quad \text{Factor out } 2x - 3.$$

  b) For this polynomial we have to rearrange the terms and factor out a common factor with a negative sign:

$$ax + 3y - 3x - ay = ax - 3x - ay + 3y \quad \text{Rearrange the terms.}$$
$$= x(a - 3) - y(a - 3) \quad \text{Factor out } x \text{ and } -y.$$
$$= (x - y)(a - 3) \quad \text{Factor out } a - 3.$$

> Now do Exercises 19–28

⟨2⟩ Factoring a Difference of Two Squares

In Section 4.7, you learned that the product of a sum and a difference is a difference of two squares:

$$(a + b)(a - b) = a^2 - ab + ab - b^2 = a^2 - b^2$$

So a difference of two squares can be factored as a product of a sum and a difference, using the following rule.

Factoring a Difference of Two Squares

For any real numbers a and b,

$$a^2 - b^2 = (a + b)(a - b).$$

Note that the square of an integer is a perfect square. For example, 64 is a perfect square because $64 = 8^2$. The square of a monomial in which the coefficient is an integer is also called a **perfect square** or simply a **square.** For example, $9m^2$ is a perfect square because $9m^2 = (3m)^2$.

EXAMPLE **4**

Factoring a difference of two squares

Factor each polynomial.

a) $y^2 - 81$ b) $9m^2 - 16$ c) $4x^2 - 9y^2$

Solution

a) Because $81 = 9^2$, the binomial $y^2 - 81$ is a difference of two squares:

$$y^2 - 81 = y^2 - 9^2 \qquad \text{Rewrite as a difference of two squares.}$$
$$= (y + 9)(y - 9) \quad \text{Factor.}$$

Check by multiplying.

b) Because $9m^2 = (3m)^2$ and $16 = 4^2$, the binomial $9m^2 - 16$ is a difference of two squares:

$$9m^2 - 16 = (3m)^2 - 4^2 \qquad \text{Rewrite as a difference of two squares.}$$
$$= (3m + 4)(3m - 4) \quad \text{Factor.}$$

Check by multiplying.

c) Because $4x^2 = (2x)^2$ and $9y^2 = (3y)^2$, the binomial $4x^2 - 9y^2$ is a difference of two squares:

$$4x^2 - 9y^2 = (2x + 3y)(2x - 3y)$$

Now do Exercises 29–42

CAUTION Don't confuse a difference of two squares $a^2 - b^2$ with a sum of two squares $a^2 + b^2$. The sum $a^2 + b^2$ is not one of the special products and it can't be factored.

⟨3⟩ **Factoring a Perfect Square Trinomial**

In Section 4.7 you learned how to square a binomial using the rule

$$(a + b)^2 = a^2 + 2ab + b^2.$$

You can reverse this rule to factor a trinomial such as $x^2 + 6x + 9$. Notice that

$$x^2 + 6x + 9 = \underset{\underset{a^2}{\uparrow}}{x^2} + \underset{2ab}{\underbrace{2 \cdot x \cdot 3}} + \underset{\underset{b^2}{\uparrow}}{3^2}.$$

So if $a = x$ and $b = 3$, then $x^2 + 6x + 9$ fits the form $a^2 + 2ab + b^2$, and

$$x^2 + 6x + 9 = (x + 3)^2.$$

A trinomial that is of the form $a^2 + 2ab + b^2$ or $a^2 - 2ab + b^2$ is called a **perfect square trinomial.** A perfect square trinomial is the square of a binomial. Perfect square trinomials will be used in solving quadratic equations by completing the square in Chapter 10. Perfect square trinomials can be identified using the following strategy.

Strategy for Identifying a Perfect Square Trinomial

A trinomial is a perfect square trinomial if

1. the first and last terms are of the form a^2 and b^2 (perfect squares), and

2. the middle term is $2ab$ or $-2ab$.

EXAMPLE 5

Identifying the special products
Determine whether each binomial is a difference of two squares and whether each trinomial is a perfect square trinomial.

a) $x^2 - 14x + 49$ **b)** $4x^2 - 81$
c) $4a^2 + 24a + 25$ **d)** $9y^2 - 24y - 16$

Solution

a) The first term is x^2, and the last term is 7^2. The middle term, $-14x$, is $-2 \cdot x \cdot 7$. So this trinomial is a perfect square trinomial.

b) Both terms of $4x^2 - 81$ are perfect squares, $(2x)^2$ and 9^2. So $4x^2 - 81$ is a difference of two squares.

c) The first term of $4a^2 + 24a + 25$ is $(2a)^2$ and the last term is 5^2. However, $2 \cdot 2a \cdot 5$ is $20a$. Because the middle term is $24a$, this trinomial is not a perfect square trinomial.

d) The first and last terms in a perfect square trinomial are both positive. Because the last term in $9y^2 - 24y - 16$ is negative, the trinomial is not a perfect square trinomial.

Now do Exercises 43–54

Note that the middle term in a perfect square trinomial may have a positive or a negative coefficient, while the first and last terms must be positive. Any perfect square trinomial can be factored as the square of a binomial by using the following rule.

Factoring Perfect Square Trinomials

For any real numbers a and b,

$$a^2 + 2ab + b^2 = (a + b)^2$$
$$a^2 - 2ab + b^2 = (a - b)^2.$$

EXAMPLE **6**

Factoring perfect square trinomials

Factor.

    a) $x^2 - 4x + 4$ **b)** $a^2 + 16a + 64$ **c)** $4x^2 - 12x + 9$

Solution

    a) The first term is x^2, and the last term is 2^2. Because the middle term is $-2 \cdot 2 \cdot x$, or $-4x$, this polynomial is a perfect square trinomial:

$$x^2 - 4x + 4 = (x - 2)^2$$

        Check by expanding $(x - 2)^2$.

    b) $a^2 + 16a + 64 = (a + 8)^2$

        Check by expanding $(a + 8)^2$.

    c) The first term is $(2x)^2$, and the last term is 3^2. Because $-2 \cdot 2x \cdot 3 = -12x$, the polynomial is a perfect square trinomial. So

$$4x^2 - 12x + 9 = (2x - 3)^2.$$

        Check by expanding $(2x - 3)^2$.

> Now do Exercises 55–72

⟨4⟩ Factoring Completely

To factor a polynomial means to write it as a product of simpler polynomials. A polynomial that can't be factored using integers is called a **prime** or **irreducible polynomial.** The polynomials $3x$, $w + 1$, and $4m - 5$ are prime polynomials. Note that $4m - 5 = 4\left(m - \frac{5}{4}\right)$, but $4m - 5$ is a prime polynomial because it can't be factored using integers only.

    A polynomial is **factored completely** when it is written as a product of prime polynomials. So $(y - 8)(y + 1)$ is a complete factorization. When factoring polynomials, we usually do not factor integers that occur as common factors. So $6x(x - 7)$ is considered to be factored completely even though 6 could be factored.

    Some polynomials have a factor common to all terms. To factor such polynomials completely, it is simpler to factor out the greatest common factor (GCF) and then factor the remaining polynomial. Example 7 illustrates factoring completely.

EXAMPLE **7**

Factoring completely

Factor each polynomial completely.

    a) $2x^3 - 50x$ **b)** $8x^2y - 32xy + 32y$ **c)** $2x^3 - 3x^2 - 2x + 3$

Solution

    a) The greatest common factor of $2x^3$ and $50x$ is $2x$:

$$2x^3 - 50x = 2x(x^2 - 25) \quad \text{Check this step by multiplying.}$$
$$= 2x(x + 5)(x - 5) \quad \text{Difference of two squares}$$

    b) $8x^2y - 32xy + 32y = 8y(x^2 - 4x + 4) \quad$ Check this step by multiplying.
$$= 8y(x - 2)^2 \qquad\qquad \text{Perfect square trinomial}$$

    c) We can factor out x^2 from the first two terms and 1 from the last two terms:

$$2x^3 - 3x^2 - 2x + 3 = x^2(2x - 3) + 1(-2x + 3)$$

However, we didn't get a common binomial. We can get a common binomial if we factor out -1 from the last two terms:

$$2x^3 - 3x^2 - 2x + 3 = x^2(2x - 3) - 1(2x - 3) \qquad \text{Factor out } x^2 \text{ and } -1.$$
$$= (x^2 - 1)(2x - 3) \qquad \text{Factor out } 2x - 3.$$
$$= (x + 1)(x - 1)(2x - 3) \qquad \text{Difference of two squares}$$

> Now do Exercises 73–98

Remember that factoring reverses multiplication and *every step of factoring can be checked by multiplication.*

Warm-Ups ▼

Fill in the blank.

1. A _____ is the square of an integer or an algebraic expression.
2. A _____ is the product of a sum and a difference.
3. A trinomial of the form $a^2 + 2ab + b^2$ is a _____ trinomial.
4. A _____ polynomial is one that can't be factored.
5. A polynomial is _____ when it is written as a product of prime polynomials.

True or false?

6. We always factor out the GCF first.
7. The polynomial $x^2 + 16$ is a difference of two squares.
8. The polynomial $x^2 - 8x + 16$ is a perfect square trinomial.
9. The polynomial $9x^2 + 21x + 49$ is a perfect square trinomial.
10. The polynomial $16y + 1$ is a prime polynomial.
11. The polynomial $4x^2 - 4$ is factored completely as $4(x^2 - 1)$.

5.2   Exercises

‹ **Study Tips** ›

- As you study a chapter, make a list of topics and questions that you would put on the test, if you were to write it.
- Write about what you read in the text. Sum things up in your own words.

‹1› **Factoring by Grouping**

Factor by grouping. See Example 1.

1. $bx + by + cx + cy$
2. $3x + 3z + ax + az$
3. $ab + b^2 + a + b$
4. $2x^2 + x + 2x + 1$
5. $wm + 3w + m + 3$
6. $ay + y + 3a + 3$
7. $6x^2 + 10x + 3xw + 5w$
8. $5ax + 2ay + 5xy + 2y^2$
9. $x^2 + 3x + 4x + 12$
10. $y^2 + 2y + 6y + 12$

Factor by grouping. See Example 2.

11. $mn + n + n^2 + m$
12. $2x^3 + y + x + 2x^2y$
13. $10 + wm + 5m + 2w$
14. $2a + 3b + 6 + ab$
15. $xa + ay + 3y + 3x$
16. $x^3 + ax + 3a + 3x^2$

17. $a^3 + w^2 + aw + a^2w$
18. $a^4 + y + ay + a^3$

Factor by grouping. See Example 3.

19. $w^2 - w - bw + b$
20. $x^2 - 2x - mx + 2m$
21. $w^2 + aw - w - a$
22. $ap + 3a - p - 3$
23. $m^2 + mx - x - m$
24. $-6n - 6b + b^2 + bn$
25. $x^2 + 7x - 5x - 35$
26. $y^2 + 3y - 8y - 24$
27. $2x^2 + 14x - 5x - 35$
28. $2y^2 + 3y - 16y - 24$

⟨2⟩ **Factoring a Difference of Two Squares**

Factor each polynomial. See Example 4.

29. $a^2 - 4$
30. $h^2 - 9$
31. $x^2 - 49$
32. $y^2 - 36$
33. $a^2 - 121$
34. $w^2 - 81$
35. $y^2 - 9x^2$
36. $16x^2 - y^2$
37. $25a^2 - 49b^2$
38. $9a^2 - 64b^2$
39. $121m^2 - 1$
40. $144n^2 - 1$
41. $9w^2 - 25c^2$
42. $144w^2 - 121a^2$

⟨3⟩ **Factoring a Perfect Square Trinomial**

Determine whether each polynomial is a difference of two squares, a perfect square trinomial, or neither of these. See Example 5. See the Strategy for Identifying Perfect Square Trinomials box on page 334.

43. $x^2 - 20x + 100$
44. $x^2 - 10x - 25$
45. $y^2 - 40$
46. $a^2 - 49$
47. $4y^2 + 12y + 9$
48. $9a^2 - 30a - 25$
49. $x^2 - 8x + 64$
50. $x^2 + 4x + 4$
51. $9y^2 - 25c^2$
52. $9x^2 + 4$
53. $9a^2 + 6ab + b^2$
54. $4x^2 - 4xy + y^2$

Factor each perfect square trinomial. See Example 6.

55. $x^2 + 2x + 1$
56. $y^2 + 4y + 4$
57. $a^2 + 6a + 9$
58. $w^2 + 10w + 25$
59. $x^2 + 12x + 36$
60. $y^2 + 14y + 49$
61. $a^2 - 4a + 4$
62. $b^2 - 6b + 9$
63. $4w^2 + 4w + 1$
64. $9m^2 + 6m + 1$
65. $16x^2 - 8x + 1$
66. $25y^2 - 10y + 1$
67. $4t^2 + 20t + 25$
68. $9y^2 - 12y + 4$
69. $9w^2 + 42w + 49$
70. $144x^2 + 24x + 1$
71. $n^2 + 2nt + t^2$
72. $x^2 - 2xy + y^2$

⟨4⟩ **Factoring Completely**

Factor each polynomial completely. See Example 7.

73. $5x^2 - 125$
74. $3y^2 - 27$
75. $-2x^2 + 18$
76. $-5y^2 + 20$
77. $a^3 - ab^2$
78. $x^2y - y$
79. $3x^2 + 6x + 3$
80. $12a^2 + 36a + 27$
81. $-5y^2 + 50y - 125$
82. $-2a^2 - 16a - 32$
83. $x^3 - 2x^2y + xy^2$
84. $x^3y + 2x^2y^2 + xy^3$
85. $-3x^2 + 3y^2$
86. $-8a^2 + 8b^2$
87. $2ax^2 - 98a$
88. $32x^2y - 2y^3$
89. $w^3 - w - w^2 + 1$
90. $x^3 + x^2 - x - 1$
91. $x^3 + x^2 - 4x - 4$
92. $a^2m - b^2n + a^2n - b^2m$
93. $3ab^2 - 18ab + 27a$
94. $-2a^2b + 8ab - 8b$
95. $-4m^3 + 24m^2n - 36mn^2$
96. $10a^3 - 20a^2b + 10ab^2$
97. $x^2a - b + bx^2 - a$
98. $wx^2 - 75 - 25w + 3x^2$

Miscellaneous

Factor each polynomial completely.

99. $6a^3y + 24a^2y^2 + 24ay^3$

100. $8b^5c - 8b^4c^2 + 2b^3c^3$

101. $24a^3y - 6ay^3$

102. $27b^3c - 12bc^3$

103. $2a^3y^2 - 6a^2y$

104. $9x^3y - 18x^2y^2$

105. $ab + 2bw - 4aw - 8w^2$

106. $3am - 6n - an + 18m$

107. $(a - b) - b(a - b)$

108. $(a + b)w - (a + b)$

109. $(4x^2 - 1)2x - (4x^2 - 1)$

110. $(a^2 - 9)a + 3(a^2 - 9)$

Applications

Use factoring to solve each problem.

111. *Skydiving.* The height in feet above the earth for a skydiver t seconds after jumping from an airplane at 6400 ft is approximated by the formula $h(t) = -16t^2 + 6400$, provided $t < 5$.

a) Rewrite the formula with the right-hand side factored completely.

b) Use the result of part (a) to find $h(2)$.

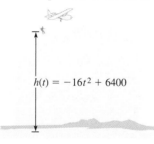

$h(t) = -16t^2 + 6400$

Figure for Exercise 111

112. *Demand for pools.* Tropical Pools sells an aboveground model for p dollars each. The monthly revenue for this model is given by the formula

$$R(p) = -0.08p^2 + 300p.$$

Revenue is the product of the price p and the demand (quantity sold).

a) Factor out the price on the right-hand side of the formula.

b) Write a formula $D(p)$ for the monthly demand.

c) Find $D(3000)$.

d) Use the accompanying graph to estimate the price at which the revenue is maximized. Approximately how many pools will be sold monthly at this price?

e) What is the approximate maximum revenue?

f) Use the accompanying graph to estimate the price at which the revenue is zero.

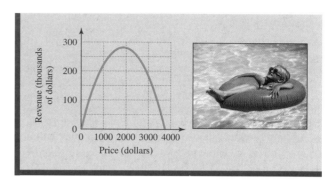

Figure for Exercise 112

113. *Volume of a tank.* The volume in cubic inches for a fish tank with a square base and height x is given by the formula

$$V(x) = x^3 - 6x^2 + 9x.$$

a) Rewrite the formula with the right-hand side factored completely.

b) Find an expression for the length of a side of the square base.

Figure for Exercise 113

Getting More Involved

114. *Discussion*

For what real number k does $3x^2 - k$ factor as $3(x - 2)(x + 2)$?

115. *Writing*

Explain in your own words how to factor a four-term polynomial by grouping.

116. *Writing*

Explain how you know that $x^2 + 1$ is a prime polynomial.

5.3　Factoring the Trinomial $ax^2 + bx + c$ with $a = 1$

In This Section

⟨1⟩ Factoring $ax^2 + bx + c$ with $a = 1$

⟨2⟩ Factoring with Two Variables

⟨3⟩ Factoring Completely

In this section, we will factor the type of trinomials that result from multiplying two different binomials. We will do this only for trinomials in which the coefficient of x^2, the leading coefficient, is 1. Factoring trinomials with a leading coefficient not equal to 1 will be done in Section 5.4.

⟨1⟩ Factoring $ax^2 + bx + c$ with $a = 1$

To find the product of the binomials $x + m$ and $x + n$, where x is the variable and m and n are constants, we use the distributive property as follows:

$$(x + m)(x + n) = (x + m)x + (x + m)n \quad \text{Distributive property}$$
$$= x^2 + mx + nx + mn \quad \text{Distributive property}$$
$$= x^2 + (m + n)x + mn \quad \text{Combine like terms.}$$

Notice that in the trinomial the coefficient of x is the sum $m + n$ and the constant term is the product mn. This observation is the key to factoring the trinomial $ax^2 + bx + c$ with $a = 1$. We first find two numbers that have a product of c (the constant term) and a sum of b (the coefficient of x). Then reverse the steps that we used in finding the product $(x + m)(x + n)$. We summarize these ideas with the following strategy.

Strategy for Factoring $x^2 + bx + c$ by Grouping

To factor $x^2 + bx + c$:

1. Find two integers that have a product of c and a sum equal to b.

2. Replace bx by the sum of two terms whose coefficients are the two numbers found in (1).

3. Factor the resulting four-term polynomial by grouping.

EXAMPLE　**1**

Factoring trinomials
Factor.

　　a) $x^2 + 5x + 6$　　　　　**b)** $x^2 + 8x + 12$　　　　　**c)** $a^2 - 9a + 20$

Solution

　a) To factor $x^2 + 5x + 6$, we need two integers that have a product of 6 and a sum of 5. If the product is positive and the sum is positive, then both integers must be positive. We can list all of the possibilities:

Product	Sum
$6 = 1 \cdot 6$	$1 + 6 - 7$
$6 = 2 \cdot 3$	$2 + 3 = 5$

The only integers that have a product of 6 and a sum of 5 are 2 and 3. Now replace $5x$ with $2x + 3x$ and factor by grouping:

$$
\begin{aligned}
x^2 + 5x + 6 &= x^2 + 2x + 3x + 6 && \text{Replace } 5x \text{ by } 2x + 3x. \\
&= x(x + 2) + 3(x + 2) && \text{Factor out } x \text{ and } 3. \\
&= (x + 3)(x + 2) && \text{Factor out } x + 2.
\end{aligned}
$$

Check by FOIL: $(x + 3)(x + 2) = x^2 + 5x + 6$.

b) To factor $x^2 + 8x + 12$, we need two integers that have a product of 12 and a sum of 8. Since the product and sum are both positive, both integers are positive.

Product	**Sum**
$12 = 1 \cdot 12$	$1 + 12 = 13$
$12 = 2 \cdot 6$	$2 + 6 = 8$
$12 = 3 \cdot 4$	$3 + 4 = 7$

The only integers that have a product of 12 and a sum of 8 are 2 and 6. Now replace $8x$ by $2x + 6x$ and factor by grouping:

$$
\begin{aligned}
x^2 + 8x + 12 &= x^2 + 2x + 6x + 12 && \text{Replace } 8x \text{ by } 2x + 6x. \\
&= x(x + 2) + 6(x + 2) && \text{Factor out } x \text{ and } 6. \\
&= (x + 6)(x + 2) && \text{Factor out } x + 2.
\end{aligned}
$$

Check by FOIL: $(x + 6)(x + 2) = x^2 + 8x + 12$.

c) To factor $a^2 - 9a + 20$, we need two integers that have a product of 20 and a sum of -9. Since the product is positive and the sum is negative, both integers must be negative.

Product	**Sum**
$20 = (-1)(-20)$	$-1 + (-20) = -21$
$20 = (-2)(-10)$	$-2 + (-10) = -12$
$20 = (-4)(-5)$	$-4 + (-5) = -9$

Only -4 and -5 have a product of 20 and a sum of -9. Now replace $-9a$ by $-4a + (-5a)$ or $-4a - 5a$ and factor by grouping:

$$
\begin{aligned}
a^2 - 9a + 20 &= a^2 - 4a - 5a + 20 && \text{Replace } -9a \text{ by } -4a - 5a. \\
&= a(a - 4) - 5(a - 4) && \text{Factor out } a \text{ and } -5. \\
&= (a - 5)(a - 4) && \text{Factor out } a - 4.
\end{aligned}
$$

Check by FOIL: $(a - 5)(a - 4) = a^2 - 9a + 20$.

Now do Exercises 1–14

We usually do not write out all of the steps shown in Example 1. We saw prior to Example 1 that

$$
x^2 + (m + n)x + mn = (x + m)(x + n).
$$

So once you know m and n, you can simply write the factors, as shown in Example 2.

E X A M P L E   **2**

Factoring trinomials more efficiently

Factor.

a) $x^2 + 5x + 4$

b) $y^2 + 6y - 16$

c) $w^2 - 5w - 24$

Solution

a) To factor $x^2 + 5x + 4$ we need two integers with a product of 4 and a sum of 5. The only possibilities for a product of 4 are

$$(1)(4), (-1)(-4), (2)(2), \quad \text{and} \quad (-2)(-2).$$

Only 1 and 4 have a sum of 5. So,

$$x^2 + 5x + 4 = (x + 1)(x + 4).$$

Check by using FOIL on $(x + 1)(x + 4)$ to get $x^2 + 5x + 4$.

b) To factor $y^2 + 6y - 16$ we need two integers with a product of -16 and a sum of 6. The only possibilities for a product of -16 are

$$(-1)(16), (1)(-16), (-2)(8), (2)(-8), \quad \text{and} \quad (-4)(4).$$

Only -2 and 8 have a sum of 6. So,

$$y^2 + 6y - 16 = (y + 8)(y - 2).$$

Check by using FOIL on $(y + 8)(y - 2)$ to get $y^2 + 6y - 16$.

c) To factor $w^2 - 5w - 24$ we need two integers with a product of -24 and a sum of -5. The only possibilities for a product of -24 are

$$(-1)(24), (1)(-24), (-2)(12), (2)(-12), (-3)(8), (3)(-8), (-4)(6), \text{ and } (4)(-6).$$

Only -8 and 3 have a sum of -5. So,

$$w^2 - 5w - 24 = (w - 8)(w + 3).$$

Check by using FOIL on $(w - 8)(w + 3)$ to get $w^2 - 5w - 24$.

> Now do Exercises 15–22

Polynomials are easiest to factor when they are in the form $ax^2 + bx + c$. So if a polynomial can be rewritten into that form, rewrite it before attempting to factor it. In Example 3, we factor polynomials that need to be rewritten.

E X A M P L E   **3**

Factoring trinomials

Factor.

a) $2x - 8 + x^2$

b) $-36 + t^2 - 9t$

Solution

a) Before factoring, write the trinomial as $x^2 + 2x - 8$. Now, to get a product of -8 and a sum of 2, use -2 and 4:

$$2x - 8 + x^2 = x^2 + 2x - 8 \qquad \text{Write in } ax^2 + bx + c \text{ form.}$$
$$= (x + 4)(x - 2) \quad \text{Factor and check by multiplying.}$$

b) Before factoring, write the trinomial as $t^2 - 9t - 36$. Now, to get a product of -36 and a sum of -9, use -12 and 3:

$$-36 + t^2 - 9t = t^2 - 9t - 36 \qquad \text{Write in } ax^2 + bx + c \text{ form.}$$
$$= (t - 12)(t + 3) \quad \text{Factor and check by multiplying.}$$

> Now do Exercises 23–24

To factor $x^2 + bx + c$, we search through all pairs of integers that have a product of c until we find a pair that has a sum of b. If there is no such pair of integers, then the polynomial cannot be factored and it is a prime polynomial. Before you can conclude that a polynomial is prime, be sure that you have tried *all* possibilities.

EXAMPLE **4**

Prime polynomials

Factor.

a) $x^2 + 7x - 6$

b) $x^2 + 9$

Solution

a) Because the last term is -6, we want a positive integer and a negative integer that have a product of -6 and a sum of 7. Check all possible pairs of integers:

Product	Sum
$-6 = (-1)(6)$	$-1 + 6 = 5$
$-6 = (1)(-6)$	$1 + (-6) = -5$
$-6 = (2)(-3)$	$2 + (-3) = -1$
$-6 = (-2)(3)$	$-2 + 3 = 1$

None of these possible factors of -6 have a sum of 7, so we can be certain that $x^2 + 7x - 6$ cannot be factored. It is a prime polynomial.

b) Because the x-term is missing in $x^2 + 9$, its coefficient is 0. That is, $x^2 + 9 = x^2 + 0x + 9$. So we seek two positive integers or two negative integers that have a product of 9 and a sum of 0. Check all possibilities:

Product	Sum
$9 = (1)(9)$	$1 + 9 = 10$
$9 = (-1)(-9)$	$-1 + (-9) = -10$
$9 = (3)(3)$	$3 + 3 = 6$
$9 = (-3)(-3)$	$-3 + (-3) = -6$

None of these pairs of integers have a sum of 0, so we can conclude that $x^2 + 9$ is a prime polynomial. Note that $x^2 + 9$ does not factor as $(x + 3)^2$ because $(x + 3)^2$ has a middle term: $(x + 3)^2 = x^2 + 6x + 9$.

> Now do Exercises 25–52

‹ **Helpful Hint** ›

Don't confuse $a^2 + b^2$ with the difference of two squares $a^2 - b^2$ which is not a prime polynomial:

$$a^2 - b^2 = (a + b)(a - b)$$

The prime polynomial $x^2 + 9$ in Example 4(b) is a sum of two squares. There are many other sums of squares that are prime. For example,

$$x^2 + 1, \quad a^2 + 4, \quad b^2 + 9, \quad \text{and} \quad 4y^2 + 25$$

are prime. However, not every sum of two squares is prime. For example, $4x^2 + 16$ is a sum of two squares that is not prime because $4x^2 + 16 = 4(x^2 + 4)$.

Sum of Two Squares

The sum of two squares $a^2 + b^2$ is prime, but not every sum of two squares is prime.

⟨2⟩ Factoring with Two Variables

In Example 5, we factor polynomials that have two variables using the same technique that we used for one variable.

EXAMPLE 5

Polynomials with two variables

Factor.

    a) $x^2 + 2xy - 8y^2$ **b)** $a^2 - 7ab + 10b^2$ **c)** $1 - 2xy - 8x^2y^2$

Solution

  a) To factor $x^2 + 2xy - 8y^2$ we need two integers with a product of -8 and a sum of 2. The only possibilities for a product of -8 are

$$(-1)(8), (1)(-8), (-2)(4), \text{ and } (2)(-4).$$

Only -2 and 4 have a sum of 2. Since $(-2y)(4y) = -8y^2$, we have

$$x^2 + 2xy - 8y^2 = (x - 2y)(x + 4y).$$

Check by using FOIL on $(x - 2y)(x + 4y)$ to get $x^2 + 2xy - 8y^2$.

  b) To factor $a^2 - 7ab + 10b^2$ we need two integers with a product of 10 and a sum of -7. The only possibilities for a product of 10 are

$$(-1)(-10), (1)(10), (-2)(-5), \text{ and } (2)(5).$$

Only -2 and -5 have a sum of -7. Since $(-2b)(-5b) = 10b^2$, we have

$$a^2 - 7ab + 10b^2 = (a - 5b)(a - 2b).$$

Check by using FOIL on $(a - 2b)(a - 5b)$ to get $a^2 - 7ab + 10b^2$.

  c) As in part (a), we need two integers with a product of -8 and a sum of -2. The integers are -4 and 2. Since 1 factors as $1 \cdot 1$ and $-8x^2y^2 = (-4xy)(2xy)$, we have

$$1 - 2xy - 8x^2y^2 = (1 + 2xy)(1 - 4xy).$$

Check by using FOIL.

> Now do Exercises 53–64

⟨3⟩ Factoring Completely

In Section 5.2 you learned that binomials such as $3x - 5$ (with no common factor) are prime polynomials. In Example 4 of this section we saw a trinomial that is a prime polynomial. There are infinitely many prime trinomials. When factoring a polynomial completely, we could have a factor that is a prime trinomial.

EXAMPLE 6

Factoring completely

Factor each polynomial completely.

   a) $x^3 - 6x^2 - 16x$ **b)** $4x^3 + 4x^2 + 4x$

Solution

a) $x^3 - 6x^2 - 16x = x(x^2 - 6x - 16)$ Factor out the GCF.

                     $= x(x - 8)(x + 2)$ Factor $x^2 - 6x - 16$.

b) First factor out $4x$, the greatest common factor:

$$4x^3 + 4x^2 + 4x = 4x(x^2 + x + 1)$$

To factor $x^2 + x + 1$, we would need two integers with a product of 1 and a sum of 1. Because there are no such integers, $x^2 + x + 1$ is prime, and the factorization is complete.

Now do Exercises 65–106

Warm-Ups ▼

Fill in the blank.

1. If there are no two integers that have a _____ of c and a ____ of b, then $x^2 + bx + c$ is prime.
2. We can check all factoring by _____ the factors.
3. The sum of two squares $a^2 + b^2$ is _____.
4. Always factor out the ____ first.

True or false?

5. $x^2 - 6x + 9 = (x - 3)^2$
6. $x^2 + 6x + 9 = (x + 3)^2$
7. $x^2 + 10x + 9 = (x - 9)(x - 1)$
8. $x^2 + 8x - 9 = (x - 1)(x + 9)$
9. $x^2 - 10xy + 9y^2 = (x - y)(x - 9y)$
10. $x^2 + 1 = (x + 1)(x + 1)$
11. $x^2 + x + 1 = (x + 1)(x + 1)$

5.3 Exercises

⟨ **Study Tips** ⟩

- Put important facts on note cards. Work on memorizing the note cards when you have a few spare minutes.
- Post some note cards on your refrigerator door. Make this course a part of your life.

⟨1⟩ **Factoring $ax^2 + bx + c$ with $a = 1$**

Factor each trinomial. Write out all of the steps as shown in Example 1. See the Strategy for Factoring $x^2 + bx + c$ by Grouping on page 339.

1. $x^2 + 4x + 3$
2. $y^2 + 6y + 5$
3. $x^2 + 9x + 18$
4. $w^2 + 6w + 8$
5. $a^2 + 7a + 10$
6. $b^2 + 7b + 12$
7. $a^2 - 7a + 12$
8. $m^2 - 9m + 14$

9. $b^2 - 5b - 6$

10. $a^2 + 5a - 6$

11. $x^2 + 3x - 10$

12. $x^2 - x - 12$

13. $x^2 + 5x - 24$

14. $a^2 - 5a - 50$

Factor each polynomial. If the polynomial is prime, say so. See Examples 2–4.

15. $y^2 + 7y + 10$

16. $x^2 + 8x + 15$

17. $a^2 - 6a + 8$

18. $b^2 - 8b + 15$

19. $m^2 - 10m + 16$

20. $m^2 - 17m + 16$

21. $w^2 + 9w - 10$

22. $m^2 + 6m - 16$

23. $w^2 - 8 - 2w$

24. $-16 + m^2 - 6m$

25. $a^2 - 2a - 12$

26. $x^2 + 3x + 3$

27. $15m - 16 + m^2$

28. $3y + y^2 - 10$

29. $a^2 - 4a + 12$

30. $y^2 - 6y - 8$

31. $z^2 - 25$

32. $p^2 - 1$

33. $h^2 + 49$

34. $q^2 + 4$

35. $m^2 + 12m + 20$

36. $m^2 + 21m + 20$

37. $t^2 - 3t + 10$

38. $x^2 - 5x - 3$

39. $m^2 - 18 - 17m$

40. $h^2 - 36 + 5h$

41. $m^2 - 23m + 24$

42. $m^2 + 23m + 24$

43. $5t - 24 + t^2$

44. $t^2 - 24 - 10t$

45. $t^2 - 2t - 24$

46. $t^2 + 14t + 24$

47. $t^2 - 10t - 200$

48. $t^2 + 30t + 200$

49. $x^2 - 5x - 150$

50. $x^2 - 25x + 150$

51. $13y + 30 + y^2$

52. $18z + 45 + z^2$

⟨2⟩ Factoring with Two Variables

Factor each polynomial. See Example 5.

53. $x^2 + 5ax + 6a^2$

54. $a^2 + 7ab + 10b^2$

55. $x^2 - 4xy - 12y^2$

56. $y^2 + yt - 12t^2$

57. $x^2 - 13xy + 12y^2$

58. $h^2 - 9hs + 9s^2$

59. $x^2 + 4xz - 33z^2$

60. $x^2 - 5xs - 24s^2$

61. $1 + 3ab - 28a^2b^2$

62. $1 - xy - 20x^2y^2$

63. $15a^2b^2 + 8ab + 1$

64. $12m^2n^2 - 8mn + 1$

⟨3⟩ Factoring Completely

Factor each polynomial completely. Use the methods discussed in Sections 5.1 through 5.3. If the polynomial is prime say so. See Example 6.

65. $5x^3 + 5x$

66. $b^3 + 49b$

67. $w^2 - 8w$

68. $x^4 - x^3$

69. $2w^2 - 162$

70. $6w^4 - 54w^2$

71. $-2b^2 - 98$

72. $-a^3 - 100a$

73. $x^3 - 2x^2 - 9x + 18$

74. $x^3 + 7x^2 - x - 7$

75. $4r^2 + 9$

76. $t^2 + 4z^2$

77. $x^2w^2 + 9x^2$

78. $a^4b + a^2b^3$

79. $w^2 - 18w + 81$

80. $w^2 + 30w + 81$

81. $6w^2 - 12w - 18$

82. $9w - w^3$

83. $3y^2 + 75$

84. $5x^2 + 500$

85. $ax + ay + cx + cy$

86. $y^3 + y^2 - 4y - 4$

87. $-2x^2 - 10x - 12$

88. $-a^3 - 2a^2 - a$

 89. $32x^2 - 2x^4$

90. $20w^2 + 100w + 40$

91. $3w^2 + 27w + 54$

92. $w^3 - 3w^2 - 18w$

93. $18w^2 + w^3 + 36w$

94. $18a^2 + 3a^3 + 36a$

95. $9y^2 + 1 + 6y$

96. $2a^2 + 1 + 3a$

97. $8vw^2 + 32vw + 32v$

 98. $3h^2t + 6ht + 3t$

99. $6x^3y + 30x^2y^2 + 36xy^3$

100. $3x^3y^2 - 3x^2y^2 + 3xy^2$

101. $5 + 8w + 3w^2$

102. $-3 + 2y + 21y^2$

103. $-3y^3 + 6y^2 - 3y$

104. $-4w^3 - 16w^2 + 20w$

105. $a^3 + ab + 3b + 3a^2$

106. $ac + xc + aw^2 + xw^2$

Applications

Use factoring to solve each problem.

107. *Area of a deck.* The area in square feet for a rectangular deck is given by $A(x) = x^2 + 6x + 8$.

    a) Find $A(6)$.

    b) If the width of the deck is $x + 2$ feet, then what is the length?

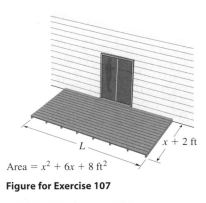

Area = $x^2 + 6x + 8$ ft^2

Figure for Exercise 107

108. *Area of a sail.* The area in square meters for a triangular sail is given by $A(x) = x^2 + 5x + 6$.

    a) Find $A(5)$.

    b) If the height of the sail is $x + 3$ meters, then what is the length of the base of the sail?

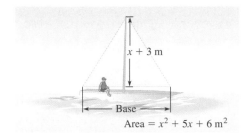

$x + 3$ m

Base

Area = $x^2 + 5x + 6$ m^2

Figure for Exercise 108

109. *Volume of a cube.* Hector designed a cubic box with volume x^3 cubic feet. After increasing the dimensions of the bottom, the box has a volume of $x^3 + 8x^2 + 15x$ cubic feet. If each of the dimensions of the bottom was increased by a whole number of feet, then how much was each increase?

110. *Volume of a container.* A cubic shipping container had a volume of a^3 cubic meters. The height was decreased by a whole number of meters and the width was increased by a whole number of meters so that the volume of the container is now $a^3 + 2a^2 - 3a$ cubic meters. By how many meters were the height and width changed?

Getting More Involved

111. *Discussion*

Which of the following products is not equivalent to the others? Explain your answer.

    a) $(2x - 4)(x + 3)$ **b)** $(x - 2)(2x + 6)$

    c) $2(x - 2)(x + 3)$ **d)** $(2x - 4)(2x + 6)$

112. *Discussion*

When asked to factor completely a certain polynomial, four students gave the following answers. Only one student gave the correct answer. Which one must it be? Explain your answer.

    a) $3(x^2 - 2x - 15)$ **b)** $(3x - 5)(5x - 15)$

    c) $3(x - 5)(x - 3)$ **d)** $(3x - 15)(x - 3)$

Mid-Chapter **Quiz** │ **Sections 5.1 through 5.3** │ **Chapter 5**

Find the prime factorization of each integer.

1. 48   **2.** 140

Find the greatest common factor for each group of integers.

3. 36, 45   **4.** 60, 144, 240

Factor each expression by factoring out the greatest common factor.

5. $8w - 6y$ **6.** $12x^3 - 30x^2$

7. $15ab^3 - 25a^2b^2 + 35a^3b$

Factor each expression.

8. $(x + 3)x - (x + 3)5$

9. $m(m - 9) - 6(m - 9)$

Factor completely.

10. $4y^2 - 9w^2$

11. $4h^2 + 12h + 9$

12. $w^2 - 16w + 64$

13. $10x^3 - 250x$

14. $-6x^2 - 36x - 54$

15. $aw - 3w + 6a - 18$

16. $bx - 5b - 6x + 30$

17. $ax^2 - a + x^2 - 1$

18. $x^3 - 5x - 4x^2$

19. $2x^3 + 18x$

20. $a^2 - 12as + 32s^2$

5.4   Factoring the Trinomial $ax^2 + bx + c$ with $a \neq 1$

In This Section

⟨1⟩ **The *ac* Method**

⟨2⟩ **Trial and Error**

⟨3⟩ **Factoring Completely**

In Section 5.3, we used grouping to factor trinomials with a leading coefficient of 1. In this section we will also use grouping to factor trinomials with a leading coefficient that is not equal to 1.

⟨1⟩   **The *ac* Method**

The first step in factoring $ax^2 + bx + c$ with $a = 1$ is to find two numbers with a product of c and a sum of b. If $a \neq 1$, then the first step is to find two numbers with a product of ac and a sum of b. This method is called the **ac method.** The strategy for factoring by the *ac* method follows. Note that this strategy works whether or not the leading coefficient is 1.

Strategy for Factoring $ax^2 + bx + c$ by the *ac* Method
To factor the trinomial $ax^2 + bx + c$:
1. Find two numbers that have a product equal to ac and a sum equal to b.
2. Replace bx by the sum of two terms whose coefficients are the two numbers found in (1).
3. Factor the resulting four-term polynomial by grouping.

E X A M P L E   **1**

The *ac* method

Factor each trinomial.

a) $2x^2 + 7x + 6$

b) $2x^2 + x - 6$

c) $10x^2 + 13x - 3$

Solution

a) In $2x^2 + 7x + 6$ we have $a = 2$, $b = 7$, and $c = 6$. So,

$$ac = 2 \cdot 6 = 12.$$

Now we need two integers with a product of 12 and a sum of 7. The pairs of integers with a product of 12 are 1 and 12, 2 and 6, and 3 and 4. Only 3 and 4 have a sum of 7. Replace $7x$ by $3x + 4x$ and factor by grouping:

$$2x^2 + 7x + 6 = 2x^2 + 3x + 4x + 6 \qquad \text{Replace } 7x \text{ by } 3x + 4x.$$
$$= (2x + 3)x + (2x + 3)2 \qquad \text{Factor out the common factors.}$$
$$= (2x + 3)(x + 2) \qquad \text{Factor out } 2x + 3.$$

Check by FOIL.

b) In $2x^2 + x - 6$ we have $a = 2$, $b = 1$, and $c = -6$. So,

$$ac = 2(-6) = -12.$$

Now we need two integers with a product of -12 and a sum of 1. We can list the possible pairs of integers with a product of -12 as follows:

$$1 \text{ and } -12 \qquad 2 \text{ and } -6 \qquad 3 \text{ and } -4$$
$$-1 \text{ and } 12 \qquad -2 \text{ and } 6 \qquad -3 \text{ and } 4$$

Only -3 and 4 have a sum of 1. Replace x by $-3x + 4x$ and factor by grouping:

$$2x^2 + x - 6 = 2x^2 - 3x + 4x - 6 \qquad \text{Replace } x \text{ by } -3x + 4x.$$
$$= (2x - 3)x + (2x - 3)2 \qquad \text{Factor out the common factors.}$$
$$= (2x - 3)(x + 2) \qquad \text{Factor out } 2x - 3.$$

Check by FOIL.

c) Because $ac = 10(-3) = -30$, we need two integers with a product of -30 and a sum of 13. The product is negative, so the integers must have opposite signs. We can list all pairs of factors of -30 as follows:

$$1 \text{ and } -30 \qquad 2 \text{ and } -15 \qquad 3 \text{ and } -10 \qquad 5 \text{ and } -6$$
$$-1 \text{ and } 30 \qquad -2 \text{ and } 15 \qquad -3 \text{ and } 10 \qquad -5 \text{ and } 6$$

The only pair that has a sum of 13 is -2 and 15:

$$10x^2 + 13x - 3 = 10x^2 - 2x + 15x - 3 \qquad \text{Replace } 13x \text{ by } -2x + 15x.$$
$$= (5x - 1)2x + (5x - 1)3 \qquad \text{Factor out the common factors.}$$
$$= (5x - 1)(2x + 3) \qquad \text{Factor out } 5x - 1.$$

Check by FOIL.

Now do Exercises 1–38

E X A M P L E   **2**

Factoring a trinomial in two variables by the *ac* method

Factor $8x^2 - 14xy + 3y^2$

Solution

Since $a = 8$, $b = -14$, and $c = 3$, we have $ac = 24$. Two numbers with a product of 24 and a sum of -14 must both be negative. The possible pairs with a product of 24 follow:

$$-1 \text{ and } -24 \qquad -3 \text{ and } -8$$
$$-2 \text{ and } -12 \qquad -4 \text{ and } -6$$

Only -2 and -12 have a sum of -14. Replace $-14xy$ by $-2xy - 12xy$ and factor by grouping:

$$8x^2 - 14xy + 3y^2 = 8x^2 - 2xy - 12xy + 3y^2$$
$$= (4x - y)2x + (4x - y)(-3y)$$
$$= (4x - y)(2x - 3y)$$

Check by FOIL.

Now do Exercises 39–44

⟨2⟩ Trial and Error

After you have gained some experience at factoring by the *ac* method, you can often find the factors without going through the steps of grouping. For example, consider the polynomial

$$3x^2 + 7x - 6.$$

The factors of $3x^2$ can only be $3x$ and x. The factors of 6 could be 2 and 3 or 1 and 6. We can list all of the possibilities that give the correct first and last terms, without regard to the signs:

$$(3x \quad 3)(x \quad 2) \qquad (3x \quad 2)(x \quad 3) \qquad (3x \quad 6)(x \quad 1) \qquad (3x \quad 1)(x \quad 6)$$

Because the factors of -6 have unlike signs, one binomial factor is a sum and the other binomial is a difference. Now we try some products to see if we get a middle term of $7x$:

$$(3x + 3)(x - 2) = 3x^2 - 3x - 6 \quad \text{Incorrect}$$
$$(3x - 3)(x + 2) = 3x^2 + 3x - 6 \quad \text{Incorrect}$$

⟨ **Helpful Hint** ⟩

If the trinomial has no common factor, then neither binomial factor can have a common factor.

Actually, there is no need to try $(3x \quad 3)(x \quad 2)$ or $(3x \quad 6)(x \quad 1)$ because each contains a binomial with a common factor. A common factor in the binomial causes a common factor in the product. But $3x^2 + 7x - 6$ has no common factor. So the factors must come from either $(3x \quad 2)(x \quad 3)$ or $(3x \quad 1)(x \quad 6)$. So we try again:

$$(3x + 2)(x - 3) = 3x^2 - 7x - 6 \quad \text{Incorrect}$$
$$(3x - 2)(x + 3) = 3x^2 + 7x - 6 \quad \text{Correct}$$

Even though there may be many possibilities in some factoring problems, it is often possible to find the correct factors without writing down every possibility. We can use a bit of guesswork in factoring trinomials. *Try* whichever possibility you think might work. *Check* it by multiplying. If it is not right, then *try again*. That is why this method is called **trial and error.**

EXAMPLE   3

Trial and error

Factor each trinomial using trial and error.

a) $2x^2 + 5x - 3$ **b)** $3x^2 - 11x + 6$

Solution

a) Because $2x^2$ factors only as $2x \cdot x$ and 3 factors only as $1 \cdot 3$, there are only two possible ways to get the correct first and last terms, without regard to the signs:

$$(2x \quad 1)(x \quad 3) \quad \text{and} \quad (2x \quad 3)(x \quad 1)$$

Because the last term of the trinomial is negative, one of the missing signs must be $+$, and the other must be $-$. The trinomial is factored correctly as

$$2x^2 + 5x - 3 = (2x - 1)(x + 3).$$

Check by using FOIL.

b) There are four possible ways to factor $3x^2 - 11x + 6$:

$$(3x \quad 1)(x \quad 6) \qquad (3x \quad 2)(x \quad 3)$$
$$(3x \quad 6)(x \quad 1) \qquad (3x \quad 3)(x \quad 2)$$

The first binomials of $(3x \quad 6)(x \quad 1)$ and $(3x \quad 3)(x \quad 2)$ have a common factor of 3. Since there is no common factor in $3x^2 - 11x + 6$, we can rule out both of these possibilities. Since the last term in $3x^2 - 11x + 6$ is positive and the middle term is negative, both signs in the factors must be negative. So the correct factorization is either $(3x - 1)(x - 6)$ or $(3x - 2)(x - 3)$. By using FOIL we can verify that $(3x - 2)(x - 3) = 3x^2 - 11x + 6$. So the polynomial is factored correctly as

$$3x^2 - 11x + 6 = (3x - 2)(x - 3).$$

> **Now do Exercises 45–64**

‹ **Helpful Hint** ›

The ac method is more systematic than trial and error. However, trial and error can be faster and easier, especially if your first or second trial is correct.

Factoring by trial and error is not just guessing. In fact, if the trinomial has a positive leading coefficient, we can determine in advance whether its factors are sums or differences.

Using Signs in Trial and Error

1. If the signs of the terms of a trinomial are $+ \; + \; +$, then both factors are sums: $x^2 + 5x + 6 = (x + 2)(x + 3)$.

2. If the signs are $+ \; - \; +$, then both factors are differences: $x^2 - 5x + 6 = (x - 2)(x - 3)$.

3. If the signs are $+ \; + \; -$ or $+ \; - \; -$, then one factor is a sum and the other is a difference: $x^2 + x - 6 = (x + 3)(x - 2)$ and $x^2 - x - 6 = (x - 3)(x + 2)$.

In Example 4 we factor a trinomial that has two variables.

EXAMPLE 4

Factoring a trinomial with two variables by trial and error
Factor $6x^2 - 7xy + 2y^2$.

Solution
We list the possible ways to factor the trinomial:

$$(3x \quad 2y)(2x \quad y) \quad\quad (3x \quad y)(2x \quad 2y) \quad\quad (6x \quad 2y)(x \quad y) \quad\quad (6x \quad y)(x \quad 2y)$$

Note that there is a common factor 2 in $(2x \quad 2y)$ and in $(6x \quad 2y)$. Since there is no common factor of 2 in the original trinomial, the second and third possibilities will not work. Because the last term of the trinomial is positive and the middle term is negative, both factors must contain subtraction symbols. Only the first possibility will give a middle term of $-7xy$ when subtraction symbols are used in both factors. So,

$$6x^2 - 7xy + 2y^2 = (3x - 2y)(2x - y).$$

> Now do Exercises 65–74

⟨3⟩ Factoring Completely

You can use the latest factoring technique along with the techniques that you learned earlier to factor polynomials completely. Remember always to first factor out the greatest common factor (if it is not 1).

EXAMPLE 5

Factoring completely
Factor each polynomial completely.

    a) $4x^3 + 14x^2 + 6x$

    b) $12x^2y + 6xy + 6y$

Solution
    a) $4x^3 + 14x^2 + 6x = 2x(2x^2 + 7x + 3)$ Factor out the GCF, $2x$.

$$= 2x(2x + 1)(x + 3) \quad \text{Factor } 2x^2 + 7x + 3.$$

    Check by multiplying.

    b) $12x^2y + 6xy + 6y = 6y(2x^2 + x + 1)$ Factor out the GCF, $6y$.

    To factor $2x^2 + x + 1$ by the ac method, we need two numbers with a product of 2 and a sum of 1. Because there are no such numbers, $2x^2 + x + 1$ is prime and the factorization is complete.

> Now do Exercises 75–84

Our first step in factoring is to factor out the greatest common factor (if it is not 1). If the first term of a polynomial has a negative coefficient, then it is better to factor out the opposite of the GCF so that the resulting polynomial will have a positive leading coefficient.

E X A M P L E   **6**

Factoring out the opposite of the GCF

Factor each polynomial completely.

a) $-18x^3 + 51x^2 - 15x$

b) $-3a^2 + 2a + 21$

Solution

a) The GCF is $3x$. Because the first term has a negative coefficient, we factor out $-3x$:

$$-18x^3 + 51x^2 - 15x = -3x(6x^2 - 17x + 5) \quad \text{Factor out } -3x.$$
$$= -3x(3x - 1)(2x - 5) \quad \text{Factor } 6x^2 - 17x + 5.$$

b) The GCF for $-3a^2 + 2a + 21$ is 1. Because the first term has a negative coefficient, factor out -1:

$$-3a^2 + 2a + 21 = -1(3a^2 - 2a - 21) \quad \text{Factor out } -1.$$
$$= -1(3a + 7)(a - 3) \quad \text{Factor } 3a^2 - 2a - 21.$$

> Now do Exercises 85–100

Warm-Ups ▼

Fill in the blank.

1. If there are no two integers that have a _____ of ac and a ____ of b, then $ax^2 + bx + c$ is prime.

2. In the _____ method we make educated guesses at the factors and then check by FOIL.

True or false?

3. $2x^2 + 3x + 1 = (2x + 1)(x + 1)$

4. $2x^2 + 5x + 3 = (2x + 1)(x + 3)$

5. $3x^2 + 10x + 3 = (3x + 1)(x + 3)$

6. $2x^2 - 7x - 9 = (2x - 9)(x + 1)$

7. $2x^2 - 16x - 9 = (2x - 9)(2x + 1)$

8. $12x^2 - 13x + 3 = (3x - 1)(4x - 3)$

5.4 Exercises

‹ 1 › **The *ac* Method**

Find the following. See Example 1.

1. Two integers that have a product of 12 and a sum of 7

2. Two integers that have a product of 20 and a sum of 12

3. Two integers that have a product of 30 and a sum of -17

4. Two integers that have a product of 36 and a sum of -20

5. Two integers that have a product of -12 and a sum of -4

6. Two integers that have a product of -8 and a sum of 7

Each of the following trinomials is in the form $ax^2 + bx + c$. For each trinomial, find two integers that have a product of ac and a sum of b. Do not factor the trinomials. See Example 1.

7. $6x^2 + 7x + 2$

8. $5x^2 + 17x + 6$

9. $6y^2 - 11y + 3$

10. $6z^2 - 19z + 10$

11. $12w^2 + w - 1$

12. $15t^2 - 17t - 4$

Factor each trinomial using the ac method. See Example 1. See the Strategy for Factoring $ax^2 + bx + c$ by the ac Method box on page 347.

13. $2x^2 + 3x + 1$

14. $2x^2 + 11x + 5$

15. $2x^2 + 9x + 4$

16. $2h^2 + 7h + 3$

17. $3t^2 + 7t + 2$

18. $3t^2 + 8t + 5$

19. $2x^2 + 5x - 3$

20. $3x^2 - x - 2$

21. $6x^2 + 7x - 3$

22. $21x^2 + 2x - 3$

23. $3x^2 - 5x + 4$

24. $6x^2 - 5x + 3$

25. $2x^2 - 7x + 6$

26. $3a^2 - 14a + 15$

27. $5b^2 - 13b + 6$

28. $7y^2 + 16y - 15$

29. $4y^2 - 11y - 3$

30. $35x^2 - 2x - 1$

31. $3x^2 + 2x + 1$

32. $6x^2 - 4x - 5$

33. $8x^2 - 2x - 1$

34. $8x^2 - 10x - 3$

35. $9t^2 - 9t + 2$

36. $9t^2 + 5t - 4$

37. $15x^2 + 13x + 2$

38. $15x^2 - 7x - 2$

Use the ac method to factor each trinomial. See Example 2.

39. $4a^2 + 16ab + 15b^2$

40. $10x^2 + 17xy + 3y^2$

41. $6m^2 - 7mn - 5n^2$

42. $3a^2 + 2ab - 21b^2$

43. $3x^2 - 8xy + 5y^2$

44. $3m^2 - 13mn + 12n^2$

⟨2⟩ Trial and Error

Factor each trinomial using trial and error. See Examples 3 and 4.

45. $5a^2 + 6a + 1$

46. $7b^2 + 8b + 1$

47. $6x^2 + 5x + 1$

48. $15y^2 + 8y + 1$

49. $5a^2 + 11a + 2$

50. $3y^2 + 10y + 7$

51. $4w^2 + 8w + 3$

52. $6z^2 + 13z + 5$

53. $15x^2 - x - 2$

54. $15x^2 + 13x - 2$

55. $8x^2 - 6x + 1$

56. $8x^2 - 22x + 5$

57. $15x^2 - 31x + 2$

58. $15x^2 + 31x + 2$

59. $4x^2 - 4x + 3$

60. $4x^2 + 12x - 5$

61. $2x^2 + 18x - 90$

62. $3x^2 + 11x + 10$

63. $3x^2 + x - 10$

64. $3x^2 - 17x + 10$

65. $10x^2 - 3xy - y^2$

66. $8x^2 - 2xy - y^2$

67. $42a^2 - 13ab + b^2$

68. $10a^2 - 27ab + 5b^2$

Complete the factoring.

69. $3x^2 + 7x + 2 = (x + 2)(\qquad)$

70. $2x^2 - x - 15 = (x - 3)(\qquad)$

71. $5x^2 + 11x + 2 = (5x + 1)(\qquad)$

72. $4x^2 - 19x - 5 = (4x + 1)(\qquad)$

73. $6a^2 - 17a + 5 = (3a - 1)(\qquad)$

74. $4b^2 - 16b + 15 = (2b - 5)(\qquad)$

⟨3⟩ Factoring Completely

Factor each polynomial completely. See Examples 5 and 6.

75. $81w^3 - w$

76. $81w^3 - w^2$

77. $4w^2 + 2w - 30$

78. $2x^2 - 28x + 98$

79. $27 + 12x^2 + 36x$

80. $24y + 12y^2 + 12$

81. $6w^2 - 11w - 35$

82. $8y^2 - 14y - 15$

83. $3x^2z - 3zx - 18z$

84. $a^2b + 2ab - 15b$

85. $9x^3 - 21x^2 + 18x$

86. $-8x^3 + 4x^2 - 2x$

87. $a^2 + 2ab - 15b^2$

88. $a^2b^2 - 2a^2b - 15a^2$

89. $2x^2y^2 + xy^2 + 3y^2$

90. $18x^2 - 6x + 6$

91. $-6t^3 - t^2 + 2t$

92. $-36t^2 - 6t + 12$

93. $12t^4 - 2t^3 - 4t^2$

94. $12t^3 + 14t^2 + 4t$

95. $4x^2y - 8xy^2 + 3y^3$

96. $9x^2 + 24xy - 9y^2$

97. $-4w^2 + 7w - 3$

98. $-30w^2 + w + 1$

99. $-12a^3 + 22a^2b - 6ab^2$

100. $-36a^2b + 21ab^2 - 3b^3$

Applications

Solve each problem.

101. *Height of a ball.* If a ball is thrown straight upward at 40 feet per second from a rooftop 24 feet above the ground, then its height in feet above the ground t seconds after it is thrown is given by

$$h(t) = -16t^2 + 40t + 24.$$

a) Find $h(0)$, $h(1)$, $h(2)$, and $h(3)$.

b) Rewrite the formula with the polynomial factored completely.

c) Find $h(3)$ using the result of part (b).

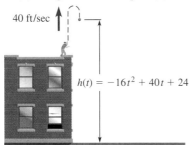

40 ft/sec

$h(t) = -16t^2 + 40t + 24$

Figure for Exercise 101

102. *Worker efficiency.* In a study of worker efficiency at Wong Laboratories it was found that the number of components assembled per hour by the average worker t hours after starting work could be modeled by the formula

$$N(t) = -3t^3 + 23t^2 + 8t.$$

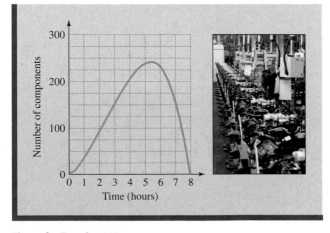

Figure for Exercise 102

a) Rewrite the formula by factoring the right-hand side completely.

b) Use the factored version of the formula to find $N(3)$.

c) Use the accompanying graph to estimate the time at which the workers are most efficient.

d) Use the accompanying graph to estimate the maximum number of components assembled per hour during an 8-hour shift.

Getting More Involved

103. *Exploration*

Find all positive and negative integers b for which each polynomial can be factored.

a) $x^2 + bx + 3$ **b)** $3x^2 + bx + 5$

c) $2x^2 + bx - 15$

104. *Exploration*

Find two integers c (positive or negative) for which each polynomial can be factored. Many answers are possible.

a) $x^2 + x + c$

b) $x^2 - 2x + c$

c) $2x^2 - 3x + c$

105. *Cooperative learning*

Working in groups, cut two large squares, three rectangles, and one small square out of paper that are exactly the same size as shown in the accompanying figure. Then try to place the six figures next to one another so that they form a large rectangle. Do not overlap the pieces or leave any gaps. Explain how factoring $2x^2 + 3x + 1$ can help you solve this puzzle.

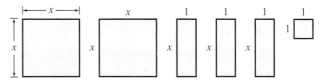

Figure for Exercise 105

106. *Cooperative learning*

Working in groups, cut four squares and eight rectangles out of paper as in Exercise 105 to illustrate the trinomial $4x^2 + 7x + 3$. Select one group to demonstrate how to arrange the 12 pieces to form a large rectangle. Have another group explain how factoring the trinomial can help you solve this puzzle.

5.5   Difference and Sum of Cubes and a Strategy

In Sections 5.1 to 5.4, we established the general idea of factoring and some special cases. In this section we will see two more special cases. We will then summarize all of the factoring that we have done with a factoring strategy.

⟨1⟩ Factoring a Difference or Sum of Two Cubes

We can use division to discover that $a - b$ is a factor of $a^3 - b^3$ (a difference of two cubes) and $a + b$ is a factor of $a^3 + b^3$ (a sum of two cubes):

$$
\begin{array}{r}
a^2 + ab + b^2 \\
a - b\,\overline{)a^3 + 0a^2b + 0ab^2 - b^3} \\
\underline{a^3 - \;\;a^2b} \\
a^2b + 0ab^2 \\
\underline{a^2b - \;\;ab^2} \\
ab^2 - b^3 \\
\underline{ab^2 - b^3} \\
0
\end{array}
\qquad
\begin{array}{r}
a^2 - ab + b^2 \\
a + b\,\overline{)a^3 + 0a^2b + 0ab^2 + b^3} \\
\underline{a^3 + \;\;a^2b} \\
-a^2b + 0ab^2 \\
\underline{-a^2b - \;\;ab^2} \\
ab^2 + b^3 \\
\underline{ab^2 + b^3} \\
0
\end{array}
$$

In each division the remainder is 0. So in each case the dividend is equal to the divisor times the quotient. These results give us two new factoring rules.

Factoring a Difference or Sum of Two Cubes

$$a^3 - b^3 = (a - b)(a^2 + ab + b^2)$$
$$a^3 + b^3 = (a + b)(a^2 - ab + b^2)$$

Use the following strategy to factor a difference or sum of two cubes.

Strategy for Factoring $a^3 - b^3$ or $a^3 + b^3$

1. The first factor is the original polynomial without the exponents, and the middle term in the second factor has the opposite sign from the first factor:

$$a^3 - b^3 = (a - b)(a^2 + ab + b^2) \qquad a^3 + b^3 = (a + b)(a^2 - ab + b^2)$$
$$\quad\;\uparrow\quad\;\uparrow \qquad\qquad\qquad\qquad\qquad \uparrow\quad\;\uparrow$$
$$\text{opposite signs} \qquad\qquad\qquad\qquad \text{opposite signs}$$

2. Recall the two perfect square trinomials $a^2 + 2ab + b^2$ and $a^2 - 2ab + b^2$. The second factor is *almost* a perfect square trinomial. Just delete the 2.

It is helpful also to compare the differences and sums of squares and cubes:

$$a^2 - b^2 = (a - b)(a + b) \qquad\qquad a^2 + b^2 \;\; \text{Prime}$$
$$a^3 - b^3 = (a - b)(a^2 + ab + b^2) \qquad a^3 + b^3 = (a + b)(a^2 - ab + b^2)$$

The factors $a^2 + ab + b^2$ and $a^2 - ab + b^2$ are prime. They can't be factored. The perfect square trinomials $a^2 + 2ab + b^2$ and $a^2 - 2ab + b^2$, which are almost the same, are not prime. They can be factored:

$$a^2 + 2ab + b^2 = (a + b)^2 \qquad \text{and} \qquad a^2 - 2ab + b^2 = (a - b)^2.$$

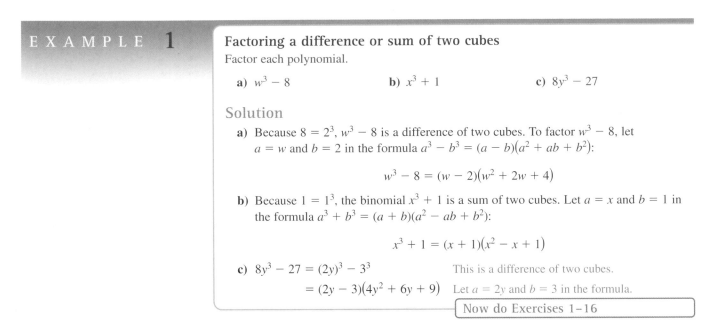

E X A M P L E   **1**

Factoring a difference or sum of two cubes

Factor each polynomial.

  a) $w^3 - 8$ **b)** $x^3 + 1$ **c)** $8y^3 - 27$

Solution

  a) Because $8 = 2^3$, $w^3 - 8$ is a difference of two cubes. To factor $w^3 - 8$, let
$a = w$ and $b = 2$ in the formula $a^3 - b^3 = (a - b)(a^2 + ab + b^2)$:

$$w^3 - 8 = (w - 2)(w^2 + 2w + 4)$$

  b) Because $1 = 1^3$, the binomial $x^3 + 1$ is a sum of two cubes. Let $a = x$ and $b = 1$ in
the formula $a^3 + b^3 = (a + b)(a^2 - ab + b^2)$:

$$x^3 + 1 = (x + 1)(x^2 - x + 1)$$

  c) $8y^3 - 27 = (2y)^3 - 3^3$ This is a difference of two cubes.

$\qquad\qquad = (2y - 3)(4y^2 + 6y + 9)$ Let $a = 2y$ and $b = 3$ in the formula.

> Now do Exercises 1–16

In Example 1, we used the first three perfect cubes, 1, 8, and 27. You should verify
that 1, 8, 27, 64, 125, 216, 343, 512, 729, and 1000 are the first 10 perfect cubes.

CAUTION   The polynomial $(a - b)^3$ is not equivalent to $a^3 - b^3$ because if $a = 2$
and $b = 1$, then

$$(a - b)^3 = (2 - 1)^3 = 1^3 = 1$$

and

$$a^3 - b^3 = 2^3 - 1^3 = 8 - 1 = 7.$$

Likewise, $(a + b)^3$ is not equivalent to $a^3 + b^3$.

⟨2⟩ Factoring a Difference of Two Fourth Powers

A difference of two fourth powers of the form $a^4 - b^4$ is also a difference of two
squares, $(a^2)^2 - (b^2)^2$. It can be factored by the rule for factoring a difference of two
squares:

$\quad a^4 - b^4 = (a^2)^2 - (b^2)^2$ Write as a difference of two squares.

$\qquad\qquad = (a^2 - b^2)(a^2 + b^2)$ Difference of two squares

$\qquad\qquad = (a - b)(a + b)(a^2 + b^2)$ Factor completely.

Note that the sum of two squares $a^2 + b^2$ is prime and cannot be factored.

E X A M P L E   **2**

Factoring a difference of two fourth powers

Factor each polynomial completely.

  a) $x^4 - 16$ **b)** $81m^4 - n^4$

Solution

a) $x^4 - 16 = (x^2)^2 - 4^2$ Write as a difference of two squares.

$= (x^2 - 4)(x^2 + 4)$ Difference of two squares

$= (x - 2)(x + 2)(x^2 + 4)$ Factor completely.

b) $81m^4 - n^4 = (9m^2)^2 - (n^2)^2$ Write as a difference of two squares.

$= (9m^2 - n^2)(9m^2 + n^2)$ Factor.

$= (3m - n)(3m + n)(9m^2 + n^2)$ Factor completely.

Now do Exercises 17–24

CAUTION A difference of two squares or cubes can be factored, and a sum of two cubes can be factored. But the sums of two squares $x^2 + 4$ and $9m^2 + n^2$ in Example 2 are prime.

⟨3⟩ The Factoring Strategy

The following is a summary of the ideas that we use to factor a polynomial completely.

> ## Strategy for Factoring Polynomials Completely
>
> 1. Factor out the GCF (with a negative coefficient if necessary).
> 2. When factoring a binomial, check to see whether it is a difference of two squares, a difference of two cubes, or a sum of two cubes. *A sum of two squares does not factor.*
> 3. When factoring a trinomial, check to see whether it is a perfect square trinomial.
> 4. If the polynomial has four terms, try factoring by grouping.
> 5. When factoring a trinomial that is not a perfect square, use the *ac* method or the trial-and-error method.
> 6. Check to see whether any of the factors can be factored again.

We will use the factoring strategy in Example 3.

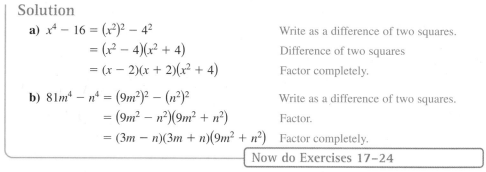

EXAMPLE **3**

Factoring polynomials

Factor each polynomial completely.

a) $2a^2b - 24ab + 72b$ **b)** $3x^3 + 6x^2 - 75x - 150$

c) $-3x^4 - 15x^3 + 72x^2$ **d)** $60y^3 - 85y^2 - 25y$

Solution

a) $2a^2b - 24ab + 72b = 2b(a^2 - 12a + 36)$ First factor out the GCF, $2b$.

$= 2b(a - 6)^2$ Factor the perfect square trinomial.

b) $3x^3 + 6x^2 - 75x - 150 = 3[x^3 + 2x^2 - 25x - 50]$ Factor out the GCF, 3.

$= 3[x^2(x + 2) - 25(x + 2)]$ Factor out common factors.

$= 3(x^2 - 25)(x + 2)$ Factor by grouping.

$= 3(x + 5)(x - 5)(x + 2)$ Factor the difference of two squares.

c) Factor out $-3x^2$ to get $-3x^4 - 15x^3 + 72x^2 = -3x^2(x^2 + 5x - 24)$. To factor the trinomial, find two numbers with a product of -24 and a sum of 5. For a product of 24 we have $1 \cdot 24$, $2 \cdot 12$, $3 \cdot 8$, and $4 \cdot 6$. To get a sum of 5 and a product of -24 choose 8 and -3:

$$-3x^4 - 15x^3 + 72x^2 = -3x^2(x^2 + 5x - 24)$$
$$= -3x^2(x - 3)(x + 8)$$

d) Factor out $5y$ to get $60y^3 - 85y^2 - 25y = 5y(12y^2 - 17y - 5)$. By the *ac* method we need two numbers that have a product of -60 (*ac*) and a sum of -17. The numbers are -20 and 3. Now factor by grouping:

$$\begin{aligned} 60y^3 - 85y^2 - 25y &= 5y(12y^2 - 17y - 5) && \text{Factor out } 5y. \\ &= 5y(12y^2 - 20y + 3y - 5) && -17y = -20y + 3y \\ &= 5y[4y(3y - 5) + 1(3y - 5)] && \text{Factor by grouping.} \\ &= 5y(3y - 5)(4y + 1) && \text{Factor out } 3y - 5. \end{aligned}$$

> Now do Exercises 25–92

Warm-Ups ▼

Fill in the blank.

1. If there is no _____ , then the dividend is the divisor times the quotient.
2. The binomial $a^3 + b^3$ is a ____ of two cubes.
3. The binomial $a^3 - b^3$ is a _____ of two cubes.
4. If $a^3 - b^3$ is divided by $a - b$, then the remainder is ____ .

True or false?

5. For any real number x, $x^2 - 4 = (x - 2)^2$.
6. The trinomial $4x^2 + 6x + 9$ is a perfect square trinomial.
7. The binomial $4y^2 + 25$ is prime.
8. If the GCF is not 1, then you should factor it out first.
9. You can factor $y^2 - 5y - my + 5m$ by grouping.
10. You can factor $x^2 + ax - 3x + 3a$ by grouping.

5.5 Exercises

‹ **Study Tips** ›

- If you have a choice, sit at the front of the class. It is easier to stay alert when you are at the front.
- If you miss what is going on in class, you miss what your instructor feels is important and most likely to appear on tests and quizzes.

‹ 1 › **Factoring a Difference or Sum of Two Cubes**

Factor each difference or sum of cubes. See Example 1.

1. $m^3 - 1$
2. $z^3 - 27$
3. $x^3 + 8$
4. $y^3 + 27$
5. $a^3 + 125$
6. $b^3 - 216$

VIDEO

7. $c^3 - 343$

8. $d^3 + 1000$

9. $8w^3 + 1$

10. $125m^3 + 1$

11. $8t^3 - 27$

VIDEO **12.** $125n^3 - 8$

13. $x^3 - y^3$

14. $m^3 + n^3$

15. $8t^3 + y^3$

16. $u^3 - 125v^3$

⟨2⟩ Factoring a Difference of Two Fourth Powers

Factor each polynomial completely. See Example 2.

17. $x^4 - y^4$

18. $m^4 - n^4$

19. $x^4 - 1$

20. $a^4 - 81$

VIDEO **21.** $16b^4 - 1$

22. $625b^4 - 1$

23. $a^4 - 81b^4$

24. $16a^4 - m^4$

⟨3⟩ The Factoring Strategy

Factor each polynomial completely. If a polynomial is prime, say so. See Example 3. See the Strategy for Factoring Polynomials Completely box on page 357.

25. $2x^2 - 18$

26. $3x^3 - 12x$

27. $a^2 + 4$

28. $x^2 + y^2$

29. $4x^2 + 8x - 60$

30. $3x^2 + 18x + 27$

31. $x^3 + 4x^2 + 4x$

32. $a^3 - 5a^2 + 6a$

33. $5max^2 + 20ma$

34. $3bmw^2 - 12bm$

35. $2x^2 - 3x - 1$

36. $3x^2 - 8x - 5$

37. $9x^2 + 6x + 1$

38. $9x^2 + 6x + 3$

39. $9m^2 + 1$

40. $4b^2 + 25$

41. $w^4 - z^4$

42. $y^4 - 1$

43. $6x^2y + xy - 2y$

44. $5x^2y^2 - xy^2 - 6y^2$

45. $y^2 + 10y - 25$

46. $x^2 - 20x + 25$

47. $48a^2 - 24a + 3$

48. $8b^2 + 24b + 18$

49. $16m^2 - 4m - 2$

50. $32a^2 + 4a - 6$

51. $s^4 - 16t^4$

52. $81 - q^4$

53. $9a^2 + 24a + 16$

54. $3x^2 - 18x - 48$

VIDEO **55.** $24x^2 - 26x + 6$

56. $4x^2 - 6x - 12$

57. $3m^2 + 27$

58. $5a^2 + 20b^2$

59. $3a^2 - 27a$

60. $a^2 - 25a$

61. $8 - 2x^2$

62. $x^3 + 6x^2 + 9x$

63. $w^2 + 4t^2$

64. $9x^2 + 4y^2$

65. $6x^3 - 5x^2 + 12x$

66. $x^3 + 2x^2 - x - 2$

67. $a^3b - 4ab$

68. $2m^2 - 1800$

69. $x^3 + 2x^2 - 4x - 8$

70. $-2x^3 - 50x$

71. $-7m^3n - 28mn^3$

72. $x^3 - x^2 - x + 1$

73. $2x^3 + 16$

VIDEO **74.** $m^2a + 2ma^2 + a^3$

75. $2w^4 - 16w$

76. $m^4n + mn^4$

77. $3a^2w - 18aw + 27w$

78. $8a^3 + 4a$

79. $5x^2 - 500$

80. $25x^2 - 16y^2$

81. $2m + 2n - wm - wn$

82. $aw - 5b - bw + 5a$

83. $3x^4 + 3x$

84. $3a^5 - 81a^2$

85. $4w^2 + 4w - 4$

86. $4w^2 + 8w - 5$

87. $a^4 + 7a^3 - 30a^2$

88. $2y^5 + 3y^4 - 20y^3$

89. $4aw^3 - 12aw^2 + 9aw$

90. $9bn^3 + 15bn^2 - 14bn$

91. $t^2 + 6t + 9$

92. $t^3 + 12t^2 + 36t$

Getting More Involved

93. *Discussion*

Are there any values for a and b for which $(a + b)^3 = a^3 + b^3$? Find a pair of values for a and b for which $(a + b)^3 \neq a^3 + b^3$. Is $(a + b)^3$ equivalent to $a^3 + b^3$? Explain your answers.

94. *Writing*

Explain why $a^2 + ab + b^2$ and $a^2 - ab + b^2$ are prime polynomials.

95. *Discussion*

The polynomial $a^6 + 1$ is a sum of two squares and a sum of two cubes. You can't factor it as a sum of two squares, but you can factor any sum of two cubes. Factor $a^6 + 1$.

96. *Discussion*

Factor $a^6 + b^6$ and $a^6 - b^6$ completely.

Extra Factoring Exercises

Factor each polynomial completely.

97. $3w^2 + 30w + 75$

98. $4z^2 + 16z + 16$

99. $81 - b^2$

100. $9 - 4p^2$

101. $w^2 - 8w$

102. $6z^2 + 12z$

103. $3x^2 + 6x - 105$

104. $6m^2 - 36m - 96$

105. $ax - 5a + 4x - 20$

106. $w^2 + 3w - 3c - cw$

107. $12x^2 - 7x - 12$

108. $8x^2 - 6x - 27$

109. $-9x^2 - 15x + 6$

110. $-8x^2 - 4x + 40$

111. $w^3 - 27$

112. $y^3 + 1$

113. $y^3 + y^2 + y + 1$

114. $a^3 + 2a^2 + 4a + 8$

115. $m^4 - 81$

116. $t^4 - 256$

117. $a^2 - 2ab - 8b^2$

118. $x^2 - xy - 12y^2$

119. $m^3y + 6m^2y^2 + 9my^3$

120. $w^4a - 10w^3a^2 + 25w^2a^3$

121. $x^4 + 2x^3 + 4x^2$

122. $y^5 - 6y^4 - 9y^3$

123. $y^7 - y^3$

124. $a^6 - 16a^2$

125. $x^2 - 18x + 72$

126. $m^2 - 17m + 72$

127. $-6a^3 + 5a^2 + 4a$

128. $-12x^2 + 15x + 18$

129. $x^4 - 8x$

130. $a^4 + ab^3$

131. $16t^2 - 24tx + 9x^2$

132. $9y^2 + 30yz + 25z^2$

5.6 **Solving Quadratic Equations by Factoring**

The techniques of factoring can be used to solve equations involving polynomials. These equations cannot be solved by the other methods that you have learned. After you learn to solve equations by factoring, you will use this technique to solve some new types of problems.

⟨1⟩ **The Zero Factor Property**

In this chapter you learned to factor polynomials such as $x^2 + x - 6$. The equation $x^2 + x - 6 = 0$ is called a *quadratic equation*.

> **Quadratic Equation**
>
> If a, b, and c are real numbers with $a \neq 0$, then
> $$ax^2 + bx + c = 0$$
> is called a **quadratic equation.**

A quadratic equation always has a second-degree term because it is specified in the definition that a is not zero. The main idea used to solve quadratic equations, the **zero factor property,** is simply a fact about multiplication by zero.

> **The Zero Factor Property**
>
> The equation $a \cdot b = 0$ is equivalent to
> $$a = 0 \qquad \text{or} \qquad b = 0.$$

We will use the zero factor property most often to solve quadratic equations that have two factors, as shown in Example 1. However, this property holds for more than two factors as well. If a product of any number of factors is zero, then at least one of the factors is zero.

The following strategy gives the steps to follow when solving a quadratic equation by factoring. Of course, this method applies only to quadratic equations in which the quadratic polynomial can be factored. Methods that can be used for solving all quadratic equations are presented in Chapter 10.

> **Strategy for Solving an Equation by Factoring**
>
> **1.** Rewrite the equation with 0 on one side.
> **2.** Factor the other side completely.
> **3.** Use the zero factor property to get simple linear equations.
> **4.** Solve the linear equations.
> **5.** Check the answer in the original equation.
> **6.** State the solution(s) to the original equation.

EXAMPLE   **1**

Using the zero factor property

Solve $x^2 + x - 6 = 0$.

Solution

First factor the polynomial on the left-hand side:

$$x^2 + x - 6 = 0$$
$$(x + 3)(x - 2) = 0 \quad \text{Factor the left-hand side.}$$
$$x + 3 = 0 \quad \text{or} \quad x - 2 = 0 \quad \text{Zero factor property}$$
$$x = -3 \quad \text{or} \quad x = 2 \quad \text{Solve each equation.}$$

We now check that -3 and 2 satisfy the original equation.

For $x = -3$:

$$x^2 + x - 6 = (-3)^2 + (-3) - 6$$
$$= 9 - 3 - 6$$
$$= 0$$

For $x = 2$:

$$x^2 + x - 6 = (2)^2 + (2) - 6$$
$$= 4 + 2 - 6$$
$$= 0$$

The solutions to $x^2 + x - 6 = 0$ are -3 and 2. Checking -3 and 2 in the factored form of the equation $(x + 3)(x - 2) = 0$ will help you understand the zero factor property:

$$(-3 + 3)(-3 - 2) = (0)(-5) = 0$$
$$(2 + 3)(2 - 2) = (5)(0) = 0$$

For each solution to the equation, one of the factors is zero and the other is not zero. All it takes to get a product of zero is one of the factors being zero.

Now do Exercises 1–12

A sentence such as $x = -3$ or $x = 2$, which is made up of two or more equations connected with the word "or," is called a **compound equation.** In Example 2, we again solve a quadratic equation by using the zero factor property to write a compound equation.

EXAMPLE   **2**

Using the zero factor property

Solve the equation $3x^2 = -3x$.

Solution

First rewrite the equation with 0 on the right-hand side:

$$3x^2 = -3x$$
$$3x^2 + 3x = 0 \quad \text{Add } 3x \text{ to each side.}$$
$$3x(x + 1) = 0 \quad \text{Factor the left-hand side.}$$
$$3x = 0 \quad \text{or} \quad x + 1 = 0 \quad \text{Zero factor property}$$
$$x = 0 \quad \text{or} \quad x = -1 \quad \text{Solve each equation.}$$

Check 0 and -1 in the original equation $3x^2 = -3x$.

For $x = 0$:

$$3(0)^2 = -3(0)$$
$$0 = 0$$

For $x = -1$:

$$3(-1)^2 = -3(-1)$$
$$3 = 3$$

There are two solutions to the original equation, 0 and -1.

Now do Exercises 13–20

CAUTION   If in Example 2 you divide each side of $3x^2 = -3x$ by $3x$, you would get $x = -1$ but not the solution $x = 0$. For this reason we usually do not divide each side of an equation by a variable.

E X A M P L E   **3**

Using the zero factor property

Solve $(2x + 1)(x - 1) = 14$.

Solution

To write the equation with 0 on the right-hand side, multiply the binomials on the left and then subtract 14 from each side:

$$\begin{aligned}
(2x + 1)(x - 1) &= 14 && \text{Original equation}\\
2x^2 - x - 1 &= 14 && \text{Multiply the binomials.}\\
2x^2 - x - 15 &= 0 && \text{Subtract 14 from each side.}\\
(2x + 5)(x - 3) &= 0 && \text{Factor.}\\
2x + 5 = 0 \quad\text{or}\quad x - 3 &= 0 && \text{Zero factor property}\\
2x = -5 \quad\text{or}\quad x &= 3\\
x = -\frac{5}{2} \quad\text{or}\quad x &= 3
\end{aligned}$$

Check $-\frac{5}{2}$ and 3 in the original equation:

$$\left(2 \cdot -\frac{5}{2} + 1\right)\left(-\frac{5}{2} - 1\right) = (-5 + 1)\left(-\frac{5}{2} - \frac{2}{2}\right)$$

$$= (-4)\left(-\frac{7}{2}\right)$$

$$= 14$$

$$(2 \cdot 3 + 1)(3 - 1) = (7)(2)$$

$$= 14$$

So the solutions are $-\frac{5}{2}$ and 3.

Now do Exercises 21–26

CAUTION   In Example 3, we started with a product equal to 14. Because $1 \cdot 14 = 14$, $2 \cdot 7 = 14$, $\frac{1}{2} \cdot 28 = 14$, $\frac{1}{3} \cdot 42 = 14$, and so on, we cannot make any conclusion about the factors that have a product of 14. If the product of two factors is zero, then we can conclude that one or the other factor is zero.

If a perfect square trinomial occurs in a quadratic equation with 0 on one side, then there are two identical factors of the trinomial. In this case it is not necessary to set both factors equal to zero. The solution can be found from one factor.

EXAMPLE   **4**

An equation with a repeated factor

Solve $5x^2 - 30x + 45 = 0$.

Solution

Notice that the trinomial on the left-hand side has a common factor:

$$5x^2 - 30x + 45 = 0$$
$$5(x^2 - 6x + 9) = 0 \quad \text{Factor out the GCF.}$$
$$5(x - 3)^2 = 0 \quad \text{Factor the perfect square trinomial.}$$
$$(x - 3)^2 = 0 \quad \text{Divide each side by 5.}$$
$$x - 3 = 0 \quad \text{Zero factor property}$$
$$x = 3$$

Even though $x - 3$ occurs twice as a factor, it is not necessary to write $x - 3 = 0$ or $x - 3 = 0$. If $x = 3$ in $5x^2 - 30x + 45 = 0$, we get

$$5 \cdot 3^2 - 30 \cdot 3 + 45 = 0,$$

which is correct. So the only solution to the equation is 3.

Now do Exercises 27–30

CAUTION   Do not include 5 in the solution to Example 4. Dividing by 5 eliminates it. Instead of dividing by 5 we could have applied the zero factor property to $5(x - 3)^2 = 0$. Since 5 is not 0, we must have $(x - 3)^2 = 0$ or $x - 3 = 0$.

If the left-hand side of the equation has more than two factors, we can write an equivalent equation by setting each factor equal to zero.

EXAMPLE   **5**

An equation with three solutions

Solve $2x^3 - x^2 - 8x + 4 = 0$.

Solution

We can factor the four-term polynomial by grouping:

$$2x^3 - x^2 - 8x + 4 = 0$$
$$x^2(2x - 1) - 4(2x - 1) = 0 \quad \text{Factor out the common factors.}$$
$$(x^2 - 4)(2x - 1) = 0 \quad \text{Factor out } 2x - 1.$$
$$(x - 2)(x + 2)(2x - 1) = 0 \quad \text{Difference of two squares}$$
$$x - 2 = 0 \quad \text{or} \quad x + 2 = 0 \quad \text{or} \quad 2x - 1 = 0 \quad \text{Zero factor property}$$
$$x = 2 \quad \text{or} \quad x = -2 \quad \text{or} \quad x = \frac{1}{2} \quad \text{Solve each equation.}$$

To check let $x = -2, \frac{1}{2}$, and 2 in $2x^3 - x^2 - 8x + 4 = 0$:

$$2(-2)^3 - (-2)^2 - 8(-2) + 4 = 0$$
$$2\left(\frac{1}{2}\right)^3 - \left(\frac{1}{2}\right)^2 - 8\left(\frac{1}{2}\right) + 4 = 0$$
$$2(2)^3 - 2^2 - 8(2) + 4 = 0$$

Since all of these equations are correct, the solutions are $-2, \frac{1}{2}$, and 2.

Now do Exercises 31–38

‹ **Helpful Hint** ›

Compare the number of solutions in Examples 1 through 5 to the degree of the polynomial. The number of real solutions to any polynomial equation is less than or equal to the degree of the polynomial. This fact is known as the fundamental theorem of algebra.

⟨2⟩ **Fractions and Decimals**

If the coefficients in an equation are not integers, we might be able to convert them into integers. Fractions can be eliminated by multiplying each side of the equation by the least common denominator (LCD). To eliminate decimals multiply each side by the smallest power of 10 that will eliminate all of the decimals.

EXAMPLE 6

Converting to Integers

Solve.

a) $\dfrac{1}{12}x^2 + \dfrac{1}{6}x - 2 = 0$ b) $0.02x^2 - 0.19x = 0.1$

Solution

a) The LCD for 6 and 12 is 12. So multiply each side of the equation by 12:

$$\dfrac{1}{12}x^2 + \dfrac{1}{6}x - 2 = 0 \qquad \text{Original equation}$$

$$12\left(\dfrac{1}{12}x^2 + \dfrac{1}{6}x - 2\right) = 12(0) \qquad \text{Multiply each side by 12.}$$

$$x^2 + 2x - 24 = 0 \qquad \text{Simplify.}$$

$$(x + 6)(x - 4) = 0 \qquad \text{Factor.}$$

$$x + 6 = 0 \quad \text{or} \quad x - 4 = 0 \qquad \text{Zero factor property}$$

$$x = -6 \quad \text{or} \qquad x = 4$$

Check:

$$\dfrac{1}{12}(-6)^2 + \dfrac{1}{6}(-6) - 2 = 3 - 1 - 2 = 0$$

$$\dfrac{1}{12}(4)^2 + \dfrac{1}{6}(4) - 2 = \dfrac{4}{3} + \dfrac{2}{3} - 2 = 0$$

The solutions are -6 and 4.

b) Multiply each side by 100 to eliminate the decimals:

$$0.02x^2 - 0.19x = 0.1 \qquad \text{Original equation}$$

$$100(0.02x^2 - 0.19x) = 100(0.1) \qquad \text{Multiply each side by 100.}$$

$$2x^2 - 19x = 10 \qquad \text{Simplify.}$$

$$2x^2 - 19x - 10 = 0 \qquad \text{Get 0 on one side.}$$

$$(2x + 1)(x - 10) = 0 \qquad \text{Factor.}$$

$$2x + 1 = 0 \quad \text{or} \quad x - 10 = 0 \qquad \text{Zero factor property}$$

$$x = -\dfrac{1}{2} \quad \text{or} \qquad x = 10$$

The solutions are $-\dfrac{1}{2}$ and 10. You might want to use a calculator to check.

> **Now do Exercises 39–46**

CAUTION You can multiply each side of the equation in Example 6(a) by 12 to clear the fractions and get an equivalent equation, but multiplying the polynomial $\dfrac{1}{12}x^2 + \dfrac{1}{6}x - 2$ by 12 to clear the fractions is not allowed. It would result in an expression that is not equivalent to the original.

Note that all of the equations in this section can be solved by factoring. However, we can have equations involving prime polynomials. Such equations cannot be solved by factoring but can be solved by the methods in Chapter 10.

⟨3⟩ Applications

There are many problems that can be solved by equations like those we have just discussed.

E X A M P L E   7

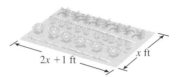

Figure 5.1

Area of a garden

Merida's garden has a rectangular shape with a length that is 1 foot longer than twice the width. If the area of the garden is 55 square feet, then what are the dimensions of the garden?

Solution

If x represents the width of the garden, then $2x + 1$ represents the length. See Fig. 5.1. Because the area of a rectangle is the length times the width, we can write the equation

$$x(2x + 1) = 55.$$

We must have zero on the right-hand side of the equation to use the zero factor property. So we rewrite the equation and then factor:

$$2x^2 + x - 55 = 0$$
$$(2x + 11)(x - 5) = 0 \quad \text{Factor.}$$
$$2x + 11 = 0 \quad \text{or} \quad x - 5 = 0 \quad \text{Zero factor property}$$
$$x = -\frac{11}{2} \quad \text{or} \quad x = 5$$

The width is certainly not $-\frac{11}{2}$. So we use $x = 5$ to get the length:

$$2x + 1 = 2(5) + 1 = 11$$

We check by multiplying 11 feet and 5 feet to get the area of 55 square feet. So the width is 5 ft, and the length is 11 ft.

Now do Exercises 65–66

⟨ **Helpful Hint** ⟩

To prove the Pythagorean theorem start with two identical squares with sides of length $a + b$, and partition them as shown.

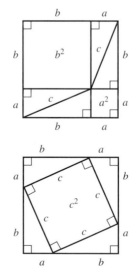

There are eight identical triangles in the diagram. Erasing four of them from each original square will leave smaller squares with areas a^2, b^2, and c^2. Since the original squares had equal areas, the remaining areas must be equal. So $a^2 + b^2 = c^2$.

The **Pythagorean theorem** was one of the earliest theorems known to ancient civilizations. It is named for the Greek mathematician and philosopher Pythagoras. Builders from ancient to modern times have used the theorem to guarantee they had right angles when laying out foundations. The Pythagorean theorem says that in any right triangle the sum of the squares of the lengths of the legs is equal to the square of the length of the hypotenuse.

The Pythagorean Theorem

The triangle shown in Fig. 5.2 is a right triangle if and only if

$$a^2 + b^2 = c^2.$$

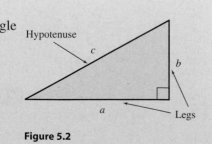

Figure 5.2

If you do an Internet search, you can find sites that have many different proofs to this theorem. One proof is shown in the Helpful Hint in the margin.

E X A M P L E   **8**

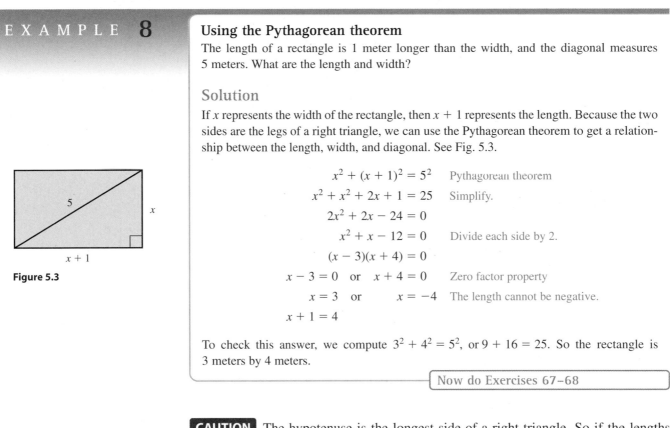

Figure 5.3

Using the Pythagorean theorem

The length of a rectangle is 1 meter longer than the width, and the diagonal measures 5 meters. What are the length and width?

Solution

If x represents the width of the rectangle, then $x + 1$ represents the length. Because the two sides are the legs of a right triangle, we can use the Pythagorean theorem to get a relationship between the length, width, and diagonal. See Fig. 5.3.

$$x^2 + (x + 1)^2 = 5^2 \qquad \text{Pythagorean theorem}$$
$$x^2 + x^2 + 2x + 1 = 25 \qquad \text{Simplify.}$$
$$2x^2 + 2x - 24 = 0$$
$$x^2 + x - 12 = 0 \qquad \text{Divide each side by 2.}$$
$$(x - 3)(x + 4) = 0$$
$$x - 3 = 0 \quad \text{or} \quad x + 4 = 0 \qquad \text{Zero factor property}$$
$$x = 3 \quad \text{or} \qquad x = -4 \qquad \text{The length cannot be negative.}$$
$$x + 1 = 4$$

To check this answer, we compute $3^2 + 4^2 = 5^2$, or $9 + 16 = 25$. So the rectangle is 3 meters by 4 meters.

Now do Exercises 67–68

CAUTION   The hypotenuse is the longest side of a right triangle. So if the lengths of the sides of a right triangle are 5 meters, 12 meters, and 13 meters, then the length of the hypotenuse is 13 meters, and $5^2 + 12^2 = 13^2$.

Warm-Ups ▼

Fill in the blank.

1. A _____ equation has the form $ax^2 + bx + c = 0$ where $a \neq 0$.
2. A _____ equation is two equations connected with the word "or."
3. The _____ property says that if $ab = 0$, then $a = 0$ or $b = 0$.
4. Some quadratic equations can be solved by _____.
5. We do not usually _____ each side of an equation by a variable.
6. The _____ theorem says that a triangle is a right triangle if and only if the sum of the squares of the legs is equal to the square of the hypotenuse.

True or false?

7. The equation $x(x + 2) = 3$ is equivalent to $x = 3$ or $x + 2 = 3$.
8. Equations solved by factoring always have two different solutions.
9. The equation $ad = 0$ is equivalent to $a = 0$ or $d = 0$.
10. The solution set to $(x - 1)(x + 4) = 0$ is $\{1, -4\}$.
11. If a, b, and c are the sides of any triangle, then $a^2 + b^2 = c^2$.
12. The solution set to $3(x - 4)(x - 5) = 0$ is $\{3, 4, 5\}$.

‹1› The Zero Factor Property

Solve by factoring. See Example 1. See the Strategy for Solving an Equation by Factoring box on page 361.

1. $(x + 5)(x + 4) = 0$

2. $(a + 6)(a + 5) = 0$

3. $(2x + 5)(3x - 4) = 0$

4. $(3k - 8)(4k + 3) = 0$

5. $x^2 + 3x + 2 = 0$

6. $x^2 + 7x + 12 = 0$

7. $w^2 - 9w + 14 = 0$

8. $t^2 + 6t - 27 = 0$

9. $y^2 - 2y - 24 = 0$

10. $q^2 + 3q - 18 = 0$

11. $2m^2 + m - 1 = 0$

12. $2h^2 - h - 3 = 0$

Solve each equation. See Examples 2 and 3.

13. $x^2 = x$

14. $w^2 = 2w$

15. $m^2 = -7m$

16. $h^2 = -5h$

17. $a^2 + a = 20$

18. $p^2 + p = 42$

19. $2x^2 + 5x = 3$

20. $3x^2 - 10x = -7$

21. $(x + 2)(x + 6) = 12$

22. $(x + 2)(x - 6) = 20$

23. $(a + 3)(2a - 1) = 15$

24. $(b - 3)(3b + 4) = 10$

25. $2(4 - 5h) = 3h^2$

26. $2w(4w + 1) = 1$

Solve each equation. See Examples 4 and 5.

27. $2x^2 + 50 = 20x$

28. $3x^2 + 48 = 24x$

29. $4m^2 - 12m + 9 = 0$

30. $25y^2 + 20y + 4 = 0$

31. $x^3 - 9x = 0$

32. $25x - x^3 = 0$

33. $w^3 + 4w^2 - 4w = 16$

34. $a^3 + 2a^2 - a = 2$

35. $n^3 - 3n^2 + 3 = n$

36. $w^3 + w^2 - 25w = 25$

37. $6y^3 - y^2 - 2y = 0$

38. $12m^3 - 13m^2 + 3m = 0$

‹2› Fractions and Decimals

Solve each equation. See Example 6.

39. $\frac{1}{6}x^2 - \frac{5}{6}x - 1 = 0$

40. $\frac{1}{10}x^2 + \frac{3}{10}x - 1 = 0$

41. $\frac{1}{9}x^2 + \frac{2}{3}x - 3 = 0$

42. $\frac{1}{10}x^2 - \frac{3}{2}x + 5 = 0$

43. $0.01x^2 + 0.08x = 0.2$

44. $0.01x^2 - 0.07x = -0.1$

45. $0.3x^2 + 0.7x - 2 = 0$

46. $0.1x^2 + 0.7x + 1 = 0$

Miscellaneous

Solve each equation.

47. $x^2 - 16 = 0$

48. $x^2 - 36 = 0$

49. $4x^2 = 9$

50. $25x^2 = 1$

51. $a^3 = a$

52. $x^3 = 4x$

53. $3x^2 + 15x + 18 = 0$

54. $-2x^2 - 2x + 24 = 0$

55. $z^2 + \dfrac{11}{2}z = -6$

56. $m^2 + \dfrac{8}{3}m = 1$

57. $(t - 3)(t + 5) = 9$

58. $3x(2x + 1) = 18$

59. $(x - 2)^2 + x^2 = 10$

60. $(x - 3)^2 + (x + 2)^2 = 17$

61. $\dfrac{1}{16}x^2 + \dfrac{1}{8}x = \dfrac{1}{2}$

 **62.** $\dfrac{1}{18}h^2 - \dfrac{1}{2}h + 1 = 0$

63. $a^3 + 3a^2 - 25a = 75$

64. $m^4 + m^3 = 100m^2 + 100m$

⟨3⟩ Applications

Solve each problem. See Examples 7 and 8.

65. *Dimensions of a rectangle.* The perimeter of a rectangle is 34 feet, and the diagonal is 13 feet long. What are the length and width of the rectangle?

66. *Address book.* The perimeter of the cover of an address book is 14 inches, and the diagonal measures 5 inches. What are the length and width of the cover?

ADDRESS
BOOK
5 in.

Figure for Exercise 66

67. *Violla's bathroom.* The length of Violla's bathroom is 2 feet longer than twice the width. If the diagonal measures 13 feet, then what are the length and width?

 68. *Rectangular stage.* One side of a rectangular stage is 2 meters longer than the other. If the diagonal is 10 meters, then what are the lengths of the sides?

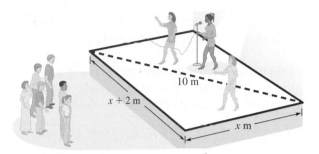

10 m

$x + 2$ m

x m

Figure for Exercise 68

69. *Consecutive integers.* The sum of the squares of two consecutive integers is 13. Find the integers.

70. *Consecutive integers.* The sum of the squares of two consecutive even integers is 52. Find the integers.

71. *Two numbers.* The sum of two numbers is 11, and their product is 30. Find the numbers.

72. *Missing ages.* Molly's age is twice Anita's. If the sum of the squares of their ages is 80, then what are their ages?

73. *Three even integers.* The sum of the squares of three consecutive even integers is 116. Find the integers.

74. *Two odd integers.* The product of two consecutive odd integers is 63. Find the integers.

75. *Consecutive integers.* The product of two consecutive integers is 5 more than their sum. Find the integers.

76. *Consecutive even integers.* If the product of two consecutive even integers is 34 larger than their sum, then what are the integers?

77. *Two integers.* Two integers differ by 5. If the sum of their squares is 53, then what are the integers?

78. *Two negative integers.* Two negative integers have a sum of -10. If the sum of their squares is 68, then what are the integers?

79. *Lucy's kids.* The sum of the squares of the ages of Lucy's two kids is 100. If the boy is two years older than the girl, then what are their ages?

80. *Sheri's kids.* The sum of the squares of the ages of Sheri's three kids is 114. If the twin girls are three years younger than the boy, then what are their ages?

81. *Area of a rectangle.* The area of a rectangle is 72 square feet. If the length is 6 feet longer than the width, then what are the length and the width?

82. *Area of a triangle.* The base of a triangle is 4 inches longer than the height. If its area is 70 square inches, then what are the base and the height?

83. *Legs of a right triangle.* The hypotenuse of a right triangle is 15 meters. If one leg is 3 meters longer than the other, then what are the lengths of the legs?

84. *Legs of a right triangle.* If the longer leg of a right triangle is 1 cm longer than the shorter leg and the hypotenuse is 5 cm, then what are the lengths of the legs?

85. *Skydiving.* If there were no air resistance, then the height (in feet) above the earth for a skydiver t seconds after jumping from an airplane at 10,000 feet would be given by

$$h(t) = -16t^2 + 10{,}000.$$

a) Find the time that it would take to fall to earth with no air resistance; that is, find t for which $h(t) = 0$. A skydiver actually gets about twice as much free fall time due to air resistance.

b) Use the accompanying graph to determine whether the skydiver (with no air resistance) falls farther in the first 5 seconds or the last 5 seconds of the fall.

c) Is the skydiver's velocity increasing or decreasing as she falls?

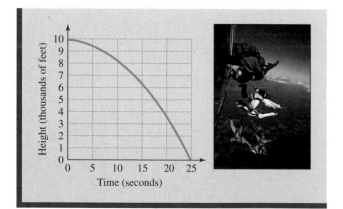

Figure for Exercise 85

86. *Skydiving.* If a skydiver jumps from an airplane at a height of 8256 feet, then for the first five seconds, her height above the earth is approximated by the formula $h(t) = -16t^2 + 8256$. How many seconds does it take her to reach 8000 feet?

87. *Throwing a sandbag.* A balloonist throws a sandbag downward at 24 feet per second from an altitude of 720 feet. Its height (in feet) above the ground after t seconds is given by $S(t) = -16t^2 - 24t + 720$.

a) Find $S(1)$.

b) What is the height of the sandbag 2 seconds after it is thrown?

c) How long does it take for the sandbag to reach the ground? [On the ground, $S(t) = 0$.]

88. *Throwing a wrench.* An angry construction worker throws his wrench downward from a height of 128 feet with an initial velocity of 32 feet per second. The height of the wrench above the ground after t seconds is given by $S(t) = -16t^2 - 32t + 128$.

a) What is the height of the wrench after 1 second?

b) How long does it take for the wrench to reach the ground?

89. *Glass prism.* One end of a glass prism is in the shape of a triangle with a height that is 1 inch longer than twice the base. If the area of the triangle is 39 square inches, then how long are the base and height?

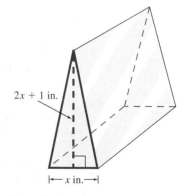

Figure for Exercise 89

90. *Areas of two circles.* The radius of a circle is 1 meter longer than the radius of another circle. If their areas differ by 5π square meters, then what is the radius of each?

91. *Changing area.* Last year Otto's garden was square. This year he plans to make it smaller by shortening one side 5 feet and the other 8 feet. If the area of the smaller garden will be 180 square feet, then what was the size of Otto's garden last year?

92. *Dimensions of a box.* Rosita's Christmas present from Carlos is in a box that has a width that is 3 inches shorter than the height. The length of the base is 5 inches longer than the height. If the area of the base is 84 square inches, then what is the height of the package?

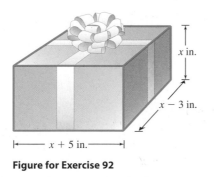

Figure for Exercise 92

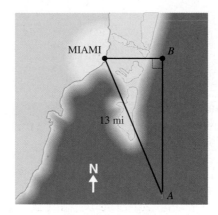

Figure for Exercise 97

93. *Flying a kite.* Imelda and Gordon have designed a new kite. While Imelda is flying the kite, Gordon is standing directly below it. The kite is designed so that its altitude is always 20 feet larger than the distance between Imelda and Gordon. What is the altitude of the kite when it is 100 feet from Imelda?

94. *Avoiding a collision.* A car is traveling on a road that is perpendicular to a railroad track. When the car is 30 meters from the crossing, the car's new collision detector warns the driver that there is a train 50 meters from the car and heading toward the same crossing. How far is the train from the crossing?

95. *Carpeting two rooms.* Virginia is buying carpet for two square rooms. One room is 3 yards wider than the other. If she needs 45 square yards of carpet, then what are the dimensions of each room?

96. *Winter wheat.* While finding the amount of seed needed to plant his three square wheat fields, Hank observed that the side of one field was 1 kilometer longer than the side of the smallest field and that the side of the largest field was 3 kilometers longer than the side of the smallest field. If the total area of the three fields is 38 square kilometers, then what is the area of each field?

97. *Sailing to Miami.* At point A the captain of a ship determined that the distance to Miami was 13 miles. If she sailed north to point B and then west to Miami, the distance would be 17 miles. If the distance from point A to point B is greater than the distance from point B to Miami, then how far is it from point A to point B?

98. *Buried treasure.* Ahmed has half of a treasure map, which indicates that the treasure is buried in the desert $2x + 6$ paces from Castle Rock. Vanessa has the other half of the map. Her half indicates that to find the treasure, one must get to Castle Rock, walk x paces to

the north, and then walk $2x + 4$ paces to the east. If they share their information, then they can find x and save a lot of digging. What is x?

99. *Broken Bamboo I.* A 10 chi high bamboo stalk is broken by the wind. The top touches the ground 3 chi from its base as shown in the accompanying figure. At what height did the stalk break? This problem appeared in a book by Chinese mathematician Yang Hui in 1261.

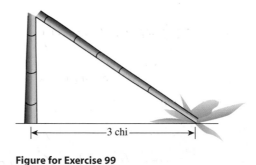

Figure for Exercise 99

100. *Broken Bamboo II.* A section of bamboo that is 5 chi in length is broken from a stalk of bamboo of unknown height. If the broken section touches the ground 3 chi from the base as in Exercise 99, then what was the original height of the bamboo stalk?

101. *Emerging markets.* Catarina's investment of $16,000 in an emerging market fund grew to $25,000 in two years. Find the average annual rate of return by solving the equation $16{,}000(1 + r)^2 = 25{,}000$.

102. *Venture capital.* Henry invested $12,000 in a new restaurant. When the restaurant was sold two years later, he received $27,000. Find his average annual return by solving the equation $12{,}000(1 + r)^2 = 27{,}000$.

Chapter 5 Wrap-Up

Summary

Factoring		Examples
Prime number	A positive integer larger than 1 that has no integral factors other than 1 and itself	2, 3, 5, 7, 11
Prime polynomial	A polynomial that cannot be factored is prime.	$x^2 + 3$ and $x^2 - x + 5$ are prime.
Strategy for finding the GCF for monomials	1. Find the GCF for the coefficients of the monomials. 2. Form the product of the GCF of the coefficients and each variable that is common to all of the monomials, where the exponent on each variable equals the smallest power of that variable in any of the monomials.	$12x^3yz,\ 8x^2y^3$ $GCF = 4x^2y$
Factoring out the GCF	Use the distributive property to factor out the GCF from all terms of a polynomial.	$2x^3 - 4x = 2x(x^2 - 2)$

Special Cases		Examples
Difference of two squares	$a^2 - b^2 = (a + b)(a - b)$	$m^2 - 9 = (m - 3)(m + 3)$
Perfect square trinomial	$a^2 + 2ab + b^2 = (a + b)^2$ $a^2 - 2ab + b^2 = (a - b)^2$	$x^2 + 6x + 9 = (x + 3)^2$ $4h^2 - 12h + 9 = (2h - 3)^2$
Difference or sum of two cubes	$a^3 - b^3 = (a - b)(a^2 + ab + b^2)$ $a^3 + b^3 = (a + b)(a^2 - ab + b^2)$	$t^3 - 8 = (t - 2)(t^2 + 2t + 4)$ $p^3 + 1 = (p + 1)(p^2 - p + 1)$

Factoring Polynomials		Examples
Factoring by grouping	Factor out common factors from groups of terms.	$6x + 6w + ax + aw$ $= 6(x + w) + a(x + w)$ $= (6 + a)(x + w)$
Strategy for factoring $ax^2 + bx + c$ by the ac method	1. Find two numbers that have a product equal to ac and a sum equal to b. 2. Replace bx by two terms using the two new numbers as coefficients. 3. Factor the resulting four-term polynomial by grouping.	$6x^2 + 17x + 12$ $= 6x^2 + 9x + 8x + 12$ $= (2x + 3)3x + (2x + 3)4$ $= (2x + 3)(3x + 4)$

| Factoring by trial and error | Try possible factors of the trinomial and check by using FOIL. If incorrect, try again. | $2x^2 + 5x - 12 = (2x - 3)(x + 4)$ |

Strategy for factoring polynomials completely	1. First factor out the greatest common factor.	$x^4 - 4x^2 = x^2(x^2 - 4)$
	2. When factoring a binomial, check to see whether it is a difference of two squares, a difference of two cubes, or a sum of two cubes. The sum of two squares (with no common factor) is prime.	$x^2 - 4 = (x + 2)(x - 2)$ $x^3 - 8 = (x - 2)(x^2 + 2x + 4)$ $x^3 + 8 = (x + 2)(x^2 - 2x + 4)$ $x^2 + 4$ is prime.
	3. When factoring a trinomial, check to see whether it is a perfect square trinomial.	$x^2 + 6x + 9 = (x + 3)^2$ $x^2 - 6x + 9 = (x - 3)^2$
	4. If the polynomial has four terms, try factoring by grouping.	$x^2 + bx + 2x + 2b = x(x + b) + 2(x + b)$ $\qquad\qquad\qquad\quad = (x + 2)(x + b)$
	5. When factoring a trinomial that is not a perfect square, use the *ac* method or trial and error.	$x^2 + 7x + 12 = (x + 3)(x + 4)$
	6. Check to see whether any factors can be factored again.	$x^4 - 4x^2 = x^2(x^2 - 4)$ $\qquad\qquad\;\; = x^2(x + 2)(x - 2)$

Solving Equations

Examples

| Zero factor property | The equation $a \cdot b = 0$ is equivalent to $\qquad a = 0 \quad$ or $\quad b = 0.$ | $x(x - 1) = 0$
 $x = 0 \quad$ or $\quad x - 1 = 0$ |

Strategy for solving an equation by factoring	1. Rewrite the equation with 0 on the right-hand side.	$x^2 + 3x = 18$ $x^2 + 3x - 18 = 0$
	2. Factor the left-hand side completely.	$(x + 6)(x - 3) = 0$
	3. Set each factor equal to zero to get linear equations.	$x + 6 = 0 \quad$ or $\quad x - 3 = 0$ $\qquad x = -6 \quad$ or $\qquad x = 3$
	4. Solve the linear equations.	
	5. Check the answers in the original equation.	$(-6)^2 + 3(-6) = 18, 3^2 + 3(3) = 18$
	6. State the solution(s) to the original equation.	The solutions are -6 and 3.

Enriching Your Mathematical Word Power

Fill in the blank.

1. A _____ number is an integer greater than 1 that has no integral factors other than itself and 1.

2. An integer larger than 1 that is not prime is _____.

3. A polynomial that has no factors is a _____ polynomial.

4. Writing a polynomial as a product is _____.

5. Writing a polynomial as a product of primes is factoring _____.

6. The largest integer that is a factor of two or more integers is the _____ common factor.

7. The square of a monomial in which the coefficient is an integer is a _____ square.

8. The trinomial $a^2 + 2ab + b^2$ is a perfect _____ trinomial.

9. The polynomial $a^3 + b^3$ is a _____ of two cubes.

10. The polynomial $a^3 - b^3$ is a _____ of two cubes.

11. A _____ equation is an equation of the form $ax^2 + bx + c = 0$.

12. According to the _____ factor property, if $ab = 0$ then $a = 0$ or $b = 0$.

13. The _____ theorem indicates that a triangle is a right triangle if and only if the sum of the squares of the legs is equal to the square of the hypotenuse.

Review Exercises

5.1 Factoring Out Common Factors
Find the prime factorization for each integer.

1. 144
2. 121
3. 58
4. 76
5. 150
6. 200

Find the greatest common factor for each group.

7. 36, 90
8. 30, 42, 78
9. $8x$, $12x^2$
10. $6a^2b$, $9ab^2$, $15a^2b^2$

Complete the factorization of each binomial.

11. $3x + 6 = 3($ $)$
12. $7x^2 + x = x($ $)$
13. $2a - 20 = -2($ $)$
14. $a^2 - a = -a($ $)$

Factor each polynomial by factoring out the GCF.

15. $2a - a^2$
16. $9 - 3b$
17. $6x^2y^2 - 9x^5y$
18. $a^3b^5 + a^3b^2$
19. $3x^2y - 12xy - 9y^2$
20. $2a^2 - 4ab^2 - ab$

5.2 Special Products and Grouping
Factor each polynomial completely.

21. $y^2 + y + by + b$
22. $ac + mc + aw^2 + mw^2$
23. $w^2 + 2a - 2w - aw$
24. $a^2 + 3x - ax - 3a$
25. $abc - 3 + c - 3ab$
26. $mnx - 5 + 5nx - m$
27. $y^2 - 400$
28. $4m^2 - 9$
29. $w^2 - 8w + 16$
30. $t^2 + 20t + 100$
31. $4y^2 + 20y + 25$
32. $2a^2 - 4a - 2$
33. $r^2 - 4r + 4$
34. $3m^2 - 75$
35. $8t^3 - 24t^2 + 18t$
36. $t^2 - 9w^2$
37. $x^2 + 12xy + 36y^2$
38. $9y^2 - 12xy + 4x^2$
39. $x^2 + 5x - xy - 5y$
40. $x^2 + xy + ax + ay$

5.3 Factoring the Trinomial $ax^2 + bx + c$ with $a = 1$
Factor each polynomial.

41. $b^2 + 5b - 24$
42. $a^2 - 2a - 35$
43. $r^2 - 4r - 60$
44. $x^2 + 13x + 40$
45. $y^2 - 6y - 55$
46. $a^2 + 6a - 40$
47. $u^2 + 26u + 120$
48. $v^2 - 22v - 75$

Factor completely.

49. $3t^3 + 12t^2$
50. $-4m^4 - 36m^2$
51. $5w^3 + 25w^2 + 25w$
52. $-3t^3 + 3t^2 - 6t$
53. $2a^3b + 3a^2b^2 + ab^3$
54. $6x^2y^2 - xy^3 - y^4$
55. $9x^3 - xy^2$
56. $h^4 - 100h^2$

5.4 Factoring the Trinomial $ax^2 + bx + c$ with $a \neq 1$
Factor each polynomial completely.

57. $14t^2 + t - 3$
58. $15x^2 - 22x - 5$
59. $6x^2 - 19x - 7$
60. $2x^2 - x - 10$
61. $6p^2 + 5p - 4$
62. $3p^2 + 2p - 5$
63. $-30p^3 + 8p^2 + 8p$
64. $-6q^2 - 40q - 50$
65. $6x^2 - 29xy - 5y^2$
66. $10a^2 + ab - 2b^2$
67. $32x^2 + 16xy + 2y^2$
68. $8a^2 + 40ab + 50b^2$

5.5 Difference and Sum of Cubes and a Strategy
Factor completely.

69. $5x^3 + 40x$
70. $w^2 + 6w + 9$
71. $9x^2 + 3x - 2$
72. $ax^3 + ax$
73. $n^2 + 64$
74. $4t^2 + h^2$
75. $x^3 + 2x^2 - x - 2$
76. $16x^2 - 2x - 3$
77. $x^2y - 16xy^2$
78. $-3x^2 + 27$
79. $w^2 + 4w + 5$
80. $2n^2 + 3n - 1$
81. $a^2 + 2a + 1$
82. $-2w^2 - 12w - 18$
83. $x^3 - x^2 + x - 1$
84. $9x^2y^2 - 9y^2$

85. $a^2 + ab + 2a + 2b$ **86.** $4m^2 + 20m + 25$

87. $-2x^2 + 16x - 24$ **88.** $6x^2 + 21x - 45$

89. $m^3 - 1000$ **90.** $8p^3 + 1$

91. $p^4 - q^4$ **92.** $z^4 - 81$

93. $a^3 + 3a^2 + a + 3$ **94.** $y^3 - 5y^2 + 8y - 40$

5.6 Solving Quadratic Equations by Factoring
Solve each equation.

95. $x^3 = 5x^2$ **96.** $2m^2 + 10m = -12$

97. $(a - 2)(a - 3) = 6$ **98.** $(w - 2)(w + 3) = 50$

99. $2m^2 - 9m - 5 = 0$ **100.** $12x^2 + 5x - 3 = 0$

101. $m^3 + 4m^2 - 9m = 36$ **102.** $w^3 + 5w^2 - w = 5$

103. $(x + 3)^2 + x^2 = 5$ **104.** $(h - 2)^2 + (h + 1)^2 = 9$

105. $p^2 + \dfrac{1}{4}p - \dfrac{1}{8} = 0$ **106.** $t^2 + 1 = \dfrac{13}{6}t$

107. $0.1x^2 + 0.01 = 0.07x$ **108.** $0.2y^2 + 0.03y = 0.02$

Applications
Solve each problem.

109. *Positive numbers.* Two positive numbers differ by 6, and their squares differ by 96. Find the numbers.

110. *Consecutive integers.* Find three consecutive integers such that the sum of their squares is 77.

111. *Dimensions of a notebook.* The perimeter of a notebook is 28 inches, and the diagonal measures 10 inches. What are the length and width of the notebook?

112. *Two numbers.* The sum of two numbers is 8.5, and their product is 18. Find the numbers.

113. *Poiseuille's law.* According to the nineteenth-century physician Poiseuille, the velocity (in centimeters per second) of blood r centimeters from the center of an artery of radius R centimeters is given by $v = kR^2 - kr^2$, where k is a constant. Rewrite the formula by factoring the right-hand side completely.

114. *Racquetball.* The volume of rubber (in cubic centimeters) in a hollow rubber ball used in racquetball is given by

$$V = \frac{4}{3}\pi R^3 - \frac{4}{3}\pi r^3,$$

where the inside radius is r centimeters and the outside radius is R centimeters.

a) Rewrite the formula by factoring the right-hand side completely.

b) The accompanying graph shows the relationship between r and V when $R = 3$. Use the graph to estimate the value of r for which $V = 100$ cm³.

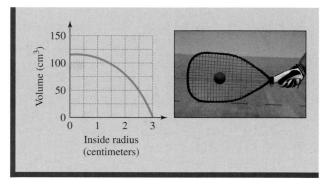

Figure for Exercise 114

115. *Leaning ladder.* A 10-foot ladder is placed against a building so that the distance from the bottom of the ladder to the building is 2 feet less than the distance from the top of the ladder to the ground. What is the distance from the bottom of the ladder to the building?

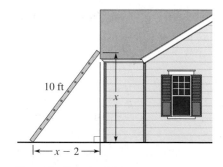

Figure for Exercise 115

116. *Towering antenna.* A guy wire of length 50 feet is attached to the ground and to the top of an antenna. The height of the antenna is 10 feet larger than the distance from the base of the antenna to the point where the guy wire is attached to the ground. What is the height of the antenna?

Chapter 5 Test

Give the prime factorization for each integer.

1. 66

2. 336

Find the greatest common factor (GCF) for each group.

3. 48, 80

4. 42, 66, 78

5. $6y^2$, $15y^3$

6. $12a^2b$, $18ab^2$, $24a^3b^3$

Factor each polynomial completely.

7. $5x^2 - 10x$

8. $6x^2y^2 + 12xy^2 + 12y^2$

9. $3a^3b - 3ab^3$

10. $a^2 + 2a - 24$

11. $4b^2 - 28b + 49$

12. $3m^3 + 27m$

13. $ax - ay + bx - by$

14. $ax - 2a - 5x + 10$

15. $6b^2 - 7b - 5$

16. $m^2 + 4mn + 4n^2$

17. $2a^2 - 13a + 15$

18. $z^3 + 9z^2 + 18z$

19. $x^3 + 125$

20. $a^4 - ab^3$

Solve each equation.

21. $x^2 + 6x + 9 = 0$

22. $2x^2 + 5x - 12 = 0$

23. $3x^3 = 12x$

24. $(2x - 1)(3x + 5) = 5$

25. $\dfrac{1}{8}x^2 - \dfrac{3}{4}x + 1 = 0$

26. $0.3x^2 - 1.7x + 1 = 0$

Write a complete solution to each problem.

27. If the length of a rectangle is 3 feet longer than the width and the diagonal is 15 feet, then what are the length and width?

28. The sum of two numbers is 4, and their product is -32. Find the numbers.

29. A ball is dropped from a height of 64 feet. Its height above the earth in feet is given by $h(t) = -16t^2 + 64$, where t is the number of seconds after it is dropped.

   a) Find $h(1)$.

   b) How long does it take the ball to fall to the earth?

*Making*Connections | A Review of Chapters 1–5

Simplify each expression.

1. $\dfrac{91 - 17}{17 - 91}$

2. $\dfrac{4 - 18}{-6 - 1}$

3. $5 - 2(7 - 3)$

4. $3^2 - 4(6)(-2)$

5. $2^5 - 2^4$

6. $0.07(37) + 0.07(63)$

Perform the indicated operations.

7. $x \cdot 2x$

8. $x + 2x$

9. $\dfrac{6 + 2x}{2}$

10. $\dfrac{6 \cdot 2x}{2}$

11. $2 \cdot 3y \cdot 4z$

12. $2(3y + 4z)$

13. $2 - (3 - 4z)$

14. $-(x - 3) - 2(5 - x)$

15. $-2(3x - 4)$

16. $5x - 2(3x - 4)$

17. $(5x - 2)(3x - 4)$

18. $(5x - 2)(3x^2 - 4x + 1)$

19. $\dfrac{9x^2 - 6x}{3x}$

20. $\dfrac{9x - 6}{3} - \dfrac{7x - 14}{7}$

Find the solution set to each equation.

21. $2x - 3 = 0$

22. $2x + 1 = 0$

23. $(x - 3)(x + 5) = 0$

24. $(2x - 3)(2x + 1) = 0$

25. $3x(x - 3) = 0$

26. $x^2 = x$

27. $3x - 3x = 0$

28. $3x - 3x = 1$

29. $0.01x - x + 14.9 = 0.5x$

30. $0.05x + 0.04(x - 40) = 2$

31. $2x^2 = 18$

32. $2x^2 + 7x - 15 = 0$

Simplify each expression. Write answers without negative exponents. All variables represent nonzero real numbers.

33. $t^8 \div t^2$

34. $t^8 \cdot t^2$

35. $t^2 \div t^8$

36. $(t^8)^2$

37. $\dfrac{8t^8}{2t^2}$

38. $\dfrac{3y^{-5}}{9y^2}$

39. $\dfrac{6x^{-6}}{15x^{-8}}$

40. $\dfrac{\left(4w^{-3}\right)^2}{24w^{-6}}$

41. $(-2x^{-3}y^2)^3$

42. $\left(\dfrac{x^{-2}}{3y^3}\right)^{-2}$

43. $-3^2 + \left(\dfrac{1}{2}\right)^{-2}$

44. $-4^0 + \left(-\dfrac{1}{3}\right)^{-3}$

Solve each inequality. State the solution set in interval notation and sketch its graph.

45. $2x - 5 > 3x + 4$

46. $4 - 5x \le -11$

47. $-\dfrac{2}{3}x + 3 < -5$

48. $0.05(x - 120) - 24 < 0$

Factor each expression completely.

49. $4p^3 + 12p^2 + 32p$

50. $3m^4 - 12m^3 + 9m^2$

51. $-12a^2 + 12a - 3$

52. $-2b^2 + 8$

53. $ab + qb + a + q$

54. $2am + 2bm - 3an - 3bn$

55. $-7x^3 + 7$

56. $2a^3 + 54$

Solve each problem.

57. The area of a rectangular garden is 750 square yards and the length is 30 yards. What is the width?

58. The perimeter of a rectangular canvas is 66 inches and its length is 19 inches. Find the width.

59. The area of a rectangular balcony is 66 square feet. If the length is 5 feet more than the width, then what is the length?

60. A craft shop charges five cents per square inch for a rectangular piece of copper. If the width is 3 inches less than the length and the charge is $5.40, then what is the width?

61. The diagonal measure of a small television screen is 1 inch greater than the length and 2 inches greater than the width. Find the length and width.

62. *Another ace.* Professional tennis players can serve a tennis ball at speeds over 120 mph into a rectangular region that has a perimeter of 69 feet and an area of 283.5 square feet. Find the length and width of the service region.

Photo for Exercise 62

CriticalThinking | For Individual or Group Work | Chapter 5

These exercises can be solved by a variety of techniques, which may or may not require algebra. So be creative and think critically. Explain all answers. Answers are in the Instructor's Edition of this text.

1. *Counting cubes.* What is the total number of cubes that are in each of the following diagrams?

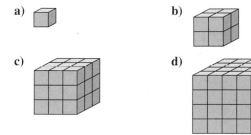

a) b)

c) d)

2. *More cubes.* Imagine a large cube that is made up of 125 small cubes like those in Exercise 1. What is the total number of cubes that could be found in this arrangement?

3. *Timely coincidence.* Starting at 8 A.M. determine the number of times in the next 24 hours for which the hour and minute hands on a clock coincide.

Photo for Exercise 3

4. *Chess board.* There are 64 squares on a square chess board. How many squares are neither diagonal squares nor edge squares?

Photo for Exercise 4

5. *Last digit.* Find the last digit in 3^{9999}.

6. *Reconciling remainders.* Find a positive integer smaller than 500 that has a remainder of 3 when divided by 5, a remainder of 6 when divided by 9, and a remainder of 8 when divided by 11.

7. *Exact sum.* Find this sum exactly:

$$\frac{1}{2} + \frac{1}{2^2} + \frac{1}{2^3} + \frac{1}{2^4} + \cdots + \frac{1}{2^{19}}$$

8. *Ten-digit number.* Find a 10-digit number whose first digit is the number of 1's in the 10-digit number, whose second digit is the number of 2's in the 10-digit number, whose third digit is the number of 3's in the 10-digit number, and so on. The ninth digit must be the number of nines in the 10-digit number and the tenth digit must be the number of zeros in the 10-digit number.

Rational Expressions

Advanced technical developments have made sports equipment faster, lighter, and more responsive to the human body. Behind the more flexible skis, lighter bats, and comfortable athletic shoes lies the science of biomechanics, which is the study of human movement and the factors that influence it.

Designing and testing an athletic shoe go hand in hand. While a shoe is being designed, it is tested in a multitude of ways, including long-term wear, rear foot stability, and strength of materials. Testing basketball shoes usually includes an evaluation of the force applied to the ground by the foot during running, jumping, and landing. Many biomechanics laboratories have a special platform that can measure the force exerted when a player cuts from side to side, as well as the force against the bottom of the shoe. Force exerted in landing from a layup shot can be as high as 14 times the weight of the body. Side-to-side force is usually about 1 to 2 body weights in a cutting movement.

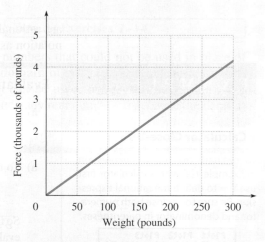

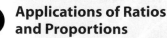
In Exercises 53 and 54 of Section 6.7 you will see how designers of athletic shoes use proportions to find the amount of force on the foot and soles of shoes for activities such as running and jumping.

| EXAMPLE | 8 |

Factoring out the opposite of a common factor

Reduce $\frac{-3w - 3w^2}{w^2 - 1}$ to lowest terms.

Solution

We can factor $3w$ or $-3w$ from the numerator. If we factor out $-3w$, we get a common factor in the numerator and denominator:

$$\frac{-3w - 3w^2}{w^2 - 1} = \frac{-3w(1 + w)}{(w - 1)(w + 1)} \quad \text{Factor.}$$
$$= \frac{-3w}{w - 1} \quad \text{Since } 1 + w = w + 1, \text{ we divide out } w + 1.$$
$$= \frac{3w}{1 - w} \quad \text{Multiply numerator and denominator by } -1.$$

The last step is not absolutely necessary, but we usually perform it to express the answer with one less negative sign.

> Now do Exercises 75–84

The main points to remember for reducing rational expressions are summarized in the following reducing strategy.

Strategy for Reducing Rational Expressions

1. Factor the numerator and denominator completely. Factor out a common factor with a negative sign if necessary.
2. Divide out all common factors. Use the quotient rule if the common factors are powers.

⟨6⟩ Writing Rational Expressions

Rational expressions occur in applications involving rates. For uniform motion, rate is distance divided by time, $R = \frac{D}{T}$. For example, if you drive 500 miles in 10 hours, your rate is $\frac{500}{10}$ or 50 mph. If you drive 500 miles in x hours, your rate is $\frac{500}{x}$ mph. In work problems, rate is work divided by time, $R = \frac{W}{T}$. For example, if you lay 400 tiles in 4 hours, your rate is $\frac{400}{4}$ or 100 tiles/hour. If you lay 400 tiles in x hours, your rate is $\frac{400}{x}$ tiles/hour.

| EXAMPLE | 9 |

Writing rational expressions

Answer each question with a rational expression.

a) If a trucker drives 500 miles in $x + 1$ hours, then what is his average speed?
b) If a wholesaler buys 100 pounds of shrimp for x dollars, then what is the price per pound?
c) If a painter completes an entire house in $2x$ hours, then at what rate is she painting?

Solution

a) Because $R = \frac{D}{T}$, he is averaging $\frac{500}{x + 1}$ mph.
b) At x dollars for 100 pounds, the wholesaler is paying $\frac{x}{100}$ dollars per pound or $\frac{x}{100}$ dollars/pound.
c) By completing 1 house in $2x$ hours, her rate is $\frac{1}{2x}$ house/hour.

> Now do Exercises 107–112

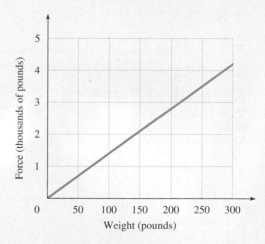

Rational Expressions

Advanced technical developments have made sports equipment faster, lighter, and more responsive to the human body. Behind the more flexible skis, lighter bats, and comfortable athletic shoes lies the science of biomechanics, which is the study of human movement and the factors that influence it.

Designing and testing an athletic shoe go hand in hand. While a shoe is being designed, it is tested in a multitude of ways, including long-term wear, rear foot stability, and strength of materials. Testing basketball shoes usually includes an evaluation of the force applied to the ground by the foot during running, jumping, and landing. Many biomechanics laboratories have a special platform that can measure the force exerted when a player cuts from side to side, as well as the force against the bottom of the shoe. Force exerted in landing from a lay-up shot can be as high as 14 times the weight of the body. Side-to-side force is usually about 1 to 2 body weights in a cutting movement.

 6.1 Reducing Rational Expressions

 6.2 Multiplication and Division

6.3 Finding the Least Common Denominator

6.4 Addition and Subtraction

6.5 Complex Fractions

6.6 Solving Equations with Rational Expressions

6.7 Applications of Ratios and Proportions

 6.8 Applications of Rational Expressions

In Exercises 53 and 54 of Section 6.7 you will see how designers of athletic shoes use proportions to find the amount of force on the foot and soles of shoes for activities such as running and jumping.

6.1 Reducing Rational Expressions

Rational expressions in algebra are similar to the rational numbers in arithmetic. In this section, you will learn the basic ideas of rational expressions.

⟨1⟩ Rational Expressions and Functions

A rational number is the ratio of two integers with the denominator not equal to 0. For example,

$$\frac{3}{4}, \quad \frac{-9}{-6}, \quad \frac{-7}{1}, \quad \text{and} \quad \frac{0}{2}$$

are rational numbers. Of course, we usually write the last three of these rational numbers in their simpler forms $\frac{3}{2}$, -7, and 0. A **rational expression** is the ratio of two polynomials with the denominator not equal to 0. Because an integer is a monomial, a rational number is also a rational expression. As with rational numbers, if the denominator is 1, it can be omitted. Some examples of rational expressions are

$$\frac{x^2 - 1}{x + 8}, \quad \frac{3a^2 + 5a - 3}{a - 9}, \quad \frac{3}{7}, \quad \text{and} \quad 9x.$$

A rational expression involving a variable has no value unless we assign a value to the variable. If the value of a rational expression is used to determine the value of a second variable, then we have a **rational function.** For example,

$$y = \frac{x^2 - 1}{x + 8} \quad \text{and} \quad w = \frac{3a^2 + 5a - 3}{a - 9}$$

are rational functions. We can evaluate a rational expression with or without function notation as we did for polynomials in Chapter 5.

EXAMPLE 1

Evaluating a rational expression

a) Find the value of $\frac{4x - 1}{x + 2}$ for $x = -3$.   **b)** If $R(x) = \frac{3x + 2}{2x - 1}$, find $R(4)$.

Solution

a) To find the value of $\frac{4x - 1}{x + 2}$ for $x = -3$, replace x by -3 in the rational expression:

$$\frac{4(-3) - 1}{-3 + 2} = \frac{-13}{-1} = 13$$

So the value of the rational expression is 13. The Calculator Close-Up shows how to evaluate the expression with a graphing calculator using a variable. With a scientific or graphing calculator you could also evaluate the expression by entering $(4(-3) - 1)/(-3 + 2)$. Be sure to enclose the numerator and denominator in parentheses.

b) $R(4)$ is the value of the rational expression when $x = 4$. To find $R(4)$, replace x by 4 in $R(x) = \frac{3x + 2}{2x - 1}$:

$$R(4) = \frac{3(4) + 2}{2(4) - 1}$$

$$R(4) = \frac{14}{7} = 2$$

So the value of the rational expression is 2 when $x = 4$, or $R(4) = 2$ (read "R of 4 is 2").

Now do Exercises 1–6

⟨ **Calculator Close-Up** ⟩

To evaluate the rational expression in Example 1(a) with a calculator, first use Y = to define the rational expression. Be sure to enclose both numerator and denominator in parentheses.

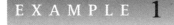

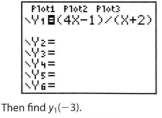
```
Plot1  Plot2  Plot3
\Y1◘(4X-1)/(X+2)
\Y2=
\Y3=
\Y4=
\Y5=
\Y6=
```

Then find $y_1(-3)$.

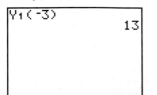
```
Y1(-3)
              13
```

An expression such as $\frac{5}{0}$ is **undefined** because the definition of rational numbers does not allow zero in the denominator. When a variable occurs in a denominator, any real number can be used for the variable *except* numbers that make the expression undefined.

EXAMPLE   2

Ruling out values for x

Which numbers cannot be used in place of x in each rational expression?

a) $\dfrac{x^2 - 1}{x + 8}$ b) $\dfrac{x + 2}{2x + 1}$ c) $\dfrac{x + 5}{x^2 - 4}$

Solution

a) The denominator is 0 if $x + 8 = 0$, or $x = -8$. So -8 cannot be used in place of x. (All real numbers except -8 can be used in place of x.)

b) The denominator is zero if $2x + 1 = 0$, or $x = -\frac{1}{2}$. So we cannot use $-\frac{1}{2}$ in place of x. $\left(\text{All real numbers except } -\frac{1}{2} \text{ can be used in place of } x.\right)$

c) The denominator is zero if $x^2 - 4 = 0$. Solve this equation:

$$x^2 - 4 = 0$$
$$(x - 2)(x + 2) = 0 \quad \text{Factor.}$$
$$x - 2 = 0 \quad \text{or} \quad x + 2 = 0 \quad \text{Zero factor property}$$
$$x = 2 \quad \text{or} \quad x = -2$$

So 2 and -2 cannot be used in place of x. (All real numbers except 2 and -2 can be used in place of x.)

| Now do Exercises 7–14 |

In Example 2 we determined the real numbers that could not be used in place of the variable in a rational expression. The **domain** of any algebraic expression in one variable is the set of all real numbers that *can* be used in place of the variable. For rational expressions, the domain must exclude any real numbers that cause the denominator to be zero.

EXAMPLE   3

Domain

Find the domain of each expression.

a) $\dfrac{x^2 - 9}{x + 3}$ b) $\dfrac{x}{x^2 - x - 6}$ c) $\dfrac{x - 5}{4}$

Solution

a) The denominator is 0 if $x + 3 = 0$, or $x = -3$. So -3 can't be used for x. The domain is the set of all real numbers except -3, which is written in set notation as $\{x \mid x \neq -3\}$.

b) The denominator is 0 if $x^2 - x - 6 = 0$:

$$x^2 - x - 6 = 0$$
$$(x - 3)(x + 2) = 0$$
$$x - 3 = 0 \quad \text{or} \quad x + 2 = 0$$
$$x = 3 \quad \text{or} \quad x = -2$$

So -2 and 3 can't be used in place of x. The domain is the set of all real numbers except -2 and 3, which is written as $\{x \mid x \neq -2 \text{ and } x \neq 3\}$.

c) Since the denominator is 4, the denominator can't be 0 no matter what number is used for x. The domain is the set of all real numbers, R.

> Now do Exercises 15–22

Note that if a rational expression is used to define a function, then the domain of the rational expression is also called the **domain of the function.** For example, the domain of the function $y = \frac{x^2 - 9}{x + 3}$ is the set of all real numbers except -3 or $\{x \mid x \neq -3\}$.

When dealing with rational expressions in this book, we will generally assume that the variables represent numbers for which the denominator is not zero.

⟨2⟩ Reducing to Lowest Terms

Rational expressions are a generalization of rational numbers. The operations that we perform on rational numbers can be performed on rational expressions in exactly the same manner.

Each rational number can be written in infinitely many equivalent forms. For example,

$$\frac{3}{5} = \frac{6}{10} = \frac{9}{15} = \frac{12}{20} = \frac{15}{25} = \cdots.$$

⟨ **Helpful Hint** ⟩

How would you fill in the blank in $\frac{3}{5} = \frac{}{10}$? Most students learn to divide 5 into 10 to get 2, and then multiply 3 by 2 to get 6. In algebra, it is better to multiply the numerator and denominator of $\frac{3}{5}$ by 2, as shown here.

Each equivalent form of $\frac{3}{5}$ is obtained from $\frac{3}{5}$ by multiplying both numerator and denominator by the same nonzero number. This is equivalent to multiplying the fraction by 1, which does not change its value. For example,

$$\frac{3}{5} = \frac{3}{5} \cdot 1 = \frac{3}{5} \cdot \frac{2}{2} = \frac{6}{10} \quad \text{and} \quad \frac{3}{5} = \frac{3 \cdot 3}{5 \cdot 3} = \frac{9}{15}.$$

If we start with $\frac{6}{10}$ and convert it into $\frac{3}{5}$, we say that we are **reducing $\frac{6}{10}$ to lowest terms.** We reduce by dividing the numerator and denominator by the common factor 2:

$$\frac{6}{10} = \frac{\cancel{2} \cdot 3}{\cancel{2} \cdot 5} = \frac{3}{5}$$

A rational number is expressed in lowest terms when the numerator and the denominator have no common factors other than 1.

CAUTION   We can reduce fractions only by dividing the numerator and the denominator by a common factor. Although it is true that

$$\frac{6}{10} = \frac{2 + 4}{2 + 8},$$

we cannot eliminate the 2's, because they are not factors. Removing them from the sums in the numerator and denominator would not result in $\frac{3}{5}$.

Reducing Fractions

If $a \neq 0$ and $c \neq 0$, then

$$\frac{ab}{ac} = \frac{b}{c}.$$

To reduce rational expressions to lowest terms, we use exactly the same procedure as with fractions:

> ### Reducing Rational Expressions
> **1.** Factor the numerator and denominator completely.
> **2.** Divide the numerator and denominator by the greatest common factor.

Dividing the numerator and denominator by the GCF is often referred to as **dividing out** or **canceling** the GCF.

EXAMPLE 4

Reducing

Reduce to lowest terms.

a) $\dfrac{30}{42}$ b) $\dfrac{x^2 - 9}{6x + 18}$ c) $\dfrac{3x^2 + 9x + 6}{2x^2 - 8}$

Solution

a) $\dfrac{30}{42} = \dfrac{2 \cdot 3 \cdot 5}{2 \cdot 3 \cdot 7}$ Factor.

$= \dfrac{5}{7}$ Divide out the GCF: $2 \cdot 3$ or 6.

b) Since $\dfrac{9}{18} = \dfrac{9 \cdot 1}{9 \cdot 2} = \dfrac{1}{2}$, it is tempting to apply that fact here. However, 9 is not a common factor of the numerator and denominator of $\dfrac{x^2 - 9}{6x + 18}$, as it is in $\dfrac{9}{18}$. You must factor the numerator and denominator completely before reducing.

$$\frac{x^2 - 9}{6x + 18} = \frac{(x - 3)(x + 3)}{6(x + 3)}$$ Factor.

$$= \frac{x - 3}{6}$$ Divide out the GCF: $x + 3$.

This reduction is valid for all real numbers except -3, because that is the domain of the original expression. If $x = -3$, then $x + 3 = 0$ and we would be dividing out 0 from the numerator and denominator, which is prohibited in the rule for reducing fractions.

c) $\dfrac{3x^2 + 9x + 6}{2x^2 - 8} = \dfrac{3(x + 2)(x + 1)}{2(x + 2)(x - 2)}$ Factor completely.

$= \dfrac{3x + 3}{2(x - 2)}$ Divide out the GCF: $x + 2$.

This reduction is valid for all real numbers except -2 and 2, because that is the domain of the original expression.

> **Now do Exercises 23–46**

CAUTION In reducing, you can divide out or cancel common factors only. You cannot cancel x from $\dfrac{x + 3}{x + 2}$, because it is not a factor of either $x + 3$ or $x + 2$. But x is a common factor in $\dfrac{3x}{2x}$, and $\dfrac{3x}{2x} = \dfrac{3}{2}$.

Note that there are four ways to write the answer to Example 3(c) depending on whether the numerator and denominator are factored. Since

$$\frac{3x + 3}{2(x - 2)} = \frac{3(x + 1)}{2(x - 2)} = \frac{3(x + 1)}{2x - 4} = \frac{3x + 3}{2x - 4},$$

any of these four rational expressions is correct. We usually give such answers with the denominator factored and the numerator not factored. With the denominator factored you can easily spot the values for x that will cause an undefined expression.

⟨3⟩ **Reducing with the Quotient Rule for Exponents**

To reduce rational expressions involving exponential expressions, we use the quotient rule for exponents from Chapter 4. We restate it here for reference.

Quotient Rule for Exponents

If $a \neq 0$, and m and n are any integers, then

$$\frac{a^m}{a^n} = a^{m-n}.$$

EXAMPLE 5

Using the quotient rule in reducing

Reduce to lowest terms.

a) $\dfrac{3a^{15}}{6a^7}$

b) $\dfrac{6x^4y^2}{4xy^5}$

Solution

a) $\dfrac{3a^{15}}{6a^7} = \dfrac{\cancel{3}a^{15}}{\cancel{3} \cdot 2a^7}$ Factor.

$= \dfrac{a^{15-7}}{2}$ Quotient rule for exponents

$= \dfrac{a^8}{2}$

b) $\dfrac{6x^4y^2}{4xy^5} = \dfrac{\cancel{2} \cdot 3x^4y^2}{\cancel{2} \cdot 2xy^5}$ Factor.

$= \dfrac{3x^{4-1}y^{2-5}}{2}$ Quotient rule for exponents

$= \dfrac{3x^3y^{-3}}{2} = \dfrac{3x^3}{2y^3}$

> Now do Exercises 47–58

The essential part of reducing is getting a complete factorization for the numerator and denominator. To get a complete factorization, you must use the techniques for factoring from Chapter 5. If there are large integers in the numerator and denominator, you can use the technique shown in Section 5.1 to get a prime factorization of each integer.

EXAMPLE 6

Reducing expressions involving large integers

Reduce $\frac{420}{616}$ to lowest terms.

Solution

Use the method of Section 5.1 to get a prime factorization of 420 and 616:

$\begin{array}{r} 2\overline{)420} \\ 2\overline{)210} \\ 3\overline{)105} \\ 5\overline{)35} \\ 7 \end{array}$ $\begin{array}{r} 2\overline{)616} \\ 2\overline{)308} \\ 2\overline{)154} \\ 7\overline{)77} \\ 11 \end{array}$

The complete factorization for 420 is $2^2 \cdot 3 \cdot 5 \cdot 7$, and the complete factorization for 616 is $2^3 \cdot 7 \cdot 11$. To reduce the fraction, we divide out the common factors:

$$\frac{420}{616} = \frac{2^2 \cdot 3 \cdot 5 \cdot 7}{2^3 \cdot 7 \cdot 11}$$

$$= \frac{3 \cdot 5}{2 \cdot 11}$$

$$= \frac{15}{22}$$

Now do Exercises 59–66

⟨4⟩ Dividing $a - b$ by $b - a$

In Section 4.5 you learned that $a - b = -(b - a) = -1(b - a)$. So if $a - b$ is divided by $b - a$, the quotient is -1:

$$\frac{a - b}{b - a} = \frac{-1(b - a)}{b - a}$$

$$= -1$$

We will use this fact in Example 7.

EXAMPLE **7**

Expressions with $a - b$ and $b - a$

Reduce to lowest terms.

a) $\dfrac{5x - 5y}{4y - 4x}$

b) $\dfrac{m^2 - n^2}{n - m}$

Solution

a) Factor out 5 from the numerator and 4 from the denominator and use $\frac{x - y}{y - x} = -1$:

$$\frac{5x - 5y}{4y - 4x} = \frac{5(x - y)}{4(y - x)} = \frac{5}{4}(-1) = -\frac{5}{4}$$

Another way is to factor out -5 from the numerator and 4 from the denominator and then use $\frac{y - x}{y - x} = 1$:

$$\frac{5x - 5y}{4y - 4x} = \frac{-5(y - x)}{4(y - x)} = \frac{-5}{4}(1) = -\frac{5}{4}$$

b) 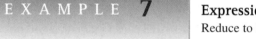 $\dfrac{m^2 - n^2}{n - m} = \dfrac{(m - n)(m + n)}{n - m}$ Factor.

$$= -1(m + n) \qquad \frac{m - n}{n - m} = -1$$

$$= -m - n$$

Now do Exercises 67–74

CAUTION We can reduce $\frac{a - b}{b - a}$ to -1, but we cannot reduce $\frac{a - b}{a + b}$. There is no factor that is common to the numerator and denominator of $\frac{a - b}{a + b}$ or $\frac{a + b}{a - b}$.

⟨5⟩ Factoring Out the Opposite of a Common Factor

If we can factor out a common factor, we can also factor out the opposite of that common factor. For example, from $-3x - 6y$ we can factor out the common factor 3 or the common factor -3:

$$-3x - 6y = 3(-x - 2y) \qquad \text{or} \qquad -3x - 6y = -3(x + 2y)$$

To reduce an expression, it is sometimes necessary to factor out the opposite of a common factor.

EXAMPLE **8**

Factoring out the opposite of a common factor

Reduce $\dfrac{-3w - 3w^2}{w^2 - 1}$ to lowest terms.

Solution

We can factor $3w$ or $-3w$ from the numerator. If we factor out $-3w$, we get a common factor in the numerator and denominator:

$$\frac{-3w - 3w^2}{w^2 - 1} = \frac{-3w(1 + w)}{(w - 1)(w + 1)} \qquad \text{Factor.}$$

$$= \frac{-3w}{w - 1} \qquad \text{Since } 1 + w = w + 1, \text{ we divide out } w + 1.$$

$$= \frac{3w}{1 - w} \qquad \text{Multiply numerator and denominator by } -1.$$

The last step is not absolutely necessary, but we usually perform it to express the answer with one less negative sign.

> Now do Exercises 75–84

The main points to remember for reducing rational expressions are summarized in the following reducing strategy.

Strategy for Reducing Rational Expressions

1. Factor the numerator and denominator completely. Factor out a common factor with a negative sign if necessary.

2. Divide out all common factors. Use the quotient rule if the common factors are powers.

⟨6⟩ Writing Rational Expressions

Rational expressions occur in applications involving rates. For uniform motion, rate is distance divided by time, $R = \dfrac{D}{T}$. For example, if you drive 500 miles in 10 hours, your rate is $\dfrac{500}{10}$ or 50 mph. If you drive 500 miles in x hours, your rate is $\dfrac{500}{x}$ mph. In work problems, rate is work divided by time, $R = \dfrac{W}{T}$. For example, if you lay 400 tiles in 4 hours, your rate is $\dfrac{400}{4}$ or 100 tiles/hour. If you lay 400 tiles in x hours, your rate is $\dfrac{400}{x}$ tiles/hour.

EXAMPLE **9**

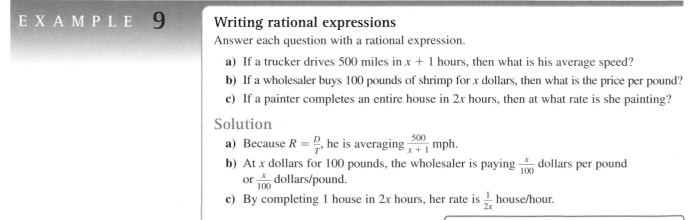

Writing rational expressions

Answer each question with a rational expression.

a) If a trucker drives 500 miles in $x + 1$ hours, then what is his average speed?

b) If a wholesaler buys 100 pounds of shrimp for x dollars, then what is the price per pound?

c) If a painter completes an entire house in $2x$ hours, then at what rate is she painting?

Solution

a) Because $R = \dfrac{D}{T}$, he is averaging $\dfrac{500}{x + 1}$ mph.

b) At x dollars for 100 pounds, the wholesaler is paying $\dfrac{x}{100}$ dollars per pound or $\dfrac{x}{100}$ dollars/pound.

c) By completing 1 house in $2x$ hours, her rate is $\dfrac{1}{2x}$ house/hour.

> Now do Exercises 107–112

Warm-Ups ▼

Fill in the blank.

1. A rational number is a ratio of two _____ with the denominator not 0.
2. A rational expression is a ratio of two _____ with the denominator not 0.
3. A rational expression is reduced to lowest terms by _____ the numerator and denominator by the GCF.
4. The _____ rule is used in reducing a ratio of monomials.
5. The expressions $a - b$ and $b - a$ are _____ .
6. If a rational expression is used to determine y from x, then y is a _____ function of x.

True or false?

7. A complete factorization of 3003 is $2 \cdot 3 \cdot 7 \cdot 11 \cdot 13$.
8. A complete factorization of 120 is $2^3 \cdot 3 \cdot 5$.
9. We can't replace x by -1 or 3 in $\dfrac{x + 1}{x - 3}$.
10. For any real number x, $\dfrac{2x}{2} = x$.
11. Reducing $\dfrac{a^2 + b^2}{a + b}$ to lowest terms yields $a + b$.

Exercises 6.1

‹ **Study Tips** ›

• If you must miss class, let your instructor know. Be sure to get notes from a reliable classmate.
• Take good notes in class for yourself and your classmates. You never know when a classmate will ask to see your notes.

‹1› Rational Expressions and Functions

Evaluate each rational expression. See Example 1.

1. Evaluate $\frac{3x - 3}{x + 5}$ for $x = -2$.

2. Evaluate $\frac{3x + 1}{4x - 4}$ for $x = 5$.

3. If $R(x) = \frac{2x + 9}{x}$, find $R(3)$.

4. If $R(x) = \frac{-20x - 2}{x - 8}$, find $R(-1)$.

5. If $R(x) = \frac{x - 5}{x + 3}$, find $R(2)$, $R(-4)$, $R(-3.02)$, and $R(-2.96)$. Note how a small difference in x (-3.02 to -2.96) can make a big difference in $R(x)$.

6. If $R(x) = \frac{x^2 - 2x - 3}{x - 2}$, find $R(3)$, $R(5)$, $R(2.05)$, and $R(1.999)$.

Which numbers cannot be used in place of the variable in each rational expression? See Example 2.

7. $\dfrac{x}{x + 1}$

8. $\dfrac{3x}{x - 7}$

9. $\dfrac{7a}{3a - 5}$

10. $\dfrac{84}{3 - 2a}$

11. $\dfrac{2x + 3}{x^2 - 16}$

12. $\dfrac{2y + 1}{y^2 - y - 6}$

13. $\dfrac{p - 1}{2}$

14. $\dfrac{m + 31}{5}$

Find the domain of each rational expression. See Example 3.

15. $\dfrac{x^2 + x}{x - 2}$

16. $\dfrac{x + 4}{x - 5}$

17. $\dfrac{x}{x^2 + 5x + 6}$

18. $\dfrac{x^2 + 2}{x^2 - x - 12}$

19. $\dfrac{x^2 - 4}{2}$

20. $\dfrac{x^2 - 3x}{9}$

21. $\dfrac{x - 5}{x}$

22. $\dfrac{x^2 - 3}{x + 9}$

⟨2⟩ Reducing to Lowest Terms

Reduce each rational expression to lowest terms. Assume that the variables represent only numbers for which the denominators are nonzero. See Example 4.

23. $\dfrac{6}{27}$

24. $\dfrac{14}{21}$

25. $\dfrac{42}{90}$

26. $\dfrac{42}{54}$

27. $\dfrac{36a}{90}$

28. $\dfrac{56y}{40}$

29. $\dfrac{78}{30w}$

30. $\dfrac{68}{44y}$

31. $\dfrac{6x + 2}{6}$

32. $\dfrac{2w + 2}{2w}$

33. $\dfrac{2x + 4y}{6y + 3x}$

34. $\dfrac{5x - 10a}{10x - 20a}$

35. $\dfrac{3b - 9}{6b - 15}$

36. $\dfrac{3m + 9w}{3m - 6w}$

37. $\dfrac{w^2 - 49}{w + 7}$

38. $\dfrac{a^2 - b^2}{a - b}$

39. $\dfrac{a^2 - 1}{a^2 + 2a + 1}$

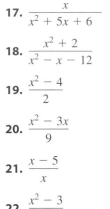**40.** $\dfrac{x^2 - y^2}{x^2 + 2xy + y^2}$

41. $\dfrac{2x^2 + 4x + 2}{4x^2 - 4}$

42. $\dfrac{2x^2 + 10x + 12}{3x^2 - 27}$

43. $\dfrac{3x^2 + 18x + 27}{21x + 63}$

44. $\dfrac{x^3 - 3x^2 - 4x}{x^2 - 4x}$

45. $\dfrac{2a^3 + 16}{4a + 8}$

46. $\dfrac{w^3 - 27}{w^2 - 3w}$

⟨3⟩ Reducing with the Quotient Rule for Exponents

Reduce each expression to lowest terms. Assume that all variables represent nonzero real numbers, and use only positive exponents in your answers. See Example 5.

47. $\dfrac{x^{10}}{x^7}$

48. $\dfrac{y^8}{y^5}$

49. $\dfrac{z^3}{z^8}$

50. $\dfrac{w^9}{w^{12}}$

51. $\dfrac{4x^7}{-2x^5}$

52. $\dfrac{-6y^3}{3y^9}$

53. $\dfrac{-12m^9n^{18}}{8m^6n^{16}}$

54. $\dfrac{-9u^9v^{19}}{6u^9v^{14}}$

55. $\dfrac{6b^{10}c^4}{-8b^{10}c^7}$

56. $\dfrac{9x^{20}y}{-6x^{25}y^3}$

57. $\dfrac{30a^3bc}{18a^7b^{17}}$

58. $\dfrac{15m^{10}n^3}{24m^{12}np}$

Reduce each expression to lowest terms. Assume that all variables represent nonzero real numbers, and use only positive exponents in your answers. See Example 6.

59. $\dfrac{210}{264}$

60. $\dfrac{616}{660}$

61. $\dfrac{231}{168}$

62. $\dfrac{936}{624}$

63. $\dfrac{630x^5}{300x^9}$

64. $\dfrac{96y^2}{108y^5}$

65. $\dfrac{924a^{23}}{448a^{19}}$

66. $\dfrac{270b^{75}}{165b^{12}}$

⟨4⟩ Dividing a − b by b − a

Reduce each expression to lowest terms. See Example 7.

67. $\dfrac{3a - 2b}{2b - 3a}$

68. $\dfrac{5m - 6n}{6n - 5m}$

69. $\dfrac{h^2 - t^2}{t - h}$

70. $\dfrac{r^2 - s^2}{s - r}$

71. $\dfrac{2g - 6h}{9h^2 - g^2}$

72. $\dfrac{5a - 10b}{4b^2 - a^2}$

73. $\dfrac{x^2 - x - 6}{9 - x^2}$

74. $\dfrac{1 - a^2}{a^2 + a - 2}$

⟨5⟩ Factoring Out the Opposite of a Common Factor

Reduce each expression to lowest terms. See Example 8.

75. $\dfrac{-x - 6}{x + 6}$

76. $\dfrac{-5x - 20}{3x + 12}$

77. $\dfrac{-2y - 6y^2}{3 + 9y}$

78. $\dfrac{y^2 - 16}{-8 - 2y}$

79. $\dfrac{-3x - 6}{3x - 6}$

80. $\dfrac{8 - 4x}{-8x - 16}$

81. $\dfrac{-12a - 6}{2a^2 + 7a + 3}$

82. $\dfrac{-2b^2 - 6b - 4}{b^2 - 1}$

83. $\dfrac{a^3 - b^3}{2b^2 - 2ab}$

84. $\dfrac{x^3 - 1}{x - x^2}$

Reduce each expression to lowest terms. See the Strategy for Reducing Rational Expressions box on page 388.

85. $\dfrac{2x^{12}}{4x^8}$

86. $\dfrac{4x^2}{2x^9}$

87. $\dfrac{2x + 4}{4x}$

88. $\dfrac{2x + 4x^2}{4x}$

89. $\dfrac{a - 4}{4 - a}$

90. $\dfrac{2b - 4}{2b + 4}$

91. $\dfrac{2c - 4}{4 - c^2}$

92. $\dfrac{-2t - 4}{4 - t^2}$

93. $\dfrac{x^2 + 4x + 4}{x^2 - 4}$

94. $\dfrac{3x - 6}{x^2 - 4x + 4}$

95. $\dfrac{-2x - 4}{x^2 + 5x + 6}$

96. $\dfrac{-2x - 8}{x^2 + 2x - 8}$

97. $\dfrac{2q^8 + q^7}{2q^6 + q^5}$

98. $\dfrac{8s^{12}}{12s^6 - 16s^5}$

99. $\dfrac{u^2 - 6u - 16}{u^2 - 16u + 64}$

100. $\dfrac{v^2 + 3v - 18}{v^2 + 12v + 36}$

101. $\dfrac{a^3 - 8}{2a - 4}$

102. $\dfrac{4w^2 - 12w + 36}{2w^3 + 54}$

103. $\dfrac{y^3 - 2y^2 - 4y + 8}{y^2 - 4y + 4}$

104. $\dfrac{mx + 3x + my + 3y}{m^2 - 3m - 18}$

105. $\dfrac{2x + 2w - ax - aw}{x^3 - xw^2}$

106. $\dfrac{x^2 + ax - 4x - 4a}{x^2 - 16}$

⟨6⟩ Writing Rational Expressions

Answer each question with a rational expression. Be sure to include the units. See Example 9.

107. If Sergio drove 300 miles at $x + 10$ miles per hour, then how many hours did he drive?

108. If Carrie walked 40 miles in x hours, then how fast did she walk?

109. If $x + 4$ pounds of peaches cost \$4.50, then what is the cost per pound?

110. If nine pounds of pears cost x dollars, then what is the price per pound?

111. If Ayesha can clean the entire swimming pool in x hours, then how much of the pool does she clean per hour?

112. If Ramon can mow the entire lawn in $x - 3$ hours, then how much of the lawn does he mow per hour?

Applications

Solve each problem.

113. *Annual reports.* The Crest Meat Company found that the cost per report for printing x annual reports at Peppy Printing is given by the formula

$$C(x) = \frac{150 + 0.60x}{x},$$

where $C(x)$ is in dollars.

a) Use the accompanying graph to estimate the cost per report for printing 1000 reports.

b) Use the formula to find $C(1000)$, $C(5000)$, and $C(10,000)$.

c) What happens to the cost per report as the number of reports gets very large?

given by the formula,

$$C(p) = \frac{500,000}{100 - p}.$$

a) Use the accompanying graph to estimate the cost for removing 90% and 95% of the toxic chemicals.

b) Use the formula to find $C(99.5)$ and $C(99.9)$.

c) What happens to the cost as the percentage of pollutants removed approaches 100%?

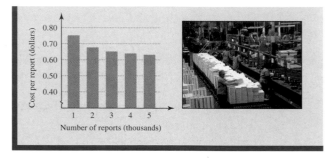

Figure for Exercise 113

114. *Toxic pollutants.* The annual cost in dollars for removing $p\%$ of the toxic chemicals from a town's water supply is

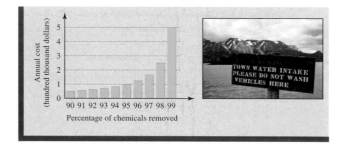

Figure for Exercise 114

6.2   Multiplication and Division

In Section 6.1, you learned to reduce rational expressions in the same way that we reduce rational numbers. In this section, we will multiply and divide rational expressions using the same procedures that we use for rational numbers.

⟨1⟩   Multiplication of Rational Numbers

Two rational numbers are multiplied by multiplying their numerators and multiplying their denominators.

> **Multiplication of Rational Numbers**
> If $b \neq 0$ and $d \neq 0$, then
>
> $$\frac{a}{b} \cdot \frac{c}{d} = \frac{ac}{bd}.$$

EXAMPLE   1

‹ **Helpful Hint** ›

Did you know that the line separating the numerator and denominator in a fraction is called the *vinculum*?

Multiplying rational numbers

Find the product $\frac{6}{7} \cdot \frac{14}{15}$.

Solution

The product is found by multiplying the numerators and multiplying the denominators:

$$\frac{6}{7} \cdot \frac{14}{15} = \frac{84}{105}$$

$$= \frac{21 \cdot 4}{21 \cdot 5} \qquad \text{Factor the numerator and denominator}$$

$$= \frac{4}{5} \qquad \text{Divide out the GCF 21.}$$

The reducing that we did after multiplying is easier to do before multiplying. First factor all terms, reduce, and then multiply:

$$\frac{6}{7} \cdot \frac{14}{15} = \frac{2 \cdot \cancel{3}}{\cancel{7}} \cdot \frac{2 \cdot \cancel{7}}{\cancel{3} \cdot 5}$$

$$= \frac{4}{5}$$

Now do Exercises 1–8

‹2› Multiplication of Rational Expressions

Rational expressions are multiplied just like rational numbers: factor, reduce, and then multiply. A rational number cannot have zero in its denominator and neither can a rational expression. Since a rational expression can have variables in its denominator, the results obtained in Examples 2 and 3 are valid only for values of the variable(s) that would not cause a denominator to be 0.

EXAMPLE   2

Multiplying rational expressions

Find the indicated products.

a) $\dfrac{9x}{5y} \cdot \dfrac{10y}{3xy}$

b) $\dfrac{-8xy^4}{3z^3} \cdot \dfrac{15z}{2x^5y^3}$

Solution

a) $\dfrac{9x}{5y} \cdot \dfrac{10y}{3xy} = \dfrac{3 \cdot \cancel{3}\cancel{x}}{\cancel{5}\cancel{y}} \cdot \dfrac{2 \cdot \cancel{5}\cancel{y}}{\cancel{3}\cancel{x}y}$ Factor.

$$= \frac{6}{y}$$

b) $\dfrac{-8xy^4}{3z^3} \cdot \dfrac{15z}{2x^5y^3} = \dfrac{-2 \cdot 2 \cdot \cancel{2}xy^4}{\cancel{3}z^3} \cdot \dfrac{\cancel{3} \cdot 5z}{\cancel{2}x^5y^3}$ Factor.

$$= \frac{-20xy^4z}{z^3x^5y^3} \qquad \text{Reduce.}$$

$$= \frac{-20y}{z^2x^4} \qquad \text{Quotient rule}$$

Now do Exercises 9–18

EXAMPLE   3

Multiplying rational expressions
Find the indicated products.

a) $\dfrac{2x - 2y}{4} \cdot \dfrac{2x}{x^2 - y^2}$

b) $\dfrac{x^2 + 7x + 12}{2x + 6} \cdot \dfrac{x}{x^2 - 16}$

c) $\dfrac{a + b}{6a} \cdot \dfrac{8a^2}{a^2 + 2ab + b^2}$

Solution

a) $\dfrac{2x - 2y}{4} \cdot \dfrac{2x}{x^2 - y^2} = \dfrac{2(x - y)}{2 \cdot 2} \cdot \dfrac{2 \cdot x}{(x - y)(x + y)}$ Factor.

$= \dfrac{x}{x + y}$ Reduce.

b) $\dfrac{x^2 + 7x + 12}{2x + 6} \cdot \dfrac{x}{x^2 - 16} = \dfrac{(x + 3)(x + 4)}{2(x + 3)} \cdot \dfrac{x}{(x - 4)(x + 4)}$ Factor.

$= \dfrac{x}{2(x - 4)}$ Reduce.

c) $\dfrac{a + b}{6a} \cdot \dfrac{8a^2}{a^2 + 2ab + b^2} = \dfrac{a + b}{2 \cdot 3a} \cdot \dfrac{2 \cdot 4a^2}{(a + b)^2}$ Factor.

$= \dfrac{4a}{3(a + b)}$ Reduce.

Now do Exercises 19–26

⟨3⟩ Division of Rational Numbers

By the definition of division, a quotient is found by multiplying the dividend by the reciprocal of the divisor. If the divisor is a rational number $\dfrac{c}{d}$, its reciprocal is simply $\dfrac{d}{c}$.

Division of Rational Numbers

If $b \neq 0$, $c \neq 0$, and $d \neq 0$, then

$$\frac{a}{b} \div \frac{c}{d} = \frac{a}{b} \cdot \frac{d}{c}.$$

EXAMPLE   4

Dividing rational numbers
Find each quotient.

a) $5 \div \dfrac{1}{2}$

b) $\dfrac{6}{7} \div \dfrac{3}{14}$

Solution

a) $5 \div \dfrac{1}{2} = 5 \cdot 2 = 10$

b) $\dfrac{6}{7} \div \dfrac{3}{14} = \dfrac{6}{7} \cdot \dfrac{14}{3} = \dfrac{2 \cdot 3}{7} \cdot \dfrac{2 \cdot 7}{3} = 4$

Now do Exercises 27–34

⟨4⟩ Division of Rational Expressions

We divide rational expressions in the same way we divide rational numbers: Invert the divisor and multiply.

EXAMPLE 5	**Dividing rational expressions**

Find each quotient.

a) $\dfrac{5}{3x} \div \dfrac{5}{6x}$ **b)** $\dfrac{x^7}{2} \div (2x^2)$ **c)** $\dfrac{4 - x^2}{x^2 + x} \div \dfrac{x - 2}{x^2 - 1}$

Solution

a) $\dfrac{5}{3x} \div \dfrac{5}{6x} = \dfrac{5}{3x} \cdot \dfrac{6x}{5}$ Invert the divisor and multiply.

$= \dfrac{\cancel{5}}{\cancel{3x}} \cdot \dfrac{2 \cdot \cancel{3x}}{\cancel{5}}$ Factor.

$= 2$ Divide out the common factors.

b) $\dfrac{x^7}{2} \div (2x^2) = \dfrac{x^7}{2} \cdot \dfrac{1}{2x^2}$ Invert and multiply.

$= \dfrac{x^5}{4}$ Quotient rule

c) $\dfrac{4 - x^2}{x^2 + x} \div \dfrac{x - 2}{x^2 - 1} = \dfrac{4 - x^2}{x^2 + x} \cdot \dfrac{x^2 - 1}{x - 2}$ Invert and multiply.

$= \dfrac{\overset{-1}{\cancel{(2 - x)}}(2 + x)}{x\cancel{(x + 1)}} \cdot \dfrac{\cancel{(x + 1)}(x - 1)}{\cancel{x - 2}}$ Factor.

$= \dfrac{-1(2 + x)(x - 1)}{x}$ $\dfrac{2 - x}{x - 2} = -1$

$= \dfrac{-1(x^2 + x - 2)}{x}$ Simplify.

$= \dfrac{-x^2 - x + 2}{x}$

> **Now do Exercises 35–48**

We sometimes write division of rational expressions using the fraction bar. For example, we can write

$$\dfrac{a + b}{3} \div \dfrac{1}{6} \quad \text{as} \quad \dfrac{\dfrac{a + b}{3}}{\dfrac{1}{6}}.$$

No matter how division is expressed, we invert the divisor and multiply.

EXAMPLE 6	**Division expressed with a fraction bar**

Find each quotient.

a) $\dfrac{\dfrac{a + b}{3}}{\dfrac{1}{6}}$ **b)** $\dfrac{\dfrac{x^2 - 1}{2}}{\dfrac{x - 1}{3}}$ **c)** $\dfrac{\dfrac{a^2 + 5}{3}}{2}$

Solution

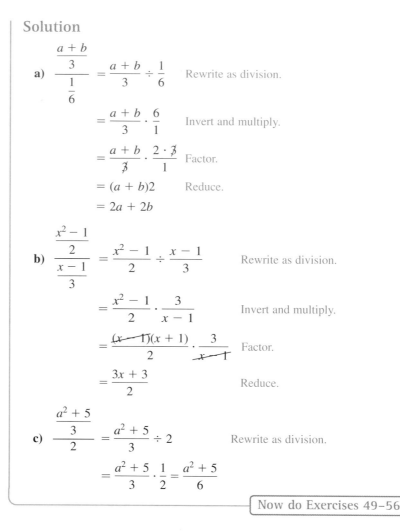

a) $\dfrac{\dfrac{a+b}{3}}{\dfrac{1}{6}} = \dfrac{a+b}{3} \div \dfrac{1}{6}$ Rewrite as division.

$= \dfrac{a+b}{3} \cdot \dfrac{6}{1}$ Invert and multiply.

$= \dfrac{a+b}{\cancel{3}} \cdot \dfrac{2 \cdot \cancel{3}}{1}$ Factor.

$= (a+b)2$ Reduce.

$= 2a + 2b$

b) $\dfrac{\dfrac{x^2-1}{2}}{\dfrac{x-1}{3}} = \dfrac{x^2-1}{2} \div \dfrac{x-1}{3}$ Rewrite as division.

$= \dfrac{x^2-1}{2} \cdot \dfrac{3}{x-1}$ Invert and multiply.

$= \dfrac{\cancel{(x-1)}(x+1)}{2} \cdot \dfrac{3}{\cancel{x-1}}$ Factor.

$= \dfrac{3x+3}{2}$ Reduce.

c) $\dfrac{\dfrac{a^2+5}{3}}{2} = \dfrac{a^2+5}{3} \div 2$ Rewrite as division.

$= \dfrac{a^2+5}{3} \cdot \dfrac{1}{2} = \dfrac{a^2+5}{6}$

⟨ **Helpful Hint** ⟩

In Section 6.5 you will see another technique for finding the quotients in Example 6.

> Now do Exercises 49–56

⟨5⟩ Applications

We saw in Section 6.1 that rational expressions can be used to represent rates. Note that there are several ways to write rates. For example, miles per hour is written mph, mi/hr, or $\frac{mi}{hr}$. The last way is best when doing operations with rates because it helps us reconcile our answers. Notice how hours "cancels" when we multiply miles per hour and hours in Example 7, giving an answer in miles, as it should be.

E X A M P L E   **7**

Using rational expressions with uniform motion

Shasta drove 200 miles on I-10 in x hours before lunch.

a) Write a rational expression for her average speed before lunch.

b) She drives for 3 hours after lunch at the same average speed. Write a rational expression for her distance after lunch.

Solution

a) Because $R = \dfrac{D}{T}$, her rate before lunch is $\dfrac{200 \text{ miles}}{x \text{ hours}}$ or $\dfrac{200}{x}$ mph.

b) Because $D = R \cdot T$, her distance after lunch is the product of $\frac{200}{x}$ mph (her rate) and 3 hours (her time):

$$D = \frac{200 \text{ mi}}{x \text{ hr}} \cdot 3 \text{ hr} = \frac{600}{x} \text{ mi}$$

> Now do Exercises 77–78

The amount of work completed is the product of rate and time, $W = R \cdot T$. So if a machine washes cars at the rate of 12 per hour and it works for 3 hours, the amount of work completed is 36 cars washed. Note that the rate is given by $R = \frac{W}{T}$.

E X A M P L E   **8**

Using rational expressions with work

It takes x minutes to fill a bathtub.

   a) Write a rational expression for the rate at which the tub is filling.
   b) Write a rational expression for the portion of the tub that is filled in 10 minutes.

Solution

   a) The work completed in this situation is 1 tub being filled. Because $R = \frac{W}{T}$, the rate at which the tub is filling is $\frac{1 \text{ tub}}{x \text{ min}}$ or $\frac{1}{x}$ tub/min.

   b) Because $W = R \cdot T$, the work completed in 10 minutes or the portion of the tub that is filled in 10 minutes is the product of $\frac{1}{x}$ tub/min (the rate) and 10 minutes (the time):

$$W = \frac{1 \text{ tub}}{x \text{ min}} \cdot 10 \text{ min} = \frac{10}{x} \text{ tub}$$

> Now do Exercises 79–80

Warm-Ups ▼

Fill in the blank.

  1. _____ expressions are multiplied by multiplying their numerators and multiplying their denominators.

  2. _____ can be done before multiplying rational expressions.

  3. To _____ rational expressions, invert the divisor and multiply.

True or false?

  4. One-half of one-fourth is one-sixth.

  5. $\dfrac{2}{3} \cdot \dfrac{5}{7} = \dfrac{10}{21}$

  6. The product of $\dfrac{x-7}{3}$ and $\dfrac{6}{7-x}$ is -2.

  7. Dividing by 2 is equivalent to multiplying by $\dfrac{1}{2}$.

  8. For any real number a, $\dfrac{a}{3} \div 3 = \dfrac{a}{9}$.

  9. $\dfrac{2}{3} \div \dfrac{1}{2} = \dfrac{4}{3}$

‹ 1 › **Multiplication of Rational Numbers**

Perform the indicated operation. See Example 1.

1. $\dfrac{2}{3} \cdot \dfrac{5}{6}$

2. $\dfrac{3}{4} \cdot \dfrac{2}{5}$

3. $\dfrac{8}{15} \cdot \dfrac{35}{24}$

4. $\dfrac{3}{4} \cdot \dfrac{8}{21}$

5. $\dfrac{12}{17} \cdot \dfrac{51}{10}$

6. $\dfrac{25}{48} \cdot \dfrac{56}{35}$

7. $24 \cdot \dfrac{7}{20}$

8. $\dfrac{3}{10} \cdot 35$

‹ 2 › **Multiplication of Rational Expressions**

Perform the indicated operation. See Example 2.

9. $\dfrac{2x}{3} \cdot \dfrac{5}{4x}$

10. $\dfrac{3y}{7} \cdot \dfrac{21}{2y}$

11. $\dfrac{5x^2}{6} \cdot \dfrac{3}{x}$

12. $\dfrac{9x}{10} \cdot \dfrac{5}{x^2}$

13. $\dfrac{5a}{12b} \cdot \dfrac{3ab}{55a}$

14. $\dfrac{3m}{7p} \cdot \dfrac{35p}{6mp}$

15. $\dfrac{-2x^6}{7a^5} \cdot \dfrac{21a^2}{6x}$

16. $\dfrac{5z^3w}{-9y^3} \cdot \dfrac{-6y^5}{20z^9}$

17. $\dfrac{15t^3y^5}{20w^7} \cdot 24t^5w^3y^2$

18. $22x^2y^3z \cdot \dfrac{6x^5}{33y^3z^4}$

Perform the indicated operation. See Example 3.

19. $\dfrac{2x + 2y}{7} \cdot \dfrac{15}{6x + 6y}$

20. $\dfrac{3}{a^2 + a} \cdot \dfrac{2a + 2}{6}$

21. $\dfrac{3a + 3b}{15} \cdot \dfrac{10a}{a^2 - b^2}$

22. $\dfrac{b^3 + b}{5} \cdot \dfrac{10}{b^2 + b}$

23. $(x^2 - 6x + 9) \cdot \dfrac{3}{x - 3}$

24. $\dfrac{12}{4x + 10} \cdot (4x^2 + 20x + 25)$

25. $\dfrac{16a + 8}{5a^2 + 5} \cdot \dfrac{2a^2 + a - 1}{4a^2 - 1}$

26. $\dfrac{6x - 18}{2x^2 - 5x - 3} \cdot \dfrac{4x^2 + 4x + 1}{6x + 3}$

‹ 3 › **Division of Rational Numbers**

Perform the indicated operation. See Example 4.

27. $\dfrac{1}{4} \div \dfrac{1}{2}$

28. $\dfrac{1}{6} \div \dfrac{1}{2}$

29. $12 \div \dfrac{2}{5}$

30. $32 \div \dfrac{1}{4}$

31. $\dfrac{5}{7} \div \dfrac{15}{14}$

32. $\dfrac{3}{4} \div \dfrac{15}{2}$

33. $\dfrac{40}{3} \div 12$

34. $\dfrac{22}{9} \div 9$

‹ 4 › **Division of Rational Expressions**

Perform the indicated operation. See Example 5.

35. $\dfrac{x^2}{4} \div \dfrac{x}{2}$

36. $\dfrac{3}{2a^2} \div \dfrac{6}{2a}$

37. $\dfrac{5x^2}{3} \div \dfrac{10x}{21}$

38. $\dfrac{4u^2}{3v} \div \dfrac{14u}{15v^6}$

39. $\dfrac{8m^3}{n^4} \div (12mn^2)$

40. $\dfrac{2p^4}{3q^3} \div (4pq^5)$

41. $\dfrac{y - 6}{2} \div \dfrac{6 - y}{6}$

42. $\dfrac{4 - a}{5} \div \dfrac{a^2 - 16}{3}$

43. $\dfrac{x^2 + 4x + 4}{8} \div \dfrac{(x + 2)^3}{16}$

44. $\dfrac{a^2 + 2a + 1}{3} \div \dfrac{a^2 - 1}{a}$

45. $\dfrac{t^2 + 3t - 10}{t^2 - 25} \div (4t - 8)$

46. $\dfrac{w^2 - 7w + 12}{w^2 - 4w} \div (w^2 - 9)$

47. $(2x^2 - 3x - 5) \div \dfrac{2x - 5}{x - 1}$

48. $(6y^2 - y - 2) \div \dfrac{2y + 1}{3y - 2}$

Perform the indicated operation. See Example 6.

49. $\dfrac{\dfrac{x - 2y}{5}}{\dfrac{1}{10}}$ **50.** $\dfrac{\dfrac{3m + 6n}{8}}{\dfrac{3}{4}}$

51. $\dfrac{\dfrac{x^2 - 4}{12}}{\dfrac{x - 2}{6}}$ **52.** $\dfrac{\dfrac{6a^2 + 6}{5}}{\dfrac{6a + 6}{5}}$

53. $\dfrac{\dfrac{x^2 + 9}{3}}{5}$ **54.** $\dfrac{\dfrac{1}{a - 3}}{4}$

55. $\dfrac{\dfrac{x^2 - y^2}{x - y}}{9}$ **56.** $\dfrac{\dfrac{x^2 + 6x + 8}{x + 2}}{x + 1}$

Miscellaneous

Perform the indicated operation.

57. $\dfrac{x - 1}{3} \cdot \dfrac{9}{1 - x}$ **58.** $\dfrac{2x - 2y}{3} \cdot \dfrac{1}{y - x}$

59. $\dfrac{3a + 3b}{a} \cdot \dfrac{1}{3}$ **60.** $\dfrac{a - b}{2b - 2a} \cdot \dfrac{2}{5}$

61. $\dfrac{\dfrac{b}{a}}{\dfrac{1}{2}}$ **62.** $\dfrac{\dfrac{2g}{3h}}{\dfrac{1}{h}}$

63. $\dfrac{6y}{3} \div (2x)$ **64.** $\dfrac{8x}{9} \div (18x)$

65. $\dfrac{a^3 b^4}{-2ab^2} \cdot \dfrac{a^5 b^7}{ab}$ **66.** $\dfrac{-2a^2}{3a^2} \cdot \dfrac{20a}{15a^3}$

67. $\dfrac{2mn^4}{6mn^2} \div \dfrac{3m^5 n^7}{m^2 n^4}$

68. $\dfrac{rt^2}{rt^2} \div \dfrac{rt^2}{r^3 t^2}$

69. $\dfrac{3x^2 + 16x + 5}{x} \cdot \dfrac{x^2}{9x^2 - 1}$

70. $\dfrac{x^2 + 6x + 5}{x} \cdot \dfrac{x^4}{3x + 3}$

71. $\dfrac{a^2 - 2a + 4}{a^2 - 4} \cdot \dfrac{(a + 2)^3}{2a + 4}$

72. $\dfrac{w^2 - 1}{(w - 1)^2} \cdot \dfrac{w - 1}{w^2 + 2w + 1}$

73. $\dfrac{2x^2 + 19x - 10}{x^2 - 100} \div \dfrac{4x^2 - 1}{2x^2 - 19x - 10}$

74. $\dfrac{x^3 - 1}{x^2 + 1} \div \dfrac{9x^2 + 9x + 9}{x^2 - x}$

75. $\dfrac{9 + 6m + m^2}{9 - 6m + m^2} \cdot \dfrac{m^2 - 9}{m^2 + mk + 3m + 3k}$

76. $\dfrac{3x + 3w + bx + bw}{x^2 - w^2} \cdot \dfrac{6 - 2b}{9 - b^2}$

⟨5⟩ Applications

Solve each problem. Answers could be rational expressions. Be sure to give your answers with appropriate units. See Examples 7 and 8.

77. *Marathon run.* Florence ran 26.2 miles in x hours in the Boston Marathon.
 a) Write a rational expression for her average speed.

 b) She runs at the same average speed for $\frac{1}{2}$ hour in the Cripple Creek Fun Run. Write a rational expression for her distance at Cripple Creek.

78. *Driving marathon.* Felix drove 800 miles in x hours on Monday.
 a) Write a rational expression for his average speed.

 b) On Tuesday he drove for 6 hours at the same average speed. Write a rational expression for his distance on Tuesday.

79. *Filling the tank.* Chantal filled her empty gas tank in x minutes.
  a) Write a rational expression for the rate at which she filled her tank.

  b) Write a rational expression for the portion of the tank that is filled in 2 minutes.

80. *Magazine sales.* Henry sold 120 magazine subscriptions in x days.
  a) Write a rational expression for the rate at which he sold the subscriptions.

  b) Suppose that he continues to sell at the same rate for 5 more days. Write a rational expression for the number of magazines sold in those 5 days.

81. *Area of a rectangle.* If the length of a rectangular flag is x meters and its width is $\frac{5}{x}$ meters, then what is the area of the rectangle?

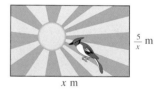

Figure for Exercise 81

82. *Area of a triangle.* If the base of a triangle is $8x + 16$ yards and its height is $\frac{1}{x+2}$ yards, then what is the area of the triangle?

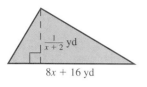

Figure for Exercise 82

Getting More Involved

83. *Discussion*

Evaluate each expression.

  a) One-half of $\frac{1}{4}$ **b)** One-third of 4

  c) One-half of $\frac{4x}{3}$ **d)** One-half of $\frac{3x}{2}$

84. *Exploration*

Let $R = \dfrac{6x^2 + 23x + 20}{24x^2 + 29x - 4}$ and $H = \dfrac{2x + 5}{8x - 1}$.

  a) Find R when $x = 2$ and $x = 3$. Find H when $x = 2$ and $x = 3$.

  b) How are these values of R and H related and why?

6.3   **Finding the Least Common Denominator**

In This Section

⟨1⟩ **Building Up the Denominator**

⟨2⟩ **Finding the Least Common Denominator**

⟨3⟩ **Converting to the LCD**

Every rational expression can be written in infinitely many equivalent forms. Because we can add or subtract only fractions with identical denominators, we must be able to change the denominator of a fraction. You have already learned how to change the denominator of a fraction by reducing. In this section, you will learn the opposite of reducing, which is called **building up the denominator.**

⟨1⟩ **Building Up the Denominator**

To convert the fraction $\frac{2}{3}$ into an equivalent fraction with a denominator of 21, we factor 21 as $21 = 3 \cdot 7$. Because $\frac{2}{3}$ already has a 3 in the denominator, multiply the numerator and denominator of $\frac{2}{3}$ by the missing factor 7 to get a denominator of 21:

$$\frac{2}{3} = \frac{2}{3} \cdot \frac{7}{7} = \frac{14}{21}$$

For rational expressions the process is the same. To convert the rational expression

$$\frac{5}{x+3}$$

into an equivalent rational expression with a denominator of $x^2 - x - 12$, first factor $x^2 - x - 12$:

$$x^2 - x - 12 = (x+3)(x-4)$$

From the factorization we can see that the denominator $x + 3$ needs only a factor of $x - 4$ to have the required denominator. So multiply the numerator and denominator by the missing factor $x - 4$:

$$\frac{5}{x+3} = \frac{5(x-4)}{(x+3)(x-4)} = \frac{5x-20}{x^2-x-12}$$

E X A M P L E   **1**

Building up the denominator

Build each rational expression into an equivalent rational expression with the indicated denominator.

a) $3 = \dfrac{?}{12}$ **b)** $\dfrac{3}{w} = \dfrac{?}{wx}$ **c)** $\dfrac{2}{3y^3} = \dfrac{?}{12y^8}$

Solution

a) Because $3 = \frac{3}{1}$, we get a denominator of 12 by multiplying the numerator and denominator by 12:

$$3 = \frac{3}{1} = \frac{3 \cdot 12}{1 \cdot 12} = \frac{36}{12}$$

b) Multiply the numerator and denominator by x:

$$\frac{3}{w} = \frac{3 \cdot x}{w \cdot x} = \frac{3x}{wx}$$

c) Note that $12y^8 = 3y^3 \cdot 4y^5$. So to build $3y^3$ up to $12y^8$ multiply by $4y^5$:

$$\frac{2}{3y^3} = \frac{2 \cdot 4y^5}{3y^3 \cdot 4y^5} = \frac{8y^5}{12y^8}$$

> Now do Exercises 1–20

In Example 2 we must factor the original denominator before building up the denominator.

E X A M P L E   **2**

Building up the denominator

Build each rational expression into an equivalent rational expression with the indicated denominator.

a) $\dfrac{7}{3x-3y} = \dfrac{?}{6y-6x}$ **b)** $\dfrac{x-2}{x+2} = \dfrac{?}{x^2+8x+12}$

‹ **Helpful Hint** ›

Notice that reducing and building up are exactly the opposite of each other. In reducing you remove a factor that is common to the numerator and denominator, and in building up you put a common factor into the numerator and denominator.

Solution

a) Because $3x - 3y = 3(x - y)$, we factor -6 out of $6y - 6x$. This will give a factor of $x - y$ in each denominator:

$$3x - 3y = 3(x - y)$$
$$6y - 6x = -6(x - y) = -2 \cdot 3(x - y)$$

To get the required denominator, we multiply the numerator and denominator by -2 only:

$$\frac{7}{3x - 3y} = \frac{7(-2)}{(3x - 3y)(-2)}$$

$$= \frac{-14}{6y - 6x}$$

b) Because $x^2 + 8x + 12 = (x + 2)(x + 6)$, we multiply the numerator and denominator by $x + 6$, the missing factor:

$$\frac{x - 2}{x + 2} = \frac{(x - 2)(x + 6)}{(x + 2)(x + 6)}$$

$$= \frac{x^2 + 4x - 12}{x^2 + 8x + 12}$$

> Now do Exercises 21–32

CAUTION   When building up a denominator, *both* the numerator and the denominator must be multiplied by the appropriate expression.

‹2› Finding the Least Common Denominator

We can use the idea of building up the denominator to convert two fractions with different denominators into fractions with identical denominators. For example,

$$\frac{5}{6} \quad \text{and} \quad \frac{1}{4}$$

can both be converted into fractions with a denominator of 12, since $12 = 2 \cdot 6$ and $12 = 3 \cdot 4$:

$$\frac{5}{6} = \frac{5 \cdot 2}{6 \cdot 2} = \frac{10}{12} \qquad \frac{1}{4} = \frac{1 \cdot 3}{4 \cdot 3} = \frac{3}{12}$$

The smallest number that is a multiple of all of the denominators is called the **least common denominator (LCD).** The LCD for the denominators 6 and 4 is 12.

To find the LCD in a systematic way, we look at a complete factorization of each denominator. Consider the denominators 24 and 30:

$$24 = 2 \cdot 2 \cdot 2 \cdot 3 = 2^3 \cdot 3$$
$$30 = 2 \cdot 3 \cdot 5$$

Any multiple of 24 must have three 2's in its factorization, and any multiple of 30 must have one 2 as a factor. So a number with three 2's in its factorization will have enough to be a multiple of both 24 and 30. The LCD must also have one 3 and one 5 in its factorization. *We use each factor the maximum number of times it appears in either factorization.* So the LCD is $2^3 \cdot 3 \cdot 5$:

$$2^3 \cdot 3 \cdot 5 = \overbrace{2 \cdot 2 \cdot \underbrace{2 \cdot 3 \cdot 5}_{30}}^{24} = 120$$

If we omitted any one of the factors in $2 \cdot 2 \cdot 2 \cdot 3 \cdot 5$, we would not have a multiple of both 24 and 30. That is what makes 120 the *least* common denominator. To find the LCD for two polynomials, we use the same strategy.

Strategy for Finding the LCD for Polynomials

1. Factor each denominator completely. Use exponent notation for repeated factors.

2. Write the product of all of the different factors that appear in the denominators.

3. On each factor, use the highest power that appears on that factor in any of the denominators.

E X A M P L E   **3**

Finding the LCD

If the given expressions were used as denominators of rational expressions, then what would be the LCD for each group of denominators?

   a) 20, 50   **b)** x^3yz^2, x^5y^2z, xyz^5 **c)** $a^2 + 5a + 6, a^2 + 4a + 4$

Solution

a) First factor each number completely:

$$20 = 2^2 \cdot 5 \qquad 50 = 2 \cdot 5^2$$

The highest power of 2 is 2, and the highest power of 5 is 2. So the LCD of 20 and 50 is $2^2 \cdot 5^2$, or 100.

b) The expressions x^3yz^2, x^5y^2z, and xyz^5 are already factored. For the LCD, use the highest power of each variable. So the LCD is $x^5y^2z^5$.

c) First factor each polynomial.

$$a^2 + 5a + 6 = (a + 2)(a + 3) \qquad a^2 + 4a + 4 = (a + 2)^2$$

The highest power of $(a + 3)$ is 1, and the highest power of $(a + 2)$ is 2. So the LCD is $(a + 3)(a + 2)^2$.

> Now do Exercises 33–46

⟨3⟩ Converting to the LCD

When adding or subtracting rational expressions, we must convert the expressions into expressions with identical denominators. To keep the computations as simple as possible, we use the least common denominator.

E X A M P L E   **4**

Converting to the LCD

Find the LCD for the rational expressions, and convert each expression into an equivalent rational expression with the LCD as the denominator.

   a) $\dfrac{4}{9xy}, \dfrac{2}{15xz}$ **b)** $\dfrac{5}{6x^2}, \dfrac{1}{8x^3y}, \dfrac{3}{4y^2}$

Solution

a) Factor each denominator completely:

$$9xy = 3^2xy \qquad 15xz = 3 \cdot 5xz$$

The LCD is $3^2 \cdot 5xyz$. Now convert each expression into an expression with this denominator. We must multiply the numerator and denominator of the first rational expression by $5z$ and the second by $3y$:

$$\left.\begin{array}{l} \dfrac{4}{9xy} = \dfrac{4 \cdot 5z}{9xy \cdot 5z} = \dfrac{20z}{45xyz} \\[3mm] \dfrac{2}{15xz} = \dfrac{2 \cdot 3y}{15xz \cdot 3y} = \dfrac{6y}{45xyz} \end{array}\right\} \text{Same denominator}$$

b) Factor each denominator completely:

$$6x^2 = 2 \cdot 3x^2 \qquad 8x^3y = 2^3x^3y \qquad 4y^2 = 2^2y^2$$

The LCD is $2^3 \cdot 3 \cdot x^3y^2$ or $24x^3y^2$. Now convert each expression into an expression with this denominator:

$$\dfrac{5}{6x^2} = \dfrac{5 \cdot 4xy^2}{6x^2 \cdot 4xy^2} = \dfrac{20xy^2}{24x^3y^2}$$

$$\dfrac{1}{8x^3y} = \dfrac{1 \cdot 3y}{8x^3y \cdot 3y} = \dfrac{3y}{24x^3y^2}$$

$$\dfrac{3}{4y^2} = \dfrac{3 \cdot 6x^3}{4y^2 \cdot 6x^3} = \dfrac{18x^3}{24x^3y^2}$$

Now do Exercises 47–58

EXAMPLE 5

Converting to the LCD
Find the LCD for the rational expressions

$$\dfrac{5x}{x^2 - 4} \quad \text{and} \quad \dfrac{3}{x^2 + x - 6}$$

and convert each into an equivalent rational expression with that denominator.

Solution
First factor the denominators:

$$x^2 - 4 = (x - 2)(x + 2)$$
$$x^2 + x - 6 = (x - 2)(x + 3)$$

The LCD is $(x - 2)(x + 2)(x + 3)$. Now we multiply the numerator and denominator of the first rational expression by $(x + 3)$ and those of the second rational expression by $(x + 2)$. Because each denominator already has one factor of $(x - 2)$, there is no reason to multiply by $(x - 2)$. We multiply each denominator by the factors in the LCD that are missing from that denominator:

$$\left.\begin{array}{l} \dfrac{5x}{x^2 - 4} = \dfrac{5x(x + 3)}{(x - 2)(x + 2)(x + 3)} = \dfrac{5x^2 + 15x}{(x - 2)(x + 2)(x + 3)} \\[3mm] \dfrac{3}{x^2 + x - 6} = \dfrac{3(x + 2)}{(x - 2)(x + 3)(x + 2)} = \dfrac{3x + 6}{(x - 2)(x + 2)(x + 3)} \end{array}\right\} \begin{array}{l} \text{Same} \\ \text{denominator} \end{array}$$

Now do Exercises 59–70

Warm-Ups ▼

Fill in the blank.

1. To _____ the denominator of a fraction, we multiply the numerator and denominator by the same nonzero real number.

2. The _____ is the smallest number that is a multiple of all denominators.

3. The LCD is the product of every factor that appears in the factorizations, raised to the _____ power that appears on the factor.

True or false?

4. $\dfrac{2}{3} = \dfrac{2 \cdot 5}{3 \cdot 5}$

5. $\dfrac{2}{3} = \dfrac{2 + 5}{3 + 5}$

6. The LCD for the denominators $2^5 \cdot 3$ and $2^4 \cdot 3^2$ is $2^5 \cdot 3^2$.

7. The LCD for $\dfrac{1}{6}$ and $\dfrac{1}{10}$ is 60.

8. The LCD for $\dfrac{1}{x-2}$ and $\dfrac{1}{x+2}$ is $x^2 - 4$.

9. The LCD for $\dfrac{1}{a^2-1}$ and $\dfrac{1}{a-1}$ is $a^2 - 1$.

Exercises 6.3

‹ **Study Tips** ›

- Try changing subjects or tasks every hour when you study. The brain does not easily assimilate the same material hour after hour.
- You will learn more from working on a subject one hour per day than seven hours on Saturday.

‹ 1 › **Building Up the Denominator**

Build each rational expression into an equivalent rational expression with the indicated denominator. See Example 1.

1. $\dfrac{1}{3} = \dfrac{?}{27}$

2. $\dfrac{2}{5} = \dfrac{?}{35}$

3. $\dfrac{3}{4} = \dfrac{?}{16}$

4. $\dfrac{3}{7} = \dfrac{?}{28}$

5. $1 = \dfrac{?}{7}$

6. $1 = \dfrac{?}{3x}$

7. $2 = \dfrac{?}{6}$

8. $5 = \dfrac{?}{12}$

9. $\dfrac{5}{x} = \dfrac{?}{ax}$

10. $\dfrac{x}{3} = \dfrac{?}{3x}$

11. $7 = \dfrac{?}{2x}$

12. $6 = \dfrac{?}{4y}$

13. $\dfrac{5}{b} = \dfrac{?}{3bt}$

14. $\dfrac{7}{2ay} = \dfrac{?}{2ayz}$

15. $\dfrac{-9z}{2aw} = \dfrac{?}{8awz}$

16. $\dfrac{-7yt}{3x} = \dfrac{?}{18xyt}$

17. $\dfrac{2}{3a} = \dfrac{?}{15a^3}$

18. $\dfrac{7b}{12c^5} = \dfrac{?}{36c^8}$

19. $\dfrac{4}{5xy^2} = \dfrac{?}{10x^2y^5}$

20. $\dfrac{5y^2}{8x^3z} = \dfrac{?}{24x^5z^3}$

Build each rational expression into an equivalent rational expression with the indicated denominator. See Example 2.

21. $\dfrac{5}{x+3} = \dfrac{?}{2x+6}$

22. $\dfrac{4}{a-5} = \dfrac{?}{3a-15}$

23. $\dfrac{5}{2x+2} = \dfrac{?}{-8x-8}$

24. $\dfrac{3}{m-n} = \dfrac{?}{2n-2m}$

25. $\dfrac{8a}{5b^2-5b} = \dfrac{?}{20b^2-20b^3}$

26. $\dfrac{5x}{-6x-9} = \dfrac{?}{18x^2+27x}$

27. $\dfrac{3}{x+2} = \dfrac{?}{x^2-4}$

28. $\dfrac{a}{a+3} = \dfrac{?}{a^2-9}$

29. $\dfrac{3x}{x+1} = \dfrac{?}{x^2+2x+1}$

30. $\dfrac{-7x}{2x-3} = \dfrac{?}{4x^2-12x+9}$

31. $\dfrac{y-6}{y-4} = \dfrac{?}{y^2+y-20}$

32. $\dfrac{z-6}{z+3} = \dfrac{?}{z^2-2z-15}$

⟨2⟩ Finding the Least Common Denominator

If the given expressions were used as denominators of rational expressions, then what would be the LCD for each group of denominators? See Example 3. See the Strategy for Finding the LCD for Polynomials box on page 403.

33. 12, 16

34. 28, 42

35. 12, 18, 20

36. 24, 40, 48

37. $6a^2$, $15a$

38. $18x^2$, $20xy$

39. $2a^4b$, $3ab^6$, $4a^3b^2$

40. $4m^3nw$, $6mn^5w^8$, $9m^6nw$

41. x^2-16, $x^2+8x+16$

42. x^2-9, x^2+6x+9

43. x, $x+2$, $x-2$

44. y, $y-5$, $y+2$

45. x^2-4x, x^2-16, $2x$

46. y, y^2-3y, $3y$

⟨3⟩ Converting to the LCD

Find the LCD for the given rational expressions, and convert each rational expression into an equivalent rational expression with the LCD as the denominator. See Example 4.

47. $\dfrac{1}{6}, \dfrac{3}{8}$

48. $\dfrac{5}{12}, \dfrac{3}{20}$

49. $\dfrac{1}{2x}, \dfrac{5}{6x}$

50. $\dfrac{3}{5x}, \dfrac{1}{10x}$

51. $\dfrac{2}{3a}, \dfrac{1}{2b}$

52. $\dfrac{y}{4x}, \dfrac{x}{6y}$

53. $\dfrac{3}{84a}, \dfrac{5}{63b}$

54. $\dfrac{4b}{75a}, \dfrac{6}{105ab}$

55. $\dfrac{1}{3x^2}, \dfrac{3}{2x^5}$

56. $\dfrac{3}{8a^3b^9}, \dfrac{5}{6a^2c}$

57. $\dfrac{x}{9y^5z}, \dfrac{y}{12x^3}, \dfrac{1}{6x^2y}$

58. $\dfrac{5}{12a^6b}, \dfrac{3b}{14a^3}, \dfrac{1}{2ab^3}$

Find the LCD for the given rational expressions, and convert each rational expression into an equivalent rational expression with the LCD as the denominator. See Example 5.

59. $\dfrac{2x}{x-3}, \dfrac{5x}{x+2}$

60. $\dfrac{2a}{a-5}, \dfrac{3a}{a+2}$

61. $\dfrac{4}{a-6}, \dfrac{5}{6-a}$

62. $\dfrac{4}{x-y}, \dfrac{5x}{2y-2x}$

63. $\dfrac{x}{x^2-9}, \dfrac{5x}{x^2-6x+9}$

64. $\dfrac{5x}{x^2-1}, \dfrac{-4}{x^2-2x+1}$

65. $\dfrac{w+2}{w^2-2w-15}, \dfrac{-2w}{w^2-4w-5}$

66. $\dfrac{z-1}{z^2+6z+8}, \dfrac{z+1}{z^2+5z+6}$

67. $\dfrac{-5}{6x-12}, \dfrac{x}{x^2-4}, \dfrac{3}{2x+4}$

68. $\dfrac{3}{4b^2-9}, \dfrac{2b}{2b+3}, \dfrac{-5}{2b^2-3b}$

69. $\dfrac{2}{2q^2-5q-3}, \dfrac{3}{2q^2+9q+4}, \dfrac{4}{q^2+q-12}$

70. $\dfrac{-3}{2p^2+7p-15}, \dfrac{p}{2p^2-11p+12}, \dfrac{2}{p^2+p-20}$

Getting More Involved

71. *Discussion*

Why do we learn how to convert two rational expressions into equivalent rational expressions with the same denominator?

72. *Discussion*

Which expression is the LCD for

$$\dfrac{3x-1}{2^2 \cdot 3 \cdot x^2(x+2)} \quad \text{and} \quad \dfrac{2x+7}{2 \cdot 3^2 \cdot x(x+2)^2}?$$

a) $2 \cdot 3 \cdot x(x+2)$ **b)** $36x(x+2)$

c) $36x^2(x+2)^2$ **d)** $2^3 \cdot 3^3 x^3 (x+2)^2$

6.4 Addition and Subtraction

In This Section

⟨1⟩ **Addition and Subtraction of Rational Numbers**

⟨2⟩ **Addition and Subtraction of Rational Expressions**

⟨3⟩ **Applications**

In Section 6.3, you learned how to find the LCD and build up the denominators of rational expressions. In this section, we will use that knowledge to add and subtract rational expressions with different denominators.

⟨1⟩ Addition and Subtraction of Rational Numbers

We can add or subtract rational numbers (or fractions) only with identical denominators according to the following definition.

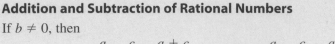

Addition and Subtraction of Rational Numbers

If $b \neq 0$, then

$$\frac{a}{b} + \frac{c}{b} = \frac{a+c}{b} \qquad \text{and} \qquad \frac{a}{b} - \frac{c}{b} = \frac{a-c}{b}.$$

E X A M P L E **1**

Adding or subtracting fractions with the same denominator

Perform the indicated operations. Reduce answers to lowest terms.

a) $\dfrac{1}{12} + \dfrac{7}{12}$
b) $\dfrac{1}{4} - \dfrac{3}{4}$

Solution

a) $\dfrac{1}{12} + \dfrac{7}{12} = \dfrac{8}{12} = \dfrac{\cancel{4} \cdot 2}{\cancel{4} \cdot 3} = \dfrac{2}{3}$
b) $\dfrac{1}{4} - \dfrac{3}{4} = \dfrac{-2}{4} = -\dfrac{1}{2}$

> Now do Exercises 1–8

If the rational numbers have different denominators, we must convert them to equivalent rational numbers that have identical denominators and then add or subtract. Of course, it is most efficient to use the least common denominator (LCD), as in Example 2.

E X A M P L E **2**

Adding or subtracting fractions with different denominators

Find each sum or difference.

a) $\dfrac{3}{20} + \dfrac{7}{12}$
b) $\dfrac{1}{6} - \dfrac{4}{15}$

⟨ **Helpful Hint** ⟩

Note how all of the operations with rational expressions are performed according to the rules for fractions. So keep thinking of how you perform operations with fractions, and you will improve your skills with fractions and with rational expressions.

Solution

a) Because $20 = 2^2 \cdot 5$ and $12 = 2^2 \cdot 3$, the LCD is $2^2 \cdot 3 \cdot 5$, or 60. Convert each fraction to an equivalent fraction with a denominator of 60:

$$\frac{3}{20} + \frac{7}{12} = \frac{3 \cdot 3}{20 \cdot 3} + \frac{7 \cdot 5}{12 \cdot 5} \qquad \text{Build up the denominators.}$$

$$= \frac{9}{60} + \frac{35}{60} \qquad \text{Simplify numerators and denominators.}$$

$$= \frac{44}{60} \qquad \text{Add the fractions.}$$

$$= \frac{4 \cdot 11}{4 \cdot 15} \qquad \text{Factor.}$$

$$= \frac{11}{15} \qquad \text{Reduce.}$$

b) Because $6 = 2 \cdot 3$ and $15 = 3 \cdot 5$, the LCD is $2 \cdot 3 \cdot 5$ or 30:

$$\frac{1}{6} - \frac{4}{15} = \frac{1}{2 \cdot 3} - \frac{4}{3 \cdot 5} \qquad \text{Factor the denominators.}$$

$$= \frac{1 \cdot 5}{2 \cdot 3 \cdot 5} - \frac{4 \cdot 2}{3 \cdot 5 \cdot 2} \qquad \text{Build up the denominators.}$$

$$= \frac{5}{30} - \frac{8}{30} \qquad \text{Simplify the numerators and denominators.}$$

$$= \frac{-3}{30} \qquad \text{Subtract.}$$

$$= \frac{-1 \cdot 3}{10 \cdot 3} \qquad \text{Factor.}$$

$$= -\frac{1}{10} \qquad \text{Reduce.}$$

Now do Exercises 9–18

⟨2⟩ Addition and Subtraction of Rational Expressions

Rational expressions are added or subtracted just like rational numbers. We can add or subtract only when we have identical denominators. All answers should be reduced to lowest terms. Remember to factor first when reducing, and then divide out any common factors.

EXAMPLE 3

Rational expressions with the same denominator
Perform the indicated operations and reduce answers to lowest terms.

a) $\dfrac{2}{3y} + \dfrac{4}{3y}$ **b)** $\dfrac{2x}{x+2} + \dfrac{4}{x+2}$ **c)** $\dfrac{x^2 + 2x}{(x-1)(x+3)} - \dfrac{2x+1}{(x-1)(x+3)}$

Solution

a) $\dfrac{2}{3y} + \dfrac{4}{3y} = \dfrac{6}{3y}$ Add the fractions.

$\qquad\qquad\quad = \dfrac{2}{y}$ Reduce.

b) $\dfrac{2x}{x+2} + \dfrac{4}{x+2} = \dfrac{2x+4}{x+2}$ Add the fractions.

$\qquad\qquad\qquad\quad = \dfrac{2(x+2)}{x+2}$ Factor the numerator.

$\qquad\qquad\qquad\quad = 2$ Reduce.

c) $\dfrac{x^2 + 2x}{(x-1)(x+3)} - \dfrac{2x+1}{(x-1)(x+3)} = \dfrac{x^2 + 2x - (2x+1)}{(x-1)(x+3)}$ Subtract the fractions.

$\qquad\qquad\qquad\qquad\qquad = \dfrac{x^2 + 2x - 2x - 1}{(x-1)(x+3)}$ Remove parentheses.

$\qquad\qquad\qquad\qquad\qquad = \dfrac{x^2 - 1}{(x-1)(x+3)}$ Combine like terms.

$\qquad\qquad\qquad\qquad\qquad = \dfrac{(x-1)(x+1)}{(x-1)(x+3)}$ Factor.

$\qquad\qquad\qquad\qquad\qquad = \dfrac{x+1}{x+3}$ Reduce.

Now do Exercises 19–30

> **CAUTION** When subtracting a numerator containing more than one term, be sure to enclose it in parentheses, as in Example 3(c). Because that numerator is a binomial, the sign of each of its terms must be changed for the subtraction.

In Example 4, the rational expressions have different denominators.

E X A M P L E **4**

Rational expressions with different denominators
Perform the indicated operations.

a) $\dfrac{5}{2x} + \dfrac{2}{3}$

b) $\dfrac{4}{x^3y} + \dfrac{2}{xy^3}$

c) $\dfrac{a+1}{6} - \dfrac{a-2}{8}$

‹ **Helpful Hint** ›

You can remind yourself of the difference between addition and multiplication of fractions with a simple example: If you and your spouse each own 1/7 of Microsoft, then together you own 2/7 of Microsoft. If you own 1/7 of Microsoft, and give 1/7 of your stock to your child, then your child owns 1/49 of Microsoft.

Solution

a) The LCD for $2x$ and 3 is $6x$:

$$\frac{5}{2x} + \frac{2}{3} = \frac{5 \cdot 3}{2x \cdot 3} + \frac{2 \cdot 2x}{3 \cdot 2x} \qquad \text{Build up both denominators to } 6x.$$

$$= \frac{15}{6x} + \frac{4x}{6x} \qquad \text{Simplify numerators and denominators.}$$

$$= \frac{15 + 4x}{6x} \qquad \text{Add the rational expressions.}$$

b) The LCD is x^3y^3.

$$\frac{4}{x^3y} + \frac{2}{xy^3} = \frac{4 \cdot y^2}{x^3y \cdot y^2} + \frac{2 \cdot x^2}{xy^3 \cdot x^2} \qquad \text{Build up both denominators to the LCD.}$$

$$= \frac{4y^2}{x^3y^3} + \frac{2x^2}{x^3y^3} \qquad \text{Simplify numerators and denominators.}$$

$$= \frac{4y^2 + 2x^2}{x^3y^3} \qquad \text{Add the rational expressions.}$$

c) Because $6 = 2 \cdot 3$ and $8 = 2^3$, the LCD is $2^3 \cdot 3$, or 24:

$$\frac{a+1}{6} - \frac{a-2}{8} = \frac{(a+1)4}{6 \cdot 4} - \frac{(a-2)3}{8 \cdot 3} \qquad \text{Build up both denominators to the LCD 24.}$$

$$= \frac{4a+4}{24} - \frac{3a-6}{24} \qquad \text{Simplify numerators and denominators.}$$

$$= \frac{4a+4-(3a-6)}{24} \qquad \text{Subtract the rational expressions.}$$

$$= \frac{4a+4-3a+6}{24} \qquad \text{Remove the parentheses.}$$

$$= \frac{a+10}{24} \qquad \text{Combine like terms.}$$

Now do Exercises 31–46

EXAMPLE 5

Rational expressions with different denominators
Perform the indicated operations:

a) $\dfrac{1}{x^2 - 9} + \dfrac{2}{x^2 + 3x}$

b) $\dfrac{4}{5 - a} - \dfrac{2}{a - 5}$

⟨ **Helpful Hint** ⟩

Once the denominators are factored as in Example 5(a), you can simply look at each denominator and ask, "What factor does the other denominator(s) have that is missing from this one?" Then use the missing factor to build up the denominator. Repeat until all denominators are identical, and you will have the LCD.

Solution

a) $\dfrac{1}{x^2 - 9} + \dfrac{2}{x^2 + 3x} = \underbrace{\dfrac{1}{(x - 3)(x + 3)}}_{\text{Needs } x} + \underbrace{\dfrac{2}{x(x + 3)}}_{\text{Needs } x - 3}$ The LCD is $x(x - 3)(x + 3)$.

$= \dfrac{1 \cdot x}{(x - 3)(x + 3)x} + \dfrac{2(x - 3)}{x(x + 3)(x - 3)}$

$= \dfrac{x}{x(x - 3)(x + 3)} + \dfrac{2x - 6}{x(x - 3)(x + 3)}$

$= \dfrac{3x - 6}{x(x - 3)(x + 3)}$ We usually leave the denominator in factored form.

b) Because $-1(5 - a) = a - 5$, we can get identical denominators by multiplying only the first expression by -1 in the numerator and denominator:

$\dfrac{4}{5 - a} - \dfrac{2}{a - 5} = \dfrac{4(-1)}{(5 - a)(-1)} - \dfrac{2}{a - 5}$

$= \dfrac{-4}{a - 5} - \dfrac{2}{a - 5}$

$= \dfrac{-6}{a - 5}$ $-4 - 2 = -6$

$= -\dfrac{6}{a - 5}$

Now do Exercises 47–64

In Example 6, we combine three rational expressions by addition and subtraction.

EXAMPLE 6

Rational expressions with different denominators
Perform the indicated operations.

$$\dfrac{x + 1}{x^2 + 2x} + \dfrac{2x + 1}{6x + 12} - \dfrac{1}{6}$$

Solution
The LCD for $x(x + 2)$, $6(x + 2)$, and 6 is $6x(x + 2)$.

$\dfrac{x + 1}{x^2 + 2x} + \dfrac{2x + 1}{6x + 12} - \dfrac{1}{6} = \dfrac{x + 1}{x(x + 2)} + \dfrac{2x + 1}{6(x + 2)} - \dfrac{1}{6}$ Factor denominators.

$= \dfrac{6(x + 1)}{6x(x + 2)} + \dfrac{x(2x + 1)}{6x(x + 2)} - \dfrac{1x(x + 2)}{6x(x + 2)}$ Build up to the LCD.

$$= \frac{6x + 6}{6x(x + 2)} + \frac{2x^2 + x}{6x(x + 2)} - \frac{x^2 + 2x}{6x(x + 2)} \quad \text{Simplify numerators.}$$

$$= \frac{6x + 6 + 2x^2 + x - x^2 - 2x}{6x(x + 2)} \quad \text{Combine the numerators.}$$

$$= \frac{x^2 + 5x + 6}{6x(x + 2)} \quad \text{Combine like terms.}$$

$$= \frac{(x + 3)(x + 2)}{6x(x + 2)} \quad \text{Factor.}$$

$$= \frac{x + 3}{6x} \quad \text{Reduce.}$$

> Now do Exercises 65–70

⟨3⟩ Applications

We have seen how rational expressions can occur in problems involving rates. In Example 7, we see an applied situation in which we add rational expressions.

E X A M P L E 7

Adding work

Harry takes twice as long as Lucy to proofread a manuscript. Write a rational expression for the amount of work they do in 3 hours working together on a manuscript.

Solution

Let $x =$ the number of hours it would take Lucy to complete the manuscript alone and $2x =$ the number of hours it would take Harry to complete the manuscript alone. Make a table showing rate, time, and work completed:

	Rate	Time	Work
Lucy	$\dfrac{1}{x} \dfrac{\text{msp}}{\text{hr}}$	3 hr	$\dfrac{3}{x}$ msp
Harry	$\dfrac{1}{2x} \dfrac{\text{msp}}{\text{hr}}$	3 hr	$\dfrac{3}{2x}$ msp

Now find the sum of each person's work.

$$\frac{3}{x} + \frac{3}{2x} = \frac{2 \cdot 3}{2 \cdot x} + \frac{3}{2x}$$

$$= \frac{6}{2x} + \frac{3}{2x}$$

$$= \frac{9}{2x}$$

So in 3 hours working together they will complete $\frac{9}{2x}$ of the manuscript.

> Now do Exercises 81–86

Warm-Ups ▼

Fill in the blank.

1. We can _____ rational expressions only if they have identical denominators.

2. We can _____ any two rational expressions so that their denominators are identical.

True or false?

3. $\dfrac{1}{2} + \dfrac{1}{3} = \dfrac{2}{5}$

4. $\dfrac{7}{12} - \dfrac{1}{12} = \dfrac{1}{2}$

5. $\dfrac{3}{5} + \dfrac{4}{3} = \dfrac{29}{15}$

6. $\dfrac{4}{5} - \dfrac{5}{7} = \dfrac{3}{35}$

7. $\dfrac{5}{20} + \dfrac{3}{4} = 1$

8. For any nonzero value of x, $\dfrac{2}{x} + 1 = \dfrac{3}{x}$.

9. For any nonzero value of a, $1 + \dfrac{1}{a} = \dfrac{a+1}{a}$.

10. For any value of a, $a - \dfrac{1}{4} = \dfrac{4a-1}{4}$.

Exercises

6.4

⟨ **Study Tips** ⟩

- When studying for a midterm or final, review the material in the order it was originally presented. This strategy will help you to see connections between the ideas.
- Studying the oldest material first will give top priority to material that you might have forgotten.

⟨1⟩ **Addition and Subtraction of Rational Numbers**

Perform the indicated operation. Reduce each answer to lowest terms. See Example 1.

1. $\dfrac{1}{10} + \dfrac{1}{10}$

2. $\dfrac{1}{8} + \dfrac{3}{8}$

3. $\dfrac{7}{8} - \dfrac{1}{8}$

4. $\dfrac{4}{9} - \dfrac{1}{9}$

5. $\dfrac{1}{6} - \dfrac{5}{6}$

6. $-\dfrac{3}{8} - \dfrac{7}{8}$

7. $-\dfrac{7}{8} + \dfrac{1}{8}$

8. $-\dfrac{9}{20} + \left(-\dfrac{3}{20}\right)$

Perform the indicated operation. Reduce each answer to lowest terms. See Example 2.

9. $\dfrac{1}{3} + \dfrac{2}{9}$

10. $\dfrac{1}{4} + \dfrac{5}{6}$

11. $\dfrac{7}{10} + \dfrac{5}{6}$

12. $\dfrac{5}{6} + \dfrac{3}{10}$

13. $\dfrac{7}{16} + \dfrac{5}{18}$

14. $\dfrac{7}{6} + \dfrac{4}{15}$

15. $\dfrac{1}{8} - \dfrac{9}{10}$

16. $\dfrac{2}{15} - \dfrac{5}{12}$

17. $-\dfrac{1}{6} - \left(-\dfrac{3}{8}\right)$

18. $-\dfrac{1}{5} - \left(-\dfrac{1}{7}\right)$

⟨2⟩ Addition and Subtraction of Rational Expressions

Perform the indicated operation. Reduce each answer to lowest terms. See Example 3.

19. $\dfrac{1}{2x} + \dfrac{1}{2x}$

20. $\dfrac{1}{3y} + \dfrac{2}{3y}$

21. $\dfrac{3}{2w} + \dfrac{7}{2w}$

22. $\dfrac{5x}{3y} + \dfrac{7x}{3y}$

23. $\dfrac{3a}{a+5} + \dfrac{15}{a+5}$

24. $\dfrac{a+7}{a-4} + \dfrac{9-5a}{a-4}$

25. $\dfrac{q-1}{q-4} - \dfrac{3q-9}{q-4}$

26. $\dfrac{3-a}{3} - \dfrac{a-5}{3}$

27. $\dfrac{4h-3}{h(h+1)} - \dfrac{h-6}{h(h+1)}$

28. $\dfrac{2t-9}{t(t-3)} - \dfrac{t-9}{t(t-3)}$

29. $\dfrac{x^2-x-5}{(x+1)(x+2)} + \dfrac{1-2x}{(x+1)(x+2)}$

30. $\dfrac{2x-5}{(x-2)(x+6)} + \dfrac{x^2-2x+1}{(x-2)(x+6)}$

Perform the indicated operation. Reduce each answer to lowest terms. See Example 4.

31. $\dfrac{1}{a} + \dfrac{1}{2a}$

32. $\dfrac{1}{3w} + \dfrac{2}{w}$

33. $\dfrac{x}{3} + \dfrac{x}{2}$

34. $\dfrac{y}{4} + \dfrac{y}{2}$

35. $\dfrac{m}{5} + m$

36. $\dfrac{y}{4} + 2y$

37. $\dfrac{1}{x} + \dfrac{2}{y}$

38. $\dfrac{2}{a} + \dfrac{3}{b}$

39. $\dfrac{3}{2a} + \dfrac{1}{5a}$

40. $\dfrac{5}{6y} - \dfrac{3}{8y}$

41. $\dfrac{w-3}{9} - \dfrac{w-4}{12}$

42. $\dfrac{y+4}{10} - \dfrac{y-2}{14}$

43. $\dfrac{b^2}{4a} - c$

44. $y + \dfrac{3}{7b}$

45. $\dfrac{2}{wz^2} + \dfrac{3}{w^2z}$

46. $\dfrac{1}{a^5b} - \dfrac{5}{ab^3}$

Perform the indicated operation. Reduce each answer to lowest terms. See Examples 5 and 6.

47. $\dfrac{1}{x} + \dfrac{1}{x+2}$

48. $\dfrac{1}{y} + \dfrac{2}{y+1}$

49. $\dfrac{2}{x+1} - \dfrac{3}{x}$

50. $\dfrac{1}{a-1} - \dfrac{2}{a}$

51. $\dfrac{2}{a-b} + \dfrac{1}{a+b}$

52. $\dfrac{3}{x+1} + \dfrac{2}{x-1}$

53. $\dfrac{3}{x^2+x} - \dfrac{4}{5x+5}$

54. $\dfrac{3}{a^2+3a} - \dfrac{2}{5a+15}$

55. $\dfrac{2a}{a^2-9} + \dfrac{a}{a-3}$

56. $\dfrac{x}{x^2-1} + \dfrac{3}{x-1}$

57. $\dfrac{4}{a-b} + \dfrac{4}{b-a}$

58. $\dfrac{2}{x-3} + \dfrac{3}{3-x}$

59. $\dfrac{3}{2a-2} - \dfrac{2}{1-a}$

60. $\dfrac{5}{2x-4} - \dfrac{3}{2-x}$

61. $\dfrac{1}{x^2-4} - \dfrac{3}{x^2-3x-10}$

62. $\dfrac{2x}{x^2-9} + \dfrac{3x}{x^2+4x+3}$

63. $\dfrac{3}{x^2+x-2} + \dfrac{4}{x^2+2x-3}$

64. $\dfrac{x-1}{x^2-x-12} + \dfrac{x+4}{x^2+5x+6}$

65. $\dfrac{1}{a} + \dfrac{1}{b} + \dfrac{1}{c}$

66. $\dfrac{1}{x} + \dfrac{1}{x^2} + \dfrac{1}{x^3}$

67. $\dfrac{2}{x} - \dfrac{1}{x-1} + \dfrac{1}{x+2}$

68. $\dfrac{1}{a} - \dfrac{2}{a+1} + \dfrac{3}{a-1}$

69. $\dfrac{5}{3a-9} - \dfrac{3}{2a} + \dfrac{4}{a^2-3a}$

70. $\dfrac{3}{4c+2} - \dfrac{c-4}{2c^2+c} - \dfrac{5}{6c}$

Match each expression in (a)–(f) with the equivalent expression in (A)–(F).

71. a) $\dfrac{1}{y} + 2$ **b)** $\dfrac{1}{y} + \dfrac{2}{y}$ **c)** $\dfrac{1}{y} + \dfrac{1}{2}$

    d) $\dfrac{1}{y} + \dfrac{1}{2y}$ **e)** $\dfrac{2}{y} + 1$ **f)** $\dfrac{y}{2} + 1$

    A) $\dfrac{3}{y}$ **B)** $\dfrac{3}{2y}$ **C)** $\dfrac{y+2}{2}$

    D) $\dfrac{y+2}{y}$ **E)** $\dfrac{y+2}{2y}$ **F)** $\dfrac{2y+1}{y}$

72. a) $\dfrac{1}{x} - x$ **b)** $\dfrac{1}{x} - \dfrac{1}{x^2}$ **c)** $\dfrac{1}{x} - 1$

    d) $\dfrac{1}{x^2} - x$ **e)** $x - \dfrac{1}{x}$ **f)** $\dfrac{1}{x^2} - \dfrac{1}{x}$

    A) $\dfrac{1-x^3}{x^2}$ **B)** $\dfrac{1-x}{x}$ **C)** $\dfrac{1-x^2}{x}$

    D) $\dfrac{1-x}{x^2}$ **E)** $\dfrac{x^2-1}{x}$ **F)** $\dfrac{x-1}{x^2}$

Perform the indicated operation. Reduce each answer to lowest terms.

73. $\dfrac{3}{2p} - \dfrac{1}{2p+8}$

74. $\dfrac{3}{2y} - \dfrac{3}{2y+4}$

75. $\dfrac{3}{a^2+3a+2} + \dfrac{3}{a^2+5a+6}$

76. $\dfrac{4}{w^2+w} + \dfrac{12}{w^2-3w}$

77. $\dfrac{2}{b^2+4b+3} - \dfrac{1}{b^2+5b+6}$

78. $\dfrac{9}{m^2-m-2} - \dfrac{6}{m^2-1}$

79. $\dfrac{3}{2t} - \dfrac{2}{t+2} - \dfrac{3}{t^2+2t}$

80. $\dfrac{4}{3n} + \dfrac{2}{n+1} + \dfrac{2}{n^2+n}$

⟨ 3 ⟩ **Applications**

Solve each problem. See Example 7.

81. Perimeter of a rectangle. Suppose that the length of a rectangle is $\frac{3}{x}$ feet and its width is $\frac{5}{2x}$ feet. Find a rational expression for the perimeter of the rectangle.

82. Perimeter of a triangle. The lengths of the sides of a triangle are $\frac{1}{x}, \frac{1}{2x},$ and $\frac{2}{3x}$ meters. Find a rational expression for the perimeter of the triangle.

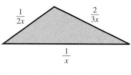

Figure for Exercise 82

83. Traveling time. Janet drove 120 miles at x mph before 6:00 A.M. After 6:00 A.M., she increased her speed by 5 mph and drove 195 additional miles. Use the fact that $T = \frac{D}{R}$ to complete the following table.

	Rate	Time	Distance
Before	$x \dfrac{\text{mi}}{\text{hr}}$		120 mi
After	$x+5 \dfrac{\text{mi}}{\text{hr}}$		195 mi

Write a rational expression for her total traveling time. Evaluate the expression for $x = 60$.

84. Traveling time. After leaving Moose Jaw, Hanson drove 200 kilometers at x km/hr and then decreased his speed by 20 km/hr and drove 240 additional kilometers. Make a table like the one in Exercise 83. Write a rational expression for his total traveling time. Evaluate the expression for $x = 100$.

85. House painting. Kent can paint a certain house by himself in x days. His helper Keith can paint the same house by himself in $x + 3$ days. Suppose that they work together on the job for 2 days. To complete the table on the next page, use the fact that the work completed is the product of the

	Rate	Time	Work
Kent	$\dfrac{1 \text{ job}}{x \text{ day}}$	2 days	
Keith	$\dfrac{1 \text{ job}}{x+3 \text{ day}}$	2 days	

rate and the time. Write a rational expression for the fraction of the house that they complete by working together for 2 days. Evaluate the expression for $x = 6$.

86. *Barn painting.* Melanie can paint a certain barn by herself in x days. Her helper Melissa can paint the same barn by herself in $2x$ days. Write a rational expression for the fraction of the barn that they complete in one day by working together. Evaluate the expression for $x = 5$.

Photo for Exercise 86

Getting More Involved

87. *Writing*

Write a step-by-step procedure for adding rational expressions.

88. *Writing*

Explain why fractions must have the same denominator to be added. Use real-life examples.

Math *at Work*   Gravity on the Moon

Hundreds of years before humans even considered traveling beyond the earth, Isaac Newton established the laws of gravity. So when Neil Armstrong made the first human step onto the moon in 1969, he knew what amount of gravitational force to expect. Let's see how he knew.

Newton's equation for the force of gravity between two objects is $F = G\frac{m_1 m_2}{d^2}$, where m_1 and m_2 are the masses of the objects (in kilograms), d is the distance (in meters) between the centers of the two objects, and G is the gravitational constant 6.67×10^{-11}. To find the force of gravity for Armstrong on earth, use 5.98×10^{24} kg for the mass of the earth, 6.378×10^6 m for the radius of the earth, and 80 kg for Armstrong's mass. We get

$$F = 6.67 \times 10^{-11} \cdot \frac{5.98 \times 10^{24} \text{ kg} \cdot 80 \text{ kg}}{(6.378 \times 10^6 \text{ m})^2} \approx 784 \text{ Newtons.}$$

To find the force of gravity for Armstrong on the moon, use 7.34×10^{22} kg for the mass of the moon and 1.737×10^6 m for the radius of the moon. We get

$$F = 6.67 \times 10^{-11} \cdot \frac{7.34 \times 10^{22} \text{ kg} \cdot 80 \text{ kg}}{(1.737 \times 10^6 \text{ m})^2} \approx 130 \text{ Newtons.}$$

So the force of gravity for Armstrong on the moon was about one-sixth of the force of gravity for Armstrong on earth. Fortunately, the moon is smaller than the earth. Walking on a planet much larger than the earth would present a real problem in terms of gravitational force.

Mid-Chapter **Quiz** | Sections 6.1 through 6.4 | **Chapter 6**

Reduce to lowest terms.

1. $\dfrac{36}{84}$

2. $\dfrac{8x - 2}{8}$

3. $\dfrac{w^2 - 1}{2w + 2}$

4. $\dfrac{2a^2 - 10a + 12}{6 - 3a}$

Perform the indicated operation.

5. $\dfrac{6}{7} \cdot \dfrac{21}{10}$

6. $\dfrac{3xy^2}{5z} \cdot \dfrac{8x^2z^3}{8y^4}$

7. $\dfrac{a^2 - 9}{2a + 4} \cdot \dfrac{5a + 10}{2a - 6}$

8. $\dfrac{b^2}{3} \div \dfrac{b^6}{21}$

9. $\dfrac{5}{9} \div \dfrac{25}{33}$

10. $\dfrac{3x - 9}{8} \div \dfrac{x^2 - 6x + 9}{12}$

11. $\dfrac{s^2}{3} + \dfrac{s^2}{21}$

12. $\dfrac{m^2 - 8m + 7}{2m} \div (m - 7)$

13. $\dfrac{5}{6} - \dfrac{5}{21}$

14. $\dfrac{4}{ab^3} + \dfrac{5}{a^2b}$

15. $\dfrac{3x}{x + 1} + \dfrac{x}{x^2 + 2x + 1}$

16. $\dfrac{y}{y + 5} - \dfrac{y}{y + 2}$

17. $\dfrac{1}{a} + \dfrac{1}{b} + \dfrac{1}{c}$

Miscellaneous.

18. What numbers(s) can't be used in place of x in $\dfrac{3x - 6}{2x + 1}$?

19. Find the value of $\dfrac{3x - 6}{2x + 1}$ when $x = -2$.

20. Find $R(-1)$ if $R(x) = \dfrac{6x^2 + 3}{5x - 1}$.

6.5   Complex Fractions

In This Section

⟨1⟩ Complex Fractions

⟨2⟩ Using the LCD to Simplify
     Complex Fractions

⟨3⟩ Applications

In this section, we will use the idea of least common denominator to simplify complex fractions. Also we will see how complex fractions can arise in applications.

⟨1⟩ Complex Fractions

A **complex fraction** is a fraction having rational expressions in the numerator, denominator, or both. Consider the following complex fraction:

$$\dfrac{\dfrac{1}{2} + \dfrac{2}{3}}{\dfrac{1}{4} - \dfrac{5}{8}}$$

← Numerator of complex fraction

← Denominator of complex fraction

Since the fraction bar is a grouping symbol, we can compute the value of the numerator, the value of the denominator, and then divide them, as shown in Example 1.

EXAMPLE   1

Simplifying complex fractions
Simplify.

a) $\dfrac{\dfrac{1}{2} + \dfrac{2}{3}}{\dfrac{1}{4} - \dfrac{5}{8}}$

b) $\dfrac{4 - \dfrac{2}{5}}{\dfrac{1}{10} + 3}$

Solution

a) Combine the fractions in the numerator:

$$\frac{1}{2} + \frac{2}{3} = \frac{1 \cdot 3}{2 \cdot 3} + \frac{2 \cdot 2}{3 \cdot 2} = \frac{3}{6} + \frac{4}{6} = \frac{7}{6}$$

Combine the fractions in the denominator as follows:

$$\frac{1}{4} - \frac{5}{8} = \frac{1 \cdot 2}{4 \cdot 2} - \frac{5}{8} = \frac{2}{8} - \frac{5}{8} = -\frac{3}{8}$$

Now divide the numerator by the denominator:

$$\frac{\dfrac{1}{2} + \dfrac{2}{3}}{\dfrac{1}{4} - \dfrac{5}{8}} = \frac{\dfrac{7}{6}}{-\dfrac{3}{8}} = \frac{7}{6} \div \left(-\frac{3}{8}\right) = \frac{7}{6} \cdot \left(-\frac{8}{3}\right) = -\frac{56}{18} = -\frac{28}{9}$$

b) $\dfrac{4 - \dfrac{2}{5}}{\dfrac{1}{10} + 3} = \dfrac{\dfrac{20}{5} - \dfrac{2}{5}}{\dfrac{1}{10} + \dfrac{30}{10}} = \dfrac{\dfrac{18}{5}}{\dfrac{31}{10}} = \dfrac{18}{5} \div \dfrac{31}{10} = \dfrac{18}{5} \cdot \dfrac{10}{31} = \dfrac{36}{31}$

> Now do Exercises 1–12

⟨2⟩ Using the LCD to Simplify Complex Fractions

A complex fraction can be simplified by performing the operations in the numerator and denominator, and then dividing the results, as shown in Example 1. However, there is a better method. All of the fractions in the complex fraction can be eliminated in one step by multiplying by the LCD of all of the single fractions. The strategy for this method is detailed in the following box and illustrated in Example 2.

Strategy for Simplifying a Complex Fraction

1. Find the LCD for all the denominators in the complex fraction.
2. Multiply both the numerator and the denominator of the complex fraction by the LCD. Use the distributive property if necessary.
3. Combine like terms if possible.
4. Reduce to lowest terms when possible.

EXAMPLE   2

Using the LCD to simplify a complex fraction
Use the LCD to simplify

$$\frac{\dfrac{1}{2} + \dfrac{2}{3}}{\dfrac{1}{4} - \dfrac{5}{8}}.$$

You can check Example 2 with a calculator as shown here.

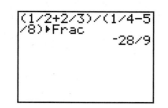

Solution

The LCD of 2, 3, 4, and 8 is 24. Now multiply the numerator and denominator of the complex fraction by the LCD:

$$\frac{\dfrac{1}{2}+\dfrac{2}{3}}{\dfrac{1}{4}-\dfrac{5}{8}} = \frac{\left(\dfrac{1}{2}+\dfrac{2}{3}\right)24}{\left(\dfrac{1}{4}-\dfrac{5}{8}\right)24}$$ Multiply the numerator and denominator by the LCD.

$$= \frac{\dfrac{1}{2}\cdot 24 + \dfrac{2}{3}\cdot 24}{\dfrac{1}{4}\cdot 24 - \dfrac{5}{8}\cdot 24}$$ Distributive property

$$= \frac{12+16}{6-15}$$ Simplify.

$$= \frac{28}{-9}$$

$$= -\frac{28}{9}$$

Now do Exercises 13–20

CAUTION We simplify a complex fraction by multiplying the numerator and denominator of the *complex fraction* by the LCD. Do not multiply the numerator and denominator of each fraction in the complex fraction by the LCD.

In Example 3 we simplify a complex fraction involving variables.

E X A M P L E   **3**

A complex fraction with variables
Simplify

$$\frac{2-\dfrac{1}{x}}{\dfrac{1}{x^2}-\dfrac{1}{2}}.$$

When students see addition or subtraction in a complex fraction, they often convert all fractions to the same denominator. This is not wrong, but it is not necessary. Simply multiplying every fraction by the LCD eliminates the denominators of the original fractions.

Solution

The LCD of the denominators x, x^2, and 2 is $2x^2$:

$$\frac{2-\dfrac{1}{x}}{\dfrac{1}{x^2}-\dfrac{1}{2}} = \frac{\left(2-\dfrac{1}{x}\right)(2x^2)}{\left(\dfrac{1}{x^2}-\dfrac{1}{2}\right)(2x^2)}$$ Multiply the numerator and denominator by $2x^2$.

$$= \frac{2\cdot 2x^2 - \dfrac{1}{x}\cdot 2x^2}{\dfrac{1}{x^2}\cdot 2x^2 - \dfrac{1}{2}\cdot 2x^2}$$ Distributive property

$$= \frac{4x^2 - 2x}{2 - x^2} \qquad \text{Simplify.}$$

The numerator of this answer can be factored, but the rational expression cannot be reduced.

Now do Exercises 21–30

E X A M P L E   **4**

Simplifying a complex fraction

Simplify

$$\frac{\dfrac{1}{x - 2} - \dfrac{2}{x + 2}}{\dfrac{3}{2 - x} + \dfrac{4}{x + 2}}.$$

Solution

Because $x - 2$ and $2 - x$ are opposites, we can use $(x - 2)(x + 2)$ as the LCD. Multiply the numerator and denominator by $(x - 2)(x + 2)$:

$$\frac{\dfrac{1}{x - 2} - \dfrac{2}{x + 2}}{\dfrac{3}{2 - x} + \dfrac{4}{x + 2}} = \frac{\dfrac{1}{x - 2}(x - 2)(x + 2) - \dfrac{2}{x + 2}(x - 2)(x + 2)}{\dfrac{3}{2 - x}(x - 2)(x + 2) + \dfrac{4}{x + 2}(x - 2)(x + 2)}$$

$$= \frac{x + 2 - 2(x - 2)}{3(-1)(x + 2) + 4(x - 2)} \qquad \frac{x - 2}{2 - x} = -1$$

$$= \frac{x + 2 - 2x + 4}{-3x - 6 + 4x - 8} \qquad \text{Distributive property}$$

$$= \frac{-x + 6}{x - 14} \qquad \text{Combine like terms.}$$

Now do Exercises 31–46

⟨3⟩ Applications

As their name suggests, complex fractions arise in some fairly complex situations.

E X A M P L E   **5**

Fast-food workers

A survey of college students found that $\frac{1}{2}$ of the female students had jobs and $\frac{2}{3}$ of the male students had jobs. It was also found that $\frac{1}{4}$ of the female students worked in fast-food restaurants and $\frac{1}{6}$ of the male students worked in fast-food restaurants. If equal numbers of male and female students were surveyed, then what fraction of the working students worked in fast-food restaurants?

Solution

Let x represent the number of males surveyed. The number of females surveyed is also x. The total number of students working in fast-food restaurants is

$$\frac{1}{4}x + \frac{1}{6}x.$$

The total number of working students in the survey is

$$\frac{1}{2}x + \frac{2}{3}x.$$

So the fraction of working students who work in fast-food restaurants is

$$\frac{\frac{1}{4}x + \frac{1}{6}x}{\frac{1}{2}x + \frac{2}{3}x}.$$

The LCD of the denominators 2, 3, 4, and 6 is 12. Multiply the numerator and denominator by 12 to eliminate the fractions as follows:

$$\frac{\frac{1}{4}x + \frac{1}{6}x}{\frac{1}{2}x + \frac{2}{3}x} = \frac{\left(\frac{1}{4}x + \frac{1}{6}x\right)12}{\left(\frac{1}{2}x + \frac{2}{3}x\right)12}$$ Multiply numerator and denominator by 12.

$$= \frac{3x + 2x}{6x + 8x}$$ Distributive property

$$= \frac{5x}{14x}$$ Combine like terms.

$$= \frac{5}{14}$$ Reduce.

So $\frac{5}{14}$ (or about 36%) of the working students work in fast-food restaurants.

> **Now do Exercises 61–62**

Warm-Ups ▼

Fill in the blank.

1. A _____ fraction has fractions in its numerator, denominator, or both.

2. To simplify a complex fraction, you can multiply its _____ and _____ by the LCD of all of the fractions.

True or false?

3. The LCD for the denominator 4, x, 6, and x^2 is $12x^3$.

4. The LCD for the denominator $a - b$, $2b - 2a$, and 6 is $6a - 6b$.

5. To simplify $\dfrac{\frac{1}{2} + \frac{1}{3}}{\frac{1}{4} - \frac{1}{6}}$, we multiply the numerator and denominator by 12.

6. $\dfrac{\left(\frac{1}{2} + \frac{1}{3}\right)12}{\left(\frac{1}{4} - \frac{1}{6}\right)12} = \dfrac{6 + 4}{3 - 2}$

7. $\dfrac{\frac{1}{2} + \frac{1}{3}}{\frac{1}{4} - \frac{1}{6}} = \dfrac{\frac{5}{6}}{\frac{1}{12}}$

8. For any real number x, $\dfrac{x - \frac{1}{2}}{x + \frac{1}{3}} = \dfrac{2x - 1}{3x + 1}$.

Exercises

‹ 1 › **Complex Fractions**

Simplify each complex fraction. See Example 1.

1. $\dfrac{\frac{1}{2}+\frac{1}{4}}{\frac{1}{2}+\frac{3}{4}}$

2. $\dfrac{\frac{1}{3}+\frac{5}{6}}{\frac{2}{3}+\frac{1}{6}}$

3. $\dfrac{\frac{1}{2}+\frac{1}{3}}{\frac{1}{4}-\frac{1}{2}}$

4. $\dfrac{\frac{1}{3}-\frac{1}{4}}{\frac{1}{3}+\frac{1}{6}}$

5. $\dfrac{\frac{2}{5}+\frac{5}{6}-\frac{1}{2}}{\frac{1}{2}-\frac{2}{3}+\frac{1}{15}}$

6. $\dfrac{\frac{2}{5}-\frac{2}{9}-\frac{1}{3}}{\frac{1}{3}+\frac{1}{5}+\frac{2}{15}}$

7. $\dfrac{1+\frac{1}{2}}{2+\frac{1}{4}}$

8. $\dfrac{\frac{1}{3}+1}{\frac{1}{6}+2}$

9. $\dfrac{3+\frac{1}{2}}{5-\frac{3}{4}}$

10. $\dfrac{1+\frac{1}{12}}{1-\frac{1}{12}}$

11. $\dfrac{1-\frac{1}{6}+\frac{2}{3}}{1+\frac{1}{15}-\frac{3}{10}}$

12. $\dfrac{3-\frac{2}{9}-\frac{1}{6}}{\frac{5}{18}-\frac{1}{3}-2}$

‹ 2 › **Using the LCD to Simplify Complex Fractions**

Simplify each complex fraction. See Examples 2 and 3. See the Strategy for Simplifying a Complex Fraction box on page 418.

13. $\dfrac{\frac{1}{2}+\frac{1}{3}}{\frac{1}{2}-\frac{1}{4}}$

14. $\dfrac{\frac{1}{4}-\frac{1}{3}}{\frac{1}{4}+\frac{1}{6}}$

15. $\dfrac{\frac{2}{5}+\frac{1}{10}}{\frac{1}{5}+\frac{1}{4}}$

16. $\dfrac{\frac{3}{10}+\frac{4}{5}}{\frac{1}{2}+\frac{3}{4}}$

17. $\dfrac{1+\frac{2}{3}+\frac{1}{2}}{2-\frac{1}{3}-\frac{3}{2}}$

18. $\dfrac{3-\frac{3}{5}+\frac{1}{10}}{2+\frac{6}{5}-\frac{3}{10}}$

19. $\dfrac{\frac{2}{3}+\frac{5}{6}+\frac{1}{2}}{\frac{1}{6}-\frac{1}{3}-\frac{1}{2}}$

20. $\dfrac{\frac{2}{5}-\frac{3}{2}+\frac{7}{10}}{\frac{1}{5}+\frac{1}{2}-\frac{1}{10}}$

21. $\dfrac{\frac{1}{a}+\frac{1}{b}}{\frac{2}{a}+\frac{2}{b}}$

22. $\dfrac{\frac{1}{x}+\frac{1}{y}}{\frac{3}{x}+\frac{3}{y}}$

23. $\dfrac{\frac{1}{a}+\frac{3}{b}}{\frac{1}{b}-\frac{3}{a}}$

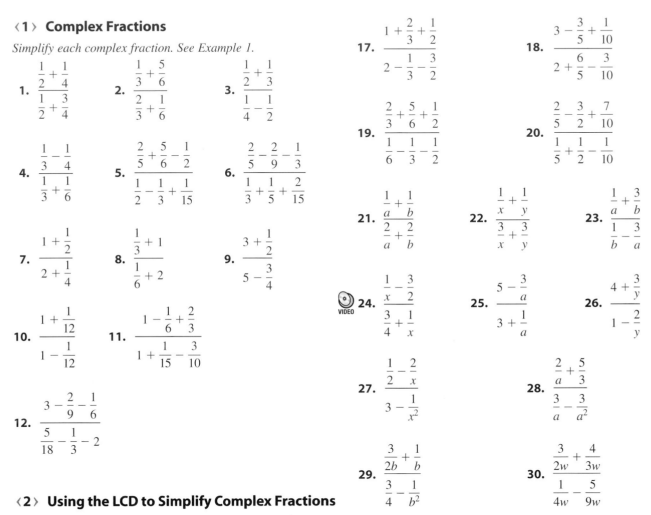 24. $\dfrac{\frac{1}{x}-\frac{3}{2}}{\frac{3}{4}+\frac{1}{x}}$

25. $\dfrac{5-\frac{3}{a}}{3+\frac{1}{a}}$

26. $\dfrac{4+\frac{3}{y}}{1-\frac{2}{y}}$

27. $\dfrac{\frac{1}{2}-\frac{2}{x}}{3-\frac{1}{x^2}}$

28. $\dfrac{\frac{2}{a}+\frac{5}{3}}{\frac{3}{a}-\frac{3}{a^2}}$

29. $\dfrac{\frac{3}{2b}+\frac{1}{b}}{\frac{3}{4}-\frac{1}{b^2}}$

30. $\dfrac{\frac{3}{2w}+\frac{4}{3w}}{\frac{1}{4w}-\frac{5}{9w}}$

Simplify each complex fraction. See Example 4.

31. $\dfrac{\frac{1}{x+1}+1}{\frac{3}{x+1}+3}$

32. $\dfrac{\frac{2}{x+3}+1}{\frac{4}{x+3}+2}$

33. $\dfrac{1-\frac{3}{y+1}}{3+\frac{1}{y+1}}$

34. $\dfrac{2-\frac{1}{a-3}}{3-\frac{1}{a-3}}$

35. $\dfrac{x + \dfrac{4}{x-2}}{x - \dfrac{x+1}{x-2}}$

36. $\dfrac{x - \dfrac{x-6}{x-1}}{x - \dfrac{x+15}{x-1}}$

37. $\dfrac{\dfrac{1}{3-x} - 5}{\dfrac{1}{x-3} - 2}$

38. $\dfrac{\dfrac{2}{x-5} - x}{\dfrac{3x}{5-x} - 1}$

39. $\dfrac{1 - \dfrac{5}{a-1}}{3 - \dfrac{2}{1-a}}$

40. $\dfrac{\dfrac{1}{3} - \dfrac{2}{9-x}}{\dfrac{1}{6} - \dfrac{1}{x-9}}$

41. $\dfrac{\dfrac{1}{m-3} - \dfrac{4}{m}}{\dfrac{3}{m-3} + \dfrac{1}{m}}$

42. $\dfrac{\dfrac{1}{y+3} - \dfrac{4}{y}}{\dfrac{1}{y} - \dfrac{2}{y+3}}$

43. $\dfrac{\dfrac{2}{w-1} - \dfrac{3}{w+1}}{\dfrac{4}{w+1} + \dfrac{5}{w-1}}$

44. $\dfrac{\dfrac{1}{x+2} - \dfrac{3}{x+3}}{\dfrac{2}{x+3} + \dfrac{3}{x+2}}$

45. $\dfrac{\dfrac{1}{a-b} - \dfrac{1}{a+b}}{\dfrac{1}{b-a} + \dfrac{1}{b+a}}$

46. $\dfrac{\dfrac{1}{2+x} - \dfrac{1}{2-x}}{\dfrac{1}{x+2} - \dfrac{1}{x-2}}$

Simplify each complex fraction. Reduce each answer to lowest terms.

47. $\dfrac{1 - \dfrac{4}{a^2}}{1 + \dfrac{2}{a} - \dfrac{8}{a^2}}$

48. $\dfrac{\dfrac{1}{3} + \dfrac{1}{y}}{\dfrac{y}{3} - \dfrac{3}{y}}$

49. $\dfrac{\dfrac{1}{2} + \dfrac{1}{4x}}{\dfrac{x}{3} - \dfrac{1}{12x}}$

50. $\dfrac{\dfrac{1}{9} + \dfrac{1}{3x}}{\dfrac{x}{9} - \dfrac{1}{x}}$

51. $\dfrac{\dfrac{1}{3} - \dfrac{5}{3x} + \dfrac{2}{x^2}}{\dfrac{1}{3} - \dfrac{3}{x^2}}$

52. $\dfrac{\dfrac{1}{2} - \dfrac{3}{2x} + \dfrac{1}{x^2}}{\dfrac{1}{2} - \dfrac{1}{2x^2}}$

53. $\dfrac{\dfrac{2x-9}{6}}{\dfrac{2x-3}{9}}$

54. $\dfrac{\dfrac{a-5}{12}}{\dfrac{a+2}{15}}$

55. $\dfrac{\dfrac{2x-4y}{xy^2}}{\dfrac{3x-6y}{x^3y}}$

56. $\dfrac{\dfrac{ab+b^2}{4ab^5}}{\dfrac{a+b}{6a^2b^4}}$

57. $\dfrac{\dfrac{a^2+2a-24}{a+1}}{\dfrac{a^2-a-12}{(a+1)^2}}$

58. $\dfrac{\dfrac{y^2-3y-18}{y^2-4}}{\dfrac{y^2+5y+6}{y-2}}$

59. $\dfrac{\dfrac{x}{x+1}}{\dfrac{1}{x^2-1} - \dfrac{1}{x-1}}$

60. $\dfrac{\dfrac{a}{a^2-b^2}}{\dfrac{1}{a+b} + \dfrac{1}{a-b}}$

⟨3⟩ Applications

Solve each problem. See Example 5.

61. *Sophomore math.* A survey of college sophomores showed that $\frac{5}{6}$ of the males were taking a mathematics class and $\frac{3}{4}$ of the females were taking a mathematics class. One-third of the males were enrolled in calculus, and $\frac{1}{5}$ of the females were enrolled in calculus. If just as many males as females were surveyed, then what fraction of the surveyed students taking mathematics were enrolled in calculus? Rework this problem assuming that the number of females in the survey was twice the number of males.

62. *Commuting students.* At a well-known university, $\frac{1}{4}$ of the undergraduate students commute, and $\frac{1}{3}$ of the graduate students commute. One-tenth of the undergraduate students drive more than 40 miles daily, and $\frac{1}{6}$ of the graduate students drive more than 40 miles daily. If there are twice as many undergraduate students as there are graduate students, then what fraction of the commuters drive more than 40 miles daily?

Photo for Exercise 62

Getting More Involved

63. *Exploration*

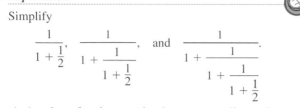

Simplify

$$\frac{1}{1+\frac{1}{2}}, \quad \frac{1}{1+\dfrac{1}{1+\frac{1}{2}}}, \quad \text{and} \quad \frac{1}{1+\dfrac{1}{1+\dfrac{1}{1+\frac{1}{2}}}}.$$

a) Are these fractions getting larger or smaller as the fractions become more complex?

b) Continuing the pattern, find the next two complex fractions and simplify them.

c) Now what can you say about the values of all five complex fractions?

64. *Discussion*

A complex fraction can be simplified by writing the numerator and denominator as single fractions and then dividing them or by multiplying the numerator and denominator by the LCD. Simplify the complex fraction

$$\frac{\dfrac{4}{xy^2}-\dfrac{6}{xy}}{\dfrac{2}{x^2}+\dfrac{4}{x^2y}}$$

by using each of these methods. Compare the number of steps used in each method, and determine which method requires fewer steps.

6.6	**Solving Equations with Rational Expressions**

In This Section

⟨1⟩ **Equations with Rational Expressions**

⟨2⟩ **Extraneous Solutions**

Many problems in algebra can be solved by using equations involving rational expressions. In this section you will learn how to solve equations that involve rational expressions, and in Sections 6.7 and 6.8 you will solve problems using these equations.

⟨1⟩ Equations with Rational Expressions

We solved some equations involving fractions in Section 2.3. In that section, the equations had only integers in the denominators. Our first step in solving those equations was to multiply by the LCD to eliminate all of the denominators.

EXAMPLE   1

Integers in the denominators

Solve $\frac{1}{2} - \frac{x-2}{3} = \frac{1}{6}$.

Solution

The LCD for 2, 3, and 6 is 6. Multiply each side of the equation by 6:

$$\frac{1}{2} - \frac{x-2}{3} = \frac{1}{6} \qquad \text{Original equation}$$

$$6\left(\frac{1}{2} - \frac{x-2}{3}\right) = 6 \cdot \frac{1}{6} \qquad \text{Multiply each side by 6.}$$

$$6 \cdot \frac{1}{2} - \cancel{6}^{2} \cdot \frac{x-2}{\cancel{3}} = \cancel{6} \cdot \frac{1}{\cancel{6}} \qquad \text{Distributive property}$$

$$3 - 2(x - 2) = 1 \qquad \text{Simplify.}$$

$$3 - 2x + 4 = 1 \qquad \text{Distributive property}$$

$$-2x = -6 \qquad \text{Subtract 7 from each side.}$$

$$x = 3 \qquad \text{Divide each side by } -2.$$

⟨ Helpful Hint ⟩

Note that it is not necessary to convert each fraction into an equivalent fraction with a common denominator here. Since we can multiply both sides of an equation by any expression we choose, we choose to multiply by the LCD. This tactic eliminates the fractions in one step.

‹ **Helpful Hint** ›

Always check your solution in the original equation by calculating the value of the left-hand side and the value of the right-hand side. If they are the same, your solution is correct.

Check $x = 3$ in the original equation:

$$\frac{1}{2} - \frac{3-2}{3} = \frac{1}{2} - \frac{1}{3} = \frac{3}{6} - \frac{2}{6} = \frac{1}{6}$$

Since the right-hand side of the equation is $\frac{1}{6}$, you can be sure that the solution to the equation is 3.

> Now do Exercises 1–12

CAUTION When a numerator contains a binomial, as in Example 1, the numerator must be enclosed in parentheses when the denominator is eliminated.

To solve an equation involving rational expressions, we usually multiply each side of the equation by the LCD for all the denominators involved, just as we do for an equation with fractions.

E X A M P L E   2

Variables in the denominators

Solve $\frac{1}{x} + \frac{1}{6} = \frac{1}{4}$.

Solution

We multiply each side of the equation by $12x$, the LCD for 4, 6, and x:

$$\frac{1}{x} + \frac{1}{6} = \frac{1}{4} \qquad \text{Original equation}$$

$$12x\left(\frac{1}{x} + \frac{1}{6}\right) = 12x\left(\frac{1}{4}\right) \qquad \text{Multiply each side by } 12x.$$

$$12x \cdot \frac{1}{x} + \overset{2}{12}x \cdot \frac{1}{6} = \overset{3}{12}x \cdot \frac{1}{4} \qquad \text{Distributive property}$$

$$12 + 2x = 3x \qquad \text{Simplify.}$$

$$12 = x \qquad \text{Subtract } 2x \text{ from each side.}$$

Check that 12 satisfies the original equation:

$$\frac{1}{12} + \frac{1}{6} = \frac{1}{12} + \frac{2}{12} = \frac{3}{12} = \frac{1}{4}$$

The solution to the equation is 12.

> Now do Exercises 13–24

E X A M P L E   3

An equation with two solutions

Solve the equation $\frac{100}{x} + \frac{100}{x+5} = 9$.

Solution

The LCD for the denominators x and $x + 5$ is $x(x + 5)$:

$$\frac{100}{x} + \frac{100}{x + 5} = 9 \qquad \text{Original equation}$$

$$x(x + 5)\frac{100}{x} + x(x + 5)\frac{100}{x + 5} = x(x + 5)9 \qquad \begin{array}{l}\text{Multiply each side by}\\ x(x + 5).\end{array}$$

$$(x + 5)100 + x(100) = \left(x^2 + 5x\right)9 \qquad \begin{array}{l}\text{All denominators are}\\ \text{eliminated.}\end{array}$$

$$100x + 500 + 100x = 9x^2 + 45x \qquad \text{Simplify.}$$

$$500 + 200x = 9x^2 + 45x$$

$$0 = 9x^2 - 155x - 500 \qquad \text{Get 0 on one side.}$$

$$0 = (9x + 25)(x - 20) \qquad \text{Factor.}$$

$$9x + 25 = 0 \qquad \text{or} \qquad x - 20 = 0 \qquad \text{Zero factor property}$$

$$x = -\frac{25}{9} \qquad \text{or} \qquad x = 20$$

A check will show that both $-\frac{25}{9}$ and 20 satisfy the original equation.

> Now do Exercises 25–32

⟨2⟩ Extraneous Solutions

In a rational expression, we can replace the variable only by real numbers that do not cause the denominator to be 0. When solving equations involving rational expressions, we must check every solution to see whether it causes 0 to appear in a denominator. If a number causes the denominator to be 0, then it cannot be a solution to the equation. A number that appears to be a solution but causes 0 in a denominator is called an **extraneous solution.** Since a solution to an equation is sometimes called a **root** to the equation, an extraneous solution is also called an **extraneous root.**

E X A M P L E 4

An equation with an extraneous solution

Solve the equation $\frac{1}{x - 2} = \frac{x}{2x - 4} + 1$.

Solution

Because the denominator $2x - 4$ factors as $2(x - 2)$, the LCD is $2(x - 2)$.

$$2(x - 2)\frac{1}{x - 2} = 2(x - 2)\frac{x}{2(x - 2)} + 2(x - 2) \cdot 1 \qquad \begin{array}{l}\text{Multiply each side of the}\\ \text{original equation by } 2(x - 2).\end{array}$$

$$2 = x + 2x - 4 \qquad \text{Simplify.}$$

$$2 = 3x - 4$$

$$6 = 3x$$

$$2 = x$$

Check 2 in the original equation:

$$\frac{1}{2 - 2} = \frac{2}{2 \cdot 2 - 4} + 1$$

The denominator $2 - 2$ is 0. So 2 does not satisfy the equation, and it is an extraneous solution. The equation has no solutions.

> Now do Exercises 33–36

      If the denominators of the rational expressions in an equation are not too complicated, you can tell at a glance which numbers cannot be solutions. For example, the equation $\frac{2}{x} + \frac{3}{x-1} = \frac{x-2}{x+5}$ could not have 0, 1, or -5 as a solution. Any solution to this equation must be in the domain of all three of the rational expressions in the equation.

E X A M P L E   5

Another extraneous solution

Solve the equation $\frac{1}{x} + \frac{1}{x-3} = \frac{x-2}{x-3}$.

Solution

The LCD for the denominators x and $x - 3$ is $x(x - 3)$:

$$\frac{1}{x} + \frac{1}{x-3} = \frac{x-2}{x-3} \qquad \text{Original equation}$$

$$x(x-3)\cdot\frac{1}{x} + x(x-3)\cdot\frac{1}{x-3} = x(x-3)\cdot\frac{x-2}{x-3} \qquad \text{Multiply each side by } x(x-3).$$

$$x - 3 + x = x(x - 2)$$

$$2x - 3 = x^2 - 2x$$

$$0 = x^2 - 4x + 3$$

$$0 = (x - 3)(x - 1)$$

$$x - 3 = 0 \qquad \text{or} \qquad x - 1 = 0$$

$$x = 3 \qquad \text{or} \qquad x = 1$$

If $x = 3$, then the denominator $x - 3$ has a value of 0. If $x = 1$, the original equation is satisfied. The only solution to the equation is 1.

> Now do Exercises 37–40

CAUTION   Always be sure to check your answers in the original equation to determine whether they are extraneous solutions.

Warm-Ups ▼

Fill in the blank.

1. The usual first step in solving an equation involving rational expressions is to multiply by the _____.

2. An _____ solution is a number that appears to be a solution but does not check in the original equation.

True or false?

3. To solve $x^2 = 8x$, we divide each side by x.

4. An extraneous solution is an irrational number.

5. Both 0 and 2 satisfy $\frac{3}{x} + \frac{5}{x-2} = \frac{2}{3}$.

6. To solve $\frac{3}{x} + \frac{5}{x-2} = \frac{2}{3}$, multiply each side by $3x^2 - 6x$.

7. To solve $\frac{1}{x-1} + 2 = \frac{1}{x+1}$, multiply each side by $x^2 - 1$.

8. The solution set to $\frac{1}{x-1} + 2 = \frac{1}{x+1}$ is $\{-1, 1\}$.

9. The solution set to $\frac{1}{x} + \frac{1}{2} = \frac{3}{x}$ is $\{4\}$.

Exercises

> ‹ **Study Tips** ›
>
> • The last couple of weeks of the semester is not the time to slack off. This is the time to double your efforts.
> • Make a schedule and plan every hour of your time.

‹1› Equations with Rational Expressions

Solve each equation. See Example 1.

1. $\dfrac{x}{2} + 1 = \dfrac{x}{4}$

2. $\dfrac{x}{3} + 2 = \dfrac{x}{6}$

3. $\dfrac{x}{3} - 5 = \dfrac{x}{2} - 7$

4. $\dfrac{x}{3} - \dfrac{x}{2} = \dfrac{x}{5} - 11$

5. $\dfrac{y}{5} - \dfrac{2}{3} = \dfrac{y}{6} + \dfrac{1}{3}$

6. $\dfrac{z}{6} + \dfrac{5}{4} = \dfrac{z}{2} - \dfrac{3}{4}$

7. $\dfrac{3}{4} - \dfrac{t-4}{3} = \dfrac{t}{12}$

8. $\dfrac{4}{5} - \dfrac{v-1}{10} = \dfrac{v-5}{30}$

9. $\dfrac{x}{3} + \dfrac{x+1}{4} = \dfrac{x+15}{12}$

10. $\dfrac{x}{8} + \dfrac{x+4}{12} = \dfrac{6x+5}{24}$

11. $\dfrac{1}{5} - \dfrac{w+10}{15} = \dfrac{1}{10} - \dfrac{w+1}{6}$

VIDEO **12.** $\dfrac{q}{5} - \dfrac{q-1}{2} = \dfrac{13}{20} - \dfrac{q+1}{4}$

Solve each equation. See Example 2.

13. $\dfrac{1}{x} + \dfrac{1}{2} = 3$

14. $\dfrac{2}{x} + \dfrac{3}{4} = 5$

15. $\dfrac{1}{x} + \dfrac{2}{x} = 7$

16. $\dfrac{5}{x} + \dfrac{6}{x} = 12$

17. $\dfrac{1}{x} + \dfrac{1}{2} = \dfrac{3}{4}$

18. $\dfrac{3}{x} + \dfrac{1}{4} = \dfrac{5}{8}$

19. $\dfrac{2}{3x} + \dfrac{1}{2x} = \dfrac{7}{24}$

20. $\dfrac{1}{6x} - \dfrac{1}{8x} = \dfrac{1}{72}$

21. $\dfrac{1}{2} + \dfrac{a-2}{a} = \dfrac{a+2}{2a}$

22. $\dfrac{1}{b} + \dfrac{1}{5} = \dfrac{b-1}{5b} + \dfrac{3}{10}$

23. $\dfrac{1}{3} - \dfrac{k+3}{6k} = \dfrac{1}{3k} - \dfrac{k-1}{2k}$

24. $\dfrac{3}{p} - \dfrac{p+3}{3p} = \dfrac{2p-1}{2p} - \dfrac{5}{6}$

Solve each equation. See Example 3.

25. $\dfrac{x}{2} = \dfrac{5}{x+3}$

26. $\dfrac{x}{3} = \dfrac{4}{x+1}$

VIDEO **27.** $\dfrac{x}{x+1} = \dfrac{6}{x+7}$

28. $\dfrac{x}{x+3} = \dfrac{2}{x-3}$

29. $\dfrac{2}{x+1} = \dfrac{1}{x} + \dfrac{1}{6}$

30. $\dfrac{1}{w+1} - \dfrac{1}{2w} = \dfrac{3}{40}$

31. $\dfrac{a-1}{a^2-4} + \dfrac{1}{a-2} = \dfrac{a+4}{a+2}$

32. $\dfrac{b+17}{b^2-1} - \dfrac{1}{b+1} = \dfrac{b-2}{b-1}$

‹2› Extraneous Solutions

Solve each equation. Watch for extraneous solutions. See Examples 4 and 5.

33. $\dfrac{1}{x-1} + \dfrac{2}{x} = \dfrac{x}{x-1}$

34. $\dfrac{4}{x} + \dfrac{3}{x-3} = \dfrac{x}{x-3} - \dfrac{1}{3}$

35. $\dfrac{5}{x+2} + \dfrac{2}{x-3} = \dfrac{x-1}{x-3}$

36. $\dfrac{6}{y-2} + \dfrac{7}{y-8} = \dfrac{y-1}{y-8}$

37. $1 + \dfrac{3y}{y-2} = \dfrac{6}{y-2}$

VIDEO **38.** $\dfrac{5}{y-3} = \dfrac{y+7}{2y-6} + 1$

39. $\dfrac{z}{z+1} - \dfrac{1}{z+2} = \dfrac{2z+5}{z^2+3z+2}$

40. $\dfrac{z}{z-2} - \dfrac{1}{z+5} = \dfrac{7}{z^2+3z-10}$

Miscellaneous

Solve each equation.

41. $\dfrac{a}{4} = \dfrac{5}{2}$

42. $\dfrac{y}{3} = \dfrac{6}{5}$

43. $\dfrac{w}{6} = \dfrac{3w}{11}$

44. $\dfrac{2m}{3} = \dfrac{3m}{2}$

45. $\dfrac{5}{x} = \dfrac{x}{5}$

46. $\dfrac{-3}{x} = \dfrac{x}{-3}$

47. $\dfrac{x-3}{5} = \dfrac{x-3}{x}$

48. $\dfrac{a+4}{2} = \dfrac{a+4}{a}$

49. $\dfrac{1}{x+2} = \dfrac{x}{x+2}$

50. $\dfrac{-3}{w+2} = \dfrac{w}{w+2}$

51. $\dfrac{1}{2x-4} + \dfrac{1}{x-2} = \dfrac{3}{2}$

52. $\dfrac{7}{3x-9} - \dfrac{1}{x-3} = \dfrac{4}{3}$

53. $\dfrac{3}{a^2-a-6} = \dfrac{2}{a^2-4}$

54. $\dfrac{8}{a^2+a-6} = \dfrac{6}{a^2-9}$

55. $\dfrac{4}{c-2} - \dfrac{1}{2-c} = \dfrac{25}{c+6}$

56. $\dfrac{3}{x+1} - \dfrac{1}{1-x} = \dfrac{10}{x^2-1}$

57. $\dfrac{1}{x^2-9} + \dfrac{3}{x+3} = \dfrac{4}{x-3}$

58. $\dfrac{3}{x-2} - \dfrac{5}{x+3} = \dfrac{1}{x^2+x-6}$

59. $\dfrac{3}{2x+4} - \dfrac{1}{x+2} = \dfrac{1}{3x+1}$

60. $\dfrac{5}{2m+6} - \dfrac{1}{m+1} = \dfrac{1}{m+3}$

61. $\dfrac{2t-1}{3t+3} + \dfrac{3t-1}{6t+6} = \dfrac{t}{t+1}$

62. $\dfrac{4w-1}{3w+6} - \dfrac{w-1}{3} = \dfrac{w-1}{w+2}$

Applications

Solve each problem.

63. *Lens equation.* The focal length f for a camera lens is related to the object distance o and the image distance i by the formula

$$\frac{1}{f} = \frac{1}{o} + \frac{1}{i}.$$

See the accompanying figure. The image is in focus at distance i from the lens. For an object that is 600 mm from a 50-mm lens, use $f = 50$ mm and $o = 600$ mm to find i.

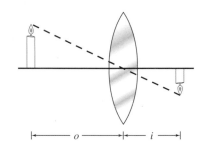

Figure for Exercise 63

64. *Telephoto lens.* Use the formula from Exercise 63 to find the image distance i for an object that is 2,000,000 mm from a 250-mm telephoto lens.

Photo for Exercise 64

6.7   Applications of Ratios and Proportions

In This Section

〈1〉 Ratios
〈2〉 Proportions

In this section, we will use the ideas of rational expressions in ratio and proportion problems. We will solve proportions in the same way we solved equations in Section 6.6.

〈1〉 Ratios

In Chapter 1 we defined a rational number as the *ratio of two integers*. We will now give a more general definition of ratio. If a and b are any real numbers (not just integers), with $b \neq 0$, then the expression $\dfrac{a}{b}$ is called the **ratio of a and b** or the **ratio of a to b.**

The ratio of a to b is also written as $a:b$. A ratio is a comparison of two numbers. Some examples of ratios are

$$\frac{3}{4}, \quad \frac{4.2}{2.1}, \quad \frac{\frac{1}{4}}{\frac{1}{2}}, \quad \frac{3.6}{5}, \quad \text{and} \quad \frac{100}{1}.$$

Ratios are treated just like fractions. We can reduce ratios, and we can build them up. We generally express ratios as ratios of integers. When possible, we will convert a ratio into an equivalent ratio of integers in lowest terms.

E X A M P L E   1

Finding equivalent ratios

Find an equivalent ratio of integers in lowest terms for each ratio.

a) $\dfrac{4.2}{2.1}$　　　　b) $\dfrac{\frac{1}{4}}{\frac{1}{2}}$　　　　c) $\dfrac{3.6}{5}$

Solution

a) Because both the numerator and the denominator have one decimal place, we will multiply the numerator and denominator by 10 to eliminate the decimals:

$$\frac{4.2}{2.1} = \frac{4.2(10)}{2.1(10)} = \frac{42}{21} = \frac{21 \cdot 2}{21 \cdot 1} = \frac{2}{1} \quad \text{Do not omit the 1 in a ratio.}$$

So the ratio of 4.2 to 2.1 is equivalent to the ratio 2 to 1.

b) This ratio is a complex fraction. We can simplify this expression using the LCD method as shown in Section 6.5. Multiply the numerator and denominator of this ratio by 4:

$$\frac{\frac{1}{4}}{\frac{1}{2}} = \frac{\frac{1}{4} \cdot 4}{\frac{1}{2} \cdot 4} = \frac{1}{2}$$

c) We can get a ratio of integers if we multiply the numerator and denominator by 10.

$$\frac{3.6}{5} = \frac{3.6(10)}{5(10)} = \frac{36}{50}$$

$$= \frac{18}{25} \quad \text{Reduce to lowest terms.}$$

> Now do Exercises 1–16

In Example 2, a ratio is used to compare quantities.

E X A M P L E   2

Nitrogen to potash

In a 50-pound bag of lawn fertilizer there are 8 pounds of nitrogen and 12 pounds of potash. What is the ratio of nitrogen to potash?

Solution

The nitrogen and potash occur in this fertilizer in the ratio of 8 pounds to 12 pounds:

$$\frac{8}{12} = \frac{2 \cdot \cancel{4}}{3 \cdot \cancel{4}} = \frac{2}{3}$$

So the ratio of nitrogen to potash is 2 to 3.

> Now do Exercises 17–18

EXAMPLE 3

Males to females

In a class of 50 students, there were exactly 20 male students. What was the ratio of males to females in this class?

Solution

Because there were 20 males in the class of 50, there were 30 females. The ratio of males to females was 20 to 30, or 2 to 3.

> Now do Exercises 19–20

Ratios give us a means of comparing the size of two quantities. For this reason *the numbers compared in a ratio should be expressed in the same units*. For example, if one dog is 24 inches high and another is 1 foot high, then the ratio of their heights is 2 to 1, not 24 to 1.

EXAMPLE 4

Quantities with different units

What is the ratio of length to width for a poster with a length of 30 inches and a width of 2 feet?

Solution

Because the width is 2 feet, or 24 inches, the ratio of length to width is 30 to 24. Reduce as follows:

$$\frac{30}{24} = \frac{5 \cdot 6}{4 \cdot 6} = \frac{5}{4}$$

So the ratio of length to width is 5 to 4.

> Now do Exercises 21–24

⟨2⟩ Proportions

A **proportion** is any statement expressing the equality of two ratios. The statement

$$\frac{a}{b} = \frac{c}{d} \qquad \text{or} \qquad a{:}b = c{:}d$$

is a proportion. In any proportion the numbers in the positions of a and d shown here are called the **extremes.** The numbers in the positions of b and c as shown are called the **means.** In the proportion

$$\frac{30}{24} = \frac{5}{4},$$

the means are 24 and 5, and the extremes are 30 and 4.

If we multiply each side of the proportion

$$\frac{a}{b} = \frac{c}{d}$$

by the LCD, bd, we get

$$\frac{a}{b} \cdot bd = \frac{c}{d} \cdot bd$$

or

$$a \cdot d = b \cdot c.$$

We can express this result by saying *that the product of the extremes is equal to the product of the means.* We call this fact the **extremes-means property** or **cross-multiplying.**

Extremes-Means Property (Cross-Multiplying)

Suppose a, b, c, and d are real numbers with $b \neq 0$ and $d \neq 0$. If

$$\frac{a}{b} = \frac{c}{d}, \quad \text{then} \quad ad = bc.$$

We use the extremes-means property to solve proportions.

E X A M P L E   5

Using the extremes-means property

Solve the proportion $\frac{3}{x} = \frac{5}{x+5}$ for x.

Solution

Instead of multiplying each side by the LCD, we use the extremes-means property:

$$\frac{3}{x} = \frac{5}{x+5} \qquad \text{Original proportion}$$

$$3(x+5) = 5x \qquad \text{Extremes-means property}$$

$$3x + 15 = 5x \qquad \text{Distributive property}$$

$$15 = 2x$$

$$\frac{15}{2} = x$$

Check:

$$\frac{3}{\frac{15}{2}} = 3 \cdot \frac{2}{15} = \frac{2}{5}$$

$$\frac{5}{\frac{15}{2} + 5} = \frac{5}{\frac{25}{2}} = 5 \cdot \frac{2}{25} = \frac{2}{5}$$

So $\frac{15}{2}$ is the solution to the equation or the solution to the proportion.

Now do Exercises 25–38

E X A M P L E   **6**

Solving a proportion

The ratio of men to women at Brighton City College is 2 to 3. If there are 894 men, then how many women are there?

Solution

Because the ratio of men to women is 2 to 3, we have

$$\frac{\text{Number of men}}{\text{Number of women}} = \frac{2}{3}.$$

If x represents the number of women, then we have the following proportion:

$$\frac{894}{x} = \frac{2}{3}$$

$$2x = 2682 \quad \text{Extremes-means property}$$

$$x = 1341$$

The number of women is 1341.

Now do Exercises 39–42

Note that any proportion can be solved by multiplying each side by the LCD as we did when we solved other equations involving rational expressions. The extremes-means property gives us a shortcut for solving proportions.

E X A M P L E   **7**

Solving a proportion

In a conservative portfolio the ratio of the amount invested in bonds to the amount invested in stocks should be 3 to 1. A conservative investor invested $2850 more in bonds than she did in stocks. How much did she invest in each category?

Solution

Because the ratio of the amount invested in bonds to the amount invested in stocks is 3 to 1, we have

$$\frac{\text{Amount invested in bonds}}{\text{Amount invested in stocks}} = \frac{3}{1}.$$

If x represents the amount invested in stocks and $x + 2850$ represents the amount invested in bonds, then we can write and solve the following proportion:

$$\frac{x + 2850}{x} = \frac{3}{1}$$

$$3x = x + 2850 \quad \text{Extremes-means property}$$

$$2x = 2850$$

$$x = 1425$$

$$x + 2850 = 4275$$

So she invested $4275 in bonds and $1425 in stocks. As a check, note that these amounts are in the ratio of 3 to 1.

Now do Exercises 43–46

Example 8 shows how conversions from one unit of measurement to another can be done by using proportions.

Converting measurements

There are 3 feet in 1 yard. How many feet are there in 12 yards?

Solution

Let x represent the number of feet in 12 yards. There are two proportions that we can write to solve the problem:

$$\frac{3 \text{ feet}}{x \text{ feet}} = \frac{1 \text{ yard}}{12 \text{ yards}} \qquad \frac{3 \text{ feet}}{1 \text{ yard}} = \frac{x \text{ feet}}{12 \text{ yards}}$$

The ratios in the second proportion violate the rule of comparing only measurements that are expressed in the same units. Note that each side of the second proportion is actually the ratio 1 to 1, since 3 feet = 1 yard and x feet = 12 yards. For doing conversions we can use ratios like this to compare measurements in different units. Applying the extremes-means property to either proportion gives

$$3 \cdot 12 = x \cdot 1,$$

or

$$x = 36.$$

So there are 36 feet in 12 yards.

Now do Exercises 47–50

Warm-Ups ▼

Fill in the blank.

1. A _____ is a comparison of two numbers.

2. A _____ is an equation that expresses the equality of two ratios.

3. In $\dfrac{a}{b} = \dfrac{c}{d}$, b and c are the _____.

4. In $\dfrac{a}{b} = \dfrac{c}{d}$, a and d are the _____.

5. The _____ property says that if $\dfrac{a}{b} = \dfrac{c}{d}$, then $ad = bc$.

True or false?

6. The ratio of 40 men to 30 women can be expressed as the ratio 4 to 3.

7. The ratio of 3 feet to 2 yards can be expressed as the ratio 3 to 2.

8. The ratio of 1.5 to 2 is equivalent to the ratio 3 to 4.

9. The product of the extremes is equal to the product of the means.

10. If $\dfrac{2}{x} = \dfrac{3}{5}$, then $5x = 6$.

11. If 4 of the 12 members of the supreme council are women, then the ratio of men to women is 1 to 3.

‹ 1 › **Ratios**

For each ratio, find an equivalent ratio of integers in lowest terms. See Example 1.

1. $\dfrac{4}{6}$

2. $\dfrac{10}{20}$

3. $\dfrac{200}{150}$

4. $\dfrac{1000}{200}$

5. $\dfrac{2.5}{3.5}$

6. $\dfrac{4.8}{1.2}$

7. $\dfrac{0.32}{0.6}$

8. $\dfrac{0.05}{0.8}$

9. $\dfrac{35}{10}$

10. $\dfrac{88}{33}$

11. $\dfrac{4.5}{7}$

12. $\dfrac{3}{2.5}$

13. $\dfrac{\frac{1}{2}}{\frac{1}{5}}$

14. $\dfrac{\frac{2}{3}}{\frac{3}{4}}$

15. $\dfrac{5}{\frac{1}{3}}$

16. $\dfrac{4}{\frac{1}{4}}$

Find a ratio for each of the following, and write it as a ratio of integers in lowest terms. See Examples 2–4.

17. *Men and women.* Find the ratio of men to women in a bowling league containing 12 men and 8 women.

18. *Coffee drinkers.* Among 100 coffee drinkers, 36 said that they preferred their coffee black and the rest did not prefer their coffee black. Find the ratio of those who prefer black coffee to those who prefer nonblack coffee.

Photo for Exercise 18

19. *Smokers.* A life insurance company found that among its last 200 claims, there were six dozen smokers. What is the ratio of smokers to nonsmokers in this group of claimants?

20. *Hits and misses.* A woman threw 60 darts and hit the target a dozen times. What is her ratio of hits to misses?

21. *Violence and kindness.* While watching television for one week, a consumer group counted 1240 acts of violence and 40 acts of kindness. What is the violence to kindness ratio for television, according to this group?

22. *Length to width.* What is the ratio of length to width for the rectangle shown?

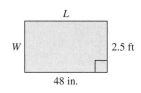

Figure for Exercise 22

23. *Rise to run.* What is the ratio of rise to run for the stairway shown in the figure?

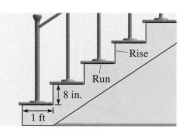

Figure for Exercise 23

24. *Rise and run.* If the rise is $\frac{3}{2}$ and the run is 5, then what is the ratio of the rise to the run?

‹ 2 › **Proportions**

Solve each proportion. See Example 5.

25. $\dfrac{4}{x} = \dfrac{2}{3}$

26. $\dfrac{9}{x} = \dfrac{3}{2}$

27. $\dfrac{a}{2} = \dfrac{-1}{5}$

28. $\dfrac{b}{3} = \dfrac{-3}{4}$

29. $-\dfrac{5}{9} = \dfrac{3}{x}$

30. $-\dfrac{3}{4} = \dfrac{5}{x}$

31. $\dfrac{x-2}{5} = \dfrac{x}{7}$

32. $\dfrac{4}{x+1} = \dfrac{2}{x}$

33. $\dfrac{10}{x} = \dfrac{34}{x+12}$

 **34.** $\dfrac{x}{3} = \dfrac{x+1}{2}$

35. $\dfrac{a}{a+1} = \dfrac{a+3}{a}$

36. $\dfrac{c+3}{c-1} = \dfrac{c+2}{c-3}$

 **37.** $\dfrac{m-1}{m-2} = \dfrac{m-3}{m+4}$

38. $\dfrac{h}{h-3} = \dfrac{h}{h-9}$

Use a proportion to solve each problem. See Examples 6–8.

39. *New shows and reruns.* The ratio of new shows to reruns on cable TV is 2 to 27. If Frank counted only eight new shows one evening, then how many reruns were there?

40. *Fast food.* If four out of five doctors prefer fast food, then at a convention of 445 doctors, how many prefer fast food?

41. *Voting.* If 220 out of 500 voters surveyed said that they would vote for the incumbent, then how many votes could the incumbent expect out of the 400,000 voters in the state?

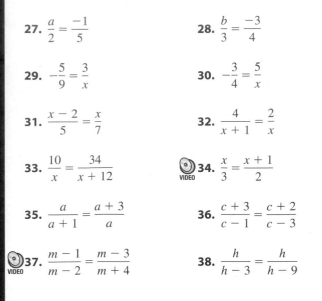

Photo for Exercise 41

 **42. *New product.*** A taste test with 200 randomly selected people found that only three of them said that they would buy a box of new Sweet Wheats cereal. How many boxes could the manufacturer expect to sell in a country of 280 million people?

43. *Basketball blowout.* As the final buzzer signaled the end of the basketball game, the Lions were 34 points ahead of the Tigers. If the Lions scored 5 points for every 3 scored by the Tigers, then what was the final score?

44. *The golden ratio.* The ancient Greeks thought that the most pleasing shape for a rectangle was one for which the ratio of the length to the width was approximately 8 to 5, the golden ratio. If the length of a rectangular painting is 2 ft longer than its width, then for what dimensions would the length and width have the golden ratio?

45. *Automobile sales.* The ratio of sports cars to luxury cars sold in Wentworth one month was 3 to 2. If 20 more sports cars were sold than luxury cars, then how many of each were sold that month?

46. *Foxes and rabbits.* The ratio of foxes to rabbits in the Deerfield Forest Preserve is 2 to 9. If there are 35 fewer foxes than rabbits, then how many of each are there?

 **47. *Inches and feet.*** If there are 12 inches in 1 foot, then how many inches are there in 7 feet?

48. *Feet and yards.* If there are 3 feet in 1 yard, then how many yards are there in 28 feet?

49. *Minutes and hours.* If there are 60 minutes in 1 hour, then how many minutes are there in 0.25 hour?

50. *Meters and kilometers.* If there are 1000 meters in 1 kilometer, then how many meters are there in 2.33 kilometers?

51. *Miles and hours.* If Alonzo travels 230 miles in 3 hours, then how many miles does he travel in 7 hours?

52. *Hiking time.* If Evangelica can hike 19 miles in 2 days on the Appalachian Trail, then how many days will it take her to hike 63 miles?

53. *Force on basketball shoes.* The force exerted on shoe soles in a jump shot is proportional to the weight of the person jumping. If a 70-pound boy exerts a force of 980 pounds on his shoe soles when he returns to the court after a jump, then what force does a 6 ft 8 in. professional ball player weighing 280 pounds exert on the soles of his shoes when he returns to the court after a jump? Use the accompanying graph to estimate the force for a 150-pound player.

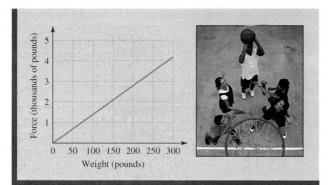

Figure for Exercise 53

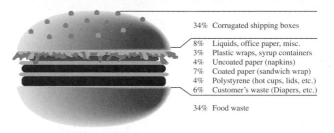

WASTE GENERATION AT A FAST-FOOD RESTAURANT

34%	Corrugated shipping boxes
8%	Liquids, office paper, misc.
3%	Plastic wraps, syrup containers
4%	Uncoated paper (napkins)
7%	Coated paper (sandwich wrap)
4%	Polystyrene (hot cups, lids, etc.)
6%	Customer's waste (Diapers, etc.)
34%	Food waste

Figure for Exercises 57 and 58

54. Force on running shoes. The ratio of the force on the shoe soles to the weight of a runner is 3 to 1. What force does a 130-pound jogger exert on the soles of her shoes?

55. Capture-recapture. To estimate the number of trout in Trout Lake, rangers used the capture-recapture method. They caught, tagged, and released 200 trout. One week later, they caught a sample of 150 trout and found that 5 of them were tagged. Assuming that the ratio of tagged trout to the total number of trout in the lake is the same as the ratio of tagged trout in the sample to the number of trout in the sample, find the number of trout in the lake.

56. Bear population. To estimate the size of the bear population on the Keweenaw Peninsula, conservationists captured, tagged, and released 50 bears. One year later, a random sample of 100 bears included only 2 tagged bears. What is the conservationist's estimate of the size of the bear population?

57. Fast-food waste. The accompanying figure shows the typical distribution of waste at a fast-food restaurant (U.S. Environmental Protection Agency, www.epa.gov).

   a) What is the ratio of customer waste to food waste?

   b) If a typical McDonald's generates 67 more pounds of food waste than customer waste per day, then how many pounds of customer waste does it generate?

58. Corrugated waste. Use the accompanying figure to find the ratio of waste from corrugated shipping boxes to waste not from corrugated shipping boxes. If a typical McDonald's generates 81 pounds of waste per day from corrugated shipping boxes, then how many pounds of

waste per day does it generate that is not from corrugated shipping boxes?

59. Mascara needs. In determining warehouse needs for a particular mascara for a chain of 2000 stores, Mike Pittman first determines a need B based on sales figures for the past 52 weeks. He then determines the actual need A from the equation $\frac{A}{B} = k$, where

$$k = 1 + V + C + X - D.$$

He uses $V = 0.22$ if there is a national TV ad and $V = 0$ if not, $C = 0.26$ if there is a national coupon and $C = 0$ if not, $X = 0.36$ if there is a chain-specific ad and $X = 0$ if not, and $D = 0.29$ if there is a special display in the chain and $D = 0$ if not. (D is subtracted because less product is needed in the warehouse when more is on display in the store.) If $B = 4200$ units and there is a special display and a national coupon but no national TV ad and no chain-specific ad, then what is the value of A?

Getting More Involved

60. Discussion

Which of the following equations is not a proportion? Explain.

   a) $\frac{1}{2} = \frac{1}{2}$ **b)** $\frac{x}{x+2} = \frac{4}{5}$

   c) $\frac{x}{4} = \frac{9}{x}$ **d)** $\frac{8}{x+2} - 1 = \frac{5}{x+2}$

61. Discussion

Find all of the errors in the following solution to an equation.

$$\frac{7}{x} = \frac{8}{x+3} + 1$$

$$7(x + 3) = 8x + 1$$

$$7x + 3 = 8x$$

$$-x = -3$$

$$x = 3$$

6.8   Applications of Rational Expressions

In This Section

⟨1⟩ **Formulas**
⟨2⟩ **Uniform Motion Problems**
⟨3⟩ **Work Problems**
⟨4⟩ **More Rate Problems**

In this section, we will study additional applications of rational expressions.

⟨1⟩ Formulas

Many formulas involve rational expressions. When solving a formula of this type for a certain variable, we usually multiply each side by the LCD to eliminate the denominators.

EXAMPLE **1**

An equation of a line

The equation for the line through $(-2, 4)$ with slope $\frac{3}{2}$ can be written as

$$\frac{y - 4}{x + 2} = \frac{3}{2}.$$

We studied equations of this type in Chapter 3. Solve this equation for y.

Solution

To isolate y on the left-hand side of the equation, we multiply each side by $x + 2$.

$$\frac{y - 4}{x + 2} = \frac{3}{2} \qquad \text{Original equation}$$

$$(x + 2) \cdot \frac{y - 4}{x + 2} = (x + 2) \cdot \frac{3}{2} \qquad \text{Multiply by } x + 2.$$

$$y - 4 = \frac{3}{2}x + 3 \qquad \text{Simplify.}$$

$$y = \frac{3}{2}x + 7 \qquad \text{Add 4 to each side.}$$

Because the original equation is a proportion, we could have used the extremes-means property to solve it for y.

> Now do Exercises 1–10

EXAMPLE **2**

Distance, rate, and time

Solve the formula $\frac{D}{T} = R$ for T.

Solution

Because the only denominator is T, we multiply each side by T:

$$\frac{D}{T} = R \qquad \text{Original formula}$$

$$T \cdot \frac{D}{T} = T \cdot R \qquad \text{Multiply each side by } T.$$

$$D = TR$$

$$\frac{D}{R} = \frac{TR}{R} \qquad \text{Divide each side by } R.$$

$$\frac{D}{R} = T \qquad \text{Simplify.}$$

The formula solved for T is $T = \frac{D}{R}$.

> Now do Exercises 11–16

In Example 3, different subscripts are used on a variable to indicate that they are different variables. Think of R_1 as the first resistance, R_2 as the second resistance, and R as a combined resistance.

EXAMPLE 3

Total resistance

The formula

$$\frac{1}{R} = \frac{1}{R_1} + \frac{1}{R_2}$$

(from physics) expresses the relationship between different amounts of resistance in a parallel circuit. Solve it for R_2.

Solution

The LCD for R, R_1, and R_2 is RR_1R_2:

$$\frac{1}{R} = \frac{1}{R_1} + \frac{1}{R_2}$$ Original formula

$$RR_1R_2 \cdot \frac{1}{R} = RR_1R_2 \cdot \frac{1}{R_1} + RR_1R_2 \cdot \frac{1}{R_2}$$ Multiply each side by the LCD, RR_1R_2.

$$R_1R_2 = RR_2 + RR_1$$ All denominators are eliminated.

$$R_1R_2 - RR_2 = RR_1$$ Get all terms involving R_2 onto the left side.

$$R_2(R_1 - R) = RR_1$$ Factor out R_2.

$$R_2 = \frac{RR_1}{R_1 - R}$$ Divide each side by $R_1 - R$.

Now do Exercises 17–24

EXAMPLE 4

Finding the value of a variable

In the formula of Example 1, find x if $y = -3$.

Solution

Substitute $y = -3$ into the formula, then solve for x:

$$\frac{y - 4}{x + 2} = \frac{3}{2}$$ Original formula

$$\frac{-3 - 4}{x + 2} = \frac{3}{2}$$ Replace y by -3.

$$\frac{-7}{x + 2} = \frac{3}{2}$$ Simplify.

$$3x + 6 = -14$$ Extremes-means property

$$3x = -20$$

$$x = -\frac{20}{3}$$

Now do Exercises 25–34

⟨2⟩ Uniform Motion Problems

In uniform motion problems we use the formula $D = RT$. In some problems in which the time is unknown, we can use the formula $T = \frac{D}{R}$ to get an equation involving rational expressions.

EXAMPLE 5

Driving to Florida

Susan drove 1500 miles to Daytona Beach for spring break. On the way back she averaged 10 miles per hour less, and the drive back took her 5 hours longer. Find Susan's average speed on the way to Daytona Beach.

Solution

If x represents her average speed going there, then $x - 10$ is her average speed for the return trip. See Fig. 6.1. We use the formula $T = \dfrac{D}{R}$ to make the following table.

	D	R	T	
Going	1500	x	$\dfrac{1500}{x}$	← Shorter time
Returning	1500	$x - 10$	$\dfrac{1500}{x - 10}$	← Longer time

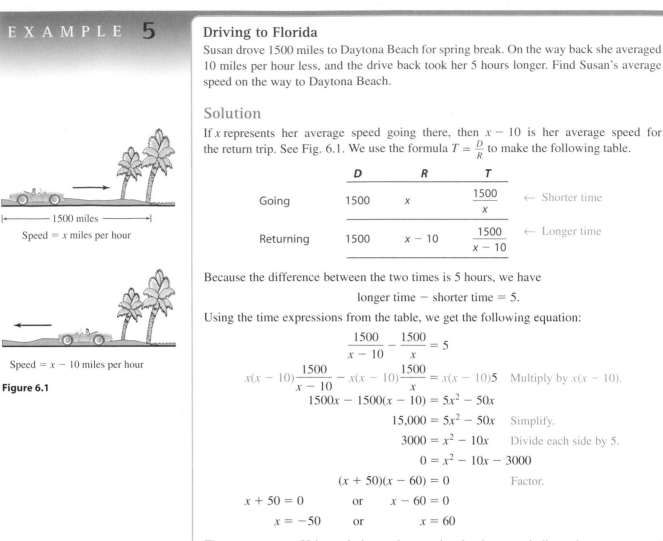

Because the difference between the two times is 5 hours, we have

$$\text{longer time} - \text{shorter time} = 5.$$

Using the time expressions from the table, we get the following equation:

$$\frac{1500}{x - 10} - \frac{1500}{x} = 5$$

$$x(x - 10)\frac{1500}{x - 10} - x(x - 10)\frac{1500}{x} = x(x - 10)5 \quad \text{Multiply by } x(x - 10).$$

$$1500x - 1500(x - 10) = 5x^2 - 50x$$

$$15{,}000 = 5x^2 - 50x \quad \text{Simplify.}$$

$$3000 = x^2 - 10x \quad \text{Divide each side by 5.}$$

$$0 = x^2 - 10x - 3000$$

$$(x + 50)(x - 60) = 0 \quad \text{Factor.}$$

$$x + 50 = 0 \quad \text{or} \quad x - 60 = 0$$

$$x = -50 \quad \text{or} \quad x = 60$$

The answer $x = -50$ is a solution to the equation, but it cannot indicate the average speed of the car. Her average speed going to Daytona Beach was 60 mph.

> Now do Exercises 35–40

〈3〉 Work Problems

If you can complete a job in 3 hours, then you are working at the rate of $\frac{1}{3}$ of the job per hour. If you work for 2 hours at the rate of $\frac{1}{3}$ of the job per hour, then you will complete $\frac{2}{3}$ of the job. The product of the rate and time is the amount of work completed. For problems involving work, we will always assume that the work is done at a constant rate. So if a job takes x hours to complete, then the rate is $\frac{1}{x}$ of the job per hour.

‹ Helpful Hint ›

Notice that a work rate is the same as a slope from Chapter 3. The only difference is that the work rates here can contain a variable.

EXAMPLE 6

Shoveling snow

After a heavy snowfall, Brian can shovel all of the driveway in 30 minutes. If his younger brother Allen helps, the job takes only 20 minutes. How long would it take Allen to do the job by himself?

Figure for image (left margin):
1500 miles
Speed = x miles per hour

Speed = $x - 10$ miles per hour

Figure 6.1

‹ **Helpful Hint** ›

The secret to work problems is remembering that the individual rates or the amounts of work can be added when people work together. If your painting rate is 1/10 of the house per day and your helper's rate is 1/5 of the house per day, then your rate together will be 3/10 of the house per day. In 2 days you will paint 2/10 of the house and your helper will paint 2/5 of the house for a total of 3/5 of the house completed.

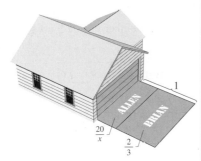

Figure 6.2

Solution

Let x represent the number of minutes it would take Allen to do the job by himself. Brian's rate for shoveling is $\frac{1}{30}$ of the driveway per minute, and Allen's rate for shoveling is $\frac{1}{x}$ of the driveway per minute. We organize all of the information in a table like the table in Example 5.

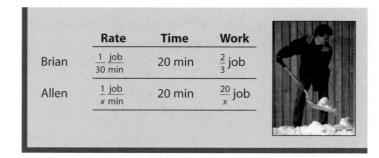

	Rate	Time	Work
Brian	$\frac{1 \text{ job}}{30 \text{ min}}$	20 min	$\frac{2}{3}$ job
Allen	$\frac{1 \text{ job}}{x \text{ min}}$	20 min	$\frac{20}{x}$ job

If Brian works for 20 min at the rate $\frac{1}{30}$ of the job per minute, then he does $\frac{20}{30}$ or $\frac{2}{3}$ of the job, as shown in Fig. 6.2. The amount of work that each boy does is a fraction of the whole job. So the expressions for work in the last column of the table have a sum of 1:

$$\frac{2}{3} + \frac{20}{x} = 1$$

$$3x \cdot \frac{2}{3} + 3x \cdot \frac{20}{x} = 3x \cdot 1 \quad \text{Multiply each side by } 3x.$$

$$2x + 60 = 3x$$

$$60 = x$$

If it takes Allen 60 min to do the job by himself, then he works at the rate of $\frac{1}{60}$ of the job per minute. In 20 minutes he does $\frac{1}{3}$ of the job while Brian does $\frac{2}{3}$. So it would take Allen 60 minutes to shovel the driveway by himself.

> Now do Exercises 41–46

Notice the similarities between the uniform motion problem in Example 5 and the work problem in Example 6. In both cases, it is beneficial to make a table. We use $D = R \cdot T$ in uniform motion problems and $W = R \cdot T$ in work problems. The main points to remember when solving work problems are summarized in the following strategy.

Strategy for Solving Work Problems

1. If a job is completed in x hours, then the rate is $\frac{1}{x}$ job/hr.
2. Make a table showing rate, time, and work completed ($W = R \cdot T$) for each person or machine.
3. The total work completed is the sum of the individual amounts of work completed.
4. If the job is completed, then the total work done is 1 job.

⟨4⟩ **More Rate Problems**

Rates are used in uniform motion and work problems. But rates also occur in other problems. If you make $400 for x hours of work, then your pay rate is $\dfrac{400}{x}$ dollars per hour. If you get $50 for selling x pounds of apples, then you are making money at the rate of $\dfrac{50}{x}$ dollars per pound.

E X A M P L E 7

Hourly rates

Dr. Watts paid $80 to her gardener and $80 to the gardener's helper for a total of 12 hours labor. If the gardener makes $10 more per hour than the helper, then how many hours did each of them work?

Solution

Let x be the number of hours for the gardener and $12 - x$ be the number of hours for the helper. Make a table as follows.

	Time	Pay	Hourly Rate
Gardener	x hours	80 dollars	$\frac{80}{x}$ dollars/hour
Helper	$12 - x$ hours	80 dollars	$\frac{80}{12 - x}$ dollars/hour

Since the gardener makes $10 more per hour, we can write the following equation.

$$\frac{80}{12 - x} + 10 = \frac{80}{x}$$

To solve the equation multiply each side by the LCD $x(12 - x)$.

$$x(12 - x)\left(\frac{80}{12 - x} + 10\right) = x(12 - x)\frac{80}{x} \qquad \text{Muliply by the LCD.}$$

$$80x + 10x(12 - x) = (12 - x)80 \qquad \text{Distributive property}$$

$$80x + 120x - 10x^2 = 960 - 80x \qquad \text{Distributive property}$$

$$-10x^2 + 280x - 960 = 0 \qquad \text{Get 0 on the right.}$$

$$x^2 - 28x + 96 = 0 \qquad \text{Divide each side by } -10.$$

$$(x - 4)(x - 24) = 0 \qquad \text{Factor.}$$

$$x - 4 = 0 \qquad \text{or} \qquad x - 24 = 0$$

$$x = 4 \qquad \text{or} \qquad x = 24$$

$$12 - x = 8 \qquad\qquad\qquad 12 - x = -12$$

Since $x = 24$ hours and $12 - x = -12$ hours does not make sense, we must have 4 hours for the gardener and 8 hours for the helper. Check: The gardener worked 4 hours at $20 per hour and the helper worked 8 hours at $10 per hour. The gardener made $10 more per hour than the helper. Note that the problem could be solved also by starting with x as the hourly pay for the gardener and $x - 10$ as the hourly pay for the helper. Try it.

> **Now do Exercises 47–48**

E X A M P L E 8

Oranges and grapefruit

Tamara bought 50 pounds of fruit consisting of Florida oranges and Texas grapefruit. She paid twice as much per pound for the grapefruit as she did for the oranges. If Tamara bought $12 worth of oranges and $16 worth of grapefruit, then how many pounds of each did she buy?

x lb

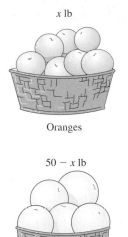

Oranges

50 − *x* lb

Grapefruit

Figure 6.3

Solution

Let *x* represent the number of pounds of oranges and 50 − *x* represent the number of pounds of grapefruit. See Fig. 6.3. Make a table.

	Rate	Quantity	Total Cost
Oranges	$\frac{12}{x}$ dollars/pound	*x* pounds	12 dollars
Grapefruit	$\frac{16}{50-x}$ dollars/pound	50 − *x* pounds	16 dollars

Since the price per pound for the grapefruit is twice that for the oranges, we have:

$$2(\text{price per pound for oranges}) = \text{price per pound for grapefruit}$$

$$2\left(\frac{12}{x}\right) = \frac{16}{50-x}$$

$$\frac{24}{x} = \frac{16}{50-x}$$

$$16x = 1200 - 24x \quad \text{Extremes-means property}$$

$$40x = 1200$$

$$x = 30$$

$$50 - x = 20$$

If Tamara purchased 20 pounds of grapefruit for $16, then she paid $0.80 per pound. If she purchased 30 pounds of oranges for $12, then she paid $0.40 per pound. Because $0.80 is twice $0.40, we can be sure that she purchased 20 pounds of grapefruit and 30 pounds of oranges.

> **Now do Exercises 49–50**

Warm-Ups ▼

True or false?

1. The formula $t = \frac{1-t}{m}$, solved for *m* is $m = \frac{1-t}{t}$.

2. To solve $\frac{1}{m} + \frac{1}{n} = \frac{1}{2}$ for *m*, we multiply each side by 2*mn*.

3. If Fiona drives 300 miles in *x* hours, then her average speed is $\frac{x}{300}$ mph.

4. If Mike drives 20 hard bargains in *x* hours, then he is driving $\frac{20}{x}$ hard bargains per hour.

5. If Fred can paint a house in *y* days, then he paints $\frac{1}{y}$ of the house per day.

6. If $\frac{1}{x}$ is 1 less than $\frac{2}{x+3}$, then $\frac{1}{x} - 1 = \frac{2}{x+3}$.

7. If *a* and *b* are nonzero and $a = \frac{m}{b}$, then $b = am$.

8. If $D = RT$, then $T = \frac{D}{R}$.

9. Solving $P + Prt = I$ for *P* yields $P = I - Prt$.

10. To solve $3R + yR = m$ for *R*, we must first factor the left side.

Exercises

> **‹ Study Tips ›**
>
> - Establish a regular routine of eating, sleeping, and exercise.
> - The ability to concentrate depends on adequate sleep, decent nutrition, and the physical well-being that comes with exercise.

‹1› Formulas

Solve each equation for y. See Example 1.

1. $\dfrac{y + 2}{x + 1} = 3$

2. $\dfrac{y + 5}{x + 2} = 6$

3. $\dfrac{y - 1}{x - 3} = 2$

4. $\dfrac{y - 2}{x - 4} = -2$

5. $\dfrac{y - 1}{x + 6} = -\dfrac{1}{2}$

6. $\dfrac{y + 5}{x - 2} = -\dfrac{1}{2}$

7. $\dfrac{y + a}{x - b} = m$

8. $\dfrac{y - h}{x + k} = a$

9. $\dfrac{y - 1}{x + 4} = -\dfrac{1}{3}$

10. $\dfrac{y - 1}{x + 3} = -\dfrac{3}{4}$

Solve each formula for the indicated variable. See Examples 2 and 3.

11. $A = \dfrac{B}{C}$ for C

12. $P = \dfrac{A}{C + D}$ for A

13. $\dfrac{1}{a} + m = \dfrac{1}{p}$ for p

14. $\dfrac{2}{f} + t = \dfrac{3}{m}$ for m

15. $F = k\dfrac{m_1 m_2}{r^2}$ for m_1

16. $F = \dfrac{mv^2}{r}$ for r

17. $\dfrac{1}{a} + \dfrac{1}{b} = \dfrac{1}{f}$ for a

18. $\dfrac{1}{R} = \dfrac{1}{R_1} + \dfrac{1}{R_2}$ for R

19. $S = \dfrac{a}{1 - r}$ for r

20. $I = \dfrac{E}{R + r}$ for R

21. $\dfrac{P_1 V_1}{T_1} = \dfrac{P_2 V_2}{T_2}$ for P_2

22. $\dfrac{P_1 V_1}{T_1} = \dfrac{P_2 V_2}{T_2}$ for T_1

23. $V = \dfrac{4}{3}\pi r^2 h$ for h

24. $h = \dfrac{S - 2\pi r^2}{2\pi r}$ for S

Find the value of the indicated variable. See Example 4.

25. In the formula of Exercise 11, if $A = 12$ and $B = 5$, find C.

26. In the formula of Exercise 12, if $A = 500$, $P = 100$, and $C = 2$, find D.

27. In the formula of Exercise 13, if $p = 6$ and $m = 4$, find a.

28. In the formula of Exercise 14, if $m = 4$ and $t = 3$, find f.

29. In the formula of Exercise 15, if $F = 32$, $r = 4$, $m_1 = 2$, and $m_2 = 6$, find k.

30. In the formula of Exercise 16, if $F = 10$, $v = 8$, and $r = 6$, find m.

31. In the formula of Exercise 17, if $f = 3$ and $a = 2$, find b.

32. In the formula of Exercise 18, if $R = 3$ and $R_1 = 5$, find R_2.

33. In the formula of Exercise 19, if $S = \dfrac{3}{2}$ and $r = \dfrac{1}{5}$, find a.

34. In the formula of Exercise 20, if $I = 15$, $E = 3$, and $R = 2$, find r.

⟨2⟩   Uniform Motion Problems

Show a complete solution to each problem. See Example 5.

35. *Fast walking.* Marcie can walk 8 miles in the same time as Frank walks 6 miles. If Marcie walks 1 mile per hour faster than Frank, then how fast does each person walk?

36. *Upstream, downstream.* Junior's boat will go 15 miles per hour in still water. If he can go 12 miles downstream in the same amount of time as it takes to go 9 miles upstream, then what is the speed of the current?

37. *Delivery routes.* Pat travels 70 miles on her milk route, and Bob travels 75 miles on his route. Pat travels 5 miles per hour slower than Bob, and her route takes her one-half hour longer than Bob's. How fast is each one traveling?

38. *Ride the peaks.* Smith bicycled 45 miles going east from Durango, and Jones bicycled 70 miles. Jones averaged 5 miles per hour more than Smith, and his trip took one-half hour longer than Smith's. How fast was each one traveling?

Photo for Exercise 38

39. *Walking and running.* Raffaele ran 8 miles and then walked 6 miles. If he ran 5 miles per hour faster than he walked and the total time was 2 hours, then how fast did he walk?

40. *Triathlon.* Luisa participated in a triathlon in which she swam 3 miles, ran 5 miles, and then bicycled 10 miles. Luisa ran twice as fast as she swam, and she cycled three times as fast as she swam. If her total time for the triathlon was 1 hour and 46 minutes, then how fast did she swim?

⟨3⟩   Work Problems

Show a complete solution to each problem. See Example 6. See the Strategy for Solving Work Problems on page 441.

41. *Fence painting.* Kiyoshi can paint a certain fence in 3 hours by himself. If Red helps, the job takes only

2 hours. How long would it take Red to paint the fence by himself?

42. *Envelope stuffing.* Every week, Linda must stuff 1000 envelopes. She can do the job by herself in 6 hours. If Laura helps, they get the job done in $5\frac{1}{2}$ hours. How long would it take Laura to do the job by herself?

43. *Garden destroying.* Mr. McGregor has discovered that a large dog can destroy his entire garden in 2 hours and that a small boy can do the same job in 1 hour. How long would it take the large dog and the small boy working together to destroy Mr. McGregor's garden?

44. *Draining the vat.* With only the small valve open, all of the liquid can be drained from a large vat in 4 hours. With only the large valve open, all of the liquid can be drained from the same vat in 2 hours. How long would it take to drain the vat with both valves open?

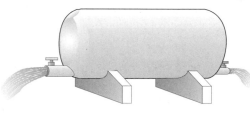

Figure for Exercise 44

45. *Cleaning sidewalks.* Edgar can blow the leaves off the sidewalks around the capitol building in 2 hours using a gasoline-powered blower. Ellen can do the same job in 8 hours using a broom. How long would it take them working together?

46. *Computer time.* It takes a computer 8 days to print all of the personalized letters for a national sweepstakes. A new computer is purchased that can do the same job in 5 days. How long would it take to do the job with both computers working on it?

⟨4⟩   More Rate Problems

Show a complete solution to each problem. See Examples 7 and 8.

47. *Repair work.* Sally received a bill for a total of 8 hours labor on the repair of her bulldozer. She paid $50 to the master mechanic and $90 to his apprentice. If the master mechanic gets $10 more per hour than his apprentice, then how many hours did each work on the bulldozer?

48. *Running backs.* In the playoff game the ball was carried by either Anderson or Brown on 21 plays. Anderson gained 36 yards, and Brown gained 54 yards. If Brown averaged twice as many yards per carry as Anderson, then on how many plays did Anderson carry the ball?

Photo for Exercise 48

49. Apples and bananas. Bertha bought 18 pounds of fruit consisting of apples and bananas. She paid $9 for the apples and $2.40 for the bananas. If the price per pound of the apples was 3 times that of the bananas, then how many pounds of each type of fruit did she buy?

50. Fuel efficiency. Last week, Joe's Electric Service used 110 gallons of gasoline in its two trucks. The large truck was driven 800 miles, and the small truck was driven 600 miles. If the small truck gets twice as many miles per gallon as the large truck, then how many gallons of gasoline did the large truck use?

Miscellaneous

Show a complete solution to each problem.

51. Small plane. It took a small plane 1 hour longer to fly 480 miles against the wind than it took the plane to fly the same distance with the wind. If the wind speed was 20 mph, then what is the speed of the plane in calm air?

52. Fast boat. A motorboat at full throttle takes two hours longer to travel 75 miles against the current than it takes to travel the same distance with the current. If the rate of the current is 5 mph, then what is the speed of the boat at full throttle in still water?

53. Light plane. At full throttle a light plane flies 275 miles against the wind in the same time as it flies 325 miles with the wind. If the plane flies at 120 mph at full throttle in still air, then what is the wind speed?

54. Big plane. A six-passenger plane cruises at 180 mph in calm air. If the plane flies 7 miles with the wind in the same amount of time as it flies 5 miles against the wind, then what is the wind speed?

55. Two cyclists. Ben and Jerry start from the same point and ride their bicycles in opposite directions. If Ben rides twice as fast as Jerry and they are 90 miles apart after four hours, then what is the speed of each rider?

56. Catching up. A sailboat leaves port and travels due south at an average speed of 9 mph. Four hours later a motorboat leaves the same port and travels due south at an average speed of 21 mph. How long will it take the motorboat to catch the sailboat?

57. Road trip. The Griswalds averaged 45 mph on their way to Las Vegas and 60 mph on the way back home using the same route. Find the distance from their home to Las Vegas if the total driving time was 70 hours.

58. Meeting cyclists. Tanya and Lebron start at the same time from opposite ends of a bicycle trail that is 81 miles long. Tanya averages 12 mph and Lebron averages 15 mph. How long does it take for them to meet?

59. Filling a fountain. Pete's fountain can be filled using a pipe or a hose. The fountain can be filled using the pipe in 6 hours or the hose in 12 hours. How long will it take to fill the fountain using both the pipe and the hose?

60. Mowing a lawn. Albert can mow a lawn in 40 minutes, while his cousin Vinnie can mow the same lawn in one hour. How long would it take to mow the lawn if Albert and Vinnie work together?

61. Printing a report. Debra plans to use two computers to print all of the copies of the annual report that are needed for the year-end meeting. The new computer can do the whole job in 2 hours while the old computer can do the whole job in 3 hours. How long will it take to get the job done using both computers simultaneously?

62. Installing a dishwasher. A plumber can install a dishwasher in 50 min. If the plumber brings his apprentice to help, the job takes 40 minutes. How long would it take the apprentice working alone to install the dishwasher?

63. Filling a tub. Using the hot and cold water faucets together, a bathtub fills in 8 minutes. Using the hot water faucet alone, the tub fills in 12 minutes. How long does it take to fill the tub using only the cold water faucet?

64. Filling a tank. A water tank has an inlet pipe and a drain pipe. A full tank can be emptied in 30 minutes if the drain is opened and an empty tank can be filled in 45 minutes with the inlet pipe opened. If both pipes are accidentally opened when the tank is full, then how long will it take to empty the tank?

Chapter

6 Wrap-Up

Summary

Rational Expressions		Examples
Rational expression	The ratio of two polynomials with the denominator not equal to 0	$\dfrac{x-1}{x-3}$ $(x \neq 3)$
Rational Function	If a rational expression is used to determine y from x, then y is a rational function of x.	$y = \dfrac{x-1}{x-3}$
Rule for reducing rational expressions	If $a \neq 0$ and $c \neq 0$, then $$\frac{ab}{ac} = \frac{b}{c}.$$ (Divide out the common factors.)	$\dfrac{8x+2}{4x} = \dfrac{2(4x+1)}{2(2x)} = \dfrac{4x+1}{2x}$ $\dfrac{x^7}{x^5} = x^2 \qquad \dfrac{x^2}{x^5} = \dfrac{1}{x^3}$

Multiplication and Division of Rational Expressions		Examples
Multiplication	If $b \neq 0$ and $d \neq 0$, then $\dfrac{a}{b} \cdot \dfrac{c}{d} = \dfrac{ac}{bd}$.	$\dfrac{3}{x^3} \cdot \dfrac{6}{x^5} = \dfrac{18}{x^8}$
Division	If $b \neq 0$, $c \neq 0$, and $d \neq 0$, then $\dfrac{a}{b} \div \dfrac{c}{d} = \dfrac{a}{b} \cdot \dfrac{d}{c}$. (Invert the divisor and multiply.)	$\dfrac{a}{x^3} \div \dfrac{5}{x^9} = \dfrac{a}{x^3} \cdot \dfrac{x^9}{5} = \dfrac{ax^6}{5}$

Addition and Subtraction of Rational Expressions		Examples
Least common denominator	The LCD of a group of denominators is the smallest number that is a multiple of all of them.	8, 12 LCD = 24
Finding the least common denominator	1. Factor each denominator completely. Use exponent notation for repeated factors. 2. Write the product of all of the different factors that appear in the denominators. 3. On each factor, use the highest power that appears on that factor in any of the denominators.	$4ab^3, 6a^2b$ $4ab^3 = 2^2ab^3$ $6a^2b = 2 \cdot 3a^2b$ LCD $= 2^2 \cdot 3a^2b^3 = 12a^2b^3$
Addition and subtraction of rational expressions	If $b \neq 0$, then $$\frac{a}{b} + \frac{c}{b} = \frac{a+c}{b} \quad \text{and} \quad \frac{a}{b} - \frac{c}{b} = \frac{a-c}{b}.$$ If the denominators are not identical, change each fraction to an equivalent fraction so that all denominators are identical.	$\dfrac{2x}{x-3} + \dfrac{7x}{x-3} = \dfrac{9x}{x-3}$ $\dfrac{2}{x} + \dfrac{1}{3x} = \dfrac{6}{3x} + \dfrac{1}{3x} = \dfrac{7}{3x}$

Complex fraction	A rational expression that has fractions in the numerator and/or the denominator	$\dfrac{\frac{1}{2}+\frac{1}{3}}{\frac{1}{3}-\frac{3}{4}}$
Simplifying complex fractions	Multiply the numerator and denominator by the LCD.	$\dfrac{\left(\frac{1}{2}+\frac{1}{3}\right)12}{\left(\frac{1}{3}-\frac{3}{4}\right)12}=\dfrac{6+4}{4-9}=-2$

Equations with Rational Expressions   **Examples**

Solving equations	Multiply each side by the LCD.	$\dfrac{1}{x}-\dfrac{1}{3}=\dfrac{1}{2x}-\dfrac{1}{6}$
		$6x\left(\dfrac{1}{x}-\dfrac{1}{3}\right)=6x\left(\dfrac{1}{2x}-\dfrac{1}{6}\right)$
		$6-2x=3-x$
Proportion	An equation expressing the equality of two ratios	$\dfrac{a}{b}=\dfrac{c}{d}$
Extremes-means property (cross-multiplying)	If $b\neq 0$ and $d\neq 0$, then $\dfrac{a}{b}=\dfrac{c}{d}$ is equivalent to $ad=bc$. Cross-multiplying is a quick way to eliminate the fractions in a proportion.	$\dfrac{2}{x-3}=\dfrac{5}{6}$ $2\cdot 6=(x-3)5$ $12=5x-15$

Enriching Your Mathematical Word Power

Fill in the blank.

1. A _____ expression is a ratio of two polynomials with the denominator not equal to zero.

2. The _____ of a rational expression is the set of all real numbers that can be used in place of the variable.

3. If a rational expression is used to determine the value of y from the value of x, then y is a rational _____ of x.

4. A rational expression is in _____ terms when the numerator and denominator have no common factors.

5. When common factors are divided out of the numerator and denominator of a rational expression, the rational expression is _____.

6. Two fractions that represent the same number are _____ fractions.

7. A _____ fraction has rational expressions in its numerator or denominator or both.

8. The opposite of reducing a fraction is _____ a fraction.

9. The smallest number that is a common multiple of a group of denominators is the ____ common denominator.

10. A number that appears to be a solution to an equation but does not satisfy the equation is an _____ root.

11. The expression a/b is the ____ of a to b.

12. A _____ is a statement expressing the equality of two rational expressions.

13. The numbers a and d in $a/b = c/d$ are the _____.

14. The numbers b and c in $a/b = c/d$ are the _____.

15. If $a/b = c/d$, then $ad = bc$ is the _____-_____ property.

Review Exercises

6.1 Reducing Rational Expressions

Find the domain of each rational expression.

1. $\dfrac{x^2}{4-x}$

2. $\dfrac{x-9}{2x+6}$

3. $\dfrac{x-5}{x^2-4x-5}$

4. $\dfrac{x+2}{x^2+6x+8}$

Reduce each rational expression to lowest terms.

5. $\dfrac{24}{28}$

6. $\dfrac{42}{18}$

7. $\dfrac{2a^3c^3}{8a^5c}$

8. $\dfrac{39x^6}{15x}$

9. $\dfrac{6w-9}{9w-12}$

10. $\dfrac{3t-6}{8-4t}$

11. $\dfrac{x^2-1}{3-3x}$

12. $\dfrac{3x^2-9x+6}{10-5x}$

6.2 Multiplication and Division

Perform the indicated operation.

13. $\dfrac{1}{6k}\cdot 3k^2$

14. $\dfrac{1}{15abc}\cdot 5a^3b^5c^2$

15. $\dfrac{2xy}{3}\div y^2$

16. $4ab\div\dfrac{1}{2a^4}$

17. $\dfrac{a^2-9}{a-2}\cdot\dfrac{a^2-4}{a+3}$

18. $\dfrac{x^2-1}{3x}\cdot\dfrac{6x}{2x-2}$

19. $\dfrac{w-2}{3w}\div\dfrac{4w-8}{6w}$

20. $\dfrac{2y+2x}{x-xy}\div\dfrac{x^2+2xy+y^2}{y^2-y}$

6.3 Finding the Least Common Denominator

Find the least common denominator for each group of denominators.

21. $36,\ 54$

22. $10,\ 15,\ 35$

23. $6ab^3,\ 8a^7b^2$

24. $20u^4v,\ 18uv^5,\ 12u^2v^3$

25. $4x,\ 6x-6$

26. $8a,\ 6a,\ 2a^2+2a$

27. $x^2-4,\ x^2-x-2$

28. $x^2-9,\ x^2+6x+9$

Convert each rational expression into an equivalent rational expression with the indicated denominator.

29. $\dfrac{5}{12}=\dfrac{?}{36}$

30. $\dfrac{2a}{15}=\dfrac{?}{45}$

31. $\dfrac{2}{3xy}=\dfrac{?}{15x^2y}$

32. $\dfrac{3z}{7x^2y}=\dfrac{?}{42x^3y^8}$

33. $\dfrac{5}{y-6}=\dfrac{?}{12-2y}$

34. $\dfrac{-3}{2-t}=\dfrac{?}{2t-4}$

35. $\dfrac{x}{x-1}=\dfrac{?}{x^2-1}$

36. $\dfrac{t}{t-3}=\dfrac{?}{t^2+2t-15}$

6.4 Addition and Subtraction

Perform the indicated operation.

37. $\dfrac{5}{36}+\dfrac{9}{28}$

38. $\dfrac{7}{30}-\dfrac{11}{42}$

39. $3-\dfrac{4}{x}$

40. $1+\dfrac{3a}{2b}$

41. $\dfrac{2}{ab^2}-\dfrac{1}{a^2b}$

42. $\dfrac{3}{4x^3}+\dfrac{5}{6x^2}$

43. $\dfrac{9a}{2a-3}+\dfrac{5}{3a-2}$

44. $\dfrac{3}{x-2}-\dfrac{5}{x+3}$

45. $\dfrac{1}{a-8}-\dfrac{2}{8-a}$

46. $\dfrac{5}{x-14}+\dfrac{4}{14-x}$

47. $\dfrac{3}{2x-4}+\dfrac{1}{x^2-4}$

48. $\dfrac{x}{x^2-2x-3}-\dfrac{3x}{x^2-9}$

6.5 Complex Fractions

Simplify each complex fraction.

49. $\dfrac{\dfrac{1}{2}-\dfrac{3}{4}}{\dfrac{2}{3}+\dfrac{1}{2}}$

50. $\dfrac{\dfrac{2}{3}+\dfrac{5}{8}}{\dfrac{1}{2}-\dfrac{3}{8}}$

51. $\dfrac{\dfrac{1}{a}+\dfrac{2}{3b}}{\dfrac{1}{2b}-\dfrac{3}{a}}$

52. $\dfrac{\dfrac{3}{xy}-\dfrac{1}{3y}}{\dfrac{1}{6x}-\dfrac{3}{5y}}$

53. $\dfrac{\dfrac{1}{x-2}-\dfrac{3}{x+3}}{\dfrac{2}{x+3}+\dfrac{1}{x-2}}$

54. $\dfrac{\dfrac{4}{a+1}+\dfrac{5}{a^2-1}}{\dfrac{1}{a^2-1}-\dfrac{3}{a-1}}$

55. $\dfrac{\dfrac{x-1}{x-3}}{\dfrac{1}{x^2-x-6}-\dfrac{4}{x+2}}$

56. $\dfrac{\dfrac{6}{a^2+5a+6}-\dfrac{8}{a+2}}{\dfrac{2}{a+3}-\dfrac{4}{a+2}}$

6.6 Solving Equations with Rational Expressions

Solve each equation.

57. $\dfrac{-2}{5}=\dfrac{3}{x}$

58. $\dfrac{3}{x}+\dfrac{5}{3x}=1$

59. $\dfrac{14}{a^2-1}+\dfrac{1}{a-1}=\dfrac{3}{a+1}$

60. $2+\dfrac{3}{y-5}=\dfrac{2y}{y-5}$

61. $z-\dfrac{3z}{2-z}=\dfrac{6}{z-2}$

62. $\dfrac{1}{x}+\dfrac{1}{3}=\dfrac{1}{2}$

6.7 Applications of Ratios and Proportions

Solve each proportion.

63. $\dfrac{3}{x}=\dfrac{2}{7}$

64. $\dfrac{4}{x}=\dfrac{x}{4}$

65. $\dfrac{2}{w-3}=\dfrac{5}{w}$

66. $\dfrac{3}{t-3}=\dfrac{5}{t+4}$

Solve each problem by using a proportion.

67. *Taxis in Times Square.* The ratio of taxis to private automobiles in Times Square at 6:00 P.M. on New Year's Eve was estimated to be 15 to 2. If there were 60 taxis, then how many private automobiles were there?

Photo for Exercise 67

68. *Student-teacher ratio.* The student-teacher ratio for Washington High was reported to be 27.5 to 1. If there are 42 teachers, then how many students are there?

69. *Water and rice.* At Wong's Chinese Restaurant the secret recipe for white rice calls for a 2 to 1 ratio of water to rice. In one batch the chef used 28 more cups of water than rice. How many cups of each did he use?

Photo for Exercise 69

70. *Oil and gas.* An outboard motor calls for a fuel mixture that has a gasoline-to-oil ratio of 50 to 1. How many pints of oil should be added to 6 gallons of gasoline?

6.8 Applications of Rational Expressions

Solve each formula for the indicated variable.

71. $\dfrac{y-b}{m}=x$ for y

72. $\dfrac{A}{h}=\dfrac{a+b}{2}$ for a

73. $F=\dfrac{mv+1}{m}$ for m

74. $m=\dfrac{r}{1+rt}$ for r

75. $\dfrac{y+1}{x-3}=4$ for y

76. $\dfrac{y-3}{x+2}=\dfrac{-1}{3}$ for y

Solve each problem.

77. Making a puzzle. Tracy, Stacy, and Fred assembled a very large puzzle together in 40 hours. If Stacy worked twice as fast as Fred, and Tracy worked just as fast as Stacy, then how long would it have taken Fred to assemble the puzzle alone?

78. Going skiing. Leon drove 270 miles to the lodge in the same time as Pat drove 330 miles to the lodge. If Pat drove 10 miles per hour faster than Leon, then how fast did each of them drive?

Photo for Exercise 78

79. Merging automobiles. When Bert and Ernie merged their automobile dealerships, Bert had 10 more cars than Ernie. While 36% of Ernie's stock consisted of new cars, only 25% of Bert's stock consisted of new cars. If they had 33 new cars on the lot after the merger, then how many cars did each one have before the merger?

80. Magazine sales. A company specializing in magazine sales over the telephone found that in 2500 phone calls, 360 resulted in sales and were made by male callers, and 480 resulted in sales and were made by female callers. If the company gets twice as many sales per call with a woman's voice than with a man's voice, then how many of the 2500 calls were made by females?

81. Distribution of waste. The accompanying figure shows the distribution of the total municipal solid waste into various categories in 2000 (U.S. Environmental Protection Agency, www.epa.gov). If the paper waste was 59.8 million tons greater than the yard waste, then what was the amount of yard waste generated?

82. Total waste. Use the information given in Exercise 81 to find the total waste generated in 2000 and the amount of food waste.

Miscellaneous

In place of each question mark, put an expression that makes each equation an identity.

83. $\dfrac{5}{x} = \dfrac{?}{2x}$

84. $\dfrac{?}{a} = \dfrac{6}{3a}$

85. $\dfrac{2}{a-5} = \dfrac{?}{5-a}$

86. $\dfrac{-1}{a-7} = \dfrac{1}{?}$

87. $3 = \dfrac{?}{x}$

88. $2a = \dfrac{?}{b}$

89. $m \div \dfrac{1}{2} = ?$

90. $5x \div \dfrac{1}{x} = ?$

91. $2a \div ? = 12a$

92. $10x \div ? = 20x^2$

93. $\dfrac{a-1}{a^2-1} = \dfrac{1}{?}$

94. $\dfrac{?}{x^2-9} = \dfrac{1}{x-3}$

95. $\dfrac{1}{a} - \dfrac{1}{5} = ?$

96. $\dfrac{3}{7} - \dfrac{2}{b} = ?$

97. $\dfrac{a}{2} - 1 = \dfrac{?}{2}$

98. $\dfrac{1}{a} - 1 = \dfrac{?}{a}$

99. $(a-b) \div (-1) = ?$

100. $(a-7) \div (7-a) = ?$

101. $\dfrac{\frac{1}{5a}}{2} = ?$

102. $\dfrac{3a}{\frac{1}{2}} = ?$

For each expression in Exercises 103–122, either perform the indicated operation or solve the equation, whichever is appropriate.

103. $\dfrac{1}{x} + \dfrac{1}{2x}$

104. $\dfrac{1}{y} + \dfrac{1}{3y} = 2$

105. $\dfrac{2}{3xy} + \dfrac{1}{6x}$

106. $\dfrac{3}{x-1} - \dfrac{3}{x}$

107. $\dfrac{5}{a-5} - \dfrac{3}{5-a}$

108. $\dfrac{2}{x-2} - \dfrac{3}{x} = \dfrac{-1}{x}$

2000 Total Waste Generation (before recycling)

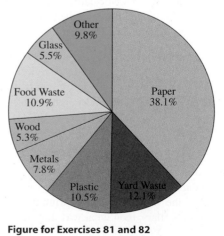

Figure for Exercises 81 and 82

109. $\dfrac{2}{x-1} - \dfrac{2}{x} = 1$

110. $\dfrac{2}{x-2} \cdot \dfrac{6x-12}{14}$

111. $\dfrac{-3}{x+2} \cdot \dfrac{5x+10}{9}$

112. $\dfrac{3}{10} = \dfrac{5}{x}$

113. $\dfrac{1}{-3} = \dfrac{-2}{x}$

114. $\dfrac{x^2-4}{x} \div \dfrac{4x-8}{x}$

115. $\dfrac{ax+am+3x+3m}{a^2-9} \div \dfrac{2x+2m}{a-3}$

116. $\dfrac{-2}{x} = \dfrac{3}{x+2}$

117. $\dfrac{2}{x^2-25} + \dfrac{1}{x^2-4x-5}$

118. $\dfrac{4}{a^2-1} + \dfrac{1}{2a+2}$

119. $\dfrac{-3}{a^2-9} - \dfrac{2}{a^2+5a+6}$

120. $\dfrac{-5}{a^2-4} - \dfrac{2}{a^2-3a+2}$

121. $\dfrac{1}{a^2-1} + \dfrac{2}{1-a} = \dfrac{3}{a+1}$

122. $3 + \dfrac{1}{x-2} = \dfrac{2x-3}{x-2}$

Chapter 6 Test

What numbers cannot be used for x in each rational expression?

1. $\dfrac{2x-1}{x^2-1}$

2. $\dfrac{5}{2-3x}$

3. $\dfrac{1}{x}$

Perform the indicated operation. Write each answer in lowest terms.

4. $\dfrac{2}{15} - \dfrac{4}{9}$

5. $\dfrac{1}{y} + 3$

6. $\dfrac{3}{a-2} - \dfrac{1}{2-a}$

7. $\dfrac{2}{x^2-4} - \dfrac{3}{x^2+x-2}$

8. $\dfrac{m^2-1}{(m-1)^2} \cdot \dfrac{2m-2}{3m+3}$

9. $\dfrac{a-b}{3} \div \dfrac{b^2-a^2}{6}$

10. $\dfrac{5a^2b}{12a} \cdot \dfrac{2a^3b}{15ab^6}$

Simplify each complex fraction.

11. $\dfrac{\frac{2}{3} + \frac{4}{5}}{\frac{2}{5} - \frac{3}{2}}$

12. $\dfrac{\frac{2}{x} + \frac{1}{x-2}}{\frac{1}{x-2} - \frac{3}{x}}$

Solve each equation.

13. $\dfrac{3}{x} = \dfrac{7}{5}$

14. $\dfrac{x}{x-1} - \dfrac{3}{x} = \dfrac{1}{2}$

15. $\dfrac{1}{x} + \dfrac{1}{6} = \dfrac{1}{4}$

Solve each formula for the indicated variable.

16. $\dfrac{y-3}{x+2} = \dfrac{-1}{5}$ for y

17. $M = \dfrac{1}{3}b(c+d)$ for c

Solve each problem.

18. If $R(x) = \dfrac{x+2}{1-x}$, then what is $R(0.9)$?

19. When all of the grocery carts escape from the supermarket, it takes Reginald 12 minutes to round them up and bring them back. Because Norman doesn't make as much per hour as Reginald, it takes Norman 18 minutes to do the same job. How long would it take them working together to complete the roundup?

20. Brenda and her husband Randy bicycled cross-country together. One morning, Brenda rode 30 miles. By traveling only 5 miles per hour faster and putting in one more hour, Randy covered twice the distance Brenda covered. What was the speed of each cyclist?

21. For a certain time period the ratio of the dollar value of exports to the dollar value of imports for the United States was 2 to 3. If the value of exports during that time period was 48 billion dollars, then what was the value of imports?

*Making*Connections | A Review of Chapters 1–6

Solve each equation.

1. $3x - 2 = 5$

2. $\dfrac{3}{5}x = -2$

3. $2(x - 2) = 4x$

4. $2(x - 2) = 2x$

5. $2(x + 3) = 6x + 6$

6. $2(3x + 4) + x^2 = 0$

7. $4x - 4x^3 = 0$

8. $\dfrac{3}{x} = \dfrac{-2}{5}$

9. $\dfrac{3}{x} = \dfrac{x}{12}$

10. $\dfrac{x}{2} = \dfrac{4}{x - 2}$

11. $\dfrac{w}{18} - \dfrac{w - 1}{9} = \dfrac{4 - w}{6}$

12. $\dfrac{x}{x + 1} + \dfrac{1}{2x + 2} = \dfrac{7}{8}$

Solve each equation for y.

13. $2x + 3y = c$

14. $\dfrac{y - 3}{x - 5} = \dfrac{1}{2}$

15. $2y = ay + c$

16. $\dfrac{A}{y} = \dfrac{C}{B}$

17. $\dfrac{A}{y} + \dfrac{1}{3} = \dfrac{B}{y}$

18. $\dfrac{A}{y} - \dfrac{1}{2} = \dfrac{1}{3}$

19. $3y - 5ay = 8$

20. $y^2 - By = 0$

21. $A = \dfrac{1}{2}h(b + y)$

22. $2(b + y) = b$

Calculate the value of $b^2 - 4ac$ for each choice of a, b, and c.

23. $a = 1, b = 2, c = -15$

24. $a = 1, b = 8, c = 12$

25. $a = 2, b = 5, c = -3$

26. $a = 6, b = 7, c = -3$

Perform each indicated operation.

27. $(3x - 5) - (5x - 3)$

28. $(2a - 5)(a - 3)$

29. $x^7 \div x^3$

30. $\dfrac{x - 3}{5} + \dfrac{x + 4}{5}$

31. $\dfrac{1}{2} \cdot \dfrac{1}{x}$

32. $\dfrac{1}{2} + \dfrac{1}{x}$

33. $\dfrac{1}{2} \div \dfrac{1}{x}$

34. $\dfrac{1}{2} - \dfrac{1}{x}$

35. $\dfrac{x - 3}{5} - \dfrac{x + 4}{5}$

36. $\dfrac{3a}{2} \div 2$

37. $(x - 8)(x + 8)$

38. $3x(x^2 - 7)$

39. $2a^5 \cdot 5a^9$

40. $x^2 \cdot x^8$

41. $(k - 6)^2$

42. $(j + 5)^2$

43. $(g - 3) \div (3 - g)$

44. $(6x^3 - 8x^2) \div (2x)$

Factor each expression completely.

45. $4x^4 + 12x^3 + 32x^2$

46. $15a^3 - 24a^2 + 9a$

47. $-12b^2 + 84b - 147$

48. $-2y^2 + 288$

49. $by + yw + 3w + 3b$

50. $2ax + 4bx - 3an - 6bn$

51. $-7b^3 - 7$

52. $2q^3 - 54$

Perform the indicated operations without using a calculator. Write each answer in scientific notation.

53. $(3 \times 10^3)(4 \times 10^4)$

54. $(3 \times 10^3)^4$

55. $\dfrac{4 \times 10^8}{8 \times 10^{15}}$

56. $(1 \times 10^3) + (1 \times 10^4)$

Solve each problem.

57. The sum of the squares of two consecutive positive even integers is 100. What are the integers?

58. The difference of the squares of two consecutive positive odd integers is 32. What are the integers?

59. *Present value.* An investor is interested in the amount or present value that she would have to invest today to receive periodic payments in the future. The present value of $1 in one year and $1 in 2 years with interest rate r compounded annually is given by the formula

$$P = \frac{1}{1 + r} + \frac{1}{(1 + r)^2}.$$

a) Rewrite the formula so that the right-hand side is a single rational expression.

b) Find P if $r = 7\%$.

c) The present value of $1 per year for the next 10 years is given by the formula

$$P = \frac{1}{1 + r} + \frac{1}{(1 + r)^2} + \frac{1}{(1 + r)^3} + \cdots + \frac{1}{(1 + r)^{10}}.$$

Use this formula to find P if $r = 5\%$.

Critical **Thinking** | For Individual or Group Work | Chapter 6

These exercises can be solved by a variety of techniques, which may or may not require algebra. So be creative and think critically. Explain all answers. Answers are in the Instructor's Edition of this text.

1. ***Equilateral triangles.*** Consider the sequence of three equilateral triangles shown in the accompanying figure.

   a) How many equilateral triangles are there in (a) of the accompanying figure?

   b) How many equilateral triangles congruent to the one in (a) can be found in (b) of the accompanying figure? How many are found in (c)?

   c) Suppose the sequence of equilateral triangles shown in (a), (b), and (c) is continued. How many equilateral triangles [congruent to the one in (a)] could be found in the *n*th such figure?

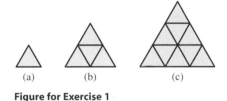

   Figure for Exercise 1

2. ***The amazing Amber.*** Amber has been amazing her friends with a math trick. Amber has a friend select a three-digit number and reverse the digits. The friend then finds the difference of the two numbers and reads the first two digits of the difference (from left to right). Amber can always tell the last digit of the difference. Explain how Amber does this.

3. ***Missing proceeds.*** Ruth and Betty sell apples at a farmers market. Ruth's apples sell at 2 for $1, while Betty's slightly smaller apples sell at 3 for $1. When Betty leaves to pick up her kids, they each have 30 apples and Ruth takes charge of both businesses. To simplify things, Ruth puts all 60 of the apples together and sells them at 5 for $2. When Betty returns, all of the apples have been sold, but they begin arguing over how to divide up the proceeds. What is the problem? Explain what went wrong.

4. ***Eyes and feet.*** A rancher has some sheep and ostriches. His young daughter observed that the animals have a total of 60 eyes and 86 feet. How many animals of each type does the rancher have?

   Photo for Exercise 4

5. ***Evaluation nightmare.*** Evaluate:

$$\frac{9{,}876{,}543{,}210}{9{,}876{,}543{,}211^2 - 9{,}876{,}543{,}210 \cdot 9{,}876{,}543{,}212}$$

6. ***Perfect squares.*** Find a positive integer such that the integer increased by 1 is a perfect square and one-half of the integer increased by 1 is a perfect square. Also find the next two larger positive integers that have this same property.

7. ***Multiplying primes.*** Find the units digit of the product of the first 500 prime numbers.

8. ***Ones and zeros.*** Find the sum of all seven-digit numbers that can be written using only ones or zeros.

Systems of Linear Equations

What determines the prices of the products that you buy? Why do prices of some products go down while the prices of others go up? Economists theorize that prices result from the collective decisions of consumers and producers. Ideally, the demand or quantity purchased by consumers depends only on the price, and price is a function of the supply. Theoretically, if the demand is greater than the supply, then prices rise and manufacturers produce more to meet the demand. As the supply of goods increases, the price comes down. The price at which the supply is equal to the demand is called the equilibrium price.

However, what happens in the real world does not always match the theory. Manufacturers cannot always control the supply, and factors other than price can affect a consumer's decision to buy. For example, droughts in Brazil decreased the supply of coffee and drove coffee prices up. Floods in California did the same to the prices of produce. With one of the most abundant wheat crops ever in 1994, cattle gained weight more quickly, increasing the supply of cattle ready for market. With supply going up, prices went down. Decreased demand for beef in Japan and Mexico drove the price of beef down further. With lower prices, consumers should be buying more beef, but increased competition from chicken and pork products, as well as health concerns, have kept consumer demand low.

The two functions that govern supply and demand form a system of equations. In this chapter you will learn how to solve systems of equations.

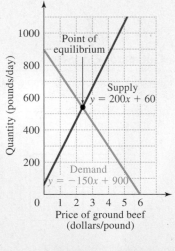

In Exercise 65 of Section 7.2 you will see an example of supply and demand equations for ground beef.

7.1   The Graphing Method

You studied linear equations in two variables in Chapter 3. In this section, you will learn to solve systems of linear equations in two variables and use systems to solve problems.

⟨1⟩ Solving a System by Graphing

Consider the linear equation $y = 2x - 1$. The graph of this equation is a straight line, and every point on the line is a solution to the equation. Now consider a second linear equation, $x + y = 2$. The graph of this equation is also a straight line, and every point on the line is a solution to this equation. Taken together, the pair of equations

$$y = 2x - 1$$
$$x + y = 2$$

is called a **system of equations.** A point that satisfies both equations is called a **solution to the system.**

E X A M P L E   1

A solution to a system

Determine whether the point $(-1, 3)$ is a solution to each system of equations.

a) $3x - y = -6$
   $x + 2y = 5$

b) $y = 2x - 1$
   $x + y = 2$

Solution

a) If we let $x = -1$ and $y = 3$ in both equations of the system, we get the following equations:

$$3(-1) - 3 = -6 \quad \text{Correct}$$
$$-1 + 2(3) = 5 \quad \text{Correct}$$

Because both of these equations are correct, $(-1, 3)$ is a solution to the system.

b) If we let $x = -1$ and $y = 3$ in both equations of the system, we get the following equations:

$$3 = 2(-1) - 1 \quad \text{Incorrect}$$
$$-1 + 3 = 2 \quad \text{Correct}$$

Because the first equation is not satisfied by $(-1, 3)$, the point $(-1, 3)$ is not a solution to the system.

Now do Exercises 1–8

⟨ **Calculator Close-Up** ⟩

Solve both equations in Example 1(a) for y to get $y = 3x + 6$ and $y = (5 - x)/2$. The graphs show that $(-1, 3)$ is on both lines.

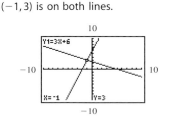

For Example 1(b), graph $y = 2x - 1$ and $y = 2 - x$ to see that $(-1, 3)$ is on one line but not the other.

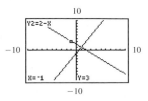

If we graph each equation of a system on the same coordinate plane, then we may be able to see the points that they have in common. Any point that is on both graphs is a solution to the system.

E X A M P L E   2

A system with only one solution

Solve the system by graphing:

$$y = x + 2$$
$$x + y = 4$$

Solution

First write the equations in slope-intercept form:

$$y = x + 2$$
$$y = -x + 4$$

Use the y-intercept and the slope to graph each line. The graph of the system is shown in Fig. 7.1. From the graph it appears that these lines intersect at $(1, 3)$. To be certain, we can check that $(1, 3)$ satisfies both equations. Let $x = 1$ and $y = 3$ in $y = x + 2$ to get

$$3 = 1 + 2.$$

Let $x = 1$ and $y = 3$ in $x + y = 4$ to get

$$1 + 3 = 4.$$

Because $(1, 3)$ satisfies both equations, the solution set to the system is $\{(1, 3)\}$.

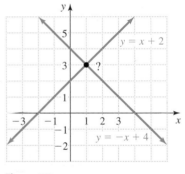

Figure 7.1

> Now do Exercises 9–16

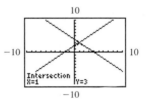
In Example 3, we graph the lines using the x- and y-intercepts.

E X A M P L E   3

A system with exactly one solution

Solve the system by graphing:

$$x - y = 6$$
$$2x + y = 6$$

Solution

We can graph these equations using their x- and y-intercepts. The intercepts for $x - y = 6$ are $(6, 0)$ and $(0, -6)$. The intercepts for $2x + y = 6$ are $(3, 0)$ and $(0, 6)$. Draw the graphs

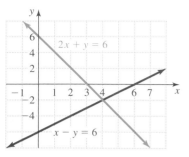

Figure 7.2

through the intercepts as shown in Fig. 7.2. The lines appear to cross at $(4, -2)$. To be certain, check $(4, -2)$ in both equations:

$$x - y = 6 \qquad\qquad\qquad\qquad 2x + y = 6$$
$$4 - (-2) = 6 \quad \text{Correct} \qquad\qquad 2 \cdot 4 + (-2) = 6 \quad \text{Correct}$$

Because both of the equations are correct, $(4, -2)$ is the solution to the system. The solution set is $\{(4, -2)\}$.

Now do Exercises 17–24

E X A M P L E   4

A system with infinitely many solutions

Solve the system by graphing:

$$4x - 2y = 6$$
$$y - 2x = -3$$

Solution

Rewrite both equations in slope-intercept form for easy graphing:

$$4x - 2y = 6 \qquad\qquad\qquad y - 2x = -3$$
$$-2y = -4x + 6 \qquad\qquad\qquad y = 2x - 3$$
$$y = 2x - 3$$

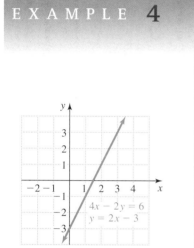

Figure 7.3

By writing the equations in slope-intercept form, we discover that they are identical. So the equations have the same graph, which is shown in Fig. 7.3. So any point on that line satisfies both of the equations, and there are infinitely many solutions to the system. The solution set consists of all points on the line $y = 2x - 3$, which is written in set notation as

$$\{(x, y) \mid y = 2x - 3\}.$$

Now do Exercises 33–36

In Example 4 we read $\{(x, y) \mid y = 2x - 3\}$ as "the set of ordered pairs (x, y) such that $y = 2x - 3$." Note that we could have used $4x - 2y = 6$ or $y - 2x = -3$ in place of $y = 2x - 3$ in set notation since these three equations are equivalent. We usually choose the simplest equation for set notation.

E X A M P L E   5

A system with no solution

Solve the system by graphing:

$$3y = 2x - 6$$
$$2x - 3y = 3$$

Solution

Write each equation in slope-intercept form to get the following system:

$$y = \frac{2}{3}x - 2$$

$$y = \frac{2}{3}x - 1$$

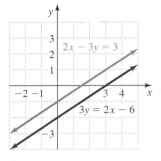

Figure 7.4

Each line has slope $\frac{2}{3}$, but they have different y-intercepts. Their graphs are shown in Fig. 7.4. Because these two lines have equal slopes, they are parallel. There is no point of intersection and no solution to the system.

> Now do Exercises 37–40

⟨2⟩ Types of Systems

A system of equations that has at least one solution is **consistent** (Examples 2, 3, and 4). A system with no solutions is **inconsistent** (Example 5). There are two types of consistent systems. A consistent system with exactly one solution is **independent** (Examples 2 and 3) and a consistent system with infinitely many solutions is **dependent** (Example 4). These ideas are summarized in Fig. 7.5.

You can classify a system as independent, dependent, or inconsistent by examining the slope-intercept form of each equation, as shown in Example 6.

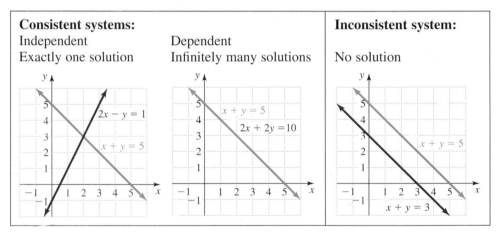

Figure 7.5

E X A M P L E   6

Types of systems

Determine whether each system is independent, dependent, or inconsistent.

a) $y = 3x - 5$
   $y = 3x + 2$

b) $y = 2x + 3$
   $y = -2x + 5$

c) $y = 5x - 1$
   $2y - 10x = -2$

Solution

a) Since $y = 3x - 5$ and $y = 3x + 2$ have the same slope and different y-intercepts, the two lines are parallel. There is no point of intersection. The system is inconsistent.

b) Since $y = 2x + 3$ and $y = -2x + 5$ have different slopes, they are not parallel. These two lines intersect at a single point. The system is independent.

c) First rewrite the second equation in slope-intercept form:

$$2y - 10x = -2$$
$$2y = 10x - 2$$
$$y = 5x - 1$$

Since the first equation is also $y = 5x - 1$, these are two different-looking equations for the same line. So every point on that line satisfies both equations. The system is dependent.

> Now do Exercises 41–54

‹ **Calculator Close-Up** ›

With a graphing calculator, you can graph both equations of a system in a single viewing window. The TRACE feature can then be used to estimate the solution to an independent system. You could also use ZOOM to "blow up" the intersection and get more accuracy. Many calculators have an intersect feature, which can find a point of intersection. First graph $y_1 = 2x - 1$ and $y_2 = 2 - x$.

From the CALC menu choose intersect.

Verify the curves (or lines) that you want to intersect by pressing ENTER. After you make a guess as to the intersection by positioning the cursor or entering a number, the calculator will find the intersection.

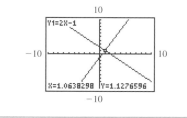

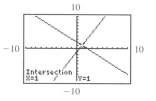

‹3› **Applications**

In a simple economic model, both supply and demand depend only on price. **Supply** is the quantity of an item that producers are willing to make or supply. **Demand** is the quantity consumers will purchase. As the price increases, producers increase the supply to take advantage of rising prices. However, as the price increases, consumer demand decreases. The **equilibrium price** is the price at which supply equals demand.

EXAMPLE 7

Supply and demand

Monthly demand for Greeny Babies (small toy frogs) is given by the equation $y = 8000 - 400x$, while monthly supply is given by the equation $y = 400x$, where x is the price in dollars. Graph the two equations, and find the equilibrium price and the demand at the equilibrium price.

Solution

The graph of $y = 8000 - 400x$ goes through $(0, 8000)$ and $(20, 0)$. The graph of $y = 400x$ goes through $(0, 0)$ and $(20, 8000)$. The two lines cross at $(10, 4000)$ as shown in Fig. 7.6. So the equilibrium price is $10, and the monthly demand is 4000 Greeny Babies.

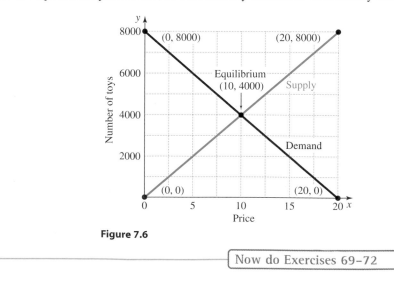

Figure 7.6

> Now do Exercises 69–72

Exercises 7.1

‹ Study Tips ›

- Working problems 1 hour per day every day of the week is better than working problems for 7 hours on one day of the week. Spread out your study time. Avoid long study sessions.
- No two students learn in exactly the same way or at the same speed. Figure out what works for you.

‹1› Solving a System by Graphing

Which of the given points is a solution to the given system? See Example 1.

1. $2x + y = 4$ $(6, 1), (3, -2), (2, 4)$
   $x - y = 5$

2. $2x - 3y = -5$ $(-1, 1), (3, 4), (2, 3)$
   $y = x + 1$

3. $6x - 2y = 4$ $(0, -2), (2, 4), (3, 7)$
   $y = 3x - 2$

4. $y = -2x + 5$ $(9, -13), (-1, 7), (0, 5)$
   $4x + 2y = 10$

5. $2x - y = 3$ $(3, 3), (5, 7), (7, 11)$
   $2x - y = 2$

6. $y = x + 5$ $(1, -2), (3, 0), (6, 3)$
   $y = x - 3$

Use the given graph to find an ordered pair that satisfies each system of equations. Check that your answer satisfies both equations of each system.

7. $y = 3x + 9$
   $2x + 3y = 5$

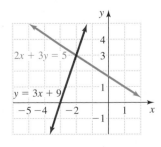

8. $x - 2y = 5$
   $y = -\dfrac{2}{3}x + 1$

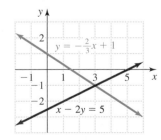

Solve each system by graphing. See Examples 2 and 3.

9. $y = 2x$
$y = -x + 6$

10. $y = 3x$
$y = -x + 4$

11. $3x - y = 1$
$2y - 3x = 1$

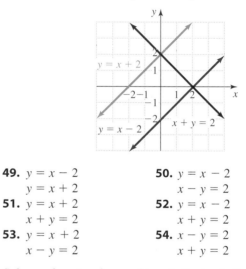 **12.** $2x + y = 3$
$x + y = 1$

13. $x - y = 5$
$x + y = -5$

14. $y + 4x = 10$
$2x - y = 2$

15. $2y + x = 4$
$2x - y = -7$

16. $2x + y = -1$
$x + y = -2$

17. $y = x$
$x + y = 0$

18. $x = 2y$
$0 = 9x - y$

19. $y = 2x - 1$
$x - 2y = -4$

20. $y = x - 1$
$2x - y = 0$

21. $x - y = 2$
$x + 3y = 6$

22. $x - y = -1$
$3x - y = 3$

23. $x - 2y = -8$
$3x - 2y = -12$

24. $x + 3y = 9$
$2x + 3y = 12$

Solve each system by graphing both equations on a graphing calculator and using the intersection feature of the calculator to find the point of intersection.

25. $y = x + 5$
$y = 9 - x$

26. $y = 2x + 1$
$y = 5 - 2x$

27. $y = 3x - 18$
$y = 32 - 2x$

28. $y = -x + 26$
$y = 2x - 34$

29. $x + y = 12$
$3x + 2y = 14$

30. $x - y = -10$
$x - 4y = 20$

31. $x + 5y = -1$
$x - 5y = 2$

32. $x + y = 0.6$
$2y + 3x = -0.5$

Solve each system by graphing. See Examples 4 and 5.

33. $x - y = 3$
$3x = 3y + 9$

34. $2x + y = 3$
$6x - 9 = -3y$

35. $4y - 2x = -16$
$x - 2y = 8$

36. $x - y = 0$
$5x = 5y$

37. $x - y = 3$
$3x = 3y + 12$

38. $2y = -3x + 6$
$2y = -3x - 2$

39. $x + y = 4$
$2y = -2x + 6$

40. $y = 3x - 5$
$y - 3x = 0$

⟨2⟩ Types of Systems

Determine whether each system is independent, dependent, or inconsistent. See Example 6.

41. $y = \dfrac{1}{2}x + 3$
$y = \dfrac{1}{2}x - 5$

42. $y = -3x - 60$
$y = \dfrac{1}{3}x - 60$

43. $y = 4x + 3$
$y = 3 + 4x$

44. $y = 5x - 4$
$y = 4 + 5x$

45. $y = \dfrac{1}{2}x + 3$
$y = -3x - 1$

46. $y = -x - 1$
$y = -1 - x$

47. $2x - 3y = 5$
$2x - 3y = 7$

48. $x + y = 1$
$2x + 2y = 2$

Use the following graph to determine whether the systems in Exercises 49–54 are independent, dependent, or inconsistent.

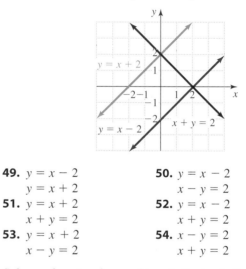

49. $y = x - 2$
$y = x + 2$

50. $y = x - 2$
$x - y = 2$

51. $y = x + 2$
$x + y = 2$

52. $y = x - 2$
$x + y = 2$

53. $y = x + 2$
$x - y = 2$

54. $x - y = 2$
$x + y = 2$

Solve each system by graphing. Indicate whether each system is independent, dependent, or inconsistent. See Examples 2–6.

55. $x - y = 3$
$3x = y + 5$

56. $3x + 2y = 6$
$2x - y = 4$

57. $x - y = 5$
$x - y = 8$

58. $y + 3x = 6$
$y - 5 = -3x$

59. $y = \dfrac{1}{3}x + 2$
$y = -\dfrac{1}{3}x$

60. $y - 4x = 4$
$y + 4x = -4$

61. $x - y = 1$
$-2y = -2x + 2$

62. $x = \frac{1}{3}y$
$y = 3x$

63. $x - y = -1$
$y = \frac{1}{2}x - 1$

64. $y = -3x + 1$
$2 - 2y = 6x$

The graphs of the following systems are given in (a) through (d). Match each system with the correct graph.

65. $5x + 4y = 7$
$x - 3y = 9$

66. $3x - 5y = -9$
$5x - 6y = -8$

67. $4x - 5y = -2$
$3y - x = -3$

68. $4x + 5y = -2$
$4y - x = 11$

a)

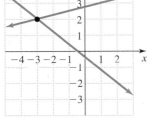

b)

c)

d)

⟨3⟩ Applications

Solve each problem by using the graphing method. See Example 7.

69. *Competing pizzas.* Mamma's Pizza charges $10 plus $2 per topping for a deep dish pizza. Papa's Pizza charges $5 plus $3 per topping for a similar pizza. The equations $C = 2n + 10$ and $C = 3n + 5$ express the cost C at each restaurant in terms of the number of toppings n.

a) Solve this system of equations by examining the accompanying graph.
b) Interpret the solution.

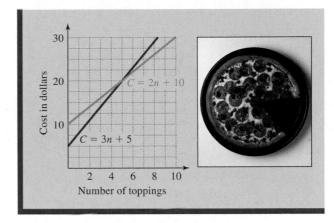

Figure for Exercise 69

70. *Equilibrium price.* A manufacturer plans to supply y units of its model 1020P CD player per month when the retail price is p dollars per player, where $y = 6p + 100$. Consumer studies show that consumer demand for the model 1020P is y units per month, where $y = -3p + 910$.

a) Fill in the missing entries in the following table.

Price	Supply	Demand
$ 0		
50		
100		
300		

b) Use the data in part (a) to graph both linear equations on the same coordinate system.
c) What is the price at which the supply is equal to the demand, the *equilibrium price?*

71. *Cost of two copiers.* An office manager figures the total cost in dollars for a certain used Xerox copier is given by $C = 800 + 0.05x$, where x is the number of copies made. She is also considering a used Panasonic copier for which the total cost is $C = 500 + 0.07x$.

a) Fill in the missing entries in the following table.

Number of Copies	Cost Xerox	Cost Panasonic
0		
5000		
10,000		
20,000		

b) Use the data from part (a) to graph both equations on the same coordinate system.

c) For what number of copies is the total cost the same for either copier?

d) If she plans to buy another copier before 10,000 copies are made, then which copier is cheaper?

72. *Flat tax proposals.* Representative Schneider has proposed a flat income tax of 15% on earnings in excess of $10,000. Under his proposal the tax T for a person earning E dollars is given by $T = 0.15(E - 10,000)$. Representative Humphries has proposed that the income tax should be 20% on earnings in excess of $20,000, or $T = 0.20(E - 20,000)$. Graph both linear equations on the same coordinate system. For what earnings would you pay the same amount of income tax under either plan? Under which plan does a rich person pay less income tax?

Getting More Involved

73. *Discussion*

If both $(-1, 3)$ and $(2, 7)$ satisfy a system of two linear equations, then what can you say about the system?

74. *Cooperative learning*

Working in groups, write an independent system of two linear equations whose solution is $(3, 5)$. Each group should then give its system to another group to solve.

75. *Cooperative learning*

Working in groups, write an inconsistent system of linear equations such that $(-2, 3)$ satisfies one equation and $(1, 4)$ satisfies the other. Each group should then give its system to another group to solve.

76. *Cooperative learning*

Suppose that $2x + 3y = 6$ is one equation of a system. Find the second equation given that $(4, 8)$ satisfies the second equation and the system is inconsistent.

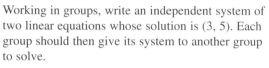

Graphing Calculator Exercises

Solve each system by graphing each pair of equations on a graphing calculator and using the calculator to estimate the point of intersection. Give the coordinates of the intersection to the nearest tenth.

77. $y = 2.5x - 6.2$

$y = -1.3x + 8.1$

78. $y = 305x + 200$

$y = -201x - 999$

79. $2.2x - 3.1y = 3.4$

$5.4x + 6.2y = 7.3$

80. $34x - 277y = 1$

$402x + 306y = 12,000$

7.2 The Substitution Method

In This Section

⟨1⟩ **Solving a System by Substitution**

⟨2⟩ **Dependent and Inconsistent Systems**

⟨3⟩ **Applications**

Solving a system by graphing is certainly limited by the accuracy of the graph. If the lines intersect at a point whose coordinates are not integers, then it is difficult to identify the solution from a graph. In this section we introduce a method for solving systems of linear equations in two variables that does not depend on a graph and is totally accurate.

⟨1⟩ Solving a System by Substitution

To solve a system by **substitution** we replace a variable in one equation by an equivalent expression for that variable (obtained from the other equation). The result should be an equation in only one variable, which we can solve by the usual techniques.

EXAMPLE **1**

Solving a system by substitution

Solve:

$$3x + 4y = 5$$
$$x = y - 1$$

Solution

Because the second equation states that $x = y - 1$, we can substitute $y - 1$ for x in the first equation:

$$3x + 4y = 5$$
$$3(y - 1) + 4y = 5 \quad \text{Replace } x \text{ with } y - 1.$$
$$3y - 3 + 4y = 5 \quad \text{Simplify.}$$
$$7y - 3 = 5$$
$$7y = 8$$
$$y = \frac{8}{7}$$

Now use the value $y = \frac{8}{7}$ in one of the original equations to find x. The simplest one to use is $x = y - 1$:

$$x = \frac{8}{7} - 1$$
$$x = \frac{1}{7}$$

Check that $\left(\frac{1}{7}, \frac{8}{7}\right)$ satisfies both equations. The solution set to the system is $\left\{\left(\frac{1}{7}, \frac{8}{7}\right)\right\}$.

Now do Exercises 1–8

⟨ **Calculator Close-Up** ⟩

To check Example 1, graph

$$y_1 = (5 - 3x)/4$$

and

$$y_2 = x + 1.$$

Use the intersect feature of your calculator to find the point of intersection.

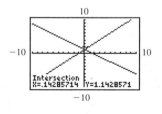

For substitution we must have one of the equations solved for x or y in terms of the other variable. In Example 1 we were given $x = y - 1$. So we replaced x with $y - 1$. In Example 2 we must rewrite one of the equations before substituting. Note how the five steps in the following strategy are used in Example 2.

Strategy for Solving a System by Substitution

1. If necessary, solve one of the equations for one variable in terms of the other. Choose the equation that is easiest to solve for x or y.
2. Substitute into the other equation to eliminate one of the variables.
3. Solve the resulting equation in one variable.
4. Insert the solution found in the last step into one of the original equations and solve for the other variable.
5. Check your solution in both equations.

E X A M P L E   2

Solving a system by substitution

Solve:

$$2x - 3y = 9$$
$$y - 4x = -8$$

Solution

(1) Solve the second equation for y:

$$y - 4x = -8$$
$$y = 4x - 8$$

(2) Substitute $4x - 8$ for y in the first equation:

$$2x - 3y = 9$$
$$2x - 3(4x - 8) = 9 \quad \text{Replace } y \text{ with } 4x - 8.$$

(3) Solve the equation for x:

$$2x - 12x + 24 = 9 \quad \text{Simplify.}$$
$$-10x + 24 = 9$$
$$-10x = -15$$
$$x = \frac{-15}{-10}$$
$$= \frac{3}{2}$$

(4) Use the value $x = \frac{3}{2}$ in $y = 4x - 8$ to find y:

$$y = 4 \cdot \frac{3}{2} - 8$$
$$= -2$$

(5) Check $x = \frac{3}{2}$ and $y = -2$ in both of the original equations:

$$2\left(\frac{3}{2}\right) - 3(-2) = 9 \quad \text{Correct}$$
$$-2 - 4\left(\frac{3}{2}\right) = -8 \quad \text{Correct}$$

Since both are correct, the solution set to the system is $\left\{\left(\frac{3}{2}, -2\right)\right\}$.

Now do Exercises 9–16

⟨2⟩ Dependent and Inconsistent Systems

Examples 3 and 4 illustrate how to solve dependent and inconsistent systems by substitution.

EXAMPLE 3

A system with infinitely many solutions
Solve:

$$2(y - x) = x + y - 1$$
$$y = 3x - 1$$

Solution

Because the second equation is solved for y, we will eliminate the variable y in the substitution. Substitute $y = 3x - 1$ into the first equation:

$$2(3x - 1 - x) = x + (3x - 1) - 1$$
$$2(2x - 1) = 4x - 2$$
$$4x - 2 = 4x - 2$$

Every real number satisfies $4x - 2 = 4x - 2$ because both sides are identical. So every real number can be used for x in the original system as long as we choose $y = 3x - 1$. The system is dependent. The graphs of these two equations are the same line. So the solution to the system is the set of all points on that line, $\{(x, y) \mid y = 3x - 1\}$.

> **Now do Exercises 17–20**

EXAMPLE 4

A system with no solution
Solve by substitution:

$$3x - 6y = 9$$
$$x = 2y + 5$$

Solution

Use $x = 2y + 5$ to replace x in the first equation:

$$3x - 6y = 9$$
$$3(2y + 5) - 6y = 9 \quad \text{Replace } x \text{ by } 2y + 5.$$
$$6y + 15 - 6y = 9 \quad \text{Simplify.}$$
$$15 = 9$$

No values for x and y will make 15 equal to 9. So there is no ordered pair that satisfies both equations. This system is inconsistent. It has no solution. The equations are the equations of parallel lines.

> **Now do Exercises 21–26**

⟨ **Calculator Close-Up** ⟩

To check Example 4, graph $y_1 = (3x - 9)/6$ and $y_2 = (x - 5)/2$. Since the lines appear to be parallel, there is no solution to the system.

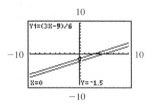

⟨ **Helpful Hint** ⟩

The purpose of Examples 3 and 4 is to show what happens when substitution is used on dependent and inconsistent systems. If we had first written the equations in slope-intercept form, we would see that the lines in Example 3 are the same and the lines in Example 4 are parallel.

When solving a system by substitution we can recognize a dependent system or an inconsistent system as follows.

Recognizing Dependent or Inconsistent Systems

Substitution in a dependent system results in an equation that is always true.
Substitution in an inconsistent system results in a false equation.

⟨3⟩ **Applications**

Many of the problems that we solved in previous chapters had two unknown quantities, but we wrote only one equation to solve the problem. For problems with two unknown quantities we can use two variables and a system of equations.

E X A M P L E   **5**

Two investments

Mrs. Robinson invested a total of $25,000 in two investments, one paying 6% and the other paying 8%. If her total income from these investments was $1790, then how much money did she invest in each?

Solution

Let x represent the amount invested at 6%, and let y represent the amount invested at 8%. The following table organizes the given information.

	Interest Rate	Amount Invested	Amount of Interest
First investment	6%	x	$0.06x$
Second investment	8%	y	$0.08y$

Write one equation describing the total of the investments, and the other equation describing the total interest:

$$x + y = 25,000 \quad \text{Total investments}$$
$$0.06x + 0.08y = 1790 \quad \text{Total interest}$$

To solve the system, we solve the first equation for y:

$$y = 25,000 - x$$

Substitute $25,000 - x$ for y in the second equation:

$$0.06x + 0.08(25,000 - x) = 1790$$
$$0.06x + 2000 - 0.08x = 1790$$
$$-0.02x + 2000 = 1790$$
$$-0.02x = -210$$
$$x = \frac{-210}{-0.02}$$
$$= 10,500$$

Let $x = 10,500$ in the equation $y = 25,000 - x$ to find y:

$$y = 25,000 - 10,500$$
$$= 14,500$$

Check these values for x and y in the original problem. Mrs. Robinson invested $10,500 at 6% and $14,500 at 8%.

⟨ **Helpful Hint** ⟩

In Chapter 2, we would have done Example 5 with one variable by letting x represent the amount invested at 6% and $25,000 - x$ represent the amount invested at 8%.

⟨ **Calculator Close-Up** ⟩

You can use a calculator to check the answers in Example 5:

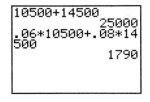

Now do Exercises 55–84

Warm-Ups ▼

Fill in the blank.

1. The disadvantage of solving a system by _____ is inaccuracy.

2. In the _____ method we eliminate a variable by substituting one equation into the other.

3. If substitution in a linear system results in a _____ equation, then the system has exactly one solution.

4. If substitution results in an identity, then the system is _____.

5. If substitution results in an _____ equation, then the system has no solution.

True or false?

6. Substituting $y = 2x$ into $x + 3y = 11$ yields $x + 6x = 11$.

7. A system of equations that has at least one solution is consistent.

8. A consistent system with infinitely many solutions is dependent.

9. An inconsistent system has no solutions.

10. No ordered pair satisfies $y = 3x - 5$ and $y = 2x - 5$.

Exercises 7.2

⟨ **Study Tips** ⟩

- Students who have difficulty with a subject often schedule a class that meets one day per week so that they do not have to see it too often. It is better to be in a class that meets more often for shorter time periods.
- Students who explain things to others often learn from it. If you must work on math alone, try explaining things to yourself.

⟨1⟩ **Solving a System by Substitution**

Solve each system by substitution. See Examples 1 and 2. See the Strategy for Solving a System by Substitution box on page 468.

1. $y = x + 2$
   $x + y = 8$

2. $y = x - 4$
   $x + y = 12$

3. $x = y - 3$
   $x + y = 11$

4. $x = y + 1$
   $x + y = 7$

5. $y = x + 3$
   $2x - 3y = -11$

VIDEO 6. $y = x - 5$
   $x + 2y = 8$

7. $x = 2y - 4$
   $2x + y = 7$

8. $x = y - 2$
   $-2x + y = -1$

9. $2x + y = 5$
   $5x + 2y = 8$

10. $5y - x = 0$
    $6x - y = 29$

11. $x + y = 0$
    $3x + 2y = -5$

12. $x - y = 6$
    $3x + 4y = -3$

13. $x + y = 1$
    $4x - 8y = -4$

14. $x - y = 2$
    $3x - 6y = 8$

15. $2x + 3y = 2$
    $4x - 9y = -1$

16. $x - 2y = 1$
    $3x + 10y = -1$

⟨2⟩ **Dependent and Inconsistent Systems**

Solve each system by substitution. Indicate whether each system is independent, dependent, or inconsistent. See Examples 1–4.

17. $21x - 35 = 7y$
    $3x - y = 5$

18. $2x + y = 3x$
    $3x - y = 2y$

19. $x - 2y = -2$
    $x + 2y = 8$

20. $y = -3x + 1$
$y = 2x + 4$

21. $x = 4 - 2y$
$4y + 2x = -8$

22. $y - 3 = 2(x - 1)$
$y = 2x + 3$

23. $y + 1 = 5(x + 1)$
$y = 5x - 1$

24. $3x - 2y = 7$

$3x + 2y = 7$

25. $2x + 5y = 5$

$3x - 5y = 6$

26. $x + 5y = 4$
$x + 5y = 4y$

Solve each system by the graphing method shown in Section 7.1, and by substitution.

27. $x + y = 5$
   $x - y = 1$

28. $x + y = 6$
   $2x - y = 3$

29. $y = x - 2$
   $y = 4 - x$

30. $y = 2x - 3$
   $y = -x + 3$

31. $y = 3x - 2$
   $y - 3x = 1$

32. $x + y = 5$
   $y = 2 - x$

Determine whether each system is independent, dependent, or inconsistent.

33. $y = -4x + 3$
   $y = -4x - 6$

34. $y = -3x - 6$
   $y = 3x - 6$

35. $y = x$
   $x = y$

36. $y = x$
   $y = x + 5$

37. $y = x$
   $y = -x$

38. $y = 3x$
   $3x - y = 0$

39. $x - y = 4$
   $x - y = 5$

40. $y = 1$
   $y + 3 = 4$

Solve each system by the substitution method.

41. $y = \dfrac{5}{2}x$
   $x + 3y = 3$

42. $6x - 3y = 3$
   $10x = y + 7$

43. $x + y = 4$
   $x - y = 5$

44. $3x - 6y = 5$
   $2y = 4x - 6$

45. $2x - 4y = 0$
   $6x + 8y = 5$

46. $-3x + 10y = 4$
   $6x - 5y = 1$

47. $3x + y = 2$
   $-x - 3y = 6$

48. $x + 3y = 2$
   $-x + y = 1$

49. $-9x + 6y = 3$
   $18x + 30y = 1$

50. $x + 6y = -2$
   $5x - 20y = 5$

51. $y = -2x$
   $3y - x = 1$

52. $y = 2x$
   $15x - 10y = -2$

53. $x = -6y + 1$
   $2y = -5x$

54. $x = -3y + 2$
   $7y = 3x$

⟨3⟩ Applications

Write a system of two equations in two unknowns for each problem. Solve each system by substitution. See Example 5.

55. *Rectangular patio.* The length of a rectangular patio is twice the width. If the perimeter is 84 feet, then what are the length and width?

56. *Rectangular lot.* The width of a rectangular lot is 50 feet less than the length. If the perimeter is 900 feet, then what are the length and width?

57. *Investing in the future.* Mrs. Miller invested $20,000 and received a total of $1600 in interest. If she invested part of the money at 10% and the remainder at 5%, then how much did she invest at each rate?

58. *Stocks and bonds.* Mr. Walker invested $30,000 in stocks and bonds and had a total return of $2880 in one year. If his stock investment returned 10% and his bond investment returned 9%, then how much did he invest in each?

59. *Gross receipts.* Two of the highest grossing movies of all time were *Titanic* and *Star Wars* with total receipts of

$1062 million (www.movieweb.com). If the gross receipts for *Titanic* exceeded the gross receipts for *Star Wars* by $140 million, then what were the gross receipts for each movie?

60. *Tennis court dimensions.* The singles court in tennis is four yards longer than it is wide. If its perimeter is 44 yards, then what are the length and width?

61. *Mowing and shoveling.* When Mr. Wilson came back from his vacation, he paid Frank $50 for mowing his lawn three times and shoveling his sidewalk two times. During Mr. Wilson's vacation last year, Frank earned $45 for mowing the lawn two times and shoveling the sidewalk three times. How much does Frank make for mowing the lawn once? How much does Frank make for shoveling the sidewalk once?

62. *Burgers and fries.* Donna ordered four burgers and one order of fries at the Hamburger Palace. However, the waiter put three burgers and two orders of fries in the bag and charged Donna the correct price for three burgers and two orders of fries, $3.15. When Donna discovered the mistake, she went back to complain. She found out that the price for four burgers and one order of fries is $3.45 and decided to keep what she had. What is the price of one burger, and what is the price of one order of fries?

63. *Racing rules.* According to NASCAR rules, no more than 52% of a car's total weight can be on any pair of tires. For optimal performance a driver of a 1150-pound car wants to have 50% of its weight on the left rear and left front tires and 48% of its weight on the left rear and right front tires. If the right front weight is determined to be 264 pounds, then what amount of weight should be on the left rear and left front? Are the NASCAR rules satisfied with this weight distribution?

64. *Weight distribution.* A driver of a 1200-pound car wants to have 50% of the car's weight on the left front and left rear tires, 48% on the left rear and right front tires, and 51% on the left rear and right rear tires. How much weight should be on each of these tires?

65. *Price of hamburger.* A grocer will supply y pounds of ground beef per day when the retail price is x dollars per pound, where $y = 200x + 60$. Consumer studies show that consumer demand for ground beef is y pounds per day, where $y = -150x + 900$. What is the price at which the supply is equal to the demand, the equilibrium price? See the accompanying figure.

Figure for Exercise 65

66. *Tweedle Dum and Dee.* Tweedle Dum said to Tweedle Dee, "The sum of my weight and twice yours is 361 pounds." Tweedle Dee said to Tweedle Dum, "Contrariwise the sum of my weight and twice yours is 362 pounds." Find the weight of each.

67. *Flying to Vegas.* Two hundred people were on a charter flight to Las Vegas. Some paid $200 for their tickets and some paid $250. If the total revenue for the flight was $44,000, then how many tickets of each type were sold?

68. *Annual concert.* A total of 150 tickets were sold for the annual concert to students and nonstudents. Student tickets were $5 and nonstudent tickets were $8. If the total revenue for the concert was $930, then how many tickets of each type were sold?

69. *Annual play.* There were twice as many tickets sold to nonstudents than to students for the annual play. Student tickets were $6 and nonstudent tickets were $11. If the total revenue for the play was $1540, then how many tickets of each type were sold?

70. *Soccer game.* There were 1000 more students at the soccer game than nonstudents. Student tickets were $8.50 and nonstudent tickets were $13.25. If the total revenue for the game was $75,925, then how many tickets of each type were sold?

71. *Mixing investments.* Helen invested $40,000 and received a total of $2300 in interest after one year. If part of the money returned 5% and the remainder 8%, then how much did she invest at each rate?

72. *Investing her bonus.* Donna invested her $33,000 bonus and received a total of $970 in interest after one year. If part of the money returned 4% and the remainder 2.25%, then how much did she invest at each rate?

73. *Mixing acid.* A chemist wants to mix a 5% acid solution with a 25% acid solution to obtain 50 liters of a 20% acid solution. How many liters of each solution should be used?

74. *Mixing fertilizer.* A farmer wants to mix a liquid fertilizer that contains 2% nitrogen with one that contains 10% nitrogen to obtain 40 gallons of a fertilizer that contains 8% nitrogen. How many gallons of each fertilizer should be used?

75. *Different interest rates.* Mrs. Brighton invested $30,000 and received a total of $2300 in interest. If she invested part of the money at 10% and the remainder at 5%, then how much did she invest at each rate?

76. *Different growth rates.* The combined population of Marysville and Springfield was 25,000 in 2000. By 2005 the population of Marysville had increased by 10%, while Springfield had increased by 9%. If the total population increased by 2380 people, then what was the population of each city in 2000?

77. *Toasters and vacations.* During one week a land developer gave away Florida vacation coupons or toasters to 100 potential customers who listened to a sales presentation. It costs the developer $6 for a toaster and $24 for a Florida vacation coupon. If his bill for prizes that week was $708, then how many of each prize did he give away?

78. *Ticket sales.* Tickets for a concert were sold to adults for $3 and to students for $2. If the total receipts were $824 and twice as many adult tickets as student tickets were sold, then how many of each were sold?

79. *Corporate taxes.* According to Bruce Harrell, CPA, the amount of federal income tax for a class C corporation is deductible on the Louisiana state tax return, and the amount of state income tax for a class C corporation is

deductible on the federal tax return. So for a state tax rate of 5% and a federal tax rate of 30%, we have

state tax = 0.05(taxable income − federal tax)

and

federal tax = 0.30(taxable income − state tax).

Find the amounts of state and federal income taxes for a class C corporation that has a taxable income of $100,000.

80. *More taxes.* Use the information given in Exercise 79 to find the amounts of state and federal income taxes for a class C corporation that has a taxable income of $300,000. Use a state tax rate of 6% and a federal tax rate of 40%.

81. *Cost accounting.* The problems presented in this exercise and Exercise 82 are encountered in cost accounting. A company has agreed to distribute 20% of its net income N to its employees as a bonus; $B = 0.20N$. If the company has an income of $120,000 before the bonus, the bonus B is deducted from the $120,000 as an expense to determine net income; $N = 120,000 − B$. Solve the system of two equations in N and B to find the amount of the bonus.

82. *Bonus and taxes.* A company has an income of $100,000 before paying taxes and a bonus. The bonus B is to be 20% of the income after deducting income taxes T but before deducting the bonus. So,

$$B = 0.20(100,000 − T).$$

Because the bonus is a deductible expense, the amount of income tax T at a 40% rate is 40% of the income after deducting the bonus. So,

$$T = 0.40(100,000 − B).$$

a) Use the accompanying graph to estimate the values of T and B that satisfy both equations.

b) Solve the system algebraically to find the bonus and the amount of tax.

83. *Textbook case.* The accompanying graph shows the cost of producing textbooks and the revenue from the sale of those textbooks.

a) What is the cost of producing 10,000 textbooks?
b) What is the revenue when 10,000 textbooks are sold?
c) For what number of textbooks is the cost equal to the revenue?
d) The cost of producing zero textbooks is called the *fixed cost*. Find the fixed cost.

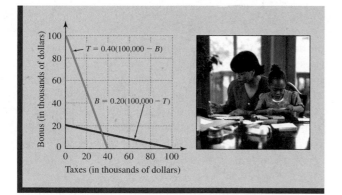

Figure for Exercise 82

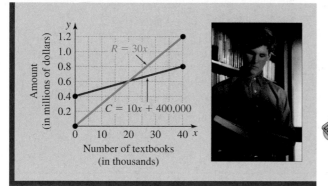

Figure for Exercise 83

84. *Free market.* The equations $S = 5000 + 200x$ and $D = 9500 - 100x$ express the supply S and the demand D, respectively, for a popular compact disc brand in terms of its price x (in dollars).

  a) Graph the equations on the same coordinate system.
  b) What happens to the supply as the price increases?
  c) What happens to the demand as the price increases?
  d) The price at which supply and demand are equal is called the *equilibrium price*. What is the equilibrium price?

Getting More Involved

85. *Discussion*

Which of the following equations is not equivalent to $2x - 3y = 6$?

  a) $3y - 2x = 6$ **b)** $y = \frac{2}{3}x - 2$

  c) $x = \frac{3}{2}y + 3$ **d)** $2(x - 5) = 3y - 4$

86. *Discussion*

Which of the following equations is inconsistent with the equation $3x + 4y = 8$?

  a) $y = \frac{3}{4}x + 2$

  b) $6x + 8y = 16$

  c) $y = -\frac{3}{4}x + 8$

  d) $3x - 4y = 8$

Graphing Calculator Exercise

87. *Life expectancy.* Since 1950, the life expectancy of a U.S. male born in year x is modeled by the formula

$$y = 0.165x - 256.7,$$

and the life expectancy of a U.S. female born in year x is modeled by

$$y = 0.186x - 290.6$$

(National Center for Health Statistics, www.cdc.gov).

  a) Find the life expectancy of a U.S. male born in 1975 and a U.S. female born in 1975.
  b) Graph both equations on your graphing calculator for $1950 < x < 2050$.
  c) Will U.S. males ever catch up with U.S. females in life expectancy?
  d) Assuming that these equations were valid before 1950, solve the system to find the year of birth for which U.S. males and females had the same life expectancy.

Math *at Work* | **Circuit Breakers**

Electricity is the flow of electrons through a circuit. It is measured in volts, amps, and watts. Volts measure the force that causes the electricity or electrons to flow. Amps measure the amount of electric current. Watts measure the amount of work done by a certain amount of current at a certain force or voltage. The basic relationship is watts = amps · volts or $W = A \cdot V$.

A circuit breaker is used as a safety device in a circuit. If the amperage exceeds a certain level, the breaker trips and prevents damage to the system. For example, suppose that 8 strings of Christmas lights each containing 25 bulbs that are 7 watts each are all plugged into one 120-volt circuit containing a 15-amp breaker. Will the breaker trip? The total wattage is $8 \cdot 25 \cdot 7$ or 1400 watts. Use $A = W/V$ to get $A = 1400/120 \approx 11.7$. So the lights will not blow a 15-amp fuse. See the accompanying figure.

While houses use standard single-phase electricity, electrical power companies may supply power for large users to transformers through three-phase lines. The power in a three-phase system is measured in volt-amps. The formula used here is volt-amps $= \sqrt{3} \cdot A \cdot V$. For example, suppose a large shopping mall has a 1,000,000 volt-amp transformer and the power company provides 25,000 volts to the mall's transformer. Will this power trip a 20-amp breaker? Because $A = $ volt-amps$/(\sqrt{3} \cdot V)$, we have $A = 1,000,000/(\sqrt{3} \cdot 25,000) \approx 23.1$ amps. So the 20-amp breaker will blow.

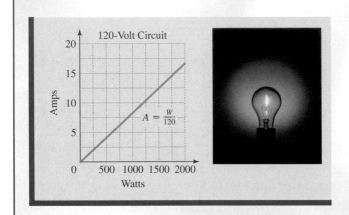

Mid-Chapter **Quiz** | **Sections 7.1 through 7.2** | **Chapter 7**

Determine whether $(1, -2)$ is in the solution set to each system.

1. $x - y = 3$
    $2x + y = 0$

2. $x + y = -1$
    $3x - y = 8$

3. $5x + 12y = -19$
    $5x + 12y = 6$

Solve by graphing.

4. $y = 2x - 4$
    $x + y = 5$

5. $x - y = 8$
    $x + y = 0$

6. $y = x + 6$
    $x - y = -6$

Solve by substitution.

7. $y = 3x - 5$
    $2x + 5y = 9$

8. $x + y = 6$
    $3x - 5y = 26$

9. $5x - y = 8$
    $35x - 6 = 7y$

Determine whether each system is independent, dependent, or inconsistent.

10. $y = \dfrac{1}{2}x - 7$
     $y = \dfrac{1}{2}x + 5$

11. $y = 5x - 12$
     $y = 3x + 7$

12. $y = \dfrac{3}{4}x + 1$
     $4y = 3x + 4$

7.3   The Addition Method

In Section 7.2, you used substitution to eliminate a variable in a system of equations. In this section, we see another method for eliminating a variable in a system of equations.

〈1〉 The Addition Method

In the **addition method** we eliminate a variable by adding the equations.

EXAMPLE 1

An independent system solved by addition

Solve the system by the addition method:

$$3x - 5y = -9$$
$$4x + 5y = 23$$

Solution

The addition property of equality allows us to add the same number to each side of an equation. We can also use the addition property of equality to add the two left sides and add the two right sides:

$$
\begin{array}{rl}
3x - 5y = -9 & \\
\underline{4x + 5y = 23} & \\
7x \quad\quad = 14 & \text{Add.} \\
x = 2 &
\end{array}
$$

The y-term was eliminated when we added the equations because the coefficients of the y-terms were opposites. Now use $x = 2$ in one of the original equations to find y. It does not matter which original equation we use. In this example we will use both equations to see that we get the same y in either case.

$$
\begin{array}{ll}
3x - 5y = -9 & \qquad 4x + 5y = 23 \\
3(2) - 5y = -9 \quad \text{Replace } x \text{ by 2.} & \qquad 4(2) + 5y = 23 \\
6 - 5y = -9 \quad \text{Solve for } y. & \qquad 8 + 5y = 23 \\
-5y = -15 & \qquad 5y = 15 \\
y = 3 & \qquad y = 3
\end{array}
$$

Because $3(2) - 5(3) = -9$ and $4(2) + 5(3) = 23$ are both true, $(2, 3)$ satisfies both equations. The solution set is $\{(2, 3)\}$.

> Now do Exercises 1–8

〈 Calculator Close-Up 〉

To check Example 1, graph

$$y_1 = (-9 - 3x)/-5$$

and

$$y_2 = (23 - 4x)/5.$$

Use the intersect feature to find the point of intersection of the two lines.

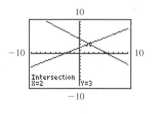

Actually the addition method can be used to eliminate any variable whose coefficients are opposites. If neither variable has coefficients that are opposites, then we use the multiplication property of equality to change the coefficients of the variables, as shown in Examples 2 and 3.

EXAMPLE **2**

Using multiplication and addition
Solve the system by the addition method:

$$2x - 3y = -13$$
$$5x - 12y = -46$$

‹ **Calculator Close-Up** ›

Check Example 2 by graphing

$$y_1 = (-13 - 2x)/(-3)$$

and

$$y_2 = (-46 - 5x)/(-12).$$

Solution
First examine the system to find the simplest way to eliminate a variable. Note that 5 is not a multiple of 2, but 12 is a multiple of 3. So if we multiply both sides of the first equation by -4, the coefficients of y will be 12 and -12, and y will be eliminated by addition:

$$(-4)(2x - 3y) = (-4)(-13) \quad \text{Multiply each side by } -4.$$
$$5x - 12y = -46$$

$$\begin{array}{r} -8x + 12y = 52 \\ 5x - 12y = -46 \qquad \text{Add.} \\ \hline -3x \qquad\quad = 6 \\ x = -2 \end{array}$$

Replace x by -2 in one of the original equations to find y:

$$2x - 3y = -13$$
$$2(-2) - 3y = -13$$
$$-4 - 3y = -13$$
$$-3y = -9$$
$$y = 3$$

Because $2(-2) - 3(3) = -13$ and $5(-2) - 12(3) = -46$ are both true, the solution set is $\{(-2, 3)\}$.

Now do Exercises 9–12

EXAMPLE **3**

Multiplying both equations before adding
Solve each system by the addition method.

a) $-2x + 3y = 6$
        $3x - 5y = -11$

b) $-2x + 3y = 0$
        $3x - 5y = 0$

Solution
a) Examine the coefficients. Since 3 is not a multiple of 2 and 5 is not a multiple of 3, we can't eliminate a variable by multiplying only one equation. However, multiplying the first equation by 3 and the second by 2 will give us $-6x$ and $6x$:

$$3(-2x + 3y) = 3(6) \qquad \text{Multiply each side by 3.}$$
$$2(3x - 5y) = 2(-11) \quad \text{Multiply each side by 2.}$$

$$\begin{array}{r} -6x + 9y = 18 \\ 6x - 10y = -22 \qquad \text{Add.} \\ \hline -y = -4 \\ y = 4 \end{array}$$

Note that we could have eliminated y by multiplying by 5 and 3. Now insert $y = 4$ into one of the original equations to find x:

$$-2x + 3(4) = 6 \quad \text{Let } y = 4 \text{ in } -2x + 3y = 6.$$
$$-2x + 12 = 6$$
$$-2x = -6$$
$$x = 3$$

Check that $(3, 4)$ satisfies both equations. The solution set is $\{(3, 4)\}$.

b) Multiplying the first equation by 3, the second by 2, and then adding will eliminate x as it did in part (a):

$$\begin{array}{llr}
-2x + 3y = 0 & \text{Multiply by 3} & -6x + 9y = 0 \\
3x - 5y = 0 & \text{Multiply by 2} & \underline{6x - 10y = 0} \\
& & -y = 0 \\
& & y = 0
\end{array}$$

If $y = 0$ in $-2x + 3y = 0$, we get $-2x = 0$ or $x = 0$. So the solution set is $\{(0, 0)\}$. Note that the graphs of these two equations intersect at the origin.

> Now do Exercises 13–18

The strategy for solving an independent system by addition follows.

Strategy for the Addition Method

1. Write both equations in the same form (usually $Ax + By = C$).
2. If necessary multiply one or both equations by the appropriate integer to obtain opposite coefficients on one of the variables.
3. Add the equations to get an equation in one variable.
4. Solve the equation in one variable.
5. Substitute the value obtained for one variable into one of the original equations to obtain the value of the other variable.
6. Check the two values in both of the original equations.

We can identify dependent and inconsistent systems in the same way that we did for the substitution method. If the result of the addition is an identity, the system is dependent and there are infinitely many solutions. If the result of the addition is a false equation, the system is inconsistent and there are no solutions. When you use addition, make sure that the equations are in the same form with the variables and equal signs aligned.

E X A M P L E   4

Solving dependent and inconsistent systems by addition
Solve each system by addition:

a) $2x - 3y = 9$
$6y = 4x - 18$

b) $-4y = 5x + 7$
$4y + 5x = 12$

Solution

a) First rewrite $6y = 4x - 18$ as $-4x + 6y = -18$ so that it is in the same form as the first equation:

$$2x - 3y = 9$$
$$-4x + 6y = -18$$

Now examine the coefficients. Multiply the first equation by 2 to get $4x - 6y = 18$ and add:

$$
\begin{array}{r}
4x - 6y = 18 \\
-4x + 6y = -18 \\
\hline
0 = 0
\end{array}
$$

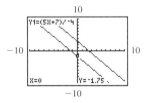

⟨ **Calculator Close-Up** ⟩

To check Example 4(b), graph

$$y_1 = (5x + 7)/{-4}$$

and

$$y_2 = (-5x + 12)/4.$$

Since the lines appear to be parallel, the graph supports the conclusion that the system is inconsistent.

Because the result of the addition is an identity, the equations are dependent and there are infinitely many solutions. The solution set is $\{(x, y) \mid 2x - 3y = 9\}$.

b) Rewrite the first equation $-4y = 5x + 7$ as $-4y - 5x = 7$ to get the same form as the second. Now add:

$$
\begin{array}{r}
-4y - 5x = 7 \\
4y + 5x = 12 \\
\hline
0 = 19
\end{array}
$$

Because the result of the addition is a false equation, the system is inconsistent. There are no solutions to the system. The solution set is the empty set, $\varnothing$.

Now do Exercises 25–30

⟨2⟩ Equations Involving Fractions or Decimals

When a system of equations involves fractions or decimals, we can use the multiplication property of equality to eliminate the fractions or decimals.

E X A M P L E   5

A system with fractions

Solve the system:

$$\frac{1}{2}x - \frac{2}{3}y = 7$$

$$\frac{2}{3}x - \frac{3}{4}y = 11$$

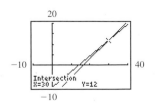

⟨ **Calculator Close-Up** ⟩

To check Example 5, graph

$$y_1 = (7 - (1/2)x)/(-2/3)$$

and

$$y_2 = (11 - (2/3)x)/(-3/4).$$

The lines appear to intersect at $(30, 12)$.

Solution

Since 2 and 3 both divide evenly into 6, multiplying the first equation by 6 will eliminate its fractions. Since 3 and 4 both divide evenly into 12, multiplying the second equation by 12 will eliminate its fractions:

$$6\left(\frac{1}{2}x - \frac{2}{3}y\right) = 6(7) \qquad \rightarrow \qquad 3x - 4y = 42$$

$$12\left(\frac{2}{3}x - \frac{3}{4}y\right) = 12(11) \qquad \rightarrow \qquad 8x - 9y = 132$$

Now examine the coefficients in the two new equations. Both equations will have to be multiplied to eliminate x or y. To eliminate x, multiply the first by -8 and the second by 3:

$$-8(3x - 4y) = -8(42) \quad \rightarrow \quad -24x + 32y = -336$$
$$3(8x - 9y) = 3(132) \quad \rightarrow \quad \underline{24x - 27y = 396}$$
$$5y = 60$$
$$y = 12$$

Substitute $y = 12$ into the first of the original equations:

$$\frac{1}{2}x - \frac{2}{3}(12) = 7$$
$$\frac{1}{2}x - 8 = 7$$
$$\frac{1}{2}x = 15$$
$$x = 30$$

Check $(30, 12)$ in the original system. The solution set is $\{(30, 12)\}$.

> Now do Exercises 31–38

Note that there are many ways to proceed in Example 5. We multiplied first to eliminate the fractions and second to eliminate a variable. That is usually the simplest approach. However, if you multiply the first equation by -48 and the second by 36, you would only have to multiply once.

EXAMPLE 6

A system with decimals

Solve the system:

$$0.05x + 0.7y = 40$$
$$x + 0.4y = 120$$

Solution

Multiplying by 10 or 100 moves the decimal point one or two places to the right, respectively. So multiplying the first equation by 100 and the second by 10 will eliminate all of the decimals:

$$100(0.05x + 0.7y) = 100(40) \quad \rightarrow \quad 5x + 70y = 4000$$
$$10(x + 0.4y) = 10(120) \quad \rightarrow \quad 10x + 4y = 1200$$

Examine the coefficients. Since 10 is a multiple of 5, we can eliminate x by multiplying the first equation by -2:

$$-2(5x + 70y) = -2(4000) \quad \rightarrow \quad -10x - 140y = -8000$$
$$10x + 4y = 1200 \quad \rightarrow \quad \underline{10x + 4y = 1200}$$
$$-136y = -6800$$
$$y = 50$$

Use $y = 50$ in $x + 0.4y = 120$ to find x:

$$x + 0.4(50) = 120$$
$$x + 20 = 120$$
$$x = 100$$

Check $(100, 50)$ in the original system. The solution set is $\{(100, 50)\}$.

> Now do Exercises 39–46

‹ **Calculator Close-Up** ›

Check Example 6 by graphing
$$y_1 = (40 - 0.05x)/(0.7)$$
and
$$y_2 = (120 - x)/(0.4).$$

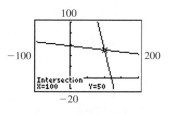

We have seen three methods for solving a system of two linear equations in two variables. For some systems the method you choose can make a difference. The following summary should help you decide which method to use.

Summary of the Methods

Graphing	It is impossible to identify the solution from a graph unless it is very simple. Graphing helps us understand the difference between independent, dependent, and inconsistent systems. Graphing works well with a graphing calculator.
Substitution	Substitution is used when one of the equations is solved for one of the variables or when it is easy to isolate one of the variables in an equation.
Addition	Addition is used when both equations are in the same form and it is easy to eliminate a variable by multiplying and adding.

⟨3⟩ Applications

Any system of two linear equations in two variables can be solved by either the addition method or substitution. In applications we use whichever method appears to be the simpler for the problem at hand.

EXAMPLE 7

Fajitas and burritos

At the Cactus Cafe the total price for four fajita dinners and three burrito dinners is $48, and the total price for three fajita dinners and two burrito dinners is $34. What is the price of each type of dinner?

⟨ **Helpful Hint** ⟩

You can see from Example 7 that the standard form $Ax + By = C$ occurs naturally in accounting. This form will occur whenever we have the price of each item and a quantity of two items and want to express the total cost.

Solution

Let x represent the price (in dollars) of a fajita dinner, and let y represent the price (in dollars) of a burrito dinner. We can write two equations to describe the given information:

$$4x + 3y = 48$$
$$3x + 2y = 34$$

Because 12 is the least common multiple of 4 and 3 (the coefficients of x), we multiply the first equation by -3 and the second by 4:

$$-3(4x + 3y) = -3(48) \quad \text{Multiply each side by } -3.$$
$$4(3x + 2y) = 4(34) \quad \text{Multiply each side by 4.}$$

$$-12x - 9y = -144$$
$$\underline{12x + 8y = 136} \qquad \text{Add.}$$
$$-y = -8$$
$$y = 8$$

To find x, use $y = 8$ in the first equation $4x + 3y = 48$:

$$4x + 3(8) = 48$$
$$4x + 24 = 48$$
$$4x = 24$$
$$x = 6$$

So the fajita dinners are $6 each, and the burrito dinners are $8 each. Check this solution in the original problem.

Now do Exercises 65–70

EXAMPLE **8**

Mixing cooking oil

Canola oil is 7% saturated fat, and corn oil is 14% saturated fat. Crisco sells a blend, Crisco Canola and Corn Oil, which is 11% saturated fat. How many gallons of each type of oil must be mixed to get 280 gallons of this blend?

Solution

Let x represent the number of gallons of canola oil, and let y represent the number of gallons of corn oil. Make a table to summarize all facts:

	Amount (gallons)	% fat	Amount of Fat (gallons)
Canola oil	x	7	$0.07x$
Corn oil	y	14	$0.14y$
Canola and Corn Oil	280	11	0.11(280) or 30.8

Since the total amount of oil is 280 gallons, we have $x + y = 280$. Since the total amount of fat is 30.8 gallons, we have $0.07x + 0.14y = 30.80$. Since we can easily solve $x + y = 280$ for y, we choose substitution to solve the system. Substitute $y = 280 - x$ into the second equation:

$$0.07x + 0.14(280 - x) = 30.80 \quad \text{Substitution}$$
$$0.07x + 39.2 - 0.14x = 30.80 \quad \text{Distributive property}$$
$$-0.07x = -8.4$$
$$x = \frac{-8.4}{-0.07} = 120$$

If $x = 120$ and $y = 280 - x$, then $y = 280 - 120 = 160$. Check that

$$0.07(120) + 0.14(160) = 30.8.$$

So it takes 120 gallons of canola oil and 160 gallons of corn oil to make 280 gallons of Crisco Canola and Corn Oil.

Now do Exercises 71–78

Warm-Ups ▼

Fill in the blank.

1. In the _____ method we eliminate a variable by adding the equations.

2. If addition in a linear system results in a _____ equation, then the system has exactly one solution.

3. If addition results in an identity, then the system is _____.

4. If addition results in an _____ equation, then the system has no solution.

5. If addition results in a _____ equation, then the two lines intersect at exactly one point.

6. If addition results in an _____, then the two lines have the same graph.

7. If addition results in an _____ equation, then the two lines are parallel.

True or false?

8. To solve $3x - y = 9$ and $3x + y = 6$ by addition we simply add the equations.

9. To solve $2x + 7y = 5$ and $3x + 2y = 8$ by addition we multiply the first equation by 3, the second by 2, and then add.

10. Both $(0, -10)$ and $(5, 0)$ satisfy $4x - 2y = 20$ and $4x + 2y = 20$.

11. The system $4x - 5y = 9$ and $-4x + 5y = -9$ has no solution.

12. Either addition or substitution could be used to solve $2x - y = 5$ and $3x + 2y = 9$

13. To eliminate fractions, multiply both sides of the equation by the least common denominator.

14. Either variable can be eliminated by the addition method.

7.3

Exercises

‹ Study Tips ›

- Don't expect to understand a topic the first time you see it. Learning mathematics takes time, patience, and repetition.
- Keep reading the text, asking questions, and working problems. Someone once said, "All math is easy once you understand it."

‹1› The Addition Method

Solve each system by addition. See Examples 1–3. See the Strategy for the Addition Method box on page 479.

1. $x - y = 1$
$x + y = 7$

2. $x + y = 7$
$x - y = 9$

3. $3x - 4y = 11$
$-3x + 2y = -7$

4. $7x - 5y = -1$
$-3x + 5y = 9$

5. $x - y = 12$
$2x + y = 3$

6. $x - 2y = -1$
$-x + 5y = 4$

7. $3x - y = 5$
$5x + y = -2$

8. $-x + 2y = 4$
$x - 5y = 1$

9. $2x - y = -5$
$3x + 2y = 3$

10. $3x + 5y = -11$
$x - 2y = 11$

11. $-3x + 5y = 1$
$9x - 3y = 5$

12. $7x - 4y = -3$
$x + 2y = 3$

13. $2x - 5y = 13$
$3x + 4y = -15$

14. $3x + 4y = -5$
$5x + 6y = -7$

15. $2x = 3y + 11$
$7x - 4y = 6$

16. $2x = 2 - y$
$3x + y = -1$

17. $x + y = 48$
$12x + 14y = 628$

18. $x + y = 13$
$22x + 36y = 356$

Use a calculator to check whether the given ordered pair satisfies both equations of the given system.

19. $(-45, 16)$
$3x + 2y = -103$
$5x - 8y = -353$

20. $(502, 388)$
$-3x + 5y = 434$
$6x - 7y = 296$

21. $(42, 99)$
$\dfrac{2}{3}x + \dfrac{5}{11}y = 73$
$-\dfrac{1}{3}x + \dfrac{2}{9}y = 9$

22. $(16.5, 25.6)$
$\dfrac{1}{3}x + \dfrac{5}{2}y = 69.5$
$\dfrac{4}{5}x - \dfrac{3}{4}y = 6$

23. $(34.56, 59.66)$
$0.02x + 0.03y = 2.481$
$0.8x + 0.9y = 81.342$

24. $(40{,}000, 120{,}000)$
$0.08x + 0.12y = 17{,}600$
$x + y = 160{,}000$

Solve each system by the addition method. Determine whether the equations are independent, dependent, or inconsistent. See Example 4.

25. $3x - 4y = 9$
$-3x + 4y = 12$

26. $x - y = 3$
$-6x + 6y = 17$

27. $5x - y = 1$
$10x - 2y = 2$

28. $4x + 3y = 2$
$-12x - 9y = -6$

29. $2x - y = 5$
$2x + y = 5$

30. $-3x + 2y = 8$
$3x + 2y = 8$

‹2› Equations Involving Fractions or Decimals

Solve each system by the addition method. See Examples 5 and 6.

31. $\dfrac{1}{4}x + \dfrac{1}{3}y = 5$
$x - y = 6$

32. $\dfrac{3x}{2} - \dfrac{2y}{3} = 10$
$\dfrac{1}{2}x + \dfrac{1}{2}y = -1$

33. $\dfrac{x}{4} - \dfrac{y}{3} = -4$

$\dfrac{x}{8} + \dfrac{y}{6} = 0$

34. $\dfrac{x}{3} - \dfrac{y}{2} = -\dfrac{5}{6}$

$\dfrac{x}{5} - \dfrac{y}{3} = -\dfrac{3}{5}$

35. $\dfrac{1}{8}x + \dfrac{1}{4}y = 5$

$\dfrac{1}{16}x + \dfrac{1}{2}y = 7$

36. $\dfrac{3}{7}x + \dfrac{5}{9}y = 27$

$\dfrac{1}{9}x + \dfrac{2}{7}y = 7$

37. $\dfrac{1}{3}x + \dfrac{1}{2}y = \dfrac{1}{3}$

$\dfrac{5}{6}x - \dfrac{3}{4}y = \dfrac{1}{6}$

38. $\dfrac{2}{3}x + \dfrac{5}{6}y = \dfrac{1}{4}$

$\dfrac{1}{5}x - \dfrac{1}{10}y = -\dfrac{1}{10}$

39. $0.05x + 0.10y = 1.30$

$x + y = 19$

40. $0.1x + 0.06y = 9$

$0.09x + 0.5y = 52.7$

41. $x + y = 1200$

$0.12x + 0.09y = 120$

42. $x - y = 100$

$0.20x + 0.06y = 150$

43. $1.5x - 2y = -0.25$

$3x + 1.5y = 6.375$

44. $3x - 2.5y = 7.125$

$2.5x - 3y = 7.3125$

45. $0.24x + 0.6y = 0.58$

$0.8x - 0.12y = 0.52$

46. $0.18x + 0.27y = 0.09$

$0.06x - 0.54y = -0.04$

Miscellaneous

Solve each system by substitution or addition, whichever is easier.

47. $y = x + 1$

$2x - 5y = -20$

48. $y = 3x - 4$

$x + y = 32$

49. $x - y = 19$

$2x + y = -13$

50. $x + y = 3$

$7x - y = 29$

51. $2y = x + 2$

$x = y - 1$

52. $2y - x = 3$

$x = 3y - 5$

53. $2y - 3x = -1$

$5y + 3x = 29$

54. $y - 5 = 2x$

$y - 9 = -2x$

55. $6x + 3y = 4$

$y = \dfrac{2}{3}x$

56. $3x - 2y = 2$

$x = \dfrac{2}{9}y$

57. $y = 3x + 1$

$x = \dfrac{1}{3}y + 5$

58. $y = -\dfrac{2}{3}x - 3$

$x = -\dfrac{3}{2}y + 9$

59. $x - y = 0$

$x + y = 2x$

60. $5x - 4y = 9$

$8y - 10x = -18$

For each system find the value of a so that the solution set to the system is $\{(2, 3)\}$.

61. $x + y = 5$

$x - y = a$

62. $2x - y = 1$

$ax + y = 13$

For each system find the values of a and b so that the solution set to the system is $\{(5, 12)\}$.

63. $y = ax + 2$

$y = bx + 17$

64. $y = 3x + a$

$y = -2x + b$

⟨**3**⟩ **Applications**

Write a system of two equations in two unknowns for each problem. Solve each system by the method of your choice. See Examples 7 and 8.

65. *Two numbers.* The sum of two numbers is 12 and their difference is 2. Find the numbers.

66. *Two more numbers.* The sum of two numbers is 11 and their difference is 6. Find the numbers.

67. *Paper size.* The length of a rectangular piece of paper is 2.5 inches greater than the width. The perimeter is 39 inches. Find the length and width.

68. *Photo size.* The length of a rectangular photo is 2 inches greater than the width. The perimeter is 20 inches. Find the length and width.

69. *Buy and sell.* Cory buys and sells baseball cards on eBay. He always buys at the same price and then sells the cards for $2 more than he buys them. One month he broke even after buying 56 cards and selling 49. Find his buying price and selling price.

70. *Jay Leno's garage.* Jay Leno's collection of cars and motorcycles totals 187. When he checks the air in the tires, he has 588 tires to check. How many cars and how many motorcycles does he own? Assume that the cars all have four tires and the motorcycles have two.

71. *Coffee and doughnuts.* On Monday, Archie paid $3.40 for three doughnuts and two coffees. On Tuesday he paid $3.60 for two doughnuts and three coffees. On Wednesday he was tired of paying the tab and went out for coffee by himself. What was his bill for one doughnut and one coffee?

b) Write a system of equations and solve it algebraically to find the exact amount of each type that should be used to obtain 50 pounds of double-dark-peanut fudge.

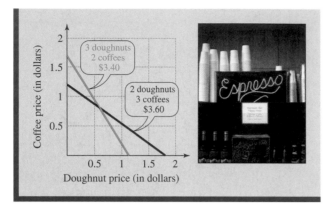

Figure for Exercise 71

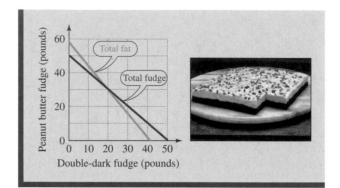

Figure for Exercise 77

72. *Books and magazines.* At Gwen's garage sale, all books were one price, and all magazines were another price. Harriet bought four books and three magazines for $1.45, and June bought two books and five magazines for $1.25. What was the price of a book and what was the price of a magazine?

73. *Boys and girls.* One-half of the boys and one-third of the girls of Freemont High attended the homecoming game, whereas one-third of the boys and one-half of the girls attended the homecoming dance. If there were 570 students at the game and 580 at the dance, then how many students are there at Freemont High?

74. *Girls and boys.* There are 385 surfers in Surf City. Two-thirds of the boys are surfers and one-twelfth of the girls are surfers. If there are two girls for every boy, then how many boys and how many girls are there in Surf City?

75. *Nickels and dimes.* Winborne has 35 coins consisting of dimes and nickels. If the value of his coins is $3.30, then how many of each type does he have?

76. *Pennies and nickels.* Wendy has 52 coins consisting of nickels and pennies. If the value of the coins is $1.20, then how many of each type does she have?

77. *Blending fudge.* The Chocolate Factory in Vancouver blends its double-dark-chocolate fudge, which is 35% fat, with its peanut butter fudge, which is 25% fat, to obtain double-dark-peanut fudge, which is 29% fat.

a) Use the accompanying graph to estimate the number of pounds of each type that must be mixed to obtain 50 pounds of double-dark-peanut fudge.

78. *Low-fat yogurt.* Ziggy's Famous Yogurt blends regular yogurt that is 3% fat with its no-fat yogurt to obtain low-fat yogurt that is 1% fat. How many pounds of regular yogurt and how many pounds of no-fat yogurt should be mixed to obtain 60 pounds of low-fat yogurt?

79. *Keystone state.* Judy averaged 42 miles per hour (mph) driving from Allentown to Harrisburg and 51 mph driving from Harrisburg to Pittsburgh. See the accompanying figure. If she drove a total of 288 miles in 6 hours, then how long did it take her to drive from Harrisburg to Pittsburgh?

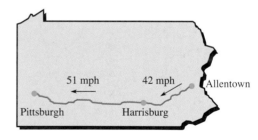

Figure for Exercise 79

80. *Empire state.* Spike averaged 45 mph driving from Rochester to Syracuse and 49 mph driving from Syracuse to Albany. If he drove a total of 237 miles in 5 hours, then how far is it from Syracuse to Albany?

81. *Probability of rain.* The probability of rain tomorrow is four times the probability that it does not rain tomorrow. The probability that it rains plus the probability that it does not rain is 1. What is the probability that it rains tomorrow?

82. *Super Bowl contender.* The probability that San Francisco plays in the next Super Bowl is nine times the probability that they do not play in the next Super Bowl. The probability that San Francisco plays in the next Super Bowl plus the probability that they do not play is 1. What is the probability that San Francisco plays in the next Super Bowl?

83. *Rectangular lot.* The width of a rectangular lot is 75% of its length. If the perimeter is 700 meters, then what are the length and width?

84. *Fence painting.* Darren and Douglas must paint the 792-foot fence that encircles their family home. Because Darren is older, he has agreed to paint 20% more than Douglas. How much of the fence will each boy paint?

Getting More Involved

85. *Discussion*

Explain how you decide whether it is easier to solve a system by substitution or addition.

86. *Exploration*

a) Write a linear equation in two variables that is satisfied by $(-3, 5)$.

b) Write another linear equation in two variables that is satisfied by $(-3, 5)$.

c) Are your equations independent or dependent?

d) Explain how to select the second equation so that it will be independent of the first.

87. *Exploration*

a) Make up a system of two linear equations in two variables such that both $(-1, 2)$ and $(4, 5)$ are in the solution set.

b) Are your equations independent or dependent?

c) Is it possible to find an independent system that is satisfied by both ordered pairs? Explain.

7.4   Systems of Linear Equations in Three Variables

In This Section

⟨1⟩ **Definition**

⟨2⟩ **Solving a System by Elimination**

⟨3⟩ **Dependent and Inconsistent Systems**

⟨4⟩ **Applications**

The techniques that you learned in Sections 7.2 and 7.3 can be extended to systems of equations in more than two variables. In this section, we use elimination of variables to solve systems of equations in three variables.

⟨1⟩ Definition

The equation $5x - 4y = 7$ is called a linear equation in two variables because its graph is a straight line. The equation $2x + 3y - 4z = 12$ is similar in form, and so it is a linear equation in three variables. An equation in three variables is graphed in a three-dimensional coordinate system. The graph of a linear equation in three variables is a plane, not a line. We will not graph equations in three variables in this text, but we can solve systems without graphing. In general, we make the following definition.

> **Linear Equation in Three Variables**
>
> If A, B, C, and D are real numbers, with A, B, and C not all zero, then
>
> $$Ax + By + Cz = D$$
>
> is called a **linear equation in three variables.**

⟨2⟩ Solving a System by Elimination

A solution to an equation in three variables is an **ordered triple** such as $(-2, 1, 5)$, where the first coordinate is the value of x, the second coordinate is the value of y, and the third coordinate is the value of z. There are infinitely many solutions to a linear equation in three variables.

The solution to a system of equations in three variables is the set of all ordered triples that satisfy all of the equations of the system. The techniques for solving a system of linear equations in three variables are similar to those used on systems of linear equations in two variables. We eliminate variables by either substitution or addition.

E X A M P L E   1

A linear system with a single solution

Solve the system:

$$(1) \qquad x + y - z = -1$$
$$(2) \qquad 2x - 2y + 3z = 8$$
$$(3) \qquad 2x - y + 2z = 9$$

Solution

We can eliminate y from Eqs. (1) and (2) by multiplying Eq. (1) by 2 and adding it to Eq. (2):

$$
\begin{array}{ll}
2x + 2y - 2z = -2 & \text{Eq. (1) multiplied by 2} \\
\underline{2x - 2y + 3z = 8} & \text{Eq. (2)} \\
(4) \quad 4x \qquad\;\; + z = 6 &
\end{array}
$$

Now we must eliminate the same variable, y, from another pair of equations. Eliminate y from Eqs. (1) and (3) by simply adding them:

$$
\begin{array}{ll}
x + y - z = -1 & \text{Eq. (1)} \\
\underline{2x - y + 2z = 9} & \text{Eq. (3)} \\
(5) \quad 3x \qquad + z = 8 &
\end{array}
$$

Equations (4) and (5) give us a system with two variables. We now solve this system. Eliminate z by multiplying Eq. (4) by -1 and adding the equations:

$$
\begin{array}{ll}
-4x - z = -6 & \text{Eq. (4) multiplied by } -1 \\
\underline{3x + z = 8} & \text{Eq. (5)} \\
-x \qquad = 2 & \\
x = -2 &
\end{array}
$$

‹ **Calculator Close-Up** ›

You can use a calculator to check that $(-2, 15, 14)$ satisfies all three equations of the original system.

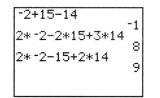

Now that we have x, we can replace x by -2 in Eq. (5) to find z:

$$
\begin{array}{ll}
3x + z = 8 & \text{Eq. (5)} \\
3(-2) + z = 8 & \\
-6 + z = 8 & \\
z = 14 &
\end{array}
$$

Now replace x by -2 and z by 14 in Eq. (1) to find y:

$$
\begin{array}{ll}
x + y - z = -1 & \text{Eq. (1)} \\
-2 + y - 14 = -1 & x = -2, z = 14 \\
y - 16 = -1 & \\
y = 15 &
\end{array}
$$

Check that $(-2, 15, 14)$ satisfies all three of the original equations. The solution set is $\{(-2, 15, 14)\}$.

Now do Exercises 1–4

Note that we could have eliminated any one of the three variables in Example 1 to get a system of two equations in two variables. We chose to eliminate y first because it was the easiest to eliminate. The strategy that we follow for solving a system of three linear equations in three variables is stated as follows:

Strategy for Solving a System in Three Variables

1. Use substitution or addition to eliminate any one of the variables from a pair of equations of the system. Look for the easiest variable to eliminate.
2. Eliminate the same variable from another pair of equations of the system.
3. Solve the resulting system of two equations in two unknowns.
4. After you have found the values of two of the variables, substitute into one of the original equations to find the value of the third variable.
5. Check the three values in all of the original equations.

In Example 2, we use a combination of addition and substitution.

EXAMPLE   2

Using addition and substitution

Solve the system:

$$
\begin{aligned}
(1) \quad & x + y \phantom{{}- 3z} = 4 \\
(2) \quad & 2x \phantom{{}+ y} - 3z = 14 \\
(3) \quad & \phantom{2x{}+{}} 2y + z = 2
\end{aligned}
$$

Solution

From Eq. (1) we get $y = 4 - x$. If we substitute $y = 4 - x$ into Eq. (3), then Eqs. (2) and (3) will be equations involving x and z only.

$$
\begin{aligned}
(3) \quad & 2y + z = 2 \\
& 2(4 - x) + z = 2 \quad \text{Replace } y \text{ by } 4 - x. \\
& 8 - 2x + z = 2 \quad \text{Simplify.} \\
(4) \quad & -2x + z = -6
\end{aligned}
$$

‹ **Helpful Hint** ›

In Example 2 we chose to eliminate y first. Try solving Example 2 by first eliminating z. Write $z = 2 - 2y$, and then substitute $2 - 2y$ for z in Eqs. (1) and (2).

Now solve the system consisting of Eqs. (2) and (4) by addition:

$$
\begin{array}{rl}
2x - 3z = 14 & \text{Eq. (2)} \\
\underline{-2x + z = -6} & \text{Eq. (4)} \\
-2z = 8 & \\
z = -4 &
\end{array}
$$

Use Eq. (3) to find y:

$$
\begin{aligned}
2y + z &= 2 \quad \text{Eq. (3)} \\
2y + (-4) &= 2 \quad \text{Let } z = -4. \\
2y &= 6 \\
y &= 3
\end{aligned}
$$

Use Eq. (1) to find x:

$$
\begin{aligned}
x + y &= 4 \quad \text{Eq. (1)} \\
x + 3 &= 4 \quad \text{Let } y = 3. \\
x &= 1
\end{aligned}
$$

Check that $(1, 3, -4)$ satisfies all three of the original equations. The solution set is $\{(1, 3, -4)\}$.

‹ **Calculator Close-Up** ›

Check that $(1, 3, -4)$ satisfies all three equations in Example 2.

```
1+3
           4
2(1)-3(-4)
           14
2(3)+(-4)
           2
```

Now do Exercises 5–20

CAUTION In solving a system in three variables it is essential to keep your work organized and neat. Writing short notes that explain your steps (as was done in the examples) will allow you to go back and check your work.

⟨3⟩ Dependent and Inconsistent Systems

The graph of any equation in three variables can be drawn on a three-dimensional coordinate system. The graph of a linear equation in three variables is a plane. To solve a system of three linear equations in three variables by graphing, we would have to draw the three planes and then identify the points that lie on all three of them. This method would be difficult even when the points have simple coordinates. So we will not attempt to solve these systems by graphing.

In Section 7.1 we classified systems of two linear equations in two variables as *consistent* if the system had at least one solution and *inconsistent* if the system had no solutions. A consistent system with exactly one solution is *independent*, and a consistent system with infinitely many solutions is *dependent*. We use the same terminology with systems of three linear equations in three variables. The only difference here is that there are more possibilities for the graphs of the dependent and inconsistent systems. Even though we don't solve systems in three variables by graphing, the graphs in Fig. 7.7 will help you to better understand these systems.

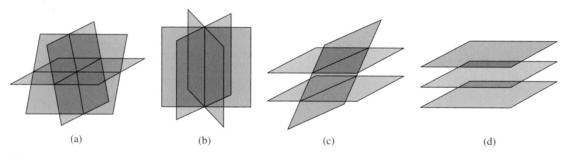

(a)   (b)   (c)   (d)

Figure 7.7

In most of the problems that we will solve, the planes intersect at a single point, as in Fig. 7.7(a). The solution set contains exactly one ordered triple, and the system is **independent.**

If the intersection of the three planes is a line or a plane, then the solution set is infinite and the system is **dependent.** There are three possibilities. All three planes could intersect along a line as shown in Fig. 7.7(b). All three planes could be the same. In which case, all points on that plane satisfy the system. We could also have two equations for the same plane with the third plane intersecting it along a line.

If there are no points in common to all three planes, then the system is **inconsistent.** The system will be inconsistent if at least two of the planes are parallel as shown in Fig. 7.7(c) and (d). There is one other configuration for an inconsistent system that is not shown here. See if you can find it.

We will not solve systems corresponding to all of the possible configurations described for the planes. Examples 3 and 4 illustrate two of these cases.

EXAMPLE   **3**

A system with infinitely many solutions
Solve the system:

$$
\begin{array}{rl}
(1) & 2x - 3y - \ \ z = 4 \\
(2) & -6x + 9y + 3z = -12 \\
(3) & 4x - 6y - 2z = 8
\end{array}
$$

Solution

We will first eliminate x from Eqs. (1) and (2). Multiply Eq. (1) by 3 and add the resulting equation to Eq. (2):

$$
\begin{array}{ll}
6x - 9y - 3z = 12 & \text{Eq. (1) multiplied by 3} \\
-6x + 9y + 3z = -12 & \text{Eq. (2)} \\
\hline
0 = 0 &
\end{array}
$$

The last statement is an identity. The identity occurred because Eq. (2) is a multiple of Eq. (1). In fact, Eq. (3) is also a multiple of Eq. (1). These equations are dependent. They are all equations for the same plane. The solution set is the set of all points on that plane,

$$
\{(x, y, z) \mid 2x - 3y - z = 4\}.
$$

> **Now do Exercises 21–22**

‹ **Helpful Hint** ›

If you recognize that multiplying Eq. (1) by -3 will produce Eq. (2), and multiplying Eq. (1) by 2 will produce Eq. (3), then you can conclude that all three equations are equivalent and there is no need to add the equations.

EXAMPLE   **4**

A system with no solutions
Solve the system:

$$
\begin{array}{rl}
(1) & x + \ \ y - \ \ z = 5 \\
(2) & 3x - 2y + \ \ z = 8 \\
(3) & 2x + 2y - 2z = 7
\end{array}
$$

Solution

We can eliminate the variable z from Eqs. (1) and (2) by adding them:

$$
\begin{array}{ll}
x + \ \ y - z = 5 & \text{Eq. (1)} \\
3x - 2y + z = 8 & \text{Eq. (2)} \\
\hline
4x - \ \ y \ \ \ \ \ = 13 &
\end{array}
$$

To eliminate z from Eqs. (1) and (3), multiply Eq. (1) by -2 and add the resulting equation to Eq. (3):

$$
\begin{array}{ll}
-2x - 2y + 2z = -10 & \text{Eq. (1) multiplied by } -2 \\
2x + 2y - 2z = 7 & \text{Eq. (3)} \\
\hline
0 = -3 &
\end{array}
$$

Because the last equation is false, the system is inconsistent. The solution set is the empty set.

> **Now do Exercises 23–34**

‹4› Applications

Problems involving three unknown quantities can often be solved by using a system of three equations in three variables.

EXAMPLE 5

Finding three unknown rents

Theresa took in a total of $1240 last week from the rental of three condominiums. She had to pay 10% of the rent from the one-bedroom condo for repairs, 20% of the rent from the two-bedroom condo for repairs, and 30% of the rent from the three-bedroom condo for repairs. If the three-bedroom condo rents for twice as much as the one-bedroom condo and her total repair bill was $276, then what is the rent for each condo?

< **Helpful Hint** >

A problem involving two unknowns can often be solved with one variable as in Chapter 2. Likewise, you can often solve a problem with three unknowns using only two variables. Solve Example 5 by letting a, b, and $2a$ be the rent for a one-bedroom, two-bedroom, and a three-bedroom condo.

Solution

Let x, y, and z represent the rent on the one-bedroom, two-bedroom, and three-bedroom condos, respectively. We can write one equation for the total rent, another equation for the total repairs, and a third equation expressing the fact that the rent for the three-bedroom condo is twice that for the one-bedroom condo:

$$x + y + z = 1240$$
$$0.1x + 0.2y + 0.3z = 276$$
$$z = 2x$$

Substitute $z = 2x$ into both of the other equations to eliminate z:

$$x + y + 2x = 1240$$
$$0.1x + 0.2y + 0.3(2x) = 276$$

$$3x + y = 1240$$
$$0.7x + 0.2y = 276$$

$-2(3x + y) = -2(1240)$	Multiply each side by -2.
$10(0.7x + 0.2y) = 10(276)$	Multiply each side by 10.

$$\begin{array}{r} -6x - 2y = -2480 \\ \underline{7x + 2y = 2760} \\ x = 280 \end{array}$$ Add.

< **Calculator Close-Up** >

Check that (280, 400, 560) satisfies all three equations in Example 5.

```
280+400+560
            1240
.1*280+.2*400+.3
*560
            276
2*280
            560
```

$z = 2(280) = 560$	Because $z = 2x$
$280 + y + 560 = 1240$	Because $x + y + z = 1240$
$y = 400$	

Check that (280, 400, 560) satisfies all three of the original equations. The condos rent for $280, $400, and $560 per week.

Now do Exercises 51–64

Warm-Ups ▼

Fill in the blank.

1. An equation of the form $Ax + By + Cz = D$ where A, B, and C are not all zero, is a _____ equation in three variables.

2. The _____ triple (a, b, c) corresponds to a point in a three-dimensional coordinate system.

3. A _____ to a linear system in three variables is an ordered triple that satisfies all of the equations.

4. To solve a linear system in three variables use _____ or _____ to eliminate variables.

5. The graph of a linear equation in three variables is a _____ in a three-dimensional coordinate system.

6. For an _____ system of three linear equations in three variables the planes intersect at a single point.

True or false?

7. The point $(1, -2, 3)$ satisfies $x + y - z = 4$.

8. The point $(4, 1, 1)$ is the only solution to $x + y - z = 4$.

9. The point $(1, -1, 2)$ satisfies $x + y + z = 2$ and $2x + y - z = -1$.

10. Two distinct planes are either parallel or intersect at a single point.

11. The equations $3x + 2y - 6z = 4$ and $-6x - 4y + 12z = -8$ are dependent.

12. The graph of $y = 2x - 3z + 4$ is a line.

13. The value of x nickels, y dimes, and z quarters is $0.05x + 0.10y + 0.25z$ cents.

14. If $x = -2$, $z = 3$, and $x + y + z = 6$, then $y = 7$.

Exercises 7.4

⟨ **Study Tips** ⟩

- Finding out what happened in class and attending class are not the same. Attend every class and be attentive.
- Don't just take notes and let your mind wander. Use class time as a learning time.

⟨2⟩ **Solving a System by Elimination**

Solve each system of equations. See Examples 1 and 2. See the Strategy for Solving a System in Three Variables box on page 489.

1. $x + y + z = 9$
$y + z = 7$
$z = 4$

2. $x + y - z = 4$
$y = 6$
$y + z = 13$

3. $x + y + z = 10$
$x - y = -1$
$x + y = 5$

4. $x + y - z = 6$
$y + z = 11$
$y - z = 3$

5. $x + y + z = 6$
$x - y + z = 2$
$x - y - z = -4$

6. $x + y + z = 0$
$x + y - z = 2$
$x - y + z = 0$

7. $x + y + z = 2$
$x + 2y - z = 6$
$2x + y - z = 5$

8. $2x - y + 3z = 14$
$x + y - 2z = -5$
$3x + y - z = 2$

9. $x - 2y + 4z = 3$
$x + 3y - 2z = 6$
$x - 4y + 3z = -5$

10. $2x + 3y + z = 13$
$-3x + 2y + z = -4$
$4x - 4y + z = 5$

11. $2x - y + z = 10$
$3x - 2y - 2z = 7$
$x - 3y - 2z = 10$

12. $x - 3y + 2z = -11$
$2x - 4y + 3z = -15$
$3x - 5y - 4z = 5$

13. $2x - 3y + z = -9$
$-2x + y - 3z = 7$
$x - y + 2z = -5$

14. $3x - 4y + z = 19$
$2x + 4y + z = 0$
$x - 2y + 5z = 17$

15. $2x - 5y + 2z = 16$
$3x + 2y - 3z = -19$
$4x - 3y + 4z = 18$

16. $-2x + 3y - 4z = 3$
$3x - 5y + 2z = 4$
$-4x + 2y - 3z = 0$

17. $x + y = 4$
$y - z = -2$
$x + y + z = 9$

18. $x + y - z = 0$
$x - y = -2$
$y + z = 10$

19.
$$x + y \quad\quad = 7$$
$$\quad\; y - z = -1$$
$$x \quad\quad + 3z = 18$$

20.
$$2x - y \quad\quad = -8$$
$$\quad\;\; y + 3z = 22$$
$$x \quad\;\; - z = -8$$

36.
$$3x - 0.4y + 9z = 1.668$$
$$0.3x + \;\; 5y - 8z = -0.972$$
$$5x - \;\; 4y - 8z = 1.8$$

Use a calculator to check whether the given ordered triple satisfies all three equations of the given system.

⟨3⟩ Dependent and Inconsistent Systems

Solve each system. See Examples 3 and 4.

21.
$$x + y - z = 2$$
$$-x - y + z = -2$$
$$2x + 2y - 2z = 4$$

22.
$$x + y + z = 1$$
$$2x + 2y + 2z = 2$$
$$4x + 4y + 4z = 4$$

23.
$$x + y - z = 2$$
$$x + y + z = 8$$
$$x + y - z = 6$$

24.
$$x + y + z = 6$$
$$2x + 2y + 2z = 9$$
$$3x + 3y + 3z = 12$$

25.
$$x + y + z = 9$$
$$x + y \quad\;\; = 5$$
$$\quad\quad\; z = 1$$

26.
$$x - y + z = 2$$
$$\quad\; y - z = 3$$
$$x \quad\quad = 4$$

27.
$$x - y + 2z = 3$$
$$2x + y - z = 5$$
$$3x - 3y + 6z = 4$$

28.
$$4x - 2y - 2z = 5$$
$$2x - y - z = 7$$
$$-4x + 2y + 2z = 6$$

29.
$$2x - 4y + 6z = 12$$
$$6x - 12y + 18z = 36$$
$$-x + 2y - 3z = -6$$

30.
$$3x - y + z = 5$$
$$9x - 3y + 3z = 15$$
$$-12x + 4y - 4z = -20$$

37. (45, 32, 12)
$$3x - 2y + z = 83$$
$$x + 5y - z = 193$$
$$5x + y - 6z = 185$$

38. (−16, 45, 19)
$$7x + 6y - 3z = 101$$
$$3x + 4y - 9z = -39$$
$$-x + 5y + 8z = 393$$

39. (244, 386, 122)
$$0.1x + 0.3y - 0.12z = 125.56$$
$$0.9x + 0.4y - 0.25z = 343.5$$
$$0.5x + 0.2y + 0.15z = 181.0$$

40. (66, 72, 84)
$$\frac{1}{2}x + \frac{1}{3}y - \frac{5}{7}z = -3$$
$$-\frac{1}{3}x - \frac{1}{4}y + \frac{5}{12}z = -5$$
$$\frac{5}{6}x - \frac{5}{4}y + \frac{5}{14}z = 5$$

Miscellaneous

Solve each system. State whether the system is independent, dependent, or inconsistent.

41.
$$x - 2y = -12$$
$$2x + 3y = 4$$

42.
$$x - 2y = 3$$
$$-2x + 4y = 6$$

43.
$$x + y = 4$$
$$2x + 2y = 8$$

44.
$$-x + y = 12$$
$$5x + 4y = -6$$

31.
$$x - y \quad\quad = 3$$
$$\quad\; y + z = 8$$
$$2x \quad\quad + 2z = 7$$

32.
$$2x - y \quad\quad = 6$$
$$\quad\; 2y + z = -4$$
$$8x \quad\quad + 2z = 3$$

33.
$$0.10x + 0.08y - 0.04z = 3$$
$$5x + 4y - 2z = 150$$
$$0.3x + 0.24y - 0.12z = 9$$

34.
$$0.06x - 0.04y + z = 6$$
$$3x - 2y + 50z = 300$$
$$0.03x - 0.02y + 0.5z = 3$$

Use a calculator to solve each system.

35.
$$3x + 2y - 0.4z = 0.1$$
$$3.7x - 0.2y + 0.05z = 0.41$$
$$-2x + 3.8y - 2.1z = -3.26$$

45.
$$x + 2y - 3z = 6$$
$$-2x - 4y + 6z = 10$$

46.
$$x - y - z = 4$$
$$2x + y + 3z = 6$$
$$2x - 2y - 2z = 10$$

47. $x - y - z = 0$
$x + y + 3z = 8$
$x + 3y + z = 10$

48. $x + y - z = 0$
$3x - y - z = -10$
$2x + y + z = 35$

49. $5x - 5y + 5z = 5$
$x - y + z = 1$
$3x - 3y + 3z = 3$

50. $4x - 7y + 3z = 5$
$8x - 14y + 6z = 10$

⟨4⟩ Applications

Solve each problem by using a system of three equations in three unknowns. See Example 5.

51. *Three cars.* The town of Springfield purchased a Chevrolet, a Ford, and a Toyota for a total of $66,000. The Ford was $2000 more than the Chevrolet and the Toyota was $2000 more than the Ford. What was the price of each car?

52. *Buying texts.* Melissa purchased an English text, a math text, and a chemistry text for a total of $276. The English text was $20 more than the math text, and the chemistry text was twice the price of the math text. What was the price of each text?

53. *Three-day drive.* In three days, Carter drove 2196 miles in 36 hours behind the wheel. The first day he averaged 64 mph, the second day 62 mph, and the third day 58 mph. If he drove 4 more hours on the third day than on the first day, then how many hours did he drive each day?

54. *Three-day trip.* In three days, Katy traveled 146 miles down the Mississippi River in her kayak with 30 hours of paddling. The first day she averaged 6 mph, the second day 5 mph, and the third day 4 mph. If her distance on the third day was equal to her distance on the first day, then for how many hours did she paddle each day?

55. *Diversification.* Ann invested a total of $12,000 in stocks, bonds, and a mutual fund. She received a 10% return on her stock investment, an 8% return on her bond investment, and a 12% return on her mutual fund. Her total return was $1230. If the total investment in stocks and

bonds equaled her mutual fund investment, then how much did she invest in each?

56. *Paranoia.* Fearful of a bank failure, Norman split his life savings of $60,000 among three banks. He received 5%, 6%, and 7% on the three deposits. In the account earning 7% interest, he deposited twice as much as in the account earning 5% interest. If his total earnings were $3760, then how much did he deposit in each account?

57. *Weighing in.* Anna, Bob, and Chris will not disclose their weights but agree to be weighed in pairs. Anna and Bob together weigh 226 pounds. Bob and Chris together weigh 210 pounds. Anna and Chris together weigh 200 pounds. How much does each student weigh?

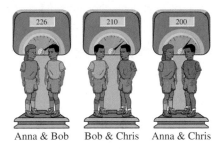

Figure for Exercise 57

58. *Big tipper.* On Monday Headley paid $1.70 for two cups of coffee and one doughnut, including the tip. On Tuesday he paid $1.65 for two doughnuts and a cup of coffee, including the tip. On Wednesday he paid $1.30 for one coffee and one doughnut, including the tip. If he always tips the same amount, then what is the amount of each item?

59. *Three coins.* Nelson paid $1.75 for his lunch with 13 coins, consisting of nickels, dimes, and quarters. If the number of dimes was twice the number of nickels, then how many of each type of coin did he use?

60. *Pocket change.* Harry has $2.25 in nickels, dimes, and quarters. If he had twice as many nickels, half as many dimes, and the same number of quarters, he would have $2.50. If he has 27 coins altogether, then how many of each does he have?

61. *Working overtime.* To make ends meet, Ms. Farnsby works three jobs. Her total income last year was $48,000. Her income from teaching was just $6000 more than her income from house painting. Royalties from her textbook sales were one-seventh of the total money she received from teaching and house painting. How much did she make from each source last year?

62. ***Lunch-box special.*** Salvador's Fruit Mart sells variety packs. The small pack contains three bananas, two apples, and one orange for $1.80. The medium pack contains four bananas, three apples, and three oranges for $3.05. The family size contains six bananas, five apples, and four oranges for $4.65. What price should Salvador charge for his lunch-box special that consists of one banana, one apple, and one orange?

63. ***Three generations.*** Edwin, his father, and his grandfather have an average age of 53. One-half of his grandfather's age, plus one-third of his father's age, plus one-fourth of Edwin's age is 65. If 4 years ago, Edwin's grandfather was four times as old as Edwin, then how old are they all now?

64. ***Three-digit number.*** The sum of the digits of a three-digit number is 11. If the digits are reversed, the new number is 46 more than five times the old number. If the hundreds digit plus twice the tens digit is equal to the units digit, then what is the number?

Getting More Involved

65. ***Exploration***

Draw diagrams showing the possible ways to position three planes in three-dimensional space.

66. ***Discussion***

Make up a system of three linear equations in three variables for which the solution set is $\{(0, 0, 0)\}$. A system with this solution set is called a *homogeneous* system. Why do you think it is given that name?

Chapter 7   Wrap-Up

Summary

Systems of Linear Equations		Examples
Methods for solving systems in two variables	Graphing: Sketch the graphs to see the solution.	The graphs of $y = x - 1$ and $x + y = 3$ intersect at $(2, 1)$.
	Substitution: Solve one equation for one variable in terms of the other, and then substitute into the other equation.	Substitution: $x + (x - 1) = 3$
	Addition: Multiply each equation as necessary to eliminate a variable upon addition of the equations.	$\begin{aligned} -x + y &= -1 \\ x + y &= 3 \\ \hline 2y &= 2 \end{aligned}$
Types of linear systems in two variables	Independent: One point in solution set The lines intersect at one point.	$y = x - 5$ $y = 2x + 3$
	Dependent: Infinite solution set The lines are the same.	$2x + 3y = 4$ $4x + 6y = 8$
	Inconsistent: Empty solution set The lines are parallel.	$2x + y = 1$ $2x + y = 5$
Linear equation in three variables	$Ax + By + Cz = D$ In a three-dimensional coordinate system the graph is a plane.	$2x - y + 3z = 5$
Linear systems in three variables	Use substitution or addition to eliminate variables in the system. The solution set may be a single point, the empty set, or an infinite set of points.	$\begin{aligned} x + y - z &= 3 \\ 2x - 3y + z &= 2 \\ x - y - 4z &= 14 \end{aligned}$

Enriching Your Mathematical Word Power

Fill in the blank.

1. A(n) _____ of equations consists of two or more equations.

2. A(n) _____ linear system is a system with exactly one solution.

3. A(n) _____ system has no solutions.

4. A(n) _____ system has infinitely many solutions.

5. In the _____ method a variable is eliminated by substituting one equation into the other.

6. In the _____ method a variable is eliminated by adding the equations.

7. A(n) _____ equation in three variables has the form $Ax + By + Cz = D$ with A, B, and C not all zero.

Review Exercises

7.1 The Graphing Method

Solve by graphing. Indicate whether each system is independent, dependent, or inconsistent.

1. $y = 2x - 1$
   $x + y = 2$

2. $y = 3x - 4$
   $y = -2x + 1$

3. $x + 2y = 4$
   $y = -\dfrac{1}{2}x + 2$

4. $2x - 3y = 12$
   $3y - 2x = -12$

5. $y = -x$
   $y = -x + 3$

6. $3x - y = 4$
   $3x - y = 0$

7.2 The Substitution Method

Solve by substitution. Indicate whether each system is independent, dependent, or inconsistent.

7. $y = 3x + 11$
   $2x + 3y = 0$

8. $x - y = 3$
   $3x - 2y = 3$

9. $x = y + 5$
   $2x - 2y = 12$

10. $3y = x + 5$
    $3x - 9y = -10$

11. $2x - y = 3$
    $6x - 9 = 3y$

12. $y = \dfrac{1}{2}x - 9$
    $3x - 6y = 54$

13. $y = \dfrac{1}{2}x - 3$
    $y = \dfrac{1}{3}x + 2$

14. $x = \dfrac{1}{8}y - 1$
    $y = \dfrac{1}{4}x + 39$

15. $x + 2y = 1$
    $8x + 6y = 4$

16. $x - 5y = 4$
    $4x + 8y = -5$

7.3 The Addition Method

Solve by addition. Indicate whether each system is independent, dependent, or inconsistent.

17. $5x - 3y = -20$
    $3x + 2y = 7$

18. $-3x + y = 3$
    $2x - 3y = 5$

19. $2(y - 5) + 4 = 3(x - 6)$
    $3x - 2y = 12$

20. $x + 3(y - 1) = 11$
    $2(x - y) + 8y = 28$

21. $3x - 4(y - 5) = x + 2$
    $2y - x = 7$

22. $4(1 - x) + y = 3$
    $3(1 - y) - 4x = -4y$

23. $\dfrac{1}{4}x + \dfrac{3}{8}y = \dfrac{3}{8}$
    $\dfrac{5}{2}x - 6y = 7$

24. $\frac{1}{3}x - \frac{1}{6}y = \frac{1}{3}$

$\frac{1}{6}x + \frac{1}{4}y = 0$

25. $0.4x + 0.06y = 11.6$

$0.8x - 0.05y = 13$

26. $0.08x + 0.7y = 37.4$

$0.06x - 0.05y = -0.7$

7.4 Systems of Linear Equations in Three Variables

Solve each system by elimination of variables.

27. $x - y + z = 4$

$-x + 2y - z = 0$

$-x + y - 3z = -16$

28. $2x - y + z = 5$

$x + y - 2z = -4$

$3x - y + 3z = 10$

29. $2x - y - z = 3$

$3x + y + 2z = 4$

$4x + 2y - z = -4$

30. $2x + 3y - 2z = -11$

$3x - 2y + 3z = 7$

$x - 4y + 4z = 14$

31. $x + y - z = 4$

$y + z = 6$

$x + 2y = 8$

32. $x - 3y + z = 5$

$2x - 4y - z = 7$

$2x - 6y + 2z = 6$

33. $x - 2y + z = 8$

$-x + 2y - z = -8$

$2x - 4y + 2z = 16$

34. $x - y + z = 1$

$2x - 2y + 2z = 2$

$-3x + 3y - 3z = -3$

Miscellaneous
Solve each system by the method of your choice.

35. $x + y = 7$

$-x + 2y = 5$

36. $-x + y = 1$

$2x - 3y = -7$

37. $2x + y = 0$

$x - 3y = 14$

38. $2x - y = 8$

$3x + 2y = -2$

39. $2x - y = 0$

$3x + y = -5$

40. $3x - 2y = 14$

$2x + 3y = -8$

41. $y = 2x - 3$

$3x - 2y = 4$

42. $y = 2x - 5$

$y = 3x - 3y$

43. $x + y - z = 0$

$x - y + 2z = 4$

$2x + y - z = 1$

44. $2x - y + 2z = 9$

$x + 3y = 5$

$3x + z = 9$

45. $x + y = 3$

$x + y + z = 0$

$x - y - z = 2$

46. $2x - y + z = 0$

$4x + 6y - 2z = 0$

$x - 2y - z = -9$

47. $y = 2x - 30$

$\frac{1}{5}x - \frac{1}{2}y = -1$

48. $3x - 5y = 4$

$y = \frac{3}{4}x - 2$

49. $2x + y = 9$

$2x - 5y = 15$

50. $3y - x = 0$

$x - 4y = -2$

51. $x - y = 0$

$2x + 3y = 35$

52. $2y = x + 6$

$-3x + 2y = -2$

53. $x + y = 40$

$0.2x + 0.8y = 23$

54. $x - y = 10$

$0.1x + 0.5y = 13$

55. $y = 2x - 5$

$y + 1 = 2(x - 2)$

56. $2x - y = 3$

$2y = 4x - 6$

57. $x - y = 5$

$2x = 2y + 14$

58. $2x - y = 4$

$2x - y = 3$

59. $y = \frac{5}{7}x$

$x = -\frac{2}{3}y$

60. $7y = 9x$

$-3x = 4y$

61. $3(y - 1) = 2(x - 3)$
$3y - 2x = -3$

62. $y = 3(x - 4)$
$3x - y = 12$

63. $y = 3x$
$y = 3x + 1$

64. $y = 3x - 4$
$y = 3x + 4$

65. $x - y = 0.1$
$2x - 3y = -0.5$

66. $y - 2x = -7.5$
$3x - 5y = 3.2$

67. $y = 2x + 4$
$3x + y = -1$

68. $3x - 2y = 6$
$3x + 2y = 6$

69. $y = -\frac{1}{2}x + 4$
$x + 2y = 8$

70. $2x - 3y = 6$
$y = \frac{2}{3}x - 2$

71. $2y - 2x = 2$
$2y - 2x = 6$

72. $3y - 3x = 9$
$x - y = 1$

73. $y = -\frac{1}{4}x$
$x + 4y = 8$

74. $y = -\frac{2}{3}x$
$2x + 3y = 5$

Use a system of equations in two or three variables to solve each problem. Solve by the method of your choice.

75. Perimeter of a rectangle. The length of a rectangular swimming pool is 15 feet longer than the width. If the perimeter is 82 feet, then what are the length and width?

76. Household income. Alkena and Hsu together earn $84,326 per year. If Alkena earns $12,468 more per year than Hsu, then how much does each of them earn per year?

77. Two-digit number. The sum of the digits in a two-digit number is 15. When the digits are reversed, the new number is 9 more than the original number. What is the original number?

78. Two-digit number. The sum of the digits in a two-digit number is 8. When the digits are reversed, the new number is 18 less than the original number. What is the original number?

79. Traveling by boat. Alonzo can travel from his camp downstream to the mouth of the river in 30 minutes. If it takes him 45 minutes to come back, then how long would it take him to go that same distance in the lake with no current?

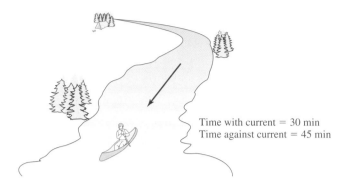

Time with current = 30 min
Time against current = 45 min

Figure for Exercise 79

80. Driving and dating. In 4 years Gasper will be old enough to drive. His parents said that he must have a driver's license for 2 years before he can date. Three years ago, Gasper's age was only one-half of the age necessary to date. How old must Gasper be to drive, and how old is he now?

81. Three solutions. A chemist has three solutions of acid that must be mixed to obtain 20 liters of a solution that is 38% acid. Solution A is 30% acid, solution B is 20% acid, and solution C is 60% acid. Because of another chemical in these solutions, the chemist must keep the ratio of solution C to solution A at 2 to 1. How many liters of each should she mix together?

82. Mixing investments. Darlene invested a total of $20,000. The part that she invested in Dell Computer stock returned 70%, and the part that she invested in U.S. Treasury bonds returned 5%. Her total return on these two investments was $9580.

a) Use the accompanying graph to estimate the amount that she put into each investment.

b) Solve a system of equations to find the exact amount that she put into each investment.

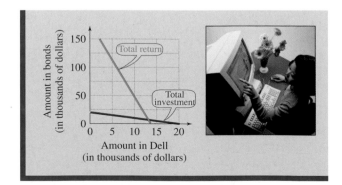

Figure for Exercise 82

83. *Beets and beans.* One serving of canned beets contains 1 gram of protein and 6 grams of carbohydrates. One serving of canned red beans contains 6 grams of protein and 20 grams of carbohydrates. How many servings of each would it take to get exactly 21 grams of protein and 78 grams of carbohydrates?

84. *Protein and carbohydrates.* One serving of Cornies breakfast cereal contains 2 grams of protein and 25 grams of carbohydrates. One serving of Oaties breakfast cereal contains 4 grams of protein and 20 grams of carbohydrates. How many servings of each would provide exactly 24 grams of protein and 210 grams of carbohydrates?

85. *Milk and a magazine.* Althia bought a gallon of milk and a magazine for a total of $4.65, excluding tax. Including the tax, the bill was $4.95. If there is a 5% sales tax on milk and an 8% sales tax on magazines, then what was the price of each item?

86. *Rectangular patio.* The length of a rectangular patio is 12 feet greater than the width. If the perimeter is 84 feet, then what are the length and width?

87. *Rectangular notepad.* The length of a rectangular notepad is 2 cm longer than twice the width. If the perimeter is 34 cm, then what are the length and width?

88. *Rectangular table.* The width of a rectangular table is 1 ft less than half of the length. If the perimeter is 28 ft, then what are the length and width?

89. *Rectangular painting.* The width of a rectangular painting is two-thirds of its length. If the perimeter is 60 in., then what are the length and width?

90. *Sum and difference.* The sum of two numbers is 10 and their difference is 3. Find the numbers.

91. *Sum and difference.* The sum of two numbers is 51 and their difference is 26. Find the numbers.

92. *Sum and difference.* The sum of two numbers is 1 and their difference is 20. Find the numbers.

93. *Sum and difference.* The sum of two numbers is 5 and their difference is 30. Find the numbers.

94. *Washing machines and refrigerators.* A truck carrying 3600 cubic feet of cargo consisting of washing machines and refrigerators was hijacked. The washing machines are worth $300 each and are shipped in 36-cubic-foot cartons. The refrigerators are worth $900 each and are shipped in 45-cubic-foot cartons. If the total value of the cargo was $51,000, then how many of each were there on the truck?

95. *Parking lot boredom.* A late-night parking lot attendant counted 50 vehicles on the lot consisting of four-wheel cars, three-wheel cars, and two-wheel motorcycles. She then counted 192 tires touching the ground and observed that the number of four-wheel cars was nine times the total of the other vehicles on the lot. How many of each type of vehicle were on the lot?

96. *Happy meals.* The total price of a hamburger, an order of fries, and Coke at a fast-food restaurant is $3.00. The price of a hamburger minus the price of an order of fries is $0.20 and the price of an order of fries minus the price of a Coke is also $0.20. Find the price of each item.

97. *Singles and doubles.* Windy's Hamburger Palace sells singles and doubles. Toward the end of the evening, Windy himself noticed that he had on hand only 32 patties and 34 slices of tomatoes. If a single takes 1 patty and 2 slices, and a double takes 2 patties and 1 slice, then how many more singles and doubles must Windy sell to use up all of his patties and tomato slices?

98. *Valuable wrenches.* Carmen has a total of 28 wrenches, all of which are either box wrenches or open-end wrenches. For insurance purposes she values the box wrenches at $3.00 each and the open-end wrenches at $2.50 each. If the value of her wrench collection is $78, then how many of each type does she have?

99. *Gary and Harry.* Gary is 5 years older than Harry. Twenty-nine years ago, Gary was twice as old as Harry. How old are they now?

100. *Acute angles.* One acute angle of a right triangle is 3° more than twice the other acute angle. What are the sizes of the acute angles?

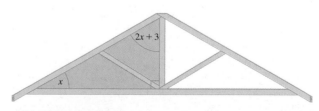

Figure for Exercise 100

101. *Equal perimeters.* A rope of length 80 feet is to be cut into two pieces. One piece will be used to form a square, and the other will be used to form an equilateral triangle. If the figures are to have equal perimeters, then what should be the length of a side of each?

Figure for Exercise 101

102. *Coffee and doughnuts.* For a cup of coffee and a doughnut, Thurrel spent $2.25, including a tip. Later he spent $4.00 for two coffees and three doughnuts, including a tip. If he always tips $1.00, then what is the price of a cup of coffee?

103. *Chlorine mixture.* A 10% chlorine solution is to be mixed with a 25% chlorine solution to obtain 30 gallons of 20% solution. How many gallons of each must be used?

104. *Safe drivers.* Emily and Camille started from the same city and drove in opposite directions on the freeway. After 3 hours, they were 354 miles apart. If they had gone in the same direction, Emily would have been 18 miles ahead of Camille. How fast did each woman drive?

105. *Weighing dogs.* Cassandra wants to determine the weights of her two dogs, Mimi and Mitzi. However, neither dog will sit on the scale by herself. Cassandra, Mimi, and Mitzi altogether weigh 175 pounds. Cassandra and Mimi together weigh 143 pounds. Cassandra and Mitzi together weigh 139 pounds. How much does each weigh individually?

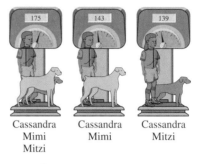

Cassandra   Cassandra   Cassandra
Mimi   Mimi   Mitzi
Mitzi

Figure for Exercise 105

106. *Nickels, dimes, and quarters.* Bernard has 41 coins consisting of nickels, dimes, and quarters, and they are worth a total of $4.00. If the number of dimes plus the number of quarters is one more than the number of nickels, then how many of each does he have?

107. *Finding three angles.* If the two acute angles of a right triangle differ by 12°, then what are the measures of the three angles of this triangle?

108. *Two acute and one obtuse.* The obtuse angle of a triangle is twice as large as the sum of the two acute angles. If the smallest angle is only one-eighth as large as the sum of the other two, then what is the measure of each angle?

Chapter 7 Test

Solve the system by graphing.

1. $x + y = 4$
   $y = 2x + 1$

Solve each system by substitution.

2. $y = 2x - 8$ **3.** $y = x - 5$
   $4x + 3y = 1$ $3x - 4(y - 2) = 28 - x$

Solve each system by the addition method.

4. $3x + 2y = 3$ **5.** $3x - y = 5$
   $4x - 3y = -13$ $-6x + 2y = 1$

Determine whether each system is independent, dependent, or inconsistent.

6. $y = 3x - 5$ **7.** $2x + 2y = 8$
   $y = 3x + 2$ $x + y = 4$

8. $y = 2x - 3$
   $y = 5x - 14$

Solve each system by the method of your choice.

9. $3x - y = 1$ **10.** $2x - y = -4$
   $x + 2y = 12$ $3x + y = -1$

11. $x + y = 0$
   $x - y + 2z = 6$
   $2x + y - z = 1$

12. $x + y - z = 2$
   $2x - y + 3z = -5$
   $x - 3y + z = 4$

13. $x - y - z = 1$
   $-x - y + 2z = -2$
   $-x - 3y + z = -5$

For each problem, write a system of equations in two or three variables. Use the method of your choice to solve each system.

14. One night the manager of the Sea Breeze Motel rented 5 singles and 12 doubles for a total of $1583. The next night he rented 9 singles and 10 doubles for a total of $1701. What is the rental charge for each type of room?

15. Jill, Karen, and Betsy studied a total of 93 hours last week. Jill's and Karen's study time totaled only one-half as much as Betsy's. If Jill studied 3 hours more than Karen, then how many hours did each one of the girls spend studying?

Making **Connections** | **A Review of Chapters 1–7**

Simplify each expression.

1. -3^4

2. $\dfrac{1}{3}(3) + 6$

3. $(-5)^2 - 4(-2)(6)$

4. $6 - (0.2)(0.3)$

5. $5(t - 3) - 6(t - 2)$

6. $0.1(x - 1) - (x - 1)$

7. $\dfrac{-9x^2 - 6x + 3}{-3}$

8. $\dfrac{4y - 6}{2} - \dfrac{3y - 9}{3}$

Factor each polynomial completely.

9. $3y^3 - 363y$

10. $2y^4 - 32$

11. $yw + 2w - 4y - 8$

12. $y^3 - 27$

13. $-3y^2 - 12y + 135$

14. $-24y^3 - 2y^2 + 12y$

15. $4a^3 - 4a^2 + 12a$

16. $2a^3b^3 + 2ab^5$

Reduce each rational expression to lowest terms.

17. $\dfrac{18x^3}{42x^4}$

18. $\dfrac{12x^8}{18x}$

19. $\dfrac{2x + 8}{2x - 14}$

20. $\dfrac{x^2 - y^2}{x^2 - xy}$

21. $\dfrac{x^2 - x - 30}{x^2 - 5x - 6}$

22. $\dfrac{2x^2 + 9x + 4}{2x^2 - x - 1}$

Perform the indicated operations.

23. $\dfrac{1}{6} + \dfrac{3}{8}$

24. $\dfrac{4}{15} + \dfrac{3}{20}$

25. $\dfrac{1}{5} - \dfrac{1}{12}$

26. $\dfrac{3}{10} - \dfrac{1}{6}$

27. $\dfrac{2}{15} \cdot \dfrac{21}{22}$

28. $\dfrac{3}{4} \cdot 88$

29. $\dfrac{2}{5} \div 4$

30. $\dfrac{9}{20} \div \dfrac{3}{10}$

31. $\dfrac{1}{3a^2} + \dfrac{5}{6a}$

32. $\dfrac{1}{2y} + y$

33. $\dfrac{3}{x^2 - 9} \cdot (3x - 9)$

34. $\dfrac{5ab^6}{7x^3y^5} \cdot \dfrac{14x^3y^7}{15ab}$

35. $\dfrac{6a}{5b} \div a$

36. $\dfrac{a^2 - 4}{a^2 + 8a + 12} \div \dfrac{2a - 4}{a^2 - 36}$

Solve each equation for y.

37. $3x - 5y = 7$

38. $Cx - Dy = W$

39. $Cy = Wy - K$

40. $A = \dfrac{1}{2}b(w - y)$

Solve each system.

41. $y = x - 5$
       $2x + 3y = 5$

42. $0.05x + 0.06y = 67$
       $x + y = 1200$

43. $3x - 15y = -51$
$x + 17 = 5y$

44. $0.07a + 0.3b = 6.70$
$7a + 30b = 67$

Find the equation of each line.

45. The line through $(0, 55)$ and $(-99, 0)$

46. The line through $(2, -3)$ and $(-4, 8)$

47. The line through $(-4, 6)$ that is parallel to $y = 5x$

48. The line through $(4, 7)$ that is perpendicular to $y = -2x + 1$

49. The line through $(3, 5)$ that is parallel to the x-axis

50. The line through $(-7, 0)$ that is perpendicular to the x-axis

Solve.

51. *Comparing copiers.* A self-employed consultant has prepared the accompanying graph to compare the total cost of purchasing and using two different copy machines.
a) Which machine has the larger purchase price?
b) What is the per copy cost for operating each machine, not including the purchase price?
c) Find the slope of each line and interpret your findings.
d) Find the equation of each line.
e) Find the number of copies for which the total cost is the same for both machines.

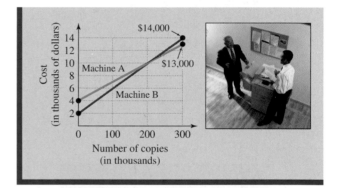

Figure for Exercise 51

*Critical*Thinking | For Individual or Group Work | Chapter 7

These exercises can be solved by a variety of techniques, which may or may not require algebra. So be creative and think critically. Explain all answers. Answers are in the Instructor's Edition of this text.

1. ***Tricky square.*** Start with a square and write any integer at each vertex. (a) At the midpoint of each side write the absolute value of the difference between the numbers at the endpoints of that side. (b) Connect the midpoints to obtain another square. Repeat parts (a) and (b) to obtain a sequence of nested squares as shown in the accompanying figure. What numbers will you always end up with?

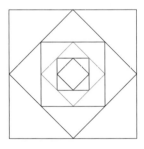

Figure for Exercise 1

2. ***Planning ahead.*** Thaddeus takes one month to build a kayak (K) and two months to build a canoe (C).

Photo for Exercise 2

While planning ahead for 1 month, he notes that there is only one thing to do that will not result in any partially finished boats. That is, build one kayak. For planning 2 months ahead there are two possibilities, KK or C. For a 3-month plan there are three possibilities, KKK or KC or CK.

a) Find the number of possibilities for a 4-month plan, a 5-month plan, and a 6-month plan by listing the possibilities. Look for a pattern.

b) Find the number of possibilities for a 7-month plan and an 8-month plan without making a list.

3. ***Five coins.*** Place five coins on a table with heads facing downward. On each move you must turn over exactly three coins. What is the minimum number of moves necessary to get all five heads facing upward?

4. ***Rotating tires.*** Helen bought a new car with four tires and a full-size spare. If she rotated the tires so that each tire would have the same amount of wear, then how many miles were on each tire when her odometer showed 40,000 miles?

5. ***Cutting pizza.*** What is the largest number of pieces of pizza you can get if you cut a circular pizza with five straight cuts? What is the largest number of pieces of pizza you can get if you cut a circular pizza with seven straight cuts?

6. ***Mysterious rectangle.*** The length of a rectangle is a two-digit number with identical digits (aa). The width of the rectangle is one-tenth of the length ($a.a$). The perimeter is numerically twice as large as the area. Find the length, width, perimeter, and area.

7. ***Finding squares.*** Evaluate the expression

$$100^2 - 99^2 + 98^2 - 97^2 + 96^2 - \cdots - 3^2 + 2^2 - 1^2$$

without using a calculator.

8. ***Five-digit sum.*** Find the sum of all five-digit numbers that are formed by using the digits 1, 2, 3, 4, and 5 once and only once.

More on Inequalities

The practice of awarding degrees originated in the universities of medieval Europe. The first known degree, a degree in civil law, was awarded in Italy at the University of Bologna in the twelfth century. The University of Paris awarded its first bachelor's degree in the thirteenth century. By the time the first colleges were opened in the American colonies, the process of awarding degrees was firmly established. At first, American schools offered only a few types of degrees. The colleges established in the colonies were primarily to train young men for the ministry. Notable were Harvard (1636; Puritan), William and Mary (1693; Anglican), Yale (1701; Congregationalist), Princeton (1746; New Lights Presbyterian), Brown (1765; Baptist), and Rutgers (1766; Dutch Reformed).

The industrial revolution sparked a demand for training in many areas. Today, approximately 1500 types of degrees are granted by academic institutions in the United States. Over 1 million Bachelor of Arts (B.A.) or of Science (B.S.) degrees are granted annually. Over one-quarter of a million Master of Arts (M.A.) or Science (M.S.) degrees are awarded annually.

The growth of bachelor's and master's degrees is modeled with linear equations in Exercise 87 of Section 8.1. In that exercise we also use inequalities and compound inequalities to discuss the growth of these degrees.

8.1   Compound Inequalities in One Variable

The inequality $a < x < b$, from Section 2.8, is one type of compound inequality in one variable. This inequality indicates that x is both greater than a and less than b. That is, x is between a and b. In this section we will study other types of compound inequalities in one variable.

〈1〉 Compound Inequalities

Inequalities involving a single inequality symbol are **simple inequalities.** If we join two simple inequalities with the connective "and" or the connective "or," we get a **compound inequality.** A compound inequality using the connective "and" is true if and only if *both* simple inequalities are true.

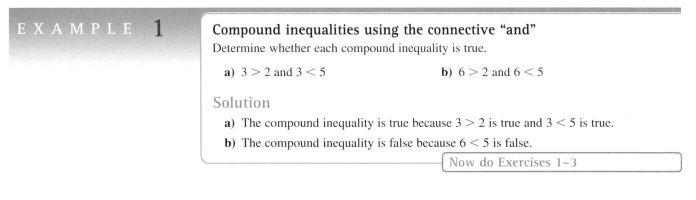

E X A M P L E   **1**

Compound inequalities using the connective "and"

Determine whether each compound inequality is true.

a) $3 > 2$ and $3 < 5$ b) $6 > 2$ and $6 < 5$

Solution

a) The compound inequality is true because $3 > 2$ is true and $3 < 5$ is true.

b) The compound inequality is false because $6 < 5$ is false.

Now do Exercises 1–3

A compound inequality using the connective "or" is true if one or the other or both of the simple inequalities are true. It is false only if both simple inequalities are false.

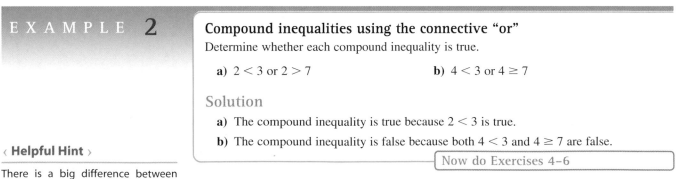

E X A M P L E   **2**

Compound inequalities using the connective "or"

Determine whether each compound inequality is true.

a) $2 < 3$ or $2 > 7$ b) $4 < 3$ or $4 \geq 7$

Solution

a) The compound inequality is true because $2 < 3$ is true.

b) The compound inequality is false because both $4 < 3$ and $4 \geq 7$ are false.

Now do Exercises 4–6

〈 **Helpful Hint** 〉

There is a big difference between "and" and "or." To get money from an automatic teller you must have a bank card *and* know a secret number (PIN). There would be a lot of problems if you could get money by having a bank card *or* knowing a PIN.

If a compound inequality involves a variable, then we are interested in the solution set to the inequality. The solution set to an "and" inequality consists of all numbers that satisfy both simple inequalities, whereas the solution set to an "or" inequality consists of all numbers that satisfy at least one of the simple inequalities.

EXAMPLE **3**   **Solutions of compound inequalities**
Determine whether 5 satisfies each compound inequality.

  a) $x < 6$ and $x < 9$ **b)** $2x - 9 \leq 5$ or $-4x \geq -12$

Solution

  a) Because $5 < 6$ and $5 < 9$ are both true, 5 satisfies the compound inequality.

  b) Because $2 \cdot 5 - 9 \leq 5$ is true, it does not matter that $-4 \cdot 5 \geq -12$ is false. So 5 satisfies the compound inequality.

  Now do Exercises 7–14

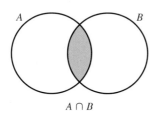

Figure 8.1

⟨2⟩ Graphing the Solution Set

If A and B are sets of numbers, then the **intersection** of A and B is the set of all numbers that are in both A and B. The intersection of A and B is denoted as $A \cap B$ (read "A intersect B"). This set is illustrated with a Venn diagram in Fig. 8.1. If $A = \{1, 2, 3\}$ and $B = \{2, 3, 4, 5\}$, then $A \cap B = \{2, 3\}$ because only 2 and 3 are in both A and B. See Appendix B for more details about sets.

The solution set to a compound inequality using the connective "and" is the intersection of the solution sets to each of the simple inequalities. Using graphs, as shown in Example 4, will help you understand compound inequalities.

EXAMPLE **4**   **Graphing compound inequalities**
Graph the solution set to the compound inequality $x > 2$ and $x < 5$.

Solution

First sketch the graph of $x > 2$ and then the graph of $x < 5$, as shown in Fig. 8.2. The intersection of these two solution sets is the portion of the number line that is shaded on both graphs, just the part between 2 and 5, not including the endpoints. In symbols, $(2, \infty) \cap (-\infty, 5) = (2, 5)$. So the solution set is the interval $(2, 5)$, and its graph is shown in Fig. 8.3. Recall from Section 2.8 that $x > 2$ and $x < 5$ is also written as $2 < x < 5$.

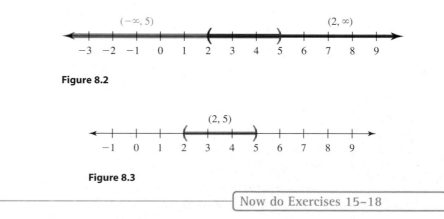

Figure 8.2

Figure 8.3

  Now do Exercises 15–18

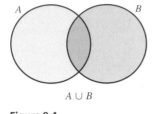

Figure 8.4

If A and B are sets of numbers, then the **union** of A and B is the set of all numbers that are in either A or B. The union of A and B is denoted as $A \cup B$ (read "A union B"). This set is illustrated with a Venn diagram in Fig. 8.4. If $A = \{1, 2, 3\}$ and $B = \{2, 3, 4, 5\}$, then $A \cup B = \{1, 2, 3, 4, 5\}$ because all of these numbers are in either A or B. Notice that the numbers in A and B are in $A \cup B$.

The solution set to a compound inequality using the connective "or" is the union of the solution sets to each of the simple inequalities.

E X A M P L E   5

Graphing compound inequalities

Graph the solution set to the compound inequality $x > 4$ or $x < -1$.

Solution

First graph the solution sets to the simple inequalities as shown in Fig. 8.5. The union of these two intervals is shown in Fig. 8.6. Since the union does not simplify to a single interval, the solution set is written using the symbol for union as $(-\infty, -1) \cup (4, \infty)$.

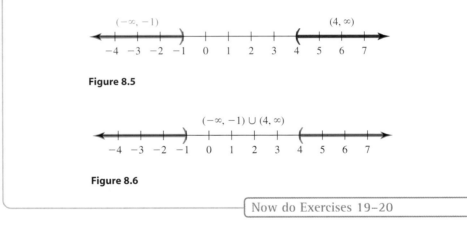

Figure 8.5

Figure 8.6

Now do Exercises 19–20

CAUTION   When graphing the intersection of two simple inequalities, do not draw too much. For the intersection, graph only numbers that satisfy *both* inequalities. Omit numbers that satisfy one but not the other inequality. Graphing a union is usually easier because we can simply draw both solution sets on the same number line.

It is not always necessary to graph the solution set to each simple inequality before graphing the solution set to the compound inequality. We can save time and work if we learn to think of the two preliminary graphs but draw only the final one.

E X A M P L E   6

Overlapping intervals

Sketch the graph and write the solution set in interval notation to each compound inequality.

   a) $x < 3$ and $x < 5$

   b) $x > 4$ or $x > 0$

Solution

a) Figure 8.7 shows $x < 3$ and $x < 5$ on the same number line. The intersection of these two intervals consists of the numbers that are less than 3. Numbers between 3 and 5 are not shaded twice and do not satisfy both inequalities. In symbols, $(-\infty, 3) \cap (-\infty, 5) = (-\infty, 3)$. So $x < 3$ and $x < 5$ is equivalent to $x < 3$. The solution set is $(-\infty, 3)$ and its graph is shown in Fig. 8.8.

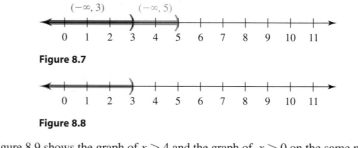

Figure 8.7

Figure 8.8

b) Figure 8.9 shows the graph of $x > 4$ and the graph of $x > 0$ on the same number line. The union of these two intervals consists of everything that is shaded in Fig. 8.9. In symbols, $(4, \infty) \cup (0, \infty) = (0, \infty)$. So $x > 4$ or $x > 0$ is equivalent to $x > 0$. The solution set to the compound inequality is $(0, \infty)$, and its graph is shown in Fig. 8.10.

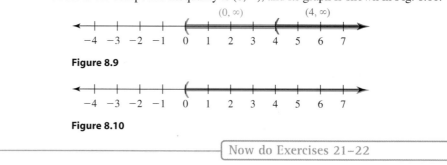

Figure 8.9

Figure 8.10

> Now do Exercises 21–22

Example 7 shows a compound inequality that has no solution and one that is satisfied by every real number.

E X A M P L E 7

All or nothing

Sketch the graph and write the solution set in interval notation to each compound inequality.

a) $x < 2$ and $x > 6$ **b)** $x < 3$ or $x > 1$

Solution

a) A number satisfies $x < 2$ and $x > 6$ if it is both less than 2 *and* greater than 6. There are no such numbers. The solution set is the empty set, $\varnothing$. In symbols, $(-\infty, 2) \cap (6, \infty) = \varnothing$.

b) To graph $x < 3$ or $x > 1$, we shade both regions on the same number line as shown in Fig. 8.11. Since the two regions cover the entire line, the solution set is the set of all real numbers $(-\infty, \infty)$. In symbols, $(-\infty, 3) \cup (1, \infty) = (-\infty, \infty)$.

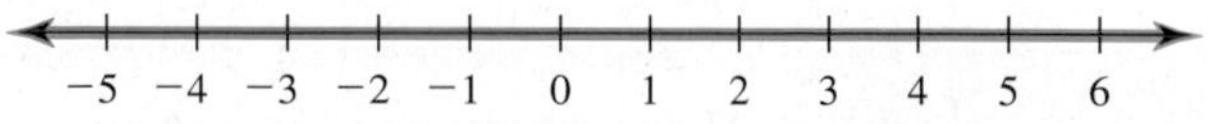

Figure 8.11

> Now do Exercises 23–28

If we start with a more complicated compound inequality, we first simplify each part of the compound inequality and then find the union or intersection.

EXAMPLE 8

‹ **Calculator Close-Up** ›

To check Example 8, press Y= and let $y_1 = x + 2$ and $y_2 = x - 6$. Now scroll through a table of values for y_1 and y_2. From the table you can see that y_1 is greater than 3 and y_2 is less than 7 precisely when x is between 1 and 13.

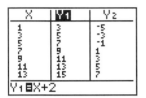

Intersection

Solve $x + 2 > 3$ and $x - 6 < 7$. Graph the solution set.

Solution

First simplify each simple inequality:

$$x + 2 - 2 > 3 - 2 \quad \text{and} \quad x - 6 + 6 < 7 + 6$$
$$x > 1 \quad \text{and} \quad x < 13$$

The intersection of these two solution sets is the set of numbers between (but not including) 1 and 13. Its graph is shown in Fig. 8.12. The solution set is written in interval notation as (1, 13). Recall from Section 2.8 that $x > 1$ and $x < 13$ is also written as $1 < x < 13$.

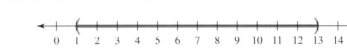

Figure 8.12

> Now do Exercises 29–32

EXAMPLE 9

‹ **Calculator Close-Up** ›

To check Example 9, press Y= and let $y_1 = 5 - 7x$ and $y_2 = 3x - 2$. Now scroll through a table of values for y_1 and y_2. From the table you can see that either $y_1 \geq 12$ or $y_2 < 7$ is true for $x < 3$. Note also that for $x \geq 3$ both $y_1 \geq 12$ and $y_2 < 7$ are incorrect. The table supports the conclusion of Example 9.

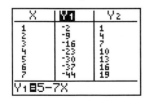

Union

Graph the solution set to the inequality

$$5 - 7x \geq 12 \quad \text{or} \quad 3x - 2 < 7.$$

Solution

First solve each of the simple inequalities:

$$5 - 7x - 5 \geq 12 - 5 \quad \text{or} \quad 3x - 2 + 2 < 7 + 2$$
$$-7x \geq 7 \quad \text{or} \quad 3x < 9$$
$$x \leq -1 \quad \text{or} \quad x < 3$$

The union of the two solution intervals is $(-\infty, 3)$. The graph is shown in Fig. 8.13.

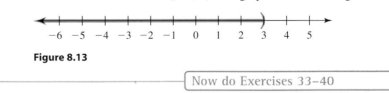

Figure 8.13

> Now do Exercises 33–40

If x is between a and b and $a < b$, then we can use the "between" notation, $a < x < b$, rather than writing $x > a$ and $x < b$. We solved compound inequalities of this type in Section 2.9. For completeness, we review that method in Examples 10 and 11.

EXAMPLE 10

Using "between" notation

Solve the inequality and graph the solution set:

$$-2 \leq 2x - 3 < 7$$

‹ **Calculator Close-Up** ›

Do not use a table on your calculator as a method of solving an inequality. Use a table to check your algebraic solution and you will get a better understanding of inequalities.

Solution

This inequality could be written as the compound inequality

$$2x - 3 \geq -2 \qquad \text{and} \qquad 2x - 3 < 7.$$

However, there is no need to rewrite the inequality because we can solve it in its original form.

$$-2 + 3 \leq 2x - 3 + 3 < 7 + 3 \qquad \text{Add 3 to each part.}$$

$$1 \leq 2x < 10$$

$$\frac{1}{2} \leq \frac{2x}{2} < \frac{10}{2} \qquad\qquad \text{Divide each part by 2.}$$

$$\frac{1}{2} \leq x < 5$$

The solution set is $\left[\frac{1}{2}, 5\right)$, and its graph is shown in Fig. 8.14.

Figure 8.14

Now do Exercises 41–44

E X A M P L E   **11**

Solving a compound inequality

Solve the inequality $-1 < 3 - 2x < 9$ and graph the solution set.

Solution

$$-1 - 3 < 3 - 2x - 3 < 9 - 3 \qquad \text{Subtract 3 from each part of the inequality.}$$

$$-4 < -2x < 6$$

$$2 > x > -3 \qquad\qquad \text{Divide each part by } -2 \text{ and reverse both inequality symbols.}$$

$$-3 < x < 2 \qquad\qquad \text{Rewrite the inequality with the smallest number on the left.}$$

The solution set is $(-3, 2)$, and its graph is shown in Fig. 8.15.

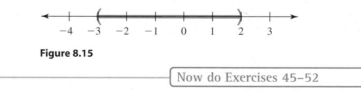

Figure 8.15

Now do Exercises 45–52

‹ **Calculator Close-Up** ›

Let $y_1 = 3 - 2x$ and make a table. Scroll through the table to see that y_1 is between -1 and 9 when x is between -3 and 2. The table supports the conclusion of Example 11.

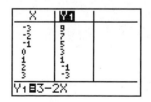

‹3› Applications

When final exams are approaching, students are often interested in finding the final exam score that would give them a certain grade for a course.

E X A M P L E   **12**

Final exam scores

Fiana made a score of 76 on her midterm exam. For her to get a B in the course, the average of her midterm exam and final exam must be between 80 and 89 inclusive. What possible scores on the final exam would give Fiana a B in the course?

⟨ **Helpful Hint** ⟩

When you use two inequality symbols as in Example 12, they must both point in the same direction. In fact, we usually have them both point to the left so that the numbers increase in size from left to right.

Solution

Let x represent her final exam score. Between 80 and 89 inclusive means that an average between 80 and 89 as well as an average of exactly 80 or 89 will get a B. So the average of the two scores must be greater than or equal to 80 and less than or equal to 89.

$$80 \le \frac{x + 76}{2} \le 89$$

$$160 \le x + 76 \le 178 \quad \text{Multiply by 2.}$$

$$160 - 76 \le x \le 178 - 76 \quad \text{Subtract 76.}$$

$$84 \le x \le 102$$

If Fiana scores between 84 and 102 inclusive, she will get a B in the course.

Now do Exercises 77–78

Warm-Ups ▼

Fill in the blank.

1. A _____ inequality consists of two simple inequalities connected by "and" or "or."
2. The compound inequality $x > 5$ ___ $x > 8$ is true only if both simple inequalities are true.
3. The compound inequality $x > 5$ ___ $x > 8$ is true if either simple inequality is true.
4. If $a < b < c$, then $a < b$ ___ $b < c$.
5. The solution set to $x > 3$ and $x < 9$ is the _____ of the intervals $(3, \infty)$ and $(-\infty, 9)$.
6. The solution set to $x > 3$ or $x < 9$ is the _____ of the intervals $(3, \infty)$ and $(-\infty, 9)$.

True or false?

7. $3 < 5$ and $3 \le 10$
8. $3 < 5$ or $3 < 10$
9. $3 > 5$ and $3 < 10$
10. $3 \ge 5$ or $3 \le 10$
11. $4 < 8$ and $4 > 2$
12. $4 < 8$ or $4 > 2$
13. $-3 < 0 < -2$
14. $(3, \infty) \cup [8, \infty) = [8, \infty)$
15. $(3, \infty) \cap [8, \infty) = [8, \infty)$
16. $(-2, \infty) \cap (-\infty, 9) = (-2, 9)$

8.1 Exercises

⟨ **Study Tips** ⟩

- When studying for a midterm or final, review the material in the order it was originally presented. This strategy will help you to see connections between the ideas.
- Studying the oldest material first will give top priority to material that you might have forgotten.

⟨ **1** ⟩ **Compound Inequalities**

Determine whether each compound inequality is true. See Examples 1 and 2.

1. $-6 < 5$ and $-6 > -3$
2. $4 \le 4$ and $-4 \le 0$
3. $1 < 5$ and $1 > -3$
4. $3 < 5$ or $0 < -3$
5. $6 < 5$ or $-4 > -3$
6. $4 \le -4$ or $0 \le 0$

Determine whether -4 *satisfies each compound inequality.*
See Example 3.

   7. $x < 5$ and $x > -3$

   8. $x > -5$ and $x < 0$

   9. $x < 5$ or $x > -3$

  10. $x < -9$ or $x > 0$

  11. $x - 3 \geq -7$ or $x + 1 > 1$

  12. $2x \leq -8$ and $5x \leq 0$

  13. $2x - 1 < -7$ or $-2x > 18$

  14. $-3x > 0$ and $3x - 4 < 11$

⟨2⟩ **Graphing the Solution Set**

Graph the solution set to each compound inequality.
See Examples 4–7.

  15. $x > -1$ and $x < 4$

  16. $x \leq 5$ and $x \geq 4$

  17. $x \leq 3$ and $x \leq 0$

  18. $x > 2$ and $x > 0$

  19. $x \geq 2$ or $x \geq 5$

  20. $x < -1$ or $x < 3$

  21. $x \leq 6$ or $x > -2$

  22. $x > -2$ and $x \leq 4$

  23. $x \leq 6$ and $x > 9$

  24. $x < 7$ or $x > 0$

  25. $x \leq 6$ or $x > 9$

  26. $x \geq 4$ and $x \leq -4$

  27. $x \geq 6$ and $x \leq 1$

  28. $x > 3$ or $x < -3$

Solve each compound inequality. Write the solution set using
interval notation and graph it. See Examples 8 and 9.

  29. $x - 3 > 7$ or $3 - x > 2$

  30. $x - 5 > 6$ or $2 - x > 4$

  31. $3 < x$ and $1 + x > 10$

  32. $-0.3x < 9$ and $0.2x > 2$

  33. $\dfrac{1}{2}x > 5$ or $-\dfrac{1}{3}x < 2$

  34. $5 < x$ or $3 - \dfrac{1}{2}x < 7$

  35. $2x - 3 \leq 5$ and $x - 1 > 0$

  36. $\dfrac{3}{4}x < 9$ and $-\dfrac{1}{3}x \leq -15$

  37. $\dfrac{1}{2}x - \dfrac{1}{3} \geq -\dfrac{1}{6}$ or $\dfrac{2}{7}x \leq \dfrac{1}{10}$

  38. $\dfrac{1}{4}x - \dfrac{1}{3} > -\dfrac{1}{5}$ and $\dfrac{1}{2}x < 2$

  39. $0.5x < 2$ and $-0.6x < -3$

  40. $0.3x < 0.6$ or $0.05x > -4$

Solve each compound inequality. Write the solution set in
interval notation and graph it. See Examples 10 and 11.

  41. $-3 < x + 1 < 3$

  42. $-4 \leq x - 4 \leq 1$

  43. $5 < 2x - 3 < 11$

  44. $-2 < 3x + 1 < 10$

45. $-1 < 5 - 3x \le 14$

46. $-1 \le 3 - 2x < 11$

47. $-3 < \dfrac{3m + 1}{2} \le 5$

48. $0 \le \dfrac{3 - 2x}{2} < 5$

49. $-2 < \dfrac{1 - 3x}{-2} < 7$

50. $-3 < \dfrac{2x - 1}{3} < 7$

51. $3 \le 3 - 5(x - 3) \le 8$

52. $2 \le 4 - \dfrac{1}{2}(x - 8) \le 10$

Write each union or intersection of intervals as a single interval if possible.

53. $(2, \infty) \cup (4, \infty)$

54. $(-3, \infty) \cup (-6, \infty)$

55. $(-\infty, 5) \cap (-\infty, 9)$

56. $(-\infty, -2) \cap (-\infty, 1)$

57. $(-\infty, 4] \cap [2, \infty)$

58. $(-\infty, 8) \cap [3, \infty)$

59. $(-\infty, 5) \cup [-3, \infty)$

60. $(-\infty, -2] \cup (2, \infty)$

61. $(3, \infty) \cap (-\infty, 3]$

62. $[-4, \infty) \cap (-\infty, -6]$

63. $(3, 5) \cap [4, 8)$

64. $[-2, 4] \cap (0, 9]$

65. $[1, 4) \cup (2, 6]$

66. $[1, 3) \cup (0, 5)$

Write either a simple or a compound inequality that has the given graph as its solution set.

67.

68.

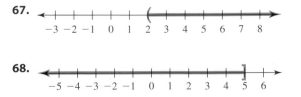

69.

70.

71.

72.

73.

74.

75.

76.

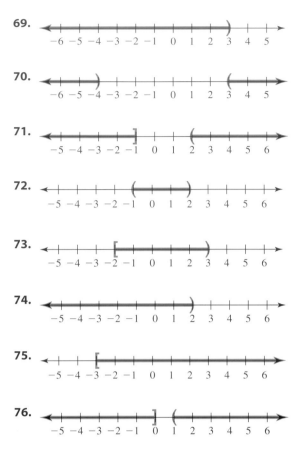

⟨3⟩ **Applications**

Solve each problem by using a compound inequality. See Example 12.

77. *Aiming for a C.* Professor Johnson gives only a midterm exam and a final exam. The semester average is computed by taking $\frac{1}{3}$ of the midterm exam score plus $\frac{2}{3}$ of the final exam score. To get a C, Beth must have a semester average between 70 and 79 inclusive. If Beth scored only 64 on the midterm, then for what range of scores on the final exam would Beth get a C?

78. *Two tests only.* Professor Davis counts his midterm as $\frac{2}{3}$ of the grade, and his final as $\frac{1}{3}$ of the grade. Jason scored only 64 on the midterm. What range of scores on the final exam would put Jason's average between 70 and 79 inclusive?

79. *Car costs.* A company uses the expression $0.0004x + 20$ to estimate the cost in cents per mile for operating a company car and the expression $20,000 - 0.2x$ to estimate the value of the car in dollars, where x is the number of miles on the odometer. If the company plans to replace any car for which the operating cost is greater than 40 cents per mile *and* the value is less than $12,000, then for what values of x is a car replaced? Use interval notation.

80. *Changing plans.* The company in Exercise 79 has changed its policy and has decided to replace any car for which the operating cost is greater than 40 cents per mile *or* the value is less than $12,000. For what values of x is a car replaced? Use interval notation.

81. *Supply and demand.* An energy minister in a small country uses the expression $20 + 0.1x$ to estimate the amount of oil in millions of barrels per day that will be supplied to his country and the expression $30 - 0.5x$ to estimate the demand for oil in millions of barrels per day, where x is the price of oil in dollars per barrel. The government must get involved if the supply is less than 22 million barrels per day *or* if the demand is less than 15 million barrels per day. For what values of x must the government get involved? Use interval notation.

82. *Predicting recession.* The country of Exercise 81 will be in recession if the supply of oil is greater than 23 million barrels per day *and* the demand is less than 14 million barrels per day. For what values of x will the country be in recession? Use interval notation.

83. *Keep on truckin'.* Abdul is shopping for a new truck in a city with an 8% sales tax. There is also an $84 title and license fee to pay. He wants to get a good truck and plans to spend at least $12,000 but no more than $15,000. What is the price range for the truck?

84. *Selling-price range.* Renee wants to sell her car through a broker who charges a commission of 10% of the selling price. The book value of the car is $14,900, but Renee still owes $13,104 on it. Although the car is in only fair condition and will not sell for more than the book value, Renee must get enough to at least pay off the loan. What is the range of the selling price?

85. *Hazardous to her health.* Trying to break her smoking habit, Jane calculates that she smokes only three full cigarettes a day, one after each meal. The rest of the time she smokes on the run and smokes only half of the cigarette. She estimates that she smokes the equivalent of 5 to 12 cigarettes per day. How many times a day does she light up on the run?

86. *Possible width.* The length of a rectangle is 20 meters longer than the width. The perimeter must be between 80 and 100 meters. What are the possible values for the width of the rectangle?

87. *Higher education.* The formulas

$$B = 16.45n + 1062.45$$

and $$M = 7.79n + 326.82$$

can be used to approximate the number of bachelor's and master's degrees in thousands, respectively, awarded per year, n years after 1990 (National Center for Educational Statistics, www.nces.ed.gov).

a) How many bachelor's degrees were awarded in 2000?

b) In what year will the number of bachelor's degrees that are awarded reach 1.4 million?

c) What is the first year in which both B is greater than 1.4 million and M is greater than 0.55 million?

d) What is the first year in which either B is greater than 1.4 million or M is greater than 0.55 million?

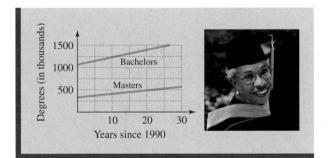

Figure for Exercise 87

88. *Senior citizens.* The number of senior citizens (65 and over) in the United States in millions n years after 1990 can be estimated by using the formula

$$s = 0.38n + 31.2$$

(U.S. Bureau of the Census, www.census.gov). See the figure on the next page. The percentage of senior citizens living below the poverty level n years after 1990 can be estimated by using the formula

$$p = -0.25n + 12.2.$$

a) How many senior citizens were there in 2000?

b) In what year will the percentage of seniors living below the poverty level reach 7%?

c) What is the first year in which we can expect both the number of seniors to be greater than 40 million and fewer than 7% living below the poverty level?

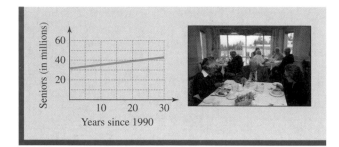

Figure for Exercise 88

Getting More Involved

89. Discussion

If $-x$ is between a and b, then what can you say about x?

90. Discussion

For which of the inequalities is the notation used correctly?

a) $-2 \le x < 3$ b) $-4 \ge x < 7$ c) $-1 \le x > 0$
d) $6 < x \le -8$ e) $5 \ge x \ge -9$

91. Discussion

In each case, write the resulting set of numbers in interval notation. Explain your answers.

a) Every number in (3, 8) is multiplied by 4.
b) Every number in $[-2, 4)$ is multiplied by -5.
c) Three is added to every number in $(-3, 6)$.
d) Every number in [3, 9] is divided by -3.

92. Discussion

Write the solution set using interval notation for each of these inequalities in terms of s and t. State any restrictions on s and t. For what values of s and t is the solution set empty?

a) $x > s$ and $x < t$
b) $x > s$ and $x > t$

Math *at Work* | Pediatric Dosing Rules

A drug is generally tested on adults, and an appropriate adult dose (*AD*) of the drug is determined. When a drug is given to a child, a doctor determines an appropriate child's dose (*CD*) using pediatric dosing rules. However, no single rule works for all children. Determining a child's dose also involves common sense and experience.

Clark's rule is based on the ratio of the child's body weight to the mean weight of an adult, 150 pounds. By Clark's rule, $CD = \frac{\text{child's weight in lbs}}{150 \text{ lbs}} \cdot AD$. A dose determined by body weight alone might be too little to be effective in a small child.

Young's rule is based on the assumption that age approximates body weight for patients over 2 years old. Of course there is a great variability of body weight of children of any given age. By Young's rule, $CD = \frac{\text{age in years}}{\text{age in years} + 12} \cdot AD$.

The area rule is often used for drugs required in radioactive imaging. It is based on the idea that (body mass)$^{2/3}$ is approximately the body surface area. For radioactive imaging the adult's body mass (*MA*) often determines *AD*. By the area rule, $CD = \frac{(MC)^{2/3}}{(MA)^{2/3}} \cdot AD$, where *MC* is the child's body mass.

Webster's rule uses age to approximate the ratio in the area rule and agrees well with the area rule until age 11 or 12. By Webster's rule $CD = \frac{\text{age} + 1}{\text{age} + 7} \cdot AD$.

Fried's rule is generally used for patients less than one year old. By Fried's rule $CD = \frac{\text{age in months}}{150} \cdot AD$.

8.2   Absolute Value Equations and Inequalities

In This Section

⟨1⟩ **Absolute Value Equations**
⟨2⟩ **Absolute Value Inequalities**
⟨3⟩ **All or Nothing**
⟨4⟩ **Applications**

In Chapter 1 we learned that absolute value measures the distance of a number from 0 on the number line. In this section we will learn to solve equations and inequalities involving absolute value.

⟨1⟩ Absolute Value Equations

Solving equations involving absolute value requires some techniques that are different from those studied in previous sections. For example, the solution set to the equation

$$|x| = 5$$

is $\{-5, 5\}$ because both 5 and -5 are five units from 0 on the number line, as shown in Fig. 8.16. So $|x| = 5$ is equivalent to the compound equation

$$x = 5 \text{ or } x = -5.$$

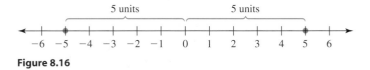

Figure 8.16

⟨ **Helpful Hint** ⟩

Some students grow up believing that the only way to solve an equation is to "do the same thing to each side." Then along come absolute value equations. For an absolute value equation we write an equivalent compound equation that is not obtained by "doing the same thing to each side."

The equation $|x| = 0$ is equivalent to the equation $x = 0$ because 0 is the only number whose distance from 0 is zero. The solution set to $|x| = 0$ is $\{0\}$.

The equation $|x| = -7$ is inconsistent because absolute value measures distance, and distance is never negative. So the solution set is empty. These ideas are summarized as follows.

Summary of Basic Absolute Value Equations

Absolute Value Equation	Equivalent Equation	Solution Set
$\lvert x \rvert = k \ (k > 0)$	$x = k \text{ or } x = -k$	$\{k, -k\}$
$\lvert x \rvert = 0$	$x = 0$	$\{0\}$
$\lvert x \rvert = k \ (k < 0)$		$\varnothing$

We can use these ideas to solve more complicated absolute value equations.

E X A M P L E   1

Absolute value equal to a positive number
Solve each equation.

    a) $|x - 7| = 2$ **b)** $|3x - 5| = 7$

‹ **Calculator Close-Up** ›

Use Y= to set y_1 = abs(x − 7). Make a table to see that y_1 has value 2 when $x = 5$ or $x = 9$. The table supports the conclusion of Example 1(a).

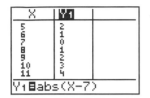

Solution

a) First rewrite $|x - 7| = 2$ without absolute value:

$$x - 7 = 2 \quad \text{or} \quad x - 7 = -2 \quad \text{Equivalent equation}$$
$$x = 9 \quad \text{or} \quad x = 5$$

The solution set is $\{5, 9\}$. The distance from 5 to 7 or from 9 to 7 is 2 units.

b) First rewrite $|3x - 5| = 7$ without absolute value:

$$3x - 5 = 7 \quad \text{or} \quad 3x - 5 = -7 \quad \text{Equivalent equation}$$
$$3x = 12 \quad \text{or} \quad 3x = -2$$
$$x = 4 \quad \text{or} \quad x = -\frac{2}{3}$$

The solution set is $\left\{-\frac{2}{3}, 4\right\}$.

> Now do Exercises 1–6

E X A M P L E   **2**

Absolute value equal to zero

Solve $|2(x - 6) + 7| = 0$.

‹ **Helpful Hint** ›

Examples 1, 2, and 3 show the three basic types of absolute value equations—absolute value equal to a positive number, zero, or a negative number. These equations have 2, 1, and no solutions, respectively.

Solution

Since 0 is the only number whose absolute value is 0, the expression within the absolute value bars must be 0.

$$2(x - 6) + 7 = 0 \quad \text{Equivalent equation}$$
$$2x - 12 + 7 = 0$$
$$2x - 5 = 0$$
$$2x = 5$$
$$x = \frac{5}{2}$$

The solution set is $\left\{\frac{5}{2}\right\}$.

> Now do Exercises 7–12

E X A M P L E   **3**

Absolute value equal to a negative number

Solve each equation.

a) $|x - 9| = -6$

b) $-5|3x - 7| + 4 = 14$

Solution

a) The equation indicates that $|x - 9| = -6$. However, the absolute value of any quantity is greater than or equal to zero. So there is no solution to the equation.

b) First subtract 4 from each side to isolate the absolute value expression:

$$-5|3x - 7| + 4 = 14 \quad \text{Original equation}$$
$$-5|3x - 7| = 10 \quad \text{Subtract 4 from each side.}$$
$$|3x - 7| = -2 \quad \text{Divide each side by } -5.$$

There is no solution because no quantity has a negative absolute value.

> Now do Exercises 13–24

The equation in Example 4 has an absolute value on both sides.

EXAMPLE **4**

Absolute value on both sides
Solve $|2x - 1| = |x + 3|$.

Solution

Two quantities have the same absolute value only if they are equal or opposites. So we can write an equivalent compound equation:

$$2x - 1 = x + 3 \quad \text{or} \quad 2x - 1 = -(x + 3)$$
$$x - 1 = 3 \quad \text{or} \quad 2x - 1 = -x - 3$$
$$x = 4 \quad \text{or} \quad 3x = -2$$
$$x = 4 \quad \text{or} \quad x = -\frac{2}{3}$$

Check 4 and $-\frac{2}{3}$ in the original equation. The solution set is $\left\{-\frac{2}{3}, 4\right\}$.

Now do Exercises 25–30

⟨2⟩ Absolute Value Inequalities

Since absolute value measures distance from 0 on the number line, $|x| > 5$ indicates that x is more than five units from 0. Any number on the number line to the right of 5 or to the left of -5 is more than five units from 0. So $|x| > 5$ is equivalent to

$$x > 5 \quad \text{or} \quad x < -5.$$

The solution set to this inequality is the union of the solution sets to the two simple inequalities. The solution set is $(-\infty, -5) \cup (5, \infty)$. The graph of $|x| > 5$ is shown in Fig. 8.17.

Figure 8.17

The inequality $|x| \leq 3$ indicates that x is less than or equal to three units from 0. Any number between -3 and 3 inclusive satisfies that condition. So $|x| \leq 3$ is equivalent to

$$-3 \leq x \leq 3.$$

The graph of $|x| \leq 3$ is shown in Fig. 8.18. These examples illustrate the basic types of absolute value inequalities.

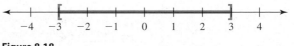

Figure 8.18

Summary of Basic Absolute Value Inequalities ($k > 0$)			
Absolute Value Inequality	**Equivalent Inequality**	**Solution Set**	**Graph of Solution Set**
$\lvert x \rvert > k$	$x > k$ or $x < -k$	$(-\infty, -k) \cup (k, \infty)$	
$\lvert x \rvert \ge k$	$x \ge k$ or $x \le -k$	$(-\infty, -k] \cup [k, \infty)$	
$\lvert x \rvert < k$	$-k < x < k$	$(-k, k)$	
$\lvert x \rvert \le k$	$-k \le x \le k$	$[-k, k]$	

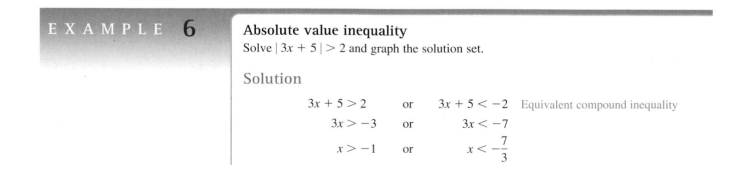

We can solve more complicated inequalities in the same manner as simple ones.

E X A M P L E   5

Absolute value inequality

Solve $\lvert x - 9 \rvert < 2$ and graph the solution set.

Solution

Because $\lvert x \rvert < k$ is equivalent to $-k < x < k$, we can rewrite $\lvert x - 9 \rvert < 2$ as follows:

$$-2 < x - 9 < 2$$
$$-2 + 9 < x - 9 + 9 < 2 + 9 \quad \text{Add 9 to each part of the inequality.}$$
$$7 < x < 11$$

The graph of the solution set $(7, 11)$ is shown in Fig. 8.19. Note that the graph consists of all real numbers that are within two units of 9.

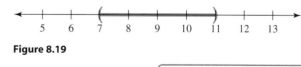

Figure 8.19

> Now do Exercises 31–32

‹ **Calculator Close-Up** ›

Use Y= to set $y_1 = \text{abs}(x - 9)$. Make a table to see that $y_1 < 2$ when x is between 7 and 11.

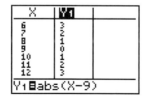

E X A M P L E   6

Absolute value inequality

Solve $\lvert 3x + 5 \rvert > 2$ and graph the solution set.

Solution

$$3x + 5 > 2 \qquad \text{or} \qquad 3x + 5 < -2 \quad \text{Equivalent compound inequality}$$
$$3x > -3 \qquad \text{or} \qquad 3x < -7$$
$$x > -1 \qquad \text{or} \qquad x < -\frac{7}{3}$$

The solution set is $\left(-\infty, -\frac{7}{3}\right) \cup (-1, \infty)$, and its graph is shown in Fig. 8.20.

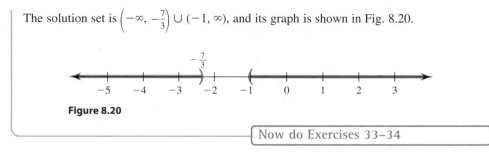

Figure 8.20

Now do Exercises 33–34

E X A M P L E **7**

Absolute value inequality

Solve $|5 - 3x| \le 6$ and graph the solution set.

Solution

$$-6 \le 5 - 3x \le 6 \quad \text{Equivalent inequality}$$

$$-11 \le -3x \le 1 \quad \text{Subtract 5 from each part.}$$

$$\frac{11}{3} \ge x \ge -\frac{1}{3} \quad \text{Divide by } -3 \text{ and reverse each inequality symbol.}$$

$$-\frac{1}{3} \le x \le \frac{11}{3} \quad \text{Write } -\frac{1}{3} \text{ on the left because it is smaller than } \frac{11}{3}.$$

The solution set is $\left[-\frac{1}{3}, \frac{11}{3}\right]$ and its graph is shown in Fig. 8.21.

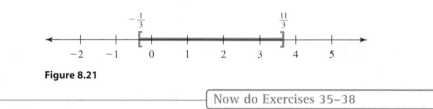

Figure 8.21

Now do Exercises 35–38

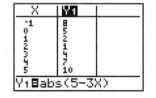
⟨**3**⟩ **All or Nothing**

The solution to an absolute value inequality can be all real numbers or no real numbers. To solve such inequalities you must remember that the absolute value of any real number is greater than or equal to zero.

E X A M P L E **8**

All real numbers

Solve $3 + |7 - 2x| \ge 3$.

Solution

Subtract 3 from each side to isolate the absolute value expression.

$$|7 - 2x| \ge 0$$

Because the absolute value of any real number is greater than or equal to 0, the solution set is R, the set of all real numbers.

Now do Exercises 59–64

EXAMPLE 9

No real numbers

Solve $|5x - 12| < -2$.

Solution

We write an equivalent inequality only when the value of k is positive. With -2 on the right-hand side, we do not write an equivalent inequality. Since the absolute value of any quantity is greater than or equal to 0, no value for x can make this absolute value less than -2. The solution set is $\varnothing$, the empty set.

Now do Exercises 65–68

⟨4⟩ Applications

A simple example will show how absolute value inequalities can be used in applications.

EXAMPLE 10

Controlling water temperature

The water temperature in a certain manufacturing process must be kept at 143°F. The computer is programmed to shut down the process if the water temperature is more than 7° away from what it is supposed to be. For what temperature readings is the process shut down?

Solution

If we let x represent the water temperature, then $x - 143$ represents the difference between the actual temperature and the desired temperature. The quantity $x - 143$ could be positive or negative. The process is shut down if the absolute value of $x - 143$ is greater than 7.

$$|x - 143| > 7$$

$$x - 143 > 7 \qquad \text{or} \qquad x - 143 < -7$$

$$x > 150 \qquad \text{or} \qquad x < 136$$

The process is shut down for temperatures greater than 150°F or less than 136°F.

Now do Exercises 77–84

Warm-Ups ▼

Fill in the blank.

1. The _____ of x is the distance from x to 0 on the number line.
2. The equation $|x| = 4$ has ___ solutions.
3. The equation $|x| = -4$ has ___ solutions.
4. The equation $|x| = 0$ has ___ solution.
5. ___ real numbers satisfy $|x| \geq 0$.
6. ___ real numbers satisfy $|x| < 0$.
7. The solution set to $|x| < 3$ is _____.
8. The inequality $|x| > 3$ is _____ to $x > 3$ or $x < -3$.

True or false?

9. If $|x| = 2$, then $x = 2$ or $x = -2$.
10. If $|x - 1| = 7$, then $x - 1 = 7$ or $x + 1 = 7$.
11. If $|x| > 5$, then $x > 5$ or $x < -5$.
12. If $|x| = -4$, then $x = -4$.
13. If $|2x - 8| = 0$, then $x = 4$.
14. If $-3 < x < 3$, then $|x| < 3$.
15. If $5 < x < 9$, then $x > 5$ and $x < 9$.
16. If x is any real number, then $|x| \geq 0$.
17. If $|x| + 1 = 5$, then $x + 1 = 5$ or $x + 1 = -5$.
18. If $|3x - 99| \leq 0$, then $x = 33$.

‹1› Absolute Value Equations

Solve each absolute value equation. See Examples 1–3 and the Summary of Basic Absolute Value Equations on page 519.

1. $|a| = 5$ **2.** $|x| = 2$ **3.** $|x - 3| = 1$

4. $|x - 5| = 2$ **5.** $|3 - x| = 6$ **6.** $|7 - x| = 6$

7. $|3x - 4| = 12$ **8.** $|5x + 2| = -3$

9. $\left|\dfrac{2}{3}x - 8\right| = 0$ **10.** $\left|3 - \dfrac{3}{4}x\right| = \dfrac{1}{4}$

11. $|6 - 0.2x| = 10$

12. $|5 - 0.1x| = 0$

13. $|7(x - 6)| = -3$

14. $|2(a + 3)| = 15$

15. $|2(x - 4) + 3| = 5$

16. $|3(x - 2) + 7| = 6$

17. $|7.3x - 5.26| = 4.215$

18. $|5.74 - 2.17x| = 10.28$

Solve each absolute value equation. See Examples 3 and 4.

19. $3 + |x| = 5$

20. $|x| - 10 = -3$

21. $2 - |x + 3| = -6$

22. $4 - 3|x - 2| = -8$

23. $5 - \dfrac{|3 - 2x|}{3} = 4$

24. $3 - \dfrac{1}{2}\left|\dfrac{1}{2}x - 4\right| = 2$

25. $|x - 5| = |2x + 1|$

26. $|w - 6| = |3 - 2w|$

27. $\left|\dfrac{5}{2} - x\right| = \left|2 - \dfrac{x}{2}\right|$

28. $\left|x - \dfrac{1}{4}\right| = \left|\dfrac{1}{2}x - \dfrac{3}{4}\right|$

29. $|x - 3| = |3 - x|$

30. $|a - 6| = |6 - a|$

‹2› Absolute Value Inequalities

Write an absolute value inequality whose solution set is shown by the graph. See Examples 5–7 and the Summary of Basic Absolute Value Inequalities on page 522.

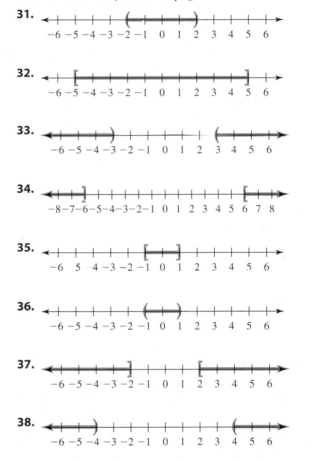

31.

32.

33.

34.

35.

36.

37.

38.

Determine whether each absolute value inequality is equivalent to the inequality following it. See Examples 5–7.

39. $|x| < 3, \; x < 3$

40. $|x| > 3, \; x > 3$

41. $|x - 3| > 1, \; x - 3 > 1 \text{ or } x - 3 < -1$

42. $|x - 3| \le 1, \; -1 \le x - 3 \le 1$

43. $|x - 3| \ge 1, \; x - 3 \ge 1 \text{ or } x - 3 \le 1$

44. $|x - 3| > 0, \; x - 3 > 0$

Solve each absolute value inequality and graph the solution set. See Examples 5–7.

45. $|x| > 6$

46. $|w| \geq 3$

47. $|t| \leq 2$

48. $|b| < 4$

49. $|2a| < 6$

50. $|3x| < 21$

51. $|x - 2| \geq 3$

52. $|x - 5| \geq 1$

53. $\dfrac{1}{5}|2x - 4| < 1$

54. $\dfrac{1}{3}|2x - 1| < 1$

55. $-2|5 - x| \geq -14$

56. $-3|6 - x| \geq -3$

57. $2|3 - 2x| - 6 \geq 18$

58. $2|5 - 2x| - 15 \geq 5$

⟨3⟩ All or Nothing

Solve each absolute value inequality and graph the solution set. See Examples 8 and 9.

59. $|x| > 0$

60. $|x - 2| > 0$

61. $|x| \leq 0$

62. $|x| < 0$

63. $|x - 5| \geq 0$

64. $|3x - 7| \geq -3$

65. $-2|3x - 7| > 6$

66. $-3|7x - 42| > 18$

67. $|2x + 3| + 6 > 0$

68. $|5 - x| + 5 > 5$

Solve each inequality. Write the solution set using interval notation.

69. $1 < |x + 2|$

70. $5 \geq |x - 4|$

71. $5 > |x| + 1$

72. $4 \leq |x| - 6$

73. $3 - 5|x| > -2$

74. $1 - 2|x| < -7$

75. $|5.67x - 3.124| < 1.68$

76. $|4.67 - 3.2x| \geq 1.43$

⟨4⟩ Applications

Solve each problem by using an absolute value equation or inequality. See Example 10.

77. *Famous battles.* In the Hundred Years' War, Henry V defeated a French army in the battle of Agincourt and Joan of Arc defeated an English army in the battle of Orleans (*The Doubleday Almanac*). Suppose you know only that these two famous battles were 14 years apart and that the battle of Agincourt occurred in 1415. Use an absolute value equation to find the possibilities for the year in which the battle of Orleans occurred.

78. *World records.* In July 1985 Steve Cram of Great Britain set a world record of 3 minutes 29.67 seconds for the 1500-meter race and a world record of 3 minutes 46.31 seconds for the 1-mile race (*The Doubleday Almanac*). Suppose you know only that these two events occurred 11 days apart and that the 1500-meter record was set on July 16. Use an absolute value equation to find the possible dates for the 1-mile record run.

79. *Weight difference.* Research at a major university has shown that identical twins generally differ by less than 6 pounds in body weight. If Kim weighs 127 pounds, then in what range is the weight of her identical twin sister Kathy?

80. **Intelligence quotient.** Jude's IQ score is more than 15 points away from Sherry's. If Sherry scored 110, then in what range is Jude's score?

81. **Approval rating.** According to a Fox News survey, the presidential approval rating is 39% plus or minus 5 percentage points.

   a) In what range is the percentage of people who approve of the president?

   b) Let x represent the actual percentage of people who approve of the president. Write an absolute value inequality for x.

82. **Time of death.** According to the coroner the time of death was 3 A.M. plus or minus 2 hours.

   a) In what range is the actual time of death?

   b) Let x represent the actual time of death. Write an absolute value inequality for x.

83. **Unidentified flying objects.** The formula

$$S = -16t^2 + v_0t + s_0$$

gives height in feet above the earth at time t seconds for an object projected into the air with an initial velocity of v_0 feet per second (ft/sec) from an initial height of s_0 feet. Two balls are tossed into the air simultaneously, one from the ground at 50 ft/sec and one from a height of 10 feet at 40 ft/sec. See the accompanying graph.

   a) Use the graph to estimate the time at which the balls are at the same height.

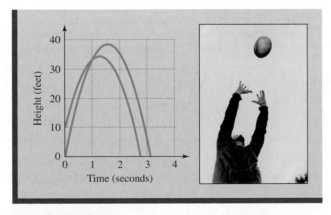

Figure for Exercise 83

   b) Find the time from part (a) algebraically.
   c) For what values of t will their heights above the ground differ by less than 5 feet (while they are both in the air)?

84. **Playing catch.** A circus clown at the top of a 60-foot platform is playing catch with another clown on the ground. The clown on the platform drops a ball at the same time as the one on the ground tosses a ball upward at 80 ft/sec. For what length of time is the distance between the balls less than or equal to 10 feet? (*Hint:* Use the formula given in Exercise 83. The initial velocity of a ball that is dropped is 0 ft/sec.) See the accompanying figure.

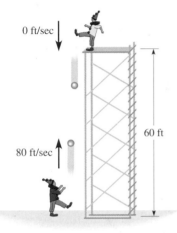

Figure for Exercise 84

Getting More Involved

85. **Discussion**

   For which real numbers m and n is each equation satisfied?

   a) $|m - n| = |n - m|$
   b) $|mn| = |m| \cdot |n|$
   c) $\left|\dfrac{m}{n}\right| = \dfrac{|m|}{|n|}$

86. **Exploration**

   a) Evaluate $|m + n|$ and $|m| + |n|$ for
      i) $m = 3$ and $n = 5$
      ii) $m = -3$ and $n = 5$
      iii) $m = 3$ and $n = -5$
      iv) $m = -3$ and $n = -5$

   b) What can you conclude about the relationship between $|m + n|$ and $|m| + |n|$?

Mid-Chapter **Quiz** | **Sections 8.1 through 8.2** | **Chapter 8**

Write the solution set to each inequality in interval notation and graph it.

1. $x > 1$ and $x < 4$

2. $x \geq 2$ or $x \geq 4$

3. $x \leq 3$ and $x \leq 5$

4. $2x - 4 \leq 6$ or $3x > -6$

5. $-3x + 1 < 7$ and $\frac{1}{2}x \leq -3$

6. $-10 \leq \dfrac{3x - 2}{2} < 5$

7. $0 < -5x - 3 < 7$

8. $|a - 3| > 4$

9. $|2w + 6| \leq 8$

10. $|2x - 7| + 5 < 3$

11. $5 - |4x| < 9$

Solve each equation.

12. $|x - 3| = 4$

13. $|w - 9| = 0$

14. $|x + 3| = |2x - 9|$

15. $4 - \left| \dfrac{1}{2}x + 1 \right| = 5$

8.3   **Compound Inequalities in Two Variables**

In This Section

⟨1⟩ **Satisfying a Compound Inequality**

⟨2⟩ **Graphing Compound Inequalities**

⟨3⟩ **Absolute Value Inequalities**

⟨4⟩ **Inequalities with No Solution**

⟨5⟩ **Applications**

A **simple inequality** in two variables involves only one inequality symbol. For example, $y > x - 3$ is a simple inequality in two variables. We graphed simple inequalities in two variables in Section 3.6. In this section we study compound inequalities in two variables.

⟨1⟩ **Satisfying a Compound Inequality**

A **compound inequality in two variables** consists of two simple inequalities joined with "and" or "or." For example, $y > x - 3$ and $y < 2 - x$ is a compound inequality in two variables. An ordered pair (or point) satisfies an "and" inequality only if it satisfies both of the simple inequalities. An ordered pair satisfies an "or" inequality if it satisfies one or the other or both inequalities.

E X A M P L E   1

Satisfying compound inequalities
Determine whether $(-2, 3)$ satisfies each compound inequality.

    a) $y > x$ and $x - y < -4$

    b) $y > x$ and $x - y > -4$

    c) $y > x$ or $x - y > -4$

Solution

a) Replacing x with -2 and y with 3 in $y > x$ and $x - y < -4$ yields $3 > -2$ and $-2 - 3 < -4$. Since both inequalities are correct, $(-2, 3)$ satisfies the compound inequality.

b) Replacing x with -2 and y with 3 in $y > x$ and $x - y > -4$ yields $3 > -2$ and $-2 - 3 > -4$. Since the second inequality is not correct, $(-2, 3)$ does not satisfy the compound inequality.

c) Replacing x with -2 and y with 3 in $y > x$ or $x - y > -4$ yields $3 > -2$ or $-2 - 3 > -4$. Since the first inequality is correct and the connecting word is "or," $(-2, 3)$ satisfies the compound inequality.

> Now do Exercises 1–6

⟨2⟩ Graphing Compound Inequalities

The solution set to a compound inequality using "and" is the intersection of the solution sets to the simple inequalities. Example 2 illustrates two methods for graphing the solution set.

EXAMPLE **2**

Graphing a compound inequality with *and*

Graph the compound inequality $y > x - 3$ and $y < -\frac{1}{2}x + 2$.

Solution

The Intersection Method

Start by graphing the lines $y = x - 3$ and $y = -\frac{1}{2}x + 2$. Points that satisfy $y > x - 3$ lie above the line $y = x - 3$, and points that satisfy $y < -\frac{1}{2}x + 2$ lie below the line $y = -\frac{1}{2}x + 2$ as shown in Fig. 8.22(a). Since the connective is "and," only points that are shaded with both colors (the intersection of the two regions) satisfy the compound inequality. The solution set to the compound inequality is shown in Fig. 8.22(b). Dashed lines are used because the inequalities are $>$ and $<$.

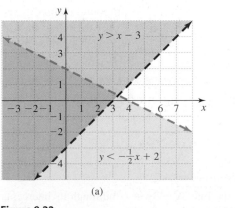

(a)

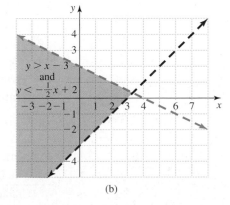

(b)

Figure 8.22

The Test Point Method

Again graph the lines, but this time select a point in each of the four regions determined by the lines as shown in Fig. 8.23(a). Test each of the four points (3, 3), (0, 0), (4, −5), and (5, 0) to see if it satisfies the compound inequality:

$$y > x - 3 \qquad \text{and} \qquad y < -\tfrac{1}{2}x + 2$$

$$3 > 3 - 3 \qquad \text{and} \qquad 3 < -\tfrac{1}{2} \cdot 3 + 2 \qquad \text{Second inequality is incorrect.}$$

$$0 > 0 - 3 \qquad \text{and} \qquad 0 < -\tfrac{1}{2} \cdot 0 + 2 \qquad \text{Both inequalities are correct.}$$

$$-5 > 4 - 3 \qquad \text{and} \qquad -5 < -\tfrac{1}{2} \cdot 4 + 2 \qquad \text{First inequality is incorrect.}$$

$$0 > 5 - 3 \qquad \text{and} \qquad 0 < -\tfrac{1}{2} \cdot 5 + 2 \qquad \text{Both inequalities are incorrect.}$$

The only point that satisfies both inequalities is (0, 0). So the solution set to the compound inequality consists of all points in the region containing (0, 0) as shown in Fig. 8.23(b).

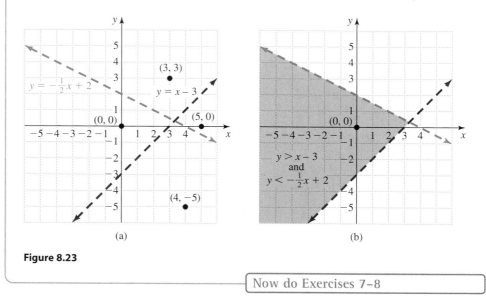

(a)   (b)

Figure 8.23

Now do Exercises 7–8

Example 3 involves a compound inequality using "or." Remember that a compound sentence with "or" is true if one, the other, or both parts of it are true. The solution set to a compound inequality with "or" is the union of the two solution sets.

E X A M P L E   3

Graphing a compound inequality with *or*
Graph the compound inequality $2x - 3y \leq -6$ or $x + 2y \geq 4$.

Solution

The Union Method

Graph the line $2x - 3y = -6$ through its intercepts (0, 2) and (−3, 0). Since (0, 0) does not satisfy this inequality, shade the region above this line as shown in Fig. 8.24(a). Graph the line $x + 2y = 4$ through $(0, 2)$ and $(4, 0)$. Since $(0, 0)$ does not satisfy this inequality, shade the region above the line as shown in Fig. 8.24(a). The union of these two solution sets consists of everything that is shaded as shown in Fig. 8.24(b). The boundary lines are solid because of the inequality symbols $\leq$ and $\geq$.

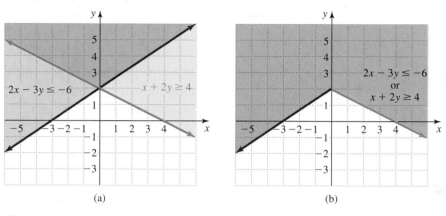

(a)   (b)

Figure 8.24

The Test Point Method

Graph the lines, and select a point in each of the four regions determined by the lines as shown in Fig. 8.25(a). Test each of the four points $(0, 0)$, $(-3, 2)$, $(0, 5)$, and $(3, 2)$ to see if it satisfies the compound inequality:

$$2x - 3y \leq -6 \qquad \text{or} \qquad x + 2y \geq 4$$
$$2(0) - 3(0) \leq -6 \qquad \text{or} \qquad 0 + 2(0) \geq 4 \qquad \text{False}$$
$$2(-3) - 3(2) \leq -6 \qquad \text{or} \qquad -3 + 2(2) \geq 4 \qquad \text{True}$$
$$2(0) - 3(5) \leq -6 \qquad \text{or} \qquad 0 + 2(5) \geq 4 \qquad \text{True}$$
$$2(3) - 3(2) \leq -6 \qquad \text{or} \qquad 3 + 2(2) \geq 4 \qquad \text{True}$$

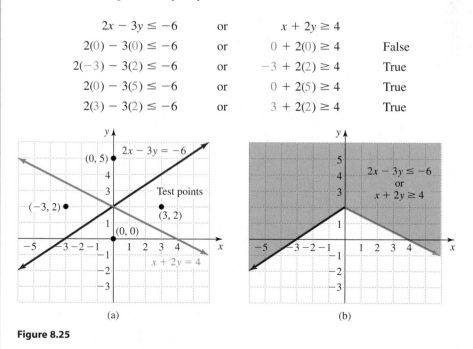

(a)   (b)

Figure 8.25

The solution set to the compound inequality consists of the three regions containing the test points that satisfy the compound inequality as shown in Fig. 8.25(b).

Now do Exercises 9–28

‹3› **Absolute Value Inequalities**

In Section 8.2 we learned that the absolute value inequality $|x| > 2$ is equivalent to the compound inequality $x < -2$ or $x > 2$. The absolute value inequality $|x| < 2$ is equivalent to the compound inequality $x > -2$ and $x < 2$. We can also write $|x| < 2$ as $-2 < x < 2$. We use these ideas with inequalities in two variables in Example 4.

E X A M P L E **4**	**Graphing absolute value inequalities**

Graph each absolute value inequality.

a) $|y - 2x| \le 3$ **b)** $|x - y| > 1$

‹ **Helpful Hint** ›

Remember that absolute value of a quantity is its distance from 0 (Section 1.1). If $|w| < 3$, then w is less than 3 units from 0:

$$-3 < w < 3$$

If $|w| > 1$, then w is more than 1 unit away from 0:

$$w > 1 \quad \text{or} \quad w < -1$$

In Example 4 we are using an expression in place of w.

Solution

a) The inequality $|y - 2x| \le 3$ is equivalent to $-3 \le y - 2x \le 3$, which is equivalent to the compound inequality

$$y - 2x \le 3 \qquad \text{and} \qquad y - 2x \ge -3.$$

First graph the lines $y - 2x = 3$ and $y - 2x = -3$ as shown in Fig. 8.26(a). These lines divide the plane into three regions. Test a point from each region in the original inequality, say $(-5, 0)$, $(0, 1)$, and $(5, 0)$:

$$|0 - 2(-5)| \le 3 \qquad |1 - 2 \cdot 0| \le 3 \qquad |0 - 2 \cdot 5| \le 3$$
$$10 \le 3 \qquad\qquad 1 \le 3 \qquad\qquad 10 \le 3$$

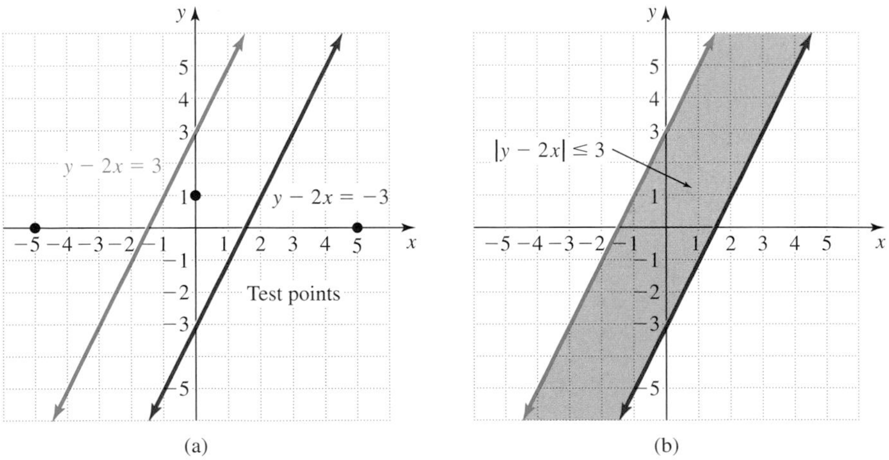

Figure 8.26

Only $(0, 1)$ satisfies the original inequality. So the region satisfying the absolute value inequality is the shaded region containing $(0, 1)$ as shown in Fig. 8.26(b). The boundary lines are solid because of the $\le$ symbol.

b) The inequality $|x - y| > 1$ is equivalent to

$$x - y > 1 \qquad \text{or} \qquad x - y < -1.$$

First graph the lines $x - y = 1$ and $x - y = -1$ as shown in Fig. 8.27(a).
Test a point from each region in the original inequality, say $(-4, 0)$, $(0, 0)$,
and $(4, 0)$:

$$|-4 - 0| > 1 \qquad |0 - 0| > 1 \qquad |4 - 0| > 1$$
$$4 > 1 \qquad\qquad 0 > 1 \qquad\qquad 4 > 1$$

Because $(-4, 0)$ and $(4, 0)$ satisfy the inequality, we shade those regions
as shown in Fig. 8.27(b). The boundary lines are dashed because of the
$>$ symbol.

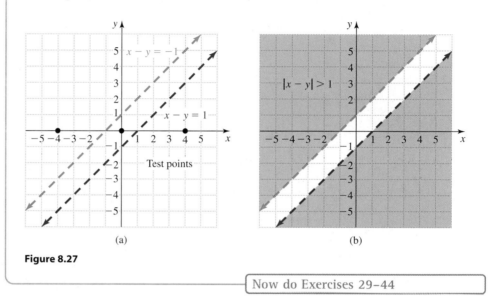

(a)   (b)

Figure 8.27

Now do Exercises 29–44

⟨4⟩ **Inequalities with No Solution**

The solution set to a compound inequality using "or" is the union of the individual
solution sets. So the solution set to an "or" inequality is not empty unless all of the
individual inequalities are inconsistent. However, the solution set to an "and" inequality
can be empty even when the solution sets to the individual inequalities are not empty.

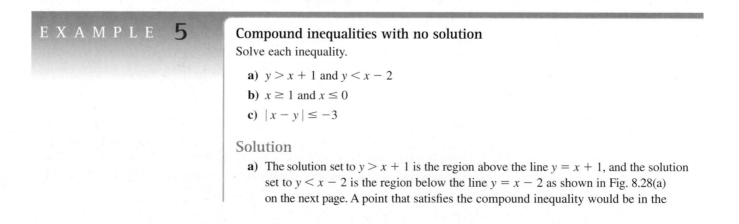

E X A M P L E **5**

Compound inequalities with no solution
Solve each inequality.

  a) $y > x + 1$ and $y < x - 2$

  b) $x \geq 1$ and $x \leq 0$

  c) $|x - y| \leq -3$

Solution

  a) The solution set to $y > x + 1$ is the region above the line $y = x + 1$, and the solution
set to $y < x - 2$ is the region below the line $y = x - 2$ as shown in Fig. 8.28(a)
on the next page. A point that satisfies the compound inequality would be in the

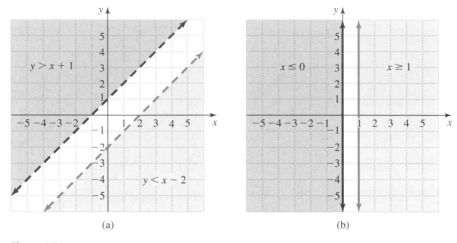

Figure 8.28

intersection of these regions. Because the lines are parallel these regions do not intersect. So the solution set to the compound inequality is the empty set $\varnothing$.

b) The solution set to $x \geq 1$ is the region on or to the right of the line $x = 1$, and the solution set to $x \leq 0$ is the region on or to the left of the line $x = 0$ as shown in Fig. 8.28(b). Because these lines are parallel these regions do not intersect and no points satisfy $x \geq 1$ and $x \leq 0$. The solution set is the empty set $\varnothing$.

c) Since the absolute value of any real number is nonnegative, there are no ordered pairs that satisfy $|x - y| \leq -3$. The solution set is the empty set, $\varnothing$.

Now do Exercises 45–60

⟨5⟩ Applications

In real situations, x and y often represent quantities or amounts, which cannot be negative. In this case our graphs are restricted to the first quadrant, where x and y are both nonnegative.

E X A M P L E   **6**

Inequalities in business

The manager of a furniture store can spend a maximum of $3000 on advertising per week. It costs $50 to run a 30-second ad on an AM radio station and $75 to run the ad on an FM station. Graph the region that shows the possible numbers of AM and FM ads that can be purchased, and identify some possibilities.

Solution

If x represents the number of AM ads and y represents the number of FM ads, then x and y must satisfy the inequality $50x + 75y \leq 3000$. Because the number of ads cannot be negative, we also have $x \geq 0$ and $y \geq 0$. So we graph only points in the first quadrant that satisfy $50x + 75y \leq 3000$. The line $50x + 75y = 3000$ goes through $(0, 40)$ and $(60, 0)$. The inequality is satisfied below this line. The region showing the possible numbers of AM ads and FM ads is shown in Fig. 8.29. We shade the entire region in Fig. 8.29, but only points in the shaded region in which both coordinates are whole numbers actually satisfy the given condition. For example, 40 AM ads and 10 FM ads could be purchased. Other possibilities are 30 AM ads and 20 FM ads, or 10 AM ads and 10 FM ads.

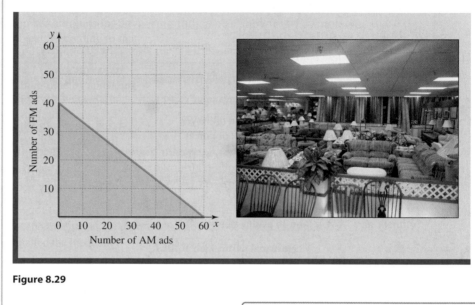

Figure 8.29

Now do Exercises 61–68

Warm-Ups ▼

Fill in the blank.

1. An inequality of the form $Ax + By \leq C$ is a _____ inequality.

2. A _____ inequality in two variables is formed by connecting two linear inequalities with "and" or "or."

3. For an "or" inequality we use the _____ of the two solution sets.

4. For an "and" inequality we use the _____ of the two solution sets.

5. A ___ point is used to determine whether all points in a region satisfy the compound inequality.

6. _____ boundary lines are used if the inequality symbols include equality.

True or false?

7. The graph of $3x - y > 2$ is the region above $3x - y = 2$.

8. The graph of $3x + y < 5$ is the region below $y = -3x + 5$.

9. The graph of $y > x + 3$ and $y < 2x - 6$ is the intersection of two regions.

10. The graph of $y \leq 2x - 3$ or $y \geq 3x + 5$ is the union of two regions.

11. The point $(2, -5)$ satisfies $y > -3x + 5$ and $y < 2x - 3$.

12. The point $(-3, 2)$ satisfies $y \leq 3x - 6$ or $y \leq x + 5$.

Exercises

‹1› **Satisfying a Compound Inequality**

Determine which of the ordered pairs $(1, 3)$, $(-2, 5)$, $(-6, -4)$, *and* $(7, -8)$ *satisfy each compound or absolute value inequality. See Example 1.*

1. $y > 5x$ and $y < -x$

2. $y > 5x$ and $y > -x$

3. $y > -x + 1$ or $y > 4x$

4. $y > -x + 1$ or $y < 4x$

5. $|x + y| < 3$

6. $|x - y| > 2$

11. $x - 4y < 0$ and $3x + 2y \geq 6$

12. $x \geq -2y$ and $x - 3y < 6$

13. $x + y \leq 5$ and $x - y \leq 3$

14. $2x - y < 3$ and $3x - y > 0$

‹2› **Graphing Compound Inequalities**

Graph each compound inequality. See Examples 2 and 3.

7. $y > x$ and $y > -2x + 3$

8. $y < x$ and $y < -3x + 2$

15. $x - 2y \leq 4$ or $2x - 3y \leq 6$

16. $4x - 3y \leq 3$ or $2x + y \geq 2$

9. $y < x + 3$ or $y > -x + 2$

10. $y \geq x - 5$ or $y \leq -2x + 1$

17. $y > 2$ and $x < 3$

18. $x \leq 5$ and $y \geq -1$

19. $y \geq x$ and $x \leq 2$ **20.** $y < x$ and $y > 0$

⟨ 3 ⟩ **Absolute Value Inequalities**

Graph the absolute value inequalities. See Example 4.

29. $|x + y| < 2$ **30.** $|2x + y| < 1$

21. $2x < y + 3$ or $y > 2 - x$ **22.** $3 - x < y + 2$ or $x > y + 5$

31. $|2x + y| \geq 1$ **32.** $|x + 2y| \geq 6$

23. $y > x - 1$ and $y < x + 3$ **24.** $y > x - 1$ and $y < 2x + 5$

33. $|y - x| > 2$ **34.** $|2y - x| > 6$

25. $0 \leq y \leq x$ and $x \leq 1$ **26.** $x \leq y \leq 1$ and $x \geq 0$

35. $|x - 2y| \leq 4$ **36.** $|x - 3y| \leq 6$

27. $1 \leq x \leq 3$ and $2 \leq y \leq 5$ **28.** $-1 < x < 1$ and $-1 < y < 1$

37. $|x| > 2$ **38.** $|x| \leq 3$

39. $|y| < 1$

40. $|y| \geq 2$

41. $|x| < 2$ and $|y| < 3$

42. $|x| \geq 3$ or $|y| \geq 1$

43. $|x - 3| < 1$ and $|y - 2| < 1$

44. $|x - 2| \geq 3$ or $|y - 5| \geq 2$

⟨4⟩ Inequalities with No Solution

Determine whether the solution set to each compound or absolute value inequality is the empty set or not. See Example 5.

45. $y > x$ and $x < 1$

46. $y > x$ and $x > 1$

47. $y < 2x - 5$ and $y > 2x + 5$

48. $y \geq 3x$ and $y \leq 3x - 1$

49. $y < 2x - 5$ or $y > 2x + 5$

50. $y \geq 3x$ or $y \leq 3x - 1$

51. $y < 2x$ and $y > 3x$

52. $y < 2x$ or $y > 3x$

VIDEO **53.** $y < x$ and $x < y$

54. $y > 3$ and $y < 1$

55. $|y + 2x| < 0$

56. $|x - 2y| < 0$

57. $|3x + 2y| \leq -4$

58. $|x - 2y| < -9$

59. $|x + y| > -4$

60. $|2x + 3y| < 4$

⟨5⟩ Applications

Solve each problem. See Example 6.

61. *Budget planning.* The Highway Patrol can spend a maximum of $120,000 on new vehicles this year. They can get a fully equipped compact car for $15,000 or a fully equipped full-size car for $20,000. Graph the region that shows the number of cars of each type that could be purchased.

62. *Allocating resources.* A furniture maker has a shop that can employ 12 workers for 40 hours per week at its maximum capacity. The shop makes tables and chairs. It takes 16 hours of labor to make a table and 8 hours of labor to make a chair. Graph the region that shows the possibilities for the number of tables and chairs that could be made in one week.

63. *More restrictions.* In Exercise 61, add the condition that the number of full-size cars must be greater than or equal to the number of compact cars. Graph the region showing the possibilities for the number of cars of each type that could be purchased.

64. *Chairs per table.* In Exercise 62, add the condition that the number of chairs must be at least four times the number of tables and at most six times the number of tables. Graph the region showing the possibilities for the number of tables and chairs that could be made in one week.

65. *Building fitness.* To achieve cardiovascular fitness, you should exercise so that your target heart rate is between 70% and 85% of its maximum rate. Your target heart rate h depends on your age a. For building fitness, you should have $h \leq 187 - 0.85a$ and $h \geq 154 - 0.70a$ (NordicTrack brochure). Graph this compound inequality for $20 \leq a \leq 75$ to see the heart rate target zone for building fitness.

66. *Waist-to-hip ratio.* A study by Dr. Aaron R. Folsom concluded that waist-to-hip ratios are a better predictor of 5-year survival than more traditional height-to-weight ratios. Dr. Folsom concluded that for good health the waist size of a woman aged 50 to 69 should be less than or equal to 80% of her hip size, $w \leq 0.80h$. Make a graph showing possible waist and hip sizes for good health for women in this age group for which hip size is no more than 50 inches.

67. *Advertising dollars.* A restaurant manager can spend at most $9000 on advertising per month and has two choices for advertising. The manager can purchase an ad in the *Daily Chronicle* (a 7-day-per-week newspaper) for $300 per day or a 30-second ad on WBTU television for $1000 each time the ad is aired. Graph the region that shows the possible number of days that an ad can be run in the newspaper and the possible number of times that an ad can be aired on television.

68. *Shipping restrictions.* The accompanying graph shows all of the possibilities for the number of refrigerators and the number of TVs that will fit into an 18-wheeler.

a) Write an inequality to describe this region.
b) Will the truck hold 71 refrigerators and 118 TVs?
c) Will the truck hold 51 refrigerators and 176 TVs?

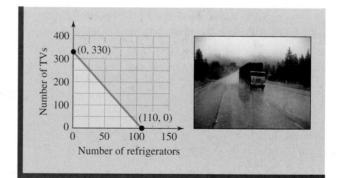

Figure for Exercise 68

Getting More Involved

69. *Writing*

Explain the difference between a compound inequality using the word "and" and a compound inequality using the word "or."

70. *Discussion*

Explain how to write an absolute value inequality as a compound inequality.

8.4   **Linear Programming**

In This Section

⟨1⟩ **Graphing the Constraints**
⟨2⟩ **Maximizing or Minimizing**

In this section we graph the solution set to a system of several linear inequalities in two variables as in Section 8.3. We then use the solution set to the inequalities to determine the maximum or minimum value of another variable. The method that we use is called **linear programming.**

⟨1⟩ Graphing the Constraints

In linear programming we have two variables that must satisfy several linear inequalities. These inequalities are called the **constraints** because they restrict the variables to only certain values. A graph in the coordinate plane is used to indicate the points that satisfy all of the constraints.

EXAMPLE 1

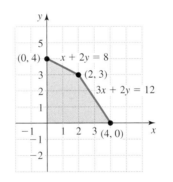

Figure 8.30

Graphing the constraints

Graph the solution set to the system of inequalities and identify each vertex of the region:

$$x \geq 0, \quad y \geq 0$$
$$3x + 2y \leq 12$$
$$x + 2y \leq 8$$

Solution

The points on or to the right of the y-axis satisfy $x \geq 0$. The points on or above the x-axis satisfy $y \geq 0$. The points on or below the line $3x + 2y = 12$ satisfy $3x + 2y \leq 12$. The points on or below the line $x + 2y = 8$ satisfy $x + 2y \leq 8$. Graph each straight line and shade the region that satisfies all four inequalities as shown in Fig. 8.30. Three of the vertices are easily identified as $(0, 0)$, $(0, 4)$, and $(4, 0)$. The fourth vertex is found by solving the system $3x + 2y = 12$ and $x + 2y = 8$. The fourth vertex is $(2, 3)$.

Now do Exercises 1–10

In linear programming the constraints usually come from physical limitations in some problem. In Example 2, we write the constraints and then graph the points in the coordinate plane that satisfy all of the constraints.

EXAMPLE 2

Writing the constraints

Jules is in the business of constructing dog houses. A small dog house requires 8 square feet (ft²) of plywood and 6 ft² of insulation. A large dog house requires 16 ft² of plywood and 3 ft² of insulation. Jules has available only 48 ft² of plywood and 18 ft² of insulation. Write the constraints on the number of small and large dog houses that he can build with the available supplies and graph the solution set to the system of constraints.

Solution

Let x represent the number of small dog houses and y represent the number of large dog houses. We have two natural constraints $x \geq 0$ and $y \geq 0$ since he cannot build a negative

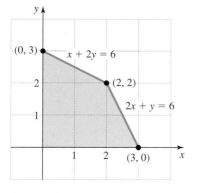

Figure 8.31

number of dog houses. Since the total plywood available for use is at most 48 ft², $8x + 16y \leq 48$. Since the total insulation available is at most 18 ft², $6x + 3y \leq 18$. Simplify the inequalities to get the following constraints:

$$x \geq 0, \quad y \geq 0$$
$$x + 2y \leq 6$$
$$2x + y \leq 6$$

The graph of the solution set to the system of inequalities is shown in Fig. 8.31.

> Now do Exercises 11–12

⟨2⟩ Maximizing or Minimizing

If a small dog house sells for $15 and a large sells for $20, then the total revenue in dollars from the sale of x small and y large dog houses is given by $R = 15x + 20y$. Since R is determined by or *is a function of* x and y, we use the function notation that was introduced in Section 2.4 and write $R(x, y)$ in place of R. The equation $R(x, y) = 15x + 20y$ is called a *linear function* of x and y. Any ordered pair within the region shown in Fig. 8.31 is a possibility for the number of dog houses of each type that could be built, and so it is the *domain* of the function R. (We will study functions in general in Chapter 11.)

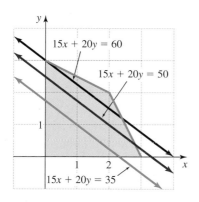

Figure 8.32

> **Linear Function of Two Variables**
>
> An equation of the form $f(x, y) = Ax + By + C$, where A, B, and C are fixed real numbers, is called a **linear function of two variables** (x and y).

Naturally, we are interested in the maximum revenue subject to the constraints on x and y. To investigate some possible revenues, replace R in $R = 15x + 20y$ with, say 35, 50, and 60. The graphs of the parallel lines $15x + 20y = 35$, $15x + 20y = 50$, and $15x + 20y = 60$ are shown in Fig. 8.32. The revenue at any point on the line $15x + 20y = 35$ is $35. We get a larger revenue on a higher revenue line (and lower revenue on a lower line). The maximum revenue possible will be on the highest revenue line that still intersects the region. Because the sides of the region are straight-line segments, the intersection of the highest (or lowest) revenue line with the region must include a vertex of the region. This is the fundamental principle behind linear programming.

> **The Principle of Linear Programming**
>
> The maximum or minimum value of a linear function subject to linear constraints occurs at a vertex of the region determined by the constraints.

E X A M P L E   **3**

Maximizing a linear function with linear constraints

A small dog house requires 8 ft² of plywood and 6 ft² of insulation. A large dog house requires 16 ft² of plywood and 3 ft² of insulation. Only 48 ft² of plywood and 18 ft² of insulation are available. If a small dog house sells for $15 and a large dog house sells for $20, then how many dog houses of each type should be built to maximize the revenue and to satisfy the constraints?

Solution

Let x be the number of small dog houses and y be the number of large dog houses. We wrote and graphed the constraints for this problem in Example 2, so we will not repeat that here. The graph in Fig. 8.31 has four vertices: $(0, 0)$, $(0, 3)$, $(3, 0)$, and $(2, 2)$. The revenue function is $R(x, y) = 15x + 20y$. Since the maximum value of this function must occur at a vertex, we evaluate the function at each vertex:

$$R(0, 0) = 15(0) + 20(0) = \$0$$
$$R(0, 3) = 15(0) + 20(3) = \$60$$
$$R(3, 0) = 15(3) + 20(0) = \$45$$
$$R(2, 2) = 15(2) + 20(2) = \$70$$

From this list we can see that the maximum revenue is \$70 when two small and two large dog houses are built. We also see that the minimum revenue is \$0 when no dog houses of either type are built.

Now do Exercises 13–32

Use the following strategy for solving linear programming problems.

Strategy for Linear Programming

Use the following steps to find the maximum or minimum value of a linear function subject to linear constraints.
1. Graph the region that satisfies all of the constraints.
2. Determine the coordinates of each vertex of the region.
3. Evaluate the function at each vertex of the region.
4. Identify which vertex gives the maximum or minimum value of the function.

In Example 4, we solve another linear programming problem.

E X A M P L E   4

Minimizing a linear function with linear constraints

One serving of food A contains 2 grams of protein and 6 grams of carbohydrates. One serving of food B contains 4 grams of protein and 3 grams of carbohydrates. A dietitian wants a meal that contains at least 12 grams of protein and at least 18 grams of carbohydrates. If the cost of food A is 9 cents per serving and the cost of food B is 20 cents per serving, then how many servings of each food would minimize the cost and satisfy the constraints?

Solution

Let x equal the number of servings of food A and y equal the number of servings of food B. If the meal is to contain at least 12 grams of protein, then $2x + 4y \geq 12$. If the meal is to contain at least 18 grams of carbohydrates, then $6x + 3y \geq 18$. Simplify each inequality and use the two natural constraints to get the following system:

$$x \geq 0, \quad y \geq 0$$
$$x + 2y \geq 6$$
$$2x + \ y \geq 6$$

The graph of the constraints is shown in Fig. 8.33. The vertices are $(0, 6)$, $(6, 0)$, and $(2, 2)$. The cost in cents for x servings of A and y servings of B is $C(x, y) = 9x + 20y$. Evaluate

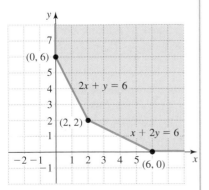

Figure 8.33

the cost at each vertex:

$$C(0, 6) = 9(0) + 20(6) = 120 \text{ cents}$$
$$C(6, 0) = 9(6) + 20(0) = 54 \text{ cents}$$
$$C(2, 2) = 9(2) + 20(2) = 58 \text{ cents}$$

The minimum cost of 54 cents is attained by using six servings of food A and no servings of food B.

> Now do Exercises 33–38

Warm-Ups ▼

Fill in the blank.

1. A _____ is an inequality that restricts the values of the variables.
2. _____ is a process used to maximize or minimize a linear function subject to linear constraints.
3. A _____ function of two variables has the form $f(x, y) = Ax + By + C$.
4. The maximum or minimum of a linear function subject to linear constraints occurs at a _____ of the region determined by the constraints.

True or false?

5. The graph of $x \geq 0$ consists of the points on or above the x-axis.
6. The graph $y \geq 0$ consists of the points on or to the right of the y-axis.
7. The graph of $x + y \leq 6$ consists of points on or below the line $x + y = 6$.
8. The graph of $2x + 3y = 30$ has x-intercept $(15, 0)$ and y-intercept $(0, 10)$.
9. The value of $R(x, y) = 3x + 5y$ at $(2, 4)$ is 26.
10. If $C(x, y) = 12x + 10y$, then $C(0, 5) = 62$.

Exercises 8.4

< Study Tips >

- Working problems 1 hour per day every day of the week is better than working problems for 7 hours on one day of the week. Spread out your study time. Avoid long study sessions.
- No two students learn in exactly the same way or at the same speed. Figure out what works for you.

⟨1⟩ Graphing the Constraints

Graph the solution set to each system of inequalities, and identify each vertex of the region. See Example 1.

1. $x \geq 0, y \geq 0$
   $x + y \leq 5$

2. $x \geq 0, y \geq 0$
   $y \leq 5, y \geq x$

VIDEO 3. $x \geq 0, y \geq 0$
   $2x + y \leq 4$
   $x + y \leq 3$

4. $x \geq 0, y \geq 0$
   $x + y \leq 4$
   $x + 2y \leq 6$

5. $x \geq 0, y \geq 0$
$2x + y \geq 3$
$x + y \geq 2$

6. $x \geq 0, y \geq 0$
$3x + 2y \geq 12$
$2x + y \geq 7$

12. *Making boats.* A company makes kayaks and canoes. Each kayak requires $80 in materials and 60 hours of labor. Each canoe requires $120 in materials and 40 hours of labor. The company has at most $12,000 available for materials and at most 4800 hours of labor. Let x represent the possible number of kayaks and y represent the possible number of canoes that can be built.

7. $x \geq 0, y \geq 0$
$x + 3y \leq 15$
$2x + y \leq 10$

8. $x \geq 0, y \geq 0$
$2x + 3y \leq 15$
$x + y \leq 7$

⟨2⟩ Maximizing or Minimizing

Let $P(x, y) = 6x + 8y$, $R(x, y) = 11x + 20y$, and $C(x, y) = 5x + 12y$. Evaluate each expression. See Example 3.

13. $P(1, 5)$

14. $P(3, 8)$

15. $R(8, 0)$

16. $R(5, 10)$

9. $x \geq 0, y \geq 0$
$x - y \geq 4$
$3x + y \geq 6$

10. $x \geq 0, y \geq 0$
$x + 3y \geq 6$
$2x + y \geq 7$

17. $C(4, 9)$

18. $C(0, 6)$

Determine the maximum value of the given linear function on the given region. See Example 3.

19. $P(x, y) = 2x + 3y$

20. $W(x, y) = 6x + 7y$

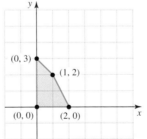

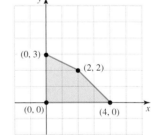

For each problem, write the constraints and graph the solution set to the system of constraints. See Example 2.

11. *Making guitars.* A company makes an acoustic and an electric guitar. Each acoustic guitar requires $100 in materials and 20 hours of labor. Each electric guitar requires $200 in materials and 15 hours of labor. The company has at most $3000 for materials and 300 hours of labor available. Let x represent the possible number of acoustic guitars and y represent the possible number of electric guitars that can be made.

21. $R(x, y) = 9x + 8y$

22. $F(x, y) = 3x + 10y$

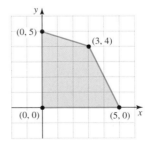

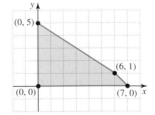

Determine the minimum value of the given function on the given region.

23. $C(x, y) = 11x + 10y$

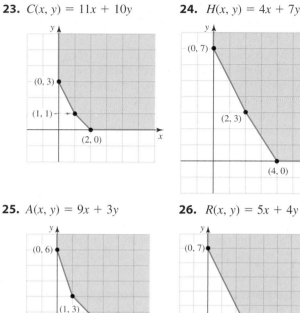

(0, 3)
(1, 1)
(2, 0)

24. $H(x, y) = 4x + 7y$

(0, 7)
(2, 3)
(4, 0)

25. $A(x, y) = 9x + 3y$

(0, 6)
(1, 3)
(4, 0)

26. $R(x, y) = 5x + 4y$

(0, 7)
(3, 1)
(6, 0)

Solve each problem. See Examples 2–4. See the Strategy for Linear Programming box on page 542.

27. *Phase I advertising.* The publicity director for Mercy Hospital is planning to bolster the hospital's image by running a TV ad and a radio ad. Due to budgetary and other constraints, the number of times that she can run the TV ad, x, and the number of times that she can run the radio ad, y, must be in the region shown in the figure. The function

$$A = 9000x + 4000y$$

gives the total number of people reached by the ads.

a) Find the total number of people reached by the ads at each vertex of the region.

b) What mix of TV and radio ads maximizes the number of people reached?

28. *Phase II advertising.* Suppose the radio station in Exercise 27 starts playing country music and the function for the total number of people changes to

$$A = 9000x + 2000y.$$

a) Find A at each vertex of the region using this function.

b) What mix of TV and radio ads maximizes the number of people reached?

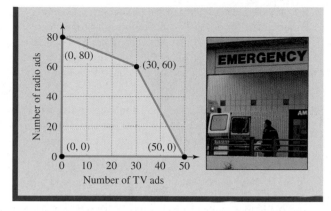

(0, 80)
(30, 60)
(0, 0)
(50, 0)
Number of radio ads
Number of TV ads

Figure for Exercises 27 and 28

29. At Burger Heaven a double contains 2 meat patties and 6 pickles, whereas a triple contains 3 meat patties and 3 pickles. Near closing time one day, only 24 meat patties and 48 pickles are available. If a double burger sells for $1.20 and a triple burger sells for $1.50, then how many of each should be made to maximize the total revenue?

30. Sam and Doris manufacture rocking chairs and porch swings in the Ozarks. Each rocker requires 3 hours of work from Sam and 2 hours from Doris. Each swing requires 2 hours of work from Sam and 2 hours from Doris. Sam cannot work more than 48 hours per week, and Doris cannot work more than 40 hours per week. If a rocker sells for $160 and a swing sells for $100, then how many of each should be made per week to maximize the revenue?

31. If a double burger sells for $1.00 and a triple burger sells for $2.00, then how many of each should be made to maximize the total revenue subject to the constraints of Exercise 29?

32. If a rocker sells for $120 and a swing sells for $100, then how many of each should be made to maximize the total revenue subject to the constraints of Exercise 30?

33. One cup of Doggie Dinner contains 20 grams of protein and 40 grams of carbohydrates. One cup of Puppy Power contains 30 grams of protein and 20 grams of carbohydrates. Susan wants her dog to get at least 200 grams of protein and 180 grams of carbohydrates per day. If Doggie Dinner costs 16 cents per cup and Puppy Power costs 20 cents per cup, then how many cups of each would satisfy the constraints and minimize the total cost?

34. Mammoth Muffler employs supervisors and helpers. According to the union contract, a supervisor does 2 brake jobs and 3 mufflers per day, whereas a helper does 6 brake jobs and 3 mufflers per day. The home office requires enough staff for at least 24 brake jobs and for at least 18 mufflers per day. If a supervisor makes $90 per day and a helper makes $100 per day, then how many of each should be employed to satisfy the constraints and to minimize the daily labor cost?

35. Suppose in Exercise 33 Doggie Dinner costs 4 cents per cup and Puppy Power costs 10 cents per cup. How many cups of each would satisfy the constraints and minimize the total cost?

36. Suppose in Exercise 34 the supervisor makes $110 per day and the helper makes $100 per day. How many of each should be employed to satisfy the constraints and to minimize the daily labor cost?

37. Anita has at most $24,000 to invest in her brother-in-law's laundromat and her nephew's car wash. Her brother-in-law has high blood pressure and heart disease, but he will pay 18%, whereas her nephew is healthier but will pay only 12%. So the amount she will invest in the car wash will be at least twice the amount that she will invest in the laundromat but not more than three times as much. How much should she invest in each to maximize her total income from the two investments?

38. Herbert assembles computers in his shop. The parts for each economy model are shipped to him in a carton with a volume of 2 cubic feet (ft^3), and the parts for each deluxe model are shipped to him in a carton with a volume of 3 ft^3. After assembly, each economy model is shipped out in a carton with a volume of 4 ft^3, and each deluxe model is shipped out in a carton with a volume of 4 ft^3. The truck that delivers the parts has a maximum capacity of 180 ft^3, and the truck that takes out the completed computers has a maximum capacity of 280 ft^3. He can receive only one shipment of parts and send out one shipment of computers per week. If his profit on an economy model is $60 and his profit on a deluxe model is $100, then how many of each should he order per week to maximize his profit?

Wrap-Up

Summary

Compound Inequalities		Examples

In one variable

Two simple inequalities in one variable connected with the word "and" or "or"

The solution set for an "and" inequality is the intersection of the solution sets.

$x > 1$ and $x < 5$

The solution set for an "or" inequality is the union of the solution sets.

$x > 3$ or $x < 1$

In two variables

Two simple inequalities in two variables connected with the word "and" or "or"

The solution set for an "and" inequality is the intersection of the solution sets.

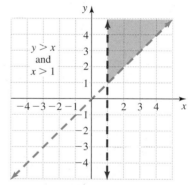

The solution set for an "or" inequality is the union of the solution sets.

Note that the graph of $x > 1$ (an inequality containing only one variable) in the rectangular coordinate system is the region to the right of the vertical line $x = 1$.

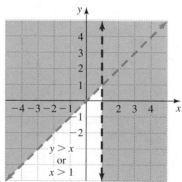

Absolute Value

	Absolute Value Equation	Equivalent Equation	Solution Set
Basic absolute value equations	$\|x\| = k$ $(k > 0)$	$x = k$ or $x = -k$	$\{k, -k\}$
	$\|x\| = 0$	$x = 0$	$\{0\}$
	$\|x\| = k$ $(k < 0)$		$\varnothing$

	Absolute Value Inequality	Equivalent Inequality	Solution Set	Graph of Solution Set
Basic absolute value inequalities ($k > 0$)	$\lvert x \rvert > k$	$x > k$ or $x < -k$	$(-\infty, -k) \cup (k, \infty)$	
	$\lvert x \rvert \geq k$	$x \geq k$ or $x \leq -k$	$(-\infty, -k] \cup [k, \infty)$	
	$\lvert x \rvert < k$	$-k < x < k$	$(-k, k)$	
	$\lvert x \rvert \leq k$	$-k \leq x \leq k$	$[-k, k]$	

Linear Programming

Use the following steps to find the maximum or minimum value of a linear function subject to linear constraints.
1. Graph the region that satisfies all of the constraints.
2. Determine the coordinates of each vertex of the region.
3. Evaluate the function at each vertex of the region.
4. Identify which vertex gives the maximum or minimum value of the function.

Enriching Your Mathematical Word Power

Fill in the blank.

1. A _____ inequality is an inequality involving only one equality symbol.
2. A _____ inequality consists of two simple inequalities joined with "and" or "or."
3. The _____ of sets A and B consists of elements that are in both A and B.
4. The _____ of sets A and B consists of elements that are either in A or B.

5. The inequality $a < x < b$ is equivalent to $a < x$ ___ $x < b$.
6. The inequality $\lvert x \rvert = k$ for $k > 0$ is equivalent to $x = k$ __ $x = -k$.
7. The inequality $\lvert x \rvert > k$ for $k > 0$ is equivalent to $x > k$ __ $x < -k$.
8. Inequalities in a linear program problem are called _____.

Review Exercises

8.1 Compound Inequalities in One Variable

Solve each compound inequality. State the solution set using interval notation and graph it.

1. $x + 2 > 3$ or $x - 6 < -10$

2. $x - 2 > 5$ or $x - 2 < -1$

3. $x > 0$ and $x - 6 < 3$

4. $x \leq 0$ and $x + 6 > 3$

5. $6 - x < 3$ or $-x < 0$

6. $-x > 0$ or $x + 2 < 7$

7. $2x < 8$ and $2(x - 3) < 6$

8. $\dfrac{1}{3}x > 2$ and $\dfrac{1}{4}x > 2$

9. $x - 6 > 2$ and $6 - x > 0$

10. $-\dfrac{1}{2}x < 6$ or $\dfrac{2}{3}x < 4$

11. $0.5x > 10$ or $0.1x < 3$

12. $0.02x > 4$ and $0.2x < 3$

13. $-2 \le \dfrac{2x - 3}{10} \le 1$

14. $-3 < \dfrac{4 - 3x}{5} < 2$

Write each union or intersection of intervals as a single interval.

15. $[1, 4) \cup (2, \infty)$

16. $(2, 5) \cup (-1, \infty)$

17. $(3, 6) \cap [2, 8]$

18. $[-1, 3] \cap [0, 8]$

19. $(-\infty, 5) \cup [5, \infty)$

20. $(-\infty, 1) \cup (0, \infty)$

21. $(-3, -1] \cap [-2, 5]$

22. $[-2, 4] \cap (4, 7]$

8.2 Absolute Value Equations and Inequalities
Solve each absolute value equation and graph the solution set.

23. $|x| + 2 = 16$

24. $\left| \dfrac{x}{2} \right| - 5 = -1$

25. $|4x - 12| = 0$

26. $|2x - 8| = 0$

27. $|x| = -5$

28. $\left| \dfrac{x}{2} - 5 \right| = -1$

29. $|2x - 1| - 3 = 0$

30. $|5 - x| - 2 = 0$

Solve each absolute value inequality and graph the solution set.

31. $|2x| \ge 8$

32. $|5x - 1| \le 14$

33. $\left| 1 - \dfrac{x}{5} \right| > \dfrac{9}{5}$

34. $\left| 1 - \dfrac{1}{6}x \right| < \dfrac{1}{2}$

35. $|x - 3| < -3$

36. $|x - 7| \le -4$

37. $|x + 4| \ge -1$

38. $|6x - 1| \ge 0$

39. $1 - \dfrac{3}{2}|x - 2| < -\dfrac{1}{2}$

40. $1 > \dfrac{1}{2}|6 - x| - \dfrac{3}{4}$

8.3 Compound Inequalities in Two Variables
Graph each compound or absolute value inequality.

41. $y > 3$ and
$y - x < 5$

42. $x + y \le 1$ or
$y \le 4$

43. $3x + 2y \geq 8$ or
$3x - 2y \leq 6$

44. $x + 8y > 8$ and
$x - 2y < 10$

45. $|x + 2y| < 10$

46. $|x - 3y| \geq 9$

47. $|x| \leq 5$

48. $|y| > 6$

49. $|y - x| > 2$

50. $|x - y| \leq 1$

Solve each problem by linear programming.

53. Find the maximum value of the function $R(x, y) = 6x + 9y$ subject to the following constraints:

$$x \geq 0, y \geq 0$$
$$2x + y \leq 6$$
$$x + 2y \leq 6$$

54. Find the minimum value of the function $C(x, y) = 9x + 10y$ subject to the following constraints:

$$x \geq 0, y \geq 0$$
$$x + y \geq 4$$
$$3x + y \geq 6$$

Miscellaneous
Solve each problem.

55. *Rockbuster video.* Stephen plans to open a video rental store in Edmonton. Industry statistics show that 45% of the rental price goes for overhead. If the maximum that anyone will pay to rent a video is $5 and Stephen wants a profit of at least $1.65 per video, then in what range should the rental price be?

56. *Working girl.* Regina makes $6.80 per hour working in the snack bar. To keep her grant, she may not earn more than $51 per week. What is the range of the number of hours per week that she may work?

57. *Skeletal remains.* Forensic scientists use the formula $h = 60.089 + 2.238F$ to predict the height h (in centimeters) for a male whose femur measures F centimeters. (See the accompanying figure.) In what range is the length of the femur for males between 150 centimeters and 180 centimeters in height? Round to the nearest tenth of a centimeter.

8.4 Linear Programming
Graph each system of inequalities and identify each vertex of the region.

51. $x \geq 0, y \geq 0$
$x + 2y \leq 6$
$x + y \leq 5$

52. $x \geq 0, y \geq 0$
$3x + 2y \geq 12$
$x + 2y \geq 8$

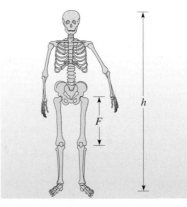

Figure for Exercise 57

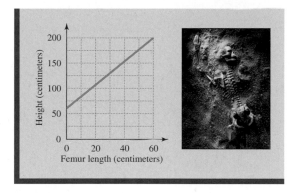

Figure for Exercise 58

58. *Female femurs.* Forensic scientists use the formula
$h = 61.412 + 2.317F$ to predict the height h in centimeters
for a female whose femur measures F centimeters.

a) Use the accompanying graph to estimate the femur
length for a female with height of 160 centimeters.

b) In what range is the length of the femur for females who
are over 170 centimeters tall?

59. *Car trouble.* Dane's car was found abandoned at mile
marker 86 on the interstate. If Dane was picked up by the
police on the interstate exactly 5 miles away, then at what
mile marker was he picked up?

60. *Comparing scores.* Scott scored 72 points on the midterm,
and Katie's score was more than 16 points away from
Scott's. What was Katie's score?

*For each graph in Exercises 61–78, write an equation or
inequality that has the solution set shown by the graph. Use
absolute value when possible.*

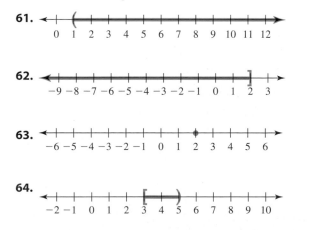

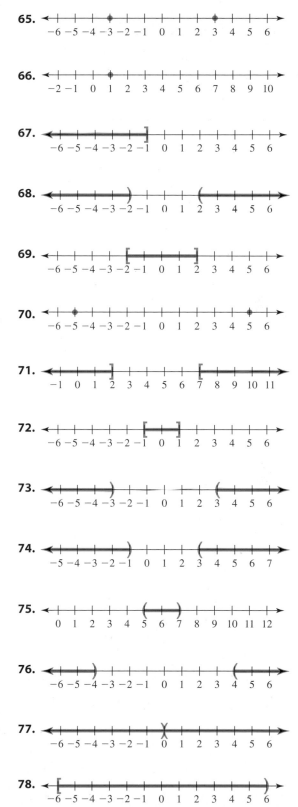

Chapter 8 Test

Write an inequality that describes each graph.

1.

2.

Write the solution set to each inequality using interval notation.

3. $x \geq 3$

4. $x > 1$ and $x \leq 6$

5. $x < 5$ or $x > 9$

6. $|x| < 3$

7. $|x| \geq 2$

Solve each inequality. State the solution set using interval notation and graph the solution set.

8. $2x + 3 > 1$

9. $|m - 6| \leq 2$

10. $2|x - 3| - 5 > 15$

11. $2 - 3(w - 1) < -2w$

12. $3x - 2 < 7$ and $-3x \leq 15$

13. $\dfrac{2}{3}y < 4$ or $y - 3 < 12$

Solve each equation or inequality.

14. $|2x - 7| = -3$

15. $x - 4 > 1$ or $x < 12$

16. $3x < 0$ and $x - 5 > 2$

17. $|2x - 5| \leq 0$

18. $|x - 3| < 0$

19. $|x - 6| > -6$

Sketch the graph of each inequality.

20. $x > 2$ and $x + y > 0$

21. $|2x + y| \geq 3$

22. $x + y > 1$ or $x - y < 2$

Solve the inequality problem.

23. Al and Brenda do the same job, but their annual salaries differ by more than $3000. Assume Al makes $28,000 per year, and write an absolute value inequality to describe this situation. What are the possibilities for Brenda's salary?

Solve the following problem by linear programming.

24. Find the maximum value of the function

$$P(x, y) = 8x + 10y$$

subject to the following constraints:

$$x \geq 0, y \geq 0$$
$$2x + 3y \leq 12$$
$$x + y \leq 5$$

*Making*Connections | A Review of Chapters 1–8

Simplify each expression.

1. $5x + 6x$

2. $5x \cdot 6x$

3. $\dfrac{6x + 2}{2}$

4. $5 - 4(2 - x)$

5. $(30 - 1)(30 + 1)$

6. $(30 + 1)^2$

7. $(30 - 1)^2$

8. $(2 + 3)^2$

9. $2^2 + 3^2$

10. $(8 - 3)(3 - 8)$

11. $(-1)(3 - 8)$

12. -2^2

13. $3x + 8 - 5(x - 1)$

14. $(-6)^2 - 4(-3)2$

15. $3^2 \cdot 2^3$

16. $4(-6) - (-5)(3)$

Solve each equation.

17. $5x + 6x = 8x$

18. $5x + 6x = 11x$

19. $5x + 6x = 0$

20. $5x + 6 = 11x$

21. $3x + 1 = 0$

22. $5 - 4(2 - x) = 1$

23. $x - 0.01x = 990$

24. $|5x + 6| = 11$

Solve each system of equations.

25. $\begin{aligned} 2x + y &= 5 \\ x - y &= 7 \end{aligned}$

26. $\begin{aligned} 3x - y &= 5 \\ y - 3x &= -5 \end{aligned}$

27. $\begin{aligned} 2x + 5y &= 16 \\ 3x - 4y &= -22 \end{aligned}$

28. $\begin{aligned} \tfrac{1}{2}x - \tfrac{2}{3}y &= -6 \\ \tfrac{3}{4}x + \tfrac{2}{5}y &= 12 \end{aligned}$

Simplify each expression. Write answers without negative exponents. All variables represent nonzero real numbers.

29. $x^8 \cdot x^{-3}$

30. $x^8 \div x^{-3}$

31. $x^{-3} \div x^{-5}$

32. $x^{-4} \cdot x^{-2}$

33. $\dfrac{2}{3^{-2}}$

34. $-1^{-1} + 2^0$

35. $(-3a^2b^3)^3$

36. $\left(\dfrac{4a^{-2}}{12a^6}\right)^{-3}$

37. $2^3 \cdot 3^2$

38. $3^2 \cdot 5^2$

Match each inequality in Exercises 39–48 with an equivalent inequality in A–J.

39. $2 - x < 5$

40. $x + 1 > x - 2$

41. $x > 2$ and $x > 5$

42. $x < -5$ or $x < -3$

43. $x < -9$ and $x > -3$

44. $x < -3$ or $x > 3$

45. $x > -3$ and $x < 3$

46. $|x + 3| > 0$

47. $y < x + 3$ and $y < x$

48. $y > x - 3$ or $y > x$

A. $y < x$ **B.** $|x| < 3$ **C.** $|x| > 3$

D. $x < -3$ **E.** $x > -3$ **F.** $x + 1 > x$

G. $x > 5$ **H.** $x + 1 < x$ **I.** $x \neq -3$

J. $y > x - 3$

Find each product.

49. $(x - 2)(x^2 + 2x + 4)$

50. $(a + 10)(a^2 - 10a + 100)$

51. $(3a - 5b)^2$

52. $(2x^2 + 3y)^2$

53. $(a - y^2)(a + y^2)$

54. $2(3m + 2)(5m - 9)$

Factor completely.

55. $y^2 + 98y - 99$

56. $8a^2 - 10a - 3$

57. $6a^3 - 36a^2 + 54a$

58. $b^3 + b^2 + 4b + 4$

59. $a^2 - 14a + 48$

60. $-8a^3 - 8$

Solve the problem.

61. ***Cost analysis.*** Diller Electronics can rent a copy machine for 5 years from American Business Supply for $75 per month plus 6 cents per copy. The same copier can be purchased for $8000, but then it costs only 2 cents per copy for supplies and maintenance. The purchased copier has no value after 5 years.

a) Use the accompanying graph to estimate the number of copies for 5 years for which the cost of renting would equal the cost of buying.

b) Write a formula for the 5-year cost under each plan.

c) Algebraically find the number of copies for which the 5-year costs would be equal.

d) If Diller makes 120,000 copies in 5 years, which plan is cheaper and by how much?

e) For what range of copies do the two plans differ by less than $500?

Figure for Exercise 61

Critical **Thinking** | **For Individual or Group Work** | **Chapter 8**

These exercises can be solved by a variety of techniques, which may or may not require algebra. So be creative and think critically. Explain all answers. Answers are in the Instructor's Edition of this text.

1. *Tennis time.* Tennis balls are sold in a cylindrical container that contains three balls. Assume that the balls just fit into the container as shown in the accompanying figure. What is the ratio of the amount of space in the container that is occupied by the balls to the amount of space that is not occupied by the balls?

Figure for Exercise 1

Photo for Exercise 5

2. *Planting trees.* A landscaper planted 7 trees so that they were arranged in 6 rows with 3 trees in each row. How did she do this?

3. *Division problem.* Start with any three-digit number and write the number twice to form a six-digit number. Divide the six-digit number by 7. Divide the answer by 11. Finally, divide the last answer by 13. What do you notice? Explain why this works.

4. *Totaling 25.* How many ways are there to add three different positive integers and get a sum of 25? Do not count rearrangements of the integers. For example, count 1, 2, and 22 as one possibility, but do not count 2, 22, and 1 as another.

5. *Temple of gloom.* The famous explorer Indiana Smith wants to cross a desert on foot. He plans to hire some men to help him carry supplies on the journey. However, the journey takes six days, but Smith and his helpers can each carry only a four-day supply of food and water. Of course every day, each man must consume a one-day supply of food and water or he will die. Devise a plan for getting Smith across the desert without anyone dying and using the minimum number of helpers.

6. *Counting zeros.* How many zeros are at the end of the number $(5^5)!$?

7. *Perfect Computers.* Of 6000 computers coming off a manufacturer's assembly line, every third computer had a hardware problem, every fourth computer had a software problem, and every tenth computer had a cosmetic defect. The remaining computers were perfect and were shipped to Wal-Mart. How many were shipped to Wal-Mart?

8. *Leap frog.* In Martha's garden is a circular pond with a diameter of 100 feet. A frog with an average leap of two and a quarter feet is sitting on a lily pad in the exact center of the pond. If the lily pads are all in the right places, then what is the minimum number of leaps required for the frog to jump out of the pond.

Radicals and Rational Exponents

Just how cold is it in Fargo, North Dakota, in winter? According to local meteorologists, the mercury hit a low of −33°F on January 18, 1994. But air temperature alone is not always a reliable indicator of how cold you feel. On the same date, the average wind velocity was 13.8 miles per hour. This dramatically affected how cold people felt when they stepped outside. High winds along with cold temperatures make exposed skin feel colder because the wind significantly speeds up the loss of body heat. Meteorologists use the terms "wind chill factor," "wind chill index," and "wind chill temperature" to take into account both air temperature and wind velocity.

Through experimentation in Antarctica, Paul A. Siple developed a formula in the 1940s that measures the wind chill from the velocity of the wind and the air temperature. His complex formula involving the square root of the velocity of the wind is still used today to calculate wind chill temperatures. Siple's formula is unlike most scientific formulas in that it is not based on theory. Siple experimented with various formulas involving wind velocity and temperature until he found a formula that seemed to predict how cold the air felt.

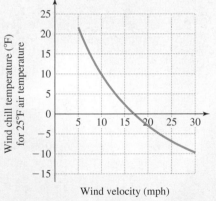

Wind velocity (mph)

Siple's formula is stated and used in Exercises 111 and 112 of Section 9.1.

9.1   Radicals

In Section 4.1, you learned the basic facts about powers. In this section, you will study roots and see how powers and roots are related.

⟨1⟩ Roots

We use the idea of roots to reverse powers. Because $3^2 = 9$ and $(-3)^2 = 9$, both 3 and -3 are square roots of 9. Because $2^4 = 16$ and $(-2)^4 = 16$, both 2 and -2 are fourth roots of 16. Because $2^3 = 8$ and $(-2)^3 = -8$, there is only one real cube root of 8 and only one real cube root of -8. The cube root of 8 is 2 and the cube root of -8 is -2.

> ### nth Roots
>
> If $a = b^n$ for a positive integer n, then b is an **nth root of a.** If $a = b^2$, then b is a **square root** of a. If $a = b^3$, then b is the **cube root** of a.

If n is a positive even integer and a is positive, then there are two real nth roots of a. We call these roots **even roots.** The positive even root of a positive number is called the **principal root.** The principal square root of 9 is 3 and the principal fourth root of 16 is 2, and these roots are even roots.

If n is a positive odd integer and a is any real number, there is only one real nth root of a. We call that root an **odd root.** Because $2^5 = 32$, the fifth root of 32 is 2 and 2 is an odd root.

We use the **radical symbol** $\sqrt{}$ to signify roots.

The parts of a radical:

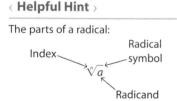

Index — Radical symbol — Radicand

> $\sqrt[n]{a}$
>
> If n is a positive *even* integer and a is positive, then $\sqrt[n]{a}$ denotes the *principal nth root of a.*
> If n is a positive *odd* integer, then $\sqrt[n]{a}$ denotes the nth root of a.
> If n is any positive integer, then $\sqrt[n]{0} = 0$.

We read $\sqrt[n]{a}$ as "the nth root of a." In the notation $\sqrt[n]{a}$, n is the **index of the radical** and a is the **radicand.** For square roots the index is omitted, and we simply write $\sqrt{a}$.

E X A M P L E   **1**

Evaluating radical expressions

Find the following roots:

    a) $\sqrt{25}$

    b) $\sqrt[3]{-27}$

    c) $\sqrt[6]{64}$

    d) $-\sqrt{4}$

Solution

a) Because $5^2 = 25$, $\sqrt{25} = 5$.

b) Because $(-3)^3 = -27$, $\sqrt[3]{-27} = -3$.

c) Because $2^6 = 64$, $\sqrt[6]{64} = 2$.

d) Because $\sqrt{4} = 2$, $-\sqrt{4} = -(\sqrt{4}) = -2$.

> Now do Exercises 1–16

CAUTION In radical notation, $\sqrt{4}$ represents the *principal square root of* 4, so $\sqrt{4} = 2$. Note that -2 is also a square root of 4, but $\sqrt{4} \neq -2$.

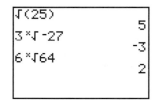

Note that even roots of negative numbers are omitted from the definition of *n*th roots because even powers of real numbers are never negative. So no real number can be an even root of a negative number. Expressions such as

$$\sqrt{-9}, \qquad \sqrt[4]{-81}, \qquad \text{and} \qquad \sqrt[6]{-64}$$

are not real numbers. Square roots of negative numbers will be discussed in Section 9.6 when we discuss the imaginary numbers.

‹2› Roots and Variables

A whole number is a perfect square if it is the square of another whole number. So 9 is a perfect square because $3^2 = 9$. Likewise, an exponential expression is a perfect square if it is the square of another exponential expression. So x^{10} is a perfect square because $(x^5)^2 = x^{10}$. The exponent in a perfect square must be divisible by 2. An exponential expression is a perfect cube if it is the cube of another exponential expression. So x^{21} is a perfect cube because $(x^7)^3 = x^{21}$. The exponent in a perfect cube must be divisible by 3. The exponent in a perfect fourth power is divisible by 4, and so on.

Perfect squares	$x^2, x^4, x^6, x^8, x^{10}, x^{12}, \ldots$	Exponent divisible by 2
Perfect cubes	$x^3, x^6, x^9, x^{12}, x^{15}, x^{18}, \ldots$	Exponent divisible by 3
Perfect fourth powers	$x^4, x^8, x^{12}, x^{16}, x^{20}, x^{24}, \ldots$	Exponent divisible by 4

To find the square root of a perfect square, divide the exponent by 2. If x is *nonnegative*, we have

$$\sqrt{x^2} = x, \quad \sqrt{x^4} = x^2, \quad \sqrt{x^6} = x^3, \quad \text{and so on.}$$

We specified that x was nonnegative because $\sqrt{x^2} = x$ and $\sqrt{x^6} = x^3$ are not correct if x is negative. If x is negative, x and x^3 are negative but the radical symbol with an even root must be a positive number. Using absolute value symbols we can say that $\sqrt{x^2} = |x|$ and $\sqrt{x^6} = |x^3|$ for any real numbers.

To find the cube root of a perfect cube, divide the exponent by 3. If x is *any real number*, we have

$$\sqrt[3]{x^3} = x, \quad \sqrt[3]{x^6} = x^2, \quad \sqrt[3]{x^9} = x^3, \quad \text{and so on.}$$

Note that both sides of each of these equations have the same sign whether x is positive or negative. For cube roots and other odd roots, we will not need absolute value symbols to make statements that are true for any real numbers. *We need absolute value symbols only when the result of an even root has an odd exponent.* For example, $\sqrt[6]{m^{30}} = |m^5|$ for any real number m.

E X A M P L E   **2**

Roots of exponential expressions

Find each root. Assume that the variables can represent any real numbers. Use absolute value symbols when necessary.

a) $\sqrt{a^2}$ b) $\sqrt{x^{22}}$ c) $\sqrt[4]{w^{40}}$ d) $\sqrt[3]{t^{18}}$ e) $\sqrt[5]{s^{30}}$

Solution

a) For a square root, divide the exponent by 2. But if a is negative, $\sqrt{a^2} = a$ is not correct, because the square root symbol represents the nonnegative square root. So if a is any real number, $\sqrt{a^2} = |a|$.

b) Divide the exponent by 2. But if x is negative, $\sqrt{x^{22}} = x^{11}$ is not correct because x^{11} is negative and $\sqrt{x^{22}}$ is positive. So if x is any real number, $\sqrt{x^{22}} = |x^{11}|$.

c) For a fourth root, divide the exponent by 4. So $\sqrt[4]{w^{40}} = w^{10}$. We don't need absolute value symbols because both sides of this equation have the same sign whether w is positive or negative.

d) For a cube root, divide the exponent by 3. So $\sqrt[3]{t^{18}} = t^6$. We don't need absolute value symbols because both sides of this equation have the same sign whether t is positive or negative.

e) For a fifth root, divide the exponent by 5. So $\sqrt[5]{s^{30}} = s^6$. We don't need absolute value symbols because both sides of this equation have the same sign whether s is positive or negative.

> Now do Exercises 17–32

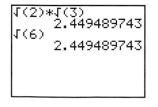
⟨3⟩ Product Rule for Radicals

Consider the expression $\sqrt{2} \cdot \sqrt{3}$. If we square this product, we get

$$(\sqrt{2} \cdot \sqrt{3})^2 = (\sqrt{2})^2(\sqrt{3})^2 \quad \text{Power of a product rule}$$
$$= 2 \cdot 3 \quad \quad (\sqrt{2})^2 = 2 \text{ and } (\sqrt{3})^2 = 3$$
$$= 6.$$

The number $\sqrt{6}$ is the unique positive number whose square is 6. Because we squared $\sqrt{2} \cdot \sqrt{3}$ and obtained 6, we must have $\sqrt{6} = \sqrt{2} \cdot \sqrt{3}$. This example illustrates the product rule for radicals.

> **Product Rule for Radicals**
>
> The nth root of a product is equal to the product of the nth roots. In symbols,
>
> $$\sqrt[n]{ab} = \sqrt[n]{a} \cdot \sqrt[n]{b},$$
>
> provided all of these roots are real numbers.

| EXAMPLE 3 | **Using the product rule for radicals to simplify** |

Simplify each radical. Assume that all variables represent nonnegative real numbers.

a) $\sqrt{4y}$ b) $\sqrt{3y^8}$ c) $\sqrt[3]{125w^2}$

Solution

a) $\sqrt{4y} = \sqrt{4} \cdot \sqrt{y}$ Product rule for radicals

      $= 2\sqrt{y}$ Simplify.

b) $\sqrt{3y^8} = \sqrt{3} \cdot \sqrt{y^8}$ Product rule for radicals

      $= \sqrt{3} \cdot y^4$ Simplify.

      $= y^4\sqrt{3}$ A radical is usually written last in a product.

c) $\sqrt[3]{125w^2} = \sqrt[3]{125} \cdot \sqrt[3]{w^2} = 5\sqrt[3]{w^2}$

Now do Exercises 33–44

In Example 4, we simplify by factoring the radicand before applying the product rule.

| EXAMPLE 4 | **Using the product rule to simplify** |

Simplify each radical.

a) $\sqrt{12}$ b) $\sqrt[3]{54}$ c) $\sqrt[4]{80}$ d) $\sqrt[5]{64}$

Solution

a) Since $12 = 4 \cdot 3$ and 4 is a perfect square, we can factor and then apply the product rule:

$$\sqrt{12} = \sqrt{4 \cdot 3} = \sqrt{4} \cdot \sqrt{3} = 2\sqrt{3}$$

b) Since $54 = 27 \cdot 2$ and 27 is a perfect cube, we can factor and then apply the product rule:

$$\sqrt[3]{54} = \sqrt[3]{27 \cdot 2} = \sqrt[3]{27} \cdot \sqrt[3]{2} = 3\sqrt[3]{2}$$

c) Since $80 = 16 \cdot 5$ and 16 is a perfect fourth power, we can factor and then apply the product rule:

$$\sqrt[4]{80} = \sqrt[4]{16 \cdot 5} = \sqrt[4]{16} \cdot \sqrt[4]{5} = 2\sqrt[4]{5}$$

d) $\sqrt[5]{64} = \sqrt[5]{32 \cdot 2} = \sqrt[5]{32} \cdot \sqrt[5]{2} = 2\sqrt[5]{2}$

Now do Exercises 45–58

In general, we simplify radical expressions of index n by using the product rule to remove any perfect nth powers from the radicand. In Example 5, we use the product rule to simplify more radicals involving variables. Remember x^n is a perfect square if n is divisible by 2, a perfect cube if n is divisible by 3, and so on.

E X A M P L E   **5**

Using the product rule to simplify

Simplify each radical. Assume that all variables represent nonnegative real numbers.

a) $\sqrt{20x^3}$ b) $\sqrt[3]{40a^8}$ c) $\sqrt[4]{48a^4b^{11}}$ d) $\sqrt[5]{w^7}$

Solution

a) Factor $20x^3$ so that all possible perfect squares are inside one radical:

$$\sqrt{20x^3} = \sqrt{4x^2 \cdot 5x} \qquad \text{Factor out perfect squares.}$$
$$= \sqrt{4x^2} \cdot \sqrt{5x} \quad \text{Product rule}$$
$$= 2x\sqrt{5x} \qquad \text{Simplify.}$$

b) Factor $40a^8$ so that all possible perfect cubes are inside one radical:

$$\sqrt[3]{40a^8} = \sqrt[3]{8a^6 \cdot 5a^2} \qquad \text{Factor out perfect cubes.}$$
$$= \sqrt[3]{8a^6} \cdot \sqrt[3]{5a^2} \quad \text{Product rule}$$
$$= 2a^2\sqrt[3]{5a^2} \qquad \text{Simplify.}$$

c) Factor $48a^4b^{11}$ so that all possible perfect fourth powers are inside one radical:

$$\sqrt[4]{48a^4b^{11}} = \sqrt[4]{16a^4b^8 \cdot 3b^3} \qquad \text{Factor out perfect fourth powers.}$$
$$= \sqrt[4]{16a^4b^8} \cdot \sqrt[4]{3b^3} \quad \text{Product rule}$$
$$= 2ab^2\sqrt[4]{3b^3} \qquad \text{Simplify.}$$

d) $\sqrt[5]{w^7} = \sqrt[5]{w^5 \cdot w^2} = \sqrt[5]{w^5} \cdot \sqrt[5]{w^2} = w\sqrt[5]{w^2}$

> Now do Exercises 59–72

‹ **Calculator Close-Up** ›

You can illustrate the quotient rule for radicals with a calculator.

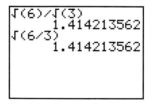

⟨4⟩ Quotient Rule for Radicals

Because $\sqrt{2} \cdot \sqrt{3} = \sqrt{6}$, we have $\sqrt{6} \div \sqrt{3} = \sqrt{2}$, or

$$\sqrt{2} = \sqrt{\frac{6}{3}} = \frac{\sqrt{6}}{\sqrt{3}}.$$

This example illustrates the quotient rule for radicals.

Quotient Rule for Radicals

The nth root of a quotient is equal to the quotient of the nth roots. In symbols,

$$\sqrt[n]{\frac{a}{b}} = \frac{\sqrt[n]{a}}{\sqrt[n]{b}},$$

provided that all of these roots are real numbers and $b \neq 0$.

E X A M P L E   **6**

Using the quotient rule for radicals

Simplify each radical. Assume that all variables represent positive real numbers.

a) $\sqrt{\dfrac{25}{9}}$ b) $\dfrac{\sqrt{15}}{\sqrt{3}}$ c) $\sqrt[3]{\dfrac{b}{125}}$ d) $\sqrt[3]{\dfrac{x^{21}}{y^6}}$

Solution

a) $\sqrt{\dfrac{25}{9}} = \dfrac{\sqrt{25}}{\sqrt{9}}$ Quotient rule for radicals

$\quad\quad = \dfrac{5}{3}$ Simplify.

b) $\dfrac{\sqrt{15}}{\sqrt{3}} = \sqrt{\dfrac{15}{3}}$ Quotient rule for radicals

$\quad\quad = \sqrt{5}$ Simplify.

c) $\sqrt[3]{\dfrac{b}{125}} = \dfrac{\sqrt[3]{b}}{\sqrt[3]{125}} = \dfrac{\sqrt[3]{b}}{5}$

d) $\sqrt[3]{\dfrac{x^{21}}{y^6}} = \dfrac{\sqrt[3]{x^{21}}}{\sqrt[3]{y^6}} = \dfrac{x^7}{y^2}$

> Now do Exercises 73–84

In Example 7, we use the product and quotient rules to simplify radical expressions.

EXAMPLE 7

Using the product and quotient rules for radicals

Simplify each radical. Assume that all variables represent positive real numbers.

a) $\sqrt{\dfrac{50}{49}}$ **b)** $\sqrt[3]{\dfrac{x^5}{8}}$ **c)** $\sqrt[4]{\dfrac{a^5}{b^8}}$

Solution

a) $\sqrt{\dfrac{50}{49}} = \dfrac{\sqrt{25} \cdot \sqrt{2}}{\sqrt{49}}$ Product and quotient rules for radicals

$\quad\quad = \dfrac{5\sqrt{2}}{7}$ Simplify.

b) $\sqrt[3]{\dfrac{x^5}{8}} = \dfrac{\sqrt[3]{x^3} \cdot \sqrt[3]{x^2}}{\sqrt[3]{8}} = \dfrac{x\sqrt[3]{x^2}}{2}$

c) $\sqrt[4]{\dfrac{a^5}{b^8}} = \dfrac{\sqrt[4]{a^4} \cdot \sqrt[4]{a}}{\sqrt[4]{b^8}} = \dfrac{a\sqrt[4]{a}}{b^2}$

> Now do Exercises 85–96

⟨5⟩ Domain of a Radical Expression or Function

The domain of any expression involving one variable is the set of all real numbers that can be used in place of the variable. For many expressions the domain of the expression is the set of all real numbers. For example, any real number can be used in place of x in the expression $2x + 3$ and its domain is the set of all real numbers, $(-\infty, \infty)$.

For a radical expression the domain depends on the radicand and whether the root is even or odd. Since every real number has an odd root, the domain of $\sqrt[3]{x}$ is $(-\infty, \infty)$.

Since there are no real even roots of negative numbers, the domain of $\sqrt{x}$ is the set of nonnegative real numbers or $[0, \infty)$.

E X A M P L E   **8**

Finding the domain of a radical expression

Find the domain of each expression. Express the answer in interval notation.

a) $\sqrt{x - 5}$ b) $\sqrt[3]{x + 7}$ c) $\sqrt[4]{2x + 6}$

Solution

a) Since the radicand in a square root must be nonnegative, $x - 5$ must be nonnegative:

$$x - 5 \geq 0$$
$$x \geq 5$$

So only values of x that are 5 or larger can be used for x. The domain is $[5, \infty)$.

b) Since any real number has a cube root, any real number can be used in place of x in $\sqrt[3]{x + 7}$. So the domain is $(-\infty, \infty)$.

c) Since the radicand in a fourth root must be nonnegative, $2x + 6$ must be nonnegative:

$$2x + 6 \geq 0$$
$$2x \geq -6$$
$$x \geq -3$$

So the domain of $\sqrt[4]{2x + 6}$ is $[-3, \infty)$.

> Now do Exercises 97–110

If a radical expression is used to determine the value of a second variable y, then we have a **radical function**. For example,

$$R(x) = \sqrt{x - 5}, \quad V(x) = \sqrt[3]{x + 7}, \quad \text{and} \quad T(x) = \sqrt[4]{2x + 6}$$

are radical functions. The **domain of a radical function** is the domain of the radical expression. Since these are the radical expressions of Example 8, the domain for $R(x)$ is $[5, \infty)$, the domain for $V(x)$ is $(-\infty, \infty)$, and the domain for $T(x)$ is $[-3, \infty)$.

Warm-Ups ▼

Fill in the blank.

1. If $b^n = a$, then b is an _____ of a.

2. If n is even and $a > 0$, then $\sqrt[n]{a}$ is the _____ nth root of a.

3. According to the _____ rule for radicals $\sqrt[n]{a} \cdot \sqrt[n]{b} = \sqrt[n]{ab}$ provided all of the roots are real.

4. According to the _____ rule for radicals $\sqrt[n]{a}/\sqrt[n]{b} = \sqrt[n]{a/b}$ provided all of the roots are real.

True or false?

5. $\sqrt{2} \cdot \sqrt{2} = 2$

6. $\sqrt[3]{2} \cdot \sqrt[3]{2} = 2$

7. $\sqrt[3]{-27} = -3$

8. $\sqrt[4]{16} = 2$

9. $\sqrt{9} = \pm 3$

10. $\sqrt{2} \cdot \sqrt{7} = \sqrt{14}$

11. $\dfrac{\sqrt{6}}{\sqrt{2}} = \sqrt{3}$

12. $\dfrac{\sqrt{10}}{2} = \sqrt{5}$

‹1› Roots

Find each root. See Example 1.

1. $\sqrt{36}$
2. $\sqrt{49}$
3. $\sqrt{100}$
4. $\sqrt{81}$
5. $-\sqrt{9}$
6. $-\sqrt{25}$
7. $\sqrt[3]{8}$
8. $\sqrt[3]{27}$
9. $\sqrt[3]{-8}$
10. $\sqrt[3]{-1}$
11. $\sqrt[5]{32}$
12. $\sqrt[4]{81}$
13. $\sqrt[3]{1000}$
14. $\sqrt[4]{16}$
15. $\sqrt[4]{-16}$
16. $\sqrt{-1}$

‹2› Roots and Variables

Find each root. See Example 2. All variables represent real numbers. Use absolute value when necessary.

17. $\sqrt{m^2}$
18. $\sqrt{m^6}$
19. $\sqrt{x^{16}}$
20. $\sqrt{y^{36}}$
21. $\sqrt[5]{y^{15}}$
22. $\sqrt[4]{m^8}$
23. $\sqrt[3]{y^{15}}$
24. $\sqrt{m^8}$
25. $\sqrt[3]{m^3}$
26. $\sqrt[4]{x^4}$
27. $\sqrt[4]{w^{12}}$
28. $\sqrt[5]{a^{30}}$
29. $\sqrt{b^{18}}$
30. $\sqrt[6]{m^{42}}$
31. $\sqrt[4]{y^{24}}$
32. $\sqrt{t^{44}}$

‹3› Product Rule for Radicals

Use the product rule for radicals to simplify each expression. See Example 3. All variables represent nonnegative real numbers.

33. $\sqrt{9y}$
34. $\sqrt{16n}$
35. $\sqrt{4a^2}$
36. $\sqrt{36n^2}$
37. $\sqrt{x^4y^2}$
38. $\sqrt{w^6t^2}$
39. $\sqrt{5m^{12}}$
40. $\sqrt{7z^{16}}$
41. $\sqrt[3]{8y}$
42. $\sqrt[3]{27z^2}$
43. $\sqrt[3]{3a^6}$
44. $\sqrt[3]{5b^9}$

Use the product rule to simplify. See Example 4.

45. $\sqrt{20}$
46. $\sqrt{18}$
47. $\sqrt{50}$
48. $\sqrt{45}$

49. $\sqrt{72}$
50. $\sqrt{98}$
51. $\sqrt[3]{40}$
52. $\sqrt[3]{24}$
53. $\sqrt[3]{81}$
54. $\sqrt[3]{250}$
55. $\sqrt[4]{48}$
56. $\sqrt[4]{32}$
57. $\sqrt[5]{96}$
58. $\sqrt[5]{2430}$

Use the product rule to simplify. See Example 5. All variables represent nonnegative real numbers.

59. $\sqrt{a^3}$
60. $\sqrt{b^5}$
61. $\sqrt{18a^6}$
62. $\sqrt{12x^8}$
63. $\sqrt{20x^5y}$
64. $\sqrt{8w^3y^3}$
65. $\sqrt[3]{24m^4}$
66. $\sqrt[3]{54ab^5}$
67. $\sqrt[4]{32a^5}$
68. $\sqrt[4]{162b^4}$
69. $\sqrt[5]{64x^6}$
70. $\sqrt[5]{96a^8}$
71. $\sqrt{48x^3y^8z^7}$
72. $\sqrt[3]{48x^3y^8z^7}$

‹4› Quotient Rule for Radicals

Simplify each radical. See Example 6. All variables represent positive real numbers.

73. $\sqrt{\dfrac{t}{4}}$
74. $\sqrt{\dfrac{w}{36}}$
75. $\sqrt{\dfrac{625}{16}}$
76. $\sqrt{\dfrac{9}{144}}$
77. $\dfrac{\sqrt{30}}{\sqrt{3}}$
78. $\dfrac{\sqrt{50}}{\sqrt{2}}$
79. $\sqrt[3]{\dfrac{t}{8}}$
80. $\sqrt[3]{\dfrac{a}{27}}$
81. $\sqrt[3]{\dfrac{-8x^6}{y^3}}$
82. $\sqrt[3]{\dfrac{-27y^{36}}{1000}}$
83. $\sqrt{\dfrac{4a^6}{9}}$
84. $\sqrt{\dfrac{9a^2}{49b^4}}$

Use the product and quotient rules to simplify. See Example 7. All variables represent positive real numbers.

85. $\sqrt{\dfrac{12}{25}}$ 　　　　　　**86.** $\sqrt{\dfrac{8}{81}}$

87. $\sqrt{\dfrac{27}{16}}$ 　　　　　　**88.** $\sqrt{\dfrac{98}{9}}$

89. $\sqrt[3]{\dfrac{a^4}{125}}$ 　　　　　　**90.** $\sqrt[3]{\dfrac{b^7}{1000}}$

91. $\sqrt[3]{\dfrac{81}{8b^3}}$ 　　　　　　**92.** $\sqrt[3]{\dfrac{a^3b^4}{125}}$

93. $\sqrt[4]{\dfrac{x^7}{y^8}}$ 　　　　　　**94.** $\sqrt[4]{\dfrac{x^5y^4}{z^{12}}}$

95. $\sqrt[4]{\dfrac{a^5}{16b^{12}}}$ 　　　　　**96.** $\sqrt[4]{\dfrac{a^7b}{81c^{16}}}$

⟨5⟩ Domain of a Radical Expression or Function

Find the domain of each radical expression. See Example 8.

97. $\sqrt{x-2}$

98. $\sqrt{2-x}$

99. $\sqrt[3]{3x-7}$

100. $\sqrt[3]{5-4x}$

101. $\sqrt[4]{9-3x}$

102. $\sqrt[4]{4x-8}$

103. $\sqrt{2x+1}$

104. $\sqrt{4x-1}$

Find the domain of each radical function.

105. $R(x) = \sqrt{x-6}$

106. $V(x) = \sqrt{7-x}$

107. $y = \sqrt[3]{x+1}$

108. $y = \sqrt[5]{3x-2}$

109. $S(x) = \sqrt[4]{9-x}$

110. $T(x) = \sqrt[4]{x-9}$

Applications

Solve each problem.

111. *Wind chill.* The wind chill temperature W (how cold the air feels) is determined by the air temperature t and

the wind velocity v. Through experimentation in Antarctica, Paul Siple developed a formula for W:

$$W = 91.4 - \frac{(10.5 + 6.7\sqrt{v} - 0.45v)(457 - 5t)}{110},$$

where W and t are in degrees Fahrenheit and v is in miles per hour (mph).

a) Find W to the nearest whole degree when $t = 25°F$ and $v = 20$ mph.

b) Use the accompanying graph to estimate W when $t = 25°F$ and $v = 30$ mph.

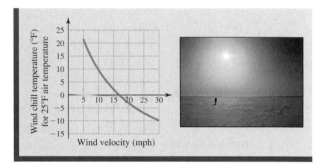

Figure for Exercise 111

112. *Comparing wind chills.* Use the formula from Exercise 111 to determine who will feel colder: a person in Minneapolis at 10°F with a 15-mph wind or a person in Chicago at 20°F with a 25-mph wind.

113. *Diving time.* The time t (in seconds) that it takes for a cliff diver to reach the water is a function of the height h (in feet) from which he dives:

$$t = \sqrt{\frac{h}{16}}$$

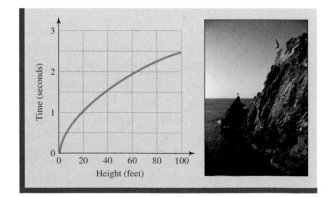

Figure for Exercise 113

a) Use the properties of radicals to simplify this formula.

b) Find the exact time (according to the formula) that it takes for a diver to hit the water when diving from a height of 40 feet.

c) Use the graph to estimate the height if a diver takes 2.5 seconds to reach the water.

114. *Sky diving.* The formula in Exercise 113 accounts for the effect of gravity only on a falling object. According to that formula, how long would it take a sky diver to reach the earth when jumping from 17,000 feet? (A sky diver can actually get about twice as much falling time by spreading out and using the air to slow the fall.)

115. *Maximum sailing speed.* To find the maximum possible speed in knots (nautical miles per hour) for a sailboat, sailors use the function $M = 1.3\sqrt{w}$, where w is the length of the waterline in feet. If the waterline for the sloop *Golden Eye* is 20 feet, then what is the maximum speed of the *Golden Eye*?

116. *America's Cup.* Since 1988 basic yacht dimensions for the America's Cup competition have satisfied the inequality

$$L + 1.25\sqrt{S} - 9.8\sqrt[3]{D} \le 16.296,$$

where L is the boat's length in meters (m), S is the sail area in square meters (m^2), and D is the displacement in cubic meters (www.sailing.com). A team of naval architects is planning to build a boat with a displacement of 21.44 cubic meters (m^3), a sail area of 320.13 m^2, and a length of 21.22 m. Does this boat satisfy the inequality? If the length and displacement of this boat cannot be changed, then how many square meters of sail area must be removed so that the boat satisfies the inequality?

117. *Landing speed.* The proper landing speed for an airplane V (in feet per second) is determined from the gross weight of the aircraft L (in pounds), the coefficient of lift C, and the wing surface area S (in square feet), by the formula

$$V = \sqrt{\frac{841L}{CS}}.$$

a) Find V (to the nearest tenth) for the Piper Cheyenne, for which $L = 8700$ lb, $C = 2.81$, and $S = 200$ ft².

b) Find V in miles per hour (to the nearest tenth).

118. *Landing speed and weight.* Because the gross weight of the Piper Cheyenne depends on how much fuel and cargo are on board, the proper landing speed (from Exercise 117) is not always the same. The formula $V = \sqrt{1.496L}$ gives the landing speed in terms of the gross weight only.

a) Find the landing speed if the gross weight is 7000 lb.

b) What gross weight corresponds to a landing speed of 115 ft/sec?

Getting More Involved

119. *Cooperative learning*

Work in a group to determine whether each equation is an identity. Explain your answers.

a) $\sqrt{x^2} = |x|$ b) $\sqrt[3]{x^3} = |x|$

c) $\sqrt{x^4} = x^2$ d) $\sqrt[4]{x^4} = |x|$

For which values of n is $\sqrt[n]{x^n} = x$ an identity?

120. *Cooperative learning*

Work in a group to determine whether each inequality is correct.

a) $\sqrt{0.9} > 0.9$

b) $\sqrt{1.01} > 1.01$

c) $\sqrt[3]{0.99} > 0.99$

d) $\sqrt[3]{1.001} > 1.001$

For which values of x and n is $\sqrt[n]{x} > x$?

121. *Discussion*

If your test scores are 80 and 100, then the arithmetic mean of your scores is 90. The geometric mean of the scores is a number h such that

$$\frac{80}{h} = \frac{h}{100}.$$

Are you better off with the arithmetic mean or the geometric mean?

<table>
<tr><td>

Math *at Work*   **Deficit and Debt**

Have you ever heard politicians talk about budget surpluses and lowering the deficit, while the national debt keeps increasing? The national debt has increased every year since 1967 and stood at $11.3 trillion in 2009. Confusing? Not if you know the definitions of these words. If the federal government spends more than it collects in taxes in a particular year, then it has a *deficit*. The amount that is overspent must be borrowed, and that adds to the *national debt,* which is the total amount that the federal government owes. Interest alone on the national debt was $676 billion in 2009 and is the second largest expense in the federal budget.

To get an idea of the size of the national debt, divide the $11.3 trillion debt in 2009 by the U.S. population of 306 million to get about $37,000 per person. The national debt went from $2.4 trillion in 1987 to $11.3 trillion in 2009. We can calculate the average annual percentage increase in the debt for these 22 years using the formula $i = \sqrt[n]{A/P} - 1$, which yields $i = \sqrt[22]{11.3/2.4} - 1 \approx 7.3\%$. With the U.S. population increasing an average of 1% per year and the debt increasing 7.3% per year, in 25 years the debt will be $11.3(1 + 0.073)^{25}$ or about $65.8 trillion while the population will increase to $306(1 + 0.01)^{25}$ or about 392 million. See the accompanying figure. So in 25 years the debt will be about $168,000 per person. Since only one person in three is a wage earner, the debt will be about one-half of a million dollars per wage earner!

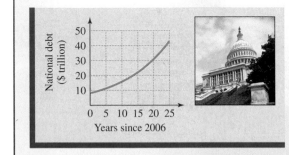

</td></tr>
</table>

9.2   Rational Exponents

In This Section

〈1〉 **Rational Exponents**

〈2〉 **Using the Rules of Exponents**

〈3〉 **Simplifying Expressions Involving Variables**

You have learned how to use exponents to express powers of numbers and radicals to express roots. In this section, you will see that roots can be expressed with exponents also. The advantage of using exponents to express roots is that the rules of exponents can be applied to the expressions.

〈1〉   Rational Exponents

Cubing and cube root are inverse operations. For example, if we start with 2 and apply both operations we get back 2: $\sqrt[3]{2^3} = 2$. If we were to use an exponent for cube root, then we must have $(2^3)^? = 2$. The only exponent that is consistent with the power of a power rule is $\frac{1}{3}$ because $(2^3)^{1/3} = 2^1 = 2$. So we make the following definition.

〈 **Calculator Close-Up** 〉

You can find the fifth root of 2 using radical notation or exponent notation. Note that the fractional exponent 1/5 must be in parentheses.

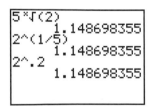

Definition of $a^{1/n}$

If n is any positive integer, then

$$a^{1/n} = \sqrt[n]{a},$$

provided that $\sqrt[n]{a}$ is a real number.

Later in this section we will see that using exponent $1/n$ for the nth root is compatible with the rules for integral exponents that we already know.

EXAMPLE 1

Radicals or exponents

Write each radical expression using exponent notation and each exponential expression using radical notation.

   a) $\sqrt[3]{35}$ **b)** $\sqrt[4]{xy}$ **c)** $5^{1/2}$ **d)** $a^{1/5}$

Solution

   a) $\sqrt[3]{35} = 35^{1/3}$ **b)** $\sqrt[4]{xy} = (xy)^{1/4}$

   c) $5^{1/2} = \sqrt{5}$ **d)** $a^{1/5} = \sqrt[5]{a}$

> Now do Exercises 1–8

In Example 2, we evaluate some exponential expressions.

EXAMPLE 2

Finding roots

Evaluate each expression.

   a) $4^{1/2}$ **b)** $(-8)^{1/3}$ **c)** $81^{1/4}$

   d) $(-9)^{1/2}$ **e)** $-9^{1/2}$

Solution

   a) $4^{1/2} = \sqrt{4} = 2$

   b) $(-8)^{1/3} = \sqrt[3]{-8} = -2$

   c) $81^{1/4} = \sqrt[4]{81} = 3$

   d) Because $(-9)^{1/2}$ or $\sqrt{-9}$ is an even root of a negative number, it is not a real number.

   e) Because the exponent in $-a^n$ is applied only to the base a (Section 1.5), we have $-9^{1/2} = -\sqrt{9} = -3$.

> Now do Exercises 9–16

We now extend the definition of exponent $1/n$ to include any rational number as an exponent. The numerator of the rational number indicates the power, and the denominator indicates the root. For example, the expression

$$8^{2/3} \quad \begin{array}{l}\longleftarrow \text{Power} \\ \longleftarrow \text{Root}\end{array}$$

represents the square of the cube root of 8. So we have

$$8^{2/3} = \left(8^{1/3}\right)^2 = (2)^2 = 4.$$

‹ **Helpful Hint** ›

Note that in $a^{m/n}$ we do not require m/n to be reduced. As long as the nth root of a is real, then the value of $a^{m/n}$ is the same whether or not m/n is in lowest terms.

Definition of $a^{m/n}$

If m and n are positive integers and $a^{1/n}$ is a real number, then

$$a^{m/n} = \left(a^{1/n}\right)^m.$$

Using radical notation, $a^{m/n} = \left(\sqrt[n]{a}\right)^m$.

By definition $a^{m/n}$ is the mth power of the nth root of a. However, $a^{m/n}$ is also equal to the nth root of the mth power of a. For example,

$$8^{2/3} = \left(8^2\right)^{1/3} = 64^{1/3} = 4.$$

Evaluating $a^{m/n}$ in Either Order

If m and n are positive integers and $a^{1/n}$ is a real number, then

$$a^{m/n} = \left(a^{1/n}\right)^m = \left(a^m\right)^{1/n}.$$

Using radical notation, $a^{m/n} = \left(\sqrt[n]{a}\right)^m = \sqrt[n]{a^m}$.

A negative rational exponent indicates a reciprocal:

Definition of $a^{-m/n}$

If m and n are positive integers, $a \neq 0$, and $a^{1/n}$ is a real number, then

$$a^{-m/n} = \frac{1}{a^{m/n}}.$$

Using radical notation, $a^{-m/n} = \dfrac{1}{\left(\sqrt[n]{a}\right)^m}$.

E X A M P L E   **3**

Radicals to exponents

Write each radical expression using exponent notation.

   a) $\sqrt[3]{x^2}$ **b)** $\dfrac{1}{\sqrt[4]{m^3}}$

Solution

   a) $\sqrt[3]{x^2} = x^{2/3}$ **b)** $\dfrac{1}{\sqrt[4]{m^3}} = \dfrac{1}{m^{3/4}} = m^{-3/4}$

> Now do Exercises 17–20

E X A M P L E   **4**

Exponents to radicals

Write each exponential expression using radicals.

   a) $5^{2/3}$ **b)** $a^{-2/5}$

Solution

   a) $5^{2/3} = \sqrt[3]{5^2}$ **b)** $a^{-2/5} = \dfrac{1}{\sqrt[5]{a^2}}$

> Now do Exercises 21–24

To evaluate an expression with a negative rational exponent, remember that the denominator indicates root, the numerator indicates power, and the negative sign indicates reciprocal:

The root, power, and reciprocal can be evaluated in any order. However, it is usually simplest to use the following strategy.

Strategy for Evaluating $a^{-m/n}$

1. Find the nth root of a.
2. Raise your result to the mth power.
3. Find the reciprocal.

For example, to evaluate $8^{-2/3}$, we find the cube root of 8 (which is 2), square 2 to get 4, then find the reciprocal of 4 to get $\frac{1}{4}$. In print $8^{-2/3}$ could be written for evaluation as $((8^{1/3})^2)^{-1}$ or $\frac{1}{(8^{1/3})^2}$.

EXAMPLE 5

Rational exponents
Evaluate each expression.

a) $27^{2/3}$ b) $4^{-3/2}$ c) $81^{-3/4}$ d) $(-8)^{-5/3}$

Solution

a) Because the exponent is 2/3, we find the cube root of 27 and then square it:
$$27^{2/3} = (27^{1/3})^2 = 3^2 = 9$$

b) Because the exponent is $-3/2$, we find the square root of 4, cube it, and find the reciprocal:
$$4^{-3/2} = \frac{1}{(4^{1/2})^3} = \frac{1}{2^3} = \frac{1}{8}$$

c) Because the exponent is $-3/4$, we find the fourth root of 81, cube it, and find the reciprocal:
$$81^{-3/4} = \frac{1}{(81^{1/4})^3} = \frac{1}{3^3} = \frac{1}{27} \quad \text{Definition of negative exponent}$$

d) $(-8)^{-5/3} = \dfrac{1}{((-8)^{1/3})^5} = \dfrac{1}{(-2)^5} = \dfrac{1}{-32} = -\dfrac{1}{32}$

> Now do Exercises 25–36

CAUTION An expression with a negative base and a negative exponent can have a positive or a negative value. For example,

$$(-8)^{-5/3} = -\frac{1}{32} \quad \text{and} \quad (-8)^{-2/3} = \frac{1}{4}.$$

⟨2⟩ Using the Rules of Exponents

All of the rules for integral exponents that you learned in Sections 4.1 and 4.2 hold for rational exponents as well. We restate those rules in the following box. Note that some expressions with rational exponents [such as $(-3)^{3/4}$] are not real numbers and the rules do not apply to such expressions.

Rules for Rational Exponents

The following rules hold for any nonzero real numbers a and b and rational numbers r and s for which the expressions represent real numbers.

1. $a^r a^s = a^{r+s}$ Product rule
2. $\dfrac{a^r}{a^s} = a^{r-s}$ Quotient rule
3. $(a^r)^s = a^{rs}$ Power of a power rule
4. $(ab)^r = a^r b^r$ Power of a product rule
5. $\left(\dfrac{a}{b}\right)^r = \dfrac{a^r}{b^r}$ Power of a quotient rule

---

We can use the product rule to add rational exponents. For example,

$$16^{1/4} \cdot 16^{1/4} = 16^{2/4}.$$

The fourth root of 16 is 2, and 2 squared is 4. So $16^{2/4} = 4$. Because we also have $16^{1/2} = 4$, we see that a rational exponent can be reduced to its lowest terms. If an exponent can be reduced, it is usually simpler to reduce the exponent before we evaluate the expression. We can simplify $16^{1/4} \cdot 16^{1/4}$ as follows:

$$16^{1/4} \cdot 16^{1/4} = 16^{2/4} = 16^{1/2} = 4$$

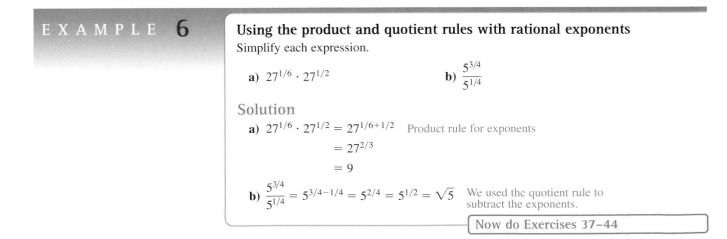

E X A M P L E   6

Using the product and quotient rules with rational exponents
Simplify each expression.

a) $27^{1/6} \cdot 27^{1/2}$

b) $\dfrac{5^{3/4}}{5^{1/4}}$

Solution

a) $27^{1/6} \cdot 27^{1/2} = 27^{1/6+1/2}$ Product rule for exponents

$$= 27^{2/3}$$

$$= 9$$

b) $\dfrac{5^{3/4}}{5^{1/4}} = 5^{3/4-1/4} = 5^{2/4} = 5^{1/2} = \sqrt{5}$ We used the quotient rule to subtract the exponents.

Now do Exercises 37–44

EXAMPLE 7

Using the power rules with rational exponents

Simplify each expression.

a) $3^{1/2} \cdot 12^{1/2}$ **b)** $\left(3^{10}\right)^{1/2}$ **c)** $\left(\dfrac{2^6}{3^9}\right)^{-1/3}$

Solution

a) Because the bases 3 and 12 are different, we cannot use the product rule to add the exponents. Instead, we use the power of a product rule to place the $1/2$ power outside the parentheses:

$$3^{1/2} \cdot 12^{1/2} = (3 \cdot 12)^{1/2} = 36^{1/2} = 6$$

b) Use the power of a power rule to multiply the exponents:

$$\left(3^{10}\right)^{1/2} = 3^5$$

c) $\left(\dfrac{2^6}{3^9}\right)^{-1/3} = \dfrac{\left(2^6\right)^{-1/3}}{\left(3^9\right)^{-1/3}}$ Power of a quotient rule

$$= \dfrac{2^{-2}}{3^{-3}}$$ Power of a power rule

$$= \dfrac{3^3}{2^2}$$ Definition of negative exponent

$$= \dfrac{27}{4}$$

> Now do Exercises 45–54

‹ **Helpful Hint** ›

We usually think of squaring and taking a square root as inverse operations, which they are as long as we stick to positive numbers. We can square 3 to get 9, and then find the square root of 9 to get 3—what we started with. We don't get back to where we began if we start with -3.

‹3› Simplifying Expressions Involving Variables

When simplifying expressions involving rational exponents and variables, we must be careful to write equivalent expressions. For example, in the equation

$$\left(x^2\right)^{1/2} = x$$

it looks as if we are correctly applying the power of a power rule. However, this statement is false if x is negative because the $1/2$ power on the left-hand side indicates the positive square root of x^2. For example, if $x = -3$, we get

$$\left[(-3)^2\right]^{1/2} = 9^{1/2} = 3,$$

which is not equal to -3. To write a simpler equivalent expression for $\left(x^2\right)^{1/2}$, we use absolute value as follows.

> **Square Root of x^2**
>
> For any real number x,
>
> $$\left(x^2\right)^{1/2} = |x| \quad \text{and} \quad \sqrt{x^2} = |x|.$$

Note that both $\left(x^2\right)^{1/2} = |x|$ and $\sqrt{x^2} = |x|$ are identities. They are true whether x is positive, negative, or zero.

It is also necessary to use absolute value when writing identities for other even roots of expressions involving variables.

E X A M P L E   **8**

Using absolute value symbols with roots

Simplify each expression. Assume the variables represent any real numbers and use absolute value symbols as necessary.

a) $\left(x^8y^4\right)^{1/4}$

b) $\left(\dfrac{x^9}{8}\right)^{1/3}$

Solution

a) Apply the power of a product rule to get the equation $\left(x^8y^4\right)^{1/4} = x^2y$. The left-hand side is nonnegative for any choices of x and y, but the right-hand side is negative when y is negative. So for any real values of x and y we have

$$\left(x^8y^4\right)^{1/4} = x^2\,|\,y\,|.$$

Note that the absolute value symbols could also be placed around the entire expression: $\left(x^8y^4\right)^{1/4} = |\,x^2y\,|$.

b) Using the power of a quotient rule, we get

$$\left(\dfrac{x^9}{8}\right)^{1/3} = \dfrac{x^3}{2}.$$

This equation is valid for every real number x, so no absolute value signs are used.

> Now do Exercises 55–64

Because there are no real even roots of negative numbers, the expressions

$$a^{1/2}, \quad x^{-3/4}, \quad \text{and} \quad y^{1/6}$$

are not real numbers if the variables have negative values. To simplify matters, we sometimes assume the variables represent only positive numbers when we are working with expressions involving variables with rational exponents. That way we do not have to be concerned with undefined expressions and absolute value.

E X A M P L E   **9**

Expressions involving variables with rational exponents

Use the rules of exponents to simplify the following. Write your answers with positive exponents. Assume all variables represent *positive* real numbers.

a) $x^{2/3}x^{4/3}$

b) $\dfrac{a^{1/2}}{a^{1/4}}$

c) $\left(x^{1/2}y^{-3}\right)^{1/2}$

d) $\left(\dfrac{x^2}{y^{1/3}}\right)^{-1/2}$

Solution

a) $x^{2/3}x^{4/3} = x^{6/3}$ Use the product rule to add the exponents.

   $\qquad\qquad = x^2$ Reduce the exponent.

b) $\dfrac{a^{1/2}}{a^{1/4}} = a^{1/2-1/4}$ Use the quotient rule to subtract the exponents.

   $\qquad = a^{1/4}$ Simplify.

c) $\left(x^{1/2}y^{-3}\right)^{1/2} = \left(x^{1/2}\right)^{1/2}\left(y^{-3}\right)^{1/2}$ Power of a product rule

   $\qquad\qquad = x^{1/4}y^{-3/2}$ Power of a power rule

   $\qquad\qquad = \dfrac{x^{1/4}}{y^{3/2}}$ Definition of negative exponent

d) Because this expression is a negative power of a quotient, we can first find the reciprocal of the quotient and then apply the power of a power rule:

$$\left(\frac{x^2}{y^{1/3}}\right)^{-1/2} = \left(\frac{y^{1/3}}{x^2}\right)^{1/2} = \frac{y^{1/6}}{x} \quad \frac{1}{3}\cdot\frac{1}{2} = \frac{1}{6}$$

Now do Exercises 65–76

Warm-Ups ▼

Fill in the blank.

1. The notation $a^{1/n}$ represents the _____ of a.

2. The notation $a^{m/n}$ represents the _____ of the nth root.

3. The expression $a^{-m/n}$ is the _____ of $a^{m/n}$.

4. The expression $a^{-m/n}$ is a real number except when n is ____ and a is _____, or when $a = 0$.

True or false?

5. $9^{1/3} = \sqrt[3]{9}$

6. $8^{5/3} = \sqrt[5]{8^3}$

7. $(-16)^{1/2} = -16^{1/2}$

8. $9^{-3/2} = \dfrac{1}{27}$

9. $6^{-1/2} = \dfrac{\sqrt{6}}{6}$

10. $2^{1/2}\cdot 2^{1/2} = 4^{1/2}$

11. $6^{1/6}\cdot 6^{1/6} = 6^{1/3}$

12. $(2^8)^{3/4} = 2^6$

Exercises   9.2

⟨ **Study Tips** ⟩

- Avoid cramming. When you have limited time to study for a test, start with class notes and homework assignments. Work one or two problems of each type.
- Don't get discouraged if you cannot work the hardest problems. Instructors often ask some relatively easy questions to see if you understand the basics.

⟨**1**⟩ **Rational Exponents**

Write each radical expression using exponent notation. See Example 1.

1. $\sqrt[4]{7}$ **2.** $\sqrt[3]{cbs}$

3. $\sqrt{5x}$ **4.** $\sqrt{3y}$

Write each exponential expression using radical notation. See Example 1.

5. $9^{1/5}$

6. $3^{1/2}$

7. $a^{1/2}$

8. $(-b)^{1/5}$

Evaluate each expression. See Example 2.

9. $25^{1/2}$ **10.** $16^{1/2}$

11. $(-125)^{1/3}$ **12.** $(-32)^{1/5}$

13. $16^{1/4}$ **14.** $8^{1/3}$

15. $(-4)^{1/2}$ **16.** $(-16)^{1/4}$

*Write each radical expression using exponent notation.
See Example 3.*

17. $\sqrt[3]{w^7}$

18. $\sqrt{a^5}$

19. $\dfrac{1}{\sqrt[3]{2^{10}}}$

20. $\sqrt[3]{\dfrac{1}{a^2}}$

*Write each exponential expression using radical notation.
See Example 4.*

21. $w^{-3/4}$

22. $6^{-5/3}$

23. $(ab)^{3/2}$

24. $(3m)^{-1/5}$

*Evaluate each expression. See Example 5. See the Strategy for
Evaluating $a^{-m/n}$ box on page 571.*

25. $125^{2/3}$

26. $1000^{2/3}$

27. $25^{3/2}$

28. $16^{3/2}$

29. $27^{-4/3}$

30. $16^{-3/4}$

31. $16^{-3/2}$

32. $25^{-3/2}$

33. $(-27)^{-1/3}$

34. $(-8)^{-4/3}$

35. $(-16)^{-1/4}$

36. $(-100)^{-3/2}$

⟨2⟩ Using the Rules of Exponents

*Use the rules of exponents to simplify each expression.
See Examples 6 and 7.*

37. $3^{1/3}3^{1/4}$

38. $2^{1/2}2^{1/3}$

39. $3^{1/3}3^{-1/3}$

40. $5^{1/4}5^{-1/4}$

41. $\dfrac{8^{1/3}}{8^{2/3}}$

42. $\dfrac{27^{-2/3}}{27^{-1/3}}$

43. $4^{3/4} \div 4^{1/4}$

44. $9^{1/4} \div 9^{3/4}$

45. $18^{1/2}2^{1/2}$

46. $8^{1/2}2^{1/2}$

47. $(2^6)^{1/3}$

48. $(3^{10})^{1/5}$

49. $(3^8)^{1/2}$

50. $(3^{-6})^{1/3}$

51. $(2^{-4})^{1/2}$

52. $(5^4)^{1/2}$

53. $\left(\dfrac{3^4}{2^6}\right)^{1/2}$

54. $\left(\dfrac{5^4}{3^6}\right)^{1/2}$

⟨3⟩ Simplifying Expressions Involving Variables

*Simplify each expression. Assume the variables represent any
real numbers and use absolute value as necessary. See
Example 8.*

55. $(x^4)^{1/4}$

56. $(y^6)^{1/6}$

57. $(a^8)^{1/2}$

58. $(b^{10})^{1/2}$

59. $(y^3)^{1/3}$

60. $(w^9)^{1/3}$

61. $(9x^6y^2)^{1/2}$

62. $(16a^8b^4)^{1/4}$

63. $\left(\dfrac{81x^{12}}{y^{20}}\right)^{1/4}$

64. $\left(\dfrac{144a^8}{9y^{18}}\right)^{1/2}$

*Simplify. Assume all variables represent positive numbers.
Write answers with positive exponents only. See Example 9.*

65. $x^{1/2}x^{1/4}$

66. $y^{1/3}y^{1/3}$

67. $(x^{1/2}y)(x^{-3/4}y^{1/?})$

68. $(a^{1/2}b^{-1/3})(ab)$

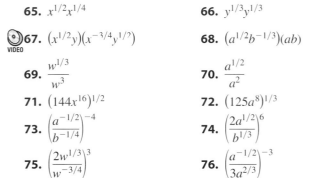

69. $\dfrac{w^{1/3}}{w^3}$

70. $\dfrac{a^{1/2}}{a^2}$

71. $(144x^{16})^{1/2}$

72. $(125a^8)^{1/3}$

73. $\left(\dfrac{a^{-1/2}}{b^{-1/4}}\right)^{-4}$

74. $\left(\dfrac{2a^{1/2}}{b^{1/3}}\right)^6$

75. $\left(\dfrac{2w^{1/3}}{w^{-3/4}}\right)^3$

76. $\left(\dfrac{a^{-1/2}}{3a^{2/3}}\right)^{-3}$

Miscellaneous

*Simplify each expression. Write your answers with positive
exponents. Assume that all variables represent positive real
numbers.*

77. $(9^2)^{1/2}$

78. $(4^{16})^{1/2}$

79. $-16^{-3/4}$

80. $-25^{-3/2}$

81. $125^{-4/3}$

82. $27^{-2/3}$

83. $2^{1/2}2^{-1/4}$

84. $9^{-1}9^{1/2}$

85. $3^{0.26}3^{0.74}$

86. $2^{1.5}2^{0.5}$

87. $3^{1/4}27^{1/4}$

88. $3^{2/3}9^{2/3}$

89. $\left(-\dfrac{8}{27}\right)^{2/3}$

90. $\left(-\dfrac{8}{27}\right)^{-1/3}$

91. $\left(-\dfrac{1}{16}\right)^{-3/4}$

92. $\left(-\dfrac{5}{9}\right)^{-7/2}$

93. $\left(\dfrac{9}{16}\right)^{-1/2}$

94. $\left(\dfrac{16}{81}\right)^{-1/4}$

95. $-\left(\dfrac{25}{36}\right)^{-3/2}$

96. $\left(-\dfrac{27}{8}\right)^{-4/3}$

97. $\left(9x^9\right)^{1/2}$

98. $\left(-27x^9\right)^{1/3}$

99. $\left(3a^{-2/3}\right)^{-3}$

100. $\left(5x^{-1/2}\right)^{-2}$

101. $\left(a^{1/2}b\right)^{1/2}\left(ab^{1/2}\right)$

102. $\left(m^{1/4}n^{1/2}\right)^2\left(m^2n^3\right)^{1/2}$

103. $\left(km^{1/2}\right)^3\left(k^3m^5\right)^{1/2}$

104. $\left(tv^{1/3}\right)^2\left(t^2v^{-3}\right)^{-1/2}$

Use a scientific calculator with a power key (x^y) to find the decimal value of each expression. Round approximate answers to four decimal places.

105. $2^{1/3}$

106. $5^{1/2}$

107. $-2^{1/2}$

108. $(-3)^{1/3}$

109. $1024^{1/10}$

110. $7776^{0.2}$

111. $\left(\dfrac{64}{15,625}\right)^{-1/6}$

112. $\left(\dfrac{32}{243}\right)^{-3/5}$

Simplify each expression. Assume a and b are positive real numbers and m and n are rational numbers.

113. $a^{m/2} \cdot a^{m/4}$

114. $b^{n/2} \cdot b^{-n/3}$

115. $\dfrac{a^{-m/5}}{a^{-m/3}}$

116. $\dfrac{b^{-n/4}}{b^{-n/3}}$

117. $\left(a^{-1/m}b^{-1/n}\right)^{-mn}$

118. $\left(a^{-m/2}b^{-n/3}\right)^{-6}$

119. $\left(\dfrac{a^{-3m}b^{-6n}}{a^{9m}}\right)^{-1/3}$

120. $\left(\dfrac{a^{-3/m}b^{6/n}}{a^{-6/m}b^{9/n}}\right)^{-1/3}$

Applications

Solve each problem. Round answers to two decimal places when necessary.

121. *Falling object.* The time in seconds t that it takes for a ball to fall to the earth from a height of h feet is given by the function

$$t(h) = 0.25h^{1/2}.$$

Find $t(1)$, $t(16)$, and $t(36)$.

122. *Sailboat speed.* The maximum speed for a sailboat in knots M is a function of the length of the waterline in feet w, given by

$$M(w) = 1.3w^{1/2}.$$

Find $M(19)$, $M(24)$, and $M(30)$ to the nearest hundredth.

123. *Diagonal of a box.* The length of the diagonal of a box D is a function of its length L, width W, and height H:

$$D = \left(L^2 + W^2 + H^2\right)^{1/2}$$

a) Find D for the box shown in the accompanying figure.

b) Find D if $L = W = H = 1$ inch.

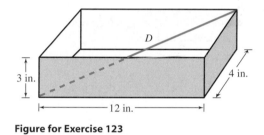

Figure for Exercise 123

124. *Radius of a sphere.* The radius of a sphere is given by the function

$$r = \left(\dfrac{0.75V}{\pi}\right)^{1/3},$$

where V is its volume. Find the radius of a spherical tank that has a volume of $\frac{32\pi}{3}$ cubic meters.

Figure for Exercise 124

125. *Maximum sail area.* According to the new International America's Cup Class Rules, the maximum sail area in square meters for a yacht in the America's Cup race is given by the function

$$S = \left(13.0368 + 7.84D^{1/3} - 0.8L\right)^2,$$

where D is the displacement in cubic meters (m^3), and L is the length in meters (m) (www.sailing.com). Find the maximum sail area for a boat that has a displacement of 18.42 m^3 and a length of 21.45 m.

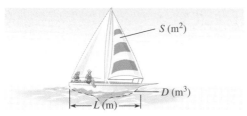

Figure for Exercise 125

126. *Orbits of the planets.* According to Kepler's third law of planetary motion, the average radius R of the orbit of a planet around the sun is determined by $R = T^{2/3}$, where T is the number of years for one orbit and R is measured in astronomical units or AUs (Windows to the Universe, www.windows.umich.edu).

a) It takes Mars 1.881 years to make one orbit of the sun. What is the average radius (in AUs) of the orbit of Mars?

b) The average radius of the orbit of Saturn is 9.5 AU. Use the accompanying graph to estimate the number of years it takes Saturn to make one orbit of the sun.

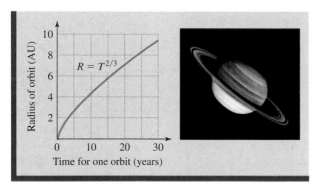

Figure for Exercise 126

127. *Top stock fund.* The average annual return r is a function of the initial investment P, the number of years n, and the amount S that it is worth after n years:

$$r = \left(\frac{S}{P}\right)^{1/n} - 1$$

An investment of $10,000 in the T. Rowe Price Latin America Fund in 2004 was worth $20,733 in 2009

(www.money.com). Find the 5-year average annual return.

128. *Top bond fund.* An investment of $10,000 in the Templeton Global Bond Fund in 2004 was worth $14,789 in 2009 (www.money.com). Use the formula from Exercise 127 to find the 5-year average annual return.

129. *Overdue loan payment.* In 1777 a wealthy Pennsylvania merchant, Jacob DeHaven, lent $450,000 to the Continental Congress to rescue the troops at Valley Forge. The loan was not repaid. In 1990 DeHaven's descendants filed suit for $141.6 billion (*New York Times,* May 27, 1990). What average annual rate of return were they using to calculate the value of the debt after 213 years? (See Exercise 127.)

130. *California growin'.* The population of California grew from 19.9 million in 1970 to 32.5 million in 2000 (U.S. Census Bureau, www.census.gov). Find the average annual rate of growth for that time period. (Use the formula from Exercise 127 with P being the initial population and S being the population n years later.)

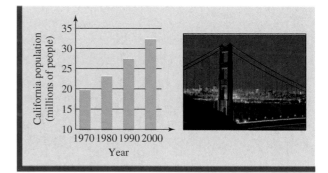

Figure for Exercise 130

Getting More Involved

131. *Discussion*

Determine whether each equation is an identity. Explain.

a) $(w^2x^2)^{1/2} = |w| \cdot |x|$

b) $(w^2x^2)^{1/2} = |wx|$

c) $(w^2x^2)^{1/2} = w|x|$

9.3   **Adding, Subtracting, and Multiplying Radicals**

In This Section

⟨1⟩ **Adding and Subtracting Radicals**

⟨2⟩ **Multiplying Radicals**

⟨3⟩ **Conjugates**

⟨4⟩ **Multiplying Radicals with Different Indices**

In this section, we will use the ideas of Section 9.1 in performing arithmetic operations with radical expressions.

⟨1⟩   **Adding and Subtracting Radicals**

To find the sum of $\sqrt{2}$ and $\sqrt{3}$, we can use a calculator to get $\sqrt{2} \approx 1.414$ and $\sqrt{3} \approx 1.732$. (The symbol $\approx$ means "is approximately equal to.") We can then add the decimal numbers and get

$$\sqrt{2} + \sqrt{3} \approx 1.414 + 1.732 = 3.146.$$

We cannot write an exact decimal form for $\sqrt{2} + \sqrt{3}$; the number 3.146 is an approximation of $\sqrt{2} + \sqrt{3}$. To represent the exact value of $\sqrt{2} + \sqrt{3}$, we just use the form $\sqrt{2} + \sqrt{3}$. This form cannot be simplified any further. However, a sum of like radicals can be simplified. **Like radicals** are radicals that have the same index and the same radicand.

To simplify the sum $3\sqrt{2} + 5\sqrt{2}$, we can use the fact that $3x + 5x = 8x$ is true for any value of x. Substituting $\sqrt{2}$ for x gives us $3\sqrt{2} + 5\sqrt{2} = 8\sqrt{2}$. So like radicals can be combined just as like terms are combined.

EXAMPLE   1

Adding and subtracting like radicals

Simplify the following expressions. Assume the variables represent positive numbers.

    a) $3\sqrt{5} + 4\sqrt{5}$ **b)** $\sqrt[4]{w} - 6\sqrt[4]{w}$

    c) $\sqrt{3} + \sqrt{5} - 4\sqrt{3} + 6\sqrt{5}$ **d)** $3\sqrt[3]{6x} + 2\sqrt[3]{x} + \sqrt[3]{6x} + \sqrt[3]{x}$

Solution

    a) $3\sqrt{5} + 4\sqrt{5} = 7\sqrt{5}$ **b)** $\sqrt[4]{w} - 6\sqrt[4]{w} = -5\sqrt[4]{w}$

    c) $\sqrt{3} + \sqrt{5} - 4\sqrt{3} + 6\sqrt{5} = -3\sqrt{3} + 7\sqrt{5}$ Only like radicals are combined.

    d) $3\sqrt[3]{6x} + 2\sqrt[3]{x} + \sqrt[3]{6x} + \sqrt[3]{x} = 4\sqrt[3]{6x} + 3\sqrt[3]{x}$

Now do Exercises 1–12

Remember that *only radicals with the same index and same radicand can be combined by addition or subtraction.* If the radicals are not in simplified form, then they must be simplified before you can determine whether they can be combined.

EXAMPLE   2

Simplifying radicals before combining

Perform the indicated operations. Assume the variables represent positive numbers.

    a) $\sqrt{8} + \sqrt{18}$ **b)** $\sqrt{2x^3} - \sqrt{4x^2} + 5\sqrt{18x^3}$

    c) $\sqrt[3]{16x^4y^3} - \sqrt[3]{54x^4y^3}$

Check that
$$\sqrt{8} + \sqrt{18} = 5\sqrt{2}.$$

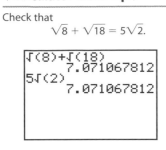

Solution

a) $\sqrt{8} + \sqrt{18} = \sqrt{4} \cdot \sqrt{2} + \sqrt{9} \cdot \sqrt{2}$

$\qquad\qquad\quad = 2\sqrt{2} + 3\sqrt{2}$ Simplify each radical.

$\qquad\qquad\quad = 5\sqrt{2}$ Add like radicals.

Note that $\sqrt{8} + \sqrt{18} \neq \sqrt{26}$.

b) $\sqrt{2x^3} - \sqrt{4x^2} + 5\sqrt{18x^3} = \sqrt{x^2} \cdot \sqrt{2x} - 2x + 5 \cdot \sqrt{9x^2} \cdot \sqrt{2x}$

$\qquad\qquad\qquad\qquad = x\sqrt{2x} - 2x + 15x\sqrt{2x}$ Simplify each radical.

$\qquad\qquad\qquad\qquad = 16x\sqrt{2x} - 2x$ Add like radicals only.

c) $\sqrt[3]{16x^4y^3} - \sqrt[3]{54x^4y^3} = \sqrt[3]{8x^3y^3} \cdot \sqrt[3]{2x} - \sqrt[3]{27x^3y^3} \cdot \sqrt[3]{2x}$

$\qquad\qquad\qquad\qquad = 2xy\sqrt[3]{2x} - 3xy\sqrt[3]{2x}$ Simplify each radical.

$\qquad\qquad\qquad\qquad = -xy\sqrt[3]{2x}$

> Now do Exercises 13–28

⟨2⟩ Multiplying Radicals

The product rule for radicals, $\sqrt[n]{a} \cdot \sqrt[n]{b} = \sqrt[n]{ab}$, allows multiplication of radicals with the same index, such as

$$\sqrt{5} \cdot \sqrt{3} = \sqrt{15}, \qquad \sqrt[3]{2} \cdot \sqrt[3]{5} = \sqrt[3]{10}, \qquad \text{and} \qquad \sqrt[5]{x^2} \cdot \sqrt[5]{x} = \sqrt[5]{x^3}.$$

CAUTION   The product rule does not allow multiplication of radicals that have different indices. We cannot use the product rule to multiply $\sqrt{2}$ and $\sqrt[3]{5}$.

E X A M P L E   3

Multiplying radicals with the same index

Multiply and simplify the following expressions. Assume the variables represent positive numbers.

a) $5\sqrt{6} \cdot 4\sqrt{3}$ $\qquad\qquad\qquad\qquad$ b) $\sqrt{3a^2} \cdot \sqrt{6a}$

c) $\sqrt[3]{4} \cdot \sqrt[3]{4}$ $\qquad\qquad\qquad\qquad$ d) $\sqrt[4]{\dfrac{x^3}{2}} \cdot \sqrt[4]{\dfrac{x^2}{8}}$

Students often write

$$\sqrt{15} \cdot \sqrt{15} = \sqrt{225} = 15.$$

Although this is correct, you should get used to the idea that

$$\sqrt{15} \cdot \sqrt{15} = 15.$$

Because of the definition of a square root, $\sqrt{a} \cdot \sqrt{a} = a$ for any positive number a.

Solution

a) $5\sqrt{6} \cdot 4\sqrt{3} = 5 \cdot 4 \cdot \sqrt{6} \cdot \sqrt{3}$

$\qquad\qquad\quad = 20\sqrt{18}$ Product rule for radicals

$\qquad\qquad\quad = 20 \cdot 3\sqrt{2}$ $\sqrt{18} = \sqrt{9} \cdot \sqrt{2} = 3\sqrt{2}$

$\qquad\qquad\quad = 60\sqrt{2}$

b) $\sqrt{3a^2} \cdot \sqrt{6a} = \sqrt{18a^3}$ Product rule for radicals

$\qquad\qquad\quad = \sqrt{9a^2} \cdot \sqrt{2a}$

$\qquad\qquad\quad = 3a\sqrt{2a}$ Simplify.

c) $\sqrt[3]{4} \cdot \sqrt[3]{4} = \sqrt[3]{16}$

$\qquad\qquad = \sqrt[3]{8} \cdot \sqrt[3]{2}$ Simplify.

$\qquad\qquad = 2\sqrt[3]{2}$

d) $\sqrt[4]{\dfrac{x^3}{2}} \cdot \sqrt[4]{\dfrac{x^2}{8}} = \sqrt[4]{\dfrac{x^5}{16}}$ Product rule for radicals

$\qquad\qquad = \dfrac{\sqrt[4]{x^4} \cdot \sqrt[4]{x}}{\sqrt[4]{16}}$ Product and quotient rules for radicals

$\qquad\qquad = \dfrac{x\sqrt[4]{x}}{2}$ Simplify.

> Now do Exercises 29–42

We find a product such as $3\sqrt{2}(4\sqrt{2} - \sqrt{3})$ by using the distributive property as we do when multiplying a monomial and a binomial. A product such as $(2\sqrt{3} + \sqrt{5})(3\sqrt{3} - 2\sqrt{5})$ can be found by using FOIL as we do for the product of two binomials.

E X A M P L E   **4**

Multiplying radicals

Multiply and simplify.

a) $3\sqrt{2}(4\sqrt{2} - \sqrt{3})$ \qquad   **b)** $\sqrt[3]{a}\left(\sqrt[3]{a} - \sqrt[3]{a^2}\right)$

c) $(2\sqrt{3} + \sqrt{5})(3\sqrt{3} - 2\sqrt{5})$ \qquad   **d)** $(3 + \sqrt{x - 9})^2$

Solution

a) $3\sqrt{2}(4\sqrt{2} - \sqrt{3}) = 3\sqrt{2} \cdot 4\sqrt{2} - 3\sqrt{2} \cdot \sqrt{3}$ Distributive property

$\qquad\qquad\qquad\qquad\qquad$ Because $\sqrt{2} \cdot \sqrt{2} = 2$

$\qquad\qquad\qquad\quad = 12 \cdot 2 - 3\sqrt{6}$ and $\sqrt{2} \cdot \sqrt{3} = \sqrt{6}$

$\qquad\qquad\qquad\quad = 24 - 3\sqrt{6}$

b) $\sqrt[3]{a}\left(\sqrt[3]{a} - \sqrt[3]{a^2}\right) = \sqrt[3]{a^2} - \sqrt[3]{a^3}$ Distributive property

$\qquad\qquad\qquad\quad = \sqrt[3]{a^2} - a$

c) $(2\sqrt{3} + \sqrt{5})(3\sqrt{3} - 2\sqrt{5})$

$\qquad\qquad\qquad$ F $\qquad\qquad$ O $\qquad\qquad$ I $\qquad\qquad$ L

$\quad = 2\sqrt{3} \cdot 3\sqrt{3} - 2\sqrt{3} \cdot 2\sqrt{5} + \sqrt{5} \cdot 3\sqrt{3} - \sqrt{5} \cdot 2\sqrt{5}$

$\quad = 18 - 4\sqrt{15} + 3\sqrt{15} - 10$

$\quad = 8 - \sqrt{15}$ Combine like radicals.

d) To square a sum, we use $(a + b)^2 = a^2 + 2ab + b^2$:

$$(3 + \sqrt{x - 9})^2 = 3^2 + 2 \cdot 3\sqrt{x - 9} + (\sqrt{x - 9})^2$$

$$= 9 + 6\sqrt{x - 9} + x - 9$$

$$= x + 6\sqrt{x - 9}$$

> Now do Exercises 43–56

CAUTION We can't simplify $\sqrt{x-9}$ in Example 4(d), because in general $\sqrt{a-b}$ $\neq \sqrt{a} - \sqrt{b}$. For example, $\sqrt{25-16} = \sqrt{9} = 3$ and $\sqrt{25} - \sqrt{16} = 1$. Find an example where $\sqrt{a+b} \neq \sqrt{a} + \sqrt{b}$.

⟨3⟩ Conjugates

Recall the special product rule $(a+b)(a-b) = a^2 - b^2$. The product of the sum $4 + \sqrt{3}$ and the difference $4 - \sqrt{3}$ can be found by using this rule:

$$(4 + \sqrt{3})(4 - \sqrt{3}) = 4^2 - (\sqrt{3})^2 = 16 - 3 = 13$$

The product of the irrational number $4 + \sqrt{3}$ and the irrational number $4 - \sqrt{3}$ is the rational number 13. For this reason the expressions $4 + \sqrt{3}$ and $4 - \sqrt{3}$ are called **conjugates** of one another. We will use conjugates in Section 9.4 to rationalize some denominators.

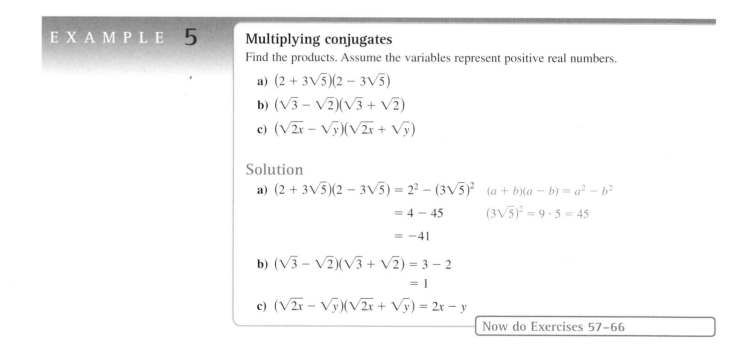

E X A M P L E   5

Multiplying conjugates

Find the products. Assume the variables represent positive real numbers.

a) $(2 + 3\sqrt{5})(2 - 3\sqrt{5})$

b) $(\sqrt{3} - \sqrt{2})(\sqrt{3} + \sqrt{2})$

c) $(\sqrt{2x} - \sqrt{y})(\sqrt{2x} + \sqrt{y})$

Solution

a) $(2 + 3\sqrt{5})(2 - 3\sqrt{5}) = 2^2 - (3\sqrt{5})^2$ $(a+b)(a-b) = a^2 - b^2$

$= 4 - 45$ $(3\sqrt{5})^2 = 9 \cdot 5 = 45$

$= -41$

b) $(\sqrt{3} - \sqrt{2})(\sqrt{3} + \sqrt{2}) = 3 - 2$

$= 1$

c) $(\sqrt{2x} - \sqrt{y})(\sqrt{2x} + \sqrt{y}) = 2x - y$

Now do Exercises 57–66

⟨4⟩ Multiplying Radicals with Different Indices

The product rule for radicals applies only to radicals with the same index. To multiply radicals with different indices we convert the radicals into exponential expressions with rational exponents. If the exponential expressions have the same base, apply the product rule for exponents $(a^m \cdot a^n = a^{m+n})$ to get a single exponential expression and then convert back to a radical [Example 6(a)]. If the bases of the exponential expression are different, get a common denominator for the rational exponents, convert back to radicals and then apply the product rule for radicals $\left(\sqrt[n]{a} \cdot \sqrt[n]{b} = \sqrt[n]{ab}\right)$ to get a single radical expression [Example 6(b)].

EXAMPLE **6**

Multiplying radicals with different indices

Write each product as a single radical expression.

a) $\sqrt[3]{2} \cdot \sqrt[4]{2}$ b) $\sqrt[3]{2} \cdot \sqrt{3}$

Solution

a) $\sqrt[3]{2} \cdot \sqrt[4]{2} = 2^{1/3} \cdot 2^{1/4}$ Write in exponential notation.

$= 2^{7/12}$ Product rule for exponents: $\frac{1}{3} + \frac{1}{4} = \frac{7}{12}$

$= \sqrt[12]{2^7}$ Write in radical notation.

$= \sqrt[12]{128}$

b) $\sqrt[3]{2} \cdot \sqrt{3} = 2^{1/3} \cdot 3^{1/2}$ Write in exponential notation.

$= 2^{2/6} \cdot 3^{3/6}$ Write the exponents with the LCD of 6.

$= \sqrt[6]{2^2} \cdot \sqrt[6]{3^3}$ Write in radical notation.

$= \sqrt[6]{2^2 \cdot 3^3}$ Product rule for radicals

$= \sqrt[6]{108}$ $2^2 \cdot 3^3 = 4 \cdot 27 = 108$

> Now do Exercises 67–74

‹ **Calculator Close-Up** ›

Check that

$$\sqrt[3]{2} \cdot \sqrt[4]{2} = \sqrt[12]{128}.$$

```
2^(1/3)*2^(1/4)
         1.498307077
128^(1/12)
         1.498307077
```

CAUTION Because the bases in $2^{1/3} \cdot 2^{1/4}$ are identical, we can add the exponents [Example 6(a)]. Because the bases in $2^{2/6} \cdot 3^{3/6}$ are not the same, we cannot add the exponents [Example 6(b)]. Instead, we write each factor as a sixth root and use the product rule for radicals.

Warm-Ups ▼

Fill in the blank.

1. _____ radicals have the same index and the same radicand.

2. The _____ property is used to combine like radicals.

3. The product rule for radicals is used to multiply radicals with the same _____.

4. The _____ of $2 - \sqrt{3}$ is $2 + \sqrt{3}$.

True or false?

5. $\sqrt{3} + \sqrt{3} = \sqrt{6}$

6. $\sqrt{8} + \sqrt{2} = 3\sqrt{2}$

7. $2\sqrt{3} \cdot 3\sqrt{3} = 6\sqrt{3}$

8. $2\sqrt{5} \cdot 3\sqrt{2} = 6\sqrt{10}$

9. $\sqrt[3]{2} \cdot \sqrt[3]{2} = 2$

10. $\sqrt{2}(\sqrt{3} - \sqrt{2}) = \sqrt{6} - 2$

11. $(\sqrt{2} + \sqrt{3})^2 = 2 + 3$

12. $(\sqrt{3} - \sqrt{2})(\sqrt{3} + \sqrt{2}) = 1$

‹1› Adding and Subtracting Radicals

All variables in the following exercises represent positive numbers. Simplify the sums and differences. Give exact answers. See Example 1.

1. $\sqrt{3} - 2\sqrt{3}$

2. $\sqrt{5} - 3\sqrt{5}$

3. $5\sqrt{7x} + 4\sqrt{7x}$

4. $3\sqrt{6a} + 7\sqrt{6a}$

5. $2\sqrt[3]{2} + 3\sqrt[3]{2}$

6. $\sqrt[3]{4} + 4\sqrt[3]{4}$

7. $\sqrt{3} - \sqrt{5} + 3\sqrt{3} - \sqrt{5}$

8. $\sqrt{2} - 5\sqrt{3} - 7\sqrt{2} + 9\sqrt{3}$

9. $\sqrt[3]{2} + \sqrt[3]{x} - \sqrt[3]{2} + 4\sqrt[3]{x}$

10. $\sqrt[3]{5y} - 4\sqrt[3]{5y} + \sqrt[3]{x} + \sqrt[3]{x}$

11. $\sqrt[3]{x} - \sqrt[3]{2x} + \sqrt[3]{x}$

12. $\sqrt[3]{ab} + \sqrt[3]{a} + 5\sqrt[3]{a} + \sqrt[3]{ab}$

Simplify each expression. Give exact answers. See Example 2.

13. $\sqrt{8} + \sqrt{28}$

14. $\sqrt{12} + \sqrt{24}$

15. $\sqrt{8} + \sqrt{18}$

16. $\sqrt{12} + \sqrt{27}$

17. $2\sqrt{45} - 3\sqrt{20}$

18. $3\sqrt{50} - 2\sqrt{32}$

19. $\sqrt{2} - \sqrt{8}$

20. $\sqrt{20} - \sqrt{125}$

21. $\sqrt{45x^3} - \sqrt{18x^2} + \sqrt{50x^2} - \sqrt{20x^3}$

22. $\sqrt{12x^5} - \sqrt{18x} - \sqrt{300x^5} + \sqrt{98x}$

23. $2\sqrt[3]{24} + \sqrt[3]{81}$

24. $5\sqrt[3]{24} + 2\sqrt[3]{375}$

25. $\sqrt[4]{48} - 2\sqrt[4]{243}$

26. $\sqrt[5]{64} + 7\sqrt[5]{2}$

27. $\sqrt[3]{54t^4y^3} - \sqrt[3]{16t^4y^3}$

28. $\sqrt[3]{2000w^2z^5} - \sqrt[3]{16w^2z^5}$

‹2› Multiplying Radicals

Simplify the products. Give exact answers. See Examples 3 and 4.

29. $\sqrt{3} \cdot \sqrt{5}$

30. $\sqrt{5} \cdot \sqrt{7}$

31. $2\sqrt{5} \cdot 3\sqrt{10}$

32. $(3\sqrt{2})(-4\sqrt{10})$

33. $2\sqrt{7a} \cdot 3\sqrt{2a}$

34. $2\sqrt{5c} \cdot 5\sqrt{5}$

35. $\sqrt[4]{9} \cdot \sqrt[4]{27}$

36. $\sqrt[3]{5} \cdot \sqrt[3]{100}$

37. $(2\sqrt{3})^2$

38. $(-4\sqrt{2})^2$

39. $\sqrt{5x^3} \cdot \sqrt{8x^4}$

40. $\sqrt{3b^3} \cdot \sqrt{6b^5}$

41. $\sqrt[4]{\dfrac{x^5}{3}} \cdot \sqrt[4]{\dfrac{x^2}{27}}$

42. $\sqrt[3]{\dfrac{a^4}{2}} \cdot \sqrt[3]{\dfrac{a^3}{4}}$

43. $2\sqrt{3}(\sqrt{6} + 3\sqrt{3})$

44. $2\sqrt{5}(\sqrt{3} + 3\sqrt{5})$

45. $\sqrt{5}(\sqrt{10} - 2)$

46. $\sqrt{6}(\sqrt{15} - 1)$

47. $\sqrt[3]{3t}(\sqrt[3]{9t} - \sqrt[3]{t^2})$

48. $\sqrt[3]{2}(\sqrt[3]{12x} - \sqrt[3]{2x})$

49. $(\sqrt{3} + 2)(\sqrt{3} - 5)$

50. $(\sqrt{5} + 2)(\sqrt{5} - 6)$

51. $(\sqrt{11} - 3)(\sqrt{11} + 3)$

52. $(\sqrt{2} + 5)(\sqrt{2} + 5)$

53. $(2\sqrt{5} - 7)(2\sqrt{5} + 4)$

54. $(2\sqrt{6} - 3)(2\sqrt{6} + 4)$

55. $(2\sqrt{3} - \sqrt{6})(\sqrt{3} + 2\sqrt{6})$

56. $(3\sqrt{3} - \sqrt{2})(\sqrt{2} + \sqrt{3})$

⟨3⟩ Conjugates

Find the product of each pair of conjugates. See Example 5.

57. $(\sqrt{3} - 2)(\sqrt{3} + 2)$

58. $(7 - \sqrt{3})(7 + \sqrt{3})$

59. $(\sqrt{5} + \sqrt{2})(\sqrt{5} - \sqrt{2})$

60. $(\sqrt{6} + \sqrt{5})(\sqrt{6} - \sqrt{5})$

61. $(2\sqrt{5} + 1)(2\sqrt{5} - 1)$

62. $(3\sqrt{2} - 4)(3\sqrt{2} + 4)$

63. $(3\sqrt{2} + \sqrt{5})(3\sqrt{2} - \sqrt{5})$

64. $(2\sqrt{3} - \sqrt{7})(2\sqrt{3} + \sqrt{7})$

65. $(5 - 3\sqrt{x})(5 + 3\sqrt{x})$

66. $(4\sqrt{y} + 3\sqrt{z})(4\sqrt{y} - 3\sqrt{z})$

⟨4⟩ Multiplying Radicals with Different Indices

Write each product as a single radical expression. See Example 6.

67. $\sqrt[3]{3} \cdot \sqrt{3}$ **68.** $\sqrt{3} \cdot \sqrt[4]{3}$

69. $\sqrt[3]{5} \cdot \sqrt[4]{5}$ **70.** $\sqrt[3]{2} \cdot \sqrt[5]{2}$

71. $\sqrt[3]{2} \cdot \sqrt{5}$ **72.** $\sqrt{6} \cdot \sqrt[3]{2}$

73. $\sqrt[3]{2} \cdot \sqrt[4]{3}$ **74.** $\sqrt[3]{3} \cdot \sqrt[4]{2}$

Miscellaneous

Simplify each expression.

75. $\sqrt{300} + \sqrt{3}$ **76.** $\sqrt{50} + \sqrt{2}$

77. $2\sqrt{5} \cdot 5\sqrt{6}$ **78.** $3\sqrt{6} \cdot 5\sqrt{10}$

79. $(3 + 2\sqrt{7})(\sqrt{7} - 2)$

80. $(2 + \sqrt{7})(\sqrt{7} - 2)$ **81.** $4\sqrt{w} \cdot 4\sqrt{w}$

82. $3\sqrt{m} \cdot 5\sqrt{m}$ **83.** $\sqrt{3x^3} \cdot \sqrt{6x^2}$

84. $\sqrt{2t^5} \cdot \sqrt{10t^4}$

85. $(2\sqrt{5} + \sqrt{2})(3\sqrt{5} - \sqrt{2})$

86. $(3\sqrt{2} - \sqrt{3})(2\sqrt{2} + 3\sqrt{3})$

87. $\dfrac{\sqrt{2}}{3} + \dfrac{\sqrt{2}}{5}$

88. $\dfrac{\sqrt{2}}{4} + \dfrac{\sqrt{3}}{5}$

89. $(5 + 2\sqrt{2})(5 - 2\sqrt{2})$

90. $(3 - 2\sqrt{7})(3 + 2\sqrt{7})$

91. $(3 + \sqrt{x})^2$

92. $(1 - \sqrt{x})^2$

93. $(5\sqrt{x} - 3)^2$

94. $(3\sqrt{a} + 2)^2$

95. $(1 + \sqrt{x + 2})^2$

96. $(\sqrt{x - 1} + 1)^2$

97. $\sqrt{4w} - \sqrt{9w}$

98. $10\sqrt{m} - \sqrt{16m}$

99. $2\sqrt{a^3} + 3\sqrt{a^3} - 2a\sqrt{4a}$

100. $5\sqrt{w^2y} - 7\sqrt{w^2y} + 6\sqrt{w^2y}$

101. $\sqrt{x^5} + 2x\sqrt{x^3}$

102. $\sqrt{8x^3} + \sqrt{50x^3} - x\sqrt{2x}$

103. $\sqrt[3]{-16x^4} + 5x\sqrt[3]{54x}$

104. $\sqrt[3]{3x^5y^7} - \sqrt[3]{24x^5y^7}$

105. $\sqrt[3]{2x} \cdot \sqrt{2x}$ **106.** $\sqrt[3]{2m} \cdot \sqrt[4]{2n}$

Applications

Solve each problem.

107. *Area of a rectangle.* Find the exact area of a rectangle that has a length of $\sqrt{6}$ feet and a width of $\sqrt{3}$ feet.

108. *Volume of a cube.* Find the exact volume of a cube with sides of length $\sqrt{3}$ meters.

109. *Area of a trapezoid.* Find the exact area of a trapezoid with a height of $\sqrt{6}$ feet and bases of $\sqrt{3}$ feet and $\sqrt{12}$ feet.

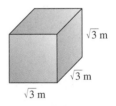

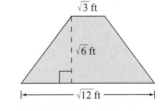

Figure for Exercise 108   **Figure for Exercise 109**

110. *Area of a triangle.* Find the exact area of a triangle with a base of $\sqrt{30}$ meters and a height of $\sqrt{6}$ meters.

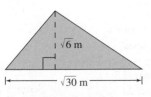

Figure for Exercise 110

Getting More Involved

111. *Discussion*

Is $\sqrt{a} + \sqrt{b} = \sqrt{a + b}$ for all values of a and b?

112. *Discussion*

Which of the following equations are identities? Explain your answers.

a) $\sqrt{9x} = 3\sqrt{x}$

b) $\sqrt{9 + x} = 3 + \sqrt{x}$

c) $\sqrt{x - 4} = \sqrt{x} - 2$

d) $\sqrt{\dfrac{x}{4}} = \dfrac{\sqrt{x}}{2}$

113. *Exploration*

Because 3 is the square of $\sqrt{3}$, a binomial such as $y^2 - 3$ is a difference of two squares.

a) Factor $y^2 - 3$ and $2a^2 - 7$ using radicals.

b) Use factoring with radicals to solve the equation $x^2 - 8 = 0$.

c) Assuming a is a positive real number, solve the equation $x^2 - a = 0$.

Mid-Chapter **Quiz** | Sections 9.1 through 9.3 | Chapter 9

Simplify each radical expression.

1. $\sqrt{64}$

2. $\sqrt[3]{-27}$

3. $\sqrt{120}$

4. $\sqrt[3]{56}$

5. $\sqrt{12x^7}$

6. $\sqrt[3]{24a^3 b^{13}}$

7. $\sqrt{\dfrac{w}{16}}$

8. $\sqrt{\dfrac{8x^3}{9}}$

9. $81^{1/2}$

10. $100^{3/2}$

11. $-16^{3/2}$

12. $\left(\dfrac{5^{1/3}}{5^{-2/3}}\right)^{-3}$

Perform the indicated operations.

13. $2\sqrt{3} - 5\sqrt{6} - 4\sqrt{3} + \sqrt{6}$ **14.** $9\sqrt{20} - 3\sqrt{45}$

15. $3\sqrt{10} \cdot 2\sqrt{14}$

16. $(8 + \sqrt{10})(8 - \sqrt{10})$

17. $\sqrt[3]{8x^5} + \sqrt[3]{27x^5}$

Miscellaneous.

18. Find the domain of the expression $\sqrt{6 - 3x}$.

19. Find the solution set to $\sqrt{x^2} = x$.

20. Find the solution set to $(x^4)^{1/4} = |x|$.

21. Write the product $\sqrt{2} \cdot \sqrt[3]{2}$ as a single radical expression.

22. Suppose that $h(t) = 5t^{2/3}$. Find $h(8)$.

9.4 Quotients, Powers, and Rationalizing Denominators

In This Section

⟨1⟩ **Rationalizing the Denominator**

⟨2⟩ **Simplifying Radicals**

⟨3⟩ **Dividing Radicals**

⟨4⟩ **Rationalizing Denominators Using Conjugates**

⟨5⟩ **Powers of Radical Expressions**

In this section, we will continue studying operations with radicals. We will first learn how to rationalize denominators, and then we will find quotients and powers with radicals.

⟨1⟩ Rationalizing the Denominator

Square roots such as $\sqrt{2}$, $\sqrt{3}$, and $\sqrt{5}$ are irrational numbers. If roots of this type appear in the denominator of a fraction, it is customary to rewrite the fraction with a rational number in the denominator, or **rationalize** it. We rationalize a denominator by multiplying both the numerator and denominator by another radical that makes the denominator rational.

You can find products of radicals in two ways. By definition, $\sqrt{2}$ is the positive number that you multiply by itself to get 2. So,

$$\sqrt{2} \cdot \sqrt{2} = 2.$$

By the product rule, $\sqrt{2} \cdot \sqrt{2} = \sqrt{4} = 2$. Note that $\sqrt[3]{2} \cdot \sqrt[3]{2} = \sqrt[3]{4}$ by the product rule, but $\sqrt[3]{4} \neq 2$. By definition of a cube root,

$$\sqrt[3]{2} \cdot \sqrt[3]{2} \cdot \sqrt[3]{2} = 2.$$

E X A M P L E   1

Rationalizing the denominator
Rewrite each expression with a rational denominator.

a) $\dfrac{\sqrt{3}}{\sqrt{5}}$ **b)** $\dfrac{3}{\sqrt[3]{2}}$

‹ **Helpful Hint** ›

If you are going to compute the value of a radical expression with a calculator, it does not matter if the denominator is rational. However, rationalizing the denominator provides another opportunity to practice building up the denominator of a fraction and multiplying radicals.

Solution

a) Because $\sqrt{5} \cdot \sqrt{5} = 5$, multiplying both the numerator and denominator by $\sqrt{5}$ will rationalize the denominator:

$$\frac{\sqrt{3}}{\sqrt{5}} = \frac{\sqrt{3}}{\sqrt{5}} \cdot \frac{\sqrt{5}}{\sqrt{5}} = \frac{\sqrt{15}}{5} \qquad \text{By the product rule, } \sqrt{3} \cdot \sqrt{5} = \sqrt{15}.$$

b) We must build up the denominator to be the cube root of a perfect cube. So we multiply by $\sqrt[3]{4}$ to get $\sqrt[3]{4} \cdot \sqrt[3]{2} = \sqrt[3]{8}$:

$$\frac{3}{\sqrt[3]{2}} = \frac{3}{\sqrt[3]{2}} \cdot \frac{\sqrt[3]{4}}{\sqrt[3]{4}} = \frac{3\sqrt[3]{4}}{\sqrt[3]{8}} = \frac{3\sqrt[3]{4}}{2}$$

> Now do Exercises 1–8

CAUTION   To rationalize a denominator with a single square root, you simply multiply by that square root. If the denominator has a cube root, you build the denominator to a cube root of a perfect cube, as in Example 1(b). For a fourth root you build to a fourth root of a perfect fourth power, and so on.

‹2› Simplifying Radicals

When simplifying a radical expression, we have three specific conditions to satisfy. First, we use the product rule to factor out perfect nth powers from the radicand in nth roots. That is, we factor out perfect squares in square roots, perfect cubes in cube roots, and so on. For example,

$$\sqrt{72} = \sqrt{36} \cdot \sqrt{2} = 6\sqrt{2} \qquad \text{and} \qquad \sqrt[3]{24} = \sqrt[3]{8} \cdot \sqrt[3]{3} = 2\sqrt[3]{3}.$$

Second, we use the quotient rule to remove all fractions from inside a radical. For example,

$$\sqrt{\frac{2}{3}} = \frac{\sqrt{2}}{\sqrt{3}}.$$

Third, we remove radicals from denominators by rationalizing the denominator:

$$\sqrt{\frac{2}{3}} = \frac{\sqrt{2} \cdot \sqrt{3}}{\sqrt{3} \cdot \sqrt{3}} = \frac{\sqrt{6}}{3}.$$

A radical expression that satisfies the following three conditions is in *simplified radical form*.

Simplified Radical Form for Radicals of Index *n*

A radical expression of index *n* is in **simplified radical form** if it has

    1. *no* perfect *n*th powers as factors of the radicand,
    2. *no* fractions inside the radical, and
    3. *no* radicals in the denominator.

E X A M P L E   2   **Writing radical expressions in simplified radical form**
Simplify.

    a) $\dfrac{\sqrt{10}}{\sqrt{6}}$

    b) $\sqrt[3]{\dfrac{5}{9}}$

Solution

    a) To rationalize the denominator, multiply the numerator and denominator by $\sqrt{6}$:

$$\frac{\sqrt{10}}{\sqrt{6}} = \frac{\sqrt{10}}{\sqrt{6}} \cdot \frac{\sqrt{6}}{\sqrt{6}} \qquad \text{Rationalize the denominator.}$$

$$= \frac{\sqrt{60}}{6}$$

$$= \frac{\sqrt{4}\sqrt{15}}{6} \qquad \text{Remove the perfect square from } \sqrt{60}.$$

$$= \frac{2\sqrt{15}}{6}$$

$$= \frac{\sqrt{15}}{3} \qquad \text{Reduce } \frac{2}{6} \text{ to } \frac{1}{3}. \text{ Note that } \sqrt{15} \div 3 \neq \sqrt{5}.$$

b) To rationalize the denominator, build up the denominator to a cube root of a perfect cube. Because $\sqrt[3]{9} \cdot \sqrt[3]{3} = \sqrt[3]{27} = 3$, we multiply by $\sqrt[3]{3}$:

$$\sqrt[3]{\frac{5}{9}} = \frac{\sqrt[3]{5}}{\sqrt[3]{9}} \qquad \text{Quotient rule for radicals}$$

$$= \frac{\sqrt[3]{5}}{\sqrt[3]{9}} \cdot \frac{\sqrt[3]{3}}{\sqrt[3]{3}} \qquad \text{Rationalize the denominator.}$$

$$= \frac{\sqrt[3]{15}}{\sqrt[3]{27}}$$

$$= \frac{\sqrt[3]{15}}{3}$$

> **Now do Exercises 9–18**

EXAMPLE 3

Rationalizing the denominator with variables

Simplify each expression. Assume all variables represent positive real numbers.

a) $\sqrt{\dfrac{a}{b}}$ **b)** $\sqrt{\dfrac{x^3}{y^5}}$ **c)** $\sqrt[3]{\dfrac{x}{y}}$

Solution

a) $\sqrt{\dfrac{a}{b}} = \dfrac{\sqrt{a}}{\sqrt{b}}$ Quotient rule for radicals

$$= \frac{\sqrt{a} \cdot \sqrt{b}}{\sqrt{b} \cdot \sqrt{b}} \qquad \text{Rationalize the denominator.}$$

$$= \frac{\sqrt{ab}}{b}$$

b) $\sqrt{\dfrac{x^3}{y^5}} = \dfrac{\sqrt{x^3}}{\sqrt{y^5}}$ Quotient rule for radicals

$$= \frac{\sqrt{x^2} \cdot \sqrt{x}}{\sqrt{y^4} \cdot \sqrt{y}} \qquad \text{Product rule for radicals}$$

$$= \frac{x\sqrt{x}}{y^2\sqrt{y}} \qquad \text{Simplify.}$$

$$= \frac{x\sqrt{x} \cdot \sqrt{y}}{y^2\sqrt{y} \cdot \sqrt{y}} \qquad \text{Rationalize the denominator.}$$

$$= \frac{x\sqrt{xy}}{y^2 \cdot y} = \frac{x\sqrt{xy}}{y^3}$$

c) Multiply by $\sqrt[3]{y^2}$ to rationalize the denominator:

$$\sqrt[3]{\frac{x}{y}} = \frac{\sqrt[3]{x}}{\sqrt[3]{y}} = \frac{\sqrt[3]{x}}{\sqrt[3]{y}} \cdot \frac{\sqrt[3]{y^2}}{\sqrt[3]{y^2}} = \frac{\sqrt[3]{xy^2}}{\sqrt[3]{y^3}} = \frac{\sqrt[3]{xy^2}}{y}$$

> **Now do Exercises 19–28**

⟨3⟩ Dividing Radicals

In Section 9.3 you learned how to add, subtract, and multiply radical expressions. To divide two radical expressions, simply write the quotient as a ratio and then simplify. In general, we have

$$\sqrt[n]{a} \div \sqrt[n]{b} = \frac{\sqrt[n]{a}}{\sqrt[n]{b}} = \sqrt[n]{\frac{a}{b}},$$

provided that all expressions represent real numbers. Note that the quotient rule is applied only to radicals that have the same index.

EXAMPLE **4**

Dividing radicals with the same index

Divide and simplify. Assume the variables represent positive numbers.

a) $\sqrt{10} \div \sqrt{5}$ **b)** $(3\sqrt{2}) \div (2\sqrt{3})$ **c)** $\sqrt[3]{10x^2} \div \sqrt[3]{5x}$

Solution

a) $\sqrt{10} \div \sqrt{5} = \dfrac{\sqrt{10}}{\sqrt{5}}$ $a \div b = \frac{a}{b}$, provided that $b \neq 0$.

$\qquad\qquad = \sqrt{\dfrac{10}{5}}$ Quotient rule for radicals

$\qquad\qquad = \sqrt{2}$ Reduce.

b) $(3\sqrt{2}) \div (2\sqrt{3}) = \dfrac{3\sqrt{2}}{2\sqrt{3}}$

$\qquad\qquad\qquad = \dfrac{3\sqrt{2}}{2\sqrt{3}} \cdot \dfrac{\sqrt{3}}{\sqrt{3}}$ Rationalize the denominator.

$\qquad\qquad\qquad = \dfrac{3\sqrt{6}}{2 \cdot 3}$

$\qquad\qquad\qquad = \dfrac{\sqrt{6}}{2}$ Note that $\sqrt{6} \div 2 \neq \sqrt{3}$.

c) $\sqrt[3]{10x^2} \div \sqrt[3]{5x} = \dfrac{\sqrt[3]{10x^2}}{\sqrt[3]{5x}}$

$\qquad\qquad\qquad = \sqrt[3]{\dfrac{10x^2}{5x}}$ Quotient rule for radicals

$\qquad\qquad\qquad = \sqrt[3]{2x}$ Reduce.

Now do Exercises 29–36

Note that in Example 4(a) we applied the quotient rule to get $\sqrt{10} \div \sqrt{5} = \sqrt{2}$. In Example 4(b) we did not use the quotient rule because 2 is not evenly divisible by 3. Instead, we rationalized the denominator to get the result in simplified form.

When working with radicals it is usually best to write them in simplified radical form before doing any operations with the radicals.

EXAMPLE **5**

Simplifying before dividing

Divide and simplify. Assume the variables represent positive numbers.

  a) $\sqrt{12} \div \sqrt{72x}$ **b)** $\sqrt[4]{16a} \div \sqrt[4]{a^5}$

Solution

  a) $\sqrt{12} \div \sqrt{72x} = \dfrac{\sqrt{4} \cdot \sqrt{3}}{\sqrt{36} \cdot \sqrt{2x}}$ Factor out perfect squares.

$\qquad\qquad\qquad = \dfrac{2\sqrt{3}}{6\sqrt{2x}}$ Simplify.

$\qquad\qquad\qquad = \dfrac{\sqrt{3} \cdot \sqrt{2x}}{3\sqrt{2x} \cdot \sqrt{2x}}$ Reduce $\frac{2}{6}$ to $\frac{1}{3}$ and rationalize.

$\qquad\qquad\qquad = \dfrac{\sqrt{6x}}{6x}$ Multiply the radicals.

  b) $\sqrt[4]{16a} \div \sqrt[4]{a^5} = \dfrac{\sqrt[4]{16} \cdot \sqrt[4]{a}}{\sqrt[4]{a^4} \cdot \sqrt[4]{a}}$ Factor out perfect fourth powers.

$\qquad\qquad\qquad = \dfrac{\sqrt[4]{16}}{\sqrt[4]{a^4}}$ Reduce.

$\qquad\qquad\qquad = \dfrac{2}{a}$ Simplify the radicals.

> Now do Exercises 37–44

In Chapter 10 it will be necessary to simplify expressions of the type found in Example 6.

EXAMPLE **6**

Simplifying radical expressions

Simplify.

  a) $\dfrac{4 - \sqrt{12}}{4}$ **b)** $\dfrac{-6 + \sqrt{20}}{-2}$

‹ **Helpful Hint** ›

The expressions in Example 6 are the types of expressions that you must simplify when learning the quadratic formula in Chapter 10.

Solution

  a) First write $\sqrt{12}$ in simplified form. Then simplify the expression.

$\qquad \dfrac{4 - \sqrt{12}}{4} = \dfrac{4 - 2\sqrt{3}}{4}$ Simplify $\sqrt{12}$.

$\qquad\qquad\quad = \dfrac{2(2 - \sqrt{3})}{2 \cdot 2}$ Factor.

$\qquad\qquad\quad = \dfrac{2 - \sqrt{3}}{2}$ Divide out the common factor.

  b) $\dfrac{-6 + \sqrt{20}}{-2} = \dfrac{-6 + 2\sqrt{5}}{-2}$

$\qquad\qquad\quad = \dfrac{-2(3 - \sqrt{5})}{-2}$

$\qquad\qquad\quad = 3 - \sqrt{5}$

> Now do Exercises 45–48

CAUTION To simplify the expressions in Example 6, you must simplify the radical, factor the numerator, and then divide out the common factors. You cannot simply "cancel" the 4's in $\dfrac{4 - \sqrt{12}}{4}$ or the 2's in $\dfrac{2 - \sqrt{3}}{2}$ because they are not common factors.

⟨4⟩ Rationalizing Denominators Using Conjugates

A simplified expression involving radicals does not have radicals in the denominator. If an expression such as $4 - \sqrt{3}$ appears in a denominator, we can multiply both the numerator and denominator by its conjugate $4 + \sqrt{3}$ to get a rational number in the denominator.

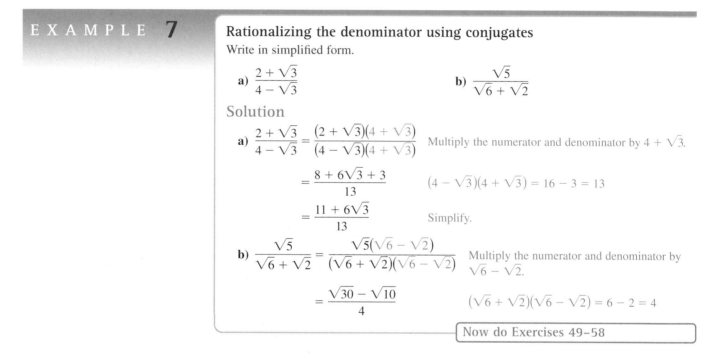

EXAMPLE 7

Rationalizing the denominator using conjugates

Write in simplified form.

a) $\dfrac{2 + \sqrt{3}}{4 - \sqrt{3}}$

b) $\dfrac{\sqrt{5}}{\sqrt{6} + \sqrt{2}}$

Solution

a) $\dfrac{2 + \sqrt{3}}{4 - \sqrt{3}} = \dfrac{(2 + \sqrt{3})(4 + \sqrt{3})}{(4 - \sqrt{3})(4 + \sqrt{3})}$ Multiply the numerator and denominator by $4 + \sqrt{3}$.

$= \dfrac{8 + 6\sqrt{3} + 3}{13}$ $(4 - \sqrt{3})(4 + \sqrt{3}) = 16 - 3 = 13$

$= \dfrac{11 + 6\sqrt{3}}{13}$ Simplify.

b) $\dfrac{\sqrt{5}}{\sqrt{6} + \sqrt{2}} = \dfrac{\sqrt{5}(\sqrt{6} - \sqrt{2})}{(\sqrt{6} + \sqrt{2})(\sqrt{6} - \sqrt{2})}$ Multiply the numerator and denominator by $\sqrt{6} - \sqrt{2}$.

$= \dfrac{\sqrt{30} - \sqrt{10}}{4}$ $(\sqrt{6} + \sqrt{2})(\sqrt{6} - \sqrt{2}) = 6 - 2 = 4$

Now do Exercises 49–58

⟨5⟩ Powers of Radical Expressions

In Example 8, we use the power of a product rule $[(ab)^n = a^n b^n]$ and the power of a power rule $[(a^m)^n = a^{mn}]$ with radical expressions. We also use the fact that a root and a power can be found in either order. That is, $\left(\sqrt[n]{a}\right)^m = \sqrt[n]{a^m}$.

EXAMPLE 8

Finding powers of rational expressions

Simplify. Assume the variables represent positive numbers.

a) $(5\sqrt{2})^3$ b) $(2\sqrt{x^3})^4$ c) $(3w\sqrt[3]{2w})^3$ d) $(2t\sqrt[4]{3t})^3$

Solution

a) $(5\sqrt{2})^3 = 5^3(\sqrt{2})^3$ Power of a product rule

$= 125\sqrt{8}$ $(\sqrt{2})^3 = \sqrt{2^3} = \sqrt{8}$

$= 125 \cdot 2\sqrt{2}$ $\sqrt{8} = \sqrt{4}\sqrt{2} = 2\sqrt{2}$

$= 250\sqrt{2}$

b) $(2\sqrt{x^3})^4 = 2^4(\sqrt{x^3})^4$ Power of a product rule

$\qquad\qquad = 2^4\sqrt{(x^3)^4}$ $(\sqrt[n]{a})^m = \sqrt[n]{a^m}$

$\qquad\qquad = 16\sqrt{x^{12}}$ $(a^m)^n = a^{mn}$

$\qquad\qquad = 16x^6$

c) $(3w\sqrt[3]{2w})^3 = 3^3w^3(\sqrt[3]{2w})^3$

$\qquad\qquad\qquad = 27w^3(2w)$

$\qquad\qquad\qquad = 54w^4$

d) $(2t\sqrt[4]{3t})^3 = 2^3t^3(\sqrt[4]{3t})^3 = 8t^3\sqrt[4]{27t^3}$

> Now do Exercises 59–70

Warm-Ups ▼

Fill in the blank.

1. The numbers $\sqrt{2}$, $\sqrt{3}$, and $\sqrt{5}$ are _____ numbers.

2. Writing a fraction with an irrational denominator as an equivalent fraction with a rational denominator is _____ the denominator.

3. A simplified square root expression has no perfect _____ as factors of the radicand.

4. A simplified radical expression has no fractions inside the _____.

5. A simplified radical expression has no radicals in the _____.

True or false?

6. $\dfrac{\sqrt{6}}{\sqrt{2}} = \sqrt{3}$

7. $\dfrac{2}{\sqrt{2}} = \sqrt{2}$

8. $\dfrac{4 - \sqrt{10}}{2} = 2 - \sqrt{10}$

9. $\dfrac{4 - \sqrt{10}}{2} = 2 - \sqrt{5}$

10. $\dfrac{1}{\sqrt{3}} = \dfrac{\sqrt{3}}{3}$

11. $(2\sqrt{4})^2 = 16$

12. $(3\sqrt{5})^3 = 27\sqrt{125}$

Exercises

9.4

‹ **Study Tips** ›

• Personal issues can have a tremendous effect on your progress in any course. If you need help, get it.
• Most schools have counseling centers that can help you to overcome personal issues that are affecting your studies.

‹ 1 › **Rationalizing the Denominator**

All variables in the following exercises represent positive numbers.
Rewrite each expression with a rational denominator. See Example 1.

1. $\dfrac{2}{\sqrt{5}}$

2. $\dfrac{5}{\sqrt{3}}$

3. $\dfrac{\sqrt{3}}{\sqrt{7}}$

4. $\dfrac{\sqrt{6}}{\sqrt{5}}$

5. $\dfrac{1}{\sqrt[3]{4}}$

6. $\dfrac{7}{\sqrt[3]{3}}$

7. $\dfrac{\sqrt[3]{6}}{\sqrt[3]{5}}$

8. $\dfrac{\sqrt[4]{2}}{\sqrt[4]{27}}$

⟨2⟩ Simplifying Radicals

Write each radical expression in simplified radical form. See Example 2.

9. $\dfrac{\sqrt{5}}{\sqrt{12}}$

10. $\dfrac{\sqrt{7}}{\sqrt{18}}$

11. $\dfrac{\sqrt{3}}{\sqrt{12}}$

12. $\dfrac{\sqrt{2}}{\sqrt{18}}$

13. $\sqrt{\dfrac{1}{2}}$

14. $\sqrt{\dfrac{3}{8}}$

15. $\sqrt[3]{\dfrac{2}{3}}$

16. $\sqrt[3]{\dfrac{3}{5}}$

17. $\sqrt[3]{\dfrac{7}{4}}$

18. $\sqrt[4]{\dfrac{1}{5}}$

Simplify. See Example 3.

19. $\sqrt{\dfrac{x}{y}}$

20. $\sqrt{\dfrac{x^2}{a}}$

21. $\sqrt{\dfrac{a^3}{b^7}}$

22. $\sqrt{\dfrac{w^5}{y^3}}$

23. $\sqrt{\dfrac{a}{3b}}$

24. $\sqrt{\dfrac{5x}{2y}}$

25. $\sqrt[3]{\dfrac{a}{b}}$

26. $\sqrt[3]{\dfrac{4a}{b}}$

27. $\sqrt[3]{\dfrac{5}{2b^2}}$

28. $\sqrt[3]{\dfrac{3}{4a^2}}$

⟨3⟩ Dividing Radicals

Divide and simplify. See Examples 4 and 5.

29. $\sqrt{15} \div \sqrt{5}$

30. $\sqrt{14} \div \sqrt{7}$

31. $\sqrt{3} \div \sqrt{5}$

32. $\sqrt{5} \div \sqrt{7}$

33. $(3\sqrt{3}) \div (5\sqrt{6})$

34. $(2\sqrt{2}) \div (4\sqrt{10})$

35. $(2\sqrt{3}) \div (3\sqrt{6})$

36. $(5\sqrt{12}) \div (4\sqrt{6})$

37. $\sqrt{24a^2} \div \sqrt{72a}$

38. $\sqrt{32x^3} \div \sqrt{48x^2}$

39. $\sqrt[3]{20} \div \sqrt[3]{2}$

40. $\sqrt[3]{8x^7} \div \sqrt[3]{2x}$

41. $\sqrt[4]{48} \div \sqrt[4]{3}$

42. $\sqrt[4]{4a^{10}} \div \sqrt[4]{2a^2}$

43. $\sqrt[4]{16w} \div \sqrt[4]{w^5}$

44. $\sqrt[4]{81b^5} \div \sqrt[4]{b}$

Simplify. See Example 6.

45. $\dfrac{6 + \sqrt{45}}{3}$

46. $\dfrac{10 + \sqrt{50}}{5}$

47. $\dfrac{-2 + \sqrt{12}}{-2}$

48. $\dfrac{-6 + \sqrt{72}}{-6}$

⟨4⟩ Rationalizing Denominators Using Conjugates

Simplify each expression by rationalizing the denominator. See Example 7.

49. $\dfrac{4}{2 + \sqrt{8}}$

50. $\dfrac{6}{3 - \sqrt{18}}$

51. $\dfrac{3}{\sqrt{11} - \sqrt{5}}$

52. $\dfrac{6}{\sqrt{5} - \sqrt{14}}$

53. $\dfrac{1 + \sqrt{2}}{\sqrt{3} - 1}$

54. $\dfrac{2 - \sqrt{3}}{\sqrt{2} + \sqrt{6}}$

55. $\dfrac{\sqrt{2}}{\sqrt{6} + \sqrt{3}}$

56. $\dfrac{5}{\sqrt{7} - \sqrt{5}}$

57. $\dfrac{2\sqrt{3}}{3\sqrt{2} - \sqrt{5}}$

58. $\dfrac{3\sqrt{5}}{5\sqrt{2} + \sqrt{6}}$

⟨5⟩ Powers of Radical Expressions

Simplify. See Example 8.

59. $(2\sqrt{2})^5$

60. $(3\sqrt{3})^4$

61. $(\sqrt{x})^5$

62. $(2\sqrt{y})^3$

63. $(-3\sqrt{x^3})^3$

64. $(-2\sqrt{x^3})^4$

65. $(2x\sqrt[3]{x^2})^3$

66. $(2y\sqrt[3]{4y})^3$

67. $(-2\sqrt[3]{5})^2$

68. $(-3\sqrt[3]{4})^2$

69. $(\sqrt[3]{x^2})^6$

70. $(2\sqrt[4]{y^3})^3$

Miscellaneous

Simplify.

71. $\dfrac{\sqrt{3}}{\sqrt{2}} + \dfrac{2}{\sqrt{2}}$

72. $\dfrac{2}{\sqrt{7}} + \dfrac{5}{\sqrt{7}}$

73. $\dfrac{\sqrt{3}}{\sqrt{2}} + \dfrac{3\sqrt{6}}{2}$

74. $\dfrac{\sqrt{3}}{2\sqrt{2}} + \dfrac{\sqrt{5}}{3\sqrt{2}}$

75. $\dfrac{\sqrt{6}}{2} \cdot \dfrac{1}{\sqrt{3}}$

76. $\dfrac{\sqrt{6}}{\sqrt{7}} \cdot \dfrac{\sqrt{14}}{\sqrt{3}}$

77. $\dfrac{8 - \sqrt{32}}{20}$

78. $\dfrac{4 - \sqrt{28}}{6}$

79. $\dfrac{5 + \sqrt{75}}{10}$

80. $\dfrac{3 + \sqrt{18}}{6}$

81. $\sqrt{a}(\sqrt{a} - 3)$

82. $3\sqrt{m}(2\sqrt{m} - 6)$

83. $4\sqrt{a}(a + \sqrt{a})$

84. $\sqrt{3ab}(\sqrt{3a} + \sqrt{3})$

85. $(2\sqrt{3m})^2$

86. $(-3\sqrt{4y})^2$

87. $\left(-2\sqrt{xy^2z}\right)^2$

88. $(5a\sqrt{ab})^2$

89. $\sqrt[3]{m}\left(\sqrt[3]{m^2} - \sqrt[3]{m^5}\right)$

90. $\sqrt[4]{w}\left(\sqrt[4]{w^3} - \sqrt[4]{w^7}\right)$

91. $\sqrt[3]{8x^4} + \sqrt[3]{27x^4}$

92. $\sqrt[3]{16a^4} + a\sqrt[3]{2a}$

93. $\left(2m\sqrt[4]{2m^2}\right)^3$

94. $\left(-2t\sqrt[6]{2t^2}\right)^5$

95. $\dfrac{x - 9}{\sqrt{x} - 3}$

96. $\dfrac{x - y}{\sqrt{x} - \sqrt{y}}$

97. $\dfrac{3\sqrt{k}}{\sqrt{k} + \sqrt{7}}$

98. $\dfrac{\sqrt{hk}}{\sqrt{h} + 3\sqrt{k}}$

99. $\dfrac{5}{\sqrt{2} - 1} + \dfrac{3}{\sqrt{2} + 1}$

100. $\dfrac{\sqrt{3}}{\sqrt{6} - 1} - \dfrac{\sqrt{3}}{\sqrt{6} + 1}$

101. $\dfrac{1}{\sqrt{2}} + \dfrac{1}{\sqrt{3}}$

102. $\dfrac{4}{2\sqrt{3}} + \dfrac{1}{\sqrt{5}}$

103. $\dfrac{3}{\sqrt{2} - 1} + \dfrac{4}{\sqrt{2} + 1}$

104. $\dfrac{3}{\sqrt{5} - \sqrt{3}} - \dfrac{2}{\sqrt{5} + \sqrt{3}}$

105. $\dfrac{\sqrt{x}}{\sqrt{x} + 2} + \dfrac{3\sqrt{x}}{\sqrt{x} - 2}$

106. $\dfrac{\sqrt{5}}{3 - \sqrt{y}} - \dfrac{\sqrt{5y}}{3 + \sqrt{y}}$

107. $\dfrac{1}{\sqrt{x}} + \dfrac{1}{1 - \sqrt{x}}$

108. $\dfrac{\sqrt{x}}{\sqrt{x} - 3} + \dfrac{5}{\sqrt{x}}$

Getting More Involved

109. *Exploration*

A polynomial is prime if it cannot be factored by using integers, but many prime polynomials can be factored if we use radicals.

a) Find the product $(x - \sqrt[3]{2})(x^2 + \sqrt[3]{2}x + \sqrt[3]{4})$.

b) Factor $x^3 + 5$ using radicals.

c) Find the product

$$(\sqrt[3]{5} - \sqrt[3]{2})(\sqrt[3]{25} + \sqrt[3]{10} + \sqrt[3]{4}).$$

d) Use radicals to factor $a + b$ as a sum of two cubes and $a - b$ as a difference of two cubes.

110. *Discussion*

Which one of the following expressions is not equivalent to the others?

a) $(\sqrt[3]{x})^4$ **b)** $\sqrt[4]{x^3}$ **c)** $\sqrt[3]{x^4}$

d) $x^{4/3}$ **e)** $(x^{1/3})^4$

9.5   Solving Equations with Radicals and Exponents

In This Section

⟨1⟩ The Odd-Root Property
⟨2⟩ The Even-Root Property
⟨3⟩ Equations Involving Radicals
⟨4⟩ Equations Involving Rational Exponents
⟨5⟩ Applications

One of our goals in algebra is to keep increasing our knowledge of solving equations because the solutions to equations can give us the answers to various applied questions. In this section, we will apply our knowledge of radicals and exponents to solving some new types of equations.

⟨1⟩ The Odd-Root Property

Because $(-2)^3 = -8$ and $2^3 = 8$, the equation $x^3 = 8$ is equivalent to $x = 2$. The equation $x^3 = -8$ is equivalent to $x = -2$. Because there is only one real odd root of each real number, there is a simple rule for writing an equivalent equation in this situation.

Odd-Root Property

If n is an odd positive integer,

$$x^n = k \qquad \text{is equivalent to} \qquad x = \sqrt[n]{k}$$

for any real number k.

Note that $x^n = k$ is equivalent to $x = \sqrt[n]{k}$ means that these two equations have the same *real* solutions. So $x^3 = 1$ and $x = \sqrt[3]{1}$ each have only one real solution.

E X A M P L E   1

Using the odd-root property
Solve each equation.

  a) $x^3 = 27$ **b)** $x^5 + 32 = 0$ **c)** $(x - 2)^3 = 24$

Solution

  a) $x^3 = 27$
    $x = \sqrt[3]{27}$ Odd-root property
    $x = 3$
  Check 3 in the original equation. The solution set is $\{3\}$.

  b) $x^5 + 32 = 0$
        $x^5 = -32$ Isolate the variable.
        $x = \sqrt[5]{-32}$ Odd-root property
        $x = -2$
  Check -2 in the original equation. The solution set is $\{-2\}$.

  c) $(x - 2)^3 = 24$
        $x - 2 = \sqrt[3]{24}$ Odd-root property
        $x = 2 + 2\sqrt[3]{3}$ $\sqrt[3]{24} = \sqrt[3]{8} \cdot \sqrt[3]{3} = 2\sqrt[3]{3}$
  Check. The solution set is $\{2 + 2\sqrt[3]{3}\}$.

  Now do Exercises 1–8

⟨2⟩ The Even-Root Property

In solving the equation $x^2 = 4$, you might be tempted to write $x = 2$ as an equivalent equation. But $x = 2$ is not equivalent to $x^2 = 4$ because $2^2 = 4$ and $(-2)^2 = 4$. So the solution set to $x^2 = 4$ is $\{-2, 2\}$. The equation $x^2 = 4$ is equivalent to the compound sentence $x = 2$ or $x = -2$, which we can abbreviate as $x = \pm 2$. The equation $x = \pm 2$ is read "x equals positive or negative 2."

Equations involving other even powers are handled like the squares. Because $2^4 = 16$ and $(-2)^4 = 16$, the equation $x^4 = 16$ is equivalent to $x = \pm 2$. So $x^4 = 16$ has two real solutions. Note that $x^4 = -16$ has no real solutions. The equation $x^6 = 5$ is equivalent to $x = \pm\sqrt[6]{5}$. We can now state a general rule.

Even-Root Property

Suppose n is a positive even integer.

If $k > 0$, then $x^n = k$ is equivalent to $x = \pm\sqrt[n]{k}$.

If $k = 0$, then $x^n = k$ is equivalent to $x = 0$.

If $k < 0$, then $x^n = k$ has no real solution.

Note that $x^n = k$ for $k > 0$ is equivalent to $x = \pm\sqrt[n]{k}$ means that these two equations have the same *real* solutions.

EXAMPLE **2**

Using the even-root property

Solve each equation.

a) $x^2 = 10$ **b)** $w^8 = 0$ **c)** $x^4 = -4$

⟨ **Helpful Hint** ⟩

We do not say, "take the square root of each side." We are not doing the same thing to each side of $x^2 = 9$ when we write $x = \pm 3$. This is the third time that we have seen a rule for obtaining an equivalent equation without "doing the same thing to each side." (What were the other two?) Because there is only one odd root of every real number, you can actually take an odd root of each side.

Solution

a) $x^2 = 10$

    $x = \pm\sqrt{10}$ Even-root property

   The solution set is $\{-\sqrt{10}, \sqrt{10}\}$, or $\{\pm\sqrt{10}\}$.

b) $w^8 = 0$

    $w = 0$ Even-root property

   The solution set is $\{0\}$.

c) By the even-root property, $x^4 = -4$ has no real solution. (The fourth power of any real number is nonnegative.)

> Now do Exercises 9–14

Whether an equation has a solution depends on the domain of the variable. For example, $2x = 5$ has no solution in the set of integers and $x^2 = -9$ has no solution in the set of real numbers. We can say that the solution set to both of these equations is the empty set, $\varnothing$, as long as the domain of the variable is clear. In Section 9.6 we introduce a new set of numbers, the *imaginary numbers*, in which $x^2 = -9$ will have two solutions. So in this section it is best to say that $x^2 = -9$ has no real solution, because in Section 9.6 its solution set will *not* be $\varnothing$. An equation such as $x = x + 1$ never has a solution, and so saying that its solution set is $\varnothing$ is clear.

In Example 3, the even-root property is used to solve some equations that are a bit more complicated than those of Example 2.

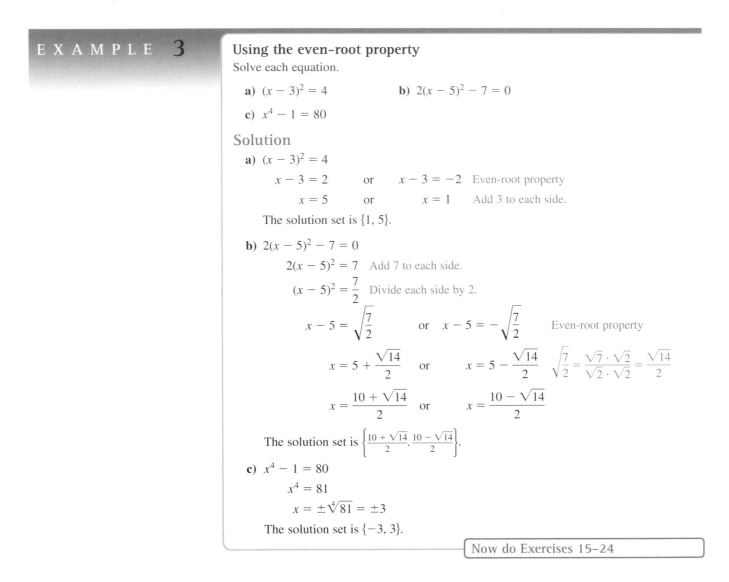

EXAMPLE 3

Using the even-root property

Solve each equation.

a) $(x - 3)^2 = 4$ b) $2(x - 5)^2 - 7 = 0$

c) $x^4 - 1 = 80$

Solution

a) $(x - 3)^2 = 4$

$\qquad x - 3 = 2 \quad$ or $\quad x - 3 = -2 \quad$ Even-root property

$\qquad\qquad x = 5 \quad$ or $\qquad x = 1 \quad$ Add 3 to each side.

The solution set is $\{1, 5\}$.

b) $2(x - 5)^2 - 7 = 0$

$\qquad 2(x - 5)^2 = 7 \quad$ Add 7 to each side.

$\qquad (x - 5)^2 = \dfrac{7}{2} \quad$ Divide each side by 2.

$\qquad x - 5 = \sqrt{\dfrac{7}{2}} \quad$ or $\quad x - 5 = -\sqrt{\dfrac{7}{2}} \quad$ Even-root property

$\qquad x = 5 + \dfrac{\sqrt{14}}{2} \quad$ or $\quad x = 5 - \dfrac{\sqrt{14}}{2} \qquad \sqrt{\dfrac{7}{2}} = \dfrac{\sqrt{7} \cdot \sqrt{2}}{\sqrt{2} \cdot \sqrt{2}} = \dfrac{\sqrt{14}}{2}$

$\qquad x = \dfrac{10 + \sqrt{14}}{2} \quad$ or $\quad x = \dfrac{10 - \sqrt{14}}{2}$

The solution set is $\left\{\dfrac{10 + \sqrt{14}}{2}, \dfrac{10 - \sqrt{14}}{2}\right\}$.

c) $x^4 - 1 = 80$

$\qquad x^4 = 81$

$\qquad x = \pm\sqrt[4]{81} = \pm 3$

The solution set is $\{-3, 3\}$.

Now do Exercises 15–24

In Chapter 5 we solved quadratic equations by factoring. The quadratic equations that we encounter in this chapter can be solved by using the even-root property as in parts (a) and (b) of Example 3. In Chapter 10 you will learn general methods for solving any quadratic equation.

⟨3⟩ Equations Involving Radicals

If we start with the equation $x = 3$ and square both sides, we get $x^2 = 9$, which has the solution set $\{-3, 3\}$. But the solution set to $x = 3$ is $\{3\}$. Squaring both sides produced an equation with more solutions than the original. We call the extra solutions **extraneous solutions.** The same problem can occur when we raise each side to any even power. Note that we don't always get extraneous solutions. We *might* get one or more of them.

Raising each side to an odd power does not cause extraneous solutions. For example, if we cube each side of $x = 3$ we get $x^3 = 27$. The solution set to both equations is $\{3\}$. Likewise, $x = -3$ and $x^3 = -27$ both have solution set $\{-3\}$.

> **Raising each side of an equation to a power**
>
> If n is odd, then $a = b$ and $a^n = b^n$ are equivalent equations.
>
> If n is even, then $a = b$ and $a^n = b^n$ may not be equivalent. However, the solution set to $a^n = b^n$ contains all of the solutions to $a = b$.

It has always been important to check solutions any time you solve an equation. When raising each side to a power, it is even more important. We use these ideas most often with equations involving radicals as shown in Example 4.

EXAMPLE   4

Raising each side to a power to eliminate radicals

Solve each equation.

    a) $\sqrt{2x - 3} - 5 = 0$ b) $\sqrt[3]{3x + 5} = \sqrt[3]{x - 1}$ c) $\sqrt{3x + 18} = x$

‹ Calculator Close-Up ›

If 14 satisfies the equation

$$\sqrt{2x - 3} - 5 = 0,$$

then (14, 0) is an x-intercept for the graph of

$$y = \sqrt{2x - 3} - 5.$$

So the calculator graph shown here provides visual support for the conclusion that 14 is the only solution to the equation.

Solution

  a) Eliminate the square root by raising each side to the power 2:

$$\sqrt{2x - 3} - 5 = 0 \qquad \text{Original equation}$$
$$\sqrt{2x - 3} = 5 \qquad \text{Isolate the radical.}$$
$$(\sqrt{2x - 3})^2 = 5^2 \qquad \text{Square both sides.}$$
$$2x - 3 = 25$$
$$2x = 28$$
$$x = 14$$

Check by evaluating $x = 14$ in the original equation:

$$\sqrt{2(14) - 3} - 5 = 0$$
$$\sqrt{28 - 3} - 5 = 0$$
$$\sqrt{25} - 5 = 0$$
$$0 = 0$$

The solution set is $\{14\}$.

  b) $\sqrt[3]{3x + 5} = \sqrt[3]{x - 1}$ Original equation
$$(\sqrt[3]{3x + 5})^3 = (\sqrt[3]{x - 1})^3 \qquad \text{Cube each side.}$$
$$3x + 5 = x - 1$$
$$2x = -6$$
$$x = -3$$

Check $x = -3$ in the original equation:

$$\sqrt[3]{3(-3) + 5} = \sqrt[3]{-3 - 1}$$
$$\sqrt[3]{-4} = \sqrt[3]{-4}$$

Note that $\sqrt[3]{-4}$ is a real number. The solution set is $\{-3\}$. In this example, we checked for arithmetic mistakes. There was no possibility of extraneous solutions here because we raised each side to an odd power.

c) $\sqrt{3x + 18} = x$ Original equation

$(\sqrt{3x + 18})^2 = x^2$ Square both sides.

$3x + 18 = x^2$ Simplify.

$-x^2 + 3x + 18 = 0$ Subtract x^2 from each side
                                            to get zero on one side.

$x^2 - 3x - 18 = 0$ Multiply each side by -1
                                            for easier factoring.

$(x - 6)(x + 3) = 0$ Factor.

$x - 6 = 0$ or $x + 3 = 0$ Zero factor property

$x = 6$ or $x = -3$

Because we squared both sides, we must check for extraneous solutions. If $x = -3$ in the original equation $\sqrt{3x + 18} = x$, we get

$$\sqrt{3(-3) + 18} = -3$$
$$\sqrt{9} = -3$$
$$3 = -3,$$

which is not correct. If $x = 6$ in the original equation, we get

$$\sqrt{3(6) + 18} = 6,$$

which is correct. The solution set is $\{6\}$.

Now do Exercises 25–44

< Calculator Close-Up >

The graphs of
$$y_1 = \sqrt{3x + 18}$$
and $y_2 = x$ provide visual support that 6 is the only value of x for which x and $\sqrt{3x + 18}$ are equal.

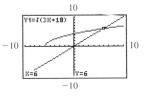

In Example 5, the radicals are not eliminated after squaring both sides of the equation. In this case, we must square both sides a second time. Note that we square the side with two terms the same way we square a binomial.

E X A M P L E   **5**

Squaring both sides twice

Solve $\sqrt{5x - 1} - \sqrt{x + 2} = 1$.

Solution

It is easier to square both sides if the two radicals are not on the same side.

$\sqrt{5x - 1} - \sqrt{x + 2} = 1$ Original equation

$\sqrt{5x - 1} = 1 + \sqrt{x + 2}$ Add $\sqrt{x + 2}$ to each side.

$(\sqrt{5x - 1})^2 = (1 + \sqrt{x + 2})^2$ Square both sides.

$5x - 1 = 1 + 2\sqrt{x + 2} + x + 2$ Square the right side like
                                            a binomial.

$5x - 1 = 3 + x + 2\sqrt{x + 2}$ Combine like terms on
                                            the right side.

$4x - 4 = 2\sqrt{x + 2}$ Isolate the square root.

$2x - 2 = \sqrt{x + 2}$ Divide each side by 2.

$(2x - 2)^2 = (\sqrt{x + 2})^2$ Square both sides.

$4x^2 - 8x + 4 = x + 2$ Square the binomial on
                                            the left side.

$4x^2 - 9x + 2 = 0$

$(4x - 1)(x - 2) = 0$

$4x - 1 = 0$ or $x - 2 = 0$

$x = \dfrac{1}{4}$ or $x = 2$

Check to see whether $\sqrt{5x-1} - \sqrt{x+2} = 1$ for $x = \frac{1}{4}$ and for $x = 2$:

$$\sqrt{5 \cdot \frac{1}{4} - 1} - \sqrt{\frac{1}{4} + 2} = \sqrt{\frac{1}{4}} - \sqrt{\frac{9}{4}} = \frac{1}{2} - \frac{3}{2} = -1$$

$$\sqrt{5 \cdot 2 - 1} - \sqrt{2+2} = \sqrt{9} - \sqrt{4} = 3 - 2 = 1$$

So the original equation is not satisfied for $x = \frac{1}{4}$ but is satisfied for $x = 2$. Since 2 is the only solution to the equation, the solution set is $\{2\}$.

> Now do Exercises 45–60

⟨4⟩ Equations Involving Rational Exponents

Equations involving rational exponents can be solved by combining the methods that you just learned for eliminating radicals and integral exponents. For equations involving rational exponents, always eliminate the root first and the power second.

EXAMPLE 6

Eliminating the root, then the power

Solve each equation.

a) $x^{2/3} = 4$

b) $(w-1)^{-2/5} = 4$

Solution

a) Because the exponent 2/3 indicates a cube root, raise each side to the power 3:

$x^{2/3} = 4$	Original equation
$(x^{2/3})^3 = 4^3$	Cube each side.
$x^2 = 64$	Multiply the exponents: $\frac{2}{3} \cdot 3 = 2.$
$x = 8$ or $x = -8$	Even-root property

All of the equations are equivalent. Check 8 and -8 in the original equation. The solution set is $\{-8, 8\}$.

b)

$(w-1)^{-2/5} = 4$	Original equation
$[(w-1)^{-2/5}]^{-5} = 4^{-5}$	Raise each side to the power -5 to eliminate the negative exponent.
$(w-1)^2 = \dfrac{1}{1024}$	Multiply the exponents: $-\frac{2}{5}(-5) = 2.$
$w - 1 = \pm\sqrt{\dfrac{1}{1024}}$	Even-root property
$w - 1 = \dfrac{1}{32}$ or $w - 1 = -\dfrac{1}{32}$	
$w = \dfrac{33}{32}$ or $w = \dfrac{31}{32}$	

Check the values in the original equation. The solution set is $\left\{\frac{31}{32}, \frac{33}{32}\right\}$.

> Now do Exercises 61–72

⟨ Helpful Hint ⟩

Note how we eliminate the root first by raising each side to an integer power, and then apply the even-root property to get two solutions in Example 6(a). A common mistake is to raise each side to the 3/2 power and get $x = 4^{3/2} = 8$. If you do not use the even-root property, you can easily miss the solution -8.

⟨ Calculator Close-Up ⟩

Check that 31/32 and 33/32 satisfy the original equation.

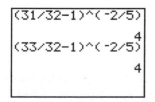

```
(31/32-1)^(-2/5)
                4
(33/32-1)^(-2/5)
                4
```

An equation with a rational exponent might not have a real solution because all even powers of real numbers are nonnegative.

E X A M P L E   7

An equation with no solution

Solve $(2t - 3)^{-2/3} = -1$.

Solution

Raise each side to the power -3 to eliminate the root and the negative sign in the exponent:

$$(2t - 3)^{-2/3} = -1 \qquad \text{Original equation}$$

$$\left[(2t - 3)^{-2/3}\right]^{-3} = (-1)^{-3} \qquad \text{Raise each side to the } -3 \text{ power.}$$

$$(2t - 3)^2 = -1 \qquad \text{Multiply the exponents: } -\frac{2}{3}(-3) = 2.$$

By the even-root property this equation has no real solution. The square of every real number is nonnegative.

> Now do Exercises 73–74

The three most important rules for solving equations with exponents and radicals are restated here.

Strategy for Solving Equations with Exponents and Radicals

1. In raising each side of an equation to an even power, we can create an equation that gives extraneous solutions. We must check all possible solutions in the original equation.

2. When applying the even-root property, remember that there is a positive and a negative even root for any positive real number.

3. For equations with rational exponents, raise each side to a positive or negative integral power first and then apply the even- or odd-root property. (Positive fraction—raise to a positive power; negative fraction—raise to a negative power.)

⟨5⟩ Applications

The square of the hypotenuse of any right triangle is equal to the sum of the squares of the legs (the Pythagorean theorem). In Example 8 we use this fact and the even-root property to find a distance on a baseball diamond.

E X A M P L E   8

Diagonal of a baseball diamond

A baseball diamond is actually a square, 90 feet on each side. What is the distance from third base to first base?

Solution

First make a sketch as in Fig. 9.1. The distance x from third base to first base is the length of the diagonal of the square shown in Fig. 9.1. The Pythagorean theorem can be applied to

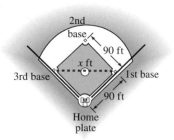

Figure 9.1

the right triangle formed from the diagonal and two sides of the square. The sum of the squares of the sides is equal to the diagonal squared:

$$x^2 = 90^2 + 90^2$$
$$x^2 = 8100 + 8100$$
$$x^2 = 16{,}200$$
$$x = \pm\sqrt{16{,}200} = \pm 90\sqrt{2}$$

The length of the diagonal of a square must be positive, so we disregard the negative solution. Checking the answer in the original equation verifies that the *exact* length of the diagonal is $90\sqrt{2}$ feet.

> Now do Exercises 95–110

Warm-Ups ▼

Fill in the blank.

1. If n is an _____ positive integer, then $x^n = k$ is equivalent to $x = \sqrt[n]{k}$ for any real number k.

2. If n is an _____ positive integer, then $x^n = k$ is equivalent to $x = \pm\sqrt[n]{k}$ for $k > 0$.

3. An _____ solution is a solution that appears when solving an equation but does not satisfy the original equation.

4. Raising each side of an equation to an _____ power can produce an extraneous solution.

True or false?

5. The equations $x^2 = 4$ and $x = 2$ are equivalent.

6. The equation $x^2 = -25$ has no real solution.

7. The equation $x^2 = 0$ has no solution.

8. The equation $x^3 = 8$ is equivalent to $x = 2$.

9. Squaring both sides of $\sqrt{x} = -7$ yields an equation with an extraneous solution.

10. The equations $x^2 - 6 = 0$ and $x = \pm\sqrt{6}$ are equivalent.

Exercises 9.5

‹ **Study Tips** ›

• Try changing subjects or tasks every hour when you study. The brain does not easily assimilate the same material hour after hour.
• You will learn more from working on a subject 1 hour per day than 7 hours on Saturday.

‹1› The Odd-Root Property

Solve each equation. See Example 1.

1. $x^3 = -1000$

2. $y^3 = 125$

3. $32m^5 - 1 = 0$

4. $243a^5 + 1 = 0$

5. $(y - 3)^3 = -8$

6. $(x - 1)^3 = -1$

7. $\dfrac{1}{2}x^3 + 4 = 0$

8. $3(x - 9)^7 = 0$

‹2› The Even-Root Property

Find all real solutions to each equation. See Examples 2 and 3.

9. $x^2 = 25$

10. $x^2 = 36$

11. $x^2 - 20 = 0$

12. $a^2 - 40 = 0$

13. $x^2 + 9 = 0$

14. $w^2 + 49 = 0$

15. $(x - 3)^2 = 16$

16. $(a - 2)^2 = 25$

17. $(x + 1)^2 - 8 = 0$

18. $(w + 3)^2 - 12 = 0$

19. $\frac{1}{2}x^2 = 5$

20. $\frac{1}{3}x^2 = 6$

21. $(y - 3)^4 = 0$

22. $(2x - 3)^6 = 0$

23. $2x^6 = 128$

24. $3y^4 = 48$

⟨3⟩ Equations Involving Radicals

Solve each equation and check for extraneous solutions. See Example 4.

25. $\sqrt{x - 3} - 3 = 4$

26. $\sqrt{a - 1} - 5 = 1$

27. $2\sqrt{w + 4} = 5$

28. $3\sqrt{w + 1} = 6$

29. $\sqrt[3]{2x + 3} = \sqrt[3]{x + 12}$

30. $\sqrt[3]{a + 3} = \sqrt[3]{2a - 7}$

31. $\sqrt{2t - 4} = \sqrt{t - 1}$

32. $\sqrt{w - 3} = \sqrt{4w - 15}$

33. $\sqrt{4x^2 + x - 3} = 2x$

34. $\sqrt{x^2 - 5x + 2} = x$

35. $\sqrt{x^2 + 2x - 6} = 3$

36. $\sqrt{x^2 - x - 4} = 4$

37. $\sqrt{2x^2 - 1} = x$

38. $\sqrt{2x^2 - 3x - 10} = x$

39. $\sqrt{2x^2 + 5x + 6} = x$

40. $\sqrt{2x^2 + 6x + 9} = x$

41. $\sqrt{x - 1} = x - 1$

42. $\sqrt{2x - 1} = 2x - 1$

43. $x + \sqrt{x - 9} = 9$

44. $\sqrt{3x - 1} + 3x = 1$

Solve each equation and check for extraneous solutions. See Example 5.

45. $\sqrt{x} + \sqrt{x - 3} = 3$

46. $\sqrt{x} + \sqrt{x + 3} = 3$

47. $\sqrt{x + 2} + \sqrt{x - 1} = 3$

48. $\sqrt{x} + \sqrt{x - 5} = 5$

49. $\sqrt{x + 3} - \sqrt{x - 2} = 1$

50. $\sqrt{2x + 1} - \sqrt{x} = 1$

51. $\sqrt{3x + 1} - \sqrt{2x - 1} = 1$

52. $\sqrt{4x + 1} - \sqrt{3x - 2} = 1$

53. $\sqrt{2x + 2} - \sqrt{x - 3} = 2$

54. $\sqrt{3x} - \sqrt{x - 2} = 4$

55. $\sqrt{4 - x} - \sqrt{x + 6} = 2$

56. $\sqrt{6 - x} - \sqrt{x - 2} = 2$

57. $\sqrt{x - 5} - \sqrt{x} = 3$

58. $\sqrt{2x} - \sqrt{2x - 12} = 6$

59. $\sqrt{3x + 1} + \sqrt{2x + 4} = 3$

60. $\sqrt{2x + 5} + \sqrt{x + 2} = 1$

⟨4⟩ Equations Involving Rational Exponents

Solve each equation. See Examples 6 and 7.

61. $x^{2/3} = 3$

62. $a^{2/3} = 2$

63. $y^{-2/3} = 9$

64. $w^{-2/3} = 4$

65. $w^{1/3} = 8$

66. $a^{1/3} = 27$

67. $t^{-1/2} = 9$

68. $w^{-1/4} = \frac{1}{2}$

69. $(3a - 1)^{-2/5} = 1$

70. $(r - 1)^{-2/3} = 1$

71. $(t - 1)^{-2/3} = 2$

72. $(w + 3)^{-1/3} = \frac{1}{3}$

73. $(x - 3)^{2/3} = -4$

74. $(x + 2)^{3/2} = -1$

Miscellaneous

Solve each equation. See the Strategy for Solving Equations with Exponents and Radicals box on page 602.

75. $2x^2 + 3 = 7$

76. $3x^2 - 5 = 16$

77. $\sqrt[3]{2w + 3} - \sqrt[3]{w - 2}$

78. $\sqrt[3]{2 - w} = \sqrt[3]{2w - 28}$

79. $(w + 1)^{2/3} = -3$

80. $(x - 2)^{4/3} = -2$

81. $(a + 1)^{1/3} = -2$

82. $(a - 1)^{1/3} = -3$

83. $(4y - 5)^7 = 0$

84. $(5x)^9 = 0$

85. $\sqrt{5x^2 + 4x + 1} - x = 0$

86. $3 + \sqrt{x^2 - 8x} = 0$

87. $\sqrt{4x^2} = x + 2$

88. $\sqrt{9x^2} = x + 6$

89. $(t + 2)^4 = 32$

90. $(w + 1)^4 = 48$

91. $\sqrt{x^2 - 3x} = x$

92. $\sqrt[4]{4x^4 - 48} = -x$

93. $x^{-3} = 8$

94. $x^{-2} = 4$

⟨5⟩ Applications

Solve each problem by writing an equation and solving it. Find the exact answer and simplify it using the rules for radicals. See Example 8.

95. *Side of a square.* Find the length of the side of a square whose diagonal is 8 feet.

96. *Diagonal of a patio.* Find the length of the diagonal of a square patio with an area of 40 square meters.

97. *Side of a sign.* Find the length of the side of a square sign whose area is 50 square feet.

98. *Side of a cube.* Find the length of the side of a cubic box whose volume is 80 cubic feet.

99. *Diagonal of a rectangle.* If the sides of a rectangle are 30 feet and 40 feet in length, find the length of the diagonal of the rectangle.

100. *Diagonal of a sign.* What is the length of the diagonal of a rectangular billboard whose sides are 5 meters and 12 meters?

101. *A 30-60-90 triangle.* In a 30°-60°-90° triangle, the side opposite the 30° angle is half the length of the hypotenuse. See the accompanying figure.
   a) Find the length of the hypotenuse if the side opposite the 30° angle is 1.
   b) Find the length of the side opposite 60° if the side opposite 30° is 1.
   c) Find the length of the side opposite 60° if the length of the hypotenuse is 1.

102. *An isosceles right triangle.* An isosceles right triangle has two 45° angles. The sides opposite those angles are equal in length. See the accompanying figure.
   a) Find the length of the hypotenuse if the length of each of the equal sides is 1.
   b) Find the length of each of the equal sides if the length of the hypotenuse is 1.

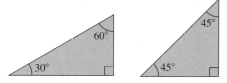

Figure for Exercises 101 and 102

103. *Sailboat stability.* To be considered safe for ocean sailing, the capsize screening value C should be less than 2 (www.sailing.com). For a boat with a beam (or width) b in feet and displacement d in pounds, C is determined by the function

$$C = 4d^{-1/3}b.$$

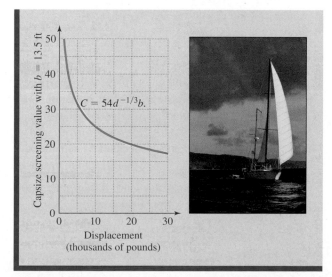

Figure for Exercise 103

a) Find the capsize screening value for the Tartan 4100, which has a displacement of 23,245 pounds and a beam of 13.5 feet.

b) Solve this formula for d.

c) The accompanying graph shows C in terms of d for the Tartan 4100 ($b = 13.5$). For what displacement is the Tartan 4100 safe for ocean sailing?

104. *Sailboat speed.* The sail area-displacement ratio S provides a measure of the sail power available to drive a boat. For a boat with a displacement of d pounds and a sail area of A square feet S is determined by the function
$$S = 16Ad^{-2/3}.$$

a) Find S to the nearest tenth for the Tartan 4100, which has a sail area of 810 square feet and a displacement of 23,245 pounds.

b) Write d in terms of A and S.

105. *Diagonal of a side.* Find the length of the diagonal of a side of a cubic packing crate whose volume is 2 cubic meters.

106. *Volume of a cube.* Find the volume of a cube on which the diagonal of a side measures 2 feet.

107. *Length of a road.* An architect designs a public park in the shape of a trapezoid. Find the length of the diagonal road marked a in the figure.

108. *Length of a boundary.* Find the length of the border of the park marked b in the trapezoid shown in the figure.

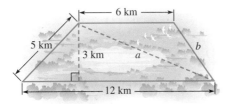

Figure for Exercises 107 and 108

109. *Average annual return.* In Exercise 127 of Section 9.2, the function
$$r = \left(\frac{S}{P}\right)^{1/n} - 1$$
was used to find the average annual return for an investment.

a) Write S in terms of r, P, and n.

b) Write P in terms of r, S, and n.

110. *Surface area of a cube.* The function $A = 6V^{2/3}$ gives the surface area of a cube in terms of its volume V. What

is the volume of a cube with surface area 12 square feet?

111. *Kepler's third law.* According to Kepler's third law of planetary motion, the ratio $\frac{T^2}{R^3}$ has the same value for every planet in our solar system. R is the average radius of the orbit of the planet measured in astronomical units (AU), and T is the number of years it takes for one complete orbit of the sun. Jupiter orbits the sun in 11.86 years with an average radius of 5.2 AU, whereas Saturn orbits the sun in 29.46 years. Find the average radius of the orbit of Saturn. (One AU is the distance from the earth to the sun.)

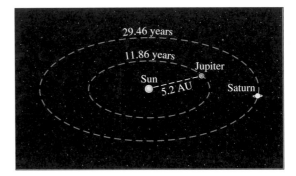

Figure for Exercise 111

112. *Orbit of Venus.* If the average radius of the orbit of Venus is 0.723 AU, then how many years does it take for Venus to complete one orbit of the sun? Use the information in Exercise 111.

Use a calculator to find approximate solutions to the following equations. Round your answers to three decimal places.

113. $x^2 = 3.24$

114. $(x + 4)^3 = 7.51$

115. $\sqrt{x - 2} = 1.73$

116. $\sqrt[3]{x - 5} = 3.7$

117. $x^{2/3} = 8.86$

118. $(x - 1)^{-3/4} = 7.065$

Getting More Involved

119. *Cooperative learning*

Work in a small group to write a formula that gives the side of a cube in terms of the volume of the cube and explain the formula to the other groups.

120. *Cooperative learning*

Work in a small group to write a formula that gives the side of a square in terms of the diagonal of the square and explain the formula to the other groups.

9.6   Complex Numbers

In Chapter 1, we discussed the real numbers and the various subsets of the real numbers. In this section, we define a set of numbers that has the real numbers as a subset. This new set of numbers is the set of *complex numbers*. Although it is hard to imagine numbers beyond the real numbers, the complex numbers are used to model some very real phenomena in physics and electrical engineering. These applications are beyond the scope of this text, but if you want a better understanding of them, search the Internet for "applications of complex numbers."

⟨1⟩ Definition

The equation $2x = 1$ has no solution in the set of integers, but in the set of rational numbers, $2x = 1$ has a solution. The situation is similar for the equation $x^2 = -4$. It has no solution in the set of real numbers because the square of every real number is nonnegative. However, in the set of complex numbers $x^2 = -4$ has two solutions. The complex numbers were developed so that equations such as $x^2 = -4$ would have solutions.

The complex numbers are based on the symbol $\sqrt{-1}$. In the real number system this symbol has no meaning. In the set of complex numbers this symbol is given meaning. We call it i. We make the definition that

$$i = \sqrt{-1} \quad \text{and} \quad i^2 = -1.$$

> **Complex Numbers**
>
> The set of **complex numbers** is the set of all numbers of the form
> $$a + bi,$$
> where a and b are real numbers, $i = \sqrt{-1}$, and $i^2 = -1$.

In the complex number $a + bi$, a is called the **real part** and b is called the **imaginary part.** If $b \neq 0$, the number $a + bi$ is called an **imaginary number.** If $b = 0$, then the complex number $a + 0i$ is the real number a.

In dealing with complex numbers, we treat $a + bi$ as if it were a binomial, with i being a variable. Thus, we would write $2 + (-3)i$ as $2 - 3i$. We agree that $2 + i3$, $3i + 2$, and $i3 + 2$ are just different ways of writing $2 + 3i$ (the standard form). Some examples of complex numbers are

$$-3 - 5i, \quad \frac{2}{3} - \frac{3}{4}i, \quad 1 + i\sqrt{2}, \quad 9 + 0i, \quad \text{and} \quad 0 + 7i.$$

For simplicity we write only $7i$ for $0 + 7i$. The complex number $9 + 0i$ is the real number 9, and $0 + 0i$ is the real number 0. Any complex number with $b = 0$ is a real number. For any real number a,

$$a + 0i = a.$$

‹ **Helpful Hint** ›
Note that a complex number does not have to have an *i* in it. All real numbers are complex numbers. So 1, 2, and 3 are complex numbers.

The set of real numbers is a subset of the set of complex numbers. See Fig. 9.2.

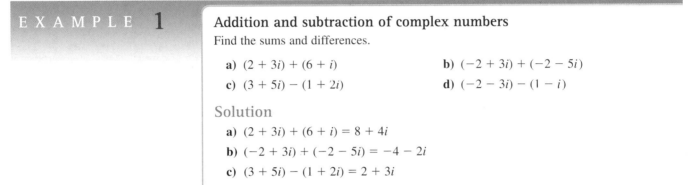

Complex numbers	
Real numbers	Imaginary numbers
$3, \pi, \frac{5}{2}, 0, -9, \sqrt{2}$	$i, 2 + 3i, \sqrt{-5}, -3 - 8i$

Figure 9.2

‹2› Addition, Subtraction, and Multiplication

Addition and subtraction of complex numbers are performed as if the complex numbers were algebraic expressions with *i* being a variable.

EXAMPLE 1

Addition and subtraction of complex numbers
Find the sums and differences.

a) $(2 + 3i) + (6 + i)$ b) $(-2 + 3i) + (-2 - 5i)$

c) $(3 + 5i) - (1 + 2i)$ d) $(-2 - 3i) - (1 - i)$

Solution

a) $(2 + 3i) + (6 + i) = 8 + 4i$

b) $(-2 + 3i) + (-2 - 5i) = -4 - 2i$

c) $(3 + 5i) - (1 + 2i) = 2 + 3i$

d) $(-2 - 3i) - (1 - i) = -3 - 2i$

Now do Exercises 1–8

Informally, we add and subtract complex numbers as in Example 1. Formally, we use the following symbolic definition. We include this definition for completeness, but you don't need to memorize it. Just add or subtract as in Example 1.

Addition and Subtraction of Complex Numbers
The sum and difference of $a + bi$ and $c + di$ are defined as follows:

$$(a + bi) + (c + di) = (a + c) + (b + d)i$$
$$(a + bi) - (c + di) = (a - c) + (b - d)i$$

Complex numbers are multiplied as if they were algebraic expressions. Whenever i^2 appears, we replace it by -1.

EXAMPLE 2

Products of complex numbers
Find each product.

a) $2i(1 + i)$ b) $(2 + 3i)(4 + 5i)$ c) $(3 + i)(3 - i)$

Solution

a) $2i(1 + i) = 2i + 2i^2$ Distributive property

            $= 2i + 2(-1)$ $i^2 = -1$

            $= -2 + 2i$

Many graphing calculators can perform operations with complex numbers.

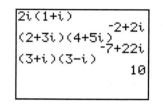

b) Use the FOIL method to find the product:

$$(2 + 3i)(4 + 5i) = 8 + 10i + 12i + 15i^2$$
$$= 8 + 22i + 15(-1) \quad \text{Replace } i^2 \text{ by } -1.$$
$$= 8 + 22i - 15$$
$$= -7 + 22i$$

c) This product is the product of a sum and a difference.

$$(3 + i)(3 - i) = 9 - 3i + 3i - i^2$$
$$= 9 - (-1) \quad i^2 = -1$$
$$= 10$$

Now do Exercises 9–26

For completeness we give the following symbolic definition of multiplication of complex numbers. However, it is simpler to find products as we did in Example 2 than to use this definition.

Multiplication of Complex Numbers

The complex numbers $a + bi$ and $c + di$ are multiplied as follows:

$$(a + bi)(c + di) = (ac - bd) + (ad + bc)i$$

We can find powers of i using the fact that $i^2 = -1$. For example,

$$i^3 = i^2 \cdot i = -1 \cdot i = -i.$$

The value of i^4 is found from the value of i^3:

$$i^4 = i^3 \cdot i = -i \cdot i = -i^2 = 1$$

Using $i^2 = -1$, $i^3 = -i$, and $i^4 = 1$, you can actually find any power of i by factoring out all of the fourth powers. For example,

$$i^{13} = (i^4)^3 \cdot i = (1)^3 \cdot i = i \quad \text{and} \quad i^{18} = (i^4)^4 \cdot i^2 = (1)^4 \cdot i^2 = -1.$$

E X A M P L E   **3**

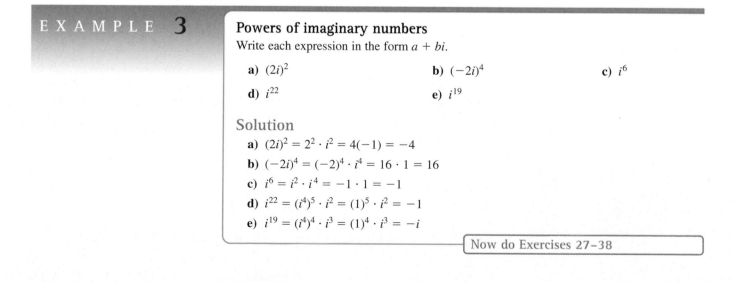

Powers of imaginary numbers

Write each expression in the form $a + bi$.

a) $(2i)^2$ **b)** $(-2i)^4$ **c)** i^6

d) i^{22} **e)** i^{19}

Solution

a) $(2i)^2 = 2^2 \cdot i^2 = 4(-1) = -4$

b) $(-2i)^4 = (-2)^4 \cdot i^4 = 16 \cdot 1 = 16$

c) $i^6 = i^2 \cdot i^4 = -1 \cdot 1 = -1$

d) $i^{22} = (i^4)^5 \cdot i^2 = (1)^5 \cdot i^2 = -1$

e) $i^{19} = (i^4)^4 \cdot i^3 = (1)^4 \cdot i^3 = -i$

Now do Exercises 27–38

⟨3⟩ Division of Complex Numbers

To divide a complex number by a real number, divide each term by the real number, just as we would divide a binomial by a number. For example,

$$\frac{4 + 6i}{2} = \frac{2(2 + 3i)}{2}$$

$$= 2 + 3i.$$

⟨ **Helpful Hint** ⟩

Here is that word "conjugate" again. It is generally used to refer to two things that go together in some way.

To understand division by a complex number, we first look at imaginary numbers that have a real product. The product of the two imaginary numbers in Example 2(c) is a real number:

$$(3 + i)(3 - i) = 10$$

We say that $3 + i$ and $3 - i$ are complex conjugates of each other.

> **Complex Conjugates**
>
> The complex numbers $a + bi$ and $a - bi$ are called **complex conjugates** of one another. Their product is the real number $a^2 + b^2$.

EXAMPLE 4

Products of conjugates

Find the product of the given complex number and its conjugate.

a) $2 + 3i$ **b)** $5 - 4i$

Solution

a) The conjugate of $2 + 3i$ is $2 - 3i$.

$$(2 + 3i)(2 - 3i) = 4 - 9i^2$$
$$= 4 + 9$$
$$= 13$$

b) The conjugate of $5 - 4i$ is $5 + 4i$.

$$(5 - 4i)(5 + 4i) = 25 + 16$$
$$= 41$$

Now do Exercises 39–46

We use complex conjugates to divide complex numbers. The process is the same as rationalizing the denominator. We multiply the numerator and denominator of the quotient by the complex conjugate of the denominator. If we were to use $\sqrt{-1}$ instead of i, then Example 5 here would look just like Example 7 in Section 9.4.

EXAMPLE 5

Dividing complex numbers

Find each quotient. Write the answer in the form $a + bi$.

a) $\dfrac{5}{3 - 4i}$ **b)** $\dfrac{3 - i}{2 + i}$ **c)** $\dfrac{3 + 2i}{i}$

Solution

a) Multiply the numerator and denominator by $3 + 4i$, the conjugate of $3 - 4i$:

$$\frac{5}{3 - 4i} = \frac{5(3 + 4i)}{(3 - 4i)(3 + 4i)}$$

$$= \frac{15 + 20i}{9 - 16i^2}$$

$$= \frac{15 + 20i}{25} \qquad 9 - 16i^2 = 9 - 16(-1) = 25$$

$$= \frac{15}{25} + \frac{20}{25}i$$

$$= \frac{3}{5} + \frac{4}{5}i$$

b) Multiply the numerator and denominator by $2 - i$, the conjugate of $2 + i$:

$$\frac{3 - i}{2 + i} = \frac{(3 - i)(2 - i)}{(2 + i)(2 - i)}$$

$$= \frac{6 - 5i + i^2}{4 - i^2}$$

$$= \frac{6 - 5i - 1}{4 - (-1)}$$

$$= \frac{5 - 5i}{5}$$

$$= 1 - i$$

c) Multiply the numerator and denominator by $-i$, the conjugate of i:

$$\frac{3 + 2i}{i} = \frac{(3 + 2i)(-i)}{i(-i)}$$

$$= \frac{-3i - 2i^2}{-i^2}$$

$$= \frac{-3i + 2}{1}$$

$$= 2 - 3i$$

> Now do Exercises 47–58

The symbolic definition of division of complex numbers follows.

Division of Complex Numbers

We divide the complex number $a + bi$ by the complex number $c + di$ as follows:

$$\frac{a + bi}{c + di} = \frac{(a + bi)(c - di)}{(c + di)(c - di)}$$

⟨4⟩ Square Roots of Negative Numbers

In the complex number system, negative numbers have two square roots. Because $i^2 = -1$ and $(-i)^2 = -1$, both i and $-i$ are square roots of -1. Because $(2i)^2 = -4$ and $(-2i)^2 = -4$, both $2i$ and $-2i$ are square roots of -4. We use the radical symbol only for the square root that has the positive coefficient, as in $\sqrt{-4} = 2i$.

Square Root of a Negative Number

For any positive real number b,

$$\sqrt{-b} = i\sqrt{b}.$$

For example, $\sqrt{-9} = i\sqrt{9} = 3i$ and $\sqrt{-7} = i\sqrt{7}$. Note that the expression $\sqrt{7}i$ could easily be mistaken for the expression $\sqrt{7i}$, where i is under the radical. For this reason, when the coefficient of i is a radical, we write i preceding the radical.

Note that the product rule $(\sqrt{a} \cdot \sqrt{b} = \sqrt{ab})$ does not apply to negative numbers. For example, $\sqrt{-2} \cdot \sqrt{-3} \neq \sqrt{6}$:

$$\sqrt{-2} \cdot \sqrt{-3} = i\sqrt{2} \cdot i\sqrt{3} = i^2\sqrt{6} = -\sqrt{6}$$

Square roots of negative numbers must be written in terms of i before operations are performed.

E X A M P L E   6

Square roots of negative numbers

Write each expression in the form $a + bi$, where a and b are real numbers.

a) $3 + \sqrt{-9}$ **b)** $\sqrt{-12} + \sqrt{-27}$

c) $\dfrac{-1 - \sqrt{-18}}{3}$ **d)** $\sqrt{-4} \cdot \sqrt{-9}$

Solution

a) $3 + \sqrt{-9} = 3 + i\sqrt{9}$
$$= 3 + 3i$$

b) $\sqrt{-12} + \sqrt{-27} = i\sqrt{12} + i\sqrt{27}$
$$= 2i\sqrt{3} + 3i\sqrt{3} \qquad \begin{array}{l} \sqrt{12} = \sqrt{4}\,\sqrt{3} = 2\sqrt{3} \\ \sqrt{27} = \sqrt{9}\,\sqrt{3} = 3\sqrt{3} \end{array}$$
$$= 5i\sqrt{3}$$

c) $\dfrac{-1 - \sqrt{-18}}{3} = \dfrac{-1 - i\sqrt{18}}{3}$
$$= \dfrac{-1 - 3i\sqrt{2}}{3}$$
$$= -\dfrac{1}{3} - i\sqrt{2}$$

d) $\sqrt{-4} \cdot \sqrt{-9} = i\sqrt{4} \cdot i\sqrt{9} = 2i \cdot 3i = 6i^2 = -6$

> Now do Exercises 59–78

⟨5⟩ Imaginary Solutions to Equations

In the complex number system the even-root property can be restated so that $x^2 = k$ is equivalent to $x = \pm\sqrt{k}$ for any $k \neq 0$. So an equation such as $x^2 = -9$ that has no real solutions has two imaginary solutions in the complex numbers.

E X A M P L E   7

Imaginary solutions to equations

Find the imaginary solutions to each equation.

a) $x^2 = -9$ **b)** $3x^2 + 2 = 0$

Solution

a) First apply the even-root property:

$$x^2 = -9$$
$$x = \pm\sqrt{-9} \quad \text{Even-root property}$$
$$= \pm i\sqrt{9}$$
$$= \pm 3i$$

Check these solutions in the original equation:

$$(3i)^2 = 9i^2 = 9(-1) = -9$$
$$(-3i)^2 = 9i^2 = -9$$

The solution set is $\{\pm 3i\}$.

b) First solve the equation for x^2:

$$3x^2 + 2 = 0$$
$$x^2 = -\frac{2}{3}$$
$$x = \pm\sqrt{-\frac{2}{3}} = \pm i\sqrt{\frac{2}{3}} = \pm i\frac{\sqrt{6}}{3}$$

Check these solutions in the original equation. The solution set is $\left\{\pm i\frac{\sqrt{6}}{3}\right\}$.

> Now do Exercises 79–86

The basic facts about complex numbers are listed in the following box.

Complex Numbers

1. Definition of i: $i = \sqrt{-1}$, and $i^2 = -1$.
2. A complex number has the form $a + bi$, where a and b are real numbers.
3. The complex number $a + 0i$ is the real number a.
4. If b is a positive real number, then $\sqrt{-b} = i\sqrt{b}$.
5. The numbers $a + bi$ and $a - bi$ are called complex conjugates of each other. Their product is the real number $a^2 + b^2$.
6. Add, subtract, and multiply complex numbers as if they were algebraic expressions with i being the variable, and replace i^2 by -1.
7. Divide complex numbers by multiplying the numerator and denominator by the conjugate of the denominator.
8. In the complex number system, $x^2 = k$ for any real number k is equivalent to $x = \pm\sqrt{k}$.

Warm-Ups ▼

Fill in the blank.

1. A _____ number is a number of the form $a + bi$ where a and b are real numbers.

2. An _____ number is a complex number in which $b \neq 0$.

3. The union of the real numbers and the imaginary numbers is the set of _____ numbers.

4. The _____ of $a + bi$ is $a - bi$.

5. The _____ of $a + bi$ and $a - bi$ is $a^2 + b^2$.

6. If $b = 0$, then $a + bi$ is a ____ number.

7. The set of real numbers is a _____ of the set of complex numbers.

True or false?

8. $\sqrt{-1} = i$

9. $2 - \sqrt{-6} = 2 - 6i$

10. $\sqrt{-9} = \pm 3$

11. $2 - 3i - (4 - 2i) = -2 - i$

12. $i^4 = 1$

13. $i^3 = i$

14. $i^{48} = 1$

15. $(2 - i)(2 + i) = 5$

16. If $x^2 = -9$, then $x = \pm 3i$.

9.6 Exercises

⟨ **Study Tips** ⟩

- When studying for a midterm or final, review the material in the order it was originally presented. This strategy will help you to see connections between the ideas.
- Studying the oldest material first will give top priority to material that you might have forgotten.

⟨ **2** ⟩ **Addition, Subtraction, and Multiplication**

Find the indicated sums and differences of complex numbers. See Example 1.

1. $(2 + 3i) + (-4 + 5i)$

2. $(-1 + 6i) + (5 - 4i)$

3. $(2 - 3i) - (6 - 7i)$

4. $(2 - 3i) - (6 - 2i)$

5. $(-1 + i) + (-1 - i)$

6. $(-5 + i) + (-5 - i)$

7. $(-2 - 3i) - (6 - i)$

8. $(-6 + 4i) - (2 - i)$

Find each product. Express each answer in the form $a + bi$. See Example 2.

9. $3(2 + 5i)$

10. $4(1 - 3i)$

11. $2i(i - 5)$

12. $3i(2 - 6i)$

13. $-4i(3 - i)$

14. $-5i(2 + 3i)$

15. $(2 + 3i)(4 + 6i)$

16. $(2 + i)(3 + 4i)$

17. $(-1 + i)(2 - i)$

18. $(3 - 2i)(2 - 5i)$

19. $(-1 - 2i)(2 + i)$

20. $(1 - 3i)(1 + 3i)$

21. $(5 - 2i)(5 + 2i)$

22. $(4 + 3i)(4 + 3i)$

23. $(1 - i)(1 + i)$

24. $(2 + 6i)(2 - 6i)$

25. $(4 + 2i)(4 - 2i)$

26. $(4 - i)(4 + i)$

Find the indicated powers of complex numbers. See Example 3.

27. $(3i)^2$

28. $(5i)^2$

29. $(-5i)^2$

30. $(-9i)^2$

31. $(2i)^4$

32. $(-2i)^3$

33. i^9

34. i^{12}

35. i^{18}

36. i^{33}

37. i^{25}

38. i^{31}

⟨ **3** ⟩ **Division of Complex Numbers**

Find the product of the given complex number and its conjugate. See Example 4.

39. $3 + 5i$

40. $3 + i$

41. $1 - 2i$

42. $4 - 6i$

43. $-2 + i$

44. $-3 - 2i$

45. $2 - i\sqrt{3}$

46. $\sqrt{5} - 4i$

Find each quotient. Express each answer in the form a + bi.
See Example 5.

47. $\dfrac{3}{4+i}$

48. $\dfrac{6}{7-2i}$

49. $\dfrac{2+i}{3-2i}$

50. $\dfrac{3+5i}{2-i}$

51. $\dfrac{4+3i}{i}$

52. $\dfrac{5-6i}{3i}$

53. $\dfrac{2+6i}{2}$

54. $\dfrac{9-3i}{-6}$

55. $\dfrac{1+i}{3i-2}$

56. $\dfrac{2+i}{i+5}$

57. $\dfrac{6}{3i}$

58. $\dfrac{8}{-2i}$

⟨4⟩ Square Roots of Negative Numbers

Write each expression in the form a + bi, where a and b are
real numbers. See Example 6.

59. $\sqrt{-25}$

60. $\sqrt{-81}$

61. $2 + \sqrt{-4}$

62. $3 + \sqrt{-9}$

63. $2\sqrt{-9} + 5$

64. $3\sqrt{-16} + 2$

65. $7 - \sqrt{-6}$

66. $\sqrt{-5} + 3$

67. $\sqrt{-8} + \sqrt{-18}$

68. $2\sqrt{-20} - \sqrt{-45}$

69. $\dfrac{2 + \sqrt{-12}}{2}$

70. $\dfrac{-6 - \sqrt{-18}}{3}$

71. $\dfrac{-4 - \sqrt{-24}}{4}$

72. $\dfrac{8 + \sqrt{-20}}{-4}$

73. $\sqrt{-2} \cdot \sqrt{-6}$

74. $\sqrt{-3} \cdot \sqrt{-15}$

75. $\sqrt{-3} \cdot \sqrt{-27}$

76. $\sqrt{-3} \cdot \sqrt{-7}$

77. $\dfrac{\sqrt{8}}{\sqrt{-4}}$

78. $\dfrac{\sqrt{6}}{\sqrt{-2}}$

⟨5⟩ Imaginary Solutions to Equations

Find the imaginary solutions to each equation. See Example 7.

79. $x^2 = -36$

80. $x^2 + 4 = 0$

81. $x^2 = -12$

82. $x^2 = -25$

83. $2x^2 + 5 = 0$

84. $3x^2 + 4 = 0$

85. $3x^2 + 6 = 0$

86. $x^2 + 1 = 0$

Miscellaneous

Write each expression in the form a + bi, where a and b are
real numbers.

87. $(2 - 3i)(3 + 4i)$

88. $(2 - 3i)(2 + 3i)$

89. $(2 - 3i) + (3 + 4i)$

90. $(3 - 5i) - (2 - 7i)$

91. $\dfrac{2 - 3i}{3 + 4i}$

92. $\dfrac{-3i}{3 - 6i}$

93. $i(2 - 3i)$

94. $-3i(4i - 1)$

95. $(-3i)^2$

96. $(-2i)^6$

97. $\sqrt{-12} + \sqrt{-3}$

98. $\sqrt{-49} - \sqrt{-25}$

99. $(2 - 3i)^2$

100. $(5 + 3i)^2$

101. $\dfrac{-4 + \sqrt{-32}}{2}$

102. $\dfrac{-2 - \sqrt{-27}}{-6}$

Getting More Involved

103. *Writing*

Explain why $2 - i$ is a solution to

$$x^2 - 4x + 5 = 0.$$

104. *Cooperative learning*

Work with a group to verify that $-1 + i\sqrt{3}$ and
$-1 - i\sqrt{3}$ satisfy the equation

$$x^3 - 8 = 0.$$

In the complex number system there are three cube roots
of 8. What are they?

105. *Discussion*

What is wrong with using the product rule for
radicals to get

$$\sqrt{-4} \cdot \sqrt{-4} = \sqrt{(-4)(-4)} = \sqrt{16} = 4?$$

What is the correct product?

9 Wrap-Up

Summary

Powers and Roots		Examples
nth roots	If $a = b^n$ for a positive integer n, then b is an nth root of a.	2 and -2 are fourth roots of 16.
Principal root	The positive even root of a positive number	The principal fourth root of 16 is 2.
Radical notation	If n is a positive even integer and a is positive, then the symbol $\sqrt[n]{a}$ denotes the principal nth root of a. If n is a positive odd integer, then the symbol $\sqrt[n]{a}$ denotes the nth root of a. If n is any positive integer, then $\sqrt[n]{0} = 0$. If x is any real number, then use absolute value when an even root of an exponential expression has an odd exponent.	$\sqrt[4]{16} = 2$ $\sqrt[4]{16} \neq -2$ $\sqrt[3]{-8} = -2, \sqrt[3]{8} = 2$ $\sqrt[5]{0} = 0, \sqrt[6]{0} = 0$ $\sqrt{x^2} = |x|, \sqrt{x^6} = |x^3|$ $\sqrt{x^4} = x^2, \sqrt[4]{x^4} = |x|$
Domain of a radical expression	The set of all real numbers that can be used in place of the variable in the radical expression	$\sqrt{x}$, domain $[0, \infty)$ $\sqrt[3]{x-1}$, domain $(-\infty, \infty)$ $\sqrt[4]{x-5}$, domain $[5, \infty)$
Domain of a radical function	The domain of the radical expression that defines the function	$G(x) = \sqrt{x}$ Domain $[0, \infty)$
Definition of $a^{1/n}$	If n is any positive integer, then $a^{1/n} = \sqrt[n]{a}$, provided that $\sqrt[n]{a}$ is a real number.	$8^{1/3} = \sqrt[3]{8} = 2$ $(-4)^{1/2}$ is not real.
Definition of $a^{m/n}$	If m and n are positive integers, then $a^{m/n} = (a^{1/n})^m$, provided that $a^{1/n}$ is a real number.	$8^{2/3} = (8^{1/3})^2 = 2^2 = 4$ $(-16)^{3/4}$ is not real.
Definition of $a^{-m/n}$	If m and n are positive integers and $a \neq 0$, then $a^{-m/n} = \frac{1}{a^{m/n}}$, provided that $a^{1/n}$ is a real number.	$8^{-2/3} = \frac{1}{8^{2/3}} = \frac{1}{4}$

Rules for Radicals		Examples
Product rule for radicals	Provided that all roots are real, $\sqrt[n]{ab} = \sqrt[n]{a} \cdot \sqrt[n]{b}$.	$\sqrt{2} \cdot \sqrt{3} = \sqrt{6}$ $\sqrt{4x} = 2\sqrt{x}$

Quotient rule for radicals	Provided that all roots are real and $b \neq 0$, $$\sqrt[n]{\frac{a}{b}} = \frac{\sqrt[n]{a}}{\sqrt[n]{b}}.$$	$$\sqrt{\frac{5}{9}} = \frac{\sqrt{5}}{3}$$ $$\sqrt{10} \div \sqrt{5} = \sqrt{2}$$
Simplified radical form for radicals of index n	A simplified radical of index n has 1. *no* perfect nth powers as factors of the radicand, 2. *no* fractions inside the radical, and 3. *no* radicals in the denominator.	$$\sqrt{20} = \sqrt{4 \cdot 5} = 2\sqrt{5}$$ $$\sqrt{\frac{3}{2}} = \frac{\sqrt{3}}{\sqrt{2}}$$ $$\frac{\sqrt{3}}{\sqrt{2}} = \frac{\sqrt{3}}{\sqrt{2}} \cdot \frac{\sqrt{2}}{\sqrt{2}} = \frac{\sqrt{6}}{2}$$

Rules for Rational Exponents

Examples

If a and b are nonzero real numbers and r and s are rational numbers, then the following rules hold, provided all expressions represent real numbers.

Product rule	$a^r \cdot a^s = a^{r+s}$	$3^{1/4} \cdot 3^{1/2} = 3^{3/4}$
Quotient rule	$\dfrac{a^r}{a^s} = a^{r-s}$	$\dfrac{x^{3/4}}{x^{1/4}} = x^{1/2}$
Power of a power rule	$(a^r)^s = a^{rs}$	$(2^{1/2})^{-1/2} = 2^{-1/4}$ $(x^{3/4})^4 = x^3$
Power of a product rule	$(ab)^r = a^r b^r$	$(a^2 b^6)^{1/2} = ab^3$
Power of a quotient rule	$\left(\dfrac{a}{b}\right)^r = \dfrac{a^r}{b^r}$	$\left(\dfrac{8}{x^6}\right)^{2/3} = \dfrac{4}{x^4}$

Equations

Examples

Equations with radicals and exponents	1. In raising each side of an equation to an even power, we can create an equation that gives extraneous solutions. We must check. 2. When applying the even-root property, remember that there is a positive and a negative root. 3. For equations with rational exponents, raise each side to a positive or a negative power first and then apply the even- or odd-root property.	$\sqrt{x} = -3$ $x = 9$ $x^2 = 36$ $x = \pm 6$ $x^{-2/3} = 4$ $(x^{-2/3})^{-3} = 4^{-3}$ $x^2 = \dfrac{1}{64}$ $x = \pm\dfrac{1}{8}$

Complex Numbers

Complex Numbers		Examples
Complex numbers	Numbers of the form $a + bi$, where a and b are real numbers: $i = \sqrt{-1}$, $i^2 = -1$	$2 + 3i$ $-6i$ $\sqrt{2} + i$
Complex conjugates	Complex numbers of the form $a + bi$ and $a - bi$: Their product is the real number $a^2 + b^2$.	$(2 + 3i)(2 - 3i) = 2^2 + 3^2$ $= 13$
Complex number operations	Add, subtract, and multiply as algebraic expressions with i being the variable. Simplify using $i^2 = -1$.	$(2 + 5i) + (4 - 2i) = 6 + 3i$ $(2 + 5i) - (4 - 2i) = -2 + 7i$ $(2 + 5i)(4 - 2i) = 18 + 16i$
	Divide complex numbers by multiplying the numerator and denominator by the conjugate of the denominator.	$(2 + 5i) \div (4 - 2i)$ $= \dfrac{(2 + 5i)(4 + 2i)}{(4 - 2i)(4 + 2i)}$ $= \dfrac{-2 + 24i}{20} = -\dfrac{1}{10} + \dfrac{6}{5}i$
Square root of a negative number	For any positive real number b, $\sqrt{-b} = i\sqrt{b}$.	$\sqrt{-9} = i\sqrt{9} = 3i$
Imaginary solutions to equations	In the complex number system, $x^2 = k$ for any real k is equivalent to $x = \pm\sqrt{k}$.	$x^2 = -25$ $x = \pm\sqrt{-25} = \pm5i$

Enriching Your Mathematical Word Power

Fill in the blank.

1. A number b such that $b^n = a$ is the nth _____ of a.
2. The expression a^2 is the _____ of a.
3. A number b such that $b^3 = a$ is the _____ root of a.
4. If $a > 0$ and n is even, then $\sqrt[n]{a}$ is the _____ nth root of a.
5. If n is an odd number and $b^n = a$, then b is an _____ root of a.
6. The number n in $\sqrt[n]{a}$ is the _____.
7. The number a in $\sqrt[n]{a}$ is the _____.
8. Radicals with the same radicand and the same index are _____ radicals.
9. The set of real numbers that can be used in place of the variable in an algebraic expression is the _____ of the expression.
10. If an exponent is one of the numbers in the set $\{\ldots, -3, -2, -1, 0, 1, 2, 3, \ldots\}$, then it is an _____ exponent.
11. A _____ number has the form $a + bi$ where a and b are real numbers.
12. In $a + bi$, i is the _____ unit.
13. The complex numbers $a + bi$ and $a - bi$ are complex _____.
14. A complex number in which $b \neq 0$ is an _____ number.

Review Exercises

9.1 Radicals
Find each root. Variables represent any real numbers. Use absolute value when necessary.

1. $\sqrt{81}$
2. $\sqrt[4]{16}$
3. $\sqrt[3]{27}$
4. $\sqrt[5]{32}$
5. $\sqrt[3]{-27}$
6. $\sqrt[5]{-32}$
7. $\sqrt{100}$
8. $\sqrt[3]{1000}$
9. $\sqrt{y^2}$
10. $\sqrt[4]{y^{12}}$
11. $\sqrt[3]{a^3}$
12. $\sqrt[5]{b^{10}}$

13. $\sqrt[4]{n^{24}}$

14. $\sqrt{m^{32}}$

15. $\sqrt[4]{n^{20}}$

16. $\sqrt{m^{30}}$

Simplify each radical expression. Assume all variables represent positive real numbers.

17. $\sqrt{72}$

18. $\sqrt{48}$

19. $\sqrt{x^{12}}$

20. $\sqrt{a^{10}}$

21. $\sqrt[3]{x^6}$

22. $\sqrt[3]{a^9}$

23. $\sqrt{2x^9}$

24. $\sqrt{3a^7}$

25. $\sqrt{8w^5}$

26. $\sqrt{20n^{25}}$

27. $\sqrt[3]{16x^4}$

28. $\sqrt[3]{54b^5}$

29. $\sqrt[4]{a^9b^5}$

30. $\sqrt[4]{32m^{11}}$

31. $\sqrt{\dfrac{x^3}{16}}$

32. $\sqrt{\dfrac{12a^3}{25}}$

Find the domain of each radical expression. Use interval notation.

33. $\sqrt{2x - 5}$

34. $\sqrt{3x + 12}$

35. $\sqrt[3]{7x - 1}$

36. $\sqrt[3]{9 - 2x}$

37. $\sqrt[4]{-3x + 1}$

38. $\sqrt[4]{-5x - 1}$

39. $\sqrt{\dfrac{1}{2}x + 1}$

40. $\sqrt{\dfrac{2}{3}x - 2}$

Find the domain of each radical function.

41. $T(x) = \sqrt{x + 5}$

42. $W(x) = \sqrt{6 - 2x}$

43. $y = \sqrt[4]{20 - x}$

44. $y = \sqrt[4]{3x}$

45. $S(x) = \sqrt[3]{17x - 12}$

46. $T(x) = \sqrt[5]{9 - 5x}$

9.2 Rational Exponents

Simplify the expressions involving rational exponents. Assume all variables represent positive real numbers. Write your answers with positive exponents.

47. $(-27)^{-2/3}$

48. $-25^{3/2}$

49. $\left(2^6\right)^{1/3}$

50. $\left(5^2\right)^{1/2}$

51. $100^{-3/2}$

52. $1000^{-2/3}$

53. $\dfrac{3x^{-1/2}}{3^{-2}x^{-1}}$

54. $\dfrac{\left(x^2y^{-3}z\right)^{1/2}}{x^{1/2}yz^{-1/2}}$

55. $\left(a^{1/2}b\right)^3\left(ab^{1/4}\right)^2$

56. $\left(t^{-1/2}\right)^{-2}\left(t^{-2}v^2\right)$

57. $\left(x^{1/2}y^{1/4}\right)\left(x^{1/4}y\right)$

58. $\left(a^{1/3}b^{1/6}\right)^2\left(a^{1/3}b^{2/3}\right)$

9.3 Adding, Subtracting, and Multiplying Radicals

Perform the operations and simplify. Assume the variables represent positive real numbers.

59. $\sqrt{13} \cdot \sqrt{13}$

60. $\sqrt[3]{14} \cdot \sqrt[3]{14} \cdot \sqrt[3]{14}$

61. $\sqrt{27} + \sqrt{45} - \sqrt{75}$

62. $\sqrt{12} - \sqrt{50} + \sqrt{72}$

63. $3\sqrt{2}\left(5\sqrt{2} - 7\sqrt{3}\right)$

64. $-2\sqrt{a}\left(\sqrt{a} - \sqrt{ab^6}\right)$

65. $(2 - \sqrt{3})(3 + \sqrt{2})$

66. $(2\sqrt{x} - \sqrt{y})(\sqrt{x} + \sqrt{y})$

9.4 Quotients, Powers, and Rationalizing Denominators

Perform the operations and simplify. Assume the variables represent positive real numbers.

67. $5 \div \sqrt{2}$

68. $(10\sqrt{6}) \div (2\sqrt{2})$

69. $\sqrt{\dfrac{2}{5}}$

70. $\sqrt{\dfrac{1}{6}}$

71. $\sqrt[3]{\dfrac{2}{3}}$

72. $\sqrt[3]{\dfrac{1}{9}}$

73. $\dfrac{2}{\sqrt{3x}}$

74. $\dfrac{3}{\sqrt{2y}}$

75. $\dfrac{\sqrt{10y^3}}{\sqrt{6}}$

76. $\dfrac{\sqrt{5x^5}}{\sqrt{8}}$

77. $\dfrac{3}{\sqrt[3]{2a}}$

78. $\dfrac{a}{\sqrt[3]{a^2}}$

79. $\dfrac{5}{\sqrt[4]{3x^2}}$

80. $\dfrac{b}{\sqrt[4]{a^2b^3}}$

81. $(\sqrt{3})^4$

82. $(-2\sqrt{x})^9$

83. $\dfrac{2 - \sqrt{8}}{2}$

84. $\dfrac{-3 - \sqrt{18}}{-6}$

85. $\dfrac{\sqrt{6}}{1 - \sqrt{3}}$

86. $\dfrac{\sqrt{15}}{2 + \sqrt{5}}$

87. $\dfrac{2\sqrt{3}}{3\sqrt{6} - \sqrt{12}}$

88. $\dfrac{-\sqrt{xy}}{3\sqrt{x} + \sqrt{xy}}$

89. $\left(2w\sqrt[3]{2w^2}\right)^6$ **90.** $\left(m\sqrt[4]{m^3}\right)^8$

9.5 Solving Equations with Radicals and Exponents
Find all real solutions to each equation.

91. $x^2 = 16$

92. $w^2 = 100$

93. $(a - 5)^2 = 4$

94. $(m - 7)^2 = 25$

95. $(a + 1)^2 = 5$

96. $(x + 5)^2 = 3$

97. $(m + 1)^2 = -8$

98. $(w + 4)^2 = 16$

99. $\sqrt{m - 1} = 3$

100. $3\sqrt{x + 5} = 12$

101. $\sqrt[3]{2x + 9} = 3$

102. $\sqrt[4]{2x - 1} = 2$

103. $w^{2/3} = 4$

104. $m^{-4/3} = 16$

105. $(m + 1)^{1/3} = 5$

106. $(w - 3)^{-2/3} = 4$

107. $\sqrt{x - 3} = \sqrt{x + 2} - 1$

108. $\sqrt{x^2 + 3x + 6} = 4$

109. $\sqrt{5x - x^2} = \sqrt{6}$

110. $\sqrt{x + 4} - 2\sqrt{x - 1} = -1$

111. $\sqrt{x + 7} - 2\sqrt{x} = -2$

112. $\sqrt{x} - \sqrt{x - 1} = 1$

113. $2\sqrt{x} - \sqrt{x - 3} = 3$

114. $1 + \sqrt{x + 7} = \sqrt{2x + 7}$

9.6 Complex Numbers
Perform the indicated operations. Write answers in the form a + bi.

115. $(2 - 3i)(-5 + 5i)$

116. $(2 + i)(5 - 2i)$

117. $(2 + i) + (5 - 4i)$

118. $(2 + i) + (3 - 6i)$

119. $(1 - i) - (2 - 3i)$

120. $(3 - 2i) - (1 - i)$

121. $\dfrac{6 + 3i}{3}$

122. $\dfrac{8 + 12i}{4}$

123. $\dfrac{4 - \sqrt{-12}}{2}$

124. $\dfrac{6 + \sqrt{-18}}{3}$

125. $\dfrac{2 - 3i}{4 + i}$

126. $\dfrac{3 + i}{2 - 3i}$

127. $(-2i)^4$

128. $(-2i)^5$

129. i^{14}

130. i^{21}

Find the imaginary solutions to each equation.

131. $x^2 + 100 = 0$

132. $25a^2 + 3 = 0$

133. $2b^2 + 9 = 0$

134. $3y^2 + 8 = 0$

Miscellaneous

Determine whether each equation is true or false and explain your answer. An equation involving variables should be marked true only if it is an identity. Do not use a calculator.

135. $2^3 \cdot 3^2 = 6^5$

136. $16^{1/4} = 4^{1/2}$

137. $(\sqrt{2})^3 = 2\sqrt{2}$

138. $\sqrt[3]{9} = 3$

139. $8^{200} \cdot 8^{200} = 64^{200}$

140. $\sqrt{295} \cdot \sqrt{295} = 295$

141. $4^{1/2} = \sqrt{2}$

142. $\sqrt{a^2} = |a|$

143. $5^2 \cdot 5^2 = 25^4$

144. $\sqrt{6} \div \sqrt{2} = \sqrt{3}$

145. $\sqrt{w^{10}} = w^5$

146. $\sqrt{a^{16}} = a^4$

147. $\sqrt{x^6} = x^3$

148. $\sqrt[6]{16} = \sqrt[3]{4}$

149. $\sqrt{x^8} = x^4$

150. $\sqrt[9]{2^6} = 2^{2/3}$

151. $\sqrt{16} = 2$

152. $2^{1/2} \cdot 2^{1/4} = 2^{3/4}$

153. $2^{600} = 4^{300}$

154. $\sqrt{2} \cdot \sqrt[4]{2} = \sqrt[6]{2}$

155. $\dfrac{2 + \sqrt{6}}{2} = 1 + \sqrt{6}$

156. $\dfrac{4 + 2\sqrt{3}}{2} = 2 + \sqrt{3}$

157. $\sqrt{\dfrac{4}{6}} = \dfrac{2}{3}$

158. $8^{200} \cdot 8^{200} = 8^{400}$

159. $81^{2/4} = 81^{1/2}$

160. $(-64)^{2/6} = (-64)^{1/3}$

161. $(a^4 b^2)^{1/2} = |a^2 b|$

162. $\left(\dfrac{a^2}{b^6}\right)^{1/2} = \dfrac{|a|}{b^3}$

Solve each problem.

163. *Falling objects.* Neglecting air resistance, the number of feet s that an object falls from rest during t seconds is given by the formula $s = 16t^2$. How long would it take an object to reach the earth if it is dropped from 12,000 feet?

164. *Timber.* Anne is pulling on a 60-foot rope attached to the top of a 48-foot tree while Walter is cutting the tree at its base. How far from the base of the tree is Anne standing?

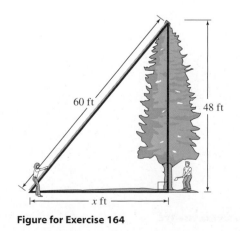

60 ft   48 ft

x ft

Figure for Exercise 164

165. *Dropping a rock.* The time in seconds t that it takes for a rock to fall to the earth from a height of h feet is given by the function

$$t(h) = 0.25h^{1/2}.$$

a) Find $t(100)$.

b) If it takes 4 seconds for a rock to reach the ground when dropped from the top of a tall building, then what is the height of the building?

166. *Skid marks.* Under certain conditions the speed S in miles per hour prior to an accident is determined from the length L in feet of the skid marks by the function

$$S(L) = (20L)^{1/2}.$$

a) Find $S(80)$.

b) Find the length of the skid marks for a car traveling 70 mph.

167. *Guy wire.* If a guy wire of length 40 feet is attached to an antenna at a height of 30 feet, then how far from the base of the antenna is the wire attached to the ground?

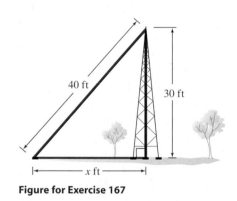

40 ft   30 ft

x ft

Figure for Exercise 167

168. *Touchdown.* Suppose at the kickoff of a football game, the receiver catches the football at the left side of the goal line and runs for a touchdown diagonally across the field. How many yards would he run? (A football field is 100 yards long and 160 feet wide.)

169. *Long guy wires.* The manufacturer of an antenna recommends that guy wires from the top of the antenna to the ground be attached to the ground at a distance from the base equal to the height of the antenna. How long would the guy wires be for a 200-foot antenna?

170. *Height of a post.* Betty observed that the lamp post in front of her house casts a shadow of length 8 feet when the angle of inclination of the sun is 60 degrees. How tall is the lamp post? (In a 30-60-90 right triangle, the side opposite 30 is one-half the length of the hypotenuse.)

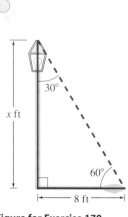

Figure for Exercise 170

171. *Manufacturing a box.* A cubic box has a volume of 40 cubic feet. The amount of recycled cardboard that it takes to make the six-sided box is 10% larger than the surface area of the box. Find the exact amount of recycled cardboard used in manufacturing the box.

172. *Shipping parts.* A cubic box with a volume of 32 cubic feet is to be used to ship some machine parts. All of the parts are small except for a long, straight steel connecting rod. What is the maximum length of a connecting rod that will fit into this box?

173. *Health care costs.* The total annual cost of health care in the United States grew from $993.3 billion in 1995 to $2151.1 billion in 2009 (Statistical Abstract of the United States, www.census.gov).

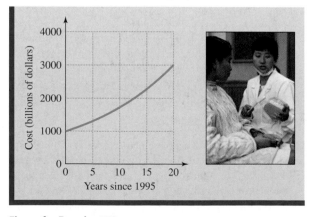

Figure for Exercise 173

a) Find the average annual rate of growth r for that period by solving $2151.1 = 993.3(1 + r)^{14}$.

b) Estimate the total annual cost of health care in 2015 by reading the accompanying graph.

174. *Population growth.* The formula $P = P_0(1 + r)^n$ gives the population P at the end of an n-year time period, where P_0 is the initial population and r is the average annual growth rate. The U.S. population grew from 248.7 million in 1990 to 307.8 million in 2009 (U.S. Census Bureau). Find r for that period.

175. *Landing speed.* Aircraft engineers determine the proper landing speed V (in feet per second) for an airplane from the formula

$$V = \sqrt{\frac{841L}{CS}},$$

where L is the gross weight of the aircraft in pounds, C is the coefficient of lift, and S is the wing surface area in square feet. Rewrite the formula so that the expression on the right-hand side is in simplified radical form.

176. *Spillway capacity.* Civil engineers use the formula

$$Q = 3.32LH^{3/2}$$

to find the maximum discharge that the dam (a broad-crested weir) shown in the figure can pass before the water breaches its abutments (*Standard Handbook for Civil Engineers,* 1968). In the formula, Q is the discharge in cubic feet per second, L is the length of the spillway in feet, and H is the depth of the spillway. Find Q given that $L = 60$ feet and $H = 5$ feet. Find H given that $Q = 3000$ cubic feet per second and $L = 70$ feet.

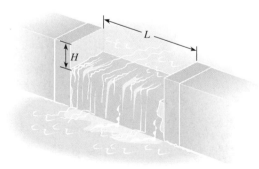

Figure for Exercise 176

Chapter 9 Test

Find each root. Variables represent any real numbers. Use absolute value when necessary.

1. $\sqrt{36}$

2. $\sqrt[3]{-125}$

3. $\sqrt{t^2}$

4. $\sqrt[3]{p^3}$

5. $\sqrt[4]{w^8}$

6. $\sqrt[4]{w^{12}}$

Simplify each expression. Assume all variables represent positive numbers.

7. $8^{2/3}$

8. $4^{-3/2}$

9. $\sqrt{21} \div \sqrt{7}$

10. $2\sqrt{5} \cdot 3\sqrt{5}$

11. $\sqrt{20} + \sqrt{5}$

12. $\sqrt{5} + \dfrac{1}{\sqrt{5}}$

13. $2^{1/2} \cdot 2^{1/2}$

14. $\sqrt{72}$

15. $\sqrt{\dfrac{5}{12}}$

16. $\dfrac{6 + \sqrt{18}}{6}$

17. $(2\sqrt{3} + 1)(\sqrt{3} - 2)$

18. $\sqrt[4]{32a^5y^8}$

19. $\dfrac{1}{\sqrt[3]{2x^2}}$

20. $\sqrt{\dfrac{8a^9}{b^3}}$

21. $\sqrt[3]{-27x^9}$

22. $\sqrt{20m^3}$

23. $x^{1/2} \cdot x^{1/4}$

24. $\left(9y^4x^{1/2}\right)^{1/2}$

25. $\sqrt[3]{40x^7}$

26. $\left(4 + \sqrt{3}\right)^2$

Find the domain of each radical expression. Use interval notation.

27. $\sqrt{4 - x}$

28. $\sqrt[3]{5x - 3}$

Rationalize the denominator and simplify.

29. $\dfrac{2}{5 - \sqrt{3}}$

30. $\dfrac{\sqrt{6}}{4\sqrt{3} + \sqrt{2}}$

Write each expression in the form a + bi.

31. $(3 - 2i)(4 + 5i)$

32. $i^4 - i^5$

33. $\dfrac{3 - i}{1 + 2i}$

34. $\dfrac{-6 + \sqrt{-12}}{8}$

Find all real or imaginary solutions to each equation.

35. $(x - 2)^2 = 49$

36. $2\sqrt{x + 4} = 3$

37. $w^{2/3} = 4$

38. $9y^2 + 16 = 0$

39. $\sqrt{2x^2 + x - 12} = x$

40. $\sqrt{x - 1} + \sqrt{x + 4} = 5$

Show a complete solution to each problem.

41. Find the exact length of the side of a square whose diagonal is 3 feet.

42. Two positive numbers differ by 11, and their square roots differ by 1. Find the numbers.

43. If the perimeter of a rectangle is 20 feet and the diagonal is $2\sqrt{13}$ feet, then what are the length and width?

44. The average radius R of the orbit of a planet around the sun is determined by $R = T^{2/3}$, where T is the number of years for one orbit and R is measured in astronomical units (AU). If it takes Pluto 248.530 years to make one orbit of the sun, then what is the average radius of the orbit of Pluto? If the average radius of the orbit of Neptune is 30.08 AU, then how many years does it take Neptune to complete one orbit of the sun?

45. The maximum speed for a sailboat in knots M is a function of the length of the waterline in feet w, given by $M(w) = 1.3\sqrt{w}$.

   a) Find $M(16)$ and $M(25)$.

   b) Find the length of the waterline if the maximum speed is 9.1 knots.

*Making*Connections | A Review of Chapters 1–9

Evaluate each expression

1. $3 + 2\sqrt{14 - 2 \cdot 5}$

2. $4 - 3|5 - 2 \cdot 4|$

3. $5 - 2(6 - 2 \cdot 4^2)$

4. $\sqrt{13^2 - 12^2} + 6$

5. $\sqrt[3]{6^2 - 3^2} - 2^5$

6. $\sqrt[3]{4(7 + 3^2)} - 2^3$

7. $(4 + 3^2) \div |5 - 2 \cdot 9|$

8. $\sqrt{9 + 16} - |9 - 16|$

9. $\sqrt{(-30)^2 - 4 \cdot 9 \cdot 25}$

10. $\sqrt{(-23)^2 - 4 \cdot 12 \cdot 5}$

Which elements of

$$\left\{ -5, -\sqrt[3]{2}, -\frac{1}{9}, 0, 1, \sqrt{4}, 2.99, \pi, \frac{33}{4}, 2 + 3i \right\}$$

are members of these sets?

11. Whole numbers

12. Natural numbers

13. Integers

14. Rational numbers

15. Irrational numbers

16. Real numbers

17. Imaginary numbers

18. Complex numbers

Fill in the blank.

19. Zero is the _____ identity.

20. One is the _____ identity.

21. According to the _____ property of addition, $a + b = b + a$ for all real numbers a and b.

22. According to the _____ property of multiplication, $ab = ba$ for all real numbers a and b.

23. According to the _____ property of addition, $a + (b + c) = (a + b) + c$ for all real numbers a, b, and c.

24. According to the _____ property, $a(b + c) = ab + ac$ for all real numbers a, b, and c.

25. Every real number a has a(n) _____ inverse $-a$ such that $a + (-a) = 0$.

26. Every nonzero real number a has a(n) _____ inverse $\dfrac{1}{a}$ such that $a \cdot \dfrac{1}{a} = 1$.

Find all real solutions to each equation or inequality. For the inequalities, also sketch the graph of the solution set.

27. $3(x - 2) + 5 = 7 - 4(x + 3)$

28. $\sqrt{6x + 7} = 4$

29. $|2x + 5| > 1$

30. $8x^3 - 27 = 0$

31. $2x - 3 > 3x - 4$

32. $\sqrt{2x - 3} - \sqrt{3x + 4} = 0$

33. $\dfrac{w}{3} + \dfrac{w - 4}{2} = \dfrac{11}{2}$

34. $2(x + 7) - 4 = x - (10 - x)$

35. $(x + 7)^2 = 25$

36. $a^{-1/2} = 4$

37. $x - 3 > 2$ or $x < 2x + 6$

38. $a^{-2/3} = 16$

39. $3x^2 - 1 = 0$

40. $5 - 2(x - 2) = 3x - 5(x - 2) - 1$

41. $|3x - 4| < 5$

42. $3x - 1 = 0$

43. $\sqrt{y - 1} = 9$

44. $|5(x - 2) + 1| = 3$

45. $0.06x - 0.04(x - 20) = 2.8$

46. $|3x - 1| > -2$

47. $\dfrac{3\sqrt{2}}{x} = \dfrac{\sqrt{3}}{4\sqrt{5}}$

48. $\dfrac{\sqrt{x} - 4}{x} = \dfrac{1}{\sqrt{x} + 5}$

49. $\dfrac{3\sqrt{2} + 4}{\sqrt{2}} = \dfrac{x\sqrt{18}}{3\sqrt{2} + 2}$

50. $\dfrac{x}{2\sqrt{5} - \sqrt{2}} = \dfrac{2\sqrt{5} + \sqrt{2}}{x}$

51. $\dfrac{\sqrt{2x} - 5}{x} = \dfrac{-3}{\sqrt{2x} + 5}$

52. $\dfrac{\sqrt{6}+2}{x} = \dfrac{2}{\sqrt{6}+4}$

53. $\dfrac{x-1}{\sqrt{6}} = \dfrac{\sqrt{6}}{x}$

54. $\dfrac{x+3}{\sqrt{10}} = \dfrac{\sqrt{10}}{x}$

55. $\dfrac{1}{x} - \dfrac{1}{x-1} = -\dfrac{1}{6}$

56. $\dfrac{1}{x^2 - 2x} + \dfrac{1}{x} = \dfrac{2}{3}$

The expression $\dfrac{-b + \sqrt{b^2 - 4ac}}{2a}$ *will be used in Chapter 10 to solve quadratic equations. Evaluate it for the given values of a, b, and c.*

57. $a = 1, b = 2, c = -15$

58. $a = 1, b = 8, c = 12$

59. $a = 2, b = 5, c = -3$

60. $a = 6, b = 7, c = -3$

Solve each problem.

61. *Popping corn.* If 1 gram of popcorn with moisture content $x\%$ is popped in a hot-air popper, then the volume of

popped corn v (in cubic centimeters) that results is modeled by the formula

$$v = -94.8 + 21.4x - 0.761x^2.$$

a) Use the formula to find the volume that results when 1 gram of popcorn with moisture content 11% is popped.

b) Use the accompanying graph to estimate the moisture content that will produce the maximum volume of popped corn.

c) Use the graph to estimate the maximum possible volume for popping 1 gram of popcorn in a hot-air popper.

Figure for Exercise 61

Critical **Thinking** | **For Individual or Group Work** | **Chapter 9**

These exercises can be solved by a variety of techniques, which may or may not require algebra. So be creative and think critically. Explain all answers. Answers are in the Instructor's Edition of this text.

1. *Wagon wheel.* A wagon wheel is placed against a wall as shown in the accompanying figure. One point on the edge of the wheel is 5 inches from the ground and 10 inches from the wall. What is the radius of the wheel?

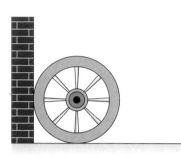

Figure for Exercise 1

2. *Comparing jobs.* Bob has two job offers with a starting salary of $100,000 per year and monthly paychecks. The Atlanta employer will raise his annual salary by $2000 at the end of every year, while the Chicago employer will raise his annual salary by $1000 at the end of every six months.

   a) Which job is the better deal?

   b) How much more will Bob have made at the end of 10 years with the better deal?

3. *Floor tiles.* A square floor is tiled using 121 square floor tiles. Only whole tiles are used. How many tiles are neither diagonal tiles nor edge tiles?

4. *Counting days.* If the first day of this century was January 1, 2000, then how many days are there in this century? (A year is a leap year if it is divisible by 4, unless it's divisible by 100, in which case it isn't, unless it's divisible by 400, in which case it is.)

5. *Planting trees.* How can you plant 10 trees in five rows with four trees in each row?

6. *Counting rectangles.* How many rectangles of any size are there on an 8 by 8 checker board?

7. *Chime time.* The clock in the bell tower at Webster College chimes every hour on the hour: once at 1 o'clock, twice at 2 o'clock, and so on. The clock takes 5 seconds to chime at 4 o'clock and 15 seconds to chime at 10 o'clock. The time needed to chime 1 o'clock is negligible. What is the total number of seconds needed for the clock to do all of its chiming in a 24-hour period starting at 1 P.M.?

Photo for Exercise 7

8. *Arranging digits.* In how many ways can you arrange the digits 8, 7, 6, and 3 to form a four-digit number divisible by 9, using each digit once and only once?

Quadratic Equations, Functions, and Inequalities

Is it possible to measure beauty? For thousands of years artists and philosophers have been challenged to answer this question. The seventeenth-century philosopher John Locke said, "Beauty consists of a certain composition of color and figure causing delight in the beholder." Over the centuries many architects, sculptors, and painters have searched for beauty in their work by exploring numerical patterns in various art forms.

Today many artists and architects still use the concepts of beauty given to us by the ancient Greeks. One principle, called the Golden Rectangle, concerns the most pleasing proportions of a rectangle. The Golden Rectangle appears in nature as well as in many cultures. Examples of it can be seen in Leonardo da Vinci's *Proportions of the Human Figure* as well as in Indonesian temples and Chinese pagodas. Perhaps one of the best-known examples of the Golden Rectangle is in the façade and floor plan of the Parthenon, built in Athens in the fifth century B.C.

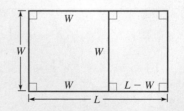

In Exercise 89 of Section 10.3 we will see that the principle of the Golden Rectangle is based on a proportion that we can solve using the quadratic formula.

10.1   Factoring and Completing the Square

Factoring and the even-root property were used to solve quadratic equations in Chapters 5, 6, and 9. In this section we first review those methods. Then you will learn the method of completing the square, which can be used to solve any quadratic equation.

⟨1⟩ Review of Factoring

A quadratic equation has the form $ax^2 + bx + c = 0$, where a, b, and c are real numbers with $a \neq 0$. In Section 5.6 we solved quadratic equations by factoring and then applying the zero factor property.

Zero Factor Property

The equation $ab = 0$ is equivalent to the compound equation

$$a = 0 \qquad \text{or} \qquad b = 0.$$

Of course we can only use the factoring method when we can factor the quadratic polynomial. To solve a quadratic equation by factoring we use the following strategy.

Strategy for Solving Quadratic Equations by Factoring

1. Write the equation with 0 on one side.
2. Factor the other side.
3. Use the zero factor property to set each factor equal to zero.
4. Solve the simpler equations.
5. Check the answers in the original equation.

EXAMPLE 1

Solving a quadratic equation by factoring
Solve $3x^2 - 4x = 15$ by factoring.

Solution

Subtract 15 from each side to get 0 on the right-hand side:

$$3x^2 - 4x - 15 = 0$$
$$(3x + 5)(x - 3) = 0 \quad \text{Factor the left-hand side.}$$

$$3x + 5 = 0 \qquad \text{or} \qquad x - 3 = 0 \quad \text{Zero factor property}$$
$$3x = -5 \qquad \text{or} \qquad x = 3$$
$$x = -\frac{5}{3}$$

The solution set is $\left\{-\frac{5}{3}, 3\right\}$. Check the solutions in the original equation.

Now do Exercises 1–10

⟨2⟩ Review of the Even-Root Property

In Chapter 9 we solved some simple quadratic equations by using the even-root property, which we restate as follows:

> **Even-Root Property**
>
> Suppose n is a positive even integer.
>
> If $k > 0$, then $x^n = k$ is equivalent to $x = \pm\sqrt[n]{k}$.
> If $k = 0$, then $x^n = k$ is equivalent to $x = 0$.
> If $k < 0$, then $x^n = k$ has no real solution.

By the even-root property $x^2 = 4$ is equivalent to $x = \pm 2$, $x^2 = 0$ is equivalent to $x = 0$, and $x^2 = -4$ has no real solutions.

E X A M P L E   2

Solving a quadratic equation by the even-root property
Solve $(a - 1)^2 = 9$.

Solution

By the even-root property $x^2 = k$ is equivalent to $x = \pm\sqrt{k}$.

$$(a - 1)^2 = 9$$
$$a - 1 = \pm\sqrt{9} \quad \text{Even-root property}$$

$$a - 1 = 3 \quad \text{or} \quad a - 1 = -3$$
$$a = 4 \quad \text{or} \quad a = -2$$

Check these solutions in the original equation. The solution set is $\{-2, 4\}$.

Now do Exercises 11–20

⟨ **Helpful Hint** ⟩

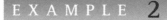

The area of an x by x square and two x by 3 rectangles is $x^2 + 6x$. The area needed to "complete the square" in this figure is 9:

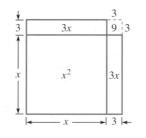

⟨3⟩ Completing the Square

We cannot solve every quadratic by factoring because not all quadratic polynomials can be factored. However, we can write any quadratic equation in the form of Example 2 and then apply the even-root property to solve it. This method is called **completing the square.**

The essential part of completing the square is to recognize a perfect square trinomial when given its first two terms. For example, if we are given $x^2 + 6x$, how do we recognize that these are the first two terms of the perfect square trinomial $x^2 + 6x + 9$? To answer this question, recall that $x^2 + 6x + 9$ is a perfect square trinomial because it is the square of the binomial $x + 3$:

$$(x + 3)^2 = x^2 + 2 \cdot 3x + 3^2 = x^2 + 6x + 9$$

Notice that the 6 comes from multiplying 3 by 2 and the 9 comes from squaring the 3. So to find the missing 9 in $x^2 + 6x$, divide 6 by 2 to get 3, and then square 3 to get 9. This procedure can be used to find the last term in any perfect square trinomial in which the coefficient of x^2 is 1.

> **Rule for Finding the Last Term**
>
> The last term of a perfect square trinomial is the square of one-half of the coefficient of the middle term. In symbols, the perfect square trinomial whose first two terms are $x^2 + bx$ is $x^2 + bx + \left(\frac{b}{2}\right)^2$.

E X A M P L E   **3**

Finding the last term

Find the perfect square trinomial whose first two terms are given.

a) $x^2 + 8x$ b) $x^2 - 5x$ c) $x^2 + \frac{4}{7}x$ d) $x^2 - \frac{3}{2}x$

Solution

a) One-half of 8 is 4, and 4 squared is 16. So the perfect square trinomial is

$$x^2 + 8x + 16.$$

b) One-half of -5 is $-\frac{5}{2}$, and $-\frac{5}{2}$ squared is $\frac{25}{4}$. So the perfect square trinomial is

$$x^2 - 5x + \frac{25}{4}.$$

c) Since $\frac{1}{2} \cdot \frac{4}{7} = \frac{2}{7}$ and $\frac{2}{7}$ squared is $\frac{4}{49}$, the perfect square trinomial is

$$x^2 + \frac{4}{7}x + \frac{4}{49}.$$

d) Since $\frac{1}{2}\left(-\frac{3}{2}\right) = -\frac{3}{4}$ and $\left(-\frac{3}{4}\right)^2 = \frac{9}{16}$, the perfect square trinomial is

$$x^2 - \frac{3}{2}x + \frac{9}{16}.$$

> Now do Exercises 21–28

CAUTION The rule for finding the last term applies only to perfect square trinomials with $a = 1$. A trinomial such as $9x^2 + 6x + 1$ is a perfect square trinomial because it is $(3x + 1)^2$, but the last term is certainly not the square of one-half the coefficient of the middle term.

Another essential step in completing the square is to write the perfect square trinomial as the square of a binomial. Recall that

$$a^2 + 2ab + b^2 = (a + b)^2$$

and

$$a^2 - 2ab + b^2 = (a - b)^2.$$

E X A M P L E   **4**

Factoring perfect square trinomials

Factor each trinomial.

a) $x^2 + 12x + 36$ b) $y^2 - 7y + \frac{49}{4}$

c) $z^2 - \frac{4}{3}z + \frac{4}{9}$

Solution

a) The trinomial $x^2 + 12x + 36$ is of the form $a^2 + 2ab + b^2$ with $a = x$ and $b = 6$. So,

$$x^2 + 12x + 36 = (x + 6)^2.$$

Check by squaring $x + 6$.

b) The trinomial $y^2 - 7y + \frac{49}{4}$ is of the form $a^2 - 2ab + b^2$ with $a = y$ and $b = \frac{7}{2}$. So,

$$y^2 - 7y + \frac{49}{4} = \left(y - \frac{7}{2}\right)^2.$$

Check by squaring $y - \frac{7}{2}$.

c) The trinomial $z^2 - \frac{4}{3}z + \frac{4}{9}$ is of the form $a^2 - 2ab + b^2$ with $a = z$ and $b = -\frac{2}{3}$. So,

$$z^2 - \frac{4}{3}z + \frac{4}{9} = \left(z - \frac{2}{3}\right)^2.$$

Now do Exercises 29–36

In Example 5, we use the skills that we learned in Examples 2, 3, and 4 to solve the quadratic equation $ax^2 + bx + c = 0$ with $a = 1$ by the method of completing the square. This method works only if $a = 1$ because the method for completing the square developed in Examples 2, 3, and 4 works only for $a = 1$.

E X A M P L E   **5**

Completing the square with $a = 1$

Solve $x^2 + 6x + 5 = 0$ by completing the square.

Solution

The perfect square trinomial whose first two terms are $x^2 + 6x$ is

$$x^2 + 6x + 9.$$

So we move 5 to the right-hand side of the equation, and then add 9 to each side to create a perfect square on the left side:

$x^2 + 6x \qquad = -5$	Subtract 5 from each side.
$x^2 + 6x + 9 = -5 + 9$	Add 9 to each side to get a perfect square trinomial.
$(x + 3)^2 = 4$	Factor the left-hand side.
$x + 3 = \pm\sqrt{4}$	Even-root property

$$x + 3 = 2 \qquad \text{or} \qquad x + 3 = -2$$
$$x = -1 \qquad \text{or} \qquad x = -5$$

Check in the original equation:

$$(-1)^2 + 6(-1) + 5 = 0$$

and

$$(-5)^2 + 6(-5) + 5 = 0$$

The solution set is $\{-1, -5\}$.

Now do Exercises 37–44

⟨ **Calculator Close-Up** ⟩

The solutions to

$$x^2 + 6x + 5 = 0$$

correspond to the x-intercepts for the graph of

$$y = x^2 + 6x + 5.$$

So we can check our solutions by graphing and using the TRACE feature as shown here.

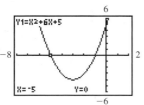

CAUTION All of the perfect square trinomials that we have used so far had a leading coefficient of 1. If $a \neq 1$, then we must divide each side of the equation by a to get an equation with a leading coefficient of 1.

The strategy for solving a quadratic equation by completing the square is stated in the following box.

Strategy for Solving Quadratic Equations by Completing the Square

1. If $a \neq 1$, then divide each side of the equation by a.
2. Get only the x^2- and the x-terms on the left-hand side.
3. Add to each side the square of $\frac{1}{2}$ the coefficient of x.
4. Factor the left-hand side as the square of a binomial.
5. Apply the even-root property.
6. Solve for x.
7. Simplify.

E X A M P L E   6

Completing the square with $a \neq 1$

Solve $2x^2 + 3x - 2 = 0$ by completing the square.

Solution

For completing the square, the coefficient of x^2 must be 1. So we first divide each side of the equation by 2:

$$\frac{2x^2 + 3x - 2}{2} = \frac{0}{2} \qquad \text{Divide each side by 2.}$$

$$x^2 + \frac{3}{2}x - 1 = 0 \qquad \text{Simplify.}$$

$$x^2 + \frac{3}{2}x = 1 \qquad \text{Get only } x^2\text{- and } x\text{-terms on the left-hand side.}$$

$$x^2 + \frac{3}{2}x + \frac{9}{16} = 1 + \frac{9}{16} \qquad \text{One-half of } \frac{3}{2} \text{ is } \frac{3}{4}, \text{ and } \left(\frac{3}{4}\right)^2 = \frac{9}{16}.$$

$$\left(x + \frac{3}{4}\right)^2 = \frac{25}{16} \qquad \text{Factor the left-hand side.}$$

$$x + \frac{3}{4} = \pm\sqrt{\frac{25}{16}} \qquad \text{Even-root property}$$

$$x + \frac{3}{4} = \frac{5}{4} \qquad \text{or} \qquad x + \frac{3}{4} = -\frac{5}{4}$$

$$x = \frac{2}{4} = \frac{1}{2} \qquad \text{or} \qquad x = -\frac{8}{4} = -2$$

‹ **Calculator Close-Up** ›

Note that the x-intercepts for the graph of
$$y = 2x^2 + 3x - 2$$
are $(-2, 0)$ and $\left(\frac{1}{2}, 0\right)$:

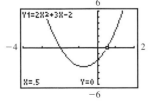

Check these values in the original equation. The solution set is $\left\{-2, \frac{1}{2}\right\}$.

Now do Exercises 45–46

In Examples 5 and 6, the solutions were rational numbers, and the equations could have been solved by factoring. In Example 7, the solutions are irrational numbers, and factoring will not work.

EXAMPLE **7**

A quadratic equation with irrational solutions

Solve $x^2 - 3x - 6 = 0$ by completing the square.

Solution

Because $a = 1$, we first get the x^2- and x-terms on the left-hand side:

$$x^2 - 3x - 6 = 0$$

$$x^2 - 3x \quad = 6 \qquad \text{Add 6 to each side.}$$

$$x^2 - 3x + \frac{9}{4} = 6 + \frac{9}{4} \qquad \text{One-half of } -3 \text{ is } -\frac{3}{2}, \text{ and } \left(-\frac{3}{2}\right)^2 = \frac{9}{4}.$$

$$\left(x - \frac{3}{2}\right)^2 = \frac{33}{4} \qquad 6 + \frac{9}{4} = \frac{24}{4} + \frac{9}{4} = \frac{33}{4}$$

$$x - \frac{3}{2} = \pm\sqrt{\frac{33}{4}} \qquad \text{Even-root property}$$

$$x = \frac{3}{2} \pm \frac{\sqrt{33}}{2} \qquad \text{Add } \frac{3}{2} \text{ to each side.}$$

$$x = \frac{3 \pm \sqrt{33}}{2}$$

The solution set is $\left\{\frac{3 + \sqrt{33}}{2}, \frac{3 - \sqrt{33}}{2}\right\}$.

> **Now do Exercises 47–56**

⟨4⟩ Radicals and Rational Expressions

Examples 8 and 9 show equations that are not originally in the form of quadratic equations. However, after simplifying these equations, we get quadratic equations. Even though completing the square can be used on any quadratic equation, factoring and the square root property are usually easier and we can use them when applicable. In Examples 8 and 9, we will use the most appropriate method.

EXAMPLE **8**

An equation containing a radical

Solve $x + 3 = \sqrt{153 - x}$.

Solution

Square both sides of the equation to eliminate the radical:

$$x + 3 = \sqrt{153 - x} \qquad \text{The original equation}$$

$$(x + 3)^2 = \left(\sqrt{153 - x}\right)^2 \qquad \text{Square each side.}$$

$$x^2 + 6x + 9 = 153 - x \qquad \text{Simplify.}$$

$$x^2 + 7x - 144 = 0$$

$$(x - 9)(x + 16) = 0 \qquad \text{Factor.}$$

$$x - 9 = 0 \quad \text{or} \quad x + 16 = 0 \qquad \text{Zero factor property}$$

$$x = 9 \quad \text{or} \qquad x = -16$$

You can provide graphical support for the solution to Example 8 by graphing
$$y_1 = x + 3$$
and
$$y_2 = \sqrt{153 - x}.$$
It appears that the only point of intersection occurs when $x = 9$.

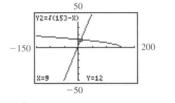

Because we squared each side of the original equation, we must check for extraneous roots. Let $x = 9$ in the original equation:

$$9 + 3 = \sqrt{153 - 9}$$
$$12 = \sqrt{144} \quad \text{Correct}$$

Let $x = -16$ in the original equation:

$$-16 + 3 = \sqrt{153 - (-16)}$$
$$-13 = \sqrt{169} \quad \text{Incorrect because } \sqrt{169} = 13$$

Because -16 is an extraneous root, the solution set is $\{9\}$.

Now do Exercises 57–60

E X A M P L E **9**

An equation containing rational expressions

Solve $\dfrac{1}{x} + \dfrac{3}{x-2} = \dfrac{5}{8}$.

Solution

The least common denominator (LCD) for x, $x - 2$, and 8 is $8x(x - 2)$.

$$\frac{1}{x} + \frac{3}{x-2} = \frac{5}{8}$$

$$8x(x-2)\frac{1}{x} + 8x(x-2)\frac{3}{x-2} = 8x(x-2)\frac{5}{8} \quad \text{Multiply each side by the LCD.}$$

$$8x - 16 + 24x = 5x^2 - 10x$$

$$32x - 16 = 5x^2 - 10x$$

$$-5x^2 + 42x - 16 = 0$$

$$5x^2 - 42x + 16 = 0 \qquad \text{Multiply each side by } -1$$
$$\text{for easier factoring.}$$

$$(5x - 2)(x - 8) = 0 \qquad \text{Factor.}$$

$$5x - 2 = 0 \quad \text{or} \quad x - 8 = 0$$

$$x = \frac{2}{5} \quad \text{or} \quad x = 8$$

Check these values in the original equation. The solution set is $\left\{\frac{2}{5}, 8\right\}$.

Now do Exercises 61–64

‹5› Imaginary Solutions

In Chapter 9, we found imaginary solutions to quadratic equations using the even-root property. We can get imaginary solutions also by completing the square.

E X A M P L E **10**

An equation with imaginary solutions

Find the complex solutions to $x^2 - 4x + 12 = 0$.

Solution

Because the quadratic polynomial cannot be factored, we solve the equation by completing the square.

< **Calculator Close-Up** >

The answer key (ANS) can be used to check imaginary answers as shown here.

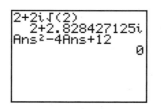

$$x^2 - 4x + 12 = 0 \qquad \text{The original equation}$$

$$x^2 - 4x = -12 \qquad \text{Subtract 12 from each side.}$$

$$x^2 - 4x + 4 = -12 + 4 \qquad \text{One-half of } -4 \text{ is } -2, \text{ and } (-2)^2 = 4.$$

$$(x - 2)^2 = -8$$

$$x - 2 = \pm\sqrt{-8} \qquad \text{Even-root property}$$

$$x = 2 \pm i\sqrt{8}$$

$$= 2 \pm 2i\sqrt{2}$$

Check these values in the original equation. The solution set is $\{2 \pm 2i\sqrt{2}\}$.

Now do Exercises 65–74

Warm-Ups ▼

Fill in the blank.

1. In this section quadratic equations are solved by _____, the _____ property, and _____ the square.

2. If $b = 0$ in $ax^2 + bx + c = 0$, then the equation can be solved by the _____.

3. The last term of a perfect square trinomial is the square of one-half the coefficient of the _____ term.

4. If the leading coefficient is not 1, then the first step in completing the square is to divide both sides of the equation by the _____.

True or false?

5. Every quadratic equation can be solved by factoring.

6. All quadratic equations have two distinct complex solutions.

7. The trinomial $x^2 + \frac{3}{2}x + \frac{9}{16}$ is a perfect square trinomial.

8. Every quadratic equation can be solved by completing the square.

9. $(x - 3)^2 = 12$ is equivalent to $x - 3 = 2\sqrt{3}$.

10. $(2x - 3)(3x + 5) = 0$ is equivalent to $x = \frac{3}{2}$ or $x = \frac{5}{3}$.

11. $x^2 = 8$ is equivalent to $x = \pm 2\sqrt{2}$.

12. To complete the square for $x^2 - 3x = 4$, add $\frac{9}{4}$ to each side.

Exercises

10.1

< **Study Tips** >

- Stay calm and confident. Take breaks when you study. Get 6 to 8 hours of sleep every night.
- Keep reminding yourself that working hard throughout the semester will really pay off in the end.

< 1 > **Review of Factoring**

Solve by factoring. See Example 1. See the Strategy for Solving Quadratic Equations by Factoring box on page 628.

1. $x^2 - x - 6 = 0$

2. $x^2 + 6x + 8 = 0$

3. $a^2 + 2a = 15$

4. $w^2 - 2w = 15$

5. $2x^2 - x - 3 = 0$

6. $6x^2 - x - 15 = 0$

7. $y^2 + 14y + 49 = 0$

8. $a^2 - 6a + 9 = 0$

9. $a^2 - 16 = 0$ **10.** $4w^2 - 25 = 0$

33. $z^2 - \dfrac{4}{7}z + \dfrac{4}{49}$ **34.** $m^2 - \dfrac{6}{5}m + \dfrac{9}{25}$

⟨2⟩ Review of the Even-Root Property

Use the even-root property to solve each equation.
See Example 2.

35. $t^2 + \dfrac{3}{5}t + \dfrac{9}{100}$ **36.** $h^2 + \dfrac{3}{2}h + \dfrac{9}{16}$

11. $x^2 = 81$ **12.** $x^2 = \dfrac{9}{4}$

13. $x^2 = \dfrac{16}{9}$ **14.** $a^2 = 32$

Solve by completing the square. See Examples 5–7. See the
Strategy for Solving Quadratic Equations by Completing the
Square box on page 632. Use your calculator to check.

37. $x^2 - 2x - 15 = 0$

15. $(x - 3)^2 = 16$ **16.** $(x + 5)^2 = 4$

38. $x^2 - 6x - 7 = 0$

17. $(z + 1)^2 = 5$ **18.** $(a - 2)^2 = 8$

39. $2x^2 - 4x = 70$

40. $3x^2 - 6x = 24$

19. $\left(w - \dfrac{3}{2}\right)^2 = \dfrac{7}{4}$ **20.** $\left(w + \dfrac{2}{3}\right)^2 = \dfrac{5}{9}$

41. $w^2 - w - 20 = 0$

42. $y^2 - 3y - 10 = 0$

43. $q^2 + 5q = 14$

44. $z^2 + z = 2$

⟨3⟩ Completing the Square

Find the perfect square trinomial whose first two terms are
given. See Example 3.

45. $2h^2 - h - 3 = 0$

46. $2m^2 - m - 15 = 0$

21. $x^2 + 2x$ **22.** $m^2 + 14m$

47. $x^2 + 4x = 6$

48. $x^2 + 6x - 8 = 0$

23. $x^2 - 3x$ **24.** $w^2 - 5w$

49. $x^2 + 8x - 4 = 0$

50. $x^2 + 10x - 3 = 0$

25. $y^2 + \dfrac{1}{4}y$ **26.** $z^2 + \dfrac{3}{2}z$

51. $x^2 + 5x + 5 = 0$

52. $x^2 - 7x + 4 = 0$

27. $x^2 + \dfrac{2}{3}x$ **28.** $p^2 + \dfrac{6}{5}p$

53. $4x^2 - 4x - 1 = 0$

54. $4x^2 + 4x - 2 = 0$

Factor each perfect square trinomial. See Example 4.

29. $x^2 + 8x + 16$ **30.** $x^2 - 10x + 25$

31. $y^2 - 5y + \dfrac{25}{4}$ **32.** $w^2 + w + \dfrac{1}{4}$

55. $2x^2 + 3x - 4 = 0$

56. $2x^2 + 5x - 1 = 0$

77. $4x^2 + 25 = 0$

78. $5w^2 - 3 = 0$

⟨4⟩ Radicals and Rational Expressions

Solve each equation by an appropriate method.
See Examples 8 and 9.

57. $\sqrt{2x + 1} = x - 1$ **58.** $\sqrt{2x - 4} = x - 14$

79. $\left(p + \dfrac{1}{2}\right)^2 = \dfrac{9}{4}$

80. $\left(y - \dfrac{2}{3}\right)^2 = \dfrac{4}{9}$

59. $w = \dfrac{\sqrt{w + 1}}{2}$ **60.** $y - 1 = \dfrac{\sqrt{y + 1}}{2}$

81. $5t^2 + 4t - 3 = 0$

82. $3v^2 + 4v - 1 = 0$

61. $\dfrac{t}{t - 2} = \dfrac{2t - 3}{t}$ **62.** $\dfrac{z}{z + 3} = \dfrac{3z}{5z - 1}$

83. $m^2 + 2m - 24 = 0$

84. $q^2 + 6q - 7 = 0$

63. $\dfrac{2}{x^2} + \dfrac{4}{x} + 1 = 0$

85. $(x - 2)^2 = -9$

64. $\dfrac{1}{x^2} + \dfrac{3}{x} + 1 = 0$

86. $(2x - 1)^2 = -4$

87. $-x^2 + x + 6 = 0$

⟨5⟩ Imaginary Solutions

Use completing the square to find the imaginary solutions to
each equation. See Example 10.

88. $-x^2 + x + 12 = 0$

65. $x^2 + 2x + 5 = 0$ **66.** $x^2 + 4x + 5 = 0$

89. $x^2 - 6x + 10 = 0$

67. $x^2 - 6x + 11 = 0$ **68.** $x^2 - 8x + 19 = 0$

90. $x^2 - 8x + 17 = 0$

91. $2x - 5 = \sqrt{7x + 7}$

69. $x^2 = -\dfrac{1}{2}$ **70.** $x^2 = -\dfrac{1}{8}$

92. $\sqrt{7x + 29} = x + 3$

93. $\dfrac{1}{x} + \dfrac{1}{x - 1} = \dfrac{1}{4}$

71. $x^2 + 12 = 0$ **72.** $-3x^2 - 21 = 0$

94. $\dfrac{1}{x} - \dfrac{2}{1 - x} = \dfrac{1}{2}$

73. $5z^2 - 4z + 1 = 0$ **74.** $2w^2 - 3w + 2 = 0$

Find the real solutions to each equation by examining the
graphs on page 638.

Miscellaneous

Find all real or imaginary solutions to each equation.
Use the method of your choice.

95. $x^2 + 2x - 15 = 0$

96. $100x^2 + 20x - 3 = 0$

75. $x^2 = -121$

76. $w^2 = -225$

97. $x^2 + 4x + 15 = 0$

98. $100x^2 - 60x + 9 = 0$

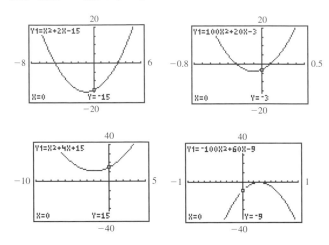

Applications

Solve each problem.

99. Approach speed. The formula $1211.1L = CA^2S$ is used to determine the approach speed for landing an aircraft, where L is the gross weight of the aircraft in pounds, C is the coefficient of lift, S is the surface area of the wings in square feet (ft^2), and A is approach speed in feet per second. Find A for the Piper Cheyenne, which has a gross weight of 8700 lb, a coefficient of lift of 2.81, and a wing surface area of 200 ft^2.

100. Time to swing. The period T (time in seconds for one complete cycle) of a simple pendulum is related to the length L (in feet) of the pendulum by the formula $8T^2 = \pi^2L$. If a child is on a swing with a 10-foot chain, then how long does it take to complete one cycle of the swing?

101. Time for a swim. Tropical Pools figures that its monthly revenue in dollars on the sale of x aboveground pools is given by $R = 1500x - 3x^2$, where x is less than 25. What number of pools sold would provide a revenue of $17,568?

102. Pole vaulting. In 1981 Vladimir Poliakov (USSR) set a world record of 19 ft $\frac{3}{4}$ in. for the pole vault (www.polevault.com). To reach that height, Poliakov obtained a speed of approximately 36 feet per second on the runway. The formula $h = -16t^2 + 36t$ gives his height t seconds after leaving the ground.

   a) Use the formula to find the exact values of t for which his height was 18 feet.
   b) Use the accompanying graph to estimate the value of t for which he was at his maximum height.
   c) Approximately how long was he in the air?

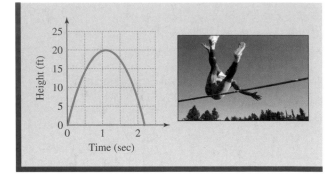

Figure for Exercise 102

Getting More Involved

103. Discussion

Which of the following equations is not a quadratic equation? Explain your answer.

   a) $\pi x^2 - \sqrt{5}x - 1 = 0$ **b)** $3x^2 - 1 = 0$
   c) $4x + 5 = 0$ **d)** $0.009x^2 = 0$

104. Exploration

Solve $x^2 - 4x + k = 0$ for $k = 0, 4, 5,$ and 10.

   a) When does the equation have only one solution?
   b) For what values of k are the solutions real?
   c) For what values of k are the solutions imaginary?

105. Cooperative learning

Write a quadratic equation of each of the following types, and then trade your equations with those of a classmate. Solve the equations and verify that they are of the required types.

   a) a single rational solution   **b)** two rational solutions
   c) two irrational solutions   **d)** two imaginary solutions

106. Exploration

In Section 10.2 we will solve $ax^2 + bx + c = 0$ for x by completing the square. Try it now without looking ahead.

Graphing Calculator Exercises

For each equation, find approximate solutions rounded to two decimal places.

107. $x^2 - 7.3x + 12.5 = 0$
108. $1.2x^2 - \pi x + \sqrt{2} = 0$
109. $2x - 3 = \sqrt{20 - x}$
110. $x^2 - 1.3x = 22.3 - x^2$

Math *at* Work   **Financial Matters**

In the United States, over 1 million new homes are sold annually, with a median price of about $200,000. Over 17 million new cars are sold each year with a median price over $20,000. Americans are constantly saving and borrowing. Nearly everyone will need to know a monthly payment or what their savings will total over time. The answers to these questions are in the following table.

What $P Left at Compound Interest Will Grow to	What $R Deposited Periodically Will Grow to	Periodic Payment That Will Pay off a Loan of $P
$P(1 + i)^{nt}$	$R\dfrac{(1 + i)^{nt} - 1}{i}$	$P\dfrac{i}{1 - (1 + i)^{-nt}}$

In each case, n is the number of periods per year, r is the annual percentage rate (APR), t is the number of years, and i is the interest rate per period $\left(i = \dfrac{r}{n}\right)$. For periodic payments or deposits these expressions apply only if the compounding period equals the payment period. So let's see what these expressions do.

A person inherits $10,000 and lets it grow at 4% APR compounded daily for 20 years. Use the first expression with $n = 365$, $i = \dfrac{0.04}{365}$, and $t = 20$ to get $10,000\left(1 + \dfrac{0.04}{365}\right)^{365\cdot20}$ or $22,254.43, which is the amount after 20 years.

20-year $200,000 mortgage

More often, people save money with periodic deposits. Suppose you deposit $100 per month at 4% compounded monthly for 20 years. Use the second expression with $R = 100$, $i = \dfrac{0.04}{12}$, $n = 12$, and $t = 20$ to get $100\dfrac{(1 + 0.04/12)^{12\cdot20} - 1}{0.04/12}$ or $36,677.46, which is the amount after 20 years.

Suppose that you get a 20-year $200,000 mortgage at 7% APR compounded monthly to buy an average house. Try using the third expression to calculate the monthly payment of $1550.60. See the accompanying figure.

10.2 The Quadratic Formula

In This Section

Completing the square from Section 10.1 can be used to solve any quadratic equation. Here we apply this method to the general quadratic equation to get a formula for the solutions to any quadratic equation.

⟨1⟩ Developing the Formula

Start with the general form of the quadratic equation,

$$ax^2 + bx + c = 0.$$

Assume a is positive for now, and divide each side by a:

$$\frac{ax^2 + bx + c}{a} = \frac{0}{a}$$

$$x^2 + \frac{b}{a}x + \frac{c}{a} = 0$$

$$x^2 + \frac{b}{a}x = -\frac{c}{a} \qquad \text{Subtract } \tfrac{c}{a} \text{ from each side.}$$

One-half of $\frac{b}{a}$ is $\frac{b}{2a}$, and $\frac{b}{2a}$ squared is $\frac{b^2}{4a^2}$:

$$x^2 + \frac{b}{a}x + \frac{b^2}{4a^2} = -\frac{c}{a} + \frac{b^2}{4a^2}$$

Factor the left-hand side and get a common denominator for the right-hand side:

$$\left(x + \frac{b}{2a}\right)^2 = \frac{b^2}{4a^2} - \frac{4ac}{4a^2} \qquad\qquad \frac{c(4a)}{a(4a)} = \frac{4ac}{4a^2}$$

$$\left(x + \frac{b}{2a}\right)^2 = \frac{b^2 - 4ac}{4a^2}$$

$$x + \frac{b}{2a} = \pm\sqrt{\frac{b^2 - 4ac}{4a^2}} \qquad \text{Even-root property}$$

$$x = \frac{-b}{2a} \pm \frac{\sqrt{b^2 - 4ac}}{2a} \qquad \text{Because } a > 0, \sqrt{4a^2} = 2a.$$

$$x = \frac{-b \pm \sqrt{b^2 - 4ac}}{2a}$$

We assumed a was positive so that $\sqrt{4a^2} = 2a$ would be correct. If a is negative, then $\sqrt{4a^2} = -2a$, and we get

$$x = \frac{-b}{2a} \pm \frac{\sqrt{b^2 - 4ac}}{-2a}.$$

However, the negative sign can be omitted in $-2a$ because of the $\pm$ symbol preceding it. For example, the results of $5 \pm (-3)$ and 5 ± 3 are the same. So when a is negative, we get the same formula as when a is positive. It is called the **quadratic formula.**

The Quadratic Formula

The solution to $ax^2 + bx + c = 0$, with $a \neq 0$, is given by the formula

$$x = \frac{-b \pm \sqrt{b^2 - 4ac}}{2a}.$$

⟨2⟩ Using the Formula

The quadratic formula solves any quadratic equation. Simply identify a, b, and c and insert those numbers into the formula. Note that if b is positive then $-b$ (the opposite of b) is a negative number. If b is negative, then $-b$ is a *positive* number.

EXAMPLE 1

Two rational solutions

Solve $x^2 + 2x - 15 = 0$ using the quadratic formula.

Solution

To use the formula, we first identify the values of a, b, and c:

$$\underset{\underset{a}{\uparrow}}{1x^2} + \underset{\underset{b}{\uparrow}}{2x} - \underset{\underset{c}{\uparrow}}{15} = 0$$

The coefficient of x^2 is 1, so $a = 1$. The coefficient of $2x$ is 2, so $b = 2$. The constant term is -15, so $c = -15$. Substitute these values into the quadratic formula:

$$x = \frac{-2 \pm \sqrt{2^2 - 4(1)(-15)}}{2(1)}$$

$$= \frac{-2 \pm \sqrt{4 + 60}}{2}$$

$$= \frac{-2 \pm \sqrt{64}}{2}$$

$$= \frac{-2 \pm 8}{2}$$

$$x = \frac{-2 + 8}{2} = 3 \quad \text{or} \quad x = \frac{-2 - 8}{2} = -5$$

Check 3 and -5 in the original equation. The solution set is $\{-5, 3\}$.

Now do Exercises 1–8

⟨ Calculator Close-Up ⟩

Note that the two solutions to

$$x^2 + 2x - 15 = 0$$

correspond to the two x-intercepts for the graph of

$$y = x^2 + 2x - 15.$$

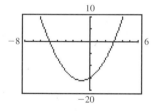

CAUTION To identify a, b, and c for the quadratic formula, the equation must be in the standard form $ax^2 + bx + c = 0$. If it is not in that form, then you must first rewrite the equation.

EXAMPLE 2

One rational solution

Solve $4x^2 = 12x - 9$ by using the quadratic formula.

Solution

Rewrite the equation in the form $ax^2 + bx + c = 0$ before identifying a, b, and c:

$$4x^2 - 12x + 9 = 0$$

Note that the single solution to

$$4x^2 - 12x + 9 = 0$$

corresponds to the single x-intercept for the graph of

$$y = 4x^2 - 12x + 9.$$

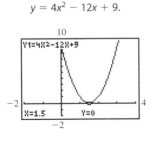

In this form we get $a = 4$, $b = -12$, and $c = 9$.

$$x = \frac{12 \pm \sqrt{(-12)^2 - 4(4)(9)}}{2(4)} \quad \text{Because } b = -12, -b = 12.$$

$$= \frac{12 \pm \sqrt{144 - 144}}{8}$$

$$= \frac{12 \pm 0}{8}$$

$$= \frac{12}{8}$$

$$= \frac{3}{2}$$

Check $\frac{3}{2}$ in the original equation. The solution set is $\left\{\frac{3}{2}\right\}$.

> Now do Exercises 9–14

Because the solutions to the equations in Examples 1 and 2 were rational numbers, these equations could have been solved by factoring. In Example 3, the solutions are irrational.

E X A M P L E 3

Two irrational solutions

Solve $\frac{1}{3}x^2 + x + \frac{1}{2} = 0$.

Solution

We could use $a = \frac{1}{3}$, $b = 1$, and $c = \frac{1}{2}$ in the quadratic formula, but it is easier to use the formula with integers. So we first multiply each side of the equation by 6, the least common denominator. Multiplying by 6 yields

$$2x^2 + 6x + 3 = 0.$$

Now let $a = 2$, $b = 6$, and $c = 3$ in the quadratic formula:

$$x = \frac{-6 \pm \sqrt{(6)^2 - 4(2)(3)}}{2(2)}$$

$$= \frac{-6 \pm \sqrt{36 - 24}}{4}$$

$$= \frac{-6 \pm \sqrt{12}}{4}$$

$$= \frac{-6 \pm 2\sqrt{3}}{4}$$

$$= \frac{2(-3 \pm \sqrt{3})}{2 \cdot 2}$$

$$= \frac{-3 \pm \sqrt{3}}{2}$$

The two irrational solutions to

$$2x^2 + 6x + 3 = 0$$

correspond to the two x-intercepts for the graph of

$$y = 2x^2 + 6x + 3.$$

Check these values in the original equation. The solution set is $\left\{\frac{-3 \pm \sqrt{3}}{2}\right\}$.

> Now do Exercises 15–20

EXAMPLE **4**

< **Calculator Close-Up** >

Because $x^2 + x + 5 = 0$ has no real solutions, the graph of

$$y = x^2 + x + 5$$

has no x-intercepts.

Two imaginary solutions, no real solutions

Find the complex solutions to $x^2 + x + 5 = 0$.

Solution

Let $a = 1$, $b = 1$, and $c = 5$ in the quadratic formula:

$$x = \frac{-1 \pm \sqrt{(1)^2 - 4(1)(5)}}{2(1)}$$

$$= \frac{-1 \pm \sqrt{-19}}{2}$$

$$= \frac{-1 \pm i\sqrt{19}}{2}$$

Check these values in the original equation. The solution set is $\left\{\frac{-1 \pm i\sqrt{19}}{2}\right\}$. There are no real solutions to the equation.

Now do Exercises 21–26

You have learned to solve quadratic equations by four different methods: the even-root property, factoring, completing the square, and the quadratic formula. The even-root property and factoring arc limited to certain special equations, but you should use those methods when possible. Any quadratic equation can be solved by completing the square or using the quadratic formula. Because the quadratic formula is usually faster, it is used more often than completing the square. However, completing the square is an important skill to learn. It will be used in the study of conic sections later in this text.

Summary of Methods for Solving $ax^2 + bx + c = 0$		
Method	**Comments**	**Examples**
Even-root property	Use when $b = 0$.	$(x - 2)^2 = 8$ $x - 2 = \pm\sqrt{8}$
Factoring	Use when the polynomial can be factored.	$x^2 + 5x + 6 = 0$ $(x + 2)(x + 3) = 0$
Quadratic formula	Solves any quadratic equation	$x^2 + 5x + 3 = 0$ $x = \dfrac{-5 \pm \sqrt{25 - 4(3)}}{2}$
Completing the square	Solves any quadratic equation, but quadratic formula is faster	$x^2 - 6x + 7 = 0$ $x^2 - 6x + 9 = -7 + 9$ $(x - 3)^2 = 2$

〈3〉 **Number of Solutions**

The quadratic equations in Examples 1 and 3 had two real solutions each. In each of those examples, the value of $b^2 - 4ac$ was positive. In Example 2, the quadratic equation had only one solution because the value of $b^2 - 4ac$ was zero. In Example 4, the quadratic equation had no real solutions because $b^2 - 4ac$ was negative. Because $b^2 - 4ac$ determines the kind and number of solutions to a quadratic equation, it is called the **discriminant.**

> ### Number of Solutions to a Quadratic Equation
> The quadratic equation $ax^2 + bx + c = 0$ with $a \neq 0$ has
> *two* real solutions if $b^2 - 4ac > 0$,
> *one* real solution if $b^2 - 4ac = 0$, and
> *no* real solutions (two imaginary solutions) if $b^2 - 4ac < 0$.

EXAMPLE 5

Using the discriminant
Use the discriminant to determine the number of real solutions to each quadratic equation.

   a) $x^2 - 3x - 5 = 0$ **b)** $x^2 = 3x - 9$ **c)** $4x^2 - 12x + 9 = 0$

Solution
   a) For $x^2 - 3x - 5 = 0$, use $a = 1$, $b = -3$, and $c = -5$ in $b^2 - 4ac$:

$$b^2 - 4ac = (-3)^2 - 4(1)(-5) = 9 + 20 = 29$$

Because the discriminant is positive, there are two real solutions to this quadratic equation.

   b) Rewrite $x^2 = 3x - 9$ as $x^2 - 3x + 9 = 0$. Then use $a = 1$, $b = -3$, and $c = 9$ in $b^2 - 4ac$:

$$b^2 - 4ac = (-3)^2 - 4(1)(9) = 9 - 36 = -27$$

Because the discriminant is negative, the equation has no real solutions. It has two imaginary solutions.

   c) For $4x^2 - 12x + 9 = 0$, use $a = 4$, $b = -12$, and $c = 9$ in $b^2 - 4ac$:

$$b^2 - 4ac = (-12)^2 - 4(4)(9) = 144 - 144 = 0$$

Because the discriminant is zero, there is only one real solution to this quadratic equation.

Now do Exercises 27–42

⟨4⟩ Applications
With the quadratic formula we can easily solve problems whose solutions are irrational numbers. When the solutions are irrational numbers, we usually use a calculator to find rational approximations and to check.

EXAMPLE 6

Area of a tabletop
The area of a rectangular tabletop is 6 square feet. If the width is 2 feet shorter than the length, then what are the dimensions?

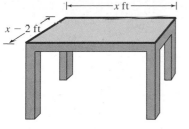

Figure 10.1

Solution

Let x be the length and $x - 2$ be the width, as shown in Fig. 10.1. Because the area is 6 square feet and $A = LW$, we can write the equation

$$x(x - 2) = 6$$

or

$$x^2 - 2x - 6 = 0.$$

Because this equation cannot be factored, we use the quadratic formula with $a = 1$, $b = -2$, and $c = -6$:

$$x = \frac{2 \pm \sqrt{(-2)^2 - 4(1)(-6)}}{2(1)}$$

$$= \frac{2 \pm \sqrt{28}}{2} = \frac{2 \pm 2\sqrt{7}}{2} = 1 \pm \sqrt{7}$$

Because $1 - \sqrt{7}$ is a negative number, it cannot be the length of a tabletop. If $x = 1 + \sqrt{7}$, then $x - 2 = 1 + \sqrt{7} - 2 = \sqrt{7} - 1$. Checking the product of $\sqrt{7} + 1$ and $\sqrt{7} - 1$, we get

$$\left(\sqrt{7} + 1\right)\left(\sqrt{7} - 1\right) = 7 - 1 = 6.$$

The exact length is $\sqrt{7} + 1$ feet, and the width is $\sqrt{7} - 1$ feet. Using a calculator, we find that the approximate length is 3.65 feet and the approximate width is 1.65 feet.

> Now do Exercises 71–90

Warm-Ups ▼

Fill in the blank.

1. The _____ formula can be used to solve any quadratic equation.

2. The _____ is $b^2 - 4ac$.

3. In the _____ number system every quadratic equation has at least one solution.

4. If $b^2 - 4ac = 0$, then the quadratic equation has ___ real solution.

5. If $b^2 - 4ac > 0$, then the quadratic equation has ___ real solutions.

6. If $b^2 - 4ac < 0$, then the quadratic equation has ___ imaginary solutions.

True or false?

7. Completing the square is used to develop the quadratic formula.

8. The quadratic formula will not work on $x^2 - 3 = 0$.

9. If $a = 2$, $b = -3$, and $c = -4$, then $b^2 - 4ac = 41$.

10. If $x^2 + 4x - 5 = 0$, then $x = \dfrac{-4 \pm \sqrt{16 - 4(-5)}}{2}$.

11. If $3x^2 - 5x + 9 = 0$, then $x = \dfrac{5 \pm \sqrt{25 - 4(3)(9)}}{2}$.

12. If $mx^2 + nx + p = 0$ and $m \neq 0$, then

$$x = \frac{-n \pm \sqrt{n^2 - 4mp}}{2m}.$$

‹ **2** › **Using the Formula**

Solve each equation by using the quadratic formula.
See Example 1.

1. $x^2 - 3x + 2 = 0$ **2.** $x^2 - 7x + 12 = 0$
3. $x^2 + 5x + 6 = 0$ **4.** $x^2 + 4x + 3 = 0$

5. $y^2 + y = 6$ **6.** $m^2 + 2m = 8$
7. $-6z^2 + 7z + 3 = 0$ **8.** $-8q^2 - 2q + 1 = 0$

Solve each equation by using the quadratic formula.
See Example 2.

9. $4x^2 - 4x + 1 = 0$ **10.** $4x^2 - 12x + 9 = 0$

11. $-9x^2 + 6x - 1 = 0$ **12.** $-9x^2 + 24x - 16 = 0$

13. $9 + 24x + 16x^2 = 0$ **14.** $4 + 20x = -25x^2$

Solve each equation by using the quadratic formula.
See Example 3.

15. $v^2 + 8v + 6 = 0$ **16.** $p^2 + 6p + 4 = 0$

17. $-x^2 - 5x + 1 = 0$ **18.** $-x^2 - 3x + 5 = 0$

19. $\frac{1}{3}t^2 - t + \frac{1}{6} = 0$ **20.** $\frac{3}{4}x^2 - 2x + \frac{1}{2} = 0$

Solve each equation by using the quadratic formula.
See Example 4.

21. $2t^2 - 6t + 5 = 0$ **22.** $2y^2 + 1 = 2y$

23. $-2x^2 + 3x = 6$ **24.** $-3x^2 - 2x - 5 = 0$

25. $\frac{1}{2}x^2 + 13 = 5x$ **26.** $\frac{1}{4}x^2 + \frac{17}{4} = 2x$

‹ **3** › **Number of Solutions**

Find $b^2 - 4ac$ and the number of real solutions to each
equation. See Example 5.

27. $x^2 - 6x + 2 = 0$ **28.** $x^2 + 6x + 9 = 0$
29. $-2x^2 + 5x - 6 = 0$ **30.** $-x^2 + 3x - 4 = 0$

31. $4m^2 + 25 = 20m$ **32.** $v^2 = 3v + 5$

33. $y^2 - \frac{1}{2}y + \frac{1}{4} = 0$ **34.** $\frac{1}{2}w^2 - \frac{1}{3}w + \frac{1}{4} = 0$

35. $-3t^2 + 5t + 6 = 0$ **36.** $9m^2 + 16 = 24m$
37. $9 - 24z + 16z^2 = 0$ **38.** $12 - 7x + x^2 = 0$
39. $5x^2 - 7 = 0$ **40.** $-6x^2 - 5 = 0$
41. $x^2 = x$ **42.** $-3x^2 + 7x = 0$

Miscellaneous

Solve by the method of your choice. See the Summary of
Methods for Solving $ax^2 + bx + c = 0$ on page 643.

43. $\frac{1}{4}y^2 + y = 1$ **44.** $\frac{1}{2}x^2 + x = 1$

45. $\frac{1}{3}x^2 + \frac{1}{2}x = \frac{1}{3}$ **46.** $\frac{4}{9}w^2 + 1 = \frac{5}{3}w$

47. $3y^2 + 2y - 4 = 0$ **48.** $2y^2 - 3y - 6 = 0$

49. $\frac{w}{w - 2} = \frac{w}{w - 3}$ **50.** $\frac{y}{3y - 4} = \frac{2}{y + 4}$

51. $\frac{9(3x - 5)^2}{4} = 1$ **52.** $\frac{25(2x + 1)^2}{9} = 0$

53. $25 - \frac{1}{3}x^2 = 0$ **54.** $\frac{49}{2} - \frac{1}{4}x^2 = 0$

55. $1 + \frac{20}{x^2} = \frac{8}{x}$ **56.** $\frac{34}{x^2} = \frac{6}{x} - 1$

57. $(x - 8)(x + 4) = -42$ **58.** $(x - 10)(x - 2) = -20$

59. $y = \frac{3(2y + 5)}{8(y - 1)}$ **60.** $z = \frac{7z - 4}{12(z - 1)}$

Use the quadratic formula and a calculator to solve each equation. Round answers to three decimal places and check your answers.

61. $x^2 + 3.2x - 5.7 = 0$

62. $x^2 + 7.15x + 3.24 = 0$

63. $x^2 - 7.4x + 13.69 = 0$

64. $1.44x^2 + 5.52x + 5.29 = 0$

65. $1.85x^2 + 6.72x + 3.6 = 0$

66. $3.67x^2 + 4.35x - 2.13 = 0$

67. $3x^2 + 14,379x + 243 = 0$

68. $x^2 + 12,347x + 6741 = 0$

69. $x^2 + 0.00075x - 0.0062 = 0$

70. $4.3x^2 - 9.86x - 3.75 = 0$

⟨4⟩ Applications

Find the exact solution(s) to each problem. If the solution(s) are irrational, then also find approximate solution(s) to the nearest tenth. See Example 6.

71. *Missing numbers.* Find two positive real numbers that differ by 1 and have a product of 16.

72. *Missing numbers.* Find two positive real numbers that differ by 2 and have a product of 10.

73. *More missing numbers.* Find two real numbers that have a sum of 6 and a product of 4.

74. *More missing numbers.* Find two real numbers that have a sum of 8 and a product of 2.

75. *Bulletin board.* The length of a bulletin board is 1 foot more than the width. The diagonal has a length of $\sqrt{3}$ feet (ft). Find the length and width of the bulletin board.

76. *Diagonal brace.* The width of a rectangular gate is 2 meters (m) larger than its height. The diagonal brace measures $\sqrt{6}$ m. Find the width and height.

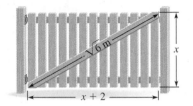

Figure for Exercise 76

77. *Area of a rectangle.* The length of a rectangle is 4 ft longer than the width, and its area is 10 square feet (ft²). Find the length and width.

78. *Diagonal of a square.* The diagonal of a square is 2 m longer than a side. Find the length of a side.

If an object is given an initial velocity of v_0 feet per second from a height of s_0 feet, then its height S after t seconds is given by the formula $S = -16t^2 + v_0 t + s_0$.

79. *Projected pine cone.* If a pine cone is projected upward at a velocity of 16 ft/sec from the top of a 96-foot pine tree, then how long does it take to reach the earth?

80. *Falling pine cone.* If a pine cone falls from the top of a 96-foot pine tree, then how long does it take to reach the earth?

81. *Tossing a ball.* A ball is tossed into the air at 10 ft/sec from a height of 5 feet. How long does it take to reach the earth?

82. *Time in the air.* A ball is tossed into the air from a height of 12 feet at 16 ft/sec. How long does it take to reach the earth?

83. *Penny tossing.* If a penny is thrown downward at 30 ft/sec from the bridge at Royal Gorge, Colorado, how long does it take to reach the Arkansas River 1000 ft below?

84. *Foul ball.* Suppose Charlie O'Brian of the Braves hits a baseball straight upward at 150 ft/sec from a height of 5 ft.

     a) Use the formula to determine how long it takes the ball to return to the earth.

b) Use the accompanying graph to estimate the maximum height reached by the ball.

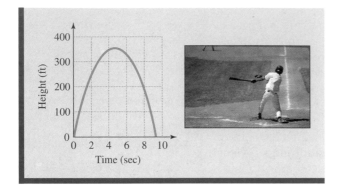

Figure for Exercise 84

Solve each problem.

85. Kitchen countertop. A 30 in. by 40 in. countertop for a work island is to be covered with green ceramic tiles, except for a border of uniform width as shown in the figure. If the area covered by the green tiles is 704 square inches (in.^{2}), then how wide is the border?

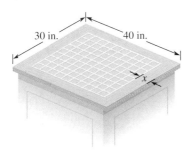

Figure for Exercise 85

86. Recovering an investment. The manager at Cream of the Crop bought a load of watermelons for $200. She priced the melons so that she would make $1.50 profit on each melon. When all but 30 had been sold, the manager had recovered her initial investment. How many did she buy originally?

87. Baby shower. A group of office workers plans to share equally the $100 cost of giving a baby shower for a coworker. If they can get six more people to share the cost, then the cost per person will decrease by $15. How many people are in the original group?

88. Sharing cost. The members of a flying club plan to share equally the cost of a $200,000 airplane. The members want to find five more people to join the club so that the cost per

person will decrease by $2000. How many members are currently in the club?

89. Farmer's delight. The manager of Farmer's Delight bought a load of watermelons for $750 and priced the watermelons so that he would make a profit of $2 on each melon. When all but 100 of the melons had been sold, he broke even. How many did he buy originally?

90. Traveling club. The members of a traveling club plan to share equally the cost of a $150,000 motorhome. If they can find 10 more people to join and share the cost, then the cost per person will decrease by $1250. How many members are there originally in the club?

Getting More Involved

91. Discussion

Find the solutions to $6x^2 + 5x - 4 = 0$. Is the sum of your solutions equal to $-\frac{b}{a}$? Explain why the sum of the solutions to any quadratic equation is $-\frac{b}{a}$. (*Hint:* Use the quadratic formula.)

92. Discussion

Use the result of Exercise 91 to check whether $\left\{\frac{2}{3}, \frac{1}{3}\right\}$ is the solution set to $9x^2 - 3x - 2 = 0$. If this solution set is not correct, then what is the correct solution set?

93. Discussion

What is the product of the two solutions to $6x^2 + 5x - 4 = 0$? Explain why the product of the solutions to any quadratic equation is $\frac{c}{a}$.

94. Discussion

Use the result of Exercise 93 to check whether $\left\{\frac{9}{2}, -2\right\}$ is the solution set to $2x^2 - 13x + 18 = 0$. If this solution set is not correct, then what is the correct solution set?

Graphing Calculator Exercises

Determine the number of real solutions to each equation by examining the calculator graph of $y = ax^2 + bx + c$. Use the discriminant to check your conclusions.

95. $x^2 - 6.33x + 3.7 = 0$

96. $1.8x^2 + 2.4x - 895 = 0$

97. $4x^2 - 67.1x + 344 = 0$

98. $-2x^2 - 403 = 0$

99. $-x^2 + 30x - 226 = 0$

100. $16x^2 - 648x + 6562 = 0$

10.3 More on Quadratic Equations

In this section, we use the ideas and methods of the previous sections to explore additional topics involving quadratic equations.

〈1〉 Writing a Quadratic Equation with Given Solutions

Not every quadratic equation can be solved by factoring, but the factoring method can be used (in reverse) to write a quadratic equation with given solutions.

E X A M P L E   1

Writing a quadratic given the solutions
Write a quadratic equation that has each given pair of solutions.

  a) 4, −6   **b)** $-\sqrt{2}, \sqrt{2}$ **c)** $-3i, 3i$

Solution

  a) Reverse the factoring method using solutions 4 and −6:

$$x = 4 \qquad \text{or} \qquad x = -6$$
$$x - 4 = 0 \qquad \text{or} \qquad x + 6 = 0$$
$$(x - 4)(x + 6) = 0 \quad \text{Zero factor property}$$
$$x^2 + 2x - 24 = 0 \quad \text{Multiply the factors.}$$

  b) Reverse the factoring method using solutions $-\sqrt{2}$ and $\sqrt{2}$:

$$x = -\sqrt{2} \qquad \text{or} \qquad x = \sqrt{2}$$
$$x + \sqrt{2} = 0 \qquad \text{or} \qquad x - \sqrt{2} = 0$$
$$(x + \sqrt{2})(x - \sqrt{2}) = 0 \quad \text{Zero factor property}$$
$$x^2 - 2 = 0 \quad \text{Multiply the factors.}$$

  c) Reverse the factoring method using solutions $-3i$ and $3i$:

$$x = -3i \qquad \text{or} \qquad x = 3i$$
$$x + 3i = 0 \qquad \text{or} \qquad x - 3i = 0$$
$$(x + 3i)(x - 3i) = 0 \quad \text{Zero factor property}$$
$$x^2 - 9i^2 = 0 \quad \text{Multiply the factors.}$$
$$x^2 + 9 = 0 \quad \text{Note: } i^2 = -1$$

Now do Exercises 1–12

〈 **Calculator Close-Up** 〉

The graph of $y = x^2 + 2x - 24$ supports the conclusion in Example 1(a) because the graph crosses the x-axis at $(4, 0)$ and $(-6, 0)$.

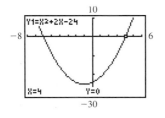

The process in Example 1 can be shortened somewhat if we observe the correspondence between the solutions to the equation and the factors.

> **Correspondence Between Solutions and Factors**
>
> If a and b are solutions to a quadratic equation, then the equation is equivalent to
> $$(x - a)(x - b) = 0.$$

So if 2 and −3 are solutions to a quadratic equation, then the quadratic equation is $(x - 2)(x + 3) = 0$ or $x^2 + x - 6 = 0$. If the solutions are fractions, it is not necessary to use fractions in the factors. For example, if $\frac{2}{3}$ is a solution, then $3x - 2$ is a factor because $3x - 2 = 0$ is equivalent to $x = \frac{2}{3}$. If $-\frac{1}{5}$ is a solution, then $5x + 1$ is a factor because $5x + 1 = 0$ is equivalent to $x = -\frac{1}{5}$. So if $\frac{2}{3}$ and $-\frac{1}{5}$ are solutions to a quadratic equation, then the equation is $(3x - 2)(5x + 1) = 0$ or $15x^2 - 7x - 2 = 0$.

⟨2⟩ Using the Discriminant in Factoring

The quadratic formula $x = \frac{-b \pm \sqrt{b^2 - 4ac}}{2a}$ gives the solutions to the quadratic equation $ax^2 + bx + c = 0$. If a, b, and c are integers and $b^2 - 4ac$ is a perfect square, then $\sqrt{b^2 - 4ac}$ is a whole number and the quadratic formula produces solutions that are rational. The quadratic equations with rational solutions are precisely the ones that we solve by factoring. So we can use the discriminant $b^2 - 4ac$ to determine whether a quadratic polynomial is prime.

Identifying Prime Quadratic Polynomials Using $b^2 - 4ac$

Let $ax^2 + bx + c$ be a quadratic polynomial with integral coefficients having a greatest common factor of 1. The quadratic polynomial is prime if and only if the discriminant $b^2 - 4ac$ is *not* a perfect square.

EXAMPLE 2

Using the discriminant
Use the discriminant to determine whether each polynomial can be factored.

a) $6x^2 + x - 15$ **b)** $5x^2 - 3x + 2$

Solution

a) Use $a = 6$, $b = 1$, and $c = -15$ to find $b^2 - 4ac$:

$$b^2 - 4ac = 1^2 - 4(6)(-15) = 361$$

Because $\sqrt{361} = 19$, $6x^2 + x - 15$ can be factored. Using the *ac* method, we get

$$6x^2 + x - 15 = (2x - 3)(3x + 5).$$

b) Use $a = 5$, $b = -3$, and $c = 2$ to find $b^2 - 4ac$:

$$b^2 - 4ac = (-3)^2 - 4(5)(2) = -31$$

Because the discriminant is not a perfect square, $5x^2 - 3x + 2$ is prime.

> Now do Exercises 13–24

⟨3⟩ Equations Quadratic in Form

An equation in which an expression appears in place of x in $ax^2 + bx + c = 0$ is called **quadratic in form**. So,

$$3(x - 7)^2 + (x - 7) + 8 = 0,$$
$$-2(x^2 + 3)^2 - (x^2 + 3) + 1 = 0, \quad \text{and}$$
$$-7x^4 + 5x^2 - 6 = 0$$

are quadratic in form. Note that the last equation is quadratic in form because it could be written as $-7(x^2)^2 + 5(x^2) - 6$, where x^2 is used in place of x. To solve an equation that is quadratic in form we replace the expression with a single variable and then solve the resulting quadratic equation, as shown in Example 3.

E X A M P L E   **3**

An equation quadratic in form
Solve $(x + 15)^2 - 3(x + 15) - 18 = 0$.

Solution

Note that $x + 15$ and $(x + 15)^2$ both appear in the equation. Let $a = x + 15$ and substitute a for $x + 15$ in the equation:

$$(x + 15)^2 - 3(x + 15) - 18 = 0$$
$$a^2 - 3a - 18 = 0$$
$$(a - 6)(a + 3) = 0 \qquad \text{Factor.}$$

$$a - 6 = 0 \qquad \text{or} \qquad a + 3 = 0$$
$$a = 6 \qquad \text{or} \qquad a = -3$$
$$x + 15 = 6 \qquad \text{or} \qquad x + 15 = -3 \quad \text{Replace } a \text{ by } x + 15.$$
$$x = -9 \qquad \text{or} \qquad x = -18$$

Check in the original equation. The solution set is $\{-18, -9\}$.

 Now do Exercises 25–30

In Example 4, we have a fourth-degree equation that is quadratic in form. Note that the fourth-degree equation has four solutions.

E X A M P L E   **4**

A fourth-degree equation
Solve $x^4 - 6x^2 + 8 = 0$.

Solution

Note that x^4 is the square of x^2. If we let $w = x^2$, then $w^2 = x^4$. Substitute these expressions into the original equation.

$$x^4 - 6x^2 + 8 = 0$$
$$w^2 - 6w + 8 = 0 \qquad \text{Replace } x^4 \text{ by } w^2 \text{ and } x^2 \text{ by } w.$$
$$(w - 2)(w - 4) = 0 \qquad \text{Factor.}$$

$$w - 2 = 0 \qquad \text{or} \qquad w - 4 = 0$$
$$w = 2 \qquad \text{or} \qquad w = 4$$
$$x^2 = 2 \qquad \text{or} \qquad x^2 = 4 \qquad \text{Substitute } x^2 \text{ for } w.$$
$$x = \pm\sqrt{2} \qquad \text{or} \qquad x = \pm 2 \qquad \text{Even-root property}$$

Check. The solution set is $\{-2, -\sqrt{2}, \sqrt{2}, 2\}$.

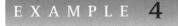 Now do Exercises 31–38

‹ **Helpful Hint** ›

The fundamental theorem of algebra says that the number of solutions to a polynomial equation is less than or equal to the degree of the polynomial. This famous theorem was proved by Carl Friedrich Gauss when he was a young man.

CAUTION   If you replace x^2 by w, do not quit when you find the values of w. If the variable in the original equation is x, then you must solve for x.

EXAMPLE 5

A quadratic within a quadratic
Solve $(x^2 + 2x)^2 - 11(x^2 + 2x) + 24 = 0$.

Solution

Note that $x^2 + 2x$ and $(x^2 + 2x)^2$ appear in the equation. Let $a = x^2 + 2x$ and substitute.

$$a^2 - 11a + 24 = 0$$
$$(a - 8)(a - 3) = 0 \quad \text{Factor.}$$

$a - 8 = 0$ or $a - 3 = 0$

$\quad\quad a = 8$ or $\quad\quad a = 3$

$x^2 + 2x = 8$ or $x^2 + 2x = 3$ Replace a by $x^2 + 2x$.

$x^2 + 2x - 8 = 0$ or $x^2 + 2x - 3 = 0$

$(x - 2)(x + 4) = 0$ or $(x + 3)(x - 1) = 0$

$x - 2 = 0$ or $x + 4 = 0$ or $x + 3 = 0$ or $x - 1 = 0$

$\quad x = 2$ or $\quad x = -4$ or $\quad x = -3$ or $\quad x = 1$

Check. The solution set is $\{-4, -3, 1, 2\}$.

> Now do Exercises 39–44

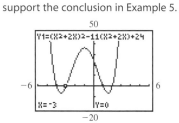

< **Calculator Close-Up** >

The four x-intercepts on the graph of
$y = (x^2 + 2x)^2 - 11(x^2 + 2x) + 24$
support the conclusion in Example 5.

Example 6 involves a fractional exponent. To identify this type of equation as quadratic in form, recall how to square an expression with a fractional exponent. For example, $(x^{1/2})^2 = x$, $(x^{1/4})^2 = x^{1/2}$, and $(x^{1/3})^2 = x^{2/3}$.

EXAMPLE 6

A fractional exponent
Solve $x - 9x^{1/2} + 14 = 0$.

Solution

Note that the square of $x^{1/2}$ is x. Let $w = x^{1/2}$; then $w^2 = (x^{1/2})^2 = x$. Now substitute w and w^2 into the original equation:

$$w^2 - 9w + 14 = 0$$
$$(w - 7)(w - 2) = 0$$

$w - 7 = 0$ or $w - 2 = 0$

$\quad\quad w = 7$ or $\quad\quad w = 2$

$x^{1/2} = 7$ or $x^{1/2} = 2$ Replace w by $x^{1/2}$.

$\quad x = 49$ or $\quad x = 4$ Square each side.

Because we squared each side, we must check for extraneous roots. First evaluate $x - 9x^{1/2} + 14$ for $x = 49$:

$$49 - 9 \cdot 49^{1/2} + 14 = 49 - 9 \cdot 7 + 14 = 0$$

Now evaluate $x - 9x^{1/2} + 14$ for $x = 4$:

$$4 - 9 \cdot 4^{1/2} + 14 = 4 - 9 \cdot 2 + 14 = 0$$

Because each solution checks, the solution set is $\{4, 49\}$.

> Now do Exercises 45–52

CAUTION An equation of quadratic form with variable x must have a power of x and its square. Equations such as $x^4 - 5x^3 + 6 = 0$ or $x^{1/2} - 3x^{1/3} - 18 = 0$ are not quadratic in form and cannot be solved by substitution.

‹4› Applications

Applied problems often result in quadratic equations that cannot be factored. For such equations we use the quadratic formula to find exact solutions and a calculator to find decimal approximations for the exact solutions.

EXAMPLE **7**

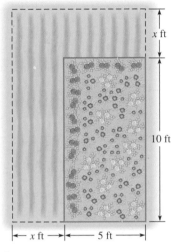

Figure 10.2

Changing area

Marvin's flower bed is rectangular in shape with a length of 10 feet and a width of 5 feet (ft). He wants to increase the length and width by the same amount to obtain a flower bed with an area of 75 square feet (ft^2). What should the amount of increase be?

Solution

Let x be the amount of increase. The length and width of the new flower bed are $x + 10$ ft and $x + 5$ ft, respectively, as shown in Fig. 10.2. Because the area is to be 75 ft^2, we have

$$(x + 10)(x + 5) = 75.$$

Write this equation in the form $ax^2 + bx + c = 0$:

$$x^2 + 15x + 50 = 75$$

$$x^2 + 15x - 25 = 0 \quad \text{Get 0 on the right.}$$

$$x = \frac{-15 \pm \sqrt{225 - 4(1)(-25)}}{2(1)}$$

$$= \frac{-15 \pm \sqrt{325}}{2} = \frac{-15 \pm 5\sqrt{13}}{2}$$

Because the value of x must be positive, the exact increase is

$$\frac{-15 + 5\sqrt{13}}{2} \text{ feet.}$$

Using a calculator, we can find that x is approximately 1.51 ft. If $x = 1.51$ ft, then the new length is 11.51 ft, and the new width is 6.51 ft. The area of a rectangle with these dimensions is 74.93 ft^2. Of course, the approximate dimensions do not give an area of exactly 75 ft^2.

> Now do Exercises 79–86

EXAMPLE **8**

Mowing the lawn

It takes Carla 1 hour longer to mow the lawn than it takes Sharon to mow the lawn. If they can mow the lawn in 5 hours working together, then how long would it take each girl by herself?

Solution

If Sharon can mow the lawn by herself in x hours, then she works at the rate of $\frac{1}{x}$ of the lawn per hour. If Carla can mow the lawn by herself in $x + 1$ hours, then she works at the rate of $\frac{1}{x+1}$ of the lawn per hour. We can use a table to list all of the important quantities.

	Rate	Time	Work
Sharon	$\frac{1}{x} \frac{\text{lawn}}{\text{hr}}$	5 hr	$\frac{5}{x}$ lawn
Carla	$\frac{1}{x+1} \frac{\text{lawn}}{\text{hr}}$	5 hr	$\frac{5}{x+1}$ lawn

Note that the equation concerns the portion of the job done by each girl. We could have written an equation about the rates at which the two girls work. Because they can finish the lawn together in 5 hours, they are mowing together at the rate of $\frac{1}{5}$ lawn per hour. So,

$$\frac{1}{x} + \frac{1}{x+1} = \frac{1}{5}.$$

Because they complete the lawn in 5 hours, the portion of the lawn done by Sharon and the portion done by Carla have a sum of 1:

$$\frac{5}{x} + \frac{5}{x+1} = 1$$

$$x(x+1)\,\frac{5}{x} + x(x+1)\,\frac{5}{x+1} = x(x+1)1 \quad \text{Multiply by the LCD.}$$

$$5x + 5 + 5x = x^2 + x$$

$$10x + 5 = x^2 + x$$

$$-x^2 + 9x + 5 = 0$$

$$x^2 - 9x - 5 = 0$$

$$x = \frac{9 \pm \sqrt{(-9)^2 - 4(1)(-5)}}{2(1)}$$

$$= \frac{9 \pm \sqrt{101}}{2}$$

Using a calculator, we find that $\frac{9 - \sqrt{101}}{2}$ is negative. So Sharon's time alone is

$$\frac{9 + \sqrt{101}}{2} \text{ hours.}$$

To find Carla's time alone, we add 1 hour to Sharon's time:

$$\frac{9 + \sqrt{101}}{2} + 1 = \frac{9 + \sqrt{101}}{2} + \frac{2}{2} = \frac{11 + \sqrt{101}}{2} \text{ hours}$$

Sharon's time alone is approximately 9.525 hours, and Carla's time alone is approximately 10.525 hours.

> Now do Exercises 87–90

Warm-Ups ▼

Fill in the blank.

1. If d is a solution to a quadratic equation, then $x - d$ is a _____ of the quadratic polynomial.

2. If $b^2 - 4ac$ is not a perfect square, then $ax^2 + bx + c$ is a _____ polynomial.

3. An equation that is quadratic after a substitution is called _____ in form.

4. If m and n are _____ to a quadratic equation, then $(x - m)(x - n) = 0$ is a quadratic equation with those solutions.

True or false?

5. The equation $(x - 4)(x + 5) = 0$ is a quadratic equation with solutions 4 and −5.

6. If $w = x^{1/6}$, then $w^2 = x^{1/3}$.

7. If $y = 2^{1/2}$, then $y^2 = 2^{1/4}$.

8. To solve $x^4 - 5x^2 + 6 = 0$ by substitution, let $w = x^2$.

9. To solve $x^5 - 3x^3 - 10 = 0$ by substitution, let $w = x^3$.

10. If Ann's boat goes 10 mph in still water, then against a 5-mph current it will go 2 mph.

11. If Elvia drives 300 miles in x hours, then her rate is $\dfrac{300}{x}$ mph.

12. If John paints a 100-foot fence in x hours, then his rate is $\dfrac{100}{x}$ of the fence per hour.

‹ 1 › Writing a Quadratic Equation with Given Solutions

For each given pair of numbers find a quadratic equation with integral coefficients that has the numbers as its solutions. See Example 1.

1. $3, -7$
2. $-8, 2$
3. $4, 1$
4. $3, 2$
5. $\sqrt{5}, -\sqrt{5}$
6. $-\sqrt{7}, \sqrt{7}$
7. $4i, -4i$
8. $-3i, 3i$
9. $i\sqrt{2}, -i\sqrt{2}$
10. $3i\sqrt{2}, -3i\sqrt{2}$
11. $\dfrac{1}{2}, \dfrac{1}{3}$
12. $-\dfrac{1}{5}, -\dfrac{1}{2}$

‹ 2 › Using the Discriminant in Factoring

Use the discriminant to determine whether each quadratic polynomial can be factored, and then factor the ones that are not prime. See Example 2.

13. $x^2 + 9$
14. $x^2 - 9$
15. $2x^2 - x + 4$
16. $2x^2 + 3x - 5$
17. $2x^2 + 6x - 5$
18. $3x^2 + 5x - 1$
19. $6x^2 + 19x - 36$
20. $8x^2 + 6x - 27$
21. $4x^2 - 5x - 12$
22. $4x^2 - 27x + 45$
23. $8x^2 - 18x - 45$
24. $6x^2 + 9x - 16$

‹ 3 › Equations Quadratic in Form

Find all real solutions to each equation. See Example 3.

25. $(x - 1)^2 - 2(x - 1) - 8 = 0$
26. $(m + 3)^2 + 5(m + 3) - 14 = 0$
27. $(2a - 1)^2 + 2(2a - 1) - 8 = 0$
28. $(3a + 2)^2 - 3(3a + 2) = 10$
29. $(w - 1)^2 + 5(w - 1) + 5 = 0$
30. $(2x - 1)^2 - 4(2x - 1) + 2 = 0$

Find all real solutions to each equation. See Example 4.

31. $x^4 - 13x^2 + 36 = 0$
32. $x^4 - 20x^2 + 64 = 0$
33. $x^6 - 28x^3 + 27 = 0$
34. $x^6 - 3x^3 - 4 = 0$
35. $x^4 - 14x^2 + 45 = 0$
36. $x^4 + 2x^2 = 15$
37. $x^6 + 7x^3 = 8$
38. $a^6 + 6a^3 = 16$

Find all real solutions to each equation. See Example 5.

39. $(x^2 + 1)^2 - 11(x^2 + 1) = -10$
40. $(x^2 + 2)^2 - 11(x^2 + 2) = -30$
41. $(x^2 + 2x)^2 - 7(x^2 + 2x) + 12 = 0$
42. $(x^2 + 3x)^2 + (x^2 + 3x) - 20 = 0$
43. $(y^2 + y)^2 - 8(y^2 + y) + 12 = 0$
44. $(w^2 - 2w)^2 + 24 = 11(w^2 - 2w)$

Find all real solutions to each equation. See Example 6.

45. $x - 3x^{1/2} + 2 = 0$
46. $x^{1/2} - 3x^{1/4} + 2 = 0$
47. $x^{2/3} + 4x^{1/3} + 3 = 0$
48. $x^{2/3} - 3x^{1/3} - 10 = 0$
49. $x^{1/2} - 5x^{1/4} + 6 = 0$
50. $2x - 5\sqrt{x} + 2 = 0$
51. $2x - 5x^{1/2} - 3 = 0$
52. $x^{1/4} + 2 = x^{1/2}$

Find all real solutions to each equation.

53. $x^{-2} + x^{-1} - 6 = 0$
54. $x^{-2} - 2x^{-1} = 8$
55. $x^{1/6} - x^{1/3} + 2 = 0$
56. $x^{2/3} - x^{1/3} - 20 = 0$
57. $\left(\dfrac{1}{y - 1}\right)^2 + \left(\dfrac{1}{y - 1}\right) = 6$
58. $\left(\dfrac{1}{w + 1}\right)^2 - 2\left(\dfrac{1}{w + 1}\right) - 24 = 0$
59. $2x^2 - 3 - 6\sqrt{2x^2 - 3} + 8 = 0$
60. $x^2 + x + \sqrt{x^2 + x} - 2 = 0$

61. $x^{-2} - 2x^{-1} - 1 = 0$

62. $x^{-2} - 6x^{-1} + 6 = 0$

Miscellaneous

Find all real and imaginary solutions to each equation.

63. $w^2 + 4 = 0$ **64.** $w^2 + 9 = 0$

65. $a^4 + 6a^2 + 8 = 0$

66. $b^4 + 13b^2 + 36 = 0$

67. $m^4 - 16 = 0$

68. $t^4 - 4 = 0$

69. $16b^4 - 1 = 0$ **70.** $b^4 - 81 = 0$

71. $x^3 + 1 = 0$

72. $x^3 - 1 = 0$

73. $x^3 + 8 = 0$

74. $x^3 - 27 = 0$

75. $a^{-2} - 2a^{-1} + 5 = 0$

76. $b^{-2} - 4b^{-1} + 6 = 0$

77. $(2x - 1)^2 - 2(2x - 1) + 5 = 0$

78. $(4x - 1)^2 - 6(4x - 1) + 25 = 0$

⟨4⟩ Applications

Find the exact solution to each problem. If the exact solution is an irrational number, then also find an approximate decimal solution. See Examples 7 and 8.

79. *Country singers.* Harry and Gary are traveling to Nashville to make their fortunes. Harry leaves on the train at 8:00 A.M. and Gary travels by car, starting at 9:00 A.M. To complete the 300-mile trip and arrive at the same time as Harry, Gary travels 10 miles per hour (mph) faster than the train. At what time will they both arrive in Nashville?

80. *Gone fishing.* Debbie traveled by boat 5 miles upstream to fish in her favorite spot. Because of the 4-mph current, it took her 20 minutes longer to get there than to return. How fast will her boat go in still water?

81. *Cross-country cycling.* Erin was traveling across the desert on her bicycle. Before lunch she traveled 60 miles (mi); after lunch she traveled 46 mi. She put in 1 hour more after lunch than before lunch, but her speed was

4 mph slower than before. What was her speed before lunch and after lunch?

Photo for Exercise 81

82. *Extreme hardship.* Kim starts to walk 3 mi to school at 7:30 A.M. with a temperature of 0°F. Her brother Bryan starts at 7:45 A.M. on his bicycle, traveling 10 mph faster than Kim. If they get to school at the same time, then how fast is each one traveling?

83. *American pie.* John takes 3 hours longer than Andrew to peel 500 pounds (lb) of apples. If together they can peel 500 lb of apples in 8 hours, then how long would it take each one working alone?

84. *On the half shell.* It takes Brent 1 hour longer than Calvin to shuck a sack of oysters. If together they shuck a sack of oysters in 45 minutes, then how long would it take each one working alone?

85. *The growing garden.* Eric's garden is 20 ft by 30 ft. He wants to increase the length and width by the same amount to have a 1000-ft² garden. What should be the new dimensions of the garden?

86. *Open-top box.* Thomas is going to make an open-top box by cutting equal squares from the four corners of an 11 inch by 14 inch sheet of cardboard and folding up the sides. If the area of the base is to be 80 square inches, then what size square should be cut from each corner?

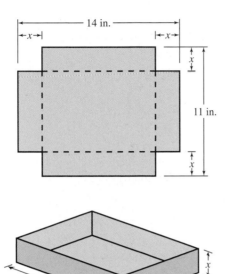

Figure for Exercise 86

87. *Pumping the pool.* It takes pump A 2 hours less time than pump B to empty a certain swimming pool. Pump A is started at 8:00 A.M., and pump B is started at 11:00 A.M. If the pool is still half full at 5:00 P.M., then how long would it take pump A working alone?

88. *Time off for lunch.* It usually takes Eva 3 hours longer to do the monthly payroll than it takes Cicely. They start working on it together at 9:00 A.M. and at 5:00 P.M. they have 90% of it done. If Eva took a 2-hour lunch break while Cicely had none, then how much longer will it take for them to finish the payroll working together?

89. *Golden Rectangle.* One principle used by the ancient Greeks to get shapes that are pleasing to the eye in art and architecture was the Golden Rectangle. If a square is removed from one end of a Golden Rectangle, as shown in the figure, the sides of the remaining rectangle are proportional to the original rectangle. So the length and width of the original rectangle satisfy

$$\frac{L}{W} = \frac{W}{L - W}.$$

If the length of a Golden Rectangle is 10 meters, then what is its width?

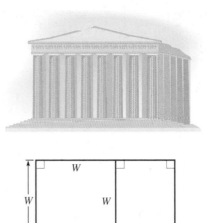

Figure for Exercise 89

90. *Golden painting.* An artist wants her painting to be in the shape of a Golden Rectangle. If the length of the painting is 36 inches, then what should be the width? See Exercise 89.

Getting More Involved

91. *Exploration*

  a) Given that $P(x) = x^4 + 6x^2 - 27$, find $P(3i)$, $P(-3i)$, $P(\sqrt{3})$, and $P(-\sqrt{3})$.

  b) What can you conclude about the values $3i$, $-3i$, $\sqrt{3}$, and $-\sqrt{3}$ and their relationship to each other?

92. *Cooperative learning*

  Work with a group to write a quadratic equation that has each given pair of solutions.

  a) $3 + \sqrt{5}, 3 - \sqrt{5}$ **b)** $4 - 2i, 4 + 2i$

  c) $\dfrac{1 + i\sqrt{3}}{2}, \dfrac{1 - i\sqrt{3}}{2}$

Graphing Calculator Exercises

Solve each equation by locating the x-intercepts on a calculator graph. Round approximate answers to two decimal places.

93. $(5x - 7)^2 - (5x - 7) - 6 = 0$

94. $x^4 - 116x^2 + 1600 = 0$

95. $(x^2 + 3x)^2 - 7(x^2 + 3x) + 9 = 0$

96. $x^2 - 3x^{1/2} - 12 = 0$

Mid-Chapter Quiz | Sections 10.1 through 10.3 | Chapter 10

Solve each equation by factoring.

1. $x^2 - 4x - 32 = 0$

2. $6x^2 - 5x + 1 = 0$

Solve using the even-root property.

3. $x^2 = \dfrac{16}{25}$

4. $(w + 3)^2 = 6$

Solve by completing the square.

5. $x^2 - 4x = 1$

6. $2z^2 + z = 1$

Solve by using the quadratic formula.

7. $2x^2 - 5x + 2 = 0$

8. $2h^2 + 4h + 1 = 0$

Solve by any method.

9. $(x + 7)^2 - 5(x + 7) + 6 = 0$

10. $x^4 - 6x^2 + 5 = 0$

11. $x - 2x^{1/2} - 8 = 0$

12. $\sqrt{2x - 3} = \dfrac{x - 4}{2}$

Miscellaneous.

13. Find the imaginary solutions to $x^2 - 10x + 26 = 0$.

14. Find the discriminant for the equation $3x^2 - x + 5 = 0$.

15. Find a quadratic equation that has -3 and 8 as its solutions.

10.4   Graphing Quadratic Functions

In This Section

⟨1⟩ **Finding Ordered Pairs**

⟨2⟩ **Graphing Quadratic Functions**

⟨3⟩ **The Vertex and Intercepts**

⟨4⟩ **Applications**

An equation of the form $y = mx + b$ is a linear function. Its graph is a straight line. An equation of the form $y = ax^2 + bx + c$ (with $a \neq 0$) is a **quadratic function.** We will see in this section that all quadratic functions have similar graphs that are in the shape of a *parabola.* Note that a linear function is a first-degree polynomial function and a quadratic function is a second-degree polynomial function.

⟨1⟩   Finding Ordered Pairs

It is straightforward to calculate y when given x for an equation of the form $y = ax^2 + bx + c$. However, if we are given y and want to find x, then we must use methods for solving quadratic equations.

EXAMPLE   1

Finding ordered pairs

Complete each ordered pair so that it satisfies the given equation.

    a) $(2, \quad), (\quad, 0), y = x^2 - x - 6$

    b) $(0, \quad), (\quad, 20), y = -16x^2 + 48x + 84$

Solution

    a) If $x = 2$, then $y = 2^2 - 2 - 6 = -4$. So the ordered pair is $(2, -4)$. To find x when $y = 0$, replace y by 0 and solve the resulting quadratic equation:

$$x^2 - x - 6 = 0$$

$$(x - 3)(x + 2) = 0$$

$$x - 3 = 0 \quad \text{or} \quad x + 2 = 0$$
$$x = 3 \quad \text{or} \quad x = -2$$

The ordered pairs are $(-2, 0)$ and $(3, 0)$.

b) If $x = 0$, then $y = -16 \cdot 0^2 + 48 \cdot 0 + 84 = 84$. The ordered pair is $(0, 84)$. To find x when $y = 20$, replace y by 20 and solve the equation for x:

$$-16x^2 + 48x + 84 = 20$$

$$-16x^2 + 48x + 64 = 0 \qquad \text{Subtract 20 from each side.}$$

$$x^2 - 3x - 4 = 0 \qquad \text{Divide each side by } -16.$$

$$(x - 4)(x + 1) = 0 \qquad \text{Factor.}$$

$$x - 4 = 0 \quad \text{or} \quad x + 1 = 0 \qquad \text{Zero factor property}$$

$$x = 4 \quad \text{or} \quad x = -1$$

The ordered pairs are $(-1, 20)$ and $(4, 20)$.

> Now do Exercises 1–4

⟨2⟩ Graphing Quadratic Functions

All equations of the form $y = ax^2 + bx + c$ with $a \neq 0$ have graphs that are similar in shape. The graph of any equation of this form is called a **parabola.** Note that any real number can be used in place of x.

E X A M P L E 2

The simplest parabola

Make a table of ordered pairs that satisfy $y = x^2$, and then sketch the graph of $y = x^2$.

Solution

Make a table of values for x and y:

x	-2	-1	0	1	2
$y = x^2$	4	1	0	1	4

Plot the ordered pairs from the table, and draw a parabola through the points as shown in Fig. 10.3.

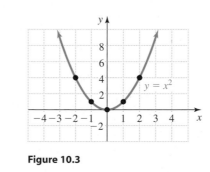

Figure 10.3

⟨ Calculator Close-Up ⟩

This close-up view of $y = x^2$ shows how rounded the curve is at the bottom. When drawing a parabola by hand, be sure to draw it smoothly.

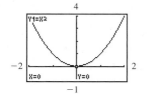

> Now do Exercises 5–14

The parabola in Example 2 is said to **open upward.** In Example 3 we see a parabola that **opens downward.** If $a > 0$ in the equation $y = ax^2 + bx + c$, then the parabola opens upward. If $a < 0$, then the parabola opens downward.

Note the symmetry of the parabola in Fig. 10.3. If the paper was folded along the y-axis, the two sides of the parabola would come together. The point $(-1, 1)$ would match up with $(1, 1)$, the point $(-2, 4)$ would match up with $(2, 4)$, and so on.

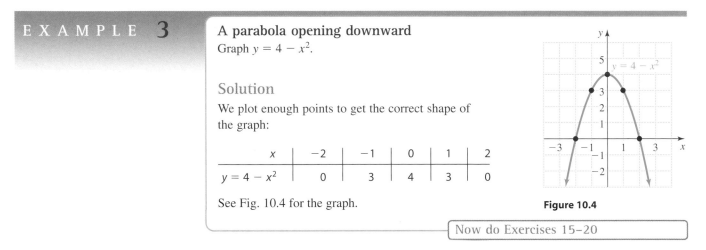

E X A M P L E   3

A parabola opening downward
Graph $y = 4 - x^2$.

Solution

We plot enough points to get the correct shape of the graph:

x	-2	-1	0	1	2
$y = 4 - x^2$	0	3	4	3	0

See Fig. 10.4 for the graph.

Figure 10.4

Now do Exercises 15–20

⟨3⟩ The Vertex and Intercepts

The lowest point on a parabola that opens upward or the highest point on a parabola that opens downward is called the **vertex.** The y-coordinate of the vertex is the **minimum value** of y if the parabola opens upward, and it is the **maximum value** of y if the parabola opens downward. For $y = x^2$ the vertex is $(0, 0)$, and 0 is the minimum value of y. For $y = 4 - x^2$ the vertex is $(0, 4)$, and 4 is the maximum value of y.

If $y = ax^2 + bx + c$ has x-intercepts, they can be found by solving $ax^2 + bx + c = 0$ by the quadratic formula. The vertex is midway between the x-intercepts as shown in Fig. 10.5. Note that in the quadratic formula

$$x = \frac{-b \pm \sqrt{b^2 - 4ac}}{2a},$$

$\sqrt{b^2 - 4ac}$ is added and subtracted from the numerator of $\frac{-b}{2a}$. So $\left(\frac{-b}{2a}, 0\right)$ is the point midway between the x-intercepts and the vertex has the same x-coordinate. Even if the parabola has no x-intercepts, the x-coordinate of the vertex is still $\frac{-b}{2a}$.

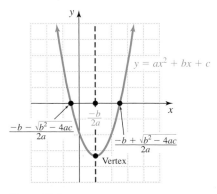

Figure 10.5

> **Vertex of a Parabola**
>
> The x-coordinate of the vertex of $y = ax^2 + bx + c$ is $\dfrac{-b}{2a}$, provided that $a \neq 0$.

When you graph a parabola, you should always locate the vertex because it is the point at which the graph "turns around." With the vertex and several nearby points you can see the correct shape of the parabola.

Using function notation we can write $f(x) = ax^2 + bx + c$ rather than $y = ax^2 + bx + c$. With this notation, the coordinates of the vertex are

$$x = \frac{-b}{2a} \quad \text{and} \quad y = f\left(\frac{-b}{2a}\right).$$

Note that in this context we are thinking of f as the name of the function rather than as a variable. We are keeping x and y as the variables and using the function called f to find y for any given x.

E X A M P L E   **4**

Using the vertex in graphing a parabola

Find the vertex and graph $f(x) = -x^2 - x + 2$.

Solution

First find the x-coordinate of the vertex:

$$x = \frac{-b}{2a} = \frac{-(-1)}{2(-1)} = \frac{1}{-2} = -\frac{1}{2}$$

Now find $f\left(-\dfrac{1}{2}\right)$:

$$f\left(-\frac{1}{2}\right) = -\left(-\frac{1}{2}\right)^2 - \left(-\frac{1}{2}\right) + 2 = -\frac{1}{4} + \frac{1}{2} + 2 = \frac{9}{4}$$

The vertex is $\left(-\dfrac{1}{2}, \dfrac{9}{4}\right)$. Now find a few points on either side of the vertex:

x	-2	-1	$-\dfrac{1}{2}$	0	1
$f(x) = -x^2 - x + 2$	0	2	$\dfrac{9}{4}$	2	0

Sketch a parabola through these points as in Fig. 10.6.

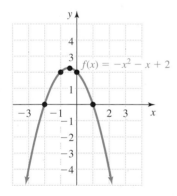

$f(x) = -x^2 - x + 2$

Figure 10.6

Now do Exercises 21–28

The y-intercept for the parabola $y = ax^2 + bx + c$ is the point that has 0 as its x-coordinate. If we let $x = 0$, we get $y = a(0)^2 + b(0) + c = c$. So the y-intercept is $(0, c)$. To find the x-intercepts let $y = 0$ and solve $ax^2 + bx + c = 0$. A parabola may have 0, 1, or 2 x-intercepts depending on the number of solutions to this equation.

> **Finding Intercepts**
>
> The y-intercept for $y = ax^2 + bx + c$ is $(c, 0)$.
> To find the x-intercepts solve $ax^2 + bx + c = 0$.

EXAMPLE **5**

Using the intercepts in graphing a parabola

Find the vertex and intercepts, and sketch the graph of each parabola.

a) $f(x) = x^2 - 2x - 8$

b) $s = -16t^2 + 64t$

Solution

a) Use $x = \frac{-b}{2a}$ to get $x = 1$ as the x-coordinate of the vertex. If $x = 1$, then

$$f(1) = 1^2 - 2 \cdot 1 - 8$$
$$= -9.$$

So the vertex is $(1, -9)$. If $x = 0$, then

$$f(0) = 0^2 - 2 \cdot 0 - 8$$
$$= -8.$$

The y-intercept is $(0, -8)$. To find the x-intercepts, replace $f(x)$ by 0:

$$x^2 - 2x - 8 = 0$$
$$(x - 4)(x + 2) = 0$$
$$x - 4 = 0 \quad \text{or} \quad x + 2 = 0$$
$$x = 4 \quad \text{or} \quad x = -2$$

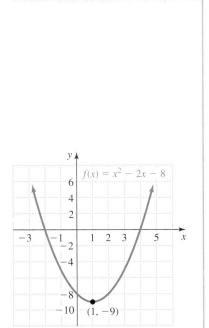

Figure 10.7

The x-intercepts are $(-2, 0)$ and $(4, 0)$. The graph is shown in Fig. 10.7.

b) Because s is expressed in terms of t in the equation $s = -16t^2 + 64t$, the independent variable is t and the dependent variable is s. Since we always put the independent variable first in an ordered pair, the ordered pairs are written in the form (t, s). To find the vertex use $t = -\frac{b}{2a}$ to get

$$t = \frac{-64}{2(-16)} = 2.$$

If $t = 2$, then

$$s = -16 \cdot 2^2 + 64 \cdot 2$$
$$= 64.$$

So the vertex is $(2, 64)$. If $t = 0$, then

$$s = -16 \cdot 0^2 + 64 \cdot 0$$
$$= 0.$$

So the s-intercept is $(0, 0)$. To find the t-intercepts, replace s by 0:

$$-16t^2 + 64t = 0$$
$$-16t(t - 4) = 0$$
$$-16t = 0 \quad \text{or} \quad t - 4 = 0$$
$$t = 0 \quad \text{or} \quad t = 4$$

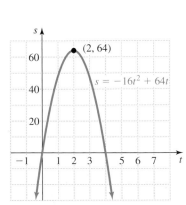

Figure 10.8

The t-intercepts are $(0, 0)$ and $(4, 0)$. The graph is shown in Fig. 10.8.

Now do Exercises 29–44

‹ **Calculator Close-Up** ›

You can find the vertex of a parabola with a calculator by using either the maximum or minimum feature. First graph the parabola as shown.

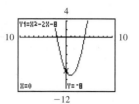

Because this parabola opens upward, the y-coordinate of the vertex is the minimum

y-coordinate on the graph. Press CALC and choose minimum.

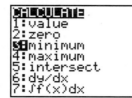

The calculator will ask for a left bound, a right bound, and a guess. For the left bound choose a point to the left of the vertex by

moving the cursor to the point and pressing ENTER. For the right bound choose a point to the right of the vertex. For the guess choose a point close to the vertex.

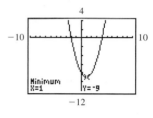

‹4› Applications

In applications we are often interested in finding the maximum or minimum value of a variable. If the graph of a parabola opens downward, then the maximum value of the dependent variable is the second coordinate of the vertex. If the parabola opens upward, then the minimum value of the dependent variable is the second coordinate of the vertex.

E X A M P L E 6

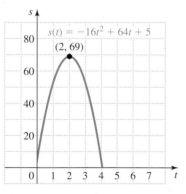

Figure 10.9

Finding the maximum height

If a projectile is launched with an initial velocity of v_0 feet per second from an initial height of s_0 feet, then its height $s(t)$ in feet is determined by $s(t) = -16t^2 + v_0t + s_0$, where t is the time in seconds. If a ball is tossed upward with velocity 64 feet per second from a height of 5 feet, then what is the maximum height reached by the ball?

Solution

The height $s(t)$ of the ball for any time t is given by $s(t) = -16t^2 + 64t + 5$. Because the maximum height occurs at the vertex of the parabola, we use $t = \frac{-b}{2a}$ to find the vertex:

$$t = \frac{-64}{2(-16)} = 2$$

Now use $t = 2$ to find the second coordinate of the vertex:

$$s(2) = -16(2)^2 + 64(2) + 5 = 69$$

The maximum height reached by the ball is 69 feet. See Fig. 10.9.

> Now do Exercises 53–61

Note that the graph in Fig. 10.9 shows the height of the ball as a function of time. It does not show the path of the ball. The ball in Example 6 is tossed straight upward and falls straight downward. The path of a projectile is discussed in trigonometry. It turns out that a ball thrown nonvertically travels through the air on a parabolic path, which depends on the velocity and the angle at which it is thrown.

Warm-Ups ▼

Fill in the blank.

1. The graph of $y = ax^2 + bx + c$ with $a \neq 0$ is a _____.
2. The graph of $y = ax^2 + bx + c$ with $a > 0$ opens _____.
3. The graph of $y = ax^2 + bx + c$ with $a < 0$ opens _____.
4. The _____ is the highest point on a parabola that opens downward or the lowest point on a parabola that opens upward.
5. The x-coordinate of the vertex for $y = ax^2 + bx + c$ is _____.
6. To find the _____ of the vertex, evaluate $y = ax^2 + bx + c$ with $x = -b/(2a)$.

True or false?

7. The ordered pair $(-2, -1)$ satisfies $f(x) = x^2 - 5$.
8. The y-intercept for $g(x) = x^2 - 3x + 9$ is $(9, 0)$.
9. The x-intercepts for $y = x^2 - 5$ are $\left(\sqrt{5}, 0\right)$ and $\left(-\sqrt{5}, 0\right)$.
10. The graph of $f(x) = x^2 - 12$ opens upward.
11. The graph of $y = 4 + x^2$ opens downward.
12. The parabola $y = x^2 + 1$ has no x-intercepts.
13. The y-intercept for $g(x) = ax^2 + bx + c$ is $(0, c)$.
14. If $w = -2t^2 + 9$, then the maximum value of w is 9.

10.4 Exercises

‹ Study Tips ›

- Be sure to ask your instructor what to expect on the final exam. Will it be the same format as other tests?
- If there are any sample final exams available, use them as a guide for your studying.

‹1› Finding Ordered Pairs

Complete each ordered pair so that it satisfies the given equation. See Example 1.

1. $y = x^2 - x - 12$ $(3, \quad), (\quad, 0)$

2. $y = -\dfrac{1}{2}x^2 - x + 1$ $(0, \quad), (\quad, -3)$

3. $y = -16x^2 + 32x$ $(4, \quad), (\quad, 0)$

4. $y = x^2 + 4x + 5$ $(-2, \quad), (\quad, 2)$

‹2› Graphing Quadratic Functions

Determine whether the graph of each parabola opens upward or downward. See Examples 2 and 3.

5. $y = x^2 + 5$

6. $y = 2x^2 + x - 1$

7. $y = -3x^2 + 4x + 2$

8. $y = -x^2 + 3$

9. $y = (-2x + 3)^2$

10. $y = (5 - x)^2$

Graph each parabola. See Examples 2 and 3.

11. $y = x^2 + 2$

12. $y = x^2 - 4$

13. $y = \dfrac{1}{2}x^2 - 4$

14. $y = \dfrac{1}{3}x^2 - 6$

15. $y = -2x^2 + 5$

16. $y = -x^2 - 1$

17. $y = -\dfrac{1}{3}x^2 + 5$

18. $y = -\dfrac{1}{2}x^2 + 3$

19. $y = (x - 2)^2$

20. $y = (x + 3)^2$

⟨3⟩ **The Vertex and Intercepts**

Find the vertex for the graph of each parabola. See Example 4.

21. $f(x) = x^2 - 9$ **22.** $f(x) = x^2 + 12$

23. $y = x^2 - 4x + 1$ **24.** $y = x^2 + 8x - 3$

25. $f(x) = -2x^2 + 20x + 1$ **26.** $f(x) = -3x^2 + 18x - 7$

27. $y = x^2 - x + 1$ **28.** $y = 3x^2 - 2x + 1$

Find all intercepts for the graph of each parabola. See Example 5.

29. $f(x) = 16 - x^2$ **30.** $f(x) = x^2 - 9$

31. $y = x^2 - 2x - 15$ **32.** $y = x^2 - x - 6$

33. $f(x) = -4x^2 + 12x - 9$ **34.** $f(x) = -2x^2 - x + 3$

Find the vertex and intercepts for each parabola. Sketch the graph. See Examples 4 and 5.

35. $f(x) = x^2 - x - 2$

36. $f(x) = x^2 + 2x - 3$

37. $g(x) = x^2 + 2x - 8$

Find the maximum or minimum value of y.

45. $y = x^2 - 8$ **46.** $y = 33 - x^2$

47. $y = -3x^2 + 14$ **48.** $y = 6 + 5x^2$

49. $y = x^2 + 2x + 3$ **50.** $y = x^2 - 2x + 5$

38. $g(x) = x^2 + x - 6$

51. $y = -2x^2 - 4x$ **52.** $y = -3x^2 + 24x$
VIDEO

⟨**4**⟩ **Applications**

Solve each problem. See Example 6.

53. *Maximum height.* If a baseball is projected upward
VIDEO   from ground level with an initial velocity of 64 feet per
second, then its height in feet is given by the function

39. $y = -x^2 - 4x - 3$

$$s(t) = -16t^2 + 64t$$

where *t* is time in seconds. Graph this parabola for $0 \le t \le 4$.
What is the maximum height reached by the ball?

40. $y = -x^2 - 5x - 4$

54. *Maximum height.* If a soccer ball is kicked straight up
from the ground with an initial velocity of 32 feet per
second, then its height above the earth in feet is given by the
function $s(t) = -16t^2 + 32t$ where *t* is time in seconds.
Graph this parabola for $0 \le t \le 2$. What is the maximum
height reached by the ball?

41. $h(x) = -x^2 + 3x + 4$

42. $h(x) = -x^2 - 2x + 8$ **43.** $a = b^2 - 6b - 16$

55. *Minimum cost.* It costs Acme Manufacturing *C* dollars
per hour to operate its golf ball division. An analyst has
determined that *C* is related to the number of golf balls
produced per hour, *x*, by the function $C = 0.009x^2 -$
$1.8x + 100$. What number of balls per hour should Acme
44. $v = -u^2 - 8u + 9$ produce to minimize the cost per hour of manufacturing
these golf balls?

56. *Maximum profit.* A chain store manager has been told by
the main office that daily profit, *P*, is related to the number
of clerks working that day, *x*, according to the function

$P = -25x^2 + 300x$. What number of clerks will maximize the profit, and what is the maximum possible profit?

57. Maximum area. Jason plans to fence a rectangular area with 100 meters of fencing. He has written the function $A = w(50 - w)$ to express the area in terms of the width w. What is the maximum possible area that he can enclose with his fencing?

Photo for Exercise 57

58. Minimizing cost. A company uses the function $C(x) = 0.02x^2 - 3.4x + 150$ to model the unit cost in dollars for producing x stabilizer bars. For what number of bars is the unit cost at its minimum? What is the unit cost at that level of production?

59. Air pollution. The amount of nitrogen dioxide A in parts per million (ppm) that was present in the air in the city of Homer on a certain day in June is modeled by the function

$$A(t) = -2t^2 + 32t + 12,$$

where t is the number of hours after 6:00 A.M. Use this function to find the time at which the nitrogen dioxide level was at its maximum.

60. Stabilization ratio. The stabilization ratio (births/deaths) for South and Central America can be modeled by the function

$$y = -0.0012x^2 + 0.074x + 2.69,$$

where y is the number of births divided by the number of deaths in the year 1950 + x (World Resources Institute, www.wri.org).

a) Use the graph to estimate the year in which the stabilization ratio was at its maximum.
b) Use the function to find the year in which the stabilization ratio was at its maximum.

c) What was the maximum stabilization ratio from part (b)?
d) What is the significance of a stabilization ratio of 1?

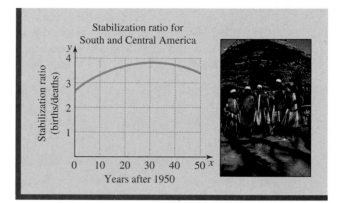

Figure for Exercise 60

61. Suspension bridge. The cable of the suspension bridge shown in the figure hangs in the shape of a parabola with equation $y = 0.0375x^2$, where x and y are in meters. What is the height of each tower above the roadway? What is the length z for the cable bracing the tower?

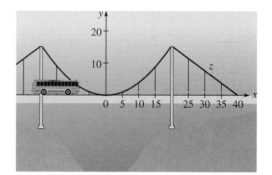

Figure for Exercise 61

Getting More Involved

62. Exploration

a) Write the equation $y = 3(x - 2)^2 + 6$ in the form $y = ax^2 + bx + c$, and find the vertex of the parabola using the formula $x = \frac{-b}{2a}$.
b) Repeat part (a) with the equations $y = -4(x - 5)^2 - 9$ and $y = 3(x + 2)^2 - 6$.
c) What is the vertex for a parabola that is written in the form $y = a(x - h)^2 + k$? Explain your answer.

Graphing Calculator Exercises

63. Graph $y = x^2$, $y = \frac{1}{2}x^2$, and $y = 2x^2$ on the same coordinate system. What can you say about the graph of $y = ax^2$ for $a > 0$?

64. Graph $y = x^2$, $y = (x - 3)^2$, and $y = (x + 3)^2$ on the same coordinate system. How does the graph of $y = (x - h)^2$ compare to the graph of $y = x^2$?

65. The equation $x = y^2$ is equivalent to $y = \pm\sqrt{x}$. Graph both $y = \sqrt{x}$ and $y = -\sqrt{x}$ on a graphing calculator. How does the graph of $x = y^2$ compare to the graph of $y = x^2$?

66. Graph each of the following equations by solving for y.
  a) $x = y^2 - 1$ **b)** $x = -y^2$

  c) $x^2 + y^2 = 4$

67. Graph each parabola using a viewing window that contains the vertex and all intercepts. Answers may vary.
  a) $y = 100x^2 - 30x + 2$

  b) $y = x^2 - 110x + 3000$

  c) $y = 999x - 10 - 10x^2$

68. Determine the approximate vertex and x-intercepts for each parabola.
  a) $y = 3.2x^2 - 5.4x + 1.6$
  b) $y = -1.09x^2 + 13x + 7.5$

10.5   Quadratic Inequalities

In This Section

⟨1⟩ Solving Quadratic Inequalities Graphically

⟨2⟩ Solving Quadratic Inequalities with the Test-Point Method

⟨3⟩ Applications

In this section, we solve inequalities involving quadratic polynomials. We use two methods, which are based on the graphs of the corresponding quadratic functions.

⟨1⟩ Solving Quadratic Inequalities Graphically

An inequality involving a quadratic polynomial is called a *quadratic inequality*.

> **Quadratic Inequality**
>
> A **quadratic inequality** has one of the following forms:
>
> $$ax^2 + bx + c > 0, \qquad\qquad ax^2 + bx + c \geq 0,$$
> $$ax^2 + bx + c < 0, \qquad\qquad ax^2 + bx + c \leq 0,$$
>
> where a, b, and c are real numbers with $a \neq 0$.

To solve $ax^2 + bx + c > 0$ we can examine the graph of the corresponding quadratic function $y = ax^2 + bx + c$. The values of x that satisfy the inequality are the same as the values of x for which $y > 0$ on the graph of $y = ax^2 + bx + c$. We use the following strategy for the **graphical method.**

Strategy for the Graphical Method

1. Rewrite the inequality (if necessary) so that 0 is on the right side and a quadratic polynomial is on the left side.
2. Find the roots to the quadratic polynomial.
3. Plot the x-intercepts using the roots found in step 2, and graph the parabola passing through the x-intercepts.
4. Read the solution set to the inequality from the graph.

E X A M P L E   1

Solving quadratic inequalities graphically

Solve each quadratic inequality. Write the solution set in interval notation and graph it.

a) $x^2 + 3x > 10$ **b)** $x^2 - 2x - 1 \leq 0$ **c)** $-x^2 + 3 \geq 0$

Solution

a) Rewrite the inequality as $x^2 + 3x - 10 > 0$. Then find the roots to the quadratic polynomial:

$$x^2 + 3x - 10 = 0$$
$$(x + 5)(x - 2) = 0$$
$$x + 5 = 0 \qquad \text{or} \qquad x - 2 = 0$$
$$x = -5 \qquad \text{or} \qquad x = 2$$

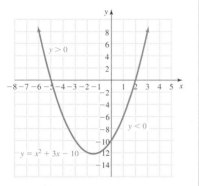

Figure 10.10

The graph of $y = x^2 + 3x - 10$ is a parabola that opens upward with x-intercepts at $(-5, 0)$ and $(2, 0)$ as shown in Fig. 10.10. The y-coordinates on the parabola are negative between the intercepts and positive to the left and right of the intercepts. Since $y = x^2 + 3x - 10$, whenever y is positive $x^2 + 3x - 10$ is positive and the inequality is satisfied. So the solution set to the inequality is $(-\infty, -5) \cup (2, \infty)$. The graph of the solution set is shown in Fig. 10.11.

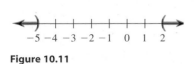

Figure 10.11

b) Find the roots to the quadratic polynomial using the quadratic formula:

$$x^2 - 2x - 1 = 0$$
$$x = \frac{-(-2) \pm \sqrt{(-2)^2 - 4(1)(-1)}}{2(1)}$$
$$= \frac{2 \pm \sqrt{8}}{2} = \frac{2 \pm 2\sqrt{2}}{2} = 1 \pm \sqrt{2}$$

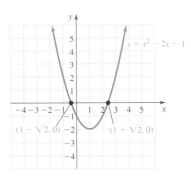

Figure 10.12

Figure 10.13

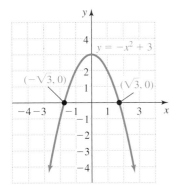

Figure 10.14

The graph of $y = x^2 - 2x - 1$ is a parabola that opens upward with x-intercepts at $(1 - \sqrt{2}, 0)$ and $(1 + \sqrt{2}, 0)$ as shown in Fig. 10.12. The y-coordinates on the parabola are negative between the intercepts, positive to the left and right of the intercepts, and zero at the intercepts. Because the inequality symbol is $\leq$, the solution set includes the roots to the polynomial. So the solution set to the inequality is $\left[1 - \sqrt{2}, 1 + \sqrt{2}\right]$. The graph of the solution set is shown in Fig. 10.13.

c) Find the roots to the quadratic polynomial:

$$-x^2 + 3 = 0$$
$$-x^2 = -3$$
$$x^2 = 3$$
$$x = \pm\sqrt{3}$$

The graph of $y = -x^2 + 3$ is a parabola that opens downward with x-intercepts at $(-\sqrt{3}, 0)$ and $(\sqrt{3}, 0)$ as shown in Fig. 10.14. The y-coordinates on the parabola are greater than or equal to zero whenever x is in the interval $\left[-\sqrt{3}, \sqrt{3}\right]$, and that is the solution set to $-x^2 + 3 \geq 0$. The graph of the solution set is shown in Fig. 10.15.

$$\begin{array}{c} \longleftarrow \quad [\underline{\quad\quad\quad}] \quad \longrightarrow \\ -\sqrt{3} \qquad \sqrt{3} \end{array}$$

Figure 10.15

Now do Exercises 1–18

The graphs of the corresponding quadratic polynomials in Example 1 all had two x-intercepts. In Example 2 the graphs have fewer than two x-intercepts.

E X A M P L E   2

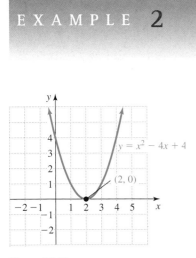

Figure 10.16

Solving quadratic inequalities graphically

Solve each quadratic inequality. Write the solution set in interval notation and graph it.

a) $x^2 - 4x + 4 \leq 0$ **b)** $x^2 + 2x + 3 > 0$ **c)** $-x^2 - 4 \geq 0$

Solution

a) Find the roots to the quadratic polynomial:

$$x^2 - 4x + 4 = 0$$
$$(x - 2)^2 = 0$$
$$x - 2 = 0$$
$$x = 2$$

The graph of $y = x^2 - 4x + 4$ is a parabola that opens upward with an x-intercept at $(2, 0)$ as shown in Fig. 10.16. The y-coordinates on the parabola are positive except when $x = 2$. At $x = 2$ the y-coordinate is zero. So there is only one value for x that satisfies $x^2 - 4x + 4 \leq 0$, and that is $x = 2$. So the solution set to the inequality is $\{2\}$. The graph of the solution set is shown in Fig. 10.17.

Figure 10.17

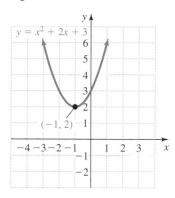

Figure 10.18

Figure 10.19

b) Find the roots to the quadratic polynomial using the quadratic formula:

$$x^2 + 2x + 3 = 0$$

$$x = \frac{-2 \pm \sqrt{2^2 - 4(1)(3)}}{2(1)} = \frac{-2 \pm \sqrt{-8}}{2}$$

Since the radical contains a negative number, there are no real solutions to the equation and no x-intercepts. The graph of $y = x^2 + 2x + 3$ is a parabola that opens upward from its vertex $(-1, 2)$ as shown in Fig. 10.18. Since all y-coordinates on this graph are positive for any value of x, the solution set to $x^2 + 2x + 3 > 0$ is the set of all real numbers, $(-\infty, \infty)$. The graph of the solution set is shown in Fig. 10.19.

c) Find the roots to the quadratic polynomial:

$$-x^2 - 4 = 0$$
$$-x^2 = 4$$
$$x^2 = -4$$
$$x = \pm\sqrt{-4} = \pm 2i$$

Since there are no real solutions to this equation, there are no x-intercepts for the graph of $y = -x^2 - 4$. The graph of $y = -x^2 - 4$ opens downward from its vertex $(0, -4)$ as shown in Fig. 10.20. The y-coordinates on the parabola are negative for every value of x. So there are no values of x that would make $-x^2 - 4 \geq 0$ and the solution set for the inequality is the empty set, $\varnothing$.

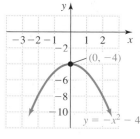

Figure 10.20

Now do Exercises 19–30

⟨2⟩ Solving Quadratic Inequalities with the Test-Point Method

The **test-point method** is a variation of the graphical method, but we don't graph the parabola. We have seen that the y-coordinates on a parabola can change sign only at an x-intercept. So we find the x-intercepts (if there are any) and then test points in the intervals determined by the intercepts to see if they satisfy the inequality. Here is the strategy.

Strategy for the Test-Point Method

1. Rewrite the inequality (if necessary) with 0 on the right.
2. Solve the quadratic equation that results from replacing the inequality symbol with the equals symbol.
3. Locate the solutions to the quadratic equation on a number line.
4. Select one test point in each interval determined by the solutions.
5. Check to see whether each test point satisfies the original inequality.
6. Write the solution set using interval notation.

EXAMPLE **3**

Solving quadratic inequalities with the test-point method

Solve each quadratic inequality. Write the solution set in interval notation and graph it.

a) $x^2 - x - 6 \leq 0$ **b)** $x^2 - 4x > 6$ **c)** $x^2 + 6x + 10 \geq 0$

Solution

a) We can solve the quadratic equation $x^2 - x - 6 = 0$ by factoring:

$$(x - 3)(x + 2) = 0$$
$$x - 3 = 0 \quad \text{or} \quad x + 2 = 0$$
$$x = 3 \quad \text{or} \quad x = -2$$

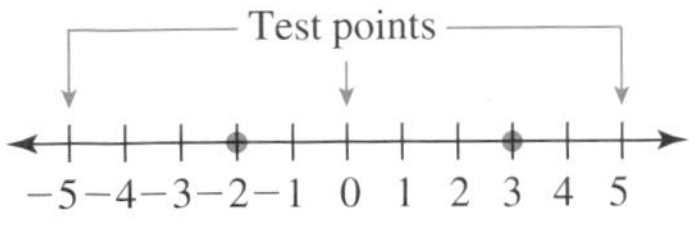

Locate -2 and 3 on the number line, and select three test points as shown in Fig. 10.21. We have chosen the points $-5, 0,$ and 5. Now test $-5, 0,$ and 5 in the original inequality $x^2 - x - 6 \leq 0$:

$$(-5)^2 - (-5) - 6 \leq 0 \qquad \text{Incorrect}$$
$$0^2 - 0 - 6 \leq 0 \qquad \text{Correct}$$
$$5^2 - 5 - 6 \leq 0 \qquad \text{Incorrect}$$

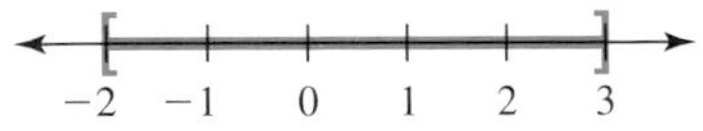

Figure 10.21

Of the three test points, only 0 satisfies the inequality. So the solution set is the interval containing 0. Since the inequality symbol includes equality, -2 and 3 are included in the solution set $[-2, 3]$. The graph of the solution set is shown in Fig. 10.22.

Figure 10.22

b) First rewrite the inequality as $x^2 - 4x - 6 > 0$. Then solve $x^2 - 4x - 6 = 0$ using the quadratic formula:

$$x = \frac{-(-4) \pm \sqrt{(-4)^2 - 4(1)(-6)}}{2(1)} = \frac{4 \pm \sqrt{40}}{2} = 2 \pm \sqrt{10}$$

Now $2 - \sqrt{10} \approx -1.2$ and $2 + \sqrt{10} \approx 5.2$. Plot these points on a number line, and select three test points as shown in Fig. 10.23. We have chosen the points $-2, 0,$ and 7. Now test $-2, 0,$ and 7 in the original inequality $x^2 - 4x > 6$:

$$(-2)^2 - 4(-2) > 6 \qquad \text{Correct}$$
$$0^2 - 4(0) > 6 \qquad \text{Incorrect}$$
$$7^2 - 4(7) > 6 \qquad \text{Correct}$$

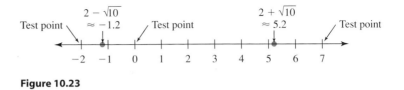

Figure 10.23

So the inequality is satisfied on the intervals containing -2 and 7. The solution set is $\left(-\infty, 2 - \sqrt{10}\right) \cup \left(2 + \sqrt{10}, \infty\right)$, and its graph is shown in Fig. 10.24.

Figure 10.24

c) Solve $x^2 + 6x + 10 = 0$ using the quadratic formula:

$$x = \frac{-6 \pm \sqrt{6^2 - 4(1)(10)}}{2(1)} = \frac{-6 \pm \sqrt{-4}}{2}$$

Since there is a negative number inside the radical, there are no real solutions to the equation. So there are no points to plot on the number line. There is just one interval to consider, and that is $(-\infty, \infty)$. We can select any point in $(-\infty, \infty)$ as a test point. Let's try 2:

$$2^2 + 6(2) + 10 \geq 0 \qquad \text{Correct}$$

Since the inequality is satisfied for $x = 2$, it is satisfied for every real number and the solution set is $(-\infty, \infty)$. The graph is shown in Fig. 10.25.

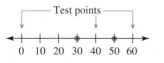

Figure 10.25

| Now do Exercises 31–44 |

If there are no solutions to the quadratic equation, then the quadratic polynomial does not change sign. The solution set is either all real numbers or no real numbers. If the inequality in Example 3(c) was not satisfied when $x = 2$, then the solution set would have been the empty set. In fact, the solution set to $x^2 + 6x + 10 \leq 0$ is the empty set, $\varnothing$.

⟨3⟩ Applications

Example 4 shows how a quadratic inequality can be used to solve a problem.

E X A M P L E 4

Making a profit

Charlene's daily profit P (in dollars) for selling x magazine subscriptions is determined by the formula

$$P = -x^2 + 80x - 1500.$$

For what values of x is her profit positive?

Solution

The profit is positive whenever $-x^2 + 80x - 1500 > 0$. Find the solutions to the corresponding quadratic equation:

$$-x^2 + 80x - 1500 = 0$$
$$x^2 - 80x + 1500 = 0$$
$$(x - 30)(x - 50) = 0$$
$$x - 30 = 0 \qquad \text{or} \qquad x - 50 = 0$$
$$x = 30 \qquad \text{or} \qquad x = 50$$

Figure 10.26

Locate 30 and 50 on a number line as shown in Fig. 10.26, and select 0, 40, and 60 as test points. Check the test points in the original inequality $-x^2 + 80x - 1500 > 0$:

$$-0^2 + 80(0) - 1500 > 0 \qquad \text{Incorrect}$$
$$-40^2 + 80(40) - 1500 > 0 \qquad \text{Correct}$$
$$-60^2 + 80(60) - 1500 > 0 \qquad \text{Incorrect}$$

Since 40 satisfies the inequality, every point between 30 and 50 also satisfies the inequality. So for a positive profit, she must sell between 30 and 50 magazine subscriptions.

| Now do Exercises 65–70 |

Warm-Ups ▼

Fill in the blank.

1. A _____ inequality has the form $ax^2 - bx + c > 0$.
2. A quadratic inequality can be solved by the _____ method or the _____ method.

True or false?

3. The solution set to $x^2 > 4$ is $(2, \infty)$.
4. To solve $x^2 + x - 2 < 0$ by graphing, we graph $y = x^2 + x - 2$.

5. In solving quadratic inequalities, we must get 0 on one side.
6. To solve $(x - 3)(x + 5) \geq 0$ using test points, the test points could be -5, 0, and 3.
7. We can't solve quadratic inequalities that do not factor.

8. The parabola $y = x^2 + 4$ has no x-intercepts.

10.5

Exercises

⟨ **Study Tips** ⟩

- Keep track of your time for one entire week. Account for every half hour.
- You should be sleeping 50 to 60 hours per week and studying 1 to 2 hours for every credit hour you are taking. For a 3-credit-hour class, you should be studying 3 to 6 hours per week.

⟨1⟩ **Solving Quadratic Inequalities Graphically**

Use the graphical method to solve each inequality. State the solution set using interval notation and graph it. See Example 1. See the Strategy for the Graphical Method on page 669.

1. $x^2 + x - 6 < 0$

2. $x^2 - 3x - 4 \geq 0$

3. $z^2 - 16 < 0$

4. $y^2 - 4 > 0$

5. $x^2 - 2x - 8 \leq 0$

6. $x^2 + x - 12 \leq 0$

7. $2u^2 + 5u \geq 12$

8. $2v^2 + 7v < 4$

9. $4x^2 - 8x \geq 0$

10. $x^2 + x > 0$

11. $5x - 10x^2 < 0$

12. $3x - x^2 > 0$

13. $x^2 - 5 > 0$

25. $25x^2 + 10x + 1 > 0$

26. $16x^2 - 16x + 4 > 0$

14. $x^2 - 3 < 0$

27. $x^2 + 5x + 12 \geq 0$

28. $x^2 + 3x + 9 > 0$

15. $x^2 - 2x - 5 \leq 0$

29. $2x^2 + 5x + 5 < 0$

30. $-3x^2 + x - 6 \geq 0$

16. $x^2 - 2x - 4 > 0$

⟨2⟩ Solving Quadratic Inequalities with the Test-Point Method

Use the test-point method to solve each inequality. State the solution set using interval notation and graph it. See Example 3. See the Strategy for the Test-Point Method on page 671.

17. $2x^2 - 6x + 3 > 0$

31. $x^2 - 4x - 12 > 0$

18. $2x^2 - 8x + 3 < 0$

32. $x^2 + 7x - 18 \leq 0$

33. $x^2 + 3x < 40$

Use the graphical method to solve each inequality. State the solution set using interval notation and graph it. See Example 2.

34. $x^2 - 15x > 16$

19. $x^2 + 6x + 9 \geq 0$

20. $x^2 + 10x + 25 \geq 0$

35. $x^2 + 8x + 17 \geq 0$

21. $x^2 + 4 < 4x$

22. $x^2 < 8x - 16$

36. $x^2 + 10x + 27 \leq 0$

23. $4x^2 - 20x + 25 \leq 0$

37. $9x - 4x^2 > x$

24. $9x^2 + 12x + 4 \leq 0$

38. $5x - x^2 \leq x$

39. $x^2 + 4 > 6x$

40. $x^2 - 2 \le 4x$

41. $-5x^2 + 2x \le 4$

42. $3x - 5 \le 3x^2$

43. $y^2 - 3y - 9 \le 0$

44. $z^2 - 5z - 7 < 0$

Miscellaneous

Solve each inequality. State the solution set using interval notation when possible.

45. $x^2 > 0$

46. $x^2 \ge 0$

47. $x^2 + 4 \ge 0$

48. $x^2 + 1 \le 0$

49. $x^2 \le 9$

50. $x^2 \ge 36$

51. $16 - x^2 > 0$

52. $9 - x^2 < 0$

53. $x^2 - 4x \ge 0$

54. $4x^2 - 9 > 0$

55. $3(2w^2 - 5) < w$

56. $6(y^2 - 2) + y < 0$

57. $z^2 \ge 4(z + 3)$

58. $t^2 < 3(2t - 3)$

59. $(q + 4)^2 > 10q + 31$

60. $(2p + 4)(p - 1) < (p + 2)^2$

61. $\frac{1}{2}x^2 \ge 4 - x$

62. $\frac{1}{2}x^2 \le x + 12$

63. $0.23x^2 + 6.5x + 4.3 < 0$

64. $0.65x^2 + 3.2x + 5.1 > 0$

⟨3⟩ **Applications**

Solve each problem by using a quadratic inequality. See Example 4.

65. *Positive profit.* The monthly profit P (in dollars) that Big Jim makes on the sale of x mobile homes is determined by the formula $P = x^2 + 5x - 50$. For what values of x is his profit positive?

66. *Profitable fruitcakes.* Sharon's revenue R (in dollars) on the sale of x fruitcakes is determined by the formula $R = 50x - x^2$. Her cost C (in dollars) for producing x fruitcakes is given by the formula $C = 2x + 40$. For what values of x is Sharon's profit positive? (Profit = revenue − cost.)

If an object is given an initial velocity straight upward of v_0 feet per second from a height of s_0 feet, then its altitude S after t seconds is given by the formula

$$S = -16t^2 + v_0 t + s_0.$$

67. *Flying high.* An arrow is shot straight upward with a velocity of 96 feet per second (ft/sec) from an altitude of 6 feet. For how many seconds is this arrow more than 86 feet high?

68. *Putting the shot.* In 1978 Udo Beyer (East Germany) set a world record in the shot-put of 72 ft 8 in. If Beyer had projected the shot straight upward with a velocity of 30 ft/sec from a height of 5 ft, then for what values of t would the shot be under 15 ft high?

If a projectile is fired at a 45° angle from a height of s_0 feet with initial velocity v_0 ft/sec, then its altitude S in feet after t seconds is given by

$$S = -16t^2 + \frac{v_0}{\sqrt{2}}t + s_0.$$

69. ***Siege and garrison artillery.*** An 8-inch mortar used in the Civil War fired a 44.5-lb projectile from ground level a distance of 3600 ft when aimed at a 45° angle (Harold R. Peterson, *Notes on Ordinance of the American Civil War*). The accompanying graph shows the altitude of the projectile when it is fired with a velocity of $240\sqrt{2}$ ft/sec.

a) Use the graph to estimate the maximum altitude reached by the projectile.

b) Use the graph to estimate approximately how long the altitude of the projectile was greater than 864 ft.

c) Use the formula to determine the length of time for which the projectile had an altitude of more than 864 ft.

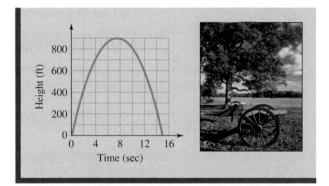

Figure for Exercise 69

 **70.** ***Seacoast artillery.*** The 13-inch mortar used in the Civil War fired a 220-lb projectile a distance of 12,975 ft when aimed at a 45° angle. If the 13-inch mortar was fired from a hill 100 ft above sea level with an initial velocity of 644 ft/sec, then for how long was the projectile more than 800 ft above sea level?

Figure for Exercise 70

Getting More Involved

71. ***Cooperative learning***

Work in a small group to solve each inequality for x, given that h and k are real numbers with $h < k$.

a) $(x - h)(x - k) < 0$

b) $(x - h)(x - k) > 0$

c) $(x + h)(x + k) < 0$

d) $(x + h)(x + k) \geq 0$

72. ***Cooperative learning***

Work in a small group to solve $ax^2 + bx + c > 0$ for x in each case.

a) $b^2 - 4ac = 0$ and $a > 0$

b) $b^2 - 4ac = 0$ and $a < 0$

c) $b^2 - 4ac < 0$ and $a > 0$

d) $b^2 - 4ac < 0$ and $a < 0$

e) $b^2 - 4ac > 0$ and $a > 0$

f) $b^2 - 4ac > 0$ and $a < 0$

10

Chapter

Wrap-Up

Summary

Quadratic Equations		Examples
Quadratic equation	An equation of the form $ax^2 + bx + c = 0$, where a, b, and c are real numbers, with $a \neq 0$	$x^2 = 11$ $(x - 5)^2 = 99$ $x^2 + 3x - 20 = 0$
Methods for solving quadratic equations	Factoring: Factor the quadratic polynomial, and then set each factor equal to 0.	$x^2 + x - 6 = 0$ $(x + 3)(x - 2) = 0$ $x + 3 = 0$ or $x - 2 = 0$
	The even-root property: If $x^2 = k$ ($k > 0$), then $x = \pm\sqrt{k}$. If $x^2 = 0$, then $x = 0$. There are no real solutions to $x^2 = k$ for $k < 0$.	$(x - 5)^2 = 10$ $x - 5 = \pm\sqrt{10}$
	Completing the square: Take one-half of middle term, square it, and then add it to each side.	$x^2 + 6x = -4$ $x^2 + 6x + 9 = -4 + 9$ $(x + 3)^2 = 5$
	Quadratic formula: If $ax^2 + bx + c = 0$ with $a \neq 0$, then $$x = \frac{-b \pm \sqrt{b^2 - 4ac}}{2a}.$$	$2x^2 + 3x - 5 = 0$ $$x = \frac{-3 \pm \sqrt{3^2 - 4(2)(-5)}}{2(2)}$$
Number of solutions	Determined by the discriminant $b^2 - 4ac$: $b^2 - 4ac > 0$ 2 real solutions	$x^2 + 6x - 12 = 0$ $6^2 - 4(1)(-12) > 0$
	$b^2 - 4ac = 0$ 1 real solution	$x^2 + 10x + 25 = 0$ $10^2 - 4(1)(25) = 0$
	$b^2 - 4ac < 0$ no real solutions, 2 imaginary solutions	$x^2 + 2x + 20 = 0$ $2^2 - 4(1)(20) < 0$
Writing equations	To write an equation with given solutions, reverse the steps in solving an equation by factoring.	$x = 2$ or $x = -3$ $(x - 2)(x + 3) = 0$ $x^2 + x - 6 = 0$
Factoring	The quadratic polynomial $ax^2 + bx + c$ (with integral coefficients) can be factored if and only if $b^2 - 4ac$ is a perfect square.	$2x^2 - 11x + 12$ $b^2 - 4ac = 25$ $(2x - 3)(x - 4)$
Equations quadratic in form	Use substitution to convert to a quadratic.	$x^4 + 3x^2 - 10 = 0$ Let $a = x^2$. $a^2 + 3a - 10 = 0$

Graphing Quadratic Functions		**Examples**
Quadratic function	A function of the form $y = ax^2 + bx + c$ where $a \neq 0$	
Parabola	The graph of a quadratic function is a parabola.	

Properties of parabolas	If $a > 0$, then the parabola opens upward.	$y = x^2 + 2x - 8$
	If $a < 0$, then the parabola opens downward.	Opens upward
	The first coordinate of the vertex is $\frac{-b}{2a}$.	$x = \frac{-b}{2a} = \frac{-2}{2(1)} = -1$
	The second coordinate of the vertex is the minimum y-value if $a > 0$ or the maximum y-value if $a < 0$.	Vertex: $(-1, -9)$ Minimum y-value: -9
	The x-intercepts are found by solving $ax^2 + bx + c = 0$.	x-intercepts: $(-4, 0), (2, 0)$
	Let $x = 0$ to find the y-intercept.	y-intercept: $(0, -8)$

Quadratic Inequalities		**Examples**
Quadratic inequality	An inequality involving a quadratic polynomial	$2x^2 - 7x + 6 \geq 0$ $x^2 - 4x - 5 < 0$

The Graphical Method	Graph the corresponding parabola, and determine the solution from the graph.	To solve $x^2 - 5x + 6 < 0$ graph $y = x^2 - 5x + 6$. Solution set: $(2, 3)$

The Test-Point Method	Test points in the intervals on the number line that are determined by the roots to the quadratic polynomial.	To solve $x^2 - 4 > 0$ plot -2 and 2 on a number line. Test $-5, 0,$ and 5 in the inequality. Solution set: $(-\infty, -2) \cup (2, \infty)$

Enriching Your Mathematical Word Power

Fill in the blank.

1. A _____ equation has the form $ax^2 + bx + c = 0$ where $a \neq 0$.

2. A _____ function has the form $y = ax^2 + bx + c$ where $a \neq 0$.

3. The trinomial $a^2 + 2ab + b^2$ is a _____ square trinomial.

4. Finding the third term of a perfect square trinomial is _____ the square.

5. The equation $x = \dfrac{-b \pm \sqrt{b^2 - 4ac}}{2a}$ is the _____ formula.

6. The expression $b^2 - 4ac$ is the _____.

7. The graph of $y = ax^2 + bx + c$ with $a \neq 0$ is
a _____.

8. An equation that is quadratic after a substitution is quadratic in _____.

9. The inequality $ax^2 + bx + c > 0$ with $a \neq 0$ is a _____ inequality.

10. A number that is used to check if an inequality is satisfied is a ___ point.

Review Exercises

10.1 Factoring and Completing the Square
Solve by factoring.

1. $x^2 - 2x - 15 = 0$

2. $x^2 - 2x - 24 = 0$

3. $2x^2 + x = 15$

4. $2x^2 + 7x = 4$

5. $w^2 - 25 = 0$

6. $a^2 - 121 = 0$

7. $4x^2 - 12x + 9 = 0$

8. $x^2 - 12x + 36 = 0$

Solve by using the even-root property.

9. $x^2 = 12$

10. $x^2 = 20$

11. $(x - 1)^2 = 9$

12. $(x + 4)^2 = 4$

13. $(x - 2)^2 = \dfrac{3}{4}$

14. $(x - 3)^2 = \dfrac{1}{4}$

15. $4x^2 = 9$

16. $2x^2 = 3$

Solve by completing the square.

17. $x^2 - 6x + 8 = 0$

18. $x^2 + 4x + 3 = 0$

19. $x^2 - 5x + 6 = 0$

20. $x^2 - x - 6 = 0$

21. $2x^2 - 7x + 3 = 0$

22. $2x^2 - x = 6$

23. $x^2 + 4x + 1 = 0$

24. $x^2 + 2x - 2 = 0$

10.2 The Quadratic Formula
Solve by the quadratic formula.

25. $x^2 - 3x - 10 = 0$

26. $x^2 - 5x - 6 = 0$

27. $6x^2 - 7x = 3$

28. $6x^2 = x + 2$

29. $x^2 + 4x + 2 = 0$

30. $x^2 + 6x = 2$

31. $3x^2 + 1 = 5x$

32. $2x^2 + 3x - 1 = 0$

Find the value of the discriminant and the number of real solutions to each equation.

33. $25x^2 - 20x + 4 = 0$

34. $16x^2 + 1 = 8x$

35. $x^2 - 3x + 7 = 0$

36. $3x^2 - x + 8 = 0$

37. $2x^2 + 1 = 5x$

38. $-3x^2 + 6x - 2 = 0$

Find the complex solutions to the quadratic equations.

39. $2x^2 - 4x + 3 = 0$

40. $2x^2 - 6x + 5 = 0$

41. $2x^2 + 3 = 3x$

42. $x^2 + x + 1 = 0$

43. $3x^2 + 2x + 2 = 0$

44. $x^2 + 2 = 2x$

45. $\dfrac{1}{2}x^2 + 3x + 8 = 0$

46. $\dfrac{1}{2}x^2 - 5x + 13 = 0$

10.3 More on Quadratic Equations
Use the discriminant to determine whether each quadratic polynomial can be factored, and then factor the ones that are not prime.

47. $8x^2 - 10x - 3$

48. $18x^2 + 9x - 2$

49. $4x^2 \quad 5x + 2$

50. $6x^2 - 7x - 4$

51. $8y^2 + 10y - 25$

52. $25z^2 - 15z - 18$

Write a quadratic equation that has each given pair of solutions.

53. $-3, -6$

54. $4, -9$

55. $-5\sqrt{2}, 5\sqrt{2}$

56. $-2i\sqrt{3}, 2i\sqrt{3}$

Find all real solutions to each equation.

57. $x^6 + 7x^3 - 8 = 0$

58. $8x^6 + 63x^3 - 8 = 0$

59. $x^4 - 13x^2 + 36 = 0$

60. $x^4 + 7x^2 + 12 = 0$

61. $(x^2 + 3x)^2 - 28(x^2 + 3x) + 180 = 0$

62. $(x^2 + 1)^2 - 8(x^2 + 1) + 15 = 0$

63. $x^2 - 6x + 6\sqrt{x^2 - 6x} - 40 = 0$

64. $x^2 - 3x - 3\sqrt{x^2 - 3x} + 2 = 0$

65. $t^{-2} + 5t^{-1} - 36 = 0$

66. $a^{-2} + a^{-1} - 6 = 0$

67. $w - 13\sqrt{w} + 36 = 0$

68. $4a - 5\sqrt{a} + 1 = 0$

10.4 Graphing Quadratic Functions

Find the vertex and intercepts for each parabola, and sketch its graph.

69. $f(x) = x^2 - 6x$

70. $f(x) = x^2 + 4x$

71. $g(x) = x^2 - 4x - 12$

72. $g(x) = x^2 + 2x - 24$

73. $h(x) = -2x^2 + 8x$

74. $h(x) = -3x^2 + 6x$

75. $y = -x^2 + 2x + 3$

76. $y = -x^2 - 3x - 2$

Determine whether each equation has a maximum or minimum y-value and find it.

77. $f(x) = x^2 + 4x + 1$

78. $f(x) = x^2 - 6x + 2$

79. $y = -2x^2 - x + 4$

80. $y = -3x^2 + 2x + 7$

10.5 Quadratic Inequalities

Solve each inequality. State the solution set using interval notation and graph it.

81. $a^2 + a > 6$

82. $x^2 - 5x + 6 > 0$

83. $x^2 - x - 20 \leq 0$

84. $a^2 + 2a \leq 15$

85. $w^2 - w < 0$

86. $x - x^2 \leq 0$

87. $2x^2 + 5x \geq 3$

88. $3x^2 - 4 < x$

89. $x^2 + 2x + 4 \geq 0$

90. $10x - x^2 \leq 28$

91. $x^2 - 10x + 25 \leq 0$

92. $4x \geq 4x^2 + 1$

93. $x^2 - 2x + 10 < 0$

94. $-x^2 + 4x - 5 > 0$

Miscellaneous

Find all real or imaginary solutions to each equation.

95. $144x^2 - 120x + 25 = 0$

96. $49x^2 + 9 = 42x$

97. $(2x + 3)^2 + 7 = 12$

98. $6x = -\dfrac{19x + 25}{x + 1}$

99. $1 + \dfrac{20}{9x^2} = \dfrac{8}{3x}$

100. $\dfrac{x - 1}{x + 2} = \dfrac{2x - 3}{x + 4}$

101. $\sqrt{3x^2 + 7x - 30} = x$

102. $\dfrac{x^4}{3} = x^2 + 6$

103. $2(2x + 1)^2 + 5(2x + 1) = 3$

104. $(w^2 - 1)^2 + 2(w^2 - 1) = 15$

105. $x^{1/2} - 15x^{1/4} + 50 = 0$

106. $x^{-2} - 9x^{-1} + 18 = 0$

Find exact and approximate solutions to each problem.

107. *Missing numbers.* Find two positive real numbers that differ by 4 and have a product of 4.

108. *One on one.* Find two positive real numbers that differ by 1 and have a product of 1.

109. *Big screen TV.* On a 19-inch diagonal measure television picture screen, the height is 4 inches less than the width. Find the height and width.

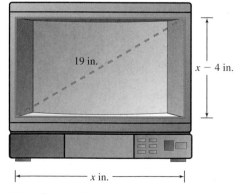

19 in.

$x - 4$ in.

x in.

Figure for Exercise 109

110. *Boxing match.* A boxing ring is in the shape of a square, 20 ft on each side. How far apart are the fighters when they are in opposite corners of the ring?

111. *Students for a Clean Environment.* A group of environmentalists plans to print a message on an 8 inch by 10 inch paper. If the typed message requires 24 square inches of paper and the group wants an equal border on all sides, then how wide should the border be?

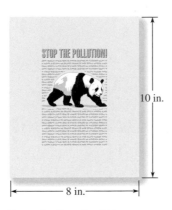

10 in.

8 in.

Figure for Exercise 111

112. *Winston works faster.* Winston can mow his dad's lawn in 1 hour less than it takes his brother Willie. If they take 2 hours to mow it when working together, then how long would it take Winston working alone?

113. *Ping Pong.* The table used for table tennis is 4 ft longer than it is wide and has an area of 45 ft^2. What are the dimensions of the table?

Figure for Exercise 113

114. *Swimming pool design.* An architect has designed a motel pool within a rectangular area that is fenced on three sides as shown in the figure. If she uses 60 yards of fencing to enclose an area of 352 square yards, then what are the dimensions marked L and W in the figure? Assume L is greater than W.

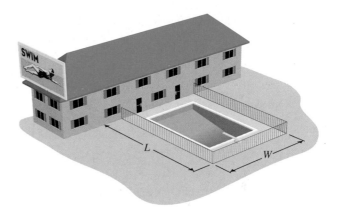

L W

Figure for Exercise 114

115. *Minimizing cost.* The unit cost in dollars for manufacturing n starters is given by $C(n) = 0.004n^2 - 3.2n + 660$. What is the unit cost when 390 starters are manufactured? For what number of starters is the unit cost at a minimum?

116. *Maximizing profit.* The total profit (in dollars) for sales of x rowing machines is given by $P(x) = -0.2x^2 + 300x - 200$. What is the profit if 500 are sold? For what value of x will the profit be at a maximum?

117. *Decathlon champion.* For 1989 and 1990 Dave Johnson had the highest decathlon score in the world. When Johnson reached a speed of 32 ft/sec on the pole vault runway, his height above the ground t seconds after leaving the ground was given by $h = -16t^2 + 32t$. (The elasticity of the pole converts the horizontal speed into vertical speed.) Find the value of t for which his height was 12 ft.

118. *Time of flight.* Use the information from Exercise 117 to determine how long Johnson was in the air. For how long was he more than 14 ft in the air?

119. *Golden ratio.* The ancient Greeks believed that a rectangle had the most pleasing shape when the ratio of its length to width was the *golden ratio*. To find the golden ratio remove a 1 by 1 square from a 1 by x rectangle as shown in the diagram. The ratio of the length to width of the small rectangle that remains should be equal to the ratio of the length to width of the original rectangle. So,

$$\frac{x}{1} = \frac{1}{x - 1}.$$

Find x (the golden ratio) to three decimal places.

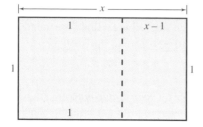

Figure for Exercise 119

Chapter 10 Test

Calculate the value of $b^2 - 4ac$, and state how many real solutions each equation has.

1. $2x^2 - 3x + 2 = 0$

2. $-3x^2 + 5x - 1 = 0$

3. $4x^2 - 4x + 1 = 0$

Solve by using the quadratic formula.

4. $2x^2 + 5x - 3 = 0$

5. $x^2 + 6x + 6 = 0$

Solve by completing the square.

6. $x^2 + 10x + 25 = 0$

7. $2x^2 + x - 6 = 0$

Solve by any method.

8. $x(x + 1) = 12$

9. $a^4 - 5a^2 + 4 = 0$

10. $x - 2 - 8\sqrt{x - 2} + 15 = 0$

Find the complex solutions to the quadratic equations.

11. $x^2 + 36 = 0$

12. $x^2 + 6x + 10 = 0$

13. $3x^2 - x + 1 = 0$

Graph each parabola. Identify the vertex, intercepts, and the maximum or minimum y-value.

14. $f(x) = 16 - x^2$

15. $g(x) = x^2 - 3x$

Write a quadratic equation that has each given pair of solutions.

16. $-4, 6$

17. $-5i, 5i$

Solve each inequality. State and graph the solution set.

18. $w^2 + 3w < 18$

19. $x^2 + 2x \geq 1$

20. $x^2 - 6x + 13 > 0$

21. $x - x^2 \geq 4$

Find the exact solution to each problem.

22. The length of a rectangle is 2 ft longer than the width. If the area is 16 ft², then what are the length and width?

23. A new computer can process a company's monthly payroll in 1 hour less time than the old computer. To really save time, the manager used both computers and finished the payroll in 3 hours. How long would it take the new computer to do the payroll by itself?

24. The height in feet for a ball thrown upward at 48 feet per second is given by $s(t) = -16t^2 + 48t$, where t is the time in seconds after the ball is tossed. What is the maximum height that the ball will reach?

*Making*Connections | A Review of Chapters 1–10

Evaluate each expression.

1. $\sqrt{\dfrac{16}{81}}$

2. $\sqrt[3]{\dfrac{8}{27}}$

3. $(100 + 21)^{1/2}$

4. $(4 + 4)^{2/3}$

5. $8^{1/6} \cdot 8^{1/2}$

6. $\dfrac{32^{1/10}}{32^{-3/10}}$

7. $\left(4^{1/2} + 36^{1/2}\right)^{-2/3}$

8. $\left(3^{-2} + 5^{-2} - 9^{-1}\right)^{3/2}$

Factor completely.

9. $y^2 + 97y - 300$

10. $20y^2 - 7y - 3$

11. $6a^3 - 60a^2 + 150a$

12. $b^3 + 2b^2 + 4b + 8$

13. $ab - ay^2 - by^2 + y^4$

14. $-2m^3 - 16$

Solve each equation.

15. $2x - 15 = 0$

16. $2x^2 - 15 = 0$

17. $2x^2 + x - 15 = 0$

18. $2x^2 + 4x - 15 = 0$

19. $|4x + 11| = 3$

20. $|4x^2 + 11x| = 3$

21. $\sqrt{x} = x - 6$

22. $(2x - 5)^{2/3} = 4$

Solve each inequality. State the solution set using interval notation.

23. $1 - 2x < 5 - x$

24. $(1 - 2x)(5 - x) \le 0$

25. $x^2 \ge x$

26. $5x^2 + 3 \le 0$

27. $3x - 1 < 5$ and $-3 \le x$

28. $x - 3 < 1$ or $2x \ge 8$

Solve each equation for y.

29. $2x - 3y = 9$

30. $\dfrac{y - 3}{x + 2} = -\dfrac{1}{2}$

31. $3y^2 + cy + d = 0$

32. $my^2 - ny = w$

33. $\dfrac{1}{3}x - \dfrac{2}{5}y = \dfrac{5}{6}$

34. $y - 3 = -\dfrac{2}{3}(x - 4)$

Let $m = \dfrac{y_2 - y_1}{x_2 - x_1}$. *Find the value of m for each of the following choices of* $x_1, x_2, y_1,$ *and* y_2.

35. $x_1 = 2, x_2 = 5, y_1 = 3, y_2 = 7$

36. $x_1 = -3, x_2 = 4, y_1 = 5, y_2 = -6$

37. $x_1 = 0.3, x_2 = 0.5, y_1 = 0.8, y_2 = 0.4$

38. $x_1 = \dfrac{1}{2}, x_2 = \dfrac{1}{3}, y_1 = \dfrac{3}{5}, y_2 = -\dfrac{4}{3}$

Solve each problem.

39. *Ticket prices.* If the price of a concert ticket goes up, then the number sold will go down, as shown in the figure. If you use the formula $n = 48{,}000 - 400p$ to predict the number sold depending on the price p, then how many will be sold at \$20 per ticket? How many will be sold at \$25 per ticket? Use the bar graph to estimate the price if 35,000 tickets were sold.

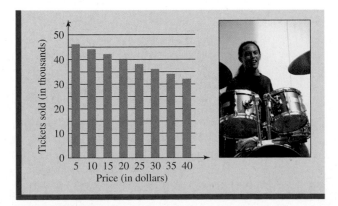

Figure for Exercise 39

40. *Increasing revenue.* Even though the number of tickets sold for a concert decreases with increasing price, the revenue generated does not necessarily decrease. Use the formula $R = p(48,000 - 400p)$ to determine the revenue when the price is $20 and when the price is $25. What price would produce a revenue of $1.28 million? Use the graph to find the price that determines the maximum revenue.

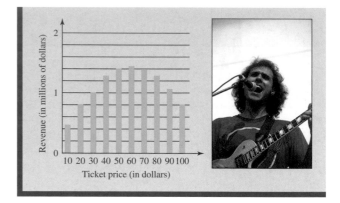

Figure for Exercise 40

Critical Thinking | **For Individual or Group Work** | **Chapter 10**

These exercises can be solved by a variety of techniques, which may or may not require algebra. So be creative and think critically. Explain all answers. Answers are in the Instructor's Edition of this text.

1. Ant parade. An ant marches from point A to point B on the cylindrical garbage can shown in the accompanying figure. The can is 1 foot in diameter and 2 feet high. If the ant makes two complete revolutions of the can in a perfect spiral, then exactly how far did he travel?

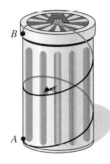

Figure for Exercise 1

2. Connecting points. Draw a circle and pick any three points on the circle.

a) How many line segments can be drawn connecting these points?

b) How many line segments can be drawn connecting four points on a circle? Five points? Six points?

c) How many line segments can be drawn connecting n points on a circle?

3. Summing the digits. Find the sum of the digits in the standard form of the number $2^{2005} \cdot 5^{2007}$.

4. Consecutive odd numbers. Find three consecutive odd whole numbers such that the sum of their squares is a four-digit whole number whose digits are all the same.

5. Reversible prime numbers. The prime number 13 has an interesting property. When its digits are reversed, the new number 31 is also prime. Find the sum of all prime numbers greater than 10 yet less than 125 that have this property.

6. Circles and squares. Start with a square piece of paper. Draw the largest possible circle inside the square. Cut out the circle and keep it. Now draw the largest possible square inside the circle. Cut out the square and keep it. What is the ratio of the area of the original square to the area of the final square? If you repeat this process six more times, then what is the ratio of the area of the original square to the area of the final square?

7. Perpendicular hands. What are the first two times (to the nearest second) after 12 noon for which the minute hand and hour hand of a clock are perpendicular to each other?

Photo for Exercise 7

8. Going broke. Albert and Zelda agreed to play a game. If heads appeared on the toss of an ordinary coin, Zelda had to double the amount of money that Albert had. If the result was tails, then Albert had to pay Zelda $24. As it turned out, the coin came up heads, tails, heads, tails, heads, tails. Then Albert was broke. How much money did Albert start with?

Chapter

11

Functions

Working in a world of numbers, designers of racing boats blend art with science to design attractive boats that are also fast and safe. If the sail area is increased, the boat will go faster but will be less stable in open seas. If the displacement is increased, the boat will be more stable but slower. Increasing length increases speed but reduces stability. To make yacht racing both competitive and safe, racing boats must satisfy complex systems of rules, many of which involve mathematical formulas.

After the 1988 mismatch between Dennis Conner's catamaran and New Zealander Michael Fay's 133-foot monohull, an international group of yacht designers rewrote the America's Cup rules to ensure the fairness of the race. In addition to hundreds of pages of other rules, every yacht must satisfy the basic inequality

$$\frac{L + 1.25\sqrt{S} - 9.8\sqrt[3]{D}}{0.679} \le 24.000,$$

which balances the length L, the sail area S, and the displacement D.

 11.1 Functions and Relations

 11.2 Graphs of Functions and Relations

11.3 Transformations of Graphs

11.4 Graphs of Polynomial Functions

11.5 Graphs of Rational Functions

11.6 Combining Functions

 11.7 Inverse Functions

In the 1979 Fastnet Race, 15 sailors lost their lives. After *Exide Challenger*'s carbon-fiber keel snapped off, Tony Bullimore spent 4 days inside the overturned hull before being rescued by the Australian navy. Yacht racing is a dangerous sport. To determine the general performance and safety of a yacht, designers calculate the displacement-length ratio, the sail area-displacement ratio, the ballast-displacement ratio, and the capsize screening value.

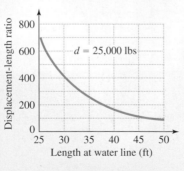

In Exercises 83 and 84 of Section 11.6 we will see how composition of functions is used to define the displacement-length ratio and the sail area-displacement ratio.

11.1   Functions and Relations

We have been using the language of functions and function notation since we first studied formulas in Chapter 2. In this section we will review what we have already studied about functions and delve further into this important concept.

⟨1⟩   The Concept of a Function

If the value of the variable y is determined by the value of the variable x, then **y is a function of x.** So "is a function of" means "is uniquely determined by." But what does uniquely determined mean? According to the dictionary "determine" means "to settle conclusively." There can be no ambiguity. There is only one y for any x.

The x-value is thought of as *input* and the y-value as *output*. If y is a function of x, then there is only one output for any input. For example, after a shopper places an order on the Internet, the shopper is asked to *input* a ZIP code so that the shipping cost (*output*) can be determined. For that order the shipping cost is a function of ZIP code. Note that many different ZIP codes can correspond to the same output. If any ZIP code caused the computer to output more than one shipping cost, then shipping cost is not a function of ZIP code. The shopper is confused and probably cancels the order. See Fig. 11.1.

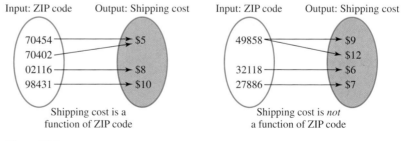

Figure 11.1

E X A M P L E   1

Deciding if y is a function of x

In each case determine whether y is a function of x.

a) Consider all possible circles. Let y represent the area of a circle and x represent its radius.

b) Consider all possible first-class letters mailed today in the United States. Let y represent the weight of a letter and x represent the amount of postage on the letter.

c) Consider all students at Pasadena City College. Let y represent the weight of a student to the nearest pound and x represent the height of the same student to the nearest inch.

d) Consider all possible rectangles. Let y represent the area of a rectangle and x represent the width.

e) Consider all cars sold at Bill Hood Ford this year where the sales tax rate is 9%. Let y represent the amount of sales tax and x represent the selling price of the car.

Solution

a) Can the area of a circle be determined from its radius? The well-known formula $A = \pi r^2$ (or in this case $y = \pi x^2$) indicates exactly how to determine the area if the radius is known. So there is only one area for any given radius and y is a function of x.

b) Can the weight of a letter be determined if the amount of postage on the letter is known? There are certainly letters that have the same amount of postage and different weights. Since the weight cannot be determined conclusively from the postage, the weight is *not* a function of the postage and y is *not* a function of x.

c) Can the weight of a student be determined from the height of the student? Imagine that we have a list containing the weights and heights for all students. There will certainly be two 5 ft 9 in. students with different weights. So weight cannot be determined from the height and y is *not* a function of x.

d) Can the area of a rectangle be determined from the width? Among all possible rectangles there are infinitely many rectangles with width 1 ft and different areas. So the area is not determined by the width and y is *not* a function of x.

e) Can the amount of sales tax be determined from the price of the car? The formula $y = 0.09x$ is used to determine the amount of tax. For example, the tax on every $20,000 car is $1800. So y is a function of x.

> Now do Exercises 1–8

⟨2⟩ Functions Expressed by Formulas

If you get a speeding ticket in St. John's Parish, Louisiana, there is a rule that is used to determine your fine. You can mail to the judge $153 plus $1 for every mile per hour over 80 miles per hour, but if your speed is over 90 miles per hour you must appear before the judge. Since there is no ambiguity, the amount of the fine is a function of your speed.

> **Function (as a Rule)**
>
> A function is a rule by which any allowable value of one variable (the **independent variable**) determines a *unique* value of a second variable (the **dependent variable**).

There are many ways to express a rule. A rule can be expressed verbally (as in the speeding ticket), with a formula, a table, or a graph. Of course, in mathematics we prefer the preciseness that formulas or equations provide. Since a formula such as $A = \pi r^2$ gives us a rule for obtaining the value of the dependent variable A from the value of the independent variable r, we say that this formula is a function. Formulas are used to describe or **model** relationships between variables.

EXAMPLE   2

Writing a formula for a function

A carpet layer charges $25 plus $4 per square yard for installing carpet. Write the total charge C as a function of the number n of square yards of carpet installed.

Solution

At $4 per square yard, n square yards installed cost $4n$ dollars. If we include the $25 charge, then the total cost is $4n + 25$ dollars. Thus, the equation

$$C = 4n + 25$$

expresses C as a function of n.

> Now do Exercises 9–12

Any formula that has the form $y = mx + b$ with $m \neq 0$ is a **linear function.** If $m = 0$, then the formula has the form $y = b$ and is called a **constant function.** So in Example 2, the charge is a linear function of the number of square yards installed and $C = 4n + 25$ is a linear function.

EXAMPLE   3

A function in geometry

Express the area of a circle as a function of its diameter.

Solution

The area of a circle is given by $A = \pi r^2$. Because the radius of a circle is one-half of the diameter, we have $r = \frac{d}{2}$. Now replace r by $\frac{d}{2}$ in the formula $A = \pi r^2$:

$$A = \pi \left(\frac{d}{2} \right)^2$$

$$= \frac{\pi d^2}{4}$$

So $A = \frac{\pi}{4} d^2$ expresses the area of a circle as a function of its diameter.

> Now do Exercises 13–18

⟨3⟩ Functions Expressed by Tables

Tables are often used to provide a rule for pairing the value of one variable with the value of another. For a table to define a function, each value of the independent variable must correspond to only one value of the dependent variable.

EXAMPLE   4

Functions defined by tables

Determine whether each table expresses y as a function of x.

a)

Weight (lbs) x	Cost ($) y
0 to 10	4.60
11 to 30	12.75
31 to 79	32.90
80 to 99	55.82

b)

Weight (lbs) x	Cost ($) y
0 to 15	4.60
10 to 30	12.75
31 to 79	32.90
80 to 99	55.82

c)

x	y
1	1
−1	1
2	2
−2	2
3	3

Solution

a) For each allowable weight, this table gives a unique cost. So the cost is a function of the weight and y is a function of x.

b) Using this table a weight of say 12 pounds would correspond to a cost of \$4.60 and also to \$12.75. Either the table has an error or perhaps there is some other factor that is being used to determine cost. In any case the weight does not determine a unique cost and y is *not* a function of x.

c) In this table every allowable value for x corresponds to a unique y-value, so y is a function of x. Note that different values of x corresponding to the same y-value are permitted in a function.

> Now do Exercises 19–26

⟨4⟩ Functions Expressed by Ordered Pairs

A computer at your grocery store determines the price of each item by searching a long list of ordered pairs in which the first coordinate is the universal product code and the second coordinate is the price of the item with that code. For each product code there is a unique price. This process certainly satisfies the rule definition of a function. Since the set of ordered pairs is the essential part of this rule, we say that the set of ordered pairs is a function.

> **Function (as a Set of Ordered Pairs)**
>
> A function is a set of ordered pairs of real numbers such that no two ordered pairs have the same first coordinates and different second coordinates.

Note the importance of the phrase "no two ordered pairs have the same first coordinates and different second coordinates." Imagine the problems at the grocery store if the computer gave two different prices for the same universal product code. Note also that the product code is an identification number and it cannot be used in calculations. So the computer can use a function defined by a formula to determine the amount of tax, but it cannot use a formula to determine the price from the product code.

> Any set of ordered pairs is called a **relation.** A function is a special relation.

EXAMPLE 5

Relations given as lists of ordered pairs

Determine whether each relation is a function. (Determine whether y is a function of x.)

a) $\{(1, 2), (1, 5), (3, 7)\}$ **b)** $\{(4, 5), (3, 5), (2, 6), (1, 7)\}$

Solution

a) This relation is not a function because $(1, 2)$ and $(1, 5)$ have the same first coordinate but different second coordinates.

b) This relation is a function. Note that the same second coordinate with different first coordinates is permitted in a function.

> Now do Exercises 27–34

The solution set to any equation in x and y is the set of ordered pairs that satisfy the equation. For example, the solution set to $x = y^2$ is expressed in set-builder notation as $\{(x, y) \mid x = y^2\}$. Since any set of ordered pairs is a relation, this solution set is a relation. We can use the definition of a function to determine whether the solution set is a function. For simplicity we often refer to an equation in x and y as a relation or a function.

EXAMPLE **6**

Relations given as equations

Determine whether each relation is a function. (Determine whether y is a function of x.)

a) $x = y^2$

b) $y = 2x$

c) $x = |y|$

⟨ **Helpful Hint** ⟩

To determine whether an equation expresses y as a function of x, always select a number for x (the independent variable) and then see if there is more than one corresponding value for y (the dependent variable). If there is more than one corresponding y-value, then y is not a function of x.

Solution

a) Is it possible to find two ordered pairs with the same first coordinate and different second coordinates that satisfy $x = y^2$? Since $(1, 1)$ and $(1, -1)$ both satisfy $x = y^2$, this relation is not a function.

b) The equation $y = 2x$ indicates that the y-coordinate is always twice the x-coordinate. Ordered pairs such as $(0, 0)$, $(2, 4)$, and $(3, 6)$ satisfy $y = 2x$. It is not possible to find two ordered pairs with the same first coordinate and different second coordinates. So $y = 2x$ is a function.

c) The equation $x = |y|$ is satisfied by ordered pairs such as $(2, 2)$ and $(2, -2)$ because $2 = |2|$ and $2 = |-2|$ are both correct. So this relation is not a function.

Now do Exercises 35–62

⟨5⟩ The Vertical-Line Test

Since every graph illustrates a set of ordered pairs, every graph is a relation. To determine whether a graph is a function, we must see whether there are two (or more) ordered pairs on the graph that have the same first coordinate and different second coordinates. Two points with the same first coordinate lie on a vertical line that crosses the graph.

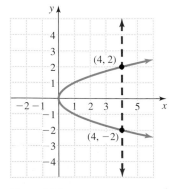

Figure 11.2

The Vertical-Line Test

A graph is the graph of a function if and only if there is no vertical line that crosses the graph more than once.

If there is a vertical line that crosses a graph twice (or more) as in Fig. 11.2, then we have two points with the same x-coordinate and different y-coordinates, and the graph is not the graph of a function. If you mentally consider every possible vertical line and none of them crosses the graph more than once, then you can conclude that the graph is the graph of a function.

EXAMPLE **7**

Using the vertical-line test

Which of these graphs are graphs of functions?

a)

b)

c)

Solution

Neither (a) nor (c) is the graph of a function, since we can draw vertical lines that cross these graphs twice. The graph (b) is the graph of a function, since no vertical line crosses it more than once.

Now do Exercises 63–68

The vertical-line test illustrates the visual difference between a set of ordered pairs that is a function and one that is not. Because graphs are not precise and not always complete, the vertical-line test might be inconclusive.

⟨6⟩ Domain and Range

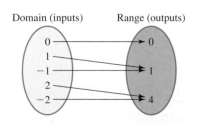

Domain (inputs)   Range (outputs)

Figure 11.3

A relation (or function) is a set of ordered pairs. The set of all first coordinates of the ordered pairs is the **domain** of the relation (or function). The set of all second coordinates of the ordered pairs is the **range** of the relation (or function). A function is a rule that pairs each member of the domain (the inputs) with a unique member of the range (the outputs). See Fig. 11.3. If a function is given as a table or a list of ordered pairs, then the domain and range are determined by simply reading them from the table or list. More often, a relation or function is given by an equation, with no domain stated. In this case, *the domain consists of all real numbers that, when substituted for the independent variable, produce real numbers for the dependent variable.*

EXAMPLE   **8**   **Identifying the domain and range**
Determine the domain and range of each relation.

a) $\{(2, 5), (2, 7), (4, 3)\}$ 　　　b) $y = 2x$ 　　　c) $y = \sqrt{x - 1}$

Solution

a) The domain is the set of first coordinates, $\{2, 4\}$. The range is the set of second coordinates, $\{3, 5, 7\}$.

b) Since any real number can be used in place of x in $y = 2x$, the domain is $(-\infty, \infty)$. Since any real number can be used in place of y in $y = 2x$, the range is also $(-\infty, \infty)$.

c) Since the square root of a negative number is not a real number, we must have $x - 1 \geq 0$ or $x \geq 1$. So the domain is the interval $[1, \infty)$. Since the square root of a nonnegative real number is a nonnegative real number, we must have $y \geq 0$. So the range is the interval $[0, \infty)$.

> Now do Exercises 69–80

⟨7⟩ Function Notation

If y is a function of x, we can use the notation $f(x)$ to represent y. The expression $f(x)$ is read as "f of x." The notation $f(x)$ is called **function notation.** So if x is the independent variable, then either y or $f(x)$ is the dependent variable. For example, the function $y = 2x + 3$ can be written as

$$f(x) = 2x + 3.$$

We use y and $f(x)$ interchangeably. We can think of f as the name of the function. We may use letters other than f. For example $g(x) = 2x + 3$ is the same function as $f(x) = 2x + 3$. The ordered pairs for each function are identical. Note that $f(x)$ does not mean f times x. The expression $f(x)$ represents the second coordinate when the first coordinate is x.

If $f(x) = 2x + 3$, then $f(4) = 2(4) + 3 = 11$. So the second coordinate is 11 if the first coordinate is 4. The ordered pair $(4, 11)$ is an ordered pair in the function f. Figure 11.4 illustrates this situation.

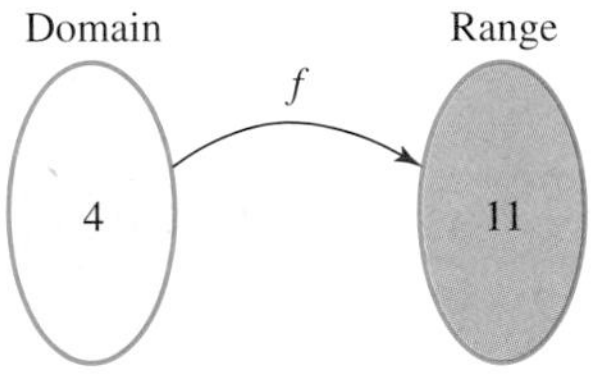

Domain 　　　 Range
　　　f
4 　　　　11

Figure 11.4

EXAMPLE   **9**   **Using function notation**
Let $f(x) = 3x - 2$ and $g(x) = x^2 - x$. Evaluate each expression.

a) $f(-5)$ 　　　　　b) $g(-5)$ 　　　　　c) $f(0) + g(3)$

Solution

a) Replace x by -5 in the equation defining the function f:
$$f(x) = 3x - 2$$
$$f(-5) = 3(-5) - 2$$
$$= -17$$
So $f(-5) = -17$.

b) Replace x by -5 in the equation defining the function g:
$$g(x) = x^2 - x$$
$$g(-5) = (-5)^2 - (-5) = 30$$
So $g(-5) = 30$.

c) Since $f(0) = 3(0) - 2 = -2$ and $g(3) = 3^2 - 3 = 6$, we have $f(0) + g(3) = -2 + 6 = 4$.

> Now do Exercises 81–96

CAUTION The notation $f(x)$ does not mean f times x.

E X A M P L E 10

An application of function notation

To determine the cost of an in-home repair, a computer technician uses the linear function $C(n) = 40n + 30$, where n is the time in hours and $C(n)$ is the cost in dollars. Find $C(2)$ and $C(4)$.

Solution

Replace n with 2 to get

$$C(2) = 40(2) + 30 = 110.$$

Replace n with 4 to get

$$C(4) = 40(4) + 30 = 190.$$

So for 2 hours the cost is \$110 and for 4 hours the cost is \$190.

> Now do Exercises 97–104

‹ **Calculator Close-Up** ›

A graphing calculator has function notation built in. To find $C(2)$ and $C(4)$ with a graphing calculator, enter $y_1 = 40x + 30$ as shown here:

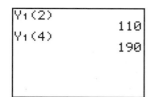

To find $C(2)$ and $C(4)$, enter $y_1(2)$ and $y_1(4)$ as shown here:

```
Y1(2)
                    110
Y1(4)
                    190
```

In this section we studied functions of one variable. However, a variable can be a function of another variable or a function of many other variables. For example, your grade on the next test is not a function of the number of hours that you study for it. Your grade is a function of many variables: study time, sleep time, work time, your mother's IQ, and so on. Even though study time alone does not determine your grade, it is the variable that has the most influence on your grade.

Warm-Ups ▼

Fill in the blank.

1. A set of ordered pairs is a _____.

2. A _____ is a set of ordered pairs in which no two have the same first coordinate and different second coordinates.

3. If m is a _____ of w, then m is uniquely determined by w.

4. The _____ of a relation is the set of all first coordinates of the ordered pairs.

5. The _____ of a relation is the set of all second coordinates of the ordered pairs.

6. In function notation $f(x)$ is used for the _____ variable.

True or false?

7. The set $\{(1, 2), (3, 0), (9, 0)\}$ is a function.

8. The set $\{(2, 1), (0, 3), (0, 9)\}$ is a function.

9. The diameter of a circle is a function of the radius.

10. The equation $y = x^2$ is a function.

11. Every relation is a function.

12. The domain of $\{(2, 1), (0, 3), (0, 9)\}$ is $\{0, 2\}$.

13. The range of $\{(2, 1), (0, 3), (0, 9)\}$ is $\{0, 1, 2, 3, 9\}$.

14. The domain of $f(x) = \sqrt{x}$ is $[0, \infty)$.

15. If $h(x) = x^2 - 3$, the $h(-2) = 1$.

Exercises

‹ **Study Tips** ›

• Instructors love to help students who are eager to learn.
• Show a genuine interest in the subject when you ask questions and you will get a good response from your instructor.

‹1› The Concept of a Function

In each situation determine whether y is a function of x. Explain your answer. See Example 1.

1. Consider all gas stations in your area. Let x represent the price per gallon of regular unleaded gasoline and y represent the number of gallons that you can get for $10.

2. Consider all items at Sears. Let x represent the universal product code for an item and y represent the price of that item.

3. Consider all students taking algebra at your school. Let x represent the number of hours (to the nearest hour) a student spent studying for the first test and y represent the student's score on the test.

4. Consider all students taking algebra at your school. Let x represent a student's height to the nearest inch and y represent the student's IQ.

5. Consider the air temperature at noon today in every town in the United States. Let x represent the Celsius temperature for a town and y represent the Fahrenheit temperature.

6. Consider all first-class letters mailed within the United States today. Let x represent the weight of a letter and y represent the amount of postage on the letter.

7. Consider all items for sale at the nearest Wal-Mart. Let x represent the cost of an item and y represent the universal product code for the item.

8. Consider all packages shipped by UPS. Let x represent the weight of a package and y represent the cost of shipping that package.

‹2› Functions Expressed by Formulas

Write a formula that describes the function. See Examples 2 and 3.

9. A small pizza costs $5.00 plus 50 cents for each topping. Express the total cost C as a function of the number of toppings t.

10. A developer prices condominiums in Florida at $20,000 plus $40 per square foot of living area. Express the cost C as a function of the number of square feet of living area s.

11. The sales tax rate on groceries in Mayberry is 9%. Express the total cost T (including tax) as a function of the total price of the groceries S.

12. With a GM MasterCard, 5% of the amount charged is credited toward a rebate on the purchase of a new car. Express the rebate R as a function of the amount charged A.

13. Express the circumference of a circle as a function of its radius.

14. Express the circumference of a circle as a function of its diameter.

15. Express the perimeter P of a square as a function of the length s of a side.

16. Express the perimeter P of a rectangle with width 10 ft as a function of its length L.

17. Express the area A of a triangle with a base of 10 m as a function of its height h.

18. Express the area A of a trapezoid with bases 12 cm and 10 cm as a function of its height h.

‹3› Functions Expressed by Tables

Determine whether each table expresses the second variable as a function of the first variable. See Example 4.

19.

x	y
1	1
4	2
9	3
16	4
25	5
36	6
49	8

20.

x	y
2	4
3	9
4	16
5	25
8	36
9	49
10	100

21.

t	v
2	2
−2	2
3	3
−3	3
4	4
−4	4
5	5

22.

s	W
5	17
6	17
−1	17
−2	17
−3	17
7	17
8	17

23.

a	P
2	2
2	−2
3	3
3	−3
4	4
4	−4
5	5

24.

n	r
17	5
17	6
17	−1
17	−2
17	−3
17	−4
17	−5

25.

b	q
1970	0.14
1972	0.18
1974	0.18
1976	0.22
1978	0.25
1980	0.28

26.

c	h
345	0.3
350	0.4
355	0.5
360	0.6
365	0.7
370	0.8
380	0.9

⟨4⟩   **Functions Expressed by Ordered Pairs**

Determine whether each relation is a function. See Example 5.

27. $\{(2, 4), (3, 4), (4, 5)\}$

28. $\{(2, -5), (2, 5), (3, 10)\}$

29. $\{(-2, 4), (-2, 6), (3, 6)\}$

30. $\{(3, 6), (6, 3)\}$

31. $\{(\pi, -1), (\pi, 1)\}$

32. $\{(-0.3, -0.3), (-0.2, 0), (-0.3, 1)\}$

33. $\left\{\left(\dfrac{1}{2}, \dfrac{1}{2}\right)\right\}$

34. $\left\{\left(\dfrac{1}{3}, 7\right), \left(-\dfrac{1}{3}, 7\right), \left(\dfrac{1}{6}, 7\right)\right\}$

Find two ordered pairs that satisfy each equation and have the same x-coordinate but different y-coordinates. Answers may vary. See Example 6.

35. $x = 2y^2$ **36.** $x^2 = y^2$

37. $x = |2y|$ **38.** $|x| = |y|$

39. $x^2 + y^2 = 1$ **40.** $x^2 + y^2 = 4$

41. $x = y^4$ **42.** $x^4 = y^4$

43. $x - 2 = |y|$ **44.** $x + 5 = |y|$

Determine whether each relation is a function. See Example 6.

45. $y = x^2$ **46.** $y = x^2 + 3$
47. $x = |y| + 1$ **48.** $|x| = |y + 1|$
49. $y = x$ **50.** $x = y + 4$
51. $x = y^4 + 1$ **52.** $x^4 = y^2$
53. $y = \sqrt{x}$ **54.** $x = \sqrt{y}$
55. $|x| = |2y|$ **56.** $|4x| = |2y|$
57. $x^2 + y^2 = 9$ **58.** $x^2 + y^4 = 1$
59. $x = 2\sqrt{y}$ **60.** $y = \sqrt{x - 5}$
61. $x + 5 = |y|$ **62.** $x - 2 = |y|$

⟨5⟩   **The Vertical-Line Test**

Use the vertical-line test to determine which of the graphs are graphs of functions. See Example 7.

63.

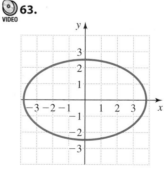

64.

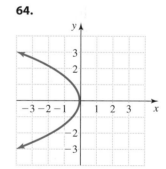

65.

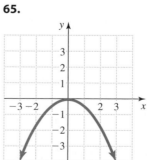

66.

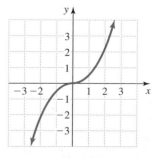

67.   **68.**

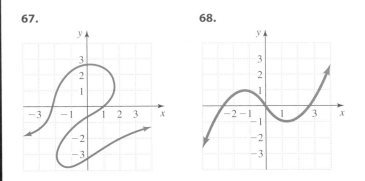

‹6› Domain and Range

Determine the domain and range of each relation. See Example 8.

69. $\{(4, 1), (7, 1)\}$

70. $\{(0, 2), (3, 5)\}$

71. $\{(2, 3), (2, 5), (2, 7)\}$

72. $\{(3, 1), (5, 1), (4, 1)\}$

73. $y = x + 1$

74. $y = 3x + 1$

75. $y = 5 - x$

76. $y = -2x + 1$

77. $y = \sqrt{x - 2}$

78. $y = \sqrt{x + 4}$

79. $y = \sqrt{2x}$

80. $y = \sqrt{2x - 4}$

‹7› Function Notation

Let $f(x) = 3x - 2$, $g(x) = -x^2 + 3x - 2$, and $h(x) = |x + 2|$. Evaluate each expression. See Example 9.

81. $f(0)$ **82.** $f(1)$

83. $f(4)$ **84.** $f(100)$

85. $g(-2)$ **86.** $g(-3)$

87. $h(-3)$ **88.** $h(-19)$

89. $h(-4.236)$ **90.** $h(-1.99)$

91. $f(2) + g(3)$ **92.** $f(1) - g(0)$

93. $\dfrac{g(2)}{h(-3)}$ **94.** $\dfrac{h(-10)}{f(2)}$

95. $f(-1) \cdot h(-4)$ **96.** $h(0) \cdot g(0)$

Solve each problem. See Example 10.

97. Height. If a ball is dropped from the top of a 256-ft building, then the formula
$$h(t) = 256 - 16t^2$$

expresses its height $h(t)$ in feet as a function of the time t in seconds.

a) Find $h(2)$, the height of the ball 2 seconds after it is dropped.

b) Find $h(4)$.

98. Velocity. If a ball is dropped from a height of 256 ft, then the formula
$$v(t) = -32t$$

expresses its velocity $v(t)$ in feet per second as a function of time t in seconds.

a) Find $v(0)$, the velocity of the ball at time $t = 0$.

b) Find $v(4)$.

99. Area of a square. Find a formula that expresses the area of a square A as a function of the length of its side s.

100. Perimeter of a square. Find a formula that expresses the perimeter of a square P as a function of the length of its side s.

101. Cost of fabric. If a certain fabric is priced at $3.98 per yard, express the cost $C(x)$ as a function of the number of yards x. Find $C(3)$.

102. Earned income. If Mildred earns $14.50 per hour, express her total pay $P(h)$ as a function of the number of hours worked h. Find $P(40)$.

103. Cost of pizza. A pizza parlor charges $14.95 for a pizza plus $0.50 for each topping. Express the total cost of a pizza $C(n)$ in dollars as a function of the number of toppings n. Find $C(6)$.

104. Cost of gravel. A gravel dealer charges $50 plus $30 per cubic yard for delivering a truckload of gravel. Express the total cost $C(n)$ in dollars as a function of the number of cubic yards delivered n. Find $C(12)$.

Getting More Involved

105. Writing

Consider $y = x + 2$ and $y > x + 2$. Explain why one of these relations is a function and the other is not.

106. Writing

Consider the graphs of $y = 2$ and $x = 3$ in the rectangular coordinate system. Explain why one of these relations is a function and the other is not.

11.2 Graphs of Functions and Relations

Functions were introduced in Section 11.1. In this section, we will study the graphs of several types of functions. We graphed linear functions in Chapter 3 and quadratic functions in Chapter 10, but for completeness we will review them here.

⟨1⟩ Linear and Constant Functions

Linear functions get their name from the fact that their graphs are straight lines.

> **Linear Function**
>
> A **linear function** is a function of the form
> $$f(x) = mx + b,$$
> where m and b are real numbers with $m \neq 0$.

The graph of the linear function $f(x) = mx + b$ is exactly the same as the graph of the linear equation $y = mx + b$. If $m = 0$, then we get $f(x) = b$, which is called a **constant function.** If $m = 1$ and $b = 0$, then we get the function $f(x) = x$, which is called the **identity function.** When we graph a function given in function notation, we usually label the vertical axis as $f(x)$ rather than y.

E X A M P L E   1

Graphing a constant function

Graph $f(x) = 3$, and state the domain and range.

Solution

The graph of $f(x) = 3$ is the same as the graph of $y = 3$, which is the horizontal line in Fig. 11.5. Since any real number can be used for x in $f(x) = 3$ and since the line in Fig. 11.5 extends without bounds to the left and right, the domain is the set of all real numbers, $(-\infty, \infty)$. Since the only y-coordinate for $f(x) = 3$ is 3, the range is {3}.

> Now do Exercises 1–2

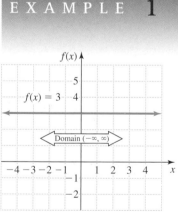

Figure 11.5

The domain and range of a function can be determined from the formula or the graph. However, the graph is usually very helpful for understanding domain and range.

E X A M P L E   2

Graphing a linear function

Graph the function $f(x) = 3x - 4$, and state the domain and range.

Solution

The y-intercept is $(0, -4)$ and the slope of the line is 3. We can use the y-intercept and the slope to draw the graph in Fig. 11.6 on the next page. Since any real number can be used for x in $f(x) = 3x - 4$, and since the line in Fig. 11.6 extends without bounds to the left and right, the domain is the set of all real numbers, $(-\infty, \infty)$. Since the graph extends without bounds upward and downward, the range is the set of all real numbers, $(-\infty, \infty)$.

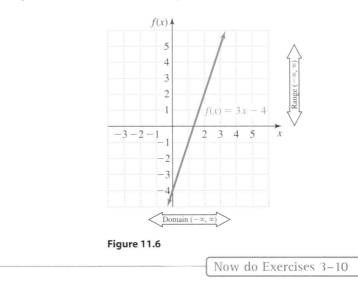

Figure 11.6

Now do Exercises 3–10

⟨2⟩ Absolute Value Functions

The equation $y = |x|$ defines a function because every value of x determines a unique value of y. We call this function the absolute value function.

> **Absolute Value Function**
>
> The **absolute value function** is the function defined by
>
> $$f(x) = |x|.$$

To graph the absolute value function, we simply plot enough ordered pairs of the function to see what the graph looks like.

EXAMPLE **3**

⟨ **Helpful Hint** ⟩

The most important feature of an absolute value function is its V-shape. If we had plotted only points in the first quadrant, we would not have seen the V-shape. So for an absolute value function we always plot enough points to see the V-shape.

The absolute value function

Graph $f(x) = |x|$, and state the domain and range.

Solution

To graph this function, we find points that satisfy the equation $f(x) = |x|$.

x	-2	-1	0	1	2
$f(x) = |x|$	2	1	0	1	2

Plotting these points, we see that they lie along the V-shaped graph shown in Fig. 11.7. Since any real number can be used for x in $f(x) = |x|$ and since the graph extends without bounds to the left and right, the domain is $(-\infty, \infty)$. Because $|x|$ is never negative, the

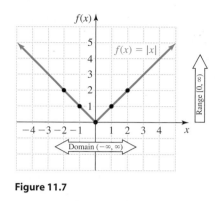

Figure 11.7

graph does not go below the *x*-axis. So the range is the set of nonnegative real numbers, $[0, \infty)$.

> Now do Exercises 11–12

Many functions involving absolute value have graphs that are V-shaped, as in Fig. 11.7. To graph functions involving absolute value, we must choose points that determine the correct shape and location of the V-shaped graph.

E X A M P L E   4

Other functions involving absolute value
Graph each function, and state the domain and range.

   a) $f(x) = |x| - 2$ **b)** $g(x) = |2x - 6|$

Solution

a) Choose values for *x* and find $f(x)$.

x	-2	-1	0	1	2
$f(x) = |x| - 2$	0	-1	-2	-1	0

Plot these points and draw a V-shaped graph through them as shown in Fig. 11.8. The domain is $(-\infty, \infty)$, and the range is $[-2, \infty)$.

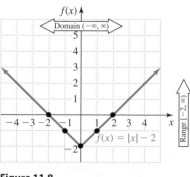
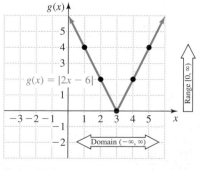

Figure 11.8

Figure 11.9

b) Make a table of values for x and $g(x)$.

x	1	2	3	4	5
$g(x) = \lvert 2x - 6 \rvert$	4	2	0	2	4

Draw the graph as shown in Fig. 11.9. The domain is $(-\infty, \infty)$, and the range is $[0, \infty)$.

Now do Exercises 13–20

⟨3⟩ Quadratic Functions

A function defined by a second-degree polynomial is a *quadratic function*.

> **Quadratic Function**
>
> A **quadratic function** is a function of the form
> $$f(x) = ax^2 + bx + c,$$
> where a, b, and c are real numbers, with $a \neq 0$.

In Chapter 10 we learned that the graph of any quadratic function is a parabola, which opens upward or downward. The vertex of a parabola is the lowest point on a parabola that opens upward or the highest point on a parabola that opens downward. Parabolas will be discussed again when we study conic sections later in this text.

EXAMPLE **5**

A quadratic function
Graph the function $g(x) = 4 - x^2$, and state the domain and range.

Solution
We plot enough points to get the correct shape of the graph.

x	-2	-1	0	1	2
$g(x) = 4 - x^2$	0	3	4	3	0

See Fig. 11.10 for the graph. The domain is $(-\infty, \infty)$. From the graph we see that the largest y-coordinate is 4. So the range is $(-\infty, 4]$.

⟨ **Calculator Close-Up** ⟩

You can find the vertex of a parabola with a calculator. For example, graph
$$y = -x^2 - x + 2.$$
Then use the maximum feature, which is found in the CALC menu. For the left bound pick a point to the left of the vertex; for the right bound pick a point to the right of the vertex; and for the guess pick a point near the vertex.

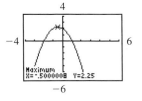

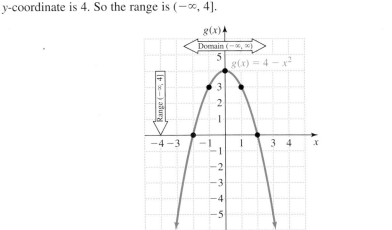

Figure 11.10

Now do Exercises 21–28

⟨4⟩ **Square-Root Functions**

We define the square-root function as follows.

Square-Root Function

The **square-root function** is the function defined by

$$f(x) = \sqrt{x}.$$

Because squaring and square root are inverse operations, the graph of $f(x) = \sqrt{x}$ is related to the graph of $f(x) = x^2$. Recall that there are two square roots of every positive real number, but the radical symbol represents only the positive root. That is why we get only half of a parabola, as shown in Example 6.

EXAMPLE   6

Square-root functions

Graph each equation, and state the domain and range.

  a) $y = \sqrt{x}$ **b)** $y = \sqrt{x + 3}$

Solution

  a) The graph of the equation $y = \sqrt{x}$ and the graph of the function $f(x) = \sqrt{x}$ are the same. Because $\sqrt{x}$ is a real number only if $x \geq 0$, the domain of this function is the set of nonnegative real numbers. The following ordered pairs are on the graph:

x	0	1	4	9
$y = \sqrt{x}$	0	1	2	3

The graph goes through these ordered pairs as shown in Fig. 11.11. Note that x is chosen from the nonnegative numbers. The domain is $[0, \infty)$ and the range is $[0, \infty)$.

  b) Note that $\sqrt{x + 3}$ is a real number only if $x + 3 \geq 0$, or $x \geq -3$. So we make a table of ordered pairs in which $x \geq -3$:

x	-3	-2	1	6
$y = \sqrt{x + 3}$	0	1	2	3

The graph goes through these ordered pairs as shown in Fig. 11.12. The domain is $[-3, \infty)$ and the range is $[0, \infty)$.

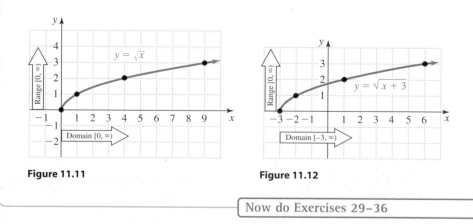

Figure 11.11   **Figure 11.12**

Now do Exercises 29–36

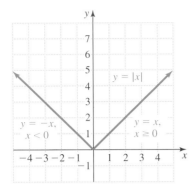

Figure 11.13

⟨5⟩ Piecewise Functions

Most of our functions are defined by a single formula, but functions can be defined by different formulas for different regions of the domain. Such functions are called **piecewise functions.** The simplest example of a piecewise function is the absolute value function. The graph of $f(x) = |x|$ is the straight line $y = x$ to the right of the y-axis and the straight line $y = -x$ to the left of the y-axis, as shown in Fig. 11.13. So $f(x) = |x|$ could be written as

$$f(x) = \begin{cases} x & \text{for} \quad x \geq 0 \\ -x & \text{for} \quad x < 0 \end{cases}.$$

In Example 7, we graph some piecewise functions.

EXAMPLE 7

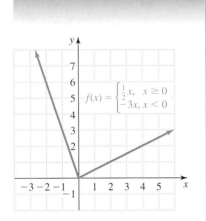

Figure 11.14

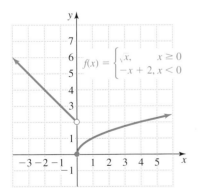

Figure 11.15

Graphing piecewise functions

Graph each function.

a) $f(x) = \begin{cases} \frac{1}{2}x & \text{for} \quad x \geq 0 \\ -3x & \text{for} \quad x < 0 \end{cases}$ **b)** $f(x) = \begin{cases} \sqrt{x} & \text{for} \quad x \geq 0 \\ -x + 2 & \text{for} \quad x < 0 \end{cases}$

Solution

a) For $x \geq 0$, we graph the line $y = \frac{1}{2}x$. For $x < 0$, we graph the line $y = -3x$. Make a table of ordered pairs for each.

$x \, (x \geq 0)$	0	2	4	6
$y = \frac{1}{2}x$	0	1	2	3

$x \, (x < 0)$	-0.1	-1	-2	-3
$y = -3x$	0.3	3	6	9

Plot these ordered pairs and draw the lines as shown in Fig. 11.14. Note that both lines "start" at the origin and neither line extends below the x-axis.

b) For $x \geq 0$, we graph the curve $y = \sqrt{x}$. For $x < 0$, we graph the line $y = -x + 2$. Make a table of ordered pairs for each.

$x \, (x \geq 0)$	0	1	4	9
$y = \sqrt{x}$	0	1	2	3

$x \, (x < 0)$	-0.1	-1	-2	-3
$y = -x + 2$	2.1	3	4	5

Plot these ordered pairs and sketch the graph, as shown in Fig. 11.15. Note that the line comes right up to the point $(0, 2)$ but does not include it. So the point is shown with a hollow circle. The point $(0, 0)$ is included on the curve. So it is shown with a solid circle.

Now do Exercises 37–44

⟨6⟩ Graphing Relations

A function is a set of ordered pairs in which no two have the same first coordinate and different second coordinates. A relation is any set of ordered pairs. The domain of a relation is the set of x-coordinates of the ordered pairs and the range of a relation is the set of y-coordinates of the ordered pairs. In Example 8, we graph the relation $x = y^2$. Note that this relation is not a function because ordered pairs such as $(4, 2)$ and $(4, -2)$ satisfy $x = y^2$.

EXAMPLE **8**

Graphing relations that are not functions

Graph each relation, and state the domain and range.

a) $x = y^2$

b) $x = |y - 3|$

Solution

a) Because the equation $x = y^2$ expresses x in terms of y, it is easier to choose the y-coordinate first and then find the x-coordinate:

$x = y^2$	4	1	0	1	4
y	-2	-1	0	1	2

Figure 11.16 shows the graph. The domain is $[0, \infty)$ and the range is $(-\infty, \infty)$.

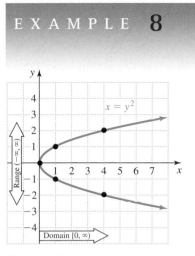

Figure 11.16

b) Again we select values for y first and find the corresponding x-coordinates:

$x = |y - 3|$	2	1	0	1	2
y	1	2	3	4	5

Plot these points as shown in Fig. 11.17. The domain is $[0, \infty)$ and the range is $(-\infty, \infty)$.

> Now do Exercises 45–56

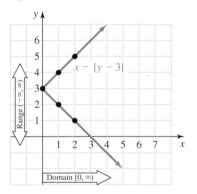

Figure 11.17

Note that $y = x^2$ is a function and $x = y^2$ is not a function because we have agreed that x is *always* the independent variable and y the dependent variable. You can determine the y-coordinate from x in $y = x^2$, but y can't be determined from x in $x = y^2$. If we use variables other than x and y, then we must know which is the independent variable to decide if the equation is a function. For example, if $W = a^2$ and a is the independent variable, then W is a function of a.

Warm-Ups ▼

Fill in the blank.

1. A _____ function has the form $f(x) = mx + b$ where $m \neq 0$.

2. A _____ function has the form $f(x) = k$ where k is a real number.

3. The _____ function is $f(x) = x$.

4. The graph of a quadratic function is a _____.

5. An _____ function has a V-shaped graph.

True or false?

6. The graph of a function is a picture of all of the ordered pairs of the function.

7. The domain of $f(x) = 3$ is $(-\infty, \infty)$.

8. The range of a quadratic function is $(-\infty, \infty)$.

9. The y-axis and the $f(x)$-axis are the same.

10. The domain of $x = y^2$ is $[0, \infty)$.

11. The domain of $f(x) = \sqrt{x - 1}$ is $(1, \infty)$.

12. The domain of a quadratic function is $(-\infty, \infty)$.

⟨1⟩ **Linear and Constant Functions**

Graph each function, and state its domain and range.
See Examples 1 and 2.

1. $h(x) = -2$

2. $f(x) = 4$

3. $f(x) = 2x - 1$

4. $g(x) = x + 2$

5. $g(x) = \dfrac{1}{2}x + 2$

6. $h(x) = \dfrac{2}{3}x - 4$

7. $y = -\dfrac{2}{3}x + 3$

8. $y = -\dfrac{3}{4}x + 4$

9. $y = -0.3x + 6.5$

10. $y = 0.25x - 0.5$

⟨2⟩ **Absolute Value Functions**

Graph each absolute value function and state its domain and range. See Examples 3 and 4.

11. $f(x) = |x| + 1$

12. $g(x) = |x| - 3$

13. $h(x) = |x + 1|$

14. $f(x) = |x - 2|$

15. $g(x) = |3x|$

16. $h(x) = |-2x|$

17. $f(x) = |2x - 1|$

18. $y = |2x - 3|$

19. $f(x) = |x - 2| + 1$ **20.** $y = |x - 1| + 2$

⟨4⟩ Square-Root Functions

Graph each square-root function, and state its domain and range. See Example 6.

29. $g(x) = 2\sqrt{x}$ **30.** $g(x) = \sqrt{x} - 1$

⟨3⟩ Quadratic Functions

Graph each quadratic function, and state its domain and range. See Example 5.

21. $y = x^2$ **22.** $y = -x^2$

31. $f(x) = \sqrt{x - 1}$ **32.** $f(x) = \sqrt{x + 1}$

33. $h(x) = -\sqrt{x}$ **34.** $h(x) = -\sqrt{x - 1}$

23. $g(x) = x^2 + 2$ **24.** $f(x) = x^2 - 4$

35. $y = \sqrt{x} + 2$ **36.** $y = 2\sqrt{x} + 1$

25. $f(x) = 2x^2$ **26.** $h(x) = -3x^2$

⟨5⟩ Piecewise Functions

Graph each piecewise function. See Example 7.

27. $y = 6 - x^2$ **28.** $y = -2x^2 + 3$

37. $f(x) = \begin{cases} x & \text{for } x \geq 0 \\ -4x & \text{for } x < 0 \end{cases}$ **38.** $f(x) = \begin{cases} 3x + 1 & \text{for } x \geq 0 \\ -x + 1 & \text{for } x < 0 \end{cases}$

39. $f(x) = \begin{cases} 2 & \text{for } x > 1 \\ -2 & \text{for } x \le 1 \end{cases}$ **40.** $f(x) = \begin{cases} 3 & \text{for } x > -2 \\ -4 & \text{for } x \le -2 \end{cases}$

47. $x = -y^2$

48. $x = 1 - y^2$

41. $f(x) = \begin{cases} \sqrt{x} & \text{for } x > 1 \\ x + 3 & \text{for } x \le 1 \end{cases}$ **42.** $f(x) = \begin{cases} \sqrt{x - 2} & \text{for } x \ge 3 \\ 6 - x & \text{for } x < 3 \end{cases}$

49. $x = 5$

50. $x = -3$

43. $f(x) = \begin{cases} \sqrt{x} & \text{for } 0 \le x \le 4 \\ x - 4 & \text{for } x > 4 \end{cases}$

51. $x + 9 = y^2$

52. $x + 3 = |y|$

44. $f(x) = \begin{cases} \sqrt{x + 1} & \text{for } -1 \le x \le 3 \\ x - 5 & \text{for } x > 3 \end{cases}$

53. $x = \sqrt{y}$

54. $x = -\sqrt{y}$

⟨6⟩ Graphing Relations

Graph each relation, and state its domain and range.
See Example 8.

45. $x = |y|$ **46.** $x = -|y|$

55. $x = (y - 1)^2$

56. $x = (y + 2)^2$

Miscellaneous

Graph each function, and state the domain and range.

57. $f(x) = 1 - |x|$ **58.** $h(x) = \sqrt{x - 3}$

59. $y = (x - 3)^2 - 1$ **60.** $y = x^2 - 2x - 3$

61. $y = |x + 3| + 1$ **62.** $f(x) = -2x + 4$

63. $y = \sqrt{x} - 3$ **64.** $y = 2|x|$

65. $y = 3x - 5$ **66.** $g(x) = (x + 2)^2$

67. $y = -x^2 + 4x - 4$ **68.** $y = -2|x - 1| + 4$

Classify each function as either a linear, constant, quadratic, square-root, or absolute value function.

69. $f(x) = \sqrt{x - 3}$
70. $f(x) = |x| + 5$
71. $f(x) = 4$
72. $f(x) = 4x - 7$
73. $f(x) = 4x^2 - 7$
74. $f(x) = -3$
75. $f(x) = 5 + \sqrt{x}$
76. $f(x) = |x - 99|$
77. $f(x) = 99x - 100$
78. $f(x) = -5x^2 + 8x + 2$

Graphing Calculator Exercises

79. Graph the function $f(x) = \sqrt{x^2}$, and explain what this graph illustrates.

80. Graph the function $f(x) = \frac{1}{x}$, and state the domain and range.

81. Graph $y = x^2$, $y = \frac{1}{2}x^2$, and $y = 2x^2$ on the same coordinate system. What can you say about the graph of $y = ax^2$ for $a > 0$?

82. Graph $y = x^2$, $y = x^2 + 2$, and $y = x^2 - 3$ on the same screen. What can you say about the position of $y = x^2 + k$ relative to $y = x^2$?

83. Graph $y = x^2$, $y = (x + 5)^2$, and $y = (x - 2)^2$ on the same screen. What can you say about the position of $y = (x - h)^2$ relative to $y = x^2$?

84. You can graph the relation $x = y^2$ by graphing the two functions $y = \sqrt{x}$ and $y = -\sqrt{x}$. Try it and explain why this works.

85. Graph $y = (x - 3)^2$, $y = |x - 3|$, and $y = \sqrt{x - 3}$ on the same coordinate system. How does the graph of $y = f(x - h)$ compare to the graph of $y = f(x)$?

11.3 Transformations of Graphs

If a, h, and k are real numbers with $a \neq 0$, then the graph of $y = af(x - h) + k$ is a **transformation** of the graph of $y = f(x)$. All of the transformations of a function form a **family of functions.** For example, all functions of the form $y = a(x - h)^2 + k$ form the **square or quadratic family** because they are transformations of $y = x^2$. The **absolute-value family** consists of functions of the form $y = a|x - h| + k$, and the **square-root family** consists of functions of the form $y = a\sqrt{x - h} + k$. Understanding families of functions makes graphing easier because all of the functions in a family have similar graphs. The graph of any function in the square family is a **parabola.** We will now see what effect each of the numbers a, h, and k has on the graph of the original function $y = f(x)$.

⟨1⟩ Horizontal Translation

According to the order of operations, to find y in $y = af(x - h) + k$, we first subtract h from x, evaluate $f(x - h)$, multiply by a, and then add k. The order is important here, and we look at the effects of these numbers in the order h, a, and k.

Consider the graphs of $f(x) = \sqrt{x}$, $g(x) = \sqrt{x - 2}$, and $h(x) = \sqrt{x + 6}$ shown in Fig. 11.18. In the expression $\sqrt{x - 2}$, subtracting 2 is the first operation to perform. So every point on the graph of g is exactly two units to the right of a corresponding point on the graph of f. (We must start with a larger value of x to get the same y-coordinate because we first subtract 2.) Every point on the graph of h is exactly six units to the left of a corresponding point on the graph of f.

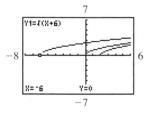

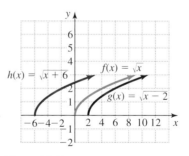

Figure 11.18

Translating to the Right or Left

If $h > 0$, then the graph of $y = f(x - h)$ is a **translation to the right** of the graph of $y = f(x)$.

If $h < 0$, then the graph of $y = f(x - h)$ is a **translation to the left** of the graph of $y = f(x)$.

E X A M P L E   **1**

Horizontal Translation

Sketch the graph of each function and state the domain and range.

a) $f(x) = (x - 2)^2$ **b)** $f(x) = |x + 3|$

Solution

a) The graph of $f(x) = (x - 2)^2$ is a translation two units to the right of the familiar graph of $f(x) = x^2$. Calculate a few ordered pairs for accuracy. The points $(2, 0)$, $(0, 4)$, and $(4, 4)$ are on the graph in Fig. 11.19. Since any real number can be used in place of x in $(x - 2)^2$, the domain is $(-\infty, \infty)$. Since the graph extends upward from $(2, 0)$, the range is $[0, \infty)$.

b) The graph of $f(x) = |x + 3|$ is a translation three units to the left of the familiar graph of $f(x) = |x|$. The points $(0, 3)$, $(-3, 0)$, and $(-6, 3)$ are on the graph in Fig. 11.20. Since any real number can be used in place of x in $|x + 3|$, the domain is $(-\infty, \infty)$. Since the graph extends upward from $(-3, 0)$, the range is $[0, \infty)$.

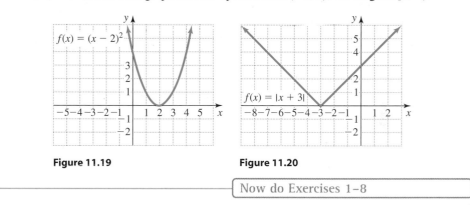

Figure 11.19   **Figure 11.20**

Now do Exercises 1–8

⟨2⟩ Stretching and Shrinking

Consider the graphs of $f(x) = x^2$, $g(x) = 2x^2$, and $h(x) = \frac{1}{2}x^2$ shown in Fig. 11.21. Every point on $g(x) = 2x^2$ corresponds to a point directly below on the graph of $f(x) = x^2$. The y-coordinate on g is exactly twice as large as the corresponding y-coordinate on f. This situation occurs because, in the expression $2x^2$, multiplying by 2 is the last operation performed. Every point on h corresponds to a point directly above on f, where the y-coordinate on h is half as large as the y-coordinate on f. The factor 2 has stretched the graph of f to form the graph of g, and the factor $\frac{1}{2}$ has shrunk the graph of f to form the graph of h.

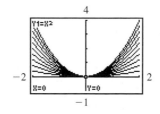

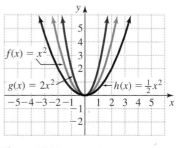

Figure 11.21

Stretching and Shrinking

If $a > 1$, then the graph of $y = af(x)$ is obtained by **stretching** the graph of $y = f(x)$. If $0 < a < 1$, then the graph of $y = af(x)$ is obtained by **shrinking** the graph of $y = f(x)$.

11-26

Note that the last operation to be performed in stretching or shrinking is multiplication by a. Whereas the function $g(x) = 2\sqrt{x}$ is obtained by stretching $f(x) = \sqrt{x}$ by a factor of 2, $h(x) = \sqrt{2x}$ is not.

EXAMPLE 2

Stretching and shrinking

Graph the functions $f(x) = \sqrt{x}$, $g(x) = 2\sqrt{x}$, and $h(x) = \frac{1}{2}\sqrt{x}$ on the same coordinate system.

Solution

The graph of g is obtained by stretching the graph of f, and the graph of h is obtained by shrinking the graph of f. The graph of f includes the points $(0, 0)$, $(1, 1)$, and $(4, 2)$. The graph of g includes the points $(0, 0)$, $(1, 2)$, and $(4, 4)$. The graph of h includes the points $(0, 0)$, $(1, 0.5)$, and $(4, 1)$. The graphs are shown in Fig. 11.22.

‹ **Calculator Close-Up** ›

The following calculator screen shows the curves $y = \sqrt{x}$, $y = 2\sqrt{x}$, $y = 3\sqrt{x}$, and so on, through $y = 10\sqrt{x}$.

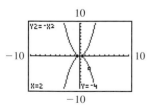

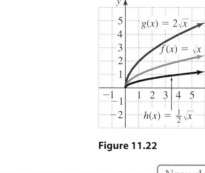

Figure 11.22

Now do Exercises 9–16

‹ **Calculator Close-Up** ›

With a graphing calculator, you can quickly see the result of modifying the formula for a function. If you have a graphing calculator, use it to graph the functions in the examples. Experimenting with it will help you to understand the ideas in this section.

‹3› Reflecting

Consider the graphs of $f(x) = x^2$ and $g(x) = -x^2$ shown in Fig. 11.23. Notice that the graph of g is a mirror image of the graph of f. For any value of x we compute the y-coordinate of an ordered pair of f by squaring x. For an ordered pair of g we square first and then find the opposite because of the order of operations. This gives a correspondence between the ordered pairs of f and the ordered pairs of g. For every ordered pair on the graph of f there is a corresponding ordered pair directly below it on the graph of g, and these ordered pairs are the same distance from the x-axis. We say that the graph of g is obtained by reflecting the graph of f in the x-axis or that g is a reflection of the graph of f.

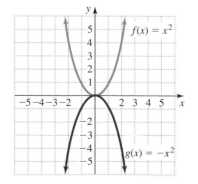

Figure 11.23

Reflection

The graph of $y = -f(x)$ is a **reflection** in the x-axis of the graph of $y = f(x)$.

EXAMPLE **3**

Reflection

Sketch the graphs of each pair of functions on the same coordinate system.

a) $f(x) = \sqrt{x}, \; g(x) = -\sqrt{x}$ **b)** $f(x) = |x|, \; g(x) = -|x|$

Solution

In each case the graph of g is a reflection in the x-axis of the graph of f. Recall that we graphed the square-root function and the absolute value function in Section 11.2. Figures 11.24 and 11.25 show the graphs for these functions.

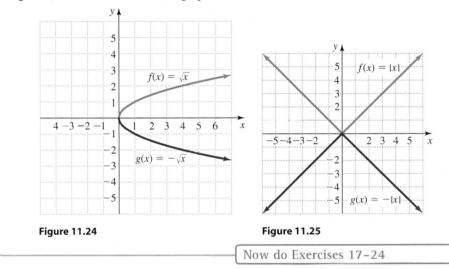

Figure 11.24 **Figure 11.25**

Now do Exercises 17–24

‹**4**› **Vertical Translation**

Consider the graphs of the functions $f(x) = \sqrt{x}$, $g(x) = \sqrt{x} + 2$, and $h(x) = \sqrt{x} - 6$ shown in Fig. 11.26. In the expression $\sqrt{x} + 2$, adding 2 is the last operation to perform. So every point on the graph of g is exactly two units above a corresponding point on the graph of f, and g has the same shape as the graph of f. Every point on the graph of h is exactly six units below a corresponding point on the graph of f. The graph of g is an upward translation of the graph of f, and the graph of h is a downward translation of the graph of f.

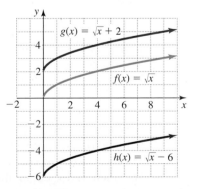

Figure 11.26

Translating Upward or Downward

If $k > 0$, then the graph of $y = f(x) + k$ is an **upward translation** of the graph of $y = f(x)$.

If $k < 0$, then the graph of $y = f(x) + k$ is a **downward translation** of the graph of $y = f(x)$.

E X A M P L E   **4**

Vertical translation

Graph the function $f(x) = |x| - 6$, and state the domain and range.

Solution

The graph of $f(x) = |x| - 6$ is a translation six units downward of the familiar graph of $f(x) = |x|$. Calculate a few ordered pairs for accuracy. The ordered pairs $(0, -6)$, $(1, -5)$, and $(-1, -5)$ are on the graph in Fig. 11.27. Since any real number can be used in place of x in $|x| - 6$, the domain is $(-\infty, \infty)$. Since the graph extends upward from $(0, -6)$, the range is $[-6, \infty)$.

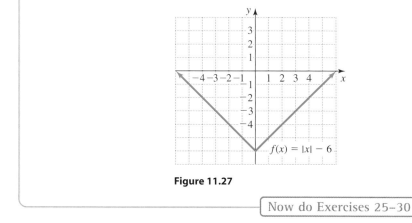

Figure 11.27

Now do Exercises 25–30

⟨5⟩ Multiple Transformations

When graphing a function containing more than one transformation, perform the transformations in the following order:

Strategy for graphing $y = af(x - h) + k$

To graph $y = af(x - h) + k$, start with the graph of $y = f(x)$ and perform

1. Horizontal translation   (right for $h > 0$ and left for $h < 0$)
2. Stretching/shrinking   (stretch for $a > 1$ and shrink for $0 < a < 1$)
3. Reflection   (reflect in x-axis for $a < 0$ or $y = -f(x)$)
4. Vertical translation   (up for $k > 0$ and down for $k < 0$).

Note that the order in which you reflect, stretch, or shrink does not matter. It does matter that you do vertical translation last. For example, if $y = x^2$ is reflected in the x-axis and then moved up two units, the equation is $y = -x^2 + 2$. If it is moved up two units and then reflected in the x-axis, the equation is $y = -(x^2 + 2)$ or $y = -x^2 - 2$. A change in the order produces different functions.

E X A M P L E   5

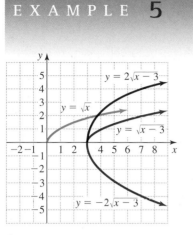

Figure 11.28

‹ **Calculator Close-Up** ›

You can check Example 5 by graph-ing $y = -2\sqrt{x-3}$ with a graphing calculator.

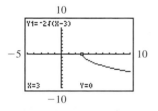

A multiple transformation of $y = \sqrt{x}$

Graph the function $y = -2\sqrt{x-3}$, and state the domain and range.

Solution

Start with the graph of $y = \sqrt{x}$ through $(0, 0)$, $(1, 1)$, and $(4, 2)$, as shown in Fig. 11.28. Translate it three units to the right to get the graph of $y = \sqrt{x-3}$. Stretch this graph by a factor of two to get the graph of $y = 2\sqrt{x-3}$ shown in Fig. 11.28. Now reflect in the x-axis to get the graph of $y = -2\sqrt{x-3}$. To get an accurate graph calculate a few points on the final graph as follows:

x	3	4	7
$y = -2\sqrt{x-3}$	0	-2	-4

Since $x - 3$ must be nonnegative in the expression $-2\sqrt{x-3}$, we must have $x - 3 \geq 0$ and $x \geq 3$. So the domain is $[3, \infty)$. Since the graph extends downward from the point $(3, 0)$, the range is $(-\infty, 0]$.

Now do Exercises 31–32

The graph of $y = x^2$ is a parabola opening upward with vertex $(0, 0)$. The graph of a function of the form $y = a(x - h)^2 + k$ is a transformation of $y = x^2$ and is also a parabola. It opens upward if $a > 0$ and downward if $a < 0$. Its vertex is (h, k). In Example 6, we graph a multiple transformation of $y = x^2$.

E X A M P L E   6

A multiple transformation of the parabola $y = x^2$

Graph the function $y = -2(x + 3)^2 + 4$, and state the domain and range.

Solution

Think of the parabola $y = x^2$ through $(-1, 1)$, $(0, 0)$, and $(1, 1)$. To get the graph of $y = -2(x + 3)^2 + 4$, translate it three units to the left, stretch by a factor of two, reflect in the x-axis, and finally translate upward four units. The graph is a stretched parabola open-ing downward from the vertex $(-3, 4)$ as shown in Fig. 11.29. To get an accurate graph

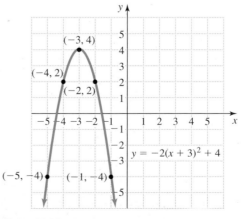

Figure 11.29

calculate a few points around the vertex as follows:

x	-5	-4	-3	-2	-1
$y = -2(x + 3)^2 + 4$	-4	2	4	2	-4

Since any real number can be used for x in $-2(x + 3)^2 + 4$, the domain is $(-\infty, \infty)$. Since the graph extends downward from $(-3, 4)$, the range is $(-\infty, 4]$.

Now do Exercises 33–34

Understanding transformations helps us to see the location of the graph of a function. To get an accurate graph we must still calculate ordered pairs that satisfy the equation. However, if we know where to expect the graph, it is easier to choose appropriate ordered pairs.

E X A M P L E   7

A multiple transformation of the absolute value function $y = |x|$

Graph the function $y = \frac{1}{2}|x - 4| - 1$, and state the domain and range.

Solution

Think of the V-shaped graph of $y = |x|$ through $(-1, 1)$, $(0, 0)$, and $(1, 1)$. To get the graph of $y = \frac{1}{2}|x - 4| - 1$, translate $y = |x|$ to the right four units, shrink by a factor of $\frac{1}{2}$, and finally translate downward one unit. The graph is shown in Fig. 11.30. To get an accurate graph, calculate a few points around the lowest point on the V-shaped graph as follows:

x	2	4	6
$y = \frac{1}{2}|x - 4| - 1$	0	-1	0

Since any real number can be used for x in $\frac{1}{2}|x - 4| - 1$, the domain is $(-\infty, \infty)$. Since the graph extends upward from $(4, -1)$, the range is $[-1, \infty)$.

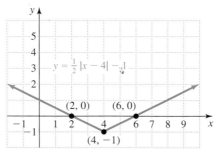

Figure 11.30

Now do Exercises 35–48

Warm-Ups ▼

Fill in the blank.

1. The graph of $y = -f(x)$ is a _____ of the graph of $y = f(x)$.

2. The graph of $y = f(x) + k$ for $k > 0$ is a(n) _____ of the graph of $y = f(x)$.

3. The graph of $y = f(x) + k$ for $k < 0$ is a(n) _____ of the graph of $y = f(x)$.

4. The graph of $y = f(x - h)$ for $h > 0$ is a translation to the ____ of the graph of $y = f(x)$.

5. The graph of $y = f(x - h)$ for $h < 0$ is a translation to the ____ of the graph of $y = f(x)$.

6. The graph of $y = af(x)$ is a _____ of the graph of $y = f(x)$ if $a > 1$.

7. The graph of $y = af(x)$ is a _____ of the graph of $y = f(x)$ if $0 < a < 1$.

True or false?

8. The graph of $f(x) = -x^2$ is a reflection in the x-axis of the graph of $f(x) = x^2$.

9. The graph of $y = |x - 3|$ lies 3 units to the left of the graph of $y = |x|$.

10. The graph of $y = |x| - 3$ lies 3 units below the graph of $y = |x|$.

11. The graph of $f(x) = 2$ is a reflection in the x-axis of the graph of $f(x) = -2$.

12. The graph of $y = -2x^2$ can be obtained by stretching and reflecting the graph of $y = x^2$.

13. The graph of $y = \sqrt{x - 3} + 5$ has the same shape as the graph of $y = \sqrt{x}$.

Exercises

11.3

‹ **Study Tips** ›

- When you take notes, leave space. Go back later and fill in details and make corrections.
- You can even leave enough space to work another problem of the same type in your notes.

‹ 1 › **Horizontal Translation**

Graph each function and state the domain and range.
See Example 1.

1. $y = x + 3$

2. $y = x - 1$

3. $f(x) = (x - 3)^2$

4. $f(x) = (x + 1)^2$

5. $f(x) = \sqrt{x - 1}$

6. $f(x) = \sqrt{x + 6}$

VIDEO **7.** $f(x) = |x + 2|$ **8.** $f(x) = |x - 4|$

⟨**3**⟩ **Reflecting**

Sketch the graphs of each pair of functions on the same coordinate system. See Example 3.

17. $f(x) = \sqrt{2x}$,   **18.** $y = x$, $y = -x$

   $g(x) = -\sqrt{2x}$

⟨**2**⟩ **Stretching and Shrinking**

Use stretching and shrinking to graph each function, and state the domain and range. See Example 2.

9. $f(x) = 3x^2$ **10.** $f(x) = \dfrac{1}{3}x^2$

VIDEO **19.** $f(x) = x^2 + 1$,   **20.** $f(x) = |x| + 1$,

   $g(x) = -(x^2 + 1)$ $g(x) = -|x| - 1$

11. $y = \dfrac{1}{5}x$ **12.** $y = 5x$

21. $y = \sqrt{x - 2}$,   **22.** $y = |x - 1|$,

   $y = -\sqrt{x - 2}$ $y = -|x - 1|$

VIDEO **13.** $f(x) = 3\sqrt{x}$ **14.** $f(x) = \dfrac{1}{3}\sqrt{x}$

23. $f(x) = x - 3$,   **24.** $f(x) = x^2 - 2$,

   $g(x) = 3 - x$ $g(x) = 2 - x^2$

15. $y = \dfrac{1}{4}|x|$ **16.** $y = 4|x|$

〈4〉 Vertical Translation

Graph each function, and state the domain and range.
See Example 4.

25. $y = \sqrt{x} + 1$ **26.** $y = \sqrt{x} - 3$

27. $f(x) = x^2 - 4$ **28.** $f(x) = x^2 + 2$

29. $y = |x| + 2$ **30.** $y = |x| - 4$

33. $f(x) = (x + 3)^2 - 5$ **34.** $f(x) = -2x^2$

35. $y = -|x + 3|$ **36.** $y = |x - 2| + 1$

37. $y = -\sqrt{x + 1} - 2$ **38.** $y = -3\sqrt{x + 4} + 6$

39. $y = -2|x - 3| + 4$ **40.** $y = 3|x - 1| + 2$

〈5〉 Multiple Transformations

Sketch the graph of each function, and state the domain and
range. See Examples 5–7. See the Strategy for graphing
$y = af(x - h) + k$ *on page 716.*

31. $y = \sqrt{x - 2} + 1$ **32.** $y = -\sqrt{x + 3}$

41. $y = -2x + 3$ **42.** $y = 3x - 1$

43. $y = 2(x + 3)^2 + 1$ **44.** $y = 2(x + 1)^2 - 2$

45. $y = -2(x - 4)^2 + 2$ **46.** $y = -2(x - 1)^2 + 3$

47. $y = -3(x - 1)^2 + 6$ **48.** $y = 3(x + 2)^2 - 6$

Match each function with its graph a–h.

49. $y = 2 + \sqrt{x}$ **50.** $y = \sqrt{2 + x}$

51. $y = 2\sqrt{x}$ **52.** $y = \sqrt{\dfrac{x}{2}}$

53. $y = \dfrac{1}{2}\sqrt{x}$ **54.** $y = 2 + \sqrt{x - 2}$

55. $y = -2\sqrt{x}$

56. $y = \sqrt{-x}$

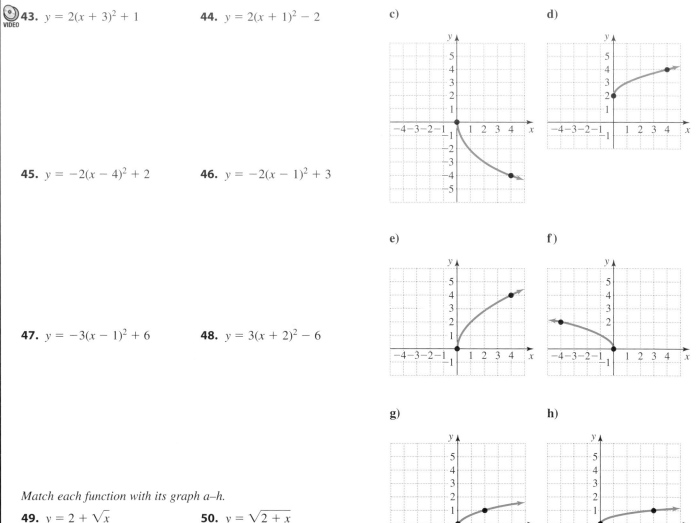

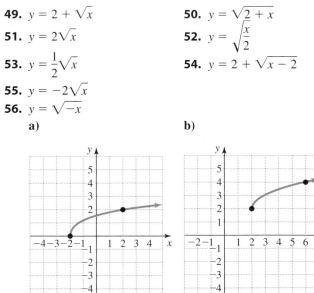

Getting More Involved

57. If the graph of $y = x^2$ is translated eight units upward, then what is the equation of the curve at that location?

58. If the graph of $y = x^2$ is translated six units to the right, then what is the equation of the curve at that location?

59. If the graph of $y = \sqrt{x}$ is translated five units to the left, then what is the equation of the curve at that location?

60. If the graph of $y = \sqrt{x}$ is translated four units downward, then what is the equation of the curve at that location?

61. If the graph of $y = |x|$ is translated three units to the left and then five units upward, then what is the equation of the curve at that location?

62. If the graph of $y = |x|$ is translated four units downward and then nine units to the right, then what is the equation of the curve at that location?

Graphing Calculator Exercises

63. Graph $f(x) = |x|$ and $g(x) = |x - 20| + 30$ on the same screen of your calculator. What transformations will transform the graph of f into the graph of g?

64. Graph $f(x) = (x + 3)^2$, $g(x) = x^2 + 3^2$, and $h(x) = x^2 + 6x + 9$ on the same screen of your calculator.

    a) Which two of these functions has the same graph? Why are they the same?

    b) Is it true that $(x + 3)^2 = x^2 + 9$ for all real numbers x?

    c) Describe each graph in terms of a transformation of the graph of $y = x^2$.

Math *at* Work | Sailboat Design

Mention sailing and your mind drifts to exotic locations, azure seas with soothing tropical breezes, crystal-clear waters, and dazzling white sand. But sailboat designers live in a world of computers, numbers, and formulas. Some of the measurements and formulas used to describe the sailing characteristics and stability of sailboats are the maximum hull speed formula, the sail area-displacement ratio, and the motion-comfort ratio.

    To estimate the theoretical maximum hull speed (M) in knots, designers use the formula $M = 1.34\sqrt{LWL}$, where LWL is the loaded waterline length (the length of the hull at the waterline). See the accompanying figure.

    Sail area-displacement ratio r indicates how fast the boat is in light wind. It is given by $r = \frac{A}{D^{2/3}}$, where A is the sail area in square feet and D is the displacement in cubic feet. Values of r range from 10 to 15 for cruisers and above 24 for high-performance racers.

    The motion-comfort ratio MCR, created by boat designer Ted Brewer, predicts the speed of the upward and downward motion of the boat as it encounters waves. The faster the motion, the more uncomfortable the passengers. If D is the displacement in pounds, LWL the loaded waterline length in feet, LOA the length overall, and B is the beam (width) in feet, then

$$MCR = \frac{D}{\frac{2}{3}B^{3/4}\left(\frac{7}{10}LWL + \frac{1}{3}LOA\right)}.$$

As the displacement increases, MCR increases. As the length and beam increases, MCR decreases. MCR should be in the low 30's for a boat with an LOA of 42 feet.

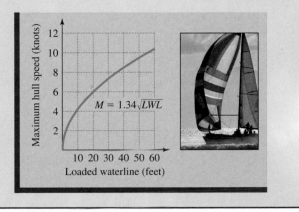

Mid-Chapter **Quiz** | Sections 11.1 through 11.3 | Chapter 11

Determine whether y is a function of x using the ordered pairs given in each table.

1.

x	-1	0	3	4	3
y	4	2	6	8	5

2.

x	2	4	6	8	10
y	1	2	3	4	5

Determine whether each set of ordered pairs is a function.

3. $\{(99, 0), (76, 0), (44, 0)\}$

4. $\{(1/2, 6), (1/3, 12), (1/4, -9)\}$

Determine whether y is a function of x for each relation.

5. $y = x^2 + 100$

6. $x = y^2 + 100$

Find the domain and range of each relation.

7. $y = \sqrt{x - 3}$

8. $\{(1, 2), (3, 4), (20, 30), (40, 30)\}$

Graph each relation, and state its domain and range.

9. $f(x) = 2x - 4$ **10.** $g(x) = |2x - 4|$

11. $h(x) = -x^2 + 3$ **12.** $y = \sqrt{x + 6} - 5$

13. $y = \begin{cases} x + 1 & \text{for } x \geq 0 \\ 2 & \text{for } x < 0 \end{cases}$ **14.** $x = y^2 - 2$

15. $y = -2(x - 1)^2 + 3$

Miscellaneous.

16. Find $f(3)$ if $f(x) = -x^2 + 9$.

17. Find $g(-4)$ if $g(x) = -2|x + 4| - 9$.

18. If the graph of $y = x^2$ is translated 2 units to the left and 4 units upward, then what is the equation of the curve in its final position?

19. If the graph of $y = |x|$ is stretched by a factor of 2, translated 3 units to the left, and reflected in the x-axis, then what is the equation of the curve in its final position?

20. If the graph of $y = \sqrt{x}$ is shrunk by a factor of $\frac{1}{2}$, translated 9 units to the left and 5 units downward, then what is the equation of the curve in its final position?

11.4   **Graphs of Polynomial Functions**

In This Section

We have already graphed constant functions, linear functions, and quadratic functions, which are polynomial functions of degree 0, 1, and 2, respectively. In this section we will graph some polynomial functions with degrees that are greater than 2.

⟨1⟩   **Cubic Functions**

A third-degree polynomial function is called a **cubic function.** The most basic third-degree polynomial function is $f(x) = x^3$, which is called the **cubing function.** We can graph it by plotting some ordered pairs that satisfy the equation $f(x) = x^3$.

E X A M P L E **1**

The cubing function

Graph the function $f(x) = x^3$, and identify the intercepts.

Solution

Make a table of ordered pairs as follows:

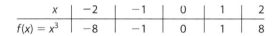

x	−2	−1	0	1	2
$f(x) = x^3$	−8	−1	0	1	8

Plot these ordered pairs, and sketch a smooth curve through them as shown in Fig. 11.31. The x-intercept and the y-intercept are both at the origin, $(0, 0)$.

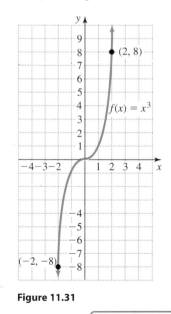

Figure 11.31

Now do Exercises 1–2

In general, the x-intercepts for a polynomial function can be difficult to find, but we will consider only polynomial functions for which the x-intercepts can be found by factoring. Example 2 shows a third-degree polynomial function that has three x-intercepts.

E X A M P L E   **2**

A cubic function with three *x*-intercepts

Graph the function $f(x) = x^3 - 4x$, and identify the intercepts.

Solution

The *y*-intercept is found by replacing *x* with 0. Since $f(0) = 0^3 - 4(0) = 0$, the *y*-intercept is (0, 0). The *x*-intercepts are found by replacing *y* or $f(x)$ with 0 and then solving for *x*:

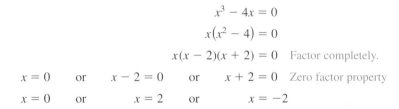

$$x^3 - 4x = 0$$
$$x(x^2 - 4) = 0$$
$$x(x - 2)(x + 2) = 0 \quad \text{Factor completely.}$$

| $x = 0$ | or | $x - 2 = 0$ | or | $x + 2 = 0$ | Zero factor property |
| $x = 0$ | or | $x = 2$ | or | $x = -2$ | |

The *x*-intercepts are $(-2, 0)$, $(0, 0)$, and $(2, 0)$. Now make a table that includes those values for *x*:

x	-3	-1	1	3
$f(x) = x^3 - 4x$	-15	3	-3	15

Plot these ordered pairs, and sketch a smooth curve through them as shown in Fig. 11.32.

Now do Exercises 3–10

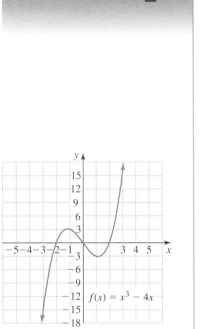

Figure 11.32

⟨2⟩ Quartic Functions

A fourth-degree polynomial function is called a **quartic function.** The most basic fourth-degree polynomial function is $f(x) = x^4$. We can graph it by plotting some ordered pairs that satisfy the equation $f(x) = x^4$.

E X A M P L E   **3**

The most basic fourth-degree polynomial function

Graph $f(x) = x^4$, and identify the intercepts.

Solution

Since $f(0) = 0^4 = 0$, the *y*-intercept is (0, 0). Since $x^4 = 0$ is satisfied only if $x = 0$, the only *x*-intercept is also (0, 0). Make a table of ordered pairs as follows:

x	-2	-1	1	2
$f(x) = x^4$	16	1	1	16

Plot these ordered pairs, and sketch a smooth curve through them as shown in Fig. 11.33. The shape of $f(x) = x^4$ is similar to a parabola, except it is "flatter" on the bottom.

Now do Exercises 11–12

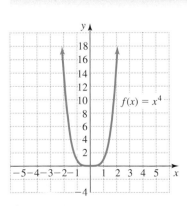

Figure 11.33

Example 4 shows a fourth-degree polynomial function that has four *x*-intercepts.

EXAMPLE 4

A fourth-degree polynomial function with four *x*-intercepts

Graph $f(x) = x^4 - 10x^2 + 9$, and identify the intercepts.

Solution

To find the *y*-intercept replace *x* with 0. Since $f(0) = 0^4 - 10(0^2) + 9 = 9$, the *y*-intercept is (0, 9). To find the *x*-intercepts replace *y* or $f(x)$ with 0 and then solve for *x*:

$$x^4 - 10x^2 + 9 = 0$$
$$(x^2 - 1)(x^2 - 9) = 0$$
$$(x - 1)(x + 1)(x - 3)(x + 3) = 0 \quad \text{Factor completely.}$$
$$x - 1 = 0 \quad \text{or} \quad x + 1 = 0 \quad \text{or} \quad x - 3 = 0 \quad \text{or} \quad x + 3 = 0$$
$$x = 1 \quad \text{or} \quad x = -1 \quad \text{or} \quad x = 3 \quad \text{or} \quad x = -3$$

The four *x*-intercepts are (±1, 0) and (±3, 0). Now make a table that includes those values for *x*:

x	-4	-2	0	2	4
$x^4 - 10x^2 + 9$	105	-15	9	-15	105

Plot these ordered pairs, and sketch a smooth curve through them as shown in Fig. 11.34.

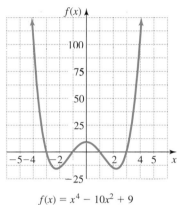

$$f(x) = x^4 - 10x^2 + 9$$

Figure 11.34

Now do Exercises 13–20

⟨3⟩ **Symmetry**

Consider the graph of the quadratic function $f(x) = x^2$ shown in Fig. 11.35. Notice that both (2, 4) and (-2, 4) are on the graph. In fact, $f(x) = f(-x)$ for any value of *x*. We get the same *y*-coordinate whether we evaluate the function at a number or its opposite. This fact causes the graph to be symmetric about the *y*-axis. If we folded the paper along the *y*-axis, the two halves of the graph would coincide.

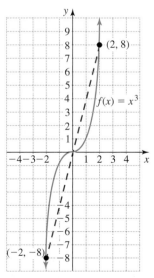

Figure 11.36

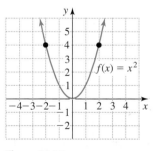

Figure 11.35

Symmetric about the *y*-Axis

If $f(x)$ is a function such that $f(x) = f(-x)$ for any value of *x* in its domain, then the graph of the function is said to be **symmetric about the *y*-axis.**

Consider the graph of $f(x) = x^3$ shown in Fig. 11.36. It is not symmetric about the *y*-axis like the graph of $f(x) = x^2$, but it has a different kind of symmetry. On the graph of $f(x) = x^3$ we find the points $(2, 8)$ and $(-2, -8)$. In this case $f(x)$ and $f(-x)$ are not equal, but $f(-x) = -f(x)$. Notice that the points $(2, 8)$ and $(-2, -8)$ are the same distance from the origin and lie on a line through the origin.

Symmetric about the Origin

If $f(x)$ is a function such that $f(-x) = -f(x)$ for any value of x in its domain, then the graph of the function is said to be **symmetric about the origin.**

E X A M P L E   **5**

Determining the symmetry of a graph

Discuss the symmetry of the graph of each polynomial function.

    a) $f(x) = 5x^3 - x$ **b)** $f(x) = 2x^4 - 3x^2$ **c)** $f(x) = x^2 - 3x + 6$

Solution

    a) Since $f(-x) = 5(-x)^3 - (-x) = -5x^3 + x$, we have $f(-x) = -f(x)$. So the graph is symmetric about the origin.

    b) Since $f(-x) = 2(-x)^4 - 3(-x)^2 = 2x^4 - 3x^2$, we have $f(x) = f(-x)$. So the graph is symmetric about the y-axis.

    c) In this case $f(-x) = (-x)^2 - 3(-x) + 6 = x^2 + 3x + 6$. So $f(-x) \neq f(x)$ and $f(-x) \neq -f(x)$. This graph has neither type of symmetry.

Now do Exercises 21–38

‹ **Calculator Close-Up** ›

We can use graphs to check the conclusions about symmetry that were arrived at algebraically in Example 5. The graph of $f(x) = 5x^3 - x$ appears to be symmetric about the origin.

The graph of $f(x) = 2x^4 - 3x^2$ appears to be symmetric about the y-axis.

The graph of $f(x) = x^2 - 3x + 6$ does not appear to have either type of symmetry.

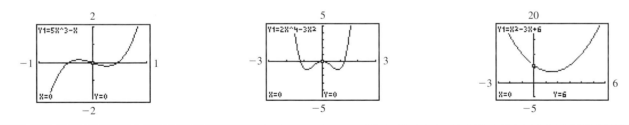

‹ **4** › **Behavior at the x-Intercepts**

The graphs of $y = x$, $y = x^2$, $y = x^3$, and $y = x^4$ all have the same x-intercept $(0, 0)$. But they have two different types of behavior at that x-intercept. The graphs of $y = x$ and $y = x^3$ cross the x-axis at $(0, 0)$, whereas the graphs of $y = x^2$ and $y = x^4$ touch but do not cross the x-axis at $(0, 0)$. The reason for this behavior is the power of the factor x. If a nonzero number is raised to an odd power, the result has the same sign as the original number. But if the power is even, the result is positive. Since $y = x$ and $y = x^3$ have odd powers, the y-coordinates are positive to the right of $(0, 0)$ and negative to the left of $(0, 0)$, and the graph crosses the x-axis at $(0, 0)$. Since the exponents in $y = x^2$ and $y = x^4$ are even, the y-coordinates are positive on either side of $(0, 0)$, and the graphs touch but do not cross the x-axis at $(0, 0)$. In general, we have the following theorem.

> **Behavior at the *x*-Intercepts**
>
> Suppose that $x - c$ is a factor of a polynomial function. The graph of the function crosses the *x*-axis at $(c, 0)$ if $x - c$ occurs an odd number of times and touches but does not cross the *x*-axis if $x - c$ occurs an even number of times.

Since factoring can get difficult for higher-degree polynomials, we will often discuss functions that are given in factored form as in Example 6.

EXAMPLE **6**

Behavior at the *x*-intercepts

Find the *x*-intercepts, and discuss the behavior of the graph of each polynomial function at its *x*-intercepts.

a) $f(x) = (x - 1)^2(x - 3)$ **b)** $y = x^3 + 2x^2 - x - 2$

Solution

a) Replace $f(x)$ with 0 to find the *x*-intercepts:

$$(x - 1)^2(x - 3) = 0$$
$$(x - 1)^2 = 0 \quad \text{or} \quad x - 3 = 0$$
$$x - 1 = 0 \quad \text{or} \quad x = 3$$
$$x = 1$$

The *x*-intercepts are $(1, 0)$ and $(3, 0)$. Since the factor corresponding to $(1, 0)$ is $x - 1$ and its power is even, the graph touches but does not cross the *x*-axis at $(1, 0)$. Since the factor corresponding to $(3, 0)$ is $x - 3$ and its power is odd, the graph crosses the *x*-axis at $(3, 0)$.

b) Replace y with 0 to find the *x*-intercepts:

$$x^3 + 2x^2 - x - 2 = 0$$
$$x^2(x + 2) - 1(x + 2) = 0 \quad \text{Factor by grouping.}$$
$$(x^2 - 1)(x + 2) = 0 \quad \text{Factor out } x + 2.$$
$$(x - 1)(x + 1)(x + 2) = 0 \quad \text{Factor completely.}$$
$$x - 1 = 0 \quad \text{or} \quad x + 1 = 0 \quad \text{or} \quad x + 2 = 0$$
$$x = 1 \quad \text{or} \quad x = -1 \quad \text{or} \quad x = -2$$

The *x*-intercepts are $(1, 0)$, $(-1, 0)$, and $(-2, 0)$. Since each factor occurs with the power of one and one is odd, the graph crosses the *x*-axis at each of the three *x*-intercepts.

> Now do Exercises 39–52

‹ **Calculator Close-Up** ›

The graphs of the functions in Example 6 support the conclusions that were made about the behavior at the *x*-intercepts.

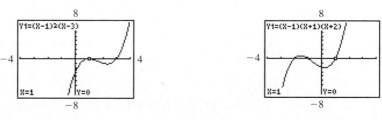

⟨5⟩ Transformations

In Section 11.3 we learned how changes in the formula defining a function can transform the graph of the function. In Example 7, we perform some transformations on $f(x) = x^3$ and $f(x) = x^4$.

EXAMPLE 7

Transformations of graphs

Write the equation of each curve in its final position.

a) The graph of $f(x) = x^3$ is translated 3 units to the right and 2 units downward.

b) The graph of $f(x) = x^4$ is translated 4 units to the left and reflected in the x-axis.

Solution

a) To move the graph 3 units to the right, replace x with $x - 3$ to get $f(x) = (x - 3)^3$. To move the graph 2 units downward, subtract 2. So $f(x) = (x - 3)^3 - 2$ is the equation for the graph in its final position.

b) To move the graph 4 units to the left, replace x with $x + 4$ to get $f(x) = (x + 4)^4$. To reflect in the x-axis, multiply by -1. So $f(x) = -(x + 4)^4$ is the equation for the graph in its final position.

Now do Exercises 53–60

⟨ **Calculator Close-Up** ⟩

The graph of $f(x) = (x - 3)^3 - 2$ shows that it is a translation 3 units to the right and 2 units downward of $f(x) = x^3$.

The graph of $f(x) = -(x + 4)^4$ shows that it is a translation 4 units to the left and a reflection of $f(x) = x^4$.

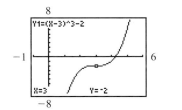

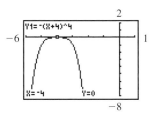

⟨6⟩ Solving Polynomial Inequalities

An inequality such as $x^3 - 3x > 0$ is a **polynomial inequality.** The graphical method that we used for quadratic inequalities in Section 10.5 can also be used with polynomial inequalities. We can read the solution to the inequality from the graph of $y = x^3 - 3x$ provided we know all of the x-intercepts. Any value of x for which $y > 0$ on the graph is a solution to the inequality.

EXAMPLE 8

Solving a polynomial inequality with the graphical method

Solve each polynomial inequality. Write the solution set in interval notation, and graph it.

a) $x^3 - 3x > 0$ **b)** $(x - 1)(x + 2)(x - 3) \leq 0$

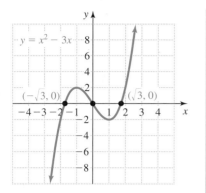

Figure 11.37

Figure 11.38

Solution

a) To solve $x^3 - 3x > 0$, graph $y = x^3 - 3x$. We can determine the solution set to the inequality from the graph if we know the x-intercepts. So first find them by solving $x^3 - 3x = 0$:

$$x^3 - 3x = 0$$
$$x(x^2 - 3) = 0$$

$x = 0$ or $x^2 - 3 = 0$

$x = 0$ or $x^2 = 3$

$x = 0$ or $x = \pm\sqrt{3}$

The x-intercepts are $(-\sqrt{3}, 0)$, $(0, 0)$, and $(\sqrt{3}, 0)$. The graph in Fig. 11.37 crosses the x-axis at each intercept. The inequality is satisfied for any x that corresponds to a positive y-coordinate on this graph. So the solution set to the inequality is $(-\sqrt{3}, 0) \cup (\sqrt{3}, \infty)$ and its graph is shown in Fig. 11.38.

b) To solve $(x - 1)(x + 2)(x - 3) \le 0$, graph $y = (x - 1)(x + 2)(x - 3)$. To find the x-intercepts we solve $(x - 1)(x + 2)(x - 3) = 0$. The x-intercepts are $(1, 0)$, $(-2, 0)$, and $(3, 0)$. The graph in Fig. 11.39 crosses the x-axis at each intercept.

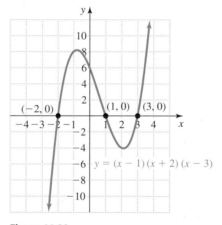

Figure 11.39

The inequality is satisfied for any x that corresponds to a negative or zero y-coordinate on this graph. So the solution set is $(-\infty, -2] \cup [1, 3]$ and its graph is shown in Fig. 11.40.

Figure 11.40

> **Now do Exercises 87–96**

The **test-point method** that we used for quadratic inequalities in Section 10.5 can be used also with polynomial inequalities. For this method we first find all of the roots to the polynomial, as in the graphical method. Instead of graphing the polynomial function, we plot the roots on a number line and then test a point in each interval determined by the roots.

E X A M P L E   **9**

Solving a polynomial inequality with the test-point method
Solve each polynomial inequality. Write the solution set in interval notation and graph it.

a) $x^4 - 16 \le 0$ **b)** $x^4 - 3x^3 - 18x^2 \ge 0$

Solution

a) Find the roots to $x^4 - 16 = 0$:

$$x^4 - 16 = 0$$
$$(x^2 - 4)(x^2 + 4) = 0$$
$$x^2 - 4 = 0 \quad \text{or} \quad x^2 + 4 = 0$$
$$x^2 = 4 \qquad\qquad x^2 = -4$$
$$x = \pm 2$$

The only real solutions to the equation are -2 and 2. Locate these numbers on a number line as in Fig. 11.41.

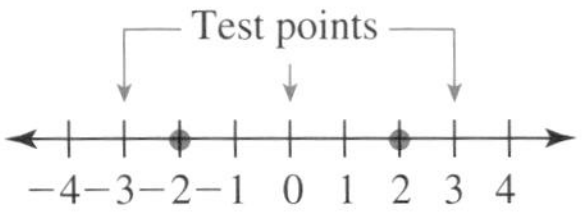

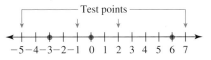

Figure 11.41

Select the test points -3, 0, and 3. Test them in the original inequality $x^4 - 16 \leq 0$.

$$(-3)^4 - 16 \leq 0 \qquad\qquad \text{Incorrect}$$
$$(0)^4 - 16 \leq 0 \qquad\qquad \text{Correct}$$
$$(3)^4 - 16 \leq 0 \qquad\qquad \text{Incorrect}$$

Since 0 is the only test point that satisfies the inequality, the interval containing 0 is the solution set to the inequality. Because of the $\leq$ symbol we include the endpoints. The solution set is $[-2, 2]$, and its graph is shown in Fig. 11.42.

Figure 11.42

b) Find the roots to $x^4 - 3x^3 - 18x^2 = 0$:

$$x^2(x^2 - 3x - 18) = 0$$
$$x^2(x - 6)(x + 3) = 0$$
$$x^2 = 0 \quad \text{or} \quad x - 6 = 0 \quad \text{or} \quad x + 3 = 0$$
$$x = 0 \quad \text{or} \qquad x = 6 \quad \text{or} \qquad x = -3$$

Now locate -3, 0, and 6 on a number line as in Fig. 11.43.

Figure 11.43

Select the test points -5, -1, 2 and 7. Use a calculator to test them in the original inequality $x^4 - 3x^3 - 18x^2 \geq 0$:

$$(-5)^4 - 3(-5)^3 - 18(-5)^2 \leq 0 \qquad\qquad \text{Incorrect}$$
$$(-1)^4 - 3(-1)^3 - 18(-1)^2 \leq 0 \qquad\qquad \text{Correct}$$
$$(2)^4 - 3(2)^3 - 18(2)^2 \leq 0 \qquad\qquad \text{Correct}$$
$$(7)^4 - 3(7)^3 - 18(7)^2 \leq 0 \qquad\qquad \text{Incorrect}$$

The inequality is satisfied on the intervals containing -1 and 2, which are $[-3, 0]$ and $[0, 6]$. Since the symbol is $\leq$, the endpoints of the intervals are included. The solution set is $[-3, 0] \cup [0, 6]$, which is simplified to $[-3, 6]$. The graph of the solution set is shown in Fig. 11.44.

Figure 11.44

Now do Exercises 97–104

Warm-Ups ▼

Fill in the blank.

1. The graph of $y = f(x)$ is symmetric about the _____ if $f(x) = f(-x)$ for all x in the domain of f.

2. The graph of $y = f(x)$ is symmetric about the _____ if $f(-x) = -f(x)$ for all x in the domain of f.

3. The graph of a polynomial function $P(x)$ crosses the x-axis at c if $x - c$ occurs an _____ number of times in the prime factorization of P.

4. The graph of a polynomial function $P(x)$ touches but does not cross the x-axis at c if $x - c$ occurs an _____ number of times in the prime factorization of P.

True or false?

5. The graph of $f(x) = x^3 - x$ is symmetric about the y-axis.

6. The graph of $y = 2x - 1$ is symmetric about the origin.

7. If $f(x) = 3x$, then $f(x) = f(-x)$ for any real number x.

8. If $f(x) = 3x^4 - 5x^3 + 2x^2 - 6x + 7$, then $f(-x) = 3x^4 + 5x^3 + 2x^2 + 6x + 7$.

9. There is only one x-intercept for the graph of $f(x) = x^2 - 4x + 4$.

10. The graph of $y = (x - 1)^2(x + 4)^4$ does not cross the x-axis at either of its intercepts.

Exercises 11.4

‹ Study Tips ›

• Always study math with a pencil and paper. Just sitting back and reading the text rarely works.
• A good way to study the examples in the text is to cover the solution with a piece of paper and see how much of the solution you can write on your own.

‹1› Cubic Functions

Graph each function, and identify the x- and y-intercepts. See Examples 1 and 2.

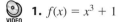

1. $f(x) = x^3 + 1$

2. $f(x) = x^3 - 1$

3. $f(x) = x^3 - 9x$

4. $f(x) = x^3 - x$

5. $f(x) = -x^3 - 4x^2$ **6.** $f(x) = -x^3 + 3x^2$ **13.** $f(x) = x^4 - 4x^2$ **14.** $f(x) = x^4 - 9x^2$

15. $f(x) = x^4 - 5x^2 + 4$ **16.** $f(x) = x^4 - 20x^2 + 64$

7. $f(x) = x^3 + x^2 - 4x - 4$ **8.** $f(x) = x^3 + 2x^2 - 9x - 18$

17. $f(x) = x^4 + x^3 - 4x^2 - 4x$

9. $f(x) = x^3 - 3x^2 - 9x + 27$ **10.** $f(x) = x^3 - 2x^2 - 4x + 8$

18. $f(x) = x^4 + 2x^3 - 9x^2 - 18x$

⟨2⟩ **Quartic Functions**

Graph each function, and identify the x- and y-intercepts. See Examples 3 and 4.

11. $f(x) = x^4 - 1$ **12.** $f(x) = x^4 + 3$

19. $f(x) = x^4 - 3x^3 - 9x^2 + 27x$

20. $f(x) = x^4 - 2x^3 - 4x^2 + 8x$

44. $f(x) = (x + 1)(x - 3)(x + 9)^2$

45. $f(x) = x^3 + 6x^2 - x - 6$

46. $f(x) = x^3 + 5x^2 - 4x - 20$

47. $f(x) = -x^3 + 5x^2$

48. $f(x) = -x^3 - 9x^2$

49. $f(x) = x^4 - 5x^3$

⟨3⟩ Symmetry

*Discuss the symmetry of the graph of each polynomial function.
See Example 5.*

50. $f(x) = x^4 + x^3$

21. $f(x) = 2x$

22. $f(x) = -x$

51. $f(x) = x^4 + 6x^3 + 9x^2$

23. $f(x) = -2x^2$ [VIDEO]

52. $f(x) = x^4 - 4x^3 + 4x^2$

24. $f(x) = x^2 + 1$

25. $f(x) = 2x^3$

26. $f(x) = -x^3$

27. $f(x) = x^4$

⟨5⟩ Transformations

*Write the equation of each curve in its final position.
See Example 7.*

28. $f(x) = x^4 - x$

29. $f(x) = x^3 - 5x + 1$ [VIDEO]

53. The graph of $f(x) = x^3$ is translated 5 units to the right and
4 units downward.

30. $f(x) = 5x^3 + 7x$

31. $f(x) = 6x^6 - 3x^2 - x$

54. The graph of $f(x) = x^3$ is translated 2 units to the right and
1 unit upward.

32. $f(x) = x^6 - x^4 + x^2 - 8$

33. $f(x) = (x - 3)^2$

55. The graph of $f(x) = x^3$ is translated 6 units to the left and
3 units upward.

34. $f(x) = 3(x + 2)^2$

35. $f(x) = (x^2 - 5)^3$

56. The graph of $f(x) = x^3$ is translated 4 units to the left and
7 units downward.

36. $f(x) = (x^2 + 1)^2$

37. $f(x) = x$

57. The graph of $f(x) = x^3$ is reflected in the x-axis.

38. $f(x) = -3x$

58. The graph of $f(x) = x^3$ is reflected in the x-axis and then
translated 1 unit upward.

⟨4⟩ Behavior at the *x*-Intercepts

*Find the x-intercepts, and discuss the behavior of the graph of
each polynomial function at its x-intercepts. Check your
answers with a graphing calculator if you have one. See
Example 6.*

59. The graph of $f(x) = x^4$ is translated 3 units to the right and
then reflected in the x-axis.

60. The graph of $f(x) = x^4$ is translated 5 units to the left and
then reflected in the x-axis.

Miscellaneous

Match each polynomial function with its graph a–h.

39. $f(x) = (x - 2)^2(x - 8)$

40. $f(x) = (x + 3)^2(x - 5)$ [VIDEO]

61. $f(x) = -2x + 3$

62. $f(x) = -2x^2 + 3$

63. $f(x) = -2x^3 + 3$

64. $f(x) = -2x^2 + 4x + 3$

41. $f(x) = (x - 1)^2(x + 4)^2$

65. $f(x) = -x^4 + 3$

66. $f(x) = x^3 - 3x^2$

42. $f(x) = (x + 4)^2(x + 6)^2$

67. $f(x) = x^3 + 3x^2 - x - 3$

43. $f(x) = (x - 1)(x + 4)(x - 7)^2$

68. $f(x) = \dfrac{1}{2}x^4 - 3$

a)

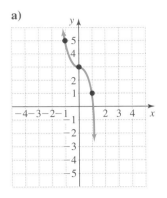

b)

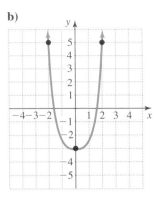

Sketch the graph of each polynomial function.

69. $f(x) = 2x - 6$

70. $f(x) = -3x + 3$

c)

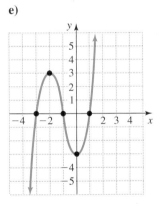

d)

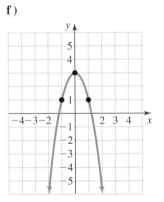

71. $f(x) = -x^2$

72. $f(x) = x^2 - 3$

73. $f(x) = x^3 - 2x^2$

74. $f(x) = x^3 - 4x$

e)

f)

75. $f(x) = (x - 1)^2(x + 1)^2$

76. $f(x) = (x + 2)^2(x - 1)$

g)

h)

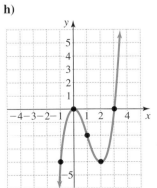

77. $f(x) = (x - 1)^2(x - 3)$

78. $f(x) = x^3 + 2x^2 - 3x$

79. $f(x) = x^4 - 4x^3 + 4x^2$ **80.** $f(x) = -x^4 + 6x^3 - 9x^2$ **86.** $f(x) = (x - 20)^2(x + 30)^2x^2$

Graphing Calculator Exercises

Sketch the graph of each polynomial function. First graph the function on a calculator and use the calculator graph as a guide.

81. $f(x) = x - 20$ **82.** $f(x) = (x - 20)^2$

83. $f(x) = (x - 20)^2(x + 30)$

84. $f(x) = (x - 20)^2(x + 30)^2$

85. $f(x) = (x - 20)^2(x + 30)^2x$

⟨6⟩ Solving Polynomial Inequalities

Solve each polynomial inequality using the graphical method. State the solution set using interval notation and graph it. See Example 8.

87. $x^3 - 4x > 0$

88. $x^3 - 16x < 0$

89. $(x - 3)(x - 5)(x + 2) \leq 0$

90. $(x + 4)(x + 1)(x - 6) \geq 0$

91. $x^3 - 2x^2 \geq 0$

92. $-x^3 + 5x^2 \leq 0$

93. $x^4 - 4x^3 + 4x^2 \geq 0$

94. $-x^4 + 6x^3 - 9x^2 \geq 0$

95. $(x - 1)^2(x + 1)^2 \leq 0$

96. $(x + 2)^2(x - 1) \leq 0$

Solve each polynomial inequality using the test-point method. State the solution set using interval notation, and graph it. See Example 9.

97. $x^4 - 81 > 0$

98. $x^4 - 1 < 0$

99. $x^4 - x^3 - 6x^2 \leq 0$

100. $x^4 + 2x^3 - 8x^2 \geq 0$

101. $x^3 + 6x^2 - 4x - 24 > 0$

102. $x^3 + 5x^2 - 9x - 45 < 0$

103. $x^4 - 10x^2 + 9 \leq 0$

104. $x^4 - 18x^2 + 32 \geq 0$

Getting More Involved

In each case, find a polynomial function $f(x)$ whose graph behaves in the required manner. Answers may vary.

105. The graph has only one x-intercept at $(3, 0)$ and crosses the x-axis there.

106. The graph has only one x-intercept at $(3, 0)$ but does not cross the x-axis there.

107. The graph has only two x-intercepts at $(-2, 0)$ and $(1, 0)$. It crosses the x-axis at $(-2, 0)$ but does not cross at $(1, 0)$.

108. The graph has only two x-intercepts at $(5, 0)$ and $(-6, 0)$. It does not cross the x-axis at either x-intercept.

11.5 Graphs of Rational Functions

In This Section

⟨1⟩ Rational Functions

⟨2⟩ Asymptotes

⟨3⟩ Sketching the Graphs

⟨4⟩ Rational Inequalities

We first studied rational expressions in Chapter 6. In this section we will study functions that are defined by rational expressions.

⟨1⟩ Rational Functions

A rational expression was defined in Chapter 6 as a ratio of two polynomials. If a ratio of two polynomials is used to define a function, then the function is called a rational function.

> **Rational Function**
>
> If $P(x)$ and $Q(x)$ are polynomials with no common factor and $f(x) = \frac{P(x)}{Q(x)}$ for $Q(x) \neq 0$, then $f(x)$ is called a **rational function.**

The domain of a rational function is the set of all real numbers except those that cause the denominator to have a value of 0.

EXAMPLE 1

Domain of a rational function

Find the domain of each rational function.

a) $f(x) = \dfrac{x - 3}{x - 1}$

b) $g(x) = \dfrac{2x - 3}{x^2 - 4}$

Solution

a) Since $x - 1 = 0$ only for $x = 1$, the domain of f is the set of all real numbers except 1, $(-\infty, 1) \cup (1, \infty)$.

b) Since $x^2 - 4 = 0$ for $x = \pm 2$, the domain of g is the set of all real numbers excluding 2 and -2, $(-\infty, -2) \cup (-2, 2) \cup (2, \infty)$.

Now do Exercises 1–6

‹ **Calculator Close-Up** ›

If the viewing window is too large, a rational function will appear to touch its asymptotes.

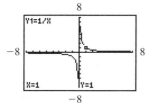

Because the asymptotes are an important feature of a rational function, we should draw it so that it approaches but does not touch its asymptotes.

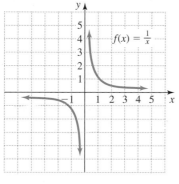

Figure 11.45

‹2› Asymptotes

Consider the simplest rational function $f(x) = 1/x$. Its domain does not include 0, but 0 is an important number for the graph of this function. The behavior of the graph of f when x is very close to 0 is what interests us. For this function the y-coordinate is the reciprocal of the x-coordinate. When the x-coordinate is close to 0, the y-coordinate is far from 0. Consider the following tables of ordered pairs that satisfy $f(x) = 1/x$:

<table>
<tr><td colspan="2" align="center">$x > 0$</td><td colspan="2" align="center">$x < 0$</td></tr>
<tr><td>x</td><td>y</td><td>x</td><td>y</td></tr>
<tr><td>0.1</td><td>10</td><td>−0.1</td><td>−10</td></tr>
<tr><td>0.01</td><td>100</td><td>−0.01</td><td>−100</td></tr>
<tr><td>0.001</td><td>1000</td><td>−0.001</td><td>−1000</td></tr>
<tr><td>0.0001</td><td>10,000</td><td>−0.0001</td><td>−10,000</td></tr>
</table>

As x gets closer and closer to 0 from above 0, the value of y gets larger and larger. We say that y goes to positive infinity. As x gets closer and closer to 0 from below 0, the values of y are negative but $|y|$ gets larger and larger. We say that y goes to negative infinity. The graph of f gets closer and closer to the vertical line $x = 0$, and so $x = 0$ is called a **vertical asymptote.** On the other hand, as $|x|$ gets larger and larger, y gets closer and closer to 0. The graph approaches the x-axis as x goes to infinity, and so the x-axis is a **horizontal asymptote** for the graph of f. See Fig. 11.45 for the graph of $f(x) = 1/x$.

In general, a rational function has a vertical asymptote for every number excluded from the domain of the function. The horizontal asymptotes are determined by the behavior of the function when $|x|$ is large.

E X A M P L E **2**

‹ **Calculator Close-Up** ›

The graph for Example 2(a) should consist of three separate pieces, but in connected mode the calculator connects the separate pieces. Even though the calculator does not draw a very good graph of this function, it does support the conclusion that the horizontal asymptote is the x-axis and the vertical asymptotes are $x = -1$ and $x = 1$.

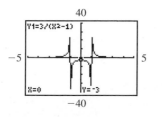

Horizontal and vertical asymptotes

Find the horizontal and vertical asymptotes for each rational function.

a) $f(x) = \dfrac{3}{x^2 - 1}$

b) $g(x) = \dfrac{x}{x^2 - 4}$

c) $h(x) = \dfrac{2x + 1}{x + 3}$

Solution

a) The denominator $x^2 - 1$ has a value of 0 if $x = \pm 1$. So the lines $x = 1$ and $x = -1$ are vertical asymptotes. If $|x|$ is very large, the value of $\dfrac{3}{x^2 - 1}$ is approximately 0. So the x-axis is a horizontal asymptote.

b) The denominator $x^2 - 4$ has a value of 0 if $x = \pm 2$. So the lines $x = 2$ and $x = -2$ are vertical asymptotes. If $|x|$ is very large, the value of $\dfrac{x}{x^2 - 4}$ is approximately 0. So the x-axis is a horizontal asymptote.

c) The denominator $x + 3$ has a value of 0 if $x = -3$. So the line $x = -3$ is a vertical asymptote. If $|x|$ is very large, the value of $h(x)$ is not approximately 0.

To understand the value of $h(x)$, we change the form of the rational expression by using long division:

$$\begin{array}{r} 2 \\ x + 3\overline{)2x + 1} \\ 2x + 6 \\ \hline -5 \end{array}$$

Writing the rational expression as quotient $+ \frac{\text{remainder}}{\text{divisor}}$, we get $h(x) = \frac{2x + 1}{x + 3} = 2 + \frac{-5}{x + 3}$. If $|x|$ is very large, $\frac{-5}{x + 3}$ is approximately 0, and so the y-coordinate is approximately 2. The line $y = 2$ is a horizontal asymptote.

> Now do Exercises 7–12

Example 2 illustrates two important facts about horizontal asymptotes. If the degree of the numerator is less than the degree of the denominator, then the x-axis is the horizontal asymptote. For example, $y = \frac{x - 4}{x^2 - 7}$ has the x-axis as a horizontal asymptote. If the degree of the numerator is equal to the degree of the denominator, then the ratio of the leading coefficients determines the horizontal asymptote. For example, $y = \frac{2x - 7}{3x - 5}$ has $y = \frac{2}{3}$ as its horizontal asymptote. The remaining case is when the degree of the numerator is greater than the degree of the denominator. This case is discussed next.

Each rational function of Example 2 had one horizontal asymptote and a vertical asymptote for each number that caused the denominator to be 0. The horizontal asymptote $y = 0$ occurs because, as $|x|$ gets larger and larger, the y-coordinate gets closer and closer to 0. Some rational functions have a nonhorizontal line for an asymptote. An asymptote that is neither horizontal nor vertical is called an **oblique asymptote** or **slant asymptote.**

EXAMPLE 3

Finding an oblique asymptote

Determine all of the asymptotes for

$$g(x) = \frac{2x^2 + 3x - 5}{x + 2}.$$

Solution

If $x + 2 = 0$, then $x = -2$. So the line $x = -2$ is a vertical asymptote. Use long division to rewrite the function as quotient $+ \frac{\text{remainder}}{\text{divisor}}$:

$$g(x) = \frac{2x^2 + 3x - 5}{x + 2} = 2x - 1 + \frac{-3}{x + 2}$$

If $|x|$ is large, the value of $\frac{-3}{x + 2}$ is approximately 0. So when $|x|$ is large, the value of $g(x)$ is approximately $2x - 1$. The line $y = 2x - 1$ is an oblique asymptote for the graph of g.

> Now do Exercises 13–14

We can summarize this discussion of asymptotes with the following strategy for finding asymptotes for a rational function.

Strategy for Finding Asymptotes for a Rational Function

Suppose $f(x) = \frac{P(x)}{Q(x)}$ is a rational function with the degree of $Q(x)$ at least 1.

1. Solve the equation $Q(x) = 0$. The graph of f has a vertical asymptote corresponding to each solution to the equation.
2. If the degree of $P(x)$ is less than the degree of $Q(x)$, then the x-axis is a horizontal asymptote.
3. If the degree of $P(x)$ is equal to the degree of $Q(x)$, then find the ratio of the leading coefficients. The horizontal line through that ratio is the horizontal asymptote.
4. If the degree of $P(x)$ is one larger than the degree of $Q(x)$, then use division to rewrite the function as

$$\text{quotient} + \frac{\text{remainder}}{\text{divisor}}.$$

The equation formed by setting y equal to the quotient gives us an oblique asymptote.

⟨3⟩ Sketching the Graphs

We now use asymptotes to help us sketch the graphs of some rational functions.

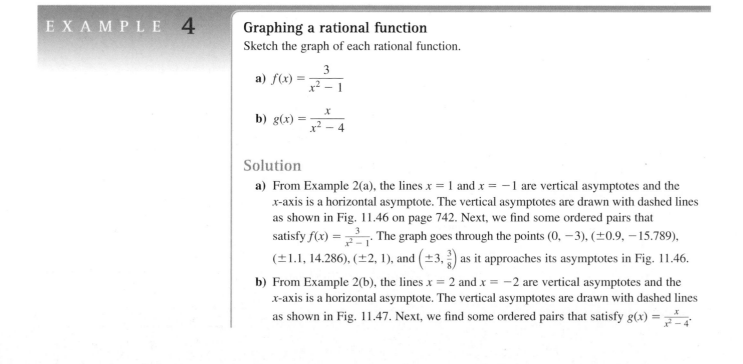

EXAMPLE **4**

Graphing a rational function
Sketch the graph of each rational function.

a) $f(x) = \dfrac{3}{x^2 - 1}$

b) $g(x) = \dfrac{x}{x^2 - 4}$

Solution

a) From Example 2(a), the lines $x = 1$ and $x = -1$ are vertical asymptotes and the x-axis is a horizontal asymptote. The vertical asymptotes are drawn with dashed lines as shown in Fig. 11.46 on page 742. Next, we find some ordered pairs that satisfy $f(x) = \frac{3}{x^2 - 1}$. The graph goes through the points $(0, -3)$, $(\pm 0.9, -15.789)$, $(\pm 1.1, 14.286)$, $(\pm 2, 1)$, and $\left(\pm 3, \frac{3}{8}\right)$ as it approaches its asymptotes in Fig. 11.46.

b) From Example 2(b), the lines $x = 2$ and $x = -2$ are vertical asymptotes and the x-axis is a horizontal asymptote. The vertical asymptotes are drawn with dashed lines as shown in Fig. 11.47. Next, we find some ordered pairs that satisfy $g(x) = \frac{x}{x^2 - 4}$.

This calculator graph supports the graph drawn in Fig. 11.47. Remember that the calculator graph can be misleading. The vertical lines drawn by the calculator are not part of the graph of the function.

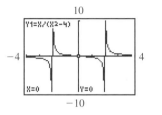

The graph goes through the points $(0, 0)$, $\left(1, -\frac{1}{3}\right)$, $(1.9, -4.872)$, $(2.1, 5.122)$, $\left(3, \frac{3}{5}\right)$, and $\left(4, \frac{1}{3}\right)$ as it approaches its asymptotes in Fig. 11.47.

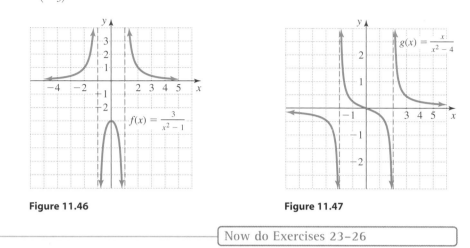

Figure 11.46

Figure 11.47

Now do Exercises 23–26

E X A M P L E   **5**

Graphing a rational function

Sketch the graph of each rational function.

a) $h(x) = \dfrac{2x + 1}{x + 3}$

b) $g(x) = \dfrac{2x^2 + 3x - 5}{x + 2}$

Solution

a) Draw the vertical asymptote $x = -3$ and the horizontal asymptote $y = 2$ from Example 2(c) as dashed lines. The points $(-2, -3)$, $\left(0, \frac{1}{3}\right)$, $\left(-\frac{1}{2}, 0\right)$, $(7, 1.5)$, $(-4, 7)$, and $(-13, 2.5)$ are on the graph shown in Fig. 11.48.

This calculator graph supports the graph drawn in Fig. 11.48. Note that if x is -3, there is no y-coordinate because $x = -3$ is the vertical asymptote.

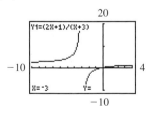

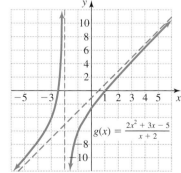

Figure 11.48

Figure 11.49

b) Draw the vertical asymptote $x = -2$ and the oblique asymptote $y = 2x - 1$ from Example 3 as dashed lines. The points $(-1, -6)$, $\left(0, -\frac{5}{2}\right)$, $(1, 0)$, $(4, 6.5)$, and $(-2.5, 0)$ are on the graph shown in Fig. 11.49.

Now do Exercises 27–32

‹4› Rational Inequalities

Inequalities involving rational expressions are **rational inequalities.** We can solve rational inequalities using the graphical method or the test-point method as we did for quadratic inequalities in Section 10.5 and polynomial inequalities in Section 11.4.

EXAMPLE 6

Solving a rational inequality graphically

Solve each rational inequality. Write the solution set in interval notation, and graph it.

a) $\dfrac{x-1}{x+2} > 0$ **b)** $\dfrac{x+4}{x-2} \le 3$

Solution

a) To solve the rational inequality, graph $y = \frac{x-1}{x+2}$. The vertical asymptote is $x = -2$. The x-intercept is $(1, 0)$, and the y-intercept is $\left(0, -\frac{1}{2}\right)$. The graph is shown in Fig. 11.50. The values of x that satisfy the inequality are the same values of x for which $y > 0$ on the graph in the figure. The y-coordinates in the figure are positive for x in the interval $(-\infty, -2)$ and also in the interval $(1, \infty)$. So the solution set is $(-\infty, -2) \cup (1, \infty)$ and the graph of the solution set is shown in Fig. 11.51.

b) First rewrite the inequality so that 0 is on the right:

$$\frac{x+4}{x-2} \le 3$$

$$\frac{x+4}{x-2} - 3 \le 0 \qquad \text{Subtract 3 from each side.}$$

$$\frac{x+4}{x-2} - \frac{3(x-2)}{x-2} \le 0 \qquad \text{Get a common denominator.}$$

$$\frac{-2x+10}{x-2} \le 0 \qquad \text{Get a single rational expression.}$$

Now graph $y = \frac{-2x+10}{x-2}$. The vertical asymptote is $x = 2$. Find the x-intercept by solving $-2x + 10 = 0$. The x-intercept is $(5, 0)$. The y-intercept is $(0, -5)$. The graph is shown in Fig. 11.52. From the graph we see that the y-coordinates are less than 0 for x in the interval $(-\infty, 2)$ and less than or equal to zero for x in the interval $[5, \infty)$. So the solution set is $(-\infty, 2) \cup [5, \infty)$. The graph of the solution set is shown in Fig. 11.53.

> **Now do Exercises 47–58**

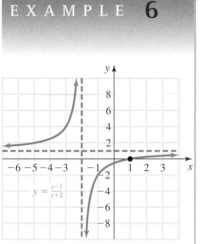

Figure 11.50

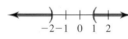

Figure 11.51

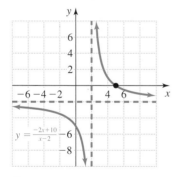

Figure 11.52

Figure 11.53

When solving an equation involving rational expressions, we multiply each side by the least common denominator. When solving a rational inequality, we do not use that technique. If we multiply each side of an inequality by a negative number, then the inequality symbol is reversed. But it is not reversed if we multiply by a positive number. If the LCD involves a variable, then the value of the LCD could be positive or negative. We won't know what to do with the inequality symbol if we multiply by an LCD containing a variable.

Example 7 illustrates the **test-point method.** The key fact here is that the y-coordinates on the graph of a rational function can change sign only at an x-intercept or a vertical asymptote. The x-intercepts are found by setting the numerator equal to zero, and the vertical asymptotes are found by setting the denominator equal to zero. We locate these x-values on a number line and then test a point from each interval that is determined by them.

E X A M P L E   **7**

Solving a rational inequality with test points

Solve each rational inequality. Write the solution set in interval notation, and graph it.

a) $\dfrac{4 - x}{2x - 1} \le 0$

b) $\dfrac{x - 1}{x + 2} > \dfrac{x}{x - 4}$

c) $\dfrac{x^2 + 2}{x^2 + 2x + 4} \ge 0$

Solution

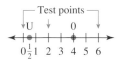

Figure 11.54

a) First solve $4 - x = 0$ to get $x = 4$. Then solve $2x - 1 = 0$ to get $x = \frac{1}{2}$. Plot $\frac{1}{2}$ and 4 on a number line as shown in Fig. 11.54. We put a 0 above 4 and a U above $\frac{1}{2}$ because the value of the rational expression is 0 when $x = 4$ and undefined if $x = \frac{1}{2}$. Select three test points, say 0, 2, and 6. Now try each point in the original inequality $\frac{4 - x}{2x - 1} \le 0$:

$$\frac{4 - 0}{2(0) - 1} \le 0 \qquad \text{Correct}$$

$$\frac{4 - 2}{2(2) - 1} \le 0 \qquad \text{Incorrect}$$

$$\frac{4 - 6}{2(6) - 1} \le 0 \qquad \text{Correct}$$

Figure 11.55

Since 0 and 6 satisfy the inequality, the solution set consists of the intervals containing 0 and 6. Since the inequality symbol is $\le$, we include 4 in the solution set. Note that $\frac{1}{2}$ does not satisfy the inequality. The solution set is the interval $\left(-\infty, \frac{1}{2}\right) \cup [4, \infty)$. The graph of the solution set is shown in Fig. 11.55.

b) First rewrite the inequality with 0 on the right.

$$\frac{x - 1}{x + 2} > \frac{x}{x - 4}$$

$$\frac{x - 1}{x + 2} - \frac{x}{x - 4} > 0$$

$$\frac{(x - 1)(x - 4)}{(x + 2)(x - 4)} - \frac{x(x + 2)}{(x - 4)(x + 2)} > 0$$

$$\frac{x^2 - 5x + 4 - (x^2 + 2x)}{(x + 2)(x - 4)} > 0$$

$$\frac{-7x + 4}{(x + 2)(x - 4)} > 0$$

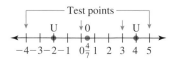

Figure 11.56

Now solve $-7x + 4 = 0$ to get $x = \frac{4}{7}$. The denominator is zero if $x = -2$ or $x = 4$. Plot -2, $\frac{4}{7}$, and 4 on a number line as in Fig. 11.56. Select a test point in each interval determined by these three numbers. We have chosen -4, 0, 3, and 5. Evaluate the original inequality at the test points:

$$\frac{-4 - 1}{-4 + 2} > \frac{-4}{-4 - 4} \qquad \text{Correct}$$

$$\frac{0 - 1}{0 + 2} > \frac{0}{0 - 4} \qquad \text{Incorrect}$$

$$\frac{3-1}{3+2} > \frac{3}{3-4} \qquad \text{Correct}$$

$$\frac{5-1}{5+2} > \frac{5}{5-4} \qquad \text{Incorrect}$$

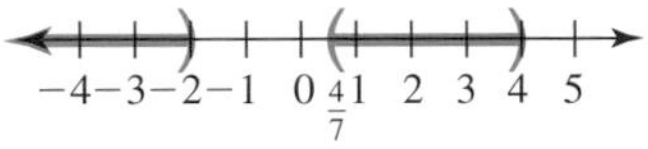

Figure 11.57

Since -4 and 3 satisfy the inequality, the solution set consists of the intervals containing -4 and 3. Since the inequality symbol is $>$, no endpoints are included in the intervals. The solution set is the interval $(-\infty, -2) \cup \left(\frac{4}{7}, 4\right)$. The graph of the solution set is shown in Fig. 11.57.

c) If $x^2 + 2 = 0$, then $x^2 = -2$ and there is no real solution to this equation. If $x^2 + 2x + 4 = 0$, then

$$x = \frac{-2 \pm \sqrt{2^2 - 4(1)(4)}}{2(1)} = \frac{-2 \pm \sqrt{-12}}{2}$$

and again there is no real solution. So the solution set is either all real numbers or the empty set. To decide, test 0 in $\dfrac{x^2 + 2}{x^2 + 2x + 4} \geq 0$:

$$\frac{0^2 + 2}{0^2 + 2(0)x + 4} \geq 0 \qquad \text{Correct}$$

Since the inequality is satisfied at the test point, it is satisfied for all real numbers. The solution set is $(-\infty, \infty)$, and the graph of the solution set is shown in Fig. 11.58.

```
 ←——┼——┼——┼——┼——┼——→
   −2−1  0  1  2
```

Figure 11.58

> **Now do Exercises 59–74**

Warm-Ups ▼

Fill in the blank.

1. A _____ function has the form $f(x) = P(x)/Q(x)$ where $P(x)$ and $Q(x)$ are polynomials with no common factor and $Q(x) \neq 0$.

2. The _____ of a rational function is all real numbers except those that cause the denominator to be 0.

3. A _____ asymptote is a vertical line that is approached by the graph of a rational function.

4. A _____ asymptote is a horizontal line approached by the graph of a rational function.

5. An _____ or _____ asymptote is a nonvertical nonhorizontal line approached by the graph of a rational function.

True or false?

6. The domain of $f(x) = \dfrac{1}{x - 9}$ is $x = 9$.

7. The domain of $f(x) = \dfrac{x - 1}{x - 2}$ is $(-\infty, -2) \cup (-2, 1) \cup (1, \infty)$.

8. The line $x = 2$ is the only vertical asymptote for the graph of $f(x) = \dfrac{1}{x^2 - 4}$.

9. The x-axis is a horizontal asymptote for the graph of $f(x) = \dfrac{x^2 - 3x + 5}{x^3 - 9x}$.

10. The line $y = 2x - 5$ is an asymptote for the graph of $f(x) = 2x - 5 + \dfrac{1}{x}$.

11. The graph of $f(x) = \dfrac{x^2}{x^2 - 9}$ is symmetric about the y-axis.

12. The solution set to $\dfrac{4}{x - 3} \geq 0$ is $(3, \infty)$.

11.5 Exercises

⟨1⟩ Rational Functions

Find the domain of each rational function. See Example 1.

1. $f(x) = \dfrac{2}{x-1}$

2. $f(x) = \dfrac{-2}{x+3}$

3. $f(x) = \dfrac{x^2-1}{x}$

4. $f(x) - \dfrac{-2x+3}{x^2}$

5. $f(x) = \dfrac{5}{x^2-16}$

6. $f(x) = \dfrac{x+12}{x^2-x-6}$

⟨2⟩ Asymptotes

Determine all asymptotes for the graph of each rational function. See Examples 2 and 3 and the Strategy for Finding Asymptotes for a Rational Function on page 741.

7. $f(x) = \dfrac{7}{x+4}$

8. $f(x) = \dfrac{-8}{x-9}$

9. $f(x) = \dfrac{1}{x^2-16}$

10. $f(x) = \dfrac{-2}{x^2-5x+6}$

11. $f(x) = \dfrac{5x}{x-7}$

12. $f(x) = \dfrac{3x+8}{x-2}$

13. $f(x) = \dfrac{2x^2}{x-3}$

14. $f(x) = \dfrac{3x^2+2}{x+1}$

⟨3⟩ Sketching the Graphs

Match each rational function with its graph a–h.

15. $f(x) = -\dfrac{2}{x}$

16. $f(x) = -\dfrac{1}{x-2}$

17. $f(x) = \dfrac{x}{x-2}$

18. $f(x) = \dfrac{x-2}{x}$

19. $f(x) = \dfrac{1}{x^2-2x}$

20. $f(x) = \dfrac{x^2}{x^2-4}$

21. $f(x) = -\dfrac{x+4}{2}$

22. $f(x) = \dfrac{x^2+2x+1}{x}$

a)
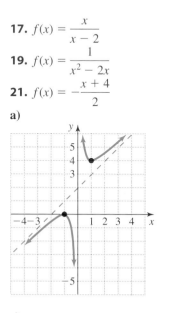

b)
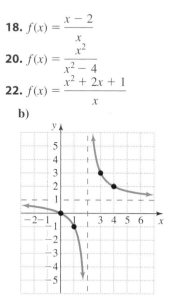

c)
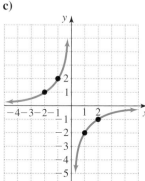

d)
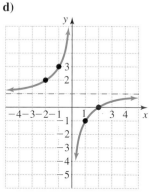

e)
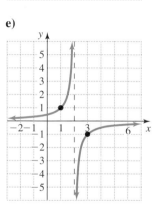

f)

g)

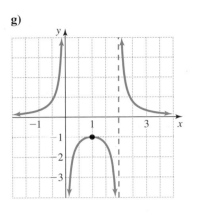

27. $f(x) = \dfrac{2x - 1}{x + 3}$

28. $f(x) = \dfrac{5 - 2x}{x - 2}$

29. $f(x) = \dfrac{x^2 - 3x + 1}{x}$

30. $f(x) = \dfrac{x^3 + 1}{x^2}$

h)

31. $f(x) = \dfrac{3x^2 - 2x}{x - 1}$

32. $f(x) = \dfrac{-x^2 + 5x - 5}{x - 3}$

Determine all asymptotes, and sketch the graph of each function. See Examples 4 and 5.

23. $f(x) = \dfrac{2}{x + 4}$

24. $f(x) = \dfrac{-3}{x - 1}$

Find all asymptotes, x-intercepts, and y-intercepts for the graph of each rational function, and sketch the graph of the function.

33. $f(x) = \dfrac{1}{x^2}$

34. $f(x) = \dfrac{2}{x^2 - 4x + 4}$

25. $f(x) = \dfrac{x}{x^2 - 9}$

26. $f(x) = \dfrac{-2}{x^2 + x - 2}$

35. $f(x) = \dfrac{2x - 3}{x^2 + x - 6}$

36. $f(x) = \dfrac{x}{x^2 + 4x + 4}$

37. $f(x) = \dfrac{x + 1}{x^2}$

38. $f(x) = \dfrac{x - 1}{x^2}$

39. $f(x) = \dfrac{2x - 1}{x^3 - 9x}$

40. $f(x) = \dfrac{2x^2 + 1}{x^3 - x}$

41. $f(x) = \dfrac{x}{x^2 - 1}$

42. $f(x) = \dfrac{x}{x^2 + x - 2}$

43. $f(x) = \dfrac{2}{x^2 + 1}$

44. $f(x) = \dfrac{x}{x^2 + 1}$

45. $f(x) = \dfrac{x^2}{x + 1}$

46. $f(x) = \dfrac{x^2}{x - 1}$

⟨4⟩ **Rational Inequalities**

Solve each rational inequality using the graphical method. State the solution set using interval notation, and graph it. See Example 6.

47. $\dfrac{1}{x} > 0$

48. $\dfrac{1}{x} \le 0$

49. $\dfrac{x}{x-3} > 0$

50. $\dfrac{a}{a+2} > 0$

51. $\dfrac{x+2}{x} \le 0$

52. $\dfrac{w-6}{w} \le 0$

53. $\dfrac{t-3}{t+6} > 0$

54. $\dfrac{x-2}{2x+5} < 0$

55. $\dfrac{x}{x+2} > -1$

56. $\dfrac{x+3}{x} \le -2$

57. $\dfrac{x-3}{x+5} \ge 2$

58. $\dfrac{x+2}{x-6} \le 3$

Solve each rational inequality using the test-point method. State the solution set using interval notation, and graph it. See Example 7.

59. $\dfrac{x-4}{x+2} \ge 0$

60. $\dfrac{x-3}{x+5} < 0$

61. $\dfrac{x-2}{x+3} < 1$

62. $\dfrac{x-3}{x+4} > 2$

63. $\dfrac{3}{x+2} > \dfrac{1}{x+1}$

64. $\dfrac{1}{x+1} < \dfrac{1}{x-1}$

65. $\dfrac{2}{x-5} > \dfrac{1}{x+4}$

66. $\dfrac{3}{x+2} > \dfrac{2}{x-1}$

67. $\dfrac{m}{m-5} + \dfrac{3}{m-1} > 0$

68. $\dfrac{p}{p-16} + \dfrac{2}{p-6} \le 0$

69. $\dfrac{x}{x-3} \le \dfrac{-8}{x-6}$

70. $\dfrac{x}{x+20} > \dfrac{2}{x+8}$

71. $\dfrac{x^2}{x^2+4} \ge 0$

72. $\dfrac{x^2+2x+3}{x^2} > 0$

73. $\dfrac{x^2}{x^2+9} < 0$

74. $\dfrac{x^2+4x+5}{x^2+1} \le 0$

Applications

Solve each problem.

75. *Oscillating modulators.* The number of oscillating modulators produced by a factory in t hours is given by the polynomial function $n(t) = t^2 + 6t$ for $t \geq 1$. The cost in dollars of operating the factory for t hours is given by the function $c(t) = 36t + 500$ for $t \geq 1$. The average cost per modulator is given by the rational function $f(t) = \frac{36t + 500}{t^2 + 6t}$ for $t \geq 1$. Graph the function f. What is the average cost per modulator at time $t = 20$ and time $t = 30$? What can you conclude about the average cost per modulator after a long period of time?

76. *Nonoscillating modulators.* The number of nonoscillating modulators produced by a factory in t hours is given by the polynomial function $n(t) = 16t$ for $t \geq 1$. The cost in dollars of operating the factory for t hours is given by the function $c(t) = 64t + 500$ for $t \geq 1$. The average cost per modulator is given by the rational function $f(t) = \frac{64t + 500}{16t}$ for $t \geq 1$. Graph the function f. What is the average cost per modulator at time $t = 10$ and $t = 20$? What can you conclude about the average cost per modulator after a long period of time?

77. *Average cost of an SUV.* Mercedes-Benz spent $700 million to design its new SUV (Motor Trend, www.motortrend.com). If it costs $25,000 to manufacture each SUV, then the average cost per vehicle in dollars when x vehicles are manufactured is given by the rational function

$$A(x) = \frac{25,000x + 700,000,000}{x}.$$

a) What is the horizontal asymptote for the graph of this function?

b) What is the average cost per vehicle when 50,000 vehicles are made?

c) For what number of vehicles is the average cost $30,000?

d) Graph this function for x ranging from 0 to 100,000.

78. *Average cost of a pill.* Assuming Pfizer spent a typical $350 million to develop its latest miracle drug and $0.10 each to make the pills, then the average cost per pill in dollars when x pills are made is given by the rational function

$$A(x) = \frac{0.10x + 350,000,000}{x}.$$

a) What is the horizontal asymptote for the graph of this function?

b) What is the average cost per pill when 100 million pills are made?

c) For what number of pills is the average cost per pill $2?

Photo for Exercise 78

d) Graph this function for x ranging from 0 to 100 million.   **84.** $f(x) = x^3$, $g(x) = x^3 + 1/x^2$

Getting More Involved

In each case find a rational function whose graph has the required asymptotes. Answers may vary.

85. The graph has the x-axis as a horizontal asymptote and the y-axis as a vertical asymptote.

Graphing Calculator Exercises

Sketch the graph of each pair of functions in the same coordinate system. What do you observe in each case?

79. $f(x) = x^2$, $g(x) = x^2 + 1/x$

80. $f(x) = x^2$, $g(x) = x^2 + 1/x^2$

81. $f(x) = |x|$, $g(x) = |x| + 1/x$

82. $f(x) = |x|$, $g(x) = |x| + 1/x^2$

83. $f(x) = \sqrt{x}$, $g(x) = \sqrt{x} + 1/x$

86. The graph has the x-axis as a horizontal asymptote and the line $x = 2$ as a vertical asymptote.

87. The graph has the x-axis as a horizontal asymptote and lines $x = 3$ and $x = -1$ as vertical asymptotes.

88. The graph has the line $y = 2$ as a horizontal asymptote and the line $x = 1$ as a vertical asymptote.

11.6 **Combining Functions**

In this section you will learn how to combine functions to obtain new functions.

In This Section

⟨1⟩ **Basic Operations with Functions**

⟨2⟩ **Composition**

⟨1⟩ Basic Operations with Functions

An entrepreneur plans to rent a stand at a farmers market for \$25 per day to sell strawberries. If she buys x flats of berries for \$5 per flat and sells them for \$9 per flat, then her daily cost in dollars can be written as a function of x:

$$C(x) = 5x + 25$$

Assuming she sells as many flats as she buys, her revenue in dollars is also a function of x:

$$R(x) = 9x$$

Because profit is revenue minus cost, we can find a function for the profit by subtracting the functions for cost and revenue:

$$\begin{aligned}
P(x) &= R(x) - C(x) \\
&= 9x - (5x + 25) \\
&= 4x - 25
\end{aligned}$$

The function $P(x) = 4x - 25$ expresses the daily profit as a function of x. Since $P(6) = -1$ and $P(7) = 3$, the profit is negative if 6 or fewer flats are sold and positive if 7 or more flats are sold.

In the example of the entrepreneur we subtracted two functions to find a new function. In other cases we may use addition, multiplication, or division to combine two functions. For any two given functions we can define the sum, difference, product, and quotient functions as follows.

> **Sum, Difference, Product, and Quotient Functions**
> Given two functions f and g, the functions $f + g$, $f - g$, $f \cdot g$, and $\frac{f}{g}$ are defined as follows:
>
> Sum function: $\quad (f + g)(x) = f(x) + g(x)$
> Difference function: $\quad (f - g)(x) = f(x) - g(x)$
> Product function: $\quad (f \cdot g)(x) = f(x) \cdot g(x)$
>
> Quotient function: $\quad \left(\dfrac{f}{g}\right)(x) = \dfrac{f(x)}{g(x)} \quad$ provided that $g(x) \neq 0$

The domain of the function $f + g$, $f - g$, $f \cdot g$, or $\frac{f}{g}$ is the intersection of the domain of f and the domain of g. For the function $\frac{f}{g}$ we also rule out any values of x for which $g(x) = 0$.

EXAMPLE 1

Operations with functions
Let $f(x) = 4x - 12$ and $g(x) = x - 3$. Find the following.

a) $(f + g)(x)$
b) $(f - g)(x)$
c) $(f \cdot g)(x)$
d) $\left(\dfrac{f}{g}\right)(x)$

⟨ Helpful Hint ⟩

Note that we use $f + g$, $f - g$, $f \cdot g$, and f/g to name these functions only because there is no application in mind here. We generally use a single letter to name functions after they are combined as we did when using P for the profit function rather than $R - C$.

Solution

a) $(f + g)(x) = f(x) + g(x)$
$= 4x - 12 + x - 3$
$= 5x - 15$

b) $(f - g)(x) = f(x) - g(x)$
$= 4x - 12 - (x - 3)$
$= 3x - 9$

c) $(f \cdot g)(x) = f(x) \cdot g(x)$
$= (4x - 12)(x - 3)$
$= 4x^2 - 24x + 36$

d) $\left(\dfrac{f}{g}\right)(x) = \dfrac{f(x)}{g(x)} = \dfrac{4x - 12}{x - 3} = \dfrac{4(x - 3)}{x - 3} = 4 \quad$ for $x \neq 3$.

> Now do Exercises 1–4

EXAMPLE 2

Evaluating a sum function
Let $f(x) = 4x - 12$ and $g(x) = x - 3$. Find $(f + g)(2)$.

Solution

In Example 1(a) we found a general formula for the function $f + g$, namely, $(f + g)(x) = 5x - 15$. If we replace x by 2, we get

$$(f + g)(2) = 5(2) - 15$$
$$= -5.$$

We can also find $(f + g)(2)$ by evaluating each function separately and then adding the results. Because $f(2) = -4$ and $g(2) = -1$, we get

$$(f + g)(2) = f(2) + g(2)$$
$$= -4 + (-1)$$
$$= -5.$$

Now do Exercises 5–12

‹2› Composition

A salesperson's monthly salary is a function of the number of cars he sells: \$1000 plus \$50 for each car sold. If we let S be his salary and n be the number of cars sold, then S in dollars is a function of n:

$$S = 1000 + 50n$$

Each month the dealer contributes \$100 plus 5% of his salary to a profit-sharing plan. If P represents the amount put into profit sharing, then P (in dollars) is a function of S:

$$P = 100 + 0.05S$$

Now P is a function of S, and S is a function of n. Is P a function of n? The value of n certainly determines the value of P. In fact, we can write a formula for P in terms of n by substituting one formula into the other:

$$P = 100 + 0.05S$$
$$= 100 + 0.05(1000 + 50n) \quad \text{Substitute } S = 1000 + 50n.$$
$$= 100 + 50 + 2.5n \quad\quad\quad \text{Distributive property}$$
$$= 150 + 2.5n$$

Now P is written as a function of n, bypassing S. We call this idea **composition of functions.**

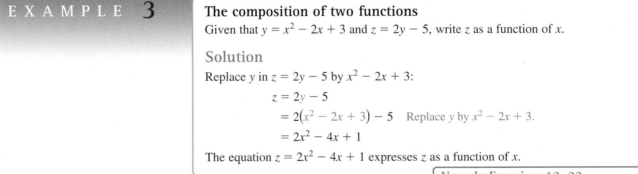

EXAMPLE   3

The composition of two functions

Given that $y = x^2 - 2x + 3$ and $z = 2y - 5$, write z as a function of x.

Solution

Replace y in $z = 2y - 5$ by $x^2 - 2x + 3$:

$$z = 2y - 5$$
$$= 2(x^2 - 2x + 3) - 5 \quad \text{Replace } y \text{ by } x^2 - 2x + 3.$$
$$= 2x^2 - 4x + 1$$

The equation $z = 2x^2 - 4x + 1$ expresses z as a function of x.

Now do Exercises 13–22

A composition of functions is simply one function followed by another. The output of the first function is the input for the second. For example, let $f(x) = x - 3$ and $g(x) = x^2$. If we start with 5, then $f(5) = 5 - 3 = 2$. Now use 2 as the input for g, $g(2) = 2^2 = 4$. So $g(f(5)) = 4$. The function that pairs 5 with 4 is called the *composition* of g and f, and we write $(g \circ f)(5) = 4$. Since we subtracted 3 first and then squared, a formula for $g \circ f$ is $(g \circ f)(x) = (x - 3)^2$. If we apply g first and then f, we get a different function, $(f \circ g)(x) = x^2 - 3$, the composition of f and g.

A composition of functions can be viewed as two function machines where the output of the first is the input of the second.

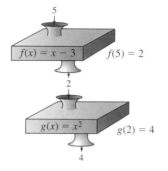

Composition of Functions

The **composition** of f and g is denoted $f \circ g$ and is defined by the equation

$$(f \circ g)(x) = f(g(x)),$$

provided that $g(x)$ is in the domain of f.

The notation $f \circ g$ is read as "the composition of f and g" or "f compose g." The diagram in Fig. 11.59 shows a function g pairing numbers in its domain with numbers in its range. If the range of g is contained in or equal to the domain of f, then f pairs the second coordinates of g with numbers in the range of f. The composition function $f \circ g$ is a rule for pairing numbers in the domain of g directly with numbers in the range of f, bypassing the middle set. The domain of the function $f \circ g$ is the domain of g (or a subset of it), and the range of $f \circ g$ is the range of f (or a subset of it).

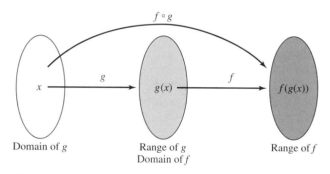

Figure 11.59

CAUTION The order in which functions are written is important in composition. For the function $f \circ g$ the function f is applied to $g(x)$. For the function $g \circ f$ the function g is applied to $f(x)$. The function closest to the variable x is applied first.

E X A M P L E   4

Evaluating compositions

Let $f(x) = 3x - 2$ and $g(x) = x^2 + 2x$. Evaluate each of the following expressions.

a) $g(f(3))$ **b)** $f(g(-4))$ **c)** $(g \circ f)(2)$ **d)** $(f \circ g)(2)$

Set $y_1 = 3x - 2$ and $y_2 = x^2 + 2x$. You can find the composition for Examples 4(c) and 4(d) by evaluating $y_2(y_1(2))$ and $y_1(y_2(2))$. Note that the order in which you evaluate the functions is critical.

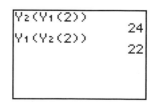

Solution

a) Because $f(3) = 3(3) - 2 = 7$, we have

$$g(f(3)) = g(7) = 7^2 + 2 \cdot 7 = 63.$$

So $g(f(3)) = 63$.

b) Because $g(-4) = (-4)^2 + 2(-4) = 8$, we have

$$f(g(-4)) = f(8) = 3(8) - 2 = 22.$$

So $f(g(-4)) = 22$.

c) Because $(g \circ f)(2) = g(f(2))$ we first find $f(2)$:

$$f(2) = 3(2) - 2 = 4$$

Because $f(2) = 4$, we have

$$(g \circ f)(2) = g(f(2)) = g(4) = 4^2 + 2(4) = 24.$$

So $(g \circ f)(2) = 24$.

d) Because $(f \circ g)(2) = f(g(2))$, we first find $g(2)$:

$$g(2) = 2^2 + 2(2) = 8$$

Because $g(2) = 8$, we have

$$(f \circ g)(2) = f(g(2)) = f(8) = 3(8) - 2 = 22.$$

So $(f \circ g)(2) = 22$.

Now do Exercises 23–36

In Example 4, we found specific values of compositions of two functions. In Example 5, we find a general formula for the two functions from Example 4.

EXAMPLE 5

Finding formulas for compositions

Let $f(x) = 3x - 2$ and $g(x) = x^2 + 2x$. Find the following.

a) $(g \circ f)(x)$ **b)** $(f \circ g)(x)$

Solution

a) Since $f(x) = 3x - 2$, we replace $f(x)$ with $3x - 2$:

$$(g \circ f)(x) = g(f(x))$$
$$= g(3x - 2) \qquad \text{Replace } f(x) \text{ with } 3x - 2.$$
$$= (3x - 2)^2 + 2(3x - 2) \qquad \text{Replace } x \text{ in } g(x) = x^2 + 2x \text{ with } 3x - 2.$$
$$= 9x^2 - 12x + 4 + 6x - 4 \qquad \text{Simplify.}$$
$$= 9x^2 - 6x$$

So $(g \circ f)(x) = 9x^2 - 6x$.

b) Since $g(x) = x^2 + 2x$, we replace $g(x)$ with $x^2 + 2x$:

$$(f \circ g)(x) = f(g(x)) \qquad \text{Definition of composition}$$
$$= f(x^2 + 2x) \qquad \text{Replace } g(x) \text{ with } x^2 + 2x.$$
$$= 3(x^2 + 2x) - 2 \qquad \text{Replace } x \text{ in } f(x) = 3x - 2 \text{ with } x^2 + 2x.$$
$$= 3x^2 + 6x - 2 \qquad \text{Simplify.}$$

So $(f \circ g)(x) = 3x^2 + 6x - 2$.

Now do Exercises 37–46

Notice that in Example 4(c) and (d), $(g \circ f)(2) \neq (f \circ g)(2)$. In Example 5(a) and (b) we see that $(g \circ f)(x)$ and $(f \circ g)(x)$ have different formulas defining them. In general, $f \circ g \neq g \circ f$. However, in Section 11.7 we will see some functions for which the composition in either order results in the same function.

It is often useful to view a complicated function as a composition of simpler functions. For example, the function $Q(x) = (x - 3)^2$ consists of two operations, subtracting 3 and squaring. So Q can be described as a composition of the functions $f(x) = x - 3$ and $g(x) = x^2$. To check this, we find $(g \circ f)(x)$:

$$(g \circ f)(x) = g(f(x))$$
$$= g(x - 3)$$
$$= (x - 3)^2$$

We can express the fact that Q is the same as the composition function $g \circ f$ by writing $Q = g \circ f$ or $Q(x) = (g \circ f)(x)$.

EXAMPLE 6

Expressing a function as a composition of simpler functions

Let $f(x) = x - 2$, $g(x) = 3x$, and $h(x) = \sqrt{x}$. Write each of the following functions as a composition, using f, g, and h.

a) $F(x) = \sqrt{x - 2}$ **b)** $H(x) = x - 4$ **c)** $K(x) = 3x - 6$

Solution

a) The function F consists of first subtracting 2 from x and then taking the square root of that result. So $F = h \circ f$. Check this result by finding $(h \circ f)(x)$:

$$(h \circ f)(x) = h(f(x)) = h(x - 2) = \sqrt{x - 2}$$

b) Subtracting 4 from x can be accomplished by subtracting 2 from x and then subtracting 2 from that result. So $H = f \circ f$. Check by finding $(f \circ f)(x)$:

$$(f \circ f)(x) = f(f(x)) = f(x - 2) = x - 2 - 2 = x - 4$$

c) Notice that $K(x) = 3(x - 2)$. The function K consists of subtracting 2 from x and then multiplying the result by 3. So $K = g \circ f$. Check by finding $(g \circ f)(x)$:

$$(g \circ f)(x) = g(f(x)) = g(x - 2) = 3(x - 2) = 3x - 6$$

Now do Exercises 47–56

CAUTION In Example 6(a) we have $F = h \circ f$ because in F we subtract 2 before taking the square root. If we had the function $G(x) = \sqrt{x} - 2$, we would take the square root before subtracting 2. So $G = f \circ h$. Notice how important the order of operations is here.

In Example 7, we see functions for which the composition is the identity function. Each function undoes what the other function does. We will study functions of this type further in Section 11.7.

EXAMPLE 7

Composition of functions

Show that $(f \circ g)(x) = x$ for each pair of functions.

a) $f(x) = 2x - 1$ and $g(x) = \dfrac{x + 1}{2}$

b) $f(x) = x^3 + 5$ and $g(x) = (x - 5)^{1/3}$

Solution

a) $(f \circ g)(x) = f(g(x)) = f\left(\dfrac{x + 1}{2}\right)$

$$= 2\left(\dfrac{x + 1}{2}\right) - 1$$
$$= x + 1 - 1$$
$$= x$$

b) $(f \circ g)(x) = f(g(x)) = f\left((x - 5)^{1/3}\right)$

$$= \left((x - 5)^{1/3}\right)^3 + 5$$
$$= x - 5 + 5$$
$$= x$$

Now do Exercises 57–64

Warm-Ups ▼

Fill in the blank.

1. The function $(f + g)(x)$ is the _____ of the functions $f(x)$ and $g(x)$.

2. The function $(f - g)(x)$ is the _____ of the functions $f(x)$ and $g(x)$.

3. The function $(f \cdot g)(x)$ is the _____ of the functions $f(x)$ and $g(x)$.

4. The function $(f/g)(x)$ is the _____ of the functions $f(x)$ and $g(x)$.

5. The function $(f \circ g)(x)$ is the _____ of the functions $f(x)$ and $g(x)$.

True or false?

6. For the composition of f and g, the function f is evaluated after g.

7. If $f(x) = x - 2$ and $g(x) = x + 3$, then $(f - g)(x) = -5$.

8. If $f(x) = x^2 - 2x$ and $g(x) = 3x + 9$, then $(f + g)(x) = x^2 + x + 9$.

9. If $f(x) = x^2$ and $g(x) = x + 2$, then $(f \circ g)(x) = x^2 + 2$.

10. If $f(x) = 3x$ and $g(x) = \dfrac{x}{3}$, then $(f \circ g)(x) = x$.

11. The function $f \circ g$ and $g \circ f$ are always the same.

12. If $F(x) = (x - 1)^2$, $h(x) = x - 1$, and $g(x) = x^2$, then $F = g \circ h$.

Exercises

11.6

⟨ **Study Tips** ⟩

- Stay alert for the entire class period. The first 20 minutes are the easiest, and the last 20 minutes the hardest.
- Think of how much time you will have to spend outside of class figuring out what happened during the last 20 minutes in which you were daydreaming.

⟨1⟩ Basic Operations with Functions

Let $f(x) = 4x - 3$ and $g(x) = x^2 - 2x$. Find the following.
See Examples 1 and 2.

1. $(f + g)(x)$
2. $(f - g)(x)$

3. $(f \cdot g)(x)$
4. $\left(\dfrac{f}{g}\right)(x)$

5. $(f + g)(3)$
6. $(f + g)(2)$
7. $(f - g)(-3)$
8. $(f - g)(-2)$
9. $(f \cdot g)(-1)$
10. $(f \cdot g)(-2)$

11. $\left(\dfrac{f}{g}\right)(4)$
12. $\left(\dfrac{f}{g}\right)(-2)$

⟨2⟩ Composition

Use the two functions to write y as a function of x.
See Example 3.

13. $y = 2a, a = 3x$
14. $y = w^2, w = 5x$
15. $y = 3a - 2, a = 2x - 6$
16. $y = 2c + 3, c = -3x + 4$
17. $y = 2d + 1, d = \dfrac{x + 1}{2}$

18. $y = -3d + 2, d = \dfrac{2 - x}{3}$

19. $y = m^2 - 1, m = x + 1$

20. $y = n^2 - 3n + 1, n = x + 2$

21. $y = \dfrac{a - 3}{a + 2}, a = \dfrac{2x + 3}{1 - x}$

22. $y = \dfrac{w + 2}{w - 5}, w = \dfrac{5x + 2}{x - 1}$

Let $f(x) = 2x - 3$, $g(x) = x^2 + 3x$, and $h(x) = \dfrac{x + 3}{2}$. Find the following. See Examples 4 and 5.

23. $(g \circ f)(1)$ **24.** $(f \circ g)(-2)$

25. $(f \circ g)(1)$ **26.** $(g \circ f)(-2)$

27. $(f \circ f)(4)$ **28.** $(h \circ h)(3)$

29. $(h \circ f)(5)$ **30.** $(f \circ h)(0)$

31. $(f \circ h)(5)$ **32.** $(h \circ f)(0)$

33. $(g \circ h)(-1)$ **34.** $(h \circ g)(-1)$

35. $(f \circ g)(2.36)$ **36.** $(h \circ f)(23.761)$

37. $(g \circ f)(x)$ **38.** $(g \circ h)(x)$

39. $(f \circ g)(x)$ **40.** $(h \circ g)(x)$

41. $(h \circ f)(x)$ **42.** $(f \circ h)(x)$

43. $(f \circ f)(x)$ **44.** $(g \circ g)(x)$

45. $(h \circ h)(x)$ **46.** $(f \circ f \circ f)(x)$

Let $f(x) = \sqrt{x}$, $g(x) = x^2$, and $h(x) = x - 3$. Write each of the following functions as a composition using f, g, or h. See Example 6.

47. $F(x) = \sqrt{x - 3}$ **48.** $N(x) = \sqrt{x} - 3$

49. $G(x) = x^2 - 6x + 9$ **50.** $P(x) = x$ for $x \geq 0$

51. $H(x) = x^2 - 3$ **52.** $M(x) = x^{1/4}$

53. $J(x) = x - 6$ **54.** $R(x) = \sqrt{x^2 - 3}$

55. $K(x) = x^4$ **56.** $Q(x) = \sqrt{x^2 - 6x + 9}$

Show that $(f \circ g)(x) = x$ and $(g \circ f)(x) = x$ for each given pair of functions. See Example 7.

57. $f(x) = 3x + 5, g(x) = \dfrac{x - 5}{3}$

58. $f(x) = 3x - 7, g(x) = \dfrac{x + 7}{3}$

59. $f(x) = x^3 - 9, g(x) = \sqrt[3]{x + 9}$

60. $f(x) = x^3 + 1, g(x) = \sqrt[3]{x - 1}$

61. $f(x) = \dfrac{x - 1}{x + 1}, g(x) = \dfrac{x + 1}{1 - x}$

62. $f(x) = \dfrac{x + 1}{x - 3}, g(x) = \dfrac{3x + 1}{x - 1}$

63. $f(x) = \dfrac{1}{x}, g(x) = \dfrac{1}{x}$

64. $f(x) = 2x^3, g(x) = \left(\dfrac{x}{2}\right)^{1/3}$

Miscellaneous

Let $f(x) = x^2$ and $g(x) = x + 5$. Determine whether each of these statements is true or false.

65. $f(3) = 9$

66. $g(3) = 8$

67. $(f + g)(4) = 21$

68. $(f - g)(0) = 5$

69. $(f \cdot g)(3) = 72$

70. $(f/g)(0) = 5$

71. $(f \circ g)(2) = 14$

72. $(g \circ f)(7) = 54$

73. $f(g(x)) = x^2 + 25$

74. $(g \circ f)(x) = x^2 + 5$

75. If $h(x) = x^2 + 10x + 25$, then $h = f \circ g$.

76. If $p(x) = x^2 + 5$, then $p = g \circ f$.

Applications

Solve each problem.

77. Area. A square gate in a wood fence has a diagonal brace with a length of 10 feet.

    a) Find the area of the square gate.

    b) Write a formula for the area of a square as a function of the length of its diagonal.

78. Perimeter. Write a formula for the perimeter of a square as a function of its area.

79. Profit function. A plastic bag manufacturer has determined that the company can sell as many bags as it can produce each month. If it produces x thousand bags in a month, the revenue is $R(x) = x^2 - 10x + 30$ dollars, and the cost is $C(x) = 2x^2 - 30x + 200$ dollars. Use the fact that profit is revenue minus cost to write the profit as a function of x.

80. *Area of a sign.* A sign is in the shape of a square with a semicircle of radius x adjoining one side and a semicircle of diameter x removed from the opposite side. If the sides of the square are length $2x$, then write the area of the sign as a function of x.

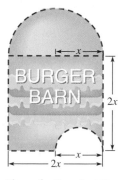

Figure for Exercise 80

81. *Junk food expenditures.* Suppose the average family spends 25% of its income on food, $F = 0.25I$, and 10% of each food dollar on junk food, $J = 0.10F$. Write J as a function of I.

82. *Area of an inscribed circle.* A pipe of radius r must pass through a square hole of area M as shown in the figure. Write the cross-sectional area of the pipe A as a function of M.

Figure for Exercise 82

83. *Displacement-length ratio.* To find the displacement-length ratio D for a sailboat, first find x, where $x = (L/100)^3$ and L is the length at the water line in feet (www.sailing.com). Next find D, where $D = (d/2240)/x$ and d is the displacement in pounds.

   a) For the Pacific Seacraft 40, $L = 30$ ft 3 in. and $d = 24{,}665$ pounds. Find D.

   b) For a boat with a displacement of 25,000 pounds, write D as a function of L.

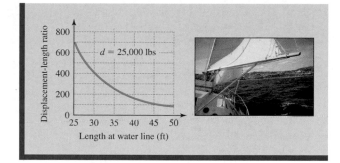

Figure for Exercise 83

   c) The graph for the function in part (b) is shown in the accompanying figure. For a fixed displacement, does the displacement-length ratio increase or decrease as the length increases?

84. *Sail area-displacement ratio.* To find the sail area-displacement ratio S, first find y, where $y = (d/64)^{2/3}$ and d is the displacement in pounds. Next find S, where $S = A/y$ and A is the sail area in square feet.

   a) For the Pacific Seacraft 40, $A = 846$ square feet (ft^2) and $d = 24{,}665$ pounds. Find S.

   b) For a boat with a sail area of 900 ft^2, write S as a function of d.

   c) For a fixed sail area, does S increase or decrease as the displacement increases?

Getting More Involved

85. *Discussion*

Let $f(x) = \sqrt{x} - 4$ and $g(x) = \sqrt{x}$. Find the domains of f, g, and $g \circ f$.

86. *Discussion*

Let $f(x) = \sqrt{x - 4}$ and $g(x) = \sqrt{x - 8}$. Find the domains of f, g, and $f + g$.

Graphing Calculator Exercises

87. Graph $y_1 = x$, $y_2 = \sqrt{x}$, and $y_3 = x + \sqrt{x}$ in the same screen. Find the domain and range of $y_3 = x + \sqrt{x}$ by examining its graph. (On some graphing calculators you can enter y_3 as $y_3 = y_1 + y_2$.)

88. Graph $y_1 = |x|$, $y_2 = |x - 3|$, and $y_3 = |x| + |x - 3|$. Find the domain and range of $y_3 = |x| + |x - 3|$ by examining its graph.

11.7   Inverse Functions

In Section 11.6, we introduced the idea of a pair of functions such that $(f \circ g)(x) = x$ and $(g \circ f)(x) = x$. Each function reverses what the other function does. In this section we explore that idea further.

⟨1⟩ Inverse of a Function

You can buy a 6-, 7-, or 8-foot conference table in the K-LOG Catalog for $299, $329, or $349, respectively. The set

$$f = \{(6, 299), (7, 329), (8, 349)\}$$

gives the price as a function of the length. We use the letter f as a name for this set or function, just as we use the letter f as a name for a function in the function notation. In the function f, lengths in the domain $\{6, 7, 8\}$ are paired with prices in the range $\{299, 329, 349\}$. The **inverse** of the function f, denoted f^{-1}, is a function whose ordered pairs are obtained from f by interchanging the x- and y-coordinates:

$$f^{-1} = \{(299, 6), (329, 7), (349, 8)\}$$

We read f^{-1} as "f inverse." The domain of f^{-1} is $\{299, 329, 349\}$, and the range of f^{-1} is $\{6, 7, 8\}$. The inverse function reverses what the function does: it pairs prices in the range of f with lengths in the domain of f. For example, to find the cost of a 7-foot table, we use the function f to get $f(7) = 329$. To find the length of a table costing $349, we use the function f^{-1} to get $f^{-1}(349) = 8$. Of course, we could find the length of a $349 table by looking at the function f, but f^{-1} is a function whose input is price and whose output is length. In general, *the domain of f^{-1} is the range of f, and the range of f^{-1} is the domain of f.* See Fig. 11.60.

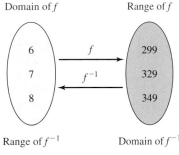

Domain of f Range of f

Range of f^{-1} Domain of f^{-1}

Figure 11.60

CAUTION   The -1 in f^{-1} is not read as an exponent. It does not mean $\dfrac{1}{f}$.

The cost per ink cartridge is a function of the number of boxes of ink cartridges purchased:

$$g = \{(1, 4.85), (2, 4.60), (3, 4.60), (4, 4.35)\}$$

If we interchange the first and second coordinates in the ordered pairs of this function, we get

$$\{(4.85, 1), (4.60, 2), (4.60, 3), (4.35, 4)\}.$$

This set of ordered pairs is not a function because it contains ordered pairs with the same first coordinates and different second coordinates. So g does not have an inverse function. A function is **invertible** if you obtain a function when the coordinates of all ordered pairs are reversed. So f is invertible and g is not invertible.

Any function that pairs more than one number in the domain with the same number in the range is not invertible, because the set is not a function when the ordered pairs are reversed. So we turn our attention to functions where each member of the domain corresponds to one member of the range and vice versa.

⟨ **Helpful Hint** ⟩

Consider the universal product codes (UPC) and the prices for all of the items in your favorite grocery store. The price of an item is a function of the UPC because every UPC determines a price. This function is not invertible because you cannot determine the UPC from a given price.

One-to-One Function

If a function is such that no two ordered pairs have different x-coordinates and the same y-coordinate, then the function is called a **one-to-one** function.

In a one-to-one function each member of the domain corresponds to just one member of the range, and each member of the range corresponds to just one member of the domain. *Functions that are one-to-one are invertible functions.*

Inverse Function

The inverse of a one-to-one function f is the function f^{-1}, which is obtained from f by interchanging the coordinates in each ordered pair of f.

EXAMPLE 1

Identifying invertible functions

Determine whether each function is invertible. If it is invertible, then find the inverse function.

  a) $f = \{(2, 4), (-2, 4), (3, 9)\}$

  b) $g = \left\{\left(2, \dfrac{1}{2}\right), \left(5, \dfrac{1}{5}\right), \left(7, \dfrac{1}{7}\right)\right\}$

  c) $h = \{(3, 5), (7, 9)\}$

Solution

  a)   Since $(2, 4)$ and $(-2, 4)$ have the same y-coordinate, this function is not one-to-one, and it is not invertible.

  b)   This function is one-to-one, and so it is invertible.

$$g^{-1} = \left\{\left(\frac{1}{2}, 2\right), \left(\frac{1}{5}, 5\right), \left(\frac{1}{7}, 7\right)\right\}$$

  c)   This function is invertible, and $h^{-1} = \{(5, 3), (9, 7)\}$.

                                                    Now do Exercises 1–10

You learned to use the vertical-line test in Section 11.1 to determine whether a graph is the graph of a function. The **horizontal-line test** is a similar visual test for determining whether a function is invertible. If a horizontal line crosses a graph two (or more) times, as in Fig. 11.61, then there are two points on the graph, say (x_1, y) and (x_2, y), that have different x-coordinates and the same y-coordinate. So the function is not one-to-one, and the function is not invertible.

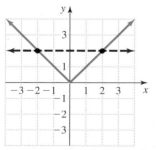

Figure 11.61

Horizontal-Line Test

A function is invertible if and only if no horizontal line crosses its graph more than once.

E X A M P L E   2

Using the horizontal–line test

Determine whether each function is invertible by examining its graph.

a)

b)

< **Helpful Hint** >

Tests such as the vertical-line test and the horizontal-line test are certainly not accurate in all cases. We discuss these tests to get a visual idea of what graphs of functions and invertible functions look like.

Solution

a) This function is not invertible because a horizontal line can be drawn so that it crosses the graph at $(2, 4)$ and $(-2, 4)$.

b) This function is invertible because every horizontal line that crosses the graph crosses it only once.

Now do Exercises 11–14

⟨2⟩ Identifying Inverse Functions

Consider the one-to-one function $f(x) = 3x$. The inverse function must reverse the ordered pairs of the function. Because division by 3 undoes multiplication by 3, we could guess that $g(x) = \frac{x}{3}$ is the inverse function. To verify our guess, we can use the following rule for determining whether two given functions are inverses of each other.

> **Identifying Inverse Functions**
>
> Functions f and g are inverses of each other if and only if
>
> $(g \circ f)(x) = x$ for every number x in the domain of f and
>
> $(f \circ g)(x) = x$ for every number x in the domain of g.

In Example 3, we verify that $f(x) = 3x$ and $g(x) = \frac{x}{3}$ are inverses.

E X A M P L E   3

Identifying inverse functions

Determine whether the functions f and g are inverses of each other.

a) $f(x) = 3x$ and $g(x) = \dfrac{x}{3}$ **b)** $f(x) = 2x - 1$ and $g(x) = \dfrac{1}{2}x + 1$

c) $f(x) = x^2$ and $g(x) = \sqrt{x}$

Solution

a) Find $g \circ f$ and $f \circ g$:

$$(g \circ f)(x) = g(f(x)) = g(3x) = \frac{3x}{3} = x$$

$$(f \circ g)(x) = f(g(x)) = f\left(\frac{x}{3}\right) = 3 \cdot \frac{x}{3} = x$$

Because each of these equations is true for any real number x, f and g are inverses of each other. We write $g = f^{-1}$ or $f^{-1}(x) = \frac{x}{3}$.

b) Find the composition of g and f:

$$(g \circ f)(x) = g(f(x))$$
$$= g(2x - 1) = \frac{1}{2}(2x - 1) + 1 = x + \frac{1}{2}$$

So f and g are not inverses of each other.

c) If x is any real number, we can write

$$(g \circ f)(x) = g(f(x))$$
$$= g(x^2) = \sqrt{x^2} = |x|.$$

The domain of f is $(-\infty, \infty)$, and $|x| \neq x$ if x is negative. So g and f are not inverses of each other. Note that $f(x) = x^2$ is not a one-to-one function, since both $(3, 9)$ and $(-3, 9)$ are ordered pairs of this function. Thus, $f(x) = x^2$ does not have an inverse.

Now do Exercises 15–22

⟨3⟩ Switch-and-Solve Strategy

If an invertible function is defined by a list of ordered pairs, as in Example 1, then the inverse function is found by simply interchanging the coordinates in the ordered pairs. If an invertible function is defined by a formula, then the inverse function must reverse or undo what the function does. Because the inverse function interchanges the roles of x and y, we interchange x and y in the formula and then solve the new formula for y to undo what the original function did. The steps to follow in this **switch-and-solve** strategy are given in the following box and illustrated in Examples 4 and 5.

Strategy for Finding f^{-1} by Switch-and-Solve

1. Replace $f(x)$ by y.
2. Interchange x and y.
3. Solve the equation for y.
4. Replace y by $f^{-1}(x)$.

EXAMPLE **4**

The switch-and-solve strategy
Find the inverse of $h(x) = 2x + 1$.

Solution

First write the function as $y = 2x + 1$, and then interchange x and y:

$$y = 2x + 1$$
$$x = 2y + 1 \qquad \text{Interchange } x \text{ and } y.$$
$$x - 1 = 2y \qquad \text{Solve for } y.$$
$$\frac{x - 1}{2} = y$$
$$h^{-1}(x) = \frac{x - 1}{2} \qquad \text{Replace } y \text{ by } h^{-1}(x).$$

We can verify that h and h^{-1} are inverses by using composition:

$$\left(h^{-1} \circ h\right)(x) = h^{-1}(h(x)) = h^{-1}(2x + 1) = \frac{2x + 1 - 1}{2} = \frac{2x}{2} = x$$

$$\left(h \circ h^{-1}\right)(x) = h(h^{-1}(x)) = h\left(\frac{x - 1}{2}\right) = 2 \cdot \frac{x - 1}{2} + 1 = x - 1 + 1 = x$$

Now do Exercises 23–36

E X A M P L E   **5**

The switch-and-solve strategy

If $f(x) = \dfrac{x + 1}{x - 3}$, find $f^{-1}(x)$.

Solution

Replace $f(x)$ by y, interchange x and y, and then solve for y:

$$y = \frac{x + 1}{x - 3} \qquad \text{Use } y \text{ in place of } f(x).$$

$$x = \frac{y + 1}{y - 3} \qquad \text{Switch } x \text{ and } y.$$

$$x(y - 3) = y + 1 \qquad \text{Multiply each side by } y - 3.$$

$$xy - 3x = y + 1 \qquad \text{Distributive property}$$

$$xy - y = 3x + 1$$

$$y(x - 1) = 3x + 1 \qquad \text{Factor out } y.$$

$$y = \frac{3x + 1}{x - 1} \qquad \text{Divide each side by } x - 1.$$

$$f^{-1}(x) = \frac{3x + 1}{x - 1} \qquad \text{Replace } y \text{ by } f^{-1}(x).$$

To check, compute $(f \circ f^{-1})(x)$:

$$(f \circ f^{-1})(x) = f\left(\frac{3x + 1}{x - 1}\right) = \frac{\dfrac{3x + 1}{x - 1} + 1}{\dfrac{3x + 1}{x - 1} - 3} = \frac{(x - 1)\left(\dfrac{3x + 1}{x - 1} + 1\right)}{(x - 1)\left(\dfrac{3x + 1}{x - 1} - 3\right)}$$

$$= \frac{3x + 1 + 1(x - 1)}{3x + 1 - 3(x - 1)} = \frac{4x}{4} = x$$

You should check that $(f^{-1} \circ f)(x) = x$.

Now do Exercises 37–40

If we use the switch-and-solve strategy to find the inverse of $f(x) = x^3$, then we get $f^{-1}(x) = x^{1/3}$. For $h(x) = 6x$ we have $h^{-1}(x) = \frac{x}{6}$. The inverse of $k(x) = x - 9$ is $k^{-1}(x) = x + 9$. For each of these functions there is an appropriate operation of arithmetic that undoes what the function does.

If a function involves two operations, the inverse function undoes those operations in the opposite order from which the function does them. For example, the function $g(x) = 3x - 5$ multiplies x by 3 and then subtracts 5 from that result. To undo these operations, we add 5 and then divide the result by 3. So,

$$g^{-1}(x) = \frac{x+5}{3}.$$

Note that $g^{-1}(x) \neq \frac{x}{3} + 5$.

⟨4⟩ Even Roots or Even Powers

We need to use special care in finding inverses for functions that involve even roots or even powers. We saw in Example 3(c) that $f(x) = x^2$ is not the inverse of $g(x) = \sqrt{x}$. However, because $g(x) = \sqrt{x}$ is a one-to-one function, it has an inverse. The domain of g is $[0, \infty)$, and the range is $[0, \infty)$. So the inverse of g must have domain $[0, \infty)$ and range $[0, \infty)$. See Fig. 11.62. The only reason that $f(x) = x^2$ is not the inverse of g is that it has the wrong domain. So to write the inverse function, we must use the appropriate domain:

$$g^{-1}(x) = x^2 \qquad \text{for} \quad x \geq 0$$

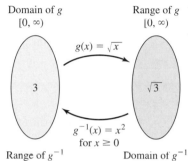

Domain of g
[0, ∞)

Range of g
[0, ∞)

$g(x) = \sqrt{x}$

3

$\sqrt{3}$

$g^{-1}(x) = x^2$
for $x \geq 0$

Range of g^{-1}

Domain of g^{-1}

Figure 11.62

Note that by restricting the domain of g^{-1} to $[0, \infty)$, g^{-1} is one-to-one. With this restriction it is true that $(g \circ g^{-1})(x) = x$ and $(g^{-1} \circ g)(x) = x$ for every nonnegative number x.

E X A M P L E 6

Inverse of a function with an even exponent

Find the inverse of the function $f(x) = (x - 3)^2$ for $x \geq 3$.

Solution

Because of the restriction $x \geq 3$, f is a one-to-one function with domain $[3, \infty)$ and range $[0, \infty)$. The domain of the inverse function is $[0, \infty)$, and its range is $[3, \infty)$. Use the switch-and-solve strategy to find the formula for the inverse:

$$y = (x - 3)^2$$
$$x = (y - 3)^2$$
$$y - 3 = \pm\sqrt{x}$$
$$y = 3 \pm \sqrt{x}$$

Because the inverse function must have range $[3, \infty)$, we use the formula $f^{-1}(x) = 3 + \sqrt{x}$. Because the domain of f^{-1} is assumed to be $[0, \infty)$, no restriction is required on x.

Now do Exercises 41–48

⟨5⟩ Graphs of f and f⁻¹

Consider $f(x) = x^2$ for $x \geq 0$ and $f^{-1}(x) = \sqrt{x}$. Their graphs are shown in Fig. 11.63 on page 766. Notice the symmetry. If we folded the paper along the line $y = x$, the two graphs would coincide.

If a point (a, b) is on the graph of the function f, then (b, a) must be on the graph of $f^{-1}(x)$. See Fig. 11.64 on page 766. The points (a, b) and (b, a) lie on opposite sides

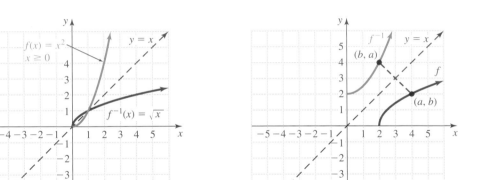

Figure 11.63   **Figure 11.64**

of the diagonal line $y = x$ and are the same distance from it. For this reason the graphs of f and f^{-1} are symmetric with respect to the line $y = x$.

E X A M P L E **7**

Inverses and their graphs

Find the inverse of the function $f(x) = \sqrt{x-1}$, and graph f and f^{-1} on the same pair of axes.

Solution

To find f^{-1}, first switch x and y in the formula $y = \sqrt{x-1}$:

$$x = \sqrt{y-1}$$
$$x^2 = y - 1 \quad \text{Square both sides.}$$
$$x^2 + 1 = y$$

Because the range of f is the set of nonnegative real numbers $[0, \infty)$, we must restrict the domain of f^{-1} to be $[0, \infty)$. Thus, $f^{-1}(x) = x^2 + 1$ for $x \geq 0$. The two graphs are shown in Fig. 11.65.

> Now do Exercises 49–58

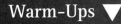

Figure 11.65

Warm-Ups ▼

Fill in the blank.

1. The _____ of a function is a function with the same ordered pairs except that the coordinates are reversed.
2. The domain of f^{-1} is the _____ of f.
3. The range of f^{-1} is the _____ of f.
4. A function is _____ if no two ordered pairs have the same second coordinates with different first coordinates.
5. The graphs of f and f^{-1} are _____ with respect to the line $y = x$.
6. If a _____ line can be drawn so that it crosses the graph of a function more than once, then the function is not one-to-one.

True or false?

7. The inverse of $\{(1, 3), (2, 5)\}$ is $\{(3, 1), (2, 5)\}$.
8. The function $f(x) = 3$ is one-to-one.
9. Only one-to-one functions are invertible.
10. The function $f(x) = x^4$ is invertible.
11. If $f(x) = -x$, then $f^{-1}(x) = -x$.
12. If h is invertible and $h(7) = -95$, then $h^{-1}(-95) = 7$.
13. If $f(x) = 4x + 5$, then $f^{-1}(x) = \dfrac{x-5}{4}$.
14. If $g(x) = 3x - 6$, then $g^{-1}(x) = \dfrac{1}{3}x + 2$.

‹1› Inverse of a Function

Determine whether each function is invertible. If it is invertible, then find the inverse. See Example 1.

1. $\{(1, 3), (2, 9)\}$
2. $\{(0, 5), (-2, 0)\}$
3. $\{(-3, 3), (-2, 2), (0, 0), (2, 2)\}$
4. $\{(1, 1), (2, 8), (3, 27)\}$
5. $\{(16, 4), (9, 3), (0, 0)\}$
6. $\{(-1, 1), (-3, 81), (3, 81)\}$
7. $\{(0, 5), (5, 0), (6, 0)\}$
8. $\{(3, -3), (-2, 2), (1, -1)\}$
9. $\{(0, 0), (2, 2), (9, 9)\}$
10. $\{(9, 1), (2, 1), (7, 1), (0, 1)\}$

Determine whether each function is invertible by examining the graph of the function. See Example 2.

11.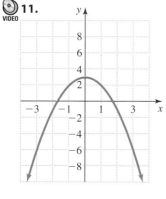

12.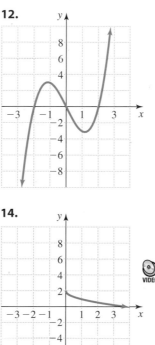

13.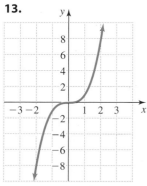

14.

‹2› Identifying Inverse Functions

Determine whether each pair of functions f and g are inverses of each other. See Example 3.

15. $f(x) = 2x$ and $g(x) = 0.5x$
16. $f(x) = 3x$ and $g(x) = 0.33x$
17. $f(x) = 2x - 10$ and $g(x) = \frac{1}{2}x + 5$
18. $f(x) = 3x + 7$ and $g(x) = \frac{x - 7}{3}$
19. $f(x) = -x$ and $g(x) = -x$
20. $f(x) = \frac{1}{x}$ and $g(x) = \frac{1}{x}$
21. $f(x) = x^4$ and $g(x) = x^{1/4}$
22. $f(x) = |2x|$ and $g(x) = \left|\frac{x}{2}\right|$

‹3› Switch-and-Solve Strategy

Find f^{-1}. Check that $(f \circ f^{-1})(x) = x$ and $(f^{-1} \circ f)(x) = x$. See Examples 4 and 5. See the Strategy for Finding f^{-1} by Switch-and-Solve box on page 763.

23. $f(x) = 5x$
24. $h(x) = -3x$

25. $g(x) = x - 9$
26. $j(x) = x + 7$

27. $k(x) = 5x - 9$
28. $r(x) = 2x - 8$

29. $m(x) = \frac{2}{x}$
30. $s(x) = \frac{-1}{x}$

31. $f(x) = \sqrt[3]{x - 4}$
32. $f(x) = \sqrt[3]{x + 2}$

33. $f(x) = \frac{3}{x - 4}$
34. $f(x) = \frac{2}{x + 1}$

35. $f(x) = \sqrt[3]{3x + 7}$
36. $f(x) = \sqrt[3]{7 - 5x}$

37. $f(x) = \dfrac{x+1}{x-2}$ **38.** $f(x) = \dfrac{1-x}{x+3}$

51. $f(x) = x^2 - 1$ for $x \geq 0$

39. $f(x) = \dfrac{x+1}{3x-4}$ **40.** $g(x) = \dfrac{3x+5}{2x-3}$

52. $f(x) = x^2 + 3$ for $x \geq 0$

⟨4⟩ Even Roots or Even Powers

Find the inverse of each function. See Example 6.

41. $p(x) = \sqrt[4]{x}$

42. $v(x) = \sqrt[6]{x}$

43. $f(x) = (x-2)^2$ for $x \geq 2$

44. $g(x) = (x+5)^2$ for $x \geq -5$

45. $f(x) = x^2 + 3$ for $x \geq 0$

46. $f(x) = x^2 - 5$ for $x \geq 0$ **53.** $f(x) = 5x$

47. $f(x) = \sqrt{x+2}$

48. $f(x) = \sqrt{x-4}$

⟨5⟩ Graphs of *f* and *f*⁻¹

Find the inverse of each function, and graph f and f^{-1} on the same pair of axes. See Example 7.

49. $f(x) = 2x + 3$ **54.** $f(x) = \dfrac{x}{4}$

50. $f(x) = -3x + 2$ **55.** $f(x) = x^3$

56. $f(x) = 2x^3$

57. $f(x) = \sqrt{x - 2}$

58. $f(x) = \sqrt{x + 3}$

Miscellaneous

Find the inverse of each function.

59. $f(x) = 2x$

60. $f(x) = x - 1$

61. $f(x) = 2x - 1$

62. $f(x) = 2(x - 1)$

63. $f(x) = \sqrt[3]{x}$

64. $f(x) = 2\sqrt[3]{x}$

65. $f(x) = \sqrt[3]{x - 1}$

66. $f(x) = \sqrt[3]{2x - 1}$

67. $f(x) = 2\sqrt[3]{x} - 1$

68. $f(x) = 2\sqrt[3]{x - 1}$

For each pair of functions, find $(f^{-1} \circ f)(x)$.

69. $f(x) = x^3 - 1$ and $f^{-1}(x) = \sqrt[3]{x + 1}$

70. $f(x) = 2x^3 + 1$ and $f^{-1}(x) = \sqrt[3]{\dfrac{x - 1}{2}}$

71. $f(x) = \dfrac{1}{2}x - 3$ and $f^{-1}(x) = 2x + 6$

72. $f(x) = 3x - 9$ and $f^{-1}(x) = \dfrac{1}{3}x + 3$

73. $f(x) = \dfrac{1}{x} + 2$ and $f^{-1}(x) = \dfrac{1}{x - 2}$

74. $f(x) = 4 - \dfrac{1}{x}$ and $f^{-1}(x) = \dfrac{1}{4 - x}$

75. $f(x) = \dfrac{x + 1}{x - 2}$ and $f^{-1}(x) = \dfrac{2x + 1}{x - 1}$

76. $f(x) = \dfrac{3x - 2}{x + 2}$ and $f^{-1}(x) = \dfrac{2x + 2}{3 - x}$

Applications

Solve each problem.

77. *Accident reconstruction.* The distance that it takes a car to stop is a function of the speed and the drag factor. The drag factor is a measure of the resistance between the tire and the road surface. The formula $S = \sqrt{30LD}$ is used to determine the minimum speed S [in miles per hour (mph)] for a car that has left skid marks of length L feet (ft) on a surface with drag factor D.

a) Find the minimum speed for a car that has left skid marks of length 50 ft where the drag factor is 0.75.

b) Does the drag factor increase or decrease for a road surface when it gets wet?

c) Write L as a function of S for a road surface with drag factor 1 and graph the function.

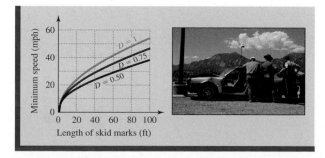

Figure for Exercise 77

78. *Area of a circle.* Let x be the radius of a circle and $h(x)$ be the area of the circle. Write a formula for $h(x)$ in terms of x. What does x represent in the notation $h^{-1}(x)$? Write a formula for $h^{-1}(x)$.

79. *Vehicle cost.* At Bill Hood Ford in Hammond a sales tax of 9% of the selling price x and a \$125 title and license fee are added to the selling price to get the total cost of a vehicle. Find the function $T(x)$ that the dealer uses to get the total cost as a function of the selling price x. Citizens National Bank will not include sales tax or fees in a loan. Find the function $T^{-1}(x)$ that the bank can use to get the selling price as a function of the total cost x.

80. *Carpeting cost.* At the Windrush Trace apartment complex all living rooms are square, but the length of x feet may vary. The cost of carpeting a living room is \$18 per square yard plus a \$50 installation fee. Find the function $C(x)$ that gives the total cost of carpeting a living room of length x. The manager has an invoice for the total cost of a living room carpeting job but does not know in which apartment it was done. Find the function $C^{-1}(x)$ that gives the length of a living room as a function of the total cost of the carpeting job x.

Getting More Involved

81. *Discussion*

Let $f(x) = x^n$ where n is a positive integer. For which values of n is f an invertible function? Explain.

82. *Discussion*

Suppose f is a function with range $(-\infty, \infty)$ and g is a function with domain $(0, \infty)$. Is it possible that g and f are inverse functions? Explain.

Graphing Calculator Exercises

83. Most graphing calculators can form compositions of functions. Let $f(x) = x^2$ and $g(x) = \sqrt{x}$. To graph the composition $g \circ f$, let $y_1 = x^2$ and $y_2 = \sqrt{y_1}$. The graph of y_2 is the graph of $g \circ f$. Use the graph of y_2 to determine whether f and g are inverse functions.

84. Let $y_1 = x^3 - 4$, $y_2 = \sqrt[3]{x + 4}$, and $y_3 = \sqrt[3]{y_1 + 4}$. The function y_3 is the composition of the first two functions. Graph all three functions on the same screen. What do the graphs indicate about the relationship between y_1 and y_2?

Chapter 11 Wrap-Up

Summary

Relations and Functions		**Examples**
Relation	Any set of ordered pairs of real numbers	$\{(1, 2), (1, 3)\}$
Function	A relation in which no two ordered pairs have the same first coordinate and different second coordinates. If y is a function of x, then y is uniquely determined by x. A function may be defined by a table, a listing of ordered pairs, or an equation.	$\{(1, 2), (3, 5), (4, 5)\}$
Domain	The set of first coordinates of the ordered pairs	Function: $y = x^2$, Domain: $(-\infty, \infty)$
Range	The set of second coordinates of the ordered pairs.	Function: $y = x^2$, Range: $[0, \infty)$
Function notation	If y is a function of x, the expression $f(x)$ is used in place of y.	$y = 2x + 3$ $f(x) = 2x + 3$
Vertical-line test	If a graph can be crossed more than once by a vertical line, then it is not the graph of a function.	
Linear function	A function of the form $f(x) = mx + b$ with $m \neq 0$	$f(x) = 3x - 7$ $f(x) = -2x + 5$
Constant function	A function of the form $f(x) = b$, where b is a real number	$f(x) = 2$

Types of Functions		**Examples**
Linear function	$y = mx + b$ or $f(x) = mx + b$ for $m \neq 0$ Domain $(-\infty, \infty)$, range $(-\infty, \infty)$ If $m = 0$, $y = b$ is a constant function. Domain $(-\infty, \infty)$, range $\{b\}$	$f(x) = 2x - 3$
Absolute value function	$y = |x|$ or $f(x) = |x|$ Domain $(-\infty, \infty)$, range $[0, \infty)$	$f(x) = |x + 5|$
Quadratic function	$f(x) = ax^2 + bx + c$ for $a \neq 0$	$f(x) = x^2 - 4x + 3$
Square-root function	$f(x) = \sqrt{x}$ Domain $[0, \infty)$, range $[0, \infty)$	$f(x) = \sqrt{x - 4}$

Transformations of Graphs

Reflecting	The graph of $y = -f(x)$ is a reflection in the x-axis of the graph of $y = f(x)$.	The graph of $y = -x^2$ is a reflection of the graph of $y = x^2$.

Translating	The graph of $y = f(x) + k$ is
	  k units above $y = f(x)$ if $k > 0$ or
	  $|k|$ units below $y = f(x)$ if $k < 0$.
		The graph of $y = x^2 + 3$ is three units above $y = x^2$, and $y = x^2 - 3$ is three units below $y = x^2$.

	The graph of $y = f(x - h)$ is
	  h units to the right of $y = f(x)$ if $h > 0$ or
	  $|h|$ units to the left of $y = f(x)$ if $h < 0$.
		The graph of $y = (x - 3)^2$ is three units to the right of $y = x^2$, and $y = (x + 3)^2$ is three units to the left.

Stretching and shrinking	The graph of $y = af(x)$ is obtained by stretching (if $a > 1$) or shrinking (if $0 < a < 1$) the graph of $y = f(x)$.	The graph of $y = 5x^2$ is obtained by stretching $y = x^2$, and $y = 0.1x^2$ is obtained by shrinking $y = x^2$.

Polynomial Functions

Examples

Polynomial function	A function defined by a polynomial	$P(x) = x^3 - x^2 - 12x + 5$
Symmetric about the y-axis	If $f(x)$ is a function such that $f(x) = f(-x)$ for any value of x in its domain, then the graph of f is symmetric about the y-axis.	If $f(x) = x^4 - x^2$, then $f(-x) = (-x)^4 - (-x)^2 = x^4 - x^2$, and $f(x) = f(-x)$.
Symmetric about the origin	If $f(x)$ is a function such that $f(-x) = -f(x)$ for any value of x in its domain, then the graph of f is symmetric about the origin.	If $f(x) = x^3 + x$, then $f(-x) = (-x)^3 + (-x) = -x^3 - x$, and $f(-x) = -f(x)$.
Behavior at the x-intercepts	The graph of a polynomial function crosses the x-axis at $(c, 0)$ if $(x - c)$ has an odd exponent. The graph touches but does not cross the x-axis if $(x - c)$ has an even exponent.	Graph of $f(x) = (x - 3)^2(x + 5)$ touches but does not cross x-axis at $(3, 0)$ and crosses x-axis at $(-5, 0)$.
Polynomial inequality	An inequality involving a polynomial	$x^3 - x > 0$
Methods for solving	Use either the graphical method or the test-point method.	$x^3 - x > 0$ **Solution set:** $(-1, 0) \cup (1, \infty)$

Rational Functions

Examples

Rational function	If $P(x)$ and $Q(x)$ are polynomials with no common factor and $f(x) = \dfrac{P(x)}{Q(x)}$ for $Q(x) \neq 0$, then $f(x)$ is a rational function.	$f(x) = \dfrac{x^2 - 1}{3x - 2}, f(x) = \dfrac{1}{x - 3}$

Finding asymptotes for a rational function $f(x) = \dfrac{P(x)}{Q(x)}$	1. The graph of f has a vertical asymptote for each solution to the equation $Q(x) = 0$. 2. If the degree of $P(x)$ is less than the degree of $Q(x)$, then the x-axis is a horizontal asymptote. 3. If the degree of $P(x)$ is equal to the degree of $Q(x)$, then the horizontal asymptote is determined by the ratio of the leading coefficients. 4. If the degree of $P(x)$ is one larger than the degree of $Q(x)$, then use long division to find the quotient of $P(x)$ and $Q(x)$.	$f(x) = \dfrac{1}{x - 2}$ Vertical: $x = 2$ Horizontal: x-axis $f(x) = \dfrac{x}{x - 2}$ Vertical: $x = 2$ Horizontal: $y = 1$ $f(x) = \dfrac{2x^2 + 3x - 5}{x + 2} = 2x - 1 + \dfrac{-3}{x + 2}$ Vertical: $x = -2$ Oblique: $y = 2x - 1$
Rational inequality	An inequality involving a rational expression	$\dfrac{1}{x - 2} \leq 0, \dfrac{x + 3}{x - 9} > 1, \dfrac{2}{x} < \dfrac{x}{2}$
Methods for solving	Use either the graphical method or the test-point method.	$\dfrac{1}{x - 2} \leq 0$ **Solution set:** $(-\infty, 2)$

Combining Functions

Examples

Sum	$(f + g)(x) = f(x) + g(x)$	For $f(x) = x^2$ and $g(x) = x + 1$ $(f + g)(x) = x^2 + x + 1$
Difference	$(f - g)(x) = f(x) - g(x)$	$(f - g)(x) = x^2 - x - 1$
Product	$(f \cdot g)(x) = f(x) \cdot g(x)$	$(f \cdot g)(x) = x^3 + x^2$
Quotient	$\left(\dfrac{f}{g}\right)(x) = \dfrac{f(x)}{g(x)}$	$\left(\dfrac{f}{g}\right)(x) = \dfrac{x^2}{x + 1}$
Composition of functions	$(g \circ f)(x) = g(f(x))$ $(f \circ g)(x) = f(g(x))$	$(g \circ f)(x) = g(x^2) = x^2 + 1$ $(f \circ g)(x) = f(x + 1)$ $\qquad = x^2 + 2x + 1$

Inverse Functions

Examples

| One-to-one function | A function in which no two ordered pairs have different x-coordinates and the same y-coordinate | $f = \{(2, 20), (3, 30)\}$ |
| Inverse function | The inverse of a one-to-one function f is the function f^{-1}, which is obtained from f by interchanging the coordinates in each ordered pair of f. The domain of f^{-1} is the range of f, and the range of f^{-1} is the domain of f. | $f^{-1} = \{(20, 2), (30, 3)\}$ |

Horizontal-line test	If there is a horizontal line that crosses the graph of a function more than once, then the function is not invertible.	
Function notation for inverse	Two functions f and g are inverses of each other if and only if both of the following conditions are met. 1. $(g \circ f)(x) = x$ for every number x in the domain of f. 2. $(f \circ g)(x) = x$ for every number x in the domain of g.	$f(x) = x^3 + 1$ $f^{-1}(x) = \sqrt[3]{x - 1}$
Switch-and-solve strategy for finding f^{-1}	1. Replace $f(x)$ by y. 2. Interchange x and y. 3. Solve for y. 4. Replace y by $f^{-1}(x)$.	$y = x^3 + 1$ $x = y^3 + 1$ $x - 1 = y^3$ $y = \sqrt[3]{x - 1}$ $f^{-1}(x) = \sqrt[3]{x - 1}$
Graphs of f and f^{-1}	Graphs of inverse functions are symmetric with respect to the line $y = x$.	

Enriching Your Mathematical Word Power

Fill in the blank.

1. Any set of ordered pairs is a _____.

2. A _____ is a set of ordered pairs in which no two have the same first coordinate and different second coordinates.

3. The set of first coordinates of a relation is the _____.

4. The set of second coordinates of a relation is the ____.

5. The notation in which $f(x)$ is used as the dependent variable is _____ notation.

6. A line that is approached by a curve is an _____.

7. An _____ asymptote is neither horizontal nor vertical.

8. The _____ of f and g is the function $f \circ g$ where $(f \circ g)(x) = f(g(x))$.

9. A function in which no two ordered pairs have the same second coordinate and different first coordinates is a _____ function.

10. The _____ line test is a visual method for determining whether a graph is the graph of a function.

11. The _____ line test is a visual method for determining whether a function is one-to-one.

12. The graph of $y = -f(x)$ is a _____ in the x-axis of the graph of $y = f(x)$.

13. The graph of $y = f(x) + c$ for $c > 0$ is an upward _____ of the graph of $y = f(x)$.

14. If $f(-x) = f(x)$, then the graph of f is _____ about the y-axis.

15. If $f(-x) = -f(x)$, then the graph of f is symmetric about the ____.

16. A ratio of two polynomial functions is a _____ function.

Review Exercises

11.1 Functions and Relations

Determine whether each relation is a function.

1. $\{(5, 7), (5, 10), (5, 3)\}$

2. $\{(1, 3), (4, 7), (1, 6)\}$

3. $\{(1, 1), (2, 1), (3, 3)\}$

4. $\{(2, 4), (4, 6), (6, 8)\}$

5. $y = x^2$

6. $x^2 = 1 + y^2$

7. $x = y^4$

8. $y = \sqrt{x - 1}$

Determine the domain and range of each relation.

9. $\{(3, 5), (4, 9), (5, 1)\}$

10. $\{(2, 6), (6, 7), (8, 9)\}$

11. $y = x + 1$

12. $y = 2x - 3$

13. $y = \sqrt{x + 5}$

14. $y = \sqrt{x - 1}$

Let $f(x) = 2x - 5$ and $g(x) = x^2 + x - 6$. Evaluate each expression.

15. $f(0)$ **16.** $f(-3)$

17. $g(0)$ **18.** $g(-2)$

19. $g\left(\dfrac{1}{2}\right)$ **20.** $g\left(-\dfrac{1}{2}\right)$

11.2 Graphs of Functions and Relations

Graph each function, and state the domain and range.

21. $f(x) = 3x - 4$

22. $y = 0.3x$

23. $h(x) = |x| - 2$

24. $y = |x - 2|$

25. $y = x^2 - 2x + 1$

26. $g(x) = x^2 - 2x - 15$

27. $k(x) = \sqrt{x} + 2$

28. $y = \sqrt{x - 2}$

29. $y = 30 - x^2$

30. $y = 4 - x^2$

31. $f(x) = \begin{cases} \sqrt{x+4} & \text{for } -4 \le x \le 0 \\ x+2 & \text{for } x > 0 \end{cases}$

32. $f(x) = \begin{cases} \sqrt{x+1} & \text{for } -1 \le x \le 3 \\ x-1 & \text{for } x > 3 \end{cases}$

Graph each relation, and state its domain and range.

33. $x = 2$

34. $x = y^2 - 1$

35. $x = |y| + 1$

36. $x = \sqrt{y-1}$

11.3 Transformations of Graphs

Sketch the graph of each function, and state the domain and range.

37. $y = \sqrt{x}$

38. $y = -\sqrt{x}$

39. $y = -2\sqrt{x}$

40. $y = 2\sqrt{x}$

41. $y = \sqrt{x-2}$

42. $y = \sqrt{x+2}$

43. $y = \frac{1}{2}\sqrt{x}$

44. $y = \sqrt{x - 1} + 2$

45. $y = -\sqrt{x + 1} + 3$

46. $y = 3\sqrt{x + 4} - 5$

11.4 Graphs of Polynomial Functions

Graph each function, and identify the x- and y-intercepts.

47. $f(x) = x^3 - 25x$

48. $f(x) = x^3 + 2x^2 - 4x - 8$

49. $f(x) = (x^2 - 4)(x - 1)$

50. $f(x) = (x^2 - 3x - 4)(x + 3)$

51. $f(x) = x^4 - 10x^2 + 9$

52. $f(x) = x^4 - 4x^3$

Find the x-intercepts, and discuss the behavior of the graph of each polynomial function at its x-intercepts.

53. $f(x) = x^2 - 6x + 9$

54. $f(x) = x^2 - 3x - 18$

55. $f(x) = (x - 3)(x + 5)(x - 4)^2$

56. $f(x) = (x - 1)^2(x + 7)$

57. $f(x) = x^3 - 8x^2 - 9x + 72$

58. $f(x) = x^4 - 29x^2 + 100$

Solve each polynomial inequality. State the solution set using interval notation and graph it.

59. $\frac{1}{2}x^3 - 2x \le 0$

60. $x^3 - 6x^2 + 8x \ge 0$

61. $x^4 - 2x^2 \geq 0$

77. $f(x) = \dfrac{2x - 1}{x - 1}$

78. $f(x) = \dfrac{-x - 1}{x}$

62. $-x^4 + 2x^2 - 1 \geq 0$

63. $x^4 - 2x^2 \leq 0$

64. $-x^4 + 2x^2 - 1 \leq 0$

65. $x^3 - x^2 - 9x + 9 < 0$

79. $f(x) = \dfrac{x^2 - 2x + 1}{x - 2}$

80. $f(x) = \dfrac{-x^2 + x + 2}{x - 1}$

66. $x^3 - 4x^2 - 16x + 64 > 0$

67. $x^3 - x^2 - 9x + 9 \geq 0$

68. $x^3 - 4x^2 - 16x + 64 \leq 0$

11.5 Graphs of Rational Functions

Find the domain of each rational function.

69. $f(x) = \dfrac{x^2 - 1}{2x + 3}$

70. $f(x) = \dfrac{3x + 2}{x^2 - x - 12}$

71. $f(x) = \dfrac{1}{x^2 + 9}$

72. $f(x) = \dfrac{x - 4}{x^2 - 9}$

Find all asymptotes for each rational function, and sketch the graph of the function.

73. $f(x) = \dfrac{2}{x - 3}$

74. $f(x) = \dfrac{-1}{x + 1}$

75. $f(x) = \dfrac{x}{x^2 - 4}$

76. $f(x) = \dfrac{x^2}{x^2 - 4}$

Solve each rational inequality. State the solution set using interval notation, and graph it.

81. $\dfrac{x + 3}{x - 4} \geq 0$

82. $\dfrac{x + 5}{x - 7} \leq 0$

83. $\dfrac{x^2 - 4}{x} > 0$

84. $\dfrac{x^2}{x^2 - 16} < 0$

85. $\dfrac{x}{x + 5} < 2$

86. $\dfrac{x + 4}{x - 1} \leq 4$

87. $\dfrac{x^2 + 4}{x} \leq 0$

88. $\dfrac{x^2}{x^2 + 8} \geq 0$

89. $\dfrac{2x}{x + 5} \leq \dfrac{x}{x - 2}$

90. $\dfrac{x}{x - 6} \geq \dfrac{x}{x + 3}$

11.6 Combining Functions
Let $f(x) = 3x + 5$, $g(x) = x^2 - 2x$, and $h(x) = \frac{x-5}{3}$. Find the following.

120. $f(x) = 2 - x^2$ for $x \geq 0$

91. $f(-3)$ **92.** $h(-4)$

93. $(h \circ f)(\sqrt{2})$ **94.** $(f \circ h)(\pi)$

95. $(g \circ f)(2)$ **96.** $(g \circ f)(x)$

97. $(f + g)(3)$ **98.** $(f - g)(x)$

99. $(f \cdot g)(x)$ **100.** $\left(\dfrac{f}{g}\right)(1)$

121. $f(x) = \dfrac{x^3}{2}$

101. $(f \circ f)(0)$ **102.** $(f \circ f)(x)$

Let $f(x) = |x|$, $g(x) = x + 2$, and $h(x) = x^2$. Write each of the following functions as a composition of functions, using f, g, or h.

103. $F(x) = |x + 2|$ **104.** $G(x) = |x| + 2$

105. $H(x) = x^2 + 2$ **106.** $K(x) = x^2 + 4x + 4$

122. $f(x) = -\dfrac{1}{4}x$

107. $I(x) = x + 4$ **108.** $J(x) = x^4 + 2$

11.7 Inverse Functions
Determine whether each function is invertible. If it is invertible, find the inverse.

109. $\{(-2, 4), (2, 4)\}$ **110.** $\{(1, 1), (3, 3)\}$

Miscellaneous
Sketch the graph of each function.

111. $f(x) = 8x$ **112.** $i(x) = -\dfrac{x}{3}$

123. $f(x) = 3$ **124.** $f(x) - 2x - 3$

113. $g(x) = 13x - 6$ **114.** $h(x) = \sqrt[4]{x - 6}$

115. $j(x) = \dfrac{x + 1}{x - 1}$ **116.** $k(x) = |x| + 7$

117. $m(x) = (x - 1)^2$ **118.** $n(x) = \dfrac{3}{x}$

125. $f(x) = x^2 - 3$ **126.** $f(x) = 3 - x^2$

Find the inverse of each function, and graph f and f^{-1} on the same pair of axes.

119. $f(x) = 3x - 1$

127. $f(x) = \dfrac{1}{x^2 - 3}$

128. $f(x) = \dfrac{x}{(x - 1)(x + 2)}$

129. $f(x) = x(x - 1)(x + 2)$

130. $f(x) = x^3 - 4x^2 + 4x$

Solve each inequality. State the solution set using interval notation.

131. $4 - 2x > 0$

132. $2x - 3 > 0$

133. $x^2 - 3 \le 0$

134. $3 - x^2 \le 0$

135. $\dfrac{1}{x^2 - 3} < 0$

136. $\dfrac{x}{(x - 1)(x + 2)} \ge 0$

137. $x(x - 1)(x + 2) < 0$

138. $x^3 - 4x^2 + 4x \ge 0$

139. $\dfrac{x - 4}{x + 3} \le 0$

140. $\dfrac{2x - 1}{x + 5} \ge 0$

141. $(x - 2)(x + 1)(x - 5) \ge 0$

142. $(x - 1)(x + 2)(2x - 5) < 0$

143. $x^3 + 3x^2 - x - 3 < 0$

144. $x^3 + 5x^2 - 4x - 20 \ge 0$

 **Graphing Calculator Exercises**

Match the given inequalities with their solution sets (a through d) by examining a table or a graph.

145. $x^2 - 2x - 8 < 0$ **a.** $(-2, 2) \cup (8, \infty)$

146. $x^2 - 3x > 54$ **b.** $(2, 4)$

147. $\dfrac{x}{x - 2} > 2$ **c.** $(-2, 4)$

148. $\dfrac{3}{x - 2} < \dfrac{5}{x + 2}$ **d.** $(-\infty, -6) \cup (9, \infty)$

Solve each problem.

149. *Inscribed square.* Given that B is the area of a square inscribed in a circle of radius r and area A, write B as a function of A.

150. *Area of a window.* A window is in the shape of a square of side s, with a semicircle of diameter s above it. Write a function that expresses the total area of the window as a function of s.

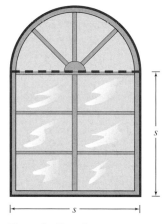

Figure for Exercise 150

151. *Composition of functions.* Given that $a = 3k + 2$ and $k = 5w - 6$, write a as a function of w.

152. *Volume of a cylinder.* The volume of a cylinder with a fixed height of 10 centimeters (cm) is given by $V = 10\pi r^2$, where r is the radius of the circular base. Write the volume as a function of the area of the base, A.

153. *Square formulas.* Write the area of a square A as a function of the length of a side of the square s. Write the length of a side of a square as a function of the area.

154. *Circle formulas.* Write the area of a circle A as a function of the radius of the circle r. Write the radius of a circle as a function of the area of the circle. Write the area as a function of the diameter d.

Chapter 11 Test

Solve each problem.

1. Determine whether $\{(0, 5), (9, 5), (4, 5)\}$ is a function.

2. Let $f(x) = -2x + 5$. Find $f(-3)$.

3. Find the domain and range of the function $y = \sqrt{x - 7}$.

4. A mail-order firm charges its customers a shipping and handling fee of $3.00 plus $0.50 per pound for each order shipped. Express the shipping and handling fee S as a function of the weight of the order n.

5. If a ball is tossed into the air from a height of 6 feet with a velocity of 32 feet per second, then its altitude at time t (in seconds) can be described by the function

$$A(t) = -16t^2 + 32t + 6.$$

Find the altitude of the ball at 2 seconds.

Sketch the graph of each function or relation, and state the domain and range.

6. $f(x) = -\dfrac{2}{3}x + 1$

7. $y = |x| - 4$

8. $g(x) = x^2 + 2x - 8$ **9.** $x = y^2$

10. $f(x) = \begin{cases} \sqrt{x} & \text{for } x \geq 0 \\ -x - 3 & \text{for } x < 0 \end{cases}$ **11.** $y = -|x - 2|$

12. $y = \sqrt{x + 5} - 2$

Graph each function. Identify all intercepts.

13. $f(x) = (x + 2)(x - 2)^2$

14. $f(x) = \dfrac{1}{x^2 - 4x + 4}$

15. $f(x) = \dfrac{2x - 3}{x - 2}$

16. $f(x) = x^3 - x^2 - 4x + 4$

Solve each inequality. State the solution set using interval notation.

17. $x^3 + 4x^2 - 32x > 0$

18. $\dfrac{x}{x^2 - 25} \le 0$

Let $f(x) = -2x + 5$ and $g(x) = x^2 + 4$. Find the following.

19. $f(-3)$

20. $(g \circ f)(-3)$

21. $f^{-1}(11)$

22. $f^{-1}(x)$

23. $(g + f)(x)$

24. $(f \cdot g)(1)$

25. $(f^{-1} \circ f)(1776)$

26. $(f/g)(2)$

27. $(f \circ g)(x)$

28. $(g \circ f)(x)$

Let $f(x) = x - 7$ and $g(x) = x^2$. Write each of the following functions as a composition of functions using f and g.

29. $H(x) = x^2 - 7$

30. $W(x) = x^2 - 14x + 49$

Determine whether each function is invertible. If it is invertible, find the inverse.

31. $\{(2, 3), (4, 3), (1, 5)\}$

32. $\{(2, 3), (3, 4), (4, 5)\}$

Find the inverse of each function.

33. $f(x) = x - 5$

34. $f(x) = 3x - 5$

35. $f(x) = \sqrt[3]{x} + 9$

36. $f(x) = \dfrac{2x + 1}{x - 1}$

*Making*Connections | A Review of Chapters 1–11

Simplify each expression.

1. $125^{-2/3}$

2. $\left(\dfrac{8}{27}\right)^{-1/3}$

3. $\sqrt{18} - \sqrt{8}$

4. $x^5 \cdot x^3$

5. $16^{1/4}$

6. $\dfrac{x^{12}}{x^3}$

Find the real solution set to each equation.

7. $x^2 = 9$

8. $x^2 = 8$

9. $x^2 = x$

10. $x^2 - 4x - 6 = 0$

11. $x^{1/4} = 3$

12. $x^{1/6} = -2$

13. $|x| = 8$

14. $|5x - 4| = 21$

15. $x^3 = 8$

16. $(3x - 2)^3 = 27$

17. $\sqrt{2x - 3} = 9$

18. $\sqrt{x - 2} = x - 8$

Sketch the graph of each set.

19. $\{(x, y) \mid y = 5\}$

20. $\{(x, y) \mid y = 2x - 5\}$

21. $\{(x, y) \mid x = 5\}$

22. $\{(x, y) \mid 3y = x\}$

23. $\{(x, y) \mid y = 5x^2\}$

24. $\{(x, y) \mid y = -2x^2\}$

Find the missing coordinates in each ordered pair so that the ordered pair satisfies the given equation.

25. $(2, \ ), (3, \ ), (\ , 2), (\ , 16), \quad 2^x = y$

26. $\left(\dfrac{1}{2}, \ \right), (-1, \ ), (\ , 16), (\ , 1), \quad 4^x = y$

Find the domain of each expression.

27. $\sqrt{x}$

28. $\sqrt{6 - 2x}$

29. $\dfrac{5x - 3}{x^2 + 1}$

30. $\dfrac{x - 3}{x^2 - 10x + 9}$

Solve each system of equations, and state whether the system is independent, dependent, or inconsistent.

31. $4x - 9y = -1$
$2x + 12y = 5$

32. $10x + 20y = 143$
$y = x + 5.2$

33. $3x - 9y = 6$
$y = \dfrac{1}{3}x - \dfrac{2}{3}$

34. $x = 5y + 12$
$y = \dfrac{1}{5}x - 7$

35. $x + y - z = -5$
$x - 2y + z = 11$
$3x - y - z = 3$

36. $2x + y + 3z = 2$
$x + y + z = 5$
$3x + 2y + 4z = 6$

Perform the indicated operations.

37. $\dfrac{11}{15} - \dfrac{5}{12}$

38. $\dfrac{12}{35} \cdot \dfrac{10}{21}$

39. $\dfrac{4}{9} \div \dfrac{14}{15}$

40. $\dfrac{2}{3x^2} + \dfrac{1}{6x}$

41. $\dfrac{5xy^3}{6a^3b^3} \cdot \dfrac{2a^3b^7}{5xy^5}$

42. $\dfrac{12a^2 - 48}{a^2 + a - 6} \div \dfrac{9a + 18}{a^2 - 9}$

Solve each problem.

43. *Capital cost and operating cost.* To decide when to replace company cars, an accountant looks at two cost components: capital cost and operating cost. The capital cost C (the difference between the original cost and the salvage value) for a certain car is $3000 plus $0.12 for each mile that the car is driven.

    a) Write the capital cost C as a linear function of x, the number of miles that the car is driven.

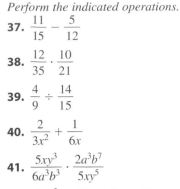

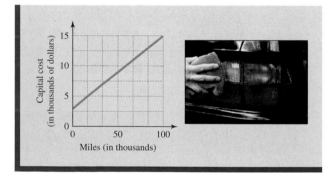

Figure for Exercise 43(a)

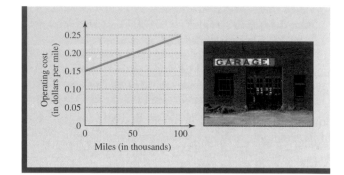

Figure for Exercise 43(b)

    b) The operating cost P is $0.15 per mile initially and increases linearly to $0.25 per mile when the car reaches 100,000 miles. Write P as a function of x, the number of miles that the car is driven.

44. *Total cost.* The accountant in Exercise 43 uses the function $T = \dfrac{C}{x} + P$ to find the total cost per mile.

    a) Find T for $x = 20{,}000$, $30{,}000$, and $90{,}000$.

    b) Sketch a graph of the total cost function.

    c) The accountant has decided to replace the car when T reaches $0.38 for the second time. At what mileage will the car be replaced?

    d) For what values of x is T less than or equal to $0.38?

*Critical*Thinking | **For Individual or Group Work** | **Chapter 11**

These exercises can be solved by a variety of techniques, which may or may not require algebra. So be creative and think critically. Explain all answers. Answers are in the Instructor's Edition of this text.

1. *Knight moves.* Draw a 3 by 3 chess board on paper, and place two pennies (P) and two nickels (N) in the corners as shown in (a) of the figure. Move the Ns to the positions of the Ps and the Ps to the position of the Ns using the moves that a knight can make in chess (one space vertically followed by two spaces horizontally or one space horizontally followed by two spaces vertically). If you allow a P or an N to make more than one move on a given turn, then it takes six turns. Try it. Find the minimum number of turns required to interchange the coins starting with the arrangement in (b).

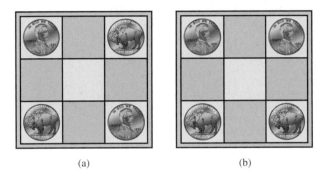

(a)   (b)

Figure for Exercise 1

2. *Friedman numbers.* A Friedman number is a positive integer that can be written in some nontrivial way using its own digits together with the elementary operations $(+, -, \cdot, \div,$ exponents, and grouping symbols). For example, $25 = 5^2$, and $126 = 21 \cdot 6$. The only two-digit Friedman number is 25. Show that 121 and 125 are Friedman numbers. There are 13 three-digit Friedman numbers. Find the other 10 three-digit Friedman numbers.

3. *Large Friedman numbers.* Show that 123,456,789 and 987,654,321 are Friedman numbers.

4. *Year numbers.* Using all four of the digits in the current year and only those digits, write expressions for the integers from 1 through 100. You may use grouping symbols, and the operations of addition, subtraction, multiplication, division, powers, roots, and factorial, but no two-digit numbers or decimal points. For example if the year is 2005, then $5^0 + 0 \cdot 2 = 1$, $5^0 + 2^0 = 2$, and so on. See how far you can go. Vary the problem by trying another year (say 1776), or allowing decimal points, or two-digit numbers.

5. *Real numbers.* Two real numbers have a sum of 200 and a product of 50. What is the sum of their reciprocals?

6. *Telling time.* Find the first time after 11 A.M. for which the minute hand and hour hand of a clock form a perfect right angle. Find the time to the nearest tenth of a second. The answer is not 11:10.

7. *Identity crisis.* Determine the value of a that will make this equation an identity.

$$\frac{1}{x-1} + \frac{2}{1-x} + \frac{3}{x-1} + \frac{4}{1-x} + \frac{5}{x-1} + \frac{6}{1-x}$$
$$+ \frac{7}{x-1} + \frac{8}{1-x} + \frac{9}{x-1} + \frac{10}{1-x} = \frac{a}{x-1}$$

8. *Difference of two squares.* Let $a = 7^{6006} + 7^{-6006}$ and $b = 7^{6006} - 7^{-6006}$. Find $a^2 - b^2$.

12

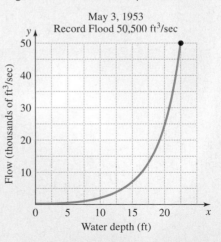

Exponential and Logarithmic Functions

Water is one of the essentials of life, yet it is something that most of us take for granted. Among other things, the U.S. Geological Survey (U.S.G.S.) studies freshwater. For over 50 years the Water Resources Division of the U.S.G.S. has been gathering basic data about the flow of both freshwater and saltwater from streams and groundwater surfaces. This division collects, compiles, analyzes, verifies, organizes, and publishes data gathered from groundwater data collection networks in each of the 50 states, Puerto Rico, and the Trust Territories. Records of stream flow, groundwater levels, and water quality provide hydrological information needed by local, state, and federal agencies as well as the private sector.

 Exponential Functions and Their Applications

 Logarithmic Functions and Their Applications

12.3 **Properties of Logarithms**

12.4 **Solving Equations and Applications**

There are many instances of the importance of the data collected by the U.S.G.S. For example, before 1987 the Tangipahoa River in Louisiana was used extensively for swimming and boating. In 1987 data gathered by the U.S.G.S. showed that fecal coliform levels in the river exceeded safe levels. Consequently, Louisiana banned recreational use of the river. Other studies by the Water Resources Division include the results of pollutants on salt marsh environments and the effect that salting highways in winter has on our drinking water supply.

In Exercises 87 and 88 of Section 12.2 you will see how data from the U.S.G.S. is used in a logarithmic function to measure water quality.

12.1   Exponential Functions and Their Applications

We have studied functions such as

$$f(x) = x^2, \qquad g(x) = x^3, \qquad \text{and} \qquad h(x) = x^{1/2}.$$

For these functions the variable is the base. In this section, we discuss functions that have a variable as an exponent. These functions are called *exponential functions.*

⟨1⟩ Exponential Functions

Some examples of exponential functions are

$$f(x) = 2^x, \qquad f(x) = \left(\frac{1}{2}\right)^x, \qquad \text{and} \qquad f(x) = 3^x.$$

> **Exponential Function**
>
> An **exponential function** is a function of the form
> $$f(x) = a^x,$$
> where $a > 0$, $a \neq 1$, and x is a real number.

We rule out the base 1 in the definition because $f(x) = 1^x$ is the same as the constant function $f(x) = 1$. Zero is not used as a base because $0^x = 0$ for any positive x and nonpositive powers of 0 are undefined. Negative numbers are not used as bases because an expression such as $(-4)^x$ is not a real number if $x = \frac{1}{2}$.

E X A M P L E   1

Evaluating exponential functions

Let $f(x) = 2^x$, $g(x) = \left(\frac{1}{4}\right)^{1-x}$, and $h(x) = -3^x$. Find the following:

a) $f\left(\dfrac{3}{2}\right)$ **b)** $f(-3)$ **c)** $g(3)$ **d)** $h(2)$

Solution

a) $f\left(\dfrac{3}{2}\right) = 2^{3/2} = \sqrt{2^3} = \sqrt{8} = 2\sqrt{2}$

b) $f(-3) = 2^{-3} = \dfrac{1}{2^3} = \dfrac{1}{8}$

c) $g(3) = \left(\dfrac{1}{4}\right)^{1-3} = \left(\dfrac{1}{4}\right)^{-2} = 4^2 = 16$

d) $h(2) = -3^2 = -9$ Note that $-3^2 \neq (-3)^2$.

> Now do Exercises 1–12

For many applications of exponential functions we use base 10 or another base called e. The number e is an irrational number that is approximately 2.718. We will see how e is used in compound interest in Example 9 of this section. Base 10 will be used in Section 12.2. Base 10 is called the **common base,** and base e is called the **natural base.**

Base 10 and base e

Let $f(x) = 10^x$ and $g(x) = e^x$. Find the following, and round approximate answers to four decimal places:

    a) $f(3)$ **b)** $f(1.51)$ **c)** $g(0)$ **d)** $g(2)$

‹ Calculator Close-Up ›

Most graphing calculators have keys for the functions 10^x and e^x.

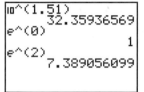

Solution

    a) $f(3) = 10^3 = 1000$

    b) $f(1.51) - 10^{1.51} \approx 32.3594$ Use the 10^x key on a calculator.

    c) $g(0) = e^0 = 1$

    d) $g(2) = e^2 \approx 7.3891$ Use the e^x key on a calculator.

> Now do Exercises 13–20

In the definition of an exponential function no restrictions were placed on the exponent x because the domain of an exponential function is the set of all real numbers. So both rational and irrational numbers can be used as the exponent. We have been using rational numbers for exponents since Chapter 9, but we have not yet seen an irrational number as an exponent. Even though we do not formally define irrational exponents in this text, an irrational number such as π can be used as an exponent, and you can evaluate an expression such as 2^π by using a calculator. Try it:

$$2^\pi \approx 8.824977827$$

Domain

The domain of an exponential function is the set of all real numbers.

‹2› Graphing Exponential Functions

Even though the domain of an exponential function is the set of all real numbers, we can graph an exponential function by evaluating it for just a few integers.

Exponential functions with base greater than 1

Sketch the graph of each function.

    a) $f(x) = 2^x$ **b)** $g(x) = 3^x$

Solution

    a) We first make a table of ordered pairs that satisfy $f(x) = 2^x$:

x	-2	-1	0	1	2	3
$f(x) = 2^x$	$\frac{1}{4}$	$\frac{1}{2}$	1	2	4	8

As x increases, 2^x increases: $2^4 = 16$, $2^5 = 32$, $2^6 = 64$, and so on. As x decreases, the powers of 2 are getting closer and closer to 0, but always remain positive: $2^{-3} = \frac{1}{8}$, $2^{-4} = \frac{1}{16}$, $2^{-5} = \frac{1}{32}$, and so on. So as x decreases, the graph approaches but does not touch the x-axis. Because the domain of the function is $(-\infty, \infty)$

The graph of $f(x) = 2^x$ on a calculator appears to touch the *x*-axis. When drawing this graph by hand, make sure that it does not touch the *x*-axis. Use zoom to see that the curve is always above the *x*-axis.

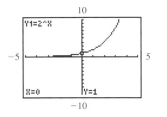

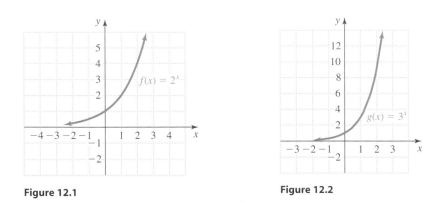

Figure 12.1

Figure 12.2

we draw the graph in Fig. 12.1 as a smooth curve through the points in the table. Since the powers of 2 are always positive, the range is $(0, \infty)$.

b) Make a table of ordered pairs that satisfy $g(x) = 3^x$:

x	-2	-1	0	1	2	3
$g(x) = 3^x$	$\frac{1}{9}$	$\frac{1}{3}$	1	3	9	27

Draw a smooth curve through the points indicated in the table. As x increases, 3^x increases. As x decreases, 3^x gets closer and closer to 0, but does not reach 0. So the graph shown in Fig. 12.2 approaches but does not touch the *x*-axis. The domain is $(-\infty, \infty)$ and the range is $(0, \infty)$.

> Now do Exercises 25–26

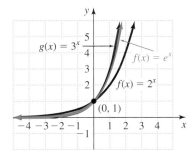

Figure 12.3

The curves in Figs. 12.1 and 12.2 are said to approach the *x*-axis **asymptotically,** and the *x*-axis is called an **asymptote** for the curves. Every exponential function has a horizontal asymptote.

Because $e \approx 2.718$, the graph of $f(x) = e^x$ lies between the graphs of $f(x) = 2^x$ and $g(x) = 3^x$, as shown in Fig. 12.3. Note that all three functions have the same domain and range and the same *y*-intercept. We summarize these ideas as follows:

$f(x) = a^x$ with $a > 1$

1. The *y*-intercept of the curve is $(0, 1)$.

2. The domain is $(-\infty, \infty)$ and the range is $(0, \infty)$.

3. The *x*-axis is an asymptote for the curve.

4. The *y*-values increase as we go from left to right on the curve.

E X A M P L E   **4**   **Exponential functions with base between 0 and 1**
Graph each function.

a) $f(x) = \left(\dfrac{1}{2}\right)^x$

b) $f(x) = 4^{-x}$

The graph of $y = (1/2)^x$ is a reflection of the graph of $y = 2^x$.

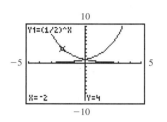

Solution

a) First make a table of ordered pairs that satisfy $f(x) = \left(\frac{1}{2}\right)^x$:

x	-2	-1	0	1	2	3
$f(x) = \left(\frac{1}{2}\right)^x$	4	2	1	$\frac{1}{2}$	$\frac{1}{4}$	$\frac{1}{8}$

As x increases, $\left(\frac{1}{2}\right)^x$ decreases, getting closer and closer to 0. Draw a smooth curve through these points as shown in Fig. 12.4.

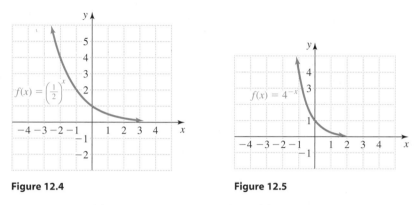

Figure 12.4   **Figure 12.5**

b) Because $4^{-x} = \left(\frac{1}{4}\right)^x$, we make a table for $f(x) = \left(\frac{1}{4}\right)^x$:

x	-2	-1	0	1	2	3
$f(x) = \left(\frac{1}{4}\right)^x$	16	4	1	$\frac{1}{4}$	$\frac{1}{16}$	$\frac{1}{64}$

As x increases, $\left(\frac{1}{4}\right)^x$, or 4^{-x}, decreases, getting closer and closer to 0. Draw a smooth curve through these points as shown in Fig. 12.5.

> **Now do Exercises 27–30**

Notice the similarities and differences between the exponential functions with $a > 1$ and those with base between 0 and 1. The main points are summarized as follows:

> **$f(x) = a^x$ with $0 < a < 1$**
>
> 1. The y-intercept of the curve is $(0, 1)$.
> 2. The domain is $(-\infty, \infty)$ and the range is $(0, \infty)$.
> 3. The x-axis is an asymptote for the curve.
> 4. The y-values decrease as we go from left to right on the curve.

CAUTION An exponential function can be written in more than one form. For example, $f(x) = \left(\frac{1}{2}\right)^x$ is the same as $f(x) = \frac{1}{2^x}$, or $f(x) = 2^{-x}$.

‹3› Transformations of Exponential Functions

We discussed transformation of functions in Section 11.3. In Example 5, we will graph some transformations of $f(x) = a^x$. Any transformation of an exponential function can be called an exponential function also.

EXAMPLE 5

Transformations of $f(x) = a^x$

Use transformations to graph each exponential function.

a) $f(x) = -2^x$ **b)** $f(x) = \frac{1}{3} \cdot 2^x + 1$ **c)** $f(x) = 2^{x-3} - 4$

Solution

a) The graph of $f(x) = -2^x$ is a reflection in the x-axis of the graph of $f(x) = 2^x$. Calculate a few ordered pairs for accuracy:

x	-1	0	1	2
$y = -2^x$	$-\frac{1}{2}$	-1	-2	-4

Plot these ordered pairs, and draw a curve through them as shown in Fig. 12.6.

b) To graph $f(x) = \frac{1}{3} \cdot 2^x + 1$, shrink the graph of $y = 2^x$ by a factor of $\frac{1}{3}$ and translate it upward one unit. Calculate a few ordered pairs for accuracy:

x	-1	0	1	2
$y = \frac{1}{3} \cdot 2^x + 1$	$\frac{7}{6}$	$\frac{4}{3}$	$\frac{5}{3}$	$\frac{7}{3}$

Plot these ordered pairs, and draw a curve through them as shown in Fig. 12.7.

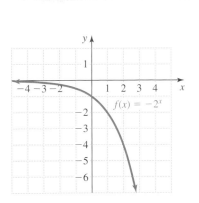

Figure 12.6

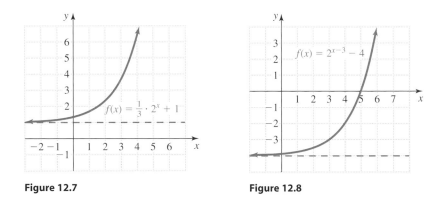

Figure 12.7   **Figure 12.8**

c) To graph $f(x) = 2^{x-3} - 4$ move $f(x) = 2^x$ to the right 3 units and down 4 units. Calculate a few ordered pairs for accuracy:

x	2	3	4	5
$y = 2^{x-3} - 4$	-3.5	-3	-2	0

Plot these ordered pairs, and draw a curve through them as shown in Fig. 12.8.

> Now do Exercises 35–46

⟨4⟩ Exponential Equations

In Chapter 11, we used the horizontal-line test to determine whether a function is one-to-one. Because no horizontal line can cross the graph of an exponential function more than once, exponential functions are one-to-one functions. For an exponential function one-to-one means that *if two exponential expressions with the same base are equal, then the exponents are equal.* If $2^x = 2^y$ then $x = y$.

One-to-One Property of Exponential Functions

For $a > 0$ and $a \neq 1$,

$$\text{if } \quad a^m = a^n, \quad \text{then} \quad m = n.$$

In Example 6, we use the one-to-one property to solve equations involving exponential functions.

E X A M P L E　6

Using the one-to-one property

Solve each equation.

a) $2^{2x-1} = 8$　　　　　**b)** $9^{|x|} = 3$　　　　　**c)** $\dfrac{1}{8} = 4^x$

Solution

a) Because 8 is 2^3, we can write each side as a power of the same base, 2:

$$2^{2x-1} = 8 \quad \text{Original equation}$$
$$2^{2x-1} = 2^3 \quad \text{Write each side as a power of the same base.}$$
$$2x - 1 = 3 \quad \text{One-to-one property}$$
$$2x = 4$$
$$x = 2$$

Check: $2^{2 \cdot 2 - 1} = 2^3 = 8$. The solution set is $\{2\}$.

b) Because $9 = 3^2$, we can write each side as a power of 3:

$$9^{|x|} = 3 \quad \text{Original equation}$$
$$(3^2)^{|x|} = 3^1$$
$$3^{2|x|} = 3^1 \quad \text{Power of a power rule}$$
$$2|x| = 1 \quad \text{One-to-one property}$$
$$|x| = \frac{1}{2}$$
$$x = \pm\frac{1}{2} \quad \text{Since } \left|-\frac{1}{2}\right| = \left|\frac{1}{2}\right| = \frac{1}{2}, \text{ there}$$
$$\text{are two solutions to } |x| = \frac{1}{2}.$$

Check $x = \pm\frac{1}{2}$ in the original equation. The solution set is $\left\{-\frac{1}{2}, \frac{1}{2}\right\}$.

c) Because $\frac{1}{8} = 2^{-3}$ and $4 = 2^2$, we can write each side as a power of 2:

$$\frac{1}{8} = 4^x \quad \text{Original equation}$$
$$2^{-3} = (2^2)^x \quad \text{Write each side as a power of 2.}$$
$$2^{-3} = 2^{2x} \quad \text{Power of a power rule}$$
$$2x = -3 \quad \text{One-to-one property}$$
$$x = -\frac{3}{2}$$

Check $x = -\frac{3}{2}$ in the original equation. The solution set is $\left\{-\frac{3}{2}\right\}$.

> Now do Exercises 47–60

‹ **Calculator Close-Up** ›

You can see the solution to $2^{2x-1} = 8$ by graphing $y_1 = 2^{2x-1}$ and $y_2 = 8$. The x-coordinate of the point of intersection is the solution to the equation.

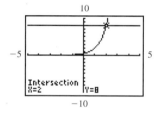

‹ **Calculator Close-Up** ›

The equation $9^{|x|} = 3$ has two solutions because the graphs of $y_1 = 9^{|x|}$ and $y_2 = 3$ intersect twice.

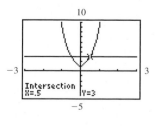

The one-to-one property is also used to find the first coordinate when given the second coordinate of an exponential function.

E X A M P L E   **7**

Finding the x-coordinate in an exponential function

Let $f(x) = 2^x$ and $g(x) = \left(\frac{1}{2}\right)^{1-x}$. Find x if

a) $f(x) = 32$

b) $g(x) = 8$

Solution

a) Because $f(x) = 2^x$ and $f(x) = 32$, we can find x by solving $2^x = 32$:

$$2^x = 32$$

$$2^x = 2^5 \quad \text{Write both sides as a power of the same base.}$$

$$x = 5 \quad \text{One-to-one property}$$

b) Because $g(x) = \left(\frac{1}{2}\right)^{1-x}$ and $g(x) = 8$, we can find x by solving $\left(\frac{1}{2}\right)^{1-x} = 8$:

$$\left(\frac{1}{2}\right)^{1-x} = 8$$

$$\left(2^{-1}\right)^{1-x} = 2^3 \quad \text{Because } \tfrac{1}{2} = 2^{-1} \text{ and } 8 = 2^3$$

$$2^{x-1} = 2^3 \quad \text{Power of a power rule}$$

$$x - 1 = 3 \quad \text{One-to-one property}$$

$$x = 4$$

Now do Exercises 61–72

〈**5**〉 **Applications**

The simple interest formula $A = P + Prt$ gives the amount A after t years for a principal P invested at simple interest rate r. If an investment is earning **compound interest,** then interest is periodically paid into the account and the interest that is paid also earns interest. To compute the amount of an account earning compound interest, the simple interest formula is used repeatedly. For example, if an account earns 6% compounded quarterly and the amount at the beginning of the first quarter is $5000, we apply the simple interest formula with $P = \$5000$, $r = 0.06$, and $t = \frac{1}{4}$ to find the amount in the account at the end of the first quarter:

$$A = P + Prt$$

$$= P(1 + rt) \qquad \text{Factor.}$$

$$= 5000\left(1 + 0.06 \cdot \frac{1}{4}\right) \qquad \text{Substitute.}$$

$$= 5000(1.015)$$

$$= \$5075$$

To repeat this computation for another quarter, we multiply $5075 by 1.015. If A represents the amount in the account at the end of n quarters, we can write A as an exponential function of n:

$$A = \$5000(1.015)^n$$

In general, the amount A is given by the following formula.

> **Compound Interest Formula**
>
> If P represents the principal, i the interest rate per period, n the number of periods, and A the amount at the end of n periods, then
>
> $$A = P(1 + i)^n.$$

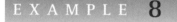

E X A M P L E   8

Compound interest formula

If $350 is deposited in an account paying 12% compounded monthly, then how much is in the account at the end of 6 years and 6 months?

Solution

Interest is paid 12 times per year, so the account earns $\frac{1}{12}$ of 12%, or 1% each month, for 78 months. So $i = 0.01$, $n = 78$, and $P = \$350$:

$$A = P(1 + i)^n$$

$$A = \$350(1.01)^{78}$$

$$\approx \$760.56$$

Now do Exercises 77–82

⟨ **Calculator Close-Up** ⟩

Graph $y = 350(1.01)^x$ to see the growth of the $350 deposit in Example 8 over time. After 360 months, it is worth $12,582.37.

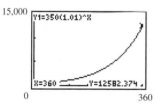

If we shorten the length of the time period (yearly, quarterly, monthly, daily, hourly, etc.), the number of periods n increases while the interest rate for the period decreases. As n increases, the amount A also increases but will not exceed a certain amount. That certain amount is the amount obtained from *continuous compounding* of the interest. It is shown in more advanced courses that the following formula gives the amount when interest is compounded continuously.

⟨ **Helpful Hint** ⟩

Compare Examples 8 and 9 to see the difference between compounded monthly and compounded continuously. Although there is not much difference to an individual investor, there could be a large difference to the bank. Rework Examples 8 and 9 using $50 million as the deposit.

> **Continuous-Compounding Formula**
>
> If P is the principal or beginning balance, r is the annual percentage rate compounded continuously, t is the time in years, and A is the amount or ending balance, then
>
> $$A = Pe^{rt}.$$

CAUTION The value of t in the continuous-compounding formula must be in years. For example, if the time is 1 year and 3 months, then $t = 1.25$ years. If the time is 3 years and 145 days, then

$$t = 3 + \frac{145}{365}$$

$$\approx 3.3973 \text{ years.}$$

E X A M P L E   **9**

Continuous-compounding formula

If $350 is deposited in an account paying 12% compounded continuously, then how much is in the account after 6 years and 6 months?

⟨ **Calculator Close-Up** ⟩

Graph $y = 350e^{0.12x}$ to see the growth of the $350 deposit in Example 9 over time. After 30 years, it is worth $12,809.38.

Solution

Use $r = 12\%$, $t = 6.5$ years, and $P = \$350$ in the formula for compounding interest continuously:

$$A = Pe^{rt}$$

$$= 350e^{(0.12)(6.5)}$$

$$= 350e^{0.78}$$

$$\approx \$763.52 \quad \text{Use the } e^x \text{ key on a scientific calculator.}$$

15,000

Y1=350e^(.12X)

X=30   Y=12809.382

0   30

Note that compounding continuously amounts to a few dollars more than compounding monthly did in Example 8.

Now do Exercises 83–90

Warm-Ups ▼

Fill in the blank.

1. An _____ function has the form $f(x) = a^x$, where $a > 0$ and $a \neq 1$.

2. The _____ of an exponential function is all real numbers.

3. Base e is the _____ base.

4. Base 10 is the _____ base.

5. The _____ property states that if $a^m = a^n$, then $m = n$.

6. The formula $A = P(1 + i)^n$ is for _____ interest.

7. The formula $A = Pe^{rt}$ is used when interest is compounded _____.

True or false?

8. If $f(x) = 4^x$, then $f\left(-\frac{1}{2}\right) = -2$.

9. If $f(x) = \left(\frac{1}{3}\right)^x$, then $f(-1) = 3$.

10. The functions $f(x) = \left(\frac{1}{2}\right)^x$ and $f(x) = 2^{-x}$ have the same graph.

11. The function $f(x) = 2^x$ is invertible.

12. The graph of $y = 2^x$ has an x-intercept.

13. The y-intercept for $f(x) = e^x$ is (0, 1).

14. The expression $2^{\sqrt{2}}$ is undefined.

‹1› Exponential Functions

Let $f(x) = 4^x$, $g(x) = \left(\frac{1}{3}\right)^{x+1}$, and $h(x) = -2^x$. Find the following. See Example 1.

1. $f(2)$ **2.** $f(-1)$

3. $f\left(\frac{1}{2}\right)$ **4.** $f\left(-\frac{3}{2}\right)$

5. $g(-2)$ **6.** $g(1)$

7. $g(0)$ **8.** $g(-3)$

9. $h(0)$ **10.** $h(3)$

11. $h(-2)$ **12.** $h(-4)$

Let $h(x) = 10^x$ and $j(x) = e^x$. Find the following. Use a calculator as necessary, and round approximate answers to three decimal places. See Example 2.

13. $h(0)$ **14.** $h(-1)$

15. $h(2)$ **16.** $h(3.4)$

17. $j(1)$ **18.** $j(3.5)$

19. $j(-2)$ **20.** $j(0)$

Fill in the missing entries in each table.

21.

x	−2	−1	0	1	2
4^x					

22.

x	−2	−1	0	1	2
5^x					

23.

x	−2	−1	0	1	2
$\left(\frac{1}{3}\right)^x$					

24.

x	−2	−1	0	1	2
$\left(\frac{1}{5}\right)^x$					

‹2› Graphing Exponential Functions

Sketch the graph of each function. See Examples 3 and 4.

25. $f(x) = 4^x$ **26.** $g(x) = 5^x$

27. $h(x) = \left(\frac{1}{3}\right)^x$ **28.** $i(x) = \left(\frac{1}{5}\right)^x$

29. $y = 10^x$ **30.** $y = (0.1)^x$

Fill in the missing entries in each table.

31.

x	−4	−3	−2	−1	0
10^{x+2}					

32.

x	−2	−1	$-\frac{1}{2}$	0	1
3^{2x+1}					

33.

x	−2	−1	0	1	2
-2^x					

34.

x	0	1	2	3	4
-2^{x-2}					

45. $f(x) = e^{-x} + 2$

46. $f(x) = e^{-x} - 1$

⟨3⟩ Transformations of Exponential Functions

Use transformations to help you sketch the graph of each function. See Example 5.

35. $f(x) = -3^x$

36. $f(x) = -10^x$

37. $f(x) = \dfrac{1}{2} \cdot 3^x$

38. $f(x) = -2 \cdot 3^x$

39. $f(x) = -3^x + 2$

40. $f(x) = -3^x - 4$

41. $f(x) = 3^{x-2} + 1$

42. $f(x) = 3^{x+1} - 2$

43. $f(x) = 10^x + 2$

44. $f(x) = -10^x + 3$

⟨4⟩ Exponential Equations

Solve each equation. See Example 6.

47. $2^x = 64$

48. $3^x = 9$

49. $10^x = 0.001$

50. $10^{2x} = 0.1$

51. $2^x = \dfrac{1}{4}$

52. $3^x = \dfrac{1}{9}$

VIDEO 53. $\left(\dfrac{2}{3}\right)^{x-1} = \dfrac{9}{4}$

54. $\left(\dfrac{1}{4}\right)^{3x} = 16$

55. $5^{-x} = 25$

56. $10^{-x} = 0.01$

57. $-2^{1-x} = -8$

58. $-3^{2-x} = -81$

59. $10^{|x|} = 1000$

60. $3^{|2x-5|} = 81$

Let $f(x) = 2^x$, $g(x) = \left(\dfrac{1}{3}\right)^x$, and $h(x) = 4^{2x-1}$. Find x in each case. See Example 7.

61. $f(x) = 4$

62. $f(x) = \dfrac{1}{4}$

63. $f(x) = 4^{2/3}$

64. $f(x) = 1$

65. $g(x) = 9$

66. $g(x) = \dfrac{1}{9}$

67. $g(x) = 1$

68. $g(x) = \sqrt{3}$

VIDEO 69. $h(x) = 16$

70. $h(x) = \dfrac{1}{2}$

71. $h(x) = 1$

72. $h(x) = \sqrt{2}$

Fill in the missing entries in each table.

73.

x	-5			0		4
2^x			$\dfrac{1}{8}$		2	

74.

x	-4			0		3
3^x			$\dfrac{1}{9}$		3	

75.

x		-2		1	
$\left(\dfrac{1}{2}\right)^x$	8		1		$\dfrac{1}{32}$

76.

x		-1		2	
$\left(\dfrac{1}{10}\right)^x$	100		1		$\dfrac{1}{1000}$

⟨5⟩ Applications

Solve each problem. See Example 8.

77. *Compounding quarterly.* If $6000 is deposited in an account paying 5% compounded quarterly, then what amount will be in the account after 10 years?

78. *Compounding quarterly.* If $400 is deposited in an account paying 10% compounded quarterly, then what amount will be in the account after 7 years?

79. *Bond fund.* Fidelity's Municipal Income Fund (www.fidelity.com) returned an average of 4.72% annually from 1999 to 2009.

   a) How much was an investment of $10,000 in this fund in 1999 worth in 2009 at 4.72% compounded annually?

   b) Use the accompanying graph to estimate the year in which the $10,000 investment in 1999 would be worth $20,000 if it continued to return 4.72% annually.

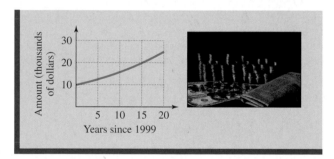

Figure for Exercise 79

80. *Slow growth.* Fidelity's Contrafund returned an average of 1.98% annually from 1999 to 2009. How much was an investment of $10,000 in this fund in 1999 worth in 2009?

81. *Depreciating knowledge.* The value of a certain textbook seems to decrease according to the formula $V = 45 \cdot 2^{-0.9t}$, where V is the value in dollars and t is the age of the book in years. What is the book worth when it is new? What is it worth when it is 2 years old?

82. *Mosquito abatement.* In a Minnesota swamp in the springtime the number of mosquitoes per acre appears to grow according to the formula $N = 10^{0.1t\,+2}$, where t is the number of days since the last frost. What is the size of the mosquito population at times $t = 10$, $t = 20$, and $t = 30$?

Solve each problem. See Example 9.

83. *Compounding continuously.* If $500 is deposited in an account paying 7% compounded continuously, then how much will be in the account after 3 years?

84. *Compounding continuously.* If $7000 is deposited in an account paying 8% compounded continuously, then what will it amount to after 4 years?

85. *One year's interest.* How much interest will be earned the first year on $80,000 on deposit in an account paying 7.5% compounded continuously?

86. *Partial year.* If $7500 is deposited in an account paying 6.75% compounded continuously, then how much will be in the account after 5 years and 215 days?

87. *Radioactive decay.* The number of grams of a certain radioactive substance present at time t is given by the formula $A = 300 \cdot e^{-0.06t}$, where t is the number of years. Find the amount present at time $t = 0$. Find the amount present after 20 years. Use the graph below to estimate the number of years that it takes for one-half of the substance to decay. Will the substance ever decay completely?

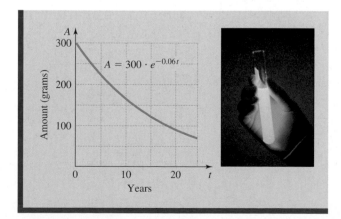

Figure for Exercise 87

88. *Population growth.* The population of a certain country appears to be growing according to the formula $P = 20 \cdot e^{0.1t}$, where P is the population in millions and t is the number of years since 1990. What was the population in 1990? What will the population be in the year 2010?

89. *Man overboard.* The difference in temperature between a warm human body (98.6°F) and a cold ocean (48.6°F) is given by the function $D = 50e^{-0.03t}$, where D is in degrees Fahrenheit and t is time in minutes. What is the difference between the body and the ocean for $t = 0$? What is the difference for $t = 15$? What is the ocean temperature at $t = 15$? What is the temperature of the human body at $t = 15$?

90. *Cooking a turkey.* The difference in temperature between a hot oven (350°F) and a cold turkey (38°F) is given by the function $D = 312e^{-0.12t}$, where D is in degrees Fahrenheit and t is time in hours. What is the difference between the turkey and the oven for $t = 0$? What is the difference for $t = 4$? What is the oven temperature at $t = 4$? What is the temperature of the turkey at $t = 4$?

Getting More Involved

91. *Exploration*

An approximate value for e can be found by adding the terms in the following infinite sum:

$$1 + \frac{1}{1} + \frac{1}{2 \cdot 1} + \frac{1}{3 \cdot 2 \cdot 1} + \frac{1}{4 \cdot 3 \cdot 2 \cdot 1} + \cdots$$

Use a calculator to find the sum of the first four terms. Find the difference between the sum of the first four terms and e. (For e, use all of the digits that your calculator gives for e^1.) What is the difference between e and the sum of the first eight terms?

Graphing Calculator Exercises

92. Graph $y_1 = 2^x$, $y_2 = e^x$, and $y_3 = 3^x$ on the same coordinate system. Which point do all three graphs have in common?

93. Graph $y_1 = 3^x$, $y_2 = 3^{x-1}$, and $y_3 = 3^{x-2}$ on the same coordinate system. What can you say about the graph of $y = 3^{x-h}$ for any real number h?

12.2 Logarithmic Functions and Their Applications

In Section 12.1, you learned that exponential functions are one-to-one functions. Because they are one-to-one functions, they have inverse functions. In this section we study the inverses of the exponential functions.

In This Section

⟨1⟩ **Logarithmic Functions**

⟨2⟩ **Graphing Logarithmic Functions**

⟨3⟩ **Logarithmic Equations**

⟨4⟩ **Applications**

⟨1⟩ Logarithmic Functions

We define $\log_a(x)$ as *the exponent that is used on the base a to obtain the result x.* Read the expression $\log_a(x)$ as "the base a logarithm of x." The expression $\log_a(x)$ is called a **logarithm.** If the *exponent* 3 is used on the *base* 2, then the *result* is 8 ($2^3 = 8$). So,

$$\log_2(8) = 3.$$

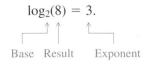

Base Result   Exponent

Because $5^2 = 25$, the exponent used to obtain 25 with base 5 is 2 and $\log_5(25) = 2$. Because $2^{-5} = \frac{1}{32}$, the exponent used to obtain $\frac{1}{32}$ with base 2 is -5 and $\log_2\left(\frac{1}{32}\right) = -5$. From these examples, we see that the definition of $\log_a(x)$ can also be stated as follows:

Definition of $\log_a(x)$

For any $a > 0$ and $a \neq 1$,

$$y = \log_a(x) \qquad \text{if and only if} \qquad a^y = x.$$

Note that the base of a logarithm must be a positive number and it cannot be 1.

EXAMPLE 1

Using the definition of logarithm

Write each logarithmic equation as an exponential equation and each exponential equation as a logarithmic equation.

a) $\log_5(125) = 3$

b) $6 = \log_{1/4}(x)$

c) $\left(\dfrac{1}{2}\right)^m = 8$

d) $7 = 3^z$

Solution

a) "The base-5 logarithm of 125 equals 3" means that 3 is the exponent on 5 that produces 125. So $5^3 = 125$.

b) The equation $6 = \log_{1/4}(x)$ is equivalent to $\left(\frac{1}{4}\right)^6 = x$ by the definition of logarithm.

c) The equation $\left(\frac{1}{2}\right)^m = 8$ is equivalent to $\log_{1/2}(8) = m$.

d) The equation $7 = 3^z$ is equivalent to $\log_3(7) = z$.

> Now do Exercises 1–12

The inverse of the base-a exponential function $f(x) = a^x$ is the **base-a logarithmic function** $f^{-1}(x) = \log_a(x)$. For example, $f(x) = 2^x$ and $f^{-1}(x) = \log_2(x)$ are inverse functions, as shown in Fig. 12.9. Each function undoes the other.

$$f(5) = 2^5 = 32 \quad \text{and} \quad f^{-1}(32) = \log_2(32) = 5.$$

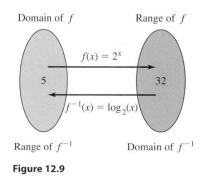

Figure 12.9

To evaluate logarithmic functions remember that a logarithm is an exponent: $\log_a(x)$ is the exponent that is used on the base a to obtain x.

EXAMPLE 2

Finding logarithms
Evaluate each logarithm.

a) $\log_5(25)$

b) $\log_2\left(\dfrac{1}{8}\right)$

c) $\log_{1/2}(4)$

d) $\log_{10}(0.001)$

e) $\log_9(3)$

< **Helpful Hint** >

When we write $C(x) = 12x$, we may think of C as a variable and write $C = 12x$, or we may think of C as the name of a function, the cost function. In $y = \log_a(x)$ we are thinking of $\log_a$ only as the name of the function that pairs an x-value with a y-value.

Solution

a) The number $\log_5(25)$ is the exponent that is used on the base 5 to obtain 25. Because $25 = 5^2$, we have $\log_5(25) = 2$.

b) The number $\log_2\left(\dfrac{1}{8}\right)$ is the power of 2 that gives us $\dfrac{1}{8}$. Because $\dfrac{1}{8} = 2^{-3}$, we have $\log_2\left(\dfrac{1}{8}\right) = -3$.

c) The number $\log_{1/2}(4)$ is the power of $\dfrac{1}{2}$ that produces 4. Because $4 = \left(\dfrac{1}{2}\right)^{-2}$, we have $\log_{1/2}(4) = -2$.

d) Because $0.001 = 10^{-3}$, we have $\log_{10}(0.001) = -3$.

e) Because $9^{1/2} = 3$, we have $\log_9(3) = \dfrac{1}{2}$.

Now do Exercises 13-22

There are two bases for logarithms that are used more frequently than the others: They are 10 and e. The base-10 logarithm is called the **common logarithm** and is usually written as $\log(x)$. The base-e logarithm is called the **natural logarithm** and is usually written as $\ln(x)$. Most scientific calculators have function keys for $\log(x)$ and $\ln(x)$. The simplest way to obtain a common or natural logarithm is to use a scientific calculator.

In Example 3, we find natural and common logarithms of certain numbers without a calculator.

EXAMPLE 3

Finding common and natural logarithms
Evaluate each logarithm.

a) $\log(1000)$

b) $\ln(e)$

c) $\log\left(\dfrac{1}{10}\right)$

< **Calculator Close-Up** >

A graphing calculator has keys for the common logarithm (LOG) and the natural logarithm (LN).

```
log(1000)
              3
ln(e)
              1
log(1/10)
             -1
```

Solution

a) Because $10^3 = 1000$, we have $\log(1000) = 3$.

b) Because $e^1 = e$, we have $\ln(e) = 1$.

c) Because $10^{-1} = \dfrac{1}{10}$, we have $\log\left(\dfrac{1}{10}\right) = -1$.

Now do Exercises 23-34

The domain of the exponential function $y = 2^x$ is $(-\infty, \infty)$, and its range is $(0, \infty)$. Because the logarithmic function $y = \log_2(x)$ is the inverse of $y = 2^x$, the domain of $y = \log_2(x)$ is $(0, \infty)$, and its range is $(-\infty, \infty)$.

CAUTION The domain of $y = \log_a(x)$ for $a > 0$ and $a \neq 1$ is $(0, \infty)$. So expressions such as $\log_2(-4)$, $\log_{1/3}(0)$, and $\ln(-1)$ are undefined, because -4, 0, and -1 are not in the domain $(0, \infty)$.

⟨2⟩ Graphing Logarithmic Functions

In Chapter 11, we saw that the graphs of a function and its inverse function are symmetric about the line $y = x$. Because the logarithm functions are inverses of exponential functions, their graphs are also symmetric about $y = x$.

EXAMPLE **4**

⟨ **Calculator Close-Up** ⟩

The graphs of $y = \ln(x)$ and $y = e^x$ are symmetric with respect to the line $y = x$. Logarithmic functions with bases other than e and 10 will be graphed on a calculator in Section 12.4.

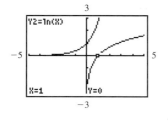

A logarithmic function with base greater than 1

Sketch the graph of $g(x) = \log_2(x)$, and compare it to the graph of $y = 2^x$.

Solution

Make a table of ordered pairs for $g(x) = \log_2(x)$ using positive numbers for x:

x	$\frac{1}{4}$	$\frac{1}{2}$	1	2	4	8
$g(x) = \log_2(x)$	-2	-1	0	1	2	3

Draw a curve through these points as shown in Fig. 12.10. The graph of the inverse function $y = 2^x$ is also shown in Fig. 12.10 for comparison. Note the symmetry of the two curves about the line $y = x$.

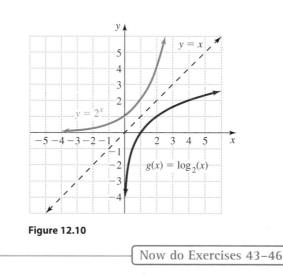

Figure 12.10

Now do Exercises 43–46

All logarithm functions with bases greater than 1 have graphs that are similar to the one in Fig. 12.11. In general, these functions have the following characteristics.

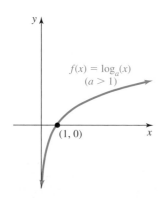

Figure 12.11

$f(x) = \log_a(x)$ with $a > 1$

1. The x-intercept of the curve is $(1, 0)$.

2. The domain is $(0, \infty)$ and the range is $(-\infty, \infty)$.

3. The y-axis is an asymptote for the curve.

4. The y-values increase as we go from left to right on the curve.

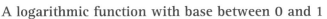

EXAMPLE 5

A logarithmic function with base between 0 and 1

Sketch the graph of $f(x) = \log_{1/2}(x)$, and compare it to the graph of $y = \left(\frac{1}{2}\right)^x$.

Solution

Make a table of ordered pairs for $f(x) = \log_{1/2}(x)$ using positive numbers for x:

x	$\frac{1}{4}$	$\frac{1}{2}$	1	2	4	8
$f(x) = \log_{1/2}(x)$	2	1	0	-1	-2	-3

The curve through these points is shown in Fig. 12.12. The graph of the inverse function $y = \left(\frac{1}{2}\right)^x$ is also shown in Fig. 12.12 for comparison. Note the symmetry with respect to the line $y = x$.

Now do Exercises 47–50

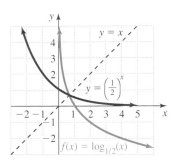

Figure 12.12

All logarithm functions with bases between 0 and 1 have graphs that are similar to the one in Fig. 12.13. In general, these functions have the following characteristics.

$f(x) = \log_a(x)$ with $0 < a < 1$

1. The x-intercept of the curve is $(1, 0)$.
2. The domain is $(0, \infty)$ and the range is $(-\infty, \infty)$.
3. The y-axis is an asymptote for the curve.
4. The y-values decrease as we go from left to right on the curve.

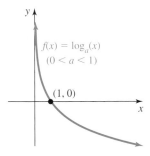

Figure 12.13

Figures 12.11 and 12.13 illustrate the fact that $y = \log_a(x)$ and $y = a^x$ are inverse functions for any base a. For any given exponential or logarithmic function the inverse function can be easily obtained from the definition of logarithm.

EXAMPLE 6

Inverses of logarithmic and exponential functions

Find the inverse of each function.

a) $f(x) = 10^x$ **b)** $g(x) = \log_3(x)$

Solution

a) To find any inverse function we switch the roles of x and y. So $y = 10^x$ becomes $x = 10^y$. Now $x = 10^y$ is equivalent to $y = \log_{10}(x)$. So the inverse of $f(x) = 10^x$ is $y = \log(x)$ or $f^{-1}(x) = \log(x)$.

b) In $g(x) = \log_3(x)$ or $y = \log_3(x)$ we switch x and y to get $x = \log_3(y)$. Now $x = \log_3(y)$ is equivalent to $y = 3^x$. So the inverse of $g(x) = \log_3(x)$ is $y = 3^x$ or $g^{-1}(x) = 3^x$.

Now do Exercises 51–56

⟨3⟩ Logarithmic Equations

In Section 12.1, we learned that the exponential functions are one-to-one functions. Because logarithmic functions are inverses of exponential functions, they are one-to-one functions also. For a base-a logarithmic function *one-to-one means that if the base-a logarithms of two numbers are equal, then the numbers are equal.*

> **One-to-One Property of Logarithms**
>
> For $a > 0$ and $a \neq 1$,
>
> $$\text{if} \quad \log_a(m) = \log_a(n), \quad \text{then} \quad m = n.$$

The one-to-one property of logarithms and the definition of logarithms are the two basic tools that we use to solve equations involving logarithms. We use these tools in Example 7.

EXAMPLE 7

Logarithmic equations

Solve each equation.

a) $\log_3(x) = -2$ 　　　　b) $\log_x(8) = -3$ 　　　　c) $\log(x^2) = \log(4)$

Solution

a) Use the definition of logarithms to rewrite the logarithmic equation as an equivalent exponential equation:

$$\log_3(x) = -2$$
$$3^{-2} = x \quad \text{Definition of logarithm}$$
$$\frac{1}{9} = x$$

Because $3^{-2} = \frac{1}{9}$ or $\log_3\left(\frac{1}{9}\right) = -2$, the solution set is $\left\{\frac{1}{9}\right\}$.

b) Use the definition of logarithms to rewrite the logarithmic equation as an equivalent exponential equation:

$$\log_x(8) = -3$$
$$x^{-3} = 8 \quad\quad\quad\quad \text{Definition of logarithm}$$
$$\left(x^{-3}\right)^{-1} = 8^{-1} \quad\quad \text{Raise each side to the } -1 \text{ power.}$$
$$x^3 = \frac{1}{8}$$
$$x = \sqrt[3]{\frac{1}{8}} = \frac{1}{2} \quad \text{Odd-root property}$$

Because $\left(\frac{1}{2}\right)^{-3} = 2^3 = 8$ or $\log_{1/2}(8) = -3$, the solution set is $\left\{\frac{1}{2}\right\}$.

c) To write an equation equivalent to $\log(x^2) = \log(4)$, we use the one-to-one property of logarithms:

$$\log(x^2) = \log(4)$$
$$x^2 = 4 \quad \text{One-to-one property of logarithms}$$
$$x = \pm 2 \quad \text{Even-root property}$$

If $x = \pm 2$, then $x^2 = 4$ and $\log(4) = \log(4)$. The solution set is $\{-2, 2\}$.

> Now do Exercises 57–68

CAUTION If we have equality of two logarithms with the same base, we use the one-to-one property to eliminate the logarithms. If we have an equation with only one logarithm, such as $\log_a(x) = y$, we use the definition of logarithm to write $a^y = x$ and to eliminate the logarithm.

⟨4⟩ **Applications**

The definition of logarithm indicates that $y = \log_a(x)$ if and only if $a^y = x$. If the base is e, then the definition indicates that

$$y = \ln(x) \quad \text{if and only if} \quad e^y = x.$$

In Example 8, we use the definition of logarithm to solve a problem involving the continuous-compounding formula

$$A = Pe^{rt},$$

where A is the amount after t years of an investment of P dollars at annual percentage rate r compounded continuously.

EXAMPLE 8

Finding the time with continuous compounding

How long does it take for $80 to grow to $240 at 12% annual percentage rate compounded continuously?

Solution

Use $r = 0.12$, $P = \$80$, and $A = \$240$ in the formula $A = Pe^{rt}$ to get $240 = 80e^{0.12t}$. Now use the definition of logarithm to solve for t:

$$240 = 80e^{0.12t}$$

$$3 = e^{0.12t} \quad \text{Divide each side by 80.}$$

$$0.12t = \ln(3) \quad \begin{array}{l}\text{Definition of logarithm:} \\ y = e^x \text{ means } x = \ln(y)\end{array}$$

$$t = \frac{\ln(3)}{0.12} \quad \text{Divide each side by 0.12.}$$

$$t \approx 9.155$$

The time is approximately 9.155 years. Multiply 365 by 0.155 to get approximately 57 days. So the time is 9 years and 57 days to the nearest day.

Now do Exercises 79–90

Note that we can also use the technique of Example 8 to solve a continuous-compounding problem in which the rate is the only unknown quantity.

Warm-Ups ▼

Fill in the blank.

1. The inverse for an exponential function is a _____ function.

2. A _____ logarithm is a base-10 logarithm.

3. A _____ logarithm is a base-e logarithm.

4. The _____ of $f(x) = \log_a(x)$ is $(0, \infty)$.

5. The _____ property states that if $\log_a(m) = \log_a(n)$, then $m = n$.

6. The graphs of $f(x) = 2^x$ and $g(x) = \log_2(x)$ are _____ about the line $y = x$.

7. The expression $\log_a(x)$ is the _____ that is used on base a to obtain x.

True or false?

8. The equation $a^3 = 2$ is equivalent to $\log_a(2) = 3$.

9. If (a, b) satisfies $y = 8^x$, then (a, b) satisfies $y = \log_8(x)$.

10. The inverse of $y = 5^x$ is $y = \log_5(x)$.

11. If $f(x) = \ln(x)$, then $f^{-1}(x) = e^x$.

12. $\log_{25}(5) = 2$

13. $\log(-10) = 1$

14. $\log_{1/2}(32) = -5$

15. $10^{\log(19)} = 19$

‹1› Logarithmic Functions

Write each exponential equation as a logarithmic equation and each logarithmic equation as an exponential equation. See Example 1.

1. $\log_2(8) = 3$ **2.** $\log_{10}(10) = 1$

3. $10^2 = 100$ **4.** $5^3 = 125$

5. $y = \log_5(x)$ **6.** $m = \log_b(N)$

7. $2^a = b$ **8.** $a^3 = c$

9. $\log_3(x) = 10$ **10.** $\log_c(t) = 4$

11. $e^3 = x$ **12.** $m = e^x$

Evaluate each logarithm. See Examples 2 and 3.

13. $\log_2(4)$ **14.** $\log_2(1)$

15. $\log_2(16)$ **16.** $\log_4(16)$

17. $\log_2(64)$ **18.** $\log_8(64)$

19. $\log_4(64)$ **20.** $\log_{64}(64)$

21. $\log_2\left(\dfrac{1}{4}\right)$ **22.** $\log_2\left(\dfrac{1}{8}\right)$

23. $\log(100)$ **24.** $\log(1)$

25. $\log(0.01)$ **26.** $\log(10{,}000)$

27. $\log_{1/3}\left(\dfrac{1}{3}\right)$ **28.** $\log_{1/3}\left(\dfrac{1}{9}\right)$

29. $\log_{1/3}(27)$ **30.** $\log_{1/3}(1)$

31. $\log_{25}(5)$ **32.** $\log_{16}(4)$

33. $\ln(e^2)$ **34.** $\ln\left(\dfrac{1}{e}\right)$

Use a calculator to evaluate each logarithm. Round answers to four decimal places.

35. $\log(5)$ **36.** $\log(0.03)$
37. $\ln(6.238)$ **38.** $\ln(0.23)$

Fill in the missing entries in each table.

39.

x	$\dfrac{1}{9}$	$\dfrac{1}{3}$	1	3	9
$\log_3(x)$					

40.

x	$\dfrac{1}{100}$	$\dfrac{1}{10}$	1	10	100
$\log_{10}(x)$					

41.

x	16	4	1	$\dfrac{1}{4}$	$\dfrac{1}{16}$
$\log_{1/4}(x)$					

42.

x	9	3	1	$\dfrac{1}{3}$	$\dfrac{1}{9}$
$\log_{1/3}(x)$					

‹2› Graphing Logarithmic Functions

Sketch the graph of each function. See Examples 4 and 5.

43. $f(x) = \log_3(x)$ **44.** $g(x) = \log_{10}(x)$

45. $y = \log_4(x)$ **46.** $y = \log_5(x)$

47. $h(x) = \log_{1/4}(x)$ **48.** $y = \log_{1/3}(x)$

49. $y = \log_{1/5}(x)$ **50.** $y = \log_{1/6}(x)$

Find the inverse of each function. See Example 6.

51. $f(x) = 6^x$ **52.** $f(x) = 4^x$

53. $f(x) = \ln(x)$ **54.** $f(x) = \log(x)$

55. $f(x) = \log_{1/2}(x)$ **56.** $f(x) = \log_{1/4}(x)$

⟨3⟩ Logarithmic Equations

Solve each equation. See Example 7.

57. $x = \left(\dfrac{1}{2}\right)^{-2}$ **58.** $x = 16^{-1/2}$

59. $5 = 25^x$ **60.** $0.1 = 10^x$

61. $\log(x) = -3$ **62.** $\log(x) = 5$

63. $\log_x(36) = 2$ **64.** $\log_x(100) = 2$

65. $\log_x(5) = -1$ **66.** $\log_x(16) = -2$

⊙67. $\log(x^2) = \log(9)$ **68.** $\ln(2x - 3) = \ln(x + 1)$
VIDEO

Use a calculator to solve each equation. Round answers to four decimal places.

69. $3 = 10^x$ **70.** $10^x = 0.03$

71. $10^x = \dfrac{1}{2}$ **72.** $75 = 10^x$

73. $e^x = 7.2$ **74.** $e^{3x} = 0.4$

Fill in the missing entries in each table.

75.

x	$\frac{1}{4}$		1		16
$\log_2(x)$		-1		2	

76.

x	$\frac{1}{125}$		1		625
$\log_5(x)$		-2		1	

77.

x		4		$\frac{1}{2}$	
$\log_{1/2}(x)$	-4		0		2

78.

x		6		$\frac{1}{36}$	
$\log_{1/6}(x)$	-2		0		3

⟨4⟩ Applications

Solve each problem. See Example 8. Use a calculator as necessary.

79. **Double your money.** How long does it take $5000 to grow to $10,000 at 12% compounded continuously?

80. **Half the rate.** How long does it take $5000 to grow to $10,000 at 6% compounded continuously?

81. **Earning interest.** How long does it take to earn $1000 in interest on a deposit of $6000 at 8% compounded continuously?

82. **Lottery winnings.** How long does it take to earn $1000 interest on a deposit of one million dollars at 9% compounded continuously?

83. **Investing.** An investment of $10,000 in Bonavista Energy in 2000 grew to $15,431 in 2009.

  a) Assuming that the investment grew continuously, what was the annual growth rate?

  b) If Bonavista Energy continued to grow continuously at the rate from part (a), then what would the investment be worth in 2020?

84. **Investing.** An investment of $10,000 in Baytex Energy in 2002 was worth $18,125 in 2009.

  a) Assuming that the investment grew continuously, what was the annual rate?

  b) If Baytex Energy continued to grow continuously at the rate from part (a), then what would the investment be worth in 2015?

In chemistry the pH of a solution is defined by

$$pH = -\log_{10}[H+],$$

where H+ is the hydrogen ion concentration of the solution in moles per liter. Distilled water has a pH of approximately 7. A solution with a pH under 7 is called an acid, and one with a pH over 7 is called a base.

85. **Tomato juice.** Tomato juice has a hydrogen ion concentration of $10^{-4.1}$ mole per liter (mol/L). Find the pH of tomato juice.

86. **Stomach acid.** The gastric juices in your stomach have a hydrogen ion concentration of 10^{-1} mol/L. Find the pH of your gastric juices.

87. **Neuse River pH.** The hydrogen ion concentration of a water sample from the Neuse River at New Bern, North Carolina, was 1.58×10^{-7} mol/L (www.nc.usgs.gov). What was the pH of this water sample?

88. **Roanoke River pH.** The hydrogen ion concentration of a water sample from the Roanoke River at Janesville, North Carolina, was 1.995×10^{-7} mol/L (www.nc.usgs.gov). What was the pH of this water sample?

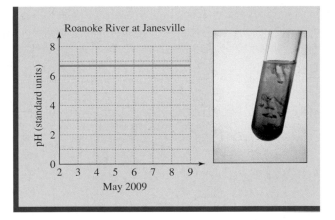

Figure for Exercise 88

Solve each problem.

89. **Sound level.** The level of sound in decibels (dB) is given by the formula

$$L = 10 \cdot \log(I \times 10^{12}),$$

where I is the intensity of the sound in watts per square meter. If the intensity of the sound at a rock concert is 0.001 watt per square meter at a distance of 75 meters from the stage, then what is the level of the sound at this point in the audience?

90. **Logistic growth.** If a rancher has one cow with a contagious disease in a herd of 1000, then the time in days t for n of the cows to become infected is modeled by

$$t = -5 \cdot \ln\left(\frac{1000 - n}{999n}\right).$$

Find the number of days that it takes for the disease to spread to 100, 200, 998, and 999 cows. This model, called a *logistic growth model,* describes how a disease can spread very rapidly at first and then very slowly as nearly all of the population has become infected. See the accompanying figure.

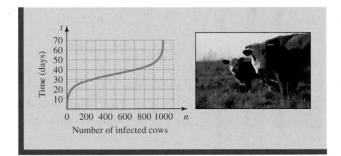

Figure for Exercise 90

Getting More Involved

91. **Discussion**

Use the switch-and-solve method from Chapter 11 to find the inverse of the function $f(x) = 5 + \log_2(x - 3)$. State the domain and range of the inverse function.

92. **Discussion**

Find the inverse of the function $f(x) = 2 + e^{x+4}$. State the domain and range of the inverse function.

Graphing Calculator Exercises

93. **Composition of inverses.** Graph the functions $y = \ln(e^x)$ and $y = e^{\ln(x)}$. Explain the similarities and differences between the graphs.

94. **The population bomb.** The population of the earth is growing continuously with an annual rate of about 1.6%. If the present population is 6 billion, then the function $y = 6e^{0.016x}$ gives the population in billions x years from now. Graph this function for $0 \le x \le 200$. What will the population be in 100 years and in 200 years?

Math *at Work* | Drug Administration

When a drug is taken continuously or intermittently, plasma concentrations of the drug increase. Over time, the rate of increase slows and eventually reaches a plateau. As concentration increases, the rate of elimination increases until a point is reached at which the amount of drug being eliminated from the body equals the amount being administered (steady state).

The time to reach steady state depends on the half-life of the drug. The half-life of a drug is the time it takes for the plasma concentration to be reduced by one-half. See the accompanying figure. The basic rule is that after administering a drug for a period equal to the half-life of the drug, the plasma concentration will be halfway between the starting concentration and steady state. This rule holds for any starting concentration. Mathematically, steady state is a limit and it is never reached. It is usually assumed that when a drug reaches 90% or more of steady state it is at steady state. It takes 3.3 half-lives of drug administration to reach 90% of steady state.

The half-life $t_{1/2}$ of a drug depends on the patient and is calculated from two plasma levels separated by a time interval. The first plasma level or peak (P) is measured after the drug has been fully distributed. The second plasma level or trough (T) is measured at some interval later (t). From P, T, and t, the elimination constant k is found by $k = \frac{\ln(P) - \ln(T)}{t}$. The half-life is then found using $t_{1/2} = \frac{\ln(2)}{k}$. When the dosing interval is much longer than the half-life, there is more time for elimination between doses and accumulation is small. When the dosing interval is much shorter than the half-life, there is little time for elimination and more accumulation of the drug.

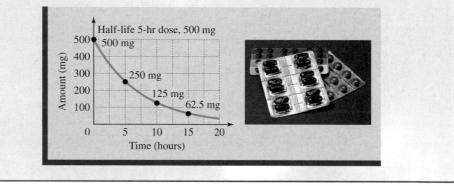

Mid-Chapter **Quiz** | Sections 12.1 through 12.2 | **Chapter 12**

Fill in the missing entries in each table.

1.

x	-2		0		3
2^x		$\frac{1}{2}$		4	

2.

x	-4		1		4
$\left(\frac{1}{2}\right)^x$		4		$\frac{1}{4}$	

3.

x	16	1	$\frac{1}{4}$	$\frac{1}{16}$
$\log_4(x)$				

4.

x	32	16	1	$\frac{1}{8}$
$\log_{1/2}(x)$				

Solve each equation.

5. $3^x = 81$

6. $\left(\frac{1}{2}\right)^{3x} = 8$

7. $10^x = 0.01$

8. $\log_x(64) = 2$

9. $\log(x) = 4$

Find the inverse of each function.   **16.** $y = \log_2(x)$
10. $f(x) = 10^x$ **11.** $g(x) = \log_8(x)$

Rewrite the exponential equation as a logarithmic equation and the logarithmic equation as an exponential equation.
12. $M = \log_5(W)$ **13.** $a^3 = y$

Graph each function, and state its domain and range.
14. $f(x) = 2^x + 1$
                                                                   Miscellaneous.
                                                                   17. Use a calculator to find $g(-2.3)$ to three decimal places if $g(x) = e^x$.

                                                                   18. If \$4000 is invested at 6% compounded quarterly, then what amount will be in the account after 5 years?

                                                                   19. If \$8000 is invested at 4.3% compounded continuously, then what will it amount to after $12\frac{1}{2}$ years?

15. $g(x) = \left(\dfrac{1}{2}\right)^x - 3$

                                                                   20. How long to the nearest day does it take for \$5000 to grow to \$8000 at 5% compounded continuously?

12.3   Properties of Logarithms

In This Section

⟨1⟩ **The Inverse Properties**

⟨2⟩ **The Product Rule for Logarithms**

⟨3⟩ **The Quotient Rule for Logarithms**

⟨4⟩ **The Power Rule for Logarithms**

⟨5⟩ **Using the Properties**

The properties of logarithms are very similar to the properties of exponents because *logarithms are exponents.* In this section, we use the properties of exponents to write some properties of logarithms. The properties will be used in solving logarithmic equations in Section 12.4.

⟨1⟩   The Inverse Properties

An exponential function and logarithmic function with the same base are inverses of each other. For example, the logarithm of 32 base 2 is 5 and the fifth power of 2 is 32. In symbols, we have

$$2^{\log_2(32)} = 2^5 = 32.$$

If we raise 3 to the fourth power, we get 81; and if we find the base-3 logarithm of 81, we get 4. In symbols, we have

$$\log_3(3^4) = \log_3(81) = 4.$$

We can state the inverse relationship between exponential and logarithm functions in general with the following inverse properties:

> **The Inverse Properties**
> **1.** $\log_a(a^x) = x$ for any real number x.
> **2.** $a^{\log_a(x)} = x$ for any positive real number x.

E X A M P L E   1

Using the inverse properties
Simplify each expression.

 a) $\ln(e^5)$ **b)** $2^{\log_2(8)}$

Solution

 a) Using the first inverse property, we get $\ln(e^5) = 5$.

 b) Using the second inverse property, we get $2^{\log_2(8)} = 8$.

> Now do Exercises 1–8

‹ **Calculator Close-Up** ›

You can illustrate the product rule for logarithms with a graphing calculator.

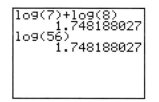

‹2› The Product Rule for Logarithms

Using the product rule for exponents and the inverse property $a^{\log_a(x)} = x$ we have

$$a^{\log_a M + \log_a N} = a^{\log_a M} a^{\log_a N} = M \cdot N.$$

By the definition of logarithm, that power of a that produces $M \cdot N$ is the base a logarithm of $M \cdot N$. So,

$$\log_a(M \cdot N) = \log_a M + \log_a N.$$

This last equation is called the **product rule for logarithms.** It says that *the logarithm of a product of two numbers is equal to the sum of their logarithms,* provided all logarithms are defined and all have the same base.

> **The Product Rule for Logarithms**
> For $M > 0$ and $N > 0$,
> $$\log_a(M \cdot N) = \log_a M + \log_a N.$$

E X A M P L E   2

Using the product rule for logarithms
Write each expression as a single logarithm.

 a) $\log_2(7) + \log_2(5)$ **b)** $\ln(\sqrt{2}) + \ln(\sqrt{3})$

Solution

 a) $\log_2(7) + \log_2(5) = \log_2(35)$ Product rule for logarithms

 b) $\ln(\sqrt{2}) + \ln(\sqrt{3}) = \ln(\sqrt{6})$ Product rule for logarithms

> Now do Exercises 9–20

⟨3⟩ The Quotient Rule for Logarithms

Using the quotient rule for exponents and the inverse property $a^{\log_a(x)} = x$ we have

$$a^{\log_a M - \log_a N} = \frac{a^{\log_a M}}{a^{\log_a N}} = \frac{M}{N}.$$

You can illustrate the quotient rule for logarithms with a graphing calculator.

```
ln(99/2)
            3.90197267
ln(99)-ln(2)
            3.90197267
```

By the definition of logarithm, the power of a that produces $\frac{M}{N}$ is the base a logarithm of $\frac{M}{N}$. So,

$$\log_a\left(\frac{M}{N}\right) = \log_a M - \log_a N.$$

This last equation is called the **quotient rule for logarithms.** It says that *the logarithm of a quotient of two numbers is equal to the difference of their logarithms,* provided all logarithms are defined and all have the same base.

The Quotient Rule for Logarithms

For $M > 0$ and $N > 0$,

$$\log_a\left(\frac{M}{N}\right) = \log_a M - \log_a N.$$

E X A M P L E   3

Using the quotient rule for logarithms
Write each expression as a single logarithm.

a) $\log_2(3) - \log_2(7)$

b) $\ln(w^8) - \ln(w^2)$

Solution

a) $\log_2(3) - \log_2(7) = \log_2\left(\dfrac{3}{7}\right)$ Quotient rule for logarithms

b) $\ln(w^8) - \ln(w^2) = \ln\left(\dfrac{w^8}{w^2}\right)$ Quotient rule for logarithms

$\phantom{\textbf{b)} \ln(w^8) - \ln(w^2)} = \ln(w^6)$ Quotient rule for exponents

> Now do Exercises 21–32

You can illustrate the power rule for logarithms with a graphing calculator.

```
log(11^13)
            13.53810491
13*log(11)
            13.53810491
```

⟨4⟩ The Power Rule for Logarithms

Using the power rule for exponents and the inverse property $a^{\log_a(x)} = x$ we have

$$a^{N\cdot \log_a M} = \left(a^{\log_a M}\right)^N = M^N.$$

By the definition of logarithm, the power of a that produces M^N is the base a logarithm of M^N. So,

$$\log_a(M^N) = N \cdot \log_a M.$$

This last equation is called the **power rule for logarithms.** It says that *the logarithm of a power of a number is equal to the power times the logarithm of the number,* provided all logarithms are defined.

The Power Rule for Logarithms

For $M > 0$,

$$\log_a(M^N) = N \cdot \log_a M.$$

E X A M P L E   **4**

Using the power rule for logarithms

Rewrite each logarithm in terms of $\log(2)$.

a) $\log(2^{10})$ b) $\log(\sqrt{2})$ c) $\log\left(\dfrac{1}{2}\right)$

Solution

a) $\log(2^{10}) = 10 \cdot \log(2)$ Power rule for logarithms

b) $\log(\sqrt{2}) = \log(2^{1/2})$ Write $\sqrt{2}$ as a power of 2.

$\qquad\qquad = \dfrac{1}{2}\log(2)$ Power rule for logarithms

c) $\log\left(\dfrac{1}{2}\right) = \log(2^{-1})$ Write $\frac{1}{2}$ as a power of 2.

$\qquad\qquad = -1 \cdot \log(2)$ Power rule for logarithms

$\qquad\qquad = -\log(2)$

Now do Exercises 33–38

⟨5⟩ **Using the Properties**

We have already seen many properties of logarithms. There are three properties that we have not yet formally stated. Because $a^1 = a$ and $a^0 = 1$, we have $\log_a(a) = 1$ and $\log_a(1) = 0$ for any positive number a. If we apply the quotient rule to $\log_a(1/N)$, we get

$$\log_a\left(\frac{1}{N}\right) = \log_a(1) - \log_a(N) = 0 - \log_a(N) = -\log_a(N).$$

So $\log_a\left(\frac{1}{N}\right) = -\log_a(N)$. These three new properties along with all of the other properties of logarithms are summarized as follows.

Properties of Logarithms

If M, N, and a are positive numbers, $a \neq 1$, then

1. $\log_a(a) = 1$

2. $\log_a(1) = 0$

3. $\log_a(a^x) = x$ for any real number x.   Inverse properties

4. $a^{\log_a(x)} = x$ for any positive real number x.   Inverse properties

5. $\log_a(MN) = \log_a(M) + \log_a(N)$ Product rule

6. $\log_a\left(\dfrac{M}{N}\right) = \log_a(M) - \log_a(N)$ Quotient rule

7. $\log_a\left(\dfrac{1}{N}\right) = -\log_a(N)$

8. $\log_a(M^N) = N \cdot \log_a(M)$.   Power rule

We have already seen several ways in which to use the properties of logarithms. In Examples 5, 6, and 7 we see more uses of the properties. First we use the rules of logarithms to write the logarithm of a complicated expression in terms of logarithms of simpler expressions.

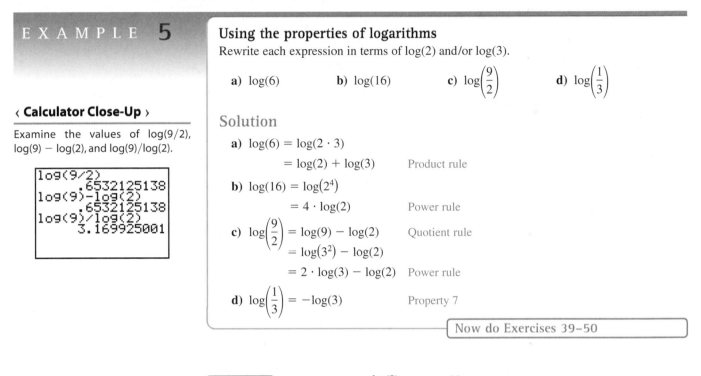

E X A M P L E   **5**

Using the properties of logarithms

Rewrite each expression in terms of log(2) and/or log(3).

a) $\log(6)$ b) $\log(16)$ c) $\log\left(\dfrac{9}{2}\right)$ d) $\log\left(\dfrac{1}{3}\right)$

‹ **Calculator Close-Up** ›

Examine the values of log(9/2), log(9) − log(2), and log(9)/log(2).

```
log(9/2)
      .6532125138
log(9)-log(2)
      .6532125138
log(9)/log(2)
      3.169925001
```

Solution

a) $\log(6) = \log(2 \cdot 3)$

$\qquad = \log(2) + \log(3)$ Product rule

b) $\log(16) = \log(2^4)$

$\qquad = 4 \cdot \log(2)$ Power rule

c) $\log\left(\dfrac{9}{2}\right) = \log(9) - \log(2)$ Quotient rule

$\qquad = \log(3^2) - \log(2)$

$\qquad = 2 \cdot \log(3) - \log(2)$ Power rule

d) $\log\left(\dfrac{1}{3}\right) = -\log(3)$ Property 7

Now do Exercises 39–50

CAUTION Do not confuse $\dfrac{\log(9)}{\log(2)}$ with $\log\left(\dfrac{9}{2}\right)$. We can use the quotient rule to write $\log\left(\dfrac{9}{2}\right) = \log(9) - \log(2)$, but $\dfrac{\log(9)}{\log(2)} \neq \log(9) - \log(2)$. The expression $\dfrac{\log(9)}{\log(2)}$ means $\log(9) \div \log(2)$. Use your calculator to verify these two statements.

The properties of logarithms can be used to combine several logarithms into a single logarithm (as in Examples 2 and 3) or to write a logarithm of a complicated expression in terms of logarithms of simpler expressions.

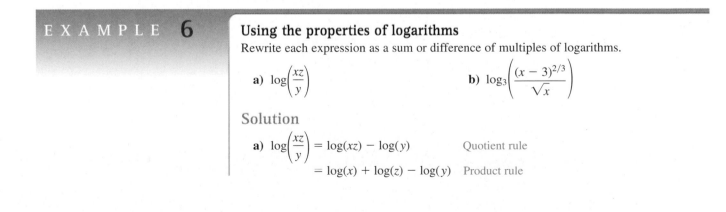

E X A M P L E   **6**

Using the properties of logarithms

Rewrite each expression as a sum or difference of multiples of logarithms.

a) $\log\left(\dfrac{xz}{y}\right)$ b) $\log_3\left(\dfrac{(x-3)^{2/3}}{\sqrt{x}}\right)$

Solution

a) $\log\left(\dfrac{xz}{y}\right) = \log(xz) - \log(y)$ Quotient rule

$\qquad = \log(x) + \log(z) - \log(y)$ Product rule

b) $\log_3\left(\dfrac{(x-3)^{2/3}}{\sqrt{x}}\right) = \log_3\left((x-3)^{2/3}\right) - \log_3\left(x^{1/2}\right)$ Quotient rule

$$= \frac{2}{3}\log_3(x-3) - \frac{1}{2}\log_3(x) \qquad \text{Power rule}$$

Now do Exercises 51–62

In Example 7, we use the properties of logarithms to convert expressions involving several logarithms into a single logarithm. The skills we are learning here will be used to solve logarithmic equations in Section 12.4.

E X A M P L E 7

Combining logarithms

Rewrite each expression as a single logarithm.

a) $\dfrac{1}{2}\log(x) - 2 \cdot \log(x+1)$ **b)** $3 \cdot \log(y) + \dfrac{1}{2}\log(z) - \log(x)$

Solution

a) $\dfrac{1}{2}\log(x) - 2 \cdot \log(x+1) = \log\left(x^{1/2}\right) - \log\left((x+1)^2\right)$ Power rule

$$= \log\left(\frac{\sqrt{x}}{(x+1)^2}\right) \qquad \text{Quotient rule}$$

b) $3 \cdot \log(y) + \dfrac{1}{2}\log(z) - \log(x) = \log\left(y^3\right) + \log\left(\sqrt{z}\right) - \log(x)$ Power rule

$$= \log\left(y^3 \cdot \sqrt{z}\right) - \log(x) \qquad \text{Product rule}$$

$$= \log\left(\frac{y^3\sqrt{z}}{x}\right) \qquad \text{Quotient rule}$$

Now do Exercises 63–74

Warm-Ups ▼

Fill in the blank.

1. The _____ rule for logarithms states that $\log_a(MN) = \log_a(M) + \log_a(N)$.

2. The _____ rule for logarithms states that $\log_a(M/N) = \log_a(M) - \log_a(N)$.

3. The _____ rule for logarithms states that $\log_a(M^N) = N \cdot \log_a(M)$.

4. The _____ properties of logarithms state that $\log_a(a^x) = x$ and $a^{\log_a(x)} = x$.

True or false?

5. $\log_2\left(\dfrac{x^2}{8}\right) = \log_2(x^2) - 3$

6. $\dfrac{\log(100)}{\log(10)} = \log(100) - \log(10)$

7. $\dfrac{\log(100)}{10} = \log(10)$

8. $\ln\left(\sqrt{2}\right) = \dfrac{\ln(2)}{2}$

9. $\ln(1) = e$

10. $\ln(2) + \ln(3) - \ln(7) = \ln\left(\dfrac{6}{7}\right)$

11. $\ln(8) = 3 \cdot \ln(2)$

12. $e^{\ln(7)} = 7$

‹ 1 › **The Inverse Properties**

Simplify each expression. See Example 1.

1. $\log_2(2^{10})$ **2.** $\ln(e^9)$

3. $5^{\log_5(19)}$ **4.** $10^{\log(2.3)}$

5. $\log(10^8)$ **6.** $\log_4(4^5)$

7. $e^{\ln(4.3)}$ **8.** $3^{\log_3(5.5)}$

‹ 2 › **The Product Rule for Logarithms**

Assume all variables involved in logarithms represent numbers for which the logarithms are defined. Write each expression as a single logarithm and simplify. See Example 2.

9. $\log(3) + \log(7)$

10. $\ln(5) + \ln(4)$

11. $\log_3(\sqrt{5}) + \log_3(\sqrt{x})$

12. $\ln(\sqrt{x}) + \ln(\sqrt{y})$

13. $\log(x^2) + \log(x^3)$

14. $\ln(a^3) + \ln(a^5)$

15. $\ln(2) + \ln(3) + \ln(5)$

16. $\log_2(x) + \log_2(y) + \log_2(z)$

17. $\log(x) + \log(x + 3)$

18. $\ln(x - 1) + \ln(x + 1)$

19. $\log_2(x - 3) + \log_2(x + 2)$

20. $\log_3(x - 5) + \log_3(x - 4)$

‹ 3 › **The Quotient Rule for Logarithms**

Write each expression as a single logarithm. See Example 3.

21. $\log_2(8) - \log_2(2)$

22. $\ln(3) - \ln(6)$

23. $\log_2(x^6) - \log_2(x^2)$

24. $\ln(w^9) - \ln(w^3)$

25. $\log(\sqrt{10}) - \log(\sqrt{2})$

26. $\log_3(\sqrt{6}) - \log_3(\sqrt{3})$

27. $\ln(4h - 8) - \ln(4)$

28. $\log(3x - 6) - \log(3)$

29. $\log_2(w^2 - 4) - \log_2(w + 2)$

30. $\log_3(k^2 - 9) - \log_3(k - 3)$

31. $\ln(x^2 + x - 6) - \ln(x + 3)$

32. $\ln(t^2 - t - 12) - \ln(t - 4)$

‹ 4 › **The Power Rule for Logarithms**

Write each expression in terms of $\log(3)$. See Example 4.

33. $\log(27)$ **34.** $\log\left(\dfrac{1}{9}\right)$

35. $\log(\sqrt{3})$ **36.** $\log(\sqrt[4]{3})$

37. $\log(3^x)$ **38.** $\log(3^{-99})$

‹ 5 › **Using the Properties**

Rewrite each expression in terms of $\log(3)$ and/or $\log(5)$. See Example 5.

39. $\log(15)$ **40.** $\log(9)$

41. $\log\left(\dfrac{5}{3}\right)$ **42.** $\log\left(\dfrac{3}{5}\right)$

43. $\log(25)$ **44.** $\log\left(\dfrac{1}{27}\right)$

45. $\log(75)$ **46.** $\log(0.6)$

47. $\log\left(\dfrac{1}{3}\right)$ **48.** $\log(45)$

49. $\log(0.2)$ **50.** $\log\left(\dfrac{9}{25}\right)$

Rewrite each expression as a sum or a difference of multiples of logarithms. See Example 6.

51. $\log(xyz)$

52. $\log(3y)$

53. $\log_2(8x)$

54. $\log_2(16y)$

55. $\ln\left(\dfrac{x}{y}\right)$

56. $\ln\left(\dfrac{z}{3}\right)$

57. $\log(10x^2)$

58. $\log(100\sqrt{x})$

59. $\log_5\left(\dfrac{(x-3)^2}{\sqrt{w}}\right)$

60. $\log_3\left(\dfrac{(y+6)^3}{y-5}\right)$

61. $\ln\left(\dfrac{yz\sqrt{x}}{w}\right)$

62. $\ln\left(\dfrac{(x-1)\sqrt{w}}{x^3}\right)$

Rewrite each expression as a single logarithm. See Example 7.

63. $\log(x) + \log(x-1)$

64. $\log_2(x-2) + \log_2(5)$

65. $\ln(3x-6) - \ln(x-2)$

66. $\log_3(x^2 - 1) - \log_3(x-1)$

67. $\ln(x) - \ln(w) + \ln(z)$

68. $\ln(x) - \ln(3) - \ln(7)$

69. $3 \cdot \ln(y) + 2 \cdot \ln(x) - \ln(w)$

70. $5 \cdot \ln(r) + 3 \cdot \ln(t) - 4 \cdot \ln(s)$

71. $\dfrac{1}{2}\log(x-3) - \dfrac{2}{3}\log(x+1)$

72. $\dfrac{1}{2}\log(y-4) + \dfrac{1}{2}\log(y+4)$

73. $\dfrac{2}{3}\log_2(x-1) - \dfrac{1}{4}\log_2(x+2)$

74. $\dfrac{1}{2}\log_3(y+3) + 6 \cdot \log_3(y)$

Determine whether each equation is true or false.

75. $\log(56) = \log(7) \cdot \log(8)$

76. $\log\left(\dfrac{5}{9}\right) = \dfrac{\log(5)}{\log(9)}$

77. $\log_2(4^2) = (\log_2(4))^2$

78. $\ln(4^2) = (\ln(4))^2$

79. $\ln(25) = 2 \cdot \ln(5)$

80. $\ln(3e) = 1 + \ln(3)$

81. $\dfrac{\log_2(64)}{\log_2(8)} = \log_2(8)$

82. $\dfrac{\log_2(16)}{\log_2(4)} = \log_2(4)$

83. $\log\left(\dfrac{1}{3}\right) = -\log(3)$

84. $\log_2(8 \cdot 2^{59}) = 62$

85. $\log_2(16^5) = 20$

86. $\log_2\left(\dfrac{5}{2}\right) = \log_2(5) - 1$

87. $\log(10^3) = 3$

88. $\log_3(3^7) = 7$

89. $\log(100 + 3) = 2 + \log(3)$

90. $\dfrac{\log_7(32)}{\log_7(8)} = \dfrac{5}{3}$

Applications

Solve each problem.

91. *Richter scale.* The Richter scale rating of an earthquake is given by the formula $r = \log(I) - \log(I_0)$, where I is the *intensity* of the earthquake and I_0 is the intensity of a small "benchmark" earthquake. Use the appropriate property of logarithms to rewrite this formula using a single logarithm. Find r if $I = 100 \cdot I_0$.

92. *Diversity index.* The U.S.G.S. measures the quality of a water sample by using the diversity index d, given by

$$d = -[p_1 \cdot \log_2(p_1) + p_2 \cdot \log_2(p_2) + \cdots + p_n \cdot \log_2(p_n)],$$

where n is the number of different taxons (biological classifications) represented in the sample and p_1 through p_n are the percentages of organisms in each of the n taxons. The value of d ranges from 0 when all organisms in the water sample are the same to some positive number when all organisms in the sample are different. If two-thirds of the organisms in a water sample are in one taxon and one-third of the organisms are in a second taxon, then $n = 2$ and

$$d = -\left[\dfrac{2}{3}\log_2\left(\dfrac{2}{3}\right) + \dfrac{1}{3}\log_2\left(\dfrac{1}{3}\right)\right].$$

Use the properties of logarithms to write the expression on the right-hand side as $\log_2\left(\dfrac{3\sqrt[3]{2}}{2}\right)$. (In Section 12.4 you will learn how to evaluate a base-2 logarithm using a calculator.)

Getting More Involved

93. *Discussion*

Which of the following equations is an identity? Explain.

a) $\ln(3x) = \ln(3) \cdot \ln(x)$

b) $\ln(3x) = \ln(3) + \ln(x)$

c) $\ln(3x) = 3 \cdot \ln(x)$

d) $\ln(3x) = \ln(x^3)$

94. *Discussion*

Which of the following expressions is not equal to $\log(5^{2/3})$? Explain.

a) $\dfrac{2}{3}\log(5)$

b) $\dfrac{\log(5) + \log(5)}{3}$

c) $\left(\log(5)\right)^{2/3}$

d) $\dfrac{1}{3}\log(25)$

Graphing Calculator Exercises

95. Graph the functions $y_1 = \ln(\sqrt{x})$ and $y_2 = 0.5 \cdot \ln(x)$ on the same screen. Explain your results.

96. Graph the functions $y_1 = \log(x)$, $y_2 = \log(10x)$, $y_3 = \log(100x)$, and $y_4 = \log(1000x)$ using the viewing window $-2 \le x \le 5$ and $-2 \le y \le 5$. Why do these curves appear as they do?

97. Graph the function $y = \log(e^x)$. Explain why the graph is a straight line. What is its slope?

12.4 Solving Equations and Applications

In This Section

⟨1⟩ **Logarithmic Equations**

⟨2⟩ **Exponential Equations**

⟨3⟩ **Changing the Base**

⟨4⟩ **Strategy for Solving Equations**

⟨5⟩ **Applications**

We solved some equations involving exponents and logarithms in Sections 12.1 and 12.2. In this section, we use the properties of exponents and logarithms to solve more complex equations.

⟨1⟩ Logarithmic Equations

The main tool that we have for solving logarithmic equations is the definition of logarithms: $y = \log_a(x)$ if and only if $a^y = x$. We can use the definition to rewrite any equation that has only one logarithm as an equivalent exponential equation.

EXAMPLE **1**

A logarithmic equation with only one logarithm

Solve $\log(x + 3) = 2$.

Solution

Write the equivalent exponential equation:

$$\log(x + 3) = 2 \qquad \text{Original equation}$$
$$10^2 = x + 3 \qquad \text{Definition of logarithm}$$
$$100 = x + 3$$
$$97 = x$$

Check: $\log(97 + 3) = \log(100) = 2$. The solution set is $\{97\}$.

Now do Exercises 1–8

In Example 2, we use the product rule for logarithms to write a sum of two logarithms as a single logarithm.

E X A M P L E   **2**

Using the product rule to solve an equation
Solve $\log_2(x + 3) + \log_2(x - 3) = 4$.

Solution
Rewrite the sum of the logarithms as the logarithm of a product:

$$\log_2(x + 3) + \log_2(x - 3) = 4 \qquad \text{Original equation}$$
$$\log_2[(x + 3)(x - 3)] = 4 \qquad \text{Product rule}$$
$$\log_2[x^2 - 9] = 4 \qquad \text{Multiply the binomials.}$$
$$x^2 - 9 = 2^4 \qquad \text{Definition of logarithm}$$
$$x^2 - 9 = 16$$
$$x^2 = 25$$
$$x = \pm 5 \qquad \text{Even-root property}$$

To check, first let $x = -5$ in the original equation:

$$\log_2(-5 + 3) + \log_2(-5 - 3) = 4$$
$$\log_2(-2) + \log_2(-8) = 4 \qquad \text{Incorrect}$$

Because the domain of any logarithm function is the set of positive real numbers, these logarithms are undefined. Now check $x = 5$ in the original equation:

$$\log_2(5 + 3) + \log_2(5 - 3) = 4$$
$$\log_2(8) + \log_2(2) = 4$$
$$3 + 1 = 4 \qquad \text{Correct}$$

The solution set is $\{5\}$.

> Now do Exercises 9–16

CAUTION   Always check that your solutions to a logarithmic equation do not produce undefined logarithms in the original equation.

E X A M P L E   **3**

Using the one-to-one property of logarithms
Solve $\log(x) + \log(x - 1) = \log(8x - 12) - \log(2)$.

‹ Calculator Close-Up ›

Graph

$$y_1 = \log(x) + \log(x - 1)$$

and

$$y_2 = \log(8x - 12) - \log(2)$$

to see the two solutions to the equation in Example 3.

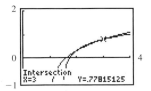

Solution

Apply the product rule to the left-hand side and the quotient rule to the right-hand side to get a single logarithm on each side:

$$\log(x) + \log(x - 1) = \log(8x - 12) - \log(2)$$
$$\log[x(x - 1)] = \log\left(\frac{8x - 12}{2}\right) \qquad \text{Product rule; quotient rule}$$
$$\log(x^2 - x) = \log(4x - 6) \qquad \text{Simplify.}$$
$$x^2 - x = 4x - 6 \qquad \text{One-to-one property of logarithms}$$
$$x^2 - 5x + 6 = 0$$
$$(x - 2)(x - 3) = 0$$
$$x - 2 = 0 \qquad \text{or} \qquad x - 3 = 0$$
$$x = 2 \qquad \text{or} \qquad x = 3$$

Neither $x = 2$ nor $x = 3$ produces undefined terms in the original equation. Use a calculator to check that they both satisfy the original equation. The solution set is $\{2, 3\}$.

Now do Exercises 17–22

CAUTION The product rule, quotient rule, and power rule do not eliminate logarithms from equations. To do so, we use the definition to change $y = \log_a(x)$ into $a^y = x$ or the one-to-one property to change $\log_a(m) = \log_a(n)$ into $m = n$.

⟨2⟩ Exponential Equations

If an equation has a single exponential expression, we can write the equivalent logarithmic equation.

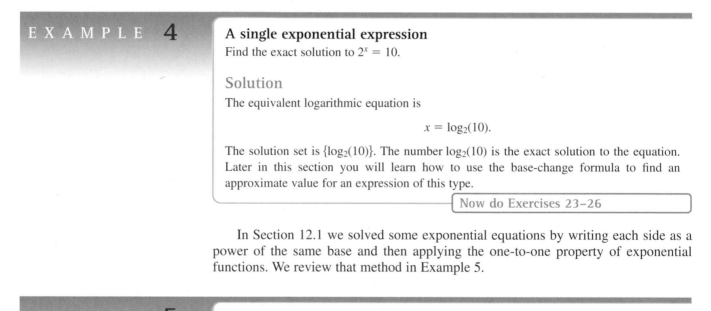

EXAMPLE 4

A single exponential expression
Find the exact solution to $2^x = 10$.

Solution

The equivalent logarithmic equation is

$$x = \log_2(10).$$

The solution set is $\{\log_2(10)\}$. The number $\log_2(10)$ is the exact solution to the equation. Later in this section you will learn how to use the base-change formula to find an approximate value for an expression of this type.

Now do Exercises 23–26

In Section 12.1 we solved some exponential equations by writing each side as a power of the same base and then applying the one-to-one property of exponential functions. We review that method in Example 5.

EXAMPLE 5

Powers of the same base
Solve $2^{(x^2)} = 4^{3x-4}$.

Solution

We can write each side as a power of the same base:

$$2^{(x^2)} = (2^2)^{3x-4} \quad \text{Because } 4 = 2^2$$
$$2^{(x^2)} = 2^{6x-8} \quad \text{Power of a power rule}$$
$$x^2 = 6x - 8 \quad \text{One-to-one property of exponential functions}$$
$$x^2 - 6x + 8 = 0$$
$$(x - 4)(x - 2) = 0$$
$$x - 4 = 0 \quad \text{or} \quad x - 2 = 0$$
$$x = 4 \quad \text{or} \quad x = 2$$

Check $x = 2$ and $x = 4$ in the original equation. The solution set is $\{2, 4\}$.

Now do Exercises 27–30

For some exponential equations we cannot write each side as a power of the same base as we did in Example 5. In this case, we take a logarithm of each side and simplify, using the rules for logarithms.

E X A M P L E   **6**

Exponential equation with two different bases

Find the exact and approximate solution to $2^{x-1} = 3^x$.

Solution

Since we want an approximate solution, we must use base 10 or base e, which are both available on a calculator. Either one will work here. We will use base 10:

$$2^{x-1} = 3^x \qquad \text{Original equation}$$

$$\log(2^{x-1}) = \log(3^x) \qquad \text{Take log of each side.}$$

$$(x - 1)\log(2) = x \cdot \log(3) \qquad \text{Power rule}$$

$$x \cdot \log(2) - \log(2) = x \cdot \log(3) \qquad \text{Distributive property}$$

$$x \cdot \log(2) - x \cdot \log(3) = \log(2) \qquad \text{Get all } x\text{-terms on one side.}$$

$$x[\log(2) - \log(3)] = \log(2) \qquad \text{Factor out } x.$$

$$x = \frac{\log (2)}{\log(2) - \log(3)} \qquad \text{Exact solution}$$

$$x \approx -1.7095 \qquad \text{Approximate solution}$$

You can use a calculator to check -1.7095 in the original equation.

> Now do Exercises 31–36

⟨3⟩ **Changing the Base**

Scientific calculators have an x^y key for computing any power of any base, in addition to the function keys for computing 10^x and e^x. For logarithms we have the keys ln and log, but there are no function keys for logarithms using other bases. To solve this problem, we develop a formula for expressing a base-a logarithm in terms of base-b logarithms.

If $y = \log_a(M)$, then $a^y = M$. Now we solve $a^y = M$ for y, using base-b logarithms:

$$a^y = M$$

$$\log_b(a^y) = \log_b(M) \quad \text{Take the base-}b\text{ logarithm of each side.}$$

$$y \cdot \log_b(a) = \log_b(M) \quad \text{Power rule}$$

$$y = \frac{\log_b(M)}{\log_b(a)} \quad \text{Divide each side by } \log_b(a).$$

Because $y = \log_a(M)$, we can write $\log_a(M)$ in terms of base-b logarithms.

⟨ **Calculator Close-Up** ⟩

The base-change formula enables you to graph logarithmic functions with bases other than e and 10. For example, to graph $y = \log_2(x)$, graph $y = \ln(x)/\ln(2)$.

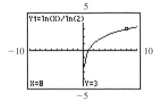

Base-Change Formula

If a and b are positive numbers not equal to 1 and M is positive, then

$$\log_a(M) = \frac{\log_b(M)}{\log_b(a)}.$$

In words, we take the logarithm with the new base and divide by the logarithm of the old base. The most important use of the base-change formula is to find base-a logarithms using a calculator. If the new base is 10 or e, then

$$\log_a(M) = \frac{\log(M)}{\log(a)} = \frac{\ln(M)}{\ln(a)}.$$

EXAMPLE 7

Using the base-change formula

Find $\log_7(99)$ to four decimal places.

Solution

Use the base-change formula with $a = 7$ and $b = 10$:

$$\log_7(99) = \frac{\log(99)}{\log(7)} \approx 2.3614$$

Check by finding $7^{2.3614}$ with your calculator. Note that we also have

$$\log_7(99) = \frac{\ln(99)}{\ln(7)} \approx 2.3614.$$

Now do Exercises 37–44

⟨4⟩ **Strategy for Solving Equations**

There is no formula that will solve every equation in this section. However, we have a strategy for solving exponential and logarithmic equations. The following list summarizes the ideas that we need for solving these equations.

Strategy for Solving Exponential and Logarithmic Equations

1. If the equation has a single logarithm or a single exponential expression, rewrite the equation using the definition $y = \log_a(x)$ if and only if $a^y = x$.

2. Use the properties of logarithms to combine logarithms as much as possible.

3. Use the one-to-one properties:
   a) If $\log_a(m) = \log_a(n)$, then $m = n$.
   b) If $a^m = a^n$, then $m = n$.

4. To get an approximate solution of an exponential equation, take the common or natural logarithm of each side of the equation.

⟨5⟩ Applications

In compound interest problems, logarithms are used to find the time it takes for money to grow to a specified amount.

E X A M P L E   **8**

Finding the time

If $500 is deposited into an account paying 8% compounded quarterly, then in how many quarters will the account have $1000 in it?

Solution

We use the compound interest formula $A = P(1 + i)^n$ with a principal of $500, an amount of $1000, and an interest rate of 2% each quarter:

$$A = P(1 + i)^n$$

$$1000 = 500(1.02)^n \quad \text{Substitute.}$$

$$2 = (1.02)^n \quad \text{Divide each side by 500.}$$

$$n = \log_{1.02}(2) \quad \text{Definition of logarithm}$$

$$= \frac{\ln(2)}{\ln(1.02)} \quad \text{Base-change formula}$$

$$\approx 35.0028 \quad \text{Use a calculator.}$$

It takes approximately 35 quarters, or 8 years and 9 months, for the initial investment to be worth $1000. Note that we could also solve $2 = (1.02)^n$ by taking the common or natural logarithm of each side. Try it.

> Now do Exercises 79–82

⟨ **Helpful Hint** ⟩

When we get $2 = (1.02)^n$, we can use the definition of log as in Example 8 or take the natural log of each side:

$$\ln(2) = \ln(1.02^n)$$
$$\ln(2) = n \cdot \ln(1.02)$$
$$n = \frac{\ln(2)}{\ln(1.02)}$$

In either way we arrive at the same solution.

Radioactive substances decay continuously over time in the same manner as money grows continuously with the continuous-compounding formula from Section 12.1. The model for radioactive decay is

$$A = A_0 e^{rt},$$

where A is the amount of the substance present at time t, r is the decay rate, and A_0 is the amount present at time $t = 0$. Note that this formula is actually the same as the continuous-compounding formula, but since the amount is decreasing, the rate r is a negative number.

E X A M P L E   **9**

Finding the rate in radioactive decay

The number of grams of a radioactive substance that is present in an old bone after t years is given by

$$A = 8e^{rt},$$

where r is the decay rate. How many grams of the radioactive substance were present when the bone was in a living organism at time $t = 0$? If it took 6300 years for the radioactive substance to decay from 8 grams to 4 grams, then what is the decay rate?

Solution

If $t = 0$, then $A = 8e^{r \cdot 0} = 8e^0 = 8 \cdot 1 = 8$. So the bone contained 8 grams of the substance when it was in a living organism. Now use $A = 4$ and $t = 6300$ in the formula $A = 8e^{rt}$ and solve for r:

$$4 = 8e^{6300r}$$

$$0.5 = e^{6300r} \quad \text{Divide each side by 8.}$$

$$6300r = \ln(0.5) \quad \text{Definition of logarithm}$$

$$r = \frac{\ln(0.5)}{6300} \quad \text{Divide each side by 6300.}$$

$$r \approx -1.1 \times 10^{-4} \text{ or } -0.00011$$

Note that the rate is negative because the substance is decaying.

Now do Exercises 83–94

Warm-Ups ▼

Fill in the blank.

1. The equations $a^y = x$ and $\log_a(x) = y$ are _____.

2. According to the _____ formula $\log_a(x) = \ln(x)/\ln(a)$.

True or false?

3. If $\log(x - 2) + \log(x + 2) = 7$, then $\log(x^2 - 4) = 7$.

4. If $\log(3x - 2) = \log(x + 2)$, then $3x - 2 = x + 2$.

5. If $e^{x-6} = e^{x^2-5x}$, then $x - 6 = x^2 - 5x$.

6. If $2^{3x-1} = 3^{5x-4}$, then $3x - 1 = 5x - 4$.

7. If $\log_2(x^2 - 2x + 5) = 3$, then $x^2 - 2x + 5 = 8$.

8. If $5^x = 23$, then $x \cdot \ln(5) = \ln(23)$.

9. $\dfrac{\ln(2)}{\ln(6)} = \dfrac{\log(2)}{\log(6)}$

10. $\log_3(5) = \dfrac{\ln(3)}{\ln(5)}$

Exercises 12.4

‹ **Study Tips** ›

• Always study math with a pencil and paper. Just sitting back and reading the text rarely works.
• A good way to study the examples in the text is to cover the solution with a piece of paper and see how much of the solution you can write on your own.

‹1› **Logarithmic Equations**

Solve each equation. See Examples 1 and 2.

1. $\log(x + 100) = 3$

2. $\log(x - 5) = 2$

3. $\log_2(x + 1) = 3$

4. $\log_3(x^2) = 4$

5. $3\log_2(x + 1) - 2 = 13$

6. $4\log_3(2x) - 1 = 7$

7. $12 + 2\ln(x) = 14$

8. $23 = 3\ln(x - 1) + 14$

9. $\log(x) + \log(5) = 1$

10. $\ln(x) + \ln(3) = 0$

11. $\log_2(x - 1) + \log_2(x + 1) = 3$

12. $\log_3(x - 4) + \log_3(x + 4) = 2$

13. $\log_2(x - 1) - \log_2(x + 2) = 2$

14. $\log_4(8x) - \log_4(x - 1) = 2$

15. $\log_2(x - 4) + \log_2(x + 2) = 4$

16. $\log_6(x + 6) + \log_6(x - 3) = 2$

Solve each equation. See Example 3.

17. $\ln(x) + \ln(x + 5) = \ln(x + 1) + \ln(x + 3)$

18. $\log(x) + \log(x + 5) = 2 \cdot \log(x + 2)$

19. $\log(x + 3) + \log(x + 4) = \log(x^3 + 13x^2) - \log(x)$

20. $\log(x^2 - 1) - \log(x - 1) = \log(6)$

21. $2 \cdot \log(x) = \log(20 - x)$

22. $2 \cdot \log(x) + \log(3) = \log(2 - 5x)$

⟨2⟩ Exponential Equations

Solve each equation. See Examples 4 and 5.

23. $3^x = 7$

24. $2^{x-1} = 5$

25. $e^{2x} = 7$

26. $e^{x+3} = 2$

27. $2^{3x+4} = 4^{x-1}$

28. $9^{2x-1} = 27^{1/2}$

29. $\left(\dfrac{1}{3}\right)^x = 3^{1+x}$

30. $4^{3x} = \left(\dfrac{1}{2}\right)^{1-x}$

Find the exact solution and approximate solution to each equation. Round approximate answers to three decimal places. See Example 6.

31. $2^x = 3^{x+5}$

32. $e^x = 10^x$

33. $5^{x+2} = 10^{x-4}$

34. $3^{2x} = 6^{x+1}$

35. $8^x = 9^{x-1}$

36. $5^{x+1} = 8^{x-1}$

⟨3⟩ Changing the Base

Use the base-change formula to find each logarithm to four decimal places. See Example 7.

37. $\log_2(3)$

38. $\log_3(5)$

39. $\log_3\left(\dfrac{1}{2}\right)$

40. $\log_5(2.56)$

41. $\log_{1/2}(4.6)$

42. $\log_{1/3}(3.5)$

43. $\log_{0.1}(0.03)$

44. $\log_{0.2}(1.06)$

⟨4⟩ Strategy for Solving Equations

For each equation, find the exact solution and an approximate solution when appropriate. Round approximate answers to three decimal places. See the Strategy for Solving Exponential and Logarithmic Equations box on page 823.

45. $x \cdot \ln(2) = \ln(7)$

46. $x \cdot \log(3) = \log(5)$

47. $3x - x \cdot \ln(2) = 1$

48. $2x + x \cdot \log(5) = \log(7)$

49. $3^x = 5$

50. $2^x = \dfrac{1}{3}$

51. $2^{x-1} = 9$

52. $10^{x-2} = 6$

53. $3^x = 20$

54. $2^x = 128$

55. $\log_3(x) + \log_3(5) = 1$

56. $\log(x) - \log(3) = \log(6)$

57. $8^x = 2^{x+1}$

58. $2^x = 5^{x+1}$

59. $\log_2(1 - x) = 2$

60. $\log_5(-x) = 3$

61. $\log_3(1 - x) + \log_3(2x + 13) = 3$

62. $\log_2(3 - x) + \log_2(x + 9) = 5$

63. $\ln(2x - 1) - \ln(x + 1) = \ln(5)$

64. $\log(x - 4) - \log(x + 5) = 1$

65. $\log_3(x - 14) - \log_3(x - 6) = 2$

66. $\log_3(7 - x^2) - \log_3(1 - x) = 1$

67. $\log(x + 1) + \log(x - 2) = 1$

68. $\log_2(x^2 - 8) - \log_2(x^2 - 5) = 2$

69. $2 \cdot \ln(x) = \ln(2) + \ln(5x - 12)$

70. $\ln(8 - x^3) - \ln(2 - x) = \ln(2x + 5)$

71. $\log_3(x^3 + 16x^2) - \log_3(x) = \log_3(36)$

72. $\ln(x) + \ln(x - 2) = \ln(x + 2) + \ln(x - 3)$

73. $\log(x) + \log(x + 5) = 2 \cdot \log(x + 2)$

74. $\log_2(x^2 - 9) - \log_2(x + 3) = \log_2(12)$

75. $\log_7(x^2 + 6x + 8) - \log_7(x + 2) = \log_7(3)$

76. $3 \cdot \log_5(x) = 2 \cdot \log_5(x)$

77. $\ln(6) + 2 \cdot \ln(x) = \ln(38x - 30) - \ln(2)$

78. $3 \cdot \ln(x + 1) = \ln(x + 1) + \ln(x^2 - x + 1)$

⟨5⟩ **Applications**

Solve each problem. See Examples 8 and 9.

79. *Finding the time.* How many months does it take for $1000 to grow to $1500 in an account paying 12% compounded monthly?

80. *Finding the time.* How many years does it take for $25 to grow to $100 in an account paying 8% compounded annually?

81. *Finding days.* How many days does it take for a deposit of $100 to grow to $105 at 3% annual percentage rate compounded daily? Round to the nearest day.

82. *Finding quarters.* How many quarters does it take for a deposit of $500 to grow to $600 at 2% annual percentage rate compounded quarterly? Round to the nearest quarter.

83. *Radioactive decay.* The number of grams of a radioactive substance that is present in an old piece of cloth after t years is given by

$$A = 10e^{-0.0001t}.$$

How many grams of the radioactive substance did the cloth contain when it was made at time $t = 0$? If the cloth now contains only 4 grams of the substance, then when was the cloth made?

84. *Finding the decay rate.* The number of grams of a radioactive substance that is present in an old log after t years is given by

$$A = 5e^{rt},$$

where r is the decay rate. How many grams of the radioactive substance were present when the log was alive at time $t = 0$? If it took 5000 years for the substance to decay from 5 grams to 2 grams, then what is the decay rate?

85. *Going with the flow.* The flow y [in cubic feet per second (ft^3/sec)] of the Tangipahoa River at Robert, Louisiana, is modeled by the exponential function $y = 114.308e^{0.265x}$, where x is the depth in feet. Find the flow when the depth is 15.8 feet.

Figure for Exercises 85 and 86

86. *Record flood.* Use the formula of Exercise 85 to find the depth of the Tangipahoa River at Robert, Louisiana, on May 3, 1953, when the flow reached an all-time record of 50,500 ft^3/sec (U.S.G.S., waterdata.usgs.gov).

87. *Above the poverty level.* In a certain country the number of people above the poverty level is currently 28 million and growing 5% annually. Assuming the population is growing continuously, the population P (in millions), t years from now, is determined by the formula $P = 28e^{0.05t}$. In how many years will there be 40 million people above the poverty level?

88. *Below the poverty level.* In the same country as in Exercise 87, the number of people below the poverty level is currently 20 million and growing 7% annually. This population (in millions), t years from now, is determined by the formula $P = 20e^{0.07t}$. In how many years will there be 40 million people below the poverty level?

89. *Fifty-fifty.* For this exercise, use the information given in Exercises 87 and 88. In how many years will the number of people above the poverty level equal the number of people below the poverty level?

90. *Golden years.* In a certain country there are currently 100 million workers and 40 million retired people. The population of workers is decreasing according to the formula $W = 100e^{-0.01t}$, where t is in years and W is in millions. The population of retired people is increasing according to the formula $R = 40e^{0.09t}$, where t is in years and R is in millions. In how many years will the number of workers equal the number of retired people?

91. *Ions for breakfast.* Orange juice has a pH of 3.7. What is the hydrogen ion concentration of orange juice? (See Exercises 85–88 of Section 12.2.)

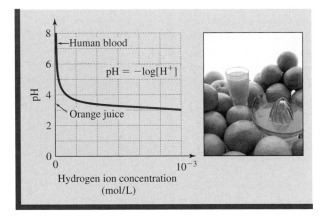

Figure for Exercises 91 and 92

92. *Ions in your veins.* Normal human blood has a pH of 7.4. What is the hydrogen ion concentration of normal human blood?

93. *Diversity index.* In Exercise 92 of Section 12.3 we expressed the diversity index d for a certain water sample as

$$d = \log_2\left(\frac{3\sqrt[3]{2}}{2}\right).$$

Use the base-change formula and a calculator to calculate the value of d. Round the answer to four decimal places.

94. *Quality water.* In a certain water sample, 5% of the organisms are in one taxon, 10% are in a second taxon, 20% are in a third taxon, 15% are in a fourth taxon, 23% are in a fifth taxon, and the rest are in a sixth taxon. Use the formula given in Exercise 92 of Section 12.3 with $n = 6$ to find the diversity index of the water sample.

Getting More Involved

95. *Exploration*

Logarithms were designed to solve equations that have variables in the exponents, but logarithms can be used to solve certain polynomial equations. Consider the following example:

$$x^5 = 88$$

$$5 \cdot \ln(x) = \ln(88)$$

$$\ln(x) = \frac{\ln(88)}{5} \approx 0.895467$$

$$x = e^{0.895467} \approx 2.4485$$

Solve $x^3 = 12$ by taking the natural logarithm of each side. Round the approximate solution to four decimal places. Solve $x^3 = 12$ without using logarithms and compare with your previous answer.

96. *Discussion*

Determine whether each logarithm is positive or negative without using a calculator. Explain your answers.

a) $\log_2(0.45)$
b) $\ln(1.01)$
c) $\log_{1/2}(4.3)$
d) $\log_{1/3}(0.44)$

Graphing Calculator Exercises

97. Graph $y_1 = 2^x$ and $y_2 = 3^{x-1}$ on the same coordinate system. Use the intersect feature of your calculator to find the point of intersection of the two curves. Round to two decimal places.

98. Bob invested $1000 at 6% compounded continuously. At the same time Paula invested $1200 at 5% compounded monthly. Write two functions that give the amounts of Bob's and Paula's investments after x years. Graph these functions on a graphing calculator. Use the intersect feature of your graphing calculator to find the approximate value of x for which the investments are equal in value.

99. Graph the functions $y_1 = \log_2(x)$ and $y_2 = 3^{x-4}$ on the same coordinate system, and use the intersect feature to find the points of intersection of the curves. Round to two decimal places. [*Hint:* To graph $y = \log_2(x)$, use the base-change formula to write the function as $y = \ln(x)/\ln(2)$.]

12

Wrap-Up

Summary

Exponential and Logarithmic Functions		**Examples**
Exponential function	A function of the form $f(x) = a^x$ for $a > 0$ and $a \neq 1$	$f(x) = 3^x$
Logarithmic function	A function of the form $f(x) = \log_a(x)$ for $a > 0$ and $a \neq 1$ $\quad y = \log_a(x)$ if and only if $a^y = x$.	$f(x) = \log_2(x)$ $\log_3(8) = x \leftrightarrow 3^x = 8$
Common logarithm	Base-10: $f(x) = \log(x)$	$\log(100) = 2$ because $100 = 10^2$.
Natural logarithm	Base-e: $f(x) = \ln(x)$ $e \approx 2.718$	$\ln(e) = 1$ because $e^1 = e$.
Inverse functions	$f(x) = a^x$ and $g(x) = \log_a(x)$ are inverse functions.	If $f(x) = e^x$, then $f^{-1}(x) = \ln(x)$.

Properties		**Examples**
M, N, and a are positive numbers with $a \neq 1$. $\quad \log_a(a) = 1 \qquad \log_a(1) = 0$		$\log_5(5) = 1$, $\log_5(1) = 0$
Inverse properties	$\log_a(a^x) = x$ for any real number x. $a^{\log_a(x)} = x$ for any positive real number x.	$\log(10^7) = 7$, $e^{\ln(3.4)} = 3.4$
Product rule	$\log_a(MN) = \log_a(M) + \log_a(N)$	$\ln(3x) = \ln(3) + \ln(x)$
Quotient rule	$\log_a\left(\dfrac{M}{N}\right) = \log_a(M) - \log_a(N)$ $\log_a\left(\dfrac{1}{N}\right) = -\log_a(N)$	$\ln\left(\dfrac{2}{3}\right) = \ln(2) - \ln(3)$ $\ln\left(\dfrac{1}{3}\right) = -\ln(3)$
Power rule	$\log_a(M^N) = N \cdot \log_a(M)$	$\log(x^3) = 3 \cdot \log(x)$
Base-change formula	$\log_a(M) = \dfrac{\log_b(M)}{\log_b(a)}$	$\log_3(5) = \dfrac{\ln(5)}{\ln(3)}$

Equations Involving Logarithms and Exponents		**Examples**
Strategy	1. If there is a single logarithm or a single exponential expression, rewrite the equation using the definition of logarithms: $y = \log_a(x)$ if and only if $a^y = x$.	$2^x = 3$ and $x = \log_2(3)$ are equivalent.

2. Use the properties of logarithms to combine logarithms as much as possible.

$\log(x) + \log(x - 3) = 1$
$\log(x^2 - 3x) = 1$

3. Use the one-to-one properties:
   a) If $\log_a(m) = \log_a(n)$, then $m = n$.
   b) If $a^m = a^n$, then $m = n$.

$\ln(x) = \ln(5 - x)$,
$x = 5 - x$
$2^{3x} = 2^{5x-7}$, $3x = 5x - 7$
$2^x = 3$, $\ln(2^x) = \ln(3)$
$x \cdot \ln(2) = \ln(3)$

4. To get an approximate solution, take the common or natural logarithm of each side of an exponential equation.

$$x = \frac{\ln(3)}{\ln(2)}$$

Enriching Your Mathematical Word Power

Fill in the blank.

1. A(n) _____ function has the form $f(x) = a^x$ where $a > 0$ and $a \neq 1$.

2. Base 10 is the _____ base.

3. Base e is the _____ base.

4. The _____ of an exponential function is $(-\infty, \infty)$.

5. With _____ interest, interest is paid periodically into the account and the interest earns interest.

6. Using $A = Pe^{rt}$ to compute the amount is _____ compounding.

7. The exponent used on base a to produce x is the base-a _____ of x.

8. A base-10 logarithm is a _____ logarithm.

9. A base-e logarithm is a _____ logarithm.

10. The base-a logarithm _____ is $f(x) = \log_a(x)$.

Review Exercises

12.1 Exponential Functions and Their Applications

Use $f(x) = 5^x$, $g(x) = 10^{x-1}$, and $h(x) = \left(\frac{1}{4}\right)^x$ for Exercises 1–28. Find the following.

1. $f(-2)$

2. $f(0)$

3. $f(3)$

4. $f(4)$

5. $g(1)$

6. $g(-1)$

7. $g(0)$

8. $g(3)$

9. $h(-1)$

10. $h(2)$

11. $h\left(\frac{1}{2}\right)$

12. $h\left(-\frac{1}{2}\right)$

Find x in each case.

13. $f(x) = 25$

14. $f(x) = -\frac{1}{125}$

15. $g(x) = 1000$

16. $g(x) = 0.001$

17. $h(x) = 32$

18. $h(x) = 8$

19. $h(x) = \frac{1}{16}$

20. $h(x) = 1$

Find the following.

21. $f(1.34)$

22. $f(-3.6)$

23. $g(3.25)$

24. $g(4.87)$

25. $h(2.82)$

26. $h(\pi)$

27. $h(\sqrt{2})$

28. $h\left(\frac{1}{3}\right)$

Sketch the graph of each function.

29. $f(x) = 5^x$

30. $g(x) = e^x$

31. $y = \left(\frac{1}{5}\right)^x$

32. $y = e^{-x}$

33. $f(x) = 3^{-x}$

34. $f(x) = -3^{x-1}$

35. $y = 1 + 2^x$

36. $y = 1 - 2^x$

12.2 Logarithmic Functions and Their Applications

Write each exponential equation as a logarithmic equation and each logarithmic equation as an exponential equation.

37. $10^m = n$ **38.** $b = a^5$

39. $h = \log_k(t)$ **40.** $\log_v(5) = u$

Let $f(x) = \log_2(x)$, $g(x) = \log(x)$, and $h(x) = \log_{1/2}(x)$. Find the following.

41. $f\left(\frac{1}{8}\right)$ **42.** $f(64)$

43. $g(0.1)$ **44.** $g(1)$

45. $g(100)$ **46.** $h\left(\frac{1}{8}\right)$

47. $h(1)$ **48.** $h(4)$

49. x, if $f(x) = 8$ **50.** x, if $g(x) = 3$

51. $f(77)$ **52.** $g(88.4)$

53. $h(33.9)$ **54.** $h(0.05)$

55. x, if $f(x) = 2.475$ **56.** x, if $g(x) = 1.426$

For each function f, find f^{-1}, and sketch the graphs of f and f^{-1} on the same set of axes.

57. $f(x) = 10^x$

58. $f(x) = \log_8(x)$

59. $f(x) = e^x$

60. $f(x) = \log_3(x)$

12.3 Properties of Logarithms

Rewrite each expression as a sum or a difference of multiples of logarithms.

61. $\log(x^2 y)$

62. $\log_3(x^2 + 2x)$

63. $\ln(16)$

64. $\log\left(\dfrac{y}{\sqrt{x}}\right)$

65. $\log_5\left(\dfrac{1}{x}\right)$

66. $\ln\left(\dfrac{xy}{z}\right)$

Rewrite each expression as a single logarithm.

67. $\dfrac{1}{2}\log(x + 2) - 2 \cdot \log(x - 1)$

68. $3 \cdot \ln(x) + 2 \cdot \ln(y) - \dfrac{1}{3}\ln(z)$

12.4 Solving Equations and Applications

Find the exact solution to each equation.

69. $\log_2(x) = 8$

70. $\log_3(x) = 0.5$

71. $\log_2(8) = x$

72. $3^x = 8$

73. $x^3 = 8$

74. $3^2 = x$

75. $\log_x(27) = 3$

76. $\log_x(9) = -\dfrac{1}{3}$

77. $x \cdot \ln(3) - x = \ln(7)$

78. $x \cdot \log(8) = x \cdot \log(4) + \log(9)$

79. $3^x = 5^{x-1}$

80. $5^{(2x^2)} = 5^{3-5x}$

81. $4^{2x} = 2^{x+1}$

82. $\log(12) = \log(x) + \log(7 - x)$

83. $\ln(x + 2) - \ln(x - 10) = \ln(2)$

84. $2 \cdot \ln(x + 3) = 3 \cdot \ln(4)$

85. $\log(x) - \log(x - 2) = 2$

86. $\log_2(x) = \log_2(x + 16) - 1$

Use a calculator to find an approximate solution to each of the following. Round your answers to four decimal places.

87. $6^x = 12$

88. $5^x = 8^{3x+2}$

89. $3^{x+1} = 5$

90. $\log_3(x) = 2.634$

Miscellaneous

Solve each problem.

91. *Compounding annually.* What does \$10,000 invested at 11.5% compounded annually amount to after 15 years?

92. *Doubling time.* How many years does it take for an investment to double at 6.5% compounded annually?

93. *Decaying substance.* The amount, A, of a certain radioactive substance remaining after t years is given by the formula $A = A_0 e^{-0.0003t}$, where A_0 is the initial amount. If we have 218 grams of this substance today, then how much of it will be left 1000 years from now?

94. *Wildlife management.* The number of white-tailed deer in the Hiawatha National Forest is believed to be growing according to the function

$$P = 517 + 10 \cdot \ln(8t + 1),$$

where t is the time in years from the year 2000.

a) What is the size of the population in 2000?

b) In what year will the population reach 600?

c) Does the population as shown on the accompanying graph appear to be growing faster during the period 2000 to 2005 or during the period 2005 to 2010?

d) What is the average rate of change of the population for each period in part (c)?

where I and E are in millions of dollars and t is the number of years after 2000.

a) What are the values of imports and exports in 2000?

b) Use the accompanying graph to estimate the year in which imports will equal exports.

c) Algebraically find the year in which imports will equal exports.

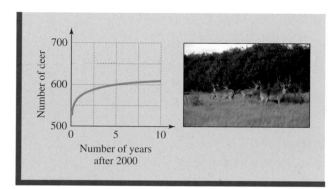

Figure for Exercise 94

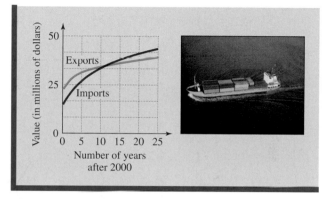

Figure for Exercise 96

95. *Comparing investments.* Melissa deposited $1000 into an account paying 5% annually; on the same day Frank deposited $900 into an account paying 7% compounded continuously. Find the number of years that it will take for the amounts in the accounts to be equal.

96. *Imports and exports.* The value of imports for a small Central American country is believed to be growing according to the function

$$I = 15 \cdot \log(16t + 33),$$

and the value of exports appears to be growing according to the function

$$E = 30 \cdot \log(t + 3),$$

97. *Finding river flow.* The U.S.G.S. measures the water height h (in feet above sea level) for the Tangipahoa River at Robert, Louisiana, and then finds the flow y [in cubic feet per second (ft^3/sec)], using the formula

$$y = 114.308e^{0.265(h-6.87)}.$$

Find the flow when the river at Robert is 20.6 ft above sea level.

98. *Finding the height.* Rewrite the formula in Exercise 97 to express h as a function of y. Use the new formula to find the water height above sea level when the flow is 10,000 ft^3/sec.

Chapter 12 Test

Let $f(x) = 5^x$ and $g(x) = \log_5(x)$. Find the following.

1. $f(2)$ **2.** $f(-1)$

3. $f(0)$ **4.** $g(125)$

5. $g(1)$ **6.** $g\left(\dfrac{1}{5}\right)$

Sketch the graph of each function.

7. $y = 2^x$

8. $f(x) = \log_2(x)$

9. $y = \left(\dfrac{1}{3}\right)^x$

10. $g(x) = \log_{1/3}(x)$

11. $f(x) = -2^x + 3$

12. $f(x) = 2^{x-3} - 1$

Suppose $\log_a(M) = 6$ and $\log_a(N) = 4$. Find the following.

13. $\log_a(MN)$

14. $\log_a\left(\dfrac{M^2}{N}\right)$

15. $\dfrac{\log_a(M)}{\log_a(N)}$

16. $\log_a(a^3 M^2)$

17. $\log_a\left(\dfrac{1}{N}\right)$

Find the exact solution to each equation.

18. $3^x = 12$

19. $\log_3(x) = \dfrac{1}{2}$

20. $5^x = 8^{x-1}$

21. $\log(x) + \log(x + 15) = 2$

22. $2 \cdot \ln(x) = \ln(3) + \ln(6 - x)$

Use a scientific calculator to find an approximate solution to each of the following. Round your answers to four decimal places.

23. Solve $20^x = 5$.

24. Solve $\log_3(x) = 2.75$.

25. The number of bacteria present in a culture at time t is given by the formula $N = 10e^{0.4t}$, where t is in hours. How many bacteria are present initially? How many are present after 24 hours?

26. How many hours does it take for the bacteria population of Problem 25 to double?

*Making*Connections | A Review of Chapters 1–12

Find the exact solution to each equation.

1. $(x - 3)^2 = 8$

2. $\log_2(x - 3) = 8$

3. $2^{x-3} = 8$

4. $2x - 3 = 8$

5. $|x - 3| = 8$

6. $\sqrt{x - 3} = 8$

7. $\log_2(x - 3) + \log_2(x) = \log_2(18)$

8. $2 \cdot \log_2(x - 3) = \log_2(5 - x)$

9. $\dfrac{1}{2}x - \dfrac{2}{3} = \dfrac{3}{4}x + \dfrac{1}{5}$

10. $3x^2 - 6x + 2 = 0$

Find the inverse of each function.

11. $f(x) = \dfrac{1}{3}x$

12. $g(x) = \log_3(x)$

13. $f(x) = 2x - 4$

14. $h(x) = \sqrt{x}$

15. $j(x) = \dfrac{1}{x}$

16. $k(x) = 5^x$

17. $m(x) = e^{x-1}$

18. $n(x) = \ln(x)$

Sketch the graph of each equation.

19. $y = 2x$

20. $y = 2^x$

21. $y = x^2$

22. $y = \log_2(x)$

23. $y = \dfrac{1}{2}x - 4$

24. $y = |2 - x|$

25. $y = 2 - x^2$

26. $y = e^2$

Simplify each expression.

27. $\dfrac{1}{25} + \dfrac{1}{5} + \dfrac{1}{4} + \dfrac{1}{2}$

28. $\dfrac{1}{100} + \dfrac{1}{50} + \dfrac{1}{5} + \dfrac{1}{4} + \dfrac{1}{2}$

29. $-6^{-1} + 5^{-1}$

30. $27^{4/3}$

31. $\sqrt[4]{625}$

32. $\sqrt[5]{-32}$

33. $\sqrt{80}$

34. $\sqrt[3]{80}$

35. $\dfrac{6\sqrt{2}}{\sqrt{3}}$

36. $\dfrac{20\sqrt{2}}{\sqrt[3]{5}}$

37. $\dfrac{\sqrt{3}\sqrt{14}}{\sqrt{21}}$

38. $\dfrac{\sqrt{8} + \sqrt{50}}{\sqrt{18} + \sqrt{32}}$

Solve each problem.

39. *Civilian labor force.* The number of workers in the civilian labor force can be modeled by the linear function

$$n(t) = 1.51t + 125.5$$

or by the exponential function

$$n(t) = 125.6e^{0.011t},$$

where t is the number of years since 1990 and $n(t)$ is in millions of workers (Bureau of Labor Statistics, www.bls.gov).

a) Graph both functions on the same coordinate system for $0 \le t \le 30$.

b) What does each model predict for the value of n in 2010?

c) What does each model predict for the value of n in the present year? Which model's prediction is closest to the actual size of the present civilian labor force?

40. *Measuring ocean depths.* In this exercise you will see how a geophysicist uses sound reflection to measure the depth of the ocean. Let v be the speed of sound through the water and d_1 be the depth of the ocean below the ship, as shown in the accompanying figure.

a) The time it takes for sound to travel from the ship at point S straight down to the ocean floor at point B_1 and back to point S is 0.270 second. Write d_1 as a function of v.

b) It takes 0.432 second for sound to travel from point S to point B_2 and then to a receiver at R, which is towed 500 meters behind the ship. Assuming $d_2 = d_3$, write d_2 as a function of v.

c) Use the Pythagorean theorem to find v. Then find the ocean depth d_1.

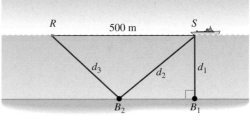

Figure for Exercise 40

Critical **Thinking** | **For Individual or Group Work** | **Chapter 12**

These exercises can be solved by a variety of techniques, which may or may not require algebra. So be creative and think critically. Explain all answers. Answers are in the Instructor's Edition of this text.

1. ***Shady crescents.*** Start with any right triangle and draw three semicircles so that each semicircle has one side of the triangle as its diameter as shown in the accompanying figure. Show that total area of the two smaller crescents is equal to the area of the largest crescent.

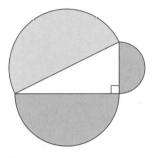

Figure for Exercise 1

2. ***Huge integer.*** The value of the expression $16^9 \cdot 5^{25}$ is an integer. How many digits does it have?

3. ***Sevens galore.*** How many seven-digit whole numbers contain the number seven at least once?

4. ***Ten-digit surprise.*** Use the digits 0 through 9 once each to construct a 10-digit number such that the first n digits (counting from the left) form a number divisible by n, for each n from 1 through 10. For example, for 3428, 3 is divisible by 1, 34 is divisible by 2, 342 is divisible by 3, and 3428 is divisible by 4, but 3428 is not a 10-digit number.

5. ***Cattle drive.*** A group of cowboys is driving a herd of cattle across the plains at a constant rate. The cowboys always keep the herd in the shape of a square that is 1 kilometer on each side. One of the cowboys starts at the left rear of the square/herd and rides his four-wheeler

Photo for Exercise 5

around the perimeter of the square at a constant rate in the same time that the herd advances 1 kilometer. How far does this cowboy travel?

6. ***Counting game.*** A teacher plays a counting game with his students. The first student says 1. The second student says 2 and 3. The third student says 4, 5, and 6. The fourth says 7, 8, 9, and 10. This pattern continues with the fifth student saying the next five counting numbers, and so on. Find a formula for the sum of the numbers said by the kth student. $\left[\text{\textit{Hint:} The sum of the first } n \text{ counting numbers is } \frac{n(n+1)}{2}.\right]$

7. ***Numerical palindrome.*** A numerical palindrome is a positive integer with at least two digits that reads the same forward or backward. For example, 55 and 343 are numerical palindromes. How many numerical palindromes are there less than 1000?

8. ***Fractional chickens.*** If 1.5 chickens lay 1.5 eggs in 1.5 days, then how many eggs do 3.5 chickens lay in 3 days?

13

Nonlinear Systems and the Conic Sections

With a cruising speed of 1540 miles per hour, the Concorde was the fastest commercial aircraft ever built. First flown in 1969, the Concorde could fly from London to New York in about 3 hours. However, the Concordes never made a profit and were all taken out of service in 2003, which ended the age of supersonic commercial air travel.

Perhaps the biggest problem for the Concorde was that it was generally prohibited from flying over land areas because of the noise. Any jet flying faster than the speed of sound creates a cone-shaped wave in the air on which there is a momentary change in air pressure. This change in air pressure causes a thunderlike sonic boom. When the jet is traveling parallel to the ground, the cone-shaped wave intersects the ground along one branch of a hyperbola. People on the ground hear the boom as the hyperbola passes them.

In this chapter, we will discuss curves, including the hyperbola, that occur when a geometric plane intersects a cone.

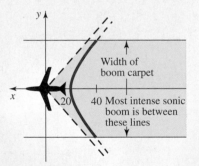

In Exercise 60 of Section 13.4 you will see how the altitude of the aircraft is related to the width of the area where the sonic boom is heard.

13.1 Nonlinear Systems of Equations

We studied systems of linear equations in Chapter 7. In this section, we turn our attention to nonlinear systems of equations.

In This Section

⟨1⟩ **Solving by Elimination**

⟨2⟩ **Applications**

⟨1⟩ Solving by Elimination

An equation whose graph is not a straight line is a **nonlinear equation.** For example,

$$y = x^2, \qquad y = \sqrt{x}, \qquad y = |x|, \qquad y = 2^x, \qquad \text{and} \qquad y = \log_2(x)$$

are nonlinear equations. A **nonlinear system** is a system of equations in which there is at least one nonlinear equation. We use the same techniques for solving nonlinear systems that we use for linear systems. Graphing the equations is used to explain the number of solutions to the system, but is generally not an accurate method for solving systems of equations. Eliminating a variable by either substitution or addition is used for solving linear or nonlinear systems.

E X A M P L E   1

A parabola and a line

Solve the system of equations, and draw the graph of each equation on the same coordinate system:

$$y = x^2 - 1$$
$$x + y = 1$$

Solution

We can eliminate y by substituting $y = x^2 - 1$ into $x + y = 1$:

$$x + y = 1$$
$$x + (x^2 - 1) = 1 \qquad \text{Substitute } x^2 - 1 \text{ for } y.$$
$$x^2 + x - 2 = 0$$
$$(x - 1)(x + 2) = 0$$
$$x - 1 = 0 \qquad \text{or} \qquad x + 2 = 0$$
$$x = 1 \qquad \text{or} \qquad x = -2$$

Replace x by 1 and -2 in $y = x^2 - 1$ to find the corresponding values of y:

$$y = (1)^2 - 1 \qquad y = (-2)^2 - 1$$
$$y = 0 \qquad\qquad y = 3$$

Check that each of the points $(1, 0)$ and $(-2, 3)$ satisfies both of the original equations. The solution set is $\{(1, 0), (-2, 3)\}$. If we solve $x + y = 1$ for y, we get $y = -x + 1$. The line $y = -x + 1$ has y-intercept $(0, 1)$ and slope -1. The graph of $y = x^2 - 1$ is a parabola with vertex $(0, -1)$. Of course, $(1, 0)$ and $(-2, 3)$ are on both graphs. The two graphs are shown in Fig. 13.1.

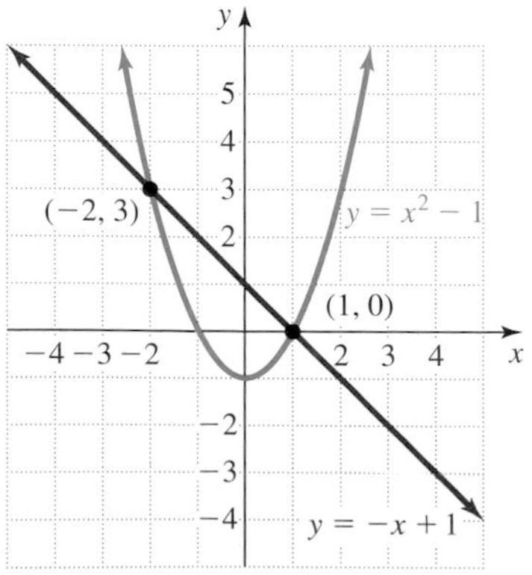

Figure 13.1

Now do Exercises 1–10

Nonlinear systems often have more than one solution, and drawing the graphs helps us to understand why. However, it is not necessary to draw the graphs to solve the system, as shown in Example 2.

EXAMPLE 2

Solving a system algebraically with substitution

Solve the system:

$$x^2 + y^2 + 2y = 3$$
$$x^2 - y = 5$$

Solution

If we substitute $y = x^2 - 5$ into the first equation to eliminate y, we will get a fourth-degree equation to solve. Instead, we can eliminate the variable x by writing $x^2 - y = 5$ as $x^2 = y + 5$. Now replace x^2 by $y + 5$ in the first equation:

$$x^2 + y^2 + 2y = 3$$
$$(y + 5) + y^2 + 2y = 3$$
$$y^2 + 3y + 5 = 3$$
$$y^2 + 3y + 2 = 0$$
$$(y + 2)(y + 1) = 0 \quad \text{Solve by factoring.}$$
$$y + 2 = 0 \quad \text{or} \quad y + 1 = 0$$
$$y = -2 \quad \text{or} \quad y = -1$$

Let $y = -2$ in the equation $x^2 = y + 5$ to find the corresponding x:

$$x^2 = -2 + 5$$
$$x^2 = 3$$
$$x = \pm\sqrt{3}$$

Now let $y = -1$ in the equation $x^2 = y + 5$ to find the corresponding x:

$$x^2 = -1 + 5$$
$$x^2 = 4$$
$$x = \pm 2$$

Check these values in the original equations. The solution set is

$$\{(\sqrt{3}, -2), (-\sqrt{3}, -2), (2, -1), (-2, -1)\}.$$

The graphs of these two equations intersect at four points.

Now do Exercises 11–20

EXAMPLE 3

Solving a system with the addition method

Solve each system:

a) $x^2 - y^2 = 5$
$x^2 + y^2 = 7$

b) $\dfrac{2}{x} + \dfrac{1}{y} = \dfrac{1}{5}$

$\dfrac{1}{x} - \dfrac{3}{y} = \dfrac{1}{3}$

Solution

a) We can eliminate y by adding the equations:

$$x^2 - y^2 = 5$$
$$\underline{x^2 + y^2 = 7}$$
$$2x^2 \quad\;\; = 12$$
$$x^2 = 6$$
$$x = \pm\sqrt{6}$$

Since $x^2 = 6$, the second equation yields $6 + y^2 = 7$, $y^2 = 1$, and $y = \pm 1$. If $x^2 = 6$ and $y^2 = 1$, then both of the original equations are satisfied. The solution set is

$$\{(\sqrt{6}, 1)(\sqrt{6}, -1), (-\sqrt{6}, 1), (-\sqrt{6}, -1)\}.$$

b) Usually with equations involving rational expressions we first multiply by the least common denominator (LCD), but this would make the given system more complicated. So we will just use the addition method to eliminate y:

$$\dfrac{6}{x} + \dfrac{3}{y} = \dfrac{3}{5} \qquad \text{Eq. (1) multiplied by 3}$$

$$\dfrac{1}{x} - \dfrac{3}{y} = \dfrac{1}{3} \qquad \text{Eq. (2)}$$

$$\overline{\dfrac{7}{x} \quad = \dfrac{14}{15}} \qquad \dfrac{3}{5} + \dfrac{1}{3} = \dfrac{14}{15}$$

$$14x = 7 \cdot 15$$

$$x = \dfrac{7 \cdot 15}{14} = \dfrac{15}{2}$$

To find y, substitute $x = \dfrac{15}{2}$ into Eq. (1):

$$\dfrac{2}{\dfrac{15}{2}} + \dfrac{1}{y} = \dfrac{1}{5}$$

$$\dfrac{4}{15} + \dfrac{1}{y} = \dfrac{1}{5} \qquad \dfrac{2}{\dfrac{15}{2}} = 2 \cdot \dfrac{2}{15} = \dfrac{4}{15}$$

$$15y \cdot \dfrac{4}{15} + 15y \cdot \dfrac{1}{y} = 15y \cdot \dfrac{1}{5} \qquad \text{Multiply each side by the LCD, } 15y.$$

$$4y + 15 = 3y$$

$$y = -15$$

Check that $x = \dfrac{15}{2}$ and $y = -15$ satisfy both original equations. The solution set is $\left\{\left(\dfrac{15}{2}, -15\right)\right\}$.

Now do Exercises 21–36

A system of nonlinear equations might involve exponential or logarithmic functions. To solve such systems, you will need to recall some facts about exponents and logarithms.

EXAMPLE 4

A system involving logarithms

Solve the system:

$$y = \log_2(x + 28)$$
$$y = 3 + \log_2(x)$$

Solution

Eliminate y by substituting $\log_2(x + 28)$ for y in the second equation:

$$\log_2(x + 28) = 3 + \log_2(x) \qquad \text{Eliminate } y.$$

$$\log_2(x + 28) - \log_2(x) = 3 \qquad\qquad \text{Subtract } \log_2(x) \text{ from each side.}$$

$$\log_2\!\left(\frac{x + 28}{x}\right) = 3 \qquad\qquad \text{Quotient rule for logarithms}$$

$$\frac{x + 28}{x} = 8 \qquad\qquad \text{Definition of logarithm}$$

$$x + 28 = 8x \qquad\qquad \text{Multiply each side by } x.$$

$$28 = 7x \qquad\qquad \text{Subtract } x \text{ from each side.}$$

$$4 = x \qquad\qquad \text{Divide each side by 7.}$$

If $x = 4$, then $y = \log_2(4 + 28) = \log_2(32) = 5$. Check $(4, 5)$ in both equations. The solution to the system is $\{(4, 5)\}$.

> Now do Exercises 37–42

⟨2⟩ Applications

Example 5 shows a geometric problem that can be solved with a system of nonlinear equations.

EXAMPLE 5

Nonlinear equations in applications

A 15-foot ladder is leaning against a wall so that the distance from the bottom of the ladder to the wall is one-half of the distance from the top of the ladder to the ground. Find the distance from the top of the ladder to the ground.

Solution

Let x be the number of feet from the bottom of the ladder to the wall and y be the number of feet from the top of the ladder to the ground (see Fig. 13.2 on the next page). We can write two equations involving x and y:

$$x^2 + y^2 = 15^2 \qquad \text{Pythagorean theorem}$$
$$y = 2x$$

Solve by substitution:

$$x^2 + (2x)^2 = 225 \qquad \text{Replace } y \text{ by } 2x.$$
$$x^2 + 4x^2 = 225$$
$$5x^2 = 225$$
$$x^2 = 45$$
$$x = \pm\sqrt{45} = \pm 3\sqrt{5}$$

⟨ **Calculator Close-Up** ⟩

To see the solutions, graph

$$y_1 = \sqrt{15^2 - x^2},$$
$$y_2 = -\sqrt{15^2 - x^2}, \text{and}$$
$$y_3 = 2x.$$

The line intersects the circle twice.

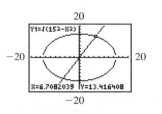

Figure 13.2

Because x represents distance, x must be positive. So $x = 3\sqrt{5}$. Because $y = 2x$, we get $y = 6\sqrt{5}$. The distance from the top of the ladder to the ground is $6\sqrt{5}$ feet.

Now do Exercises 43–46

Example 6 shows how a nonlinear system can be used to solve a problem involving work.

EXAMPLE 6

Nonlinear equations in applications

A large fish tank at the Gulf Aquarium can usually be filled in 10 minutes using pumps A and B. However, pump B can pump water in or out at the same rate. If pump B is inadvertently run in reverse, then the tank will be filled in 30 minutes. How long would it take each pump to fill the tank by itself?

‹ **Helpful Hint** ›

Note that we could write equations about the rates. Pump A's rate is $\frac{1}{a}$ tank per minute, B's rate is $\frac{1}{b}$ tank per minute, and together their rate is $\frac{1}{10}$ tank per minute or $\frac{1}{30}$ tank per minute.

$$\frac{1}{a} + \frac{1}{b} = \frac{1}{10}$$

$$\frac{1}{a} - \frac{1}{b} = \frac{1}{30}$$

Solution

Let a represent the number of minutes that it takes pump A to fill the tank alone and b represent the number of minutes it takes pump B to fill the tank alone. The rate at which pump A fills the tank is $\frac{1}{a}$ of the tank per minute, and the rate at which pump B fills the tank is $\frac{1}{b}$ of the tank per minute. Because the work completed is the product of the rate and time, we can make the following table when the pumps work together to fill the tank:

	Rate	Time	Work
Pump A	$\frac{1}{a}$ $\frac{\text{tank}}{\text{min}}$	10 min	$\frac{10}{a}$ tank
Pump B	$\frac{1}{b}$ $\frac{\text{tank}}{\text{min}}$	10 min	$\frac{10}{b}$ tank

Note that each pump fills a fraction of the tank and those fractions have a sum of 1:

$$(1) \qquad \frac{10}{a} + \frac{10}{b} = 1$$

In the 30 minutes in which pump B is working in reverse, A puts in $\frac{30}{a}$ of the tank whereas B takes out $\frac{30}{b}$ of the tank. Since the tank still gets filled, we can write the following equation:

$$(2) \qquad \frac{30}{a} - \frac{30}{b} = 1$$

Multiply Eq. (1) by 3 and add the result to Eq. (2) to eliminate b:

$$\frac{30}{a} + \frac{30}{b} = 3 \quad \text{Eq. (1) multiplied by 3}$$

$$\frac{30}{a} - \frac{30}{b} = 1 \quad \text{Eq. (2)}$$

$$\frac{60}{a} \qquad = 4$$

$$4a = 60$$

$$a = 15$$

Use $a = 15$ in Eq. (1) to find b:

$$\frac{10}{15} + \frac{10}{b} = 1$$

$$\frac{10}{b} = \frac{1}{3} \quad \text{Subtract } \tfrac{10}{15} \text{ from each side.}$$

$$b = 30$$

So pump A fills the tank in 15 minutes working alone, and pump B fills the tank in 30 minutes working alone.

Now do Exercises 47–56

Warm-Ups ▼

Fill in the blank.

1. The graph of a _____ equation is not a straight line.

2. The _____ of a nonlinear system can show us the number of solutions and the approximate value of the solutions.

3. We solve nonlinear systems using _____ and _____.

True or false?

4. The graph of $y = x^2$ is a parabola.

5. The graph of $y = |x|$ is a straight line.

6. The point $(3, -4)$ satisfies both $x^2 + y^2 = 25$ and $y = \sqrt{5x + 1}$.

7. There is no solution to the system $y = x^2 + 2$ and $y = x$.

8. The graphs of $y = \sqrt{x}$ and $y = -x - 2$ intersect at a single point.

9. If Bob paints a fence in x hours, then he paints $\dfrac{1}{x}$ of the fence per hour.

10. The area of a right triangle is one-half the product of the lengths of its legs.

Exercises

⟨ **Study Tips** ⟩

- If your instructor does not tell you what is coming tomorrow, ask.
- Read the material before it is discussed in class, and the instructor's explanation will make a lot more sense.

⟨1⟩ **Solving by Elimination**

Solve each system, and graph both equations on the same set of axes. See Example 1.

1. $y = x^2$
$x + y = 6$

2. $y = x^2 - 1$
$x + y = 11$

9. $y = -x^2 + 1$
$y = x^2$

10. $y = x^2$
$y = \sqrt{x}$

3. $y = |x|$
$2y - x = 6$

4. $y = |x|$
$3y = x + 6$

Solve each system. See Examples 2 and 3.

11. $xy = 6$
$x = 2$

12. $xy = 1$
$y = 3$

13. $xy = 1$
$y = x$

14. $y = x^2$
$y = x$

5. $y = \sqrt{2x}$
$x - y = 4$

6. $y = \sqrt{x}$
$x - y = 6$

15. $y = x^2$
$y = 2$

16. $xy = 3$
$y = x$

17. $x^2 + y^2 = 25$
$y = x^2 - 5$

18. $x^2 + y^2 = 25$
$y = x + 1$

19. $xy - 3x = 8$
$y = x + 1$

20. $xy + 2x = 9$
$x - y = 2$

7. $4x - 9y = 9$
$xy = 1$

8. $2x + 2y = 3$
$xy = -1$

21. $xy - x = 8$
$xy + 3x = -4$

22. $2xy - 3x = -1$
$xy + 5x = -7$

23. $x^2 + y^2 = 8$
$x^2 - y^2 = 2$

24. $y^2 - 2x^2 = 1$
$y^2 + 2x^2 = 5$

25. $x^2 + 2y^2 = 8$
$2x^2 - y^2 = 1$

26. $2x^2 + 3y^2 = 8$
$3x^2 + 2y^2 = 7$

27. $\dfrac{1}{x} - \dfrac{1}{y} = 5$
$\dfrac{2}{x} + \dfrac{1}{y} = -3$

28. $\dfrac{2}{x} - \dfrac{3}{y} = \dfrac{1}{2}$
$\dfrac{3}{x} + \dfrac{1}{y} = \dfrac{1}{2}$

29. $\dfrac{2}{x} - \dfrac{1}{y} = \dfrac{5}{12}$
$\dfrac{1}{x} - \dfrac{3}{y} = -\dfrac{5}{12}$

30. $\dfrac{3}{x} - \dfrac{2}{y} = 5$
$\dfrac{4}{x} + \dfrac{3}{y} = 18$

31. $x^2y = 20$
$xy + 2 = 6x$

32. $y^2x = 3$
$xy + 1 = 6x$

33. $x^2 + xy - y^2 = -11$
$x + y = 7$

34. $x^2 + xy + y^2 = 3$
$y = 2x - 5$

35. $3y - 2 = x^4$
$y = x^2$

36. $y - 3 = 2x^4$
$y = 7x^2$

Solve the following systems involving logarithmic and exponential functions. See Example 4.

37. $y = \log_2(x - 1)$
$y = 3 - \log_2(x + 1)$

38. $y = \log_3(x - 4)$
$y = 2 - \log_3(x + 4)$

39. $y = \log_2(x - 1)$
$y = 2 + \log_2(x + 2)$

40. $y = \log_4(8x)$
$y = 2 + \log_4(x - 1)$

41. $y = 2^{3x+4}$
$y = 4^{x-1}$

42. $y = 4^{3x}$
$y = \left(\dfrac{1}{2}\right)^{1-x}$

⟨2⟩ Applications

Solve each problem by using a system of two equations in two unknowns. See Examples 5 and 6.

43. *Known hypotenuse.* Find the lengths of the legs of a right triangle whose hypotenuse is $\sqrt{15}$ feet and whose area is 3 square feet.

44. *Known diagonal.* A small television is advertised to have a picture with a diagonal measure of 5 inches and a viewing area of 12 square inches (in.^{2}). What are the length and width of the screen?

Figure for Exercise 44

45. *House of seven gables.* Vincent has plans to build a house with seven gables. The plans call for an attic vent in the shape of an isosceles triangle in each gable. Because of the slope of the roof, the ratio of the height to the base of each triangle must be 1 to 4. If the vents are to provide a total ventilating area of 3500 in.^{2}, then what should be the height and base of each triangle?

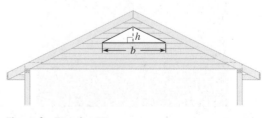

Figure for Exercise 45

46. *Known perimeter.* Find the lengths of the sides of a triangle whose perimeter is 6 feet (ft) and whose angles are 30°, 60°, and 90° (see inside the front cover of the book).

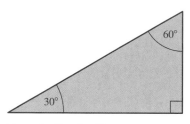

Figure for Exercise 46

47. *Filling a tank.* Pump A can either fill a tank or empty it in the same amount of time. If pump A and pump B are working together, the tank can be filled in 6 hours. When pump A was inadvertently left in the drain position while pump B was trying to fill the tank, it took 12 hours to fill the tank. How long would it take either pump working alone to fill the tank?

48. *Cleaning a house.* Roxanne either cleans the house or messes it up at the same rate. When Roxanne is cleaning with her mother, they can clean up a completely messed up house in 6 hours. If Roxanne is not cooperating, it takes her mother 9 hours to clean the house, with Roxanne continually messing it up. How long would it take her mother to clean the entire house if Roxanne were sent to her grandmother's house?

49. *Cleaning fish.* Jan and Beth work in a seafood market that processes 200 pounds of catfish every morning.

Photo for Exercise 49

On Monday, Jan started cleaning catfish at 8:00 A.M. and finished cleaning 100 pounds just as Beth arrived. Beth then took over and finished the job at 8:50 A.M. On Tuesday they both started at 8 A.M. and worked together to finish the job at 8:24 A.M. On Wednesday, Beth was sick. If Jan is the faster worker, then how long did it take Jan to complete all of the catfish by herself?

50. *Building a patio.* Richard has already formed a rectangular area for a flagstone patio, but his wife Susan is unsure of the size of the patio they want. If the width is increased by 2 ft, then the area is increased by 30 square feet (ft²). If the width is increased by 1 ft and the length by 3 ft, then the area is increased by 54 ft². What are the dimensions of the rectangle that Richard has already formed?

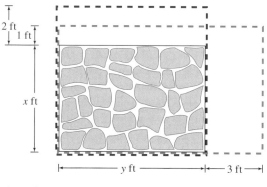

Figure for Exercise 50

51. *Fencing a rectangle.* If 34 ft of fencing are used to enclose a rectangular area of 72 ft², then what are the dimensions of the area?

52. *Real numbers.* Find two numbers that have a sum of 8 and a product of 10.

53. *Imaginary numbers.* Find two complex numbers whose sum is 8 and whose product is 20.

54. *Imaginary numbers.* Find two complex numbers whose sum is −6 and whose product is 10.

55. *Making a sign.* Rico's Sign Shop has a contract to make a sign in the shape of a square with an isosceles triangle on top of it, as shown in the figure. The contract calls for a total height of 10 ft with an area of 72 ft². How long should Rico make the side of the square and what should be the height of the triangle?

hold. It must have 184 square inches of surface area to provide enough space for all of the special offers and coupons. What should be the dimensions of the box?

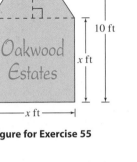

Figure for Exercise 55

56. *Designing a box.* Angelina is designing a rectangular box of 120 cubic inches that is to contain new Eaties breakfast cereal. The box must be 2 inches thick so that it is easy to

Graphing Calculator Exercises

57. Solve each system by graphing each pair of equations on a graphing calculator and using the intersect feature to estimate the point of intersection. Find the coordinates of each intersection to the nearest hundredth.

a) $y = e^x - 4$
$y = \ln(x + 3)$

b) $3^{y-1} = x$
$y = x^2$

c) $x^2 + y^2 = 4$
$y = x^3$

13.2 The Parabola

In This Section

⟨1⟩ **The Distance and Midpoint Formulas**

⟨2⟩ **The Parabola**

⟨3⟩ **Changing Forms**

⟨4⟩ **Parabolas Opening to the Right or Left**

The **conic sections** are the four curves that are obtained by intersecting a cone and a plane as in Fig. 13.3. The figure explains why the parabola, ellipse, circle, and hyperbola are called conic sections, but it does not help us find equations for the curves. To develop equations for these curves we will redefine them more precisely using distance between points. So we will first discuss the distance formula.

Parabola   Circle   Ellipse   Hyperbola

Figure 13.3

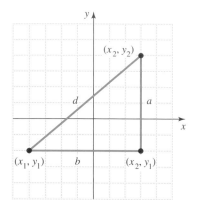

Figure 13.4

⟨1⟩ The Distance and Midpoint Formulas

Consider the points (x_1, y_1) and (x_2, y_2), as shown in Fig. 13.4. The distance between these points is the length of the hypotenuse of a right triangle as shown in the figure. The length of side a is $y_2 - y_1$, and the length of side b is $x_2 - x_1$. Using the Pythagorean theorem, we can write

$$d^2 = (x_2 - x_1)^2 + (y_2 - y_1)^2.$$

If we apply the even-root property and omit the negative square root (because the distance is positive), we can express this formula as follows.

Distance Formula

The distance d between (x_1, y_1) and (x_2, y_2) is given by the formula

$$d = \sqrt{(x_2 - x_1)^2 + (y_2 - y_1)^2}.$$

EXAMPLE 1

Using the distance formula

Find the length of the line segment with endpoints $(-8, -10)$ and $(6, -4)$.

Solution

Let $(x_1, y_1) = (-8, -10)$ and $(x_2, y_2) = (6, -4)$. Now substitute the appropriate values into the distance formula:

$$\begin{aligned}
d &= \sqrt{[6 - (-8)]^2 + [-4 - (-10)]^2} \\
&= \sqrt{(14)^2 + (6)^2} \\
&= \sqrt{196 + 36} \\
&= \sqrt{232} \\
&= \sqrt{4 \cdot 58} \\
&= 2\sqrt{58} \quad \text{Simplified form}
\end{aligned}$$

The exact length of the segment is $2\sqrt{58}$.

Now do Exercises 1–10

The **midpoint** of a line segment is a point that is on the line segment and equidistant from the endpoints. We use the notation $(\bar{x}, \bar{y})$ (read "x bar, y bar") for the midpoint of a line segment. The midpoint is found by "averaging" the x-coordinates and y-coordinates of the endpoints, in the same manner that you would average two test scores:

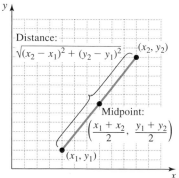

Figure 13.5

Midpoint Formula

The midpoint of the line segment with endpoints (x_1, y_1) and (x_2, y_2) is given by

$$(\bar{x}, \bar{y}) = \left(\frac{x_1 + x_2}{2}, \frac{y_1 + y_2}{2} \right).$$

The length of a line segment is the distance between its endpoints, and it is given by the distance formula. See Fig. 13.5.

EXAMPLE **2**

Finding the midpoint and length of a line segment
Find the midpoint and length of the line segment with endpoints (1, 7) and (5, 4).

Solution
Use the midpoint formula with $(x_1, y_1) = (1, 7)$ and $(x_2, y_2) = (5, 4)$:

$$(\bar{x}, \bar{y}) = \left(\frac{1+5}{2}, \frac{7+4}{2}\right) = \left(3, \frac{11}{2}\right)$$

Use the distance formula to find the length of the line segment:

$$\sqrt{(x_2 - x_1)^2 + (y_2 - y_1)^2} = \sqrt{(5-1)^2 + (4-7)^2}$$
$$= \sqrt{16 + 9}$$
$$= \sqrt{25} = 5$$

Note that $(x_1, y_1) = (5, 4)$ and $(x_2, y_2) = (1, 7)$ gives the same midpoint and length. Try it.

Now do Exercises 11–18

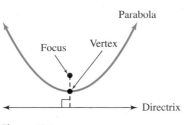

Figure 13.6

‹2› **The Parabola**

In Section 10.4 we called the graph of $y = ax^2 + bx + c$ a parabola. In this section, you will see that the following geometric definition describes the same curve as the equation.

> **Parabola**
>
> Given a line (the **directrix**) and a point not on the line (the **focus**), the set of all points in the plane that are equidistant from the point and the line is called a **parabola.**

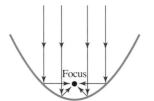

Figure 13.7

In Section 10.4 we defined the vertex as the highest point on a parabola that opens downward or the lowest point on a parabola that opens upward. We learned that $x = -b/(2a)$ gives the x-coordinate of the vertex. We can also describe the vertex of a parabola as the midpoint of the line segment that joins the focus and directrix, perpendicular to the directrix. See Fig. 13.6.

The focus of a parabola is important in applications. When parallel rays of light travel into a parabolic reflector, they are reflected toward the focus, as in Fig. 13.7. This property is used in telescopes to see the light from distant stars. If the light source is at the focus, as in a searchlight, the light is reflected off the parabola and projected outward in a narrow beam. This reflecting property is also used in camera lenses, satellite dishes, and eavesdropping devices.

To develop an equation for a parabola, given the focus and directrix, choose the point $(0, p)$, where $p > 0$, as the focus and the line $y = -p$ as the directrix, as shown in Fig. 13.8. The vertex of this parabola is $(0, 0)$. For an arbitrary point (x, y) on the parabola the distance to the directrix is the distance from (x, y) to $(x, -p)$. The distance to the focus is the distance between (x, y) and $(0, p)$. We use the fact that these distances are equal to write the equation of the parabola:

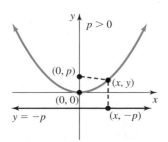

Figure 13.8

$$\sqrt{(x-0)^2 + (y-p)^2} = \sqrt{(x-x)^2 + (y-(-p))^2}$$

To simplify the equation, first remove the parentheses inside the radicals:

$$\sqrt{x^2 + y^2 - 2py + p^2} = \sqrt{y^2 + 2py + p^2}$$

$$x^2 + y^2 - 2py + p^2 = y^2 + 2py + p^2 \quad \text{Square each side.}$$

$$x^2 = 4py \qquad\qquad \text{Subtract } y^2 \text{ and } p^2 \text{ from each side.}$$

$$y = \frac{1}{4p}x^2$$

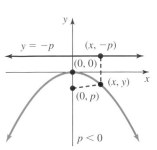

$y = -p$ $(x, -p)$
$(0, 0)$
(x, y)
$(0, p)$
$p < 0$

Figure 13.9

So the parabola with focus $(0, p)$ and directrix $y = -p$ for $p > 0$ has equation $y = \frac{1}{4p}x^2$. This equation has the form $y = ax^2 + bx + c$, where $a = \frac{1}{4p}$, $b = 0$, and $c = 0$.

If the focus is $(0, p)$ with $p < 0$ and the directrix is $y = -p$, then the parabola opens downward, as shown in Fig. 13.9. Deriving the equation using the distance formula again yields $y = \frac{1}{4p}x^2$.

The simplest parabola, $y = x^2$, has vertex $(0, 0)$. The transformation $y = a(x - h)^2 + k$ is also a parabola, and its vertex is (h, k). The focus and directrix of the transformation are found as follows:

Parabolas in the Form $y = a(x - h)^2 + k$

The graph of the equation $y = a(x - h)^2 + k$ ($a \neq 0$) is a parabola with vertex (h, k), focus $(h, k + p)$, and directrix $y = k - p$, where $a = \frac{1}{4p}$. If $a > 0$, the parabola opens upward; if $a < 0$, the parabola opens downward.

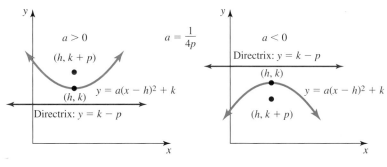

$a > 0$
$a = \frac{1}{4p}$
$(h, k + p)$
(h, k)
$y = a(x - h)^2 + k$
Directrix: $y = k - p$

$a < 0$
Directrix: $y = k - p$
(h, k)
$y = a(x - h)^2 + k$
$(h, k + p)$

Figure 13.10

Figure 13.10 shows the location of the focus and directrix for parabolas with vertex (h, k) and opening either upward or downward. Note that the location of the focus and directrix determine the value of a and the shape and opening of the parabola.

CAUTION For a parabola that opens upward, $p > 0$, and the focus $(h, k + p)$ is above the vertex (h, k). For a parabola that opens downward, $p < 0$, and the focus $(h, k + p)$ is below the vertex (h, k). In either case, the distance from the vertex to the focus and the vertex to the directrix is $|p|$.

In Example 3 we find the vertex, focus, and directrix from an equation of a parabola. In Example 4 we find the equation given the focus and directrix.

EXAMPLE **3**

Finding the vertex, focus, and directrix, given an equation

Find the vertex, focus, and directrix for the parabola $y = x^2$.

Solution

Compare $y = x^2$ to the general formula $y = a(x - h)^2 + k$. We see that $h = 0$, $k = 0$, and $a = 1$. So the vertex is $(0, 0)$. Because $a = 1$, we can use $a = \frac{1}{4p}$ to get

$$1 = \frac{1}{4p},$$

or $p = \frac{1}{4}$. Use $(h, k + p)$ to get the focus $\left(0, \frac{1}{4}\right)$. Use the equation $y = k - p$ to get $y = -\frac{1}{4}$ as the equation of the directrix. See Fig. 13.11.

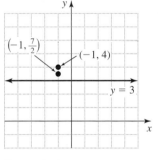

Figure 13.11

Now do Exercises 19–26

EXAMPLE **4**

Finding an equation, given a focus and directrix

Find the equation of the parabola with focus $(-1, 4)$ and directrix $y = 3$.

Solution

Because the vertex is halfway between the focus and directrix, the vertex is $\left(-1, \frac{7}{2}\right)$. See Fig. 13.12. The distance from the vertex to the focus is $\frac{1}{2}$. Because the focus is above the vertex, p is positive. So $p = \frac{1}{2}$, and $a = \frac{1}{4p} = \frac{1}{2}$.

The equation is

$$y = \frac{1}{2}(x - (-1))^2 + \frac{7}{2}.$$

Convert to $y = ax^2 + bx + c$ form as follows:

$$y = \frac{1}{2}(x + 1)^2 + \frac{7}{2}$$

$$y = \frac{1}{2}(x^2 + 2x + 1) + \frac{7}{2}$$

$$y = \frac{1}{2}x^2 + x + 4$$

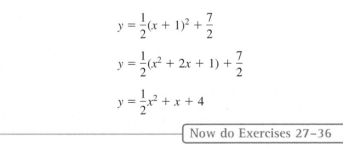

Figure 13.12

Now do Exercises 27–36

The graph of $y = x^2$ shown in Fig. 13.11 is **symmetric about the y-axis** because the two halves of the parabola would coincide if the paper were folded on the y-axis. In general, the vertical line through the vertex is the **axis of symmetry** for the

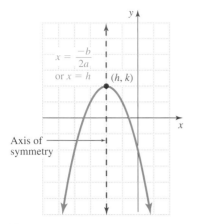

Figure 13.13

parabola. See Fig. 13.13. In the form $y = ax^2 + bx + c$, the x-coordinate of the vertex is $-b/(2a)$ and the equation of the axis of symmetry is $x = -b/(2a)$. In the form $y = a(x - h)^2 + k$, the vertex is (h, k) and the equation for the axis of symmetry is $x = h$.

⟨3⟩ Changing Forms

Since there are two forms for the equation of a parabola, it is sometimes useful to change from one form to the other. To change from $y = a(x - h)^2 + k$ to the form $y = ax^2 + bx + c$, we square the binomial and combine like terms, as in Example 4. To change from $y = ax^2 + bx + c$ to the form $y = a(x - h)^2 + k$, we complete the square, as in Example 5.

E X A M P L E   5

Converting $y = ax^2 + bx + c$ to $y = a(x - h)^2 + k$

Write $y = 2x^2 - 4x + 5$ in the form $y = a(x - h)^2 + k$, and identify the vertex, focus, directrix, and axis of symmetry of the parabola.

Solution

Use completing the square to rewrite the equation:

$$y = 2(x^2 - 2x) + 5$$
$$y = 2(x^2 - 2x + 1 - 1) + 5 \quad \text{Complete the square.}$$
$$y = 2(x^2 - 2x + 1) - 2 + 5 \quad \text{Move } 2(-1) \text{ outside the parentheses.}$$
$$y = 2(x - 1)^2 + 3$$

The vertex is $(1, 3)$. Because $a = \frac{1}{4p}$, we have

$$\frac{1}{4p} = 2,$$

and $p = \frac{1}{8}$. Because the parabola opens upward, the focus is $\frac{1}{8}$ unit above the vertex at $\left(1, 3\frac{1}{8}\right)$, or $\left(1, \frac{25}{8}\right)$, and the directrix is the horizontal line $\frac{1}{8}$ unit below the vertex, $y = 2\frac{7}{8}$ or $y = \frac{23}{8}$. The axis of symmetry is $x = 1$.

> Now do Exercises 37–44

< **Calculator Close-Up** ›

The graphs of
$$y_1 = 2x^2 - 4x + 5$$
and
$$y_2 = 2(x - 1)^2 + 3$$

appear to be identical. This supports the conclusion that the equations are equivalent.

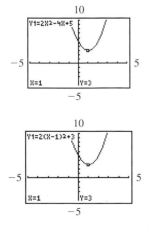

CAUTION   Be careful when you complete a square within parentheses as in Example 5. For another example, consider the equivalent equations

$$y = -3(x^2 + 4x),$$
$$y = -3(x^2 + 4x + 4 - 4),$$

and

$$y = -3(x + 2)^2 + 12.$$

EXAMPLE **6**

Finding the features of a parabola in the form $y = ax^2 + bx + c$

Find the vertex, focus, directrix, and axis of symmetry of the parabola $y = -3x^2 + 9x - 5$, and determine whether the parabola opens upward or downward.

Solution

The x-coordinate of the vertex is

$$x = \frac{-b}{2a} = \frac{-9}{2(-3)} = \frac{-9}{-6} = \frac{3}{2}.$$

To find the y-coordinate of the vertex, let $x = \frac{3}{2}$ in $y = -3x^2 + 9x - 5$:

$$y = -3\left(\frac{3}{2}\right)^2 + 9\left(\frac{3}{2}\right) - 5 = -\frac{27}{4} + \frac{27}{2} - 5 = \frac{7}{4}$$

The vertex is $\left(\frac{3}{2}, \frac{7}{4}\right)$. Because $a = -3$, the parabola opens downward. To find the focus, use $-3 = \frac{1}{4p}$ to get $p = -\frac{1}{12}$. The focus is $\frac{1}{12}$ of a unit below the vertex at $\left(\frac{3}{2}, \frac{7}{4} - \frac{1}{12}\right)$ or $\left(\frac{3}{2}, \frac{5}{3}\right)$. The directrix is the horizontal line $\frac{1}{12}$ of a unit above the vertex, $y = \frac{7}{4} + \frac{1}{12}$ or $y = \frac{11}{6}$.

The equation of the axis of symmetry is $x = \frac{3}{2}$.

> Now do Exercises 45–54

⟨ **Calculator Close-Up** ⟩

A calculator graph can be used to check the vertex and opening of a parabola.

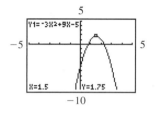

⟨4⟩ Parabolas Opening to the Right or Left

If we interchange x and y in the equation $y = a(x - h)^2 + k$, we get the equation $x = a(y - k)^2 + h$, which is a parabola opening to the right or left.

> **Parabolas in the Form $x = a(y - k)^2 + h$**
>
> The graph of $x = a(y - k)^2 + h$ $(a \neq 0)$ is a parabola with vertex (h, k), focus $(h + p, k)$, and directrix $x = h - p$, where $a = \frac{1}{4p}$. If $a > 0$, the parabola opens to the right; if $a < 0$, the parabola opens to the left.

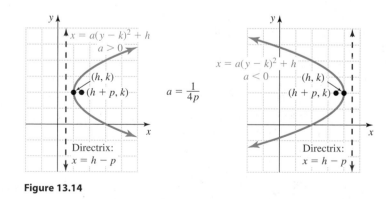

Figure 13.14

Figure 13.14 shows the location of the focus and directrix for parabolas with vertex (h, k) and opening either right or left. The location of the focus and directrix determine the value of a and the shape and opening of the parabola. Note that a and p have the same sign because $a = \dfrac{1}{4p}$.

The equation $x = ay^2 + by + c$ could be converted to the form $x = a(y - k)^2 + h$ from which the vertex, focus, and directrix could be determined. Without converting we can determine that the graph of $x = ay^2 + by + c$ opens to the right for $a > 0$ and to the left for $a < 0$. The y-coordinate of the vertex is $\dfrac{-b}{2a}$. The x-coordinate of the vertex can be determined by substituting $\dfrac{-b}{2a}$ for y in $x = ay^2 + by + c$.

E X A M P L E   **7**

Graphing a parabola opening to the right

Find the vertex, focus, and directrix for the parabola $x = \frac{1}{2}(y - 2)^2 + 1$, and sketch the graph.

Solution

In the form $x = a(y - k)^2 + h$, the vertex is (h, k). So the vertex for $x = \frac{1}{2}(y - 2)^2 + 1$ is $(1, 2)$. Since $a = \frac{1}{4p}$ and $a = \frac{1}{2}$, we have $p = \frac{1}{2}$ and the focus is $\left(\frac{3}{2}, 2\right)$. The directrix is the vertical line $x = \frac{1}{2}$. Find a few points that satisfy $x = \frac{1}{2}(y - 2)^2 + 1$ as follows:

$x = \frac{1}{2}(y - 2)^2 + 1$	3	$\frac{3}{2}$	1	$\frac{3}{2}$	3
y	0	1	2	3	4

Sketch the graph through these points, as shown in Fig. 13.15.

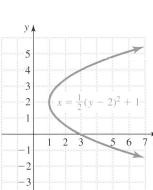

Figure 13.15

Now do Exercises 55–60

Warm-Ups ▼

Fill in the blank.

1. A _____ is the set of all points in a plane that are equidistant from a given line and a fixed point not on the line.

2. The _____ of a parabola is the fixed point in the definition.

3. The _____ is the given line in the definition.

4. The _____ of a parabola is the midpoint of the line segment joining the focus and directrix, perpendicular to the directrix.

5. The distance from the _____ or the _____ to the vertex is $|p|$, where $a = \dfrac{1}{4p}$.

6. We can convert $y = ax^2 + bx + c$ to the form $y = a(x - h)^2 + k$ by _____.

True or false?

7. If the focus of a parabola is $(0, 4)$ and the directrix is $y = 1$, then the vertex is $(0, 3)$.

8. The focus for $y = \dfrac{1}{4}x^2 + 1$ is $(0, 2)$.

9. The vertex for $y - 3 = 5(x - 4)^2$ is $(4, 3)$.

10. The parabola $y = 2x - x^2 + 9$ opens upward.

11. The vertex for $y = x^2$ is the y-intercept.

12. A parabola with vertex $(2, 3)$ and focus $(2, 4)$ has no x-intercepts.

‹1› **The Distance and Midpoint Formulas**

Find the distance between each given pair of points. See Example 1.

1. $(2, 1), (5, 5)$

2. $(3, 2), (8, 14)$

3. $(4, -3), (5, -2)$

4. $(-1, 5), (-2, 6)$

5. $(6, 5), (4, 2)$

6. $(7, 3), (5, 1)$

7. $(3, 5), (1, -3)$

8. $(6, 2), (3, -5)$

9. $(4, -2), (-3, -6)$

10. $(-2, 3), (1, -4)$

Find the midpoint and length of the line segment with the given endpoints. See Example 2.

11. $(0, 0)$ and $(6, 8)$

12. $(0, 0)$ and $(-6, 8)$

13. $(2, 5)$ and $(5, 1)$

14. $(1, 7)$ and $(5, 10)$

15. $(-2, 4)$ and $(6, -2)$

16. $(-3, 5)$ and $(3, -3)$

17. $(-1, 4)$ and $(1, 1)$

18. $(-3, -4)$ and $(-6, 1)$

‹2› **The Parabola**

Find the vertex, focus, and directrix for each parabola. See Example 3.

19. $y = 2x^2$

20. $y = \dfrac{1}{2}x^2$

21. $y = -\dfrac{1}{4}x^2$

22. $y = -\dfrac{1}{12}x^2$

23. $y = \dfrac{1}{2}(x - 3)^2 + 2$

24. $y = \dfrac{1}{4}(x + 2)^2 - 5$

25. $y = -(x + 1)^2 + 6$

26. $y = -3(x - 4)^2 + 1$

Find the equation of the parabola with the given focus and directrix. See Example 4.

27. Focus $(0, 2)$, directrix $y = -2$

28. Focus $(0, -3)$, directrix $y = 3$

29. Focus $\left(0, -\dfrac{1}{2}\right)$, directrix $y = \dfrac{1}{2}$

30. Focus $\left(0, \dfrac{1}{8}\right)$, directrix $y = -\dfrac{1}{8}$

31. Focus $(3, 2)$, directrix $y = 1$

32. Focus $(-4, 5)$, directrix $y = 4$

33. Focus $(1, -2)$, directrix $y = 2$

34. Focus $(2, -3)$, directrix $y = 1$

35. Focus $(-3, 1.25)$, directrix $y = 0.75$

36. Focus $\left(5, \dfrac{17}{8}\right)$, directrix $y = \dfrac{15}{8}$

⟨3⟩ **Changing Forms**

Write each equation in the form $y = a(x - h)^2 + k$. Identify the vertex, focus, directrix, and axis of symmetry of each parabola. See Example 5.

37. $y = x^2 - 6x + 1$

38. $y = x^2 + 4x - 7$

39. $y = 2x^2 + 12x + 5$

40. $y = 3x^2 + 6x - 7$

41. $y = -2x^2 + 16x + 1$

42. $y = -3x^2 - 6x + 7$

43. $y = 5x^2 + 40x$

44. $y = -2x^2 + 10x$

Find the vertex, focus, directrix, and axis of symmetry of each parabola (without completing the square), and determine whether the parabola opens upward or downward. See Example 6.

45. $y = x^2 - 4x + 1$

46. $y = x^2 - 6x - 7$

47. $y = -x^2 + 2x - 3$

48. $y = -x^2 + 4x + 9$

49. $y = 3x^2 - 6x + 1$

50. $y = 2x^2 + 4x - 3$

51. $y = -x^2 - 3x + 2$

52. $y = -x^2 + 3x - 1$

53. $y = 3x^2 + 5$

54. $y = -2x^2 - 6$

⟨4⟩ **Parabolas Opening to the Right or Left**

Find the vertex, focus, and directrix for each parabola. See Example 7.

55. $x = (y - 2)^2 + 3$

56. $x = (y + 3)^2 - 1$

57. $x = \dfrac{1}{4}(y - 1)^2 - 2$

58. $x = \dfrac{1}{4}(y + 1)^2 + 2$

59. $x = -\dfrac{1}{2}(y - 2)^2 + 4$

60. $x = -\dfrac{1}{2}(y + 1)^2 - 1$

Miscellaneous

Sketch the graph of each parabola.

61. $y = (x - 2)^2 + 3$　　　　**62.** $y = (x + 3)^2 - 1$

63. $y = -2(x - 1)^2 + 3$ **64.** $y = -\dfrac{1}{2}(x + 1)^2 + 5$

71. $y = x^2 - 2$
$\ y = 2x - 3$

72. $y = x^2 + x - 6$
$\ y = 7x - 15$

65. $x = (y - 2)^2 + 3$ **66.** $x = (y + 3)^2 - 1$

73. $y = x^2 + 3x - 4$
$\ y = -x^2 - 2x + 8$

67. $x = -2(y - 1)^2 + 3$ **68.** $x = -\dfrac{1}{2}(y + 1)^2 + 5$

74. $y = x^2 + 2x - 8$
$\ y = -x^2 - x + 12$

Graph both equations of each system on the same coordinate axes.
Use elimination of variables to find all points of intersection.

69. $y = -x^2 + 3$
$\ y = x^2 + 1$

75. $y = x^2 + 3x - 4$
$\ y = 2x + 2$

70. $y = x^2 - 3$
$\ y = -x^2 + 5$

76. $y = x^2 + 5x + 6$
$\ y = x + 11$

Solve each problem.

77. Find all points of intersection of the parabola $y = x^2 - 2x - 3$ and the x-axis.

78. Find all points of intersection of the parabola $y = 80x^2 - 33x + 255$ and the y-axis.

79. Find all points of intersection of the parabola $y = 0.01x^2$ and the line $y = 4$.

80. Find all points of intersection of the parabola $y = 0.02x^2$ and the line $y = x$.

81. Find all points of intersection of the parabolas $y = x^2$ and $x = y^2$.

82. Find all points of intersection of the parabolas $y = x^2$ and $y = (x - 3)^2$.

Applications

Solve each problem.

83. *Pipeline charges.* Ewing Oil paid a subcontractor $84 per yard for laying a pipe in a west Texas oil field. The pipe connects wells located at (185, 234) and (−215, −352) in the oil field coordinate system shown in the figure. The units in the figure are yards.

   a) What was the cost to the nearest dollar for this project?

   b) What is the location of the valve installed at the midpoint?

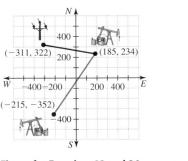

Figure for Exercises 83 and 84

84. *Electricity charges.* Texas Power installed a power line from a transformer at (−311, 322) to the well at (185, 234) as shown in the figure for $116 per yard.

   a) What was the cost to the nearest dollar for the power line?

   b) What is the location of the pole used at the midpoint?

85. *World's largest telescope.* The largest reflecting telescope in the world is the 6-meter (m) reflector on Mount Pastukhov in Russia. The accompanying figure shows a cross section of a parabolic mirror 6 m in diameter with the vertex at the origin and the focus at (0, 15). Find the equation of the parabola.

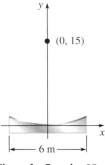

Figure for Exercise 85

86. *Arecibo observatory.* The largest radio telescope in the world uses a 1000-ft parabolic dish, suspended in a valley in Arecibo, Puerto Rico. The antenna hangs above the vertex of the dish on cables stretching from two towers. The accompanying figure shows a cross section of the parabolic dish and the towers. Assuming the vertex is at (0, 0), find the equation for the parabola. Find the

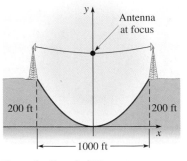

Figure for Exercise 86

distance from the vertex to the antenna located at the focus.

Getting More Involved

87. *Exploration*

Consider the parabola with focus $(p, 0)$ and directrix $x = -p$ for $p > 0$. Let (x, y) be an arbitrary point on the parabola. Write an equation expressing the fact that the distance from (x, y) to the focus is equal to the distance from (x, y) to the directrix. Rewrite the equation in the form $x = ay^2$, where $a = \dfrac{1}{4p}$.

88. *Exploration*

In general, the graph of $x = a(y - k)^2 + h$ for $a \neq 0$ is a parabola opening left or right with vertex at (h, k).

a) For which values of a does the parabola open to the right, and for which values of a does it open to the left?

b) What is the equation of its axis of symmetry?

c) Sketch the graphs $x = 2(y - 3)^2 + 1$ and $x = -(y + 1)^2 + 2$.

Graphing Calculator Exercises

89. Graph $y = x^2$ using the viewing window with $-1 \leq x \leq 1$ and $0 \leq y \leq 1$. Next graph $y = 2x^2 - 1$ using the viewing window $-2 \leq x \leq 2$ and $-1 \leq y \leq 7$. Explain what you see.

90. Graph $y = x^2$ and $y = 6x - 9$ in the viewing window $-5 \leq x \leq 5$ and $-5 \leq y \leq 20$. Does the line appear to be tangent to the parabola? Solve the system $y = x^2$ and $y = 6x - 9$ to find all points of intersection for the parabola and the line.

13.3   The Circle

In This Section

⟨1⟩ **The Equation of a Circle**

⟨2⟩ **Equations Not in Standard Form**

⟨3⟩ **Systems of Equations**

In this section, we continue the study of the conic sections with a discussion of the circle.

⟨1⟩ The Equation of a Circle

A circle is obtained by cutting a cone, as was shown in Fig. 13.3. We can also define a circle using points and distance, as we did for the parabola.

> **Circle**
>
> A **circle** is the set of all points in a plane that lie a fixed distance from a given point in the plane. The fixed distance is called the **radius,** and the given point is called the **center.**

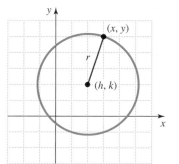

Figure 13.16

We can use the distance formula of Section 13.2 to write an equation for the circle with center (h, k) and radius r, shown in Fig. 13.16. If (x, y) is a point on the circle, its distance from the center is r. So,

$$\sqrt{(x - h)^2 + (y - k)^2} = r.$$

We square both sides of this equation to get the **standard form** for the equation of a circle.

> **Standard Equation for a Circle**
> The graph of the equation
>
> $$(x - h)^2 + (y - k)^2 = r^2$$
>
> with $r > 0$, is a circle with center (h, k) and radius r.

Note that a circle centered at the origin with radius r ($r > 0$) has the standard equation

$$x^2 + y^2 = r^2.$$

E X A M P L E   1

Finding the equation, given the center and radius
Write the equation for the circle with the given center and radius.

   a) Center $(0, 0)$, radius 2

   b) Center $(-1, 2)$, radius 4

Solution

a) The center at $(0, 0)$ means that $h = 0$ and $k = 0$ in the standard equation. So the equation is $(x - 0)^2 + (y - 0)^2 = 2^2$, or $x^2 + y^2 = 4$. The circle with radius 2 centered at the origin is shown in Fig. 13.17.

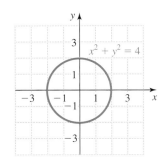

Figure 13.17

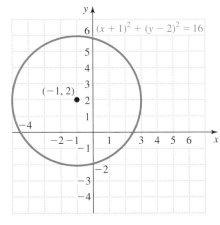

Figure 13.18

b) The center at $(-1, 2)$ means that $h = -1$ and $k = 2$. So,

$$[x - (-1)]^2 + [y - 2]^2 = 4^2.$$

Simplify this equation to get

$$(x + 1)^2 + (y - 2)^2 = 16.$$

The circle with center $(-1, 2)$ and radius 4 is shown in Fig. 13.18.

 Now do Exercises 1–12

CAUTION The equations $(x - 1)^2 + (y + 3)^2 = -9$ and $(x - 1)^2 + (y + 3)^2 = 0$ might look like equations of circles, but they are not. The first equation is not satisfied by any ordered pair of real numbers because the left-hand side is nonnegative for any x and y. The second equation is satisfied only by the point $(1, -3)$.

E X A M P L E   **2**

Finding the center and radius, given the equation

Determine the center and radius of the circle $x^2 + (y + 5)^2 = 2$.

Solution

We can write this equation as

$$(x - 0)^2 + [y - (-5)]^2 = (\sqrt{2})^2.$$

In this form we see that the center is $(0, -5)$ and the radius is $\sqrt{2}$.

Now do Exercises 13–22

E X A M P L E   **3**

Graphing a circle

Find the center and radius of $(x - 1)^2 + (y + 2)^2 = 9$, and sketch the graph.

Solution

The graph of this equation is a circle with center $(1, -2)$ and radius 3. See Fig. 13.19 for the graph.

Now do Exercises 23–32

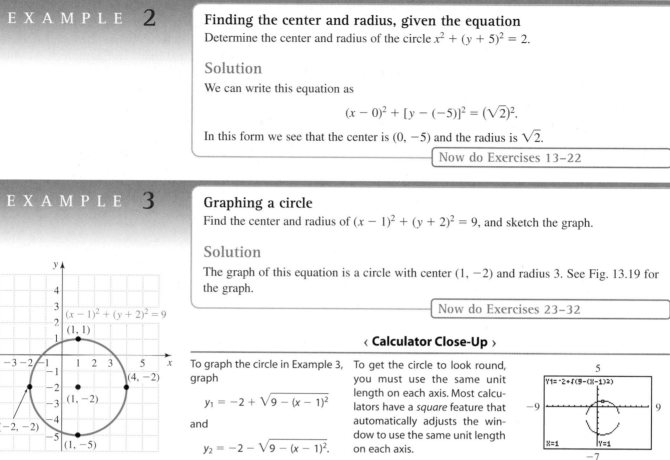

Figure 13.19

‹ **Calculator Close-Up** ›

To graph the circle in Example 3, graph

$$y_1 = -2 + \sqrt{9 - (x - 1)^2}$$

and

$$y_2 = -2 - \sqrt{9 - (x - 1)^2}.$$

To get the circle to look round, you must use the same unit length on each axis. Most calculators have a *square* feature that automatically adjusts the window to use the same unit length on each axis.

⟨2⟩ Equations Not in Standard Form

It is not easy to recognize that $x^2 - 6x + y^2 + 10y = -30$ is the equation of a circle, but it is. In Example 4, we convert this equation into the standard form for a circle by completing the squares for the variables x and y.

E X A M P L E   **4**

Converting to standard form

Find the center and radius of the circle given by the equation

$$x^2 - 6x + y^2 + 10y = -30.$$

‹ **Helpful Hint** ›

What do circles and lines have in common? They are the two simplest graphs to draw. We have compasses to make our circles look good and rulers to make our lines look good.

Solution

To complete the square for $x^2 - 6x$, we add 9, and for $y^2 + 10y$, we add 25. To get an equivalent equation, we must add on both sides:

$$x^2 - 6x \qquad + y^2 + 10y \qquad = -30$$

$$x^2 - 6x + 9 + y^2 + 10y + 25 = -30 + 9 + 25 \quad \text{Add 9 and 25 to both sides.}$$

$$(x - 3)^2 + (y + 5)^2 = 4 \qquad\qquad \text{Factor the trinomials on the left-hand side.}$$

From the standard form we see that the center is $(3, -5)$ and the radius is 2.

Now do Exercises 33–44

⟨3⟩ **Systems of Equations**

We first solved systems of nonlinear equations in two variables in Section 13.1. We found the points of intersection of two graphs without drawing the graphs. Here we will solve systems involving circles, parabolas, and lines. In Example 5, we find the points of intersection of a line and a circle.

EXAMPLE 5

Intersection of a line and a circle

Graph both equations of the system

$$(x - 3)^2 + (y + 1)^2 = 9$$

$$y = x - 1$$

on the same coordinate axes, and solve the system by elimination of variables.

Solution

The graph of the first equation is a circle with center $(3, -1)$ and radius 3. The graph of the second equation is a straight line with slope 1 and y-intercept $(0, -1)$. Both graphs are shown in Fig. 13.20. To solve the system by elimination, we substitute $y = x - 1$ into the equation of the circle:

$$(x - 3)^2 + (x - 1 + 1)^2 = 9$$

$$(x - 3)^2 + x^2 = 9$$

$$x^2 - 6x + 9 + x^2 = 9$$

$$2x^2 - 6x = 0$$

$$x^2 - 3x = 0$$

$$x(x - 3) = 0$$

$$x = 0 \quad \text{or} \quad x = 3$$

$$y = -1 \qquad y = 2 \quad \text{Because } y = x - 1$$

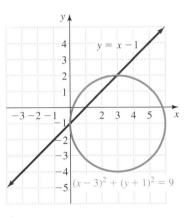

Figure 13.20

Check $(0, -1)$ and $(3, 2)$ in the original system and with the graphs in Fig. 13.20. The solution set is $\{(0, -1), (3, 2)\}$.

Now do Exercises 45–50

Warm-Ups ▼

Fill in the blank.

1. A _____ is the set of all points in a plane that lie at a fixed distance from a fixed point.

2. The _____ of a circle is the fixed point in the definition.

3. The _____ is the fixed distance in the definition of a circle.

4. The equation $(x - h)^2 + (y - k)^2 = r^2$ (for $r > 0$) is the standard equation for a _____ with _____ (h, k) and _____ r.

True or false?

5. The radius of a circle can be any nonzero real number.

6. The coordinates for the center satisfy the equation for the circle.

7. The center for $x^2 + y^2 = 4$ is the origin.

8. The radius for $x^2 + y^2 = 9$ is 9.

9. The center for $(x - 3)^2 + (y + 4)^2 = 25$ is $(3, -4)$.

10. The center for $x^2 + y^2 + 6y - 4 = 0$ is on the y-axis.

‹1› The Equation of a Circle

Write the standard equation for each circle with the given center and radius. See Example 1.

1. Center $(0, 0)$, radius 4

2. Center $(0, 0)$, radius 3

3. Center $(0, 3)$, radius 5

4. Center $(2, 0)$, radius 3

5. Center $(1, -2)$, radius 9

6. Center $(-3, 5)$, radius 4

7. Center $(0, 0)$, radius $\sqrt{3}$

8. Center $(0, 0)$, radius $\sqrt{2}$

9. Center $(-6, -3)$, radius $\dfrac{1}{2}$

10. Center $(-3, -5)$, radius $\dfrac{1}{4}$

11. Center $\left(\dfrac{1}{2}, \dfrac{1}{3}\right)$, radius 0.1

12. Center $\left(-\dfrac{1}{2}, 3\right)$, radius 0.2

Find the center and radius for each circle. See Example 2.

13. $x^2 + y^2 = 1$

14. $x^2 + (y - 1)^2 = 9$

15. $(x - 3)^2 + (y - 5)^2 = 2$

16. $(x + 3)^2 + (y - 7)^2 = 6$

17. $x^2 + \left(y - \dfrac{1}{2}\right)^2 = \dfrac{1}{2}$

18. $5x^2 + 5y^2 = 5$

19. $4x^2 + 4y^2 = 9$

20. $9x^2 + 9y^2 = 49$

21. $3 - y^2 = (x - 2)^2$

22. $9 - x^2 = (y + 1)^2$

Sketch the graph of each equation. See Example 3.

23. $x^2 + y^2 = 9$

24. $x^2 + y^2 = 16$

25. $x^2 + (y - 3)^2 = 9$

26. $(x - 4)^2 + y^2 = 16$

27. $(x + 1)^2 + (y - 1)^2 = 2$

28. $(x - 2)^2 + (y + 2)^2 = 8$

29. $(x - 4)^2 + (y + 3)^2 = 16$

30. $(x - 3)^2 + (y - 7)^2 = 25$

31. $\left(x - \dfrac{1}{2}\right)^2 + \left(y + \dfrac{1}{2}\right)^2 = \dfrac{1}{4}$

32. $\left(x + \dfrac{1}{3}\right)^2 + y^2 = \dfrac{1}{9}$

⟨2⟩ Equations Not in Standard Form

Rewrite each equation in the standard form for the equation of a circle, and identify its center and radius. See Example 4.

33. $x^2 + 4x + y^2 + 6y = 0$

34. $x^2 - 10x + y^2 + 8y = 0$

35. $x^2 - 2x + y^2 - 4y - 3 = 0$

36. $x^2 - 6x + y^2 - 2y + 9 = 0$

37. $x^2 + y^2 = 8y + 10x - 32$

38. $x^2 + y^2 = 8x - 10y$

39. $x^2 - x + y^2 + y = 0$

40. $x^2 - 3x + y^2 = 0$

41. $x^2 - 3x + y^2 - y = 1$

42. $x^2 - 5x + y^2 + 3y = 2$

43. $x^2 - \dfrac{2}{3}x + y^2 + \dfrac{3}{2}y = 0$

44. $x^2 + \dfrac{1}{3}x + y^2 - \dfrac{2}{3}y = \dfrac{1}{9}$

⟨3⟩ Systems of Equations

Graph both equations of each system on the same coordinate axes. Solve the system by elimination of variables to find all points of intersection of the graphs. See Example 5.

45. $x^2 + y^2 = 10$
$y = 3x$

46. $x^2 + y^2 = 4$
$y = x - 2$

47. $x^2 + y^2 = 9$
$y = x^2 - 3$

48. $x^2 + y^2 = 4$
$y = x^2 - 2$

49. $(x - 2)^2 + (y + 3)^2 = 4$
$y = x - 3$

50. $(x + 1)^2 + (y - 4)^2 = 17$
$y = x + 2$

Miscellaneous

Solve each problem.

51. Determine all points of intersection of the circle $(x - 1)^2 + (y - 2)^2 = 4$ with the y-axis.

52. Determine the points of intersection of the circle $x^2 + (y - 3)^2 = 25$ with the x-axis.

53. Find the radius of the circle that has center $(2, -5)$ and passes through the origin.

54. Find the radius of the circle that has center $(-2, 3)$ and passes through $(3, -1)$.

55. Determine the equation of the circle that is centered at $(2, 3)$ and passes through $(-2, -1)$.

56. Determine the equation of the circle that is centered at $(3, 4)$ and passes through the origin.

57. Find all points of intersection of the circles $x^2 + y^2 = 9$ and $(x - 5)^2 + y^2 = 9$.

58. A donkey is tied at the point $(2, -3)$ on a rope of length 12. Turnips are growing at the point $(6, 7)$. Can the donkey reach them?

59. *Volume of a flute.* The volume of air in a flute is a critical factor in determining its pitch. A cross section of a Renaissance flute in C is shown in the accompanying figure. If the length of the flute is 2874 millimeters, then what is the volume of air in the flute [to the nearest cubic millimeter (mm^3)]? (*Hint:* Use the formula for the volume of a cylinder.)

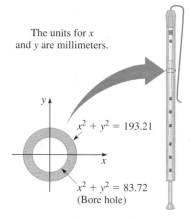

The units for x and y are millimeters.

$x^2 + y^2 = 193.21$

$x^2 + y^2 = 83.72$
(Bore hole)

Figure for Exercises 59 and 60

60. *Flute reproduction.* To make the smaller C# flute, Friedrich von Huene multiplies the length and cross-sectional area of the flute of Exercise 59 by 0.943. Find the equation for the bore hole (centered at the origin) and the volume of air in the C# flute.

Graph each equation.

61. $x^2 + y^2 = 0$

62. $x^2 - y^2 = 0$

63. $y = \sqrt{1 - x^2}$

64. $y = -\sqrt{1 - x^2}$

Getting More Involved

65. *Cooperative learning*

The equation of a circle is a special case of the general equation $Ax^2 + Bx + Cy^2 + Dy = E$, where A, B, C, D, and E are real numbers. Working in small groups, find restrictions that must be placed on A, B, C, D, and E so that the graph of this equation is a circle. What does the graph of $x^2 + y^2 = -9$ look like?

66. *Discussion*

Suppose lighthouse A is located at the origin and lighthouse B is located at coordinates $(0, 6)$. The captain of a ship has determined that the ship's distance from lighthouse A is 2 and its distance from lighthouse B is 5. What are the possible coordinates for the location of the ship?

Graphing Calculator Exercises

Graph each relation on a graphing calculator by solving for y and graphing two functions.

67. $x^2 + y^2 = 4$

68. $(x - 1)^2 + (y + 2)^2 = 1$

69. $x = y^2$

70. $x = (y + 2)^2 - 1$

71. $x = y^2 + 2y + 1$

72. $x = 4y^2 + 4y + 1$

Mid-Chapter **Quiz** │ **Sections 13.1 through 13.3** │ **Chapter 13**

Solve each system.

1. $x - y = -6$
$y = x^2$

2. $y = |x - 2|$
$y = \dfrac{1}{2}x + 2$

3. $x^2 + y^2 = 7$
$x^2 - y^2 = 1$

4. $x^2 + y^2 = 13$
$y = x^2 - 1$

Find the vertex, focus, directrix, axis of symmetry, and opening for each parabola.

5. $y = \dfrac{1}{8}x^2 - 2x + 1$

6. $y = -(x - 3)^2 + 4$

7. $x = (y - 2)^2$

Miscellaneous.

8. Find the distance between the points $(-2, 6)$ and $(-4, 5)$.

9. Find the midpoint of the line segment with endpoints $(-5, 7)$ and $(-9, -3)$.

10. Write the equation $y = 4x^2 - 8x + 2$ in the form $y = a(x - h)^2 + k$.

11. Write the standard equation for a circle with center $(-4, 5)$ and radius 10.

12. Find the center and radius of the circle $y^2 + (x - 2)^2 = 7$.

13. Write $x^2 + 4x + y^2 - 10y = 1$ in the standard form for the equation of a circle.

14. Find two numbers that have a sum of 10 and a product of 12.

13.4 The Ellipse and Hyperbola

In this section, we study the remaining two conic sections: the ellipse and the hyperbola.

In This Section

⟨1⟩ **The Ellipse**

⟨2⟩ **The Hyperbola**

Figure 13.21

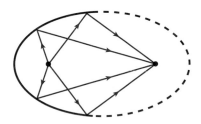

Figure 13.22

⟨1⟩ The Ellipse

An ellipse can be obtained by intersecting a plane and a cone, as was shown in Fig. 13.3. We can also give a definition of an ellipse in terms of points and distance.

> **Ellipse**
>
> An **ellipse** is the set of all points in a plane such that the sum of their distances from two fixed points is a constant. Each fixed point is called a **focus** (plural: foci).

An easy way to draw an ellipse is illustrated in Fig. 13.21. A string is attached at two fixed points, and a pencil is used to take up the slack. As the pencil is moved around the paper, the sum of the distances of the pencil point from the two fixed points remains constant. Of course, the length of the string is that constant. You may wish to try this.

Like the parabola, the ellipse also has interesting reflecting properties. All light or sound waves emitted from one focus are reflected off the ellipse to concentrate at the other focus (see Fig. 13.22). This property is used in light fixtures where a concentration of light at a point is desired or in a whispering gallery such as Statuary Hall in the U.S. Capitol Building.

The orbits of the planets around the sun and satellites around the earth are elliptical. For the orbit of the earth around the sun, the sun is at one focus. For the elliptical path of an earth satellite, the earth is at one focus and a point in space is the other focus.

Figure 13.23 shows an ellipse with foci $(c, 0)$ and $(-c, 0)$. The origin is the center of this ellipse. In general, the **center** of an ellipse is a point midway between the foci. The ellipse in Fig. 13.23 has x-intercepts at $(a, 0)$ and $(-a, 0)$ and y-intercepts at

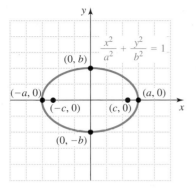

Figure 13.23

(0, b) and (0, $-b$). The distance formula can be used to write the following equation for this ellipse. (See Exercise 61.)

> ### Equation of an Ellipse Centered at the Origin
>
> An ellipse centered at (0, 0) with foci at ($\pm c$, 0) and constant sum $2a$ has equation
>
> $$\frac{x^2}{a^2} + \frac{y^2}{b^2} = 1,$$
>
> where a, b, and c are positive real numbers with $c^2 = a^2 - b^2$.

To draw a "nice-looking" ellipse, we would locate the foci and use string as shown in Fig. 13.21. We can get a rough sketch of an ellipse centered at the origin by using the x- and y-intercepts only.

E X A M P L E   1

Graphing an ellipse

Find the x- and y-intercepts for the ellipse, and sketch its graph.

$$\frac{x^2}{9} + \frac{y^2}{4} = 1$$

Solution

To find the y-intercepts, let $x = 0$ in the equation:

$$\frac{0}{9} + \frac{y^2}{4} = 1$$

$$\frac{y^2}{4} = 1$$

$$y^2 = 4$$

$$y = \pm 2$$

To find the x-intercepts, let $y = 0$. We get $x = \pm 3$. The four intercepts are (0, 2), (0, -2), (3, 0), and (-3, 0). Plot the intercepts, and draw an ellipse through them as in Fig. 13.24.

‹ **Calculator Close-Up** ›

To graph the ellipse in Example 1, graph

$$y_1 = \sqrt{4 - 4x^2/9}$$

and

$$y_2 = -y_1.$$

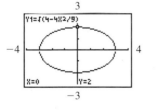

‹ **Helpful Hint** ›

When sketching ellipses or circles by hand, use your hand like a compass and rotate your paper as you draw the curve.

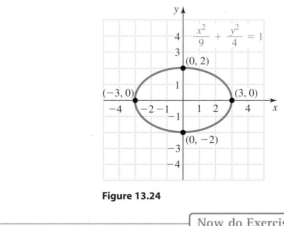

Figure 13.24

> Now do Exercises 1–14

Ellipses, like circles, may be centered at any point in the plane. To get the equation of an ellipse centered at (h, k), we replace x by $x - h$ and y by $y - k$ in the equation of the ellipse centered at the origin.

Equation of an Ellipse Centered at (h, k)

An ellipse centered at (h, k) has equation

$$\frac{(x - h)^2}{a^2} + \frac{(y - k)^2}{b^2} = 1,$$

where a and b are positive real numbers.

E X A M P L E   **2**

An ellipse with center (h, k)

Sketch the graph of the ellipse:

$$\frac{(x - 1)^2}{9} + \frac{(y + 2)^2}{4} = 1$$

Solution

The graph of this ellipse is exactly the same size and shape as the ellipse

$$\frac{x^2}{9} + \frac{y^2}{4} = 1,$$

which was graphed in Example 1. However, the center for

$$\frac{(x - 1)^2}{9} + \frac{(y + 2)^2}{4} = 1$$

is $(1, -2)$. The denominator 9 is used to determine that the ellipse passes through points that are three units to the right and three units to the left of the center: $(4, -2)$ and $(-2, -2)$. See Fig. 13.25. The denominator 4 is used to determine that the ellipse passes through points that are two units above and two units below the center: $(1, 0)$ and $(1, -4)$. We draw an ellipse using these four points, just as we did for an ellipse centered at the origin.

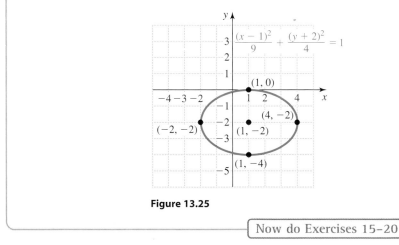

Figure 13.25

Now do Exercises 15–20

⟨2⟩ The Hyperbola

A hyperbola is the curve that occurs at the intersection of a cone and a plane, as was shown in Fig. 13.3 in Section 13.2. A hyperbola can also be defined in terms of points and distance.

Hyperbola

A **hyperbola** is the set of all points in the plane such that the difference of their distances from two fixed points (foci) is constant.

Like the parabola and the ellipse, the hyperbola also has reflecting properties. If a light ray is aimed at one focus, it is reflected off the hyperbola and goes to the other focus, as shown in Fig. 13.26. Hyperbolic mirrors are used in conjunction with parabolic mirrors in telescopes.

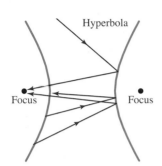

Figure 13.26

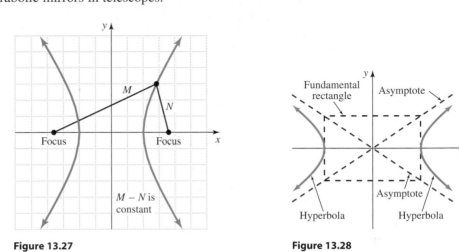

Figure 13.27

Figure 13.28

The definitions of a hyperbola and an ellipse are similar, and so are their equations. However, their graphs are very different. Figure 13.27 shows a hyperbola in which the distance from a point on the hyperbola to the closer focus is N and the distance to the farther focus is M. The value $M - N$ is the same for every point on the hyperbola.

A hyperbola has two parts called **branches.** These branches look like parabolas, but they are not parabolas. The branches of the hyperbola shown in Fig. 13.28 get closer and closer to the dashed lines, called **asymptotes,** but they never intersect them. The asymptotes are used as guidelines in sketching a hyperbola. The asymptotes are found by extending the diagonals of the **fundamental rectangle,** shown in Fig. 13.28. The key to drawing a hyperbola is getting the fundamental rectangle and extending its diagonals to get the asymptotes. You will learn how to find the fundamental rectangle from the equation of a hyperbola. The hyperbola in Fig. 13.28 opens to the left and right.

If we start with foci at $(\pm c, 0)$ and a positive number a, then we can use the definition of a hyperbola to derive the following equation of a hyperbola in which the constant difference between the distances to the foci is $2a$.

Equation of a Hyperbola Centered at (0, 0) Opening Left and Right

A hyperbola centered at $(0, 0)$ with foci $(c, 0)$ and $(-c, 0)$ and constant difference $2a$ has equation

$$\frac{x^2}{a^2} - \frac{y^2}{b^2} = 1,$$

where a, b, and c are positive real numbers such that $c^2 = a^2 + b^2$.

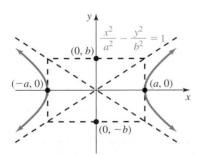

Figure 13.29

The graph of a general equation for a hyperbola is shown in Fig. 13.29. Notice that the fundamental rectangle extends to the x-intercepts along the x-axis and extends b units above and below the origin along the y-axis. Use the following procedure for graphing a hyperbola centered at the origin and opening to the left and to the right.

Strategy for Graphing a Hyperbola Centered at the Origin, Opening Left and Right

To graph the hyperbola $\frac{x^2}{a^2} - \frac{y^2}{b^2} = 1$:

1. Locate the x-intercepts at $(a, 0)$ and $(-a, 0)$.
2. Draw the fundamental rectangle through $(\pm a, 0)$ and $(0, \pm b)$.
3. Draw the extended diagonals of the rectangle to use as asymptotes.
4. Draw the hyperbola to the left and right approaching the asymptotes.

E X A M P L E   3

‹ Calculator Close-Up ›

To graph the hyperbola and its asymptotes from Example 3, graph

$$y_1 = \sqrt{x^2/4 - 9}, \quad y_2 = -y_1,$$
$$y_3 = 0.5x, \quad \text{and} \quad y_4 = -y_3.$$

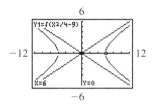

A hyperbola opening left and right

Sketch the graph of $\frac{x^2}{36} - \frac{y^2}{9} = 1$, and find the equations of its asymptotes.

Solution

The x-intercepts are $(6, 0)$ and $(-6, 0)$. Draw the fundamental rectangle through these x-intercepts and the points $(0, 3)$ and $(0, -3)$. Extend the diagonals of the fundamental rectangle to get the asymptotes. Now draw a hyperbola passing through the x-intercepts and approaching the asymptotes as shown in Fig. 13.30. From the graph in Fig. 13.30 we see that the slopes of the asymptotes are $\frac{1}{2}$ and $-\frac{1}{2}$. Because the y-intercept for both asymptotes is the origin, their equations are $y = \frac{1}{2}x$ and $y = -\frac{1}{2}x$.

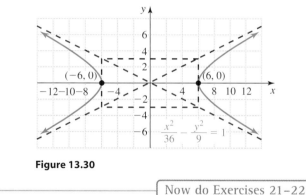

Figure 13.30

Now do Exercises 21–22

If the variables x and y are interchanged in the equation of the hyperbola, then the hyperbola opens up and down.

Equation of a Hyperbola Centered at (0, 0) Opening Up and Down

A hyperbola centered at $(0, 0)$ with foci $(0, c)$ and $(0, -c)$ and constant difference $2b$ has equation

$$\frac{y^2}{b^2} - \frac{x^2}{a^2} = 1,$$

where a, b, and c are positive real numbers such that $c^2 = a^2 + b^2$.

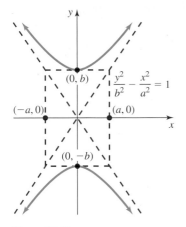

Figure 13.31

The graph of the general equation for a hyperbola opening up and down is shown in Fig. 13.31. Notice that the fundamental rectangle extends to the y-intercepts along the y-axis and extends a units to the left and right of the origin along the x-axis. The procedure for graphing a hyperbola opening up and down follows.

Strategy for Graphing a Hyperbola Centered at the Origin, Opening Up and Down

To graph the hyperbola $\frac{y^2}{b^2} - \frac{x^2}{a^2} = 1$:

1. Locate the y-intercepts at $(0, b)$ and $(0, -b)$.
2. Draw the fundamental rectangle through $(0, \pm b)$ and $(\pm a, 0)$.
3. Draw the extended diagonals of the rectangle to use as asymptotes.
4. Draw the hyperbola opening up and down approaching the asymptotes.

EXAMPLE **4**

A hyperbola opening up and down

Graph the hyperbola $\frac{y^2}{9} - \frac{x^2}{4} = 1$, and find the equations of its asymptotes.

Solution

If $y = 0$, we get

$$-\frac{x^2}{4} = 1$$

$$x^2 = -4.$$

Because this equation has no real solution, the graph has no x-intercepts. Let $x = 0$ to find the y-intercepts:

$$\frac{y^2}{9} = 1$$

$$y^2 = 9$$

$$y = \pm 3$$

The y-intercepts are $(0, 3)$ and $(0, -3)$, and the hyperbola opens up and down. From $a^2 = 4$ we get $a = 2$. So the fundamental rectangle extends to the intercepts $(0, 3)$ and $(0, -3)$ on the y-axis and to the points $(2, 0)$ and $(-2, 0)$ along the x-axis. We extend the diagonals of the rectangle and draw the graph of the hyperbola as shown in Fig. 13.32. From the graph in Fig. 13.32 we see that the asymptotes have slopes $\frac{3}{2}$ and $-\frac{3}{2}$. Because the y-intercept for both asymptotes is the origin, their equations are $y = \frac{3}{2}x$ and $y = -\frac{3}{2}x$.

Now do Exercises 23–28

‹ **Helpful Hint** ›

We could include here general formulas for the equations of the asymptotes, but that is not necessary. It is easier first to draw the asymptotes as suggested and then to figure out their equations by looking at the graph.

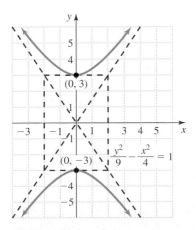

Figure 13.32

E X A M P L E **5**

A hyperbola not in standard form

Sketch the graph of the hyperbola $4x^2 - y^2 = 4$.

Solution

First write the equation in standard form. Divide each side by 4 to get

$$x^2 - \frac{y^2}{4} = 1.$$

There are no y-intercepts. If $y = 0$, then $x = \pm 1$. The hyperbola opens left and right with x-intercepts at $(1, 0)$ and $(-1, 0)$. The fundamental rectangle extends to the intercepts along the x-axis and to the points $(0, 2)$ and $(0, -2)$ along the y-axis. We extend the diagonals of the rectangle for the asymptotes and draw the graph as shown in Fig. 13.33.

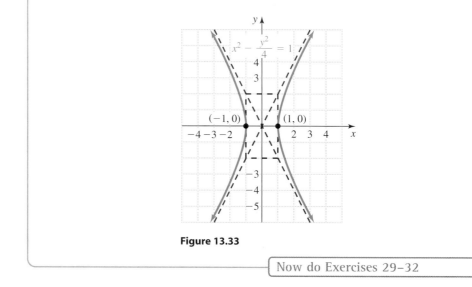

Figure 13.33

> Now do Exercises 29–32

Like circles and ellipses, hyperbolas may be centered at any point in the plane. To get the equation of a hyperbola centered at (h, k), we replace x by $x - h$ and y by $y - k$ in the equation of the hyperbola centered at the origin.

Equation of a Hyperbola Centered at (h, k)

A hyperbola centered at (h, k) has one of the following equations depending on which way it opens.

Opening left and right: Opening up and down:

$$\frac{(x - h)^2}{a^2} - \frac{(y - k)^2}{b^2} = 1 \qquad \frac{(y - k)^2}{b^2} - \frac{(x - h)^2}{a^2} = 1$$

E X A M P L E   **6**

Graphing a hyperbola centered at (h, k)

Graph the hyperbola $\frac{(x-3)^2}{16} - \frac{(y+1)^2}{4} = 1$.

Solution

This hyperbola is centered at $(3, -1)$ and opens left and right. It is a transformation of the graph of $\frac{x^2}{16} - \frac{y^2}{4} = 1$. The fundamental rectangle for $\frac{x^2}{16} - \frac{y^2}{4} = 1$ is centered at the origin and goes through $(\pm 4, 0)$ and $(0, \pm 2)$. So draw a fundamental rectangle centered at $(3, -1)$ that extends four units to the right and left and two units up and down as shown in Fig. 13.34. Draw the asymptotes through the vertices of the fundamental rectangle and the hyperbola opening to the left and right.

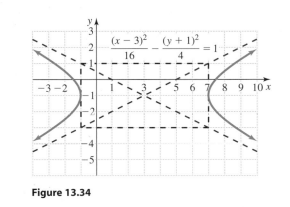

Figure 13.34

Now do Exercises 33–38

Warm-Ups ▼

Fill in the blank.

1. An _____ is the set of all points in a plane such that the sum of their distances from two fixed points is constant.

2. The ____ of an ellipse are the two fixed points in the definition.

3. The _____ of an ellipse is the point that is midway between the foci.

4. A _____ is the set of all points in a plane such that the difference of their distances from two fixed points is constant.

5. A hyperbola is made up of two separate curves called _____.

6. A hyperbola approaches two lines called _____.

True or false?

7. The graph of $\frac{x^2}{36} + \frac{y^2}{25} = 1$ is an ellipse.

8. The x-intercepts for $\frac{x^2}{36} + \frac{y^2}{25} = 1$ are $(5, 0)$ and $(-5, 0)$.

9. If the foci of an ellipse coincide, then the ellipse is a circle.

10. The graph of $\frac{x^2}{36} - \frac{y^2}{25} = 1$ is a hyperbola.

11. The x-intercepts for $\frac{x^2}{36} - \frac{y^2}{25} = 1$ are $(6, 0)$ and $(-6, 0)$.

12. The graph of $4x^2 - y^2 = 4$ is a hyperbola.

Exercises

‹ 1 › **The Ellipse**

Sketch the graph of each ellipse. See Example 1.

1. $\dfrac{x^2}{9} + \dfrac{y^2}{4} = 1$

2. $\dfrac{x^2}{9} + \dfrac{y^2}{16} = 1$

3. $\dfrac{x^2}{9} + y^2 = 1$

4. $x^2 + \dfrac{y^2}{4} = 1$

5. $\dfrac{x^2}{36} + \dfrac{y^2}{25} = 1$

6. $\dfrac{x^2}{25} + \dfrac{y^2}{49} = 1$

7. $\dfrac{x^2}{24} + \dfrac{y^2}{5} = 1$

8. $\dfrac{x^2}{6} + \dfrac{y^2}{17} = 1$

9. $9x^2 + 16y^2 = 144$

10. $9x^2 + 25y^2 = 225$

11. $25x^2 + y^2 = 25$

12. $x^2 + 16y^2 = 16$

13. $4x^2 + 9y^2 = 1$

14. $25x^2 + 16y^2 = 1$

Sketch the graph of each ellipse. See Example 2.

15. $\dfrac{(x-3)^2}{4} + \dfrac{(y-1)^2}{9} = 1$

16. $\dfrac{(x+5)^2}{49} + \dfrac{(y-2)^2}{25} = 1$

17. $\dfrac{(x+1)^2}{16} + \dfrac{(y-2)^2}{25} = 1$

18. $\dfrac{(x-3)^2}{36} + \dfrac{(y+4)^2}{64} = 1$

19. $(x - 2)^2 + \dfrac{(y + 1)^2}{36} = 1$ **20.** $\dfrac{(x + 3)^2}{9} + (y + 1)^2 = 1$

29. $9x^2 - 16y^2 = 144$ **30.** $9x^2 - 25y^2 = 225$

31. $x^2 - y^2 = 1$ **32.** $y^2 - x^2 = 1$

⟨2⟩ **The Hyperbola**

Graph each hyperbola, and write the equations of its asymptotes.
See Examples 3–5. See the Strategies for Graphing a Hyperbola
boxes on pages 872 and 873.

21. $\dfrac{x^2}{4} - \dfrac{y^2}{9} = 1$ **22.** $\dfrac{x^2}{16} - \dfrac{y^2}{9} = 1$

Sketch the graph of each hyperbola. See Example 6.

33. $\dfrac{(x - 2)^2}{4} - (y + 1)^2 = 1$ **34.** $(x + 3)^2 - \dfrac{(y - 1)^2}{4} = 1$

23. $\dfrac{y^2}{4} - \dfrac{x^2}{25} = 1$ **24.** $\dfrac{y^2}{9} - \dfrac{x^2}{16} = 1$

35. $\dfrac{(x + 1)^2}{16} - \dfrac{(y - 1)^2}{9} = 1$ **36.** $\dfrac{(x - 2)^2}{9} - \dfrac{(y + 2)^2}{16} = 1$

25. $\dfrac{x^2}{25} - y^2 = 1$ **26.** $x^2 - \dfrac{y^2}{9} = 1$

27. $x^2 - \dfrac{y^2}{25} = 1$ **28.** $\dfrac{x^2}{9} - y^2 = 1$

37. $\dfrac{(y - 2)^2}{9} - \dfrac{(x - 4)^2}{4} = 1$ **38.** $\dfrac{(y + 3)^2}{16} - \dfrac{(x + 1)^2}{9} = 1$

Miscellaneous

Determine whether the graph of each equation is a circle, parabola, ellipse, or hyperbola.

39. $y = x^2 + 1$

40. $x^2 + y^2 = 1$

41. $x^2 - y^2 = 1$

42. $4x^2 + y^2 = 1$

43. $\dfrac{x^2}{2} + y^2 = 1$

44. $x^2 - \dfrac{y^2}{9} = 1$

45. $(x - 2)^2 + (y - 4)^2 = 9$

46. $(x - 2)^2 + y = 9$

Graph both equations of each system on the same coordinate axes. Use elimination of variables to find all points of intersection.

47. $\dfrac{x^2}{4} + \dfrac{y^2}{9} = 1$

$x^2 - \dfrac{y^2}{9} = 1$

48. $x^2 - \dfrac{y^2}{4} = 1$

$\dfrac{x^2}{9} + \dfrac{y^2}{4} = 1$

49. $\dfrac{x^2}{4} + \dfrac{y^2}{16} = 1$

$x^2 + y^2 = 1$

50. $x^2 + \dfrac{y^2}{9} = 1$

$x^2 + y^2 = 4$

51. $x^2 + y^2 = 4$

$x^2 - y^2 = 1$

52. $x^2 + y^2 = 16$

$x^2 - y^2 = 4$

53. $x^2 + 9y^2 = 9$

$x^2 + y^2 = 4$

54. $x^2 + y^2 = 25$

$x^2 + 25y^2 = 25$

55. $x^2 + 9y^2 = 9$
    $y = x^2 - 1$

56. $4x^2 + y^2 = 4$
    $y = 2x^2 - 2$

57. $9x^2 - 4y^2 = 36$
    $2y = x - 2$

58. $25y^2 - 9x^2 = 225$
    $y = 3x + 3$

Applications

Solve each problem.

59. *Marine navigation.* The loran (long-range navigation) system is used by boaters to determine their location at sea. The loran unit on a boat measures the difference in time that it takes for radio signals from pairs of fixed points to reach the boat. The unit then finds the equations of two hyperbolas that pass through the location of the boat. Suppose a boat is located in the first quadrant at the intersection of $x^2 - 3y^2 = 1$ and $4y^2 - x^2 = 1$.

   a) Use the accompanying graph to approximate the location of the boat.

   b) Algebraically find the exact location of the boat.

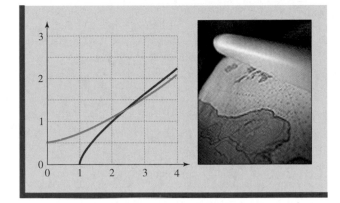

Figure for Exercise 59

60. *Sonic boom.* An aircraft traveling at supersonic speed creates a cone-shaped wave that intersects the ground along a hyperbola, as shown in the accompanying figure. A thunderlike sound is heard at any point on the hyperbola. This sonic boom travels along the ground, following the aircraft. The area where the sonic boom is most noticeable is called the *boom carpet*. The width of the boom carpet is roughly five times the altitude of the aircraft. Suppose the equation of the hyperbola in the figure is

$$\frac{x^2}{400} - \frac{y^2}{100} = 1,$$

where the units are miles and the width of the boom carpet is measured 40 miles behind the aircraft. Find the altitude of the aircraft.

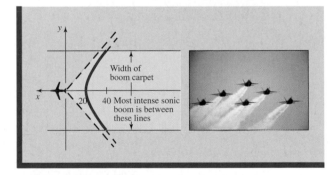

Figure for Exercise 60

Getting More Involved

61. *Cooperative learning*

Let (x, y) be an arbitrary point on an ellipse with foci $(c, 0)$ and $(-c, 0)$ for $c > 0$. The following equation expresses the fact that the distance from (x, y) to $(c, 0)$ plus the distance from (x, y) to $(-c, 0)$ is the constant value $2a$ (for $a > 0$):

$$\sqrt{(x - c)^2 + (y - 0)^2} + \sqrt{(x - (-c))^2 + (y - 0)^2} = 2a$$

Working in groups, simplify this equation. First get the radicals on opposite sides of the equation, and then square both sides twice to eliminate the square roots. Finally, let $b^2 = a^2 - c^2$ to get the equation

$$\frac{x^2}{a^2} + \frac{y^2}{b^2} = 1.$$

62. *Cooperative learning*

Let (x, y) be an arbitrary point on a hyperbola with foci $(c, 0)$ and $(-c, 0)$ for $c > 0$. The following equation expresses the fact that the distance from (x, y) to $(c, 0)$ minus the distance from (x, y) to $(-c, 0)$ is the constant value $2a$ (for $a > 0$):

$$\sqrt{(x - c)^2 + (y - 0)^2} - \sqrt{(x - (-c))^2 + (y - 0)^2} = 2a$$

Working in groups, simplify the equation. You will need to square both sides twice to eliminate the square roots. Finally, let $b^2 = c^2 - a^2$ to get the equation

$$\frac{x^2}{a^2} - \frac{y^2}{b^2} = 1.$$

Graphing Calculator Exercises

63. Graph $y_1 = \sqrt{x^2 - 1}$, $y_2 = -\sqrt{x^2 - 1}$, $y_3 = x$, and $y_4 = -x$ to get the graph of the hyperbola $x^2 - y^2 = 1$ along with its asymptotes. Use the viewing window $-3 \le x \le 3$ and $-3 \le y \le 3$. Notice how the branches of the hyperbola approach the asymptotes.

64. Graph the same four functions in Exercise 63, but use $-30 \le x \le 30$ and $-30 \le y \le 30$ as the viewing window. What happened to the hyperbola?

Math *at Work*   **Kepler's Laws**

With great patience, Danish astronomer Tycho Brahe (1546–1601) made very careful observations of the motion of the planets in the sky. Brahe tried to explain the orbits of the planets using circles. His assistant, Johannes Kepler (1571–1630), studied Tycho's tables and came up with three laws that better explained the motion of the planets. Kepler's first law went contrary to Brahe's theory and states that each planet moves around the sun in an elliptical orbit with the sun at one focus of the ellipse.

The second law states that the line joining a planet with the sun sweeps out equal areas in equal times. A planet moves faster when it is closer to the sun and slower when it is far from the sun. So the planet illustrated in the accompanying figure moves from A to B in the same time that it moves from C to D, even though the distance from A to B is greater. According to Kepler's law, the shaded areas in the figure are equal.

The third law states that the square of the period of a planet orbiting the sun is equal to the cube of the mean distance from the planet to the sun. In symbols, $P^2 = a^3$, where P is the number of earth years that it takes for the planet to orbit the sun, and a is the mean distance from the planet to the sun in astronomical units (AU). (One AU is the mean distance from the earth to the sun.) $P^2 = a^3$ can be written as $P = a^{3/2}$ or $a = P^{2/3}$ and used to find the period or the distance. For example, the period of Mars is observed to be 1.88 years. So the mean distance from Mars to the sun is $1.88^{2/3}$ or 1.53 AU. The mean distance from Pluto to the sun is observed to be 39.44 AU, so Pluto takes $39.44^{3/2}$ or 247.69 years to complete one orbit of the sun.

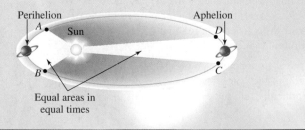

Equal areas in
equal times

13.5 Second-Degree Inequalities

In This Section

⟨1⟩ **Graphing a Second-Degree Inequality**

⟨2⟩ **Systems of Inequalities**

In this section we graph second-degree inequalities and systems of inequalities involving second-degree inequalities.

⟨1⟩ Graphing a Second-Degree Inequality

A second-degree inequality is an inequality involving squares of at least one of the variables. Changing the equal sign to an inequality symbol for any of the equations of the conic sections gives us a second-degree inequality. Second-degree inequalities are graphed in the same manner as linear inequalities.

EXAMPLE **1**

A second-degree inequality

Graph the inequality $y < x^2 + 2x - 3$.

Solution

We first graph $y = x^2 + 2x - 3$. This parabola has x-intercepts at $(1, 0)$ and $(-3, 0)$, a y-intercept at $(0, -3)$, and a vertex at $(-1, -4)$. The graph of the parabola is drawn with a dashed line, as shown in Fig. 13.35. The graph of the parabola divides the plane into two regions. Every point on one side of the parabola satisfies the inequality $y < x^2 + 2x - 3$, and every point on the other side satisfies the inequality $y > x^2 + 2x - 3$. To determine which side is which, we test a point that is not on the parabola, say $(0, 0)$. Because

$$0 < 0^2 + 2 \cdot 0 - 3$$

is false, the region not containing the origin is shaded, as in Fig. 13.35.

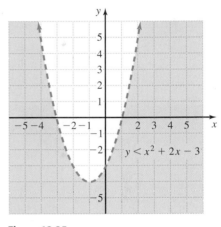

Figure 13.35

Now do Exercises 1–6

EXAMPLE **2**

A second-degree inequality

Graph the inequality $x^2 + y^2 \leq 9$.

Solution

The graph of $x^2 + y^2 = 9$ is a circle of radius 3 centered at the origin. The circle divides the plane into two regions. Every point in one region satisfies $x^2 + y^2 < 9$, and every point in the other region satisfies $x^2 + y^2 > 9$. To identify the regions, we pick a point and test it. Select $(0, 0)$. The inequality

$$0^2 + 0^2 < 9$$

is true. Because $(0, 0)$ is inside the circle, all points inside the circle satisfy $x^2 + y^2 < 9$. Points outside the circle satisfy $x^2 + y^2 > 9$. Because the inequality symbol is $\leq$, the circle is included in the solution set. So the circle is drawn as a solid curve as shown in Fig. 13.36, and the area inside the circle is shaded.

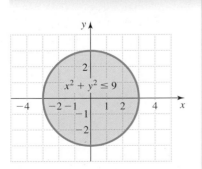

Figure 13.36

Now do Exercises 7–10

EXAMPLE **3**

A second-degree inequality

Graph the inequality $\frac{x^2}{4} - \frac{y^2}{9} > 1$.

Solution

First graph the hyperbola $\frac{x^2}{4} - \frac{y^2}{9} = 1$. Because the hyperbola shown in Fig. 13.37 divides the plane into three regions, we select a test point in each region and check to see whether it satisfies the inequality. Testing the points $(-3, 0)$, $(0, 0)$, and $(3, 0)$ gives us the inequalities

$$\frac{(-3)^2}{4} - \frac{0^2}{9} > 1, \qquad \frac{0^2}{4} - \frac{0^2}{9} > 1, \qquad \text{and} \qquad \frac{3^2}{4} - \frac{0^2}{9} > 1.$$

Because only the first and third inequalities are correct, we shade only the regions containing $(3, 0)$ and $(-3, 0)$, as shown in Fig. 13.37.

> Now do Exercises 11–22

Figure 13.37

⟨2⟩ Systems of Inequalities

A point is in the solution set to a system of inequalities if it satisfies all inequalities of the system. We graph a system of inequalities by first determining the graph of each inequality and then finding the intersection of the graphs.

EXAMPLE **4**

Systems of second-degree inequalities

Graph the system of inequalities:

$$\frac{y^2}{4} - \frac{x^2}{9} > 1 \qquad \frac{x^2}{9} + \frac{y^2}{16} < 1$$

Solution

Figure 13.38(a) shows the graph of the first inequality. Figure 13.38(b) shows the graphs of both inequalities on the same coordinate system. Points that are shaded for both inequalities in Fig. 13.38(b) satisfy the system. Figure 13.38(c) shows the graph of the system.

> Now do Exercises 27–46

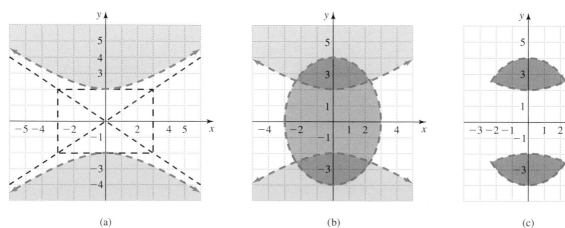

(a)   (b)   (c)

Figure 13.38

Warm-Ups ▼

Fill in the blank.

1. The graph of $x^2 + 9y^2 = 1$ is a(n) _____.
2. The graph of $x^2 - y^2 = 9$ is a(n) _____.
3. The graph of $x^2 + y^2 = 9$ is a(n) _____.
4. The graph of $x^2 + y = 9$ is a(n) _____.
5. The graph of $x + y = 9$ is a(n) ___.

True or false?

6. The point $(0, 0)$ satisfies $2x^2 - y < 3$.
7. The graph of $y > x^2 - 3x + 2$ is a parabola.
8. We can use the origin as a test point when graphing $x^2 > y$.
9. The graph of $x^2 + y^2 < 4$ is the region inside a circle of radius 2.
10. The point $(0, 4)$ satisfies $x^2 - y^2 < 1$ and $y > x^2 - 2x + 3$.

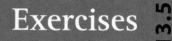

Exercises 13.5

‹ **Study Tips** ›

- Don't be discouraged by the amount of material in this text that you did not cover in this course.
- Textbooks are written for a wide audience. Most instructors skip some topics.

‹1› **Graphing a Second-Degree Inequality**

Graph each inequality. See Examples 1–3.

1. $y > x^2$
2. $y \le x^2 + 1$
3. $y < x^2 - x$
4. $y > x^2 + x$
5. $y > x^2 - x - 2$
6. $y < x^2 + x - 6$
7. $x^2 + y^2 < 9$
8. $x^2 + y^2 > 16$

9. $x^2 + 4y^2 > 4$ **10.** $4x^2 + y^2 \le 4$ **19.** $y^2 - x^2 \le 1$ **20.** $x^2 - y^2 > 1$

11. $4x^2 - 9y^2 < 36$ **12.** $25x^2 - 4y^2 > 100$ **21.** $x > y$ **22.** $x < 2y - 1$

⟨2⟩ Systems of Inequalities

Determine whether the ordered pair $(3, -4)$ *satisfies each system of inequalities.*

13. $(x - 2)^2 + (y - 3)^2 < 4$ **14.** $(x + 1)^2 + (y - 2)^2 > 1$

23. $x^2 + y^2 \le 25$
$\quad\ y \le x^2$

24. $x^2 - y^2 < 1$
$\quad\ y < x - 5$

25. $x - y > 1$
$\quad\ y > (x - 2)^2 + 3$

15. $x^2 + y^2 > 1$ **16.** $x^2 + y^2 < 25$

26. $4x^2 + y^2 \le 36$
$\quad\ x^2 + y^2 \ge 25$

Graph the solution set to each system of inequalities. See Example 4.

27. $x^2 + y^2 < 9$ **28.** $x^2 + y^2 > 1$
$\quad\ y > x$ $\quad\ x > y$

17. $4x^2 - y^2 > 4$ **18.** $x^2 - 9y^2 \le 9$

29. $x^2 - y^2 > 1$
  $x^2 + y^2 < 4$

30. $y^2 - x^2 < 1$
  $x^2 + y^2 > 9$

37. $x - y < 0$
  $y + x^2 < 1$

38. $y + 1 > x^2$
  $x + y < 2$

31. $y > x^2 + x$
  $y < 5$

32. $y > x^2 + x - 6$
  $y < x + 3$

39. $y < 5x - x^2$
  $x^2 + y^2 < 9$

40. $y < x^2 + 5x$
  $x^2 + y^2 < 16$

33. $y \geq x + 2$
  $y \leq 2 - x$

34. $y \geq 2x - 3$
  $y \leq 3 - 2x$

41. $y \geq 3$
  $x \leq 1$

42. $x > -3$
  $y < 2$

35. $4x^2 - y^2 < 4$
  $x^2 + 4y^2 > 4$

36. $x^2 - 4y^2 < 4$
  $x^2 + 4y^2 > 4$

43. $4y^2 - 9x^2 < 36$
  $x^2 + y^2 < 16$

44. $25y^2 - 16x^2 < 400$
  $x^2 + y^2 > 4$

45. $y < x^2$
        $x^2 + y^2 < 1$

46. $y > x^2$
        $4x^2 + y^2 < 4$

Photo for Exercise 47

Solve the problem.

47. *Buried treasure.* An old pirate on his deathbed gave the following description of where he had buried some treasure on a deserted island: "Starting at the large palm tree, I walked to the north and then to the east, and there I buried the treasure. I walked at least 50 paces to get to that spot, but I was not more than 50 paces, as the crow flies, from the large palm tree. I am sure that I walked farther in the northerly direction than in the easterly direction." With the large palm tree at the origin and the positive y-axis pointing to the north, graph the possible locations of the treasure.

Graphing Calculator Exercises

48. Use graphs to find an ordered pair that is in the solution set to the system of inequalities:

$$y > x^2 - 2x + 1$$
$$y < -1.1(x - 4)^2 + 5$$

Verify that your answer satisfies both inequalities.

49. Use graphs to find the solution set to the system of inequalities:

$$y > 2x^2 - 3x + 1$$
$$y < -2x^2 - 8x - 1$$

Chapter

13

Wrap-Up

Summary

Nonlinear Systems		**Examples**
Nonlinear systems in two variables	Use substitution or addition to eliminate variables. Nonlinear systems may have several points in the solution set.	$y = x^2$ $x^2 + y^2 = 4$ Substitution: $y + y^2 = 4$

The Distance and Midpoint Formulas		**Examples**
Distance formula	The distance between (x_1, y_1) and (x_2, y_2) is $\sqrt{(x_2 - x_1)^2 + (y_2 - y_1)^2}$.	Distance between $(1, -2)$ and $(3, -4)$ is $\sqrt{2^2 + (-2)^2}$ or $2\sqrt{2}$.
Midpoint formula	The midpoint of the line segment with endpoints (x_1, y_1) and (x_2, y_2) is $(\bar{x}, \bar{y}) = \left(\dfrac{x_1 + x_2}{2}, \dfrac{y_1 + y_2}{2} \right)$.	If $(x_1, y_1) = (1, -2)$ and $(x_2, y_2) = (7, 8)$, then $(\bar{x}, \bar{y}) = (4, 3)$.

Parabola		**Examples**
$y = a(x - h)^2 + k$	Opens upward for $a > 0$, downward for $a < 0$ Vertex at (h, k) To find focus and directrix, use $a = \dfrac{1}{4p}$. Distance from vertex to focus or directrix is $|p|$.	

$$y = \frac{1}{8}(x - 1)^2 - 2$$

$x = a(y - k)^2 + h$	Opens right for $a > 0$, left for $a < 0$ Vertex at (h, k) To find focus and directrix use $a = \dfrac{1}{4p}$. Distance from vertex to focus or directrix is $|p|$.	

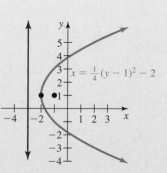

$$x = \frac{1}{4}(y - 1)^2 - 2$$

$y = ax^2 + bx + c$

Opens upward for $a > 0$, downward for $a < 0$

The x-coordinate of the vertex is $\frac{-b}{2a}$.

Find the y-coordinate of the vertex by evaluating $y = ax^2 + bx + c$ for $x = \frac{-b}{2a}$.

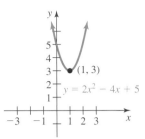

$x = ay^2 + by + c$

Opens right for $a > 0$, left for $a < 0$

The y-coordinate of the vertex is $\frac{-b}{2a}$.

Find the x-coordinate of the vertex by evaluating $x = ay^2 + by + c$ for $y = \frac{-b}{2a}$.

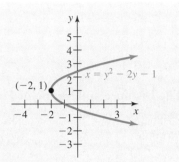

Circle

Examples

Centered at origin
$x^2 + y^2 = r^2$

Center $(0, 0)$
Radius r (for $r > 0$)

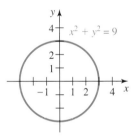

Arbitrary center
$(x - h)^2 + (y - k)^2 = r^2$

Center (h, k)
Radius r (for $r > 0$)

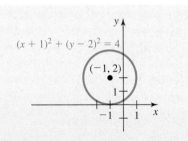

Ellipse

Examples

Centered at origin

$$\frac{x^2}{a^2} + \frac{y^2}{b^2} = 1$$

Center: $(0, 0)$
x-intercepts: $(a, 0)$ and $(-a, 0)$
y-intercepts: $(0, b)$ and $(0, -b)$
Foci: $(\pm c, 0)$ if $a^2 > b^2$ and $c^2 = a^2 - b^2$
       $(0, \pm c)$ if $b^2 > a^2$ and $c^2 = b^2 - a^2$

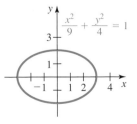

Arbitrary center   Center: (h, k)

$$\frac{(x - h)^2}{a^2} + \frac{(y - k)^2}{b^2} = 1$$

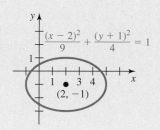

Hyperbola   **Examples**

Opening left and right   Centered at origin: $\dfrac{x^2}{a^2} - \dfrac{y^2}{b^2} = 1$

Center: $(0, 0)$

x-intercepts: $(a, 0)$ and $(-a, 0)$

y-intercepts: none

Centered at (h, k): $\dfrac{(x - h)^2}{a^2} - \dfrac{(y - k)^2}{b^2} = 1$

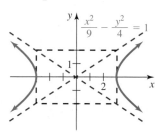

Opening up and down   Centered at origin: $\dfrac{y^2}{b^2} - \dfrac{x^2}{a^2} = 1$

Center: $(0, 0)$

x-intercepts: none

y-intercepts: $(0, b)$ and $(0, -b)$

Centered at (h, k): $\dfrac{(y - k)^2}{b^2} - \dfrac{(x - h)^2}{a^2} = 1$

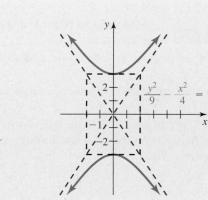

Second-Degree Inequalities   **Examples**

Solution set for   Graph the boundary curve obtained by replacing $x^2 + y^2 < 16$
a single inequality   the inequality symbol by the equal sign.
                          Use test points to determine which regions satisfy
                          the inequality.

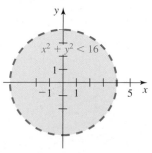

| Solution set for a system of inequalities | Graph the boundary curves. Then select a test point in each region. Shade only the regions for which the test point satisfies all inequalities of the system. | $x^2 + y^2 < 16$
 $y > x^2 - 1$ |

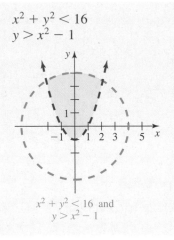

$x^2 + y^2 < 16$ and
$y > x^2 - 1$

Enriching Your Mathematical Word Power

Fill in the blank.

1. An equation whose graph is not a straight line is a(n) _____ equation.

2. A(n) _____ consists of all points in a plane that are equidistant from a point and a line.

3. The fixed point in the definition of a parabola is the _____.

4. The fixed line in the definition of a parabola is the _____.

5. The _____ of a parabola is the midpoint of the line segment that joins the focus and directrix perpendicular to the directrix.

6. A(n) _____ section is a curve obtained by intersecting a cone and a plane.

7. The ____ of symmetry is the line of symmetry of a parabola.

8. A(n) _____ consists of all points in a plane such that the sum of their distances from two fixed points is a constant.

9. A(n) _____ consists of all points in a plane that are a fixed distance from a fixed point.

10. A(n) _____ consists of all points in a plane such that the difference of their distances from two fixed points is a constant.

Review Exercises

13.1 Nonlinear Systems of Equations

Graph both equations on the same set of axes, and then determine the points of intersection of the graphs by solving the system.

1. $y = x^2$
    $y = -2x + 15$

2. $y = \sqrt{x}$
    $y = \dfrac{1}{3}x$

3. $y = 3x$
    $y = \dfrac{1}{x}$

4. $y = |x|$
    $y = -3x + 5$

Solve each system.

5. $xy = 9$
     $y = x$

6. $y = x^2$
     $y = 2x$

7. $x^2 + y^2 = 4$
     $y = \dfrac{1}{3}x^2$

8. $12y^2 - 4x^2 = 9$
     $x = y^2$

9. $x^2 + y^2 = 34$
     $y = x + 2$

10. $y = 2x + 1$
     $xy - y = 5$

11. $y = \log(x - 3)$
     $y = 1 - \log(x)$

12. $y = \left(\dfrac{1}{2}\right)^x$
     $y = 2^{x-1}$

13. $x^4 = 2(12 - y)$
     $y = x^2$

14. $x^2 + 2y^2 = 7$
     $x^2 - 2y^2 = -5$

13.2 The Parabola

Find the distance between each pair of points.

15. $(1, 1), (3, 3)$

16. $(1, 2), (4, 5)$

17. $(-4, 6), (2, -8)$

18. $(-3, -5), (5, -7)$

Find the midpoint and length of the line segment with the given endpoints.

19. $(8, -2)$ and $(2, 6)$

20. $(-9, 4)$ and $(-3, -4)$

21. $(2, -2)$ and $(3, 1)$

22. $(0, 3)$ and $(-1, -1)$

Determine the vertex, axis of symmetry, focus, and directrix for each parabola.

23. $y = x^2 + 3x - 18$

24. $y = x - x^2$

25. $y = x^2 + 3x + 2$

26. $y = -x^2 - 3x + 4$

27. $y = -\dfrac{1}{2}(x - 2)^2 + 3$

28. $y = \dfrac{1}{4}(x + 1)^2 - 2$

Write each equation in the form $y = a(x - h)^2 + k$, and identify the vertex of the parabola.

29. $y = 2x^2 - 8x + 1$

30. $y = -2x^2 - 6x - 1$

31. $y = -\dfrac{1}{2}x^2 - x + \dfrac{1}{2}$

32. $y = \dfrac{1}{4}x^2 + x - 9$

13.3 The Circle

Determine the center and radius of each circle, and sketch its graph.

33. $x^2 + y^2 = 100$

34. $x^2 + y^2 = 20$

35. $(x - 2)^2 + (y + 3)^2 = 81$ **36.** $x^2 + 2x + y^2 = 8$

Sketch the graph of each hyperbola.

47. $\dfrac{x^2}{49} - \dfrac{y^2}{36} = 1$

48. $\dfrac{y^2}{25} - \dfrac{x^2}{49} = 1$

37. $9y^2 + 9x^2 = 4$ **38.** $x^2 + 4x + y^2 - 6y - 3 = 0$

49. $4x^2 - 25y^2 = 100$

Write the standard equation for each circle with the given center and radius.

39. Center $(0, 3)$, radius 6

40. Center $(0, 0)$, radius $\sqrt{6}$

50. $6y^2 - 16x^2 = 96$

41. Center $(2, -7)$, radius 5

42. Center $\left(\frac{1}{2}, -3\right)$, radius $\frac{1}{2}$

13.4 The Ellipse and Hyperbola

Sketch the graph of each ellipse.

43. $\dfrac{x^2}{36} + \dfrac{y^2}{49} = 1$ **44.** $\dfrac{x^2}{25} + y^2 = 1$

13.5 Second-Degree Inequalities

Graph each inequality.

51. $4x - 2y > 3$

45. $25x^2 + 4y^2 = 100$ **46.** $6x^2 + 4y^2 = 24$

52. $y < x^2 - 3x$

53. $y^2 < x^2 - 1$

54. $y^2 < 1 - x^2$

55. $4x^2 + 9y^2 > 36$

56. $x^2 + y > 2x - 1$

Graph the solution set to each system of inequalities.

57. $y < 4x - x^2$
   $x^2 + y^2 < 9$

58. $x^2 - y^2 < 1$
   $y < 1$

59. $4x^2 + 9y^2 > 36$
   $x^2 + y^2 < 9$

60. $y^2 - x^2 > 4$
   $y^2 + 16x^2 < 16$

Miscellaneous

Identify each equation as the equation of a straight line, parabola, circle, hyperbola, or ellipse. Try to do these without rewriting the equations.

61. $x^2 = y^2 + 1$ **62.** $x = y + 1$

63. $x^2 = 1 - y^2$ **64.** $x^2 = y + 1$

65. $x^2 + x = 1 - y^2$ **66.** $(x - 3)^2 + (y + 2)^2 = 7$

67. $x^2 + 4x = 6y - y^2$ **68.** $4x + 6y = 1$

69. $\dfrac{x^2}{3} - \dfrac{y^2}{5} = 1$ **70.** $x^2 + \dfrac{y^2}{3} = 1$

71. $4y^2 - x^2 = 8$ **72.** $9x^2 + y = 9$

Sketch the graph of each equation.

73. $x^2 = 4 - y^2$ **74.** $x^2 = 4y^2 + 4$

75. $x^2 = 4y + 4$ **76.** $x = 4y + 4$

77. $x^2 = 4 - 4y^2$ **78.** $x^2 = 4y - y^2$

79. $x^2 = 4 - (y - 4)^2$

80. $(x - 2)^2 + (y - 4)^2 = 4$

88. Vertex $(1, 2)$ and focus $\left(1, \frac{3}{2}\right)$

89. Vertex $(0, 0)$, passing through $(3, 2)$, and opening upward

90. Vertex $(1, 3)$, passing through $(0, 0)$, and opening downward

Write the equation of the circle with the given features.

81. Centered at the origin and passing through $(3, 4)$

82. Centered at $(2, -3)$ and passing through $(-1, 4)$

83. Centered at $(-1, 5)$ with radius 6

84. Centered at $(0, -3)$ and passing through the origin

Solve each system of equations.

91. $x^2 + y^2 = 25$
$y = -x + 1$

92. $x^2 - y^2 = 1$
$x^2 + y^2 = 7$

93. $4x^2 + y^2 = 4$
$x^2 - y^2 = 21$

94. $y = x^2 + x$
$y = -x^2 + 3x + 12$

Write the equation of the parabola with the given features.

85. Focus $(1, 4)$ and directrix $y = 2$

86. Focus $(-2, 1)$ and directrix $y = 5$

87. Vertex $(0, 0)$ and focus $\left(0, \frac{1}{4}\right)$

Solve each problem.

95. *Perimeter of a rectangle.* A rectangle has a perimeter of 16 feet and an area of 12 square feet. Find its length and width.

96. *Tale of two circles.* Find the radii of two circles such that the difference in areas of the two is 10π square inches and the difference in radii of the two is 2 inches.

Chapter 13 Test

Sketch the graph of each equation.

1. $x^2 + y^2 = 25$

2. $\dfrac{x^2}{16} - \dfrac{y^2}{25} = 1$

3. $y^2 + 4x^2 = 4$

4. $y = x^2 + 4x + 4$

5. $y^2 - 4x^2 = 4$

Graph the solution set to each system of inequalities.

10. $x^2 + y^2 < 9$
$\quad\;\; x^2 - y^2 > 1$

6. $y = -x^2 - 2x + 3$

11. $y < -x^2 + x$
$\quad\;\; y < x - 4$

Solve each system of equations.

Sketch the graph of each inequality.

7. $x^2 \quad y^2 < 9$

12. $y = x^2 - 2x - 8$
$\quad\;\; y = 7 - 4x$

13. $x^2 + y^2 = 12$
$\quad\;\; y = x^2$

Solve each problem.

14. Find the distance between $(-1, 4)$ and $(1, 6)$.

15. Find the midpoint and length of the line segment with endpoints $(2, 0)$ and $(-3, -1)$.

16. Find the center and radius of the circle $x^2 + 2x + y^2 + 10y = 10$.

8. $x^2 + y^2 > 9$

17. Find the vertex, focus, and directrix of the parabola $y = x^2 + x + 3$. State the axis of symmetry and whether the parabola opens up or down.

18. Write the equation $y = \frac{1}{2}x^2 - 3x - \frac{1}{2}$ in the form $y = a(x - h)^2 + k$.

9. $y > x^2 - 9$

19. Write the equation of a circle with center $(-1, 3)$ that passes through $(2, 5)$.

20. Find the length and width of a rectangular room that has an area of 108 square feet and a perimeter of 42 ft.

*Making*Connections | A Review of Chapters 1–13

Sketch the graph of each equation.

1. $y = 9x - x^2$

2. $y = 9x$

3. $y = (x - 9)^2$

4. $y^2 = 9 - x^2$

5. $y = 9x^2$

6. $y = |9x|$

7. $4x^2 + 9y^2 = 36$

8. $4x^2 - 9y^2 = 36$

9. $y = 9 - x$

10. $y = 9^x$

Find the following products.

11. $(x + 2y)^2$

12. $(x + y)(x^2 + 2xy + y^2)$

13. $(a + b)^3$

14. $(a - 3b)^2$

15. $(2a + 1)(3a - 5)$

16. $(x - y)(x^2 + xy + y^2)$

Factor completely.

17. $a^3 + ab^2$

18. $a^3 - ab^2$

19. $2x^2 - 6x - 36$

20. $32x^2 + 8x - 60$

21. $mx^2 + 2x^2 - 9m - 18$

22. $2x^4 + 54x$

Solve each system of equations.

23. $2x - 3y = -4$
       $x + 2y = 5$

24. $x^2 + y^2 = 25$
       $x + y = 7$

25. $2x - y + z = 7$
       $x - 2y - z = 2$
       $x + y + z = 2$

26. $y = x^2$
       $y - 2x = 3$

Solve each formula for the specified variable.

27. $ax + b = 0$, for x

28. $wx^2 + dx + m = 0$, for x

29. $A = \dfrac{1}{2}h(B + b)$, for B

30. $\dfrac{1}{x} + \dfrac{1}{y} = \dfrac{1}{2}$, for x

31. $L = m + mxt$, for m

32. $y = 3a\sqrt{t}$, for t

Solve each problem.

33. Write the equation of the line in slope-intercept form that goes through the points $(2, -3)$ and $(-4, 1)$.

34. Write the equation of the line in slope-intercept form that contains the origin and is perpendicular to the line $2x - 4y = 5$.

35. Write the equation of the circle that has center $(2, 5)$ and passes through the point $(-1, -1)$.

36. Find the center and radius of the circle $x^2 + 3x + y^2 - 6y = 0$.

Perform the computations with complex numbers.

37. $2i(3 + 5i)$

38. i^6

39. $(2i - 3) + (6 - 7i)$

40. $(3 + i\sqrt{2})^2$

41. $(2 - 3i)(5 - 6i)$

42. $(3 - i) + (-6 + 4i)$

43. $(5 - 2i)(5 + 2i)$

44. $(2 - 3i) \div (2i)$

45. $(4 + 5i) \div (1 - i)$

46. $\dfrac{4 - \sqrt{-8}}{2}$

Solve.

47. *Going bananas.* Salvadore has observed that when bananas are $0.30 per pound (lb), he sells 250 lb per day, and when bananas are $0.40 per lb, he sells only 200 lb per day.

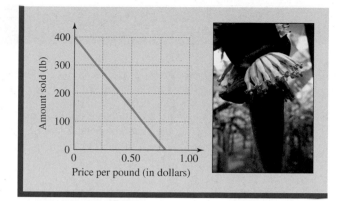

Figure for Exercise 47

a) Assume the number of pounds sold, q, is a linear function of the price per pound, x, and find that function.

b) Salvadore's daily revenue in dollars is the product of the number of pounds sold and the price per pound. Write the revenue as a function of x.

c) Graph the revenue function.

d) What price per pound maximizes his revenue?

e) What is his maximum possible revenue?

Critical **Thinking** | **For Individual or Group Work** | **Chapter 13**

These exercises can be solved by a variety of techniques, which may or may not require algebra. So be creative and think critically. Explain all answers. Answers are in the Instructor's Edition of this text.

1. ***Tiling a floor.*** Red and white floor tiles are used to make the arrangements shown in the accompanying figure. How many red tiles would appear in the 20th figure in this sequence?

Figure for Exercise 1

2. ***Rolling dice.*** A pair of dice is rolled. What is the most likely difference between the number of dots showing on the top faces?

Photo for Exercise 2

3. ***Jocks, nerds, and turkeys.*** At Ridgemont High there are 30 jocks, 20 nerds, and some turkeys. Every nerd is a turkey. One-half of the jocks are turkeys. One-half of the turkeys are nerds. No jock is a nerd. How many turkeys are there? How many turkeys are neither nerds nor jocks?

4. ***Mind reading.*** A man and a woman are on an airplane chatting about their families. The woman says that she has three children, the age of each child is a counting number, the product of their ages is 72, and the sum of their ages is the same as the flight number. The man checks his ticket for the flight number, does a bit of figuring, and says that he needs more information to determine the ages. The woman then points to the peanuts that they are munching on and says that the oldest is allergic to peanuts. The man then tells the woman the correct ages of her children. What are the ages? Explain your answer.

5. ***Five-letter takeout.*** Take out five letters from the list

   AFLIVGEELEBTRTEARS.

   The remaining letters will form a common English word. What is it?

6. ***Heads and tails.*** A bag contains three coins. One coin has heads on both sides, one has tails on both sides, and one has heads on one side and tails on the other. A single coin accidentally falls onto the floor and you observe heads on that coin, but you cannot see the other side or the other two coins in the bag. What is the probability that the other side of the coin on the floor is heads?

7. ***Adjoining ones.*** Find a positive integer such that adjoining a 1 at both ends of it increases its value by 14,789. (Adjoining a 1 at both ends of 5 would produce 151 and increase its value by 146.)

8. ***Ending digits.*** What are the last two digits (tens and ones) of 3^{1234}?

Appendix A

Geometry Review Exercises

(Answers are at the end of the answer section in this text.)

1. Find the perimeter of a triangle whose sides are 3 in., 4 in., and 5 in.

2. Find the area of a triangle whose base is 4 ft and height is 12 ft.

3. If two angles of a triangle are 30° and 90°, then what is the third angle?

4. If the area of a triangle is 36 ft² and the base is 12 ft, then what is the height?

5. If the side opposite 30° in a 30-60-90 right triangle is 10 cm, then what is the length of the hypotenuse?

6. Find the area of a trapezoid whose height is 12 cm and whose parallel sides are 4 cm and 20 cm.

7. Find the area of the right triangle that has sides of 6 ft, 8 ft, and 10 ft.

8. If a right triangle has sides of 5 ft, 12 ft, and 13 ft, then what is the length of the hypotenuse?

9. If the hypotenuse of a right triangle is 50 cm and the length of one leg is 40 cm, then what is the length of the other leg?

10. Is a triangle with sides of 5 ft, 10 ft, and 11 ft a right triangle?

11. What is the area of a triangle with sides of 7 yd, 24 yd, and 25 yd?

12. Find the perimeter of a parallelogram in which one side is 9 in. and another side is 6 in.

13. Find the area of a parallelogram which has a base of 8 ft and a height of 4 ft.

14. If one side of a rhombus is 5 km, then what is its perimeter?

15. Find the perimeter and area of a rectangle whose width is 18 in. and length is 2 ft.

16. If the width of a rectangle is 8 yd and its perimeter is 60 yd, then what is its length?

17. The radius of a circle is 4 ft. Find its area to the nearest tenth of a square foot.

18. The diameter of a circle is 12 ft. Find its circumference to the nearest tenth of a foot.

19. A right circular cone has radius 4 cm and height 9 cm. Find its volume to the nearest hundredth of a cubic centimeter.

20. A right circular cone has radius 12 ft and height 20 ft. Find its lateral surface area to the nearest hundredth of a square foot.

21. A shoe box has a length of 12 in., a width of 6 in., and a height of 4 in. Find its volume and surface area.

22. The volume of a rectangular solid is 120 cm³. If the area of its bottom is 30 cm², then what is its height?

23. What is the area and perimeter of a square in which one of the sides is 10 mi long?

24. Find the perimeter of a square whose area is 25 km².

25. Find the area of a square whose perimeter is 26 cm.

26. A sphere has a radius of 2 ft. Find its volume to the nearest thousandth of a cubic foot and its surface area to the nearest thousandth of a square foot.

27. A can of soup (right circular cylinder) has a radius of 2 in. and a height of 6 in. Find its volume to the nearest tenth of a cubic inch and total surface area to the nearest tenth of a square inch.

28. If one of two complementary angles is 34°, then what is the other angle?

29. If the perimeter of an isosceles triangle is 29 cm and one of the equal sides is 12 cm, then what is the length of the shortest side of the triangle?

30. A right triangle with sides of 6 in., 8 in., and 10 in. is similar to another right triangle that has a hypotenuse of 25 in. What are the lengths of the other two sides in the second triangle?

31. If one of two supplementary angles is 31°, then what is the other angle?

32. Find the perimeter of an equilateral triangle in which one of the sides is 4 km.

33. Find the length of a side of an equilateral triangle that has a perimeter of 30 yd.

Appendix B

Sets

Every subject has its own terminology, and **algebra** is no different. In this section we will learn the basic terms and facts about sets.

‹1› Set Notation

A **set** is a collection of objects. At home you may have a set of dishes and a set of steak knives. In algebra we generally discuss sets of numbers. For example, we refer to the numbers 1, 2, 3, 4, 5, and so on as the set of **counting numbers** or **natural numbers.** Of course, these are the numbers that we use for counting.

The objects or numbers in a set are called the **elements** or **members** of the set. To describe sets with a convenient notation, we use braces, { }, and name the sets with capital letters. For example,

$$A = \{1, 2, 3\}$$

means that set A is the set whose members are the natural numbers 1, 2, and 3. The letter N is used to represent the entire set of natural numbers.

A set that has a fixed number of elements such as $\{1, 2, 3\}$ is a **finite** set, whereas a set without a fixed number of elements such as the natural numbers is an **infinite** set. When listing the elements of a set, we use a series of three dots to indicate a continuing pattern. For example, the set of natural numbers is written as

$$N = \{1, 2, 3, \ldots\}.$$

The set of natural numbers *between* 4 and 40 can be written

$$\{5, 6, 7, 8, \ldots, 39\}.$$

Note that since the members of this set are *between* 4 and 40, it does not include 4 or 40.

Set-builder notation is another method of describing sets. In this notation we use a variable to represent the numbers in the set. A **variable** is a letter that is used to stand for some numbers. The set is then built from the variable and a description of the numbers that the variable represents. For example, the set

$$B = \{1, 2, 3, \ldots, 49\}$$

is written in set-builder notation as

$$B = \{x \mid x \text{ is a natural number less than } 50\}.$$

The set of numbers ↑↑ such that   ↑ condition for membership

This notation is read as "B is the set of numbers x such that x is a natural number less than 50." Notice that the number 50 is not a member of set B.

The symbol $\in$ is used to indicate that a specific number is a member of a set, and $\notin$ indicates that a specific number is not a member of a set. For example, the statement $1 \in B$ is read as "1 is a member of B," "1 belongs to B," "1 is in B," or "1 is an element of B." The statement $0 \notin B$ is read as "0 is not a member of B," "0 does not belong to B," "0 is not in B," or "0 is not an element of B."

Two sets are **equal** if they contain exactly the same members. Otherwise, they are said to be not equal. To indicate equal sets, we use the symbol $=$. For sets that are not equal we use the symbol $\neq$. The elements in two equal sets do not need to be written in the same order. For example, $\{3, 4, 7\} = \{3, 4, 7\}$ and $\{2, 4, 1\} = \{1, 2, 4\}$, but $\{3, 5, 6\} \neq \{3, 5, 7\}$.

EXAMPLE   **1**

Set notation

Let $A = \{1, 2, 3, 5\}$ and $B = \{x \mid x$ is an even natural number less than 10$\}$. Determine whether each statement is true or false.

a) $3 \in A$ b) $5 \in B$ c) $4 \notin A$ d) $A = N$

e) $A = \{x \mid x$ is a natural number less than 6$\}$ f) $B = \{2, 4, 6, 8\}$

Solutio n

a) True, because 3 is a member of set A.

b) False, because 5 is not an even natural number.

c) True, because 4 is not a member of set A.

d) False, because A does not contain all of the natural numbers.

e) False, because 4 is a natural number less than 6, and $4 \notin A$.

f) True, because the even counting numbers less than 10 are 2, 4, 6, and 8.

Now do Exercises 7–18

⟨2⟩ Union of Sets

Any two sets A and B can be combined to form a new set called their union that consists of all elements of A together with all elements of B.

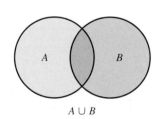

$A \cup B$

Figure B.1

> **Union of Sets**
>
> If A and B are sets, the **union** of A and B, denoted $A \cup B$, is the set of all elements that are either in A, in B, or in both. In symbols,
>
> $$A \cup B = \{x \mid x \in A \text{ or } x \in B\}.$$

In mathematics the word "or" is always used in an inclusive manner (allowing the possibility of both alternatives). The diagram in Fig. B.1 can be used to illustrate $A \cup B$. Any point that lies within circle A, circle B, or both is in $A \cup B$. Diagrams (like Fig. B.1) that are used to illustrate sets are called **Venn diagrams.**

EXAMPLE   **2**

Union of sets

Let $A = \{0, 2, 3\}$, $B = \{2, 3, 7\}$, and $C = \{7, 8\}$. List the elements in each of these sets.

a) $A \cup B$ b) $A \cup C$

⟨ Helpful Hint ⟩

To remember what "union" means think of a labor union, which is a group formed by joining together many individuals.

Solution

a) $A \cup B$ is the set of numbers that are in A, in B, or in both A and B.

$$A \cup B = \{0, 2, 3, 7\}$$

b) $A \cup C = \{0, 2, 3, 7, 8\}$

Now do Exercises 19–30

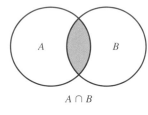

$A \cap B$

Figure B.2

⟨3⟩ **Intersection of Sets**

Another way to form a new set from two known sets is by considering only those elements that the two sets have in common. The diagram shown in Fig. B.2 illustrates the intersection of two sets A and B.

> **Intersection of Sets**
>
> If A and B are sets, the **intersection** of A and B, denoted $A \cap B$, is the set of all elements that are in both A and B. In symbols,
>
> $$A \cap B = \{x \mid x \in A \text{ and } x \in B\}.$$

⟨ **Helpful Hint** ⟩

To remember the meaning of "intersection," think of the intersection of two roads. At the intersection you are on both roads.

It is possible for two sets to have no elements in common. A set with no members is called the **empty set** and is denoted by the symbol $\varnothing$. Note that $A \cup \varnothing = A$ and $A \cap \varnothing = \varnothing$ for any set A.

CAUTION   The set $\{0\}$ is not the empty set. The set $\{0\}$ has one member, the number 0. Do not use the number 0 to represent the empty set.

E X A M P L E   **3**

Intersection of sets

Let $A = \{0, 2, 3\}$, $B = \{2, 3, 7\}$, and $C = \{7, 8\}$. List the elements in each of these sets.

   a) $A \cap B$ **b)** $B \cap C$ **c)** $A \cap C$

Solution

   a) $A \cap B$ is the set of all numbers that are in both A and B. So $A \cap B = \{2, 3\}$.

   b) $B \cap C = \{7\}$ **c)** $A \cap C = \varnothing$

Now do Exercises 19–30

E X A M P L E   **4**

Membership and equality

Let $A = \{1, 2, 3, 5\}$, $B = \{2, 3, 7, 8\}$, and $C = \{6, 7, 8, 9\}$. Place one of the symbols $=$, $\neq$, $\in$, or $\notin$ in the blank to make each statement correct.

   a) 5 _____ $A \cup B$ **b)** 5 _____ $A \cap B$

   c) $A \cup B$ _____ $\{1, 2, 3, 5, 7, 8\}$ **d)** $A \cap B$ _____ $\{2\}$

Solution

   a) $5 \in A \cup B$ because 5 is a member of A.

   b) $5 \notin A \cap B$ because 5 must belong to *both* A and B to be a member of $A \cap B$.

   c) $A \cup B = \{1, 2, 3, 5, 7, 8\}$ because the elements of A together with those of B are listed. Note that 2 and 3 are members of both sets but are listed only once.

   d) $A \cap B \neq \{2\}$ because $A \cap B = \{2, 3\}$.

Now do Exercises 31–40

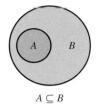

$A \subseteq B$

Figure B.3

⟨4⟩ Subsets

If every member of set A is also a member of set B, then we write $A \subseteq B$ and say that A is a **subset** of B. See Fig. B.3. For example,

$$\{2, 3\} \subseteq \{2, 3, 4\}$$

because $2 \in \{2, 3, 4\}$ and $3 \in \{2, 3, 4\}$. Note that the symbol for membership ($\in$) is used between a single element and a set, whereas the symbol for subset ($\subseteq$) is used between two sets. If A is not a subset of B, we write $A \nsubseteq B$.

CAUTION To claim that $A \nsubseteq B$, there *must* be an element of A that does *not* belong to B. For example,

$$\{1, 2\} \nsubseteq \{2, 3, 4\}$$

because 1 is a member of the first set but not of the second.

Is the empty set $\varnothing$ a subset of $\{2, 3, 4\}$? If we say that $\varnothing$ is *not* a subset of $\{2, 3, 4\}$, then there must be an element of $\varnothing$ that does not belong to $\{2, 3, 4\}$. But that cannot happen because $\varnothing$ is empty. So $\varnothing$ is a subset of $\{2, 3, 4\}$. In fact, by the same reasoning, *the empty set is a subset of every set.*

EXAMPLE **5**

Subsets

Determine whether each statement is true or false.

a) $\{1, 2, 3\}$ is a subset of the set of natural numbers.

b) The set of natural numbers is not a subset of $\{1, 2, 3\}$.

c) $\{1, 2, 3\} \nsubseteq \{2, 4, 6, 8\}$

d) $\{2, 6\} \subseteq \{1, 2, 3, 4, 5\}$

e) $\varnothing \subseteq \{2, 4, 6\}$

⟨ **Helpful Hint** ⟩

The symbols $\subseteq$ and $\subset$ are often used interchangeably. The symbol $\subseteq$ combines the subset symbol $\subset$ and the equal symbol $=$. We use it when sets are equal, $\{1, 2\} \subseteq \{1, 2\}$, and when they are not, $\{1\} \subseteq \{1, 2\}$. When sets are not equal, we could simply use $\subset$, as in $\{1\} \subset \{1, 2\}$.

Solution

a) True, because 1, 2, and 3 are natural numbers.

b) True, because 5, for example, is a natural number and $5 \notin \{1, 2, 3\}$.

c) True, because 1 is in the first set but not in the second.

d) False, because 6 is in the first set but not in the second.

e) True, because we cannot find anything in $\varnothing$ that fails to be in $\{2, 4, 6\}$.

Now do Exercises 41–52

⟨5⟩ Combining Three or More Sets

We know how to find the union and intersection of two sets. For three or more sets we use parentheses to indicate which pair of sets to combine first. In Example 6, notice that different results are obtained from different placements of the parentheses.

EXAMPLE **6**

Operations with three sets

Let $A = \{1, 2, 3, 4\}$, $B = \{2, 5, 6, 8\}$, and $C = \{4, 5, 7\}$. List the elements of each of these sets.

a) $(A \cup B) \cap C$ b) $A \cup (B \cap C)$

Solution

a) The parentheses indicate that the union of A and B is to be found first and then the result, $A \cup B$, is to be intersected with C.

$$A \cup B = \{1, 2, 3, 4, 5, 6, 8\}$$

Now examine $A \cup B$ and C to find the elements that belong to both sets:

$$A \cup B = \{1, 2, 3, 4, 5, 6, 8\}$$
$$C = \{4, 5, 7\}$$

The only numbers that are members of $A \cup B$ and C are 4 and 5. Thus

$$(A \cup B) \cap C = \{4, 5\}.$$

b) In $A \cup (B \cap C)$, first find $B \cap C$:

$$B \cap C = \{5\}$$

Now $A \cup (B \cap C)$ consist of all members of A together with 5 from $B \cap C$:

$$A \cup (B \cap C) = \{1, 2, 3, 4, 5\}$$

Now do Exercises 53–66

Exercises

Reading and Writing *After reading this section, write out the answers to these questions. Use complete sentences.*

1. What is a set?

2. What is the difference between a finite set and an infinite set?

3. What is a Venn diagram used for?

4. What is the difference between the intersection and the union of two sets?

5. What does it mean to say that set A is a subset of set B?

6. Which set is a subset of every set?

⟨1⟩ Set Notation

Using the sets A, B, C, and N, determine whether each statement is true or false. Explain. See Example 1.
$A = \{1, 3, 5, 7, 9\}$ $B = \{2, 4, 6, 8\}$
$C = \{1, 2, 3, 4, 5\}$ $N = \{1, 2, 3, \ldots\}$

7. $6 \in A$

8. $8 \in A$

9. $A \neq B$

10. $A = \{1, 3, 5, 7, \ldots\}$

11. $3 \in C$

12. $4 \notin B$

13. $A = \{1, 3, 7, 9\}$

14. $B \neq C$

15. $0 \in N$

16. $2.5 \in N$

17. $C = N$

18. $N = A$

⟨2–3⟩ Union and Intersection of Sets

Using the sets A, B, C, and N, list the elements in each set. If the set is empty write $\varnothing$. See Examples 2 and 3.
$A = \{1, 3, 5, 7, 9\}$ $B = \{2, 4, 6, 8\}$
$C = \{1, 2, 3, 4, 5\}$ $N = \{1, 2, 3, \ldots\}$

19. $A \cap B$

20. $A \cup B$

21. $A \cap C$

22. $A \cup C$

23. $B \cup C$

24. $B \cap C$

25. $A \cup \varnothing$

26. $B \cup \varnothing$

27. $A \cap \varnothing$

28. $B \cap \varnothing$

29. $A \cap N$

30. $A \cup N$

Use one of the symbols $\in$, $\notin$, $=$, $\neq$, $\cup$, or $\cap$ in each blank to make a true statement. See Example 4.
$A = \{1, 3, 5, 7, 9\}$ $B = \{2, 4, 6, 8\}$
$C = \{1, 2, 3, 4, 5\}$ $N = \{1, 2, 3, \ldots\}$

31. $A \cap B$ _____ $\varnothing$

32. $A \cap C$ _____ $\varnothing$

33. A ____ $B = \{1, 2, 3, 4, 5, 6, 7, 8, 9\}$

34. A ____ $B = \varnothing$

35. B ____ $C = \{2, 4\}$

36. B ____ $C = \{1, 2, 3, 4, 5, 6, 8\}$

37. 3 ____ $A \cap B$

38. 3 ____ $A \cap C$

39. 4 ____ $B \cap C$

40. 8 ____ $B \cup C$

⟨4⟩ Subsets

Determine whether each statement is true or false. Explain your answer. See Example 5.

$A = \{1, 3, 5, 7, 9\}$ $\quad B = \{2, 4, 6, 8\}$
$C = \{1, 2, 3, 4, 5\}$ $\quad N = \{1, 2, 3, \ldots\}$

41. $A \subseteq N$ **42.** $B \subseteq N$

43. $\{2, 3\} \subseteq C$ **44.** $C \subseteq A$

45. $B \nsubseteq C$ **46.** $C \nsubseteq A$

47. $\varnothing \subseteq B$ **48.** $\varnothing \subseteq C$

49. $A \subseteq \varnothing$ **50.** $B \subseteq \varnothing$

51. $A \cap B \subseteq C$ **52.** $B \cap C \subseteq \{2, 4, 6, 8\}$

⟨5⟩ Combining Three or More Sets

Using the sets D, E, and F, list the elements in each set. If the set is empty, write $\varnothing$. See Example 6.

$D = \{3, 5, 7\}$ $E = \{2, 4, 6, 8\}$ $F = \{1, 2, 3, 4, 5\}$

53. $D \cup E$ **54.** $D \cap E$

55. $D \cap F$ **56.** $D \cup F$

57. $E \cup F$ **58.** $E \cap F$

59. $(D \cup E) \cap F$ **60.** $(D \cup F) \cap E$

61. $D \cup (E \cap F)$ **62.** $D \cup (F \cap E)$

63. $(D \cap F) \cup (E \cap F)$ **64.** $(D \cap E) \cup (F \cap E)$

65. $(D \cup E) \cap (D \cup F)$ **66.** $(D \cup F) \cap (D \cup E)$

Miscellaneous

Use one of the symbols $\in$, $\subseteq$, $=$, $\cup$, or $\cap$ in each blank to make a true statement.

$D = \{3, 5, 7\}$ $E = \{2, 4, 6, 8\}$ $F = \{1, 2, 3, 4, 5\}$

67. D ____ $\{x \mid x \text{ is an odd natural number}\}$

68. E ____ $\{x \mid x \text{ is an even natural number smaller than 9}\}$

69. 3 ____ D **70.** $\{3\}$ ____ D

71. D ____ $E = \varnothing$ **72.** $D \cap E$ ____ D

73. $D \cap F$ ____ F **74.** $3 \notin E$ ____ F

75. $E \nsubseteq E$ ____ F **76.** $E \subseteq E$ ____ F

77. D ____ $F = F \cup D$ **78.** E ____ $F = F \cap E$

List the elements in each set.

79. $\{x \mid x \text{ is an even natural number less than 20}\}$

80. $\{x \mid x \text{ is a natural number greater than 6}\}$

81. $\{x \mid x \text{ is an odd natural number greater than 11}\}$

82. $\{x \mid x \text{ is an odd natural number less than 14}\}$

83. $\{x \mid x \text{ is an even natural number between 4 and 79}\}$

84. $\{x \mid x \text{ is an odd natural number between 12 and 57}\}$

Write each set using set-builder notation. Answers may vary.

85. $\{3, 4, 5, 6\}$

86. $\{1, 3, 5, 7\}$

87. $\{5, 7, 9, 11, \ldots\}$

88. $\{4, 5, 6, 7, \ldots\}$

89. $\{6, 8, 10, 12, \ldots, 82\}$

90. $\{9, 11, 13, 15, \ldots, 51\}$

Determine whether each statement is true or false.
$A = \{1, 2, 3, 4\}$ $B = \{3, 4, 5\}$ $C = \{3, 4\}$

91. $A = \{x \mid x \text{ is a counting number}\}$

92. The set B has an infinite number of elements.

93. The set of counting numbers less than 50 million is an infinite set.

94. $1 \in A \cap B$

95. $3 \in A \cup B$

96. $A \cap B = C$

97. $C \subseteq B$

98. $A \subseteq B$

99. $\varnothing \subseteq C$

100. $A \nsubseteq C$

Appendix C

Chapters 1–6 Diagnostic Test

Use this test to check your knowledge of Chapters 1–6. The test is arranged by chapters so that you can determine the chapters that you need to review. There is a review section for each of Chapters 1–6 in Appendix D, immediately following this test. Answers to this test and the review sections can be found at the end of the Answer Section of this text.

Chapter 1

Write each interval of real numbers in interval notation, and graph it on the number line.

1. The set of real numbers greater than 2

2. The set of real numbers less than or equal to −1

3. The set of real numbers between 0 and 1

4. The set of real numbers greater than −4 and less than or equal to −2

Evaluate each expression.

5. $\dfrac{3}{4} \cdot \dfrac{7}{9}$

6. $\dfrac{1}{4} + \dfrac{5}{6}$

7. $\dfrac{8}{9} \div 4$

8. $-4^2 - 3^3$

9. $|3 - 2^2| - |7 - 19|$

10. $\dfrac{-3 - 5}{-2 - (-1)}$

Name the property that justifies each equation.

11. $3(x + 4) = 3x + 12$

12. $x \cdot 7 = 7x$

13. $4 + (9 + y) = (4 + 9) + y$

14. $0 + 3 = 3$

Simplify each expression.

15. $5x - (3 - 8x)$

16. $x + 3 - 0.2(5x - 30)$

17. $(-3x)(-5x)$

18. $\dfrac{3x + 12}{-3}$

Chapter 2

Solve each equation and check your answer.

19. $11x - 2 = 3$

20. $4x - 5 = 12x + 11$

21. $3(x - 6) = 3x - 6$

22. $x - 0.1x = 0.9x$

Solve each equation for y.

23. $5x - 3y = 9$

24. $ay + b = 0$

25. $a = t - by$

26. $\dfrac{a}{2} + \dfrac{y}{3} = \dfrac{3a}{4}$

Solve each problem. Show all details.

27. The sum of three consecutive integers is 102. What are the integers?

28. The perimeter of a rectangular painting is 100 inches. If the width is 4 inches less than the length, then what is the width?

29. The area of a triangular piece of property is 44,000 square feet. If the base of the triangle is 400 feet, then what is the height?

30. Ivan has 400 pounds of mixed nuts that contain no peanuts. How many pounds of peanuts should he put into the mixed nuts so that 20% of the mixture is peanuts?

Solve each inequality. State the solution set using interval notation, and graph the solution set.

31. $3x - 4 \leq 11$

32. $5 - 7w > 26$

33. $-1 < 2a - 9 \le 7$ **34.** $5 < 6 - x < 6$

Chapter 3

Graph each equation in the coordinate plane, and identify all intercepts.

35. $y = \dfrac{2}{3}x - 2$ **36.** $3x - 5y = 150$

37. $y = 2$ **38.** $x = 2$

Find the slope of each line.

39. The line passing through the points $(1, 2)$ and $(3, 6)$

40. The line $y = \dfrac{1}{2}x - 4$

41. The line parallel to $2x + 3y = 9$

42. The line perpendicular to $y = -3x + 5$

Find the equation of each line in slope-intercept form when possible.

43. The line passing through the points $(0, 3)$ and $(2, 11)$

44. The line passing through the points $(-2, 4)$ and $(1, -2)$

45. The line through $(3, 5)$ that is parallel to $x = 4$

46. The line through $(0, 8)$ that is perpendicular to $y = \dfrac{1}{2}x$

Solve each variation problem.

47. The time that it takes to mow a large lawn varies inversely with the number of mowers working on the job. If it takes 30 hours with three mowers, then how long would it take with five mowers?

48. The cost of installing ceramic floor tile in a rectangular room varies jointly with the length and the width of the room. If the cost is \$810 for a 9 ft by 12 ft room, then what is the cost for a 14 ft by 18 ft room?

Graph the solution set to each inequality in the coordinate plane.

49. $3x - 4y > 12$ **50.** $y \le 3x + 2$

51. $x > -2$ **52.** $y \le 4$

Chapter 4

Perform the indicated operations.

53. $\left(x^2 - 3x + 2\right) - \left(3x^2 + 9x - 4\right)$

54. $-3x^2\left(-2x^2 - 3\right)$

55. $(x + 7)(x - 9)$

56. $(x + 2)\left(x^2 - 2x + 4\right)$

57. $\left(4w^2 - 3\right)^2$

58. $\left(-8m^7\right) \div \left(2m^2\right)$

59. $\left(-9y^3 - 6y^2 + 3y\right) \div (3y)$

60. $\left(x^3 - 2x^2 - x - 6\right) \div (x - 3)$

Use the rules of exponents to simplify each expression.
Write the answers without negative exponents.

61. $-8x^4 \cdot 4x^3$

62. $3x(5x^2)^3$

63. $\dfrac{-6x^2y^3}{-2x^{-3}y^4}$

64. $\left(\dfrac{2a^2}{a^{-3}}\right)^3$

Perform each operation without a calculator. Write the answer in scientific notation.

65. $400{,}000 \cdot 600$

66. $(9 \times 10^3)(2 \times 10^6)$

67. $(2 \times 10^{-3})^4$

68. $\dfrac{2 \times 10^9}{2000}$

Chapter 5

Factor each polynomial completely.

69. $24x^2y^3 + 18xy^5$

70. $x^2 + 2x + ax + 2a$

71. $4m^2 - 49$

72. $x^2 - 3x - 54$

73. $6t^2 - 11t - 10$

74. $4w^2 - 36w + 81$

75. $2a^3 - 6a^2 - 108a$

76. $w^3 - 27$

Solve each equation.

77. $x^2 = x$

78. $2x^3 - 8x = 0$

79. $a^2 + a - 6 = 0$

80. $(b - 2)(b + 3) = 24$

Write a complete solution to each problem.

81. The sum of two numbers is 10, and their product is 21. Find the numbers.

82. The length of a new television screen is 14 inches larger than the width, and the diagonal is 26 inches. What are the length and width?

Chapter 6

Perform the indicated operation. Write each answer in lowest terms.

83. $\dfrac{5x}{2} + \dfrac{3x}{4}$

84. $\dfrac{5}{x - 2} - \dfrac{3}{2 - x}$

85. $\dfrac{9}{x^2 - 9} + \dfrac{2x}{x - 3}$

86. $\dfrac{2}{a - 5} + \dfrac{3}{a + 4}$

87. $\dfrac{w^3}{2w - 4} \cdot \dfrac{w^2 - 4}{w}$

88. $\dfrac{5ab^2}{6a^2b^3} \div \dfrac{10a}{21b^6}$

Solve each equation.

89. $\dfrac{2}{x} = \dfrac{3}{4}$

90. $\dfrac{1}{w - 3} = \dfrac{2}{w + 5}$

91. $\dfrac{1}{x} + \dfrac{3}{7} = \dfrac{1}{3x}$

92. $\dfrac{3}{a - 1} + \dfrac{1}{a + 2} = \dfrac{17}{10}$

Solve each formula for y.

93. $\dfrac{3}{y} = \dfrac{5}{x}$

94. $a = \dfrac{1}{2}y(w - c)$

95. $\dfrac{y - 3}{x + 5} = -3$

96. $\dfrac{3}{y} + \dfrac{1}{2} = \dfrac{1}{t}$

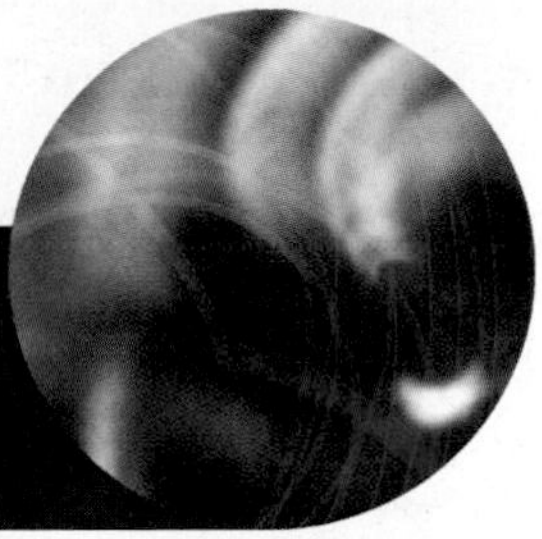

Appendix D

Chapters 1–6 Review

R.1 Real Numbers and Their Properties

This section is a review of Chapter 1 of this text. All topics in this review section are explained in greater detail in Chapter 1.

⟨1⟩ The Real Numbers

The numbers that we use in algebra are called the **real numbers.** There is a one-to-one correspondence between the set of real numbers and the points on the number line. Certain subsets of the set of real numbers are given special names.

Subsets of the Set of Real Numbers

Natural numbers	$\{1, 2, 3, \ldots\}$
Whole numbers	$\{0, 1, 2, 3, \ldots\}$
Integers	$\{\ldots, -3, -2, -1, 0, 1, 2, 3, \ldots\}$
Rational numbers	$\left\{\dfrac{a}{b} \,\middle|\, a \text{ and } b \text{ are integers, with } b \neq 0\right\}$
Irrational numbers	Real numbers that cannot be expressed as a ratio of integers

---

An **interval** of real numbers is the set of real numbers that are between two real numbers, which are called the **endpoints** of the interval. If a is less than b, then the set of real numbers between a and b, not including a or b, is written in interval notation as (a, b). If the endpoints are to be included, then we write $[a, b]$. An interval of real numbers may extend infinitely far to the right or left on the number line. In this case the infinity symbol ∞ is used as an endpoint.

E X A M P L E   1

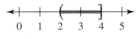

Figure R.1

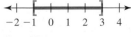

Figure R.2

Interval notation

Write each interval of real numbers in interval notation, and graph it on a number line.

a) The set of real numbers greater than 2 and less than or equal to 4

b) The set of real numbers between -1 and 3 inclusive

c) The set of real numbers greater than or equal to 0

d) The set of real numbers less than 10

Solution

a) The set of real numbers greater than 2 and less than or equal to 4 does not include 2, but does include 4. So the interval is written as $(2, 4]$ and graphed in Fig. R.1.

b) The set of real numbers between -1 and 3 inclusive includes both endpoints. So the interval is written as $[-1, 3]$ and graphed in Fig. R.2.

Figure R.3

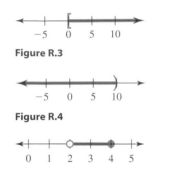

Figure R.4

c) The set of real numbers greater than or equal to 0 extends infinitely far to the right on the number line. So the interval is written as $[0, \infty)$ and graphed in Fig. R.3.

d) The set of real numbers less than 10 extends infinitely far to the left on the number line. So the interval is written as $(-\infty, 10)$ and graphed in Fig. R.4.

> **Now do Exercises 1–16**

Figure R.5

It is also common to draw the graph of an interval of real numbers using an open circle for an endpoint that does not belong to the interval and a closed circle for an endpoint that belongs to the interval. For example, the graph of $(2, 4]$ can be drawn as shown in Fig. R.5. In this text, parentheses and brackets are used so that the graphs agree with interval notation.

The **absolute value** of a real number is the number's distance from 0 on the number line. A number and its opposite have the same absolute value. For example, $|5| = 5$ and $|-5| = 5$. So the absolute value of a nonnegative number is the number, and the absolute value of a negative number is the opposite of the number. In symbols,

$$|a| = a \text{ if } a \text{ is nonnegative} \quad \text{and} \quad |a| = -a \text{ if } a \text{ is negative.}$$

E X A M P L E **2**

Absolute value
Find each absolute value.

a) $|4|$ **b)** $|-4|$ **c)** $|0|$ **d)** $|-3.9|$

Solution

a) Since 4 is 4 units from 0 on the number line, $|4| = 4$.
b) Since -4 is 4 units from 0 on the number line, $|-4| = 4$.
c) Since 0 is 0 units from 0 on the number line, $|0| = 0$.
d) Since -3.9 is negative, $|-3.9| = -(-3.9) = 3.9$.

> **Now do Exercises 17–22**

⟨2⟩ Fractions

Every fraction can be written in infinitely many equivalent forms. Consider the following equivalent forms of $\frac{2}{3}$:

$$\frac{2}{3} = \frac{4}{6} = \frac{6}{9} = \frac{8}{12} = \frac{10}{15} = \cdots$$

Note that each equivalent form of $\frac{2}{3}$ can be obtained by multiplying the numerator and denominator of $\frac{2}{3}$ by a natural number. Converting a fraction into an equivalent form with a larger denominator is called **building up** the fraction. Converting a fraction into an equivalent form with a smaller denominator is called **reducing the fraction.** A fraction that cannot be reduced is in **lowest terms.**

E X A M P L E **3**

Building up or reducing fractions
Complete each equation to make the fractions equivalent.

a) $\dfrac{3}{4} = \dfrac{?}{20}$ **b)** $\dfrac{12}{30} = \dfrac{?}{5}$

Solution

a) Since $20 = 4 \cdot 5$, we can multiply the numerator and denominator of $\frac{3}{4}$ by 5 to obtain an equivalent fraction with a denominator of 20:

$$\frac{3}{4} = \frac{3 \cdot 5}{4 \cdot 5} = \frac{15}{20}$$

Note that multiplying the numerator and denominator by 5 can also be accomplished by multiplying the fraction by the number 1 in its equivalent form $\frac{5}{5}$:

$$\frac{3}{4} = \frac{3}{4} \cdot 1 = \frac{3}{4} \cdot \frac{5}{5} = \frac{15}{20}$$

b) Since $30 = 6 \cdot 5$ and $12 = 6 \cdot 2$, we can factor the numerator and denominator. Then we divide out or cancel the common factor:

$$\frac{12}{30} = \frac{\cancel{6} \cdot 2}{\cancel{6} \cdot 5} = \frac{2}{5}$$

> Now do Exercises 23–38

In Example 4, we illustrate the four basic operations with fractions. Fractions are multiplied by multiplying their numerators and denominators. Fractions are divided by inverting the divisor and multiplying. To add or subtract fractions the fractions must have identical denominators.

E X A M P L E   4

Operations with fractions

Perform the indicated operations with fractions. Express answers in lowest terms.

a) $\dfrac{5}{6} \cdot \dfrac{3}{20}$ b) $\dfrac{2}{3} \div \dfrac{2}{5}$ c) $\dfrac{1}{3} + \dfrac{2}{7}$ d) $\dfrac{5}{6} - \dfrac{4}{15}$

Solution

a) $\dfrac{5}{6} \cdot \dfrac{3}{20} = \dfrac{15}{120} = \dfrac{\cancel{15} \cdot 1}{\cancel{15} \cdot 8} = \dfrac{1}{8}$

b) $\dfrac{2}{3} \div \dfrac{2}{5} = \dfrac{2}{3} \cdot \dfrac{5}{2} = \dfrac{5}{3}$

c) $\dfrac{1}{3} + \dfrac{2}{7} = \dfrac{1 \cdot 7}{3 \cdot 7} + \dfrac{2 \cdot 3}{7 \cdot 3} = \dfrac{7}{21} + \dfrac{6}{21} = \dfrac{13}{21}$

d) $\dfrac{5}{6} - \dfrac{4}{15} = \dfrac{5 \cdot 5}{6 \cdot 5} - \dfrac{4 \cdot 2}{15 \cdot 2} = \dfrac{25}{30} - \dfrac{8}{30} = \dfrac{17}{30}$

> Now do Exercises 39–52

⟨3⟩ Operations with Real Numbers

To find the sum of two numbers with the same sign, add their absolute values. The sum has the same sign as the original numbers. To find the sum of two numbers with unlike signs subtract their absolute values. The answer is positive if the number with the larger absolute value is positive. The answer is negative if the number with the larger absolute value is negative. The answer is zero if the original numbers have equal absolute values. All subtraction of signed numbers can be written in terms of addition according to the rule $a - b = a + (-b)$.

EXAMPLE **5**

Adding and subtracting signed numbers

Perform the indicated operations.

a) $-4 + (-5)$ b) $-5 + 8$ c) $6 + (-30)$
d) $-9 + 9$ e) $15 - 18$ f) $-3 - (-9)$

Solution

a) Since -4 and -5 have the same sign, we add their absolute values ($4 + 5 = 9$) and then give that result a negative sign. So $-4 + (-5) = -9$.

b) Since -5 and 8 have opposite signs, we subtract their absolute values ($8 - 5 = 3$). We give the result a positive sign because 8 has the larger absolute value. So $-5 + 8 = 3$.

c) Since 6 and -30 have opposite signs, we subtract their absolute values ($30 - 6 = 24$). We give the result a negative sign because -30 has the larger absolute value. So $6 + (-30) = -24$.

d) Since -9 and 9 have opposite signs and the same absolute value, their sum is 0. So $-9 + 9 = 0$.

e) Write all subtraction of signed numbers in terms of addition, and follow the rules for addition. So $15 - 18 = 15 + (-18) = -3$.

f) Write subtraction in terms of addition, and then follow the rules for addition of signed numbers. So $-3 - (-9) = -3 + 9 = 6$.

Now do Exercises 53–64

The result of multiplying two numbers is called the **product** of the numbers. To find the product of two nonzero real numbers, multiply their absolute values. The product is positive if the numbers have like signs. The product is negative if the numbers have unlike signs. If one or more of the numbers multiplied is zero, then the product is zero.

The result of dividing two numbers is called the **quotient** of the numbers. To find the quotient of two nonzero real numbers, divide their absolute values. The quotient is positive if the numbers have like signs. The quotient is negative if the numbers have unlike signs. Zero divided by any nonzero real number is zero. Division of any real number by zero is an undefined operation.

EXAMPLE **6**

Multiplying and dividing signed numbers

Perform the indicated operations.

a) $(-4)(-5)$ b) $-8 \cdot 5$ c) $-6(0)$
d) $(-9) \div (3)$ e) $-15 \div (-3)$ f) $0 \div (-9.34)$
g) $-\dfrac{1}{2} \div 0$

Solution

a) Multiply the absolute values of -4 and -5 to get $4 \cdot 5 = 20$. Since -4 and -5 have the same sign, the product is positive. So $(-4)(-5) = 20$.

b) Multiply the absolute values of -8 and 5 to get $8 \cdot 5 = 40$. Since -8 and 5 have opposite signs, the product is negative. So $-8 \cdot 5 = -40$.

c) Since the product of zero and any real number is zero, we have $-6(0) = 0$.

d) Divide the absolute values of -9 and 3 to get $9 \div 3 = 3$. Since -9 and 3 have unlike signs, the quotient is negative. So $(-9) \div (3) = -3$.

e) Divide the absolute values of -15 and -3 to get $15 \div 3 = 5$. Since -15 and -3 have like signs, the quotient is positive. So $-15 \div (-3) = 5$.

f) Zero divided by any nonzero real number is zero. So $0 \div (-9.34) = 0$.

g) Since division by zero is undefined, there is no quotient for $-\frac{1}{2} \div 0$.

Now do Exercises 65–74

⟨4⟩ Exponential Expressions and the Order of Operations

The result of writing numbers in a meaningful combination with the ordinary operations of arithmetic is called an **arithmetic expression** or simply an **expression.** To simplify the writing of repeated factors in multiplication we use exponents to indicate the number of factors that are multiplied. For example, $3 \cdot 3 \cdot 3 \cdot 3 = 3^4$. Note that in an expression such as -9^2 the exponent applies only to the 9. So $-9^2 = -(9 \cdot 9) = -81$, whereas $(-9)^2 = (-9)(-9) = 81$.

When we evaluate expressions, operations within grouping symbols are always performed first. For example, $3(2 + 5) = 3(7) = 21$. To make expressions look simpler, we often omit some or all parentheses. In this case, we follow the accepted **order of operations:** evaluate exponential expressions first, then multiplication and division, and finally addition and subtraction.

E X A M P L E 7

Evaluating arithmetic expressions

Evaluate.

a) $-4^2 + 5^3$ **b)** $(-3 + 2)(8 - 9)$ **c)** $5 \cdot 2 + 3^2$

d) $3|7 - 9| + 4$ **e)** $\dfrac{-1 - 5}{4 - (-6)}$

Solution

a) $-4^2 + 5^3 = -16 + 125 = 125 - 16$
$$= 109$$

b) $(-3 + 2)(8 - 9) = (-1)(-1)$
$$= 1$$

c) $5 \cdot 2 + 3^2 = 10 + 9$
$$= 19$$

d) $3|7 - 9| + 4 = 3|-2| + 4 = 3 \cdot 2 + 4 = 6 + 4$
$$= 10$$

e) $\dfrac{-1 - 5}{4 - (-6)} = \dfrac{-6}{10} = \dfrac{-3 \cdot 2}{5 \cdot 2}$
$$= -\dfrac{3}{5}$$

Now do Exercises 75–90

⟨5⟩ Algebraic Expressions

The result of combining numbers and variables with the ordinary operations of arithmetic in some meaningful way is called an **algebraic expression** or simply an **expression.** Expressions are named by the last operation to be performed in the expression. So $2a + b$ is a sum, $ab - xy$ is a difference, $a(x + 3)$ is a product, $\dfrac{a - 3}{b - 2}$ is a quotient, and $(a + b)^2$ is a square. An algebraic expression has a value only if a value is known for every variable in the expression.

EXAMPLE **8**

Writing and evaluating algebraic expressions

Write the algebraic expression that is described, and evaluate it for the given value(s) of the variable(s).

 a) The sum of $5x$ and 3; $x = -8$

 b) The product of $a + b$ and $a - b$; $a = -7$ and $b = 9$

 c) The difference of x^2 and y^2; $x = -2$ and $y = -5$

 d) The quotient of $x - y$ and $y - x$; $x = -2$ and $y = -5$

 e) The square of the sum $-3x + 1$; $x = -2$

Solution

 a) The sum of $5x$ and 3 is written as $5x + 3$. If $x = -8$, then
$$5x + 3 = 5(-8) + 3 = -40 + 3 = -37.$$

 b) The product of $a + b$ and $a - b$ is written as $(a + b)(a - b)$. If $a = -7$ and $b = 9$, then
$$(a + b)(a - b) = (-7 + 9)(-7 - 9) = (2)(-16) = -32.$$

 c) The difference of x^2 and y^2 is written as $x^2 - y^2$. If $x = -2$ and $y = -5$, then
$$x^2 - y^2 = (-2)^2 - (-5)^2 = 4 - 25 = -21.$$

 d) The quotient of $x - y$ and $y - x$ is written as $\frac{x-y}{y-x}$. If $x = -2$ and $y = -5$, then
$$\frac{x-y}{y-x} = \frac{-2 - (-5)}{-5 - (-2)} = \frac{3}{-3} = -1.$$

 e) The square of the sum $-3x + 1$ is written as $(-3x + 1)^2$. If $x = -2$, then
$$(-3x + 1)^2 = (-3(-2) + 1)^2 = 7^2 = 49.$$

Now do Exercises 91–104

⟨6⟩ **Properties of the Real Numbers**

The properties of the real numbers are useful in algebra. The properties are listed as follows.

Properties of the Real Numbers

For any real numbers a, b, and c the following properties are true.

Commutative property	of addition	$a + b = b + a$
	of multiplication	$ab = ba$
Associative property	of addition	$(a + b) + c = a + (b + c)$
	of multiplication	$(ab)c = a(bc)$
Distributive property	for addition	$a(b + c) = ab + ac$
	for subtraction	$a(b - c) = ab - ac$
Identity property	for addition	$a + 0 = 0 + a = a$
	for multiplication	$a \cdot 1 = 1 \cdot a = a$
Inverse property	for addition	$a + (-a) = 0$
	for multiplication	$a \cdot \dfrac{1}{a} = 1 \; (a \neq 0)$
Multiplication property of zero		$0 \cdot a = a \cdot 0 = 0$

EXAMPLE 9

Properties of the real numbers

Name the property that justifies each equation.

a) $11 \cdot 19 = 19 \cdot 11$

b) $(x + 2) + 3 = x + (2 + 3)$ for any real number x

c) $\pi a^2 + \pi b^2 = \pi(a^2 + b^2)$ for any real numbers a and b

d) $5 \cdot \dfrac{1}{5} = 1$

e) $0(499 - 365 \cdot 288) = 0$

f) $3x^2 + 0 = 3x^2$ for any real number x

Solution

a) Commutative property of multiplication

b) Associative property of addition

c) Distributive property for addition

d) Inverse property for multiplication

e) Multiplication property of zero

f) Identity property for addition

Now do Exercises 105–114

An expression containing a number or the product of a number and one or more variables raised to powers is called a **term.** The number preceding the variables in a term is called the **coefficient.** If two terms contain the same variables with the same exponents, they are called **like terms.** Using the distributive property on a sum or difference of like terms allows us to combine the like terms and simplify the expression:

$$3x + 5x = (3 + 5)x \quad \text{Distributive property}$$
$$= 8x \qquad \text{Add the coefficients.}$$

In Example 10, we use the idea of combining like terms and other properties of the real numbers to simplify expressions.

EXAMPLE 10

Using the properties of the real numbers to simplify expressions

Simplify each expression.

a) $(4x - 3) + (5x - 7)$

b) $-3a + 6 - 5(4 - 5a)$

c) $(-4b)(-7a) - (-3)(4a)$

d) $\dfrac{10y + 5}{5}$

Solution

a) $(4x - 3) + (5x - 7) = 4x + 5x - 3 - 7$ Commutative and associative
$\quad\quad\quad\quad\quad\quad\quad = 9x - 10$ properties. Combine like terms.

b) $-3a + 6 - 5(4 - 5a) = -3a + 6 - 20 + 25a$ Distributive property
$\quad\quad\quad\quad\quad\quad\quad\quad = 22a - 14$ Combine like terms.

c) $(4b)(7a) - (-3)(4a) = (4)(7)ab - (-3 \cdot 4)a$ Commutative and associative
$\quad\quad\quad\quad\quad\quad\quad\quad = 28ab + 12a$ properties. Simplify.

d) $\dfrac{10y + 5}{5} = (10y + 5) \cdot \dfrac{1}{5}$ Invert 5 and multiply.
$\quad\quad\quad\quad = 2y + 1$ Distributive property

Now do Exercises 115–126

Note that in Example 10(b) the distributive property allows us to divide 5 into both terms in the numerator of the fraction. We cannot divide the denominator into just one term of the numerator. So we cannot simplify $\frac{6a+b}{3}$ to get $2a + b$. We could write $\frac{6a+b}{3} = 2a + \frac{1}{3}b$.

R.1 Exercises

⟨1⟩ **The Real Numbers**

Write each interval of real numbers in interval notation, and graph it on a number line. See Example 1.

1. The set of real numbers between 0 and 3 inclusive

2. The set of real numbers between −2 and 5

3. The set of real numbers greater than or equal to −4 and less than 0

4. The set of real numbers greater than 3 and less than or equal to 8

5. The set of real numbers less than −1

6. The set of real numbers less than or equal to 6

7. The set of real numbers greater than or equal to 50

8. The set of real numbers greater than −10

Give a verbal description of each interval.

9. $(-2, 9)$

10. $[-4, -3]$

11. $[11, 13)$

12. $(22, 26]$

13. $(0, \infty)$

14. $[99, \infty)$

15. $(-\infty, -6]$

16. $(-\infty, 18)$

Find each absolute value. See Example 2.

17. $|-1|$

18. $|-9.35|$

19. $|0|$

20. $|5 - 5|$

21. $|50|$

22. $|6.87|$

⟨2⟩ **Fractions**

Complete each equation to make the fractions equivalent. See Example 3.

23. $\dfrac{1}{2} = \dfrac{?}{20}$

24. $\dfrac{2}{3} = \dfrac{?}{18}$

25. $\dfrac{3}{4} = \dfrac{?}{24}$

26. $\dfrac{7}{8} = \dfrac{?}{56}$

27. $\dfrac{12}{20} = \dfrac{?}{5}$

28. $\dfrac{16}{24} = \dfrac{?}{3}$

29. $\dfrac{14}{48} = \dfrac{?}{24}$

30. $\dfrac{24}{84} = \dfrac{?}{7}$

Reduce each fraction to lowest terms.

31. $\dfrac{6}{10}$

32. $\dfrac{7}{14}$

33. $\dfrac{28}{49}$

34. $\dfrac{48}{72}$

35. $\dfrac{36}{108}$

36. $\dfrac{51}{68}$

37. $\dfrac{30}{100}$

38. $\dfrac{400}{1000}$

Perform the indicated operations. Express answers in lowest terms. See Example 4.

39. $\dfrac{3}{8} \cdot \dfrac{2}{3}$

40. $\dfrac{2}{5} \cdot \dfrac{15}{26}$

41. $\dfrac{1}{4} \div \dfrac{5}{2}$

42. $\dfrac{3}{7} \div \dfrac{9}{14}$

43. $\dfrac{2}{5} \cdot 25$

44. $\dfrac{5}{8} \cdot 40$

45. $\dfrac{2}{3} \div 5$

46. $6 \div \dfrac{1}{7}$

47. $\dfrac{1}{8} + \dfrac{2}{3}$

48. $\dfrac{1}{5} + \dfrac{3}{4}$

49. $\dfrac{5}{12} - \dfrac{5}{18}$

50. $\dfrac{5}{16} - \dfrac{1}{12}$

51. $\dfrac{5}{8} + 2$

52. $\dfrac{3}{7} + 1$

⟨3⟩ Operations with Real Numbers

Perform the indicated operations. See Examples 5 and 6.

53. $-20 + (-6)$

54. $-19 + (-8)$

55. $-30 + 7$

56. $18 + (-9)$

57. $6 + (-5)$

58. $-7 + 12$

59. $-30 - 6$

60. $-15 - 12$

61. $20 - (-4)$

62. $88 - (-12)$

63. $-3 - (-5)$

64. $-9 - (-6)$

65. $(-3)(-60)$

66. $(-8)(-12)$

67. $(-7)(12)$

68. $(13)(-3)$

69. $(-30) \div (-2)$

70. $(-90) \div (-15)$

71. $-40 \div 5$

72. $100 \div (-20)$

73. $0 \div (-7)$

74. $0 \div (-2000)$

⟨4⟩ Exponential Expressions and the Order of Operations

Evaluate each arithmetic expression. See Example 7.

75. $-3^2 - 9^2$

76. $(-4)^3 - 5^2$

77. $(4 + 2^3)(1 - 4)$

78. $(4 - 5)^3(3 - 6^2)$

79. $3 + 5 \cdot 7$

80. $10 - 6 \cdot 2$

81. $2^4 - 3 \cdot 7$

82. $3 \cdot 2^5 - 5 \cdot 2^4$

83. $|3 - 9| - |5 - 8|$

84. $2|3 - 5 \cdot 4|$

85. $|-6| - 3|2 - 2^3|$

86. $|5 \cdot 4 - 10| - |-3 \cdot 2|$

87. $\dfrac{-4 - 2}{1 - 3}$

88. $\dfrac{-2^2 - 3^3}{1 - (-30)}$

89. $\dfrac{-3 \cdot 5 - 2}{1 - 3 \cdot 6}$

90. $\dfrac{4 - 2 \cdot 7}{2 - 3 \cdot 2^2}$

⟨5⟩ Algebraic Expressions

Write the algebraic expression that is described, and evaluate it for the given value(s) of the variable(s). See Example 8.

91. The sum of $5x$ and $-3y$; $x = -2$ and $y = 5$

92. The difference of a^3 and b^3; $a = -2$ and $b = 4$

93. The product of $a + b$ and $a^2 - ab + b^2$; $a = -1$ and $b = -3$

94. The quotient of $x - 7$ and $7 - x$; $x = 9$

95. The square of $2x - 3$; $x = 5$

96. The cube of $a - b$; $a = 3$ and $b = -1$

Determine whether each expression is a sum, difference, product, quotient, square, or cube.

97. $a^3 - b^3$

98. $a^2 + b^2$

99. $5a - b$

100. $5(a - b)$

101. $\dfrac{6 - a}{6a}$

102. $(5a - b)^2$

103. $(3a)^3$

104. $3 + a^3$

⟨6⟩ Properties of the Real Numbers

Name the property that justifies each equation. See Example 9.

105. $a(3) = 3a$

106. $3 + a = a + 3$

107. $5(x + 1) = 5x + 5$

108. $(w^2 + 8) + 7 = w^2 + (8 + 7)$

109. $5 \cdot 1 = 5$

110. $3 + 0 = 3$

111. $m^2 \cdot 0 = 0$

112. $6 \cdot \dfrac{1}{6} = 1$

113. $3(5x) = (3 \cdot 5)x$

114. $a + (-a) = 0$

Simplify each expression. See Example 10.

115. $(2x - 9) + (7 - 3x)$

116. $(-3x - y) + (9y - 8x)$

117. $5 + 3(4 + x)$

118. $x + 7(x + y)$

119. $6 + 7xy - 4(3 - 6xy)$

120. $4 + 3a - 5(4 - 7a)$

121. $(-2a)(5b) - 5(4ab)$

122. $(-x)(-y) - 5(-4xy)$

123. $\dfrac{3(4 - 2x)}{6}$

124. $\dfrac{2(3x - 3y)}{6}$

125. $\dfrac{44 - 2x}{-2}$

126. $\dfrac{20 + 8x}{-4}$

R.2 Linear Equations and Inequalities in One Variable

This section is a review of Chapter 2 of this text. All topics in this review section are explained in greater detail in Chapter 2.

⟨1⟩ Solving Linear Equations

An **equation** is a statement that two expressions are equal. The equations that we study in this section will contain only one variable. If the equation is correct when a number is used in place of the variable, then that number is a **solution** to the equation. The set containing all solutions to an equation is the **solution set** to the equation. Equations that have the same solution set are **equivalent equations.** To **solve** an equation means to find all solutions to the equation or to find the solution set to the equation.

A **linear equation in one variable** x is an equation of the form $ax + b = 0$, where a and b are real numbers with $a \neq 0$. Other equations that are equivalent to $ax + b = 0$ may also be called linear equations. To solve linear equations we use the properties of equality. The **addition property of equality** indicates that adding the same number to both sides of an equation does not change the solution set to the equation. The **multiplication property of equality** indicates that multiplying both sides of an equation by the same nonzero number does not change the solution set to the equation. Since subtraction and division are defined in terms of addition and multiplication, respectively, we can also subtract the same number from both sides or divide both sides by the same nonzero number.

E X A M P L E   1

Using the properties of equality to solve linear equations

Solve each equation and check.

a) $x - 5 = -13$ **b)** $\dfrac{2}{3}a = -4$

Solution

a) We can isolate the variable x by adding 5 to each side of the equation:

$$x - 5 = -13 \qquad \text{Original equation}$$
$$x - 5 + 5 = -13 + 5 \quad \text{Add 5 to each side.}$$
$$x = -8 \qquad \text{Simplify.}$$

All of the equations are equivalent, and only -8 satisfies the last equation. So -8 should be the only solution to the original equation. To check, replace x with -8:

$$x - 5 = -13$$
$$-8 - 5 = -13 \quad \text{Correct}$$

By checking, we are sure that the solution set is $\{-8\}$.

b) We can isolate a by multiplying each side of the equation by $\frac{3}{2}$:

$$\frac{2}{3}a = -4 \qquad \text{Original equation}$$

$$\frac{3}{2} \cdot \frac{2}{3}a = \frac{3}{2}(-4) \quad \text{Multiply each side by } \tfrac{3}{2}.$$

$$a = -6 \qquad \text{Simplify.}$$

Since $\frac{2}{3}(-6) = -4$ is correct, the solution set is $\{-6\}$.

Now do Exercises 1–8

In Example 2 we will solve equations that require several steps and more than one property of equality.

E X A M P L E   **2**

Using the addition and multiplication properties of equality

Solve each equation and check.

a) $3x + 5 = 9$ b) $2b - 3 = 3 + 4(b - 1)$

Solution

a) We can isolate the variable x by subtracting 5 from each side of the equation and then dividing each side by 3:

$$3x + 5 = 9 \qquad \text{Original equation}$$
$$3x + 5 - 5 = 9 - 5 \qquad \text{Subtract 5 from each side.}$$
$$3x = 4 \qquad \text{Simplify.}$$
$$\frac{3x}{3} = \frac{4}{3} \qquad \text{Divide each side by 3.}$$
$$x = \frac{4}{3} \qquad \text{Simplify.}$$

To check, replace x with $\frac{4}{3}$:

$$3x + 5 = 9$$
$$3\left(\frac{4}{3}\right) + 5 = 9 \qquad \text{Correct}$$

By checking, we are sure that the solution set is $\left\{\frac{4}{3}\right\}$.

b) Before we can apply the properties of equality we simplify the right side:

$$2b - 3 = 3 + 4(b - 1) \qquad \text{Original equation}$$
$$2b - 3 = 4b - 1 \qquad \text{Simplify.}$$
$$2b = 4b + 2 \qquad \text{Add 3 to each side.}$$
$$-2b = 2 \qquad \text{Subtract } 4b \text{ from each side.}$$
$$b = -1 \qquad \text{Divide each side by } -2.$$

Check -1 in the original equation $2b - 3 = 3 + 4(b - 1)$:

$$2(-1) - 3 = 3 + 4(-1 - 1) \qquad \text{Replace } b \text{ with } -1.$$
$$-5 = -5 \qquad \text{Correct}$$

Since -1 satisfies the original equation, we can be sure that the solution set is $\{-1\}$.

Now do Exercises 9–16

An **identity** is an equation that is satisfied for every real number for which both sides are defined. Equations such as $x + 1 = 1 + x$, $\frac{1}{x} = \frac{1}{x}$, and $2(3x) = 6x$ are identities. A **conditional equation** has at least one solution, but is not an identity. The equations that we solved in Examples 1 and 2 are conditional equations. (They are satisfied on the condition that the appropriate number is chosen to replace the variable.) An equation that has no solution is called an **inconsistent equation.** The equation $x + 1 = x + 2$ is inconsistent.

If an equation involves fractions, it is usually a good idea to multiply each side by the least common denominator to eliminate all of the fractions. If an equation involves decimals, then it is usually a good idea to multiply both sides of the equation by a power of 10 that eliminates all of the decimals. We illustrate these techniques in Example 3.

EXAMPLE 3

Equations with fractions or decimals

Solve each equation. Identify each equation as a conditional equation, an inconsistent equation, or an identity.

a) $\dfrac{1}{2}y - \dfrac{1}{3}y = y - \dfrac{5}{6}y$

b) $0.1b - 0.03 = 0.03b + 0.05$

c) $\dfrac{w}{4} - \dfrac{w}{5} = \dfrac{w}{20} + \dfrac{1}{10}$

Solution

a) Multiply each side by 6, the least common denominator:

$$\frac{1}{2}y - \frac{1}{3}y = y - \frac{5}{6}y \qquad \text{Original equation}$$

$$6\left(\frac{1}{2}y - \frac{1}{3}y\right) = 6\left(y - \frac{5}{6}y\right) \qquad \text{Multiply each side by 6.}$$

$$3y - 2y = 6y - 5y \qquad \text{Distributive property}$$

$$y = y \qquad \text{Simplify.}$$

The equation $y = y$ is satisfied by every real number. So the solution set to the original equation is the set of all real numbers, which is written symbolically as R or $(-\infty, \infty)$. The equation is an identity.

b) First multiply each side by 100 to eliminate the decimals:

$$0.1b - 0.03 = 0.03b + 0.05 \qquad \text{Original equation}$$

$$100(0.1b - 0.03) = 100(0.03b + 0.05) \qquad \text{Multiply by 100.}$$

$$10b - 3 = 3b + 5 \qquad \text{Distributive property}$$

$$10b = 3b + 8 \qquad \text{Add 3 to each side.}$$

$$7b = 8 \qquad \text{Subtract } 3b \text{ from each side.}$$

$$b = \frac{8}{7} \qquad \text{Divide each side by 7.}$$

Check in the original equation. The solution set is $\left\{\dfrac{8}{7}\right\}$, and the equation is a conditional equation.

c) Multiply each side by 20, the least common denominator:

$$\frac{w}{4} - \frac{w}{5} = \frac{w}{20} + \frac{1}{10} \qquad \text{Original equation}$$

$$20\left(\frac{w}{4} - \frac{w}{5}\right) = 20\left(\frac{w}{20} + \frac{1}{10}\right) \qquad \text{Multiply by 20.}$$

$$5w - 4w = w + 2 \qquad \text{Distributive property}$$

$$w = w + 2 \qquad \text{Simplify.}$$

$$0 = 2 \qquad \text{Subtract } w \text{ from each side.}$$

The equation $0 = 2$ is not satisfied by any real number. So the original equation has no solution and is an inconsistent equation. The solution set is the empty set, $\varnothing$.

> **Now do Exercises 17–28**

⟨2⟩ Formulas

A **formula** or **literal equation** is an equation involving two or more variables. The process of rewriting a formula for one variable in terms of the others is called **solving for a certain variable.** To solve a formula for a certain variable, we use the same techniques that we use in solving equations containing only one variable.

EXAMPLE 4

Solving for a certain variable

Solve $P = 2L + 2W$ for W.

Solution

To solve for W we can start with $2L + 2W = P$:

$$2L + 2W = P \qquad \text{Original equation}$$

$$2W = P - 2L \qquad \text{Subtract } 2L \text{ from each side.}$$

$$W = \frac{P - 2L}{2} \qquad \text{Divide each side by 2.}$$

The equation solved for W is $W = \frac{P - 2L}{2}$.

 Now do Exercises 29–42

⟨3⟩ Translating Verbal Expressions into Algebraic Expressions

The mathematical operation of addition can be indicated verbally by words such as sum, added to, more than, and increased by. Subtraction can be indicated by words such as subtracted from, less than, difference, and decreased by. Multiplication can be indicated by words such as product, twice, and a fraction or percent of. Division is indicated by ratio, quotient, and divided by.

EXAMPLE 5

Writing algebraic expressions

Translate each verbal expression into an algebraic expression.

a) The sum of a and b

b) Twelve percent of x

c) The quotient of w and 4

d) The number x decreased by 6

Solution

a) Because sum means addition, the sum of a and b is expressed as $a + b$.

b) A percent of a number is the product of the percent and the number. So twelve percent of x is expressed as $0.12x$.

c) Because quotient indicates division, the quotient of w and 4 is $\frac{w}{4}$.

d) Because decreased by indicates subtraction, the number x decreased by 6 is expressed as $x - 6$.

Now do Exercises 43–54

‹4› **Problem Solving**

In Examples 6 and 7 we apply the ideas of Example 5 to solve problems by first writing an equation that **models** or describes the problem and then solving the equation. In Example 6, we use the formula for the perimeter of a rectangle.

EXAMPLE 6

A geometric problem

The length of a rectangular patio is 1 foot larger than twice the width. If the perimeter is 92 feet, then what are the length and width?

Solution

Let W represent the width and $2W + 1$ represent the length of the patio as shown in Fig. R.6. Since the perimeter of a rectangle is twice the width plus twice the length ($P = 2W + 2L$), we can write the following equation:

$$2W + 2(2W + 1) = 92 \quad \text{2W + 2L = P}$$
$$2W + 4W + 2 = 92 \quad \text{Distributive property}$$
$$6W = 90 \quad \text{Simplify.}$$
$$W = 15 \quad \text{Divide each side by 6.}$$
$$2W + 1 = 31 \quad \text{Evaluate } 2W + 1 \text{ with } W = 15.$$

So the width is 15 feet and the length is 31 feet. Since $2(15) + 2(31) = 92$, we can be sure that the answer is correct.

Now do Exercises 55–56

W

$2W + 1$

Figure R.6

In Example 7 we will use the formula $D = RT$, which is the formula for uniform motion (motion at a constant rate).

EXAMPLE 7

A uniform-motion problem

A 44-foot-wide highway has concrete lanes and asphalt shoulders of equal width, as shown in Fig. R.7. A turtle crossing the highway travels his usual speed on the shoulders and 2 feet per hour faster on the concrete lanes. If it takes him 3 hours to cross one shoulder and 4 hours to cross the concrete lanes, then what is his usual speed and what is his speed on the concrete?

Solution

Let x represent the turtles usual speed in feet per hour and $x + 2$ represent his speed on the concrete lanes. Make a table showing rate, time, and distance for the asphalt shoulders and the concrete, using the formula $D = RT$. Note that it takes him 6 hr to cross both asphalt shoulders.

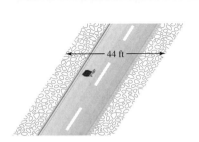

 44 ft

Figure R.7

	Rate	Time	Distance
Asphalt shoulders	x ft/hr	6 hr	$6x$ ft
Concrete lanes	$x + 2$ ft/hr	4 hr	$4(x + 2)$ ft

Since the total distance is 44 feet, we can write the following equation:

$$6x + 4(x + 2) = 44 \quad \text{Total distance is 44 feet.}$$
$$6x + 4x + 8 = 44 \quad \text{Distributive property}$$
$$10x + 8 = 44 \quad \text{Simplify.}$$
$$10x = 36 \quad \text{Subtract 8 from each side.}$$
$$x = 3.6 \quad \text{Divide each side by 10.}$$
$$x + 2 = 5.6 \quad \text{Evaluate } x + 2 \text{ with } x = 3.6.$$

So his usual speed is 3.6 ft/hr and his speed on the concrete is 5.6 ft/hr. At 3.6 ft/hr for 6 hr, his distance is 21.6 ft, and at 5.6 ft/hr for 4 hr his distance is 22.4 ft. Since 21.6 ft plus 22.4 ft is 44 ft, we can be sure that the answer is correct.

> Now do Exercises 57–58

In mixture problems the solutions might contain fat, alcohol, salt, or some other substance. We always assume that the substance in the solution neither appears nor disappears in the process. For example, if there are 3 grams of salt in one glass of water and 5 grams in another, then there are exactly 8 grams in a mixture of the two glasses of water.

EXAMPLE 8

A mixture problem

A 40-pound bag of potting soil contains 10% sand. How many pounds of sand must be added to get a mixture that is 20% sand?

Solution

Let x represent the number of pounds of sand to be added to the 40-pound bag. We can make a table as follows:

	Amount	% sand	Amount of sand
Original bag	40 lb	10%	0.10(40) lb
Sand added	x lb	100%	x lb
Mixture	$x + 40$ lb	20%	$0.20(x + 40)$ lb

Since the amount of sand in the final mixture is the sum of the sand in the original bag and the amount added, we can write the following equation:

$$0.10(40) + x = 0.20(x + 40) \quad \text{Total amounts of sand}$$
$$4 + x = 0.20x + 8 \quad \text{Distributive property}$$
$$40 + 10x = 2x + 80 \quad \text{Multiply each side by 10.}$$
$$8x = 40 \quad \text{Subtract } 2x; \text{ subtract 40.}$$
$$x = 5 \quad \text{Divide each side by 8.}$$

So to get 20% sand, 5 pounds of sand should be added. Note that the original bag contains 4 pounds of sand and that adding 5 more gives 9 pounds of sand out of 45 pounds which is 20% sand.

> Now do Exercises 59–64

〈5〉 Inequalities

The inequality symbols that we use are $<$ (less than), $\leq$ (less than or equal to), $>$ (greater than), and $\geq$ (greater than or equal to). To indicate that x is between a and b, where $a < b$, we often use the compound inequality $a < x < b$. Inequalities are solved in the same manner that we solve equations. However, to obtain an equivalent inequality when each side is multiplied or divided by a negative number the inequality symbol must be reversed.

E X A M P L E   9

Solving inequalities

Solve each inequality. State the solution set in interval notation, and graph the solution set.

a) $5x - 4 \geq 6$

b) $-3x + 5 < x - 15$

c) $-3 \leq 4x + 1 < 13$

Solution

a) To isolate x add 4 to each side and then divide each side by 5:

$$5x - 4 \geq 6 \qquad \text{Original inequality}$$

$$5x \geq 10 \qquad \text{Add 4 to each side.}$$

$$\frac{5x}{5} \geq \frac{10}{5} \qquad \text{Divide each side by 5.}$$

$$x \geq 2 \qquad \text{Simplify.}$$

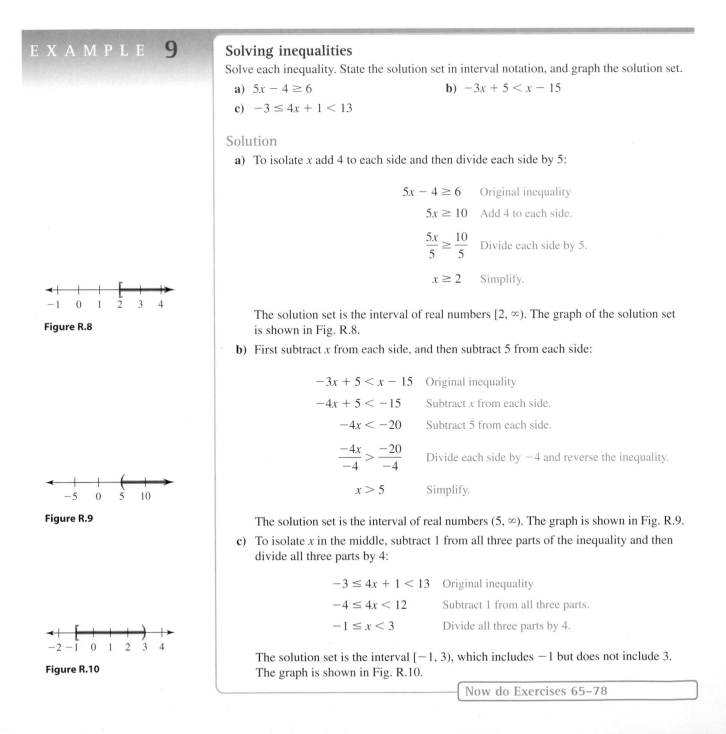

Figure R.8

The solution set is the interval of real numbers $[2, \infty)$. The graph of the solution set is shown in Fig. R.8.

b) First subtract x from each side, and then subtract 5 from each side:

$$-3x + 5 < x - 15 \qquad \text{Original inequality}$$

$$-4x + 5 < -15 \qquad \text{Subtract } x \text{ from each side.}$$

$$-4x < -20 \qquad \text{Subtract 5 from each side.}$$

$$\frac{-4x}{-4} > \frac{-20}{-4} \qquad \text{Divide each side by } -4 \text{ and reverse the inequality.}$$

$$x > 5 \qquad \text{Simplify.}$$

Figure R.9

The solution set is the interval of real numbers $(5, \infty)$. The graph is shown in Fig. R.9.

c) To isolate x in the middle, subtract 1 from all three parts of the inequality and then divide all three parts by 4:

$$-3 \leq 4x + 1 < 13 \qquad \text{Original inequality}$$

$$-4 \leq 4x < 12 \qquad \text{Subtract 1 from all three parts.}$$

$$-1 \leq x < 3 \qquad \text{Divide all three parts by 4.}$$

Figure R.10

The solution set is the interval $[-1, 3)$, which includes -1 but does not include 3. The graph is shown in Fig. R.10.

Now do Exercises 65–78

⟨1⟩ Solving Linear Equations

Solve each equation and check your answer. See Examples 1 and 2.

1. $x - 9 = -2$ **2.** $w - 8 = 7$

3. $n + 5 = -3$ **4.** $z + 4 = 21$

5. $3a = 51$ **6.** $-5b = 45$

7. $\frac{3}{4}x = -6$ **8.** $-\frac{5}{3}m = 15$

9. $4a - 1 = 49$ **10.** $3b + 2 = 0$

11. $14 - 2x = 6 - x$ **12.** $-7 - 5x = 12 - 4x$

13. $2x - 3 = 4x + 9$

14. $5 - 4x = 3 - 2x$

15. $x - 3 = 2 + 3(x + 1)$

16. $-3(x - 4) = 2x + 7$

Solve each equation. Identify each equation as a conditional equation, an inconsistent equation, or an identity. See Example 3.

17. $\frac{13}{15}x - \frac{4}{5}x = \frac{1}{5}x - \frac{2}{15}x$

18. $x + \frac{2}{3} = \frac{1}{3}(3x + 1) + \frac{1}{3}$

19. $\frac{w}{12} - \frac{w}{4} = \frac{w}{3} - 12$

20. $\frac{a}{6} - 5 = \frac{a}{15} - 2$

21. $0.05a - 0.7 = 0.12a + 0.7$

22. $0.03(z - 4) = 0.05z + 0.8$

23. $\frac{1}{8}y - \frac{1}{9}y = \frac{1}{72}y + \frac{1}{2}$

24. $\frac{1}{6}m + \frac{1}{7}m = \frac{13}{42}m - \frac{1}{21}$

25. $\frac{5}{3}t - 2\left(\frac{2}{3}t + 1\right) = 3\left(\frac{t}{3} - \frac{1}{9}\right) - \frac{7}{3}t$

26. $\frac{3}{2}v - 4\left(\frac{v}{2} + \frac{5}{2}\right) = \frac{v}{2} - (v + 10)$

27. $0.001x + 0.02 = 0.2(0.1x - 0.03)$

28. $0.2(0.3q + 0.04) = 0.005q - 0.087$

⟨2⟩ Formulas

Solve each formula for the indicated variable. See Example 4.

29. $D = RT$ for R

30. $E = mc^2$ for m

31. $K = \frac{1}{2}mv^2$ for m

32. $A = \frac{1}{2}bh$ for b

33. $P = 2L + 2W$ for L

34. $A = \frac{1}{2}h(b_1 + b_2)$ for b_2

35. $A = P + Prt$ for r

36. $2x - 3y = 6$ for y

Solve each problem.

37. *Traveling by bus.* A bus averaged 40 miles per hour while traveling from New Orleans to Memphis. If the distance is 400 miles, then how long did the bus take for the trip?

38. *Right triangle.* In a right triangle the perpendicular sides are called legs. If the area of a right triangle is 10 square meters and one leg is 4 meters, then what is the length of the other leg?

39. *Rectangular field.* If the length of a rectangular field is 45 meters and the perimeter is 150 meters, then what is the width?

40. *CD case.* If the length of a rectangular plastic CD case is 14 centimeters and the perimeter is 53 centimeters, then what is the width?

41. *Kinetic energy.* The kinetic energy K in Joules for an object of mass m kilograms with velocity v meters per second is given by $K = \frac{1}{2}mv^2$. If the kinetic energy for an object with velocity 30 meters per second is 1800 Joules, then what is the mass of the object?

42. *Upper base.* The height of a trapezoid is 4 centimeters, and its area is 40 square centimeters. If the lower base is 12 centimeters, then what is the length of the upper base?

⟨3⟩ Translating Verbal Expressions into Algebraic Expressions

Translate each verbal expression into an algebraic expression. See Example 5.

43. The sum of a^2 and b^2

44. The number x increased by 5

45. The number y decreased by 6

46. The difference between a and b

47. The product of a and b^2

48. Ten percent of x

49. The quotient of x and y

50. The number 14 divided by x

51. One-half of x

52. Two-thirds of y

53. Twice the sum of a and b

54. The square of the sum of a and b

⟨4⟩ Problem Solving
Solve each problem. See Examples 6–8.

55. *Rectangular planter.* The width of a rectangular planter is 6 inches less than its length. If the perimeter of the planter is 84 inches, then what are the length and width?

56. *Rectangular reflecting pool.* The length of a rectangular reflecting pool is 5 meters less than twice the width. If the perimeter of the pool is 170 meters, then what are the length and width?

57. *El Paso to L.A.* On Monday, Chip drove from El Paso to Phoenix in 8 hours. On Tuesday he drove from Phoenix to Los Angeles in 10 hours. If he averaged 15 miles per hour more on the first day and the total trip was 840 miles, then what was his average speed on the first day?

58. *L.A. to Portland.* On Wednesday, Chip averaged 50 miles per hour driving from Los Angeles to San Francisco. On Thursday, he continued on to Portland, averaging 64 mph. If his travel time on Wednesday was 2 hours less than his travel time on Thursday and the total trip from L.A. to Portland was 1040 miles, then what was his traveling time on Wednesday?

59. *Mixing concrete.* Concrete is a mixture of aggregate, cement, and water. A concrete truck contains 10,000 pounds of concrete that is 17% cement. How much cement must be added to the mixture to get the mixture up to 18% cement?

60. *Diluting a solution.* How many ounces of pure water must be added to 100 ounces of a saline solution that is 12% salt to get a solution that is 8% salt?

61. *Mixing alcohol.* How many liters of a 50% alcohol solution must be added to 10 liters of a 20% alcohol solution to obtain a solution that is 30% alcohol?

62. *Mixing punch.* One hundred liters of fruit punch that is 30% fruit juice is mixed with 200 liters of another fruit punch. The result is a mixture that is 20% juice. What is the percentage of fruit juice in the 200 liters of punch?

63. *Catching a speeder.* A police officer was parked on the shoulder of a highway when he was passed by a speeder. It took the officer 2 minutes to get his car started. He then averaged 100 miles per hour for 12 minutes to catch the speeder. How fast was the speeder traveling?

64. *Catching up.* At 7 A.M. the Garcias left the campground and headed east at 80 kilometers per hour. At 7:20 the Andersons left the same campground and headed east on the same road at 100 kilometers per hour. At what time will the Andersons catch up with the Garcias?

⟨5⟩ Inequalities
Solve each inequality. State the solution set using interval notation, and graph the solution set. See Example 9.

65. $3x - 1 \geq 14$

66. $2x + 5 \leq 17$

67. $4 - 3y < 0$

68. $5 - t > 0$

69. $-\dfrac{1}{2}n + 6 < 7$

70. $-\dfrac{3}{4}m - 1 > 5$

71. $5x + 7 < 2x - 8$

72. $6w - 9 > w + 31$

73. $-2z + 3 < z - 6$

74. $-5x - 8 < 2x + 13$

75. $-1 \leq 2b + 3 < 19$

76. $1 < 5a - 4 \leq 21$

77. $-5 < 3 - 2w < 31$

78. $4 \leq 1 - x \leq 5$

R.3   Linear Equations and Inequalities in Two Variables

This section is a review of Chapter 3 of this text. All topics in this review section are explained in greater detail in Chapter 3.

⟨1⟩ Graphing Lines in the Coordinate Plane

A **linear equation in two variables** is an equation of the form $Ax + By = C$, where A, B, and C are real numbers, with A and B not both equal to zero. The graph of a linear equation in two variables is a straight line in the rectangular coordinate system. The graph is a picture of the set of all ordered pairs that satisfy the equation. If $A = 0$ and $B \neq 0$, then the graph is a horizontal line. If $B = 0$ and $A \neq 0$, then the graph is a vertical line. A point at which a line crosses the x-axis is called the **x-intercept.** A point at which a line crosses the y-axis is called the **y-intercept.**

E X A M P L E   1

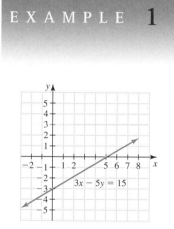

Figure R.11

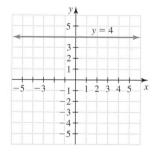

Figure R.12

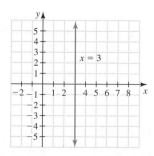

Figure R.13

Graphing linear equations using the intercepts

Graph each equation and identify all intercepts.

   a) $3x - 5y = 15$ **b)** $y = 4$ **c)** $x = 3$

Solution

a) To find the y-intercept let $x = 0$ in $3x - 5y = 15$:

$$3(0) - 5y = 15$$
$$y = -3$$

The y-intercept is $(0, -3)$. To find the x-intercept let $y = 0$ in $3x - 5y = 15$:

$$3x - 5(0) = 15$$
$$x = 5$$

The x-intercept is $(5, 0)$. Now let $x = 10$ in $3x - 5y = 15$:

$$3(10) - 5y = 15$$
$$-5y = -15$$
$$y = 3$$

So the line goes through the intercepts and $(10, 3)$. Plot these three points, and draw a line through them as shown in Fig. R.11.

b) Since the coefficient of x is zero, the graph is a horizontal line with y-intercept $(0, 4)$. Note that any number can be used for x as long as we choose $y = 4$. So the ordered pairs $(-1, 4)$, $(1, 4)$, and $(2, 4)$ also satisfy the equation. Plot these points, and draw a line through them as shown in Fig. R.12.

c) Since the coefficient of y is zero, the graph is a vertical line with x-intercept $(3, 0)$. The graph also goes through $(3, -1)$ and $(3, 2)$. Plot these points, and draw a line through them as shown in Fig. R.13.

> Now do Exercises 1–10

⟨2⟩ Slope

The **slope** of a line is the number obtained by dividing the change in y-coordinate by the change in x-coordinate for any two points on a line. The change in y-coordinate and the change in x-coordinate are also called the **rise** and the **run,** respectively. The slope of the line containing the points (x_1, y_1) and (x_2, y_2) is given by

$$m = \frac{\text{change in } y\text{-coordinate}}{\text{change in } x\text{-coordinate}} = \frac{\text{rise}}{\text{run}} = \frac{y_2 - y_1}{x_2 - x_1},$$

provided that $x_2 - x_1 \neq 0$. If $x_2 - x_1 = 0$, then the line is a vertical line and the slope of the line is not defined. Parallel lines have the same slope. If m_1 and m_2 are the slopes of two perpendicular lines, then $m_1 = -\dfrac{1}{m_2}$.

E X A M P L E   2

Finding slopes

Find the slope of each line.

    a) The line through $(-3, 5)$ and $(-1, -2)$

    b) The line through $(0, 2)$ and $(5, 2)$

    c) The line through $(3, 0)$ and $(3, 6)$

    d) A line parallel to the line through $(-1, 2)$ and $(3, 4)$

    e) A line perpendicular to the line through $(0, 6)$ and $(2, 0)$

Solution

    a) Use $(-3, 5)$ and $(-1, -2)$ in the formula $m = \frac{y_2 - y_1}{x_2 - x_1}$:

$$m = \frac{-2 - 5}{-1 - (-3)} = \frac{-7}{2} = -\frac{7}{2}$$

    b) Use $(0, 2)$ and $(5, 2)$ in the formula $m = \frac{y_2 - y_1}{x_2 - x_1}$:

$$m = \frac{2 - 2}{5 - 0} = 0$$

    c) The line through $(3, 0)$ and $(3, 6)$ is a vertical line and does not have slope.

    d) Use $(-1, 2)$ and $(3, 4)$ in the formula $m = \frac{y_2 - y_1}{x_2 - x_1}$:

$$m = \frac{4 - 2}{3 - (-1)} = \frac{2}{4} = \frac{1}{2}$$

    Any line parallel to the line through $(-1, 2)$ and $(3, 4)$ also has slope $\frac{1}{2}$.

    e) Use $(0, 6)$ and $(2, 0)$ in the formula $m = \frac{y_2 - y_1}{x_2 - x_1}$:

$$m = \frac{0 - 6}{2 - 0} = -3$$

    Any line perpendicular to the line through $(0, 6)$ and $(2, 0)$ has slope $\frac{1}{3}$.

> Now do Exercises 11–24

⟨3⟩ Equations of Lines in Slope-Intercept Form

The equation of the line with y-intercept $(0, b)$ and slope m is

$$y = mx + b.$$

The form $y = mx + b$ is called **slope-intercept form.** Of course, lines that do not have slope (vertical lines) cannot be written in this form. Every line has an equation in **standard form,** $Ax + By = C$, where A and B are not both zero. We can use the y-intercept and the slope to graph a line.

E X A M P L E   3

Using y-intercept and slope to graph a line

Identify the slope and y-intercept for each line, and then graph the line.

    a) $y = \dfrac{1}{2}x - 2$ **b)** $2x + 3y = 6$ **c)** $y = 6$

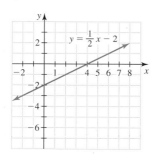

Figure R.14

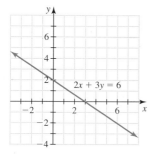

Figure R.15

Solution

a) The slope is $\frac{1}{2}$ and the y-intercept is $(0, -2)$. Since $\frac{1}{2} = \frac{\text{rise}}{\text{run}}$, start at $(0, -2)$ and move 1 unit upward and 2 units to the right to obtain a second point on the line, $(2, -1)$. Again rise 1 and run 2 to obtain a third point on the line, $(4, 0)$. Draw a line through these points as shown in Fig. R.14.

b) First solve $2x + 3y = 6$ for y:

$$2x + 3y = 6$$
$$3y = -2x + 6$$
$$y = -\frac{2}{3}x + 2$$

The slope is $-\frac{2}{3}$ and the y-intercept is $(0, 2)$. Start at $(0, 2)$ and move 2 units downward and 3 units to the right to obtain a second point on the line, $(3, 0)$. From $(3, 0)$ again move 2 units down and 3 units to the right to obtain a third point on the line, $(6, -2)$. Draw a line through these points as shown in Fig. R.15.

c) For $y = 6$ the slope is 0 and the y-intercept is $(0, 6)$. So the graph is a horizontal line through $(0, 6)$ as shown in Fig. R.16.

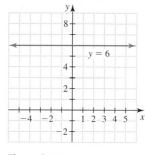

Figure R.16

> Now do Exercises 25–34

If we can determine the y-intercept and the slope from a description of a line, then we can write its equation using the slope-intercept form.

E X A M P L E   4

Writing the equation for a line using slope-intercept form

Write the equation in slope-intercept form for each line.

a) The line through $(0, 3)$ and $(4, 0)$

b) The line through $(0, \ \ 4)$ that is parallel to $y = \frac{3}{4}x - 5$

c) The line through $(0, -2)$ that is perpendicular to $2x - 5y = 3$

Solution

a) The line through $(0, 3)$ and $(4, 0)$ has slope $-\frac{3}{4}$ and y-intercept $(0, 3)$. So the equation is $y = -\frac{3}{4}x + 3$.

b) The line $y = \frac{3}{4}x - 5$ has slope $\frac{3}{4}$ and so does any line parallel to it. So the equation of the line through $(0, -4)$ that is parallel to $y = \frac{3}{4}x - 5$ is $y = \frac{3}{4}x - 4$.

c) Solve $2x - 5y = 2$ for y to determine its slope:

$$2x - 5y = 3$$
$$-5y = -2x + 3$$
$$y = \frac{2}{5}x - \frac{3}{5}$$

The slope of $2x - 5y = 3$ is $\frac{2}{5}$, and any line perpendicular to it has slope $-\frac{5}{2}$. The equation of the line through $(0, -2)$ with slope $-\frac{5}{2}$ is $y = -\frac{5}{2}x - 2$.

> Now do Exercises 35–42

‹4› **The Point-Slope Form**

The equation of the line through the point (x_1, y_1) with slope m is

$$y - y_1 = m(x - x_1).$$

This form is called the **point-slope form** for the equation of a line. To write the equation of a line with slope-intercept form you must know the slope and the y-intercept. Using the point-slope form, the point can be any point on the line.

EXAMPLE 5

Writing the equation for a line using point–slope form

Find the equation for each line. Write the answer in standard form $Ax + By = C$, where A, B, and C are integers.

a) The line through $(1, 5)$ and $(4, 2)$

b) The line through $(2, -3)$ that is parallel to $y = \frac{1}{2}x - 2$

c) The line through $(1, -4)$ that is perpendicular to $3x - y = 1$

Solution

a) The line through $(1, 5)$ and $(4, 2)$ has slope $\frac{2 - 5}{4 - 1}$ or -1. Now use one of the points, say, $(1, 5)$, and slope -1 in the point-slope form:

$$y - 5 = -1(x - 1)$$
$$y - 5 = -x + 1$$
$$x + y = 6$$

The equation of the line in standard form is $x + y = 6$. Note that this answer is not unique. Multiplying each side of $x + y = 6$ by any nonzero integer will give an equivalent equation.

b) The line $y = \frac{1}{2}x - 2$ has slope $\frac{1}{2}$ and so does any line parallel to it. So use the point $(2, -3)$ and slope $\frac{1}{2}$ in the point-slope form:

$$y - (-3) = \frac{1}{2}(x - 2) \quad \text{Point-slope form}$$

$$y + 3 = \frac{1}{2}x - 1$$

$$-\frac{1}{2}x + y = -4$$

$$x - 2y = 8 \qquad \text{Multiply each side by } -2.$$

The equation of the line in standard form is $x - 2y = 8$.

c) Solve $3x - y = 1$ for y to get $y = 3x - 1$. This line has slope 3, and any line perpendicular to it has slope $-\frac{1}{3}$. Use the point $(1, -4)$ and slope $-\frac{1}{3}$ in the point-slope form:

$$y - (-4) = -\frac{1}{3}(x - 1) \quad \text{Point-slope form}$$

$$y + 4 = -\frac{1}{3}x + \frac{1}{3}$$

$$3y + 12 = -x + 1 \qquad \text{Multiply each side by 3.}$$

$$x + 3y = -11 \qquad \text{Standard form}$$

The equation of the line in standard form is $x + 3y = -11$.

Now do Exercises 43–50

⟨5⟩ Variation

Some basic relationships between variables are expressed in terms of variation. The statement **"y varies directly as x"** or **"y is directly proportional to x"** means that $y = kx$. The statement **"y varies inversely as x"** or **"y is inversely proportional to x"** means that $y = \frac{k}{x}$. The statement **"y varies jointly as x and z"** or **"y is jointly proportional to x and z"** means that $y = kxz$. In each case, k is a nonzero constant and is called the **variation constant.**

EXAMPLE 6

Using variation terms

Solve each problem.

a) Distance varies directly with the average speed. Willy drove 200 miles with an average speed of 40 mph. Find the constant of variation.

b) The time that it takes to harvest a field of beans varies inversely with the number of pickers. If 10 pickers can harvest the field in 3 hours, then how long would it take 15 pickers?

c) The cost of waterproofing a rectangular roof varies jointly with the length and the width. If a 30-ft by 40-ft roof costs $3072, then what is the cost for a 25-ft by 50-ft roof?

Solution

a) Since distance D varies directly with the average speed R, we have $D = kR$ for some constant k. Since $D = 200$ when $R = 40$, we have $200 = k(40)$. Since 200 miles divided by 40 mph is 5 hours, the constant is 5 hours.

b) Since the time t varies inversely with the number of pickers n, we have $t = \frac{k}{n}$ for some constant k. Since $t = 3$ hr when $n = 10$ pickers, we have $3 = \frac{k}{10}$ or $k = 30$. Since 30 is obtained by multiplying hours and pickers, the units for the constant are picker-hours. It takes 30 picker-hours to harvest the field. So 1 picker can do it in 30 hours, 2 pickers in 15 hours, 3 pickers in 10 hours, and so on.

c) Since the cost C varies jointly as the length L and width W, we have $C = kLW$ for some constant k. Since $C = \$3072$ when $W = 30$ ft and $L = 40$ ft, we have $3072 = k(30)(40)$, or $k = 2.56$. Since k is obtained by dividing dollars by square feet, k is $2.56 per square foot. The cost for a 25-ft by 50-ft roof is $2.56(25)(50)$ or $3200.

Now do Exercises 51–56

⟨6⟩ Graphing Linear Inequalities in Two Variables

Linear inequalities in two variables have the same form as linear equations in two variables. If A, B, and C are real numbers with A and B not both zero, then $Ax + By < C$ is a **linear inequality in two variables.** In place of $<$ we can also use $\leq$, $>$, or $\geq$. The solution set to a linear inequality in two variables consists of infinitely many ordered pairs that lie in a region of the coordinate plane. So to graph a linear inequality, we first graph the boundary line $Ax + By = C$ and then use a test point to determine which side of the line satisfies the inequality. All points on one side of the line satisfy $Ax + By > C$, and all points on the other side satisfy $Ax + By < C$.

EXAMPLE 7

Graphing linear inequalities in two variables

Graph the solution set to each inequality in the coordinate plane.

    a) $3x - 5y > 30$ **b)** $y \le -2x + 3$ **c)** $x < 3$

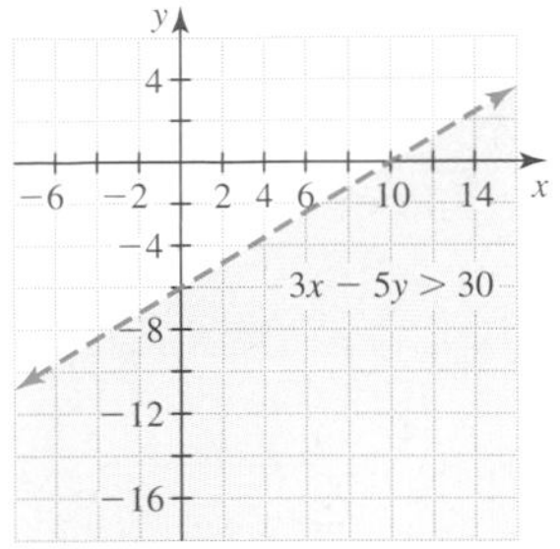

Solution

a) First graph the boundary line $3x - 5y = 30$ by using its x-intercept $(10, 0)$ and its y-intercept $(0, -6)$. Draw the line dashed because it is not included in the solution set to the inequality. Next select a test point on one side of the line, say $(1, 1)$. Since $3(1) - 5(1) > 30$ is incorrect, all points on the other side of the line must satisfy the inequality. Shade that region as shown in Fig. R.17.

b) First graph the boundary line $y = -2x + 3$ using its slope -2 and y-intercept $(0, 3)$. The line is drawn solid because it is included in the solution set to $y \le -2x + 3$. Next select a test point on one side of the line, say $(-1, 0)$. Because $0 \le -2(-1) + 3$ is correct, all points on that side of the line satisfy the inequality. Shade that region as shown in Fig. R.18.

c) First graph the vertical boundary line $x = 3$ as a dashed line. Select a test point, say $(0, 0)$. Since $0 < 3$ is correct, shade the region to the left of the line $x = 3$ as shown in Fig. R.19.

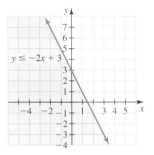

Figure R.17

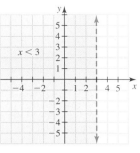

Figure R.19

Now do Exercises 57–70

Figure R.18

R.3 Exercises

⟨1⟩ Graphing Lines in the Coordinate Plane

Graph each equation and identify all intercepts. See Example 1.

  1. $3x - 4y = 12$ **2.** $x - 2y = 10$

3. $2x + y = 6$

4. $3x + 7y = 21$

5. $x = -3$

6. $x = 5$

7. $y = 2$

8. $y = -4$

9. $y = \dfrac{1}{2}x - 30$

10. $y = -\dfrac{2}{3}x + 20$

⟨2⟩ **Slope**

Find the slope of each line. See Example 2.

11. The line through $(-2, 1)$ and $(3, 6)$

12. The line through $(-1, -3)$ and $(5, 5)$

13. The line through $(-3, 3)$ and $(1, -1)$

14. The line through $(0, 0)$ and $(-5, -5)$

15. The line through $(2, 1)$ and $(2, 7)$

16. The line through $(-3, -1)$ and $(-3, 4)$

17. The line through $(4, 1)$ and $(-2, 1)$

18. The line through $(-3, 5)$ and $(3, 5)$

19. A line parallel to the line through $(1, 4)$ and $(4, 16)$

20. A line parallel to the line through $(3, 2)$ and $(-6, 2)$

21. A line perpendicular to the line through $(-1, -1)$ and $(2, 3)$

22. A line perpendicular to the line through $(-5, 8)$ and $(5, -8)$

23. A line perpendicular to the line $x = 3$

24. A line parallel to the line $y = -5$

⟨3⟩ **Equations of Lines in Slope-Intercept Form**

Identify the slope and y-intercept for each line, and then graph the line. See Example 3.

25. $y = \dfrac{1}{3}x + 1$

26. $y = \dfrac{2}{3}x - 2$

27. $y = -3x + 4$

28. $y = 2x - 5$

29. $x - y = 5$

30. $x + 2y = 4$

31. $3x - 5y = 10$ **32.** $-2x + 3y = 9$

33. $y = 4$ **34.** $y = -5$

Write the equation in slope-intercept form for each line. See Example 4.

35. The line through $(0, -2)$ and $(5, 0)$
36. The line through $(0, 5)$ and $(-3, -4)$
37. The line through $(0, 6)$ that is parallel to $y = \frac{2}{7}x + 3$
38. The line through $(0, -2)$ that is parallel to $y = -5x - 4$
39. The line through $(0, 12)$ that is perpendicular to $x - 4y = 1$
40. The line through $(0, -14)$ that is perpendicular to $3x - y = 2$
41. The line through $(0, 3)$ that is parallel to $y = -1$
42. The line through $(0, 5)$ that is perpendicular to $x = 3$

⟨4⟩ **The Point-Slope Form**
Find the equation for each line. Write the answer in standard form $Ax + By = C$, where A, B, and C are integers. See Example 5.

43. The line through $(-2, 4)$ and $(3, 7)$
44. The line through $(2, -5)$ and $(3, 9)$
45. The line through $(-1, 3)$ and $(5, 0)$
46. The line through $(2, 0)$ and $(-6, 8)$
47. The line through $(1, -4)$ that is parallel to $y = \frac{2}{3}x + 6$

48. The line through $(3, -5)$ that is parallel to $y = -\frac{1}{4}x - 9$
49. The line through $(2, -5)$ that is perpendicular to $2x + y = 5$
50. The line through $(3, -6)$ that is perpendicular to $4x - y = 2$

⟨5⟩ **Variation**
Solve each variation problem. See Example 6.

51. *Average speed.* Distance varies directly with the time. Billy drove 200 miles in 4 hours. Find the constant of variation.

52. *Hiking time.* Distance varies directly with the average speed. Cortez hiked 15 miles at 3 miles per hour. Find the constant of variation.

53. *Picking oranges.* The time that it takes to pick the entire orange grove varies inversely with the number of pickers. If 30 pickers can pick the entire grove in 14 hours, then how long would it take 40 pickers?

54. *Sharing cookies.* A box of cookies is divided among the cub scouts at the meeting. The number of cookies each scout receives varies inversely as the number of scouts in attendance. When 4 scouts are in attendance, each scout receives 12 cookies. How many cookies will each scout receive when 16 scouts are in attendance?

55. *Area of a rectangle.* The cost of wood laminate flooring for a rectangular room varies jointly as the length and width of the room. If the cost is $1148.16 for a 12-ft by 16-ft room, then what is the cost for a room that is 10 ft by 14 ft?

56. *Building bookcases.* The cost for a custom oak bookcase varies jointly with the width and height. If a bookcase that is 7 ft high and 30 in. wide costs $441, then what is the cost for a bookcase that is 32 in. wide and 6 ft high?

⟨6⟩ **Graphing Linear Inequalities in Two Variables**
Graph the solution set to each inequality in the coordinate plane. See Example 7.

57. $3x - 2y > 6$ **58.** $x - y < 5$

59. $x + 3y \leq 9$ **60.** $6x + y \geq 12$ **65.** $x < 2$ **66.** $x > -3$

61. $y \leq -x + 3$ **62.** $y \geq 2x + 1$ **67.** $x \geq -1$ **68.** $x \leq 5$

63. $y > 3x - 4$ **64.** $y < -2x + 2$ **69.** $y < 4$ **70.** $y > -2$

In This Section

R.4 Polynomials and Exponents

This section is a review of Chapter 4 of this text. All topics in this review section are explained in greater detail in Chapter 4.

⟨1⟩ Polynomials

A **polynomial** is a single term or a finite sum of terms. The **degree of a polynomial** in one variable is the highest power of the variable in the polynomial. The number preceding the variable in each term is called the **coefficient** of that variable or the coefficient of that term. A **monomial** has one term, a **binomial** has two terms, and a **trinomial** has three terms. For example, the polynomial $5x^2 - 2x + 3$ is a trinomial with degree two and the coefficient of x^2 is 5. We can also write

$$P = 5x^2 - 2x + 3 \quad \text{or} \quad P(x) = 5x^2 - 2x + 3.$$

If $x = 1$, then we can evaluate $5(1)^2 - 2(1) + 3$ to get 6. We say that the value of the polynomial is 6 when $x = 1$, or $P = 6$ when $x = 1$, or $P(1) = 6$ (read "P of 1 equals 6"). Polynomials can be added or subtracted by adding or subtracting like terms.

EXAMPLE **1**

Adding and subtracting polynomials

Perform the indicated operations.

a) $(x^2 - 6x + 3) + (-3x^2 - 5x - 4)$

b) $(-3x^3 - 4x + 2) - (-x^3 + 4x^2 - 5)$

Solution

a) $(x^2 - 6x + 3) + (-3x^2 - 5x - 4) = x^2 - 3x^2 - 6x - 5x + 3 - 4$
$$= -2x^2 - 11x - 1$$

b) $(-3x^3 - 4x + 2) - (-x^3 + 4x^2 - 5) = -3x^3 - 4x + 2 + x^3 - 4x^2 + 5$
$$= -2x^3 - 4x^2 - 4x + 7$$

> Now do Exercises 1–8

⟨2⟩ Multiplication of Polynomials

The **product rule for exponents** indicates that the exponents are added when multiplying powers of the same base. In symbols, $a^m \cdot a^n = a^{m+n}$ for any real number a and positive integers m and n. For example, $2x^3 \cdot 4x^2 = 8x^5$. We use the distributive property and the product rule for exponents to multiply polynomials.

EXAMPLE **2**

Multiplying polynomials

Find each product.

a) $3x(x^2 - 6x + 3)$ 　　　　　　　　b) $(w + 3)(w + 5)$

c) $(y - 1)(y^2 + 4y - 6)$

Solution

a) $3x(x^2 - 6x + 3) = 3x(x^2) + 3x(-6x) + 3x(3)$ 　Distributive property
$$= 3x^3 - 18x^2 + 9x \qquad\qquad \text{Multiply the monomials.}$$

b) $(w + 3)(w + 5) = (w + 3)w + (w + 3)5$ 　Distributive property
$$= w^2 + 3w + 5w + 15 \qquad \text{Distributive property}$$
$$= w^2 + 8w + 15 \qquad\qquad \text{Combine like terms.}$$

c) $(y - 1)(y^2 + 4y - 6) = (y - 1)y^2 + (y - 1)4y + (y - 1)(-6)$
$$= y^3 - y^2 + 4y^2 - 4y - 6y + 6$$
$$= y^3 + 3y^2 - 10y + 6$$

> Now do Exercises 9–24

⟨3⟩ Multiplication of Binomials

We can use the distributive property to multiply binomials as was done in Example 2(c). Because multiplication of binomials is done so frequently, we usually use the **FOIL method** instead. With FOIL we find the product of the first terms of each binomial, the product of the outer terms, the product of the inner terms, and finally the product of the last terms. In many cases, the product of the inner terms and the product of the outer terms are like terms and they can be combined.

E X A M P L E   **3**

Multiplying binomials using FOIL

Use the FOIL method to find each product.

    a) $(a + b)(c + d)$ **b)** $(2x + 3)(x - 5)$

    c) $(3a^2 - 5)(7a^2 + 2)$

Solution

    a) For $(a + b)(c + d)$ the product of the first terms is ac, the product of the outer terms is ad, the product of the inner terms is bc, and the product of the last terms is bd:

$$(a + b)(c + d) = ac + ad + bc + bd$$

In this case, there are no like terms to combine.

    b) $(2x + 3)(x - 5) = 2x \cdot x + (2x)(-5) + 3x + (3)(-5)$ FOIL

$$= 2x^2 - 10x + 3x - 15 \qquad \text{Simplify.}$$

$$= 2x^2 - 7x - 15 \qquad \text{Combine like terms.}$$

    c) $(3a^2 - 5)(7a^2 + 2) = 12a^4 - 13a^2 + 6a^2 - 10$ FOIL

$$= 12a^4 - 7a^2 - 10 \qquad \text{Combine like terms.}$$

> Now do Exercises 25–36

The idea of the FOIL method is to get the product of two binomials quickly. In Example 3 we showed more steps than are necessary. When you use the FOIL method, you should write only the steps that are necessary for you to get the correct product.

⟨**4**⟩ **Special Products**

The square of a sum, the square of a difference, and the product of a sum and a difference are called the **special products.** You can use FOIL to find these products, but it is better to learn the following rules for these products:

The Special Products

The square of a sum:	$(a + b)^2 = a^2 + 2ab + b^2$
The square of a difference:	$(a - b)^2 = a^2 - 2ab + b^2$
The product of a sum and a difference:	$(a + b)(a - b) = a^2 - b^2$

E X A M P L E   **4**

Using the special product rules

Find each product.

    a) $(x + 5)^2$ **b)** $(2x - 3)^2$ **c)** $(3w + 5)(3w - 5)$

Solution

    a) Use $(a + b)^2 = a^2 + 2ab + b^2$ with $a = x$ and $b = 5$:

$$(x + 5)^2 = x^2 + 2(x)(5) + 5^2$$

$$= x^2 + 10x + 25$$

    b) Use $(a - b)^2 = a^2 - 2ab + b^2$ with $a = 2x$ and $b = 3$:

$$(2x - 3)^2 = (2x)^2 - 2(2x)(3) + 3^2$$

$$= 4x^2 - 12x + 9$$

c) Use $(a + b)(a - b) = a^2 - b^2$ with $a = 3w$ and $b = 5$:

$$(3w + 5)(3w - 5) = (3w)^2 - 5^2$$
$$= 9w^2 - 25$$

Now do Exercises 37–48

⟨5⟩ Division of Polynomials

The **quotient rule for exponents** indicates that the exponents are subtracted when dividing powers of the same base. In symbols, $\frac{a^m}{a^n} = a^{m-n}$ for any nonzero real number a and positive integers m and n, where $m \geq n$. If $m < n$, then $\frac{a^m}{a^n} = \frac{1}{a^{n-m}}$. For example, $\frac{x^3}{x^9} = \frac{1}{x^6}$. If $a \div b = c$, then a is the **dividend**, b is the **divisor**, and c (or $a \div b$) is the **quotient**. We can use the quotient rule to divide a monomial by a monomial, but to divide polynomials with higher degrees we use a process similar to the long division process that is used to divide whole numbers.

EXAMPLE 5

Dividing polynomials

Find each quotient.

a) $\left(9x^6\right) \div \left(3x^4\right)$ **b)** $\left(12a^3 - 8a^2 + 4a\right) \div (2a)$

c) $\left(x^3 - 4x^2 + 9\right) \div (x - 3)$

Solution

a) $\left(9x^6\right) \div \left(3x^4\right) = \dfrac{9x^6}{3x^4} = \dfrac{9}{3}x^{6-4} = 3x^2$

b) $\left(12a^3 - 8a^2 + 4a\right) \div (2a) = \dfrac{12a^3 - 8a^2 + 4a}{2a}$

$$= \frac{12a^3}{2a} + \frac{-8a^2}{2a} + \frac{4a}{2a}$$

$$= 6a^2 - 4a + 2$$

c) When dividing by a binomial, use the long division process:

$$
\begin{array}{r}
x^2 - x - 3 \\
x - 3 \overline{\smash{\big)}\, x^3 - 4x^2 + 0 \cdot x + 9} \\
\underline{x^3 - 3x^2} \\
-x^2 + 0 \cdot x \\
\underline{-x^2 + 3x} \\
-3x + 9 \\
\underline{-3x + 9} \\
0
\end{array}
$$

$x^2(x - 3) = x^3 - 3x^2$

$-4x^2 - (-3x^2) = -x^2$

$-x(x - 3) = -x^2 + 3x$

Subtract: $0 \cdot x - 3x = -9x$

$-3(x - 3) = -3x + 9$

Subtract: $9 - 9 = 0$

The quotient is $x^2 - x - 3$.

Now do Exercises 49–64

If the remainder in long division is not zero, then the product of the quotient and divisor, plus the remainder, is equal to the dividend:

$$\text{dividend} = (\text{quotient})(\text{divisor}) + (\text{remainder})$$

or

$$\frac{\text{dividend}}{\text{divisor}} = \text{quotient} + \frac{\text{remainder}}{\text{divisor}}.$$

⟨6⟩ **Nonnegative Integral Exponents**

We have just seen the product rules and the quotient rule for exponents. These rules along with several other rules for exponents are stated in the following box.

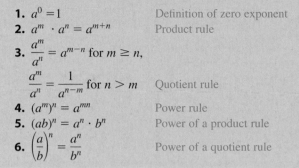

Rules for Nonnegative Integral Exponents

The following rules hold for nonzero real numbers a and b and nonnegative integers m and n.

1. $a^0 = 1$ Definition of zero exponent
2. $a^m \cdot a^n = a^{m+n}$ Product rule
3. $\dfrac{a^m}{a^n} = a^{m-n}$ for $m \geq n$,

   $\dfrac{a^m}{a^n} = \dfrac{1}{a^{n-m}}$ for $n > m$ Quotient rule
4. $(a^m)^n = a^{mn}$ Power rule
5. $(ab)^n = a^n \cdot b^n$ Power of a product rule
6. $\left(\dfrac{a}{b}\right)^n = \dfrac{a^n}{b^n}$ Power of a quotient rule

These rules are used to simplify expressions in Example 6.

EXAMPLE 6

Using the rules of exponents

Simplify each expression.

a) $\dfrac{2x^3 \cdot 3x^4}{12x^7}$ b) $(-2d^2)^3(3d^3)^4$ c) $\left(\dfrac{2a^3}{4b^2}\right)^2$

Solution

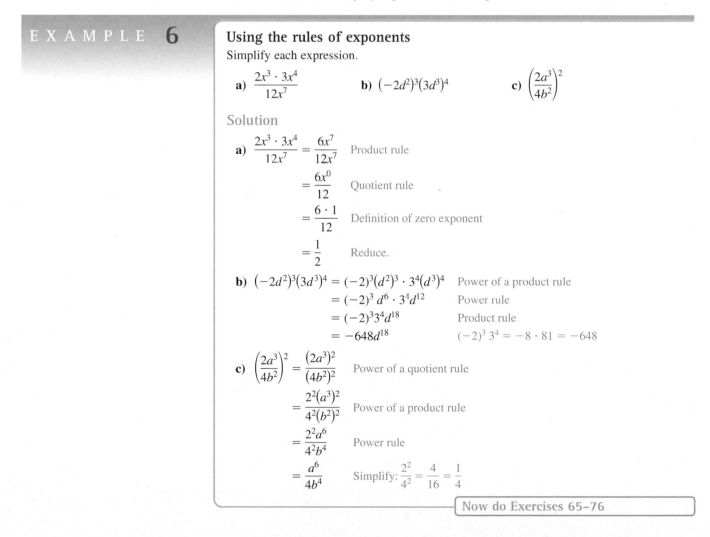

a) $\dfrac{2x^3 \cdot 3x^4}{12x^7} = \dfrac{6x^7}{12x^7}$ Product rule

$= \dfrac{6x^0}{12}$ Quotient rule

$= \dfrac{6 \cdot 1}{12}$ Definition of zero exponent

$= \dfrac{1}{2}$ Reduce.

b) $(-2d^2)^3(3d^3)^4 = (-2)^3(d^2)^3 \cdot 3^4(d^3)^4$ Power of a product rule

$= (-2)^3 d^6 \cdot 3^4 d^{12}$ Power rule

$= (-2)^3 3^4 d^{18}$ Product rule

$= -648d^{18}$ $(-2)^3 3^4 = -8 \cdot 81 = -648$

c) $\left(\dfrac{2a^3}{4b^2}\right)^2 = \dfrac{(2a^3)^2}{(4b^2)^2}$ Power of a quotient rule

$= \dfrac{2^2(a^3)^2}{4^2(b^2)^2}$ Power of a product rule

$= \dfrac{2^2 a^6}{4^2 b^4}$ Power rule

$= \dfrac{a^6}{4b^4}$ Simplify: $\dfrac{2^2}{4^2} = \dfrac{4}{16} = \dfrac{1}{4}$

Now do Exercises 65–76

⟨7⟩ **Negative Exponents and Scientific Notation**

A negative exponent is defined as a reciprocal. If a is a nonzero real number and n is a positive integer, then $a^{-n} = \frac{1}{a^n}$. So $2^{-3} = \frac{1}{2^3}$. All of the rules for positive exponents that we stated previously also hold for negative exponents. So we will not restate them here. In addition, there are a few new rules for negative exponents.

Rules for Negative Exponents

If a and b are nonzero real numbers and n is a positive integer, then

$$a^{-n} = \left(\frac{1}{a}\right)^n, \quad a^{-1} = \frac{1}{a}, \quad \frac{1}{a^{-n}} = a^n, \quad \left(\frac{a}{b}\right)^{-n} = \left(\frac{b}{a}\right)^n.$$

Using the rules for negative exponents we have $3^{-2} = \left(\frac{1}{3}\right)^2$, $3^{-1} = \frac{1}{3}$, $\frac{1}{3^{-2}} = 3^2$, and $\left(\frac{2}{3}\right)^{-3} = \left(\frac{3}{2}\right)^3$.

E X A M P L E   7

Using the rules for integral exponents

Simplify each expression. Write the answer with positive exponents only.

a) $\dfrac{-2x^{-3} \cdot 5x^4}{20x^{-6}}$
  b) $\left(-2a^{-2}\right)^{-3}\left(3a^3\right)^{-4}$
  c) $\left(\dfrac{a^{-3}}{b^2}\right)^{-2}$

Solution

a) $\dfrac{-2x^{-3} \cdot 5x^4}{20x^{-6}} = \dfrac{-10x^1}{20x^{-6}}$ Product rule

$$= \dfrac{-10x^{-1-(-6)}}{20} \qquad \text{Quotient rule}$$

$$= -\dfrac{1}{2}x^5 \qquad \text{Simplify.}$$

b) $\left(-2a^{-2}\right)^{-3}\left(3a^3\right)^{-4} = (-2)^{-3}\left(a^{-2}\right)^{-3} \cdot 3^{-4}\left(a^3\right)^{-4}$ Power of a product rule

$$= -\dfrac{1}{8}a^6 \cdot \dfrac{1}{81}a^{-12} \qquad \text{Power rule}$$

$$= -\dfrac{1}{648}a^{-6} \qquad \text{Product rule}$$

$$= -\dfrac{1}{648a^6} \qquad \text{Definition of negative exponent}$$

c) $\left(\dfrac{a^{-3}}{b^2}\right)^{-2} = \dfrac{\left(a^{-3}\right)^{-2}}{\left(b^2\right)^{-2}}$ Power of a quotient rule

$$= \dfrac{a^6}{b^{-4}} \qquad \text{Power rule}$$

$$= a^6 b^4 \qquad \text{Definition of negative exponent}$$

Now do Exercises 77–88

A number in scientific notation is written as a product of a number between 1 and 10 and a power of 10. In scientific notation there is one digit to the left of the decimal point. For example 3.5×10^3 and 2.36×10^{-4} are numbers in scientific notation. To convert these numbers to standard notation the decimal point is moved to the right for a positive power of 10 or to the left for a negative power of 10. So $3.5 \times 10^3 = 3500$ and $2.36 \times 10^{-4} = 0.000236$. To convert from standard notation to scientific notation, the process is reversed. When computing with numbers in scientific notation, we use the rules of exponents.

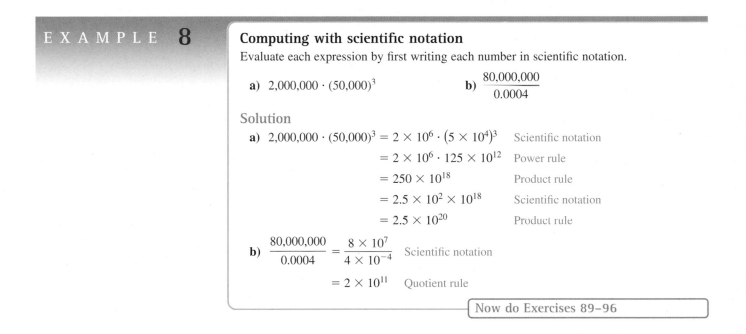

E X A M P L E   8

Computing with scientific notation

Evaluate each expression by first writing each number in scientific notation.

a) $2{,}000{,}000 \cdot (50{,}000)^3$

b) $\dfrac{80{,}000{,}000}{0.0004}$

Solution

a) $2{,}000{,}000 \cdot (50{,}000)^3 = 2 \times 10^6 \cdot \left(5 \times 10^4\right)^3$ Scientific notation

$= 2 \times 10^6 \cdot 125 \times 10^{12}$ Power rule

$= 250 \times 10^{18}$ Product rule

$= 2.5 \times 10^2 \times 10^{18}$ Scientific notation

$= 2.5 \times 10^{20}$ Product rule

b) $\dfrac{80{,}000{,}000}{0.0004} = \dfrac{8 \times 10^7}{4 \times 10^{-4}}$ Scientific notation

$= 2 \times 10^{11}$ Quotient rule

Now do Exercises 89–96

Exercises   R.4

⟨1⟩ **Polynomials**

Perform the indicated operation. See Example 1.

1. $\left(x^2 + 2x\right) + \left(x^3 - 5x\right)$

2. $\left(x^2 - 3x\right) + \left(-2x^3 - 9x\right)$

3. $\left(-w^2 + 5w - 1\right) + \left(2w^2 - w - 5\right)$

4. $\left(-2a^2 + 6a - 9\right) + \left(5a^2 - 3a + 8\right)$

5. $\left(-2y^2 + 6y\right) - \left(y^2 + 5y\right)$

6. $(-5z + 7) - (-6z - 8)$

7. $\left(-3t^2 + 5t + 1\right) - \left(-t^2 + 4t - 2\right)$

8. $\left(-n^2 - 3n + 9\right) - \left(-4n^2 - 2n + 1\right)$

⟨2⟩ **Multiplication of Polynomials**

Find each product. See Example 2.

9. $2x(4x - 3)$

10. $5x(-6x + 2)$

11. $-2a\left(a^2 - 4a + 9\right)$

12. $-3b\left(2b^2 - 5b - 1\right)$

13. $6w^2\left(w^3 - w^2 + w + 3\right)$

14. $5t^3\left(2t^3 + t^2 - 8t - 3\right)$

15. $(x + 2)(x + 4)$

16. $(a + 5)(a + 7)$

17. $(2s - 3)(3s + 1)$

18. $(4t - 1)(t - 2)$

19. $(-2x^2 + 1)(-3x^2 - 5)$

20. $(-x^3 + 5x)(2x^3 - 3x)$

21. $(x - 3)(x^2 + 3x - 9)$

22. $(a - 2)(a^2 + 5a - 8)$

23. $(w + 3)(3w^2 + 5w - 2)$

24. $(m + 7)(-2m^2 + 4m - 9)$

⟨3⟩ Multiplication of Binomials

Use the FOIL method to find each product. See Example 3.

25. $(a + m)(b + n)$

26. $(x + t)(y + s)$

27. $(x + 2)(x - 6)$

28. $(x - 5)(x + 3)$

29. $(2a + 1)(3a - 4)$

30. $(3b - 7)(5b - 9)$

31. $(-2x + 1)(-5x + 7)$

32. $(-3x - 2)(-x - 6)$

33. $(2a^3 - 6)(5a^3 + 3)$

34. $(4w^3 - 5)(3w^3 - 7)$

35. $(4x^4 - x)(4x^4 + x)$

36. $(5a^4 + x^3)(5a^4 - x^3)$

⟨4⟩ Special Products

Find each product using the special product rules. See Example 4.

37. $(x + 5)^2$

38. $(y + 3)^2$

39. $(2t + 7)^2$

40. $(3w + 4)^2$

41. $(s - 2)^2$

42. $(h - 3)^2$

43. $(3y - 5)^2$

44. $(6x - 1)^2$

45. $(3q + 4)(3q - 4)$

46. $(5m + 6)(5m - 6)$

47. $(2x^2 - 3n)(2x^2 + 3n)$

48. $(5t^2 - 3m)(5t^2 + 3m)$

⟨5⟩ Division of Polynomials

Find each quotient. See Example 5.

49. $(6x^8) \div (3x^2)$

50. $(-12a^{14}) \div (-3a^2)$

51. $(-4w^5) \div (2w^4)$

52. $(20b^{12}) \div (-5b^{10})$

53. $(3x^{22}) \div (6x^{20})$

54. $(-4t^{16}) \div (8t^8)$

55. $(30x^3 - 20x^2 + 10x) \div (10x)$

56. $(25a^3 + 20a^2 + 5a) \div (-5a)$

57. $(-3x^5 - 9x^4 + 3x^3 - 6x^2) \div (-3x^2)$

58. $(-8w^4 - 6w^3 + 4w^2) \div (-2w^2)$

59. $(x^3 - 4x^2 + 3) \div (x - 1)$

60. $(2x^3 + 3x + 5) \div (x + 1)$

61. $(x^3 - 4x^2 + x + 6) \div (x - 2)$

62. $(2x^3 + 3x^2 - 5x + 12) \div (x + 3)$

63. $(2x^3 - x^2 + 3x + 2) \div (2x + 1)$

64. $(2x^3 - x^2 - 9x + 9) \div (2x - 3)$

⟨6⟩ Nonnegative Integral Exponents

Use the rules of exponents to simplify each expression. See Example 6.

65. $\dfrac{3x^5 \cdot 5x^9}{45x^{14}}$

66. $\dfrac{-2y^3 \cdot 4y^6}{-24y^9}$

67. $\dfrac{2a^2 \cdot 6a^4}{(-2a^2)^3}$

68. $\dfrac{(-2w^3)(8w^{15})}{(-2w^3)^6}$

69. $(-3a^2)^3(2a^4)^5$

70. $(-2b)^4(2b^3)^2$

71. $(-5x^3)^2(-2x^2)^3$

72. $(-5y^5)^2(-3y^3)^2$

73. $\left(\dfrac{3x^4}{6y^2}\right)^3$

74. $\left(\dfrac{-2q^3}{4p^2}\right)^3$

75. $\left(\dfrac{2ab \cdot 3a^2b}{-4b^2}\right)^2$

76. $\left(\dfrac{3ab^2 \cdot 6a^2b^3}{(2ab)^3}\right)^2$

⟨7⟩ Negative Exponents and Scientific Notation

Simplify each expression. Write the answer with positive exponents only. See Example 7.

77. $\dfrac{-3a^{-2} \cdot 4a^3}{2a^{-5}}$

78. $\dfrac{-5b^{-6} \cdot 6b^{-5}}{2b^{-20}}$

79. $\dfrac{3w^{-7} \cdot 5w^4}{30w^{-9}}$

80. $\dfrac{4t^{-5} \cdot 8t^9}{16t^{18}}$

81. $(-3x^{-1})^{-4}$

82. $(-5y^{-6})^{-2}$

83. $(-2a^2)^{-5}(a^{-3})^6$

84. $(2b^{-1})^{-2}(-3b^{-2})^{-3}$

85. $\left(\dfrac{x^{-2}}{y^3}\right)^{-3}$

86. $\left(\dfrac{a^3}{b^{-5}}\right)^{-4}$

87. $\left(\dfrac{2x^{-2} \cdot 3x^5}{15x^{-8}}\right)^{-2}$

88. $\left(\dfrac{2y^{-1} \cdot 5y^7}{20y^2}\right)^{-2}$

Evaluate each expression by first writing each number in scientific notation. See Example 8.

89. $5{,}000 \cdot (20{,}000)^4$

90. $30{,}000 \cdot (20{,}000)^5$

91. $(0.00005)^2(2000)^3$

92. $(0.0006)^2(1000)^6$

93. $\dfrac{(10{,}000)^2}{0.000002}$

94. $\dfrac{(8000)^2}{(0.00001)^3}$

95. $\dfrac{(0.002)^3(40{,}000{,}000)^3}{(10{,}000)^5}$

96. $\dfrac{(0.0005)^2(10{,}000)}{(500)^4}$

In This Section

⟨1⟩ **Factoring Out Common Factors**

⟨2⟩ **Factoring the Special Products**

⟨3⟩ **Factoring by Grouping**

⟨4⟩ **Factoring $ax^2 + bx + c$ with $a = 1$**

⟨5⟩ **Factoring $ax^2 + bx + c$ with $a \neq 1$**

⟨6⟩ **Factoring a Difference or Sum of Two Cubes**

⟨7⟩ **Factoring Completely**

⟨8⟩ **Solving Quadratic Equations by Factoring**

R.5 Factoring

This section is a review of Chapter 5 of this text. All topics in this review section are explained in greater detail in Chapter 5.

⟨1⟩ Factoring Out Common Factors

To **factor** an expression means to write the expression as a product. For example, we factor 6 by writing 6 as $2 \cdot 3$. The largest integer that is a factor of two or more integers is the **greatest common factor (GCF)** of the integers. For example, the GCF for 12 and 18 is 6. The greatest common factor for a group of monomials includes the GCF for the coefficients of the monomials and each variable that is common to all of the monomials, where the exponent on each variable is the smallest power of that variable in any of the monomials. So the GCF or $12x^2y^3$ and $18x^4y$ is $6x^2y$. The distributive property is used to factor out the greatest common factor from a polynomial.

EXAMPLE 1

Factoring out the greatest common factor

Factor each polynomial by factoring out the greatest common factor.

a) $20x + 30$

b) $12x^2y^3 - 18x^4y$

c) $9a^3 - 12a^2 + 6a$

Solution

a) The GCF for $20x$ and 30 is 10:
$$20x + 30 = 10(2x + 3)$$

b) The GCF for $12x^2y^3$ and $18x^4y$ is $6x^2y$:
$$12x^2y^3 - 18x^4y = 6x^2y(2y^2 - 3x^2)$$

c) The GCF for $9a^3$, $12a^2$, and $6a$ is $3a$:
$$9a^3 - 12a^2 + 6a = 3a(3a^2 - 4a + 2)$$

> Now do Exercises 1–18

⟨2⟩ Factoring the Special Products

We learned the rules for finding the special products in Section R.5. The same rules are used to factor the special products. The trinomials $a^2 + 2ab + b^2$ and $a^2 - 2ab + b^2$ are called perfect square trinomials because they are the squares of binomials.

Factoring the Special Products

Perfect square trinomials: $a^2 + 2ab + b^2 = (a + b)^2$
$a^2 - 2ab + b^2 = (a - b)^2$

Difference of two squares: $a^2 - b^2 = (a + b)(a - b)$

EXAMPLE   2

Factoring the special products
Factor each polynomial.

    a) $x^2 + 6x + 9$ **b)** $4s^2 - 12st + 9t^2$ **c)** $25y^2 - 16$

Solution

    a) The trinomial $x^2 + 6x + 9$ is a perfect square trinomial. To factor it, let $a = x$ and $b = 3$ in the formula $a^2 + 2ab + b^2 = (a + b)^2$:

$$x^2 + 6x + 9 = x^2 + 2 \cdot x \cdot 3 + 3^2$$
$$= (x + 3)^2$$

    b) The trinomial $4s^2 - 12st + 9t^2$ is a perfect square trinomial. To factor it let $a = 2s$ and $b = 3t$ in the formula $a^2 - 2ab + b^2 = (a - b)^2$:

$$4s^2 - 12st + 9t^2 = (2s)^2 - 2(2s)(3t) + (3t)^2$$
$$= (2s - 3t)^2$$

    c) The binomial $25y^2 - 16$ is a difference of two squares. To factor it let $a = 5y$ and $b = 4$ in the formula $a^2 - b^2 = (a + b)(a - b)$:

$$25y^2 - 16 = (5y)^2 - 4^2$$
$$= (5y + 4)(5y - 4)$$

> Now do Exercises 19–34

⟨3⟩ Factoring by Grouping

The product of two binomials can be a polynomial with four terms. For example,

$$(x + b)(x + 2) = (x + b)x + (x + b)2$$
$$= x^2 + bx + 2x + 2b.$$

We can factor certain polynomials with four terms by reversing this process.

EXAMPLE   3

Factoring four-term polynomials by grouping
Factor each polynomial by grouping.

    a) $x^2 + 3x + cx + 3c$ **b)** $2x^3 - x^2 + 2x - 1$ **c)** $a^2 - 3a + 3b - ab$

Solution

    a) First factor out a common factor from the first two terms and from the last two terms:

$$x^2 + 3x + cx + 3c = x(x + 3) + c(x + 3)$$
$$= (x + c)(x + 3)$$

    Of course, the final answer could also be $(x + 3)(x + c)$.

    b) First factor out a common factor from the first two terms and from the last two terms:

$$2x^3 - x^2 + 2x - 1 = x^2(2x - 1) + 1(2x - 1)$$
$$= \left(x^2 + 1\right)(2x - 1)$$

    Note that $x^2 + 1$ is a sum of two squares and cannot be factored. It is a prime polynomial.

c) Factor a out of the first two terms and $-b$ out of the last two terms:

$$a^2 - 3a + 3b - ab = a(a - 3) - b(a - 3)$$
$$= (a - b)(a - 3)$$

Now do Exercises 35–46

⟨4⟩ **Factoring $ax^2 + bx + c$ with $a = 1$**

To factor $ax^2 + bx + c$ with $a = 1$, find two integers with a product of c and a sum of b.

E X A M P L E **4**

Factoring $ax^2 + bx + c$ with $a = 1$

Factor each polynomial.

a) $x^2 + 6x + 8$ **b)** $b^2 - b - 20$ **c)** $a^2 - 10a + 24$

Solution

a) Two integers with a product of 8 and a sum of 6 are 4 and 2. So we replace $6x$ with $4x + 2x$ and factor by grouping:

$$x^2 + 6x + 8 = x^2 + 4x + 2x + 8$$
$$= x(x + 4) + 2(x + 4)$$
$$= (x + 2)(x + 4)$$

Check that $(x + 2)(x + 4) = x^2 + 6x + 8$ to be sure that the factorization is correct.

b) Two integers with a product of -20 and a sum of -1 are -5 and 4. It is not necessary to write all of the steps shown in part (a). We can simply write

$$b^2 - b - 20 = (b - 5)(b + 4).$$

Use the FOIL method to check.

c) Two integers with a product of 24 and a sum of -10 are -6 and -4. So,

$$a^2 - 10a + 24 = (a - 6)(a - 4).$$

Use the FOIL method to check.

Now do Exercises 47–58

⟨5⟩ **Factoring $ax^2 + bx + c$ with $a \neq 1$**

To factor $ax^2 + bx + c$ with $a \neq 1$ by the **ac method,** find two integers with a product of ac and a sum of b. Then factor by grouping, as done in Example 4(a).

E X A M P L E **5**

Factoring $ax^2 + bx + c$ with $a \neq 1$

Factor each polynomial.

a) $6x^2 + 13x + 6$ **b)** $2w^2 + 7w - 4$ **c)** $12t^2 - 17t + 6$

Solution

a) In this case $ac = 36$ and $b = 13$. Two integers with a product of 36 and a sum of 13 are 9 and 4. So replace $13x$ with $9x + 4x$ and factor by grouping:

$$6x^2 + 13x + 6 = 6x^2 + 9x + 4x + 6$$
$$= 3x(2x + 3) + 2(2x + 3)$$
$$= (3x + 2)(2x + 3)$$

Check that $(3x + 2)(2x + 3) = 6x^2 + 13x + 6$ to be sure that the factorization is correct.

b) Two integers with a product of -8 and a sum of 7 are -1 and 8. So replace $7w$ with $-1w + 8w$ and factor by grouping:

$$2w^2 + 7w - 4 = 2w^2 - 1w + 8w - 4$$
$$= w(2w - 1) + 4(2w - 1)$$
$$= (w + 4)(2w - 1)$$

Use the FOIL method to check.

c) Two integers with a product of 72 and a sum of -17 are -9 and -8. So replace $-17t$ with $-9t - 8t$ and factor by grouping:

$$12t^2 - 17t + 6 = 12t^2 - 9t - 8t + 6$$
$$= 3t(4t - 3) - 2(4t - 3)$$
$$= (3t - 2)(4t - 3)$$

Use the FOIL method to check.

Now do Exercises 59–70

Another method that is commonly used to factor $ax^2 + bx + c$ with $a \neq 1$ is called **trial and error.** This method is not systematic like the ac method. For trial and error factoring simply try a pair of possible factors and check by FOIL. If it does not check, then try again. For example, to factor $6x^2 + 13x + 6$ we might try $(6x + 1)(x + 6)$. However,

$$(6x + 1)(x + 6) = 6x^2 + 37x + 6$$

and the middle term is wrong. With trial and error we try factors that give the correct first and last terms and then use FOIL to see if the middle term is correct.

⟨6⟩ Factoring a Difference or Sum of Two Cubes

A difference or sum of two cubes can be factored using the following rules.

Factoring a Difference or Sum of Two Cubes

$$a^3 - b^3 = (a - b)(a^2 + ab + b^2)$$
$$a^3 + b^3 = (a + b)(a^2 - ab + b^2)$$

EXAMPLE **6**

Factoring a difference or sum of two cubes

Factor each polynomial.

a) $x^3 - 125$ **b)** $8y^3 - 1$ **c)** $64w^3 + 27z^3$

Solution

a) Use $a = x$ and $b = 5$ in the formula $a^3 - b^3 = (a - b)(a^2 + ab + b^2)$:

$$x^3 - 125 = x^3 - 5^3$$
$$= (x - 5)(x^2 + 5x + 25)$$

b) Use $a = 2y$ and $b = 1$ in the formula $a^3 - b^3 = (a - b)(a^2 + ab + b^2)$:

$$8y^3 - 1 = (2y)^3 - 1^3$$
$$= (2y - 1)(4y^2 + 2y + 1)$$

c) Use $a = 4w$ and $b = 3z$ in the formula $a^3 + b^3 = (a + b)(a^2 - ab + b^2)$:

$$64w^3 + 27z^3 = (4w)^3 + (3z)^3$$
$$= (4w + 3z)(16w^2 - 12wz + 9z^2)$$

Now do Exercises 71–82

⟨7⟩ Factoring Completely

A polynomial that cannot be factored is a **prime polynomial.** A polynomial is factored completely when all of the factors are prime polynomials.

EXAMPLE **7**

Factoring a polynomial completely

Factor $8x^5 - 8xy^4$ completely.

Solution

First factor out the GCF $8x$, and then factor the difference of two squares:

$$8x^5 - 8xy^4 = 8x(x^4 - y^4)$$ Factor out the GCF.
$$= 8x(x^2 - y^2)(x^2 + y^2)$$ Factor the difference of two squares.
$$= 8x(x - y)(x + y)(x^2 + y^2)$$ Factor the difference of two squares.

Even though 8 could be factored, we do not usually factor any common integers when factoring polynomials. Note that $x^2 + y^2$ is a sum of two squares and it is a prime polynomial.

Now do Exercises 83–92

⟨8⟩ Solving Quadratic Equations by Factoring

An equation of the form $ax^2 + bx + c = 0$ with $a \neq 0$ is called a **quadratic equation.** To solve a quadratic equation by factoring we use the **zero factor property:** if $ab = 0$, then either $a = 0$ or $b = 0$.

EXAMPLE   **8**

Solving a quadratic equation by factoring

Solve $2x^2 + 5x - 12 = 0$ by factoring.

Solution

First factor $2x^2 + 5x - 12$ by the *ac* method or trial and error, and then set each factor equal to zero.

$$2x^2 + 5x - 12 = 0$$
$$(2x - 3)(x + 4) = 0 \qquad \text{Factor the polynomial.}$$
$$2x - 3 = 0 \quad \text{or} \quad x + 4 = 0 \qquad \text{Zero factor property}$$
$$2x = 3 \quad \text{or} \qquad x = -4 \quad \text{Solve each linear equation.}$$
$$x = \frac{3}{2}$$

Check $\frac{3}{2}$ and -4 in the original equation:

$$2\left(\frac{3}{2}\right)^2 + 5\left(\frac{3}{2}\right) - 12 = 0 \qquad 2(-4)^2 + 5(-4) - 12 = 0$$
$$\frac{9}{2} + \frac{15}{2} - \frac{24}{2} = 0 \qquad\qquad 32 - 20 - 12 = 0$$

The solution set is $\left\{-4, \frac{3}{2}\right\}$.

Now do Exercises 93–100

R.5 Exercises

⟨1⟩ **Factoring Out Common Factors**

Factor each polynomial by factoring out the greatest common factor. See Example 1.

1. $12x + 8$

2. $18a + 30$

3. $15y^3 - 6y^2$

4. $48z^4 - 32z^3$

5. $8a^3b^2 + 20a^4b$

6. $24y^4z^3 + 36y^3z^4$

7. $12x^4 - 20x^3 - 24x^2$

8. $14y^3 - 21y^2 - 28y$

9. $2a^3b - 6a^2b + 6ab$

10. $3w^3z - 12w^2z - 9wz$

Complete the factoring of each polynomial.

11. $4x^3 - 6x^2 = (\quad)(2x^2 - 3x)$

12. $5y^4 - 10y^2 = (\quad)(y^2 - 2)$

13. $-2x^2 - 6x = (-2x)(\quad)$

14. $-3y^3 - 9y = (-3y)(\quad)$

15. $-5a^5 + 10a^2 = (\quad)(a^3 - 2)$

16. $-4b^4 - 12b^2 = (\quad)(b^2 + 3)$

17. $-w^3x - w^2x = (-w^2x)(\quad)$

18. $-zy^3 + zy^2 = (-zy^2)(\quad)$

⟨2⟩ **Factoring the Special Products**

Factor each special product. See Example 2.

19. $x^2 + 8x + 16$

20. $x^2 + 4x + 4$

21. $a^2 - 2a + 1$

22. $b^2 - 10b + 25$

23. $y^2 - 9$

24. $n^2 - 4$

25. $9x^2 + 6x + 1$

26. $25y^2 + 20y + 4$

27. $16m^2 - 40mt + 25t^2$
28. $9s^2 - 24st + 16t^2$
29. $9x^2 - 16$
30. $81a^2 - 25$
31. $64n^2 + 48n + 9$
32. $81s^2 - 18s + 1$
33. $25x^2 - 49y^2$
34. $a^2b^2 - y^2$

⟨3⟩ **Factoring by Grouping**

Factor each polynomial by grouping. See Example 3.

35. $a^2 + 6a + ab + 6b$
36. $w^2 - 3w + wx - 3x$
37. $6x^2 - 10x + 3ax - 5a$
38. $10ax + 5a + 2x + 1$
39. $3y^3 - 4y^2 + 3y - 4$
40. $6x^3 - 3x^2 + 10x - 5$
41. $8a^3 - 4a^2 + 14a - 7$
42. $5t^3 - 10t^2 + 6t - 12$
43. $ab - 2b - 3a + 6$
44. $x^2 - xy - 7x + 7y$
45. $x^3 - x^2 + 3 - 3x$
46. $ax^2 - 4x^2 + 20 - 5a$

⟨4⟩ **Factoring $ax^2 + bx + c$ with $a = 1$**

Factor each polynomial. See Example 4.

47. $x^2 + 5x + 6$
48. $x^2 + 11x + 30$
49. $w^2 + 8w + 15$
50. $u^2 + 19u + 18$
51. $v^2 - 2v - 12$
52. $m^2 - 9m - 22$
53. $t^2 - 12t - 28$
54. $q^2 - 4q - 32$
55. $b^2 - 15b + 26$
56. $p^2 - 26p + 25$
57. $c^2 - 11c + 24$
58. $n^2 - 10n + 21$

⟨5⟩ **Factoring $ax^2 + bx + c$ with $a \neq 1$**

Factor each polynomial. See Example 5.

59. $2x^2 + 7x + 6$
60. $3w^2 + 16w + 5$
61. $15t^2 + 17t + 4$
62. $6m^2 + 29m + 20$
63. $3n^2 + 16n - 12$
64. $4y^2 + 17y - 15$
65. $8m^2 + 6m - 27$

66. $18p^2 + 9p - 5$
67. $8q^2 - 14q + 3$
68. $6t^2 - 11t + 4$
69. $15z^2 - 19z + 6$
70. $10k^2 - 41k + 4$

⟨6⟩ **Factoring a Difference or Sum of Two Cubes**

Factor each polynomial. See Example 6.

71. $x^3 - 1$
72. $y^3 - 27$
73. $a^3 - 8$
74. $b^3 - 1000$
75. $125x^3 - 1$
76. $8a^3 - 125$
77. $125q^3 - 27$
78. $1000b^3 - 343$
79. $27x^3 + 64y^3$
80. $8h^3 + 125k^3$
81. $343m^3 + 8n^3$
82. $a^3b^3 + x^3y^3$

⟨7⟩ **Factoring Completely**

Factor each polynomial completely. See Example 7.

83. $2x^2 + 8x + 6$
84. $3x^2 + 6x - 45$
85. $-2x^3 - 12x^2 - 18x$
86. $-4x^4 + 40x^3 - 100x^2$
87. $3a^4 - 3b^4$
88. $w^5 - wq^4$
89. $-a^3b - 8b^4$
90. $-24x^3 + 81$
91. $a^3 + 3a^2 - 4a - 12$
92. $x^3 - 5x^2 - 9x + 45$

⟨8⟩ **Solving Quadratic Equations by Factoring**

Solve each quadratic equation. See Example 8.

93. $x^2 - 2x - 12 = 0$
94. $y^2 + y - 20 = 0$
95. $2t^2 + 5t - 3 = 0$
96. $3p^2 - 14p + 8 = 0$
97. $4m^2 - 12m + 5 = 0$
98. $15w^2 - 8w + 1 = 0$
99. $r^3 + 5r^2 + 6r = 0$
100. $2c^3 - 2c^2 - 4c = 0$

In This Section

R.6 Rational Expressions

This section is a review of Chapter 6 of this text. All topics in this review section are explained in greater detail in Chapter 6.

⟨1⟩ Reducing Rational Expressions

A **rational expression** is the ratio of two polynomials with the denominator not equal to 0. Like rational numbers, rational expressions have infinitely many equivalent forms. If a rational expression has no factors common to the numerator and denominator, then the rational expression is in **lowest terms.** A rational expression is reduced to lowest terms by dividing out or canceling the greatest common factor for the numerator and denominator.

EXAMPLE 1

Reducing rational expressions to lowest terms

Reduce each rational expression to lowest terms. Express answers with positive exponents only.

a) $\dfrac{a^2 - 25}{a^2 + 10a + 25}$ 　　　 b) $\dfrac{12s^2t^3}{18s^3t}$ 　　　 c) $\dfrac{-6a + 6b}{a^2b - 2ab^2 + b^3}$

Solution

a) Factor the numerator and denominator completely, and then divide out the GCF.

$$\frac{a^2 - 25}{a^2 + 10a + 25} = \frac{(a + 5)(a - 5)}{(a + 5)^2} = \frac{a - 5}{a + 5}$$

b) The GCF is $6s^2t$:

$$\frac{12s^2t^3}{18s^3t} = \frac{6s^2t(2t^2)}{6s^2t(3s)} = \frac{2t^2}{3s}$$

c) We can factor -6 or positive 6 from the numerator. In this case -6 is the better choice:

$$\frac{-6a + 6b}{a^2b - 2ab^2 + b^3} = \frac{-6(a - b)}{b(a - b)^2} = \frac{-6}{b(a - b)}$$

> Now do Exercises 1–12

⟨2⟩ Multiplication and Division

Rational expressions are multiplied in the same manner that rational numbers are multiplied. As with rational numbers, we can factor, reduce, and then multiply. To divide rational expressions we invert the divisor and multiply.

EXAMPLE 2

Multiplying and dividing rational expressions

Perform the indicated operations. Express the answer in lowest terms.

a) $\dfrac{9x}{10y} \cdot \dfrac{5y^2}{6x^3}$ 　　　 b) $\dfrac{a}{a^2 + 2ab + b^2} \cdot \dfrac{a^2 - b^2}{2a}$ 　　　 c) $\dfrac{20x^2y^3}{y^2 - xy} \div \dfrac{12x^4y}{x - y}$

Solution

a) Factor the numerators and denominators completely, and then divide out the common factors:

$$\frac{9x}{10y} \cdot \frac{5y^2}{6x^3} = \frac{3 \cdot 3x}{2 \cdot 5y} \cdot \frac{5y^2}{2 \cdot 3x^3}$$

$$= \frac{3y}{4x^2}$$

b) $\dfrac{a}{a^2 + 2ab + b^2} \cdot \dfrac{a^2 - b^2}{2a} = \dfrac{a}{(a+b)^2} \cdot \dfrac{(a+b)(a-b)}{2a}$

$$= \frac{a - b}{2(a + b)}$$

c) $\dfrac{20x^2y^3}{y^2 - xy} \div \dfrac{12x^4y}{x - y} = \dfrac{20x^2y^3}{y^2 - xy} \cdot \dfrac{x - y}{12x^4y}$

$$= \frac{4 \cdot 5x^2y^3(x - y)}{4 \cdot 3x^4y(-1)(x - y)}$$

$$= -\frac{5y^2}{3x^2}$$

> **Now do Exercises 13–24**

⟨3⟩ Addition and Subtraction

We can add or subtract rational expressions only if they have identical denominators. If the denominators are not identical, then we must build up each rational expression to get identical denominators. Any common denominator will work for addition or subtraction, but the least common denominator (LCD) is the most efficient.

EXAMPLE 3

Adding and subtracting rational expressions

Perform the indicated operations. Express the answer in lowest terms.

a) $\dfrac{3}{10y} + \dfrac{5}{10y}$ b) $\dfrac{a}{a + 2} - \dfrac{a - 3}{a^2 - 4}$ c) $\dfrac{y}{y^2 - y} + \dfrac{y}{y + 2}$

Solution

a) Since the denominators are identical, the rational expressions can be added without building them up:

$$\frac{3}{10y} + \frac{5}{10y} = \frac{8}{10y} = \frac{4}{5y}$$

b) Since $a^2 - 4 = (a + 2)(a - 2)$, the LCD for these denominators is $(a + 2)(a - 2)$. To get identical denominators multiply the numerator and denominator of the first rational expression by $a - 2$:

$$\frac{a}{a + 2} - \frac{a - 3}{a^2 - 4} = \frac{a(a - 2)}{(a + 2)(a - 2)} - \frac{a - 3}{(a + 2)(a - 2)}$$

$$= \frac{a^2 - 2a - (a - 3)}{(a + 2)(a - 2)}$$

$$= \frac{a^2 - 3a + 3}{(a + 2)(a - 2)}$$

c) Since $y^2 - y = y(y - 1)$, the LCD is $y(y - 1)(y + 2)$.

$$\frac{3}{y^2 - y} + \frac{y}{y + 2} = \frac{3}{y(y - 1)} + \frac{y}{y + 2}$$

$$= \frac{3(y + 2)}{y(y - 1)(y + 2)} + \frac{y \cdot y(y - 1)}{(y + 2) \cdot y(y - 1)}$$

$$= \frac{y^3 - y^2 + 3y + 6}{y(y - 1)(y + 2)}$$

Now do Exercises 25–34

⟨4⟩ Complex Fractions

A **complex fraction** is a fraction that has rational expressions in its numerator, denominator, or both. The easiest way to simplify a complex fraction is to multiply its numerator and denominator by the LCD of all of the fractions.

EXAMPLE 4

Simplifying complex fractions

Simplify. Express the answer in lowest terms.

a) $\dfrac{\dfrac{1}{3} + \dfrac{1}{4}}{\dfrac{5}{6} - \dfrac{1}{2}}$

b) $\dfrac{\dfrac{1}{5x^2} - \dfrac{2}{3x}}{\dfrac{3}{10x} - \dfrac{4}{x}}$

Solution

a) The LCD for the denominators 3, 4, 6, and 2 is 12. So multiply the numerator and denominator by 12:

$$\frac{\dfrac{1}{3} + \dfrac{1}{4}}{\dfrac{5}{6} - \dfrac{1}{2}} = \frac{12\left(\dfrac{1}{3} + \dfrac{1}{4}\right)}{12\left(\dfrac{5}{6} - \dfrac{1}{2}\right)} = \frac{4 + 3}{10 - 6} = \frac{7}{4}$$

b) The LCD for $5x^2$, $3x$, $10x$, and x is $30x^2$. So multiply the numerator and denominator by $30x^2$.

$$\frac{\dfrac{1}{5x^2} - \dfrac{2}{3x}}{\dfrac{3}{10x} - \dfrac{4}{x}} = \frac{30x^2\left(\dfrac{1}{5x^2} - \dfrac{2}{3x}\right)}{30x^2\left(\dfrac{3}{10x} - \dfrac{4}{x}\right)} = \frac{6 - 20x}{9x - 120x}$$

$$= \frac{6 - 20x}{-111x}$$

$$= \frac{20x - 6}{111x}$$

Now do Exercises 35–40

⟨5⟩ **Solving Equations with Rational Expressions**

If an equation contains rational expressions, it is usually best to eliminate the rational expressions by multiplying both sides of the equation by the LCD.

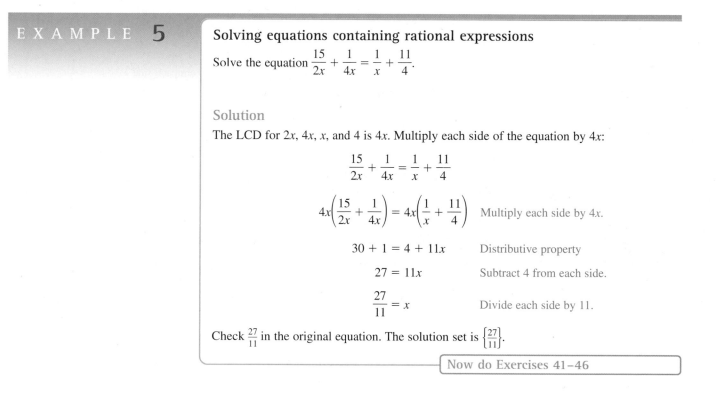

E X A M P L E 5

Solving equations containing rational expressions

Solve the equation $\dfrac{15}{2x} + \dfrac{1}{4x} = \dfrac{1}{x} + \dfrac{11}{4}$.

Solution

The LCD for $2x$, $4x$, x, and 4 is $4x$. Multiply each side of the equation by $4x$:

$$\frac{15}{2x} + \frac{1}{4x} = \frac{1}{x} + \frac{11}{4}$$

$$4x\left(\frac{15}{2x} + \frac{1}{4x}\right) = 4x\left(\frac{1}{x} + \frac{11}{4}\right) \qquad \text{Multiply each side by } 4x.$$

$$30 + 1 = 4 + 11x \qquad \text{Distributive property}$$

$$27 = 11x \qquad \text{Subtract 4 from each side.}$$

$$\frac{27}{11} = x \qquad \text{Divide each side by 11.}$$

Check $\frac{27}{11}$ in the original equation. The solution set is $\left\{\frac{27}{11}\right\}$.

Now do Exercises 41–46

⟨6⟩ **Applications of Ratios and Proportions**

If a and b are real numbers, with $b \neq 0$, then $\frac{a}{b}$ is called the **ratio of a and b** or the **ratio of a to b.** Ratios are treated just like fractions. We can reduce ratios and build them up. When possible, we usually convert ratios to ratios of integers in lowest terms. A **proportion** is a statement expressing the equality of two ratios. The equation

$$\frac{a}{b} = \frac{c}{d} \quad \text{or} \quad a : b = c : d$$

is a proportion. The numbers in the positions of a and d are called the **extremes.** The numbers in the positions of b and c are called the **means.** The **extremes-means property** indicates that *the product of the means is equal to the product of the extremes.*

E X A M P L E 6

Solving a proportion problem

The ratio of male employees to female employees at ABC Insurance is 3 to 2. If there are 20 more men than women, then how many men and how many women work at ABC?

Solution

Let x represent the number of men and $x - 20$ represent the number of women. Since the ratio of men to women is 3 to 2, we have the following proportion:

$$\frac{3}{2} = \frac{x}{x - 20}$$ The ratio of men to women is 3 to 2.

$$3(x - 20) = 2x$$ Extremes-means property

$$3x - 60 = 2x$$ Distributive property

$$3x = 2x + 60$$ Add 60 to each side.

$$x = 60$$ Subtract $2x$ from each side.

So there are 60 males and 40 females at ABC Insurance.

Now do Exercises 47–50

⟨7⟩ Applications of Rational Expressions

Many applied problems can be solved using equations that involve rational expressions.

EXAMPLE 7

Solving a uniform motion problem

Kaiser drove 600 miles from his home to Memphis. On the way back home he averaged 10 miles per hour less, and the drive back took him 2 hours longer. Find Kaiser's average speed on the way to Memphis.

Solution

Let x represent his average speed on the way to Memphis and $x - 10$ represent his average speed on the way back. Use the formula $T = \frac{D}{R}$ to make the following table:

	D	R	T
To Memphis	600 mi	x mi/hr	$\frac{600}{x}$ hr
Returning	600 mi	$x - 10$ mi/hr	$\frac{600}{x - 10}$ hr

Since the time for the return trip was 2 hours more, we have the following equation:

$$\frac{600}{x} = \frac{600}{x - 10} - 2$$

$$x(x - 10)\left(\frac{600}{x}\right) = x(x - 10)\left(\frac{600}{x - 10} - 2\right)$$ Multiply each side by the LCD.

$$600x - 6000 = 600x - 2(x)(x - 10)$$

$$600x - 6000 = -2x^2 + 620x$$

$$2x^2 - 20x - 6000 = 0$$

$$x^2 - 10x - 3000 = 0$$

$$(x - 60)(x + 50) = 0$$

$$x - 60 = 0 \quad \text{or} \quad x + 50 = 0$$

$$x = 60 \quad \text{or} \qquad x = -50$$

Since $x = -50$ is meaningless, the solution is $x = 60$. If he averaged 60 mph going to Memphis and 50 mph returning, then the time going was 600/60 or 10 hours and the time returning was 600/50 or 12 hours, which is 2 hours longer. So his average speed on the way to Memphis was 60 mph.

Now do Exercises 51–54

Exercises R.6

⟨1⟩ Reducing Rational Expressions

Reduce each rational expression to lowest terms. Express answers with positive exponents only. See Example 1.

1. $\dfrac{b^2 - 16}{b^2 + 8b + 16}$

2. $\dfrac{2x^2 - 2y^2}{2x^2 - 4xy + 2y^2}$

3. $\dfrac{4x^2 + 4x - 24}{2x^2 - 18}$

4. $\dfrac{2a^3 + 2a^2 - 40a}{a^3 + 4a^2 - 5a}$

5. $\dfrac{6x^3y^6}{8x^3y}$

6. $\dfrac{10a^3b^2}{15ab^4}$

7. $\dfrac{-20wz^9}{25w^3z^2}$

8. $\dfrac{21r^2t}{-28r^5t^3}$

9. $\dfrac{-2a - 2y}{-4a^2 + 4y^2}$

10. $\dfrac{-4a^2 - 12a + 40}{-2a + 4}$

11. $\dfrac{-3x^3 + 3y^3}{-3x^2 + 3y^2}$

12. $\dfrac{2x^2 + 10x + 12}{2x^3 + 16}$

⟨2⟩ Multiplication and Division

Perform the indicated operations. Express the answer in lowest terms. See Example 2.

13. $\dfrac{4b^2}{21a} \cdot \dfrac{35a^2}{8b^4}$

14. $\dfrac{9w^3}{5t^2} \cdot \dfrac{10t^5}{27w^8}$

15. $\dfrac{6ab^3}{40} \cdot \dfrac{25}{18a^7b}$

16. $\dfrac{3xy^3}{15xy} \cdot \dfrac{45xy^2}{18xy^9}$

17. $\dfrac{15x^3}{x^2 - 2xy + y^2} \cdot \dfrac{x^2 - y^2}{5x^7}$

18. $\dfrac{20a^6}{9a^2 + 12ab + 4b^2} \cdot \dfrac{9a^2 - 4b^2}{4a^3}$

19. $\dfrac{5x + 10}{x^2 + 5x + 6} \cdot \dfrac{x^2 + 6x + 9}{10x + 30}$

20. $\dfrac{x^2 - x - 12}{x^2 + x - 12} \cdot \dfrac{x^2 + 4x}{x^2 - 4x}$

21. $\dfrac{4a^5b^4}{a^2 - ab} \div \dfrac{24a^8b}{a^2 - b^2}$

22. $\dfrac{17x^5y^6}{x^2 - y^2} \div \dfrac{51x^5y}{x^2 + 2xy + y^2}$

23. $\dfrac{a^2 - a - 2}{a^2 + a} \div \dfrac{a^2 - 2a}{a^3 + 3a^2}$

24. $\dfrac{3w^2 - 3w - 18}{6w^2 - 18w} \div \dfrac{w + 2}{2w^2 + 2w}$

⟨3⟩ **Addition and Subtraction**

Perform the indicated operations. Express the answer in lowest terms. See Example 3.

25. $\dfrac{8}{3x} + \dfrac{4}{3x}$

26. $\dfrac{3}{5x^2y} + \dfrac{2}{5x^2y}$

27. $\dfrac{14b}{7b+1} + \dfrac{2}{7b+1}$

28. $\dfrac{2w^2+1}{w^2+4} + \dfrac{w^2+11}{w^2+4}$

29. $\dfrac{1}{x-y} - \dfrac{2x}{x^2-y^2}$

30. $\dfrac{1}{x^2-x-2} - \dfrac{1}{x-2}$

31. $\dfrac{4-3w}{2w^2-5w-3} - \dfrac{w}{2w+1}$

32. $\dfrac{t}{3t^2-t-2} - \dfrac{t}{3t+2}$

33. $\dfrac{m}{m^2+m} + \dfrac{5}{m^2+3m}$

34. $\dfrac{n}{n^2-9} + \dfrac{2}{n^2+3n}$

⟨4⟩ **Complex Fractions**

Simplify. Express the answer in lowest terms. See Example 4.

35. $\dfrac{\dfrac{1}{2}-\dfrac{1}{3}}{\dfrac{5}{4}-\dfrac{1}{6}}$

36. $\dfrac{\dfrac{3}{8}+\dfrac{2}{3}}{\dfrac{1}{2}-\dfrac{1}{4}}$

37. $\dfrac{\dfrac{1}{a}+\dfrac{2}{b}}{\dfrac{3}{ab}-\dfrac{1}{ab}}$

38. $\dfrac{\dfrac{4}{xy}-\dfrac{3}{xy}}{\dfrac{2}{x}-\dfrac{5}{y}}$

39. $\dfrac{\dfrac{1}{3t^3}-\dfrac{5}{6t}}{\dfrac{4}{9t}-\dfrac{5}{2t^2}}$

40. $\dfrac{\dfrac{2}{5m^2}+3}{\dfrac{1}{10m}-2}$

⟨5⟩ **Solving Equations with Rational Expressions**

Solve each equation. See Example 5.

41. $\dfrac{3}{x} + \dfrac{1}{2x} = \dfrac{1}{6x} + \dfrac{10}{3}$

42. $\dfrac{1}{t} + \dfrac{2}{3t} = \dfrac{3}{4t} + \dfrac{1}{6}$

43. $\dfrac{3}{x-2} - \dfrac{2}{x+2} = \dfrac{1}{x^2-4}$

44. $\dfrac{4}{y-3} + \dfrac{6}{y+1} = \dfrac{3y}{y^2-2y-3}$

45. $\dfrac{5}{a+5} + \dfrac{7}{2a-3} = \dfrac{4a}{2a^2+7a-15}$

46. $\dfrac{3}{3m+4} + \dfrac{2}{2m-1} = \dfrac{m}{6m^2+5m-4}$

⟨6⟩ **Applications of Ratios and Proportions**

Solve each problem. See Example 6.

47. *Students and teachers.* The student-teacher ratio at Bellmont High is 22.4 to 1. If there are 1904 students, then how many teachers are there?

48. *Water and oatmeal.* The recipe for hot oatmeal calls for a ratio of water to cereal of 2 to 1. If 12 cups of water are used, then how many cups of cereal should be used?

49. *Just Paws.* The ratio of dogs to cats boarded at Just Paws Kennel is 4 to 3. If there are 12 more dogs than cats, then how many dogs and how many cats are boarded at Just Paws?

50. *Cars and trucks.* At noon the ratio of trucks to cars at a rest stop in Texas was 3 to 7. If there were 12 fewer trucks than cars, then how many cars and how many trucks were at the rest stop?

⟨7⟩ **Applications of Rational Expressions**

Solve each problem. See Example 7.

51. *Driving to Dallas.* Ken drove 1400 miles from his home to Dallas. On the way back home he averaged 6 miles per hour more, and the drive back took him 3 hours less. Find Ken's average speed on the way to Dallas.

52. *Driving to San Francisco.* Amelia drove 600 miles on the first day and 400 miles on the second day of her trip to San Francisco. On the second day she averaged 10 miles per hour less and drove for 2 fewer hours. Find her average speed for each day.

53. *Sharing expenses.* A group of students can rent a motorhome and drive it to Florida for $2100. If they can get four more students to share the cost with them, then the cost per person will decrease by $400. How many students are in the original group?

54. *Sharing expenses.* A group of students can rent a small limousine for $250. If they can get 2 more couples, they can get a large limousine for $340 and pay $20 less per person. How many students are in the original group?

Answers to Selected Exercises

Chapter 1

Section 1.1 Warm-Ups
1. Integers **2.** Natural **3.** Rational **4.** Terminating, repeating
5. Irrational **6.** Real **7.** Circumference **8.** Absolute value
9. True **10.** False **11.** False **12.** True **13.** False
14. True **15.** True **16.** False **17.** True **18.** True

Section 1.1 Exercises
1. 6 **3.** 6 **5.** 0 **7.** -2 **9.** -12 **11.** -2.1

13. 1, 2, 3, 4, 5

15. 0, 1, 2, 3, 4

17. 0, 1, 2, 3, 4

19. 1, 2, 3, 4, 5, . . .

21. 1, 2, 3, 4, 5, . . .

23. True **25.** False **27.** True **29.** True **31.** True **33.** False

35. (0, 1)

37. [-2, 2]

39. (0, 5]

41. (4, ∞)

43. ($-\infty$, -1]

45. [0, ∞)

47. 6 **49.** 0 **51.** 7 **53.** 9 **55.** 45 **57.** $\frac{3}{4}$ **59.** 5.09
61. -16 **63.** $-\frac{5}{2}$ **65.** 2 **67.** 3 **69.** -9 **71.** 16
73. -4 **75.** -1.99 **77.** 74 **79.** 5.25 **81.** 40 **83.** $\frac{1}{2}$
85. -3 and 3 **87.** $-4, -3, 3, 4$ **89.** $-1, 0, 1$ **91.** [3, 8]
93. ($-30, -20$] **95.** [30, ∞) **97.** True **99.** True **101.** True
103. a) $\frac{7}{24}$ **b)** -3.115 **c)** 0.66669
   d) Add them and divide the result by 2.

105. Real: all; irrational: π, $\sqrt{3}$; rational: all except π and $\sqrt{3}$; integer: -2, $\sqrt{9}$, 6, 0; whole: $\sqrt{9}$, 6, 0; counting: $\sqrt{9}$, 6

Section 1.2 Warm-Ups
1. Equivalent **2.** Factor **3.** Common
4. Denominator, numerator **5.** Dividing **6.** True **7.** True
8. True **9.** True **10.** True **11.** False **12.** True

Section 1.2 Exercises
1. $\frac{6}{8}$ **3.** $\frac{32}{12}$ **5.** $\frac{10}{2}$ **7.** $\frac{75}{100}$ **9.** $\frac{30}{100}$ **11.** $\frac{70}{42}$ **13.** $\frac{1}{2}$
15. $\frac{2}{3}$ **17.** 3 **19.** $\frac{1}{2}$ **21.** 2 **23.** $\frac{3}{8}$ **25.** $\frac{13}{21}$ **27.** $\frac{12}{13}$
29. $\frac{10}{27}$ **31.** 5 **33.** $\frac{7}{10}$ **35.** $\frac{7}{13}$ **37.** $\frac{3}{5}$ **39.** $\frac{1}{6}$
41. 1152 in. **43.** 22.88 km **45.** 5.31 in. **47.** 402.57 g
49. 58.67 ft/sec **51.** 548.53 km/hr **53.** 3
55. $\frac{1}{15}$ **57.** 4 **59.** $\frac{4}{5}$ **61.** $\frac{3}{40}$ **63.** $\frac{1}{2}$ **65.** $\frac{1}{3}$ **67.** $\frac{1}{4}$
69. $\frac{7}{12}$ **71.** $\frac{1}{12}$ **73.** $\frac{19}{24}$ **75.** $\frac{11}{72}$ **77.** $\frac{199}{48}$ **79.** 60%, 0.6
81. $\frac{9}{100}$, 0.09 **83.** 8%, $\frac{2}{25}$ **85.** 0.75, 75% **87.** $\frac{1}{50}$, 0.02
89. $\frac{1}{100}$, 1% **91.** 3 **93.** 1 **95.** $\frac{71}{96}$ **97.** $\frac{17}{120}$
99. $\frac{65}{16}$ **101.** $\frac{69}{4}$ **103.** $\frac{13}{12}$ **105.** $\frac{1}{8}$ **107.** $\frac{3}{8}$ **109.** $\frac{3}{16}$
111. $\frac{2}{3}$ **113.** $\frac{1}{2}$ **115.** $\frac{19}{96}$
117. a) 1.3 yd^3 **b)** $36\frac{11}{24}$ ft^3 or $1\frac{227}{648}$ yd^3

121. Each daughter gets 3 km^2 ÷ 4 or a $\frac{3}{4}$ km^2 piece of the farm. Divide the farm into 12 equal squares. Give each daughter an L-shaped piece consisting of 3 of those 12 squares.

Section 1.3 Warm-Ups
1. Additive inverses, opposites **2.** Zero **3.** Subtract
4. $a + (-b)$ **5.** True **6.** True **7.** True **8.** False
9. False **10.** False **11.** True **12.** False

Section 1.3 Exercises
1. 13 **3.** -13 **5.** -8 **7.** -1.15 **9.** $-\frac{1}{2}$ **11.** 0 **13.** 0
15. 2 **17.** -6 **19.** 5.6 **21.** -2.9 **23.** $-\frac{1}{4}$ **25.** $8 + (-2)$

27. $4 + (-12)$ **29.** $-3 + 8$ **31.** $8.3 + (1.5)$ **33.** -4
35. -10 **37.** 11   **39.** -11 **41.** $-\dfrac{1}{4}$ **43.** $\dfrac{3}{4}$ **45.** 7
47. 0.93   **49.** 9.3   **51.** -5.03 **53.** 3   **55.** -9 **57.** -120
59. 78   **61.** -27 **63.** -7 **65.** -201 **67.** -322
69. -15.97 **71.** -2.92 **73.** -3.73 **75.** 3.7   **77.** $\dfrac{3}{20}$ **79.** $\dfrac{7}{24}$
81. 13   **83.** -10 **85.** 14   **87.** -4 **89.** -3 **91.** -3.49
93. -0.3422 **95.** -48.84 **97.** -8.85 **99.** $-\$8.85$
101. $-7°C$
103. When adding signed numbers, we add or subtract only positive numbers which are the absolute values of the original numbers. We then determine the appropriate sign for the answer.
105. The distance between x and y is given by either $|x - y|$ or $|y - x|$.

Section 1.4 Warm-Ups
1. Product   **2.** Absolute values   **3.** Quotient
4. Multiplication   **5.** True   **6.** True   **7.** True
8. False   **9.** True   **10.** True   **11.** False   **12.** False

Section 1.4 Exercises
1. -27 **3.** 132   **5.** $-\dfrac{1}{3}$ **7.** -0.3 **9.** 144   **11.** 0
13. -1 **15.** 3   **17.** $-\dfrac{2}{3}$ **19.** $\dfrac{5}{6}$ **21.** 0   **23.** -80
25. 0.25   **27.** 0   **29.** Undefined   **31.** Undefined   **33.** 0
35. -100 **37.** 27   **39.** -3 **41.** -4 **43.** -30 **45.** 19
47. -0.18 **49.** 0.3   **51.** -6 **53.** 1.5   **55.** 22
57. $-\dfrac{1}{3}$ **59.** -164.25 **61.** 1529.41   **63.** -12
65. -8 **67.** -6 **69.** -1 **71.** 5   **73.** 16   **75.** -8
77. 0   **79.** 0   **81.** -3.9 **83.** -40 **85.** 0.4   **87.** 0.4
89. -0.2 **91.** -7.5 **93.** $-\dfrac{1}{30}$ **95.** $-\dfrac{1}{10}$ **97.** 7.562
99. 19.35   **101.** 0   **103.** Undefined   **105.** $-\$27,778$

Mid-Chapter Quiz 1.1–1.4
1.
2.
3.
4.
5. $\dfrac{5}{8}$ **6.** $\dfrac{1}{12}$ **7.** $\dfrac{1}{10}$ **8.** $\dfrac{3}{4}$ **9.** 14   **10.** -1 **11.** -36
12. 5   **13.** -17 **14.** -100 **15.** 66   **16.** -7 **17.** $-\dfrac{3}{20}$
18. $-\dfrac{17}{42}$ **19.** $\dfrac{1}{20}$ **20.** 0   **21.** 5   **22.** 8, 8, 0   **23.** 0.25, 25%
24. 2.5 ft/sec   **25.** $\dfrac{12}{32}$ **26.** $\dfrac{2}{3}$ **27.** Undefined

Section 1.5 Warm-Ups
1. Arithmetic expression   **2.** Grouping   **3.** Exponential
4. Order of operations   **5.** False   **6.** True   **7.** False

8. False   **9.** False   **10.** False   **11.** False
12. True   **13.** True

Section 1.5 Exercises
1. -4 **3.** 1   **5.** -8 **7.** -7 **9.** -16 **11.** -4
13. 4^4 **15.** $(-5)^4$ **17.** $(-y)^3$ **19.** $\left(\dfrac{3}{7}\right)^5$ **21.** $5 \cdot 5 \cdot 5$
23. $b \cdot b$ **25.** $\left(-\dfrac{1}{2}\right)\left(-\dfrac{1}{2}\right)\left(-\dfrac{1}{2}\right)\left(-\dfrac{1}{2}\right)\left(-\dfrac{1}{2}\right)$
27. 81   **29.** 0   **31.** 625   **33.** -216 **35.** 100,000
37. -0.001 **39.** $\dfrac{1}{8}$ **41.** $\dfrac{1}{4}$ **43.** -64 **45.** -4096
47. 27   **49.** -13 **51.** 50   **53.** 10   **55.** 36   **57.** 18
59. -19 **61.** -17 **63.** -44 **65.** 18   **67.** -78 **69.** 0
71. 27   **73.** 1   **75.** 8   **77.** 7   **79.** 11   **81.** 111   **83.** 21
85. -1 **87.** -11 **89.** 9   **91.** 16   **93.** 28   **95.** 121
97. -73 **99.** 25   **101.** 0   **103.** -2 **105.** 12   **107.** 82
109. -54 **111.** -79 **113.** -24 **115.** 41.92
117. 181,806   **119.** 8.0548
121. a) \$1280   **b)** \$1275
123. a) 343.5 million   **b)** 2034
125. $(-5)^3 = -(5^3) = -5^3 = -1 \cdot 5^3$ and $-(5)^3 = 5^3$

Section 1.6 Warm-Ups
1. Algebraic expression   **2.** Sum   **3.** Product   **4.** Quotient
5. Difference   **6.** Equation   **7.** True   **8.** False   **9.** True
10. False   **11.** True   **12.** False   **13.** False   **14.** True

Section 1.6 Exercises
1. Difference   **3.** Cube   **5.** Sum   **7.** Difference
9. Product   **11.** Square   **13.** The difference of x^2 and a^2
15. The square of $x - a$ **17.** The quotient of $x - 4$ and 2
19. The difference of $\dfrac{x}{2}$ and 4   **21.** The cube of ab
23. $8 + y$ **25.** $5xz$ **27.** $8 - 7x$ **29.** $\dfrac{6}{x + 4}$
31. $(a + b)^2$ **33.** $x^3 + y^2$ **35.** $5m^2$ **37.** $(s + t)^2$
39. 3   **41.** 3   **43.** 16   **45.** -9 **47.** -3 **49.** -8
51. $-\dfrac{2}{3}$ **53.** 4   **55.** -1 **57.** 1   **59.** -4 **61.** 0
63. Yes   **65.** No   **67.** Yes   **69.** Yes   **71.** Yes
73. No   **75.** No   **77.** $5x + 3x = 8x$ **79.** $3(x + 2) = 12$
81. $\dfrac{x}{3} = 5x$ **83.** $(a + b)^2 = 9$
85. $-7, -5, -3, -1, 1$
87. $4, 8, 16; \dfrac{1}{4}, \dfrac{1}{8}, \dfrac{1}{16}; 100, 1000, 10,000; 0.01, 0.001, 0.0001$
89. 14.65   **91.** 37.12   **93.** 169.3 cm, 41 cm
95. 4, 5, 14, 16.5   **97.** 920 feet
99. For the square of the sum consider $(2 + 3)^2 = 5^2 = 25$. For the sum of the squares consider $2^2 + 3^2 = 4 + 9 = 13$. So $(2 + 3)^2 \neq 2^2 + 3^2$.

Section 1.7 Warm-Ups
1. Commutative   **2.** Distributive   **3.** Associative   **4.** Factoring
5. Additive   **6.** Multiplicative   **7.** True   **8.** False   **9.** False
10. True   **11.** False   **12.** True   **13.** True   **14.** True   **15.** True

Section 1.7 Exercises
1. $r + 9$ **3.** $3(x + 2)$ **5.** $-5x + 4$ **7.** $6x$
9. $-2(x - 4)$ **11.** $4 - 8y$ **13.** $4w^2$ **15.** $3a^2b$ **17.** $9x^3z$

19. -3 **21.** -10 **23.** -21 **25.** 0.6
27. $3x - 15$ **29.** $2a + at$ **31.** $-3w + 18$ **33.** $-20 + 4y$
35. $-a + 7$ **37.** $-t - 4$ **39.** $2(m + 6)$ **41.** $4(x - 1)$
43. $4(y - 4)$ **45.** $4(a + 2)$ **47.** $x(1 + y)$ **49.** $2(3a - b)$
51. 2 **53.** $-\dfrac{1}{5}$ **55.** $\dfrac{1}{7}$ **57.** 1 **59.** -4 **61.** $\dfrac{2}{5}$
63. Commutative property of multiplication
65. Distributive property
67. Associative property of multiplication
69. Additive inverse property
71. Commutative property of multiplication
73. Multiplicative identity property
75. Distributive property
77. Additive inverse property
79. Multiplication property of 0
81. Distributive property
83. $y + a$ **85.** $(5a)w$ **87.** $\dfrac{1}{2}(x + 1)$ **89.** $3(2x + 5)$
91. 1 **93.** 0 **95.** $\dfrac{100}{33}$
97. The perimeter is twice the sum of the length and width.
99. a) Commutative   **b)** Not commutative

Section 1.8 Warm-Ups
  1. Term   **2.** Like   **3.** Coefficient   **4.** Simplify   **5.** Sign
  6. False   **7.** True   **8.** True   **9.** True   **10.** False
11. False   **12.** False

Section 1.8 Exercises
  1. 7000   **3.** 1   **5.** 356   **7.** 350   **9.** 36   **11.** 36,000
13. 0   **15.** 98   **17.** $11w$ **19.** $3x$ **21.** $5x$ **23.** $-a$
25. $-2a$ **27.** $10 - 6t$ **29.** $8x^2$ **31.** $-4x + 2x^2$
33. $-7mw^2$ **35.** $\dfrac{5}{6}a$ **37.** $12h$ **39.** $-18b$ **41.** $-9m^2$
43. $12d^2$ **45.** y^2 **47.** $-15ab$ **49.** $-6a - 3ab$
51. $-k + k^2$ **53.** y **55.** $-3y$ **57.** y **59.** $2y^2$ **61.** $2a - 1$
63. $3x - 2$ **65.** $2x - 1$ **67.** $6c - 13$ **69.** $-7b + 1$
71. $2w - 4$ **73.** $-2x + 1$ **75.** $8 - y$ **77.** $m - 6$
79. $w - 5$ **81.** $8x + 15$ **83.** $5x - 1$ **85.** $-2a - 1$
87. $5a - 2$ **89.** $6x^2 + x - 15$ **91.** $-2b^2 - 7b + 4$
93. $3m - 18$ **95.** $-3x - 7$ **97.** $0.95x - 0.5$
99. $4x - 4$ **101.** $2y + 4$ **103.** $2y + m - 1$ **105.** 3
107. $\dfrac{7}{6}a + \dfrac{13}{6}$ **109.** $0.15x - 0.4$ **111.** $-14k + 23$
113. 45   **115.** $4x + 80$, 200 feet
117. a) $0.25x - 7625$ **b)** $12,375$ **c)** $44,000$ **d)** $310,000$
119. a) $4(2 + x) = 8 + 4x$ **b)** $4(2x) = (4 \cdot 2)x = 8x$
  c) $\dfrac{4 + x}{2} = \dfrac{1}{2}(4 + x) = 2 + \dfrac{1}{2}x$
  d) $5 - (x - 3) = 5 - x + 3 = 8 - x$

Enriching Your Mathematical Word Power
  1. Integers   **2.** Natural   **3.** Whole   **4.** Rational
  5. Irrational   **6.** Term   **7.** Like   **8.** Variable   **9.** Fraction
10. Reduced   **11.** Lowest   **12.** Additive   **13.** Order
14. Least   **15.** Absolute   **16.** Additive   **17.** Multiplicative
18. Divisor, quotient   **19.** Prime   **20.** Improper

Review Exercises
  1. 0, 1, 2, 10   **3.** $-2, 0, 1, 2, 10$ **5.** $-\sqrt{5}, \pi$
  7. True   **9.** False   **11.** False   **13.** True

15.   **17.**

19. $[4, 6]$ **21.** $[-30, \infty)$ **23.** $\dfrac{17}{24}$ **25.** 6   **27.** $\dfrac{3}{7}$ **29.** $\dfrac{14}{3}$
31. $\dfrac{13}{12}$ **33.** 2   **35.** -13 **37.** -7 **39.** -7 **41.** 11.95
43. -0.05 **45.** $-\dfrac{1}{6}$ **47.** $-\dfrac{11}{15}$ **49.** -15 **51.** 4   **53.** 5
55. $\dfrac{1}{6}$ **57.** -0.3 **59.** -0.24 **61.** 1   **63.** 66   **65.** 49
67. 41   **69.** 1   **71.** 50   **73.** -135 **75.** -2 **77.** -16
79. 16   **81.** 5   **83.** 9   **85.** 7   **87.** $-\dfrac{1}{3}$ **89.** 1   **91.** -9
93. Yes   **95.** No   **97.** Yes   **99.** No
101. Distributive property
103. Multiplicative inverse property
105. Additive identity property
107. Associative property of addition
109. Commutative property of multiplication
111. Additive inverse property
113. Multiplicative identity property
115. $-a + 12$ **117.** $6a^2 - 6a$ **119.** $-12t + 39$
121. $-0.9a - 0.57$ **123.** $-0.05x - 4$ **125.** $27x^2 + 6x + 5$
127. $-2a$ **129.** $x^2 + 4x - 3$ **131.** 0   **133.** 8   **135.** -21
137. $\dfrac{1}{2}$ **139.** -0.5 **141.** -1 **143.** $x + 2$ **145.** $4 + 2x$
147. $2x$ **149.** $-4x + 8$ **151.** $6x$ **153.** x **155.** $8x$
157. $-x^2 + 6x - 8$ **159.** $\dfrac{1}{4}x - \dfrac{3}{2}$ **161.** 3, 2, 1, 0, -1
163. 25, 125, 625; 16, -64, 256
165. a) $0.35x - 22{,}316.5$ **b)** Approximately $153,000
  c) $9,777,684

Chapter 1 Test
  1. 0, 8   **2.** $-3, 0, 8$ **3.** $-3, -\dfrac{1}{4}, 0, 8$ **4.** $-\sqrt{3}, \sqrt{5}, \pi$
  5. -21 **6.** -4 **7.** 9   **8.** -7 **9.** -0.95 **10.** -56
11. 978   **12.** 13   **13.** -1 **14.** 0   **15.** 9740   **16.** $-\dfrac{7}{24}$
17. -20 **18.** $-\dfrac{1}{6}$ **19.** -39
20.   **21.**
22. $(2, \infty)$ **23.** $[3, 9)$ **24.** Distributive property
25. Commutative property of multiplication
26. Associative property of addition
27. Additive inverse property   **28.** Multiplicative identity property
29. Multiplication property of 0   **30.** $3(x + 10)$ **31.** $7(w - 1)$
32. $6x + 6$ **33.** $4x - 2$ **34.** $7x - 3$ **35.** $0.9x + 7.5$
36. $14a^2 + 5a$ **37.** $x + 2$ **38.** $4t$ **39.** $54x^2y^2$
40. $\dfrac{3}{4}x + \dfrac{3}{2}$ **41.** 41   **42.** 5   **43.** -12 **44.** No   **45.** Yes
46. Yes   **47.** $3.66R - 0.06A + 82.205$, 168.905 cm

Chapter 2

Section 2.1 Warm-Ups
  1. Equation   **2.** Solution set   **3.** Satisfies   **4.** Equivalent
  5. Linear   **6.** Addition property of equality   **7.** True
  8. True   **9.** False   **10.** True   **11.** True   **12.** True
13. False

Section 2.1 Exercises

1. $\{1\}$ **3.** $\{9\}$ **5.** $\{1\}$ **7.** $\left\{\dfrac{2}{3}\right\}$ **9.** $\{-9\}$ **11.** $\{-19\}$

13. $\left\{\dfrac{1}{4}\right\}$ **15.** $\{0\}$ **17.** $\{5.95\}$ **19.** $\{-5\}$ **21.** $\{-4\}$ **23.** $\{3\}$

25. $\left\{\dfrac{1}{4}\right\}$ **27.** $\{-8\}$ **29.** $\{1.8\}$ **31.** $\left\{\dfrac{2}{3}\right\}$ **33.** $\left\{\dfrac{1}{2}\right\}$ **35.** $\{-5\}$

37. $\{5\}$ **39.** $\{1.25\}$ **41.** $\left\{\dfrac{1}{4}\right\}$ **43.** $\left\{\dfrac{3}{20}\right\}$ **45.** $\{-2\}$ **47.** $\{120\}$

49. $\left\{\dfrac{5}{9}\right\}$ **51.** $\left\{-\dfrac{1}{2}\right\}$ **53.** $\{-8\}$ **55.** $\left\{\dfrac{1}{3}\right\}$ **57.** $\{-3.4\}$ **59.** $\{99\}$

61. $\{-7\}$ **63.** $\{9\}$ **65.** $\{8\}$ **67.** $\{5\}$ **69.** $\{-5\}$ **71.** $\{-8\}$

73. $\{2\}$ **75.** $\left\{\dfrac{1}{6}\right\}$ **77.** $\left\{-\dfrac{1}{3}\right\}$ **79.** $\{44\}$ **81.** $\left\{\dfrac{3}{4}\right\}$ **83.** $\{7\}$

85. $\{-14\}$ **87.** $\left\{\dfrac{3}{8}\right\}$

89. a) $\dfrac{2}{3}x = 41.8$, 62.7 births per 1000 females
  b) 50 births per 1000 females
91. 2877 stocks   **93.** 3000 students

Section 2.2 Warm-Ups

1. Multiplication   **2.** Addition   **3.** Addition, multiplication
4. True   **5.** True   **6.** True   **7.** False   **8.** True
9. True   **10.** True   **11.** False

Section 2.2 Exercises

1. $\{2\}$ **3.** $\{-2\}$ **5.** $\left\{\dfrac{2}{3}\right\}$ **7.** $\{6\}$ **9.** $\{12\}$ **11.** $\left\{\dfrac{1}{2}\right\}$

13. $\left\{-\dfrac{1}{6}\right\}$ **15.** $\{4\}$ **17.** $\left\{\dfrac{5}{6}\right\}$ **19.** $\{4\}$ **21.** $\{-5\}$ **23.** $\{34\}$

25. $\{9\}$ **27.** $\{1.2\}$ **29.** $\{3\}$ **31.** $\{4\}$ **33.** $\{-3\}$ **35.** $\left\{\dfrac{1}{2}\right\}$

37. $\{30\}$ **39.** $\{6\}$ **41.** $\{-2\}$ **43.** $\{18\}$ **45.** $\{0\}$ **47.** $\left\{\dfrac{1}{6}\right\}$

49. $\{-2\}$ **51.** $\left\{\dfrac{7}{3}\right\}$ **53.** $\{1\}$ **55.** $\{-6\}$ **57.** $\{-12\}$ **59.** $\{-4\}$

61. $\{-13\}$ **63.** $\{1.7\}$ **65.** $\{2\}$ **67.** $\{4.6\}$ **69.** $\{8\}$ **71.** $\{34\}$
73. $\{6\}$ **75.** $\{0\}$ **77.** $\{-10\}$ **79.** $\{18\}$ **81.** $\{-20\}$ **83.** $\{-3\}$
85. $\{-4.3\}$ **87.** 17 hr **89.** 20°C **91.** 9 ft **93.** $14,550

Section 2.3 Warm-Ups

1. Least common denominator   **2.** Multiply   **3.** Identity
4. Conditional   **5.** Inconsistent   **6.** True   **7.** False
8. False   **9.** True   **10.** True   **11.** True

Section 2.3 Exercises

1. $\left\{\dfrac{6}{5}\right\}$ **3.** $\left\{\dfrac{2}{9}\right\}$ **5.** $\{7\}$ **7.** $\{24\}$ **9.** $\{16\}$ **11.** $\{-12\}$

13. $\{60\}$ **15.** $\{24\}$ **17.** $\left\{-\dfrac{4}{3}\right\}$ **19.** $\{90\}$ **21.** $\{6\}$ **23.** $\{-2\}$

25. $\{80\}$ **27.** $\{60\}$ **29.** $\{200\}$ **31.** $\{800\}$ **33.** $\left\{\dfrac{9}{2}\right\}$ **35.** $\{3\}$

37. $\{25\}$ **39.** $\{-2\}$ **41.** $\{-3\}$ **43.** $\{5\}$ **45.** $\{-10\}$ **47.** $\{2\}$
49. All real numbers, identity   **51.** $\varnothing$, inconsistent
53. $\{0\}$, conditional   **55.** $\varnothing$, inconsistent   **57.** $\varnothing$, inconsistent
59. $\{1\}$, conditional   **61.** $\varnothing$, inconsistent
63. All real numbers, identity   **65.** All nonzero real numbers, identity
67. All real numbers, identity   **69.** $\{-4\}$ **71.** R **73.** R **75.** $\{100\}$

77. $\left\{-\dfrac{3}{2}\right\}$ **79.** $\{30\}$ **81.** $\{6\}$ **83.** $\{0.5\}$ **85.** $\{19,608\}$

87. $128,000 **89. a)** $240,000 **b)** $250,635

Section 2.4 Warm-Ups

1. Formula, literal   **2.** Solve   **3.** Function   **4.** Perimeter
5. Area   **6.** Circumference   **7.** False   **8.** False
9. True   **10.** False   **11.** True

Section 2.4 Exercises

1. $R = \dfrac{D}{T}$ **3.** $D = \dfrac{C}{\pi}$ **5.** $P = \dfrac{I}{rt}$ **7.** $C = \dfrac{5}{9}(F - 32)$

9. $h = \dfrac{2A}{b}$ **11.** $L = \dfrac{P - 2W}{2}$ **13.** $a = 2A - b$

15. $r = \dfrac{S - P}{Pt}$ **17.** $a = \dfrac{2A - bh}{h}$ **19.** $y = -x - 9$

21. $y = -x + 6$

23. $y = 2x - 2$ **25.** $y = 3x + 4$ **27.** $y = -\dfrac{1}{2}x + 2$ **29.** $y = x - \dfrac{1}{2}$

31. $y = 3x - 14$ **33.** $y = \dfrac{1}{2}x$ **35.** $y = \dfrac{3}{2}x + 6$ **37.** $y = \dfrac{3}{2}x + \dfrac{13}{2}$

39. $y = -\dfrac{1}{4}x + \dfrac{5}{8}$ **41.** $x = \dfrac{b - a}{2}$ **43.** $x = -7a$ **45.** $x = 12 - a$

47. $x = 7ab$ **49.** 2 **51.** 7 **53.** $-\dfrac{9}{5}$ **55.** 1 **57.** 1.33

59. 60, 30, 0, -30, -60 **61.** 14, 23, 32, 104, 212 **63.** 40, 20, 10, 5, 4

65. 1, 3, 6, 10, 15 **67.** 4%, $4\dfrac{2}{3}$%, $5\dfrac{1}{3}$% **69.** 4 years

71. 14 yards, $9\dfrac{1}{3}$ yards, 7 yards **73.** 225 feet **75.** $60,500

77. $300 **79.** 20% **81.** 160 feet **83.** 24 cubic feet
85. 4 inches **87.** 8 feet **89.** 12 inches
91. a) 640 milligrams **b)** Age 4 **c)** Age 13
93. 3.75 milliliters **95.** $L = F\sqrt{S} - 2D + 5.688$

Mid-Chapter Quiz 2.1–2.4

1. $\{-21\}$ **2.** $\left\{\dfrac{2}{3}\right\}$ **3.** $\{5\}$ **4.** $\left\{\dfrac{3}{4}\right\}$ **5.** $\left\{\dfrac{9}{2}\right\}$

6. $\{7\}$ **7.** $\{4\}$ **8.** $\left\{\dfrac{1}{3}\right\}$ **9.** $\{950\}$ **10.** $\{-200\}$

11. Identity   **12.** Conditional equation   **13.** Conditional equation

14. Inconsistent equation   **15.** $x = \dfrac{c - b}{a}$ **16.** $x = \dfrac{5a - 2b}{3}$

17. $15,800 **18.** 9 yd **19.** 3 **20.** 8%

Section 2.5 Warm-Ups

1. Addition   **2.** Multiplication   **3.** Complementary
4. Supplementary   **5.** Product   **6.** Even, odd
7. True   **8.** True   **9.** True   **10.** False
11. False   **12.** False

Section 2.5 Exercises

1. $x + 3$ **3.** $x - 3$ **5.** $5x$ **7.** $0.1x$ **9.** $\dfrac{x}{3}$ **11.** $\dfrac{1}{3}x$

13. x and $x + 15$ **15.** x and $6 - x$
17. x and $x + 3$ **19.** x and $0.05x$ **21.** x and $1.30x$
23. x and $90 - x$ **25.** x and $120 - x$
27. n and $n + 2$, where n is an even integer

29. x and $x + 1$, where x is an integer
31. $x, x + 2$, and $x + 4$, where x is an odd integer
33. $x, x + 2, x + 4$, and $x + 6$, where x is an even integer
35. $3x$ miles   **37.** $0.25q$ dollars   **39.** $\dfrac{x}{20}$ hour
41. $\dfrac{x - 100}{12}$ meters per second   **43.** $5x$ square meters
45. $2w + 2(w + 3)$ inches   **47.** $150 - x$ feet   **49.** $2x + 1$ feet
51. $x(x + 5)$ square meters   **53.** $0.18(x + 1000)$
55. $\dfrac{16.50}{x}$ dollars per pound   **57.** $90 - x$ degrees
59. x is the smaller number, $x(x + 5) = 8$
61. x is the selling price, $x - 0.07x = 84{,}532$
63. x is the percent, $500x = 100$
65. x is the number of nickels, $0.05x + 0.10(x + 2) = 3.80$
67. x is the number, $x + 5 = 13$
69. x is the smallest integer, $x + (x + 1) + (x + 2) = 42$
71. x is the smaller integer, $x(x + 1) = 182$
73. x is Harriet's income, $0.12x = 3000$
75. x is the number, $0.05x = 13$
77. x is the width, $x(x + 5) = 126$
79. n is the number of nickels, $5n + 10(n - 1) = 95$
81. x is the measure of the larger angle, $x + x - 38 = 180$
83. a) $r + 0.6[220 - (30 + r)] = 144$, where r is the resting heart rate
   b) Target heart rate increases as resting heart rate increases.
85. $6 + x$ **87.** $m + 9$ **89.** $11t$ **91.** $5(x - 2)$ **93.** $m - 3m$
95. $\dfrac{h + 8}{h}$ **97.** $\dfrac{5}{y - 9}$ **99.** $\dfrac{w - 8}{2w}$ **101.** $-3v - 9$
103. $x - \dfrac{x}{7}$ **105.** $m^2 - (m + 7)$ **107.** $x + (9x - 8)$ **109.** $13n - 9$
111. $6 + \dfrac{1}{3}(x + 2)$ **113.** $\dfrac{x}{2} + x$ **115.** $x(x + 3) = 24$
117. $w(w - 4) = 24$

Section 2.6 Warm-Ups
1. Uniform   **2.** Geometric   **3.** Complementary
4. Supplementary   **5.** Even   **6.** Odd   **7.** False
8. True   **9.** True   **10.** False   **11.** True   **12.** False

Section 2.6 Exercises
1. 39, 40   **3.** 46, 47, 48   **5.** 75, 77   **7.** 47, 48, 49, 50
9. Length 50 meters, width 25 meters
11. Width 42 inches, length 46 inches
13. 13 inches   **15.** 35°   **17.** 65 miles per hour
19. 55 miles per hour   **21.** 4 hours, 2048 miles
23. Length 20 inches, width 12 inches   **25.** 5 ft, 5 ft, 3 ft
27. 20°, 40°, 120°   **29.** 20°, 80°, 80°   **31.** Raiders 32, Vikings 14
33. 3 hours, 106 miles   **35.** Crawford 1906, Wayne 1907, Stewart 1908
37. 7 ft, 7 ft, 16 ft

Section 2.7 Warm-Ups
1. Rate   **2.** Discount   **3.** Product   **4.** Table
5. Rate   **6.** True   **7.** False   **8.** True   **9.** False

Section 2.7 Exercises
1. $320   **3.** $400   **5.** $125,000   **7.** $30.24
9. 100 Fund $10,000, 101 Fund $13,000
11. Fidelity $14,000, Price $11,000

13. 30 gallons   **15.** 20 liters of 5% alcohol, 10 liters of 20% alcohol
17. 55,700 voters   **19.** $15,000   **21.** 75%   **23.** 600 students
25. 42 private rooms, 30 semiprivate rooms   **27.** 12 pounds
29. 4 nickels, 6 dimes   **31.** 800 gallons   **33.** $\frac{2}{3}$ gallon
35. Shorts $12, tops $6

Section 2.8 Warm-Ups
1. Inequality   **2.** Bracket   **3.** Parenthesis   **4.** Compound
5. Between   **6.** True   **7.** False   **8.** True
9. True   **10.** False   **11.** False

Section 2.8 Exercises
1. False   **3.** True   **5.** True   **7.** False   **9.** True   **11.** True
13. True   **15.** True

17. $(-\infty, 3]$

19. $(-2, \infty)$

21. $(-\infty, -1)$

23. $[-2, \infty)$

25. $\left[\dfrac{1}{2}, \infty\right)$

27. $(-\infty, 5.3]$

29. $(-3, 1)$

31. $[3, 7]$

33. $[-5, 0)$

35. $(40, 100]$

37. $x > 3, (3, \infty)$ **39.** $x \le 2, (-\infty, 2]$ **41.** $0 < x < 2, (0, 2)$
43. $-5 < x \le 7, (-5, 7]$ **45.** $x > -4, (-4, \infty)$
47. Yes   **49.** No   **51.** No   **53.** Yes   **55.** Yes   **57.** Yes
59. No   **61.** Yes   **63.** No   **65.** 0, 5.1   **67.** 5.1   **69.** 5.1
71. $-5.1, 0, 5.1$ **73.** $0.08p > 1500$ **75.** $p + 2p + p + 0.25 < 2.00$
77. $\dfrac{44 + 72 + s}{3} \ge 60$ **79.** $396 < 8R < 453$ **81.** $60 < 90 - x < 70$
83. a) $45 + 2(30) + 2h \le 130$ **b)** Approximately 12 in.
85. 79, moderate effort on level ground

Section 2.9 Warm-Ups
1. Equivalent   **2.** Addition   **3.** Multiplication
4. True   **5.** False   **6.** True   **7.** False
8. True   **9.** True

Section 2.9 Exercises

1. >   **3.** ≥   **5.** >   **7.** >   **9.** ≤

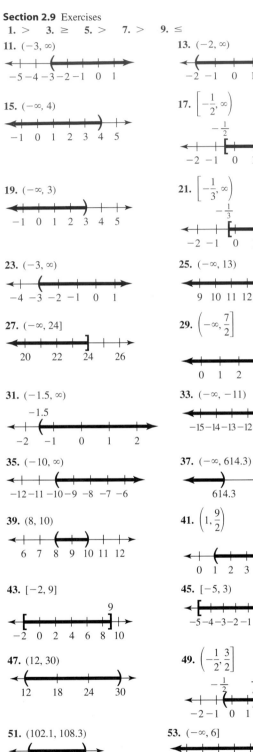

11. $(-3, \infty)$

13. $(-2, \infty)$

15. $(-\infty, 4)$

17. $\left[-\frac{1}{2}, \infty\right)$

19. $(-\infty, 3)$

21. $\left[-\frac{1}{3}, \infty\right)$

23. $(-3, \infty)$

25. $(-\infty, 13)$

27. $(-\infty, 24]$

29. $\left(-\infty, \frac{7}{2}\right]$

31. $(-1.5, \infty)$

33. $(-\infty, -11)$

35. $(-10, \infty)$

37. $(-\infty, 614.3)$

39. $(8, 10)$

41. $\left(1, \frac{9}{2}\right)$

43. $[-2, 9]$

45. $[-5, 3)$

47. $(12, 30)$

49. $\left(-\frac{1}{2}, \frac{3}{2}\right]$

51. $(102.1, 108.3)$

53. $(-\infty, 6]$

55. $(-\infty, 0)$

57. $(2, 3)$

59. At least 28 meters   **61.** Less than $9358   **63.** At most $550
65. At least 64   **67.** Between 81 and 94.5 inclusive
69. Between 49.5 and 56.625 miles per hour
71. Between 55° and 85°
73. a) Between 27 and 35 teeth inclusive   **b)** Between 23.02 in. and 24.79 in.   **c)** At least 14 teeth

Enriching Your Mathematical Word Power

1. Equation   **2.** Linear   **3.** Identity   **4.** Conditional
5. Inconsistent   **6.** Equivalent   **7.** Literal, formula
8. Function   **9.** Complementary   **10.** Supplementary
11. Uniform   **12.** Inequality   **13.** Equivalent

Review Exercises

1. {35}   **3.** {−6}   **5.** {−7}   **7.** {13}   **9.** {7}   **11.** {2}   **13.** {7}
15. {0}   **17.** {−8}   **19.** ∅, inconsistent
21. All real numbers, identity   **23.** All nonzero real numbers, identity
25. {24}, conditional   **27.** {80}, conditional   **29.** {1000}, conditional
31. $\left\{\frac{1}{4}\right\}$ **33.** $\left\{\frac{21}{8}\right\}$ **35.** $\left\{-\frac{4}{5}\right\}$ **37.** {4}   **39.** {24}   **41.** {−100}
43. $x = -\frac{b}{a}$ **45.** $x = \frac{b+2}{a}$ **47.** $x = \frac{V}{LW}$ **49.** $x = -\frac{b}{3}$
51. $y = -\frac{5}{2}x + 3$ **53.** $y = -\frac{1}{2}x + 4$ **55.** $y = -2x + 16$
57. −13   **59.** $-\frac{2}{5}$ **61.** 17   **63.** 15, 10, 5, 0, −5
65. −3, −1, 1, 3   **67.** $x + 9$, where x is the number
69. x and $x + 8$, where x is the smaller number
71. $0.65x$, where x is the number
73. $x(x + 5) = 98$, where x is the width
75. $2x = 3(x − 10)$, where x is Jim's rate
77. $x + x + 2 + x + 4 = 90$, where x is the smallest of the three even integers
79. $t + 2t + t − 10 = 180$, where t is the degree measure of an angle
81. 77, 79, 81   **83.** Betty 45 mph, Lawanda 60 mph
85. Wanda $36,000, husband $30,000   **87.** No   **89.** No
91. $x > 1$, $(1, \infty)$ **93.** $x \geq 2$, $[2, \infty)$ **95.** $-3 \leq x < 3$, $[-3, 3)$
97. $x < -1$, $(-\infty, -1)$
99. $(-1, \infty)$ **101.** $(-\infty, 3)$
103. $(-\infty, -4]$ **105.** $(-4, \infty)$
107. $(-1, 5)$ **109.** $\left(-2, \frac{1}{2}\right]$
111. $[0, 3]$ **113.** $(0, 1)$
115. $2800   **117.** $8500   **119.** $537.50   **121.** 400 movies
123. 31°   **125.** Less than 6 feet

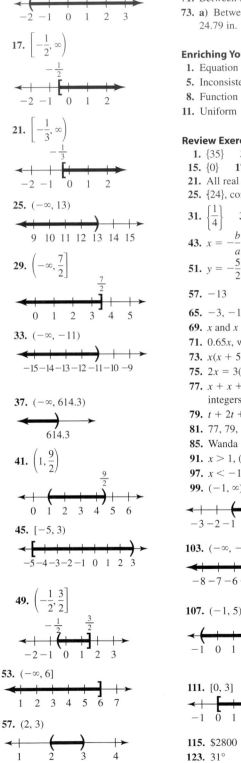

Chapter 2 Test

1. $\{-7\}$ **2.** $\{2\}$ **3.** $\{-9\}$ **4.** $\{700\}$ **5.** $\{1\}$ **6.** $\left\{\dfrac{7}{6}\right\}$ **7.** $\{2\}$

8. $\varnothing$ **9.** All real numbers   **10.** $y = \dfrac{2}{3}x - 3$ **11.** $a = \dfrac{m + w}{P}$

12. $-3 < x \le 2$, $(-3, 2]$ **13.** $x > 1$, $(1, \infty)$

14. $(19, \infty)$

17 18 19 20 21 22 23

15. $(-7, -1)$

$-7\ -6\ -5\ -4\ -3\ -2\ -1$

16. $(1, 3)$

$-1\ \ 0\ \ 1\ \ 2\ \ 3\ \ 4\ \ 5$

17. $(-6, \infty)$

$-8\ -7\ -6\ -5\ -4\ -3\ -2$

18. 14 meters   **19. a)** $A = \dfrac{1}{2}bh$ **b)** $h = \dfrac{2A}{b}$ **c)** 9 in.

20. 150 liters   **21.** At most \$2000   **22.** $30°, 60°, 90°$

Making Connections Chapters 1–2

1. $8x$ **2.** $15x^2$ **3.** $2x + 1$ **4.** $4x - 7$ **5.** $-2x + 13$

6. 60   **7.** 72   **8.** -10 **9.** $-2x^3$ **10.** -1

11. -18 **12.** -18 **13.** 5   **14.** 5   **15.** 25   **16.** 25

17. 1   **18.** 1   **19.** $(-\infty, 2)$ **20.** $(-6, \infty)$ **21.** $[5, \infty)$

22. $(-\infty, -1]$ **23.** $[2, 6]$ **24.** $(4, 8)$ **25.** $\dfrac{2}{3}$

26. $\dfrac{1}{6}$ **27.** $\dfrac{1}{9}$ **28.** $\dfrac{5}{9}$ **29.** 13   **30.** 8   **31.** $2x + 1$

32. $10x - 9$ **33.** $\left\{\dfrac{2}{3}\right\}$ **34.** $\left\{\dfrac{1}{6}\right\}$ **35.** $\left(\dfrac{2}{3}, \infty\right)$ **36.** $\left(-\infty, \dfrac{1}{6}\right]$

37. $\left\{\dfrac{1}{9}\right\}$ **38.** $\left\{\dfrac{5}{9}\right\}$ **39.** $\left[-\dfrac{1}{9}, \infty\right)$ **40.** $\left(-\infty, -\dfrac{5}{9}\right)$ **41.** $\left\{\dfrac{3}{10}\right\}$

42. $\left\{\dfrac{16}{5}\right\}$ **43.** $\left\{\dfrac{1}{2}\right\}$ **44.** $\left\{\dfrac{7}{5}\right\}$ **45.** $\{1\}$ **46.** All real numbers

47. $\{0\}$ **48.** $\{1\}$ **49.** $(0, \infty)$ **50.** $\varnothing$ **51.** $\{2\}$ **52.** $\{2\}$

53. $\left\{\dfrac{13}{2}\right\}$ **54.** $\{200\}$ **55.** $(-2, \infty)$ **56.** $[2, \infty)$

57. $[-3, 12]$ **58.** $(-41, -5)$

59. a) \$13,600   **b)** \$10,000   **c)** \$12,000

Chapter 3

Section 3.1 Warm-Ups

1. Origin   **2.** Ordered pair   **3.** x-intercept   **4.** y-intercept
5. Horizontal   **6.** Vertical   **7.** Linear   **8.** False   **9.** False
10. False   **11.** False   **12.** True   **13.** False   **14.** False

Section 3.1 Exercises

1–15. odd

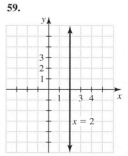

17. Quadrant II   **19.** x-axis   **21.** Quadrant III

23. Quadrant I   **25.** Quadrant II   **27.** y-axis

29. $(0, 9), (5, 24), (2, 15)$ **31.** $(0, -7), \left(\dfrac{1}{3}, -8\right), \left(-\dfrac{2}{3}, -5\right)$

33. $(0, 54.3), (10, 66.3), (0.5, 54.9)$ **35.** $(3, 0), (0, -2), (12, 6)$

37. $(5, -3), (5, 5), (5, 0)$ **39.** $(-2, 9), (0, 5), (2, 1), (4, -3), (6, -7)$

41. $(-6, 0), (-3, 1), (0, 2), (3, 3)$

43. $(-30, -200), (-20, 0), (-10, 200), (0, 400), (10, 600)$

45.

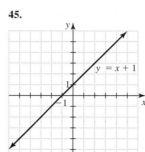

47.

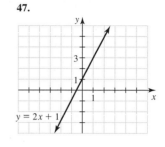

49.

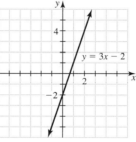

51.

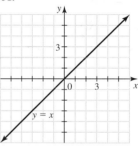

53.

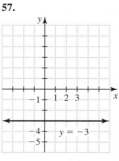

55.

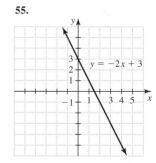

57.

59.

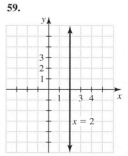

61.

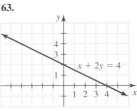

2x + y = 5

63.

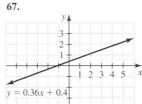

x + 2y = 4

65.

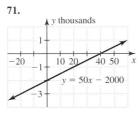

x − 3y = 6

67.

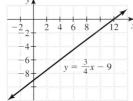

y = 0.36x + 0.4

69.

y = x + 1200

71.

y = 50x − 2000

73.

y = −400x + 2000

75. (2, 0), (0, 3)

3x + 2y = 6

77. (4, 0), (0, −1)

x − 4y = 4

79. (12, 0), (0, −9)

$y = \frac{3}{4}x - 9$

81. (2, 0), (0, 4)

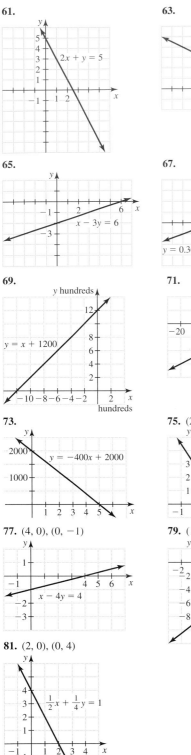

$\frac{1}{2}x + \frac{1}{4}y = 1$

83. a) $90, $190   **b)** 7 hours
85. a) 100%, 108%   **b)** age 70   **c)** $16,240 per year
87. a) $319 billion, $330.5 billion, $342 billion
    b) $353.5 billion   **c)** 2014

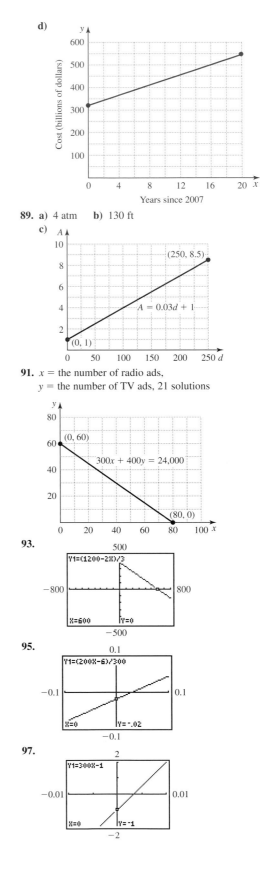

d)

Cost (billions of dollars)

Years since 2007

89. a) 4 atm   **b)** 130 ft

c)

(250, 8.5)

A = 0.03d + 1

(0, 1)

91. x = the number of radio ads,
    y = the number of TV ads, 21 solutions

(0, 60)

300x + 400y = 24,000

(80, 0)

93.

500

Y1=(1200-2X)/3

−800   800

X=600   Y=0

−500

95.

0.1

Y1=(200X-6)/300

−0.1   0.1

X=0   Y=-.02

−0.1

97.

2

Y1=300X-1

−0.01   0.01

X=0   Y=-1

−2

Section 3.2 Warm-Ups
1. Slope   2. Rise, run   3. Vertical   4. Horizontal   5. Positive
6. Negative   7. Perpendicular   8. Parallel   9. True   10. False
11. True   12. True   13. False   14. False   15. True

Section 3.2 Exercises
1. $-\dfrac{2}{3}$ 3. $\dfrac{2}{3}$ 5. $\dfrac{3}{2}$ 7. 0   9. $\dfrac{2}{5}$ 11. Undefined   13. 2
15. $\dfrac{5}{4}$ 17. $-\dfrac{5}{3}$ 19. $\dfrac{5}{7}$ 21. $-\dfrac{4}{3}$ 23. -1 25. 1
27. Undefined   29. 0   31. 3

33.
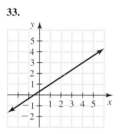

35.
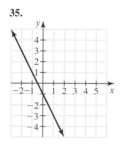

37.
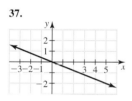

39.
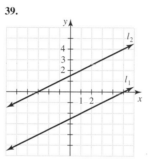

41.
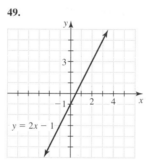

43. $-\dfrac{4}{3}$
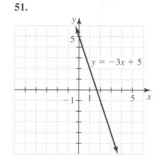

45. $\dfrac{1}{2}$
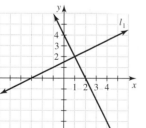

47. 1
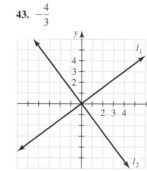

49. Parallel   51. Neither   53. Parallel   55. Perpendicular
57. a) Slope 0.1545; Cost is increasing about \$145,500 per year.
   b) \$1.918 million; yes   c) \$3.772 million
59. 1; The percentage increases 1% per year.
61. (2000, 28,100), (2003, 29,300), (2012, 32,900), (2015, 34,100)
63. Yes   65. No

Section 3.3 Warm-Ups
1. Slope-intercept   2. Slope, y-intercept   3. Standard   4. True
5. False   6. True   7. True   8. False   9. False   10. True
11. False   12. True

Section 3.3 Exercises
1. $y = \dfrac{3}{2}x + 1$ 3. $y = -2x + 2$ 5. $y = x - 2$ 7. $y = -x$
9. $y = -1$ 11. $x = -2$ 13. 3, $(0, -9)$ 15. $-\dfrac{1}{2}, (0, 3)$
17. 0, $(0, 4)$ 19. 1, $(0, 0)$ 21. $-3, (0, 0)$ 23. $-1, (0, 5)$
25. $\dfrac{1}{2}, (0, -2)$ 27. $\dfrac{2}{5}, (0, -2)$ 29. 2, $(0, 3)$
31. Undefined slope, no y-intercept   33. $x + y = 2$ 35. $x - 2y = -6$
37. $9x - 6y = 2$ 39. $6x + 10y = 7$ 41. $x = -10$ 43. $3y = 10$
45. $5x - 6y = 0$ 47. $x - 50y = -25$

49.

51.
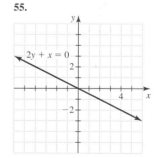

53.
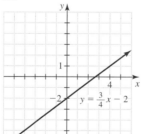

55.
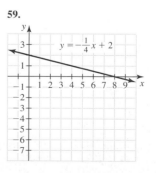

57.
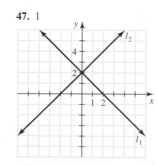

59.
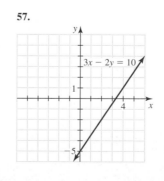

61.

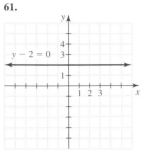

63. Parallel   **65.** Neither

4. $y = 5 - 3x$

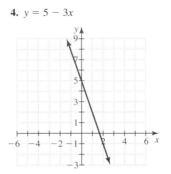

67. Parallel   **69.** Perpendicular   **71.** $y = \frac{1}{2}x - 4$ **73.** $y = 2x + 3$

75. $y = -\frac{1}{3}x + 6$ **77.** $y = -2x + 3$ **79.** $y = 3$

81. $y = -\frac{3}{2}x + 4$ **83.** $y = -\frac{4}{5}x + 4$

85. a) $80, $130, $180   **b)** 50, (0, 80)   **c)** There is an $80 fixed cost, plus $50 per hour.

87. a) $1,150,000, $1,150,200   **b)** $200   **c)** $200
    d) $1700 per mower

89. a) A slope of 1 means that the percentage of workers receiving training is going up 1% per year.   **b)** $y = x + 5$ where x is the number of years since 1982   **c)** The y-intercept (0, 5) means that 5% of the workers received training in 1982.   **d)** 33%

91. a) x = the number of packs of pansies, y = the number of packs of snapdragons

5. $x = 4$

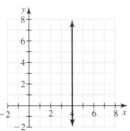

b)

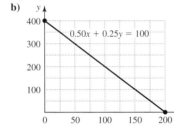

c) $y = -2x + 400$
d) -2
e) If the number of packs of pansies goes up by 1, then the number of packs of snapdragons goes down by 2.

6. $y = 2$

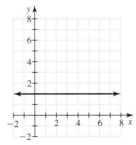

93. (2, 0), (0, 3)   **95.** $\frac{x}{9} + \frac{y}{5} = 1$

97.

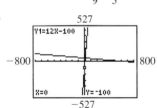

7. $2x - 3y = 6$

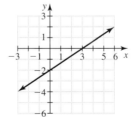

Mid-Chapter Quiz 3.1–3.3
1. $(-3, 6), (0, 4), (3, 2), (6, 0)$
2. $(-4, 0), (12, 12), (8, 9), (-12, -6)$ **3.** $y = x$

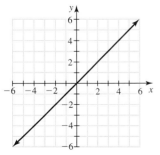

8. $y = \frac{5}{3}x - 4$

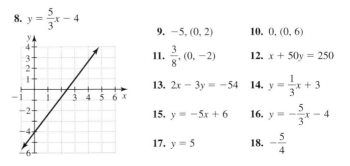

9. $-5, (0, 2)$ **10.** $0, (0, 6)$
11. $\frac{3}{8}, (0, -2)$ **12.** $x + 50y = 250$
13. $2x - 3y = -54$ **14.** $y = \frac{1}{3}x + 3$
15. $y = -5x + 6$ **16.** $y = -\frac{5}{3}x - 4$
17. $y = 5$ **18.** $-\frac{5}{4}$

19.

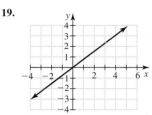

20. Perpendicular

Section 3.4 Warm-Ups

1. Point-slope   **2.** Parallel   **3.** Perpendicular   **4.** False
5. True   **6.** True   **7.** False   **8.** True   **9.** True
10. True   **11.** True   **12.** True

Section 3.4 Exercises

1. $y = -x + 1$ **3.** $y = 5x + 11$ **5.** $y = \frac{3}{4}x - 20$ **7.** $y = \frac{2}{3}x + \frac{1}{3}$

9. $y = 3x - 1$ **11.** $y = \frac{1}{2}x + 3$ **13.** $y = \frac{1}{3}x + \frac{7}{3}$

15. $y = -\frac{1}{2}x + 4$ **17.** $y = -6x - 13$ **19.** $2x - y = 7$

21. $x - 2y = 6$ **23.** $2x - 3y = 2$ **25.** $2x - y = -1$ **27.** $x - y = 0$

29. $3x - 2y = -1$ **31.** $3x + 5y = -11$ **33.** $x - y = -2$

35. $x = 2$ **37.** $y = 9$ **39.** $y = -x + 4$ **41.** $y = \frac{5}{3}x - 1$

43. $y = x + 3$ **45.** $y = -\frac{1}{3}x + 5$ **47.** $y = -\frac{2}{3}x + \frac{5}{3}$

49. $y = -2x - 5$ **51.** $y = \frac{1}{3}x + \frac{7}{3}$ **53.** $y = 2x - 1$ **55.** $y = 2$

57. $y = \frac{2}{3}x$ **59.** $y = -x$ **61.** $y = 50$ **63.** $y = -\frac{3}{5}x - 4$

65. e   **67.** f   **69.** h   **71.** g

73. a) Slope 1.625 means that the number of ATM transactions is increasing by 1.625 billion per year.   **b)** $y = 1.625x + 14.2$
c) 36.95 billion

75. a) $y = 2.8x + 45.7$ **b)** $x =$ years since 1990, $y =$ GDP in thousands of dollars   **c)** \$115,700
d)

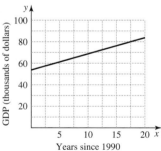

77. a) $C = 20n + 30$ or $C(n) = 20n + 30$ **b)** \$170   **c)** 12 hours

79. a) $S = 3L - \frac{41}{4}$ or $S(L) = 3L - \frac{41}{4}$ **b)** size 8.5   **c)** 7.25 inches

81. a) $v(t) = 32t + 10$ **b)** 122 ft/sec   **c)** 3 sec

83. a) $w(t) = -\frac{1}{120}t + \frac{3}{2}$ **b)** $\frac{5}{6}$ inch   **c)** 60°F

85. $A = 0.6w$, 3.6 in.   **87. a)** $a = 0.08c$ **b)** 0.24   **c)** 6.25 mg/ml

89. $2, 3, -\frac{2}{3}; 4, -5, \frac{4}{5}; \frac{1}{2}, 3, -\frac{1}{6}; 2, -\frac{1}{3}, 6$

91. a)

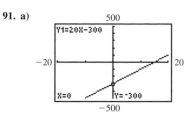

b)

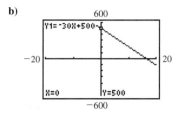

c)

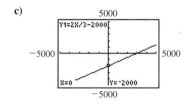

93. $-1 \le x \le 1, -1 \le y \le 1$

Section 3.5 Warm-Ups

1. Directly   **2.** Inversely   **3.** Jointly   **4.** True   **5.** True
6. False   **7.** True   **8.** False   **9.** True   **10.** True
11. True   **12.** True

Section 3.5 Exercises

1. $T = kh$ **3.** $y = \frac{k}{r}$ **5.** $R = kts$ **7.** $i = kb$ **9.** $A = kym$

11. $y = \frac{5}{3}x$ **13.** $A = \frac{6}{B}$ **15.** $m = \frac{198}{p}$ **17.** $A = 2tu$ **19.** $T = \frac{9}{2}u$

21. 25   **23.** 1   **25.** 105   **27.** 100.3 pounds   **29.** 50 minutes

31. \$17.40   **33.** 80 mph   **35.** 3 days   **37.** 1600, 12, 12

39. $\left(\frac{1}{2}, 600\right)$, $(1, 300)$, $(30, 10)$, $\left(900, \frac{1}{3}\right)$, Inversely

41. $\left(\frac{1}{3}, \frac{1}{4}\right)$, $(8, 6)$, $(12, 9)$, $(20, 15)$, Directly   **43.** Directly, $y = 3.5x$

45. Inversely, $y = \frac{20}{x}$ **47.** $(1, 65)$, $(2, 130)$, $(3, 195)$, $(4, 260)$

49. $(20, 20)$, $(40, 10)$, $(50, 8)$, $(200, 2)$ **51.** k, $(0, 0)$, no, $y = kx$

Section 3.6 Warm-Ups

1. Linear   **2.** Dashed   **3.** Solid   **4.** False   **5.** True   **6.** True
7. False   **8.** False   **9.** False   **10.** True   **11.** True

Section 3.6 Exercises

1. $(3, -1)$ **3.** $(-3, -9)$ **5.** $(3, 0)$, $(1, 3)$ **7.** $(2, 3)$, $(0, 5)$

9.

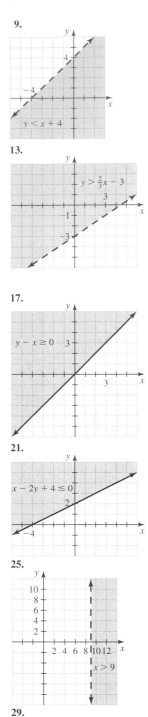

$y < x + 4$

11.

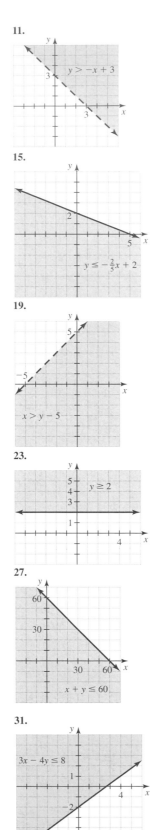

$y > -x + 3$

13.

$y > \frac{2}{3}x - 3$

15.

$y \le -\frac{2}{5}x + 2$

17.

$y - x \ge 0$

19.

$x > y - 5$

21.

$x - 2y + 4 \le 0$

23.

$y \ge 2$

25.

$x > 9$

27.

$x + y \le 60$

29.

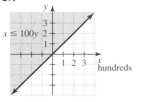

$x \le 100y$

hundreds

31.

$3x - 4y \le 8$

33.

$2x - 3y < 6$

35.

$x - 4y \le 8$

37.

$y - \frac{7}{2}x \le 7$

39.

$x - y < 5$

41.

$3x - 4y < -12$

43.

$x < 5y - 100$

45. $5x + 7y \le 770$

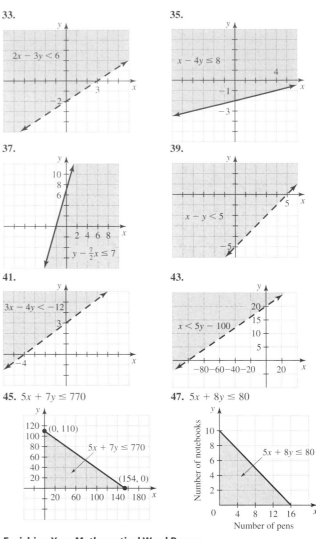

(0, 110)

$5x + 7y \le 770$

(154, 0)

47. $5x + 8y \le 80$

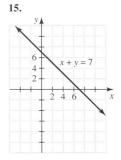

$5x + 8y \le 80$

Number of notebooks

Number of pens

Enriching Your Mathematical Word Power

1. Graph   **2.** Origin   **3.** x-coordinate   **4.** y-intercept
5. Independent   **6.** Dependent   **7.** Slope   **8.** Standard
9. Slope   **10.** Point   **11.** Linear   **12.** Function   **13.** Linear
14. Directly   **15.** Inversely   **16.** Jointly

Review Exercises

1. Quadrant II   **3.** x-axis   **5.** y-axis   **7.** Quadrant IV
9. $(0, -5), (-3, -14), (4, 7)$
11. $\left(0, -\frac{8}{3}\right), \left(3, -\frac{2}{3}\right), \left(-6, -\frac{20}{3}\right)$
13.

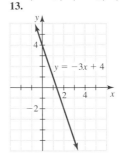

$y = -3x + 4$

15.

$x + y = 7$

17. 1   **19.** $\dfrac{3}{2}$ **21.** $\dfrac{3}{7}$ **23.** 3, $(0, -18)$ **25.** 2, $(0, -3)$

27. 2, $(0, -4)$ **29.** $y = -x + 12$ **31.** $y = \dfrac{3}{4}x - 5$

33.

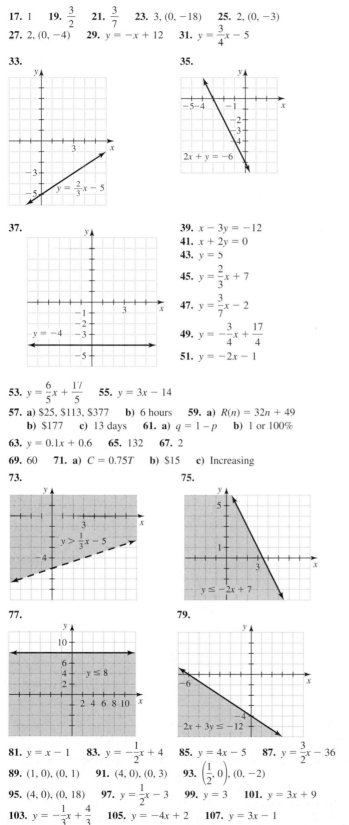

35.

37.

39. $x - 3y = -12$
41. $x + 2y = 0$
43. $y = 5$
45. $y = \dfrac{2}{3}x + 7$
47. $y = \dfrac{3}{7}x - 2$
49. $y = -\dfrac{3}{4}x + \dfrac{17}{4}$
51. $y = -2x - 1$

53. $y = \dfrac{6}{5}x + \dfrac{17}{5}$ **55.** $y = 3x - 14$

57. a) $25, $113, $377   **b)** 6 hours   **59. a)** $R(n) = 32n + 49$
  b) $177   **c)** 13 days   **61. a)** $q = 1 - p$ **b)** 1 or 100%

63. $y = 0.1x + 0.6$ **65.** 132   **67.** 2

69. 60   **71. a)** $C = 0.75T$ **b)** $15   **c)** Increasing

73.

75.

77.

79.

81. $y = x - 1$ **83.** $y = -\dfrac{1}{2}x + 4$ **85.** $y = 4x - 5$ **87.** $y = \dfrac{3}{2}x - 36$

89. $(1, 0), (0, 1)$ **91.** $(4, 0), (0, 3)$ **93.** $\left(\dfrac{1}{2}, 0\right), (0, -2)$

95. $(4, 0), (0, 18)$ **97.** $y = \dfrac{1}{2}x - 3$ **99.** $y = 3$ **101.** $y = 3x + 9$

103. $y = -\dfrac{1}{3}x + \dfrac{4}{3}$ **105.** $y = -4x + 2$ **107.** $y = 3x - 1$

Chapter 3 Test

1. Quadrant II   **2.** x-axis   **3.** Quadrant IV   **4.** y-axis

5. 1   **6.** $-\dfrac{5}{6}$ **7.** 3   **8.** 0   **9.** Undefined   **10.** $\dfrac{2}{3}$

11. $y = -\dfrac{1}{2}x + 3$ **12.** $y = \dfrac{3}{7}x - \dfrac{11}{7}$ **13.** $x - 3y = 11$

14. $5x + 3y = 27$

15.

16.

17.

18.

19.

20.

21.

22. $S = 0.75n + 2.50$
23. a) $13.25, $14.25, $17.75
  b) 7 toppings
24. a) $P(v) = 3v + 20$
  b) 80 cents   **c)** 48 ounces
25. a) 800
  b) $(0, 1000), (50, 0)$; At $0 per ticket
  1000 tickets will be sold and at $50
  per ticket zero tickets will be sold.
  c) -20 tickets/dollar; For every $1
  increase in price, 20 fewer tickets
  will be sold.

26. a) $P = kw$ **b)** $2.80

27. a) $n = \dfrac{k}{A}$ **b)** 18.75 days   **c)** Decreases

28. a) $C = kLW$ **b)** $770

Making Connections Chapters 1–3

1. -1 **2.** -34 **3.** 1   **4.** 72   **5.** -4

6. -28 **7.** $-\dfrac{7}{2}$ **8.** 0.4   **9.** $\dfrac{1}{10}$ **10.** 15   **11.** $13x$

12. $3x - 36$ **13.** $\left\{\dfrac{5}{2}\right\}$ **14.** $\left\{\dfrac{7}{3}\right\}$ **15.** $\dfrac{1}{6}$ **16.** $\dfrac{5}{12}$

17. $\{2\}$ **18.** $\{-4\}$ **19.** $2x - 4$ **20.** $x + 2$
21. $\{5\}$ **22.** $\{3\}$ **23.** $\varnothing$ **24.** All real numbers
25. $(4.5, \infty)$ **26.** $\left(-\dfrac{2}{3}, \infty\right)$ **27.** $[10, \infty)$ **28.** $[20, \infty)$
29. $\left[-\dfrac{1}{2}, \dfrac{5}{2}\right)$ **30.** $\left[0, \dfrac{2}{3}\right)$ **31.** $y = \dfrac{t - 2}{3\pi}$ **32.** $y = mx + b$
33. $y = x - 4$ **34.** $y = 6$ **35.** $y = \dfrac{4}{5}$ **36.** $y = 200$
37. IV   **38.** III and IV   **39.** $(0, 6), (2, 0)$ **40.** -3 **41.** $-\dfrac{5}{12}$
42. $-4, (0, 7)$ **43.** $\dfrac{3}{2}$ **44.** $\dfrac{1}{2}$ **45.** $y = 5x - 12$
46. $y = \dfrac{1}{2}x + 2$ **47.** $y = 5$ **48.** $y = 3x + 3$
49. $y = 2x + 11$ **50.** $y = -2x + 1$
51. a) $\dfrac{2}{15}$ **b)** $\dfrac{1}{5}$ **c)** About 13% per year
    d) \$276,000 saved, \$12,000 per year

Chapter 4

Section 4.1 Warm-Ups
1. Product   **2.** Quotient   **3.** Power   **4.** Product   **5.** Quotient
6. Zero   **7.** True   **8.** False   **9.** True   **10.** True   **11.** False
12. False   **13.** False   **14.** True

Section 4.1 Exercises
1. $27x^5$ **3.** $14a^{11}$ **5.** $-30x^4$ **7.** $27x^{17}$ **9.** $-54s^2t^2$
11. $24t^7w^8$ **13.** 1   **15.** 1   **17.** -3 **19.** 1   **21.** x^3 **23.** m^{12}
25. u^3 **27.** 1   **29.** $-3a^2$ **31.** $-4st^8$ **33.** $2x^6$ **35.** x^6
37. $2x^{12}$ **39.** t **41.** 1   **43.** $-\dfrac{1}{2}$ **45.** x^3y^6 **47.** $-8t^{15}$
49. $-8x^6y^{15}$ **51.** $a^8b^{10}c^{12}$ **53.** $\dfrac{x^3}{8}$ **55.** $\dfrac{a^{12}}{64}$ **57.** $\dfrac{16a^8}{b^{12}}$
59. $-\dfrac{x^6y^3}{8}$ **61.** 200   **63.** 1,000,000   **65.** 64   **67.** x^7
69. x^5 **71.** 1   **73.** a^{32} **75.** $a^{12}b^6$ **77.** x^3 **79.** $\dfrac{a^9}{b^{12}}$
81. $36a^{10}b^8$ **83.** \$33,502.39   **85.** \$86,357.00
87. Product rule for exponents, $P(1 + r)^{15}$

Section 4.2 Warm-Ups
1. Reciprocal   **2.** Factor   **3.** Negative   **4.** True   **5.** True
6. False   **7.** True   **8.** True   **9.** False   **10.** True

Section 4.2 Exercises
1. $\dfrac{1}{3}$ **3.** $\dfrac{1}{16}$ **5.** $-\dfrac{1}{16}$ **7.** $-\dfrac{1}{27}$ **9.** 4   **11.** $\dfrac{1}{3}$ **13.** 1250
15. $\dfrac{3b^9}{a^3}$ **17.** b^7 **19.** $\dfrac{8}{125}$ **21.** 6   **23.** $\dfrac{5}{2}$ **25.** $\dfrac{3}{2}$ **27.** 2
29. $\dfrac{10}{3}$ **31.** x^4 **33.** $\dfrac{1}{x^3}$ **35.** $\dfrac{1}{y^8}$ **37.** $-\dfrac{16}{x^4}$ **39.** $\dfrac{1}{b^6}$
41. $-\dfrac{3}{a^2}$ **43.** $\dfrac{1}{u^8}$ **45.** $-4t^2$ **47.** $\dfrac{1}{y^{12}}$ **49.** $2x^7$ **51.** b
53. $\dfrac{1}{16x^4}$ **55.** $\dfrac{y^6}{x^3}$ **57.** $\dfrac{81}{x^8}$ **59.** $\dfrac{16n^8}{m^{12}}$ **61.** $\dfrac{8b^6}{a^3}$ **63.** 1
65. -2 **67.** 2   **69.** 4   **71.** 25   **73.** 10   **75.** a^3 **77.** $\dfrac{1}{x^5}$
79. $\dfrac{w^3}{5}$ **81.** $\dfrac{b^5c^3}{2a^2}$ **83.** a^{36} **85.** \$13,940.65   **87.** \$171,928.70
89. \$10,727.41   **91. a)** $w < 0$ **b)** m is odd   **c)** $w < 0$ and m odd

Section 4.3 Warm-Ups
1. Scientific   **2.** Scientific   **3.** Standard   **4.** False   **5.** True
6. False   **7.** True   **8.** True   **9.** False   **10.** True

Section 4.3 Exercises
1. 9,860,000,000   **3.** 0.00137   **5.** 0.000001   **7.** 600,000
9. 560,000   **11.** 0.00432   **13.** 6.7   **15.** 9×10^3
17. 7.8×10^{-4} **19.** 8.5×10^{-6} **21.** 6.44×10^8
23. 5.25×10^{11} **25.** 2.3×10^7, 23,000,000
27. 1.5×10^{10}, 15,000,000,000   **29.** 1.36×10^{13}, \$13,600,000,000,000
31. 6×10^{-10} **33.** 2×10^{-38} **35.** 5×10^{27} **37.** 9×10^{24}
39. 1.25×10^{14} **41.** 2.5×10^{-33} **43.** 8.6×10^9
45. 2.1×10^2 **47.** 2.7×10^{-23} **49.** 3×10^{15}
51. 9.135×10^2 **53.** 5.715×10^{-4} **55.** 4.426×10^7
57. 1.577×10^{182} **59.** 4.910×10^{11} feet
61. 4.65×10^{-28} hours   **63.** 9.040×10^8 feet   **65.** \$39,000

Section 4.4 Warm-Ups
1. Term   **2.** Coefficient   **3.** Constant   **4.** Polynomial   **5.** Degree
6. Monomial   **7.** Binomial   **8.** Trinomial   **9.** False   **10.** False
11. True   **12.** True   **13.** False   **14.** True   **15.** True
16. True   **17.** True

Section 4.4 Exercises
1. $-3, 7$ **3.** $0, 6$ **5.** $\dfrac{1}{3}, \dfrac{7}{2}$ **7.** Monomial, 0
9. Monomial, 3   **11.** Binomial, 1   **13.** Trinomial, 10
15. Binomial, 6   **17.** Trinomial, 3   **19.** 10   **21.** 6   **23.** $\dfrac{5}{8}$
25. -85 **27.** 5   **29.** 71   **31.** -4.97665 **33.** $4x - 8$
35. $2q$ **37.** $x^2 + 3x - 2$ **39.** $x^3 + 9x - 7$ **41.** $3a^2 - 7a - 4$
43. $-3w^2 - 8w + 5$ **45.** $9.66x^2 - 1.93x - 1.49$
47. $-4x + 6$ **49.** -5 **51.** $-z^2 + 2z$ **53.** $w^5 + w^4 - w^3 - w^2$
55. $2t + 13$ **57.** $-8y + 7$ **59.** $-22.85x - 423.2$
61. $4a + 2$ **63.** $-2x + 4$ **65.** $2a$ **67.** $-5m + 7$
69. $4x^2 + 1$ **71.** $a^3 - 9a^2 + 2a + 7$ **73.** $-3x + 9$
75. $2y^3 + 4y^2 - 3y - 14$ **77.** $-3m + 3$ **79.** $-11y - 3$
81. $2x^2 - 6x + 12$ **83.** $-5z^4 - 8z^3 + 3z^2 + 7$
85. a) $P(x) = 280x - 800$ **b)** \$13,200
87. $P(x) = 6x + 3, P(4) = 27$ meters
89. $D(x) = 5x + 30$, 255 miles   **91.** 800 feet, 800 feet
93. $T(x) = 0.17x + 74.47$ dollars, \$244.47   **95.** Yes, yes, yes
97. The highest power of x is 3.

Mid-Chapter Quiz 4.1–4.4
1. -16 **2.** 16   **3.** 8   **4.** x^6 **5.** a^7 **6.** $\dfrac{1}{b^4}$ **7.** $32a^{25}b^{15}$
8. $\dfrac{8}{w^{12}}$ **9.** $\dfrac{b^6}{a^{18}}$ **10.** $-\dfrac{x^6}{27y^9}$ **11.** $9a^6b^8$ **12.** $\dfrac{1}{16y^{28}}$
13. $\dfrac{4y^6}{27x^7}$ **14.** 1.4   **15.** 5   **16.** $-3x^2 - 5x + 6$ **17.** $13x^2 - x - 6$
18. $-3x^3y$ **19.** -65 **20.** -19

Section 4.5 Warm-Ups
1. Distributive   **2.** Binomial   **3.** Opposite   **4.** Product   **5.** False
6. False   **7.** True   **8.** True   **9.** True   **10.** True

Section 4.5 Exercises
1. $27x^5$ **3.** $14a^{11}$ **5.** $-30x^4$ **7.** $27x^{17}$ **9.** $-54s^2t^2$

11. $24t^7w^8$ **13.** $25y^2$ **15.** $4x^6$ **17.** $x^2 + xy^2$ **19.** $4y^7 - 8y^3$
21. $-18y^2 + 12y$ **23.** $-3y^3 + 15y^2 - 18y$ **25.** $-xy^2 + x^3$
27. $15a^4b^3 - 5a^5b^2 - 10a^6b$ **29.** $-2t^5v^3 + 3t^3v^2 + 2t^4v^2$
31. $x^2 + 3x + 2$ **33.** $x^2 + 2x - 15$ **35.** $t^2 - 13t + 36$
37. $x^3 + 3x^2 + 4x + 2$ **39.** $6y^3 + y^2 + 7y + 6$
41. $2y^8 - 3y^6z - 5y^4z^2 + 3y^2z^3$ **43.** $u - 3t$ **45.** $-3x - y$
47. $3a^2 + a - 6$ **49.** $-3w^2 - w + 6$ **51.** $-6x^2 + 27x$
53. $-6x^2 + 27x + 2$ **55.** $-x - 7$ **57.** $36x^{12}$ **59.** $-6a^3b^{10}$
61. $25x^2 + 60x + 36$ **63.** $25x^2 - 36$ **65.** $6x^7 - 8x^4$
67. $m^3 - 1$ **69.** $3x^3 - 5x^2 - 25x + 18$
71. $A(x) = x^3 + 4x$, 140 square feet
73. $A(x) = x^2 + \frac{1}{2}x$, $A(5) = 27.5$ square feet
75. $x^2 + 5x$ or $x^2 - 5x$ **77.** $8.05x^2 + 15.93x + 6.12$ square meters
79. a) 30,000   **b)** \$300,000   **c)** $R(p) = 40,000p - 1000p^2$
     d) \$400,000, \$300,000, \$175,000

Section 4.6 Warm-Ups
 1. Distributive   **2.** FOIL   **3.** FOIL   **4.** Four   **5.** False
 6. True   **7.** True   **8.** False   **9.** True   **10.** False

Section 4.6 Exercises
 1. $x^2 + 6x + 8$ **3.** $a^2 + 5a + 4$ **5.** $x^2 + 19x + 90$
 7. $2x^2 + 7x + 3$ **9.** $a^2 - a - 6$ **11.** $2x^2 - 5x + 2$
13. $2a^2 - a - 3$ **15.** $w^2 - 60w + 500$ **17.** $y^2 + 5y - ay - 5a$
19. $5w + 5m - w^2 - mw$ **21.** $10m^2 - 9mt - 9t^2$
23. $45a^2 + 53ab + 14b^2$ **25.** $x^4 - 3x^2 - 10$ **27.** $h^6 + 10h^3 + 25$
29. $3b^6 + 14b^3 + 8$ **31.** $y^3 - 2y^2 - 3y + 6$ **33.** $6m^6 + 7m^3n^3 - 3n^4$
35. $12u^4v^2 + 10u^2v - 12$ **37.** $w^2 + 3w + 2$ **39.** $b^2 + 9b + 20$
41. $x^2 + 6x - 27$ **43.** $a^2 + 10a + 25$ **45.** $4x^2 - 4x + 1$
47. $z^2 - 100$ **49.** $a^2 + 2ab + b^2$ **51.** $a^2 - 3a + 2$
53. $2x^2 + 5x - 3$ **55.** $5t^2 - 7t + 2$ **57.** $h^2 - 16h + 63$
59. $h^2 + 14hw + 49w^2$ **61.** $4h^4 - 4h^2 + 1$ **63.** $a^3 + 4a^2 - 7a - 10$
65. $h^3 + 9h^2 + 26h + 24$ **67.** $x^3 - 2x^2 - 64x + 128$
69. $x^3 + 8x^2 - \frac{1}{4}x - 2$ **71.** $x^2 + 15x + 50$ **73.** $x^2 + x + \frac{1}{4}$
75. $8x^2 + 2x + \frac{1}{8}$ **77.** $8a^2 + a - \frac{1}{4}$ **79.** $\frac{1}{8}x^2 + \frac{1}{6}x - \frac{1}{6}$
81. $a^3 + 7a^2 + 12a$ **83.** $x^5 + 13x^4 + 42x^3$
85. $-12x^6 - 26x^5 + 10x^4$ **87.** $x^3 + 3x^2 - x - 3$
89. $9x^3 + 45x^2 - 4x - 20$ **91.** $2x + 10$
93. $A(x) = 2x^2 + 5x - 3$, 49 square feet
95. $A(x) = 4x^2 - 6x + 2$, 72 square feet
97. 12 ft^2, $3h$ ft^2, $4h$ ft^2, h^2 ft^2; $h^2 + 7h + 12$ ft^2; $(h + 3)(h + 4) = h^2 + 7h + 12$

Section 4.7 Warm-Ups
 1. Special   **2.** Difference   **3.** Square   **4.** False   **5.** True
 6. True   **7.** True   **8.** True   **9.** True

Section 4.7 Exercises
 1. $x^2 + 2x + 1$ **3.** $y^2 + 8y + 16$ **5.** $m^2 + 12m + 36$
 7. $a^2 + 18a + 81$ **9.** $9x^2 + 48x + 64$ **11.** $s^2 + 2st + t^2$
13. $4x^2 + 4xy + y^2$ **15.** $4t^2 + 12ht + 9h^2$ **17.** $p^2 - 4p + 4$
19. $a^2 - 6a + 9$ **21.** $t^2 - 2t + 1$ **23.** $9t^2 - 12t + 4$
25. $s^2 - 2st + t^2$ **27.** $9a^2 - 6ab + b^2$ **29.** $9z^2 - 30yz + 25y^2$
31. $a^2 - 25$ **33.** $y^2 - 1$ **35.** $9x^2 - 64$ **37.** $r^2 - s^2$
39. $64y^2 - 9a^2$ **41.** $25x^4 - 4$ **43.** $x^3 + 3x^2 + 3x + 1$
45. $8a^3 - 36a^2 + 54a - 27$ **47.** $a^4 - 12a^3 + 54a^2 - 108a + 81$

49. $a^4 + 4a^3b + 6a^2b^2 + 4ab^3 + b^4$ **51.** $a^2 - 400$
53. $x^2 + 15x + 56$ **55.** $16x^2 - 1$ **57.** $81y^2 - 18y + 1$
59. $6t^2 - 7t - 20$ **61.** $4t^2 - 20t + 25$ **63.** $4t^2 - 25$
65. $x^4 - 1$ **67.** $4y^6 - 36y^3 + 81$ **69.** $4x^6 + 12x^3y^2 + 9y^4$
71. $\frac{1}{4}x^2 + \frac{1}{3}x + \frac{1}{9}$ **73.** $0.04x^2 - 0.04x + 0.01$
75. $a^3 + 3a^2b + 3ab^2 + b^3$ **77.** $2.25x^2 + 11.4x + 14.44$
79. $12.25t^2 - 6.25$ **81.** $A(x) = x^2 + 6x + 9$
83. a) $A(x) = x^2 - 25$ **b)** 25 square feet
85. $A(b) = 3.14b^2 + 6.28b + 3.14$
87. $v = k(R^2 - r^2)$ **89.** $P + 2Pr + Pr^2$, \$242   **91.** \$20,230.06
93. The first is an identity and the second is a conditional equation.

Section 4.8 Warm-Ups
 1. Quotient   **2.** Dividend, divisor, quotient   **3.** Descending
 4. Less   **5.** False   **6.** False   **7.** True   **8.** True   **9.** False
10. True

Section 4.8 Exercises
 1. x^6 **3.** w^9 **5.** a^9 **7.** $3a^5$ **9.** a^6 **11.** $-4x^4$ **13.** $-y$
15. $-3x$ **17.** $-3x^3$ **19.** $x - 2$ **21.** $x^3 + 3x^2 - x$
23. $4xy - 2x + y$ **25.** $y^2 - 3xy$ **27.** $2, -1$ **29.** $x + 5, 16$
31. $x + 2, 7$ **33.** $2, -10$ **35.** $a^2 + 2a + 8, 13$ **37.** $x - 4, 4$
39. $h^2 + 3h + 9, 0$ **41.** $2x - 3, 1$ **43.** $x^2 + 1, -1$
45. $3 + \dfrac{15}{x - 5}$ **47.** $-1 + \dfrac{3}{x + 3}$ **49.** $1 - \dfrac{1}{x}$ **51.** $3 + \dfrac{1}{x}$
53. $x - 1 + \dfrac{1}{x + 1}$ **55.** $x - 2 + \dfrac{8}{x + 2}$ **57.** $x^2 + 2x + 4 + \dfrac{8}{x - 2}$
59. $x^2 + \dfrac{3}{x}$ **61.** $-3a$ **63.** $4w^5t^4$ **65.** $-a + 4$ **67.** $x - 3$
69. $-6x^2 + 2x - 3$ **71.** $t + 4$ **73.** $2w + 1$ **75.** $4x^2 - 6x + 9$
77. $t^2 - t + 3$ **79.** $v^2 - 2v + 1$ **81.** $x - 5$ meters
83. $x^8 + x^7 + x^6 + x^5 + x^4 + x^3 + x^2 + x + 1$
85. $10x \div 5x$ is not equivalent to the other two.

Enriching Your Mathematical Word Power
 1. Term   **2.** Polynomial   **3.** Degree   **4.** Leading   **5.** Monomial
 6. Binomial   **7.** Trinomial   **8.** FOIL   **9.** Principal   **10.** Amount
11. Present   **12.** Square   **13.** Difference
14. Dividend, divisor, quotient   **15.** Scientific

Review Exercises
 1. 0   **3.** $-6a^7$ **5.** $-5c^6$ **7.** b^{30} **9.** $-8x^9y^6$ **11.** $\dfrac{8a^3}{b^6}$
13. $\dfrac{8x^6y^{15}}{z^{18}}$ **15.** $\dfrac{1}{8}$ **17.** 7   **19.** $\dfrac{1}{x^3}$ **21.** a^4 **23.** $\dfrac{1}{x^{12}}$
25. $\dfrac{x^9}{8}$ **27.** $\dfrac{9}{a^2b^6}$ **29.** 8,360,000   **31.** 0.00057   **33.** 4,500,000
35. 3,561,000   **37.** 8.07×10^6 **39.** 7.09×10^{-4} **41.** 1.2×10^{12}
43. 5×10^5 **45.** 1×10^{15} **47.** 2×10^1 **49.** 1×10^2
51. $5w - 2$ **53.** $-6x + 4$ **55.** $2w^2 - 7w - 4$
57. $-2m^2 + 3m - 1$ **59.** 0   **61.** 5   **63.** $-50x^{11}$ **65.** $121a^{14}$
67. $-4x + 15$ **69.** $3x^2 - 10x + 12$ **71.** $15m^5 - 3m^3 + 6m^2$
73. $x^3 - 7x^2 + 20x - 50$ **75.** $3x^3 - 8x^2 + 16x - 8$
77. $q^2 + 2q - 48$ **79.** $2t^2 - 21t + 27$ **81.** $20y^2 - 7y - 6$
83. $6x^4 + 13x^2 + 5$ **85.** $z^2 - 49$ **87.** $y^2 + 14y + 49$
89. $w^2 - 6w + 9$ **91.** $x^4 - 9$ **93.** $9a^2 + 6a + 1$
95. $16 - 8y + y^2$ **97.** $-5x^2$ **99.** $-2a^2b^2$ **101.** $-x + 3$
103. $-3x^2 + 2x - 1$ **105.** -1 **107.** $m^3 + 2m^2 + 4m + 8$
109. $m^2 - 3m + 6, 0$ **111.** $b - 5, 15$ **113.** $2x - 1, -8$
115. $x^2 + 2x - 9, 1$ **117.** $2 + \dfrac{6}{x - 3}$ **119.** $-2 + \dfrac{2}{1 - x}$

121. $x - 1 + \dfrac{-2}{x+1}$ 123. $x - 1 + \dfrac{1}{x+1}$ 125. $x^2 + 10x + 21$
127. $t^2 - 7ty + 12y^2$ 129. 2   131. $-27h^3t^{18}$ 133. $2w^2 - 9w - 18$
135. $9u^2 - 25v^2$ 137. $9h^2 + 30h + 25$ 139. $x^3 + 9x^2 + 27x + 27$
141. $14s^5t^6$ 143. $\dfrac{k^8}{16}$ 145. $x^2 - 9x - 5$ 147. $5x^2 - x - 12$
149. $x^3 - x^2 - 19x + 4$ 151. $x + 6$
153. $P(w) = 4w + 88$, $A(w) = w^2 + 44w$, $P(50) = 288$ ft,
   $A(50) = 4700$ ft^2 155. $R(p) = -15p^2 + 600p$, \$5040, \$20
157. \$19,126.18   159. \$21,252.76

Chapter 4 Test

1. $-35x^8$ 2. $12x^5y^9$ 3. $-2ab^4$ 4. $15x^5$ 5. $\dfrac{-32a^5}{b^{10}}$ 6. $\dfrac{3a^4}{b^2}$
7. $\dfrac{3}{t^{16}}$ 8. $\dfrac{1}{w^2}$ 9. $\dfrac{s^6}{9t^4}$ 10. $\dfrac{-8y^3}{x^{18}}$ 11. 5.433×10^6
12. 6.5×10^{-6} 13. 3200   14. 0.00008   15. 3,500,000,000
16. 12,000,000,000,000   17. 4.8×10^{-1} 18. 8.1×10^{-27}
19. $7x^3 + 4x^2 + 2x - 11$ 20. $-x^2 - 9x + 2$ 21. $-2y^2 + 3y$
22. -1 23. $x^2 + x - 1$ 24. $15x^5 - 21x^4 + 12x^3 - 3x^2$
25. $x^2 + 3x - 10$ 26. $6a^2 + a - 35$ 27. $a^2 - 14a + 49$
28. $16x^2 + 24xy + 9y^2$ 29. $b^2 - 9$ 30. $9t^4 - 49$
31. $4x^4 + 5x^2 - 6$ 32. $x^3 - 3x^2 - 10x + 24$
33. $2 + \dfrac{6}{x-3}$ 34. $x - 5 + \dfrac{15}{x+2}$ 35. 13
36. 2, -4 37. $x - 2, 3$ 38. $-2x^2 + x + 15$
39. $A(x) = x^2 + 4x$, $P(x) = 4x + 8$, $A(4) = 32$ ft^2, $P(4) = 24$ ft
40. $R(q) = -150q^2 + 3000q$, \$14,400   41. \$306,209.52

Making Connections   A Review of Chapters 1–4

1. 8   2. 32   3. 41   4. -2 5. 32   6. 32   7. -144
8. 144   9. $\dfrac{5}{8}$ 10. $\dfrac{1}{9}$ 11. 64   12. 34   13. $\dfrac{5}{6}$ 14. $\dfrac{5}{36}$
15. 899   16. -1 17. $x^2 + 8x + 15$ 18. $4x + 15$ 19. $-15t^5v^7$
20. $5tv$ 21. $x^2 + 9x + 20$ 22. $x^2 + 7x + 10$ 23. $x + 3$
24. $x^3 + 13x^2 + 55x + 75$ 25. $3y - 4$ 26. $6y^2 - 4y + 1$
27. $\left\{-\dfrac{1}{2}\right\}$ 28. $\{7\}$ 29. $\left\{\dfrac{14}{3}\right\}$ 30. $\left\{\dfrac{7}{4}\right\}$ 31. $\{0\}$
32. All real numbers   33. $\left\{\dfrac{11}{15}\right\}$ 34. $\left\{\dfrac{13}{40}\right\}$ 35. $\{-40\}$
36. $\{20\}$ 37. $\varnothing$ 38. All real numbers   39. $\left(-\dfrac{1}{2}, 0\right)$
40. $(0, -7)$ 41. 2   42. $\dfrac{2}{3}$ 43. $\dfrac{14}{3}$ 44. $-\dfrac{1}{2}$
45. 200 meters   46. 30 in.   47. \$5000   48. 400
49. $\dfrac{2.25n + 100,000}{n}$; \$102.25, \$3.25, \$2.35; It averages out to 10 cents
   per disk.

Chapter 5

Section 5.1 Warm-Ups

1. Factor   2. Prime   3. Greatest common factor   4. Multiplying
5. False   6. False   7. False   8. True   9. True   10. True

Section 5.1 Exercises

1. $2 \cdot 3^2$ 3. $2^2 \cdot 13$ 5. $2 \cdot 7^2$ 7. $2^3 \cdot 3^3$
9. $2^2 \cdot 5 \cdot 23$ 11. $2^2 \cdot 3 \cdot 7 \cdot 11$ 13. 4   15. 12   17. 8
19. 4   21. 1   23. $2x$ 25. $2x$ 27. xy 29. $12ab$ 31. 1

33. $6ab$ 35. $3x$ 37. $3t$ 39. $9y^3$ 41. u^3v^2 43. $-7n^3$
45. $11xy^2z$ 47. $2(w + 2t)$ 49. $6(2x - 3y)$ 51. $x(x^2 - 6)$
53. $5a(x + y)$ 55. $h^3(h^2 + 1)$ 57. $2k^3m^4(-k^4 + 2m^2)$
59. $2x(x^2 - 3x + 4)$ 61. $6x^2t(2x^2 + 5x - 4t)$
63. $(x - 3)(a + b)$ 65. $(x - 5)(x - 1)$ 67. $(m + 1)(m + 9)$
69. $(a + b)(y + 1)^2$ 71. $8(x - y), -8(-x + y)$
73. $4x(-1 + 2x), -4x(1 - 2x)$ 75. $1(x - 5), -1(-x + 5)$
77. $1(4 - 7a), -1(-4 + 7a)$ 79. $8a^2(-3a + 2), -8a^2(3a - 2)$
81. $6x(-2x - 3), -6x(2x + 3)$ 83. $2x(-x^2 - 3x + 7), -2x(x^2 + 3x - 7)$
85. $2ab(2a^2 - 3ab - 2b^2), -2ab(-2a^2 + 3ab + 2b^2)$ 87. $x + 2$ hours
89. a) $S = 2\pi r(r + h)$ b) $S = 2\pi r^2 + 10\pi r$ c) 3 in.
91. The GCF is an algebraic expression.

Section 5.2 Warm-Ups

1. Perfect square   2. Difference of two squares   3. Perfect square
4. Prime   5. Factored completely   6. True   7. False   8. True
9. False   10. True   11. False

Section 5.2 Exercises

1. $(b + c)(x + y)$ 3. $(b + 1)(a + b)$ 5. $(w + 1)(m + 3)$
7. $(3x + 5)(2x + w)$ 9. $(x + 4)(x + 3)$ 11. $(m + n)(n + 1)$
13. $(w + 5)(m + 2)$ 15. $(a + 3)(x + y)$ 17. $(a^2 + w)(a + w)$
19. $(w - 1)(w - b)$ 21. $(w - 1)(w + a)$ 23. $(m - 1)(m + x)$
25. $(x - 5)(x + 7)$ 27. $(2x - 5)(x + 7)$ 29. $(a - 2)(a + 2)$
31. $(x - 7)(x + 7)$ 33. $(a + 11)(a - 11)$ 35. $(y + 3x)(y - 3x)$
37. $(5a + 7b)(5a - 7b)$ 39. $(11m + 1)(11m - 1)$
41. $(3w - 5c)(3w + 5c)$ 43. Perfect square trinomial
45. Neither   47. Perfect square trinomial   49. Neither
51. Difference of two squares   53. Perfect square trinomial
55. $(x + 1)^2$ 57. $(a + 3)^2$ 59. $(x + 6)^2$ 61. $(a - 2)^2$
63. $(2w + 1)^2$ 65. $(4x - 1)^2$ 67. $(2t + 5)^2$ 69. $(3w + 7)^2$
71. $(n + t)^2$ 73. $5(x - 5)(x + 5)$ 75. $-2(x - 3)(x + 3)$
77. $a(a - b)(a + b)$ 79. $3(x + 1)^2$ 81. $-5(y - 5)^2$
83. $x(x - y)^2$ 85. $-3(x - y)(x + y)$ 87. $2a(x - 7)(x + 7)$
89. $(w - 1)^2(w + 1)$ 91. $(x - 2)(x + 2)(x + 1)$
93. $3a(b - 3)^2$ 95. $-4m(m - 3n)^2$ 97. $(a + b)(x - 1)(x + 1)$
99. $6ay(a + 2y)^2$ 101. $6ay(2a - y)(2a + y)$ 103. $2a^2y(ay - 3)$
105. $(b - 4w)(a + 2w)$ 107. $(a - b)(1 - b)$ 109. $(2x - 1)^2(2x + 1)$
111. a) $h(t) = -16(t - 20)(t + 20)$ b) 6336 feet
113. a) $V(x) = x(x - 3)^2$ b) $x - 3$ inches

Section 5.3 Warm-Ups

1. Product, sum   2. Multiplying   3. Prime   4. GCF   5. True
6. True   7. False   8. True   9. True   10. False   11. False

Section 5.3 Exercises

1. $(x + 3)(x + 1)$ 3. $(x + 3)(x + 6)$ 5. $(a + 2)(a + 5)$
7. $(a - 3)(a - 4)$ 9. $(b - 6)(b + 1)$ 11. $(x - 2)(x + 5)$
13. $(x + 8)(x - 3)$ 15. $(y + 2)(y + 5)$ 17. $(a - 2)(a - 4)$
19. $(m - 8)(m - 2)$ 21. $(w + 10)(w - 1)$ 23. $(w - 4)(w + 2)$
25. Prime   27. $(m + 16)(m - 1)$ 29. Prime   31. $(z - 5)(z + 5)$
33. Prime   35. $(m + 2)(m + 10)$ 37. Prime   39. $(m - 18)(m + 1)$
41. Prime   43. $(t + 8)(t - 3)$ 45. $(t - 6)(t + 4)$
47. $(t - 20)(t + 10)$ 49. $(x - 15)(x + 10)$ 51. $(y + 3)(y + 10)$
53. $(x + 3a)(x + 2a)$ 55. $(x - 6y)(x + 2y)$ 57. $(x - 12y)(x - y)$
59. Prime   61. $(1 - 4ab)(1 + 7ab)$ 63. $(3ab + 1)(5ab + 1)$
65. $5x(x^2 + 1)$ 67. $w(w - 8)$ 69. $2(w - 9)(w + 9)$
71. $-2(b^2 + 49)$ 73. $(x + 3)(x - 3)(x - 2)$ 75. Prime
77. $x^2(w^2 + 9)$ 79. $(w - 9)^2$ 81. $6(w - 3)(w + 1)$
83. $3(y^2 + 25)$ 85. $(a + c)(x + y)$ 87. $-2(x + 2)(x + 3)$

89. $2x^2(4 - x)(4 + x)$ **91.** $3(w + 3)(w + 6)$ **93.** $w(w^2 + 18w + 36)$
95. $(3y + 1)^2$ **97.** $8v(w + 2)^2$ **99.** $6xy(x + 3y)(x + 2y)$
101. $(3w + 5)(w + 1)$ **103.** $-3y(y - 1)^2$ **105.** $(u + 3)(a^2 + b)$
107. a) 80 square feet   **b)** $x + 4$ feet   **109.** 3 feet and 5 feet   **111.** d

Mid-Chapter Quiz 5.1–5.3
  1. $2^4 \cdot 3$ **2.** $2^2 \cdot 5 \cdot 7$ **3.** 9   **4.** 12   **5.** $2(4w - 3y)$
  6. $6x^2(2x - 5)$ **7.** $5ab(3b^2 - 5ab + 7a^2)$ **8.** $(x + 3)(x - 5)$
  9. $(m - 9)(m - 6)$ **10.** $(2y + 3w)(2y - 3w)$ **11.** $(2h + 3)^2$
12. $(w - 8)^2$ **13.** $10x(x + 5)(x - 5)$ **14.** $-6(x + 3)^2$
15. $(w + 6)(a - 3)$ **16.** $(b - 6)(x - 5)$ **17.** $(a + 1)(x + 1)(x - 1)$
18. $x(x - 5)(x + 1)$ **19.** $2x(x^2 + 9)$ **20.** $(a - 8s)(a - 4s)$

Section 5.4 Warm-Ups
  1. Product, sum   **2.** Trial-and-error   **3.** True   **4.** False   **5.** True
  6. True   **7.** False   **8.** True

Section 5.4 Exercises
  1. 3 and 4   **3.** -2 and -15 **5.** -6 and 2   **7.** 3 and 4
  9. -2 and -9 **11.** -3 and 4   **13.** $(2x + 1)(x + 1)$
15. $(2x + 1)(x + 4)$ **17.** $(3t + 1)(t + 2)$ **19.** $(2x - 1)(x + 3)$
21. $(3x - 1)(2x + 3)$ **23.** Prime   **25.** $(2x - 3)(x - 2)$
27. $(5b - 3)(b - 2)$ **29.** $(4y + 1)(y - 3)$ **31.** Prime
33. $(4x + 1)(2x - 1)$ **35.** $(3t - 1)(3t - 2)$ **37.** $(5x + 1)(3x + 2)$
39. $(2a + 3b)(2a + 5b)$ **41.** $(3m - 5n)(2m + n)$
43. $(x - y)(3x - 5y)$ **45.** $(5a + 1)(a + 1)$ **47.** $(2x + 1)(3x + 1)$
49. $(5a + 1)(a + 2)$ **51.** $(2w + 3)(2w + 1)$ **53.** $(5x - 2)(3x + 1)$
55. $(4x - 1)(2x - 1)$ **57.** $(15x - 1)(x - 2)$ **59.** Prime
61. $2(x^2 + 9x - 45)$ **63.** $(3x - 5)(x + 2)$ **65.** $(5x + y)(2x - y)$
67. $(6a - b)(7a - b)$ **69.** $3x + 1$ **71.** $x + 2$ **73.** $2a - 5$
75. $w(9w - 1)(9w + 1)$ **77.** $2(2w - 5)(w + 3)$ **79.** $3(2x + 3)^2$
81. $(3w + 5)(2w - 7)$ **83.** $3z(x - 3)(x + 2)$ **85.** $3x(3x^2 - 7x + 6)$
87. $(a + 5b)(a - 3b)$ **89.** $y^2(2x^2 + x + 3)$ **91.** $-t(3t + 2)(2t - 1)$
93. $2t^2(3t - 2)(2t + 1)$ **95.** $y(2x - y)(2x - 3y)$
97. $-1(w - 1)(4w - 3)$ **99.** $-2a(2a - 3b)(3a - b)$
101. a) 24, 48, 40, 0 feet   **b)** $h(t) = -8(2t + 1)(t - 3)$ **c)** 0 feet
103. a) ± 4 **b)** $\pm 8, \pm 16$ **c)** $\pm 1, \pm 7, \pm 13, \pm 29$

Section 5.5 Warm-Ups
  1. Remainder   **2.** Sum   **3.** Difference   **4.** Zero   **5.** False
  6. False   **7.** True   **8.** True   **9.** True   **10.** False

Section 5.5 Exercises
  1. $(m - 1)(m^2 + m + 1)$ **3.** $(x + 2)(x^2 - 2x + 4)$
  5. $(a + 5)(a^2 - 5a + 25)$ **7.** $(c - 7)(c^2 + 7c + 49)$
  9. $(2w + 1)(4w^2 - 2w + 1)$ **11.** $(2t - 3)(4t^2 + 6t + 9)$
13. $(x - y)(x^2 + xy + y^2)$ **15.** $(2t + y)(4t^2 - 2ty + y^2)$
17. $(x - y)(x + y)(x^2 + y^2)$ **19.** $(x - 1)(x + 1)(x^2 + 1)$
21. $(2b - 1)(2b + 1)(4b^2 + 1)$ **23.** $(a - 3b)(a + 3b)(a^2 + 9b^2)$
25. $2(x - 3)(x + 3)$ **27.** Prime   **29.** $4(x + 5)(x - 3)$
31. $x(x + 2)^2$ **33.** $5am(x^2 + 4)$ **35.** Prime   **37.** $(3x + 1)^2$
39. Prime   **41.** $(w - z)(w + z)(w^2 + z^2)$ **43.** $y(3x + 2)(2x - 1)$
45. Prime   **47.** $3(4a - 1)^2$ **49.** $2(4m + 1)(2m - 1)$
51. $(s - 2t)(s + 2t)(s^2 + 4t^2)$ **53.** $(3a + 4)^2$ **55.** $2(3x - 1)(4x - 3)$
57. $3(m^2 + 9)$ **59.** $3a(a - 9)$ **61.** $2(2 - x)(2 + x)$ **63.** Prime
65. $x(6x^2 - 5x + 12)$ **67.** $ab(a - 2)(a + 2)$ **69.** $(x - 2)(x + 2)^2$
71. $-7mn(m^2 + 4n^2)$ **73.** $2(x + 2)(x^2 - 2x + 4)$
75. $2w(w - 2)(w^2 + 2w + 4)$ **77.** $3w(a - 3)^2$
79. $5(x - 10)(x + 10)$ **81.** $(2 - w)(m + n)$
83. $3x(x + 1)(x^2 - x + 1)$ **85.** $4(w^2 + w - 1)$

87. $a^2(a + 10)(a - 3)$ **89.** $aw(2w - 3)^2$ **91.** $(t + 3)^2$
93. $(-1 + 1)^3 = (-1)^3 + 1^3, (1 + 2)^3 \neq 1^3 + 2^3$
95. $(a^2 + 1)(a^4 - a^2 + 1)$ **97.** $3(w + 5)^2$ **99.** $(9 + b)(9 - b)$
101. $w(w - 8)$ **103.** $3(x - 5)(x + 7)$ **105.** $(x - 5)(a + 4)$
107. $(3x - 4)(4x + 3)$ **109.** $-3(3x - 1)(x + 2)$
111. $(w - 3)(w^2 + 3w + 9)$ **113.** $(y + 1)(y^2 + 1)$
115. $(m + 3)(m - 3)(m^2 + 9)$ **117.** $(a + 2b)(a - 4b)$
119. $my(m + 3y)^2$ **121.** $x^2(x^2 + 2x + 4)$
123. $y^3(y + 1)(y - 1)(y^2 + 1)$ **125.** $(x - 6)(x - 12)$
127. $-a(2a + 1)(3a - 4)$ **129.** $x(x - 2)(x^2 + 2x + 4)$
131. $(4t - 3x)^2$

Section 5.6 Warm-Ups
  1. Quadratic   **2.** Compound   **3.** Zero factor   **4.** Factoring
  5. Divide   **6.** Pythagorean   **7.** False   **8.** False   **9.** True
10. True   **11.** False   **12.** False

Section 5.6 Exercises
  1. $-4, -5$ **3.** $-\dfrac{5}{2}, \dfrac{4}{3}$ **5.** $-2, -1$ **7.** 2, 7   **9.** $-4, 6$
11. $-1, \dfrac{1}{2}$ **13.** 0, 1   **15.** $0, -7$ **17.** $-5, 4$ **19.** $\dfrac{1}{2}, -3$
21. $0, -8$ **23.** $-\dfrac{9}{2}, 2$ **25.** $\dfrac{2}{3}, -4$ **27.** 5   **29.** $\dfrac{3}{2}$ **31.** $0, -3, 3$
33. $-4, -2, 2$ **35.** $-1, 1, 3$ **37.** $-\dfrac{1}{2}, 0, \dfrac{2}{3}$ **39.** $-1, 6$
41. $-9, 3$ **43.** $-10, 2$ **45.** $-4, \dfrac{5}{3}$ **47.** $-4, 4$ **49.** $-\dfrac{3}{2}, \dfrac{3}{2}$
51. $0, -1, 1$ **53.** $-3, -2$ **55.** $-\dfrac{3}{2}, -4$ **57.** $-6, 4$ **59.** $-1, 3$
61. $-4, 2$ **63.** $-5, -3, 5$ **65.** Length 12 ft, width 5 ft
67. Width 5 ft, length 12 ft   **69.** 2 and 3, or -3 and -2 **71.** 5 and 6
73. $-8, -6, -4$, or 4, 6, 8   **75.** -2 and -1, or 3 and 4
77. -7 and -2, or 2 and 7   **79.** Boy 8, girl 6
81. Length 12 feet, width 6 feet   **83.** 9 meters and 12 meters
85. a) 25 sec   **b)** last 5 sec   **c)** increasing
87. a) 680 feet   **b)** 608 feet   **c)** 6 sec
89. Base 6 in., height 13 in.   **91.** 20 ft by 20 ft   **93.** 80 ft
95. 3 yd by 3 yd, 6 yd by 6 yd   **97.** 12 mi   **99.** 91/20 chi
101. 25%

Enriching Your Mathematical Word Power
  1. Prime   **2.** Composite   **3.** Prime   **4.** Factoring
  5. Completely   **6.** Greatest   **7.** Perfect   **8.** Square   **9.** Sum
10. Difference   **11.** Quadratic   **12.** Zero   **13.** Pythagorean

Review Exercises
  1. $2^4 \cdot 3^2$ **3.** $2 \cdot 29$ **5.** $2 \cdot 3 \cdot 5^2$ **7.** 18   **9.** $4x$ **11.** $x + 2$
13. $-a + 10$ **15.** $a(2 - a)$ **17.** $3x^2y(2y - 3x^3)$
19. $3y(x^2 - 4x - 3y)$ **21.** $(y + b)(y + 1)$ **23.** $(w - 2)(w - a)$
25. $(c - 3)(ab + 1)$ **27.** $(y - 20)(y + 20)$ **29.** $(w - 4)^2$
31. $(2y + 5)^2$ **33.** $(r - 2)^2$ **35.** $2t(2t - 3)^2$ **37.** $(x + 6y)^2$
39. $(x - y)(x + 5)$ **41.** $(b + 8)(b - 3)$ **43.** $(r - 10)(r + 6)$
45. $(y - 11)(y + 5)$ **47.** $(u + 20)(u + 6)$ **49.** $3t^2(t + 4)$
51. $5w(w^2 + 5w + 5)$ **53.** $ab(2a + b)(a + b)$
55. $x(3x - y)(3x + y)$ **57.** $(7t - 3)(2t + 1)$ **59.** $(3x + 1)(2x - 7)$
61. $(3p + 4)(2p - 1)$ **63.** $-2p(5p + 2)(3p - 2)$
65. $(6x + y)(x - 5y)$ **67.** $2(4x + y)^2$ **69.** $5x(x^2 + 8)$
71. $(3x - 1)(3x + 2)$ **73.** Prime   **75.** $(x + 2)(x - 1)(x + 1)$
77. $xy(x - 16y)$ **79.** Prime   **81.** $(a + 1)^2$ **83.** $(x^2 + 1)(x - 1)$
85. $(a + 2)(a + b)$ **87.** $-2(x - 6)(x - 2)$

89. $(m - 10)(m^2 + 10m + 100)$ **91.** $(p - q)(p + q)(p^2 + q^2)$

93. $(a^2 + 1)(a + 3)$ **95.** $0, 5$ **97.** $0, 5$ **99.** $-\dfrac{1}{2}, 5$

101. $-4, -3, 3$ **103.** $-2, -1$ **105.** $-\dfrac{1}{2}, \dfrac{1}{4}$ **107.** $\dfrac{1}{5}, \dfrac{1}{2}$

109. $5, 11$ **111.** 6 in. by 8 in.   **113.** $v = k(R - r)(R + r)$ **115.** 6 ft

Chapter 5 Test

1. $2 \cdot 3 \cdot 11$ **2.** $2^4 \cdot 3 \cdot 7$ **3.** 16 **4.** 6 **5.** $3y^2$ **6.** $6ab$
7. $5x(x - 2)$ **8.** $6y^2(x^2 + 2x + 2)$ **9.** $3ab(a - b)(a + b)$
10. $(a + 6)(a - 4)$ **11.** $(2b - 7)^2$ **12.** $3m(m^2 + 9)$
13. $(a + b)(x - y)$ **14.** $(a - 5)(x - 2)$ **15.** $(3b - 5)(2b + 1)$
16. $(m + 2n)^2$ **17.** $(2a - 3)(a - 5)$ **18.** $z(z + 3)(z + 6)$
19. $(x + 5)(x^2 - 5x + 25)$ **20.** $a(a - b)(a^2 + ab + b^2)$

21. -3 **22.** $\dfrac{3}{2}, -4$ **23.** $0, -2, 2$ **24.** $-2, \dfrac{5}{6}$ **25.** $2, 4$

26. $\dfrac{2}{3}, 5$ **27.** Length 12 ft, width 9 ft   **28.** -4 and 8

29. a) 48 feet   **b)** 2 seconds

Making Connections   A Review of Chapters 1–5

1. -1 **2.** 2 **3.** -3 **4.** 57 **5.** 16 **6.** 7 **7.** $2x^2$ **8.** $3x$
9. $3 + x$ **10.** $6x$ **11.** $24yz$ **12.** $6y + 8z$ **13.** $4z - 1$
14. $x - 7$ **15.** $-6x + 8$ **16.** $-x + 8$ **17.** $15x^2 - 26x + 8$
18. $15x^3 - 26x^2 + 13x - 2$ **19.** $3x - 2$ **20.** $2x$ **21.** $\left\{\dfrac{3}{2}\right\}$

22. $\left\{-\dfrac{1}{2}\right\}$ **23.** $\{3, -5\}$ **24.** $\left\{\dfrac{3}{2}, -\dfrac{1}{2}\right\}$ **25.** $\{0, 3\}$ **26.** $\{0, 1\}$

27. All real numbers   **28.** No solution or $\varnothing$ **29.** $\{10\}$ **30.** $\{40\}$

31. $\{-3, 3\}$ **32.** $\left\{-5, \dfrac{3}{2}\right\}$ **33.** t^6 **34.** t^{10} **35.** $\dfrac{1}{t^6}$ **36.** t^{16}

37. $4t^6$ **38.** $\dfrac{1}{3y^7}$ **39.** $\dfrac{2x^2}{5}$ **40.** $\dfrac{2}{3}$ **41.** $-\dfrac{8y^6}{x^9}$ **42.** $9x^4y^6$

43. -5 **44.** -28

45. $(-\infty, -9)$

$-13\,-12\,-11\,-10\,-9\,-8\,-7$

46. $[3, \infty)$

$1 \quad 2 \quad 3 \quad 4 \quad 5 \quad 6 \quad 7$

47. $(12, \infty)$

$10\ \ 11\ \ 12\ \ 13\ \ 14\ \ 15\ \ 16$

48. $(-\infty, 600)$

$0 \quad 200 \ \ 400 \ \ 600 \ \ 800$

49. $4p(p^2 + 3p + 8)$ **50.** $3m^2(m - 3)(m - 1)$ **51.** $-3(2a - 1)^2$
52. $-2(b + 2)(b - 2)$ **53.** $(a + q)(b + 1)$ **54.** $(a + b)(2m - 3n)$
55. $-7(x - 1)(x^2 + x + 1)$ **56.** $2(a + 3)(a^2 - 3a + 9)$
57. 25 yards   **58.** 14 inches   **59.** 11 feet   **60.** 9 inches
61. 4 inches, 3 inches   **62.** Length 21 ft, width 13.5 ft

Chapter 6

Section 6.1 Warm-Ups

1. Integers   **2.** Polynomials   **3.** Dividing   **4.** Quotient
5. Opposites   **6.** Rational   **7.** False   **8.** True   **9.** False
10. True   **11.** False

Section 6.1 Exercises

1. -3 **3.** 5 **5.** $-0.6, 9, 401, -199$ **7.** -1

9. $\dfrac{5}{3}$ **11.** $4, -4$ **13.** Any number can be used.

15. All real numbers except 2   **17.** All real numbers except -3 and -2

19. All real numbers   **21.** All real numbers except 0   **23.** $\dfrac{2}{9}$

25. $\dfrac{7}{15}$ **27.** $\dfrac{2a}{5}$ **29.** $\dfrac{13}{5w}$ **31.** $\dfrac{3x + 1}{3}$ **33.** $\dfrac{2}{3}$

35. $\dfrac{b - 3}{2b - 5}$ **37.** $w - 7$ **39.** $\dfrac{a - 1}{a + 1}$ **41.** $\dfrac{x + 1}{2(x - 1)}$ **43.** $\dfrac{x + 3}{7}$

45. $\dfrac{a^2 - 2a + 4}{2}$ **47.** x^3 **49.** $\dfrac{1}{z^5}$ **51.** $-2x^2$

53. $\dfrac{-3m^3n^2}{2}$ **55.** $\dfrac{-3}{4c^3}$ **57.** $\dfrac{5c}{3a^4b^{16}}$ **59.** $\dfrac{35}{44}$ **61.** $\dfrac{11}{8}$

63. $\dfrac{21}{10x^4}$ **65.** $\dfrac{33a^4}{16}$ **67.** -1 **69.** $-h - t$ **71.** $\dfrac{-2}{3h + g}$

73. $\dfrac{-x - 2}{x + 3}$ **75.** -1 **77.** $\dfrac{-2y}{3}$ **79.** $\dfrac{x + 2}{2 - x}$ **81.** $\dfrac{-6}{a + 3}$

83. $\dfrac{-a^2 - ab - b^2}{2b}$ **85.** $\dfrac{x^4}{2}$ **87.** $\dfrac{x + 2}{2x}$ **89.** -1 **91.** $\dfrac{-2}{c + 2}$

93. $\dfrac{x + 2}{x - 2}$ **95.** $\dfrac{-2}{x + 3}$ **97.** q^2 **99.** $\dfrac{u + 2}{u - 8}$

101. $\dfrac{a^2 + 2a + 4}{2}$ **103.** $y + 2$ **105.** $\dfrac{2 - a}{x(x - w)}$

107. $\dfrac{300}{x + 10}$ hr   **109.** $\dfrac{4.50}{x + 4}$ dollars/lb   **111.** $\dfrac{1}{x}$ pool/hr

113. a) \$0.75   **b)** \$0.75, \$0.63, \$0.615   **c)** Approaches \$0.60

Section 6.2 Warm-Ups

1. Rational   **2.** Reducing   **3.** Divide   **4.** False   **5.** True
6. True   **7.** True   **8.** True   **9.** True

Section 6.2 Exercises

1. $\dfrac{5}{9}$ **3.** $\dfrac{7}{9}$ **5.** $\dfrac{18}{5}$ **7.** $\dfrac{42}{5}$ **9.** $\dfrac{5}{6}$ **11.** $\dfrac{5x}{2}$ **13.** $\dfrac{a}{44}$

15. $\dfrac{-x^5}{a^3}$ **17.** $\dfrac{18t^8y^7}{w^4}$ **19.** $\dfrac{5}{7}$ **21.** $\dfrac{2a}{a - b}$ **23.** $3x - 9$

25. $\dfrac{8a + 8}{5(a^2 + 1)}$ **27.** $\dfrac{1}{2}$ **29.** 30 **31.** $\dfrac{2}{3}$ **33.** $\dfrac{10}{9}$ **35.** $\dfrac{x}{2}$

37. $\dfrac{7x}{2}$ **39.** $\dfrac{2m^2}{3n^6}$ **41.** -3 **43.** $\dfrac{2}{x + 2}$ **45.** $\dfrac{1}{4(t - 5)}$

47. $x^2 - 1$ **49.** $2x - 4y$ **51.** $\dfrac{x + 2}{2}$ **53.** $\dfrac{x^2 + 9}{15}$ **55.** $9x + 9y$

57. -3 **59.** $\dfrac{a + b}{a}$ **61.** $\dfrac{2b}{a}$ **63.** $\dfrac{y}{x}$ **65.** $\dfrac{-a^6b^8}{2}$ **67.** $\dfrac{1}{9m^3n}$

69. $\dfrac{x^2 + 5x}{3x - 1}$ **71.** $\dfrac{a^3 + 8}{2(a - 2)}$ **73.** 1 **75.** $\dfrac{m^2 + 6m + 9}{(m - 3)(m + k)}$

77. a) $\dfrac{26.2}{x}$ mph   **b)** $\dfrac{13.1}{x}$ miles   **79. a)** $\dfrac{1}{x}$ tank/min   **b)** $\dfrac{2}{x}$ tank

81. 5 square meters   **83. a)** $\dfrac{1}{8}$ **b)** $\dfrac{4}{3}$ **c)** $\dfrac{2x}{3}$ **d)** $\dfrac{3x}{4}$

Section 6.3 Warm-Ups

1. Build up   **2.** Least common denominator   **3.** Highest   **4.** True
5. False   **6.** True   **7.** False   **8.** True   **9.** True

Section 6.3 Exercises

1. $\dfrac{9}{27}$ **3.** $\dfrac{12}{16}$ **5.** $\dfrac{7}{7}$ **7.** $\dfrac{12}{6}$ **9.** $\dfrac{5a}{ax}$ **11.** $\dfrac{14x}{2x}$ **13.** $\dfrac{15t}{3bt}$

15. $\dfrac{-36z^2}{8awz}$ 17. $\dfrac{10a^2}{15a^3}$ 19. $\dfrac{8xy^3}{10x^2y^5}$ 21. $\dfrac{10}{2x+6}$ 23. $\dfrac{-20}{-8x-8}$

25. $\dfrac{-32ab}{20b^2-20b^3}$ 27. $\dfrac{3x-6}{x^2-4}$ 29. $\dfrac{3x^2+3x}{x^2+2x+1}$ 31. $\dfrac{y^2-y-30}{y^2+y-20}$

33. 48 35. 180 37. $30a^2$ 39. $12a^4b^6$ 41. $(x-4)(x+4)^2$

43. $x(x+2)(x-2)$ 45. $2x(x-4)(x+4)$ 47. $\dfrac{4}{24}, \dfrac{9}{24}$

49. $\dfrac{3}{6x}, \dfrac{5}{6x}$ 51. $\dfrac{4b}{6ab}, \dfrac{3a}{6ab}$ 53. $\dfrac{9b}{252ab}, \dfrac{20a}{252ab}$ 55. $\dfrac{2x^3}{6x^5}, \dfrac{9}{6x^5}$

57. $\dfrac{4x^4}{36x^3y^5z}, \dfrac{3y^6z}{36x^3y^5z}, \dfrac{6xy^4z}{36x^3y^5z}$ 59. $\dfrac{2x^2+4x}{(x-3)(x+2)}, \dfrac{5x^2-15x}{(x-3)(x+2)}$

61. $\dfrac{4}{a-6}, \dfrac{-5}{a-6}$ 63. $\dfrac{x^2-3x}{(x-3)^2(x+3)}, \dfrac{5x^2+15x}{(x-3)^2(x+3)}$

65. $\dfrac{w^2+3w+2}{(w-5)(w+3)(w+1)}, \dfrac{-2w^2-6w}{(w-5)(w+3)(w+1)}$

67. $\dfrac{-5x-10}{6(x-2)(x+2)}, \dfrac{6x}{6(x-2)(x+2)}, \dfrac{9x-18}{6(x-2)(x+2)}$

69. $\dfrac{2q+8}{(2q+1)(q-3)(q+4)}, \dfrac{3q-9}{(2q+1)(q-3)(q+4)},$

$\dfrac{8q+4}{(2q+1)(q-3)(q+4)}$

71. Identical denominators are needed for addition and subtraction.

Section 6.4 Warm-Ups
1. Add 2. Build up 3. False 4. True 5. True 6. True
7. True 8. False 9. True 10. True

Section 6.4 Exercises
1. $\dfrac{1}{5}$ 3. $\dfrac{3}{4}$ 5. $-\dfrac{2}{3}$ 7. $-\dfrac{3}{4}$ 9. $\dfrac{5}{9}$ 11. $\dfrac{23}{15}$ 13. $\dfrac{103}{144}$

15. $-\dfrac{31}{40}$ 17. $\dfrac{5}{24}$ 19. $\dfrac{1}{x}$ 21. $\dfrac{5}{w}$ 23. 3 25. -2 27. $\dfrac{3}{h}$

29. $\dfrac{x-4}{x+2}$ 31. $\dfrac{3}{2a}$ 33. $\dfrac{5x}{6}$ 35. $\dfrac{6m}{5}$ 37. $\dfrac{2x+y}{xy}$ 39. $\dfrac{17}{10a}$

41. $\dfrac{w}{36}$ 43. $\dfrac{b^2-4ac}{4a}$ 45. $\dfrac{2w+3z}{w^2z^2}$ 47. $\dfrac{2x+2}{x(x+2)}$

49. $\dfrac{-x-3}{x(x+1)}$ 51. $\dfrac{3a+b}{(a-b)(a+b)}$ 53. $\dfrac{15-4x}{5x(x+1)}$

55. $\dfrac{a^2+5a}{(a-3)(a+3)}$ 57. 0 59. $\dfrac{7}{2(a-1)}$

61. $\dfrac{-2x+1}{(x-5)(x+2)(x-2)}$ 63. $\dfrac{7x+17}{(x+2)(x-1)(x+3)}$

65. $\dfrac{bc+ac+ab}{abc}$ 67. $\dfrac{2x^2-x-4}{x(x-1)(x+2)}$ 69. $\dfrac{a+51}{6a(a-3)}$

71. a) F b) A c) E d) B e) D f) C 73. $\dfrac{p+6}{p(p+4)}$

75. $\dfrac{6}{(a+1)(a+3)}$ 77. $\dfrac{1}{(b+1)(b+2)}$ 79. $\dfrac{-1}{2(t+2)}$

81. $\dfrac{11}{x}$ feet 83. $\dfrac{120}{x}$ hr, $\dfrac{195}{x+5}$ hr, $\dfrac{315x+600}{x(x+5)}$ hours, 5 hours

85. $\dfrac{4x+6}{x(x+3)}$ job, $\dfrac{5}{9}$ job

Mid-Chapter Quiz 6.1–6.4
1. $\dfrac{3}{7}$ 2. $\dfrac{4x-1}{4}$ 3. $\dfrac{w-1}{2}$ 4. $\dfrac{-2a+6}{3}$ 5. $\dfrac{9}{5}$ 6. $\dfrac{3x^3z^2}{5y^2}$

7. $\dfrac{5a+15}{4}$ 8. $\dfrac{7}{b^4}$ 9. $\dfrac{11}{15}$ 10. $\dfrac{9}{2(x-3)}$ 11. $\dfrac{8s^2}{21}$

12. $\dfrac{m-1}{2m}$ 13. $\dfrac{25}{42}$ 14. $\dfrac{4a+5b^2}{a^2b^3}$ 15. $\dfrac{3x^2+4x}{(x+1)^2}$

16. $\dfrac{-3y}{(y+5)(y+2)}$ 17. $\dfrac{bc+ac+ab}{abc}$ 18. $-\dfrac{1}{2}$ 19. 4 20. $-\dfrac{3}{2}$

Section 6.5 Warm-Ups
1. Complex 2. Numerator, denominator 3. False 4. True
5. True 6. True 7. True 8. False

Section 6.5 Exercises
1. $\dfrac{3}{5}$ 3. $-\dfrac{10}{3}$ 5. $\dfrac{22}{7}$ 7. $\dfrac{2}{3}$ 9. $\dfrac{14}{17}$ 11. $\dfrac{45}{23}$ 13. $\dfrac{10}{3}$

15. $\dfrac{10}{9}$ 17. 13 19. -3 21. $\dfrac{1}{2}$ 23. $\dfrac{3a+b}{a-3b}$ 25. $\dfrac{5a-3}{3a+1}$

27. $\dfrac{x^2-4x}{2(3x^2-1)}$ 29. $\dfrac{10b}{3b^2-4}$ 31. $\dfrac{1}{3}$ 33. $\dfrac{y-2}{3y+4}$

35. $\dfrac{x^2-2x+4}{x^2-3x-1}$ 37. $\dfrac{5x-14}{2x-7}$ 39. $\dfrac{a-6}{3a-1}$ 41. $\dfrac{-3m+12}{4m-3}$

43. $\dfrac{-w+5}{9w+1}$ 45. -1 47. $\dfrac{a+2}{a+4}$ 49. $\dfrac{3}{2x-1}$ 51. $\dfrac{x-2}{x+3}$

53. $\dfrac{6x-27}{2(2x-3)}$ 55. $\dfrac{2x^2}{3y}$ 57. $\dfrac{a^2+7a+6}{a+3}$ 59. $1-x$

61. $\dfrac{32}{95}, \dfrac{11}{35}$ 63. a) Neither b) $\dfrac{8}{13}, \dfrac{13}{21}$ c) Converging to 0.61803

Section 6.6 Warm-Ups
1. LCD 2. Extraneous 3. False 4. False 5. False 6. True
7. True 8. False 9. True

Section 6.6 Exercises
1. -4 3. 12 5. 30 7. 5 9. 2 11. 4 13. $\dfrac{2}{5}$ 15. $\dfrac{3}{7}$

17. 4 19. 4 21. 3 23. 2 25. -5, 2 27. -3, 2 29. 2, 3
31. -3, 3 33. 2 35. No solution 37. No solution 39. 3
41. 10 43. 0 45. -5, 5 47. 3, 5 49. 1 51. 3

53. 0 55. 4 57. -20 59. 3 61. 3 63. $54\dfrac{6}{11}$ mm

Section 6.7 Warm-Ups
1. Ratio 2. Proportion 3. Means 4. Extremes
5. Extremes-means 6. True 7. False 8. True 9. True
10. False 11. False

Section 6.7 Exercises
1. $\dfrac{2}{3}$ 3. $\dfrac{4}{3}$ 5. $\dfrac{5}{7}$ 7. $\dfrac{8}{15}$ 9. $\dfrac{7}{2}$ 11. $\dfrac{9}{14}$ 13. $\dfrac{5}{2}$

15. $\dfrac{15}{1}$ 17. 3 to 2 19. 9 to 16 21. 31 to 1 23. 2 to 3 25. 6

27. $-\dfrac{2}{5}$ 29. $-\dfrac{27}{5}$ 31. 7 33. 5 35. $-\dfrac{3}{4}$ 37. $\dfrac{5}{4}$ 39. 108

41. 176,000 43. Lions 85, Tigers 51
45. 40 luxury cars, 60 sports cars 47. 84 in. 49. 15 min

51. $\dfrac{1610}{3}$ or 536.7 mi 53. 3920 lb, 2000 lb 55. 6000

57. a) 3 to 17 b) $\dfrac{201}{14}$ or 14.4 lb 59. 4074

Section 6.8 Warm-Ups

1. True   **2.** True   **3.** False   **4.** True   **5.** True   **6.** False
7. False   **8.** True   **9.** False   **10.** True

Section 6.8 Exercises

1. $y = 3x + 1$ **3.** $y = 2x - 5$ **5.** $y = -\dfrac{1}{2}x - 2$

7. $y = mx - mb - a$ **9.** $y = -\dfrac{1}{3}x - \dfrac{1}{3}$ **11.** $C = \dfrac{B}{A}$

13. $p = \dfrac{a}{1 + am}$ **15.** $m_1 = \dfrac{r^2 F}{km_2}$ **17.** $a = \dfrac{bf}{b - f}$ **19.** $r = \dfrac{S - a}{S}$

21. $P_2 = \dfrac{P_1 V_1 T_2}{T_1 V_2}$ **23.** $h = \dfrac{3V}{4\pi r^2}$ **25.** $\dfrac{5}{12}$ **27.** $-\dfrac{6}{23}$ **29.** $\dfrac{128}{3}$

31. -6 **33.** $\dfrac{6}{5}$ **35.** Marcie 4 mph, Frank 3 mph

37. Bob 25 mph, Pat 20 mph   **39.** 5 mph   **41.** 6 hours
43. 40 minutes   **45.** 1 hour 36 minutes
47. Master 2 hours, apprentice 6 hours
49. Bananas 8 pounds, apples 10 pounds   **51.** 140 mph
53. 10 mph   **55.** Ben 15 mph, Jerry 7.5 mph   **57.** 1800 miles
59. 4 hours   **61.** 1.2 hours or 1 hour 12 minutes   **63.** 24 minutes

Enriching Your Mathematical Word Power

1. Rational   **2.** Domain   **3.** Function   **4.** Lowest   **5.** Reduced
6. Equivalent   **7.** Complex   **8.** Building up   **9.** Least
10. Extraneous   **11.** Ratio   **12.** Proportion   **13.** Extremes
14. Means   **15.** Extremes, means

Review Exercises

1. All real numbers except 4

3. All real numbers except -1 and 5   **5.** $\dfrac{6}{7}$ **7.** $\dfrac{c^2}{4a^2}$

9. $\dfrac{2w - 3}{3w - 4}$ **11.** $-\dfrac{x + 1}{3}$ **13.** $\dfrac{1}{2}k$ **15.** $\dfrac{2x}{3y}$

17. $a^2 - a - 6$ **19.** $\dfrac{1}{2}$ **21.** 108   **23.** $24a^7 b^3$ **25.** $12x(x - 1)$

27. $(x + 1)(x - 2)(x + 2)$ **29.** $\dfrac{15}{36}$ **31.** $\dfrac{10x}{15x^2 y}$ **33.** $\dfrac{-10}{12 - 2y}$

35. $\dfrac{x^2 + x}{x^2 - 1}$ **37.** $\dfrac{29}{63}$ **39.** $\dfrac{3x - 4}{x}$ **41.** $\dfrac{2a - b}{a^2 b^2}$

43. $\dfrac{27a^2 - 8a - 15}{(2a - 3)(3a - 2)}$ **45.** $\dfrac{3}{a - 8}$ **47.** $\dfrac{3x + 8}{2(x + 2)(x - 2)}$

49. $-\dfrac{3}{14}$ **51.** $\dfrac{6b + 4a}{3(a - 6b)}$ **53.** $\dfrac{-2x + 9}{3x - 1}$ **55.** $\dfrac{x^2 + x - 2}{-4x + 13}$

57. $-\dfrac{15}{2}$ **59.** 9   **61.** -3 **63.** $\dfrac{21}{2}$ **65.** 5   **67.** 8

69. 56 cups water, 28 cups rice   **71.** $y = mx + b$ **73.** $m = \dfrac{1}{F - v}$

75. $y = 4x - 13$ **77.** 200 hours   **79.** Bert 60 cars, Ernie 50 cars
81. 27.83 million tons   **83.** 10   **85.** -2 **87.** $3x$ **89.** $2m$

91. $\dfrac{1}{6}$ **93.** $a + 1$ **95.** $\dfrac{5 - a}{5a}$ **97.** $a - 2$ **99.** $b - a$ **101.** $\dfrac{1}{10a}$

103. $\dfrac{3}{2x}$ **105.** $\dfrac{4 + y}{6xy}$ **107.** $\dfrac{8}{a - 5}$ **109.** $-1, 2$ **111.** $-\dfrac{5}{3}$

113. 6   **115.** $\dfrac{1}{2}$ **117.** $\dfrac{3x + 7}{(x - 5)(x + 5)(x + 1)}$

119. $\dfrac{-5a}{(a - 3)(a + 3)(a + 2)}$ **121.** $\dfrac{2}{5}$

Chapter 6 Test

1. $-1, 1$ **2.** $\dfrac{2}{3}$ **3.** 0   **4.** $-\dfrac{14}{45}$ **5.** $\dfrac{1 + 3y}{y}$ **6.** $\dfrac{4}{a - 2}$

7. $\dfrac{-x + 4}{(x + 2)(x - 2)(x - 1)}$ **8.** $\dfrac{2}{3}$ **9.** $\dfrac{-2}{a + b}$ **10.** $\dfrac{a^3}{18b^4}$ **11.** $-\dfrac{4}{3}$

12. $\dfrac{-3x + 4}{2(x - 3)}$ **13.** $\dfrac{15}{7}$ **14.** 2, 3   **15.** 12   **16.** $y = -\dfrac{1}{5}x + \dfrac{13}{5}$

17. $c = \dfrac{3M - bd}{b}$ **18.** 29   **19.** 7.2 minutes

20. Brenda 15 mph and Randy 20 mph, or Brenda 10 mph and Randy 15 mph
21. \$72 billion

Making Connections   A Review of Chapters 1–6

1. $\dfrac{7}{3}$ **2.** $-\dfrac{10}{3}$ **3.** -2 **4.** No solution   **5.** 0   **6.** $-4, -2$

7. $-1, 0, 1$ **8.** $-\dfrac{15}{2}$ **9.** $-6, 6$ **10.** $-2, 4$ **11.** 5   **12.** 3

13. $y = \dfrac{c - 2x}{3}$ **14.** $y = \dfrac{1}{2}x + \dfrac{1}{2}$ **15.** $y = \dfrac{c}{2 - a}$ **16.** $y = \dfrac{AB}{C}$

17. $y = 3B - 3A$ **18.** $y = \dfrac{6A}{5}$ **19.** $y = \dfrac{8}{3 - 5a}$

20. $y = 0$ or $y = B$ **21.** $y = \dfrac{2A - hb}{h}$ **22.** $y = -\dfrac{b}{2}$ **23.** 64

24. 16   **25.** 49   **26.** 121   **27.** $-2x - 2$ **28.** $2a^2 - 11a + 15$

29. x^4 **30.** $\dfrac{2x + 1}{5}$ **31.** $\dfrac{1}{2x}$ **32.** $\dfrac{x + 2}{2x}$ **33.** $\dfrac{x}{2}$ **34.** $\dfrac{x - 2}{2x}$

35. $-\dfrac{7}{5}$ **36.** $\dfrac{3a}{4}$ **37.** $x^2 - 64$ **38.** $3x^3 - 21x$ **39.** $10a^{14}$

40. x^{10} **41.** $k^2 - 12k + 36$ **42.** $j^2 + 10j + 25$ **43.** -1

44. $3x^2 - 4x$ **45.** $4x^2(x^2 + 3x + 8)$ **46.** $3a(5a - 3)(a - 1)$

47. $-3(2b - 7)^2$ **48.** $-2(y + 12)(y - 12)$ **49.** $(b + w)(y + 3)$

50. $(a + 2b)(2x - 3n)$ **51.** $-7(b + 1)(b^2 - b + 1)$

52. $2(q - 3)(q^2 + 3q + 9)$ **53.** 1.2×10^8 **54.** 8.1×10^{13}

55. 5×10^{-8} **56.** 1.1×10^4 **57.** 6 and 8   **58.** 7 and 9

59. a) $P = \dfrac{r + 2}{(1 + r)^2}$ b) \$1.81   c) \$7.72

Chapter 7

Section 7.1 Warm-Ups

1. System   **2.** Solution   **3.** Consistent   **4.** Independent
5. Dependent   **6.** Parallel   **7.** Coincide   **8.** True   **9.** False
10. False   **11.** True   **12.** True

Section 7.1 Exercises

1. $(3, -2)$ **3.** All three   **5.** None   **7.** $(-2, 3)$ **9.** $\{(2, 4)\}$
11. $\{(1, 2)\}$ **13.** $\{(0, -5)\}$ **15.** $\{(-2, 3)\}$ **17.** $\{(0, 0)\}$
19. $\{(2, 3)\}$ **21.** $\{(3, 1)\}$ **23.** $\{(-2, 3)\}$ **25.** $\{(2, 7)\}$
27. $\{(10, 12)\}$ **29.** $\{(-10, 22)\}$ **31.** $\{(0.5, -0.3)\}$
33. $\{(x, y) \mid x - y = 3\}$ **35.** $\{(x, y) \mid x - 2y = 8\}$ **37.** No solution
39. No solution   **41.** Inconsistent   **43.** Dependent
45. Independent   **47.** Inconsistent   **49.** Inconsistent
51. Independent   **53.** Inconsistent   **55.** $\{(1, -2)\}$, independent
57. No solution, inconsistent   **59.** $\{(-3, 1)\}$, independent
61. $\{(x, y) \mid x - y = 1\}$, dependent   **63.** $\{(-4, -3)\}$, independent
65. c   **67.** b

69. a) $\{(5, 20)\}$
b) For 5 toppings the cost is $20 at both restaurants.
71. a) 800, 500; 1050, 850; 1300, 1200; 1800, 1900
b)

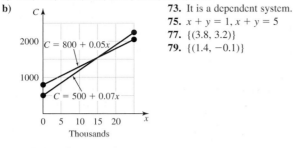

73. It is a dependent system.
75. $x + y = 1, x + y = 5$
77. $\{(3.8, 3.2)\}$
79. $\{(1.4, -0.1)\}$

c) 15,000   **d)** Panasonic

Section 7.2 Warm-Ups
1. Graphing   **2.** Substitution   **3.** Conditional   **4.** Dependent
5. Inconsistent   **6.** True   **7.** True   **8.** True   **9.** True   **10.** False

Section 7.2 Exercises
1. $\{(3, 5)\}$ **3.** $\{(4, 7)\}$ **5.** $\{(2, 5)\}$ **7.** $\{(2, 3)\}$
9. $\{(-2, 9)\}$ **11.** $\{(-5, 5)\}$ **13.** $\left\{\left(\frac{1}{3}, \frac{2}{3}\right)\right\}$ **15.** $\left\{\left(\frac{1}{2}, \frac{1}{3}\right)\right\}$
17. $\{(x, y) \mid 3x - y = 5\}$, dependent   **19.** $\left\{\left(3, \frac{5}{2}\right)\right\}$, independent
21. No solution, inconsistent   **23.** No solution, inconsistent
25. $\left\{\left(\frac{11}{5}, \frac{3}{25}\right)\right\}$, independent   **27.** $\{(3, 2)\}$ **29.** $\{(3, 1)\}$
31. No solution   **33.** Inconsistent   **35.** Dependent
37. Independent   **39.** Inconsistent   **41.** $\left\{\left(\frac{6}{17}, \frac{15}{17}\right)\right\}$ **43.** $\left\{\left(\frac{9}{2}, -\frac{1}{2}\right)\right\}$
45. $\left\{\left(\frac{1}{2}, \frac{1}{4}\right)\right\}$ **47.** $\left\{\left(\frac{3}{2}, -\frac{5}{2}\right)\right\}$ **49.** $\left\{\left(-\frac{2}{9}, \frac{1}{6}\right)\right\}$ **51.** $\left\{\left(-\frac{1}{7}, \frac{2}{7}\right)\right\}$
53. $\left\{\left(-\frac{1}{14}, \frac{5}{28}\right)\right\}$ **55.** Length 28 ft, width 14 ft
57. $12,000 at 10%, $8000 at 5%
59. *Titanic* $601 million, *Star Wars* $461 million
61. Lawn $12, sidewalk $7
63. Left rear 288 pounds, left front 287 pounds, no
65. $2.40 per pound   **67.** 120 tickets for $200, 80 tickets for $250
69. 55 tickets for $6, 110 tickets for $11
71. $30,000 at 5%, $10,000 at 8%
73. 12.5 L of 5% solution, 37.5 L of 25% solution
75. $14,000 at 5%, $16,000 at 10%   **77.** 94 toasters, 6 vacation coupons
79. State tax $3553, federal tax $28,934   **81.** $20,000
83. a) $500,000   **b)** $300,000   **c)** 20,000   **d)** $400,000   **85.** a
87. a) 69.2 years, 76.8 years
b)

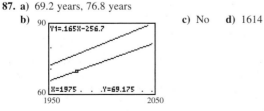

c) No   **d)** 1614

Mid-Chapter Quiz 7.1–7.2
1. Yes   **2.** No   **3.** No   **4.** $\{(3, 2)\}$ **5.** $\{(4, -4)\}$
6. $\{(x, y) \mid y = x + 6\}$ **7.** $\{(2, 1)\}$ **8.** $\{(7, -1)\}$ **9.** $\varnothing$
10. Inconsistent   **11.** Independent   **12.** Dependent

Section 7.3 Warm-Ups
1. Addition   **2.** Conditional   **3.** Dependent   **4.** Inconsistent
5. Conditional   **6.** Identity   **7.** Inconsistent   **8.** True   **9.** False
10. False   **11.** False   **12.** True   **13.** True   **14.** True

Section 7.3 Exercises
1. $\{(4, 3)\}$ **3.** $\{(1, -2)\}$ **5.** $\{(5, -7)\}$
7. $\left\{\left(\frac{3}{8}, -\frac{31}{8}\right)\right\}$ **9.** $\{(-1, 3)\}$ **11.** $\left\{\left(\frac{7}{9}, \frac{2}{3}\right)\right\}$ **13.** $\{(-1, -3)\}$
15. $\{(-2, -5)\}$ **17.** $\{(22, 26)\}$ **19.** Yes   **21.** No   **23.** Yes
25. $\varnothing$, inconsistent   **27.** $\{(x, y) \mid 5x - y = 1\}$, dependent
29. $\left\{\left(\frac{5}{2}, 0\right)\right\}$, independent   **31.** $\{(12, 6)\}$ **33.** $\{(-8, 6)\}$
35. $\{(16, 12)\}$ **37.** $\left\{\left(\frac{1}{2}, \frac{1}{3}\right)\right\}$ **39.** $\{(12, 7)\}$ **41.** $\{(400, 800)\}$
43. $\{(1.5, 1.25)\}$ **45.** $\left\{\left(\frac{3}{4}, \frac{2}{3}\right)\right\}$ **47.** $\{(5, 6)\}$ **49.** $\{(2, -17)\}$
51. $\{(0, 1)\}$ **53.** $\{(3, 4)\}$ **55.** $\left\{\left(\frac{1}{2}, \frac{1}{3}\right)\right\}$ **57.** $\varnothing$ **59.** $\{(x, y) \mid y = x\}$
61. $a = -1$ **63.** $a = 2, b = -1$ **65.** 5 and 7
67. Length 11 in., width 8.5 in.   **69.** Buys for $14, sells for $16
71. $1.40   **73.** 1380 students   **75.** 31 dimes, 4 nickels
77. a) 20 pounds chocolate, 30 pounds peanut butter
b) 20 pounds chocolate, 30 pounds peanut butter
79. 4 hours   **81.** 80%   **83.** Width 150 meters, length 200 meters

Section 7.4 Warm-Ups
1. Linear   **2.** Ordered   **3.** Solution   **4.** Addition, substitution
5. Plane   **6.** Independent   **7.** False   **8.** False   **9.** True
10. False   **11.** True   **12.** False   **13.** False   **14.** False

Section 7.4 Exercises
1. $\{(2, 3, 4)\}$ **3.** $\{(2, 3, 5)\}$ **5.** $\{(1, 2, 3)\}$ **7.** $\{(1, 2, -1)\}$
9. $\{(1, 3, 2)\}$ **11.** $\{(1, -5, 3)\}$ **13.** $\{(-1, 2, -1)\}$
15. $\{(-1, -2, 4)\}$ **17.** $\{(1, 3, 5)\}$ **19.** $\{(3, 4, 5)\}$
21. $\{(x, y, z) \mid x + y - z = 2\}$ **23.** $\varnothing$ **25.** $\varnothing$ **27.** $\varnothing$
29. $\{(x, y, z) \mid -x + 2y - 3z = -6\}$ **31.** $\varnothing$
33. $\{(x, y, z) \mid 5x + 4y - 2z - 150\}$ **35.** $\{(0.1, 0.3, 2)\}$ **37.** Yes
39. No   **41.** $\{(-4, 4)\}$, independent
43. $\{(x, y)\} \mid x + y = 4\}$, dependent   **45.** $\varnothing$, inconsistent
47. $\{(3, 2, 1)\}$, independent   **49.** $\{(x, y, z) \mid x - y + z = 1\}$, dependent
51. Chevrolet $20,000, Ford $22,000, Toyota $24,000
53. First 10 hr, second 12 hr, third 14 hr
55. $1500 stocks, $4500 bonds, $6000 mutual fund
57. Anna 108 pounds, Bob 118 pounds, Chris 92 pounds
59. 3 nickels, 6 dimes, 4 quarters
61. $24,000 teaching, $18,000 painting, $6000 royalties
63. Edwin 24, father 51, grandfather 84

Enriching Your Mathematical Word Power
1. System   **2.** Independent   **3.** Inconsistent   **4.** Dependent
5. Substitution   **6.** Addition   **7.** Linear

Review Exercises
1. $\{(1, 1)\}$, independent   **3.** $\{(x, y) \mid x + 2y = 4\}$, dependent
5. $\varnothing$, inconsistent   **7.** $\{(-3, 2)\}$, independent   **9.** $\varnothing$, inconsistent
11. $\{(x, y) \mid 2x - y = 3\}$, dependent   **13.** $\{(30, 12)\}$, independent
15. $\left\{\left(\frac{1}{5}, \frac{2}{5}\right)\right\}$, independent   **17.** $\{(-1, 5)\}$, independent

19. $\{(x, y) \mid 3x - 2y = 12\}$, dependent   **21.** $\varnothing$, inconsistent

23. $\left\{\left(2, -\dfrac{1}{3}\right)\right\}$, independent   **25.** $\{(20, 60)\}$, independent

27. $\{(2, 4, 6)\}$ **29.** $\{(1, -3, 2)\}$ **31.** $\varnothing$

33. $\{(x, y, z) \mid x - 2y + z = 8\}$ **35.** $\{(3, 4)\}$ **37.** $\{(2, -4)\}$

39. $\{(-1, -2)\}$ **41.** $\{(2, 1)\}$ **43.** $\{(1, 1, 2)\}$ **45.** $\{(1, 2, -3)\}$

47. $\{(20, 10)\}$ **49.** $\{(5, -1)\}$ **51.** $\{(7, 7)\}$ **53.** $\{(15, 25)\}$

55. $\{(x, y) \mid y = 2x - 5\}$ **57.** $\varnothing$ **59.** $\{(0, 0)\}$

61. $\{(x, y) \mid 3y - 2x = -3\}$ **63.** $\varnothing$ **65.** $\{(0.8, 0.7)\}$ **67.** $\{(-1, 2)\}$

69. $\{(x, y) \mid x + 2y = 8\}$ **71.** $\varnothing$ **73.** $\varnothing$

75. Width 13 feet, length 28 feet   **77.** 78   **79.** 36 minutes

81. 4 liters of A, 8 liters of B, 8 liters of C   **83.** Three servings of each

85. Milk $2.40, magazine $2.25   **87.** Length 12 cm, width 5 cm

89. Length 18 in., width 12 in.   **91.** 12.5 and 38.5   **93.** 17.5 and -12.5

95. 45 four-wheel cars, 2 three-wheel cars, and 3 two-wheel motorcycles

97. 12 singles, 10 doubles   **99.** Gary 39, Harry 34

101. Square 10 feet, triangle $\dfrac{40}{3}$ feet

103. 10 gallons of 10% solution, 20 gallons of 25% solution

105. Mimi 36 pounds, Mitzi 32 pounds, Cassandra 107 pounds

107. $39°, 51°, 90°$

Chapter 7 Test

1. $\{(1, 3)\}$ **2.** $\left\{\left(\dfrac{5}{2}, -3\right)\right\}$ **3.** $\{(x, y) \mid y = x - 5\}$

4. $\{(-1, 3)\}$ **5.** $\varnothing$ **6.** Inconsistent   **7.** Dependent

8. Independent   **9.** $\{(2, 5)\}$ **10.** $\{(-1, 2)\}$

11. $\{(2, -2, 1)\}$ **12.** $\{(1, -2, -3)\}$ **13.** $\{(3, 1, 1)\}$

14. Single $79, double $99

15. Jill 17 hours, Karen 14 hours, Betsy 62 hours

Making Connections   A Review of Chapters 1–7

1. -81 **2.** 7   **3.** 73   **4.** 5.94   **5.** $-t - 3$ **6.** $-0.9x + 0.9$

7. $3x^2 + 2x - 1$ **8.** y **9.** $3y(y + 11)(y - 11)$

10. $2(y + 2)(y - 2)(y^2 + 4)$ **11.** $(y + 2)(w - 4)$

12. $(y - 3)(y^2 + 3y + 9)$ **13.** $-3(y - 5)(y + 9)$

14. $-2y(3y - 2)(4y + 3)$ **15.** $4a(a^2 - a + 3)$

16. $2ab^3(a^2 + b^2)$ **17.** $\dfrac{3}{7x}$ **18.** $\dfrac{2x^7}{3}$ **19.** $\dfrac{x + 4}{x - 7}$ **20.** $\dfrac{x + y}{x}$

21. $\dfrac{x + 5}{x + 1}$ **22.** $\dfrac{x + 4}{x - 1}$ **23.** $\dfrac{13}{24}$ **24.** $\dfrac{5}{12}$ **25.** $\dfrac{7}{60}$ **26.** $\dfrac{2}{15}$

27. $\dfrac{7}{55}$ **28.** 66   **29.** $\dfrac{1}{10}$ **30.** $\dfrac{3}{2}$ **31.** $\dfrac{2 + 5a}{6a^2}$ **32.** $\dfrac{1 + 2y^2}{2y}$

33. $\dfrac{9}{x + 3}$ **34.** $\dfrac{2b^5y^2}{3}$ **35.** $\dfrac{6}{5b}$ **36.** $\dfrac{a - 6}{2}$ **37.** $y = \dfrac{3}{5}x - \dfrac{7}{5}$

38. $y = \dfrac{C}{D}x - \dfrac{W}{D}$ **39.** $y = \dfrac{K}{W - C}$ **40.** $y = \dfrac{bw - 2A}{b}$

41. $\{(4, -1)\}$ **42.** $\{(500, 700)\}$ **43.** $\{(x, y) \mid x + 17 = 5y\}$ **44.** $\varnothing$

45. $y = \dfrac{5}{9}x + 55$ **46.** $y = -\dfrac{11}{6}x + \dfrac{2}{3}$ **47.** $y = 5x + 26$

48. $y = \dfrac{1}{2}x + 5$ **49.** $y = 5$ **50.** $x = -7$

51. a) Machine A
b) Machine B $0.04 per copy, machine A $0.03 per copy
c) The slopes 0.04 and 0.03 are the per copy cost for each machine.
d) B: $y = 0.04x + 2000$, A: $y = 0.03x + 4000$
e) 200,000

Chapter 8

Section 8.1 Warm-Ups

1. Compound   **2.** And   **3.** Or   **4.** And   **5.** Intersection
6. Union   **7.** True   **8.** True   **9.** False   **10.** True   **11.** True
12. True   **13.** False   **14.** False   **15.** True   **16.** True

Section 8.1 Exercises

1. No   **3.** Yes   **5.** No   **7.** No   **9.** Yes   **11.** Yes   **13.** Yes

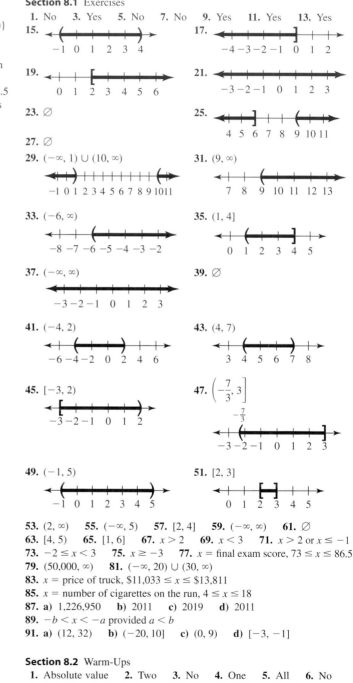

29. $(-\infty, 1) \cup (10, \infty)$ **31.** $(9, \infty)$

33. $(-6, \infty)$ **35.** $(1, 4]$

37. $(-\infty, \infty)$ **39.** $\varnothing$

41. $(-4, 2)$ **43.** $(4, 7)$

45. $[-3, 2)$ **47.** $\left(-\dfrac{7}{3}, 3\right]$

49. $(-1, 5)$ **51.** $[2, 3]$

53. $(2, \infty)$ **55.** $(-\infty, 5)$ **57.** $[2, 4]$ **59.** $(-\infty, \infty)$ **61.** $\varnothing$
63. $[4, 5)$ **65.** $[1, 6]$ **67.** $x > 2$ **69.** $x < 3$ **71.** $x > 2$ or $x \leq -1$
73. $-2 \leq x < 3$ **75.** $x \geq -3$ **77.** $x = $ final exam score, $73 \leq x \leq 86.5$
79. $(50,000, \infty)$ **81.** $(-\infty, 20) \cup (30, \infty)$
83. $x = $ price of truck, $11,033 \leq x \leq $13,811
85. $x = $ number of cigarettes on the run, $4 \leq x \leq 18$
87. a) 1,226,950   **b)** 2011   **c)** 2019   **d)** 2011
89. $-b < x < -a$ provided $a < b$
91. a) $(12, 32)$ **b)** $(-20, 10]$ **c)** $(0, 9)$ **d)** $[-3, -1]$

Section 8.2 Warm-Ups

1. Absolute value   **2.** Two   **3.** No   **4.** One   **5.** All   **6.** No
7. $(-3, 3)$ **8.** Equivalent   **9.** True   **10.** False   **11.** True
12. False   **13.** True   **14.** True   **15.** True   **16.** True   **17.** False
18. True

Section 8.2 Exercises

1. $\{-5, 5\}$ **3.** $\{2, 4\}$ **5.** $\{-3, 9\}$ **7.** $\left[-\dfrac{8}{3}, \dfrac{16}{3}\right]$ **9.** $\{12\}$

11. $\{-20, 80\}$ **13.** $\varnothing$ **15.** $\{0, 5\}$ **17.** $\{0.143, 1.298\}$

19. $\{-2, 2\}$ **21.** $\{-11, 5\}$ **23.** $\{0, 3\}$ **25.** $\left\{-6, \dfrac{4}{3}\right\}$ **27.** $\{1, 3\}$

29. $(-\infty, \infty)$ **31.** $|x| < 2$ **33.** $|x| > 3$ **35.** $|x| \le 1$

37. $|x| \ge 2$ **39.** No   **41.** Yes   **43.** No

45. $(-\infty, -6) \cup (6, \infty)$ **47.** $[-2, 2]$

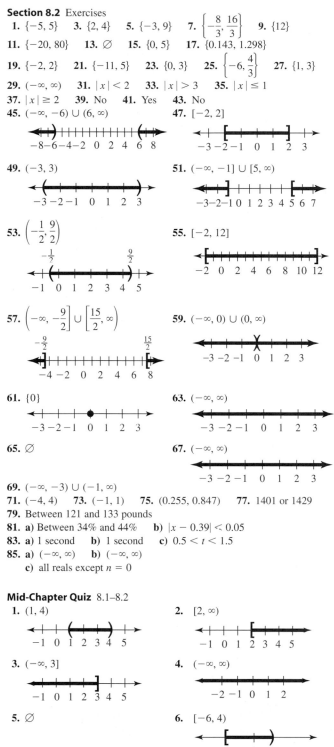

49. $(-3, 3)$ **51.** $(-\infty, -1] \cup [5, \infty)$

53. $\left(-\dfrac{1}{2}, \dfrac{9}{2}\right)$ **55.** $[-2, 12]$

57. $\left(-\infty, -\dfrac{9}{2}\right] \cup \left[\dfrac{15}{2}, \infty\right)$ **59.** $(-\infty, 0) \cup (0, \infty)$

61. $\{0\}$ **63.** $(-\infty, \infty)$

65. $\varnothing$ **67.** $(-\infty, \infty)$

69. $(-\infty, -3) \cup (-1, \infty)$
71. $(-4, 4)$ **73.** $(-1, 1)$ **75.** $(0.255, 0.847)$ **77.** 1401 or 1429
79. Between 121 and 133 pounds
81. a) Between 34% and 44%   **b)** $|x - 0.39| < 0.05$
83. a) 1 second   **b)** 1 second   **c)** $0.5 < t < 1.5$
85. a) $(-\infty, \infty)$ **b)** $(-\infty, \infty)$
  c) all reals except $n = 0$

Mid-Chapter Quiz 8.1–8.2

1. $(1, 4)$ **2.** $[2, \infty)$

3. $(-\infty, 3]$ **4.** $(-\infty, \infty)$

5. $\varnothing$ **6.** $[-6, 4)$

7. $\left(-2, -\dfrac{3}{5}\right)$ **8.** $(-\infty, -1) \cup (7, \infty)$

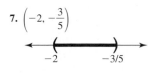
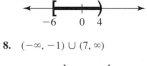

9. $[-7, 1]$ **10.** $\varnothing$

11. $(-\infty, \infty)$

12. $\{-1, 7\}$ **13.** $\{9\}$ **14.** $\{2, 12\}$ **15.** $\varnothing$

Section 8.3 Warm-Ups
1. Linear   **2.** Compound   **3.** Union   **4.** Intersection   **5.** Test
6. Solid   **7.** False   **8.** True   **9.** True   **10.** True   **11.** False
12. True

Section 8.3 Exercises
1. $(-6, -4)$ **3.** $(1, 3), (-2, 5), (-6, -4)$ **5.** $(7, -8)$
7.   **9.**

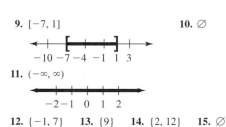

11.   **13.**

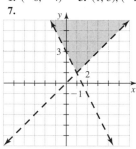
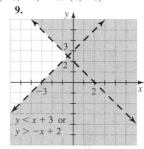

15.   **17.**

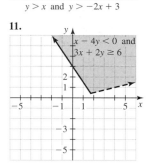
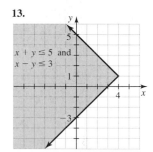

19.   **21.**

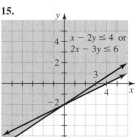
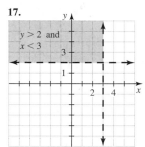
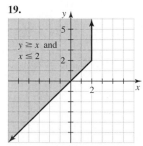
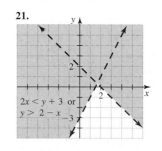
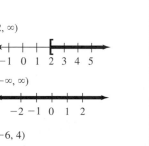

23.

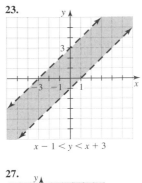

$x - 1 < y < x + 3$

25.

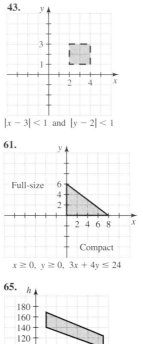

$0 \le y \le x$ and $x \le 1$

43.

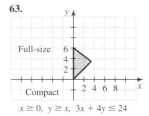

$|x - 3| < 1$ and $|y - 2| < 1$

45. Not the empty set
47. $\varnothing$
49. Not the empty set
51. Not the empty set
53. $\varnothing$
55. $\varnothing$
57. $\varnothing$
59. Not the empty set

27.

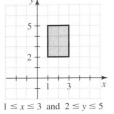

$1 \le x \le 3$ and $2 \le y \le 5$

29.

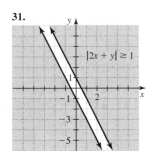

$|x + y| < 2$

61.

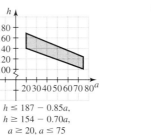

Full-size

Compact

$x \ge 0, \ y \ge 0, \ 3x + 4y \le 24$

63.

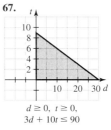

Full-size

Compact

$x \ge 0, \ y \ge x, \ 3x + 4y \le 24$

31.

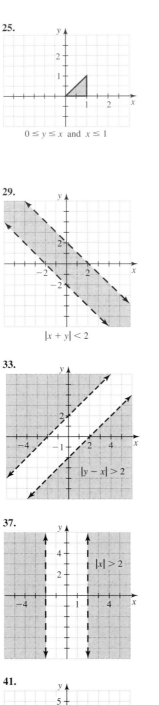

$|2x + y| \ge 1$

33.

$|y - x| > 2$

65.

$h \le 187 - 0.85a,$
$h \ge 154 - 0.70a,$
$a \ge 20, a \le 75$

67.

$d \ge 0, \ t \ge 0,$
$3d + 10t \le 90$

Section 8.4 Warm-Ups
1. Constraint   **2.** Linear programming   **3.** Linear   **4.** Vertex
5. False   **6.** False   **7.** True   **8.** True   **9.** True   **10.** False

35.

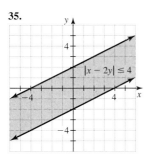

$|x - 2y| \le 4$

37.

$|x| > 2$

Section 8.4 Exercises

1.

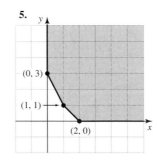

(0, 5)   (0, 0)   (5, 0)

3.

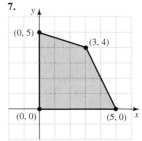

(0, 3)   (1, 2)   (0, 0)   (2, 0)

39.

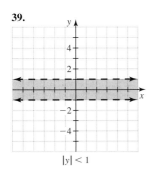

$|y| < 1$

41.

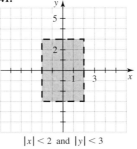

$|x| < 2$ and $|y| < 3$

5.

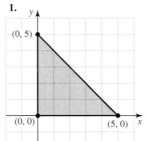

(0, 3)   (1, 1)   (2, 0)

7.

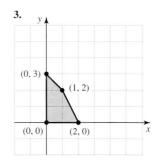

(0, 5)   (3, 4)   (0, 0)   (5, 0)

9.

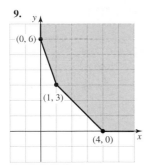

11. $x \geq 0$, $y \geq 0$, $x + 2y \leq 30$, $4x + 3y \leq 60$

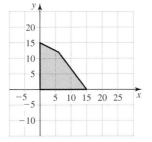

13. 46   **15.** 88   **17.** 128   **19.** 9   **21.** 59   **23.** 21   **25.** 18
27. a) 0, 320,000, 510,000, 450,000   **b)** 30 TV ads and 60 radio ads
29. 6 doubles, 4 triples   **31.** 0 doubles, 8 triples
33. 1.75 cups Doggie Dinner, 5.5 cups Puppy Power
35. 10 cups Doggie Dinner, 0 cups Puppy Power
37. Laundromat $8000, car wash $16,000

Enriching Your Mathematical Word Power
1. Simple   **2.** Compound   **3.** Intersection   **4.** Union   **5.** And
6. Or   **7.** Or   **8.** Constraints

Review Exercises
1. $(-\infty, -4) \cup (1, \infty)$

3. $(0, 9)$

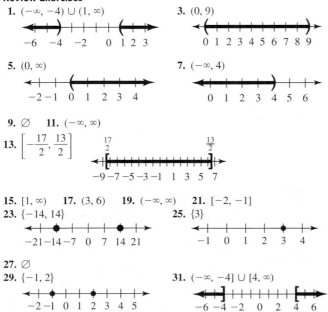

5. $(0, \infty)$

7. $(-\infty, 4)$

9. $\varnothing$ **11.** $(-\infty, \infty)$

13. $\left[-\dfrac{17}{2}, \dfrac{13}{2} \right]$

15. $[1, \infty)$ **17.** $(3, 6)$ **19.** $(-\infty, \infty)$ **21.** $[-2, -1]$
23. $\{-14, 14\}$ **25.** $\{3\}$

27. $\varnothing$
29. $\{-1, 2\}$ **31.** $(-\infty, -4] \cup [4, \infty)$

33. $(-\infty, -4) \cup (14, \infty)$ **35.** $\varnothing$

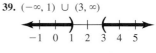

37. $(-\infty, \infty)$ **39.** $(-\infty, 1) \cup (3, \infty)$

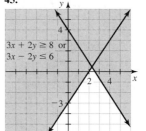

41.

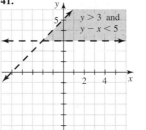

43.

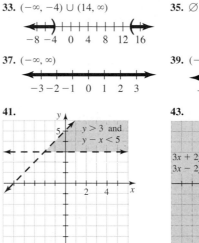

45.

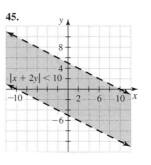

47.

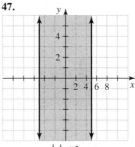

$|x| \leq 5$

49.

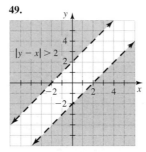

51.

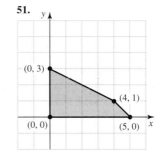

53. 30   **55.** $x =$ rental price, $3 \leq x \leq 5$ **57.** $(40.2, 53.6)$
59. 81 or 91   **61.** $x > 1$ **63.** $|x - 2| = 0$ **65.** $|x| = 3$
67. $x \leq -1$ **69.** $|x| \leq 2$ **71.** $x \leq 2$ or $x \geq 7$ **73.** $|x| > 3$
75. $5 < x < 7$ or $|x - 6| < 1$ **77.** $|x| > 0$

Chapter 8 Test
1. $-3 < x \leq 2$ **2.** $x > 1$ **3.** $[3, \infty)$ **4.** $(1, 6]$
5. $(-\infty, 5) \cup (9, \infty)$ **6.** $(-3, 3)$ **7.** $(-\infty, -2) \cup (2, \infty)$
8. $(-1, \infty)$ **9.** $[4, 8]$

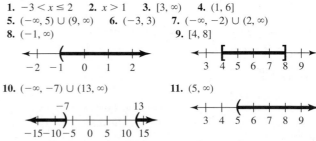

10. $(-\infty, -7) \cup (13, \infty)$ **11.** $(5, \infty)$

12. $[-5, 3)$

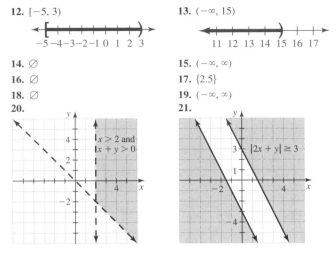

13. $(-\infty, 15)$

14. $\varnothing$

15. $(-\infty, \infty)$

16. $\varnothing$

17. $\{2.5\}$

18. $\varnothing$

19. $(-\infty, \infty)$

20.

21.

22.

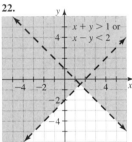

23. $|x - 28,000| > 3,000$ where x is Brenda's salary; Brenda makes more than \$31,000 or less than \$25,000.

24. 44

Making Connections   A Review of Chapters 1–8
1. $11x$ **2.** $30x^2$ **3.** $3x + 1$ **4.** $4x - 3$ **5.** 899 **6.** 961
7. 841 **8.** 25 **9.** 13 **10.** -25 **11.** 5 **12.** -4
13. $-2x + 13$ **14.** 60 **15.** 72 **16.** -9 **17.** $\{0\}$
18. R or $(-\infty, \infty)$ **19.** $\{0\}$ **20.** $\{1\}$ **21.** $\left\{-\dfrac{1}{3}\right\}$ **22.** $\{1\}$
23. $\{1000\}$ **24.** $\left\{-\dfrac{17}{5}, 1\right\}$ **25.** $\{(4, -3)\}$ **26.** $\{(x, y)\,|\,3x - y = 5\}$
27. $\{(-2, 4)\}$ **28.** $\{(8, 15)\}$ **29.** x^5 **30.** x^{11} **31.** x^2 **32.** $\dfrac{1}{x^6}$
33. 18 **34.** 0 **35.** $-27a^6b^9$ **36.** $27a^{24}$ **37.** 72 **38.** 225
39. E **40.** F **41.** G **42.** D **43.** H **44.** C **45.** B **46.** I
47. A **48.** J **49.** $x^3 - 8$ **50.** $a^3 + 1000$ **51.** $9a^2 - 30ab + 25b^2$
52. $4x^4 + 12x^2y + 9y^2$ **53.** $a^2 - y^4$ **54.** $30m^2 - 34m - 36$
55. $(y - 1)(y + 99)$ **56.** $(2a - 3)(4a + 1)$ **57.** $6a(a - 3)^2$
58. $(b^2 + 4)(b + 1)$ **59.** $(a - 8)(a - 6)$ **60.** $-8(a + 1)(a^2 - a + 1)$
61. a) 87,500 **b)** $C_r = 4500 + 0.06x$, $C_b = 8000 + 0.02x$ **c)** 87,500
d) Buying is \$1300 cheaper **e)** $(75,000, 100,000)$

Chapter 9

Section 9.1 Warm-Ups
1. nth root **2.** Principal **3.** Product **4.** Quotient **5.** True
6. False **7.** True **8.** True **9.** False **10.** True **11.** True
12. False

Section 9.1 Exercises
1. 6 **3.** 10 **5.** -3 **7.** 2 **9.** -2 **11.** 2 **13.** 10
15. Not a real number **17.** $|m|$ **19.** x^8 **21.** y^3 **23.** y^5
25. m **27.** $|w^3|$ **29.** $|b^9|$ **31.** y^6 **33.** $3\sqrt{y}$ **35.** $2a$ **37.** x^2y
39. $m^6\sqrt{5}$ **41.** $2\sqrt[3]{y}$ **43.** $a^2\sqrt[3]{3}$ **45.** $2\sqrt{5}$ **47.** $5\sqrt{2}$
49. $6\sqrt{2}$ **51.** $2\sqrt[5]{5}$ **53.** $3\sqrt[3]{3}$ **55.** $2\sqrt[4]{3}$ **57.** $2\sqrt[3]{3}$ **59.** $a\sqrt{a}$
61. $3a^3\sqrt{2}$ **63.** $2x^2\sqrt{5xy}$ **65.** $2m\sqrt[3]{3m}$ **67.** $2a\sqrt[4]{2a}$
69. $2x\sqrt[5]{2x}$ **71.** $4xy^4z^3\sqrt{3xz}$ **73.** $\dfrac{\sqrt[5]{t}}{2}$ **75.** $\dfrac{25}{4}$ **77.** $\sqrt{10}$
79. $\dfrac{\sqrt[3]{t}}{2}$ **81.** $\dfrac{-2x^2}{y}$ **83.** $\dfrac{2a^3}{3}$ **85.** $\dfrac{2\sqrt{3}}{5}$ **87.** $\dfrac{3\sqrt{3}}{4}$
89. $\dfrac{a\sqrt[3]{a}}{5}$ **91.** $\dfrac{3\sqrt{3}}{2b}$ **93.** $\dfrac{x\sqrt[3]{x^3}}{y^2}$ **95.** $\dfrac{a\sqrt[4]{a}}{2b^3}$ **97.** $[2, \infty)$
99. $(-\infty, \infty)$ **101.** $(-\infty, 3]$ **103.** $\left[-\dfrac{1}{2}, \infty\right)$ **105.** $[6, \infty)$
107. $(-\infty, \infty)$ **109.** $(-\infty, 9]$ **111. a)** $-4°F$ **b)** $-10°F$
113. a) $t = \dfrac{\sqrt{h}}{4}$ **b)** $\dfrac{\sqrt{10}}{2}$ sec **c)** 100 ft
115. 5.8 knots **117. a)** 114.1 ft/sec **b)** 77.8 mph
119. a) Yes **b)** No **c)** Yes **d)** Yes **121.** Arithmetic mean

Section 9.2 Warm-Ups
1. nth root **2.** mth power **3.** Reciprocal **4.** Even, negative
5. True **6.** False **7.** False **8.** True **9.** True **10.** True
11. True **12.** True

Section 9.2 Exercises
1. $7^{1/4}$ **3.** $(5x)^{1/2}$ **5.** $\sqrt[5]{9}$ **7.** $\sqrt{a}$ **9.** 5 **11.** -5
13. 2 **15.** Not a real number **17.** $w^{7/3}$ **19.** $2^{-10/3}$ **21.** $\sqrt[4]{\dfrac{1}{w^3}}$
23. $\sqrt{(ab)^3}$ **25.** 25 **27.** 125 **29.** $\dfrac{1}{81}$ **31.** $\dfrac{1}{64}$ **33.** $-\dfrac{1}{3}$
35. Not a real number **37.** $3^{7/12}$ **39.** 1 **41.** $\dfrac{1}{2}$ **43.** 2 **45.** 6
47. 4 **49.** 81 **51.** $\dfrac{1}{4}$ **53.** $\dfrac{9}{8}$ **55.** $|x|$ **57.** a^4 **59.** y
61. $3|x^3y|$ **63.** $3\left|\dfrac{x^3}{y^5}\right|$ **65.** $x^{3/4}$ **67.** $\dfrac{y^{3/2}}{x^{1/4}}$ **69.** $\dfrac{1}{w^{8/3}}$
71. $12x^8$ **73.** $\dfrac{a^2}{b}$ **75.** $8w^{13/4}$ **77.** 9 **79.** $-\dfrac{1}{8}$ **81.** $\dfrac{1}{625}$
83. $2^{1/4}$ **85.** 3 **87.** 3 **89.** $\dfrac{4}{9}$ **91.** Not a real number
93. $\dfrac{4}{3}$ **95.** $-\dfrac{216}{125}$ **97.** $3x^{9/2}$ **99.** $\dfrac{a^2}{27}$ **101.** $a^{5/4}b$
103. $k^{9/2}m^4$ **105.** 1.2599 **107.** -1.4142 **109.** 2 **111.** 2.5
113. $a^{3m/4}$ **115.** $a^{2m/15}$ **117.** $a^n b^m$ **119.** $a^{4m}b^{2n}$
121. 0.25 sec, 1 sec, 1.5 sec **123. a)** 13 in. **b)** $\sqrt{3}$ or 1.73 in.
125. 274.96 m² **127.** 15.7% **129.** 6.12%
131. a) Identity **b)** Identity **c)** Not an identity

Section 9.3 Warm-Ups
1. Like **2.** Distributive **3.** Index **4.** Conjugate **5.** False
6. True **7.** False **8.** True **9.** False **10.** True **11.** False
12. True

Section 9.3 Exercises
1. $-\sqrt{3}$ **3.** $9\sqrt{7x}$ **5.** $5\sqrt[3]{2}$ **7.** $4\sqrt{3} - 2\sqrt{5}$ **9.** $5\sqrt[3]{x}$
11. $2\sqrt[3]{x} - \sqrt{2x}$ **13.** $2\sqrt{2} + 2\sqrt{7}$ **15.** $5\sqrt{2}$ **17.** 0
19. $-\sqrt{2}$ **21.** $x\sqrt{5x} + 2x\sqrt{2}$ **23.** $7\sqrt[3]{3}$ **25.** $-4\sqrt[3]{3}$
27. $ty\sqrt{2t}$ **29.** $\sqrt{15}$ **31.** $30\sqrt{2}$ **33.** $6a\sqrt{14}$ **35.** $3\sqrt[4]{3}$

37. 12   **39.** $2x^3\sqrt{10x}$ **41.** $\dfrac{x\sqrt[4]{x^3}}{3}$ **43.** $6\sqrt{2}+18$

45. $5\sqrt{2}-2\sqrt{5}$ **47.** $3\sqrt[5]{t^2}-t\sqrt[5]{3}$ **49.** $7\quad 3\sqrt{3}$ **51.** 2

53. $-8-6\sqrt{5}$ **55.** $-6+9\sqrt{2}$ **57.** -1 **59.** 3   **61.** 19

63. 13   **65.** $25-9x$ **67.** $\sqrt[6]{3^5}$ **69.** $\sqrt[12]{5^7}$ **71.** $\sqrt[6]{500}$

73. $\sqrt[12]{432}$ **75.** $11\sqrt{3}$ **77.** $10\sqrt{30}$ **79.** $8-\sqrt{7}$ **81.** $16w$

83. $3x^2\sqrt{2x}$ **85.** $28+\sqrt{10}$ **87.** $\dfrac{8\sqrt{2}}{15}$ **89.** 17

91. $9+6\sqrt{x}+x$ **93.** $25x-30\sqrt{x}+9$ **95.** $x+3+2\sqrt{x+2}$

97. $-\sqrt{w}$ **99.** $a\sqrt{a}$ **101.** $3x^2\sqrt{x}$ **103.** $13x\sqrt[3]{2x}$

105. $\sqrt[6]{32x^5}$ **107.** $3\sqrt{2}$ square feet (ft²)   **109.** $\dfrac{9\sqrt{2}}{2}$ ft²   **111.** No

113. a) $(y-\sqrt{3})(y+\sqrt{3}),(\sqrt{2a}-\sqrt{7})(\sqrt{2a}+\sqrt{7})$
      b) $\{\pm 2\sqrt{2}\}$ **c)** $\{\pm\sqrt{a}\}$

Mid-Chapter Quiz 9.1–9.3
1. 8   **2.** -3 **3.** $2\sqrt{30}$ **4.** $2\sqrt[3]{7}$ **5.** $2x^3\sqrt{3x}$ **6.** $2ab^4\sqrt[3]{3b}$

7. $\dfrac{\sqrt{w}}{4}$ **8.** $\dfrac{2x\sqrt{2x}}{3}$ **9.** 9   **10.** 1000   **11.** -64 **12.** $\dfrac{1}{25}$

13. $-2\sqrt{3}-4\sqrt{6}$ **14.** $9\sqrt{5}$ **15.** $12\sqrt{35}$ **16.** 54   **17.** $5x\sqrt[3]{x^2}$
18. $(-\infty,2]$ **19.** $[0,\infty)$ **20.** $(-\infty,\infty)$ **21.** $\sqrt[4]{32}$ **22.** 20

Section 9.4 Warm-Ups
1. Irrational   **2.** Rationalizing   **3.** Squares   **4.** Radical
5. Denominator   **6.** True   **7.** True   **8.** False   **9.** False
10. True   **11.** True   **12.** True

Section 9.4 Exercises
1. $\dfrac{2\sqrt{5}}{5}$ **3.** $\dfrac{\sqrt{21}}{7}$ **5.** $\dfrac{\sqrt[4]{2}}{2}$ **7.** $\dfrac{\sqrt[4]{150}}{5}$ **9.** $\dfrac{\sqrt{15}}{6}$ **11.** $\dfrac{1}{2}$

13. $\dfrac{\sqrt{2}}{2}$ **15.** $\dfrac{\sqrt[3]{18}}{3}$ **17.** $\dfrac{\sqrt[4]{14}}{2}$ **19.** $\dfrac{\sqrt{xy}}{y}$ **21.** $\dfrac{a\sqrt{ab}}{b^4}$

23. $\dfrac{\sqrt{3ab}}{3b}$ **25.** $\dfrac{\sqrt[3]{ab^2}}{b}$ **27.** $\dfrac{\sqrt[3]{20b}}{2b}$ **29.** $\sqrt{3}$ **31.** $\dfrac{\sqrt{15}}{5}$

33. $\dfrac{3\sqrt{2}}{10}$ **35.** $\dfrac{\sqrt{2}}{3}$ **37.** $\dfrac{\sqrt[3]{3a}}{3}$ **39.** $\sqrt[3]{10}$ **41.** 2   **43.** $\dfrac{2}{w}$

45. $2+\sqrt{5}$ **47.** $1-\sqrt{3}$ **49.** $2\sqrt{2}-2$ **51.** $\dfrac{\sqrt{11}+\sqrt{5}}{2}$

53. $\dfrac{1+\sqrt{6}+\sqrt{2}+\sqrt{3}}{2}$ **55.** $\dfrac{2\sqrt{3}-\sqrt{6}}{3}$ **57.** $\dfrac{6\sqrt{6}+2\sqrt{15}}{13}$

59. $128\sqrt{2}$ **61.** $x^2\sqrt{x}$ **63.** $-27x^4\sqrt{x}$ **65.** $8x^5$ **67.** $4\sqrt[3]{25}$

69. x^4 **71.** $\dfrac{\sqrt{6}+2\sqrt{2}}{2}$ **73.** $2\sqrt{6}$ **75.** $\dfrac{\sqrt{2}}{2}$ **77.** $\dfrac{2-\sqrt{2}}{5}$

79. $\dfrac{1+\sqrt{3}}{2}$ **81.** $a-3\sqrt{a}$ **83.** $4a\sqrt{a}+4a$ **85.** $12m$

87. $4xy^2z$ **89.** $m-m^2$ **91.** $5x\sqrt[3]{x}$ **93.** $8m^4\sqrt[4]{8m^2}$

95. $\sqrt{x}+3$ **97.** $\dfrac{3k-3\sqrt{7k}}{k-7}$ **99.** $2+8\sqrt{2}$ **101.** $\dfrac{3\sqrt{2}+2\sqrt{3}}{6}$

103. $7\sqrt{2}-1$ **105.** $\dfrac{4x+4\sqrt{x}}{x-4}$ **107.** $\dfrac{x+\sqrt{x}}{x(1-x)}$

109. a) x^3-2 **b)** $(x+\sqrt[3]{5})(x^2-\sqrt[3]{5}x+\sqrt[3]{25})$ **c)** 3
      d) $(\sqrt[3]{a}+\sqrt[3]{b})(\sqrt[3]{a^2}-\sqrt[3]{ab}+\sqrt[3]{b^2})$,
          $(\sqrt[3]{a}-\sqrt[3]{b})(\sqrt[3]{a^2}+\sqrt[3]{ab}+\sqrt[3]{b^2})$

Section 9.5 Warm-Ups
1. Odd   **2.** Even   **3.** Extraneous   **4.** Even   **5.** False   **6.** True
7. False   **8.** True   **9.** True   **10.** True

Section 9.5 Exercises
1. $\{-10\}$ **3.** $\left\{\dfrac{1}{2}\right\}$ **5.** $\{1\}$ **7.** $\{-2\}$ **9.** $\{-5,5\}$

11. $\{-2\sqrt{5},2\sqrt{5}\}$ **13.** No real solution   **15.** $\{-1,7\}$

17. $\{-1-2\sqrt{2},-1+2\sqrt{2}\}$ **19.** $\{-\sqrt{10},\sqrt{10}\}$ **21.** $\{3\}$

23. $\{-2,2\}$ **25.** $\{52\}$ **27.** $\left\{\dfrac{9}{4}\right\}$ **29.** $\{9\}$ **31.** $\{3\}$ **33.** $\{3\}$

35. $\{-5,3\}$ **37.** $\{1\}$ **39.** $\varnothing$ **41.** $\{1,2\}$ **43.** $\{9\}$ **45.** $\{4\}$

47. $\{2\}$ **49.** $\{6\}$ **51.** $\{1,5\}$ **53.** $\{7\}$ **55.** $\{-5\}$ **57.** $\varnothing$

59. $\{0\}$ **61.** $\{-3\sqrt{3},3\sqrt{3}\}$ **63.** $\left\{-\dfrac{1}{27},\dfrac{1}{27}\right\}$ **65.** $\{512\}$

67. $\left\{\dfrac{1}{81}\right\}$ **69.** $\left\{0,\dfrac{2}{3}\right\}$ **71.** $\left\{\dfrac{4-\sqrt{2}}{4},\dfrac{4+\sqrt{2}}{4}\right\}$

73. No real solution   **75.** $\{-\sqrt{2},\sqrt{2}\}$ **77.** $\{-5\}$

79. No real solution   **81.** $\{-9\}$ **83.** $\left\{\dfrac{5}{4}\right\}$ **85.** $\varnothing$

87. $\left\{-\dfrac{2}{3},2\right\}$ **89.** $\{-2-2\sqrt[4]{2},-2+2\sqrt[4]{2}\}$ **91.** $\{0\}$ **93.** $\left\{\dfrac{1}{2}\right\}$

95. $4\sqrt{2}$ feet   **97.** $5\sqrt{2}$ feet   **99.** 50 feet

101. a) 2   **b)** $\sqrt{3}$ **c)** $\dfrac{\sqrt{3}}{2}$ **103. a)** 1.89   **b)** $d=\dfrac{64b^3}{C^3}$
      c) $d>19{,}683$ pounds

105. $\sqrt[6]{32}$ meters   **107.** $\sqrt{73}$ kilometers (km)
109. a) $S=P(1+r)^n$ **b)** $P=S(1+r)^{-n}$ **111.** 9.5 AU
113. $\{-1.8,1.8\}$ **115.** $\{4.993\}$ **117.** $\{-26.372,26.372\}$

Section 9.6 Warm-Ups
1. Complex   **2.** Imaginary   **3.** Complex   **4.** Conjugate
5. Product   **6.** Real   **7.** Subset   **8.** True   **9.** False   **10.** False
11. True   **12.** True   **13.** False   **14.** True   **15.** True   **16.** True

Section 9.6 Exercises
1. $-2+8i$ **3.** $-4+4i$ **5.** -2 **7.** $-8-2i$ **9.** $6+15i$
11. $-2-10i$ **13.** $-4-12i$ **15.** $-10+24i$ **17.** $-1+3i$
19. $-5i$ **21.** 29   **23.** 2   **25.** 20   **27.** -9 **29.** -25 **31.** 16
33. i **35.** -1 **37.** i **39.** 34   **41.** 5   **43.** 5   **45.** 7

47. $\dfrac{12}{17}-\dfrac{3}{17}i$ **49.** $\dfrac{4}{13}+\dfrac{7}{13}i$ **51.** $3-4i$ **53.** $1+3i$

55. $\dfrac{1}{13}-\dfrac{5}{13}i$ **57.** $-2i$ **59.** $5i$ **61.** $2+2i$ **63.** $5+6i$

65. $7-i\sqrt{6}$ **67.** $5i\sqrt{2}$ **69.** $1+i\sqrt{3}$ **71.** $-1-\dfrac{1}{2}i\sqrt{6}$

73. $-2\sqrt{3}$ **75.** -9 **77.** $-i\sqrt{2}$ **79.** $\{\pm 6i\}$ **81.** $\{\pm 2i\sqrt{3}\}$

83. $\left\{\pm\dfrac{i\sqrt{10}}{2}\right\}$ **85.** $\{\pm i\sqrt{2}\}$ **87.** $18-i$ **89.** $5+i$

91. $-\dfrac{6}{25}-\dfrac{17}{25}i$ **93.** $3+2i$ **95.** -9

97. $3i\sqrt{3}$ **99.** $-5-12i$ **101.** $-2+2i\sqrt{2}$

Enriching Your Mathematical Word Power
1. Root   **2.** Square   **3.** Cube   **4.** Principal   **5.** Odd   **6.** Index
7. Radicand   **8.** Like   **9.** Domain   **10.** Integral   **11.** Complex
12. Imaginary   **13.** Conjugates   **14.** Imaginary

Review Exercises
1. 9   **3.** 3   **5.** -3 **7.** 10   **9.** $|y|$ **11.** a **13.** n^6 **15.** $|n^5|$

17. $6\sqrt{2}$ **19.** x^6 **21.** x^2 **23.** $x^4\sqrt{2x}$ **25.** $2w^2\sqrt{2w}$

27. $2x\sqrt[3]{2x}$ **29.** $a^2b\sqrt[4]{ab}$ **31.** $\dfrac{x\sqrt{x}}{4}$ **33.** $[2.5,\infty)$ **35.** $(-\infty,\infty)$

37. $\left(-\infty, \frac{1}{3}\right]$ **39.** $[-2, \infty)$ **41.** $[-5, \infty)$ **43.** $(-\infty, 20]$

45. $(-\infty, \infty)$ **47.** $\frac{1}{9}$ **49.** 4   **51.** $\frac{1}{1000}$ **53.** $27x^{1/2}$ **55.** $a^{7/2}b^{7/2}$

57. $x^{3/4}y^{5/4}$ **59.** 13   **61.** $3\sqrt{5} - 2\sqrt{3}$ **63.** $30 - 21\sqrt{6}$

65. $6 - 3\sqrt{3} + 2\sqrt{2} - \sqrt{6}$ **67.** $\frac{5\sqrt{2}}{2}$ **69.** $\frac{\sqrt{10}}{5}$ **71.** $\frac{\sqrt[3]{18}}{3}$

73. $\frac{2\sqrt{3x}}{3x}$ **75.** $\frac{y\sqrt{15y}}{3}$ **77.** $\frac{3\sqrt[3]{4a^2}}{2a}$ **79.** $\frac{5\sqrt[3]{27x^2}}{3x}$ **81.** 9

83. $1 - \sqrt{2}$ **85.** $\frac{-\sqrt{6} - 3\sqrt{2}}{2}$ **87.** $\frac{3\sqrt{2} + 2}{7}$ **89.** $256w^{10}$

91. $\{-4, 4\}$ **93.** $\{3, 7\}$ **95.** $\{-1 - \sqrt{5}, -1 + \sqrt{5}\}$

97. No real solution   **99.** $\{10\}$ **101.** $\{9\}$ **103.** $\{-8, 8\}$ **105.** $\{124\}$

107. $\{7\}$ **109.** $\{2, 3\}$ **111.** $\{9\}$ **113.** $\{4\}$ **115.** $5 + 25i$

117. $7 - 3i$ **119.** $-1 + 2i$ **121.** $2 + i$ **123.** $2 - i\sqrt{3}$

125. $\frac{5}{17} - \frac{14}{17}i$ **127.** 16   **129.** -1 **131.** $\{\pm 10i\}$ **133.** $\left\{\pm\frac{3i\sqrt{2}}{2}\right\}$

135. False   **137.** True   **139.** True   **141.** False   **143.** False

145. False   **147.** False   **149.** True   **151.** False   **153.** True

155. False   **157.** False   **159.** True   **161.** True

163. $5\sqrt{30}$ or approximately 27.4 seconds   **165. a)** 2.5 sec   **b)** 256 ft

167. $10\sqrt{7}$ feet   **169.** $200\sqrt{2}$ feet   **171.** $26.4\sqrt[3]{25}$ ft^2

173. a) 5.7%   **b)** \$3000 billion or \$3 trillion   **175.** $V = \frac{29\sqrt{LCS}}{CS}$

Chapter 9 Test

1. 6   **2.** -5 **3.** $|t|$ **4.** p **5.** w^2 **6.** $|w^3|$ **7.** 4   **8.** $\frac{1}{8}$

9. $\sqrt{3}$ **10.** 30   **11.** $3\sqrt{5}$ **12.** $\frac{6\sqrt{5}}{5}$ **13.** 2   **14.** $6\sqrt{2}$

15. $\frac{\sqrt{15}}{6}$ **16.** $\frac{2 + \sqrt{2}}{2}$ **17.** $4 - 3\sqrt{3}$ **18.** $2ay^2\sqrt[4]{2a}$

19. $\frac{\sqrt[3]{4x}}{2x}$ **20.** $\frac{2a^4\sqrt{2ab}}{b^2}$ **21.** $-3x^3$ **22.** $2m\sqrt{5m}$ **23.** $x^{3/4}$

24. $3y^2x^{1/4}$ **25.** $2x^2\sqrt[3]{5x}$ **26.** $19 + 8\sqrt{3}$ **27.** $(-\infty, 4]$

28. $(-\infty, \infty)$ **29.** $\frac{5 + \sqrt{3}}{11}$ **30.** $\frac{6\sqrt{2} - \sqrt{3}}{23}$ **31.** $22 + 7i$

32. $1 - i$ **33.** $\frac{1}{5} - \frac{7}{5}i$ **34.** $-\frac{3}{4} + \frac{1}{4}i\sqrt{3}$ **35.** $\{-5, 9\}$ **36.** $\left\{-\frac{7}{4}\right\}$

37. $\{-8, 8\}$ **38.** $\left\{\pm\frac{4}{3}i\right\}$ **39.** $\{3\}$ **40.** $\{5\}$ **41.** $\frac{3\sqrt{2}}{2}$ feet

42. 25 and 36   **43.** Length 6 ft, width 4 ft   **44.** 39.53 AU, 164.97 years

45. a) 5.2 knots, 6.5 knots   **b)** 49 ft

Making Connections   A Review of Chapters 1–9

1. 7   **2.** -5 **3.** 57   **4.** 11   **5.** -29 **6.** -4 **7.** 1   **8.** -2

9. 0   **10.** 17   **11.** $\{0, 1, \sqrt{4}\}$ **12.** $\{1, \sqrt{4}\}$ **13.** $\{-5, 0, 1, \sqrt{4}\}$

14. $\left\{-5, -\frac{1}{9}, 0, 1, \sqrt{4}, 2.99, \frac{33}{4}\right\}$ **15.** $\{-\sqrt[3]{2}, \pi\}$

16. All except $2 + 3i$ **17.** $\{2 + 3i\}$ **18.** All   **19.** Additive

20. Multiplicative   **21.** Commutative   **22.** Commutative

23. Associative   **24.** Distributive   **25.** Additive   **26.** Multiplicative

27. $\left\{-\frac{4}{7}\right\}$ **28.** $\left\{\frac{3}{2}\right\}$

29. $(-\infty, -3) \cup (-2, \infty)$

-5 -4 -3 -2 -1 0

30. $\left\{\frac{3}{2}\right\}$ **31.** $(-\infty, 1)$ **32.** $\varnothing$

-3 -2 -1 0 1 2 3

33. $\{9\}$ **34.** $\varnothing$ **35.** $\{-12, -2\}$ **36.** $\left\{\frac{1}{16}\right\}$

37. $(-6, \infty)$

-8 -7 -6 -5 -4 -3 -2

38. $\left\{-\frac{1}{64}, \frac{1}{64}\right\}$

39. $\left\{-\frac{\sqrt{3}}{3}, \frac{\sqrt{3}}{3}\right\}$ **40.** R **41.** $\left(-\frac{1}{3}, 3\right)$

$-\frac{1}{3}$

-2 -1 0 1 2 3 4

42. $\left\{\frac{1}{3}\right\}$ **43.** $\{82\}$ **44.** $\left\{\frac{6}{5}, \frac{12}{5}\right\}$ **45.** $\{100\}$ **46.** R **47.** $\{4\sqrt{30}\}$

48. $\{400\}$ **49.** $\left\{\frac{13 + 9\sqrt{2}}{3}\right\}$ **50.** $\{-3\sqrt{2}, 3\sqrt{2}\}$ **51.** $\{5\}$

52. $\{7 + 3\sqrt{6}\}$ **53.** $\{-2, 3\}$ **54.** $\{-5, 2\}$ **55.** $\{-2, 3\}$ **56.** $\left\{\frac{1}{2}, 3\right\}$

57. 3   **58.** -2 **59.** $\frac{1}{2}$ **60.** $\frac{1}{3}$

61. a) 48.5 cm^3 **b)** 14%   **c)** 56 cm^3

Chapter 10

Section 10.1 Warm-Ups

1. Factoring, even-root, completing   **2.** Even-root property
3. Middle   **4.** Leading coefficient   **5.** False   **6.** False   **7.** True
8. True   **9.** False   **10.** False   **11.** True   **12.** True

Section 10.1 Exercises

1. $\{-2, 3\}$ **3.** $\{-5, 3\}$ **5.** $\left\{-1, \frac{3}{2}\right\}$ **7.** $\{-7\}$ **9.** $\{-4, 4\}$

11. $\{-9, 9\}$ **13.** $\left\{-\frac{4}{3}, \frac{4}{3}\right\}$ **15.** $\{-1, 7\}$ **17.** $\{-1 - \sqrt{5}, -1 + \sqrt{5}\}$

19. $\left\{\frac{3 - \sqrt{7}}{2}, \frac{3 + \sqrt{7}}{2}\right\}$ **21.** $x^2 + 2x + 1$ **23.** $x^2 - 3x + \frac{9}{4}$

25. $y^2 + \frac{1}{4}y + \frac{1}{64}$ **27.** $x^2 + \frac{2}{3}x + \frac{1}{9}$ **29.** $(x + 4)^2$ **31.** $\left(y - \frac{5}{2}\right)^2$

33. $\left(z - \frac{2}{7}\right)^2$ **35.** $\left(t + \frac{3}{10}\right)^2$ **37.** $\{-3, 5\}$ **39.** $\{-5, 7\}$

41. $\{-4, 5\}$ **43.** $\{-7, 2\}$ **45.** $\left\{-1, \frac{3}{2}\right\}$

47. $\{-2 - \sqrt{10}, -2 + \sqrt{10}\}$ **49.** $\{-4 - 2\sqrt{5}, -4 + 2\sqrt{5}\}$

51. $\left\{\frac{-5 + \sqrt{5}}{2}, \frac{-5 - \sqrt{5}}{2}\right\}$ **53.** $\left\{\frac{1 - \sqrt{2}}{2}, \frac{1 + \sqrt{2}}{2}\right\}$

55. $\left\{\frac{-3 - \sqrt{41}}{4}, \frac{-3 + \sqrt{41}}{4}\right\}$ **57.** $\{4\}$ **59.** $\left\{\frac{1 + \sqrt{17}}{8}\right\}$

61. $\{1, 6\}$ **63.** $\{-2 - \sqrt{2}, -2 + \sqrt{2}\}$ **65.** $\{-1 - 2i, -1 + 2i\}$

67. $\{3 + i\sqrt{2}, 3 - i\sqrt{2}\}$ **69.** $\left\{\pm\frac{i\sqrt{2}}{2}\right\}$ **71.** $\{-2i\sqrt{3}, 2i\sqrt{3}\}$

73. $\left\{\frac{2 \pm i}{5}\right\}$ **75.** $\{\pm 11i\}$ **77.** $\left\{-\frac{5}{2}i, \frac{5}{2}i\right\}$ **79.** $\{-2, 1\}$

81. $\left\{\frac{-2 - \sqrt{19}}{5}, \frac{-2 + \sqrt{19}}{5}\right\}$ **83.** $\{-6, 4\}$ **85.** $\{2 \pm 3i\}$

87. $\{-2, 3\}$ **89.** $\{3 - i, 3 + i\}$ **91.** $\{6\}$

93. $\left\{\frac{9 - \sqrt{65}}{2}, \frac{9 + \sqrt{65}}{2}\right\}$ **95.** $\{-5, 3\}$ **97.** $\varnothing$

99. 136.9 ft/sec   **101.** 12   **103.** c   **107.** $\{4.56, 2.74\}$ **109.** $\{3.53\}$

Section 10.2 Warm-Ups

1. Quadratic   **2.** Discriminant   **3.** Complex   **4.** One   **5.** Two
6. Two   **7.** True   **8.** False   **9.** True   **10.** True   **11.** False
12. True

Section 10.2 Exercises

1. $\{1, 2\}$ **3.** $\{-3, -2\}$ **5.** $\{-3, 2\}$ **7.** $\left\{-\frac{1}{3}, \frac{3}{2}\right\}$ **9.** $\left\{\frac{1}{2}\right\}$

11. $\left\{\frac{1}{3}\right\}$ **13.** $\left\{-\frac{3}{4}\right\}$ **15.** $\{-4 \pm \sqrt{10}\}$ **17.** $\left\{\frac{-5 \pm \sqrt{29}}{2}\right\}$

19. $\left\{\frac{3 \pm \sqrt{7}}{2}\right\}$ **21.** $\left\{\frac{3 \pm i}{2}\right\}$ **23.** $\left\{\frac{3 \pm i\sqrt{39}}{4}\right\}$ **25.** $\{5 \pm i\}$

27. 28, 2   **29.** $-23, 0$ **31.** 0, 1   **33.** $-\frac{3}{4}, 0$ **35.** 97, 2

37. 0, 1   **39.** 140, 2   **41.** 1, 2   **43.** $\{-2 \pm 2\sqrt{2}\}$ **45.** $\left\{-2, \frac{1}{2}\right\}$

47. $\left\{\frac{-1 \pm \sqrt{13}}{3}\right\}$ **49.** $\{0\}$ **51.** $\left\{\frac{13}{9}, \frac{17}{9}\right\}$ **53.** $\{\pm 5\sqrt{3}\}$

55. $\{4 \pm 2i\}$ **57.** $\{2 \pm i\sqrt{6}\}$ **59.** $\left\{-\frac{3}{4}, \frac{5}{2}\right\}$ **61.** $\{-4.474, 1.274\}$

63. $\{3.7\}$ **65.** $\{-2.979, -0.653\}$ **67.** $\{-4792.983, -0.017\}$

69. $\{-0.079, 0.078\}$ **71.** $\dfrac{1 + \sqrt{65}}{2}$ and $\dfrac{-1 + \sqrt{65}}{2}$, or 4.5 and 3.5

73. $3 + \sqrt{5}$ and $3 - \sqrt{5}$, or 5.2 and 0.8

75. $W = \dfrac{-1 + \sqrt{5}}{2} \approx 0.6$ ft, $L = \dfrac{1 + \sqrt{5}}{2} \approx 1.6$ ft

77. $W = -2 + \sqrt{14} \approx 1.7$ ft, $L = 2 + \sqrt{14} \approx 5.7$ ft

79. 3 sec   **81.** $\dfrac{5 + \sqrt{105}}{16}$ or 1.0 sec   **83.** 7.0 sec   **85.** 4 in.

87. 4   **89.** 250 melons   **95.** 2   **97.** 0   **99 .** 0

Section 10.3 Warm-Ups

1. Factor   **2.** Prime   **3.** Quadratic   **4.** Solutions   **5.** True
6. True   **7.** False   **8.** True   **9.** False   **10.** False   **11.** True
12. False

Section 10.3 Exercises

1. $x^2 + 4x - 21 = 0$ **3.** $x^2 - 5x + 4 = 0$ **5.** $x^2 - 5 = 0$
7. $x^2 + 16 = 0$ **9.** $x^2 + 2 = 0$ **11.** $6x^2 - 5x + 1 = 0$
13. Prime   **15.** Prime   **17.** Prime   **19.** $(3x - 4)(2x + 9)$ **21.** Prime

23. $(4x - 15)(2x + 3)$ **25.** $\{-1, 5\}$ **27.** $\left\{-\frac{3}{2}, \frac{3}{2}\right\}$

29. $\left\{\frac{-3 \pm \sqrt{5}}{2}\right\}$ **31.** $\{\pm 2, \pm 3\}$ **33.** $\{1, 3\}$ **35.** $\{\pm\sqrt{5}, \pm 3\}$

37. $\{-2, 1\}$ **39.** $\{0, \pm 3\}$ **41.** $\{-1 \pm \sqrt{5}, -3, 1\}$
43. $\{-3, -2, 1, 2\}$ **45.** $\{1, 4\}$ **47.** $\{-27, -1\}$ **49.** $\{16, 81\}$

51. $\{9\}$ **53.** $\left\{-\frac{1}{3}, \frac{1}{2}\right\}$ **55.** $\{64\}$ **57.** $\left\{\frac{2}{3}, \frac{3}{2}\right\}$

59. $\left\{\pm\frac{\sqrt{14}}{2}, \pm\frac{\sqrt{38}}{2}\right\}$ **61.** $\{-1 + \sqrt{2}, -1 - \sqrt{2}\}$ **63.** $\{\pm 2i\}$

65. $\{\pm i\sqrt{2}, \pm 2i\}$ **67.** $\{\pm 2, \pm 2i\}$ **69.** $\left\{\pm\frac{1}{2}, \pm\frac{i}{2}\right\}$

71. $\left\{\frac{1 \pm i\sqrt{3}}{2}, -1\right\}$ **73.** $\{1 \pm i\sqrt{3}, -2\}$ **75.** $\left\{\frac{1 \pm 2i}{5}\right\}$

77. $\{1 \pm i\}$ **79.** 2:00 P.M.
81. Before $-5 + \sqrt{265}$ or 11.3 mph, after $-9 + \sqrt{265}$ or 7.3 mph

83. Andrew $\dfrac{13 + \sqrt{265}}{2}$ or 14.6 hours, John $\dfrac{19 + \sqrt{265}}{2}$ or 17.6 hours

85. Length $5 + 5\sqrt{41}$ or 37.02 ft, width $-5 + 5\sqrt{41}$ or 27.02 ft
87. $14 + 2\sqrt{58}$ or 29.2 hours   **89.** $-5 + 5\sqrt{5}$ or 6.2 meters
93. $\{1, 2\}$ **95.** $\{-4.25, -3.49, 0.49, 1.25\}$

Mid-Chapter Quiz 10.1–10.3

1. $\{-4, 8\}$ **2.** $\left\{\frac{1}{2}, \frac{1}{3}\right\}$ **3.** $\left\{\pm\frac{4}{5}\right\}$ **4.** $\{-3 \pm \sqrt{6}\}$

5. $\{2 \pm \sqrt{5}\}$ **6.** $\left\{-1, \frac{1}{2}\right\}$ **7.** $\left\{\frac{1}{2}, 2\right\}$ **8.** $\left\{\frac{-2 \pm \sqrt{2}}{2}\right\}$

9. $\{-5, -4\}$ **10.** $\{\pm 1, \pm\sqrt{5}\}$ **11.** $\{16\}$ **12.** $\{14\}$
13. $\{5 \pm i\}$ **14.** -59 **15.** $x^2 - 5x - 24 = 0$

Section 10.4 Warm-Ups

1. Parabola   **2.** Upward   **3.** Downward   **4.** Vertex   **5.** $-b/(2a)$
6. y-coordinate   **7.** True   **8.** False   **9.** True   **10.** True
11. False   **12.** True   **13.** True   **14.** True

Section 10.4 Exercises

1. $(3, -6), (4, 0), (-3, 0)$ **3.** $(4, -128), (0, 0), (2, 0)$ **5.** Upward
7. Downward   **9.** Upward

11.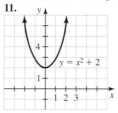

13.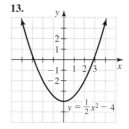

15.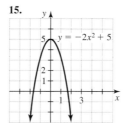

17.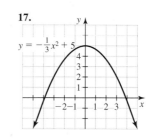

19.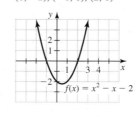

21. $(0, -9)$ **23.** $(2, -3)$

25. $(5, 51)$ **27.** $\left(\frac{1}{2}, \frac{3}{4}\right)$

29. $(0, 16), (-4, 0), (4, 0)$

31. $(0, -15), (-3, 0), (5, 0)$

33. $(0, -9), \left(\frac{3}{2}, 0\right)$

35. Vertex $\left(\frac{1}{2}, -\frac{9}{4}\right)$, intercepts $(0, -2), (-1, 0), (2, 0)$

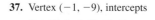

37. Vertex $(-1, -9)$, intercepts $(0, -8), (-4, 0), (2, 0)$

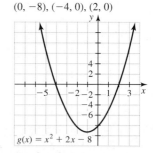

39. Vertex $(-2, 1)$, intercepts $(0, -3), (-1, 0), (-3, 0)$

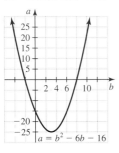

41. Vertex $\left(\dfrac{3}{2}, \dfrac{25}{4}\right)$, intercepts $(0, 4), (4, 0), (-1, 0)$

43. Vertex $(3, -25)$, intercepts $(0, -16), (8, 0), (-2, 0)$

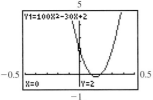

45. Minimum -8
47. Maximum 14
49. Minimum 2
51. Maximum 2
53. Maximum 64 feet

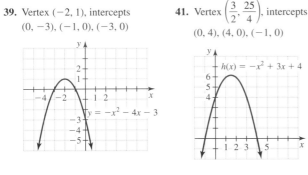

55. 100   **57.** 625 square meters   **59.** 2 P.M.   **61.** 15 meters, 25 meters
63. The graph of $y = ax^2$ gets narrower as a gets larger.
65. The graph of $y = x^2$ has the same shape as $x = y^2$.
67. Answers may vary.

a)

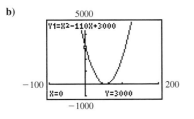

b)

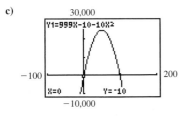

c)

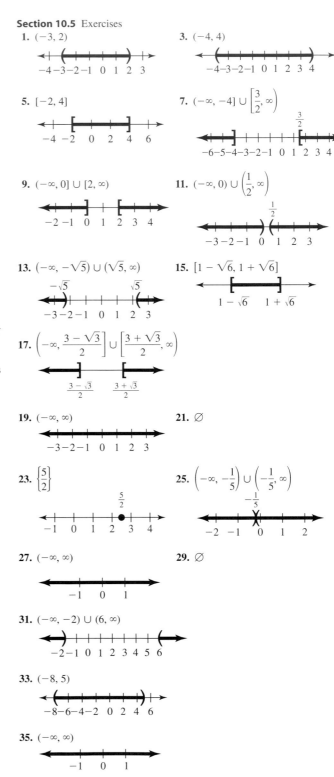

1. Quadratic   **2.** Graphical, test-point   **3.** False   **4.** True
5. True   **6.** False   **7.** False   **8.** True

Section 10.5 Exercises
1. $(-3, 2)$

3. $(-4, 4)$

5. $[-2, 4]$

7. $(-\infty, -4] \cup \left[\dfrac{3}{2}, \infty\right)$

9. $(-\infty, 0] \cup [2, \infty)$

11. $(-\infty, 0) \cup \left(\dfrac{1}{2}, \infty\right)$

13. $(-\infty, -\sqrt{5}) \cup (\sqrt{5}, \infty)$

15. $[1 - \sqrt{6}, 1 + \sqrt{6}]$

17. $\left(-\infty, \dfrac{3 - \sqrt{3}}{2}\right] \cup \left[\dfrac{3 + \sqrt{3}}{2}, \infty\right)$

19. $(-\infty, \infty)$

21. $\varnothing$

23. $\left\{\dfrac{5}{2}\right\}$

25. $\left(-\infty, -\dfrac{1}{5}\right) \cup \left(-\dfrac{1}{5}, \infty\right)$

27. $(-\infty, \infty)$

29. $\varnothing$

31. $(-\infty, -2) \cup (6, \infty)$

33. $(-8, 5)$

35. $(-\infty, \infty)$

37. $(0, 2)$

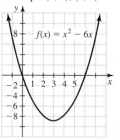

39. $\left(-\infty, 3 - \sqrt{5}\right) \cup \left(3 + \sqrt{5}, \infty\right)$

41. $(-\infty, \infty)$

43. $\left[\dfrac{3 - 3\sqrt{5}}{2}, \dfrac{3 + 3\sqrt{5}}{2}\right]$

45. $(-\infty, 0) \cup (0, \infty)$ **47.** $(-\infty, \infty)$ **49.** $[-3, 3]$ **51.** $(-4, 4)$

53. $(-\infty, 0] \cup [4, \infty)$ **55.** $\left(-\dfrac{3}{2}, \dfrac{5}{3}\right)$ **57.** $(-\infty, -2] \cup [6, \infty)$

59. $(-\infty, -3) \cup (5, \infty)$ **61.** $(-\infty, -4] \cup [2, \infty)$ **63.** $(-27.58, -0.68)$

65. Greater than 5, or 6, 7, 8, . . .   **67.** 4 seconds

69. a) 900 ft   **b)** 3 seconds   **c)** 3 seconds

71. a) (h, k) **b)** $(-\infty, h) \cup (k, \infty)$ **c)** $(-k, -h)$

     d) $(-\infty, -k] \cup [-h, \infty)$

Enriching Your Mathematical Word Power

1. Quadratic   **2.** Quadratic   **3.** Perfect   **4.** Completing

5. Quadratic   **6.** Discriminant   **7.** Parabola   **8.** Form

9. Quadratic   **10.** Test

Review Exercises

1. $\{-3, 5\}$ **3.** $\left\{-3, \dfrac{5}{2}\right\}$ **5.** $\{-5, 5\}$ **7.** $\left\{\dfrac{3}{2}\right\}$ **9.** $\{\pm 2\sqrt{3}\}$

11. $\{-2, 4\}$ **13.** $\left\{\dfrac{4 \pm \sqrt{3}}{2}\right\}$ **15.** $\left\{\pm\dfrac{3}{2}\right\}$ **17.** $\{2, 4\}$ **19.** $\{2, 3\}$

21. $\left\{\dfrac{1}{2}, 3\right\}$ **23.** $\{-2 \pm \sqrt{3}\}$ **25.** $\{-2, 5\}$ **27.** $\left\{-\dfrac{1}{3}, \dfrac{3}{2}\right\}$

29. $\{-2 \pm \sqrt{2}\}$ **31.** $\left\{\dfrac{5 \pm \sqrt{13}}{6}\right\}$ **33.** $0, 1$ **35.** $-19, 0$

37. $17, 2$ **39.** $\left\{\dfrac{2 \pm i\sqrt{2}}{2}\right\}$ **41.** $\left\{\dfrac{3 \pm i\sqrt{15}}{4}\right\}$ **43.** $\left\{\dfrac{-1 \pm i\sqrt{5}}{3}\right\}$

45. $\{-3 \pm i\sqrt{7}\}$ **47.** $(4x + 1)(2x - 3)$ **49.** Prime

51. $(4y - 5)(2y + 5)$ **53.** $x^2 + 9x + 18 = 0$ **55.** $x^2 - 50 = 0$

57. $\{-2, 1\}$ **59.** $\{\pm 2, \pm 3\}$ **61.** $\{-6, -5, 2, 3\}$ **63.** $\{-2, 8\}$

65. $\left\{-\dfrac{1}{9}, \dfrac{1}{4}\right\}$ **67.** $\{16, 81\}$

69. Vertex $(3, -9)$,   **71.** Vertex $(2, -16)$, intercepts
intercepts $(0, 0)$, $(6, 0)$ $(0, -12)$, $(-2, 0)$, and $(6, 0)$

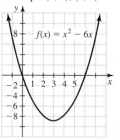
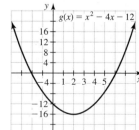

73. Vertex $(2, 8)$,   **75.** Vertex $(1, 4)$, intercepts
intercepts $(0, 0)$, $(4, 0)$ $(0, 3)$, $(-1, 0)$, $(3, 0)$

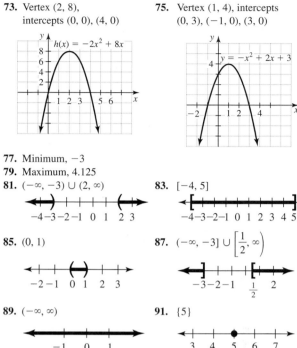

77. Minimum, -3

79. Maximum, 4.125

81. $(-\infty, -3) \cup (2, \infty)$ **83.** $[-4, 5]$

85. $(0, 1)$ **87.** $\left(-\infty, -3\right] \cup \left[\dfrac{1}{2}, \infty\right)$

89. $(-\infty, \infty)$ **91.** $\{5\}$

93. $\varnothing$ **95.** $\left\{\dfrac{5}{12}\right\}$ **97.** $\left\{\dfrac{-3 \pm \sqrt{5}}{2}\right\}$ **99.** $\left\{\dfrac{4 \pm 2i}{3}\right\}$

101. $\left\{\dfrac{5}{2}\right\}$ **103.** $\left\{-2, -\dfrac{1}{4}\right\}$ **105.** $\{625, 10,000\}$

107. $-2 + 2\sqrt{2}$ and $2 + 2\sqrt{2}$, or 0.83 and 4.83

109. Width $\dfrac{4 + \sqrt{706}}{2}$ or 15.3 inches, height $\dfrac{-4 + \sqrt{706}}{2}$ or 11.3 inches

111. 2 inches   **113.** Width 5 ft, length 9 ft   **115.** $\$20.40$, 400

117. 0.5 second and 1.5 seconds   **119.** 1.618

Chapter 10 Test

1. $-7, 0$ **2.** $13, 2$ **3.** $0, 1$ **4.** $\left\{-3, \dfrac{1}{2}\right\}$ **5.** $\{-3 \pm \sqrt{3}\}$

6. $\{-5\}$ **7.** $\left\{-2, \dfrac{3}{2}\right\}$ **8.** $\{-4, 3\}$ **9.** $\{\pm 1, \pm 2\}$ **10.** $\{11, 27\}$

11. $\{\pm 6i\}$ **12.** $\{-3 \pm i\}$ **13.** $\left\{\dfrac{1 \pm i\sqrt{11}}{6}\right\}$

14. Vertex $(0, 16)$,   **15.** Vertex $(1.5, -2.25)$,
intercepts $(0, 16)$, $(-4, 0)$, $(4, 0)$,   intercepts $(0, 0)$, $(3, 0)$,
maximum y-value 16   minimum y-value -2.25

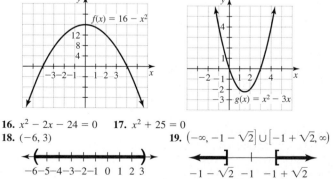

16. $x^2 - 2x - 24 = 0$ **17.** $x^2 + 25 = 0$

18. $(-6, 3)$ **19.** $\left(-\infty, -1 - \sqrt{2}\right] \cup \left[-1 + \sqrt{2}, \infty\right)$

20. $(-\infty, \infty)$ **21.** $\varnothing$

22. Width $-1 + \sqrt{17}$ ft, length $1 + \sqrt{17}$ ft   **23.** $\dfrac{5 + \sqrt{37}}{2}$ or 5.5 hours

24. 36 feet

Making Connections   A Review of Chapters 1–10

1. $\dfrac{4}{9}$ **2.** $\dfrac{2}{3}$ **3.** 11   **4.** 4   **5.** 4   **6.** 4   **7.** $\dfrac{1}{4}$ **8.** $\dfrac{1}{125}$

9. $(y - 3)(y + 100)$ **10.** $(5y - 3)(4y + 1)$ **11.** $6a(a - 5)^2$

12. $(b^2 + 4)(b + 2)$ **13.** $(a - y^2)(b - y^2)$

14. $-2(m + 2)(m^2 - 2m + 4)$

15. $\left\{\dfrac{15}{2}\right\}$ **16.** $\left\{\pm\dfrac{\sqrt{30}}{2}\right\}$ **17.** $\left\{-3, \dfrac{5}{2}\right\}$ **18.** $\left\{\dfrac{-2 \pm \sqrt{34}}{2}\right\}$

19. $\left\{-\dfrac{7}{2}, -2\right\}$ **20.** $\left\{-3, \dfrac{1}{4}, \dfrac{-11 \pm \sqrt{73}}{8}\right\}$ **21.** $\{9\}$ **22.** $\left\{-\dfrac{3}{2}, \dfrac{13}{2}\right\}$

23. $(-4, \infty)$ **24.** $\left[\dfrac{1}{2}, 5\right]$ **25.** $(-\infty, 0] \cup [1, \infty)$ **26.** $\varnothing$

27. $[-3, 2)$ **28.** $(-\infty, \infty)$ **29.** $y = \dfrac{2}{3}x - 3$

30. $y = -\dfrac{1}{2}x + 2$ **31.** $y = \dfrac{-c \pm \sqrt{c^2 - 12d}}{6}$

32. $y = \dfrac{n \pm \sqrt{n^2 + 4mw}}{2m}$ **33.** $y = \dfrac{5}{6}x - \dfrac{25}{12}$ **34.** $y = -\dfrac{2}{3}x + \dfrac{17}{3}$

35. $\dfrac{4}{3}$ **36.** $-\dfrac{11}{7}$ **37.** -2 **38.** $\dfrac{58}{5}$ **39.** 40,000, 38,000, $32.50

40. $800,000, $950,000, $40 or $80, $60

Chapter 11

Section 11.1 Warm-Ups

1. Relation   **2.** Function   **3.** Function   **4.** Domain   **5.** Range
6. Dependent   **7.** True   **8.** False   **9.** True   **10.** True
11. False   **12.** True   **13.** False   **14.** True   **15.** True

Section 11.1 Exercises

1. Yes   **3.** No   **5.** Yes   **7.** No   **9.** $C = 0.50t + 5$
11. $T = 1.09S$ **13.** $C = 2\pi r$ **15.** $P = 4s$ **17.** $A = 5h$
19. Yes   **21.** Yes   **23.** No   **25.** Yes   **27.** Yes   **29.** No
31. No   **33.** Yes   **35.** $(2, 1), (2, -1)$ **37.** $(8, 4), (8, -4)$
39. $(0, 1), (0, -1)$ **41.** $(16, 2), (16, -2)$ **43.** $(3, 1), (3, -1)$
45. Yes   **47.** No   **49.** Yes   **51.** No   **53.** Yes   **55.** No
57. No   **59.** Yes   **61.** No   **63.** No   **65.** Yes   **67.** No
69. $\{4, 7\}, \{1\}$ **71.** $\{2\}, \{3, 5, 7\}$ **73.** $(-\infty, \infty), (-\infty, \infty)$
75. $(-\infty, \infty), (-\infty, \infty)$ **77.** $[2, \infty), [0, \infty)$ **79.** $[0, \infty), [0, \infty)$
81. -2 **83.** 10   **85.** -12 **87.** 1   **89.** 2.236
91. 2   **93.** 0   **95.** -10 **97.** **a)** 192 ft   **b)** 0 ft
99. $A = s^2$ or $A(s) = s^2$ **101.** $C(x) = 3.98x$, $11.94
103. $C(n) = 14.95 + 0.50n$, $17.95

Section 11.2 Warm-Ups

1. Linear   **2.** Constant   **3.** Identity   **4.** Parabola
5. Absolute value   **6.** True   **7.** True   **8.** False   **9.** True
10. True   **11.** False   **12.** True

Section 11.2 Exercises

1. $(-\infty, \infty), \{-2\}$

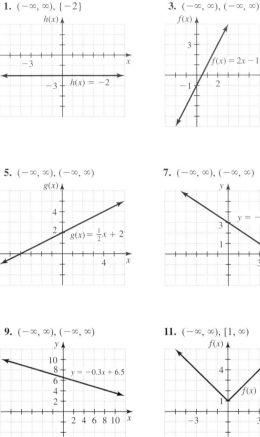

3. $(-\infty, \infty), (-\infty, \infty)$

$f(x) = 2x - 1$

5. $(-\infty, \infty), (-\infty, \infty)$

$g(x) = \frac{1}{2}x + 2$

7. $(-\infty, \infty), (-\infty, \infty)$

$y = -\frac{2}{3}x + 3$

9. $(-\infty, \infty), (-\infty, \infty)$

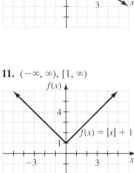

$y = -0.3x + 6.5$

11. $(-\infty, \infty), [1, \infty)$

$f(x) = |x| + 1$

13. $(-\infty, \infty), [0, \infty)$

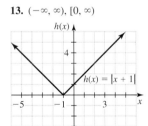

$h(x) = |x + 1|$

15. $(-\infty, \infty), [0, \infty)$

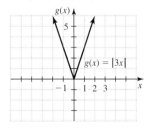

$g(x) = |3x|$

17. $(-\infty, \infty), [0, \infty)$

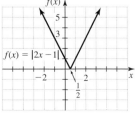

$f(x) = |2x - 1|$

19. $(-\infty, \infty), [1, \infty)$

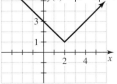

$f(x) = |x - 2| + 1$

21. $(-\infty, \infty)$, $[0, \infty)$

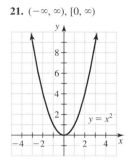

23. $(-\infty, \infty)$, $[2, \infty)$

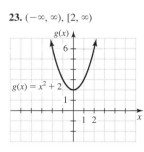

41.

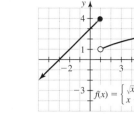

43.

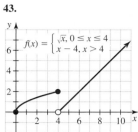

25. $(-\infty, \infty)$, $[0, \infty)$

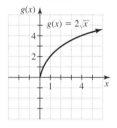

27. $(-\infty, \infty)$, $(-\infty, 6]$

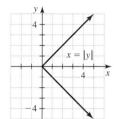

45. $[0, \infty)$, $(-\infty, \infty)$

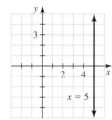

47. $(-\infty, 0]$, $(-\infty, \infty)$

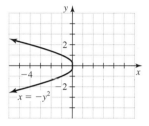

29. $[0, \infty)$, $[0, \infty)$

31. $[1, \infty)$, $[0, \infty)$

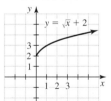

49. $\{5\}$, $(-\infty, \infty)$

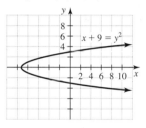

51. $[-9, \infty)$, $(-\infty, \infty)$

33. $[0, \infty)$, $(-\infty, 0]$

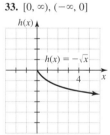

35. $[0, \infty)$, $[2, \infty)$

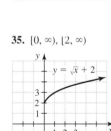

53. $[0, \infty)$, $[0, \infty)$

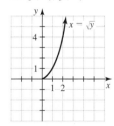

55. $[0, \infty)$, $(-\infty, \infty)$

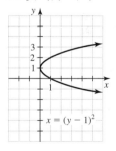

37.

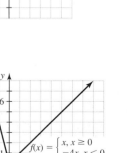

39.

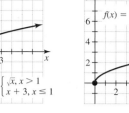

57. $(-\infty, \infty)$, $(-\infty, 1]$

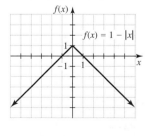

59. $(-\infty, \infty)$, $[-1, \infty)$

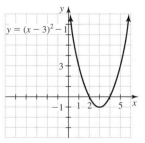

61. $(-\infty, \infty)$, $[1, \infty)$

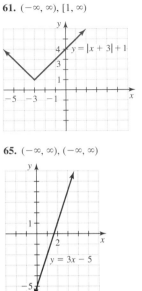

63. $[0, \infty)$, $[-3, \infty)$

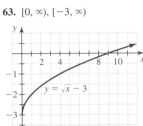

9. $(-\infty, \infty)$, $[0, \infty)$

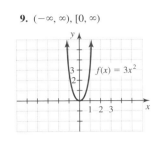

11. $(-\infty, \infty)$, $(-\infty, \infty)$

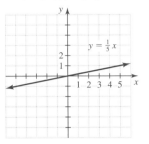

65. $(-\infty, \infty)$, $(-\infty, \infty)$

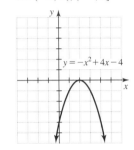

67. $(-\infty, \infty)$, $(-\infty, 0]$

13. $[0, \infty)$, $[0, \infty)$

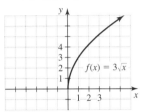

15. $(-\infty, \infty)$, $[0, \infty)$

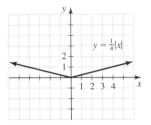

69. Square-root   **71.** Constant   **73.** Quadratic
75. Square-root   **77.** Linear
79. The graph of $f(x) = \sqrt{x^2}$ is the same as the graph of $f(x) = |x|$.
81. For large values of a the graph gets narrower and for smaller values of a the graph gets broader.
83. The graph of $y = (x - h)^2$ moves to the right for $h > 0$ and to the left for $h < 0$.
85. The graph of $y = f(x - h)$ lies to the right of the graph of $y = f(x)$ when $h > 0$.

17.

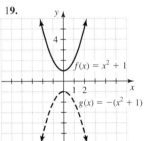

19.

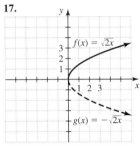

Section 11.3 Warm-Ups
1. Reflection   **2.** Upward translation   **3.** Downward translation
4. Right   **5.** Left   **6.** Stretching   **7.** Shrinking   **8.** True
9. False   **10.** True   **11.** True   **12.** True   **13.** True

Section 11.3 Exercises
1. $(-\infty, \infty)$, $(-\infty, \infty)$

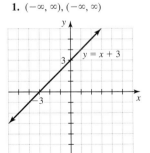

3. $(-\infty, \infty)$, $[0, \infty)$

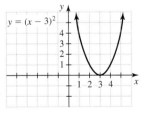

21.

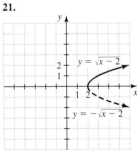

23.

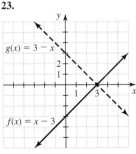

5. $[1, \infty)$, $[0, \infty)$

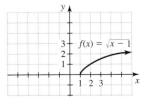

7. $(-\infty, \infty)$, $[0, \infty)$

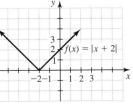

25. $[0, \infty)$, $[1, \infty)$

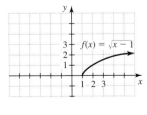

27. $(-\infty, \infty)$, $[-4, \infty)$

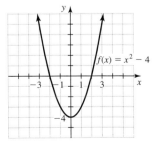

29. $(-\infty, \infty), [2, \infty)$

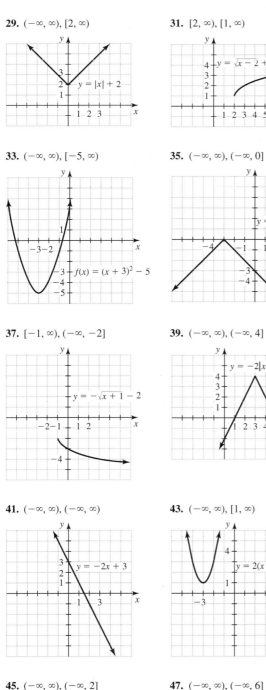

31. $[2, \infty), [1, \infty)$

$y = \sqrt{x - 2} + 1$

33. $(-\infty, \infty), [-5, \infty)$

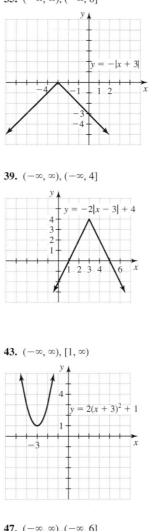

$f(x) = (x + 3)^2 - 5$

35. $(-\infty, \infty), (-\infty, 0]$

$y = -|x + 3|$

37. $[-1, \infty), (-\infty, -2]$

$y = -\sqrt{x + 1} - 2$

39. $(-\infty, \infty), (-\infty, 4]$

$y = -2|x - 3| + 4$

41. $(-\infty, \infty), (-\infty, \infty)$

$y = -2x + 3$

43. $(-\infty, \infty), [1, \infty)$

$y = 2(x + 3)^2 + 1$

45. $(-\infty, \infty), (-\infty, 2]$

$y = -2(x - 4)^2 + 2$

47. $(-\infty, \infty), (-\infty, 6]$

$y = -3(x - 1)^2 + 6$

49. d **51.** e **53.** h **55.** c **57.** $y = x^2 + 8$ **59.** $y = \sqrt{x + 5}$

61. $y = |x + 3| + 5$ **63.** Move f to the right 20 units and upward 30 units.

Mid-Chapter Quiz 11.1–11.3

1. No **2.** Yes **3.** Yes **4.** Yes **5.** Yes **6.** No

7. $[3, \infty), [0, \infty)$ **8.** $\{1, 3, 20, 40\}, \{2, 4, 30\}$

9. $(-\infty, \infty), (-\infty, \infty)$

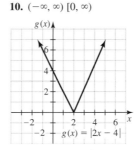

$f(x) = 2x - 4$

10. $(-\infty, \infty) [0, \infty)$

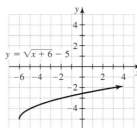

$g(x) = |2x - 4|$

11. $(-\infty, \infty), (-\infty, 3]$

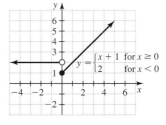

$h(x) = -x^2 + 3$

12. $[-6, \infty), [-5, \infty)$

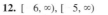

$y = \sqrt{x + 6} - 5$

13. $(-\infty, \infty), [1, \infty)$

$y = \begin{cases} x + 1 & \text{for } x \geq 0 \\ 2 & \text{for } x < 0 \end{cases}$

14. $[-2, \infty), (-\infty, \infty)$

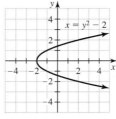

$x = y^2 - 2$

15. $(-\infty, \infty), (-\infty, 3]$

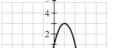

$y = -2(x - 1)^2 + 3$

16. 0 **17.** -9 **18.** $y = (x + 2)^2 + 4$ **19.** $y = -2|x + 3|$

20. $y = \dfrac{1}{2}\sqrt{x + 9} - 5$

Section 11.4 Warm-Ups
1. y-axis   **2.** Origin   **3.** Odd   **4.** Even   **5.** False   **6.** False
7. False   **8.** True   **9.** True   **10.** True

Section 11.4 Exercises
1. $(-1, 0), (0, 1)$

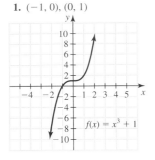

$f(x) = x^3 + 1$

3. $(-3, 0), (3, 0), (0, 0)$

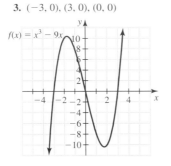

$f(x) = x^3 - 9x$

5. $(0, 0), (-4, 0)$

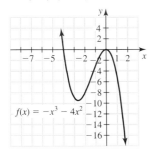

$f(x) = -x^3 - 4x^2$

7. $(-2, 0), (-1, 0), (2, 0), (0, -4)$

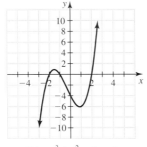

$f(x) = x^3 + x^2 - 4x - 4$

9. $(-3, 0), (3, 0), (0, 27)$

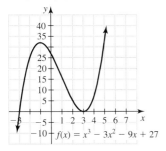

$f(x) = x^3 - 3x^2 - 9x + 27$

11. $(-1, 0), (1, 0), (0, -1)$

$f(x) = x^4 - 1$

13. $(-2, 0), (0, 0), (2, 0)$

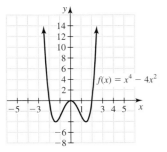

$f(x) = x^4 - 4x^2$

15. $(-2, 0), (-1, 0), (1, 0),$
$(2, 0), (0, 4)$

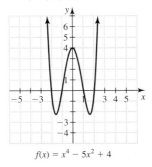

$f(x) = x^4 - 5x^2 + 4$

17. $(-2, 0), (-1, 0), (0, 0), (2, 0)$ **19.** $(-3, 0), (0, 0), (3, 0)$

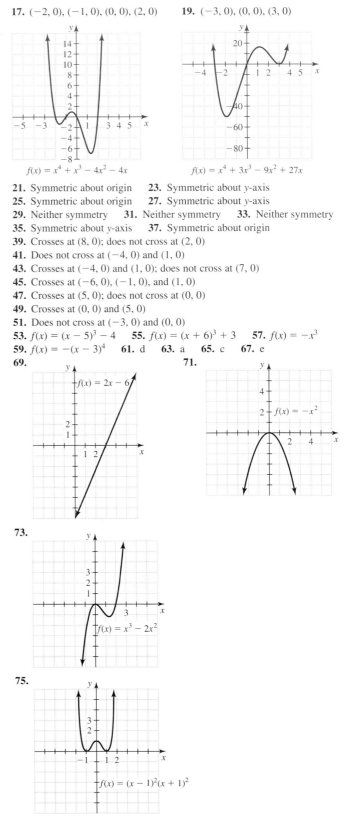

$f(x) = x^4 + x^3 - 4x^2 - 4x$ $f(x) = x^4 + 3x^3 - 9x^2 + 27x$

21. Symmetric about origin   **23.** Symmetric about y-axis
25. Symmetric about origin   **27.** Symmetric about y-axis
29. Neither symmetry   **31.** Neither symmetry   **33.** Neither symmetry
35. Symmetric about y-axis   **37.** Symmetric about origin
39. Crosses at $(8, 0)$; does not cross at $(2, 0)$
41. Does not cross at $(-4, 0)$ and $(1, 0)$
43. Crosses at $(-4, 0)$ and $(1, 0)$; does not cross at $(7, 0)$
45. Crosses at $(-6, 0), (-1, 0),$ and $(1, 0)$
47. Crosses at $(5, 0)$; does not cross at $(0, 0)$
49. Crosses at $(0, 0)$ and $(5, 0)$
51. Does not cross at $(-3, 0)$ and $(0, 0)$
53. $f(x) = (x - 5)^3 - 4$ **55.** $f(x) = (x + 6)^3 + 3$ **57.** $f(x) = -x^3$
59. $f(x) = -(x - 3)^4$ **61.** d   **63.** a   **65.** c   **67.** e
69.

$f(x) = 2x - 6$

71.

$f(x) = -x^2$

73.

$f(x) = x^3 - 2x^2$

75.

$f(x) = (x - 1)^2(x + 1)^2$

77.

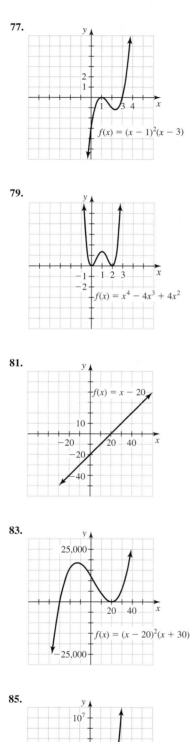

$f(x) = (x - 1)^2(x - 3)$

79.

$f(x) = x^4 - 4x^3 + 4x^2$

81.

$f(x) = x - 20$

83.

$f(x) = (x - 20)^2(x + 30)$

85.

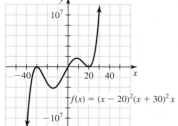

$f(x) = (x - 20)^2(x + 30)^2 x$

87. $(-2, 0) \cup (2, \infty)$ **89.** $(-\infty, -2] \cup [3, 5]$

91. $\{0\} \cup [2, \infty)$ **93.** $(-\infty, \infty)$

95. $\{-1, 1\}$ **97.** $(-\infty, -3) \cup (3, \infty)$

99. $[-2, 3]$ **101.** $(-6, -2) \cup (2, \infty)$

103. $[-3, -1] \cup [1, 3]$

105. $f(x) = x - 3$ **107.** $f(x) = (x + 2)(x - 1)^2$

Section 11.5 Warm-Ups
1. Rational   **2.** Domain   **3.** Vertical   **4.** Horizontal
5. Oblique, slant   **6.** False   **7.** False   **8.** False   **9.** True
10. True   **11.** True   **12.** True

Section 11.5 Exercises
1. $(-\infty, 1) \cup (1, \infty)$ **3.** $(-\infty, 0) \cup (0, \infty)$
5. $(-\infty, -4) \cup (-4, 4) \cup (4, \infty)$
7. Vertical: $x = -4$; horizontal: x-axis
9. Vertical: $x = 4, x = -4$; horizontal: x-axis
11. Vertical: $x = 7$; horizontal: $y = 5$
13. Vertical: $x = 3$; oblique: $y = 2x + 6$ **15.** c   **17.** b   **19.** g   **21.** f
23. $x = -4$, x-axis   **25.** $x = 3, x = -3$, x-axis

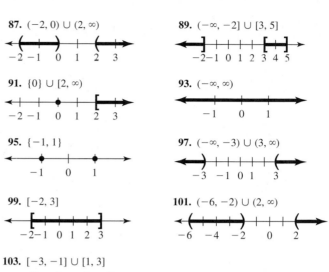

$f(x) = \dfrac{2}{x + 4}$

$f(x) = \dfrac{x}{x^2 - 9}$

27. $x = -3, y = 2$ **29.** y-axis, $y = x - 3$

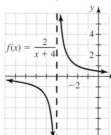

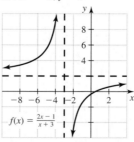

$f(x) = \dfrac{2x - 1}{x + 3}$

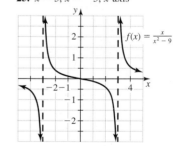

$f(x) = \dfrac{x^2 - 3x + 1}{x}$

31. $x = 1$, $y = 3x + 1$

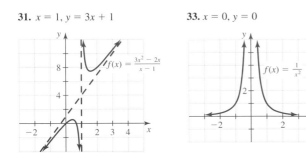

$f(x) = \frac{3x^2 - 2x}{x - 1}$

33. $x = 0$, $y = 0$

$f(x) = \frac{1}{x^2}$

35. $x = -3$, $x = 2$, $y = 0$, $\left(0, \frac{1}{2}\right)$, $\left(\frac{3}{2}, 0\right)$

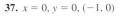
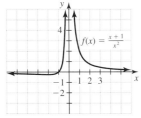

$f(x) = \frac{2x - 3}{x^2 + x - 6}$

37. $x = 0$, $y = 0$, $(-1, 0)$

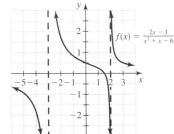

$f(x) = \frac{x + 1}{x^2}$

39. $x = 0$, $x = \pm 3$, $y = 0$, $\left(\frac{1}{2}, 0\right)$

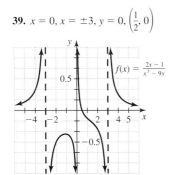

$f(x) = \frac{2x - 1}{x^3 - 9x}$

41. $x = \pm 1$, $y = 0$, $(0, 0)$

$f(x) = \frac{x}{x^2 - 1}$

43. $y = 0$, $(0, 2)$

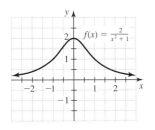

$f(x) = \frac{2}{x^2 + 1}$

45. $x = -1$, $y = x - 1$, $(0, 0)$

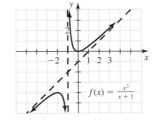

$f(x) = \frac{x^2}{x + 1}$

47. $(0, \infty)$

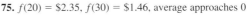

49. $(-\infty, 0) \cup (3, \infty)$

51. $[-2, 0)$

53. $(-\infty, -6) \cup (3, \infty)$

55. $(-\infty, -2) \cup (-1, \infty)$

57. $[-13, -5)$

59. $(-\infty, -2) \cup [4, \infty)$

61. $(-3, \infty)$

63. $(-2, -1) \cup \left(-\frac{1}{2}, \infty\right)$

65. $(-13, -4) \cup (5, \infty)$

67. $(-\infty, -5) \cup (1, 3) \cup (5, \infty)$

69. $[-6, 3) \cup [4, 6)$

71. $(-\infty, \infty)$

73. $\varnothing$

75. $f(20) = \$2.35$, $f(30) = \$1.46$, average approaches 0

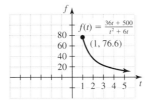

$f(t) = \frac{36t + 500}{t^2 + 6t}$

$(1, 76.6)$

77. a) $y = 25{,}000$ **b)** $\$39{,}000$ **c)** $140{,}000$

d)

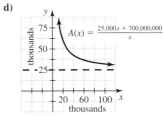

$A(x) = \frac{25{,}000x + 700{,}000{,}000}{x}$

79, 81, and 83. The graph of $f(x)$ is an asymptote for the graph of $g(x)$.

85. $f(x) = 1/x$

87. $f(x) = \dfrac{1}{(x - 3)(x + 1)}$

Section 11.6 Warm-Ups
1. Sum   **2.** Difference   **3.** Product   **4.** Quotient   **5.** Composition
6. True   **7.** True   **8.** True   **9.** False   **10.** True   **11.** False
12. True

Section 11.6 Exercises

1. $x^2 + 2x - 3$ **3.** $4x^3 - 11x^2 + 6x$ **5.** 12   **7.** -30 **9.** -21

11. $\dfrac{13}{8}$ **13.** $y = 6x$ **15.** $y = 6x - 20$ **17.** $y = x + 2$

19. $y = x^2 + 2x$ **21.** $y = x$ **23.** -2 **25.** 5   **27.** 7   **29.** 5

31. 5   **33.** 4   **35.** 22.2992   **37.** $4x^2 - 6x$ **39.** $2x^2 + 6x - 3$

41. x **43.** $4x - 9$ **45.** $\dfrac{x + 9}{4}$ **47.** $F = f \circ h$ **49.** $G = g \circ h$

51. $H = h \circ g$ **53.** $J = h \circ h$ **55.** $K = g \circ g$ **65.** True   **67.** False

69. True   **71.** False   **73.** False   **75.** True   **77. a)** 50 ft^2 **b)** $A = \dfrac{d^2}{2}$

79. $P(x) = -x^2 + 20x - 170$ **81.** $J = 0.025I$

83. a) 397.8   **b)** $D = \dfrac{1.116 \times 10^7}{L^3}$ **c)** Decreases

85. $[0, \infty), [0, \infty), [16, \infty)$ **87.** $[0, \infty), [0, \infty)$

Section 11.7 Warm-Ups

1. Inverse   **2.** Range   **3.** Domain   **4.** One-to-one   **5.** Symmetric
6. Horizontal   **7.** False   **8.** False   **9.** True   **10.** False
11. True   **12.** True   **13.** True   **14.** True

Section 11.7 Exercises

1. Yes, $\{(3, 1), (9, 2)\}$ **3.** No   **5.** Yes, $\{(4, 16), (3, 9), (0, 0)\}$
7. No   **9.** Yes, $\{(0, 0), (2, 2), (9, 9)\}$ **11.** No   **13.** Yes   **15.** Yes

17. Yes   **19.** Yes   **21.** No   **23.** $f^{-1}(x) = \dfrac{x}{5}$ **25.** $g^{-1}(x) = x + 9$

27. $k^{-1}(x) = \dfrac{x + 9}{5}$ **29.** $m^{-1}(x) = \dfrac{2}{x}$ **31.** $f^{-1}(x) = x^3 + 4$

33. $f^{-1}(x) = \dfrac{3}{x} + 4$ **35.** $f^{-1}(x) = \dfrac{x^3 - 7}{3}$ **37.** $f^{-1}(x) = \dfrac{2x + 1}{x - 1}$

39. $f^{-1}(x) = \dfrac{1 + 4x}{3x - 1}$ **41.** $p^{-1}(x) = x^4$ for $x \geq 0$ **43.** $f^{-1}(x) = 2 + \sqrt{x}$

45. $f^{-1}(x) = \sqrt{x - 3}$ **47.** $f^{-1}(x) = x^2 - 2$ for $x \geq 0$

49. $f^{-1}(x) = \dfrac{1}{2}x - \dfrac{3}{2}$ **51.** $f^{-1}(x) = \sqrt{x + 1}$

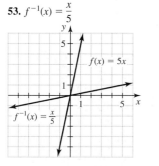

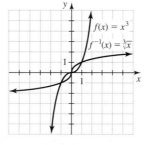

53. $f^{-1}(x) = \dfrac{x}{5}$ **55.** $f^{-1}(x) = \sqrt[3]{x}$

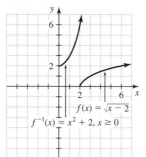

57. $f^{-1}(x) = x^2 + 2$ for $x \geq 0$

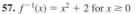

59. $f^{-1}(x) = \dfrac{x}{2}$ **61.** $f^{-1}(x) = \dfrac{x + 1}{2}$ **63.** $f^{-1}(x) = x^3$

65. $f^{-1}(x) = x^3 + 1$ **67.** $f^{-1}(x) = \left(\dfrac{x + 1}{2}\right)^3$ **69.** $(f^{-1} \circ f)(x) = x$

71. $(f^{-1} \circ f)(x) = x$ **73.** $(f^{-1} \circ f)(x) = x$ **75.** $(f^{-1} \circ f)(x) = x$

77. a) 33.5 mph   **b)** Decreases   **c)** $L = \dfrac{S^2}{30}$

79. $T(x) = 1.09x + 125$, $T^{-1}(x) = \dfrac{x - 125}{1.09}$

81. An odd positive integer   **83.** Not inverses

Enriching Your Mathematical Word Power

1. Relation   **2.** Function   **3.** Domain   **4.** Range   **5.** Function
6. Asymptote   **7.** Oblique   **8.** Composition   **9.** One-to-one
10. Vertical   **11.** Horizontal   **12.** Reflection   **13.** Translation
14. Symmetric   **15.** Origin   **16.** Rational

Review Exercises

1. No   **3.** Yes   **5.** Yes   **7.** No   **9.** $\{3, 4, 5\}, \{1, 5, 9\}$
11. $(-\infty, \infty), (-\infty, \infty)$ **13.** $[-5, \infty), [0, \infty)$

15. -5 **17.** -6 **19.** $-\dfrac{21}{4}$

21. $(-\infty, \infty), (-\infty, \infty)$ **23.** $(-\infty, \infty), [-2, \infty)$

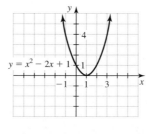

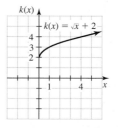

25. $(-\infty, \infty), [0, \infty)$ **27.** $[0, \infty), [2, \infty)$

29. $(-\infty, \infty)$, $(-\infty, 30]$

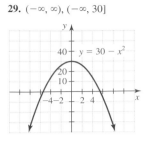

31. $[-4, \infty)$, $[0, \infty)$

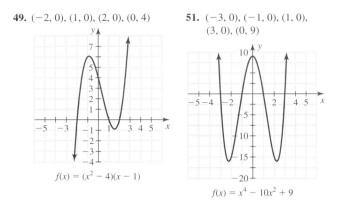

$$f(x) = \begin{cases} \sqrt{x+4}, & -4 \le x \le 0 \\ x+2, & x > 0 \end{cases}$$

49. $(-2, 0)$, $(1, 0)$, $(2, 0)$, $(0, 4)$

$$f(x) = (x^2 - 4)(x - 1)$$

51. $(-3, 0)$, $(-1, 0)$, $(1, 0)$, $(3, 0)$, $(0, 9)$

$$f(x) = x^4 - 10x^2 + 9$$

33. $\{2\}$, $(-\infty, \infty)$

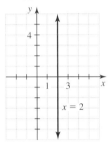

35. $[1, \infty)$, $(-\infty, \infty)$

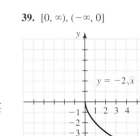

$x = |y| + 1$

53. The graph touches but does not cross the x-axis at $(3, 0)$.
55. The graph crosses the x-axis at $(3, 0)$ and $(-5, 0)$, and touches but does not cross at $(4, 0)$.
57. The graph crosses the x-axis at $(-3, 0)$, $(3, 0)$, and $(8, 0)$.

59. $(-\infty, -2] \cup [0, 2]$

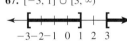

61. $(-\infty, -\sqrt{2}] \cup \{0\} \cup [\sqrt{2}, \infty)$

37. $[0, \infty)$, $[0, \infty)$

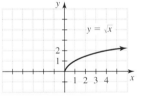

$y = \sqrt{x}$

39. $[0, \infty)$, $(-\infty, 0]$

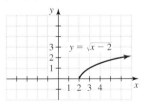

$y = -2\sqrt{x}$

63. $[-\sqrt{2}, \sqrt{2}]$

65. $(-\infty, -3) \cup (1, 3)$

67. $[-3, 1] \cup [3, \infty)$

69. $\left(-\infty, -\dfrac{3}{2}\right) \cup \left(-\dfrac{3}{2}, \infty\right)$

71. $(-\infty, \infty)$

41. $[2, \infty)$, $[0, \infty)$

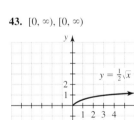

$y = \sqrt{x - 2}$

43. $[0, \infty)$, $[0, \infty)$

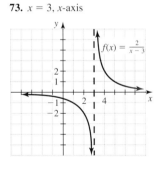

$y = \frac{1}{2}\sqrt{x}$

73. $x = 3$, x-axis

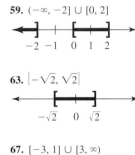

$f(x) = \dfrac{2}{x - 3}$

75. $x = 2$, $x = -2$, x-axis

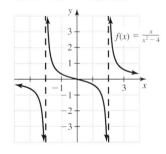

$f(x) = \dfrac{x}{x^2 - 4}$

45. $[-1, \infty)$, $(-\infty, 3]$

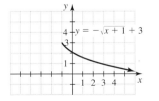

$y = -\sqrt{x + 1} + 3$

47. $(-5, 0)$, $(5, 0)$, $(0, 0)$

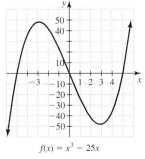

$f(x) = x^3 - 25x$

77. $x = 1$, $y = 2$

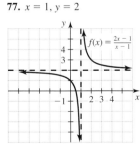

$f(x) = \dfrac{2x - 1}{x - 1}$

79. $x = 2$, $y = x$

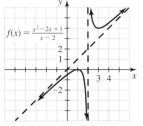

$f(x) = \dfrac{x^2 - 2x + 1}{x - 2}$

81. $(-\infty, -3] \cup (4, \infty)$ **83.** $(-2, 0) \cup (2, \infty)$

85. $(-\infty, -10) \cup (-5, \infty)$ **87.** $(-\infty, 0)$

89. $(-5, 0] \cup (2, 9]$

91. -4 **93.** $\sqrt{2}$ **95.** 99 **97.** 17
99. $3x^3 - x^2 - 10x$ **101.** 20 **103.** $F = f \circ g$
105. $H = g \circ h$ **107.** $I = g \circ g$ **109.** No
111. Yes, $f^{-1}(x) = x/8$
113. Yes, $g^{-1}(x) = \dfrac{x + 6}{13}$
115. Yes, $j^{-1}(x) = \dfrac{x + 1}{x - 1}$ **117.** No
119. $f^{-1}(x) = \dfrac{1}{3}x + \dfrac{1}{3}$ **121.** $f^{-1}(x) = \sqrt[3]{2x}$

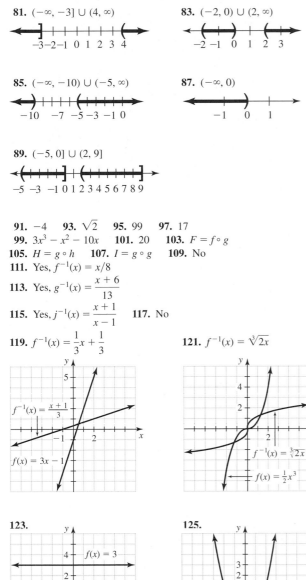

123.   **125.**

127.

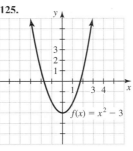

129.

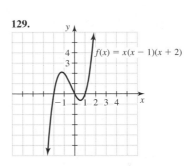

131. $(-\infty, 2)$ **133.** $\left[-\sqrt{3}, \sqrt{3}\right]$ **135.** $\left(-\sqrt{3}, \sqrt{3}\right)$
137. $(-\infty, -2) \cup (0, 1)$ **139.** $(-3, 4]$ **141.** $[-1, 2] \cup [5, \infty)$
143. $(-\infty, -3) \cup (-1, 1)$ **145.** c **147.** b

149. $B = \dfrac{2A}{\pi}$ **151.** $a = 15w - 16$
153. $A = s^2, s = \sqrt{A}$

Chapter 11 Test
1. Yes **2.** 11 **3.** $[7, \infty), [0, \infty)$ **4.** $S = 0.50n + 3$ **5.** 6 ft
6. $(-\infty, \infty), (-\infty, \infty)$ **7.** $(-\infty, \infty), [-4, \infty)$

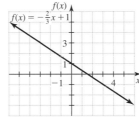   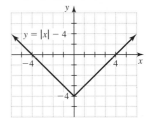

8. $(-\infty, \infty), [-9, \infty)$ **9.** $[0, \infty), (-\infty, \infty)$

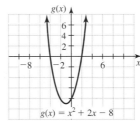   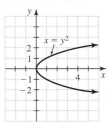

10. $(-\infty, \infty), (-3, \infty)$ **11.** $(-\infty, \infty), (-\infty, 0]$

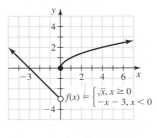   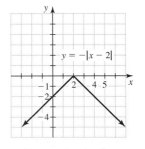

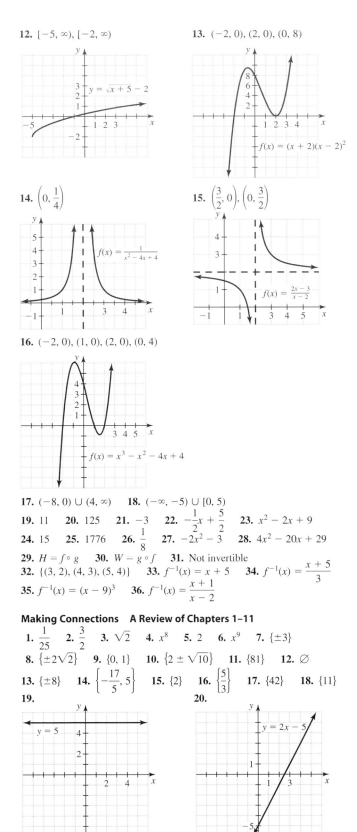

12. $[-5, \infty)$, $[-2, \infty)$

$y = \sqrt{x+5} - 2$

13. $(-2, 0)$, $(2, 0)$, $(0, 8)$

$f(x) = (x+2)(x-2)^2$

14. $\left(0, \dfrac{1}{4}\right)$

$f(x) = \dfrac{1}{x^2 - 4x + 4}$

15. $\left(\dfrac{3}{2}, 0\right)$, $\left(0, \dfrac{3}{2}\right)$

$f(x) = \dfrac{2x-3}{x-2}$

16. $(-2, 0)$, $(1, 0)$, $(2, 0)$, $(0, 4)$

$f(x) = x^3 - x^2 - 4x + 4$

17. $(-8, 0) \cup (4, \infty)$ **18.** $(-\infty, -5) \cup [0, 5)$

19. 11   **20.** 125   **21.** -3 **22.** $-\dfrac{1}{2}x + \dfrac{5}{2}$ **23.** $x^2 - 2x + 9$

24. 15   **25.** 1776   **26.** $\dfrac{1}{8}$ **27.** $-2x^2 - 3$ **28.** $4x^2 - 20x + 29$

29. $H = f \circ g$ **30.** $W = g \circ f$ **31.** Not invertible

32. $\{(3, 2), (4, 3), (5, 4)\}$ **33.** $f^{-1}(x) = x + 5$ **34.** $f^{-1}(x) = \dfrac{x+5}{3}$

35. $f^{-1}(x) = (x - 9)^3$ **36.** $f^{-1}(x) = \dfrac{x+1}{x-2}$

Making Connections   A Review of Chapters 1–11

1. $\dfrac{1}{25}$ **2.** $\dfrac{3}{2}$ **3.** $\sqrt{2}$ **4.** x^8 **5.** 2   **6.** x^9 **7.** $\{\pm 3\}$

8. $\{\pm 2\sqrt{2}\}$ **9.** $\{0, 1\}$ **10.** $\{2 \pm \sqrt{10}\}$ **11.** $\{81\}$ **12.** $\varnothing$

13. $\{\pm 8\}$ **14.** $\left\{-\dfrac{17}{5}, 5\right\}$ **15.** $\{2\}$ **16.** $\left\{\dfrac{5}{3}\right\}$ **17.** $\{42\}$ **18.** $\{11\}$

19.

$y = 5$

20.

$y = 2x - 5$

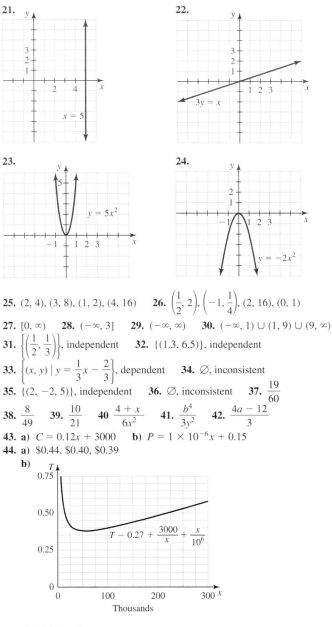

21.

$x = 5$

22.

$3y = x$

23.

$y = 5x^2$

24.

$y = -2x^2$

25. $(2, 4)$, $(3, 8)$, $(1, 2)$, $(4, 16)$ **26.** $\left(\dfrac{1}{2}, 2\right)$, $\left(-1, \dfrac{1}{4}\right)$, $(2, 16)$, $(0, 1)$

27. $[0, \infty)$ **28.** $(-\infty, 3]$ **29.** $(-\infty, \infty)$ **30.** $(-\infty, 1) \cup (1, 9) \cup (9, \infty)$

31. $\left\{\left(\dfrac{1}{2}, \dfrac{1}{3}\right)\right\}$, independent   **32.** $\{(1.3, 6.5)\}$, independent

33. $\left\{(x, y) \mid y = \dfrac{1}{3}x - \dfrac{2}{3}\right\}$, dependent   **34.** $\varnothing$, inconsistent

35. $\{(2, -2, 5)\}$, independent   **36.** $\varnothing$, inconsistent   **37.** $\dfrac{19}{60}$

38. $\dfrac{8}{49}$ **39.** $\dfrac{10}{21}$ **40** $\dfrac{4+x}{6x^2}$ **41.** $\dfrac{b^4}{3y^2}$ **42.** $\dfrac{4a - 12}{3}$

43. a) $C = 0.12x + 3000$ **b)** $P = 1 \times 10^{-6}x + 0.15$

44. a) \$0.44, \$0.40, \$0.39

b)

$T = 0.27 + \dfrac{3000}{x} + \dfrac{x}{10^6}$

Thousands

c) 60,000 miles

d) $[50,000, 60,000]$

Chapter 12

Section 12.1 Warm-Ups

1. Exponential   **2.** Domain   **3.** Natural   **4.** Common
5. One-to-one   **6.** Compound   **7.** Continuously   **8.** False
9. True   **10.** True   **11.** True   **12.** False   **13.** True   **14.** False

Section 12.1 Exercises

1. 16   **3.** 2   **5.** 3   **7.** $\dfrac{1}{3}$ **9.** -1 **11.** $-\dfrac{1}{4}$ **13.** 1   **15.** 100

17. 2.718   **19.** 0.135   **21.** $\dfrac{1}{16}, \dfrac{1}{4}, 1, 4, 16$ **23.** $9, 3, 1, \dfrac{1}{3}, \dfrac{1}{9}$

25.

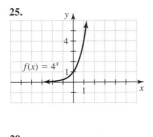

$f(x) = 4^x$

27.

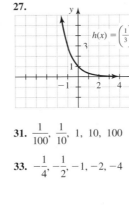

$h(x) = \left(\frac{1}{3}\right)^x$

29.

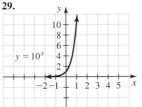

$y = 10^x$

31. $\frac{1}{100}, \frac{1}{10}, 1, 10, 100$

33. $-\frac{1}{4}, -\frac{1}{2}, -1, -2, -4$

35.

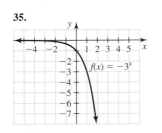

$f(x) = -3^x$

37.

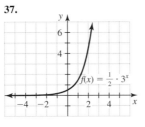

$f(x) = \frac{1}{2} \cdot 3^x$

39.

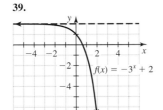

$f(x) = -3^x + 2$

41.

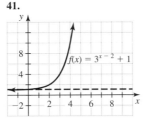

$f(x) = 3^{x-2} + 1$

43.

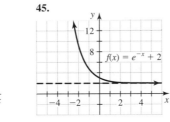

$f(x) = 10^x + 2$

45.

$f(x) = e^{-x} + 2$

47. $\{6\}$ **49.** $\{-3\}$ **51.** $\{-2\}$ **53.** $\{-1\}$ **55.** $\{-2\}$ **57.** $\{-2\}$
59. $\{-3, 3\}$ **61.** 2   **63.** $\frac{4}{3}$ **65.** -2 **67.** 0   **69.** $\frac{3}{2}$ **71.** $\frac{1}{2}$
73. $\frac{1}{32}, -3, 1, 1, 16$ **75.** $-3, 4, 0, \frac{1}{2}, 5$ **77.** $\$9861.72$
79. a) $\$15,859.75$ **b)** 2014   **81.** $\$45, \$12.92$ **83.** $\$616.84$
85. $\$6230.73$ **87.** 300 grams, 90.4 grams, 12 years, no
89. $50°F, 31.9°F, 48.6°F, 80.5°F$ **91.** $2.66666667, 0.0516, 2.8 \times 10^{-5}$
93. The graph of $y = 3^{x-h}$ lies h units to the right of $y = 3^x$ when $h > 0$
and $|h|$ units to the left of $y = 3^x$ when $h < 0$.

Section 12.2 Warm-Ups
1. Logarithmic   **2.** Common   **3.** Natural   **4.** Domain
5. One-to-one   **6.** Symmetric   **7.** Exponent   **8.** True

9. False   **10.** True   **11.** True   **12.** False   **13.** False
14. True   **15.** True

Section 12.2 Exercises
1. $2^3 = 8$ **3.** $\log(100) = 2$ **5.** $5^y = x$ **7.** $\log_2(b) = a$
9. $3^{10} = x$ **11.** $\ln(x) = 3$ **13.** 2   **15.** 4   **17.** 6   **19.** 3
21. -2 **23.** 2   **25.** -2 **27.** 1   **29.** -3 **31.** $\frac{1}{2}$ **33.** 2
35. 0.6990 **37.** 1.8307 **39.** $-2, -1, 0, 1, 2$ **41.** $-2, -1, 0, 1, 2$

43.

$f(x) = \log_3(x)$

45.

$y = \log_4(x)$

47.

$h(x) = \log_{1/4}(x)$

49.

$y = \log_{1/5}(x)$

51. $f^{-1}(x) = \log_6(x)$ **53.** $f^{-1}(x) = e^x$ **55.** $f^{-1}(x) = \left(\frac{1}{2}\right)^x$
57. $\{4\}$ **59.** $\left\{\frac{1}{2}\right\}$ **61.** $\{0.001\}$ **63.** $\{6\}$ **65.** $\left\{\frac{1}{5}\right\}$ **67.** $\{\pm 3\}$
69. $\{0.4771\}$ **71.** $\{-0.3010\}$ **73.** $\{1.9741\}$ **75.** $-2, \frac{1}{2}, 0, 4, 4$
77. $16, -2, 1, 1, \frac{1}{4}$ **79.** 5.776 years   **81.** 1.927 years
83. a) 4.82% **b)** $\$26,222$ **85.** 4.1   **87.** 6.8   **89.** 90 dB
91. $f^{-1}(x) = 2^{x-5} + 3, (-\infty, \infty), (3, \infty)$
93. $y = \ln(e^x) = x$ for $-\infty < x < \infty$, $y = e^{\ln(x)} = x$ for $0 < x < \infty$

Mid-Chapter Quiz 12.1–12.2
1. $\frac{1}{4}, -1, 1, 2, 8$ **2.** $16, -2, \frac{1}{2}, 2, \frac{1}{16}$ **3.** $2, 0, -1, -2$
4. $-5, -4, 0, 3$ **5.** $\{4\}$ **6.** $\{-1\}$ **7.** $\{-2\}$ **8.** $\{8\}$
9. $\{10,000\}$ **10.** $f^{-1}(x) = \log(x)$ **11.** $g^{-1}(x) = 8^x$
12. $5^M = W$ **13.** $\log_a(y) = 3$
14. $(-\infty, \infty), (1, \infty)$ **15.** $(-\infty, \infty), (-3, \infty)$

$f(x) = 2^x + 1$

$g(x) = \left(\frac{1}{2}\right)^x - 3$

$f(x) = 2^x + 1$

$g(x) = \left(\frac{1}{2}\right)^x - 3$

16. $(0, \infty)$, $(-\infty, \infty)$
$y = \log_2(x)$

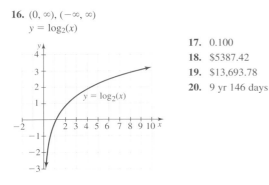

17. 0.100
18. $5387.42
19. $13,693.78
20. 9 yr 146 days

Section 12.3 Warm-Ups

1. Product   **2.** Quotient   **3.** Power   **4.** Inverse   **5.** True
6. False   **7.** False   **8.** True   **9.** False   **10.** True
11. True   **12.** True

Section 12.3 Exercises

1. 10   **3.** 19   **5.** 8   **7.** 4.3   **9.** $\log(21)$ **11.** $\log_3(\sqrt{5x})$
13. $\log(x^5)$ **15.** $\ln(30)$ **17.** $\log(x^2 + 3x)$ **19.** $\log_2(x^2 - x - 6)$
21. $\log(4)$ **23.** $\log_2(x^4)$ **25.** $\log(\sqrt{5})$ **27.** $\ln(h - 2)$

29. $\log_2(w - 2)$ **31.** $\ln(x - 2)$ **33.** $3\log(3)$ **35.** $\frac{1}{2}\log(3)$

37. $x\log(3)$ **39.** $\log(3) + \log(5)$ **41.** $\log(5) - \log(3)$
43. $2\log(5)$ **45.** $2\log(5) + \log(3)$ **47.** $-\log(3)$ **49.** $-\log(5)$
51. $\log(x) + \log(y) + \log(z)$ **53.** $3 + \log_2(x)$ **55.** $\ln(x) - \ln(y)$

57. $1 + 2\log(x)$ **59.** $2\log_5(x - 3) - \frac{1}{2}\log_5(w)$

61. $\ln(y) + \ln(z) + \frac{1}{2}\ln(x) - \ln(w)$ **63.** $\log(x^2 - x)$ **65.** $\ln(3)$

67. $\ln\left(\dfrac{xz}{w}\right)$ **69.** $\ln\left(\dfrac{x^2y^3}{w}\right)$ **71.** $\log\left(\dfrac{(x - 3)^{1/2}}{(x + 1)^{2/3}}\right)$ **73.** $\log_2\left(\dfrac{(x - 1)^{2/3}}{(x + 2)^{1/4}}\right)$

75. False   **77.** True   **79.** True   **81.** False   **83.** True   **85.** True
87. True   **89.** False   **91.** $r = \log(I/I_0)$, $r = 2$ **93.** b

95. The graphs are the same because $\ln(\sqrt{x}) = \ln(x^{1/2}) = \frac{1}{2}\ln(x)$.

97. The graph is a straight line because $\log(e^x) = x\log(e) \approx 0.434x$.
The slope is $\log(e)$ or approximately 0.434.

Section 12.4 Warm-Ups

1. Equivalent   **2.** Base-change   **3.** True   **4.** True   **5.** True
6. False   **7.** True   **8.** True   **9.** True   **10.** False

Section 12.4 Exercises

1. {900}   **3.** {7}   **5.** {31}   **7.** {e}   **9.** {2}   **11.** {3}   **13.** $\varnothing$

15. {6}   **17.** {3}   **19.** {2}   **21.** {4}   **23.** {$\log_3(7)$}   **25.** $\left\{\dfrac{\ln(7)}{2}\right\}$

27. {−6}   **29.** $\left\{-\dfrac{1}{2}\right\}$ **31.** $\dfrac{5\ln(3)}{\ln(2) - \ln(3)}$, −13.548

33. $\dfrac{4 + 2\log(5)}{1 - \log(5)}$, 17.932   **35.** $\dfrac{\ln(9)}{\ln(9) - \ln(8)}$, 18.655   **37.** 1.5850

39. −0.6309   **41.** −2.2016   **43.** 1.5229   **45.** $\dfrac{\ln(7)}{\ln(2)}$, 2.807

47. $\dfrac{1}{3 - \ln(2)}$, 0.433   **49.** $\dfrac{\ln(5)}{\ln(3)}$, 1.465   **51.** $1 + \dfrac{\ln(9)}{\ln(2)}$, 4.170

53. $\log_3(20)$, 2.727   **55.** $\dfrac{3}{5}$ **57.** $\dfrac{1}{2}$ **59.** {−3}   **61.** $\left\{-\dfrac{7}{2}, -2\right\}$

63. $\varnothing$ **65.** $\varnothing$ **67.** {4}   **69.** {4, 6}   **71.** {2}   **73.** {4}   **75.** {−1}

77. $\left[\dfrac{3}{2}, \dfrac{5}{3}\right]$ **79.** 41 months   **81.** 594 days   **83.** 10 g, 9163 years ago
85. 7524 ft³/sec   **87.** 7.1 years   **89.** 16.8 years   **91.** 2.0×10^{-4}
93. 0.9183   **95.** $\sqrt[3]{12}$ or 2.2894   **97.** (2.71, 6.54)
99. (1.03, 0.04), (4.74, 2.24)

Enriching Your Mathematical Word Power

1. Exponential   **2.** Common   **3.** Natural   **4.** Domain
5. Compound   **6.** Continuous   **7.** Logarithm   **8.** Common
9. Natural   **10.** Function

Review Exercises

1. $\dfrac{1}{25}$ **3.** 125   **5.** 1   **7.** $\dfrac{1}{10}$ **9.** 4   **11.** $\dfrac{1}{2}$ **13.** 2   **15.** 4

17. $-\dfrac{5}{2}$ **19.** 2   **21.** 8.6421   **23.** 177.828   **25.** 0.02005

27. 0.1408
29.   **31.**

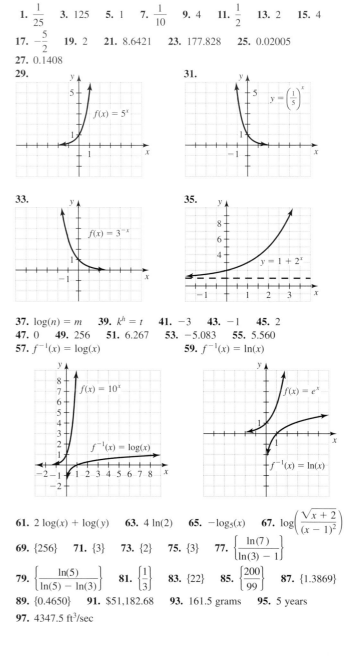

37. $\log(n) = m$ **39.** $k^h = t$ **41.** −3   **43.** −1   **45.** 2
47. 0   **49.** 256   **51.** 6.267   **53.** −5.083   **55.** 5.560
57. $f^{-1}(x) = \log(x)$ **59.** $f^{-1}(x) = \ln(x)$

61. $2\log(x) + \log(y)$ **63.** $4\ln(2)$ **65.** $-\log_5(x)$ **67.** $\log\left(\dfrac{\sqrt{x + 2}}{(x - 1)^2}\right)$

69. {256}   **71.** {3}   **73.** {2}   **75.** {3}   **77.** $\left\{\dfrac{\ln(7)}{\ln(3) - 1}\right\}$

79. $\left\{\dfrac{\ln(5)}{\ln(5) - \ln(3)}\right\}$ **81.** $\left\{\dfrac{1}{3}\right\}$ **83.** {22}   **85.** $\left\{\dfrac{200}{99}\right\}$ **87.** {1.3869}

89. {0.4650}   **91.** $51,182.68   **93.** 161.5 grams   **95.** 5 years
97. 4347.5 ft³/sec

Chapter 12 Test

1. 25   **2.** $\dfrac{1}{5}$ **3.** 1   **4.** 3   **5.** 0   **6.** -1

7.
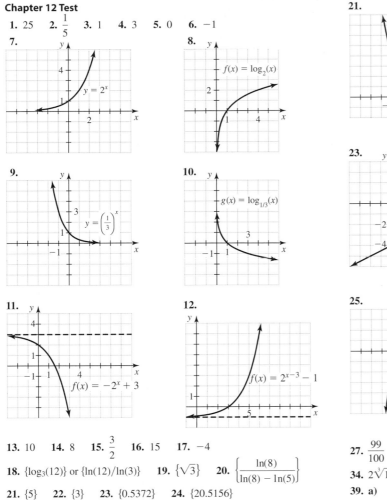

8.

9.

10.

11.

12.

21.
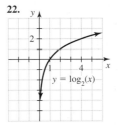

22.
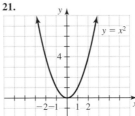

23.
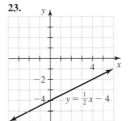

24.
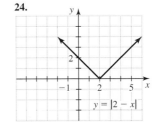

25.
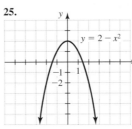

26.
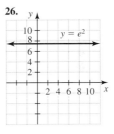

13. 10   **14.** 8   **15.** $\dfrac{3}{2}$ **16.** 15   **17.** -4

18. $\{\log_3(12)\}$ or $\{\ln(12)/\ln(3)\}$ **19.** $\{\sqrt{3}\}$ **20.** $\left\{\dfrac{\ln(8)}{\ln(8) - \ln(5)}\right\}$

21. $\{5\}$ **22.** $\{3\}$ **23.** $\{0.5372\}$ **24.** $\{20.5156\}$

25. 10; 147,648   **26.** 1.733 hours

27. $\dfrac{99}{100}$ **28.** $\dfrac{49}{50}$ **29.** $\dfrac{1}{30}$ **30.** 81   **31.** 5   **32.** -2 **33.** $4\sqrt{5}$

34. $2\sqrt[3]{10}$ **35.** $2\sqrt{6}$ **36.** $4\sqrt[3]{50}$ **37.** $\sqrt{2}$ **38.** 1

39. a)
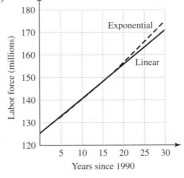

b) Linear 155.7 million, exponential 156.5 million

40. a) $d_1 = 0.135v$ **b)** $d_2 = 0.216v$

c) $v = 1482.67$ m/sec, $d_1 = 200.2$ meters

Making Connections   A Review of Chapters 1–12

1. $\left\{3 \pm 2\sqrt{2}\right\}$ **2.** $\{259\}$ **3.** $\{6\}$ **4.** $\left\{\dfrac{11}{2}\right\}$ **5.** $\{-5, 11\}$

6. $\{67\}$ **7.** $\{6\}$ **8.** $\{4\}$ **9.** $\left\{-\dfrac{52}{15}\right\}$ **10.** $\left\{\dfrac{3 \pm \sqrt{3}}{3}\right\}$

11. $f^{-1}(x) = 3x$ **12.** $g^{-1}(x) = 3^x$ **13.** $f^{-1}(x) = \dfrac{x + 4}{2}$

14. $h^{-1}(x) = x^2$ for $x \geq 0$ **15.** $j^{-1}(x) = \dfrac{1}{x}$ **16.** $k^{-1}(x) = \log_5(x)$

17. $m^{-1}(x) = 1 + \ln(x)$ **18.** $n^{-1}(x) = e^x$

19.
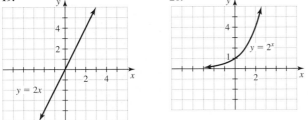

20.

Chapter 13

Section 13.1 Warm-Ups

1. Nonlinear   **2.** Graph   **3.** Substitution, addition   **4.** True

5. False   **6.** False   **7.** True   **8.** False   **9.** True   **10.** True

Section 13.1 Exercises

1. $\{(2, 4), (-3, 9)\}$

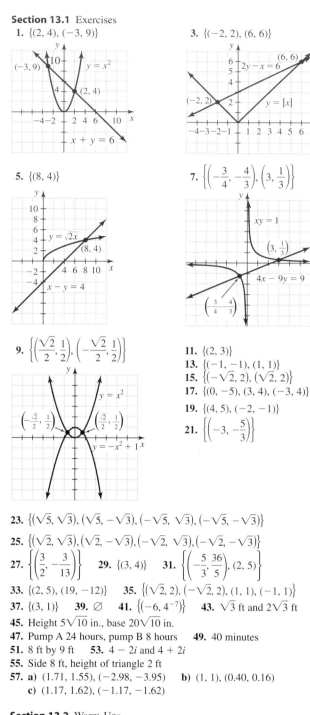

3. $\{(-2, 2), (6, 6)\}$

5. $\{(8, 4)\}$

7. $\left\{\left(-\dfrac{3}{4}, -\dfrac{4}{3}\right), \left(3, \dfrac{1}{3}\right)\right\}$

9. $\left\{\left(\dfrac{\sqrt{2}}{2}, \dfrac{1}{2}\right), \left(-\dfrac{\sqrt{2}}{2}, \dfrac{1}{2}\right)\right\}$

11. $\{(2, 3)\}$
13. $\{(-1, -1), (1, 1)\}$
15. $\{(-\sqrt{2}, 2), (\sqrt{2}, 2)\}$
17. $\{(0, -5), (3, 4), (-3, 4)\}$
19. $\{(4, 5), (-2, -1)\}$
21. $\left\{\left(-3, -\dfrac{5}{3}\right)\right\}$

23. $\{(\sqrt{5}, \sqrt{3}), (\sqrt{5}, -\sqrt{3}), (-\sqrt{5}, \sqrt{3}), (-\sqrt{5}, -\sqrt{3})\}$

25. $\{(\sqrt{2}, \sqrt{3}), (\sqrt{2}, -\sqrt{3}), (-\sqrt{2}, \sqrt{3}), (-\sqrt{2}, -\sqrt{3})\}$

27. $\left\{\left(\dfrac{3}{2}, -\dfrac{3}{13}\right)\right\}$ **29.** $\{(3, 4)\}$ **31.** $\left\{\left(-\dfrac{5}{3}, \dfrac{36}{5}\right), (2, 5)\right\}$

33. $\{(2, 5), (19, -12)\}$ **35.** $\{(\sqrt{2}, 2), (-\sqrt{2}, 2), (1, 1), (-1, 1)\}$

37. $\{(3, 1)\}$ **39.** $\varnothing$ **41.** $\{(-6, 4^{-7})\}$ **43.** $\sqrt{3}$ ft and $2\sqrt{3}$ ft

45. Height $5\sqrt{10}$ in., base $20\sqrt{10}$ in.

47. Pump A 24 hours, pump B 8 hours   **49.** 40 minutes

51. 8 ft by 9 ft   **53.** $4 - 2i$ and $4 + 2i$

55. Side 8 ft, height of triangle 2 ft

57. a) $(1.71, 1.55), (-2.98, -3.95)$ **b)** $(1, 1), (0.40, 0.16)$
   c) $(1.17, 1.62), (-1.17, -1.62)$

Section 13.2 Warm-Ups

1. Parabola   **2.** Focus   **3.** Directrix   **4.** Vertex
5. Focus, directrix   **6.** Completing the square   **7.** False
8. True   **9.** True   **10.** False   **11.** True   **12.** True

Section 13.2 Exercises

1. 5   **3.** $\sqrt{2}$ **5.** $\sqrt{13}$ **7.** $2\sqrt{17}$ **9.** $\sqrt{65}$
11. $(3, 4)$, 10   **13.** $\left(\dfrac{7}{2}, 3\right)$, 5   **15.** $(2, 1)$, 10   **17.** $\left(0, \dfrac{5}{2}\right)$, $\sqrt{13}$

19. Vertex $(0, 0)$, focus $\left(0, \dfrac{1}{8}\right)$, directrix $y = -\dfrac{1}{8}$

21. Vertex $(0, 0)$, focus $(0, -1)$, directrix $y = 1$

23. Vertex $(3, 2)$, focus $(3, 2.5)$, directrix $y = 1.5$

25. Vertex $(-1, 6)$, focus $(-1, 5.75)$, directrix $y = 6.25$

27. $y = \dfrac{1}{8}x^2$ **29.** $y = -\dfrac{1}{2}x^2$ **31.** $y = \dfrac{1}{2}x^2 - 3x + 6$

33. $y = -\dfrac{1}{8}x^2 + \dfrac{1}{4}x - \dfrac{1}{8}$ **35.** $y = x^2 + 6x + 10$

37. $y = (x - 3)^2 - 8$, vertex $(3, -8)$, focus $(3, -7.75)$,
   directrix $y = -8.25$, axis $x = 3$

39. $y = 2(x + 3)^2 - 13$, vertex $(-3, -13)$, focus $(-3, -12.875)$,
   directrix $y = -13.125$, axis $x = -3$

41. $y = -2(x - 4)^2 + 33$, vertex $(4, 33)$, focus $\left(4, 32\dfrac{7}{8}\right)$, directrix $y = 33\dfrac{1}{8}$,
   axis $x = 4$

43. $y = 5(x + 4)^2 - 80$, vertex $(-4, -80)$, focus $\left(-4, -79\dfrac{19}{20}\right)$,
   directrix $y = -80\dfrac{1}{20}$, axis $x = -4$

45. Vertex $(2, -3)$, focus $\left(2, -2\dfrac{3}{4}\right)$, directrix $y = -3\dfrac{1}{4}$, $x = 2$, upward

47. Vertex $(1, -2)$, focus $\left(1, -2\dfrac{1}{4}\right)$, directrix $y = -1\dfrac{3}{4}$, $x = 1$, downward

49. Vertex $(1, -2)$, focus $\left(1, -1\dfrac{11}{12}\right)$, directrix $y = -2\dfrac{1}{12}$, $x = 1$, upward

51. Vertex $\left(-\dfrac{3}{2}, \dfrac{17}{4}\right)$, focus $\left(-\dfrac{3}{2}, 4\right)$, directrix $y = \dfrac{9}{2}$, $x = -\dfrac{3}{2}$, downward

53. Vertex $(0, 5)$, focus $\left(0, 5\dfrac{1}{12}\right)$, directrix $y = 4\dfrac{11}{12}$, $x = 0$, upward

55. $(3, 2)$, $\left(\dfrac{13}{4}, 2\right)$, $x = \dfrac{11}{4}$ **57.** $(-2, 1)$, $(-1, 1)$, $x = -3$

59. $(4, 2)$, $\left(\dfrac{7}{2}, 2\right)$, $x = \dfrac{9}{2}$

61.

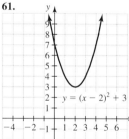

$y = (x - 2)^2 + 3$

63.

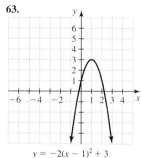

$y = -2(x - 1)^2 + 3$

65.

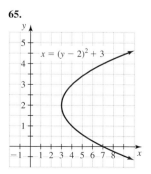

$x = (y - 2)^2 + 3$

67.

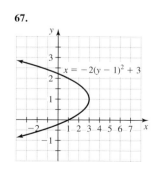

$x = -2(y - 1)^2 + 3$

69. $\{(-1, 2), (1, 2)\}$

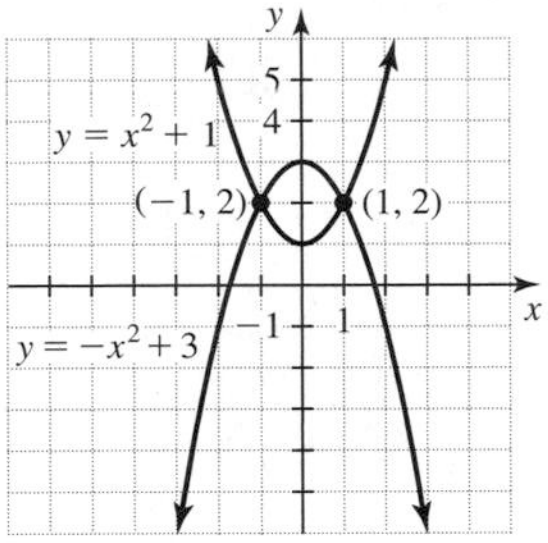

71. $\{(1, -1)\}$

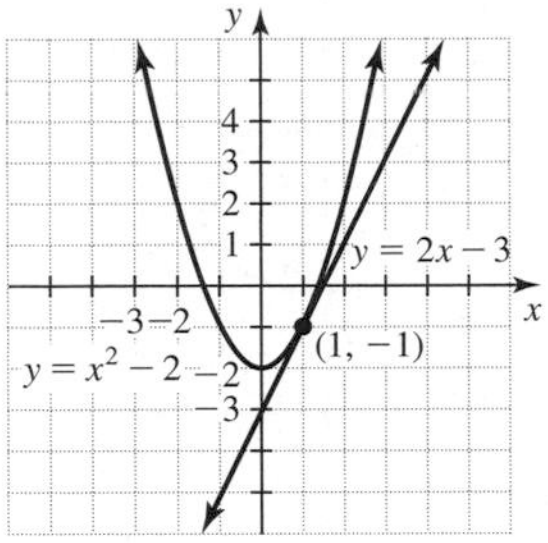

73. $\left\{\left(\frac{3}{2}, \frac{11}{4}\right), (-4, 0)\right\}$

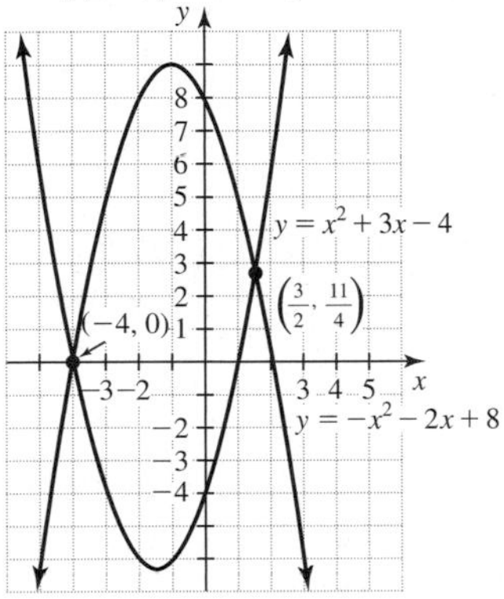

75. $\{(-3, -4), (2, 6)\}$

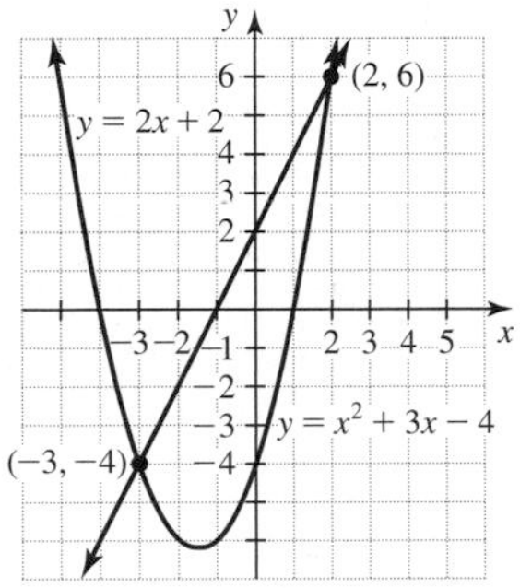

77. $(3, 0), (-1, 0)$ **79.** $(20, 4), (-20, 4)$ **81.** $(0, 0), (1, 1)$

83. a) \$59,598 **b)** $(-15, -59)$ **85.** $y = \frac{1}{60}x^2$

89. The graphs have identical shapes.

Section 13.3 Warm-Ups

1. Circle **2.** Center **3.** Radius **4.** Circle, center, radius
5. False **6.** False **7.** True **8.** False **9.** True **10.** True

Section 13.3 Exercises

1. $x^2 + y^2 = 16$ **3.** $x^2 + (y - 3)^2 = 25$ **5.** $(x - 1)^2 + (y + 2)^2 = 81$

7. $x^2 + y^2 = 3$ **9.** $(x + 6)^2 + (y + 3)^2 = \frac{1}{4}$

11. $\left(x - \frac{1}{2}\right)^2 + \left(y - \frac{1}{3}\right)^2 = 0.01$ **13.** $(0, 0), 1$ **15.** $(3, 5), \sqrt{2}$

17. $\left(0, \frac{1}{2}\right), \frac{\sqrt{2}}{2}$ **19.** $(0, 0), \frac{3}{2}$ **21.** $(2, 0), \sqrt{3}$

23.

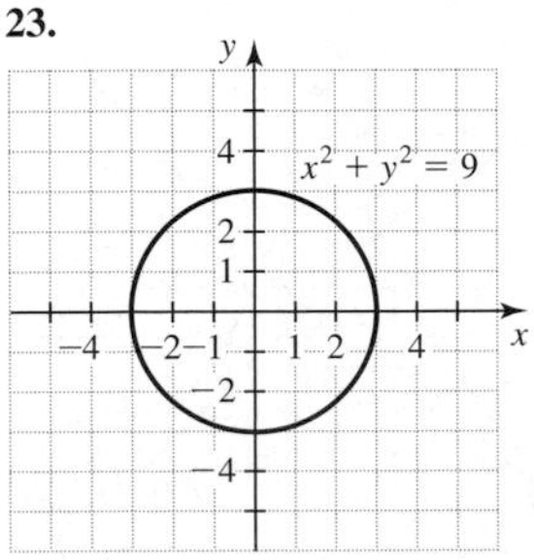

25.

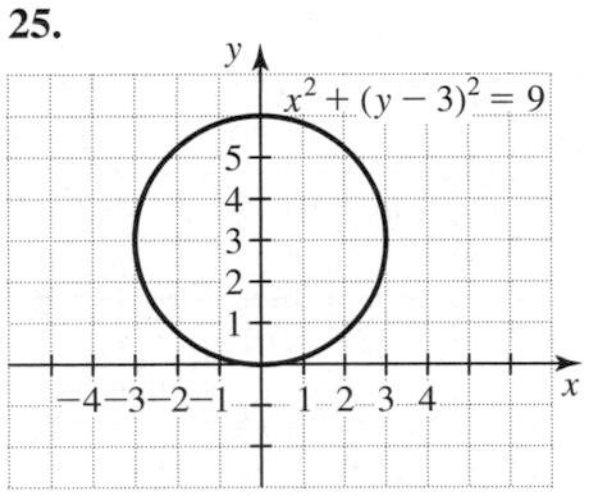

27.

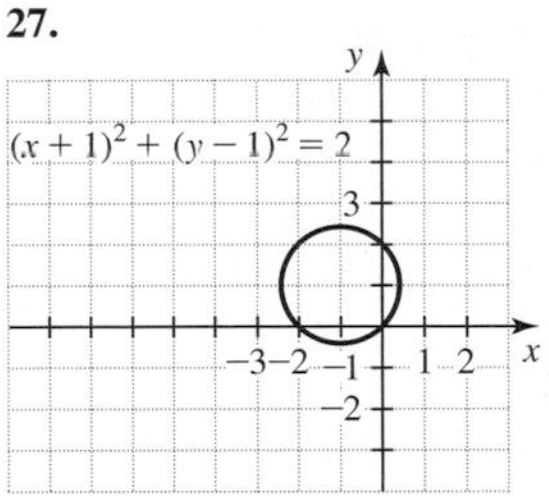

29.

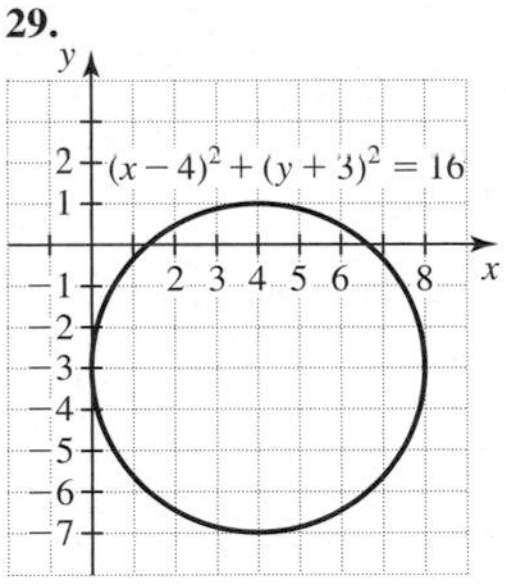

31.

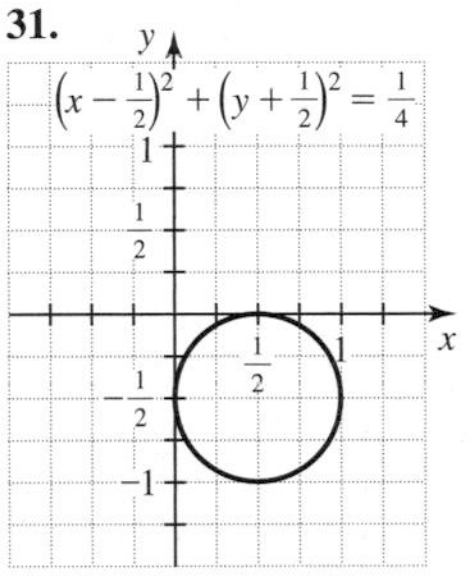

33. $(x + 2)^2 + (y + 3)^2 = 13, (-2, -3), \sqrt{13}$
35. $(x - 1)^2 + (y - 2)^2 = 8, (1, 2), 2\sqrt{2}$
37. $(x - 5)^2 + (y - 4)^2 = 9, (5, 4), 3$
39. $\left(x - \frac{1}{2}\right)^2 + \left(y + \frac{1}{2}\right)^2 = \frac{1}{2}, \left(\frac{1}{2}, -\frac{1}{2}\right), \frac{\sqrt{2}}{2}$
41. $\left(x - \frac{3}{2}\right)^2 + \left(y - \frac{1}{2}\right)^2 = \frac{7}{2}, \left(\frac{3}{2}, \frac{1}{2}\right), \frac{\sqrt{14}}{2}$

43. $\left(x - \frac{1}{3}\right)^2 + \left(y + \frac{3}{4}\right)^2 = \frac{97}{144}, \left(\frac{1}{3}, -\frac{3}{4}\right), \frac{\sqrt{97}}{12}$

45. $\{(1, 3), (-1, -3)\}$

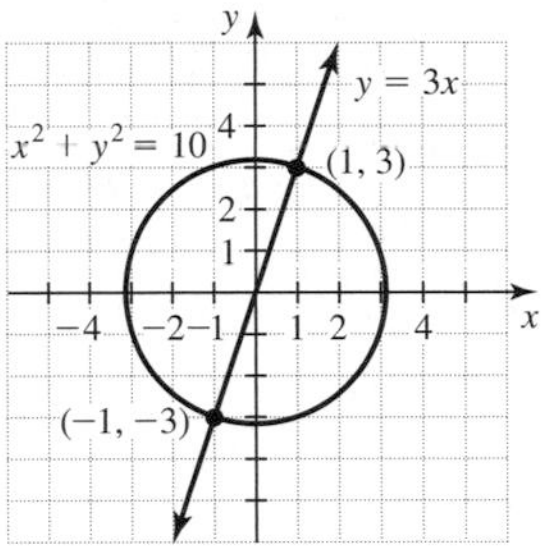

47. $\{(0, -3), (\sqrt{5}, 2), (-\sqrt{5}, 2)\}$

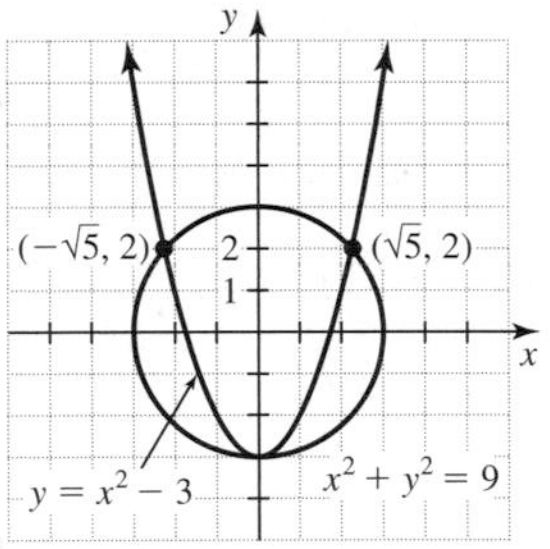

49. $\{(0, -3), (2, -1)\}$

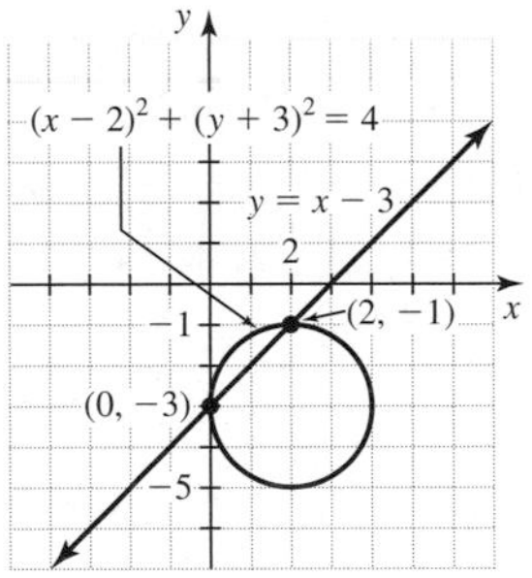

51. $(0, 2 + \sqrt{3})$ and $(0, 2 - \sqrt{3})$
53. $\sqrt{29}$
55. $(x - 2)^2 + (y - 3)^2 = 32$
57. $\left(\frac{5}{2}, -\frac{\sqrt{11}}{2}\right)$ and $\left(\frac{5}{2}, \frac{\sqrt{11}}{2}\right)$
59. 755,903 mm^3

61. $(0, 0)$ only

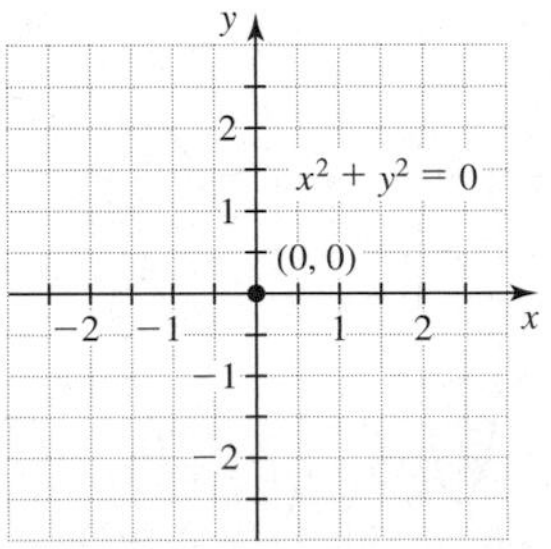

63.

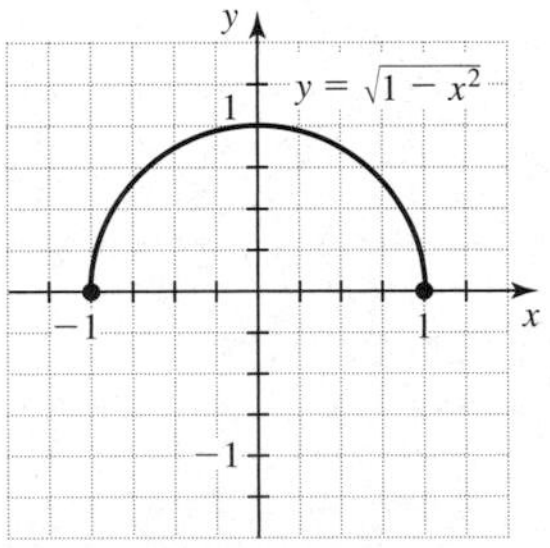

65. B and D can be any real numbers, but A must equal C,
and $4AE + B^2 + D^2 > 0$. No ordered pairs satisfy $x^2 + y^2 = -9$.
67. $y = \pm\sqrt{4 - x^2}$ **69.** $y = \pm\sqrt{x}$ **71.** $y = -1 \pm \sqrt{x}$

Mid-Chapter Quiz 13.1–13.3
1. $\{(-2, 4), (3, 9)\}$ **2.** $\{(0, 2), (8, 6)\}$
3. $\{(2, \pm\sqrt{3}), (-2, \pm\sqrt{3})\}$ **4.** $\{(\pm2, 3)\}$
5. $(8, -7), (8, -5), y = -9, x = 8$, up
6. $(3, 4), (3, 15/14), y = 17/4, x = 3$, down
7. $(0, 2), (1/4, 2), x = -1/4, y = 2$, right
8. $\sqrt{5}$ **9.** $(-7, 2)$ **10.** $y = 4(x - 1)^2 - 2$
11. $(x + 4)^2 + (y - 5)^2 = 100$ **12.** $(2, 0), \sqrt{7}$
13. $(x + 2)^2 + (y - 5)^2 = 30$
14. $5 + \sqrt{13}, 5 - \sqrt{13}$

Section 13.4 Warm-Ups
1. Ellipse   **2.** Foci   **3.** Center   **4.** Hyperbola   **5.** Branches
6. Asymptotes   **7.** True   **8.** False   **9.** True   **10.** True
11. True   **12.** True

Section 13.4 Exercises
1.
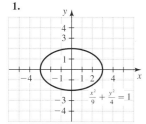
3.
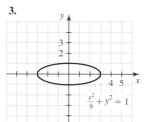
5.
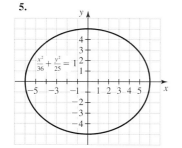
7.
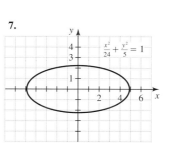
9.
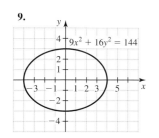
11.
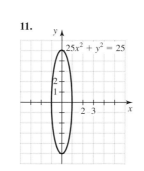

13.
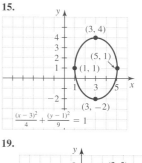
15.
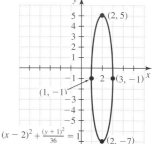
17.
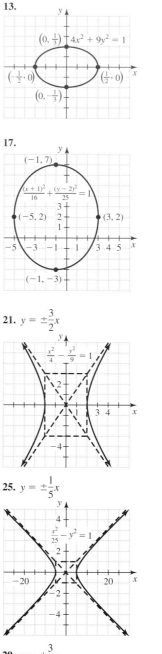
19.
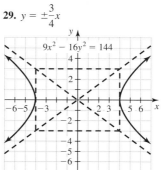

21. $y = \pm\dfrac{3}{2}x$
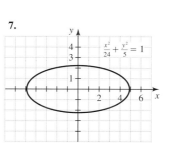
23. $y = \pm\dfrac{2}{5}x$
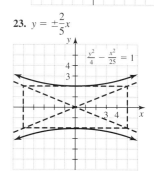
25. $y = \pm\dfrac{1}{5}x$

27. $y = \pm5x$
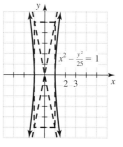
29. $y = \pm\dfrac{3}{4}x$

31. $y = \pm x$
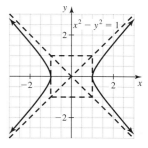

33.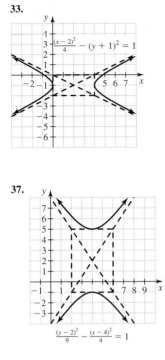

$\dfrac{(x-2)^2}{4} - (y+1)^2 = 1$

35.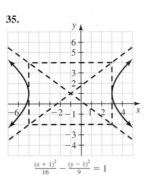

$\dfrac{(x+1)^2}{16} - \dfrac{(y-1)^2}{9} = 1$

37.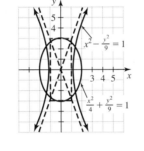

$\dfrac{(y-2)^2}{9} - \dfrac{(x-4)^2}{4} = 1$

39. Parabola
41. Hyperbola
43. Ellipse
45. Circle

47. $\left(\dfrac{2\sqrt{10}}{5}, \dfrac{3\sqrt{15}}{5}\right),$
$\left(\dfrac{2\sqrt{10}}{5}, -\dfrac{3\sqrt{15}}{5}\right),$
$\left(-\dfrac{2\sqrt{10}}{5}, \dfrac{3\sqrt{15}}{5}\right),$
$\left(-\dfrac{2\sqrt{10}}{5}, -\dfrac{3\sqrt{15}}{5}\right)$

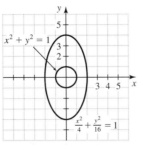

49. No points of intersection

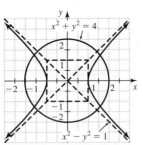

51. $\left(\dfrac{\sqrt{10}}{2}, \dfrac{\sqrt{6}}{2}\right), \left(\dfrac{\sqrt{10}}{2}, -\dfrac{\sqrt{6}}{2}\right),$
$\left(-\dfrac{\sqrt{10}}{2}, \dfrac{\sqrt{6}}{2}\right),$
$\left(-\dfrac{\sqrt{10}}{2}, -\dfrac{\sqrt{6}}{2}\right)$

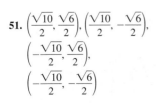

53. $\left(\dfrac{3\sqrt{6}}{4}, \dfrac{\sqrt{10}}{4}\right),$
$\left(\dfrac{3\sqrt{6}}{4}, -\dfrac{\sqrt{10}}{4}\right),$
$\left(-\dfrac{3\sqrt{6}}{4}, \dfrac{\sqrt{10}}{4}\right),$
$\left(-\dfrac{3\sqrt{6}}{4}, -\dfrac{\sqrt{10}}{4}\right)$

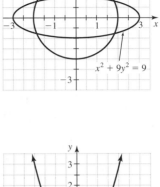

55. $\left(\dfrac{\sqrt{17}}{3}, \dfrac{8}{9}\right), \left(-\dfrac{\sqrt{17}}{3}, \dfrac{8}{9}\right),$
$(0, -1)$

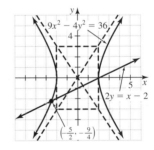

57. $(2, 0), \left(-\dfrac{5}{2}, -\dfrac{9}{4}\right)$

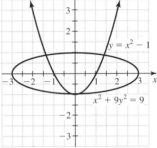

59. a) $(2.5, 1.3)$ **b)** $(\sqrt{7}, \sqrt{2})$

Section 13.5 Warm-Ups
1. Ellipse   **2.** Hyperbola   **3.** Circle   **4.** Parabola   **5.** Line
6. True   **7.** False   **8.** False   **9.** True   **10.** True

Section 13.5 Exercises

1. **3.**
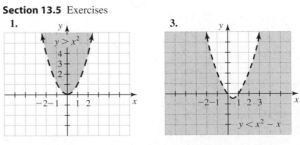

5.

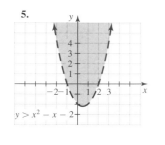

$y > x^2 - x - 2$

7.

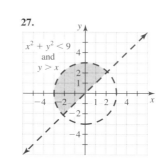

$x^2 + y^2 < 9$

27.

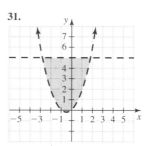

$x^2 + y^2 < 9$
and
$y > x$

29.

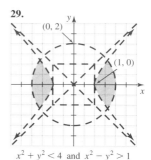

$(0, 2)$ $(1, 0)$

$x^2 + y^2 < 4$ and $x^2 - y^2 > 1$

9.

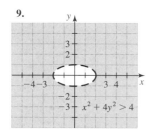

$x^2 + 4y^2 > 4$

11.

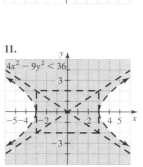

$4x^2 - 9y^2 < 36$

31.

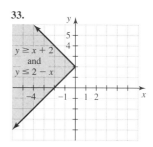

$y > x^2 + x$ and $y < 5$

33.

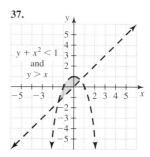

$y \geq x + 2$
and
$y \leq 2 - x$

13.

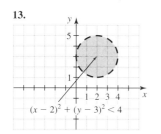

$(x - 2)^2 + (y - 3)^2 < 4$

15.

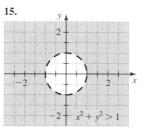

$x^2 + y^2 > 1$

35.

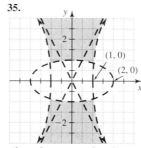

$(1, 0)$ $(2, 0)$

$x^2 + 4y^2 > 4$ and $4x^2 - y^2 < 4$

37.

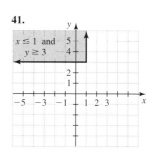

$y + x^2 < 1$
and
$y > x$

17.

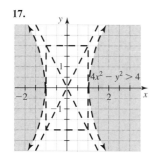

$4x^2 - y^2 > 4$

19.

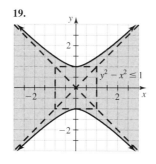

$y^2 - x^2 \leq 1$

39.

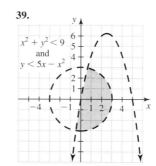

$x^2 + y^2 < 9$
and
$y < 5x - x^2$

41.

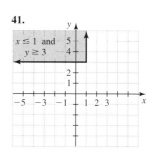

$x \leq 1$ and
$y \geq 3$

21.

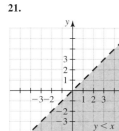

$y < x$

23. Yes

25. No

43.

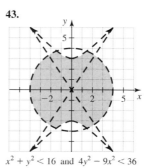

$x^2 + y^2 < 16$ and $4y^2 - 9x^2 < 36$

45.

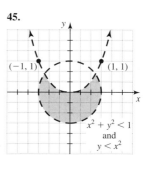

$(-1, 1)$ $(1, 1)$

$x^2 + y^2 < 1$
and
$y < x^2$

47.

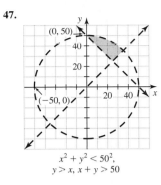

$$x^2 + y^2 < 50^2,$$
$$y > x, x + y > 50$$

49. No solution

37. $(0, 0), \dfrac{2}{3}$

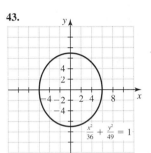

39. $x^2 + (y - 3)^2 = 36$

41. $(x - 2)^2 + (y + 7)^2 = 25$

Enriching Your Mathematical Word Power

1. Nonlinear   **2.** Parabola   **3.** Focus   **4.** Directrix   **5.** Vertex
6. Conic   **7.** Axis   **8.** Ellipse   **9.** Circle   **10.** Hyperbola

Review Exercises

1. $\{(3, 9), (-5, 25)\}$

3. $\left\{ \left(\dfrac{\sqrt{3}}{3}, \sqrt{3} \right), \left(-\dfrac{\sqrt{3}}{3}, -\sqrt{3} \right) \right\}$

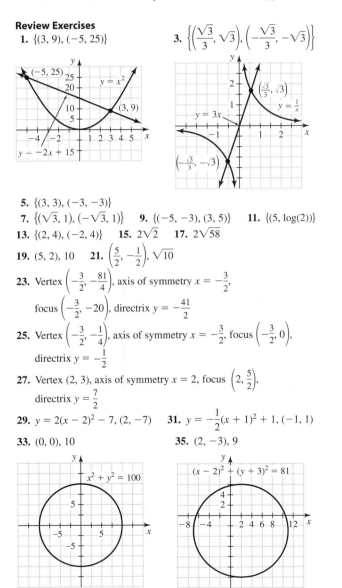

5. $\{(3, 3), (-3, -3)\}$
7. $\{(\sqrt{3}, 1), (-\sqrt{3}, 1)\}$ **9.** $\{(-5, -3), (3, 5)\}$ **11.** $\{(5, \log(2))\}$
13. $\{(2, 4), (-2, 4)\}$ **15.** $2\sqrt{2}$ **17.** $2\sqrt{58}$

19. $(5, 2), 10$ **21.** $\left(\dfrac{5}{2}, -\dfrac{1}{2} \right), \sqrt{10}$

23. Vertex $\left(-\dfrac{3}{2}, -\dfrac{81}{4} \right)$, axis of symmetry $x = -\dfrac{3}{2}$,
focus $\left(-\dfrac{3}{2}, -20 \right)$, directrix $y = -\dfrac{41}{2}$

25. Vertex $\left(-\dfrac{3}{2}, -\dfrac{1}{4} \right)$, axis of symmetry $x = -\dfrac{3}{2}$, focus $\left(-\dfrac{3}{2}, 0 \right)$,
directrix $y = -\dfrac{1}{2}$

27. Vertex $(2, 3)$, axis of symmetry $x = 2$, focus $\left(2, \dfrac{5}{2} \right)$,
directrix $y = \dfrac{7}{2}$

29. $y = 2(x - 2)^2 - 7, (2, -7)$ **31.** $y = -\dfrac{1}{2}(x + 1)^2 + 1, (-1, 1)$

33. $(0, 0), 10$ **35.** $(2, -3), 9$

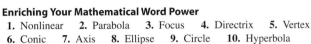

43.

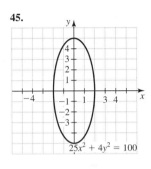

$$\dfrac{x^2}{36} + \dfrac{y^2}{49} = 1$$

45.

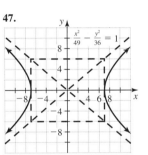

$$25x^2 + 4y^2 = 100$$

47.

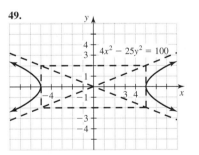

$$\dfrac{x^2}{49} - \dfrac{y^2}{36} = 1$$

49.

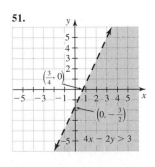

$$4x^2 - 25y^2 = 100$$

51.

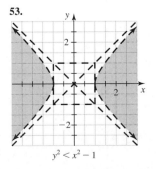

$$4x - 2y > 3$$

53.

$$y^2 < x^2 - 1$$

55.

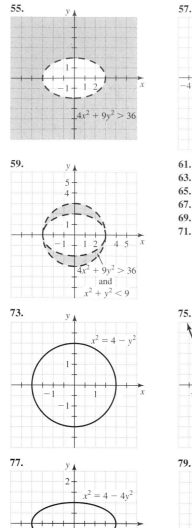

$4x^2 + 9y^2 > 36$

57.

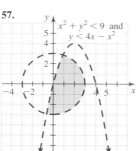

$x^2 + y^2 < 9$ and $y < 4x - x^2$

3.

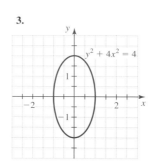

$y^2 + 4x^2 = 4$

4.

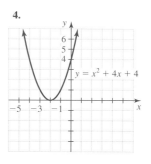

$y = x^2 + 4x + 4$

59.

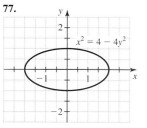

$4x^2 + 9y^2 > 36$
and
$x^2 + y^2 < 9$

61. Hyperbola
63. Circle
65. Circle
67. Circle
69. Hyperbola
71. Hyperbola

5.

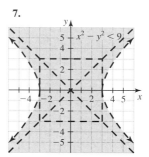

$y^2 - 4x^2 = 4$

6.

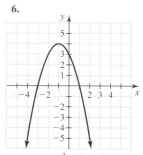

$y = -x^2 - 2x + 3$

73.

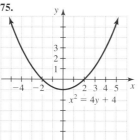

$x^2 = 4 - y^2$

75.

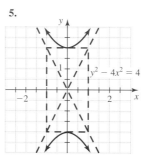

$x^2 = 4y + 4$

7.

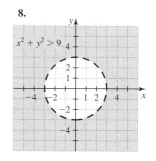

$x^2 - y^2 < 9$

8.

$x^2 + y^2 > 9$

77.

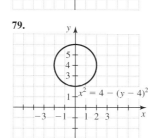

$x^2 = 4 - 4y^2$

79.

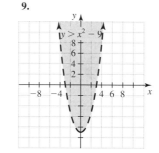

$x^2 = 4 - (y - 4)^2$

9.

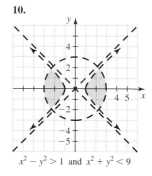

$y > x^2 - 9$

10.

$x^2 - y^2 > 1$ and $x^2 + y^2 < 9$

81. $x^2 + y^2 = 25$ **83.** $(x + 1)^2 + (y - 5)^2 = 36$

85. $y = \dfrac{1}{4}(x - 1)^2 + 3$ **87.** $y = x^2$ **89.** $y = \dfrac{2}{9}x^2$

91. $\{(4, -3), (-3, 4)\}$ **93.** $\varnothing$ **95.** 6 ft, 2 ft

Chapter 13 Test

1.

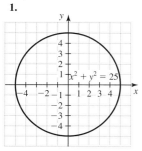

$x^2 + y^2 = 25$

2.

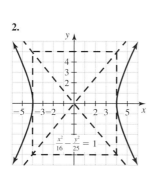

$\dfrac{x^2}{16} - \dfrac{y^2}{25} = 1$

11.

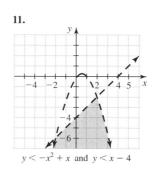

$y < -x^2 + x$ and $y < x - 4$

12. $\{(-5, 27), (3, -5)\}$
13. $\{(\sqrt{3}, 3), (-\sqrt{3}, 3)\}$
14. $2\sqrt{2}$

15. $\left(-\dfrac{1}{2}, -\dfrac{1}{2}\right), \sqrt{26}$

16. $(-1, -5), 6$

17. Vertex $\left(-\frac{1}{2}, \frac{11}{4}\right)$, focus $\left(-\frac{1}{2}, 3\right)$, directrix $y = \frac{5}{2}$, axis of symmetry $x = -\frac{1}{2}$, upward

18. $y = \frac{1}{2}(x - 3)^2 - 5$ **19.** $(x + 1)^2 + (y - 3)^2 = 13$ **20.** 12 ft, 9 ft

Making Connections   A Review of Chapters 1–13

1.
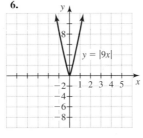

2.
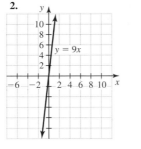

3.
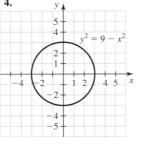

4.
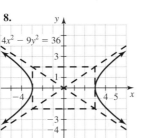

5.
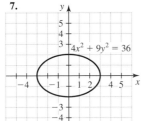

6.
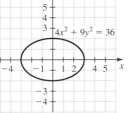

7.
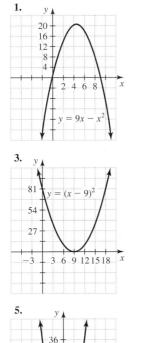

8.
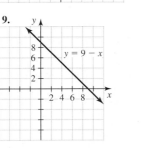

9.
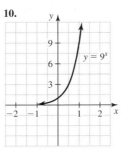

10.

11. $x^2 + 4xy + 4y^2$ **12.** $x^3 + 3x^2y + 3xy^2 + y^3$
13. $a^3 + 3a^2b + 3ab^2 + b^3$ **14.** $a^2 - 6ab + 9b^2$ **15.** $6a^2 - 7a - 5$
16. $x^3 - y^3$ **17.** $a(a^2 + b^2)$ **18.** $a(a + b)(a - b)$
19. $2(x + 3)(x - 6)$ **20.** $4(2x + 3)(4x - 5)$
21. $(m + 2)(x + 3)(x - 3)$ **22.** $2x(x + 3)(x^2 - 3x + 9)$
23. $\{(1, 2)\}$ **24.** $\{(3, 4), (4, 3)\}$ **25.** $\{(1, -2, 3)\}$
26. $\{(-1, 1), (3, 9)\}$ **27.** $x = -\frac{b}{a}$ **28.** $x = \frac{-d \pm \sqrt{d^2 - 4wm}}{2w}$
29. $B = \frac{2A - bh}{h}$ **30.** $x = \frac{2y}{y - 2}$ **31.** $m = \frac{L}{1 + xt}$
32. $t = \frac{y^2}{9a^2}$ **33.** $y = -\frac{2}{3}x - \frac{5}{3}$ **34.** $y = -2x$
35. $(x - 2)^2 + (y - 5)^2 = 45$ **36.** $\left(-\frac{3}{2}, 3\right), \frac{3\sqrt{5}}{2}$
37. $-10 + 6i$ **38.** -1 **39.** $3 - 5i$ **40.** $7 + 6i\sqrt{2}$ **41.** $-8 - 27i$
42. $-3 + 3i$ **43.** 29 **44.** $-\frac{3}{2} - i$ **45.** $-\frac{1}{2} + \frac{9}{2}i$ **46.** $2 - i\sqrt{2}$
47. a) $q = -500x + 400$ **b)** $R = -500x^2 + 400x$
c)
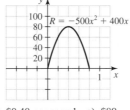

d) \$0.40 per pound   **e)** \$80

Appendix A

Geometry Review Exercises
1. 12 in.   **2.** 24 ft²   **3.** 60°   **4.** 6 ft   **5.** 20 cm   **6.** 144 cm²
7. 24 ft²   **8.** 13 ft   **9.** 30 cm   **10.** No   **11.** 84 yd²   **12.** 30 in.
13. 32 ft²   **14.** 20 km   **15.** 7 ft, 3 ft²   **16.** 22 yd   **17.** 50.3 ft²
18. 37.7 ft   **19.** 150.80 cm³   **20.** 879.29 ft²   **21.** 288 in.³, 288 in.²
22. 4 cm   **23.** 100 mi², 40 mi   **24.** 20 km   **25.** 42.25 cm²
26. 33.510 ft³, 50.265 ft²   **27.** 75.4 in.³, 100.5 in.²   **28.** 56°
29. 5 cm   **30.** 15 in. and 20 in.   **31.** 149°   **32.** 12 km   **33.** 10 yd

Appendix B

Sets
1. A set is a collection of objects.
2. A finite set has a fixed number of elements and an infinite set does not.
3. A Venn diagram is used to illustrate relationships between sets.
4. The intersection of two sets consists of elements that are in both sets, whereas the union of two sets consists of elements that are in one, in the other, or in both sets.
5. Every member of set A is also a member of set B.
6. The empty set is a subset of every set.   **7.** False   **8.** False
9. True   **10.** False   **11.** True   **12.** False   **13.** False   **14.** True
15. False   **16.** False   **17.** False   **18.** False   **19.** $\varnothing$
20. $\{1, 2, 3, 4, 5, 6, 7, 8, 9\}$ **21.** $\{1, 3, 5\}$ **22.** $\{1, 2, 3, 4, 5, 7, 9\}$
23. $\{1, 2, 3, 4, 5, 6, 8\}$ **24.** $\{2, 4\}$ **25.** A **26.** B **27.** $\varnothing$ **28.** $\varnothing$
29. A **30.** N **31.** $=$ **32.** $\neq$ **33.** $\cup$ **34.** $\cap$ **35.** $\cap$ **36.** $\cup$
37. $\not\subseteq$ **38.** $\in$ **39.** $\in$ **40.** $\in$ **41.** True   **42.** True   **43.** True
44. False   **45.** True   **46.** True   **47.** True   **48.** True   **49.** False
50. False   **51.** True   **52.** True   **53.** $\{2, 3, 4, 5, 6, 7, 8\}$ **54.** $\varnothing$

55. $\{3, 5\}$ **56.** $\{1, 2, 3, 4, 5, 7\}$ **57.** $\{1, 2, 3, 4, 5, 6, 8\}$ **58.** $\{2, 4\}$
59. $\{2, 3, 4, 5\}$ **60.** $\{2, 4\}$ **61.** $\{2, 3, 4, 5, 7\}$ **62.** $\{2, 3, 4, 5, 7\}$
63. $\{2, 3, 4, 5\}$ **64.** $\{2, 4\}$ **65.** $\{2, 3, 4, 5, 7\}$ **66.** $\{2, 3, 4, 5, 7\}$
67. $\subseteq$ **68.** $=$ **69.** $\in$ **70.** $\subseteq$ **71.** $\cap$ **72.** $\subseteq$ **73.** $\subseteq$
74. $\cap$ **75.** $\cap$ **76.** $\cup$ **77.** $\cup$ **78.** $\cap$ **79.** $\{2, 4, 6, \ldots, 18\}$
80. $\{7, 8, 9, \ldots\}$ **81.** $\{13, 15, 17, \ldots\}$ **82.** $\{1, 3, 5, \ldots, 13\}$
83. $\{6, 8, 10, \ldots, 78\}$ **84.** $\{13, 15, 17, \ldots, 55\}$
85. $\{x \mid x$ is a natural number between 2 and 7$\}$
86. $\{x \mid x$ is an odd natural number less than 8$\}$
87. $\{x \mid x$ is an odd natural number greater than 4$\}$
88. $\{x \mid x$ is a natural number greater than 3$\}$
89. $\{x \mid x$ is an even natural number between 5 and 83$\}$
90. $\{x \mid x$ is an odd natural number between 8 and 52$\}$
91. False   **92.** False   **93.** False   **94.** False   **95.** True
96. True   **97.** True   **98.** False   **99.** True   **100.** True

Appendix C

Chapters 1–6 Diagnostic Test

1. $(2, \infty)$

2. $(-\infty, -1]$

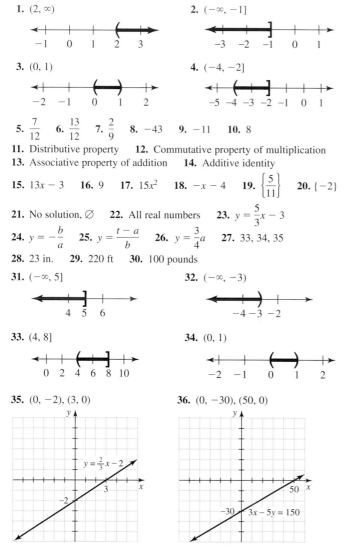

3. $(0, 1)$

4. $(-4, -2]$

5. $\dfrac{7}{12}$ **6.** $\dfrac{13}{12}$ **7.** $\dfrac{2}{9}$ **8.** -43 **9.** -11 **10.** 8

11. Distributive property   **12.** Commutative property of multiplication
13. Associative property of addition   **14.** Additive identity

15. $13x - 3$ **16.** 9 **17.** $15x^2$ **18.** $-x - 4$ **19.** $\left\{\dfrac{5}{11}\right\}$ **20.** $\{-2\}$

21. No solution, $\varnothing$ **22.** All real numbers   **23.** $y = \dfrac{5}{3}x - 3$

24. $y = -\dfrac{b}{a}$ **25.** $y = \dfrac{t - a}{b}$ **26.** $y = \dfrac{3}{4}a$ **27.** 33, 34, 35

28. 23 in.   **29.** 220 ft   **30.** 100 pounds

31. $(-\infty, 5]$

32. $(-\infty, -3)$

33. $(4, 8]$

34. $(0, 1)$

35. $(0, -2), (3, 0)$

36. $(0, -30), (50, 0)$

37. $(0, 2)$

38. $(2, 0)$

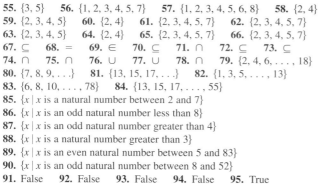

39. 2 **40.** $\dfrac{1}{2}$ **41.** $-\dfrac{2}{3}$ **42.** $\dfrac{1}{3}$ **43.** $y = 4x + 3$ **44.** $y = -2x$
45. $x = 3$ **46.** $y = -2x + 8$ **47.** 18 hours   **48.** \$1890

49.

50.

51.

52.

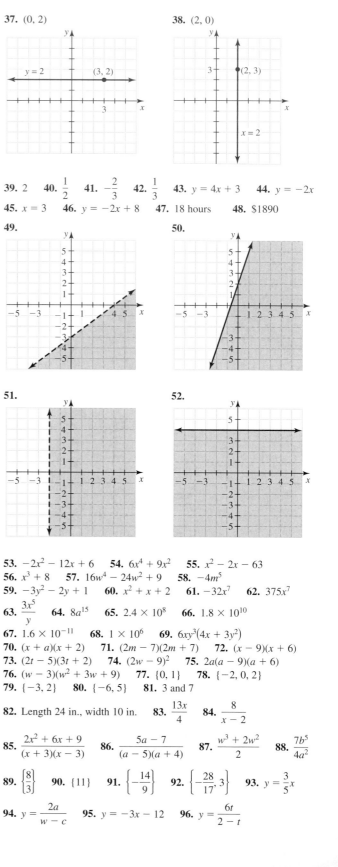

53. $-2x^2 - 12x + 6$ **54.** $6x^4 + 9x^2$ **55.** $x^2 - 2x - 63$
56. $x^3 + 8$ **57.** $16w^4 - 24w^2 + 9$ **58.** $-4m^5$
59. $-3y^2 - 2y + 1$ **60.** $x^2 + x + 2$ **61.** $-32x^7$ **62.** $375x^7$
63. $\dfrac{3x^5}{y}$ **64.** $8a^{15}$ **65.** 2.4×10^8 **66.** 1.8×10^{10}
67. 1.6×10^{-11} **68.** 1×10^6 **69.** $6xy^3(4x + 3y^2)$
70. $(x + a)(x + 2)$ **71.** $(2m - 7)(2m + 7)$ **72.** $(x - 9)(x + 6)$
73. $(2t - 5)(3t + 2)$ **74.** $(2w - 9)^2$ **75.** $2a(a - 9)(a + 6)$
76. $(w - 3)(w^2 + 3w + 9)$ **77.** $\{0, 1\}$ **78.** $\{-2, 0, 2\}$
79. $\{-3, 2\}$ **80.** $\{-6, 5\}$ **81.** 3 and 7

82. Length 24 in., width 10 in.   **83.** $\dfrac{13x}{4}$ **84.** $\dfrac{8}{x - 2}$

85. $\dfrac{2x^2 + 6x + 9}{(x + 3)(x - 3)}$ **86.** $\dfrac{5a - 7}{(a - 5)(a + 4)}$ **87.** $\dfrac{w^3 + 2w^2}{2}$ **88.** $\dfrac{7b^5}{4a^2}$

89. $\left\{\dfrac{8}{3}\right\}$ **90.** $\{11\}$ **91.** $\left\{-\dfrac{14}{9}\right\}$ **92.** $\left\{-\dfrac{28}{17}, 3\right\}$ **93.** $y = \dfrac{3}{5}x$

94. $y = \dfrac{2a}{w - c}$ **95.** $y = -3x - 12$ **96.** $y = \dfrac{6t}{2 - t}$

Appendix D

Chapters 1–6 Review
R.1 Exercises

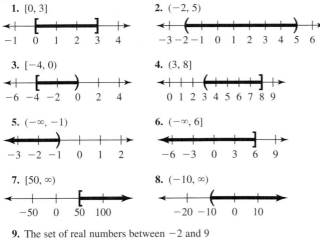

1. [0, 3]

2. (−2, 5)

3. [−4, 0)

4. (3, 8]

5. (−∞, −1)

6. (−∞, 6]

7. [50, ∞)

8. (−10, ∞)

9. The set of real numbers between −2 and 9

10. The set of real numbers between −4 and −3 inclusive

11. The set of real numbers greater than or equal to 11 and less than 13

12. The set of real numbers greater than 22 and less than or equal to 26

13. The set of real numbers greater than 0

14. The set of real numbers greater than or equal to 99

15. The set of real numbers less than or equal to −6

16. The set of real numbers less than 18

17. 1   **18.** 9.35   **19.** 0   **20.** 0   **21.** 50   **22.** 6.87   **23.** $\dfrac{10}{20}$

24. $\dfrac{12}{18}$ **25.** $\dfrac{18}{24}$ **26.** $\dfrac{49}{56}$ **27.** $\dfrac{3}{5}$ **28.** $\dfrac{2}{3}$ **29.** $\dfrac{7}{24}$ **30.** $\dfrac{2}{7}$ **31.** $\dfrac{3}{5}$

32. $\dfrac{1}{2}$ **33.** $\dfrac{4}{7}$ **34.** $\dfrac{2}{3}$ **35.** $\dfrac{1}{3}$ **36.** $\dfrac{3}{4}$ **37.** $\dfrac{3}{10}$ **38.** $\dfrac{2}{5}$ **39.** $\dfrac{1}{4}$

40. $\dfrac{3}{13}$ **41.** $\dfrac{1}{10}$ **42.** $\dfrac{2}{3}$ **43.** 10   **44.** 25   **45.** $\dfrac{2}{15}$ **46.** 42

47. $\dfrac{19}{24}$ **48.** $\dfrac{19}{20}$ **49.** $\dfrac{5}{36}$ **50.** $\dfrac{11}{48}$ **51.** $\dfrac{21}{8}$ **52.** $\dfrac{10}{7}$ **53.** −26

54. −27   **55.** −23   **56.** 9   **57.** 1   **58.** 5   **59.** −36   **60.** −27

61. 24   **62.** 100   **63.** 2   **64.** −3   **65.** 180   **66.** 96   **67.** −84

68. −39   **69.** 15   **70.** 6   **71.** −8   **72.** −5   **73.** 0   **74.** 0

75. −90   **76.** −89   **77.** −36   **78.** 33   **79.** 38   **80.** −2

81. −5   **82.** 16   **83.** 3   **84.** 34   **85.** −12   **86.** 4   **87.** 3

88. −1   **89.** 1   **90.** 1   **91.** $5x + (−3y)$, −25   **92.** $a^3 − b^3$, −72

93. $(a + b)(a^2 − ab + b^2)$, −28   **94.** $\dfrac{x − 7}{7 − x}$, −1   **95.** $(2x − 3)^2$, 49

96. $(a − b)^3$, 64   **97.** Difference   **98.** Sum   **99.** Difference

100. Product   **101.** Quotient   **102.** Square   **103.** Cube

104. Sum   **105.** Commutative property of multiplication

106. Commutative property of addition   **107.** Distributive property

108. Associative property of addition   **109.** Multiplicative identity

110. Additive identity   **111.** Multiplication property of zero

112. Multiplicative inverse property

113. Associative property of multiplication

114. Additive inverse property

115. $−x − 2$ **116.** $−11x + 8y$ **117.** $3x + 17$ **118.** $8x + 7y$

119. $31xy − 6$ **120.** $38a − 16$ **121.** $−30ab$ **122.** $21xy$

123. $2 − x$ **124.** $x − y$ **125.** $−22 + x$ **126.** $−5 − 2x$

R.2 Exercises

1. {7}   **2.** {15}   **3.** {−8}   **4.** {17}   **5.** {17}   **6.** {−9}

7. {−8}   **8.** {−9}   **9.** $\left\{\dfrac{25}{2}\right\}$ **10.** $\left\{-\dfrac{2}{3}\right\}$ **11.** {8}   **12.** {−19}

13. {−6}   **14.** {1}   **15.** {−4}   **16.** {1}   **17.** (−∞, ∞), identity

18. (−∞, ∞), identity   **19.** {24}, conditional equation

20. {30}, conditional equation   **21.** {−20}, conditional equation

22. {−46}, conditional equation   **23.** ∅, inconsistent equation

24. ∅, inconsistent equation   **25.** {1}, conditional equation

26. (−∞, ∞), identity   **27.** $\left\{\dfrac{26}{19}\right\}$, conditional equation

28. $\left\{-\dfrac{19}{11}\right\}$, conditional equation   **29.** $R = \dfrac{D}{T}$ **30.** $m = \dfrac{E}{c^2}$

31. $m = \dfrac{2K}{v^2}$ **32.** $b = \dfrac{2A}{h}$ **33.** $L = \dfrac{P − 2W}{2}$ **34.** $b_2 = \dfrac{2A − hb_1}{h}$

35. $r = \dfrac{A − P}{Pt}$ **36.** $y = \dfrac{2x − 6}{3}$ **37.** 10 hr   **38.** 5 m   **39.** 30 m

40. 12.5 cm   **41.** 4 kg   **42.** 8 cm   **43.** $a^2 + b^2$ **44.** $x + 5$

45. $y − 6$ **46.** $a − b$ **47.** ab^2 **48.** $0.10x$ **49.** $\dfrac{x}{y}$ **50.** $\dfrac{14}{x}$

51. $\dfrac{1}{2}x$ **52.** $\dfrac{2}{3}y$ **53.** $2(a + b)$ **54.** $(a + b)^2$

55. Length 24 in., width 18 in.   **56.** Length 55 m, width 30 m

57. 55 mph   **58.** 8 hr   **59.** Approximately 121.95 lb   **60.** 50 oz

61. 5 L   **62.** 15%   **63.** Approximately 85.71 mph   **64.** 8:40 A.M.

65. [5, ∞)

66. (−∞, 6]

67. $\left(\dfrac{4}{3}, \infty\right)$

68. (−∞, 5)

69. (−2, ∞)

70. (−∞, −8)

71. (−∞, −5)

72. (8, ∞)

73. (3, ∞)

74. (−3, ∞)

75. [−2, 8)

76. (1, 5]

77. (−14, 4)

78. [−4, −3]

R.3 Exercises

1. $(4, 0), (0, -3)$

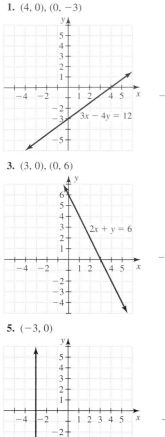

$3x - 4y = 12$

2. $(10, 0), (0, -5)$

$x - 2y = 10$

3. $(3, 0), (0, 6)$

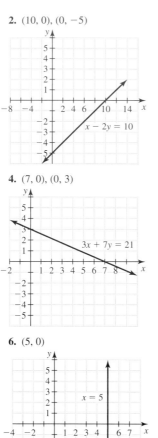

$2x + y = 6$

4. $(7, 0), (0, 3)$

$3x + 7y = 21$

5. $(-3, 0)$

$x = -3$

6. $(5, 0)$

$x = 5$

7. $(0, 2)$

$y = 2$

8. $(0, -4)$

$y = -4$

9. $(60, 0), (0, -30)$

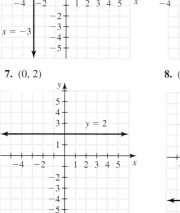

$y = \frac{1}{2}x - 30$

10. $(30, 0), (0, 20)$

$y = -\frac{2}{3}x + 20$

11. 1   **12.** $\frac{4}{3}$ **13.** -1 **14.** 1   **15.** No slope   **16.** No slope

17. 0   **18.** 0   **19.** 4   **20.** 0   **21.** $-\frac{3}{4}$ **22.** $\frac{5}{8}$ **23.** 0   **24.** 0

25. $\frac{1}{3}, (0, 1)$

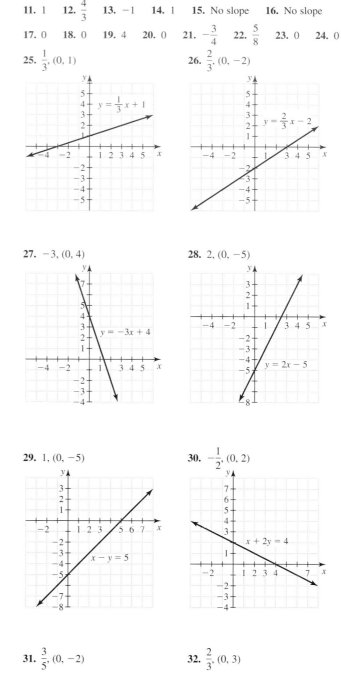

$y = \frac{1}{3}x + 1$

26. $\frac{2}{3}, (0, -2)$

$y = \frac{2}{3}x - 2$

27. $-3, (0, 4)$

$y = -3x + 4$

28. $2, (0, -5)$

$y = 2x - 5$

29. $1, (0, -5)$

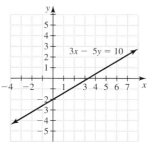

$x - y = 5$

30. $-\frac{1}{2}, (0, 2)$

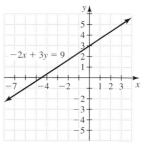

$x + 2y = 4$

31. $\frac{3}{5}, (0, -2)$

$3x - 5y = 10$

32. $\frac{2}{3}, (0, 3)$

$-2x + 3y = 9$

33. 0, (0, 4)

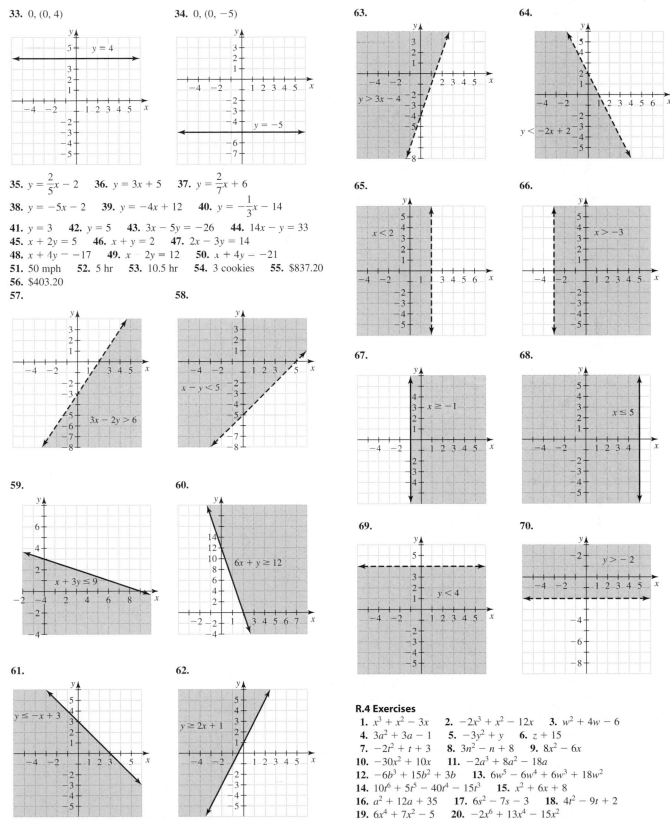

34. 0, (0, −5)

35. $y = \dfrac{2}{5}x - 2$ **36.** $y = 3x + 5$ **37.** $y = \dfrac{2}{7}x + 6$

38. $y = -5x - 2$ **39.** $y = -4x + 12$ **40.** $y = -\dfrac{1}{3}x - 14$

41. $y = 3$ **42.** $y = 5$ **43.** $3x - 5y = -26$ **44.** $14x - y = 33$
45. $x + 2y = 5$ **46.** $x + y = 2$ **47.** $2x - 3y = 14$
48. $x + 4y = -17$ **49.** $x - 2y = 12$ **50.** $x + 4y = -21$
51. 50 mph   **52.** 5 hr   **53.** 10.5 hr   **54.** 3 cookies   **55.** $837.20
56. $403.20
57.

58.

59.

60.

61.

62.

63.

64.

65.

66.

67.

68.

69.

70.

R.4 Exercises

1. $x^3 + x^2 - 3x$ **2.** $-2x^3 + x^2 - 12x$ **3.** $w^2 + 4w - 6$
4. $3a^2 + 3a - 1$ **5.** $-3y^2 + y$ **6.** $z + 15$
7. $-2t^2 + t + 3$ **8.** $3n^2 - n + 8$ **9.** $8x^2 - 6x$
10. $-30x^2 + 10x$ **11.** $-2a^3 + 8a^2 - 18a$
12. $-6b^3 + 15b^2 + 3b$ **13.** $6w^5 - 6w^4 + 6w^3 + 18w^2$
14. $10t^6 + 5t^5 - 40t^4 - 15t^3$ **15.** $x^2 + 6x + 8$
16. $a^2 + 12a + 35$ **17.** $6s^2 - 7s - 3$ **18.** $4t^2 - 9t + 2$
19. $6x^4 + 7x^2 - 5$ **20.** $-2x^6 + 13x^4 - 15x^2$

21. $x^3 - 18x + 27$ **22.** $a^3 + 3a^2 - 18a + 16$
23. $3w^3 + 14w^2 + 13w - 6$ **24.** $-2m^3 - 10m^2 + 19m - 63$
25. $ab + an + mb + mn$ **26.** $xy + sx + yt + st$
27. $x^2 - 4x - 12$ **28.** $x^2 - 2x - 15$ **29.** $6a^2 - 5a - 4$
30. $15b^2 - 62b + 63$ **31.** $10x^2 - 19x + 7$ **32.** $3x^2 + 20x + 12$
33. $10a^6 - 24a^3 - 18$ **34.** $12w^6 - 43w^3 + 35$ **35.** $16x^8 - x^2$
36. $25a^8 - x^6$ **37.** $x^2 + 10x + 25$ **38.** $y^2 + 6y + 9$
39. $4t^2 + 28t + 49$ **40.** $9w^2 + 24w + 16$ **41.** $s^2 - 4s + 4$
42. $h^2 - 6h + 9$ **43.** $9y^2 - 30y + 25$ **44.** $36x^2 - 12x + 1$
45. $9q^2 - 16$ **46.** $25m^2 - 36$ **47.** $4x^4 - 9n^2$ **48.** $25t^4 - 9m^2$
49. $2x^6$ **50.** $4a^{12}$ **51.** $-2w$ **52.** $-4b^2$ **53.** $\frac{1}{2}x^2$ **54.** $-\frac{1}{2}t^8$
55. $3x^2 - 2x + 1$ **56.** $-5a^2 - 4a - 1$ **57.** $x^3 + 3x^2 - x + 2$
58. $4w^2 + 3w - 2$ **59.** $x^2 - 3x - 3$ **60.** $2x^2 - 2x + 5$
61. $x^2 - 2x - 3$ **62.** $2x^2 - 3x + 4$ **63.** $x^2 - x + 2$ **64.** $x^2 + x - 3$
65. $\frac{1}{3}$ **66.** $\frac{1}{3}$ **67.** $-\frac{3}{2}$ **68.** $-\frac{1}{4}$ **69.** $-864a^{26}$ **70.** $64b^{10}$
71. $-200x^{12}$ **72.** $225y^{16}$ **73.** $\frac{x^{12}}{8y^6}$ **74.** $-\frac{q^9}{8p^6}$ **75.** $\frac{9a^6}{4}$ **76.** $\frac{81b^4}{16}$
77. $-6a^6$ **78.** $-15b^9$ **79.** $\frac{w^6}{2}$ **80.** $\frac{2}{t^{14}}$ **81.** $\frac{x^4}{81}$ **82.** $\frac{y^{12}}{25}$
83. $-\frac{1}{32a^{28}}$ **84.** $-\frac{b^8}{108}$ **85.** x^6y^9 **86.** $\frac{1}{a^{12}b^{20}}$ **87.** $\frac{25}{4x^{22}}$ **88.** $\frac{4}{y^8}$
89. 8×10^{20} **90.** 9.6×10^{25} **91.** 2×10^1 **92.** 3.6×10^{11}
93. 5×10^{13} **94.** 6.4×10^{22} **95.** 5.12×10^{-6} **96.** 4×10^{-14}

R.5 Exercises
1. $4(3x + 2)$ **2.** $6(3a + 5)$ **3.** $3y^2(5y - 2)$ **4.** $16z^3(3z - 2)$
5. $4a^3b(2b + 5a)$ **6.** $12y^3z^3(2y + 3z)$ **7.** $4x^2(3x^2 - 5x - 6)$
8. $7y(2y^2 - 3y - 4)$ **9.** $2ab(a^2 - 3a + 3)$ **10.** $3wz(w^2 - 4w - 3)$
11. $2x$ **12.** $5y^2$ **13.** $x + 3$ **14.** $y^2 + 3$ **15.** $-5a^2$ **16.** $-4b^2$
17. $w + 1$ **18.** $y - 1$ **19.** $(x + 4)^2$ **20.** $(x + 2)^2$ **21.** $(a - 1)^2$
22. $(b - 5)^2$ **23.** $(y + 3)(y - 3)$ **24.** $(n + 2)(n - 2)$ **25.** $(3x + 1)^2$
26. $(5y + 2)^2$ **27.** $(4m - 5t)^2$ **28.** $(3s - 4t)^2$ **29.** $(3x + 4)(3x - 4)$
30. $(9a + 5)(9a - 5)$ **31.** $(8n + 3)^2$ **32.** $(9s - 1)^2$
33. $(5x + 7y)(5x - 7y)$ **34.** $(ab + y)(ab - y)$ **35.** $(a + b)(a + 6)$
36. $(w + x)(w - 3)$ **37.** $(2x + a)(3x - 5)$ **38.** $(5a + 1)(2x + 1)$
39. $(y^2 + 1)(3y - 4)$ **40.** $(3x^2 + 5)(2x - 1)$ **41.** $(4a^2 + 7)(2a - 1)$
42. $(5t^2 + 6)(t - 2)$ **43.** $(b - 3)(a - 2)$ **44.** $(x - 7)(x - y)$
45. $(x^2 - 3)(x - 1)$ **46.** $(x^2 - 5)(a - 4)$ **47.** $(x + 2)(x + 3)$
48. $(x + 5)(x + 6)$ **49.** $(w + 3)(w + 5)$ **50.** $(u + 18)(u + 1)$
51. $(v - 6)(v + 4)$ **52.** $(m - 11)(m + 2)$ **53.** $(t - 14)(t + 2)$
54. $(q - 8)(q + 4)$ **55.** $(b - 13)(b - 2)$ **56.** $(p - 25)(p - 1)$
57. $(c - 8)(c - 3)$ **58.** $(n - 3)(n - 7)$ **59.** $(2x + 3)(x + 2)$
60. $(3w + 1)(w + 5)$ **61.** $(3t + 1)(5t + 4)$ **62.** $(6m + 5)((m + 4)$
63. $(3n - 2)(n + 6)$ **64.** $(4y - 3)(y + 5)$ **65.** $(2m - 3)(4m + 9)$

66. $(3p - 1)(6p + 5)$ **67.** $(4q - 1)(2q - 3)$ **68.** $(3t - 4)(2t - 1)$
69. $(5z - 3)(3z - 2)$ **70.** $(k - 4)(10k - 1)$ **71.** $(x - 1)(x^2 + x + 1)$
72. $(y - 3)(y^2 + 3y + 9)$ **73.** $(a - 2)(a^2 + 2a + 4)$
74. $(b - 10)(b^2 + 10b + 100)$ **75.** $(5x - 1)(25x^2 + 5x + 1)$
76. $(2a - 5)(4a^2 + 10a + 25)$ **77.** $(5q - 3)(25q^2 + 15q + 9)$
78. $(10b - 7)(100b^2 + 70b + 49)$ **79.** $(3x + 4y)(9x^2 - 12xy + 16y^2)$
80. $(2h + 5k)(4h^2 - 10hk + 25k^2)$ **81.** $(7m + 2n)(49m^2 - 14mn + 4n^2)$
82. $(ab + xy)(a^2b^2 - abxy + x^2y^2)$ **83.** $2(x + 1)(x + 3)$
84. $3(x + 5)(x - 3)$ **85.** $-2x(x + 3)^2$ **86.** $-4x^2(x - 5)^2$
87. $3(a - b)(a + b)(a^2 + b^2)$ **88.** $w(w - q)(w + q)(w^2 + q^2)$
89. $-b(a + 2b)(a^2 - 2ab + 4b^2)$ **90.** $-3(2x - 3)(4x^2 + 6x + 9)$
91. $(a - 2)(a + 2)(a + 3)$ **92.** $(x - 3)(x + 3)(x - 5)$
93. $\{-4, 6\}$ **94.** $\{-5, 4\}$ **95.** $\left\{-3, \frac{1}{2}\right\}$ **96.** $\left\{\frac{2}{3}, 4\right\}$ **97.** $\left\{\frac{1}{2}, \frac{5}{2}\right\}$
98. $\left\{\frac{1}{5}, \frac{1}{3}\right\}$ **99.** $\{-3, -2, 0\}$ **100.** $\{-1, 0, 2\}$

R.6 Exercises
1. $\frac{b - 4}{b + 4}$ **2.** $\frac{x + y}{x - y}$ **3.** $\frac{2x - 4}{x - 3}$ **4.** $\frac{2a - 8}{a - 1}$ **5.** $\frac{3y^5}{4}$
6. $\frac{2a^2}{3b^2}$ **7.** $-\frac{4z^7}{5w^2}$ **8.** $-\frac{3}{4r^3t^2}$ **9.** $\frac{1}{2(a - y)}$ **10.** $2a + 10$
11. $\frac{x^2 + xy + y^2}{x + y}$ **12.** $\frac{x + 3}{x^2 - 2x + 4}$ **13.** $\frac{5a}{6b^2}$ **14.** $\frac{2t^3}{3w^5}$
15. $\frac{5b^2}{24a^6}$ **16.** $\frac{1}{2y^5}$ **17.** $\frac{3x + 3y}{x^4(x - y)}$ **18.** $\frac{15a^4 - 10a^3b}{3a + 2b}$
19. $\frac{1}{2}$ **20.** $\frac{x + 3}{x - 3}$ **21.** $\frac{ab^3 + b^4}{6a^4}$ **22.** $\frac{xy^5 + y^6}{3(x - y)}$ **23.** $a + 3$
24. $w + 1$ **25.** $\frac{4}{x}$ **26.** $\frac{1}{x^2y}$ **27.** 2 **28.** 3 **29.** $\frac{-1}{x + y}$
30. $\frac{-x}{(x - 2)(x + 1)}$ **31.** $\frac{4 - w^2}{(2w + 1)(w - 3)}$ **32.** $\frac{2t - t^2}{(3t + 2)(t - 1)}$
33. $\frac{m^2 + 8m + 5}{m(m + 1)(m + 3)}$ **34.** $\frac{n^2 + 2n - 6}{n(n - 3)(n + 3)}$ **35.** $\frac{2}{13}$ **36.** $\frac{25}{6}$
37. $\frac{b + 2a}{2}$ **38.** $\frac{1}{2y - 5x}$ **39.** $\frac{6 - 15t^2}{8t^2 - 45t}$ **40.** $\frac{4 + 30m^2}{m - 20m^2}$
41. $\{1\}$ **42.** $\left\{\frac{11}{2}\right\}$ **43.** $\{-9\}$ **44.** $\{2\}$ **45.** $\left\{-\frac{20}{13}\right\}$ **46.** $\left\{-\frac{5}{11}\right\}$
47. 85 teachers   **48.** 6 cups of cereal   **49.** 48 dogs and 36 cats
50. 21 cars and 9 trucks   **51.** 50 mph
52. First day 60 mph and second day 50 mph or first day 50 and second day 40
53. 3 students   **54.** 4 students

Index

DEFINITIONS, RULES, AND FORMULAS

Subsets of the Real Numbers

Natural Numbers $= \{1, 2, 3, \ldots\}$

Whole Numbers $= \{0, 1, 2, 3, \ldots\}$

Integers $= \{\ldots -3, -2, -1, 0, 1, 2, 3, \ldots\}$

Rational $= \left\{ \dfrac{a}{b} \,\middle|\, a \text{ and } b \text{ are integers with } b \neq 0 \right\}$

Irrational $= \{x \mid x \text{ is not rational}\}$

Properties of the Real Numbers

For all real numbers a, b, and c

$a + b = b + a$; $a \cdot b = b \cdot a$ Commutative

$(a + b) + c = a + (b + c)$; $(ab)c = a(bc)$ Associative

$a(b + c) = ab + ac$; $a(b - c) = ab - ac$ Distributive

$a + 0 = a$; $1 \cdot a = a$ Identity

$a + (-a) = 0$; $a \cdot \dfrac{1}{a} = 1$ $(a \neq 0)$ Inverse

$a \cdot 0 = 0$ Multiplication property of 0

Absolute Value

$|a| = \begin{cases} a & \text{for } a \geq 0 \\ -a & \text{for } a < 0 \end{cases}$

Order of Operations

No parentheses or absolute value present:

1. Exponential expressions
2. Multiplication and division
3. Addition and subtraction

With parentheses or absolute value:

First evaluate within each set of parentheses or absolute value, using the order of operations.

Exponents

$a^0 = 1$

$a^{-1} = \dfrac{1}{a}$

$a^{-r} = \dfrac{1}{a^r} = \left(\dfrac{1}{a}\right)^r$

$\dfrac{1}{a^{-r}} = a^r$

$a^r a^s = a^{r+s}$

$\dfrac{a^r}{a^s} = a^{r-s}$

$(a^r)^s = a^{rs}$

$(ab)^r = a^r b^r$

$\left(\dfrac{a}{b}\right)^r = \dfrac{a^r}{b^r}$

$\left(\dfrac{a}{b}\right)^{-r} = \left(\dfrac{b}{a}\right)^r$

Roots and Radicals

$a^{1/n} = \sqrt[n]{a}$

$a^{m/n} = \left(\sqrt[n]{a}\right)^m = \sqrt[n]{a^m}$

$\sqrt[n]{ab} = \sqrt[n]{a} \cdot \sqrt[n]{b}$

$\sqrt[n]{\dfrac{a}{b}} = \dfrac{\sqrt[n]{a}}{\sqrt[n]{b}}$

Factoring

$a^2 + 2ab + b^2 = (a + b)^2$

$a^2 - 2ab + b^2 = (a - b)^2$

$a^2 - b^2 = (a + b)(a - b)$

$a^3 - b^3 = (a - b)(a^2 + ab + b^2)$

$a^3 + b^3 = (a + b)(a^2 - ab + b^2)$

Rational Expressions

$\dfrac{a}{b} + \dfrac{c}{b} = \dfrac{a + c}{b}$

$\dfrac{a}{b} - \dfrac{c}{b} = \dfrac{a - c}{b}$

$\dfrac{ac}{bc} = \dfrac{a}{b}$

$\dfrac{a}{b} + \dfrac{c}{d} = \dfrac{ad + bc}{bd}$

$\dfrac{a}{b} \cdot \dfrac{c}{d} = \dfrac{ac}{bd}$

$\dfrac{a}{b} \div \dfrac{c}{d} = \dfrac{a}{b} \cdot \dfrac{d}{c}$

If $\dfrac{a}{b} = \dfrac{c}{d}$, then $ad = bc$.